Springer Series in Bio-/Neuroinformatics

Volume 5

Series editor

Nikola K. Kasabov, Auckland University of Technology, Auckland, New Zealand
e-mail: nkasabov@aut.ac.nz

Aims & Scope

Springer Series in Bio-/Neuroinformatics publishes content on fundamental principles as well as state-of-art theories, methods and applications in the rapidly evolving fields of bioinformatics and neuroinformatics. The series is unique in that it integrates three interdependent disciplines, namely information science, bioinformatics and neuroinformatics. The series covers general informatics methods and techniques, like data mining, database management systems, machine learning, artificial neural networks, evolutionary computation, chaos theory, quantum computation, statistical methods, and image and signal processing, as well as their applications to different areas of research, including big data, functional analysis of the brain, computational systems biology, computational systems neuroscience, health informatics, personalized medicine, rehabilitation medicine, industrial biotechnology, and many more. The series publishes monographs, contributed volumes and conference proceedings, as well as advanced textbooks.

Advisory Board

More information about this series at http://www.springer.com/series/10088

Sasan Adibi
Editor

Mobile Health

A Technology Road Map

Springer

Editor
Sasan Adibi
Faculty of Science Engineering & Built Environment
School of Information Technology
Burwood, Victoria
Australia

ISSN 2193-9349 ISSN 2193-9357 (electronic)
Springer Series in Bio-/Neuroinformatics
ISBN 978-3-319-12816-0 ISBN 978-3-319-12817-7 (eBook)
DOI 10.1007/978-3-319-12817-7

Library of Congress Control Number: 2014953523

Springer Cham Heidelberg New York Dordrecht London

Printed on acid-free paper

Springer International Publishing AG Switzerland is part of Springer Science+Business Media (www.springer.com)

Foreword

The area of Mobile Health (mHealth) has been gaining attention in recent years. Due to its potential and proven impact on healthcare, more researchers and industries have become involved with it; moreover, engineers, scientists, and healthcare professionals have been collaborating to solve fundamental problems in the healthcare system, thus causing a deep systematic and progressive transformation.

Edited by a connoisseur in the field (Sasan Adibi) and written by authors with a very diverse range of expertise, this book provides the readers with technology-inspired perspectives on mHealth, and presents an important transforming force in both the healthcare industry and the field of mobile technologies. I recommend this book to all my colleagues, both in the academia and industry, who are interested in health informatics. I believe that this book will be of great benefit for a wide audience, serving to spark the interest of the general readers, and an informative and thought-provoking source for the professionals who may be more closely involved with mHealth research and developments.

Professor Majid Sarrafzadeh
IEEE Fellow, UCLA, Los Angeles, United States

There has been a tremendous interest on healthcare applications from various types of technology professionals, such as researchers, developers, integrators, and solution providers. However, there is a vast range of emerging technologies that can be gainfully utilized in healthcare. Also the healthcare sector is very complex with many disciplines and sub-disciplines. Therefore, it is difficult to provide a broad coverage of different technologies targeting various healthcare problems in one book. I congratulate Sasan Adibi for being able to cover such a wide range of technologies for different healthcare disciplines in one book. He has done an excellent job of sustained communication with so many authors to keep them motivated on improving the chapter contents based on timely reviews. This is a great achievement, also given an acceptance ratio of only about 40% showing a stringent quality control. I hope this book will become a major reference for Information and Communication Technologies (ICT) in healthcare.

Professor Pradeep Kumar Ray
University of SouthWales, Sydney, Australia

Contents

Part IV: Framework, Architecture, Software, and Hardware Systems

Part V: Cloud Applications

1. Introduction

Mobile Health (mHealth) is the intersection between Electronic Health (eHealth) and smartphone technology. The coverage of mHealth includes the acquisition, manipulation, classification, and transmission of health-related information. The end-to-end link used for transmitting health data normally initiates from biomedical sensors usually attached to the user's body. In the typical scenarios, the sensory information is collected by portable devices with relevant applications running on them. The information bits are transmitted through wireless and cloud networks. Various cloud-based stakeholders (e.g., doctors, medical labs, insurance, medical database, and family/coach) may be on the receiving ends of the end-to-end bidirectional paths.

This book is comprised of eight parts and forty five chapters covering many technological contents relevant to mHealth from various engineering perspectives, including: preventive and curative medicine, remote health monitoring, interoperability aspects, framework, architecture, software and hardware systems, cloud-based applications, radio technologies and applications, communication networks and systems, and security and privacy issues revolving around mHealth.

The aim of this introductory chapter is to discuss a few fundamental concepts and definitions that are used extensively throughout this book as well as short introductions to the covered contents.

1 Definitions

In this section, a few fundamental definitions are discussed:

Health Informatics: The body of knowledge incorporating the acquisition, retrieval, dissemination, usage, and storage, of human health information. The coverage includes the design, integration, and management of associated resources, methods, and devices to advance the understanding of healthcare practices, including: healthcare standards, healthcare systems and infrastructures, and eHealth.

Electronic Health (eHealth): Coined back in late 1990's eHealth refers to the healthcare practices assisted by communication systems and electronic processes. Various forms of eHealth systems and services may include: Electronic Health Records (EHR), Mobile Health (mHealth), Healthcare Information Systems (HIS), telemedicine, and consumer health informatics.

S. Adibi (ed.), *Mobile Health*,
Springer Series in Bio-/Neuroinformatics 5, DOI: 10.1007/978-3-319-12817-7_1

Mobile Health (mHealth): The practice of eHealth assisted by smartphones, which are used to capture, analyze, process, and transmit health-based information from sensors and other biomedical systems.

Healthcare systems: Healthcare systems and infrastructures use the organization of resources and rely on the interactions with the stakeholders, institutions, and infrastructures working towards the common objectives, which are based on providing health-related services to groups and individuals.

2 Preventive and Curative Medicine

In this four chapter part, the application of mHealth in preventive and curative medicine is covered. The focus of these chapters is the utilization of mHealth systems and devices in promoting preventive approaches and assisting healthcare staffs and patients in monitoring the treatments procedures. In these approaches, patients have the option to be monitored remotely at the comfort of their homes while the caregivers are kept informed of the ongoing medical treatments for signs of health improvements or deteriorations. A practical application of this category is the sleep technology, which is a preventive care approach in identifying sleep disorders, such as sleep-apnea. A number of chapters are tackling issues revolving around this area of preventive medicine. Another relevant application covered in another chapter is aimed at treating diabetes for long term diseases management for diabetic patients. Three other chapters cover various aspects of sleep disorders, such as sleep apnea and insomnia, which can seriously impair a patient's quality of life. Existing studies have shown that insomniacs have 350 percent higher risk of having hypertension compared to those who sleep normally. Insomnia is also a risk factor for diabetes, as well as anxiety and depression. For an accurate evaluation of insomnia, sleep measurements based on polysomnographic (PSG) signals and questionnaires are necessary.

A chapter is also dedicated to fall detection, which belongs to the preventive and curative medicine domain. A physical fall is considered a natural occurrence, which is expected to happen many times in the life span of human being from birth to death. In a perfect scenario it would be possible to predict when a fall should occur, therefore mechanisms may be in place to prevent it or deal with it. In a real-life scenario, it would be very hard to predict when a fall is going occur. Some literatures study the detection of situations where the risks point to an eminent fall after which warning can be issued immediately. The warning signals can be issued via a mobile device (either smartphones or tablets). Biofeedback sensors may also be engaged via Bluetooth connections to measure the patient physiological signals in a real-time sense, such as: electrocardiography (ECG), electromyography (EMG), electroencephalography (EEG), blood pressure, electro-dermal activity, and galvanic skin responses. In order to detect a fall, a kinematic sensor accelerometer may be included in the studies, which are typically available in most mobile devices, coupled with an algorithm devised to interact and support a fall-detection/alert system. A mobile device may use its computational power to perform a fall detection algorithm, to analyze the physiological signals received at the smartphone and in time of fall-detection, to notify the patient/caregiver immediately all at the same time.

The last chapter covers the topic of dementia and the management of the affected people through deploying mHealth mechanisms. In this chapter, mHealth sensors are explored which track and locate wandering patients suffering from dementia. The deployed system can assist caregivers by measuring dimensions of dementia-related wandering using eHealth tools. These tools analyze proximal factors as well as study background factors.

3 Remote Health Monitoring

The objective of this nine chapter part is to present a number of technologies and frameworks revolving around remote health monitoring through analyzing the individuals vital signs, which is an enabling application of mHealth. This is typically assessed through the evaluation of specific bio-physiological signals. In particular, the evaluation of such signals over time may provide doctors (or caregivers) relevant information about the evaluation of the effectiveness of the prognosis. However, this typically requires a frequent, or sometimes continuous, monitoring of the patient and the goal is to achieve an optimal involvement of human resources.

Parameters, such as blood-pressure, heart-pulse-rate, blood oxygen saturation, and respiration are examples of such bio-physiological signals. There are other bedside parameters considered in many healthcare scenarios, including those characterized by patients affected by motion disorders (e.g., Parkinson's disease or stroke) or, more simply, by patients who need a constant control of their movements (e.g., post-rehabilitation after surgery or monitoring of activities of elderly people).

Over the past decade, the progress of portable and wearable sensors has paved the way toward the use of related sensors in mHealth applications, allowing remote evaluation of different bio-physiological signals coupled with added economic savings in terms of specialized human resources and improving patient's comfort.

This part gives an overview of recently emerged physiological sensor technologies which come in the form of garment, jewelry and wearable gadgets. Some researchers are also looking into the feasibility of eye monitoring in mHealth with the use of head-mounted sensors. One chapter presents an approach regarding the dependability assessment of a mHealth monitoring system at the design time, based on an event-based formalism. The challenges in device implementation, in terms of ergonomics and overall power consumption are also considered.

Additionally, this part describes emerging concepts for mHealth supporting systems that combine physical environment (smart devices, sensors, controllers, healthcare tool), virtual environment (mobile devices enabled to access services and resources through networks anywhere and anytime) and human environment (patients and doctors may interact in real time via interfaces of smart devices within assisted living intelligent ambients). In particular, the novel trends in healthcare infrastructures and software development methodologies are discussed for remote health/vital sign monitoring systems. The presented concepts serve as practical guides for the development of complex health assistance tool expanding coverage of medical services assisted by mobile devices.

Finally, the underlying wireless technologies conveying health-related parameters from remote locations are also considered in various chapters. Low-data rate and low-power short-range wireless technologies enable local connections between medical sensors/devices and gateways or smartphones. Broadband wireless technologies relay data between local networks and core networks where electronic healthcare record servers are located. Designs, implementations, experiments and trial deployments of a number of integrated wireless communication systems and software platforms for automatic vital sign capture are also presented in these chapters.

4 Interoperability

Maintaining sound and logical relationships between healthcare practices adopted in various regions and locations in the world is a universal challenge and the goal is to bring these standards and regulations under one umbrella. This part discusses the issues revolving around interoperability aspects between various standards.

In this context, one of the interoperable eco-systems that has recently emerged is the ISO/IEEE11073 family of standards, which aims at providing seamless integration of Personal Health Devices (PHDs) into mHealth solutions. The implementation of such an approach in mobile devices poses an inherent trade-off between interoperability and processing load. One of the two chapters in this part discusses the current situation of both off-the-shelf and forthcoming X73PHD-compliant devices and the integration thereof in mHealth telemonitoring scenarios.

A number of implementations and research benefits can be derived from the use of X73PHD devices in mHealth scenarios. The mentioned chapter will also explore such added values, which include: the reduction of design costs, the shared experience within standards, marketing profits, reuse of devices, a posteriori integration, or plug-and-play. All these benefits enhance the potentials for X73PHD-compliant mHealth devices, thereby improving the delivered Quality of Service (QoS).

There are currently many standards relevant to mHealth, such as ISO/IEEE11073, X73PHD, Interchanging Healthcare Enterprise (IHE), Patient Care Device, CEN/ISO 13606, OpenEHR, and Clinical Document Architecture (CDA), and in this part, a few of them are discussed.

The second chapter argues that mHealth is enabling and accelerating three major global trends in the healthcare space: regulatory reform driven by demographic changes, industrialization of the healthcare sector and personalized medicine. This chapter covers interoperability as a key enabler of scalable mHealth.

5 Frameworks, Architecture, Software, and Hardware Systems

This six chapter part covers topics revolving around mHealth supporting frameworks, architectures, software, and hardware systems, which have undergone major technology shifts in the last decade. These technology shifts aim to support: security, open platforms, regulatory services, Quality of Care (QoC), Bing Your Own Device (BYOD), and data management.

In mobile patient monitoring systems, it is common the use wireless network technologies to transmit the data generated by the sensors, thus defining the term: application-driven wireless sensor network (WSN). The implementation of the WSN is not a trivial task, as different application scenarios have different requirements in terms of the number of variables to monitor, the number of supported users, mobility, data loss tolerance, power consumption, and coverage range among others. All these requirements must be considered when designing the WSN architecture to be used in a particular mobile patient monitoring system.

A number of frameworks will be discussed in this part, including the COHESY mode, which is a novel patient-centric collaborative healthcare system. This system features an integrated intelligent recommendation algorithm that is expected to give a new dimension to preventive and curative medicine. The proposed model helps its users to actively participate in their healthcare and prevention, thereby supporting an active life in accordance with their daily responsibilities at work, family, and friends.

Another framework discussed is ERPHA (Emergency Remote Pre-Hospital Assistance), which supports continuous monitoring of patients involved in traffic accidents; and prepares the hospital staff for the anticipated arrival of the patients to the hospitals for a more quality of care patient treatments. ERPHA deals mainly with improving communications, accessibility to information, and monitoring of patients.

6 mHealth and Cloud Computing

Cloud-ready applications have received major attentions in the past several years and cloud-based coverage can benefit mHealth services greatly by providing anytime-anywhere health and medical accessibility to the users. A number of mHealth cloud-based stakeholders, such as networked chains of clinics, hospitals, insurance (mInsurance), and pharmacy (mPharmapcy) services have recently emerged from such a deployment. In this paradigm shift, new technology implications became more challenging and require further optimizations and elaborations, such as: security, privacy, confidentiality, and integrity of information in the cloud space used for authentication, authorization, and trust enhancement of cloud users and servers.

This six chapter part covers various aspects of the concept of mHealth cloud and applications. A chapter presents a secure mHealth application which is based on hybrid cloud architecture combined with cryptographic techniques to protect data privacy, integrity and security of patients and health care givers and with role based access control to authenticate and authorize cloud users. Another chapter makes a comparative study of the existing mHealth cloud systems and constructs an energy-efficient transmission scheme in body sensor networks to find the most suitable architecture of multi-cloud offloading for mHealth application. The rest of the chapters focus on various cloud-based applications, such as: medical video communications, terahertz imaging, and multi-agent drug delivery system.

7 Radio Technologies and Applications

The main physical engine for the transmission of mHealth applications depends on the capabilities of the lowest OSI layer (physical layer), where the actual bits and bytes are transmitted via efficient modulation schemes. The main factor differentiating among different generations of mobile networks (e.g., 2G, 3G, and 4G) is the advancements in the modulation mechanisms.

This seven chapter part is focused on the radio technologies for cellular and wireless personal area networks (WPAN). A chapter is dedicated to discuss the progress of mobile networks from 2G up to 4G and beyond in relation to mHealth applications. The heart of the physical layer of modern mobile networks is based on Orthogonal Frequency-Division Multiplexing (OFDM), which is the main topic of another chapter. IEEE-based wireless access technologies (IEEE 802.16 and IEEE 802.22) are discussed in another chapter, which aims to cover telemedicine services over rural broadband networks.

Wireless Body Area Network (WBAN), which is covered under IEEE 802.15.6 is another emerging protocol for short-range interconnection between mHealth sensors and gadgets and smartphones, which is considered in another chapter. Near Field Communication (NFC) is also another short-range radio technology, which has found many applications within the mHealth domain, considered in another chapter in this part.

8 Communication Networks and Systems

Tremendous developments and worldwide deployments of cellular and wireless networks enable global and cost effective connectivity solutions, which can support many current and emerging healthcare applications. Telecommunications and network-related aspects of mHealth rely on the underlying technologies to support security and privacy, Quality of Service (QoS), radio technologies, and data management. This five chapter part is dedicated to cover such technological aspects. Moreover, the objective of this part is to provide a snapshot for some of the communication networks and wireless applications for mHealth systems. A brief review of the spectrum of these applications and the potential benefits of these efforts will be presented through the related chapters. In addition, different mHealth systems will be reviewed in order to compare their communication protocols from the QoS perspectives.

One area, which will be covered in this part, is the mHealth paradigm in supporting emerging unlicensed wireless broadband technologies. Emerging broadband wireless access technologies provide viable solutions to the deployment of mHealth applications in underserved and isolated sparsely populated areas. In such areas, the mobility support for healthcare delivery is limited by transportation availability, relatively large travelling times, and the natural geographical barriers. This part considers the use of emerging unlicensed wireless broadband technologies for the implementation of rural mHealth applications.

9 Security and Privacy

In this four chapter final part, the main focus is on security and privacy provisioning for mHealth systems and applications. Generally speaking, security and privacy have been under the main focus in healthcare industry. The challenge for the healthcare innovation is to embrace the potential of mHealth creatively within the healthcare system and not to merely replicate current technologies into a parallel wireless environment. In addressing this challenge, the complexities of securing health information along a composite clinical information pathway and in each situation of use must be defined.

The chapters within this part aim at identifying the issues in mHealth security and privacy by adopting a purely technical perspective. A list of generic security and privacy mechanisms relevant to mHealth applications, include: availability, anonymity, access control, integrity, authentication, authorization, accountability (non-repudiation), confidentiality, and each chapter address one or more of these security services in a systematic way relevant to mHealth applications.

Part I

Preventive and Curative Medicine

Chapter 2: mHealth Sensors, Techniques, and Applications for Managing Wandering Behavior of People with Dementia: A Review

Chapter 3: Ubiquitous mHealth Approach for Biofeedback Monitoring with Falls Detection Techniques and Falls Prevention Methodologies

Chapter 4: A Review of Methods to Characterize and Classify Sleep, Depression and Schizophrenia Disorders

Chapter 5: A New Era in Sleep Monitoring: The Application of Mobile Technologies in Insomnia Diagnosis

Chapter 6: Current Status and Future Trends of Wireless and Mobile Health Technologies in Sleep Medicine: Insomnia Case Study

2. mHealth Sensors, Techniques, and Applications for Managing Wandering Behavior of People with Dementia: A Review

Nhu Khue Vuong, Syin Chan, and Chiew Tong Lau

School of Computer Engineering, Nanyang Technological University, Singapore
{vuon0004,asschan,asctlau}@ntu.edu.sg

Abstract. Managing wandering behavior of people with dementia (PWD) has become increasingly imperative for these reasons: its high prevalence (60%) among PWD, its negative outcomes such as falls or elopement, and its burden on caregivers. In this chapter, we discuss the emergence of sensors, techniques, and applications for managing wandering behavior of PWD. First, we briefly present the 5Ws1H (WHO, WHAT, WHERE, WHEN, WHY, HOW) conceptual map of wandering science including stakeholders (WHO), measurements of wandering (WHAT), environments in which wandering takes place (WHERE), detection of wandering (WHEN), causes of wandering (WHY), interventions of wandering (HOW). Second, we introduce a framework that identifies specific groups of mHealth and eHealth assistive technologies for managing dementia-related wandering. Third, we review existing technological works that address these 4 domains in the 5Ws1H conceptual map: WHAT-WHERE-WHY-HOW. In particular, we explore mHealth sensors to geo-fence and prevent elopement, mHealth devices to track and locate PWD who wander, information services to assist caregivers, eHealth tools to measure dimensions of dementia-related wandering, and mHealth tools that analyze proximal factors as well as study background factors. Based on this review, we further discuss research and design issues, human factors, ethics, security and privacy that need to be considered when implementing mHealth applications for wandering management. We conclude the chapter by highlighting the future research work in this area.

Keywords: classification, dementia, dementia-related wandering, detection, wandering behavior, wandering patterns, wandering management.

1 Introduction

1.1 The Need for Managing Wandering Behavior

Wandering is generally described as a form of locomotion, most often repetitive walking or aimless meandering [1]. In clinical settings, wandering behavior

S. Adibi (ed.), *Mobile Health*,
Springer Series in Bio-/Neuroinformatics 5, DOI: 10.1007/978-3-319-12817-7_2

happens most commonly to PWD. The prevalence of PWD who wander was reported to be as high as 63% and the incidence is found to be greater in people with Alzheimer's disease than those with vascular dementia [2]. Unsafe wandering increases the risk of falls, hip fractures, sleep disturbances, getting lost, elopement and death [3-7]. In the USA, 70% of legal claims filed against nursing homes are related with wanderers' deaths [8]. Currently, there is no pharmacological treatment that can completely cure wandering. Some wanderers are given psychoactive medications to reduce wandering; however, it often leads to unpleasant side effects [9] and creates an ethical dilemma for caregivers [10]. Managing wandering of PWD is increasingly imperative for caregivers and policymakers due to the rapidly soaring number of demented persons. Statistical studies indicate that the number of PWD worldwide will quadruple to 115 million by 2050 [11] and account for nearly $640 billion in total direct and indirect costs annually [12].

In practice, family members and professional caregivers find it stressful to cope with wandering behavior and the need to multi-task in order to look after PWD. In a survey of caregivers for PWD, staff listed wandering as one of the top three most difficult behavior problems to manage [91]. The level of caregiver's burden is significant; with 27.2% of caregivers expressing feelings of burden more often than sometimes especially when this behavior occurs several times a day [92]. If the wandering patient is managed properly, staff will experience less frustration and adopt more tolerant attitudes toward the patient [91]. On the other hand, caregivers have to cope with the guilt of not doing enough and the burden of feeling personally responsible for PWD, probably due to the lack of awareness and knowledge on negative effects and consequences of wandering to the patient's well-being. Helping caregivers juggle multiple responsibilities and reducing their feelings of guilt or not being up to the task is therefore crucial for good dementia care.

In addition, managing wandering behavior will play an important role in diagnosis and assessment of dementia. First of all, the literature suggests that wandering can potentially be used as a diagnostic sign for preclinical dementia [13-15]. A five-year study of clinical records in patients with preclinical dementia [13] found that wandering is the earliest symptom followed by cognitive complaints. There is evidence to show that changes in locomotion patterns begin many years prior to the onset of dementia [14] and gait-related motor disturbances present in all subtypes of dementia, even in the early and preclinical stages [15-16]. Early detection of such changes will allow early treatment and potentially slow down the onset of dementia. This will improve the person's quality of life and delay the transition to costly care facilities.

Secondly, evidence-based research reveals that several features of wandering are associated with cognitive declination, health status, and other physiological effects in PWD. Patients diagnosed with mild cognitive impairment were found to have greater variability in walking speeds and activity levels than those without [17-18]. Increased stride time variability is also correlated negatively with cognitive performance and increased the risk of falls in community-dwelling elders [19, 20, 71]. Algase et al. [21] found that pacing patterns among PWD who wander are signals of agitation and discomfort. It is therefore essential to quantify and measure wandering in PWD in order to identify and address the decline in cognitive functions indicative of dementia or other negative health states.

Lastly, as wandering is associated with negative outcomes such as falls, getting lost, or elopements from home or care environments, wandering monitoring systems can provide timely alerts to caregivers or family members and help prevent such detrimental outcomes.

1.2 mHealth as a Promising Technology to Manage Wandering

mHealth technology assisted by mobile tracking devices and wearable sensors can be used for assessing and measuring wandering, as well as preventing negative outcomes of wandering from happening. For example, GPS-based devices and products like Em Finder [22], GPSShoes [23] let caregivers monitor a person with dementia outside the home or formal care environments. Complex systems like Lifeline 4000+ home reassurance unit [24] address the safety and security for PWD, i.e. ensure that PWD remain in designated safe zones or are easily located if they do move away from these areas. Other ambient systems with ultra wideband and radio frequency identification (UWB-RFID) technology can locate wanderers before or after elopement in indoor environments like nursing home [25]. Such technologies allow caregivers to better monitor wandering behavior of PWD.

In this chapter, we provide a thorough review of existing technological works targeted at managing wandering behavior of PWD. Specifically, we focus on mHealth sensors to geo-fence and prevent elopement, mHealth devices to track and locate PWD who wander, information services to assist caregivers, eHealth tools to measure dimensions of dementia-related wandering, and mHealth tools that analyze proximal factors and study background factors.

The objectives of this chapter are first to present a concise conceptual map of wandering science to guide technologists in developing assistive technologies. Second, it identifies main threads of technologies for managing dementia-related wandering. Third, the review summarizes achievements in each thread of technologies. Based on these summaries, we highlight what has not been achieved and addressed to point out the future works.

The rest of this chapter is organized as follows. In Section 2, we present the background of wandering science including a conceptual map of wandering and a framework of mHealth technologies for managing wandering. Section 3 reviews the current mHealth technologies for managing wandering based on the conceptual map and the framework described in Section 2. Section 4 discusses research issues and economy benefits. Section 5 concludes the chapter and points out future directions.

2 Understanding the Science of Wandering

2.1 Conceptual Map of Wandering Science

We will first introduce an overall conceptual map of wandering science. This conceptual map is synthesized and simplified from Chapters 1, 2, 5, 11, 12 and 16 of the book "Evidence-based Protocols for Managing Wandering Behaviors" (Audrey Nelson & Donna Algase) [26-31].

The purpose of introducing this overall conceptual map is three-fold. Firstly, it equips readers with the essential fundamental background on dementia-related wandering. Secondly, it serves as a guide to identify the kinds of mHealth technologies that can be developed to assist wandering management.

Thirdly, we would like to promote wandering science from the perspective of gerontologists to technologists. There are two reasons why this is important. First, technologists tend to rely on their own perception of wandering to recognize and manage it. And their perception of wandering is much simpler than the one of gerontologists. In fact, more than 90% of technological works focus on outcomes (e.g. elopement, getting lost) of wandering behavior. Terms including wandering, wandering outcomes, wandering correlates (e.g. shadowing, lingering near exits) are often used interchangeably in these studies. These technologists are likely not aware of the differences and therefore they do not cover all aspects or dimensions of wandering in designing and developing assistive technologies. Second, technologies are created to assist PWD, caregivers, and gerontologists. Thus, it is important for technologists to identify the needs of these stakeholders and base on widely utilized concept of wandering to develop assistive applications. It will then maximize the likelihood of their technologies' acceptance in reality.

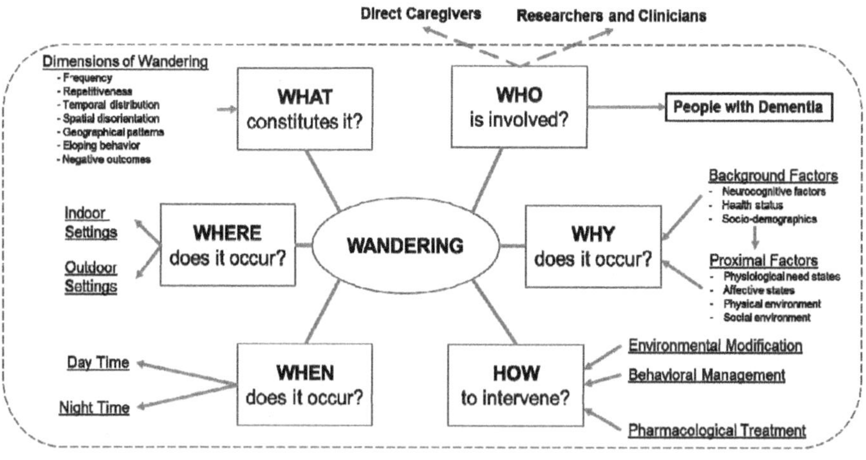

Fig. 1 Overall Conceptual Map of Dementia-Related Wandering

Fig. 1 depicts the conceptual map and briefly answers 6 interrelated questions (domains) on wandering behavior. We will not elaborate on each domain since the detailed studies as well as discussions can be found in the references [26-31]. Instead, we will provide a concise description of the domains and identify mHealth assistive technologies that can be developed in each domain.

2.1.1 WHO

Wandering directly affects PWD. It indirectly affects caregivers in the form of additional care burden and time needed to surveillance the behavior. Clinicians are challenged to monitor the occurrence and progress of wandering in the long-term care settings as well as to find effective treatments and interventions. Researchers especially gerontologists face difficulties in conducting dementia-related wandering studies due to the lack of efficient methods to observe, assess, measure, and quantify wandering.

2.1.2 WHAT

What is wandering indeed? In fact, there is the lack of a universal, widely utilized definition of wandering within the gerontology discipline and across disciplines. Table 1 summarizes 3 empirical definitions of wandering. The clinical definition [33], developed within nursing science as a nursing diagnosis almost a decade ago, has yet to be routinely seen in published wandering literature [34]. The policy-oriented definition [35] is used by policy makers and risk managers. It is meant for media reports of accidents, injuries, and deaths of wandering patients, and lawsuits related to charges of unsafe wandering in institutional settings [26]. The scientific definition [36] was proposed in 2006 at the International Consortium for Research on Wandering on the basis of analysis and synthesis of terms and definitions from 183 published papers. It is recommended to adopt this provisional consensus definition in all scientific studies regarding dementia-related wandering because it allows others to independently verify the validity of research results through observation, measurement, or testing [26].

According to the scientific definition, wandering is measured in terms of *frequency* (a sum, mean, or rate of wandering episodes per unit time); *repetitiveness* (characterized by 2 items: wanderer goes repeatedly to the same location; wanderer travels repeatedly the same route while walking); *temporal distribution* (wandering duration between lunch and dinner; between breakfast and lunch; between dinner and bedtime; between waking and breakfast); *spatial disorientation* (characterized by 5 items: wanderer cannot locate dining room/own room/bathroom without help; wanderer bumps into obstacles or other people while walking alone; wanderer walks about aimlessly; wanderer gets lost); *geographical patterns* (wandering in pacing, lapping, random patterns); *eloping behavior* (characterized by 4 items: wanderer runs off; wanderer enters unauthorized area; wanderer attempts to leave authorized area; wanderer is returned to authorized area after leaving unnoticed); *negative outcomes* (characterized by: fatigue; inadequate food intake; falls and injuries; exit attempts despite caregivers' efforts to keep wanderer in a location or home).

Table 1 Definitions of Dementia-Related Wandering

Definitions	Description
Clinical definition (NANDA, 2001) [33]	Meandering, aimless or repetitive locomotion that exposes a person to harm and is incongruent with boundaries, limits or obstacles.
Policy-oriented definition (VHA, 2002) [35]	A wandering patient is a high-risk patient who has shown a propensity to stay beyond the view or control of employees, thereby requiring a high degree of monitoring and protection to ensure the patient's safety.
Scientific definition (Algase et al, 2006) [36]	A syndrome of dementia-related locomotion behavior having a frequent, repetitive, temporally-disordered, and/or spatially disoriented nature that is manifested in lapping, random, and/or pacing patterns, some of which are associated with eloping, eloping attempts, or getting lost unless accompanied.

2.1.3 WHERE

Wandering happens in both outdoor (community-dwelling) and indoor (nursing homes, home environments, assisted living facilities or ALF) settings.

2.1.4 WHEN

Wandering usually happens in the morning hours, increases in the amount and frequency throughout the day, and reaches the peak at 5pm to 7pm [32].

2.1.5 WHY

There are two groups of factors that account for wandering in PWD. Background factors consist of neurocognitive status, general health, and socio demographic factors [27]. For example, a person may wander because he/she gradually loses memory and visual-spatial or orientation skills as dementia worsens; or because he/she is in pain due to earache, toothache or constipation; or because his/her previous jobs (deliverymen, salesperson, etc.) require him/her to travel and move around frequently. Proximal factors include personal need states, both physiological and psychological, and environmental conditions, both physical and social.

For instance, wandering might be a search for food in response to hunger, or a way to cope with boredom or loneliness. Sometimes, ambient conditions (lighting, sound, or temperature within a location) or lack of social interaction in care and home environments could also trigger wandering [28].

2.1.6 HOW

There are 4 approaches to intervene wandering.

Environmental modification involves modifying environmental features of the care settings to limit wandering. The 6 environmental features of long-term care units (suggested by research and practice literatures) that are associated with wandering behavior are: unit layout, hallways, doorways, unit exits, directional and destination signage, and ambient conditions [31]. For example: place gridlines in front of exit doors, provide a secure wandering/walking area, make exits less accessible or noticeable, use a combination of large-print signs and portrait-like photographs to aid residents to find own bedrooms [37].

Behavioral techniques are the most commonly used and productive approach to limit wandering. Examples include: promote social interaction during structured activities, introduce music sessions or reading sessions, schedule regular exercise, or apply air mat therapy as well as aromatherapy [37-38].

Technology and safety systems can be used to intervene wandering. Alarm/alert and sensory devices are used to safeguard PWD and provide alerts to caregivers or family members when PWD elope or attempt to elope [39]. Mobile locator devices are used to quickly locate wanderers [40].

Pharmacological treatment may be considered when nonpharmacologic treatments fail. In fact, behavioral symptoms of dementia like wandering can be helped with nonpharmacologic approaches and do not respond well to pharmacologic treatment. There is limited evidence-based data available specifically on efficacy of pharmaceutical options (e.g. risperidone, olanzapine, citalopram, mirtazapine) on wandering even though these options are commonly used to treat symptoms like aggression or agitation [30].

2.2 Framework of mHealth Technologies for Managing Wandering

In Fig. 2, we present a framework of mHealth technologies for managing dementia-related wandering. These technologies target the WHAT, WHERE, HOW, and WHY domains and serves 3 different application users (PWD, caregivers, clinicians and researchers) in the WHO domain. We do not exclude technologies targeting WHEN domain. Instead, we combine them with technologies targeting the WHERE domain because they are highly interdependent, i.e. it is possible to detect when wandering happens once we can detect where wandering happens. The framework consists of 3 building layers (user interface, applications & logic, sensing modalities) and 2 mediational layers (network & repository, data fusion).

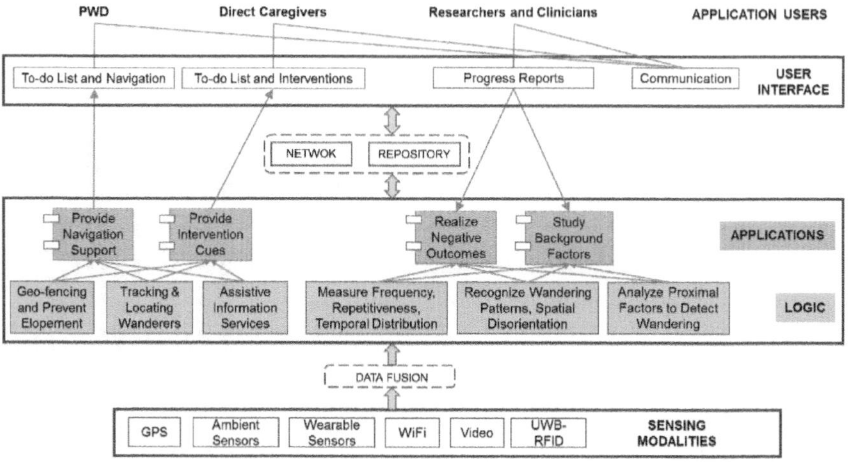

Fig. 2 Framework of mHealth Technologies for Managing Dementia-Related Wandering

The SENSING MODALITIES layer represents sensors that are deployed in assistive applications.

The DATA FUSION layer preprocesses data collected from sensors and provides inputs for the applications & logic layer.

The APPLICATIONS & LOGIC layer consists 2 sub-layers: applications and logic. The logic sub-layer consists of algorithms to: track and locate wanderers; geo-fence and prevent elopement; provide assistive information services; measure 3 dimensions of wandering including frequency, repetitiveness, temporal distribution; recognize wandering patterns and spatial disorientation; and analyze proximal factors to detect wandering. These algorithms are pivotal to creating mHealth applications to provide navigation support for PWD when they disorient and get lost, to provide intervention cues for direct caregivers when PWD wander, to assist researchers and clinicians to realize negative outcomes and study background factors associated with dementia-related wandering. The direction of arrows in Fig. 2 shows the applicability of these applications.

The NETWORK & REPOSITORY layer consists of 2 modules: network and repository. Repository module is a database that stores the lists of navigation plans to support wanderers when they get disoriented. Also, it stores intervention cues that direct caregivers can perform to limit wandering. Additionally, it records measurements of wandering dimensions and changes in proximal factors. Network module is a networking communication platform based on a client-server architecture. It is used to facilitate communication between the applications layer with the application users.

The USER INTERFACE layer consists of 4 parts. The to-do list and navigation module is used to assist PWD to navigate through the navigation plan received from the to-do list repository. The interface shows information indicating the direction where the disoriented user is expected to go. The to-do list and interventions module suggests what caregivers can perform to intervene wandering. The interface shows lists of cues or tasks received from the to-do list repository. The progress reports module enable researchers and clinicians to generate reports of possible negative outcomes and affected background factors through monitoring and analyzing wandering dimensions and proximal factors. The interface shows graphs plotting dimensions of wandering throughout the day, week, or months. It also indicates or highlights changes in these dimensions and proximal factors. The communication module offers a communication channel between PWD, caregivers, researchers and clinicians. Its interface shows assistive buttons to enable calls for help and intervention.

In this review, we focus on the applications & logic layer. From Fig. 2, we identify that the first 2 applications (navigation support and intervention cues) target the WHERE and HOW domains in the conceptual map (Fig. 1) and the last 2 applications (negative outcomes and background factors) target the WHAT and WHY domains respectively. As elaborated earlier, these applications can be developed by combining algorithms in the logic sub-layer. Therefore, in the next section, we categorize the existing technological work into the 6 groups of algorithms in the logic sub-layer. We will also review the application to study background factors separately but not the other 3 since they are addressed together with the 6 groups of algorithms.

3 Literature Review

Technologies for managing dementia-related wandering are a subset of assisted living technologies. There have been excellent surveys of assisted living technologies for elderly and dementia people in the literature [41-45], in which wandering management was sporadically reviewed. These reviews only included technologies concerning the WHERE and HOW domains of wandering. Others that evaluate the WHAT and WHY domains of wandering were not summarized and synthesized. Here we first examine the WHERE and HOW domains by synthesizing those reviews with other works not yet included, followed by expounding on the WHAT and WHY domains of these works. All systems described in this review were designed for PWD unless otherwise stated.

3.1 mHealth Sensors to Geo-fence and Prevent Elopement

The related works are categorized into 2 groups: alarm/alert and sensory systems. Both groups deploy sensors in home or care environments. Alarm/alert systems attach commercial alarms and alerts at exit doors to prevent wanderers from elopement whereas sensory systems employ a wide variety of low-cost sensors to evaluate indoor movements of PWD so as to ensure safety.

3.1.1 Alarm/Alert Systems

Alarm systems alert caregivers when the wanderer is detected to be leaving the monitored area. They may automatically lock doors or incorporate wireless technologies to trigger alarms [40]. Such systems are available in three forms. *Audible alarms* [39] sound when a monitored door is accessed by the tagged person. *Optically activated alarms* [46] activate lights via passive infrared motion detectors when a monitored door or area is accessed. *Video-activated alerts* [63] run continually and are activated by triggers and can provide real-time feedback. Audible and optically activated alarms are well-established, installable on interior and exterior doors, and nonrestrictive to locomotion. However, loud audible or even false positive alarms (e.g. when guests or caregivers access the monitored door or area) may be distressing. Video monitoring system [63] achieved 100% sensitivity but could not distinguish wanderer's elopements from routine exiting of the caregivers. In addition, the system can only cover confined area and has issues with privacy of monitored subjects.

3.1.2 Sensory Systems

Sensory systems [47-49] used different location modalities such as infrared sensors [47], optical illumination-based location sensors [48], or ultra-wide-band radio receivers with small tags for transmitters [49]. From the location information collected, subjects' movement trajectories are modeled to estimate covered distances and travel speeds [47] or anticipate dangerous situations that may happen when the wandering subjects enter adverse areas like exit doors or laundry rooms [48]. In addition, location data may be combined with other signals like the amount of social interaction and the quality of sleep [49] to monitor state changes of PWD. The limitation of these studies is they were not evaluated on large number of subjects. The evaluation of the correlation between sensory data and the development of the dementia in the inhabitants [49] has not yet been possible whereas [47] managed to get one wanderer involved in their trial for 8 months.

CareWatch [50] is a smart home system that uses motion, door opening and bed occupancy sensors to alert the caregiver of unattended home exits, particularly at night. The system has been randomly assigned to 27 homes with dementia subjects for trial. Although the final results have not yet been published, CareWatch has operated for more than 200 months of combined system time (of 27 homes) without any major failures [43]. Moreover, there have been no unattended exits during the night in homes instrumented with CareWatch.

In short, audible and optically activated alarms are commercially available. They offer quick and simple solutions to prevent wanderers from eloping. However, the problem of these solutions is false positive alarms. Video-activated and sensory systems can be further enhanced to provide intelligent solutions for managing dementia related wandering. For example, we can integrate face recognition features to the existing system [63] to distinguish wanderers from caregivers.

In general, both systems address 3 items of eloping behavior in the WHAT domain: wanderer runs off; wanderer enters unauthorized area; wanderer attempts to leave authorized area.

3.2 mHealth Devices to Track and Locate PWD

Tracking systems aim to provide the position of a wanderer at any moment and can be used to locate PWD at risk of wandering. Unlike those in Section 3.1, these systems are invasive and require PWD to wear a tag or transmitter shaped like a watch, pager, or ankle bracelet. The wanderers are then located by radio frequency (RF) range finding or global positioning systems (GPS). RF systems are mainly for indoor monitoring whereas GPS works well outdoors only. We discuss both tracking systems.

3.2.1 GPS Tracking with Mobile Locator Devices

Digital Angel, Inc. [93] has developed a commercial system which employs a GPS-enabled phone to establish a perimeter (termed a "geo-fence") around the home. Wanderers wearing the phone, who cross the geo-fence, automatically reveal their position to their caregivers through a server-based notification system of e-mail, pager, and automated telephone calls using synthesized voices. Caregivers can locate the missing wanderer within an accuracy of 30m, and coverage is dependent on cellular telephone service and GPS satellite availability.

Other well-established research works in this category share similar working mechanism with Digital Angel, i.e. combining GPS technology and mobile telephones. These works differ from one another in the proliferation of mobile phone technologies and the systems' localization accuracy. In particular, before the advent of GPS-enabled smartphones, the earliest systems used separate GPS receiver [51] and personal handy-phone system [52] to detect a person's location within 200m and 60m location error (distance between the actual and the located location) respectively. The more recent ones harnessed PDAs to locate missing persons to an accuracy of approximately 10m [53]. These were controlled studies (subjects were instructed where and how to move) with no involvement of PWD.

In other uncontrolled studies (subjects were free to move around), Miskelly [54] first tagged 4 nursing home PWD each with an electronic bracelet that buzzed a pager if the resident wandered into off-limit areas. In the second study [55], he attached GPS-integrated mobile phones to 11 dementia subjects to enable relatives or caregivers to locate them within 5m, except inside buildings or on public transport. Data and evaluation were recorded over a 6-month period. Two important results were found. First, three different types of episodes (wandering outside, wandering into a risk area, removal of a bracelet) were immediately detected by the system and the affected resident was promptly attended to [54]. Second, tracking was accurate when the users complied with wearing the phone. 90% of the location requests exactly matched the information given by the relative/caregiver. Of the location requests that were deemed failures, 95% were due to non-compliance (i.e. not wearing the phone) [55]. In short, GPS-enabled phones are effective in monitoring and locating PWD who exhibit symptoms of wandering.

3.2.2 RFID Tracking

In recent years, RFID technologies have been increasingly applied to manage wandering in institutional care settings because of three reasons. Firstly, each RFID device has a unique identification number to distinguish the subjects. Secondly, it can provide compact and real-time information on the person's location, accurate to a few tens of centimeters [25]. Thirdly, a variant of RFID called Ultra Wideband (UWB) offers advantages over passive RFID with its capability to reveal the identity and precise locations (and time) of multiple subjects moving simultaneously [25]. Therefore, it can minimize and probably remove the burden of observation on the subjects especially during night hours.

Several RFID systems have been developed and implemented in institutional care settings. The earliest prototype proposed by Ashida [56] comprised RFID tags, trigger generators, tag receivers and personal computer for data processing. A tag hooked to the subject would send the RF signal when the person is in a trigger field. The tag receiver would receive that RF signal, decode the data, and send the data including date and time, tag ID and trigger ID to a central processing computer. The information was used to monitor the subject's daily behavior. Ashida's system was further enhanced to calculate the distance that the subject moves, display the movement pattern mapping of the subject and the duration of stay at each location [57, 58].

Almudevar [59] instrumented the home of a person at risk for dementia with various RF sensors to reconstruct his or her movement trajectory within the home. Analysis of a 3-week longitudinal data suggested that the methodology was reliable enough to construct the travel trajectory of the person. Regions of high visiting frequency were identified and transitions between the regions were accurately tracked. Further replication is necessary in the homes of more subjects.

Kearns [25] introduced the use of UWB as an advanced method to track a moving transmitter (or tagged PWD) in real-time. It is similar to RFID monitoring systems but has certain advantages. UWB devices operate at lower power levels and can transmit data over periods of years without battery replacement. The large bandwidth available to UWB-RFID allows potentially very high data transmission rate (to the order of 1 Gbps) over very short range (less than 1m). It is capable of estimating a location accurate to 20 cm in real-time. Trial tests [60] concluded that the UWB-RFID technology is sufficiently reliable and precise to allow accurate measurement of lingering and shadowing behavior. However, UWB systems are costly and they must be precisely calibrated to ensure accurate localization.

To sum up, both GPS and RF based tracking systems are able to track and locate PWD in outdoor and indoor settings with high accuracy (5-10m for GPS and 30cm for RF-UWB). The technologies are mature and commercially available. They provide effective tools to address the WHERE and WHEN domains. Future research can make use of these available technologies and products to study and address more complex aspects of wandering.

3.3 Information Services to Assist Caregivers

Information services are complementary networks of hardware and software that caregivers and organizations use to collect, process, and distribute data regarding PWD who wander so as to provide interventions and improve quality of care. We have found only one of such work in the literature.

Lin's another study [62] attempted to build an information platform for care organizations. An eXtensible- Markup-Language-based dementia assessment system combining program code and assessment content was developed to provide caregivers with better flexibility and real-time response ability. With regard to dementia assessment service procedures, care personnel could obtain information concerning a dementia patient's symptoms from family members, enabling them to use the assessment system to analyze the severity of the dementia. This system facilitated the establishment of long-term case files and recording cases' structured clinical dementia ratings (SCDR), mini-mental state examination (MMSE), and daily moods. It then helped to identify easily irritable patients in order to provide additional care mechanisms.

In brief, information systems directly assisting caregivers in monitoring and managing wanderers are scarce. Future works should focus more on characterizing the HOW domain of wandering and providing assistive tools to caregivers. For example, we can document behavioral techniques to intervene wandering and implement a mobile application to teach and guide caregivers to apply such intervention techniques when PWD wander.

3.4 eHealth Tools to Measure Frequency, Repetitiveness, Temporal Distribution

There have been eHealth tools that capture 2 dimensions of wandering: frequency and temporal distribution. Frequency was measured as the rate of wandering episodes per hour. Temporal distribution was measured in different forms: walking time duration, time spent on wandering, temporal variability of wandering locomotion, duration of wandering episodes. These dimensions were used to correlate wandering and cognitive performance.

Hayes and colleagues [17-18] measured the walking time duration of 14 elderly seniors living in their own homes. Three passive infrared sensors were placed on a path in the home identified as the longest path along which the subject typically walked continuously from end-to-end. These sensors were triggered as a person walked past the monitored path. In each observation period, if the sensors are s1, s2, and s3, then each time the sensors fired in the order (s1, s2, s3) or (s3, s2, s1) without another intervening sensor firing, the subject was considered to have walked along the monitored path and the time difference in firing events was calculated as the observed sample of the walking time. Kernel density estimation was then used to approximate the distribution of walking time. They found that variability in walking time was greater among the 7 participants with mild cognitive impairment than those 7 without.

Biomechanical devices including Actillume, StepWatch, and TriTrac-R3D were used to estimate the time spent on wandering of PWD [65]. The proportion of time during which motion was detected by each device was compared to the proportion of time during which wandering was observed by human observers. StepWatch produced the closest estimate of time spent wandering (16.8%) compared to 15.4% assessed by human observers. Actillume and TriTrac-R3D detected much higher percentages, 63% and 49% respectively, reflecting oversensitivity to movement that is not wandering behavior.

Makimoto et al. [66] employed RFID tagging systems [57-58] to record locomotion data which were then used to measure temporal variability of wandering locomotion. Two indicators were calculated: the percentage of hours with locomotion and the median distance traveled per hour. They found that the former indicator was negatively correlated with the subject's cognitive ability, which did not correlate with the latter indicator. This suggests that cognitive impairment may be better reflected by the frequency of wandering rather than the total distance travelled.

Algase and associates [85] videotaped wanderers in long-term care settings. They used Noldus Observer 5.0 software and human observers to code videotapes and measure the rate and duration of wandering episodes. They found that MMSE scores were negatively correlated with overall duration of wandering and were not significantly correlated with rate-related parameters.

In conclusion, current eHealth tools and software are able to capture the frequency and temporal distribution dimensions in the WHAT domain of wandering. There is no automated tool to capture the repetitiveness dimension (i.e. wanderer goes repeatedly to the same location using the same route while traveling indoors). We suggest using camera video or RFID-UWB systems to capture the locomotion of wanderers and determine its repetitiveness by applying computational algorithms and machine learning techniques.

3.5 eHealth Tools to Recognize Wandering Patterns, Spatial Disorientation

Present eHealth tools are able to capture geographical patterns of wandering including pacing, lapping, and random. In addition, there exist systems that measure one specific aspect of spatial disorientation, i.e. getting lost.

3.5.1 Recognize Wandering Patterns

Nakaokaet al. [67] quantitatively described locomotion in PWD in terms of distance moved per day, and pacing and lapping movements. Spearman correlation coefficients were obtained to examine the relationships between spatial disorientation and relevant variables such as cognitive function and age. They found that repetitive pacing and lapping movements were observed in participants with fronto-temporal dementia. Therefore, routinized pacing is one of the predictors for fronto-temporal dementia and pathological changes in the brain could account for this behavior. However, they pointed out that the major limitation was to measure random movements and quantify pacing or lapping patterns.

In a later work, Kearns et al [68] found that fractal dimension (a measure of randomness in 14 ALF residents' movement paths continuously recorded over 30 days using UWB-RFID sensors) was significantly higher in persons with lower MMSE scores signifying cognitive impairment. The fractal dimension is computed by comparing the measurement of a line or surface at multiple scales. A truly straight line has a fractal dimension of 1, while a line that crosses a plane enough times to completely cover the surface (i.e. follows a Brownian motion) has a fractal dimension of 2.Thus when describing the fractal dimension of a line, 1 <= Fractal Dimension <= 2 [69]. It can also be applied to the 3D analysis of surfaces such that the fractal dimension ranges from 2 to 3. Fractal dimension in Kearns' studies was computed based on the 2D geographical travel paths recorded by Ubisense location systems [98].

In a subsequent study of 28 subjects [70] recorded over one month replicated the findings of Kearns [68] and found that the randomness of the paths taken by elderly ALF residents reliably differentiated PWD from those without the disorder. Most recently Kearns et al. [71] determined that randomness in the paths of 53 ALF residents' walking patterns recorded over a period of 1 year was strongly related to their imminent fall risk, establishing an important relationship between the random movement patterns in older persons and the risk of falling.

Vuong et al. [72] developed an automated algorithm to classify wandering patterns of PWD based on the Martino-Saltzman typology [73]. First, temporal information was used to segment ambulatory data into travel episodes. After that, a heuristic decision tree based algorithm was applied to classify episodes into direct, pacing, lapping, and random patterns. In a recent study [74], they deployed the classification algorithm on Android mobile devices and integrated it with a WiFi-based localization system. The integrated mobile application was able to detect wandering patterns of a person in real-time as he walked around his home environment. The application is also able to send an SMS alert to the attending caregiver once wandering is detected.

3.5.2 Detect Spatial Disorientation

Lin et al [61] proposed a framework that combines RFID, GPS, Global System for Mobile, and Geographic Information System (GIS) to construct a stray prevention system. It provides four monitoring schemes: indoor residence monitoring, outdoor activity area monitoring, emergency rescue, and remote monitoring. The service platform assists in identifying the real-time positions of missing elderly using Internet-connected devices. Chang [63] implemented a mobile location-based networking system for a specific group of subjects: people with mental illness who travel to and from work. The system was created to assist both the subjects and their job coaches. The job coaches work with the subjects to support them in learning new jobs, and more importantly remaining socially connected. The subjects may at times still require assistance to decrease the risk of getting lost. The PDA attached to the subjects reports their current location to a server through GPRS connectivity. The server tracks the subject who deviates from the pre-set route to or from the workplace, and then the matching module locates the job coaches who are within a certain distance from the lost subject. The system also includes a HELP button on the PDA that the user can press to ask for help while travelling. Based on this initial work, Chang and Wang [64] recruited 8 subjects to test the system's ability in identifying the job coaches within a distance from the target user when the subject pressed the HELP button. The experimental results showed that the precision ranges from 89 to 100% and the recall from 94 to 100% for all subjects. Precision is the proportion of relevant job coaches among all those identified by the system (which may include irrelevant ones). Recall is the proportion identified by the system among all relevant job coaches (some of which not identified by the system).

In summary, current eHealth tools and algorithms have been used to capture wandering patterns and spatial disorientation in the WHAT domain. However, these technologies are still at early stage. One challenge lies in the polymorphism of patterns (e.g. lapping in the same or opposite directions) and forms of disorientation (e.g. wanderer cannot locate the dining room or his or her own room). The lack of ground truth and benchmark datasets also makes it difficult to evaluate and validate the algorithms.

3.6 mHealth Applications That Analyze Proximal Factors to Detect Wandering

Several applications address the WHY domain of dementia-related wandering. These applications analyze proximal factors to detect the risk of wandering in PWD.

Opportunity Knocks [75] and iRoute [76] use GPS-enabled phones to learn the individual's standard travel routes in outdoor environments. Patterson, et al. [75] (Opportunity Knocks) have proposed a hierarchical Dynamic Bayesian Network model, an extended Bayesian Networks, which is able to infer common destinations such as frequently used bus stops and parking lots that users visit. It then estimates the transition probability matrices between these destinations, "trip segments and streets" and subsequently calculates the most likely trajectories. To detect abnormal events such as wandering, the approach uses two models with different transition parameters. "The first one assumes the user is behaving according to his personal historical trends and uses the learned hierarchical model for tracking. The second tracker assumes a background model of activities and uses an untrained prior model that accounts for general physical constraints but is not adjusted to the user's past routines. The trackers are run in parallel, and the probability of each model given the observations is calculated. When the user is following his ordinary routine the learned hierarchical model should have a higher probability, but when the user does something unexpected the second model should become more likely" [75].

iRoute [76] has the same objective and approach with Opportunity Knocks. However, their inference system uses the agent approach with the Belief-Desire-Intention (BDI) architecture [77]. The route prediction and matching is based on user preferences and the learned routes information stored in the BDI agent's belief base. The agent tracks the user towards the predicted route or routes until the user reaches one of the predicted destinations. However, users can have more than one predicted route at a certain time of the journey. During the travel, if the user walks off the route, then the agent tries to guide the user (using voice-activated navigation) back to the last predicted route. If the user continues moving in the wrong path, then the agent will notify either the care staff or the relatives.

iWander [78] proposes a Bayesian multi-variant model to detect the subject's probability of wandering. The variables or proximal factors include the person's standard travel routes and others like age, severity of dementia, time of the day, current weather conditions and total time spent outside. They argue that time of day greatly impacts wandering because about half of dementia patients experience night time walking of which a significant number result in the patient getting lost. Dementia patients are outside less during night hours and during bad weather. So, time of the day and weather conditions also affect the probability of wandering. Also, older adults who have more severe dementia and suffer from more advanced stage of dementia are more susceptible to wandering. Analyzing these variables against a collection of existing data can help classify behavior as normal or abnormal (wandering or not wandering). Their learning approach consists of two phases. First, the data is passed through a one class support vector machine. This filters out obvious, normal activities. After that, data is passed to a nonlinear regression function which will classify a behavior as normal or abnormal.

Similar to iWander, LaCaSa [79] also used GPS trajectories and proximal factors known as home location, close-to-home locations, and far-from-home locations to estimate the risk of wandering faced by PWD. Different with iWander, LaCaSa employed a partially observable Markov decision process for reasoning. In addition, the application also activates the appropriate response when wandering occurs, such as prompting the PWD or calling the caregiver.

Another study also tried to detect wandering locomotion including lapping and pacing patterns based on user's GPS trajectory [80]. A data-driven method is proposed to examine and count turning points in an ongoing trajectory. The angular sum of these points serves as the basis to distinguish lapping and pacing movements.

A recent medical study [81] measured the emotion expression of PWD (facial displays, vocalizations, and body movement/posture) as affective factors to predict wandering behavior. They used hierarchical linear modeling and Poisson regressions to explore the relationship of observable emotional expression and wandering. Their results revealed that positive emotional expression was associated with a 1.03 (confident interval: 1.02-1.04) increase in wandering rate, and negative emotional expression was associated with a 0.11 (confident interval: 0.83-0.95) decrease in wandering rate, even after controlling for cognitive function and subject's characteristics such as age, sex, race, and education.

In short, most of the above applications used location trajectories as proximal factors to detect wandering. The principle is daily mobility or travel behavior tends to be repetitive and follows a recurring pattern. Therefore, disoriented wandering can be detected by looking for trajectories that are not usual or optimal. We realize that detecting wandering by using proximal factors is a long-term goal, which requires the combinations of advanced sensing technology, accurate recognition algorithms, and efficient machine learning algorithms.

3.7 Applications to Study Background Factors

These are longitudinal studies which use wandering as a feature or predictor to study background factors such as neurocognitive factors or health status.

Hsiao [82] deployed a wireless sensor network system in the university hospital's elder care center for 8 months. They collected location traces and investigated the daily and long-term mobility of 4 volunteering elders. They found a recurring pattern in each elder's daily mobility. The pattern, however, differed from individual to individual. These suggested that merely quantifying how much the elders move around the facility will not be sufficient for behavioral modeling. Exact location of the elders' presence, rather, is more relevant. They concluded that long-term location tracking, not just the mere quantity of mobility, allows discovery of the moving patterns and in turn makes early detection of the elders' physical or mental problems possible.

A dataset of indoor locations recorded from 6 PWD during a year (about 900 individual daily traces organized in groups of 14 days) was used to assess the state of dementia in a real-world nursing home. By applying k-Nearest Neighbor classifier, the authors [83] were highly accurate (92%) in classifying a patient's state of well-being as positive or negative. The ground truths (positive and negative states) were documented by the attending nurses. Their results once again support the idea of using motion analysis to manage wandering behavior and well-being of PWD.

A clinical study [84] used duration, rate, general health, cognitive impairment, circadian rhythm, age, and sex factors to differentiate three types of wanderers (classic, moderate, and subclinical). They applied principal component analysis and clustering techniques on the multi-dimensional dataset of 142 PWD. They found that duration and rate parameters primarily distinguished the 3 types of wanderers. Cognitive impairment, upper and lower gastrointestinal tracts differed between groups of wanderers and from non-wanderers. Measures of circadian rhythm, age, and sex did not differentiate types.

Briefly, while there are initial efforts to study background factors that affect wandering and vice versa, the long-term effects and significant results from such studies have not been seen or achieved.

4 Discussion

4.1 Issues

Several issues need to be considered when developing mHealth applications and systems for managing dementia-related wandering.

4.1.1 Design Issues and Human Factors

mHealth applications for managing wandering aim to benefit the wanderers, the caregivers, and the clinicians/researchers. Designing mHealth applications for these stakeholders requires special attention to training and support [86].

For wanderers with dementia, due to their cognitive, perceptual, and physical limitations, such applications should not rely on their efforts. The wandering applications should be installed when the person is at the early stages of dementia. A previous study [87] observed that people with mild stage memory impairment were able to use and benefit from the GPS tracking technology to live more independently. They were able to move with greater freedom and security when using GPS systems, and some were also able to take care of the devices by themselves. Persons with later stage memory impairment could not learn to adopt the technology as it was too complex for them.

When deploying or selecting sensors for wanderers with dementia, we need to consider the following factors.

- **Size.** The size of a sensor is important. In general, a smaller size is more acceptable to the user. Small and lightweight sensors are easy to attach to the body, less intrusive, and more convenient to use.
- **Power consumption.** Most sensors are powered by batteries. With lower sensor energy consumption, the battery can last longer – ideally, over the sensor's lifetime. Some sensors can harvest power themselves; however, there may be maintenance problems. Short-range wireless energy submission can be used

also. However, the frequency and difficulty of related maintenance tasks such as charging and cleaning should be minimal.

- **Mobility.** Sensors should be able to communicate with other sensors or control devices while in motion, as they are often attached to a mobile user. There should be minimal physical interference between sensors and movement. Any potential difficulties in attaching a sensor or removing it should be eliminated.
- **Processing capability.** A sensor that performs some simple data processing (e.g. filtering or removing additive noise) before sending useful data to a data collector can help to reduce redundant data transmission.
- **Storage capability.** A sensor may possess a small amount of local storage for critical (or preprocessed) data prior to transmission. This improves robustness when the data communication channel becomes temporarily unavailable or interrupted.

Sensor radiation and compatibility with human body tissue must also be taken into account as longtime exposure to radiation from wireless devices can raise concerns about possible health hazards. Other factors may include clinical approval, durability, and accuracy of measurement, social and fashion concerns. For example, using sensors embedded in a smart phone or wearing sensors that appear as shirt buttons can help avoid unwanted attention and possible stigmatization [45].

Important factors of wandering management applications targeting caregivers, researchers and/or clinicians are interface design and functionality. Interfaces of mHealth applications tailored for caregivers should be kept as simple as possible. Some details such as large button size, quick response of touch screen, not too many steps/options to choose from, easy navigation, and nice use of pictures or metaphors, good use of colors are preferable [88]. In addition, the applications must take least time and minimal effort to learn and familiarize because caregivers already have to juggle many tasks to take care of PWD. Applications for researchers and clinicians should be adaptive to the context and their specific needs, e.g. measuring wandering dimensions, so as to maximize the usability of the applications.

4.1.2 Ethical, Privacy, and Security Issues

PWD and/or their family members must be well-informed of possible consequences of mHealth applications to protect their rights. In particular, there is an ethical and practical concern to the tagged person, i.e. the person's liberty is limited. For moderate and severe dementia patients, electronic tagging arguably satisfies an ethical principle and decreases stigma. Being lost and half dressed in the middle of the night near a dual carriageway is hugely stigmatizing, and electronic tagging may avoid this [89].

Privacy of users must be safeguarded and the involved parties should ensure confidentiality when personal data are collected [90]. Surprisingly, most nursing home residents do not share those views. When the moderator brought up the privacy issue to groups of nursing home residents, it was quickly dismissed as secondary. Residents are more concerned with the weight of tracking device placed on them, i.e. "a device should weigh no more than 113g if attached to an arm or leg, but it could weigh 227g if the device was worn on a belt" [94]. Nursing staff also support electronic monitoring of PWD because it enables them to provide better quality care and encourages PWD to live more independent lives.

While the prospective value of mHealth assistive applications far outweighs privacy concerns [94], systems should be designed with good built-in security features. It has to guard against scenarios that might lead to serious security breaches. For instance, intruders might obtain information about the wanderer's habit of leaving home at a particular hour, and use such information to break into the house, potentially even endangering the person's life. Therefore, the user's security requirements should be determined before the design stage of the applications and systems [95]. All communications should be secure and encrypted. This is especially important in the case of wireless communications which are easier to intercept [96]. Personal monitoring devices should authenticate the identity of the rightful owner using unobtrusive biometrics or possibly key physiological signs to avoid any data tampering (owner-aware devices [45]). Different levels of access authority should be granted to different users and the relevant data required or collected within the suite of mHealth applications is properly managed on a secure and password-protected database [97].

4.2 Economy Discussion

Currently, the cost of location systems and devices for managing wandering is high for personal users. Typically, the total cost of an intelligent safety phone with integrated elopement prevention is 950USD (including installation) [87]. A UWB-RFID system with 4 sensors can cost 7000USD in the USA and 16000USD in Asia markets [98]. Devices to monitor activities such as wearable inertial sensors range from 1300USD to 3500USD per item [99]. Though such cost is high for personal users, it is more affordable if group users share the systems and services together. For example, a dementia care centre purchases a system and charges its clients a fee for using the devices with the service. The cost per person will be reduced significantly. A previous study reported that the typical rental cost of the devices without other services is about 13USD to 65USD per month. These costs are not high in exchange for the following benefits:

- For **patients**, as a diagnostic sign for preclinical dementia, timely recognition of wandering provides an opportunity for early treatment of dementia and delay nursing home entry. This will realize significant cost savings because once transferred into formal care, the annual cost of care for a patient exceeds USD 32,000 in high income countries [12]. In addition, the quality of care for PWD prone to wandering will be improved. In fact, the identification of specific types of wandering helps point to possible causes or reasons for wandering. If he/she changes from a non-wanderer to a wanderer, it may reflect either advancing dementia or some undiagnosed medical event that affects cognition, perhaps causing a delirium.

- For **caregivers**, wandering management technology will help them cut down time spent on surveillance of wandering behavior. Researchers and clinicians who harness the technology are able to garner new knowledge on the antecedents and consequences of risky wandering. This knowledge can critically enable health professionals to individualize interventions and effective approaches tailored to different types of dementia and patients.
- For **societies**, long-term research outcomes will reduce the cost of care and use of public funds for institutional care. If technologies can help a caregiver spend 1 hour less per day to look after a dementia wanderer, the world will annually save USD155 billion in 2014 dollars. To governments and agencies concerned with the well-being of elderly population, particularly those who have dementia, this represents a methodological breakthrough worthy of significant investigation.

5 Case Studies

We use two case studies to illustrate some of the factors influencing wandering behavior and the different types of assistance required by a person with dementia who wanders, and to inform the design of the assistive tool. These differences highlight the need to incorporate background information and the individual and contextual information about the situation. These two case studies are adapted from [79].

5.1 Case Study 1

Choo is a 70-year-old woman who lives in a three-room flat in block 511, Bedok, Singapore. She lives with her daughter, who is 42 and works as a journalist for a local newspaper. Choo has been diagnosed with mild cognitive impairment recently. She has increasing trouble in remembering where she puts things and names of her relatives and friends. One of Choo's daily social activities is to go down to a hawker center below her block to have breakfast and coffee with her friends. She then usually goes to the wet market nearby to buy groceries for lunch and dinner. These are her greatest pleasures because they enable her to spend her free time and to interact with people as her daughter goes to work. At times, she feels disoriented on her way home after morning activities due to memory lapses. She has been brought back home a few times by other friends and neighbors. She feels increasingly inhibited to go out for morning activities because of the social stigma attached to this dependence, and her daughter feels increasingly anxious when she leaves. She has a number of friends in her neighborhood, and regularly frequents a few of the neighboring blocks, from which she can easily get back to her block.

5.2 Case Study 2

Lee worked for 32 years as a secondary school teacher. He retired in large part because he was diagnosed with Alzheimer's disease three years ago. Often Lee becomes anxious and tells his wife that he needs to go to work, and she has increasing difficulty convincing him that he has already retired. On a few occasions she has found him a block or two away from home walking in the direction to his former secondary school and with some effort she manages to convince him to get back home or walks with him until he forgets where he was going. Once, he took the early bus to the school in the morning and arrived there feeling disoriented. Fortunately, a former colleague of his found him and called his wife.

5.3 Case Studies' Discussion

In the first case, Choo is mostly aware that she is lost. She gets disoriented due to lack of attention and poor visual and spatial ability even though she might be in an area she is familiar with. In her case, the anxiety caused by feeling lost can be lessened by providing her with anchors to reality, e.g. reminding her that she was heading home or showing her a close-by known location, such as her neighboring block.

The second case is an instance of direct wandering, as Lee does not feel lost when he starts to walk to his former workplace. On the contrary, his wandering is a manifestation of his need to go to work and he will not be receptive to suggestions to return home or ask for assistance, so automatic prompting might lead to more confusion and anxiety. In this case notifying a caregiver or someone in the vicinity who can provide assistance would be more effective.

These two scenarios also show how independence and safety need to be balanced according to the profiles of PWD and the factors that lead to wandering. Lee, who is at a more advanced stage of dementia has a greater risk of getting lost and might travel larger distances. While he might have unmet needs that cause the wandering, independence is not a major issue in his quality of life as he is already under the constant care and supervision of his wife. In contrast, for Choo, being able to join breakfast with her friends and go marketing gives her satisfaction and contributes to her quality of life. The risks she faces of getting disoriented in the housing estate where she is well known do not justify her refraining from going outdoors on her own.

6 Conclusions and Future Directions

6.1 Conclusions

Table 2 summarizes the reviewed works and highlight what has been addressed and what has not been achieved. Technologies to geo-fence and prevent elopement, track and locate wanderers are mature and stable. The availability of commercial products

Table 2 Summary of reviewed works

Domain	Category of works	What has been addressed	What has not been addressed
WHAT	Tools to measure frequency, repetitiveness, temporal distribution	Measure frequency: rate of wandering episodes per hour [85] Measure temporal distribution: walking time [17,18], wandering time [65], temporal variability [66], duration of wandering episodes [85]	- Automated tools to document and incorporate all dimensions of wandering - Tools or applications to measure repetitiveness: go repeatedly to the same location, travel repeatedly the same route while walking indoors
	Tools to recognize wandering patterns and spatial disorientation	Classify pacing, lapping, random patterns [67,68,70,72,73,74] Recognize an aspect of disorientation: getting lost [61, 63, 64]	-Algorithms to address the variance in temporal, spatial, and geometrical shapes of patterns within the same wanderer, across different wanderers, between wanderers and non-wanderers - Algorithms to address other types of disorientations: wanderers cannot locate dining room/own room/bathroom without help; wanderers bump into obstacles or other people while walking alone or walk about aimlessly
	Tools to prevent elopement	Wanderer runs off, attempts to leave authorized area [39, 46, 63] Wanderer enters unauthorized area [47,48,49,50]	(Mature technologies)
	Negative outcomes	Falls [71]	Fatigue, inadequate food intake, injuries
WHERE WHEN	Track and locate PWD	GPS tracking [51,52,53,54,55, 93] RFID and/or UWB tracking [25,57,58,59]	(Mature technologies)
HOW	Information services	Dementia assessment [62]	Tools that provide intervention cues (environmental modification, behavioral techniques) to instruct caregivers
WHY	Tools that analyze proximal factors to detect wandering	Physical environment, e.g. deviation in location trajectories [75,76,78,79,80] Affective states, e.g. emotional expression [81]	Other factors that could trigger wandering including: - Physiological need states, e.g. hunger, agitation - Affective states, e.g. boredom or loneliness - Physical and social environment, e.g. ambient conditions
	Applications to study background factors	Health status, e.g. well-being [82,83] Socio-demographics, e.g. age, sex [84]	The relationship of wandering and background factors such as: - Neurocognitive factors, e.g. visual-spatial memory - Health status and socio-demographics, e.g. pain, constipation, education, previous jobs

and various tracking applications on mobile platforms is clear evidence. On the other hand, information and communication services that assist caregivers are scarce. Current eHealth tools do not support comprehensive measurement of a variety of wandering dimensions. Typically each of them is built to quantify a single, though possibly different dimension of wandering locomotion. There is still a need to provide additional assistive support and integrate these tools together. Automated technologies to recognize wandering patterns and spatial disorientation, analyze proximal factors, as well as realize negative outcomes and study background factors are limited. These technologies are still at the early stage and not many of them have been validated in large scale or longitudinal studies involving PWD from diverse geographical regions and medical backgrounds. Based on these observations, we would like to suggest future research directions in both the short- and long-term. Due to page limit, we only briefly discuss the issues and problems when developing mHealth applications for managing wandering.

6.2 Short-term Research Work

Assistive Information Services: For example, mHealth applications to provide recommended behavioral and environmental modification techniques for caregivers to target different wandering scenarios. Such applications can also contain videos, images, and scenarios showcasing different wandering behaviors of PWD. Direct caregivers can then learn to recognize wandering and provide immediate interventions. These applications will serve as educational or training tools for caregivers who have limited knowledge of wandering and its associated outcomes.

Measurements of Frequency, Repetitiveness, Temporal Distribution of Wandering: Automated tools are needed to measure these 3 dimensions. So far, there are attempts to automatically measure frequency and temporal distribution but not repetitiveness. Thus, future eHealth applications can consider using locomotion data to identify repeated locations or routes traveled by wanderers. Integrated applications to automatically measure these 3 dimensions are essential to characterize dementia-related wandering.

6.3 Long-term Research Work

Recognize Wandering Patterns and Spatial Disorientation: The core research concept on wandering aims to quantify random movement, which has been a difficult task for human observers to master. Most current algorithms to recognize wandering patterns and spatial disorientation need to improve in reliability, robustness and accuracy. The challenges lie in the polymorphism of patterns and forms of disorientation, their differences in spatial, temporal, and patterns' geometrical shapes from wanderers to wanderers, from wanderers to non-wanderers, and within a wanderer. In addition, the unavailability of ground truth labeled data as well as standard benchmark datasets make it very difficult to evaluate and validate the algorithms. Appropriate visualization tools can also be designed to enable researchers and clinicians to easily view these wandering dimensions of PWD, for example on a computer monitor with an intuitive user interface.

Analyze Proximal Factors to Detect Wandering: In general, such systems and eHealth applications require 3 subsystems. Firstly, context-aware sensing subsystems are needed to recognize relevant proximal factors. Secondly, automated algorithms to continuously monitor and detect changes in proximal factors are required. Thirdly, probabilistic models and regressions are applied to estimate the likelihood of wandering. Combinations of advanced sensing technology, accurate recognition algorithms, and efficient machine learning algorithms are potential to accomplishing this goal.

Realize Negative Outcomes and Study Background Factors: Longitudinal studies can be conducted to achieve these long-term goals. An example of such studies is we can monitor wandering and its progress to early detect deterioration in visual-spatial memory and cognition of PWD. Tools to characterize full spectrum and dimensions of wandering as well as background and proximal factors are instrumental in conducting such studies and developing the technologies. Synergistic collaboration between technologists, gerontologists, caregivers and PWD is necessary to ensure the feasibility of adoption of such technologies in healthcare settings.

References

[1] North American Nursing Diagnosis Association. Nursing diagnoses: Definitions & classification, 4th edn. Nursecom, Inc., Philadelphia

[2] Lai, C.K.Y., Arthur, D.G.: Wandering behavior in people with dementia. Journal of Advanced Nursing (J. AdvNurs) 44, 173–182 (2003)

[3] Cesari, M., et al.: Prevalence and risk factors for falls in an older community-dwelling population. Journal of Gerontology Series A: Biological Sciences and Medical Sciences 57, 722–726 (2002)

[4] Colon-Emeric, C.S., Biggs, D.P., Schenck, A.P., Lyles, K.W.: Risk factors for hip fracture in skilled nursing facilities: Who should be evaluated? Osteoporosis International 14, 484–489 (2003)

[5] Hirst, S., Metcalf, B.: Whys and whats of wandering. Geriatric Nursing 5, 237–238 (1989)

[6] McShane, R., et al.: Getting lost in dementia: A longitudinal study of a behavioral symptom. International Psychogeriatrics 10, 253–260 (1998)

[7] Rowe, M.A., Glover, J.C.: Antecedents, descriptions, and consequences of wandering in cognitively-impaired adults and the safe return (SR) program. American Journal of Alzheimer's Disease and other Dementias 16, 344–352 (2001)

[8] Rodriguez, J.: Resident falls and elopement: cost and controls. Nursing Homes 4, 16–17 (1993)

[9] Nygaard, H.: Consumption of psychotropic drugs in elderly persons registered at the home nursing center in Bergen, Norway. International Journal of Geriatric Psychiatry 7, 53–58 (1992)

[10] Siders, C., et al.: Evidence for implementing nonpharmacological interventions for wandering. Rehabilitation Nursing 29, 195–206 (2004)

[11] Wimo, A., Prince, M.: World Alzheimer Report 2010: The global economic impact of dementia. Alzheimer's Disease International (2010)

[12] Wimo, A., Winblad, B., Jonsson, L.: The worldwide societal costs of dementia: estimates for 2009. Alzheimer's & Dementia 6, 98–103 (2010)

[13] Ramakers, I.H.G.B., et al.: Symptoms of preclinical dementia in general practice up to five years before dementia diagnosis. Dementia & Geriatric Cognitive Disorde 24, 300–306 (2007)

[14] Hope, T., et al.: Wandering in dementia: A longitudinal study. International Psychogeriatrics 13, 137–147 (2001)

[15] Verghese, J., et al.: Abnormality of gait as a predictor of non-Alzheimer's dementia. New England Journal of Medicine 347, 1761–1768 (2002)

[16] Scherder, E., et al.: Gait in ageing and associated dementias; its relationship with cognition. Neurosci. Biobehav. Rev. 31, 485–497 (2007)

[17] Hayes, T.L., Pavel, M., Kaye, J.A.: An unobtrusive in-home monitoring system for detection of key motor changes preceding cognitive decline. In: 26th IEEE Conference on Engineering in Medicine and Biology, vol. 4, pp. 2480–2483 (2004)

[18] Pavel, M., Hayes, T.L., Adami, A., Jimison, H., Kaye, J.: Unobtrusive assessment of mobility. In: 28th IEEE Conference on Engineering in Medicine and Biology, vol. 1, pp. 6277–6280 (2006)

[19] Verghese, J., Holtzer, R., Lipton, R.B., Wang, C.: Quantitative gait markers and incident fall risk in older adults. Journals of Gerontology Series A: Biological and Medical Sciences 64A, 896–901 (2009)

[20] Giladi, N., Herman, T., Reider-Groswasser, I.I., Gurevich, T., Hausdorff, J.: Clinical characteristics of elderly patients with a cautious gait of unknown origin. J. Neurol. 252, 300–306 (2005)

[21] Algase, D.L., Beel-Bates, C.B., Beattie, E.R.A.: Wandering in long-term care. Annals of Long-Term Care: Clinical Care and Aging 11, 33–39 (2003)

[22] EmFinders, Emfinder, `http://www.emfinders.com/`

[23] Aetrex, GPS Shoe, `http://www.aetrex.com/aetrex-gps-shoe/` (2011)

[24] Lifeline 4000+, `http://www.tunstall.co.uk/Our-products/Telecare-solutions/`

[25] Kearns, W.D., Algase, D., Moore, D.H., Ahmed, S.: Ultra wideband radio: A novel method for measuring wandering in persons with dementia. Gerontechnology 7(1), 48–57 (2008)

[26] Algase, D.L., Moore, D.H., Gavin-Dreschnack, D., VandeWeerd, C.: Wandering definitions and terms. In: Nelson, A., Algase, D. (eds.) Evidence-Based Protocols for Managing Wandering Behaviors, pp. 3–18. Springer, New York (2007)

[27] Algase, D.L., Yao, L., Beel-Bates, C.A., Song, J.: Theoretical models of wandering. In: Nelson, A., Algase, D. (eds.) Evidence-Based Protocols for Managing Wandering Behaviors, pp. 19–52. Springer, New York (2007)

[28] Algase, D.L.: Assessment of wandering behaviors. In: Nelson, A., Algase, D. (eds.) Evidence-Based Protocols for Managing Wandering Behaviors, pp. 75–102. Springer, New York (2007)

[29] Schonfeld, L., Brown, L.M., Molinari, V.A., Algase, D.L.: Behavior management of wandering behavior: staff training issues. In: Nelson, A., Algase, D. (eds.) Evidence-Based Protocols for Managing Wandering Behaviors, pp. 195–214. Springer, New York (2007)

[30] Castle, S.C., Rutledge, M.K.: Pharmacological interventions associated with wandering. In: Nelson, A., Algase, D. (eds.) Evidence-Based Protocols for Managing Wandering Behaviors, pp. 215–250. Springer, New York (2007)

[31] Connell, B.R., Calkins, M.P.: Environmental design. In: Evidence-Based Protocols for Managing Wandering Behaviors, pp. 299–336. Springer, New York (2007)

[32] Moore, D.H., McKenzie, B.: Dementia-related wandering: management interventions. University of South Florida Geriatric Education Center – Summer Institute (June 2007)
[33] North American Nursing Diagnosis Association (NANDA). Nursing diagnosis: definitions and classification, pp. 206–207. Author, Philadelphia (2001-2002)
[34] Beattie, E.R.A., Struble, L.M.: Evidence-based practice protocols for wandering. In: Nelson, A., Algase, D. (eds.) Evidence-Based Protocols for Managing Wandering Behaviors, pp. 337–352. Springer, New York (2007)
[35] Department of Veterans Affairs, Veterans Health Administration. Managementof wandering and missing patient events. VHA DIRECTIVE 2002-013 (March 4, 2002)
[36] Algase, D.L., Moore, H.D., Vandeweerd, C., Gavin-Dreschnack, D.J.: Mapping the maze of terms and definitions in dementia-related wandering. Aging Ment. Health 11, 686–698 (2007)
[37] Hodgkinson, B., Koch, S., Nay, R., Lewis, M.: Managing the wandering behavior of people living in a residential aged care facility. Int. J. Evid. Based Healthc. 5, 406–436 (2007)
[38] Stokes, G.: Wandering – Common problems with the elderly confused. In: Holden, U.P. (ed.), p. 35. Winslow Press, Oxon England (1988)
[39] Moore, D.H., Algase, D.L., Powell-Cope, G., Applegarth, S., Beattie, E.R.A.: A framework for managing wandering and preventing elopement. American Journal of Alzheimer's Disease & Other Dementia 24(3), 208–219 (2009)
[40] Kearns, W.D., Fozard, J.L.: Technologies to manage wandering. In: Nelson, A., Algase, D. (eds.) Evidence-Based Protocols for Managing Wandering Behaviors, pp. 277–298. Springer, New York (2007)
[41] Miskelly, F.G.: Assistive technology in elderly care. Age and Ageing 30, 455–458 (2001)
[42] Meiland, F., et al.: COGKNOW: Development of an ICT device to support people with dementia. The Journal on Info. Tech. in Healthcare 5(5), 324–334 (2007)
[43] Bharucha, A.J., et al.: Intelligent assistive technology applications to dementia care: current challenges, limitations, and future challenges. Am. J. Geriat. Psychiat. 17(2), 84–104 (2009)
[44] Phua, C., et al.: State-of-the-Art assistive technology for people with dementia. In: Handbook of Research on Ambient Intelligent & Smart Environments: Trends & Perspectives, pp. 300–319 (2011)
[45] Rashidi, P., Mihailidis, A.: A survey on ambient assisted living tools for older adults. IEEE Transactions on Information Technology in Biomedicine 17(3), 579–590 (2012)
[46] Kearns, W.D., Rosenberg, D., West, L., Applegarth, S.: Attitudes and expectations of technologies to manage wandering behavior in persons with dementia. Gerontechnology 6(2), 89–101 (2007)
[47] Chan, M., Bonhomme, S., Esteve, D., Campo, E.: Individual movement trajectories in smart homes. In: ICBME, pp. 1014–1018 (2008)
[48] Taub, D.M., et al.: The Escort system: a safety monitor for people living with Alzheimer's disease. IEEE Pervasive Computing 10(2), 68–77 (2011)
[49] Hanser, F., Gruenerbl, A., Rodegast, C., Lukowicz, P.: Design and real life deployment of a pervasive monitoring system for dementia patients. In: 2nd International Conference on Pervasive Computing Technologies for Healthcare (2008)
[50] Rowe, M., Lane, S., Phipps, C.: CareWatch: a home monitoring system for use in homes of persons with cognitive impairment. Top Geriatr. Rehabil. 23, 3–8 (2007)

[51] Shimizu, K., Kawamura, K., Yamamoto, K.: Location system for dementia wandering. In: 22nd IEEE EMBS, pp. 1556–1559 (2000)
[52] Ogawa, H., Yonezawa, Y., Maki, H., Sato, H., Caldwell, W.M.: A mobile phone-based safety support system for wandering elderly persons. In: 26th IEEE EMBS, pp. 3316–3317 (2004)
[53] Chang, Y.J., Liu, H.H., Wang, T.Y.: Mobile social networks as quality of life technology for people with severe mental illness. IEEE Wireless Com. 16(3), 34–40 (2009)
[54] Miskelly, F.: A novel system of electronic tagging in patients with dementia and wandering. Age Aging 33, 304–306 (2004)
[55] Miskelly, F.: Electronic tracking of patients with dementia and wandering using mobile phone technology. Age Aging 34(5), 497–499 (2005)
[56] Ashida, N., et al.: A monitoring of wandering behavioral pattern of senile dementia with IC tag. In: 3rd APT Telemedicine Conference, pp. 80–84 (2005)
[57] Greiner, C., Makimoto, K., Suzuki, M., Yamakawa, M., Ashida, N.: Feasibility study of the integrated circuit tag monitoring system for dementia residents in Japan. Am. J. Alzheimer's Dis. 22(2), 129–136 (2007)
[58] Makimoto, K., Yamakawa, M., Ashida, N., Kang, Y., Shin, K.R.: Japan-Korea joint project on monitoring people with dementia. In: World Congress on the Internet and Medicine (2006)
[59] Almudevar, A., Leibovici, A., Tentler, A.: Home monitoring using wearable radio frequency transmitters. Artif. Intell. Med. 42, 109–120 (2008)
[60] Kearns, W.D., Fozard, J.L.: Ultra-wideband radio assessment of locomotion in elderly. Gerontechnology 7(2), 138 (2008)
[61] Lin, C.C., et al.: A wireless healthcare service system for elderly with dementia. IEEE Trans. Info. Tech. Biomed. 10(4), 696–704 (2006)
[62] Lin, C.C., et al.: A healthcare integration system for disease assessment and safety monitoring of dementia patients. IEEE Trans. Info. Tech. Biomed. 12(5), 579–586 (2008)
[63] Chen, D., Bharucha, A.J., Wactlar, H.D.: Intelligent video monitoring to improve safety of older persons. In: 29th EMBS Conference, New York, USA, pp. 3814–3817 (2007)
[64] Chang, Y., Wang, T.: Mobile location-based social networking in supported employment for people with cognitive impairments. Cybernetics and Systems 41, 245–261 (2010)
[65] Agase, D.L., Beattie, E.R.A., Leitsch, S.A., Beel-Bates, C.A.: Biomechanical activity devices to index wandering behaviour in dementia. Am. J. Alzheimer's Dis. and Other Dementias 18(2), 85–92 (2003)
[66] Makimoto, K., et al.: Temporal patterns of movements in institutionalized elderly with dementia during 12 consecutive days of observation in Seoul, Korea. Am. J. Alzheimer's Dis. 23(2), 203–206 (2008)
[67] Nakaoka, A., et al.: Pacing and lapping movements among institutionalized patients with dementia. Am. J. Alzheimer's Dis. 25(2), 167–172 (2010)
[68] Kearns, W.D., Nams, V.O., Fozard, J.L.: Tortuosity in movement paths is related to cognitive impairment. Wireless fractal estimation in assisted living facility residents. Methods of Information in Medicine 49, 592–598 (2010)
[69] Craighead, J.D.: Using fractal dimension to assess robot operator skill in a search task. Journal of Intelligent Robot. System 64, 97–118 (2011)

[70] Kearns, W.D., Fozard, J.L., Nams, V.O., Craighead, J.D.: Wireless telesurveillance system for detecting dementia. Gerontechnology 10, 90–102 (2011)

[71] Kearns, W.D., et al.: Path tortuosity in everyday movements of elderly persons increases fall prediction beyond knowledge of fall history, medication use, and standardized gait and balance assessments. Journal of the American Medical Directors Association 13(7) (2012)

[72] Vuong, N.K., Chan, S., Lau, C.T., Lau, K.M.: Feasibility study of a real-time wandering detection algorithm for dementia patients. ACM MobileHealth, 23–26 (May 2011)

[73] Martino-Saltzman, D., Blasch, B.B., Morris, R.D., McNeal, L.W.: Travel behavior of nursing home residents perceived as wanderers and nonwanderers. Gerontologist 31, 666–672 (1991)

[74] Vuong, N.K., Goh, S.G.A., Chan, S., Lau, C.T.: A Mobile-Health Application to Detect Wandering Patterns of Elderly People in Home Environment. In: IEEE International Conference in Medicine and Biology Society (2013)

[75] Patterson, D.J., et al.: Opportunity knocks: A system to provide cognitive assistance with transportation services. In: Mynatt, E.D., Siio, I. (eds.) UbiComp 2004. LNCS, vol. 3205, pp. 433–450. Springer, Heidelberg (2004)

[76] Hossain, S., Hallenborg, K., Demazeau, Y.: iRoute: Cognitive support for independent living using BDI agent deliberation. In: Corchado, J.M., Pérez, J.B., Hallenborg, K., Golinska, P., Corchuelo, R. (eds.) Trends in Practical Applications of Agents and Multiagent Systems. AISC, vol. 90, pp. 41–50. Springer, Heidelberg (2011)

[77] Bratman, M.: Intention, plans, and practical reason. Harvard Univ Press, Cambridge (1987)

[78] Sposaro, F., Danielson, J., Tyson, G.: iWander: an Android application for dementia patients. In: IEEE EMBC, pp. 3875–3878 (2010)

[79] Hoey, J., Yang, X., Grzes, M., Navarro, R., Favela, J.: Modeling and learning for LaCasa, the location and context-aware safety assistant. In: Lake Tahoe, N.V. (ed.) NIPS 2012 Workshop on Machine Learning Approaches to Mobile Context Awareness, Lake Tahoe, NV (2012)

[80] Lin, Q., Zhang, D., Huang, X., Ni, H., Zhou, X.: Detecting wandering behavior based on GPS traces for elders with dementia. In: Proc. of the 12th International Conference on Control, Automation, Robotics and Vision (2012)

[81] Lee, K.H., Algase, D.L., McConnell, E.S.: Relationship between observable emotional expression and wandering behavior of people with dementia. Int. J. Geriatr. Psychiatry (2013)

[82] Hsiao, C.C.: Towards long-term mobility tracking in NTU hospital's elder care center. In: PerCom Workshops, pp. 649–654 (2011)

[83] Grünerbl, A., Bahle, G., Lukowicz, P., Hanser, F.: Using Indoor Location to Assess the State of Dementia Patients: Results and Experience Report from a Long Term, Real World Study. In: Proceedings of the 7th International Conference on Intelligent Environmen (2011)

[84] Algase, D.L., Antonakos, C., Beattie, E.R.A., Beel-Bates, C.A., Yao, L.: Empirical derivation and validation of a wandering typology. J. Am. Geriatr. Soc. 57, 2037–2045 (2009)

[85] Algase, D.L., Antonakos, C.L., Beattie, E., Beel-Bates, C.A., Yao, L.: New parameters for daytime wandering. Res. Gerontol. Nurs. 2(1), 58–68 (2009)

[86] Heart, T., Kalderon, E.: Older adults: are they ready to adopt health-related ICT? Medical Informatics, 1–23 (2011)

[87] Perala, S., Makela, K., Salmenabo, A., Latvala, R.: Technology for elderly with memory impairment and wandering risk. e-Health Telecommunication Systems and Networks 2, 13–22 (2013)
[88] Armstrong, N., Nugent, C., Moore, G., Finlay, D.: Using smartphones to address the needs of persons with Alzheimer's disease. Ann. Telecommun. 65, 485–495 (2010)
[89] Hughes, J.C.: Electronic tagging of people with dementia who wander: Ethical considerations are possibly more important than practical benefits. BMJ 325(7369), 847–848 (2002)
[90] Jara, J., Zamora, M.A., Skarmeta, A.F.: An mHealth platform based on Android OS and 6LoWPAN to support ubiquitous access and data collection. Journal of eHealth Technology and Application 8(1), 1–8 (2010)
[91] Angiullo, L.M.: Wandering behavior in people with dementia. PhD thesis, University of Massachusetts (1997)
[92] Profiling the the dementia family carer in Singapore, `http://www.alz.org.sg/sites/alz.org.sg/files/Profiling%20the%20Dementia%20Family%20Carer%20in%20Singapore%20-%20Executive%20Summary.pdf`
[93] Digital Angel, Inc., `http://www.digitalangel.com`
[94] Kearns, W.D., Rosenberg, D., West, L., Applegarth, S.: Attitudes and expectations of technologies to manage wandering behavior in persons with dementia. Gerontechnology 6(2), 89–101 (2007)
[95] Silva, B.M.C., Rodrigues, J.J.P.C., Lopes, I.M.C., Machado, T.M.F., Zhou, L.: A Novel Cooperation Strategy for Mobile Health Applications. IEEE Journal on Selected Areas in Communications, Special Issue on Emerging Technologies in Communications - eHealth 31(9), 28–36 (2013)
[96] Jara, J., Zamora, M.A., Skarmeta, A.F.: Interconnection framework for mHealth and remote monitoring based on the Internet of Things. IEEE Journal on Selected Areas in Communications, Special Issue on Emerging Technologies in Communications - eHealth 31(9), 47–65 (2013)
[97] Silva, B.M., Rodrigues, J.J.P.C., Canelo, F., Lopes, I., Zhou, L.: A Data Encryption Solution for Mobile Health Applications in Cooperation Environments: DE4MHA. Journal of Medical Internet Research 15(4) (2013)
[98] Ubisense, Inc., `http://www.ubisense.net`
[99] APDM Movement Monitoring Solutions, Inc., `http://www.apdm.com`

Questions

Q1. What is wandering, both clinically and scientifically? List the dimensions of wandering in each definition.

Q2. Suppose that we would like to implement an mHealth location-based networking system using GPS-enabled phones (e.g. iPhones) with cellular-telephone data connections (3G or 4G) for people with mild cognitive impairment (MCI) who travel home from a day care center. The system is created to assist both caregivers and people with MCI. People with MCI are able to travel home independently. However, they may at times require assistance to decrease the risks of getting lost that may arise. Caregivers and family members need to be assured that the people with MCI travel and reach home safely. To accomplish these goals, we suggest to design a client-server architecture. The GPS-enabled phone reports the current location of the person with MCI to a server as he/she travels home. The server tracks the person who deviates from the pre-set route to his/her house, and then the matching module tries to locate the caregivers or family members who are within a certain distance from the lost person so as to offer immediate assistance.

Draw the system architecture of the described mHealth application. Focus on the four main components: the client devices, the wireless access network, the Internet and its hosts, and the server side that accommodates application-specific servers (e.g. repository, matching logic, etc.) that are required in the system.

Q3. In question 2, when the person with MCI calls for help, the matching logic needs to search for a list of k nearest caregivers or family members. Suppose that the true location of the person with MCI at the moment he/she calls for help is represented by a two dimensional vector $x(t)$. However, the true location $x(t)$ is not known due to location error produced by the GPS-enabled devices. Suppose the nearest k caregivers/family members are located within a circle of radius r around $x(t)$. We need to find the constraint on location accuracy such that if the location error is within that constraint, the set of k nearest caregivers/family members can be correctly determined. If the true location is known, determining the location error is straightforward. However, this accuracy requirement can be determined using only the estimated location $\hat{x}(t)$, without knowing the true $x(t)$.

Given the estimated location $\hat{x}(t)$, prove that the maximum location ror, $e_r(t)$, in order to ensure that the k searched caregivers/family members are indeed the k nearest caregivers/family members to the person with MCI, is constrained by:

$$e_r(t) \leq \max\{(r'_{k+1} - r'_k)/2, \Delta(t)\}$$

where r'_n is the radius of the smallest circle centered at $\hat{x}(t)$ enclosing the nearest n caregivers/family members, and $\Delta(t)$ is a lower bound on the best error that may be requested.

Q4. There are a number of factors that need to be considered when selecting sensors for tagging and monitoring PWD who wander so as to increase the user's compliance. List and discuss these desirable factors.

Q5. A tri-axis accelerometer is used to stream a wanderer's movements. Calculate the min-max bandwidth requirements given that the sampling rate ranges from 25-100 Hz.

3. Ubiquitous mHealth Approach for Biofeedback Monitoring with Falls Detection Techniques and Falls Prevention Methodologies

Edgar T. Horta, Ivo C. Lopes, and Joel J.P.C. Rodrigues*

Instituto de Telecomunicações, University of Beira Interior, Covilhã, Portugal
{edgar.horta,ivo.lopes}@it.ubi.pt, joeljr@ieee.org

Abstract. A mobile health application solution with biofeedback based on body sensors is very useful to perform a data collection for diagnosis in patients whose clinical conditions are not favorable. This system allows comfort, mobility, and efficiency in all the process of data collection providing more confidence and operability. Falls occurrence in senior people represents a high risk for their health and life. Those falls can cause fractures or injuries causing great dependence and debilitation to the elderly and even death in extreme cases. Falls can be detected by the accelerometer included in most of the available mobile devices. To reverse this tendency, it can be obtained more accurate data for patients monitoring from body sensors attached to the human body (such as, electrocardiogram, electromyography, blood volume pressure, electro dermal activity, or galvanic skin response). Then, this chapter reviews the related literature about this topic and introduces a mobile solution for falls prevention and detection, and biofeedback monitoring. The proposed system collects sensed data that is sent to a smartphone or tablet through Bluetooth. Mobile devices are used to process and display information graphically to users. The falls prevention system uses collected data from sensors in order to control and advice the patient (user) or even to give instructions to treat an abnormal condition to reduce the falls risk. In cases of symptoms even detect a possible disease. The signal processing algorithms plays a key role in the fall prevention system. In real time, these algorithms process the collected biofeedback data in order to extract relevant information from the signals and thereby warn the patient. Monitoring and processing data from sensors is realized by a smartphone that will send warnings to the user. All the process is performed in real time. The proposed system is evaluated, demonstrated, and validated through a prototype and it is ready for use.

Keywords: Falls Detection, Falls prevention, Mobile Health, Biofeedback, Health Monitoring, Body Sensor Networks, Mobile Healthcare Application.

* Corresponding author.

S. Adibi (ed.), *Mobile Health*,
Springer Series in Bio-/Neuroinformatics 5, DOI: 10.1007/978-3-319-12817-7_3

1 Introduction

Advances in technologies for healthcare along with the increasing of elderly people around the world, new challenges arise for better quality of life focusing on this senior people. It is observed that elderly people live longer with the best quality of life as possible. The continuous biofeedback monitoring has encouraged the development of technology associated with sensors and systems for physiological signal acquisition, increasingly miniaturized, and maintaining a high quality of signal, good enough for the information extraction with the minimum discomfort.

Mobile applications offer great contributions to ambient assisted living (AAL) solutions, improved with the use of electronic body sensors networks (BSNs). These sensors can be placed in a patient body and communicate with mobile devices (such as, smartphones or tablets) through short-range communications, like Bluetooth or larger-range offered by Wireless and Zigbee [1]. The emergence of mobile health (mHealth) is triggered with the greater use of tablets, smartphones, and integration of sensors in mobile devices. The information and communication technologies for the next-generation patient physiological mobile solutions offer excellent conditions to create innovative solutions and techniques for remote and real-time people monitoring with mobility support [2]. These systems can collect sensed data from patient in real time. Mobile devices can store and process incoming data and also send them to storage systems in real time, including the Internet [3], [4]. Furthermore, the patient can also have information that can be useful to him/her, allowing a detailed and on time information that can be extremely relevant for diagnosis and treatment adjustments for people from healthcare systems (physicians, nurses, and paramedics). The historical record of physiological sensed data is very useful for diagnosing further problems as the causes of falls and habits that may increase the patient risk. Thus, trying to increase patients' independence, offering a good AAL solution with mobility support, the falls detection system combined with a biofeedback monitoring, data processing, and presentation solution, using body sensors is essential for development in health systems technology area. Through a detailed analysis of the conjuncture of these physical detailed signals it is possible to control chronic diseases, cardiovascular diseases (e.g. hypertension) and others to identify risk factors normally associated with severe pathologies. With this proposal, a falls detection solution and a biofeedback monitoring activity, in real time, with mobility support was embedded on a mobile device, such as a smartphone or a tablet, including a complement of body sensors and the accelerometer already included in the mobile device. In a perfect scenario it would be possible to predict when a fall is going to occur and try to avoid it. Based on this fact, the proposed solution tries to detect situations where the risk of a fall shall be eminent and then warm the patient in order to prevent it. The patient receives a notice and can try to avoid a fall [5]. The warning messages can be sent through a smartphone or tablet. Biofeedback sensors were used by users in order to send raw data through a Bluetooth connection to a smartphone or tablet in order to measure the patient physiological parameters. This BSN can obtain more accurate data from patients monitoring signals (such as, electrocardiogram, electromyography, blood volume pressure, electro dermal activity, and galvanic skin response). This real time collected data from sensors is handled through an automatic algorithm, which is a critical piece for falls prevention [6]. For falls detection, a kinematic sensor accelerometer included in most of the available mobile devices collects data that is processed through an algorithm providing this benefit. This mobile health tool monitors the human body using a real time communication through Bluetooth with a mobile device. A device (smartphone or tablet) uses its computational power to perform a fall detection using an algorithm that also analyzes physiological signals received at the smartphone [7]. If an abnormal behavior is detected the patient is notified immediately. Furthermore, the patient caretaker and/or physician can also be notified, if needed. With the purpose of distributing information, the sensors send data to a mobile gateway. Subsequently, this gateway sends the received data to a Web server and it allows subsequent queries.

This mHealth advanced system aims to be fully focused on patients. In this sense, a methodology for falls prevention is included considering biofeedback monitoring, data processing, and presentation using data collected by body sensors. This solution is embedded on a mobile device and tries to perform user biofeedback activity, falls detection, and it will broadcast to the patient distress signals to prevent fall occurrences.

The rest of the chapter is organized as follows. Section 2 elaborates on the related work about mHealth technologies and biofeedback monitoring. The system architecture of proposed ubiquitous mHealth solution for biofeedback monitoring with falls detection and prevention is presented in Section 3 while the mobile application is described on Section 4. The performance evaluation and demonstration of the proposed system is addressed on Section 5 and the conclusion and suggestions for future work are included on Section 6.

2 Related Work

The mHealth technologies can improve the efficiency, quality, and reduce cost of healthcare services. Mobile computing technologies features have the potential to transform healthcare. Mobile health technology enables caretakers to remotely monitor patient's health and also allows individuals to manage their own health more easily and with privacy. These technologies and applications also provide patients' medical information and more easy access to electronic health records [8].

As mobile health applications evolve, more sophisticated and advanced health applications emerge, raising concerns about mobility, facility to access anytime and anywhere, and sometimes in extreme conditions. To answer AAL needs focused on elderly people affected by isolation and displacement incapacities, mobile health is presented as an excellent solution and with a high potential. This people often underestimates diseases, such as hypertension and oversights, that may cause several damage if not properly monitored for prevention and treatment awareness [9]. Through mobile health solutions, data can be collected from sensors to patient's mobile devices (e.g., smartphones or tablets) and sent to healthcare providers, enabling a remote and fast data analysis, avoiding patient's displacement. Mobile technology has being registered a progressive evolution due to the growing ubiquity of those devices in contrast to desktop computing.

Currently, the mobile devices available on the market (mainly, smartphones and tablets) already have a higher processing and storage capacity but still have many limitations. For instance, the spotlight Samsung S4 and the iPhone 5 already have multi-cores processor and high storage capacity. Due to mobile devices proliferation [10] several health applications have been developed and available to the public through online markets. They offer the possibility of monitoring users own health state, by creating and maintaining their own health records, treatment alerts, falls detection, health goals, and many others. All these types of applications are present in the most popular mobile platforms, such as Google Android OS and iOS.

The communication of patient data between different mHealth systems, caretakers and other specialists is carried to eHealth records. It enables faster and more convenient access. An eHealth record uses a structured representation of health information in an electronic format. A typical system is based on a well-structured and organized archive regarding the patient medical history in an electronic digital form [11]. This methodology presents several advantages regarding health information structure, namely the capacity to improve health provider's efficiency, reducing costs or even increasing treatments effectiveness. Moreover, it allows medical staff or patients to easily access that information in an electronic or digital way providing to a medical staff enhanced abilities to make improved decisions about possible health treatments or analysis to the patients' health information [12].

The eHealth data exchange based on mobile applications are currently gaining popularity due to the fact that users access their health records through the Web, not being constrained to spatial and temporal barriers, although it is necessary a reliable Internet connection in order to support these services. Mobile eHealth records systems rely on the utilization of Web services that enable the access to data from any personal computer, tablet, or smartphone with Internet connection.

The challenges and limitations on mHealth systems using eHealth are presented, namely, the lack of standards regarding the standardization to link telemedicine services, due to difficult operational compatibilities between mobile devices and telecommunication services. Moreover, it is stated that due to systems overly complex and with privacy problems, user acceptance tends to grow up slowly [13].

Several approaches based on tools and techniques for falls detection can be found in the literature. Generally, they are mainly concerned with health of elderly and incapacitated people, using techniques based on accelerometers movement data collection and falls detection algorithms. Considering the contributions available in [14], the differences between different approaches come from the fact that one uses a smartphone and the other is based on an embedded accelerometer, GPS receiver, and microcontroller sensor solution with Zigbee technology transceiver. Many solutions of monitoring biofeedback are also reported but all of them have in common an explanation of the importance of data acquisition in healthcare systems for surveillance of patient's physiological signals. A more static and no adaptive system showing a biofeedback monitor to handle a specific type of health problem is presented in [15]. A relatively large amount of scientific work has been performed on this topic and there is a considerable convergence among all of them, mentioning that physiological data is vital for the treatment, monitoring, and prevention of various diseases. Many of these works are quite recent as these systems should be computationally powerful, as to collect and process large amounts of data in a short time interval.

The capture and processing biofeedback signals with sensors to a smartphone became possible and increasingly used due to the development and generalized use of these devices for personal use. Several approaches for falls prevention, detection, and biofeedback monitoring may be found in the related literature, using different techniques, tools, and devices. Falls detection is a topic with abundant information where algorithms use sophisticated methods to detect falls, but it is also closely related to describe motion and movements by the patient body. This detection is performed by kinematic sensors, such as accelerometers and gyroscopes included in the most of mobile devices available in the market. In [16], a quantitative falls risk position using the kinematic sensor was presented to identify an association with different types of falls risk. Here, a wearable falls detection system with use of acceleration techniques and a falls detection algorithm is proposed. Both are mainly concerned with the health of elderly and incapacitated people. In order to develop a fall prevention system, a collection of fall characteristics to determinate and train fall events that use strategies with existing resources were reviewed [17]. A detailed methodology for electrocardiogram (ECG) sensor was included in [18]. It uses ECG beat detection and treats their data waves to heart beat recognition, R-wave detection procedure or peak detection to make diagnoses of heart diseases and prevent possible problems. An electromyography (EMG) analysis was performed to investigate the changes that occur in the parameters of the patient's electrical signals by increased risk of falling and balance disorders [19]. The galvanic skin response (GSR) was a simple method to detect arousal and analyze physical stimulus in a patient to thereby check whether patient has or not imminent risk of falling.

There was an extensive range of healthcare applications with biofeedback monitoring from critical monitoring of physiology scenarios of terminal illness up to the own patient make diagnoses by himself. The proposed solution gathers contributions from the related work available in the literature. All these systems and techniques above described offered important contributions to the solution proposed in this work.

3 System Architecture

This section presents the system architecture based on a body sensors solution and a falls detection and prevention system. Figure 1 illustrates the fall prevention and detection scenario to explain the proposed system architecture with main actions defined for communications process with the mobile application and the healthcare system.

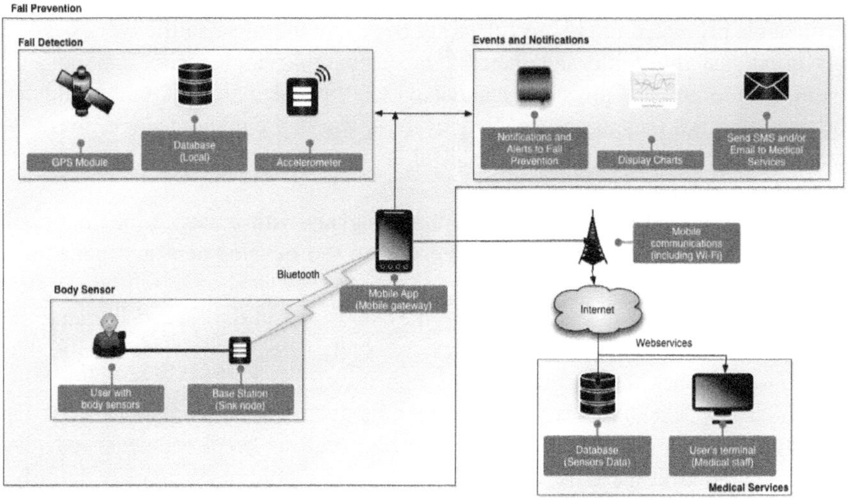

Fig. 1 Illustration of the System Architecture with all the features and interactions

The proposed system architecture considers the following four modules: the falls detection, the body sensors, medical services, and fall prevention extended system with events and notifications. The falls detection module is responsible for detecting the accelerations caused by the user daily movements and interprets the values in order to detect user falls and even patient motion recognition. In this module, the mobile gateway also receives data directly from the GPS navigation system to provide user location. The body sensors modules are responsible for collecting body parameters from the sensors placed along the user body in order to obtain the vital parameters and extra information to perform a falls detection system. The base station (sink node) sends the body sensors data through Bluetooth to a mobile device that also acts as a mobile gateway. Events and notifications module is the main responsible for the falls prevention system, this advises the user, in real time, for health problems detected. In the Web services module, all the collected data is sent to a remote database through the hypertext transfer protocol (HTTP) protocol using REST Web services. These data are also available for being consulted by user or medical staff through a browser.

The accomplishment of the system architecture begins with the user needs as well as how it fits in a real context. The mobile gateway is the key point of the architecture that depends on selected technologies that are leading and proven. It provides a system asset, user interface, and important aspects in processing and sharing information securely and effectively. It also offers a flexible integration solution for new sensors, if needed.

4 Mobile Solution

This section focuses on the system demonstration and validation in a real scenario. The mHealth solution user's interface is simple and appropriated for elderly people. The application uses large buttons, intuitive menus, and elucidative images as required for this kind of applications taking into account the users' profile. The system controls the patient physiological data, even running in background and alerting the user whenever necessary. Figure 2 presents the main page of the application where the user accesses to all the available features. Figure 3 presents a proposed questionnaire only displayed at the first time the application is used in a given mobile device.

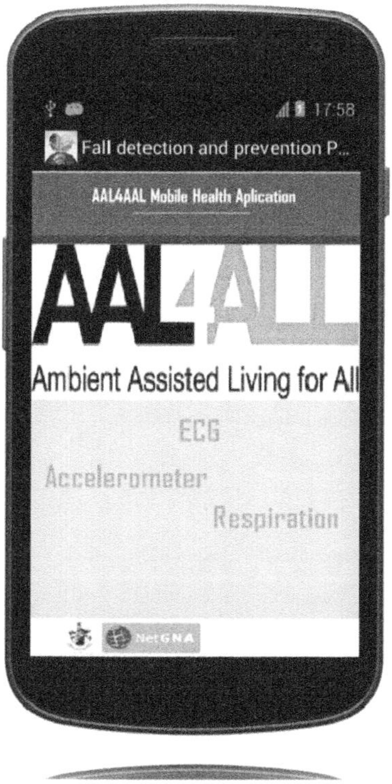

Fig. 2 Splash screen presentation

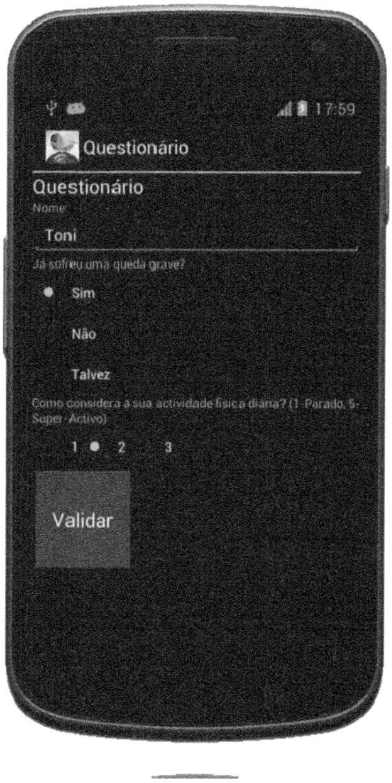

Fig. 3 First use questionnaire

4.1 Falls Detection Techniques with External Alerts

The falls detection and patient motion recognition system works by measuring a user body acceleration and position, getting acceleration values from accelerometer and then transforms the values in *G* force. If a fall is detected an automatic alert with several information about patient is sent to the caretaker or emergency.

4.1.1 Falls Detection and External Alerts

The falls detection process was deployed using paradigms of multi-threading in order to make possible the simultaneous combination of several tasks and several sensors readings. In Figure 4, the activity diagram of falls detection system, which represents the sequence of the application and its data flow is presented.

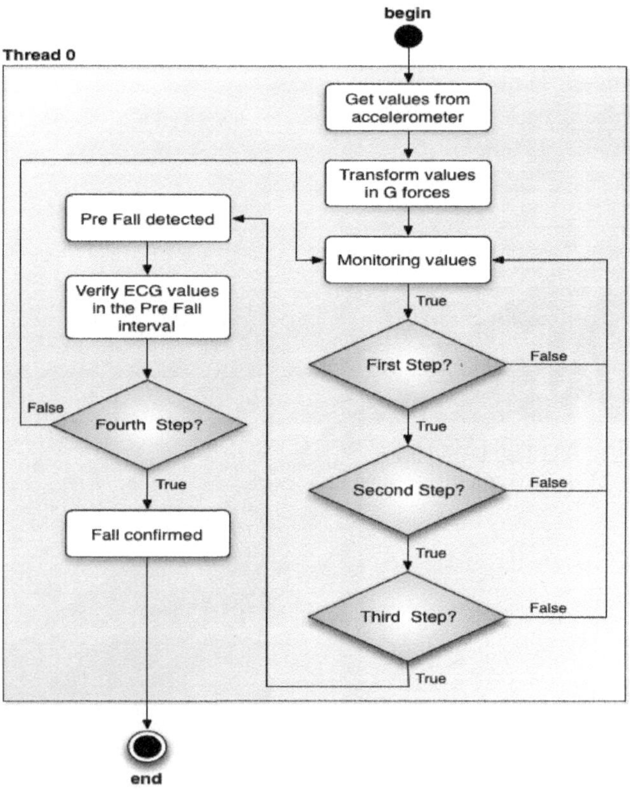

Fig. 4 Activity diagram of falls detection activity diagram

Figure 4 presents the workflow of the first thread (called *Thread 0*). The process starts from getting the accelerations values from the accelerometer where Gx, Gy, Gz, are the values of axes x, y, and z received from the accelerometer and then calculates the G force, using the equation 1. The thread will always be active and monitoring the user activity. In the first step, if G force is lower than 0.5, the process advance to a second step. In the second step, if G force is great than 2.5, it will go to the third step. In the third step, if the first and second step presents a flag as *True*, it will start counting a 10 seconds timer, if in the time interval some movement occurs, it is declared as false alarm. In the other case, if there is no activity and the verifying data from sensors (e.g. the ECG values) are too weak comparing with normal values using a threshold value or have irregularities, it is possible the user is unconscious or immobilized and need help. Then, the system will trigger the fall alert and the time interval (in milliseconds) and data from sensors are saved on the database to compare the values in the same moment of time.

$$G = \frac{\sqrt{Gx^2 \times Gy^2 \times Gz^2}}{9{,}8} \tag{1}$$

Figure 5 shows the values captured by the accelerometer in a time interval where a fall was detected and confirmed. Figure 6 shows a notification on the user mobile device highlighted with a red circle and the current location of the user mobile device is captured to send an alert to caretakers or emergency response center already registered.

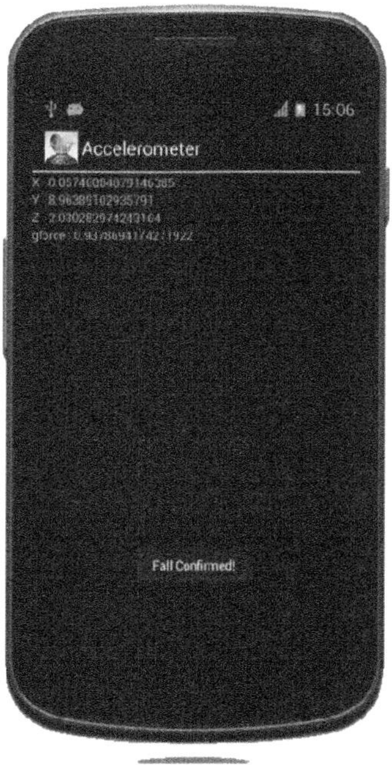

Fig. 5 Accelerometer values and toast a message fall confirmed

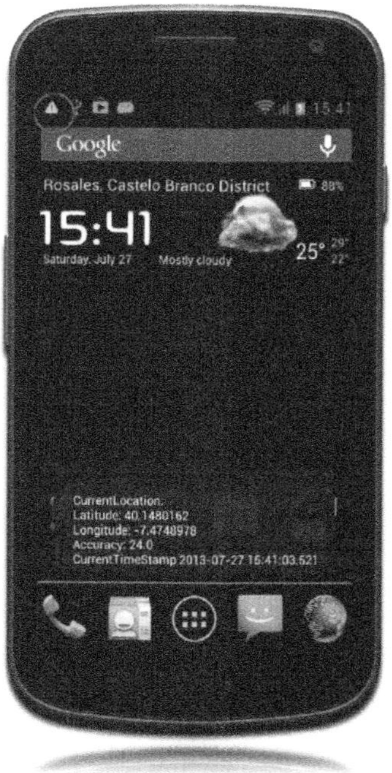

Fig. 6 User falls detection notification and current location to an external alert

Through the magnitude of the G force at a given moment and inactivity immediately after, considering the position of the device, a fall may be detected. All the collected information is stored in a local database and sent trough a Web service to the server. Nevertheless, if a false positive falling is detected and the user does not need help, the user will be able to open the notifications bar to take some actions. Next, Figure 7 shows the notification actions that can be performed at the tools bar of a patient Android mobile device.

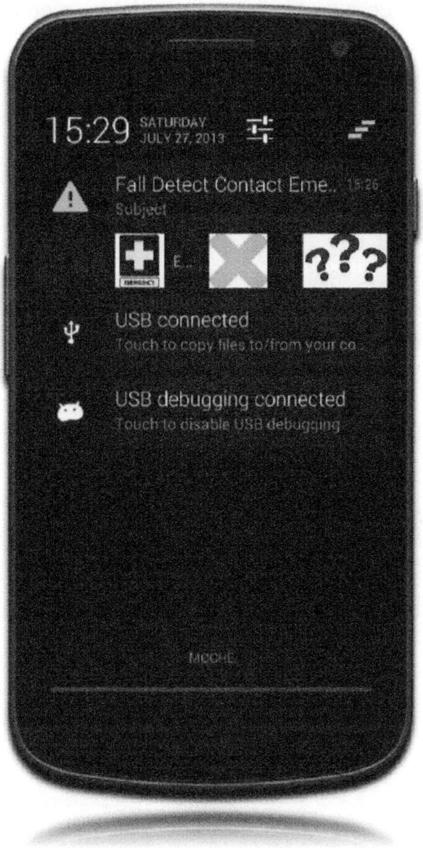

Fig. 7 Illustration of the Android tools bar falls detection notification

When the system detects a fall it automatically contacts the emergency response center or a caretaker sending alerts as an SMS or an email with the current location of patient and information about the detected problem. However, the patient may ignore and cancel the current alerts in the *green cross* button. In the *emergency* button it will contact immediately with an emergency call and in the *question mark* it is possible to read some advices.

4.1.2 Patient Motion Recognition

To improve falls detection and prevention techniques, artificial intelligence and machine learning contributions have been applied in order to detect on time a patient motion specific type. This motion detection can be an important factor in the evaluation of daily patient activity to make possible a better and quicker diagnosis. For the development of this methodology, the values of the following attributes are used: acceleration, time, and its respective motion class. The following motion classes are chosen because they are the most relevant in balance problems. Then, the considered possible classes are the following: Walking, Running, Sitting, Jumping, and Falling. So, through a decision tree algorithm presented in Figure 8, that uses the value of the acceleration and a timer, it is possible to detect in an exact time interval the corresponding motion class belongs.

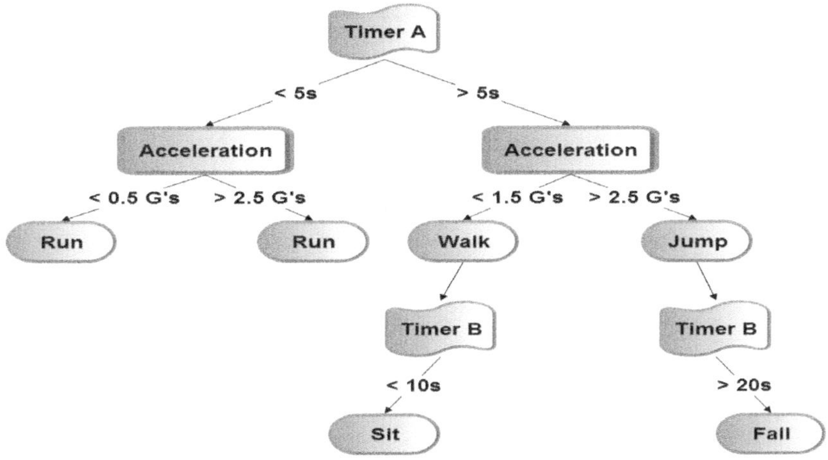

Fig. 8 Illustration of decision tree to patient motion detection

Values presented in the decision tree are examined and studied with the aid of a machine-learning tool to extract knowledge calling WEKA (Waikato Environment for Knowledge Analysis). In the tree, if the *timer A* is over a time interval, greater than 5 seconds it contains acceleration greater than 2.5 G's and if the the *timer B* counts a value greater than 20 seconds a fall is detected considering the right side of the tree, or this time instant is classified as Jump motion. Next, to confirm a motion, the *timer A* and *B* restart again with whole process.

4.2 Biofeedback Monitoring Toll

Monitoring physiological activity is a non-invasive body control technique based on the theory of electrical conductivity. The biofeedback application software presented in this chapter represents an important component for problems screening and therapy developing because it is designed for body responses monitoring, progress data compilation and reporting. For data processing and information visualization from sensors, new threads are created for the others activities, with the remaining sensors and visualization, launched in simultaneous and synchronized. Figure 9 illustrates this approach considering 2 threads, one considering Bluetooth connection, receiving, storing, and processing data (Thread 1) and other showing charts (Thread 2).

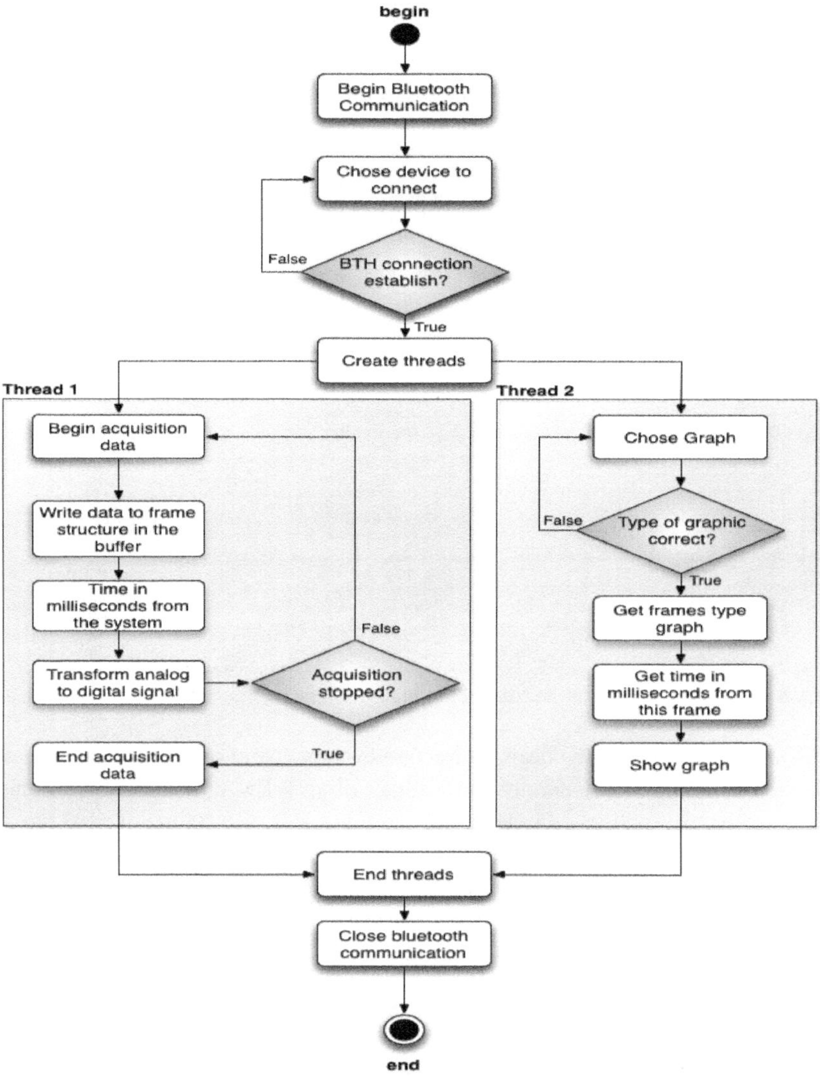

Fig. 9 Activity diagram illustrating two parallel threads, one considering Bluetooth connection, receiving, storing, and processing data (Thread 1) and other showing charts (Thread 2)

The biofeedback status is directly related to certain health issues. The biofeedback-monitoring tool is designed for working with specific monitoring equipment. This software was developed to work with available professional biofeedback tools that include most of the needed components to conduct a biofeedback session. During a biofeedback session, a patient wears sensor devices that are placed on the skin. The sensors are plugged to a sink node (also known as a base station) and through Bluetooth connection it is synchronized with a mobile device (smartphone or tablet) for data acquisition. The application features include chart-graphing options, multiple window options to display more than one body response at the same time, and storage databases to track and review progress. It is possible to control the output signal and the frequency in each pitch to do a sample frequency, in Hertz, and a final digital signal for analyses, in Volts. It also features audio notifications to users and alerts by sending a short message service (SMS) or an electronic mail (eMail) to a caretaker or an emergency response center.

In Figure 10, a smartphone presents an ECG signal and a pre-fall is detected by the accelerometer algorithm shown in the tablet computer, displaying the variation of the G force signal, emerging a warning to the user and an important value to the system when a fall is detected by the algorithm.

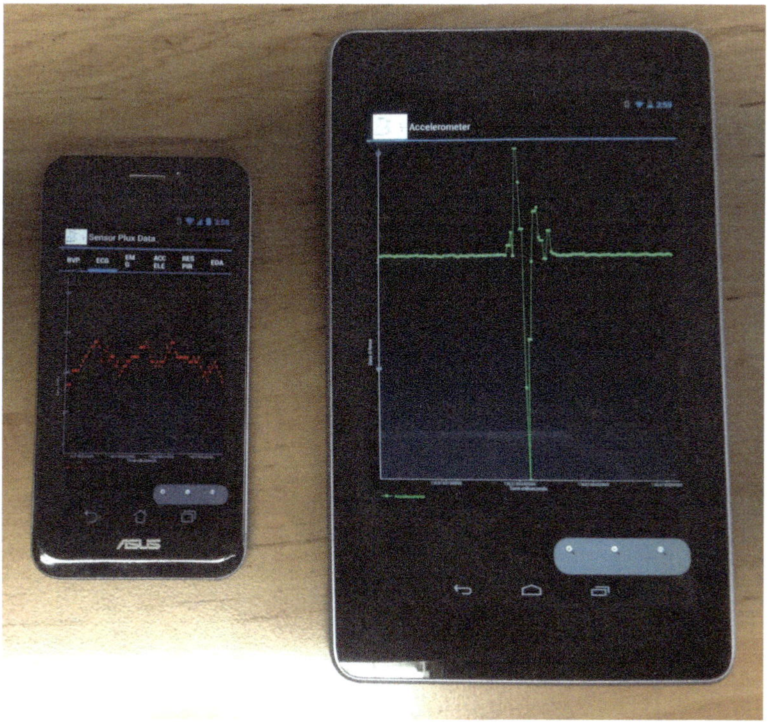

Fig. 10 Photo of a real deployment of the proposed solution with the ECG and accelerometer charts during a fall experiment in a real scenario

When a fall occurs the values from the sensors are sent and saved on the database. The importance of this time interval is to detect some happenstances, before and after, to identify anomalies in order to be analysed in more detail by specialists, who detect important scenes in the a event to discover and describe the fall occurrence. In this real scenario only the vital signs are used to detect whether values have changed to the normal rhythm of the user. The distance and the variability of the ECG waves with a threshold value were experimented returning a Boolean value whether the ECG has right signs or not. When a fall is confirmed and the patient has anomalies in vital signs the system will trigger an alert.

4.2.1 Arrangement of the Sensors in Patient Body

A good arrangement of body sensors in the patient's body is essential for proper readings. In Figure 11.A and 11.C the sensors are placed on the proper location and the base station (Figure 11.B) is synchronized with a smartphone for data acquisition.

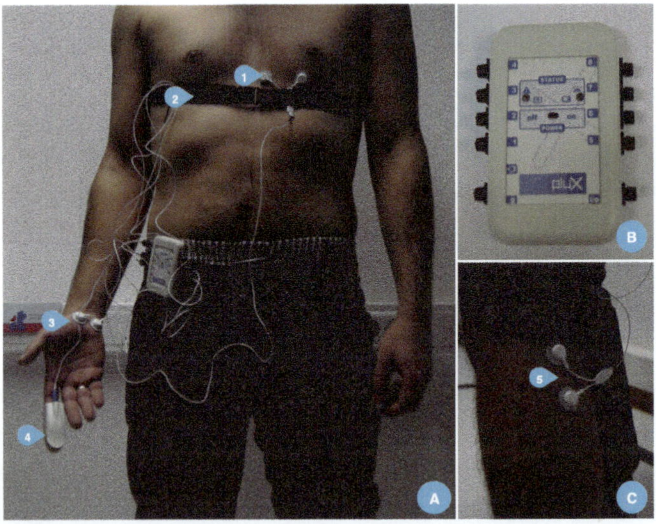

Fig. 11 Photos illustrating the body sensors placement in a real experiment: A and C - sensors placed in a human body; B - base station (sink node)

The sensors were placed on the patient body accordingly as may be seen in Figure 11.A and 11.C. The ECG (1) and Respiration (2) sensors were placed on the chest (heart area), the EDA (3) was positioned on the palm of the hand and the BVP (4) was placed on the fingertips while the EMG (5) was placed on the tibialis anterior and gastrocnemius muscles on the leg.

Modeling a BS in the human body to be prepared for a fall is not particularly easy. Sensor readings must be performed in ideal conditions of data acquisition or it can become unviable. The flexibility and robustness of the hardware used are taken into account to allow the system reliability.

4.3 Falls Prevention Methodology with Notification Alerts

Many biological processes can be measured from a person. Physiological signals measurement can offer important information about body behaviour in a given time. For biofeedback monitoring, the use of physiological and inertial sensors is proposed. They will allow a better detection of problems that can be associated with falls and possible diseases.

4.3.1 Monitoring Techniques Based on Body Sensors

With biofeedback monitoring it is possible to detect common types of physiological anomalies related to balance problems or more serious diseases. Physiological balance is related to those health problems. Through the sensors readings it was possible to detect problems as cardiac disorders using electrocardiogram, muscle activity with electromyography, and anxiety and flutter with galvanic skin response or electro dermal activity. Furthermore, through a tri-axial accelerometer, it is possible to monitor and classify movements to detect anomalous patterns preceding a crisis.

The proposed system uses the capabilities of the accelerometer to identify situations of potential fall and identify motion to improve and alert whether a patient has a bad postural control or perform fast movements, which endanger their equilibrium.

Electro cardiology monitoring is used to detect different states of problems with ECG, as illustrated in Figure 12. For that, it was used an automatic R crests detection algorithm that is based on EMD (empirical mode decomposition) and an adaptive thresholds technique in order to reduce the possibility of errors and bad interpretation of the measured signal.

The R peaks were identified by computing the relative maximum of each complete cardiac cycle where the PQRST phenomenon occurs. To detect heartbeat problems, the system count the number of P crest per minute in order to detect if heartbeat is normal, fast, or slow. To detect irregular heartbeat possible arrhythmia, the time between the P crest was measured (in milliseconds). The warning sign, red notification is a type of danger sign that indicates a hazard ahead on the health that may not be readily apparent to a patient. This notification is the most danger that a patient can have. Upon receiving this notification warning, patient should urgently consult a health care professional.

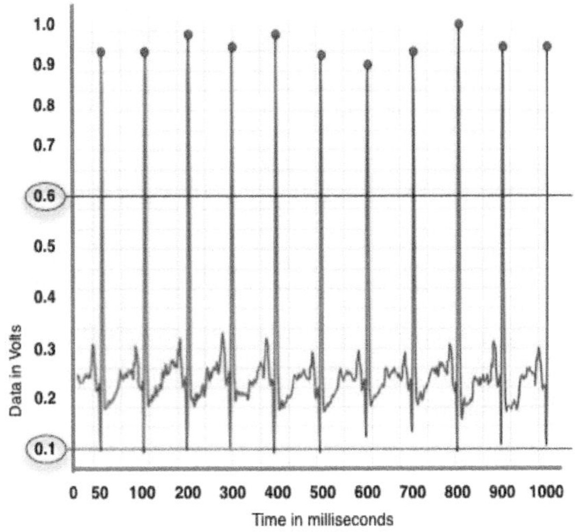

Fig. 12 Illustration of an ECG analytical signal

Galvanic skin response (GSR), also known as the electro dermal activity (EDA), is a method for measuring the electrical conductance of the skin in a patient body, which changes with its moisture level. This is interesting because the sweat glands are controlled by the sympathetic nervous system that controls emotions. Figure 13 shows a sample GSR or EDA signal. A patient can present multiple states of stress and it is possible to be measured by the wavelengths variance.

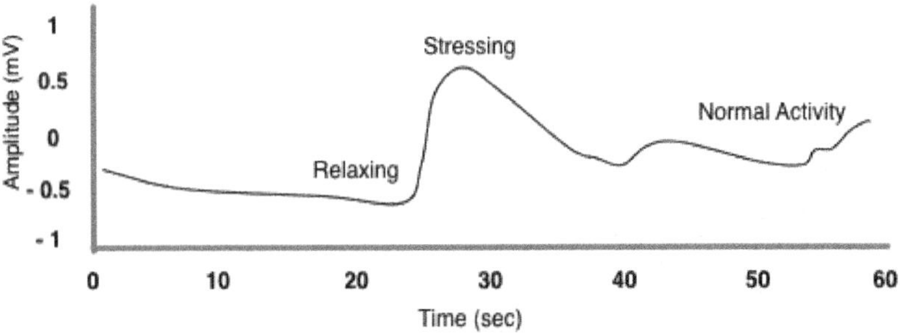

Fig. 13 Illustration of the GSR or EDA variation for diagnoses multiple states of stress

Electromyography is a technique to evaluate the recording of electrical activity produced by skeletal muscles cells when electrically or neurologically activated. The electromyography is used to diagnose neuromuscular disorders and the control muscular movements. Figure 14 illustrates the detection of balance disorders. The changes observed in EMG signals are caused by modifications of the electrical impedance of the skin-muscle system.

The different EMG wavelengths represent standards of the possible variations observed in a patient. Figure 14 presents a healthy person with a diagram showing an abnormal condition neuropathy and a diagram with myopathy, which refers to a critical condition. Skin conductance measurement is a component of polygraph devices and it is used in scientific research to capture emotional or physiological arousal. The skin electrical conductance of a person is directly correlated to his/her emotional state and it is used to detect excitement events. Therefore, the excited state of a person can be observed and it can detect psychological situations less favourable to the welfare and patient health, and when he/she may be more unprotected for fall events. The falls prevention system with biofeedback uses skin conductance to measure and present an individual stress response in real time. It aims to help the patient to control their anxiety and prevent a possible fall.

Figure 15 presents a smartphone showing the signal from an EMG sensor with tendency in a pre-fall occurrence. The pre-fall incident is marked with the red circle at the 1362609124000 millisecond of time. When it occurs the chart goes to its maximum and, immediately after, it registers one low period of values. Through it, it is possible to make diagnose of neuromuscular diseases and also muscular disorders given a poor balance in muscle legs that may cause falls.

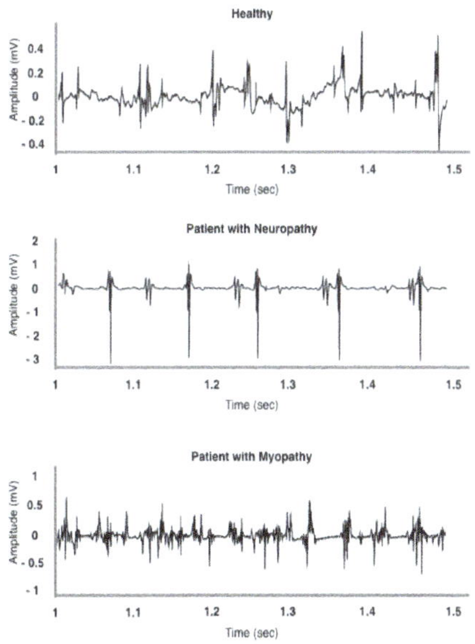

Fig. 14 EMG wavelengths variations

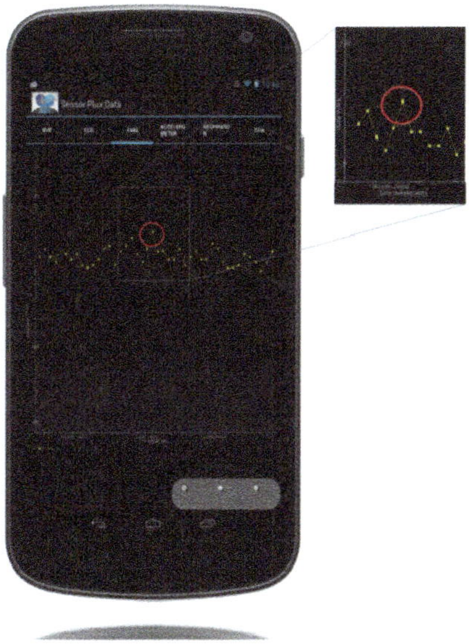

Fig. 15 The measured EMG signal

The warning signal, showing a yellow notification, is a type of danger signal that indicates balance and muscular problems interrelated with falls that may not be readily apparent for a patient. With this notification the patient should remain calm. This is the second most danger notification the patient can have. Upon receiving this warning notification, patient should try to fix the emotional or physiological arousal problems.

The activity diagram shown in Figure 16 presents the sequence of the application and its data flow. The health and movement sensors modules are launched in a cooperative state and the Android activities are handled synchronously.

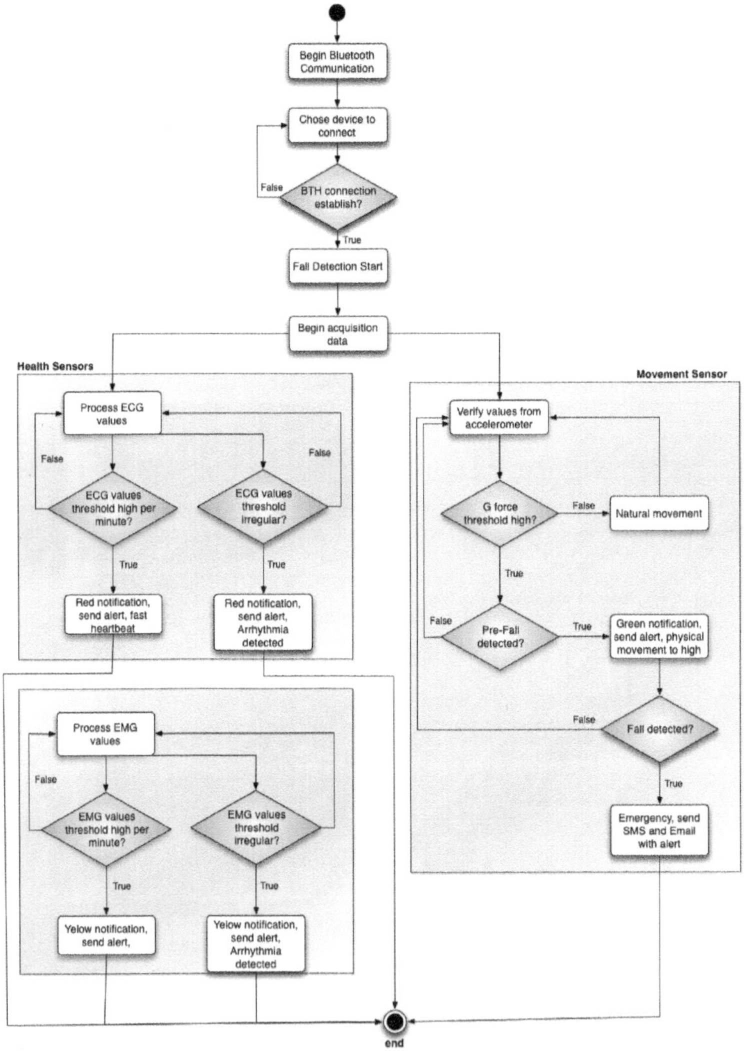

Fig. 16 Activity diagram of falls prevention with notification alerts

By the predefined threshold values and the timers, it is possible to detect real life disturbances (in real time) using body sensors (electrocardiology, electromyography, galvanic skin response, and electro dermal activity). With Android notifications it is possible to alert users for disorders, which endanger their wellbeing. A notification is displayed for each type of sensor according to the severity of the detected problem. It shows different suggestions according to the detected issue. Then, the user becomes aware and tries to solve the situation by himself/herself to prevent a further even worse situation.

4.3.2 Notification Alerts

The signal of health sensors allows the falls prevention algorithm to analyze a typical phenomenon in patients distinguishing variations as previously indicated in the falls prevention methodology. In combination with other signals (e.g. between heart rate and respiration signals), transient classification changes allow a more detailed diagnosis known as electro cardio respiration to detect only specific types of pathologies.

The notification alerts for falls prevention is the main feature to warn patients whether something is wrong with their health. As previously described, through the data collected by ECG, EMG, GSR, or EDA, respiration, accelerometer, and gyroscope, it becomes possible to identify patterns that can expose the patient to a possible fall. A different kind of warnings seen by the patient changes depending on the severity of the detected problem, as shown in Figure 17.

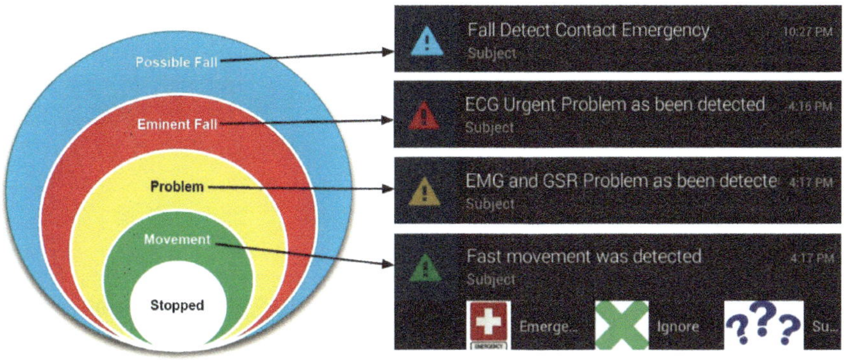

Fig. 17 Different types of event notification warnings with corresponding description associated to biofeedback sensors

The colors shown in Figure 17 represent the warning severity. When it is "Stopped" it means the patient is standing and defines the threshold values, which delimits the sequence of warnings. The patient must know these warnings in order to correct deficiencies for helping himself/herself to prevent a fall and avoid him/her entering in a danger situation.

5 Performance Evaluation and Validation

To evaluate the performance of the proposed solution several trials of experiments were performed in order to estimate the percentage of success (accuracy) of the different experimental approaches. Furthermore, a system validation and results are discussed for both approaches, concluding with an overall evaluation of the proposed solutions.

5.1 Falls Detection and ECG Experimental Analyses

The mobile application was deployed on several different devices in order to validate the proposed solution, and also in order to turn possible essays for volunteers. The application ran effortlessly on all the used devices. The experiments were approached based on several fall simulations with 3 volunteers, each one with 38 trials, in a total of 114 experiments. The mobile devices, even smartphones or tablets, obtained results for one experiment with the sensors monitoring several bio-signals during a fall experiment.

To evaluate the solution two trials were experimented to find the significant differences and improvements between them, as shown in Table 1. In order to reach a final average to the two solutions, several measurements were archived.

Table 1 Control table with comparison between Pre Fall and Pre Fall + ECG experimental results

Action	Number of experiments	Experiments with errors	Right experiments	Average
Pre Fall	114	10	104	91%
Pre Fall + ECG	114	6	108	95%

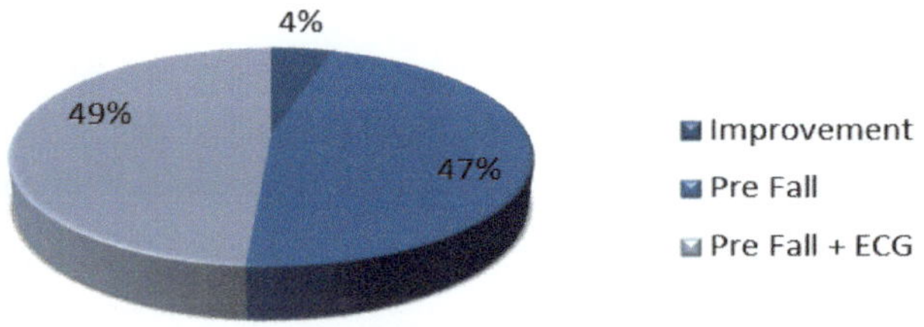

Fig. 18 Proportional area chart of the solution improvement between Pre Fall and Pre Fall + ECG

The results are also shown in Figure 18. As may be seen, the estimation of only pre-fall is about 91% and the solution with pre-fall and ECG increases about 4%. Results show a better falls detection, as expected. It is important to note that this second solution Pre Fall + ECG, brings improvements in real catastrophic scenarios, such as when the fall was detected and the user is unconscious or suffered from a cardiovascular disease or heart attack.

5.2 Falls Prevention Experimental Analyses

The presented results aim to analyze quantitatively the applicability of the system in a real scenario. Therefore, in Table 2, the data was acquired from three healthy individuals in different motions to perform the experiment and the alerts for fall prevention were considered. A total of 50 experiments were performed under the same conditions for each action. The total values line represents a sum of alert values for each motion.

Table 2 Experimental Results for falls prevention

Alerts	Number of Warnings during an Action				
	Walking (5m)	Running (10m)	Sitting	Jumping	Falling
Movement (Green)	28	32	14	46	43
Problem (Yellow)	6	21	4	41	27
Eminent Fall (Red)	0	45	0	38	36
Possible Fall (Blue)	0	2	1	1	48
Total	**34**	**100**	**19**	**125**	**154**

Table 3 Falls prevention expected Values of the experimental results

Expected Values					
Movement (Green)	30	35	15	50	50
Problem (Yellow)	8	25	5	45	30
Eminent Fall (Red)	0	45	0	45	40
Possible Fall (Blue)	0	5	0	2	50

Next, Table 3 presents the expected values and shows an estimated amount of alerts, which is considered correct for each type of motion to allow the calculation of the average.

Evaluating the result and the effectiveness of the proposed algorithm, to estimate the average value it is necessary for determining how to improve it. Figure 19 shows the average value of 50 experiments for each type of motion. It was identified for movements, which require more physical activity (e.g. Run and Jump) that the accuracy is lower. The final average evaluation of the system for all the types of motions is about 86.8% for 250 trials.

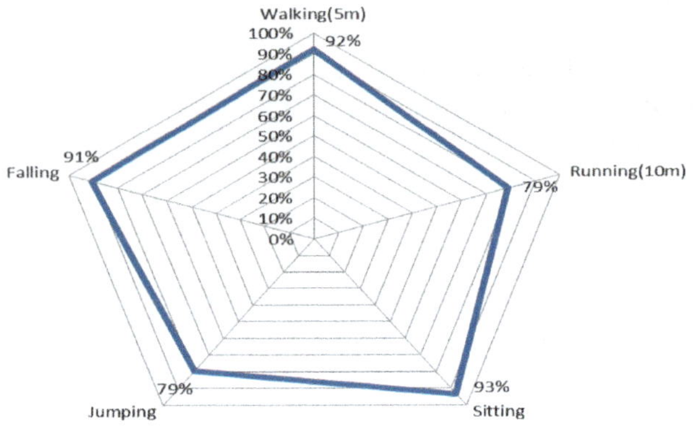

Fig. 19 Performance comparison of average values for each action (walking, running, sitting, jumping, and Falling)

6 Conclusion and Future Work

This section presents a synthesis of this chapter along with the main achievements that result from this work. It also points some directions for further work. The proposed system is prepared for being computationally powerful, performing data collection and processing a large amount of collected information. In a matter of efficiency the battery lifetime for smartphone and base-station is also considered. The system is responsible for receiving data from sensors in real time and sends them to a smartphone or tablet for processing and archiving. It should serve as a support platform for medical staff (or others) for monitoring or to identify an incident, such as a stroke, falls and other events that endanger patients' lives. The application also offers visual tools through the device screen for monitoring the related graphics using the data received from sensors that can be created and observed in real time. This tool is based on four modules as shown in the system architecture, including falls detection, data processing and visualization of information from sensors, and communication through Web services with user's terminals (Medical staff), and an external database storage so data can be accessed by medical staff, available from outside.

The increased use of mobile devices is growing exponentially. Currently, these devices offer a wide range of features that were previously only found on personal computers. Being more compact, mobile devices allow an improved ease to use and mobility, making it possible to access to the newest tools and applications like the one proposed in this chapter. However, mobile devices have the flexibility to become a better tool, to aid in everyday life tasks and much more (e.g. a medical monitoring assistive tool).

Biofeedback monitoring with falls detection is a solution that enrich the knowledge of bio-signals before, after, and during a fall process. Another attained objective of this system was the identification of fall characteristics collection for fall prevention strategies through biofeedback.

Unfortunately the technology cannot detect all the falls. Furthermore, trying to avoid a fall, it is better for the users' health and a solution to increase the accuracy rate of the mobile solution. It was shown that not all the falls are preventable but many of them can be avoided with this system. The proposed solution was evaluated and demonstrated with a real deployment and experiments, and it is ready for use.

The greatest difficulty in the developed system was the interdisciplinary character of the solution. The proposed system was efficient and reliable during rehearsals showing that it can be applied in real word presenting advantages over available solutions, especially in AAL scenarios where the continuous monitoring is a key issue. This new approach demonstrated that biofeedback with falls detection could be a useful instrument for more detailed diagnoses in order to identify the cause of a fall and to improve patient safety.

The performance evaluation and comparison with available solutions are difficult to perform but are necessary to uphold the claim that biofeedback data can be used as a method to derive the level of usability in patients. The, for further works, the following research suggestions are proposed:

- The development and update of the APPs and APIs for other mobile platforms, such as, iOS from Apple and Windows Phone.
- Other biosensors may be included on the solution. For instance, the data collected by an electroencephalography (EEG), beyond ECG, EMG, EDA, BVP, Respiration, and accelerometer can be used to accomplish the falls prevention and detection.

- The mobile devices (either smartphones or tables) may be used to collect data embedding health sensors and may become a specific device for a body health sensing smartphone.
- A performance assessment study through simulation considering different danger scenarios and evaluate the scalability of the proposed solution, may also be considered.

Acknowledgements. This work has been partially supported by the *Instituto de Telecomunicações*, Next Generation Networks and Applications Group (NetGNA), by National Funding from the FCT – *Fundação para a Ciência e Tecnologia* through the Pest-OE/EEI/LA0008/2013, and by the AAL4ALL (Ambient Assisted Living for All), project co-financed by the European Community Fund FEDER through COMPETE – *Programa Operacional Factores de Competitividade.*

References

[1] Figueiredo, C.P., Gama, O.S., Pereira, C.M., Mendes, P.M., Silva, S., Domingues, K., Hoffman, K.P.: Autonomy Suitability of Wireless Modules for Ambient Assisted Living Applications: WiFi, Zigbee, and Proprietary Devices. In: 2010 Fourth International Conference on Sensor Technologies and Applications (SENSORCOMM), July 18-25, pp. 169–172 (2010)

[2] Mulvenna, M., Carswell, W., McCullagh, P., Augusto, J.C.: Visualization of data for ambient assisted living services. IEEE Communications Magazine 49(1), 110–117 (2011)

[3] Dunkel, J., Bruns, R., Stipkovic, S.: Event-based smartphone sensor processing for ambient assisted living. In: 2013 IEEE Eleventh International Symposium Autonomous Decentralized Systems (ISADS), March 6-8, pp. 1–6 (2013)

[4] Rodrigues, J.J.P.C., Pereira, O.R.E., Neves, P.A.C.S.: Biofeedback Data Visualization for Body Sensor Networks. Journal of Network and Computer Applications 34(1), 151–158 (2011), ISSN: 1084-8045

[5] Shiozawa, N., Arai, S., Okada, S., Makikawa, M.: Gait analysis of sit-to-walk motion by using portable acceleration monitor device for fall prevention. In: 2012 IEEE International Conference on Biomedical and Health Informatics (IEEE BHI 2012), Hong Kong, China, January 5-7, pp. 563–564 (2012)

[6] Lopes, I.C., Vaidya, B., Rodrigues, J.J.P.C.: Towards an Autonomous Fall Detection and Alerting System on a Mobile and Pervasive Environment. Telecommunications Systems, Springer 52(4), 2299–2310 (2013), doi:10.1007/s11235-011-9534-0

[7] Chen, D., Feng, W., Zhang, Y., Li, X., Wang, T.: A wearable wireless fall detection system with accelerators. In: 2011 IEEE International Conference on Robotics and Biomimetics (IEEE ROBIO 2011), Karon Beach, Thailand, December 7-11, pp. 2259–2263 (2011)

[8] Mohamed, A.H.H.M., Tawfik, H., Al-Jumeily, D., Norton, L.: MoHTAM: A Technology Acceptance Model for Mobile Health Applications. In: Developments in E-systems Engineering (DeSE 2011), December 6-8, pp. 13–18 (2011)

[9] Kumar, S., Nilsen, W., Pavel, M., Srivastava, M.: Mobile Health: Revolutionizing Healthcare Through Transdisciplinary Research. IEEE Computer Society 46(1), 28–35 (2013)

[10] Rodrigues, J.J.P.C., Oliveira, M., Vaidya, B.: New Trends on Ubiquitous Mobile Multimedia Applications. Journal on Wireless Communications and Networking (EURASIP 2010) 2010(1), 517–689 (2010)

[11] ISO/TR 20514, Health informatics - Electronic health record - Definition, scope and context American National Standards Institute (ANSI), p. 34 (2005)

[12] Nosteboom, C., Bastola, D., Qureshi, S.: Cycles of Electronic Health Records Adaptation by Physicians: How Do the Positive and Negative Experiences with the EHR System Affect Physicians EHR Adaptation Process?, Maui, Hawai, USA, pp. 1530–1605 (January 2012)

[13] Kumar, B., Singh, S.P., Mohan, A.: Emerging Mobile Communication Technologies for Health. presented at the International Conference on Computer and Communication Technology (ICCCT 2010), Allahabad, India, pp. 828–832 (2010)

[14] Thammasat, E., Chaicharn, J.: A simply fall-detection algorithm using accelerometers on a smartphone. In: Biomedical Engineering International Conference (BMEiCON), December 5-7, pp. 1–4 (2012)

[15] Jang, E.H., Park, B.J., Kim, S.H., Eum, Y., Sohn, J.: Identification of the optimal emotion recognition algorithm using physiological signals. In: Engineering and Industries International Conference (ICEI 2011), November 29 - December 1, pp. 1–6 (2011)

[16] Shiozawa, N., Arai, S., Okada, S., Makikawa, M.: Gait analysis of sit-to-walk motion by using portable acceleration monitor device for fall prevention. In: 2012 IEEE International Conference on Biomedical and Health Informatics (IEEE BHI 2012), Hong Kong, China, January 5-7, pp. 563–564 (2012)

[17] Jantaraprim, P., Phukpattaranont, P., Limsakul, C., Wongkittisuksa, B.: Improving the accuracy of a fall detection algorithm using free fall characteristics. In: Electrical Engineering/ Electronics Computer Telecommunications and Information Technology International Conference (ECTI-CON), May 19-21, pp. 501–504 (2010)

[18] Li, A., Wang, S., Zheng, H., Ji, L., Wu, J.: A novel abnormal ECG beats detection method. In: Computer and Automation Engineering The 2nd International Conference (ICCAE 2010), February 26-28, pp. 47–51 (2010)

[19] Phinyomark, A., Chujit, G., Phukpattaranont, P., Limsakul, C., Hu, H.: A preliminary study assessing time-domain EMG features of classifying exercises in preventing falls in the elderly. In: 2012 IEEE 9th International Conference on Electrical Engineering/Electronics, Computer, Telecommunications and Information Technology (IEEE ECTI-CON 2012), Hua Hin, Thailand, pp. 1–4 (May 2012)

Case Studies

Problem Definition

Falling is something natural in the life of a human being from birth to death. A fall may have multiple origins, some of them can be avoided but others not. In elderly people the consequences of a fall increase with age. With the increase of elderly people around the world, new challenges arise to give greater quality of life for these senior citizens. So they can live longer and with the best quality of life as possible.

A statistical study applied in institutionalized Portuguese older adults in 2013, with 65 years old or older, being 32 males and 81 females in a total of 113 elderly. Table 4 shows falls average of this sample [1].

Table 4 Number of falls by sex and average

Last 2 years Falls		
Gender	*Number*	*Average(%)*
Male	8	7,1
Female	28	24,8
Total	36	31,9

Another statistical study performed in United States associates health problems directly with greater chances of falling. Sixteen percent of all Emergency Department visits and almost seven percent of all hospitalizations are for fall-related injuries. Table 5 shows a number of health problems per chance of falling estimation.

Table 5 Health problems per chance of falling estimation

Number of health problems	*Chance of falling*
0	
1	
2	
3	
4 or more	

In table 2, in general one elder in ten will have the risk of fall. The second line if the number of health problems known is one, the chance of falling will be two elder in ten and these critical numbers enlarge so on. Because of these statistical evidences, if the patient has health problems it will significantly increase the chances of falling [2].

This real problem in modern society is common in developed countries and the statistical facts presented before, demonstrate the seriousness of the problem. Therefore, mHealth solutions attached to health sensors may prevent or even avoid falls reducing the death rates.

Falls Detection and ECG Experimental Analysis

The mobile application was deployed on several devices in order to validate the proposed solution, and also in order to be possible essays for volunteers. The application ran effortlessly on all the used devices. The experiments were approached based on several fall simulations with 3 volunteers, each one with 38 trials, in a total of 114 trials. The mobile devices, smartphone or tablet obtained results for one experiment with the sensors monitoring several bio-signals during a fall experiment.

To evaluate the solution two trials were experimented to find the significant differences and improvements between them, as shown in Table 6. In order to reach a final average to the two solutions, several measurements were archived.

Table 6 Experimental Results Control Table

Action	*Number of trials*	*Trials errors*	*Right Trials*	*Average*
Pre Fall	114	10	104	91%
Pre Fall + ECG	114	6	108	95%

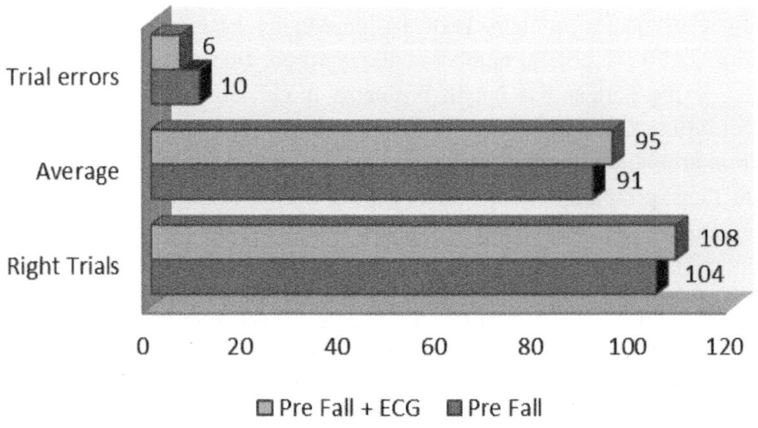

Fig. 20 Bar Chart Values Ordered

A total of 114 trials were performed under the same conditions for the two solutions. The trial errors were counted when the experimented solution did not detect a pre-fall, the system did not capture significant values from sensors or because the system presents failures and performed an exception.

In the first essay, Pre Fall results shown in Figure 20, only the accelerometer to conduct the experiment in a total of 114 trials was used. The system detected 104 falls with accuracy and only 10 of them did not detect a fall. In the second essay, Pre Fall + ECG results also shown in Figure 20, the ECG was added to the system and the results got slightly better, passing from 104 falls to 108 falls detected and only 6 trial errors.

The results shown in proportional area chart, Figure 21, represent the estimation of only pre-fall, 91% and the solution with pre-fall and ECG increases about 4%.

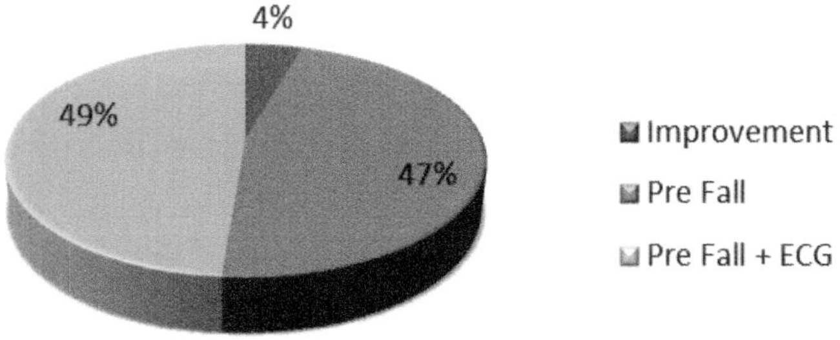

Fig. 21 Proportional Area Chart of Solution Improvement between Pre Fall and Pre Fall + ECG

Results show a better fall detection, as expected. It is important to note that this second solution Pre Fall + ECG, brings improvements in real catastrophic scenarios, such as when the fall was detected and the user is unconscious or suffered from a cardiovascular disease or heart attack.

Falls Prevention Experimental Analyses

The presented results aim to analyse quantitatively the applicability of the system in a real scenario. Therefore, in Table 4, the data was acquired from three healthy individuals in different motions to perform the experiment and the alerts for fall prevention were considered.

A total of 50 experiments were performed under the same conditions for each action. The total values line represents a sum of alert values for each motion.

Table 7 represents the expected values and shows an estimated amount of alerts, which is considered correct for each type of motion to allow calculate the average.

Table 7 Experimental Results

Alerts	*Number of Warnings during the Action*				
	Walking(5m)	Running(10m)	Sitting	Jumping	Falling
Movement (Green)	28	32	14	46	43
Problem (Yellow)	6	21	4	41	27
Eminent Fall (Red)	0	45	0	38	36
Possible Fall (Blue)	0	2	1	1	48
Total	34	100	19	125	154

Table 8 Expected Values of Experimental Results

	Expected Values				
Movement (Green)	30	35	15	50	50
Problem (Yellow)	8	25	5	45	30
Eminent Fall (Red)	0	45	0	45	40
Possible Fall (Blue)	0	5	0	2	50

Table 8 presents the average Values for each type of motion. The activities with larger accuracy are Sitting, Walking and Falling.

Table 9 Average Values for each Motion

Alerts	*Walking(5m)*	*Running(10m)*	*Sitting*	*Jumping*	*Falling*
Movement (Green)	93%	91%	93%	92%	86%
Problem (Yellow)	75%	84%	80%	91%	90%
Eminent Fall (Red)	100%	100%	100%	84%	90%
Possible Fall (Blue)	100%	40%	100%	50%	96%
Total Average	92%	79%	93%	79%	91%

During the Walk action a total of 34 warnings were triggered, 28 under the movement alert, and 6 for the problem alert. It was expected 30 for the movement alert and 8 for the problem alert. For the eminent fall and possible fall alerts it was not triggered any warning but it was not expected any warning, presenting a percentage of 92% of accuracy during the Walk action.

In the Run action, a total of 100 warnings were triggered, 32 in the movement, 21 in the problem, 45 in the eminent fall, and 2 in the possible fall. For this action, 35 warnings for the movement, 25 for the problem, 45 for the eminent fall, and 5 for the possible fall were expected. It shows about 79% of accuracy in this action.

For the Sit action a total of 19 warnings were triggered, 14 for the movement, 4 for the problem, and 1 for a possible Fall. It was expected 15 warnings for the movement and 5 for the problem. A total of 93% of accuracy was reached.

In the Jump action a total of 125 warnings were triggered, 46 for the movement, 41 for the problem, 38 for the eminent fall, and 1 for the possible fall. It was expected about 50 for the movement, 45 for the problem and eminent fall, and 2 for a possible fall. About 79% of accuracy was reached during this action.

Finally, for the Fall action, a total of 154 warnings, 43 for the movement, 27 for the problem, 36 for the eminent fall, and 48 for the possible fall were triggered. For this action, about 50 warnings for the movement, 30 for the problem, 40 for the eminent fall, and 50 for the possible fall were expected. A total of 91% of accuracy in this action was found.

Figure 22 presents a line chart to visually distinguish and evaluates the average for each type of motion and the notification received.

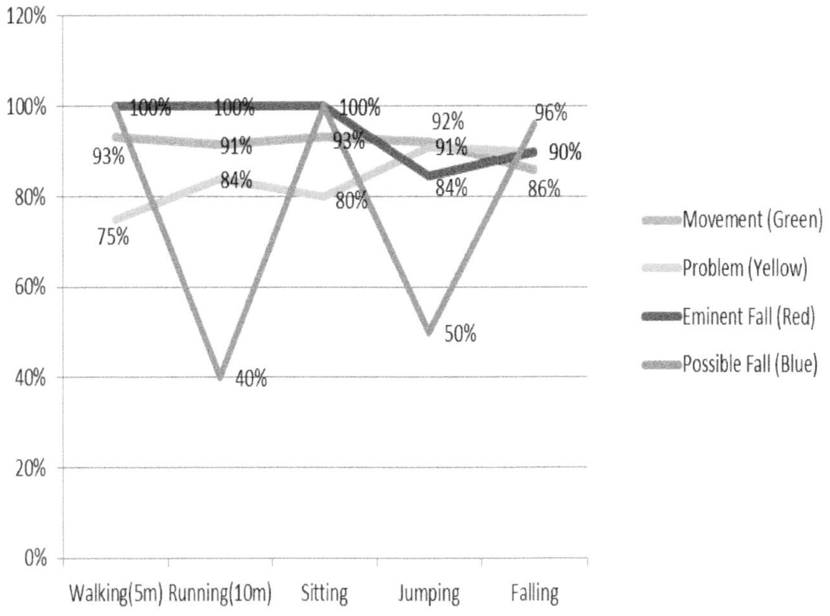

Fig. 22 Line Chart of the Average Values for each Motion

Evaluating the result and the effectiveness of the proposed algorithm, to estimate the average value is necessary to determine how to improve it. Figure 23 shows the average value of 50 experiments for each type of motion.

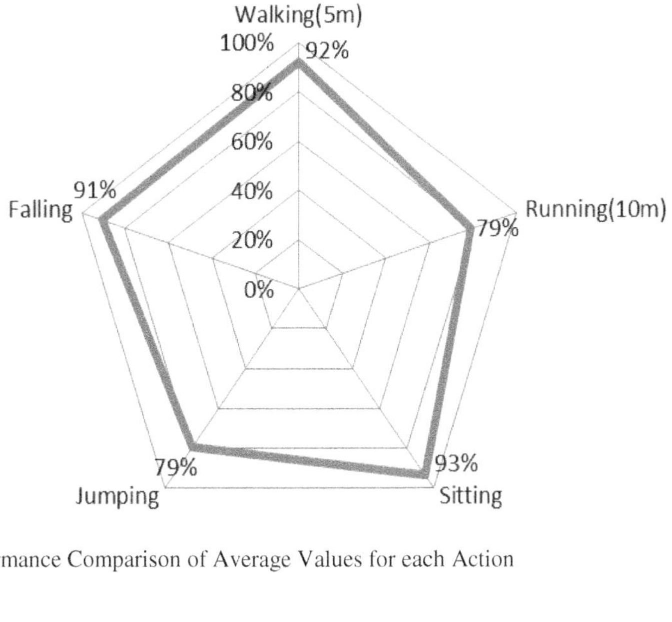

Fig. 23 Performance Comparison of Average Values for each Action

It was identified for movements which require more physical activity (e.g. Run and Jump) that the accuracy is lower, but with a nice value. The final average evaluation of the program for all types of motions is about 86.8% for 250 trials.

7 Future Work

To conclude this work, it just remains to make suggestions in terms of future developments and research directions based on current work:

- The development and update of the APPs and APIs for other mobile platforms, such as iOS from Apple and Windows Phone.

- Furthermore, the data retrieved from the other sensors for example Electroencephalography (EEG) beyond ECG, EMG, EDA, BVP, Respiration and accelerometer can be used to accomplish the falls prevention and detection.

- The mobile devices smartphones or tables may be used to collect data embed health sensors and become specific devices for example a Body Health Sensing Smartphone.

- A performance assessment study through simulation considering different danger scenarios and evaluate the scalability of the proposed solution, may also be considered.

References

[1] Almeida, P., Neves, R.: As quedas em idosos institucionalizados. Suas Características, 2013 EFDeportes.com, Revista Digital, Buenos Aires, `http://www.efdeportes.com/efd177/as-quedas-em-idososinstitucionalizados.htm` (accessed May 2013)

[2] Tinetti, M.E.: Facts about Falls - Preventable (2010), `http://www.fallprevention.org/pages/fallfacts.htm` (accessed April 2013)

Questions

Q1. Describe mobile health systems and applications?

Q2. What is biofeedback monitoring?

Q3. What kinds of biofeedback signal are used in fall prevention and detection system?

Q4. Which one is the most important physiological signal in the human body?

Q5. Which benefits can have the fall prevention and detection system to the user?

4. A Review of Methods to Characterize and Classify Sleep, Depression and Schizophrenia Disorders

Dean Cvetkovic*, Haslaile Abdullah, and Ramiro Chaparro-Vargas

School of Electrical and Computer Engineering, RMIT University,
Melbourne, VIC 3001 Australia
dean.cvetkovic@rmit.edu.au

Keywords: depression, schizophrenia, sleep, insomnia, brain and heart interaction.

1 Introduction

Why investigate depression and schizophrenia during sleep? Because studying spontaneous neural activity during sleep minimises factors related to waking such as attention, cognitive behaviour and any presence of active symptoms. Sleep architecture also activates particular neural networks and interacts with cardiovascular activity, which can be compared between depressed and schizophrenic patients, healthy and sleep disorders.

The overview of this chapter consist of the following sections: insomnia and psychiatric disorders, electrophysiological interaction and characterisation of disorders, depression in schizophrenia, schizophrenia and sleep, schizophrenia and biomarkers, author's past and current research projects and future directions.

2 Insomnia and Psychiatric Disorders

Sleep and psychiatric disorder does not only reduce the quality of life but also increases the financial burden individually and the public health care system in whole [Ozminkowski et al, 2007, Walsh et al, 1999]. In the US, almost $30 billion -$35 billion has been spent for insomnia related disorder [Chilcott L.A. and Shapiro C.M, 1996] while the overall cost of sleep disorders in Australia in 2004 was $7494 million [Hillman et al, 2006]. The estimated prevalence rate of insomnia varies greatly worldwide. In the US, the prevalence rate was approximately 25% [Ancoli Israel et al, 1999], Canada was about 9.5% [Morin et al, 2006], Japan was at 21.4% [Liu et al, 2000], Australia approximately at 7% [Lack et al, 1988] and the Europ from 4% to 22% [Chevalier et al, 1999] where in France itself the estimated prevalence was reported at 9% [Leger et al, 2000].

Experimental results have consistently shown that insomnia is a risk factor for the development of anxiety and depressive disorders [Neckelmann et al., 2007,

* Corresponding author.

S. Adibi (ed.), *Mobile Health*,
Springer Series in Bio-/Neuroinformatics 5, DOI: 10.1007/978-3-319-12817-7_4

Dombrovski et al., 2008, Lustberg et al, 2000]. The 20 and 34-year longitudinal studies have reported insomnia to be associated with risk of depression [Buysse et al, 2008, Chang et al, 1997]. The same finding was reported in the 7-year study of adolescent insomnia that increased the risk of depression and suicide [Roane et al, 2008]. In fact, insomnia can act as an independent predictor or risk factor for developing depression [Rieman et al, 2003].

Diagnostic and Statistical Manual of the American Psychiatric Association criteria for insomnia state that the sleep disturbance(s) must result in impairments of daytime functioning (or mood) or cause clinically significant distress in order to be classified as insomnia. For primary insomnia, the main symptoms are difficulty falling asleep, maintaining sleep, or a combination of both these problems. One of the diagnostic criteria for primary insomnia is: *"The disturbance does not occur exclusively during the course of another mental disorder (e.g., major depressive disorder, generalised anxiety disorder, a delirium)"* [Kennedy & Solin, 2004]. How do we measure or quantify whether sleep disturbance does or does not occur exclusively during depressive or anxiety disorder? The DSM-IV-TR diagnosis of primary insomnia subsumes several insomnia diagnoses in the *International Classifications of Sleep Disorders.* Primary insomnia is essentially the same as psychophysiological insomnia. The term psychophysiological insomnia has been used to more accurately reflect the most common causes of insomnia, which typically include inappropriate arousal and conditioning or learned factors.

In relation to secondary insomnia, people with psychiatric conditions can report a wide range of sleep-related problems [Perlis et al., 2001]. However, it can be difficult to determine whether the sleep problems are due to medications (neuroleptics), the psychiatric disorder, or both these factors together. Careful review of the person's medical records and taking a thorough history may allow the causal roles of medications and disease factors in the sleep problem(s) to be determined. Sleep disruption and insomnia are common symptoms of most psychiatric disorders. Depression can be complicated by the problem of early morning awakening and difficulty returning to sleep [Livingston et al., 1993]. About 30% of people suffering from major depression report sleep problems (e.g., insomnia, hypersomnia or dream disorders). Studies have suggested that people with depression and insomnia are particularly at risk for suicide [Fawcett et al., 1990]. The National Survey of Mental Health and Wellbeing reported within 12 months period, one in ten (8.6%) people with mental disorders reported being suicidal [Australian Bureau of Statistics (ABS), 2009].

In contrast people experiencing a manic episode usually have insomnia and sleep very little, but usually do not complain of sleeplessness. The physical and psychological symptoms associated with premenstrual syndrome (PMS) can disrupt sleep. The severity and duration of sleep disruption depends on the severity of the symptoms and the number of days that they are experienced. During pregnancy sleep can often be disrupted and daytime sleep can be a problem for some women. Insomnia can be a common problem in new mothers and should also be regarded as an indicator of possible postnatal depression.

A condition, known as sleep state misperception, is characterised by complaints of insomnia, but there is usually a marked discrepancy between reported and observed estimates of sleep duration [Edinger & Krystal, 2003]. When diagnosing primary insomnia, patient complaints of poor sleep are either related to medical problem or duration of sleep problems. If sleep is more than 7 hours per day, it may be diagnosed as sleep state misperception, or paradoxical insomnia, reported with no clear reference to corresponding PSG findings or its clinical significance. In other words, subjective estimation may be over or under estimated the amount of sleep by few hours as reported by objective normal PSG sleep duration. The psychological profile of insomniac patients, suggests that rumination may influence the sleep misperception.

There are *no clear* objective markers from PSG signals to characterize insomnia. PSG studies on insomnia patients generally show abnormalities, such as prolonged latency to sleep onset, frequent arousals, and reduced amounts of total sleep. However, objective measures of sleep do not always correlate well with the patient's experience of insomnia, which may be partially due to the fact that the function of sleep itself is still unknown, making it difficult to pinpoint which objective sleep abnormalities contribute to the clinical entity of insomnia. It is why subjective tests and interviews are necessary. American Academy of Sleep Medicine (AASM) state that PSG should not entirely be used to routinely evaluate of chronic insomnia for a number of good reasons because there is a great *discrepancy between the PSG measures and the subjective experience* of primary insomnia [Riemann, et al., 2010]. In the 2003 American Sleep Disorders Association (ASDA) Practice Parameter Report, it was stated that the use of *PSG as a diagnostic tool for insomnia is still somewhat controversial* and was not recommended for the routine evaluation of chronic insomnia or insomnia associated with psychiatric disorders,[Littner et al., 2003]. It is challenging to find the cause of insomnia, described by the ''chicken and egg'' dilemma, which makes it unclear whether the cortical state determines subcortical activity or vice versa (i.e., is it the daytime worrying that causes alterations in EEG or vice versa?). In patients with insomnia worrying may include phobia or fear of not being able to sleep and its consequences on daytime functioning. Other insomniac sufferers have certain expectations on their sleep requirements and worry when such requirements are not met [Morin et al., 2007]. This worrying is related to self-reflective, repetitive and passive focus on one's negative emotions, known as rumination. This rumination can, in turn, affect the psychophysiological state of insomniac sufferer, and lead to depression. There have been many experimental and correlational studies which have shown that rumination predicts depressive symptoms and onset of major depressive episodes [Treynor et al., 2003].

Although much of the research has been done to determine the cause of insomnia, its specific cause has not yet been demonstrated and remains elusive. In general, insomnia can be secondary to a psychiatric or sleep disorder and sometimes it can be a primary or independent. Primary insomnia is a term used to describe a subtype of insomnia that constitutes the disorder itself and is not a consequent to any other disorder. Secondary insomnia is known as comorbid insomnia, which is usually associated with new episodes of depression. Many articles have been published demonstrating insomnia as a risk factor for developing psychiatric disorders particularly depression and anxiety disorder. A study by [Taylor et al., 2005] has claimed that people with insomnia were 9.82 and 17.35 times as likely to have depression and anxiety disorder, respectively. A directional risk study of insomnia with anxiety and depressive disorders by [Johnson et al., 2006] reported that a prior insomnia was associated with onset of depression, but less association is observed from depression to insomnia. This result is in agreement with a study conducted by [Gregory, et al., 2009] which confirmed that sleep problems at age 8 predicted depression at age 10. Whereas, the depression symptoms did not predict later sleep problems. Longitudinal study across 34 years follow up found that insomnia in young men is indicative of a greater risk for subsequent clinical depression and psychiatric distress [Chang et al., 1997]. In another study, the prevalence of major depressive episode among general hospital patients with insomnia was reported to 46.9% [Rocha, et al., 2005]. In contrast, there were several studies which claimed that insomnia causally related to depression. For example, a study by [Ohayon & Roth, 2003] showed that a previous history of a mental disorder was closely related to the severity and the chronicity of current insomnia. A recent study by [Jansson-Fröjmark & Lindblom, 2008] with a one year follow up revealed that anxiety, depression and insomnia are bi-directionally related over time. The advantage of this study was its ability to show directional relationship between insomnia and depression.

3 Electrophysiological Interaction and Characterisation of Disorders

It is understood that some mental disorders such as depression, are associated with measurable physiological changes. Symptoms of depression often reflect dysregulation in the autonomic and neuroendocrine systems, which are mostly controlled by the hypothalamus. Some of the examples of physiological changes are; sleep disturbance (characterised by the loss of stage 4 sleep and difficulty staying asleep), increased resting heart rate, hyperactivity of the sympathetic nervous system, hypothalamic pituitary overactive state, temperature changes and others [Mayers & Baldwin, 2006]. Recent research has demonstrated that overnight sleeping heart rate in patients with various psychiatric disorders are distinctly different than the 'signature normal' patterns in healthy (i.e. no psychiatric illness) patients [Stampfer, 1998]. The 'categorically' different patterns of heart rate during sleep in different psychiatric disorders may be reflected in other objective parameters. It is known that there is a shortened REM latency in depression, but there are likely other EEG changes. There has been a focus on the analysis on heart rate and heart rate variability (HRV) patterns during sleep. Medications are a major confounding factor on HRV. Therefore, any recruitment of patients will need to consider medication-free and other groups. Also, the mean trend heart rate patterns would allow one to distinguish between medication effects as opposed to therapeutic change. It has been shown that the heart rate patterns throughout the wake to sleep transition (related to insomnia) in generalised anxiety and depressed patients are 'categorically' different to that found in 'melancholic' depressed patients and one would expect that this should also be reflected in EEG activity. Previous study results have revealed that the heart rate may be used as a marker to diagnose or predict whether a human is suffering from a mental illness [Iverson et al., 2005]. There has been some advancement in trying to classify various sleep and psychiatric disorders (i.e., insomnia, sleep apnoea, depression, bipolar disordered and schizophrenia) from electrophysiological EEG and ECG (objective) measures [Boostani et al., 2009]. However, the current body of knowledge lacks the proper methodology to objectively make an accurate diagnosis or prediction of any sleep or mental disorder.

It is unclear whether increased autonomic activity is causing insomnia or whether insomnia and its sleep loss triggers increased autonomic activity, prompting the investigation of ECG in the cause and effect of insomnia. In terms of EEG, PSG shows significantly reduced sleep duration and efficiency, increased arousal index, and slightly, but significantly, less REM sleep in insomniacs compared to the healthy controls. The findings revealed that the difference between healthy and insomniacs, is that EEG beta band was higher during W, S1 and REM in insomniacs, and there was a reduction in EEG delta and theta bands in both REM and NREM.

Arousal is gaining more interest to be used as a marker of sleep disruption which involved in some sleep disorders. A study in healthy men has applied a cortical arousal from EEG alpha activity to identify bouts of tachycardia-bradycardia [Togo et al., 2006]. Some arousals (shorter) may not be subjectively estimated, but may be measured objectively using EEG and ECG. Whereas, longer arousals may be better estimated subjectively. Arousals can be both objectively measured and subjectively reported. However, the problem lies with sleep fragmentation (brief arousals) where patients are not aware of their shorter period awakenings in their sleep and cannot report it subjectively. Patients are generally aware of their longer awakenings, characterised with longer periods of arousals. This heightened level of arousal commonly appeared in the people likely suffer from depression [Staner, L, 2010]. A study of EEG beta power spectral during NREM sleep and glucose metabolism has shown an increased values of these two measures which associated with arousal particularly in the prefrontal cortex and the right lateral inferior occipital cortex [Nofzinger et al., 2000]. Hyperarousals have been described as a potential key factor in the maintenance of insomnia. However, there is a need for further research to characterise the nature of this arousal and its relationship to psychological, behavioural and physiological parameters, especially the *EEG and ECG interaction.*

Jurysta et al. hypothesized that chronic insomnia will alter the relationship between autonomic cardiac activity and delta sleep EEG [Jurysta et al., 2009]. During normal sleep, thalamic, thalamocortical and cortical neurons interact and are associated with generation of Slow Wave Sleep (SWS) particularly of delta rhythms and arousals. The interaction with the cardiac cycle is observed within the cells of thalamus. In insomniacs, it was observed that the linear correlation is decreased between cardiac vagal and delta EEG in comparison to the healthy controls. In the healthy controls, the High Frequency (HF) component of HRV which reflects the cardiac parasympathetic activity was preceded by delta EEG. However, in the primary insomnia patients the time delay between these two variables was less predictable. The interaction between cardiac activity and delta sleep power has also been studied in a major depressive disorder [Jurysta et al., 2010]. The link between cardiac and delta EEG activity was not significant between the healthy and depressed patients. Besides, none of the HRV parameters differed between both groups. In contrast, a study by Udupa et al. has shown increased sympathetic and decreased parasympathetic activity in the major depression patients compared to healthy controls [Udupa et al., 2007]. Despite the fact that this study has reported the alteration between cardiac function and delta EEG in the primary insomnia and depressed patients, the understanding about this interaction has not yet been fully established and will be studied in this project.

In the late 70s, Gillin and colleagues [Gillin, 1979] were the first to have successfully separated or classified healthy, insomnia and depressive patients using EEG signals. Patients with depression have a sleep abnormality patterns which could be used as an input variable in the classification. Gilin and colleagues have suggested eight objective variables namely as total sleep time, total recording period, sleep efficiency, sleep latency, early morning awake time, awake time, REM time and REM%. Similarly, a study by Steiger et al has also acknowledged sleep EEG as a potential biomarker in the classification of depression and other psychiatric disorders particularly desinhibition of REM sleep, changes of NREM sleep and sleep continuity disturbance [Steiger et al., 2010]. Nissen et al. introduced delta sleep ratio of spectral analysis to show changes in NREM in the deep sleep stage of sleep deprived patients. A higher value of delta sleep ratio was observed in the patient [Nissen et al., 2001]. Staner et al., suggested that hyperarousal could be related to sleep disturbance and abnormal levels of arousal in the depressive insomnia patients [Staner, et al., 2003]. All night sleep EEG frequency bands were extracted during the sleep onset and first NREM from the healthy, primary insomnia and depressive insomnia patients. The results have revealed the hyperarousal only characterized the primary insomnia with higher EEG activities during sleep onset, whereas the depressive insomnia showed a lower slow wave sleep (SWS). Bonnet et al. have demonstrated an increase metabolic rate in insomnia that was associated with difficulty sleeping [Bonnet et al., 1995]. Functional neuroimaging studies have reported a similar findings that insomnia patients have shown higher whole brain metabolism which failed to decline in activity from wake to sleep and resulted in greater NREM sleep [Nofzinger et al, 2004]. The findings from all these studies suggest that symptoms of heightened physiological arousal (hyperarousal) are common characteristics that could differentiate insomnia from depressed and healthy.

In recent year, various methods have been applied to extract different features in the sleep EEG to characterize different pathological conditions. Although linear methods like spectral analysis of EEG frequency bands and wavelet have successfully classified healthy, depressed, and insomnia patients, these methods may not be sufficient enough to extract valuable information and detect abnormalities in the signals. Therefore, various methods based on non-linear analyses are necessary for the study of the complex systems. Non-linear EEG time series in the healthy and depressed patients have been studied by Leistedt et al. using detrended fluctuation analysis (DFA) [Leistedt et al., 2007]. Depressed patients have displayed less autocorrelation of DFA measures during SWS which reflect breakdown of fractal physiologic complexity during specific pathologic conditions. In this research project, linear and non-linear methods will be applied to extract the useful features from EEG and ECG that can be used as markers to separate healthy controls from depressed and insomnia patients.

4 Depression in Schizophrenia

Cotton et al. highlighted that there needs to be a better understanding of the nature and characterization of depression in the first episode schizophrenia, as depressive symptoms are often neglected in first episode schizophrenia patients [Cotton et al., 2012]. Evidence suggests that depressive symptoms initially appear during the first schizophrenic episode and continue the course of positive symptoms in schizophrenia.

5 Schizophrenia and Sleep

Previous studies have evaluated that that medicated schizophrenia patients had a reduced sleep spindles (amplitude, duration and number), compared to healthy and depressed patients during the initial NREM episode [Ferrarelli at al., 2007]. Ferrarelli and colleagues decided to investigate whether similar reduction in spindle activity was consistent in the larger schizophrenia group, throughout the whole night during all NREM stages and if this spindle impairment was related to any specific neuronal network. Ferrarelli and colleagues conducted an overnight PSG study on 49 schizophrenia, 20 medicated non-schizophrenia and 44 healthy patients [Ferrarelli at al., 2010]. The results consisted of comparing the slow wave sleep (SWS) and the mean number of sleep spindles per time (i.e. density), duration, amplitude and frequency. Findings revealed that schizophrenia patients had no changes in SWS but had a reduction in spindle power (12-16 Hz) and in slow (12-14 Hz) and fast (14-16 Hz) spindle amplitude, duration, number in the prefrontal, centro-parietal and temporal brain regions. The medications did not affect the spindle reduction during the night. Schizophrenia clearly affects the neural thalamic reticular nucleus and thalamo-reticular network, responsible for the generation of sleep spindles.

Whilst Ferrarelli and colleagues [Ferrarelli at al., 2010] found no changes in SWS, Goder and his colleagues [Goder et al., 2006] tested the hypothesis that delta power in sleep was associated with neuropsychological performance in 17 healthy patients but impaired in 16 medicated schizophrenic. The results from the neuropsychological tasks did not reveal any significant differences between patient groups. But the SWS was significantly reduced (in temporal and occipital region) in schizophrenia patients and delta was not significantly different during stage 2 and REM in comparison to healthy patients. These findings were interpreted as a dysfunction of thalamocortical and prefrontal networks in schizophrenia, linked to cognitive impairment in schizophrenia patients.

It was hypothesised that beta and gamma power bands were a suitable mechanism for describing the integration of spatially distributed brain activities [Noh et al., 2013]. Given that, both bands involved abnormalities in the oscillatory behaviour (i.e. synchronisation) related to cognitive impairments and other symptoms [Bob, 2012]. For instance, Uhlhaas reported phase synchronisation deficits in 'Gelstat' perception amongst schizophrenia patients in beta band, but not the same case in gamma band. Based upon this, it was claimed that deficits in self-monitoring and willed actions could be produced by a communication dysfunction between the frontal and temporal lobes within the corollary discharge system. Such cognitive and behavioural disturbances of schizophrenia are an expression of central nervous system's inability to adequately bind neural processes segregated across distributed areas, a condition termed as functional dysconnectivity. Hitherto, neuroimaging and electrophysiological studies cannot agree upon whether functional dysconnectivity comes from the aberrant axonal connectivity or abnormal neuronal function amongst cortical areas interwoven by a functional circuit [Zalesky et al., 2010]. Regarding additional power bands, Siekmeier documented the increment of delta and theta power band oscillatory activity in the temporal lobes. Even more, several studies argued a positive correlation between temporal lobe activity and positive schizophrenia symptoms [Bob, 2012]. The suggestion made by Rolls stated that the abnormal brain activity in schizophrenia can be modelled as a reduction in excitatory and inhibitory synaptic transmission, which can lead to rapid changes of neuronal states across spatially diverse neuronal populations. Thus, alpha oscillations in occipito-parietal and central regions manifested similar functional interaction in relation to inhibitory processes. Likewise, beta oscillations primarily managed the maintenance of the current sensorimotor cognitive state [Nikulin et al., 2012]. On the same way, [Timashev et al., 2012] research work showed that dampening in alpha activity accompanied by the amplification of low frequency in delta and theta power bands, were essential signs of schizophrenic condition.

Some additional facts and findings in regard to the studies aforementioned are deeper detailed in Table 1, including subjects, methodology, conditions and conclusions.

6 Schizophrenia and Biomarkers

Schulz and colleagues highlighted that previous research demonstrated a dysfunctional cardiac autonomic nervous system, associated with a decreased parasympathetic (vagal) and increased sympathetic activity [Schulz et al., 2012]. However, Schulz identified that respiration, responsible for the complex interaction between the brainstem and 'higher centres', was not investigated together with the cardiac activity, as part of cardiorespiratory regulation in schizophrenic, their first degree relatives and healthy patient groups. Various parameters (linear, nonlinear, indices, etc) were computed for cardiorespiratory coupling, respiratory variability and dynamics in order to characterise autonomic regulation. The results revealed a significant difference in schizophrenia patients as compared to healthy and first-degree relatives, for the respiratory variability and their dynamics, with increased complexity and reduced respiratory sinusarrhythmia (i.e. reduced cardiorespiratory coupling in schizophrenia but not for their first degree relatives).

In terms of HRV time domain measurements obtained from previous studies, it was shown that non-medicated schizophrenia patients exhibited decreased RMSSD (root mean square of successive RR differences – each RR difference is squared, summed, averaged and result of square root derived), pNN50 (the percentage of adjacent RR intervals that differ by more than 50 ms) and HF power compared to healthy subjects [Bar et al., 2007]. The SDNN (standard deviation of normal to normal RR intervals) reflects sympathetic and parasympathetic activity. The RMSSD and pNN50 correlate with HF power in HRV, more related to parasympathetic activity. These study results indicated that schizophrenia was characterised by a decrease in parasympathetic activity. The medication was not cause of this effect in HRV activity. Henry et al. study investigated HRV in acutely hospitalised manic bipolar, schizophrenia and healthy patients [Henry et al., 2010]. The HRV was analysed in time and frequency domains as well as nonlinear analyses. The results revealed that in bipolar patients, parasympathetic activity (vagal tone) was significantly reduced, and in schizophrenia patients a similar but 'not significant' reduction in the same HRV activity was evident, as compare to healthy patients. The reduction in heart rate complexity was also evident in bipolar disorder patient group when analysed using linear and nonlinear methods.

Uhlhaas and Singer claimed that in schizophrenia patients, the synchronisation of EEG beta and gamma bands was abnormal due to dysfunctional oscillations that occur during cognition, behaviour and at rest [Uhlhaas and Singer, 2010]. This synchronization abnormality was due to the impairment involving disconnectivity of cortical networks particularly in medial frontal, parietal/occipital, and left temporal [Zalesky et al., 2011]. This EEG beta and gamma synchronisation was typically present in local cortical areas [Womelsdorf et al., 2007] and the lower EEG activity (i.e. delta, theta and beta bands) synchronization was observed over and between more distant cortical areas [von Stein et al., 2000]. It is currently unknown *"to what extent impairments in local circuits contribute to long-range synchronisation impairments or whether these are two independent phenomena"* [p. 102, Uhlhaas and Singer, 2010]. *Our research study is aligned in investigating the cross-frequency synchronization between EEG high and low frequencies, related to more distant (long-range) and local cortical synchronization, respectively.* Other studies revealed that patients with schizophrenia were characterized by reduced EEG gamma and theta amplitudes in the frontal cortical area [Cho et al., 2006] mainly during cognitive tasks accompanied by reduced phase synchronization [Uhlhaas and Singer, 2010]; reduced synchronization in EEG beta and gamma bands during visuo-perceptual organization and auditory processing [Uhlhaas et al., 2006]; and reduction in coherence at EEH theta band. Some studies have investigated whether the pharmacological treatments (medication) have any influence on the particular abnormal EEG activity in patients with schizophrenia. For example, *is the abnormality in EEG beta and gamma bands due to schizophrenia disorder alone or does medication plays any part in its abnormality?* Uhlhaas and Singer recommend that more research is necessary to shed light on this, but at the moment, various studies have reported abnormalities in the amplitude and phase of EEG gamma bands in patients treated briefly with medication and who started exhibiting schizophrenic symptoms [Uhlhaas and Singer, 2010]. Thus, it is evident that even at the onset development of abnormalities in EEG activities, medication which was administered briefly, may not play any part in it. Evidence shows that several psychiatric disorders may exhibit similar abnormal EEG activities [Uhlhaas and Singer, 2010]. For example, adults with autism spectrum and schizophrenia disorders typically are characterized with changes in EEG gamma during perceptual organisation. Bipolar and schizophrenia disorders have similar impairments during auditory steady-state evoked potentials (SSEP). These findings suggest that there is a need for an improved characterization of psychiatric disorders during wake and sleep, and linked to sleep disorders as it has been discussed in this chapter.

Timashev et al. introduced a new quantitative measure of frequency-phase synchronization and spikiness factor which showed lower values and higher values respectively in the higher risk schizophrenia group as compared to the other groups [Timashev et al., 2013]. These may have corresponded to the disruption in the neural activity of cortex area in schizophrenia.

For cognitive tasks, the long-range brain regions are predominantly active, thus low EEG band oscillations can be observed. However, the relationship between cognition and EEG oscillations is still debatable topic. It is understood that the Anterior Cingulate Cortex (ACC) activity is reflected by frontal midline theta (fm-theta) frequency band (4-7 Hz) [Maron-Katz et al., 2013]. The ACC activity and fm-theta is associated with attention demanding, mental and cognitive tasks, also linked to meditation and stage 1 and REM sleep [Inanaga, 1999]. If these fm-theta oscillations can be altered, the cognition can be enhanced, and thus influence the schizophrenia and other psychiatric disorders. On the contrary, fm-theta can be induced by various mental tasks or stage 1 and REM sleep. One study found that the mean EEG theta band frequency measured over frontal midline (fm) in medicated and non-medicated schizophrenic patients was less than at parietal region when compared to healthy and neuroleptic medicated patients. Westphal and colleagues revealed that this slower fm-theta activity in schizophrenia patients was related to the 'hypofrontality hypothesis' [Westphal et al., 1990].

Dumont et al. stressed the focus given to the dynamic interaction between HRV and limited number of EEG power bands along previous studies [Dumont et. al., 2004]. His research strived to broaden the assessment of underlying relationships upon all available EEG power bands under general sleep conditions. From a general perspective, the introduction of linear and nonlinear approaches revealed the coupling between HRV normalised high frequency band (HF_{nu}) and each EEG power band across the different sleep stages. The strongest interaction or maximized coupling occurred between EEG delta and HRV HF_{nu} power bands. Having said that, this study exclusively considered healthy young adults without mental or sleep disorders, then the obtained findings were useful in the sense of sleep staging examination rather than comparative analysis amongst clinical conditions. Therefore, the authors extended their study to more defined conditions related to sympathetic and vagal cardiac activities, such as sleep apnea-hypopnea syndrome [Dumont et al. 2007]. By conducting inherited and novel linear/nonlinear investigations, the predominance of interaction between EEG delta and HRV HF_{nu} power bands was reaffirmed. Conclusively, it was argued that sleep apnoea condition did not influence significantly the functional interdependence between HF_{nu} component and all remaining sleep power bands. A complete description about the conduction of the studies is summarised in Table 1.

Critchley et al. study on healthy patients [Critchley, 2003] wanted to determine which brain areas were associated with the control of HRV during effortful cognitive and motor behaviours. The findings were that during effortful cognitive and motor behavior the dorsal ACC was influenced by associated autonomic states of cardiovascular responses (arousals). This study was performed using the event-related fMRI and simultaneous ECG with computed HRV parameters (heart rate, LF, HF). Non-linear methods have been applied to investigate the disturbance in cortical integration in schizophrenia, study based on the dynamical decoupling of brain regions in schizophrenia [Breakspear et al., 2003]. The results revealed that for schizophrenia patients dynamical interdependencies were present more often between multiple cortical sites and less often between only two sites, seen in healthy patients. The authors suggest that this relates to the "loss of balance" between independence and interdependence in large-scale cortical processes [Breakspear et al., 2003]. This larger-scale may not necessarily mean the long-range cortical synchronization related to lower EEG bands (delta, theta and alpha) and more distant cortical sites.

7 Author's Past and Current Research Projects

There is growing evidence in an alteration between cardiac and neural functions in primary insomnia and psychiatric patients. This alteration is due to abnormal levels of heart rate, heart rate variability, sleep spindles, arousal and slow-wave sleep, but the understanding of this cardio-neural interaction has not yet been fully established. The synchronisation studies upon polysomnographic (PSG) recordings are primarily oriented to reveal functional interdependence and causal relationships between neuronal regions of interest and heart rate variability (HRV) in healthy and disordered (obstructive sleep apnoea/hypopnoea, insomnia, depressed and schizophrenia) patients.

Our research project, initially undertaken over the last few years is currently progressing well and will continue into the future. The project is focusing on the development of automated tool and consists of several stages of development. For the first stage of the project, the co-author Abdullah has been investigating the interaction of electroencephalographic (EEG) and electrocardiographic (ECG) signals during sleep stages and utilize these signals to automatically classify healthy, sleep (i.e. insomnia, obstructive sleep apnoea/hypopnoea and schizophrenia) disorders. The objective of Abdullah's project was to investigate study how EEG and ECG interact (i.e. measure which activity drives each other) during sleep stages and whether this interaction differs between sleep and psychiatric disorder sleep architecture. The signal processing was conducted offline. The classification consisted of developing a computational model that was 'trained' to recognise each of the group disorders. One half of collected PSG data was used from this 'training' part of the study. The 'trained' model was then used to 'test' and classify the second half of subjects from all four groups. The final outcome from this study was to test the effectiveness of using a mathematical model to classify each of the disorders.

In the follow-up project currently conducted by another co-author Chaparro-Vargas, the objective is to utilse more advanced linear and non-linear signal processing methods in order to evaluate this EEG and ECG interaction and statistical classification in online (i.e. near real-time) mode. The online EEG and ECG interaction will be evaluated using novel synchrogram feature and online sleep staging (i.e. online hypnogram). The combined features, 'synchrohypnograms' will be the 'main engine' of the project in order to automatically classify the disorders. From this online 'synchrohypnograms', the neuronal and cardiac 'oscillators' and instantaneous coupling and decoupling phenomena in terms of amplitudes, cross frequencies and phase shifts across the different sleep stages, will be developed.

8 Future Directions

Although much of the research has been done to determine the cause of insomnia, its specific cause has not yet been demonstrated and remains elusive. It is unclear whether increased autonomic activity is causing insomnia or whether insomnia and its sleep loss triggers increased autonomic activity. This study might also help us better understand the connection of insomnia to psychiatric disorders (or vice versa). Our tool's systematic classification of patient PSG data could provide a clinician with invaluable information to treat depression and schizophrenia earlier before it worsens and becomes chronic and at the same time, treat insomnia.

Table 1 Literature review summary of subjective, objective PSG and EEG and ECG interaction and epidemiological studies on causes and effects of insomnia and psychiatric disorders. Literature review summary of epidemiological studies on causes and effects of insomnia and psychiatric disorders

Year	Disorder	Title	Author/Journal	Type of study, No. and type of participants	Methods	Findings
2013	Schizophrenia	Impaired coupling of local and global functional feedbacks underlies abnormal synchronization and negative symptoms of schizophrenia	Noh, K., et al. BMC Systems Biology	Patients: 12 M, 3 F (23.80 ± 4.6) Healthy: 11 M, 6 F (22.06 ± 2.11)	120 seconds MEG Diagnosis: DSM-IV, PANSS, DSM-IV (Non-patient version) MEG: 102 magnetometers Analysis: Time-Frequency and power analysis, Beta-gamma power switching, phase synchronization, feedback identification based on impulse reponses, Wilson-Cowan Model Statistics: Two-tailed t-test	Coupled local and global feedback (CLGF) circuits in the cortical functional network are related to abnormal synchronization and also correlated to the negative symptom of schizophrenia During rest an increase in beta band synchronization and a reduction in gamma band power compared to healthy controls
2012	Schizophrenia	Consciousness, schizophrenia and complexity	Bob, P. Cognitive Systems Research			Mental disintegration in schizophrenia could be described as a level of neural disintegration leading to more irregular neural states with higher complexity that negatively affect information integration and synchronization process in the brain
2012	Schizophrenia	Attention of long-range temporal correlations in the amplitude dynamics of alpha and beta neuronal oscillations in patients with schizophrenia	Nikulin, V., et al. Neuroimage	Patients: 16 M, 5 F (37.9 ± 5.1) Healthy: 21 M, 7 F (35.9 ± 7.1)	50 seconds EEG EEG: 21 electrodes Diagnosis: DSM-III, DSM-IV, SANS, SAPS Analysis: Instantaneous amplitude, detrended fluctuation analysis, cross-frequency correlation	Amplitude of neuronal oscillations in alpha and beta frequency ranges did not differ between patients and control subjects Long-range temporal correlations (LRTC) were strongly attenuated in patients with schizophrenia in both alpha and beta frequency ranges Cross-frequency correlation between LRTC belonging to alpha and beta oscillations was stronger for patients than

Table 1 (*Continued*)

						healthy controls
2012	Schizophrenia	Analysis of cross-correlations in electroencephalogram signals as an approach to proactive diagnosis of schizophrenia	Timashev, S. F., Physica A: Statistical Mechanics and its Applications	Subjects: 84 (12 ± 2)	60 seconds EEG EEG: 2 electrodes (F3, F4) Diagnosis: Diagnosed schizophrenia-spectrum symptoms Analysis: Flicker-noise spectroscopy, cross-correlation function, frequency-phase synchronization	Classification subjects into 4 categories corresponding to different risk levels of subjects' susceptibility to schizophrenia
2011	Schizophrenia	Disrupted Axonal Fiber Connectivity in Schizophrenia	Zalesky, A., et al. Biology Psychiatry	Patients: 47 M, 27 F (38 ± 11) Healthy: 15 M, 17 F (33 ± 13)	MRI: Preprocessing pipeline Diagnosis: Diagnostic interview for psychoses Analysis: Fibre tract network model, Network organization: nodal degree, efficiency, path length, clustering coefficient and small-worldness.	Impaired connectivity to involve a distributed network of nodes compromising medial frontal, parietal/occipital, and the left temporal lobe.
2010	Insomnia and depression	Comorbidity of insomnia and depression	Staner, L. Sleep Medicine Reviews	Review	-	Epidemiological studies show that insomnia could lead to depression.
2010		Altered interaction between cardiac vagal influence and delta sleep EEG suggests an altered neuroplasticity in patients suffering from major depressive disorder.	Jurysta, F., et al., Acta Psychiatrica Scandinavica	MDD: 10 M (38 ± 10) Healthy: 10 M (40 ± 9)	4 nights PSG MDD scored HDRS > 16 (Hamilton Depression rating Scale 24 items) EEG: Normalized delta power ECG: LF, HF, Total Power, LFnu, Hfnu, LF/HF Coherence analysis: Coherence function, gain function and phase shift Statistics: ANCOVA	Smoking and caffeine intake were related to the duration of the first three NREM-REM. MDD: Sleep latency increased, other sleep parameters not significant S1 and W duration were increased HRV: not significant Coherence: lower gain between HFnu and delta Modification in HF appear 8 and 10 min before modification of delta in MDD and healthy
2010	Sleep and depression	Wake and sleep EEG provide biomarkers in depression	Steiger, A. and Kimura, M. Journal of Psychiatric Research 4		Review article	REM sleep desinhibition, disturbed sleep continuity, lower SWS, shorter REM latency are potential biomarkers to characterize depression.
2009	Insomnia	The impact of chronic primary insomnia on the	Jurysta, F., et al.,	chronic primary insomnia : 14 M (42 ± 12)	3 nights PSG (used 2nd night) EEG: FFT on 5s data window, averaged every 20s	Insomnia: sleep efficiency was decreased, sleep latency longer

Table 1 (*Continued*)

		heart rate - EEG variability link	Clinical Neurophysiology	Healthy: 14 M (41 ± 10)-	Normalized delta power (0.5-3 Hz)=Sum delta power every 20s/mean power across full night ECG: RRI on 120s and shifting 20s ahead HRV: LFnu, HFnu, Total power HRV, LF/HF Coherence analysis: Coherence function, gain function, phase shift Statistics: ANOVA repeated measures within groups	NREM duration similar Larger mean duration REM HRV parameters not significant between groups in all sleep stages. Insomnia: decreased linear coupling between two markers. Similar gain and phase shift with the healthy. Decreased coherence represent desynchronization of both neural population.
2009	General sleep and depression	The direction of longitudinal associations between sleep problems and depression symptoms: a study of twins aged 8 and 10 years old	Gregory, A.M., et al., Sleep	Epidemiology-longitudinal 300 twin pairs aged 8-10 years old.	A cross-lagged model was used to examine the concurrent and longitudinal associations between sleep problems and depression.	Sleep problems at age 8 predicted depression at age 10 whereas depression symptoms did not predict later sleep problems.
2009	General mental	National Survey of Mental Health and Wellbeing of Australians: Summary of Results.	Australian Bureau of Statistics (ABS).	Report		20% of population had a mental disorders in any 12 months period
2008	Depression and insomnia	A bidirectional relationship between anxiety and depression, and insomnia? A prospective study in the general population	Jansson-Fröjmark, M. and Lindblom, K. Journal of Psychosomatic Research	Epidemiology-longitudinal 1498 Aged 20-60 years old	One year follow up study by posted questionnaires. Use Hospital Anxiety and Depression Scale (HADS). Sleep measures: Basic Nordic Sleep Questionnaires and Uppsala Sleep Inventory Define insomnia based on DSM and ICSD-reporting a sleep problem for 3 nights or more per week during the past 3 months Statistic: Correlation analysis for (depression-anxiety-insomnia relationship) Bidirectional analysis: Four stepwise logistics regression with odds ratio	Anxiety, depression and insomnia are bidirectional related over time. Insomnia at baseline had predictive power odds ratio=3.5 in relation to future depression.
2008	Depression and insomnia	Prevalence, course, and comorbidity of insomnia and depression in young adults	Buysse, D.J., et al., Sleep	Epidemiology-longitudinal 591 participants	20 years follow up.	Insomnia predicted future MDE and that MDE tended to predict future insomnia. Pure insomnia and insomnia comorbid with depression were strongly associated with each other longitudinally, whereas pure depression and insomnia comorbid with depression were less consistently related to each other longitudinally.
2008	Depression and insomnia	Adolescent insomnia as a risk factor for early adult	Roane, B.M. and Taylor, D.J.	Epidemiology-longitudinal 4494 adolescents, 12 to 18	6 to 7 years follow-up.	Insomnia symptoms during adolescence were a significant

Table 1 (*Continued*)

		depression and substance abuse	Sleep	years old at baseline, with 3582 young adults, 18 to 25 years.		risk factor for depression diagnosis
2008	Depression	Which symptoms predict recurrence of depression in women treated with maintenance interpersonal psychotherapy?	Dombrovski, A.Y., et al., Depress Anxiety	131 women aged 20-60.	Used Cox proportional hazards regression models	Persistent is associated with increased risk of recurrence of depression.
2007	Insomnia	The Direct and Indirect Costs of Untreated Insomnia in Adults in the United States	Ozminkowski, R.J., Wang, S. and Walsh, J.K. Sleep	Retrospective 138,820 younger adults and 75,558 elderly patients plus equal-sized, matched comparison groups	Used data from July 1, 1999 to June 30, 2003. Statistic: t-test, multiple regression (2-part regression process)	in average insomnia patients will spend $5790 within 6 months for treatment cost
2007	General sleep and sleep apnea-hypopnea	Scale-free dynamics of the synchronization between sleep EEG power bands and the high frequency component of heart rate variability in normal men and patients with sleep apnea-hypopnea syndrome	M. Dumont, F. Jurysta, J.-P. Lanquart, A. Noseda, P. van de Borne, and P. Linkowski Clinical Neurophysiology	Patients: 24 (44 ± 5) Healthy: 12 (43 ± 6)	2 overnight EEG recordings (20 second epochs) EEG: 3 electrodes (Fz-Ax, Cz-Ax, Oz-Ax) Analysis: Detrended Fluctuation Analysis	Sleep apnea-hypopnea syndrome does not affect the interdependence between the high frequency component of heart rate variability and all sleep power bands as measured by synchronization likelihood.
2007	Depression and cardio	Alteration of cardiac autonomic functions in patients with major depression: A study using heart rate variability measures	Udupa, K., et al., Journal of Affective Disorders	MDD: 26 M, 14 F (30.58 ± 7.4) Healthy: 26 M, 14 F (30.73 ± 7.1)	Subjective measures: DSM-IV TR HDRS > 16 Objectives: ECG: Lead II, 15 min basal recording , extract SDNN, RMSDD, LFnu, HFnu , LF/HF Deep breathing difference Valsava maneuver (intrathoracic pressure) Orthostatic test Statistic unpaired t-test and Pearson Correlation Coefficient	MDD: LF/HF sympathovagal balance was higher, LFnu showed increased sympathetic activity but not significant Significant negative correlation between age and parasympathetic
2007	Depression and sleep	Characterization of the sleep EEG in acutely depressed men using detrended fluctuation analysis.	Leistedt, S., et al., Clinical Neurophysiology	Healthy: 14 M (42.43 ± 5.67) Acute depressed: 10 M (41.7 ± 8.11)	Subjective measures: Pittsburgh Sleep Quality Index (PSQI). Hamilton Rating Scale for Depression (HAM-D) Meet DSM-IV-TR criteria. Objective measures: 3 nights PSG (only last 2 nights used) Extract DFA to each contiguous 20-s epoch during S2, SWS and REM. Statistics: Kruskal–Wallis test and the Spearman's rank correlation Coefficient sleep analysis, Student's	A strong linear correlation was observed between the DFA exponents and the HAM-D score during SWS. A decrease of long-range temporal correlations during stage 2 and SWS in depressive patients in comparison with the healthy controls.

Table 1 (*Continued*)

					t-test or Mann–Whitney	
2006	Insomnia	Epidemiology of insomnia: Prevalence, self-help treatments, consultations, and determinants of help-seeking behaviors	Morin, C.M., et al Sleep Medicine	Epidemiology 2001 Aged 18-91 years old	Telephone survey Used DSM-IV and ICD-10 to define insomnia Statistics: Between group comparison, Chi-squared Tests and logistic regression	9.5% of the population in Canada reported with insomnia which mainly caused by psychological distress (40%).
2006	Depression and insomnia	The association of insomnia with anxiety disorders and depression: Exploration of the direction of risk	Johnson, E.O., Roth, T. and Breslau, N. Journal of Psychiatric Research	Epidemiology 1014 participants age (13-15).	Use structured interviews of DSM-IV To examine directionality between insomnia, depression and anxiety Statistic: Cox Proportional hazards model. Factor: Sex, race/ethnicity, income, parent marital status, parent education	Prior insomnia was associated with onset of depression, prior anxiety may lead to insomnia but less association from depression to insomnia.
2006		The economic cost of sleep disorders.	Hillman, D.R., et al Sleep		Using data derived from national databases (including the Australian Institute of Health and Welfare and the Australian Bureau of Statistics).	The overall cost of sleep disorders in Australia in 2004 was $7494 million
2005	General sleep	Wake up Australia: The value of healthy sleep.	Report by Access Economics Pty Ltd for Sleep Health Australia,	Report		In Australia, it is estimated about 6% of the population are affected by sleep disorders
2005	Insomnia	Is insomnia a marker for psychiatric disorders in general hospitals?	Rocha, F.L., et al., Sleep Medicine	200 participants aged 18-65 years old	Used structured and codified questionnaires. Interviews by psychiatrist using DSM-IV Statistic: Multivariate logistics regression odds ratio	Insomnia is marker for major depressive episode in general hospital inpatients. The prevalence of MDE among patients with insomnia is 46.9%.
2005	Depression and insomnia	Epidemiology of insomnia, depression and anxiety	Taylor, D.J., et al., Sleep	722 participants aged 20-89 years old	Complete sleep diaries for two weeks. Factor: Sex, ethnicity	People with insomnia were 9.82 and 17.35 times as likely to have depression and anxiety. African American and women had higher depression level than Caucasian and men.
2004	General sleep	Interdependency between heart rate variability and sleep EEG: linear/nonlinear?	M. Dumont, F. Jurysta, J.-P. Lanquart, P.-F. Migeotte, P. van de Borne and P. Linkowski Clinical Neurophysiology	Healthy: 8 M (20.5)	2 overnight EEG recordings (20 second epochs) EEG: 3 electrodes (Fz-Ax, Cz-Ax, Oz-Ax) Analysis: Synchronization likelihood, surrogate data statistical analysis	The interaction between delta and HF_{nu}, requires a non-linear model while a more simplistic system (with less oscillators or synchronized oscillators), such as sigma and HF_{nu} interaction is described by a linear model. This study confirms that all sleep EEG power bands are coupled to HF_{nu} band of RRI in healthy young subjects. The interaction between delta, theta and alpha bands and HF_{nu}

Table 1 *(Continued)*

						is non-linear.
2003	Depression and insomnia	Primary insomnia: a risk factor to develop depression?	Riemann, D. and Voderholzer, U. Journal of Affective Disorders	Report from literature review of study conducted from 1966-2000	Investigate primary insomnia according to DSM-IV: insomnia complaints at least 1 month	Insomnia can act as independent predictor /risk factor for developing depression. Most study do not differentiate insomnia according to its subtype (eg. primary, secondary, chronic
2003	Depression and insomnia	Place of chronic insomnia in the course of depressive and anxiety disorders	Ohayon M.M. and Roth T. J Psychiatr Res.	14,915 subjects aged from 15 to 100 years old	Use Sleep EVAL system	previous history of a mental disorder is closely related to the severity and the chronicity of current insomnia
2001	Depression and sleep	Delta sleep ratio as a predictor of sleep deprivation response in major depression.	Nissen, C., et al., Journal of Psychiatric Research	Primary major depressive (SD responder): 5 M, 3 F (37.8 ± 11.5) Non responder: 5 M, 3 F (37.1 ± 13.8)	Subjective: HAMD scored > 18 SD responder: showed 50% improvement of 6 HAMD Non responder: showed <50% improvement of 6 HAMD 2 nights PSG: Sleep deprived study PSG sleep summary were calculated All night spectral analysis for frequency bands: delta, theta, alpha, sigma, beta1 and beta2. Delta sleep ratio is defined as a quotient of SWA in the first to the second NREM episode. All analysis done in log value (base e) Statistic: t-test and 2 factorial ANOVA	A high delta sleep ratio is a positive predictor for SD response.
2000	Insomnia	Prevalence of insomnia in a survey of 12 778 adults in France.	Leger, D., et al., Journal of Sleep Research	Epidemiology 12 778 subjects Aged 18-65+ years old	Questionnaires adapted from DSM-IV criteria for insomnia Factors: age, sex, profession and marital status Statistic: Student t-test, Chi-squared tests and odds ratio	9% of the population was reported with severe insomnia. Sleep complaints were higher in woman, increase significant with age, in retiree and white collar worker and in widows, divorcee
2000	General sleep	Sleep loss and daytime sleepiness in the general adult population of Japan.	Liu, X., et al Psychiatry Research	Epidemiology 3030 Aged 20+	Post questionnaire Statistic: Chi-squared test, Spearman rank correlation coefficients were computed to examine relationships between age and sleep duration, Statistic: Logistic regression analysis was applied to examine associations of sleep problems with age and gender	In Japan, the prevalence of insomnia was 21.4% and increase in elderly
2000	Depression and insomnia	Depression and insomnia: questions of cause and effect	Lustberg, Lisa Reynolds, Charles F.	Review		Chronic insomnia is a risk factor for depression and MDE. PSG revealed reduction of delta

Table 1 (*Continued*)

			Sleep Medicine Reviews			waves in NREM and decreased REM latency.
2000	Depression and sleep	Towards a neurobiology of dysfunctional arousal in depression: the relationship between beta EEG power and regional cerebral glucose metabolism during NREM sleep	Nofzinger, E.A., et al., Psychiatry Research: Neuroimaging,	Depressed: 4 M, 8 F (40.14 ± 10.5) Healthy: #1: 3 M, 6 F (32.02 ± 9.19) #2: 4 M, 8 F (40.11 ± 10.82)	Subjective measures: Depression: Meet diagnosis DSM-III-R (primary unipolar), HRSD min score 15 Healthy: Meet Research Diagnostic Criteria HRSD: score < 6 All must complete visual analog scale to describe sleep quality Objective measures: Depression: PSG sleep and NREM PET #1 PSG and NREM PET #2 PSG only PSG 3 night (used 2nd night as baseline sleep night) EEG: FFT of 4 s epochs to calculate beta power (20-32 Hz) PET: 2 nights Statistics: ANCOVA, t –test and Spearman's rho correlation to test sleep parameters	Depression: Described worse sleep quality on a baseline sleep night
1999	Insomnia	Characteristics of insomnia in the United States : Results of the 1991 National Sleep Foundation survey. I.	Ancoli Israel, et al American Academy of Sleep Medicine.	Epidemiology 1000		In US, the prevalence rate of occasional insomnia was 25% and 9% was reported for regular sleep disturbance
1999	Insomnia	Evaluation of severe insomnia in the general population: results of a European multinational survey.	Chevalier H, L.F. et al J Psychopharmacol,	Epidemiology	Face to face interview	prevalence of severe insomnia was ranged from 4% to 22% and was higher in female
1999	Insomnia	The direct economic costs of insomnia in the United States for 1995	Walsh J.K., Engelhardt C.L. Sleep	Report	Used data compiled by IMS America, Ltd and data of the 1994 National Ambulatory Medical Care Survey conducted by the National Center for Health Statistics and from the America Medical Association Center for Health Policy Research	In 1995, total direct costs in the United States for insomnia in were estimated to be $13.9 billion.
1997	Insomnia and depression	Insomnia in young men and subsequent depression. The Johns Hopkins Precursors Study	Chang, P.P., et al., American Journal of Epidemiology	1053 participants	34 years follow up. Use Cox proportional hazards analysis.	Insomnia in young men is indicative of a greater risk for subsequent clinical depression and psychiatric distress that persists for at least 30 years.
1996	Insomnia	The socioeconomic impact of insomnia. An overview.	Chilcott L.A. , Shapiro C.M. Pharmacoeconomics	overview		In 1994, the total direct, indirect and related costs of insomnia are conservatively estimated at $US30 to 35 billion annually in the US
1988	General sleep	A survey of sleeping difficulties in an Australian population	Lack, L., Miller, W. and Turner, D. Community Health Studies,	216 participants	Telephone survey in Adelaide	13-20% of population experienced various sleep disturbance with 6-7% have difficulty to initiate sleep

9 Conclusion

Most of this section of this chapter cover the comprehensive literature review on the studies published upto date based on the resting neuronal and cardiac signal interaction and synchronization during awake and sleep, in patients suffering from insomnia, depression and schizophrenia. The literature also characterises these disorders and sheds a spectrum of rays for authors and others to pursue their research towards one common cause, *and that is to assist in the diagnosis of sleep, depression and schizophrenia disorders.* The authors are currently conducting concurrent projects to better understand the connectivity between insomnia and schizophrenia by finding the reason why insomniac patients have an increase risk to develop depression that may become chronic or even lead to occurrence of first episodes of schizophrenia.

References

Neckelmann, D., Mykletun, A., Dahl, A.A.: Chronic insomnia as a risk factor for developing anxiety and depression. Sleep 30(7), 873–880 (2007)

Kennedy, G.A., Solin, P.: How to treat- insomnia- part I. Australian Doctor, 37–44 (2004)

Perlis, M.L., Sharpe, M., Smith, M.T., Greenblatt, D., Giles, D.: Behavioral Treatment of Insomnia: Treatment Outcome and the Relevance of Medical and Psychiatric Morbidity. J. Behav. Med. 24(3), 281–296 (2001)

Livingston, G., Blizard, B., Mann, A.: Does sleep disturbance predict depression in elderly people? A study in inner London. Br. J. Gen. Pract. 43(376), 445–448 (1993)

Fawcett, J., Scheftner, W.A., Fogg, L., Clark, D.C., Young, M.A., Hedeke, D., Gibbons, R.: Time-related predictors of suicide in major affective disorder. Am. J. Psychiatry 147, 1189–1194 (1990)

Riemann, D., Spiegelhalder, K., Feige, B., Voderholzer, U., Berger, M., Perlis, M., Nissen, C.: The hyperarousal model of insomnia: A review of the concept and its evidence. Sleep Medicine Reviews 14, 19–31 (2010)

Littner, M., Hirshkowitz, M., Kramer, M., Kapen, S., Anderson, W.M., Bailey, D., Berry, R.B., Davila, D., Johnson, S., Kushida, C., Loube, D.I., Wise, M., Woodson, B.T.: Practice Parameters for Using Polysomnography to Evaluate Insomnia: An Update. Sleep 26(6), 754–760 (2003)

Morin, C.M., Vallieres, A., Ivers, H.: Dysfunctional beliefs and attitudes about sleep (DBAS): validation of a belief version (DBAS-16). Sleep 30(11), 1547–1544 (2007)

Treynor, W., Gonzalez, R., Nolen-Hoeksema, S.: Rumination reconsidered: A psychometric analysis. Cognitive Therapy and Research 27(3), 247–259 (2003)

Edinger, J.D., Krystal, A.D.: Subtyping primary insomnia: is sleep state misperception a distinct clinical entity? Sleep Med. Rev. 7(3), 203–214 (2003)

Taylor, D.J., Lichstein, K.L., Durrence, H.H., Reidel, B.W., Bush, A.J.: Epidemiology of insomnia, depression, and anxiety. Sleep 28(11), 1457–1464 (2005)

Johnson, E.O., Roth, T., Breslau, N.: The association of insomnia with anxiety disorders and depression: Exploration of the direction of risk. Journal of Psychiatric Research 40(8), 700–708 (2006)

Gregory, A.M., Rijsdijk, F.V., Lau, J.Y.F., Dahl, R.E., Eley, T.C.: The Direction of Longitudinal Associations Between Sleep Problems and Depression Symptoms: A Study of Twins Aged 8 and 10 Years. Sleep 32(2), 189–199 (2009)

Chang, P.P., Ford, D.E., Mead, L.A., Cooper-Patrick, L., Klag, M.J.: Insomnia in young men and subsequent depression. The Johns Hopkins Precursors Study. Am. J. Epidemiol. 146(2), 105–114 (1997)

Rocha, F.L., Hara, C., Rodrigues, C.V., Costa, M.A., Castro e Costa, E., Fuzikawa, C., Santos, V.G.: Is insomnia a marker for psychiatric disorders in general hospitals? Sleep Medicine 6(6), 549–553 (2005)

Ohayon, M.M., Roth, T.: Place of chronic insomnia in the course of depressive and anxiety disorders. Journal of Psychiatric Research 37(1), 9–15 (2003)

Jansson-Fröjmark, M., Lindblom, K.: A bidirectional relationship between anxiety and depression, and insomnia? A prospective study in the general population. Journal of Psychosomatic Research 64(4), 443–449 (2008)

Mayers, A.G., Baldwin, D.S.: The relationship between sleep disturbance and depression. International Journal of Psychiatry in Clinical Practice 10(1), 2–16 (2006)

Stampfer, H.G.: The relationship between psychiatric illness and the circadian pattern of heart rate. Australian and New Zealand Journal of Psychiatry 32(2), 187–198 (1998)

Iverson, G.L., Gaetz, M.B., Rzempoluck, E.J., McLean, P., Linden, W., Remick, R.: A new potential marker for abnormal cardiac physiology in depression. Journal of Behavioral Medicine 28(6), 507–511 (2005)

Boostani, R., Sadatnezhad, K., Sabeti, M.: An efficient classifier to diagnose of schizophrenia based on the EEG signals. Expert Systems with Applications 36(3, Part 2), 6492–6499 (2009)

Togo, F., Cherniack, N.S., Natelson, B.H.: Electroencephalogram characteristics of autonomic arousals during sleep in healthy men. Clin. Neurophysiol. 117(12), 2597–2603 (2006)

Jurysta, F., Lanquart, J.-P., Sputaels, V., Dumont, M., Migeotte, P.-F., Leistedt, S., Linkowski, P., van de Borne, P.: The impact of chronic primary insomnia on the heart rate – EEG variability link. Clinical Neurophysiology 120, 1054–1060 (2009)

Jurysta, F., Kempenaers, C., Lancini, J., Lanquart, J.P., Van De Borne, P., Linkowski, P.: Altered interaction between cardiac vagal influence and delta sleep EEG suggests an altered neuroplasticity in patients suffering from major depressive disorder. Acta Psychiatrica Scandinavica 121, 236–239 (2010)

Udupa, K., Sathyaprabha, T.N., Thirthalli, J., Kishore, K.R., Lavekar, G.S., Raju, T.R., Gangadhar, B.N.: Alteration of cardiac autonomic functions in patients with major depression: a study using heart rate variability measures. J. Affect. Disord. 100(1-3), 137–141 (2007)

Gillin, J.C., Duncan, W., Pettigrew, K.D., Frankel, B.L., Snyder, F.: Sucessful separation of depressed, normal and insomniac subjects by EEG sleep data. Arch. Gen. Psychiatry 36(1), 85–90 (1979)

Steiger, A., Kimura, M.: Wake and sleep EEG provide biomarkers in depression. J. Psychiatr. Res. 44(4), 242–252 (2010)

Staner, L., Cornette, F., Maurice, D., Viardot, G., Le Bon, O., Haba, J., Staner, C., Luthringer, R., Muzet, A., Machner, J.-P.: Sleep microstructure around sleep onset differentiates major depressive insomnia from primary insomnia. J. Sleep Res. 12, 319–330 (2003)

Bonnet, M.H., Arand, D.L.: 24-Hour metabolic rate in insomniacs and matched normal sleepers. Sleep 18(7), 581–588 (1995)

Nofzinger, E.A., Buysse, D.J., Germain, A., Price, J.C., Miewald, J.M., Kupfer, D.J.: Functional neuroimaging evidence for hyperarousal in insomnia. Am. J. Psychiatry 161(11), 2126–2128 (2004)

Leistedt, S., Dumont, M., Lanquart, J.P., Jurysta, F., Linkowski, P.: Characterization of the sleep EEG in acutely depressed men using detrended fluctuation analysis. Clin. Neurophysiol. 118(4), 940–950 (2007)

Westphal, K.P., Grozinger, B., Diekmann, V., Scherb, W., Reeb, J., Leibing, U., Kornhuber, H.H.: Slower theta activity over the midfrontal cortex in schizophrenia patients. Acta Psychiatr. Scand. 81, 132–138 (1990)

Inanaga, K.: Frontal midline theta rhythm and mental activity. Psychiatry and Clinical Neurosciences 52(6), 555–566 (1999)

Critchley, H.D., Mathias, C.J., Josephs, O., O'Doherty, J., Zanini, S., Dewar, B.-K., Cipolotti, L., Shallice, T., Dolan, R.J.: Human cingulated cortex and autonomic control: converging neuroimaging and clinical evidence. Brain 126(10), 2139–2152 (2003), doi:10.1093/brain/awg216

Maron-Katz, A., Ben-Simon, E., Sharon, H., Gruberger, M., Cvetkovic, D.: A Neuroscientific Perspective on Meditation. In: Fábián, T.K. (ed.) Psychology of Meditations, Mind-Body Connections: Pathways of Psychosomatic Coupling under Meditation and Other Altered States of Consciousness, NOVA, Hungary, EU (2013) (in press)

Uhlhaas, P.J., Singer, W.: Abnormal Neural Oscillations and Synchrony in Schizophrenia. Nature Reviews, Neuroscience 11, 100–113 (2010)

von Stein, A., Chaing, C., Konig, P.: Top down processing mediated by interareal synchronization. Proc. Natl. Acad. Sci. 97, 14748–14753 (2000)

Womelsdorf, T., et al.: Modulation of neuronal interactions through neuronal synchronization. Science 316, 1609–1612 (2007)

Cho, R.Y., Konecky, R.O., Carter, C.S.: Impairments in frontal cortical gamma synchrony and cognitive control in schizophrenia. Proc. Natl. Acad. Sci. 103, 19878–19883 (2006)

Uhlhaas, P.A., et al.: Dysfunctional long-range coordination of neural activity during Gestalt perception in schizophrenia. J. Neurosci. 26, 8168–8175 (2006)

Cotton, S.M., Lambert, M., Schimmelmann, B.G., Mackinnon, A., Gleeson, J.F.M., Berk, M., Hides, L., Chanen, A., McGorry, P.D., Conus, P.: Depressive symptoms in first episode schizophrenia spectrum disorder. Schizophrenia Research 134, 20–26 (2012)

Breakspear, M., Terry, J.R., Friston, K.J., Harris, A.W.F., Williams, L.M., Brown, K., Brennan, J., Gordon, E.: A disturbance of non-linear interdependence in scalp EEG of subjects with first episode schizophrenia. NeuroImage 20, 466–478 (2003)

Schulz, S., Bar, K.J., Voss, A.: Respiratory variability and cardiorespiratory coupling analyses in patients suffering from schizophrenia and their healthy first-degree relatives. Biomed. Tech. 57(suppl.1), 104 (2012)

Bar, K.J., Boettger, M.K., Koschke, M., Schulz, S., Chokka, P., Yeragani, V.K., et al.: Non-linear complexity measures of heart rate variability in acute schizophrenia. Clinical Neurophysiology 118, 2009–2015 (2007)

Henry, B.L., Minassian, A., Paulus, M.P., Geyer, M.A., Perry, W.: Heart rate variability in bipolar mania and schizophrenia. Journal of Psychiatric Research 44(3), 168–176 (2010)

Ferrarelli, F., Peterson, M.J., Sarasso, S., Riedner, B.A., Murphy, M.J., Benca, R.M., Bria, P., Kalin, N.H., Tononi, G.: Thalamic dysfunction in schizophrenia suggested by whole-night deficits in slow and fast spindles. Am. J. Psychiatry 167, 1339–1348 (2010)

Ferrarelli, F., Huber, R., Peterson, M.J., Massimini, M., Murphy, M., Riedner, B.A., Watson, A., Bria, P., Tononi, G.: Reduced sleep spindle activity in schizophrenia patients. Am. J. Psychiatry 164, 483–492 (2007)

Goder, R., Aldenhoff, J.B., Boigs, M., Braun, S., Koch, J., Fritzer, G.: Delta power in sleep in relation to neuropsychological performance in healthy subjects and schizophrenia patients. Journal of Neuropsychiatry and Clinical Neurosciences 18, 529–535 (2006)

Zalesky, A., Fornito, A., Seal, M.L., Cocchi, L., Westin, C.-F., Bullmore, E.T., Egan, G.F., Pantelis, C.: Disrupted Axonal Fiber Connectivity in Schizophrenia. Biology Psychiatry 69, 80–89 (2011)

Bob, P.: Consciousness, schizophrenia and complexity. Cognitive Systems Research 13(1), 87–94 (2012)

Dumont, M., Jurysta, F., Lanquart, J.-P., Migeotte, P.-F., van de Borne, P., Linkowski, P.: Interdependency between heart rate variability and sleep EEG: linear/non-linear? Clinical Neurophysiology 115(9), 2031–2040 (2004)

Dumont, M., Jurysta, F., Lanquart, J.-P., Noseda, A., van de Borne, P., Linkowski, P.: Scale-free dynamics of the synchronization between sleep EEG power bands and the high frequency component of heart rate variability in normal men and patients with sleep apnea–hypopnea syndrome. Clinical Neurophysiology 118(12), 2752–2764 (2007)

Nikulin, V.V., Jönsson, E.G., Brismar, T.: Attenuation of long-range temporal correlations in the amplitude dynamics of alpha and beta neuronal oscillations in patients with schizophrenia. NeuroImage 61(1), 162–169 (2012)

Noh, K., Shin, K.S., Shin, D., Hwang, J.Y., Kim, J.S., Jang, J.H., Chung, C.K., Kwon, J.S., Cho, K.-H.: Impaired coupling of local and global functional feedbacks underlies abnormal synchronization and negative symptoms of schizophrenia. BMC Systems Biology 7(30), 1–13 (2013)

Timashev, S.F., Panischev, O.Y., Polyakov, Y.S., Demin, S.A., Kaplan, A.Y.: Analysis of cross-correlations in electroencephalogram signals as an approach to proactive diagnosis of schizophrenia. Physica A: Statistical Mechanics and its Applications 391(4), 1179–1194 (2012)

Questions

Q1. What is the causal relationship between insomnia and patho-physiological disorders (e.g. depression, schizophrenia, etc.)?

Q2. Which are the physiological cues characterising primary and chronic insomnia?

Q3. From a neuronal perspective, what is the relationship between schizophrenia disease and EEG spectral bands?

Q4. Explain the term functional dysconnectivity used in schizophrenia.

Q5. List the physiological changes that are associated with depression.

Q6. Differentiate the terms primary insomnia and secondary insomnia.

5. A New Era in Sleep Monitoring: The Application of Mobile Technologies in Insomnia Diagnosis

Sana Tmar-Ben Hamida*, Beena Ahmed, Dean Cvetkovic, Emil Jovanov, Gerard Kennedy, and Thomas Penzel

Electrical and Computer Engineering Department, Texas A&M University, Doha, PO Box 23874, Qatar
sana.tmar@qatar.tamu.edu

Abstract. Sleep disorders, such as insomnia can seriously impair a patient's quality of life. Existing studies have shown that insomniacs have a risk of hypertension 350 percent higher than normal sleepers. Insomnia is also a risk factor for diabetes, as well as anxiety and depression. Sleep measurements based on polysomnographic (PSG) signals and questionnaires are necessary for an accurate evaluation of insomnia; however PSG systems are uncomfortable and inconvenient as they require patients to stay overnight at sleep centers. There is an increasing interest in portable devices, which provide the opportunity for the assessment of insomnia in a native environment (e.g. patients' homes). Due to recent advances in technology, it is now possible to continuously monitor a patient's sleep at home and send their sleep data to a remote clinical back-end system for analysis and reporting. This chapter provides a systematic analysis on the sleep monitoring technologies that can be used for insomnia assessment and treatment. This study highlights the key technical challenges of sleep monitoring, describes different types of technologies and discusses their applications in insomnia assessment. An overview of some model-based signal processing for sleep staging and insomnia detection is presented. Lastly, this chapter ends with a discussion, which provides future directions for the deployment of effective in-home patient monitoring systems for insomnia diagnosis.

Keywords: Health care, sleep analysis, mobile sleep monitoring, in-home insomnia monitoring, remote physiological monitoring.

1 Introduction

Sleep is a universal need; it provides relaxation and comfort and affects the body's equilibrium (Murali, Svatikova, & Somers, 2003). Sleep deprivation can cause health problems with serious consequences to human physiology and the neurological system. Studies have shown that insomniacs have a risk of hypertension

---

* Corresponding author.

S. Adibi (ed.), *Mobile Health*,
Springer Series in Bio-/Neuroinformatics 5, DOI: 10.1007/978-3-319-12817-7_5

350 percent greater than normal sleepers. Insomnia is also a risk factor for diabetes (Murali, Svatikova, & Somers, 2003) and increases the chances of developing anxiety and depression.

Despite the prevalence of insomnia in modern society, it has received limited clinical attention due to difficulties in its accurate diagnosis and treatment. Insomnia must be evaluated in a systematic manner, with a well-established framework, so that crucial diagnostic information is not overlooked. An adequate evaluation of persistent insomnia requires detailed historical information as well as medical, psychological and psychiatric assessment (Zhang & Zhao, 2007). Sleep diaries and specialized questionnaires are necessary for an accurate evaluation of insomnia. Face-to-face clinical consultation is followed up with responses on these instruments to determine the problem and establish and inform treatment decisions. At the initial face-to-face consultation, insomniacs have an evaluation by a psychologist or medical practitioner and this can be followed by nocturnal polysomnography (PSG). Practices of good and poor sleep hygiene (keeping regular bedtime, low intake of caffeine, exercise, napping, etc.) have also been a subject of systematic focus.

Accordingly, subjective and objective measures of sleep structure need to be collected from patients and combined for a precise evaluation of sleep quality. Subjective measures are currently collected during face-to-face clinical consultations with a psychologist or medical practitioner. With an increasingly mobile society and the worldwide deployment of mobile applications, this information can be collected using an application on a smart phone or tablet device. Objective measures are obtained from PSG signals collected during several hours of sleep from on-body specific electrodes. These include recordings of electroencephalogram (EEG), electromyogram (EMG), electrocardiogram (ECG) and electrooculogram (EOG). The spectrum of these signals depends on the state of the brain and bodily activity in wakefulness and sleep states.

However, sleep studies are complicated for patients and healthy subjects particularly when they need to spend several nights sleeping in unfamiliar surroundings with limited privacy. Moreover, leads connected to different electrodes are inconvenient and uncomfortable for the sleeper, resulting in sleep activity different from their normal pattern of sleep. While various psychological experiences are often present in insomniacs, which keeps them awake at night, the unfamiliar PSG system can often add to this problem.

This has led to interest in a wearable, effective insomnia monitoring system which can be used by patients in the comfort of their homes over long periods of time. Driven by improvements in sleep recording systems and the availability of high mobile bandwidth, sleep monitoring systems are now offering a wider range of new services. This revolution makes the idea of in-home sleep monitoring practical and provides the opportunity for sleep assessment in "real-world" environments producing more ecologically valid data.

The objective of this chapter is to provide a detailed evaluation of the existing monitoring sleep systems that can be utilized for insomnia assessment and treatment. Despite, the deployment of large variety of devices developed to evaluate different sleep disorders, there is no single system that can combine subjective and objective measures for use with insomniac patients. We present in this chapter, an example of an in-home system for monitoring and screening of insomnia that allows collecting and combining efficiently different types of measurements. This system consists of four main components: (1) PSG monitoring; (2) electronic sleep diary, (3) the device coordinator; and (4) the clinical back-end system.

The outline of this chapter is as follows: Section 2 describes the physiology of sleep and the insomnia problem and reports the approach used for insomnia assessment. Section 3 reviews the actual sleep monitoring systems. These devices are classified into groups depending on the type of data collected. Section 4 derives design guidelines for the deployment of a practical in-home sleep monitoring system for an accurate insomnia assessment and presents the different system constraints and challenges that can be encountered. Section 5 reviews the signal analysis methods used to extract the required measures for insomnia diagnosis. Finally, Section 6 outlines future directions for in-home insomnia assessment and treatment methods.

2 Physiology of Sleep

2.1 Normal Sleep

In humans, sleep-wake is a periodic cycle which is regulated by different systems in the brain. There are two types of normal sleep: NREM (non-Rapid Eye Movements) and REM sleep (Rapid Eye Movements) (Danker-Hopfe, et al., 2009). The duration of the sleep cycle for adults is between 90 and 100 minutes on average (Colten & B. M. Altevogt, 2006). During a normal night of sleep, the sleep cycle is repeated 4 or 5 periods. The normal night of sleep is classified into 4 phases as shown in Figure 1.

During analysis, the PSG data is usually divided into epochs of 20s or 30s. Then, the sleep scientists visually classify the different epochs into sleep stages according to the sleep clinical standard proposed by the American Academy of Sleep Medicine (AASM) ((Danker-Hopfe, et al., 2009), (Berry, et al., 2012)). The AASM rules categorize the normal healthy sleep into four states: Movement (MT), Wake (W), NREM (non-Rapid Eye Movements) sleep and REM (Rapid Eye Movements) sleep. The NREM sleep is then classified into three stages (referred to N1 through N3).

These stages are characterized by the presence of the following rhythms in the EEG: alpha (α), beta (β), theta (θ) and delta (δ) (Danker-Hopfe, et al., 2009). Each epoch is assigned to a specific stage based on the presence of these rhythms and their properties and duration. As described in Table 1, stage W is characterized by the presence of mixed EEG frequencies (α, and β) and body movements may also appear (Carskadon & Dement, 1994). Beta (β) waves are defined as low amplitude and high frequency and are dominant during the wake stage. Stage N1 is the transition between the wake and the sleep states. It is identified by the reduction of alpha (α) rhythm and the appearance of low amplitude theta (θ) rhythms. During this stage, the respiration and cardiac rates decrease. Stage N2 is distinguished by the presence of K-complex waves (a negative high voltage transient) followed by sleep spindle bursts (their frequency range is between 11 and 16Hz) mixed with the stage N1 waves. Delta (δ) waves are characterized by a high amplitude (between 20 and 200μV) and low frequencies (below 5Hz); generally occurring during deep sleep (stage N3).

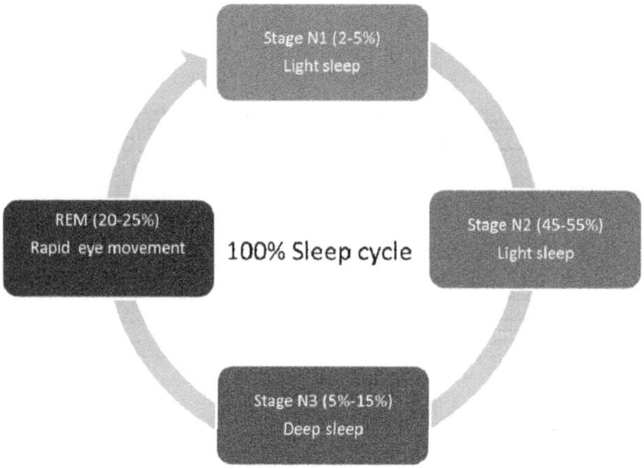

Fig. 1 A normal sleep cycle

The REM stage is known as paradoxical sleep since it is characterized by low amplitude, irregular and mixed brain waves. The brain activity at this time is similar to the wake state. The incidence ratios of delta and spindle wave decrease, rapid eyes movements appear, EOG waves are similar to those in stage W and the chin slacks.

Table 1 EEG waves and their corresponding stages

Waves	Frequency (Hz)	Amplitude (μV)	Stages
Delta	0.5-3	20-200	N3
Theta	4-7	5-100	N1,N2
Alpha	8-13	20-60	W,N1,N2 and REM
Beta	13-30	2-20	W

It is difficult to quantify the amount of sleep needed as our need for sleep depends on our age and other factors that are unique to each individual. Table 2 shows the average amount of sleep people tend to have at different stages of life. Table 2 also shows how much of the different types of sleep we have at different ages.

Table 2 Sleep structure across the lifespan

	Sleep Hours	Stage 1-2(%)	Stage 3 (%)	REM(%)
Infants	13-16	10-30	30-40	40-50
Children	8-12	40-60	20-30	20-30
Adults	6-9	45-60	15-25	15-25
Elderly	5-8	50-80	5-15	15-25

About 96% of adults sleep 6-9 hours each night. Adults, who sleep less than six hours each night, tend to show symptoms of sleep deprivation. Sleep deprivation symptoms include reduced daytime alertness and performance, negative mood changes and fatigue. The sleep/wake cycle is sensitive to a wide range of lifestyle issues, mood disturbances and medical conditions, that while not sleep disorders in themselves, may cause acute or chronic sleep disruption.

2.2 *Insomnia*

The word 'insomnia' is commonly used to describe the problem of not being able to fall asleep, not being able to stay asleep or having both of these problems at the same time. Insomnia is not a disease and can have many different causes such as stress, anxiety, depression, medical conditions, excessive caffeine, alcohol or nicotine and irregular schedules and poor sleep habits. Insomnia is a primary sleep disorder that is classified under the category of dyssomnias in the *Diagnostic and Statistical Manual* of the American Psychiatric Association (DSM-IV-R). Dyssomnias are disorders of initiating or maintaining sleep or excessive sleepiness and are characterised by disturbances in the duration, quality and/or the timing of sleep. Based on the cause and symptoms, insomnia can be classified into either primary or secondary insomnia. Primary insomnia is diagnosed when sleep disturbance is the primary or main complaint (see Table 3). Secondary insomnia is diagnosed when sleep disturbance occurs with another medical or psychiatric condition, or other types of sleep disorders (e.g., obstructive *sleep apnea and hypopnea* (OSAH)) or substance abuse disorders like alcoholism. People with psychiatric conditions can suffer from secondary insomnia, reporting a wide range of sleep-related problems (Perlis, Sharpe, Smith, Greenblatt, & Gile, 2001). The main night-time symptoms of insomnia are usually difficulty falling asleep and/or staying asleep. Daytime symptoms of insomnia usually include fatigue, sleepiness, lack of energy, irritability and other mood disturbances (e.g., depressed mood), poor motivation and concentration, and poor work performance. Daytime symptoms may also have negative effects on social functioning and can increase the risk of motor vehicle and work-related accidents.

The percentage of the population that reports "trouble with sleeping" depends on age and gender. Only about 9% of people between the ages of 20-30 years report sleep problems. In the mid-life age group (30-50 years old) up to 16% report problems, with more women than men reporting sleep problems. In people over 60 years old, about 20% report chronic problems with their sleep.

2.2.1 Primary Insomnia

The Diagnostic and Statistical Manual of the American Psychiatric Association criteria for insomnia state that the sleep disturbance(s) must result in impairments of daytime functioning (or mood) or cause clinically significant distress in order to be classified as insomnia. However, many people may have sleep difficulties that while not meeting the DSM-IV-TR criteria for insomnia, may still be defined as poor sleep in terms of the individual's sleep needs and perceptions of sleep disturbance. The diagnostic criteria for primary insomnia are shown in Table 3.

Table 3 Diagnostic criteria for primary insomnia

A.	The predominant complaint is difficulty initiating or maintaining sleep (or both) or non-restorative sleep for at least one month
B.	The sleep disturbance (or associated daytime fatigue) causes clinically significant distress or impairment in social, occupational or other important areas of functioning
C.	The sleep disturbance does not occur exclusively during the course of narcolepsy, breathing-related sleep disorders (e.g., asthma, sleep apnea), circadian rhythm sleep disorder (e.g., delayed sleep-phase insomnia) or a parasomnia (inappropriate arousal during sleep)
D.	The disturbance does not occur exclusively during the course of another mental disorder (e.g., major depressive disorder, generalised anxiety disorder, a delirium)
E.	The disturbance is not due to the direct physiological effects of a substance (e.g., a drug of abuse, a medication) or a general medical condition.

The DSM-IV-TR diagnosis of primary insomnia subsumes several insomnia diagnoses in the International Classifications of Sleep Disorders. Primary insomnia is essentially the same as psychophysiological insomnia. The term psychophysiological insomnia has been used to more accurately reflect the most common causes of insomnia, which typically include inappropriate arousal and conditioning and/or learning factors.

Another condition, sleep state misperception is characterized by complaints of insomnia, but there is usually a marked discrepancy between reported and observed estimates of sleep duration. That is, the person complains of sleeplessness, but objective measurement in a sleep laboratory show that the person actually sleeps a lot more than they think they do. Under estimating the amount of sleep one has is also common in people with longstanding primary insomnia. Sleep state misperception is a much less common problem in its own right.

2.2.2 Secondary Insomnia

Secondary insomnia is the symptom or side effect of another problem. This type of insomnia often is a symptom of an emotional, neurological, or other medical or sleep disorder. The conditions that can cause insomnia, include:

i) Psychiatric disorders

People with psychiatric conditions can report a wide range of sleep-related problems. However, it can be difficult to determine whether the sleep problems are due to medications (neuroleptics), the psychiatric disorder, or both these factors together. A careful review of the person's medical records and taking a thorough history may allow the causal roles of medications and disease factors in the sleep problem(s) to be determined. Sleep disruption and insomnia are common symptoms of some psychiatric disorders.

ii) Depression

Depression can be complicated by the problem of early morning awakening and difficulty returning to sleep. About 30% of people suffering from major depression report sleep problems (e.g. insomnia, hypersomnia or dream disorders). Studies have suggested that people with depression and insomnia are particularly at risk for suicide.

iii) Mania

People experiencing a manic episode usually have insomnia and sleep very little, but usually do not complain of sleeplessness.

iv) Schizophrenia

Insomnia, sleep fragmentation, hypersomnia, nightmares and the complete reversal of day-night sleep and activity patterns are very common problems in people suffering from schizophrenia type illnesses. These problems can be due to a number of factors including; reduced motivation and a lack of daily goals and scheduled activities, the medications used to treat the condition and the actual the condition itself.

v) Dementia

People with Alzheimer's disease often show a reversal of their day-night sleep and activity patterns. In some cases nocturnal behaviour may be associated with improved cognitive function (sundowning). Studies have shown that by increasing patients' daytime exposure to bright light, nocturnal sleep can be improved and consequently nocturnal wandering can be reduced. Bright light exposure each day particularly in the morning can be achieved by taking patients outside, making sure the blinds are raised and installing additional ceiling lighting above the day time activities areas.

vi) Post-traumatic stress disorder (PTSD)

People with post-traumatic stress disorder usually report frequent recurring nightmares and also a state of hyperarousal (difficulty initiating and maintaining sleep). Insomnia, restless sleep and trauma-related dreams are some of the primary symptoms of post-traumatic stress disorder. Many people with post-traumatic stress disorder have a delayed sleep phase staying late and sleeping during the day. The sleep problems associated with post-traumatic sleep disorder can and should be treated either in conjunction with or separately to the counselling for the trauma itself.

vii) Generalized anxiety and panic disorders

These anxiety disorders are associated with insomnia and sleep disruption when people experience nocturnal panic attacks (usually in sleep stages 2 and 3). Panic attacks may be mistaken for other parasomnias. Anxiety conditions generally make suffering from insomnia more likely because in some ways anxiety is a key driving factor in insomnia in any case. For many people who have developed poor sleep, anxiety about sleep and in particular not sleeping and the associated negative symptoms becomes a major driving factor in causing the sleep difficulties.

viii) Sleep-related eating disorder

This is a rare clinical syndrome where people experience partial awakening at night during which they eat large amounts of food. Another similar condition is nocturnal eating. In this condition the person feels an overwhelming compulsion to eat large amounts of food in the late evening. Treatment for both conditions usually requires the involvement of medical and/or psychological professionals who specialise in the area of disordered eating and mood problems. However, a sleep expert may also be helpful in diagnosing and assisting with treatment of these disorders.

ix) Medical disorders & Pain

Complaints about sleep problems or difficulties are very common in general medical patients. Sleep difficulties are often related to pain, shortness of breath, effects of medication or the medical disorder itself. Pain is one of the main factors that causes sleep difficulties in medical patients. Pain can cause problems of initiating sleep (onset insomnia) and may also be associated with frequent and/or prolonged periods of arousal (maintenance insomnia). Pain and sleep difficulties often increase as analgesic medications are metabolised during the latter part of the night. Pain may be caused by a variety of conditions such as rheumatoid arthritis, headache and cancers.

x) Breathing disorders

The treatment of insomnia caused by nocturnal shortness of breath is usually directed at the underlying cause of the respiratory problem. People with obstructive sleep apnea syndrome commonly present with snoring, daytime sleepiness and obesity, but also commonly report insomnia, which is due to the disruptive effects of obstructive (cessation of breathing) arousals. People suffering heart failure and Cheyne-Stokes respiration (central sleep apnea) often report insomnia that is due to the disruptive effects of rapid breathing cycles during the night.

2.3 Assessment of Insomnia

Assessment for insomnia involves a multi-prong approach using PSG and *multiple sleep latency tests* (MSLT) to distinguish between *obstructive sleep apnea hypopnea* (OSAH), *periodic limb movement disorder* (PLMD), *restless leg syndrome* (RLS), etc. Sleep logs, self-administered questionnaires, symptom checklists, psychological and psychiatric assessments and screening tests are also used in the assessment process to assist in developing an overall picture of the patient.

2.3.1 Polysomnography (PSG)

PSG is used in the assessment of insomnia to rule out the presence of occult primary sleep disorders and circadian rhythm disorders. Sleep staging and event scoring is used to provide information on the sleep duration and structure of the patient's sleep. A quantification of the percentage and duration of REM and stage 1, 2 and 3 sleep and quantity of sleep spindles and K-complexes present in the overnight recording allows an assessment to be made of whether or not the patient is sleeping sufficiently for their age and rule out other disorders (Chesson, et al., 2000), (American Sleep Disorders Association Task Force, 1992). PSG recordings of insomnia sufferers have been observed to have altered sleep/wake ratios and altered patterns of night-to-night variability. Polysomnography is often used to confirm or quantify a patient's subjective reporting of awakenings (for sleep state misperception disorder) (Reite, Buysse, Reynolds, & Mendelson, 1995).

2.3.2 Sleep Onset

There is little information concerning the changes that occur during the wake to sleep transition or sleep onset. Sleep onset does not occur at one time, it cannot be predicted (or at least it has not been investigated) and there are oscillations in vigilance that occur before the stable sleep. The true wake-to-sleep transition could be said to begin from stages 1 to 2 and the arrival of spindles (Ogilvie, 2001). The actual sleep onset may occur within wake, stages 1 and 2 and it is in its 'loop or cyclic mode' for certain time, until it transits into light and then deeper sleep. In diagnosing insomnia, it is imperative to develop an algorithm to measure and characterize the sleep onset parameters.

Physiologically, sleep onset occurs with: a decrease in the electromyographic (EMG) muscle tonus activity; decrease in heart rate and respiration; shift to hypometabolic parasympathetic activity; increase in rectal temperature; decrease in skin conductivity; shift in EEG activity from low amplitude, high frequency to high amplitude, low frequency. There are two alpha oscillations which characterize sleep onset process, such as alpha prior to sleep onset and alpha activity at sleep onset and during sleep (Ogilvie, Simons, Kuderian, MacDonald, & Rustenburg, 1991). Studies indicate that lower delta and higher alpha predominance is the cause of a distorted sleep onset process which is responsible for insomnia. The standard definition of sleep onset was found to coincide with an alpha power decrease (confirmed with most studies) below 28% of total EEG energy (Shinar, Akselrod, Dagan, & Baharav, 2006). This sharp decrease in alpha power at sleep onset is synchronous with a reciprocal strong increase in delta power. Though alpha activity is probably one of the main indicators of sleep onset, other EEG bands have also been used to characterize this process. ((Lubin, Johnson, & Austin, 1969); (Hori, 1985); (Merica & Fortune, 2004) and (Badia, Wright Jr, & Wauquier, 1994)).

2.3.3 Arousal and Hyperarousal

Primary insomnia is often considered a disorder of 'hyperarousal' which means that patient has an arousal that is incompatible with the falling asleep or maintenance of sleep (Riemann, et al., 2010). Arousal and hyperarousal can pose a lot of problems to the diagnostician as its significance and behavior is still an open question. An objective automated method of assessing a person's arousal level is thus challenging.

There are three types of arousals:

i) *spontaneous arousal* – an EEG arousal not associated with respiratory or limb movement event;
ii) *limb movement* – an EEG arousal that results from limb movement (e.g. periodic leg movements (PLMs));
iii) *respiratory arousal* – an EEG arousal resulting from the occurrence of an apnea/hypopnea.

Arousals are often caused by an underlying pathology like the obstructive sleep apnea hypopnea syndrome (OSAHS) or periodic limb movement syndrome (PLMS). Micro arousals can be seen either simultaneously or before and after the onset of PLM episode (Karadeniz, Ondze, & Besset, 2000). These arousals are some of the contributing factors to fragmented and disturbed sleep, leading to insomnia. Arousals are also part of the normal sleep architecture; normally occuring 20 times per hour. The cortical (EEG) arousals are often accompanied by an increase in chin EMG amplitude, leg EMG amplitude, movement, breathing rate/volume, heart rate and eye blinks. The current standard definitions of the sleep cortical arousal indexes can be defined as a sudden change in EEG theta, alpha band and/or frequencies that are greater than 16Hz but excluding sleep spindles. Autonomic arousal is another type of cortical arousal affected by the heart rate and blood pressure elevation.

3 Mobile Sleep Monitoring

Recent years have witnessed the emergence of various home-based, portable sleep monitoring technologies that integrate different functionalities: measurements, storage and wireless communications. These healthcare tools offer the ability to collect data from patients during several hours of sleep. These systems are mostly catering to patients suffering from sleep disordered breathing such as obstructive sleep apnea hypopnea (OSAH). However, these systems are not entirely wire-free. The PSG sensors plug into a central unit mounted on the patient which then wirelessly transmits data to a receiver in the control room where the PSG data can be monitored on a computer. These 'wireless' polysomnography systems allow for easy mobility since the patient is not tethered to the wiring cables of the main unit.

The recent interest in personal well-being has also led to the release of a range of non-clinical devices paired with smartphones to monitor sleep quality during the night. Most of these systems use non-contact activity sensors to track movements made during sleep to track sleep and wake periods. Due to their high level of convenience, they have gained increased popularity amongst users but have limited clinical validation. In the following sections, we present the most common devices explicitly used for sleep monitoring. The different systems are categorized into groups depending on the type of collected data.

3.1 Sleep Monitoring Based on the Brain Activity

Collecting data using traditional PSG systems requires the patient to leave his home environment, which hinders an accurate insomnia assessment. Moreover, the use of different sensors with several wires may disturb the sleep subjects. Novel alternatives have been proposed that aim to reduce the number of brain electrodes by collecting polysomnographic signals at specific emplacement (frontal, central, etc.). New studies demonstrate that the use of only a single EEG channel can be sufficient to classify epochs into sleep stages (Koley & Dey, 2012.), (Van Hese, Philips, De Koninck, Van de Walle, & Lemahieu, 2001). As example, iBrain developed by NeuroVigil (Neurovigil, The iBrain device, 2013), consists of headgear with single frontal EEG. Sleep Profiler, designed by Advanced Brain Monitoring, (Advanced Brain Monitoring, 2013) provides in-home multi-night chronic insomnia assessment based on three frontal EEG channels. It also detects the head movement positions and offers real-time monitoring. This system does not integrate the subjective measures required for accurate insomnia diagnosis. A limitation of these single channel systems is their inability to distinguish between REM and stage N1 with high accuracy. Hence, the intervention of a sleep expert is recommended to rectify this shortcoming or recording additional polysomnographic signals (as example EOG signals).

3.2 Sleep Monitoring Based on Movements

The development and miniaturization of MEMS and piezoelectric accelerometers has revolutionized systems based on movement measurements. Actigraphy monitors integrate miniature accelerometers with a sensitivity $\geq 0.01g$ in very small and unobtrusive sensors that can be worn on limbs during sleep or for prolonged periods during activities of daily living. Typical sensors include Mini-Motionlogger (Ambulatory Monitoring, Inc., 2013) and wActiSleep-BT (ActiGraph, 2013).

These sensors allow researchers and clinicians to objectively document long-term sleep, hyperactivity, daytime activity levels, fatigue, circadian rhythm, and vigilance, as well as environmental conditions (e.g. light, temperature, sound intensity). They typically provide the event-marker button to allow users to annotate typical events, such as time in and out of bed.

Wearable systems can integrate several sensors embedded in clothes (Bonato, 2009; E. Jovanov N. H., 2009) or using smart textiles (Coyle, et al., 2010). Capacitive sensing and smart clothes represent a very important trend in increased user's convenience (Coyle, et al., 2010; R. Matthews, 2007). For example, BioHarness (Zephyr-BioHarness, 2013) captures and transmits the physiological data on the wearer (vest-like device) via mobile and fixed data networks. This vest records respiration rate, heart rate, breathing rate, skin temperature, body activity, etc. The collected data is transferred to Android devices using Bluetooth.

Wireless communication within Body Area Network (BAN) or Body Sensor Network (BSN) significantly improves user's convenience and allows integration of implanted sensors. Application of specific sensing, processing and wireless communication (Yseboodt, et al., 2009; Wong, McDonagh, Omeni, Nunn, Hernandez-Silveira, & Burdett, 2009) significantly decrease power consumption, which extends battery life or facilitates use of energy harvesting (Vullers, Van Schaijk, Visser, Penders, & Van Hoof, 2010).

3.3 Sleep Monitoring Based on Bed Sensors

The user's convenience has inspired a number of approaches alternative to wearable physiological sensors. It has been demonstrated that infrared cameras and ultra-wide band radars could be used to monitor physiological parameters during sleep. However, both approaches require sensors of significant complexity and price. A growing demand for inexpensive systems to monitor elderly at their homes has motivated development of bed sensors, such as special mattresses or sensors based on ballistocardiography placed under mattress (Taylor, Grant, Knoefel, & Goubran, 2013). The sensor can extract bed occupancy information including time in bed, number of bed exits and time of first morning exit, even heart rate of the subject.

3.4 Other Monitoring Modalities

A variety of alternative approaches have been proposed which focused on the minimization of a number of monitored physiological signals or alternative methods of capturing correlates of physiological states and parameters. For example, the measurement of arousal made using a sleep belt integrating heart and respiration monitor and wrist unit with sensors measuring skin conductance and temperature (B. Grundlehner, 2009). The portable device (WatchPAT, 2013) characterizes the changes in the autonomous nervous system by monitoring peripheral arterial tone (PAT) and activity, as well as blood oxygen saturation using a finger cuff.

3.5 Current Commercial Mobile Sleep Monitoring Systems

Table 4 summarizes the existing commercial systems; most of which are designed for specialized sleep laboratories and ambulatory monitors intended for home use which employ local intelligence or wireless communication to decrease the length of wires.

Table 4 A comparison of some existing commercial sleep products

System	Hardware Features	Software (algorithms)
g.Mobilab (g.Tec)	Multi-purpose version (2 EEH, 2 EOG, 1 ECG, 1 EMG, other signals (i.e. respiration, galvanic skin response, PPG); Wireless	No sleep scoring/staging algorithms; real-time visualization and analysis under Matlab;
Crystal Monitor PSG series (Clevemed)	2 EEG, 2 EOG, 2 EMG, 1 ECG, thoracic/abdominal effort, airflow, snore, body positioning, pulse oxymetry; Wireless	Software with manual sleep scoring (no details provided)
Siesta (Compumedics)	32 EEG channels; 2 expansion ports for Oximetry, pressure transducers and 32 additional signal interface; Wireless	'Profusion PSG' software is used for sleep staging
Somte (Compumedics)	8 EEG, 2 EOG, 2 chin EMG, ECG, airflow, snore, pressure. Thoracic and abdominal RIP band, 2 limb, position, oxymetry, external position sensors, thermistor; Wireless	
Embla Titanium	34 channels Full PSG, including oximetry, nasal and differential pressure, Wireless;	'Somnologica' software with manual sleep scoring, respiration and PLM
AURA PSG LITE (Grass technologies)	16 channels (no mention of EEG, EOG, ECG, PPG, etc)	AURA PSG LITE – automatic sleep report generator, Pulse Transit Time
Easy Ambulatory PSG (Cadwell)	24 channels ambulatory system (inc. oxygen saturation, body position, nasal pressure, limb movements, respiration, EEG, etc); Non-wireless;	Easy III software is patient database and scheduler but no mention of sleep staging
Stardust II (Respironics)	Detects OSAH by measuring airflow (pressure-based), pulse rate, oxymetry, chest or abdominal effort, body positioning changes; Non-wireless;	'Sleepware' software diagnostic software for sleep scoring and apnea detection and screening;
Alice PDx (Respironics)	Measures patient airflow from nasal/oral thermistor, pulse oximetry sensor; optional 1/6/12 lead ECG; optional 4 EEG/EOG, 3 EMG; level-2 sleep type; Non-wireless	
WatchPAT (Itamar)	6 channels – PAT (signal and amplitude), pulse rate, oxygen saturation, actigraphy, snoring, body position; Non-wireless;	PAT signal; Sleep staging (awake, light, deep REM), RDI, AHI, ODI; analytical software;
SleepTrek 3 (Grass technologies)	Records oxygen saturation; pulse rate; airflow; snoring; respiratory effort (RIP); body position; Non-wireless;	'GammaST' software generates sleep report; optional FASS automated sleep staging software with arousal, sleep stages and event detection
Enobio (Starlab)	EEG, ECG and EOG monitoring; Wireless	Enabling real-time data processing; Java software; not for sleep staging
Actigraph	Sleep monitor; 3-axis accelerometer; Non-wireless	Monitors sleep onset, sleep latency, number and duration of awakenings, sleep efficiency and physical activity and light intensity
Actiwatch family (Respironics)	Sleep monitor, Non-wireless	Records long-term sleep/wake patterns, sleep onset, sleep efficiency, wake after sleep onset; circadian rhythms disorders; shift-work disturbances; subjective sleep diary score - manual entry

3.6 Needs of Insomnia Monitoring Systems

Today's home-based diagnostic devices still face many challenges; the most important issues include wearability, ease of use, unobtrusiveness, quality of recorded data, user comfort for normal night of sleep, etc. There are similar concerns with the software detection of PSG transient events that significantly influence sleep quality. The existing indices which detect and characterize these transients often require full PSG, hence they do not simplify the problem of objectively measuring and differentiating sleep disorders. The traditional approach to PSG requires monitoring of a variety of physiological signals that require special montage with multiple electrodes, such as EEG, EOG, EMG, ECG, respiration, etc. Leads connected to these electrodes are inconvenient and uncomfortable for the user, resulting in modified sleep activity. The unfamiliar PSG system can also exacerbate the problem, making it harder for insomniacs to get to sleep.

Even though the proper assessment and diagnosis of insomnia needs particular face-to-face (patient-psychologist/psychiatrist) interview and consultation, the screening can be fully automated and performed at home, eliminating the need for the patient to stay in the hospital overnight to receive the diagnostic results. PSG is mainly used to differentiate the diagnosis of sleep disorder, such as OSAH, RLS, PLMD, etc.

There are no currently available monitoring products that can effectively monitor and screen insomnia using objective and subjective measures despite the growing number of commercial sleep monitoring systems. Systems need to be developed that integrate the sensors and functionality necessary to support monitoring, screening, diagnosis and treatment of insomnia. These systems should automatically process relevant insomnia parameters from a partial PSG and combine it with subjective data to monitor and screen for insomnia.

4 A Mobile Monitoring System for Insomnia Assessment and Treatment

4.1 System Functionality

The sleep hygiene of a patient can be determined by evaluating their sleeping patterns over a 2-week period, documented daily in a sleep diary by the patient. A detailed assessment of a patient's sleep behavior allows the clinician to properly diagnose the extent of the sleeping disorder. Sleep diaries and specialized questionnaires are necessary for an accurate evaluation of insomnia. As illustrated in Figure 12, the clinician follows a multi-step methodology to validate the diagnosis of insomnia. The tools used in the assessment process include: (1) face-to-face clinical consultation, (2) sleep diaries and (3) nocturnal PSG. The results obtained using these tools affect the treatment decisions made by the clinician. In the initial face-to-face consultation, the psychologist or medical practitioner collects the background information required to complete an initial evaluation of the patient. During these consultations, the clinic incorporates various questionnaires, such as insomnia severity index (ISI), Pittsburgh sleep quality index (PSQI), dysfunctional beliefs and attitudes about sleep (DBAS) and the Epworth sleepiness scale (ESS). Patients are required to use sleep diaries or logs to record daytime

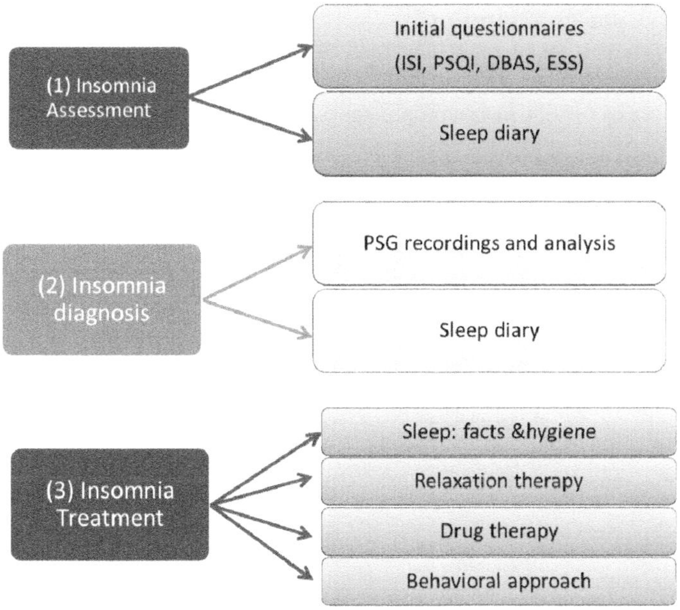

Fig. 2 Main steps for insomnia assessment and treatment

activities, pre-sleep rituals each night before and after sleeping, getting back to sleep details the and sleep disorders.

Together with nocturnal PSG recordings, these diaries help clinicians identify the cause of insomnia. The evaluation process is thus time-consuming and can be expensive.

An efficient in-home sleep monitoring system needs to provide the following functions:

i) *Recording of physiological signals*: the subject is equipped with different sensors and electrodes that constantly measure different vital signals (*e.g.* EEG, ECG, *etc.*);

ii) *Electronic diary*: consists of collecting subjective measures (*i.e.* questionnaires) from patients to provide extra information about the sleep quality (daytime activities, pre-sleep rituals, etc.);

iii) *Monitoring & Communication*: PSG signals and questionnaires are transported to the clinical back-end system using secure and efficient protocols;

iv) *Sleep analysis*: this step consists mainly of three stages: preprocessing, features extraction and classification. The aim of the first stage is to detect artifacts, eliminate noise and segment the whole signal into epochs. During the second step, features are extracted from individual epochs of recorded PSG signals. These are then used to classify each epoch into one of the sleep stages (Hasan, 1996); and finally

v) *Reporting*: it aims to provide the clinician with the required information to make a diagnosis about the patient's sleep patterns.

4.2 System Description

The architecture of in-home insomnia monitoring system is illustrated in Figure 3, and comprises of four main components:

i) *Physiological monitoring*
Different electrodes would be mounted on the patient to acquire these signals: EEG channels (frontal and central), 2 EOG channels (right and left), 1 ECG channel and 1 EMG channel. Each of these signals provides the activity of the autonomous nervous system during sleep. As shown in Figure 3, the ECG electrode is incorporated into an adjustable chest band and the EEG and EOG electrodes are fitted into adjustable head band. The signals are recorded on a central device that transmits the collected data to a medical database.

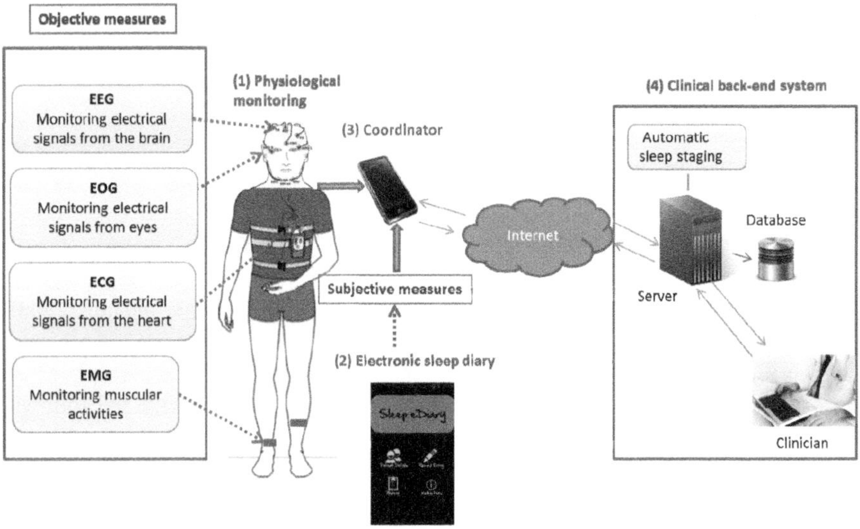

Fig. 3 In-home wearable insomnia monitoring and diagnosis system

It is essential for a sleep diary application to provide the patient relevant sleep questions and record answers using an easy and simple user-interface. The user-interface should be kept simple, to allow even the newest mobile users to successfully use the application. Hence, the sleep diary application should at least be compatible with present day software solution and provide appropriate

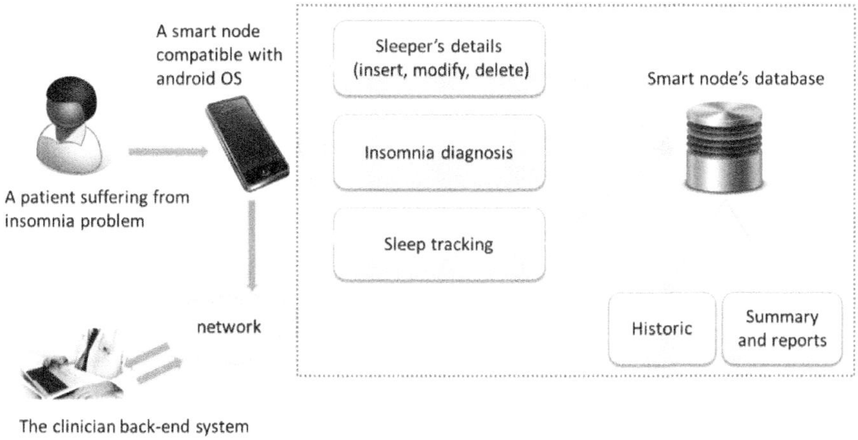

Fig. 4 Electronic sleep diary architecture

The architecture of the electronic sleep diary is illustrated in Figure 4. Lifestyle factors are collected in *'insomnia diagnosis'*, through asking relevant sleep questions which relate to insomnia diagnosis. Sleep latency (time from when going to bed to falling asleep) is a feature of sleep diary which assists in being one of the indicators of insomnia (insomniacs often overestimate sleep latency and underestimate total sleep time). In order to further ease the calculation of total sleep time, the application includes an additional tracking feature, allowing the patient to track sleep duration. A history is included to log sleep assessment data in an efficient manner. The *sleep summary report* is automatically generated which includes sleep statistics (i.e. mean sleep efficiency, mean sleep duration, range of sleep latency, number and duration of awakening etc.). These statistics will be included with the objective data used in the assessment and stored in the system database. The stored data can then be transferred individually to the clinician via email.

ii) *The Device Coordinator*

PSG signals are collected on the coordinator (or smart phone) which is equipped with various wireless technologies. The collected measures are stored and then transmitted to the clinical back-end system.

iii) *The Clinical back-end system*

The back-end system consists of a database to store the received data and a server to analyze and process these health data. It provides the clinical data used in the diagnosis of primary insomnia. It includes the sleep stage classifier which will provide information on the sleep structure for use in determining whether or not the patient is having good sleep for their age.

4.3 Challenges

The design of autonomous in-home remote sleep monitoring systems poses several intrinsic design challenges and constraints. More specifically, the following design challenges should be carefully addressed to enable the emergence of such mobile health (mHealth) solutions:

i) *Availability:* The *real-time availability* of *accurate sleep data* (*i.e.* collected patients physiological signals) is a requirement to enable timely sleep quality assessment and proper decision making;

ii) *Scalability: Scalability* of the in-home monitoring systems and communication architecture is a strong requirement to effectively enable the remote monitoring of a large number of in-home patients, and to accommodate the collection of a large amount of real-time sleep recordings.

iii) *Security and Privacy:* The security of the collected measures is a crucial challenge, and should be carefully treated. In fact, the collected patient's health information is highly confidential, and the following basic security concepts should be addressed: *authentication, availability, authorization, confidentiality, non-repudiation and integrity.* Medical staff needs to ensure each patient is assigned their own data and results with any incidents.

iv) *Reliability, resilience and quality of service (QoS):* The *reliability, resilience* and *quality of service* (QoS) of *communications* are of great importance to provide an *end-to-end connectivity* between in-home patients and their medical centers. Moreover, given the large amount of data to be transferred, *low overhead communications* are required to better accommodate available network bandwidths and data plans (*e.g.* GPRS, 3G, 4G, *etc.*).

v) *Usability:* The *usability* of the autonomous sleep monitoring system is also an important challenge. Indeed, mHealth solutions need to seamlessly exploit available networks (e.g. WiFi, GPRS, 3G, etc.) and automatically recover from communications errors, without disturbing patients or insomnia monitoring operations.

5 Signal Analysis for Insomnia Diagnosis

Subjective and objective variables need to be extracted from the collected physiological data that replicate the systematic assessment process followed by clinicians to assist in the diagnosis process.

5.1 Sleep Stage Detection and Classification

As we briefly described in the previous sections, for insomnia assessment and diagnosis, overnight PSG data collected from patients is analyzed by sleep experts and classified into the different sleep stages. The visual analysis of PSG into the sleep stages is a time-consuming and very tedious procedure. Research for automatic sleep staging is well underway. Numerous solutions have been introduced to imitate the expert identification method, and to classify automatically sleep epochs into stages. They are now being incorporated in a number of commercial sleep recordings systems.

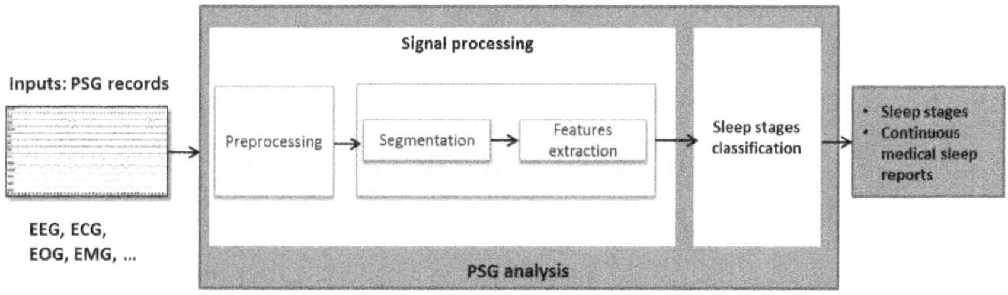

Fig. 5 Block diagram of the automated sleep staging

The automatic sleep staging procedure consists mainly of three stages: preprocessing, features extraction and classification (Penzel & Conradt, Computer based sleep recording and analysis, 2000). Figure 5 shows the block diagram of the automated sleep staging. The inputs to the system are the different PSG signals collected from overnight sleep recordings of the patient. The aim of the first stage is to detect artifacts, eliminate noise and segment the whole signal into fixed-size epochs (20 or 30s)((Thakor & Muthuswamy , 1998), (Anderer, et al., 1999)).The common artifacts in EEG signals are caused by the movements of subjects, system interference or eye movements, etc. (Bates, 1972)

During the second step, various features are extracted from individual epochs of the recorded PSG signals to detect the discriminating properties of the different characteristic waves. The extracted features may be the amplitude, the power of different waves or the number of the K-complex and sleep spindles events, etc. Classifiers are then used to combine the different features to assign each epoch to a specific sleep stage (Hasan, 1996). A number of automatic sleep staging methods are available in literature but they differ in the implementation of the above mentioned steps.

Sleep staging provides a quantification of the percentage and duration of REM and stage 1, 2 and 3 sleep as well as the quantity of sleep spindles and K-complexes present. All these variables are used in the assessment of sleep quality for insomnia diagnosis.

5.2 Sleep Onset Detection

Sleep onset is characterized by three phases, the initial process - 'alpha related change' (beginning of sleep onset and the initial transition into stage 1); intermediate process - 'theta and vertex waves' (regarded as 'middle' sleep onset) and terminal process signifying the end of wakefulness, known as 'sigma' or 'sleep spindle' process (Ogilvie R. , 2001). The intermediate process of sleep onset has not been investigated in greater detail. In fact, most automatic detection developed so far has been on the detection of sleep spindles (marker of end of sleep onset well understood), complex transients (i.e. K complexes or CAP transients), slow wave sleep (SWS), alpha, beta and gamma bands. This probably has to do with the fact that vertex and theta activity are regarded as wake transients to pre-sleep onset and are not considered to characterize the sleep onset or even have any influence over its progression. Sleep onset has been characterized from measures which indicate that alpha decreases as delta increases with a theta increase to a lesser degree (Cvetkovic & Cosic, 2008). Considering that delta or SWS is more predominant and evident ('sharper' marker) in this sleep onset progression, it has been utilized in our studies of automated detection of sleep onset and biofeedback operation.

A sleep onset index (SOI) which is measured from objective data (EEG, EOG, ECG) can assist in accurate insomnia detection. The index would need to include various feature extractions, such as measures to characterize the degree of EEG/EOG signal order/disorder, coherence and power of EEG frequency bands, various spectral edge frequencies and measures to detect micro and macro-structure of sleep EEG using a finer time scale.

5.3 Arousal Index

Currently scoring EEG micro arousals according to ASDA criteria is a manual process which requires manual and visual event scoring, but recently there has been interest in automating arousal detection. Alvarez-Estevez and Moret-Bonillo performed automated detection of EEG arousals from combined EEG and EMG signals processing of two moving windows (10 and 3 sec.) (Moret-Bonillo & Alvarez-Estevez , 2011). The feature extraction methods consisted of computing the EEG alpha and beta bands and EMG amplitude for each moving window. Various pattern classifications were conducted and artificial neural networks (ANN) were found to be most suitable. Sugi and colleagues performed their automated detection of EEG arousals (respiratory-related arousals) (Sugi, Kawana, & Nakamura, 2009) based on varying thresholds according to events of sleep apnea from EEG and EMG activity. Recently, a 'weighted-transition sleep fragmentation index', was used to characterize the quality of sleep and symptoms of disorder, consisted of capturing different sleep stage transitions and assigning different weights to them (Swarnkar, Abeyratne, Hukins, & Duce, 2009). This index has the disadvantage that it does not focus on the dynamic measure of multiple phases during the sleep onset. Mendez and colleagues applied various time frequency analyses on heart rate variability (HRV) during arousal from sleep, with and without chin EMG activity (Mendez, et al., 2008). During arousal episodes with chin activity, the energy distributions and spectral indexes from three time-frequency analyses methods revealed substantial changes in RR intervals and LF HRV band (sympathetic activity). Mendoza and co. concluded that most of the stress on the heart was when an arousal was related to muscular chin activity. De Carli and colleagues developed a methods for the automated detection of arousals from EEG and EMG activities (De Carli, Nobili, Gelcich, & Ferrill, 1999). Wavelets were employed to compute indices to discriminate arousal segments. Their results revealed sensitivity of 88% and specificity of 75%, higher and lower than human experts, respectively.

The rise of PTT measured blood pressure and heart rates are also considered sensitive markers for the arousal process (Pitson & al., 1994). The blood pressure arousals detected by PTT correlate closely with EEG micro arousals and respiratory events. However, heart rate rises are less sensitive markers for the arousal detection. In fact, Pitson suggested that PTT measured blood pressure arousals may serve as an alternative to manual scoring of EEG micros arousals according to ASDA criteria (Pitson & Stradling , 1998).

An arousal index can be used to quantify sleep fragmentation, derived as arousal number per hour of sleep. To provide accurate information about sleep quality, the arousal index (AI) needs to include features, such as: the dynamic measure of multiple phases during the sleep onset; % of NREM N1 transitional sleep; number of transient arousals; exclusion of sleep spindles; and an intrusion of alpha wave activity into SWS (deeper) sleep.

5.4 Actigraphy Index

The available commercial actigraphs have not been standardized and thus have different sensitivities and different algorithms for the assessment of sleep and wakefulness. As sleep becomes more disturbed, the actigraphy becomes less accurate. Constant turning in bed, due to the inability to sleep, causes the actigraphy to be less accurate and *overestimate* sleep and *underestimate* wake state. As an example, insomnia patients can stay inactive for prolonged periods in an attempt to fall asleep, and patients with sleep apnea have more disturbed sleep thus more body movements. Actigraphy also tends to *underestimate the frequency of leg movements during sleep*, and assessment of periodic leg movements (PLMs).

The most frequently extracted parameters include time in bed, sleep onset latency, total sleep time, mean motor activity (number of movement in one minute), number of awakenings longer than 5 minutes (NA>5), wake after sleep onset, and sleep efficiency.

5.5 Arterial Stiffness and Pulse Transit Time

The need to maximize the user's convenience and unobtrusive monitoring has motivated a variety of alternative approaches focused on the minimization of a number of monitored physiological signals or alternative methods of capturing correlates of physiological states and parameters. Characterization of changes in the autonomous nervous system requires monitoring of beat-to-beat blood pressure during daily activities. Since traditional cuff-based monitors cannot be used for ambulatory monitoring, alternative approaches, such as Pulse Wave Velocity (PWV) or Pulse Travel Time (PTT) portable sensors, have emerged. PWV measures speed of blood pulse propagation on two or more points on body (Cecelja & Chowienczyk, 2012), while pulse transit/arrival time is measured as a time from heart beat to the arrival of the wave to the tip of the finger or earlobe (Smith, Argod, Pépin, & Lévy, 1999). PTT can be used to assess activation of the sympathetic nervous system and arterial stiffness. While blood pressure monitors use inflatable finger cuffs that can be very unpleasant and disturb sleep pattern, monitors based on PTT can use light-weight reflective photoplethysmography (PPG) sensors (Shelley and Shelley, 2001), therefore providing much more realistic sleep monitoring of subjects with sleep disturbances. The PPG signal can be also used to measure heart rate and blood oxygen saturation. PTT is inversely proportional to blood pressure and correlates with intra-thoracic pressure changes while Peripheral Arterial Tonometry (PAT) determines the peripheral arterial vascular tone (Hennig & Patzak, 2013). Autonomous (sympathetic activity) arousals during sleep can be recognized by the analysis from this PAT measurement.

As shown Figure 6, the amplitude of the PPG varies with the power of the left ventricular stroke and vasomotor tone. The changes in these and other morphological details can be used to determine the hemodynamic status of the subject at any time point. Normally during sleep, BP and HR drop by 10-20% from daytime values in hypertensive patients (high blood pressure; opposite of hypertension is hypotension). This drop in blood pressure is called dipping. Other patients who do not experience this dipping during sleep, are called non-dippers and patients whose BP increases during sleep are known as reverse dippers. This dipping concept has been claimed to be linked to a restorative physiological process, with non-dipping linked to hypersensitivity, OSA and ANS dysfunction.

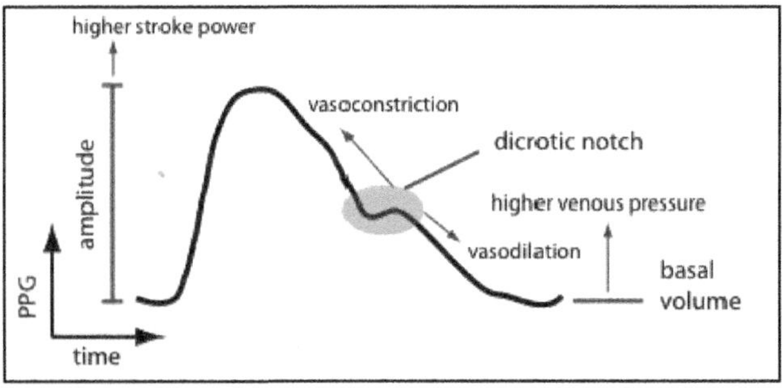

Fig. 6 PPG morphology

It is suspected that the better the sleep quality (deeper sleep, less disrupted or less fragmented sleep), the greater the dipping during sleep. Evidence also suggests that non-dippers have more arousals during sleep and more disrupted sleep. Arousals, which greatly disturb sleep, increase the sympathetic nervous system and therefore increase blood pressure, decrease PTT, and change position of the dicrotic notch in PPG. It is suggested that recurrent arousals or awakenings from sleep (which reduce sleep efficiency), increase BP, and lead to sleep fragmentation associated with low apnea-hypopnea index (AHI). The arousal in sleep is accompanied by a rapid increase in heart rate and blood pressure mediated by activation of the sympathetic nervous system and rapid parasympathetic withdrawal. On average, arousal causes a 20 mm Hg increase in systolic and 15 mm Hg increase in diastolic blood pressure, with a heart rate increase of 10 beats per minute. At times, arousals cannot be detected with EEG, but appear in sympathetic activation as part of it (Smith, Argod, Pépin, & Lévy, 1999).

6 Discussion and Conclusions

The development of home-based, automated and portable sleep monitoring technologies is on the rise, but they mostly cater to patients suffering from sleep disordered breathing such as OSAH. Even though the proper assessment and diagnosis of insomnia needs particular face-to-face (patient-psychologist/psychiatrist) interviews and consultation, the screening can be fully automated and performed at home, eliminating the need for the patient to stay in the hospital over several nights to receive the diagnostic results. PSG is mainly used to differentiate the diagnosis of sleep disorder, such as sleep-disorder breathing, OSAH, RLS, PLMD, etc. Actually, there are no objective, sleep monitoring systems that can differentiate between the sub-types of insomnia or other more general sleep disorders. Currently, there are devices that can successfully perform objective detections of sleep abnormalities, but these devices cannot distinguish insomnia from any of these abnormalities or detect co-existing (i.e. OSAH and insomnia) abnormalities. This issue poses a significant challenge in the correct diagnosis and more importantly in providing correct treatment.

An accurate insomnia diagnosis requires the collection of PSG signals and patient information by the sleep expert. In this chapter, we presented a new, cost-effective architecture of a single insomnia monitoring system that combines subjective and objective measures. As discussed, the key is to target the integration of a minimum set of sensors that can maximize the user's comfort, ease of use and compliance subject as well as provide the required measures for accurate diagnosis. This system offers the possibility to collect information from patients in their homes over long periods of time. The collected health information can be transferred to a remote clinician's back-end server for further data analysis to enable an assessment of the subject's sleep quality and assist in diagnosis. Extensive clinical studies would need to be conducted to validate the performance of the developed insomnia monitoring tool. Its assessment would also be compared to traditional monitoring techniques and statistical analysis performed to quantify its capability. This validation is necessary to pursue future commercialization options.

Home-based diagnostic devices still face many challenges; the most important issues include the availability, scalability, security, reliability, quality of recorded data, and user comfort for normal night of sleep, etc. These challenges propose different topics, which can be considered as framework for a research agenda to improve the insomnia monitoring field. Long-term PSG monitoring generates a huge amount of data, hence requiring a large bandwidth for transportation and efficient communications approaches for the transmission of collected health data to the remote clinical back-end system. There is a need to explore compression techniques that aim to obtain a temporal evolution of some significant and recurrent features. This process should be integrated into existing sleep staging systems, to allow them to become efficient, fast and have low computational cost. Another question that needs to be asked is the use of real time processing in wearable applications that allows significant features extraction to speed up the insomnia diagnosis. Moreover, such in-home sleep monitoring systems pose unprecedented threats to a patient's privacy. It is thus important to provide an efficient and secure end-to-end connectivity between in-home patients and the remote clinical servers.

Many unanswered questions remain about the design of automatic insomnia detection and the integration to the proposed cost-effective architecture. There is a need for strong cooperation between engineers and clinicians to provide an efficient automatic insomnia detection method adapted to in-home and wearable applications. Further research in this field is required to complete this analysis and to produce accurate insomnia diagnosis and prompt treatment.

References

Cvetkovic, D., Cosic, I.: Sleep Onset Estimator: Evaluation of Parameters. In: 30th Annual International Conference IEEE Engineering in Medicine and Biology Society (EMBS), Vancouver, Canada, pp. 3860–3863 (2008)

De Carli, F., Nobili, L., Gelcich, P., Ferrill, F.: A method for the automatic detection of arousals during sleep. Sleep 22(5), 561–567 (1999)

Karadeniz, D., Ondze, B., Besset, A.: EEG arousals and awakenings in relation with periodic leg movements during sleep. J. Sleep Res. 9, 273–277 (2000)

Shinar, Z., Akselrod, S., Dagan, Y., Baharav, A.: Autonomic changes during wake-sleep transition: A heart rate variability based approach. Autonomic Neuroscience: Basic and Clinical 130, 17–27 (2006)

Sugi, T., Kawana, F., Nakamura, M.: Automatic EEG arousal detection for sleep apnea syndrome. Biomedical Signal Processing and Control 4, 329–337 (2009)

Swarnkar, V., Abeyratne, U., Hukins, C., Duce, B.: A state transition-based method for quantifying EEG sleep fragmentation. Medical & Biological Engineering & Computing 47, 1053–1061 (2009)

Advanced Brain Monitoring (2013), Retrieved November 2013, from Advanced Sleep Group, `http://advancedbrainmonitoring.com/`

Ambulatory Monitoring, Inc. (2013), Retrieved November 2013, from Retrieved from `http://www.ambulatory-monitoring.com/`

Neurovigil, The iBrain device (2013), Retrieved November 2013, from `http://www.neurovigil.com/ibrain/`

WatchPAT (2013), Retrieved November 2013, from Itamar Medical: `http://www.itamar-medical.com/WatchPAT.html`

ActiGraph (2013), ActiSleep Sleep Monitor. Retrieved November 2013, from ActiGraph: `http://www.theactigraph.com/products/actisleep/`

Guilleminault, C.: American Sleep Disorders Association Task Force (Chair) EEG arousals: scoring rules and examples. Sleep 15, 173–184 (1992)

Anderer, P., Roberts, S., Schlögl, A., Gruber, G., Klösch, G., Herrmann, W., et al.: Artifact Processing in Computerized Analysis of Sleep EEG - A Review. Neuropsychobiology 40(3), 150–157 (1999)

Grundlehner, B., Brown, L.: The Design and Analysis of a Real-Time, Continuous Arousal Monitor. In: 6th Int. Workshop on Wearable and Implantable Body Sensor Networks (2009)

Badia, P., Wright Jr., K.P., Wauquier, A.: Fluctuations in single-hertz EEG activity during the transition to sleep. In: Ogilvie, R.D., Harsh, J.R. (eds.) Sleep Onset: Normal and Abnormal Processes. American Psychological Association, Washington (1994)

Bates, J.: Fundamentals of Electroencephalography. Journal of Neurology, Neurosurgery, and Psychiatry 35(1) (1972)

Berry, R.B., Budhiraja, R., Gottlieb, D.J., Gozal, D., Iber, C., Kapur, V.K., et al.: Rules for scoring respiratory events in sleep: update of the 2007 AASM Manual for the Scoring of Sleep and Associated Events. Deliberations of the Sleep Apnea Definitions Task Force of the American Academy of Sleep Medicine. J. Clin. Sleep Med. 8(5) (October 2012)

Bonato, P.: Advances in wearable technology and its medical applications. In: Proc. IEEE EMBC (2009)

Carskadon, M.A., Dement, W.C.: Normal human sleep: an overview. Principles and Practice of Sleep Medicine 4, 13–23 (1994)

Cecelja, M., Chowienczyk, P.: Role of arterial stiffness in cardiovascular disease. JRSM Cardiovascular Disease 1(4) (2012)

Chesson, A., Hartse, K., Anderson, W.M., Davila, D., Johnson, S., Littner, M., et al.: Practice Parameters for the Evaluation of Chronic Insomnia 23(2) (2000)

Colten, H., Altevogt, B.M.: Sleep Disorders and Sleep Deprivation: An Unmet Public Health Problem. Institute of Medicine (US) Committee on Sleep Medicine and Research. National Academies Press (US), Washington (DC) (2006)

Coyle, S., Lau, K.T., Moyna, N., Diamond, D., Francesco, F.D., Costanzo, D., et al.: BIOTEX—Biosensing textiles for personalised healthcare management 14(2) (2010)

Danker-Hopfe, H., Anderer, P., Zeitlhofer, J., Boeck, M., Dorn, H., Gruber, G., et al.: Interrater reliability for sleep scoring according to the Rechtschaffen & Kales and the new AASM standard. Journal of Sleep Research 18(1), 74–84 (2009)

Jovanov, E., Hanish, V.: Avatar—A multi-sensory system for real time body position monitoring. IEEE EMBC, Minneapolis (2009)

Hasan, J.: Past and future of computer-assisted sleep analysis and drowsiness assessment. Journal of Clinical Neurophysiology 13(4), 295–313 (1996)

Hennig, A., Patzak, A.: Continuous blood pressure measurement using pulse transit time. Somnologie 17, 104–110 (2013)

Hori, T.: Spatiotemporal changes of EEG activity during waking-sleeping transition period. Inter. J. Neurosci. 27, 101–114 (1985)

Koley, B., Dey, D.: An ensemble system for automatic sleep stage classification using single channel EEG signal. Computers in Biology and Medicine 42(12), 1186–1195 (2012)

Lubin, A., Johnson, L.C., Austin, M.T.: Discrimination among states of conciousness using EEG spectra. Psychophysiology 6, 122–132 (1969)

Mendez, M., Bianchi, A., Montano, N., Patruno, V., Gil, E., Mantaras, C., et al.: On arousal from sleep: time-frequency analysis. Medical & Biological Engineering & Computing 46, 341–351 (2008)

Merica, H., Fortune, R.: Physiological Review, State transitions between wake and sleep, and within the ultradian cycle, with focus on the link to neuronal activity. Sleep Medicine Reviews 8, 473–485 (2004)

Moret-Bonillo, D., Alvarez-Estevez, V.: Identification of electroenceph-alographic arousals in multichannel sleep recordings. IEEE Transactions on Biomedical Engineering 58, 54–63 (2011)

Murali, N., Svatikova, A., Somers, V.: Cardiovascular physiology and sleep. Frontiers in Bioscience: A Journal and Virtual Library 8, 636–652 (2003)

Ogilvie, R.: The process of falling asleep. Physiological Review 5(3) (2001)

Ogilvie, R., Simons, I., Kuderian, R., MacDonald, T., Rustenburg, J.: Behavioral, event-related potential, and EEG/FFT changes at sleep onset. Psychophysiology 28(1), 54–64 (1991)

Penzel, T., Conradt, R.: Computer based sleep recording and analysis. Sleep Medicine Reviews 4(2), 131–148 (2000)

Perlis, M., Sharpe, M., Smith, M., Greenblatt, D., Gile, D.: Behavioral Treatment of Insomnia: Treatment Outcome and the Relevance of Medical and Psychiatric Morbidity 24(3) (2001)

Pitson, D.J., Stradling, J.R.: Autonomic markers or arousal during sleep sleep in patients undergoing investigation for obstructive sleep apnoea, their relationship to EEG arousals, respiratory events and subjective sleepiness. J. Sleep Res. 7, 53–59 (1998)

Pitson, D., et al.: Changes in pulse transit time and pulse rate as markers of arousal from sleep in normal subjects. Clin. Sci. 87, 269–273 (1994)

Matthews, R., McDonald, N.: A Wearable Physiological Sensor Suite for Unobtrusive Monitoring of Physiological and Cognitive State. In: Proc. IEEE EMBS (2007)

Reite, M., Buysse, D., Reynolds, C., Mendelson, W.: The Use of Polysomnography in the Evaluation of Insomnia. Standards of Practice Committee, American Sleep Disorders Association, Rochester, MN, Sleep Journal 18(01), 58–70 (1995)

Riemann, D., Spiegelhalder, K., Feige, B., Voderholzer, U., Berger, M., Perlis, M., et al.: The hyperarousal model of insomnia: a review of the concept and its evidence 14(1) (2010)

Smith, R.P., Argod, J., Pépin, J., Lévy, P.A.: Pulse transit time: an appraisal of potential clinical Applications. Thorax 54, 452–457 (1999)

Taylor, M., Grant, T., Knoefel, F., Goubran, R.: Bed occupancy measurements using under mattress pressure sensors for long term monitoring of community-dwelling older adults. In: IEEE Internatinational Symposium on Medical Measurements and Applications Proceedings (MeMeA), pp. 130–134 (2013)

Thakor, J., Muthuswamy, N.V.: Spectral analysis methods for neurological signals. Journal of Neuroscience Methods 83(1), 1–14 (1998)

Van Hese, P., Philips, W., De Koninck, J., Van de Walle, R., Lemahieu, I.: Automatic detection of sleep stages using the EEG. In: Proceedings of the 23rd Annual International Conference of the IEEE Engineering in Medicine and Biology Society, vol. 2, pp. 1944–1947 (2001)

Vullers, R., Van Schaijk, R., Visser, H., Penders, J., Van Hoof, C.: Energy harvesting for autonomous wireless sensor networks. IEEE Solid-State Circuits Magazine 2(2), 29–38 (2010)

Wong, A., McDonagh, D., Omeni, O., Nunn, C., Hernandez-Silveira, M., Burdett, A.: Sensium: An Ultra-Low-Power Wireless Body Sensor Network Platform: Design & Application Challenges. In: Proc. IEEE EMBC 2009 (2009)

Yseboodt, L., Nil, M., Huisken, J., Berekovic, M., Zhao, Q., Bouwens, F., et al.: Design of 100 μW wireless sensor nodes for biomedical monitoring 57(1) (2009)

Zephyr-BioHarness. BioHarness 3 - Zephyr Technology Corporation (2013), `http://www.zephyranywhere.com/` (retrieved November 2013)

Zhang, Z., Zhao, L.: Objective and subjective measures for sleep disorders. Neuroscience Bulletin 23(4), 236–240 (2007)

Questions

Q1. How is sleep in insomniacs currently monitored in commercially available systems?

Q2. What are the requirements of an efficient remote insomnia monitoring system?

Q3. What are the challenges in in-home remote sleep monitoring systems?

Q4. What are the three steps in automatic sleep stages?

Q5. What is a sleep onset index?

Q6. What is an arousal index?

Q7. What are the three types of arousals?

Q8. What are the most frequently extracted parameters from actigraphy?

Q9. How to arousals affect the physiology of sleep?

6. Current Status and Future Trends of Wireless and Mobile Health Technologies in Sleep Medicine: Insomnia Case Study

Leila Jalali* and Philip Bigelow

School of Public Health and health Systems, University of Waterloo,
200 University Ave. West. Waterloo, ON. Canada

Abstract. This chapter presents the current status and future directions of mHealth technologies in sleep medicine. The focus is on insomnia case as one of the most common reported sleep disorders with significant consequences on daytime functionality, long-term health and economy. Recent advancements and existing wireless technologies for sleep disorder monitoring, diagnosis and treatment from the two perspectives of smartphone applications level and wireless physiological sensor level have been investigated and described. Key challenges and limitations of the existing technologies from health and medical application perspective have been explained. Moreover, the trend and future directions of mHealth and wireless physiological sensor technologies and the need for the convergence of the two levels of smartphone and physiological sensing for effective and efficient future wireless sleep medicine technology is presented.

Keywords: mHealth, sleep technology, insomnia, wireless sensing, wireless EEG, sleep monitoring, mobile apps, portable sleep monitoring.

1 Introduction

Having a sufficient amount of restorative sleep is a key element of physical and mental health. It is estimated that 50 to 70 million Americans suffer from a chronic sleep disorder that affects their daily functioning and long term health and longevity (Colten and Altevogt 2006).Insomnia is the most common sleep complaint reported to physicians (Chesson et al. 1999) and is characterized by difficulty initiating or maintaining sleep, or experiencing non-restorative sleep (International Classification of Sleep Disorder, Second Edition ICSD-2). Insomnia is a major public health problem and has a significant effect on millions of individuals as a primary disorder or is associated with other conditions. Studies have shown that one third of U.S. adults report difficulty sleeping, and this number increases to 57% in the elderly (McCall 2004, Chong et al. 2013). The number of individuals

* Corresponding author.

S. Adibi (ed.), *Mobile Health*,
Springer Series in Bio-/Neuroinformatics 5, DOI: 10.1007/978-3-319-12817-7_6

who require the use of sleep aids is increasing, and approximately 8.5 million people in the U.S. use prescription sleep medications (Chong et al. 2013). Decreased quality of life, mental health problems, weight gain and increasing risk for type 2 diabetes are just a few health consequences of insomnia (Roth 2007). Patients with insomnia are also two and half times more likely to have car accidents (Roth and Ancoli-Israel 1999, Hillman et al. 2013). The statistics on highway accidents released by the U.S. National Highway Traffic Safety Administration showed that approximately 100,000 vehicle accidents and 1500 deaths occur annually due to drowsiness while driving (Zammit et al. 1999). In terms of financial consequences, the direct cost of insomnia is estimated to be about $12 billion for health care services and $2 billion for sleep programing agents in the U.S. (Walsh and Engelhardt 1999). Indirect costs of insomnia such as workplace accidents and errors have been estimated to be about $31.1billion annually in the U.S. (Shahly et al. 2012).

The use of mobile electronic devices to support medical or public health practice is currently an active area of research and technology development. Wireless sensing and mHealth technology are powerful tools that have the potential to revolutionize diagnosis, monitoring and management of chronic insomnia. The advancements in technology in this area are progressing at the level of the sensor as well as within the handheld device (smartphone). At the physiological sensor level, wireless sensors offer improved comfort, minimal movement restriction for patients as well as possibility of home sleep monitoring. At smartphone level, health data collection, monitoring, and analysis using smartphone computational power as well as the transfer of the health data to the health care provider are the areas of focus. Currently, the application of mHealth for insomnia diagnosis and treatment covers primarily sleep monitoring and data collection by smartphones; however, it is expected that future technologies will connect the physiological wireless sensors to smartphones to offer a precise diagnosis, better monitoring and improved treatment techniques.

In addition to diagnosis and monitoring power, smartphone technology can be used for cognitive behavioral treatment which is one of the most common treatment techniques to address insomnia. Measuring sleep quality, journaling sleep, monitoring sleep cycles and recording daily habits and environmental factors through smartphone applications may provide invaluable data for sleep clinicians resulting in more efficient and effective diagnosis of sleep disorders and treatments.

This chapter starts with the description of insomnia, its etiology and health impacts. Patients' assessment and treatment will be explained. A review of advanced mHealth technologies and their applications in screening, diagnosis and treatment of insomnia will be provided. Moreover, a perspective on mHealth technology and its limitations in sleep disorders and future directions will be described.

2 Definition, Etiology and Health Consequences of Insomnia

Based on ICSD-2 criteria, insomnia is defined as difficulty initiating or maintaining sleep despite adequate opportunities, increased awakening, non-refreshing sleep and low sleep efficiency for at least one month. Three different groupings of factors play a role in insomnia. Predisposing factors are a group of factors that present before insomnia occurs. Examples include personality traits, genetic predisposition, physiological hyper-arousal and tendency to worry or ruminate. Precipitating factors are acute stressors that trigger insomnia. The most common precipitating factors are drug/medication use, medical conditions with nocturnal symptoms, environmental factors, psychiatric disorders, primary sleep disorders, stressful events and changes in usual habits. Perpetuating factors contribute to the persistence of insomnia and include excessive time in bed, irregular sleep/wake schedule, daytime napping, caffeine, nicotine, alcohol, drugs, and anxiety about daytime consequences (Lee-Chiong 2008).

A vast body of existing research supports the idea that not getting sufficient sleep can negatively affect health and wellbeing in several ways. In the short term, insomnia decreases performance and alertness, and causes memory and cognitive impairment, learning disabilities, sleepiness and fatigue (Carskadon 1990, Roehrs et al 1995, Zammit et al. 1999). Insomnia also increases occupational and automobile injuries (Nakata et al. 2005), the use of alcohol, relapses in alcohol dependence and difficulty with social relationships and occupational functioning (Drammond et al. 1998) (Simon and VonKorff 1997, Chevalier et al. 1999). In the long term, the clinical consequences of untreated insomnia are associated with numerous medical outcomes including weight gain, insulin resistance and increasing risk for type 2 diabetes (McNeil and Doucet 2013), diminished quality of life (Zammit et al. 1999), increased blood pressure (Suka et al 2003), a weaker immune system (Cover and Irwin 1994, Irwin et al.1996) and higher rates of morbidity (Kripke et al. 2002) and mortality (Pigeon 2010).

Insomnia is a risk factor for psychiatric disorders such as depression and anxiety (Thase 2005, Breslau et al. 1996). In addition 90% of people who suffer from depression experience insomnia (Tsuno et al.2005). Insomnia is also associated with the number of chronic medical conditions including chronic pain syndromes, coronary heart disease, asthma, gastrointestinal disorders, neuro-degenerative diseases, vascular disorders, chronic fatigue and endocrine and metabolic disorders (Roth 2007, Thase 2005).

3 Patient Assessment and Treatment

Insomnia is primarily diagnosed by clinical evaluations including a detailed medical exam with physical, psychiatric and sleep history (Littner et al. 2003). A sleep diary is also a useful method not only for diagnosis of insomnia but also for monitoring treatment. Standardized questionnaires such as the Insomnia Severity Index (ISI) and Epworth Sleepiness Scale (ESS) are often administered to gather self-reported information to aid in the diagnosis of insomnia and degree of sleepiness. Actigraphy can provide more objective measures of sleep behavior by detecting the amount of movement and estimation of sleep-and-wake cycles. Polysomnography (PSG) is not routinely used in the evaluation of insomnia. It is used only in cases of severe insomnia resistant to behavior therapy and medication, and to rule out other sleep disorders (Littner et al. 2003).

Although pharmacotherapy is widely used for treatment of insomnia, cognitive behavioral therapy (CBT) is considered the first line of treatment for chronic insomnia (Schutte-Rodin et al. 2008)(Wilson et al. 2010). CBT focuses on recognizing and changing patterns of thoughts and behaviors that cause or worsen sleep problems. Unlike medication, it also does not result in adverse side effects or dependency. CBT is as effective as pharmacologic therapy in the short term and has better long term outcomes (Jacobs et al.2004, Morin et al. 1999, Sivertsen et al. 2006). CBT for insomnia involves sleep education, cognitive control and psychotherapy, sleep restriction, stimulus control therapy, sleep hygiene, relaxation training, paradoxical intention, keeping a sleep diary and phototherapy (Taylor and Roane 2010). The most effective treatment varies with the individual and it often involves a combination of these techniques.

4 mHealth Technology for Diagnosis and Treatment of Insomnia

Smartphones as powerful tools in the mHealth system offer accessibility, portability, connectivity, convenience and microelectromechanical systems, and can benefit healthcare and sleep medicine by providing computational and communication opportunities. Moreover, the built-in sensors, such as accelerometers, gyroscopes, microphones and cameras in smartphones devices can be useful in advancing the practice of medicine and public health (Lee 2011). Smartphones can connect to external hardware devices, such as biofeedback sensors and monitor physiological factors, to motivate and reinforce relaxation. The PLX xWave, for example, is a portable, wearable sensor that measures and detects brainwaves, relaxation and focus levels and sends data to a smartphone to provide users with neurofeedback which has been shown to be helpful in improving sleep quality (Luxton et al. 2011).

Smartphones provide patients with the opportunity to benefit from social support regardless of time and location by making social media sites and social networking apps immediately accessible. Text messaging, an effective way to disseminate behavioral health information, is a popular, simple and inexpensive method for sending information to patients (Luxton et al. 2011).

The GPS capabilities of smartphones can be used to benefit patients, caregivers and clinicians as they aid in locating patients with cognitive impairment and can also be used to increase the application of some behavior therapies such as behavioral activation (BA)(Miskelly 2005). Their use in sleep disorders has not been described in the literature but has potential as a component in CBT treatments for insomnia.

Smartphones can provide an opportunity for telebehavioral health, to create and maintain access to rural and undeserved area for offering care to communities that clinical services are not easily available. Smartphones also can offer audiovisual education alternatives to text based material (Luxton et al. 2011). These uses have also not been described in the literature but there are potential applications as patients in rural areas could be evaluated and provided treatment remotely by specialized sleep physicians located in an urban medical center.

5 Current Status of mHealth Technology in Diagnosis and Treatment of Insomnia

In the area of sleep medicine and health, the mHealth technology has been progressing quite rapidly both at the level of handheld device and the sensor. At the smartphone level, the main focus has been on the developing different applications for sleep monitoring using available resources and sensors on the smartphone, whereas on the sensor level, the key focus has been on improving sensing technologies and adding wireless capability to the sensor nodes so that they can transfer their data to the mobile device. The following sections describe the current state of the technology in these two areas with a focus on insomnia.

5.1 Smartphone Advancements in Insomnia Monitoring

From the smartphone perspective in the case of insomnia, mobile phone applications are available for a variety of useful tasks including psycho-education, screening for disease, symptom assessment, cognitive and behavioral therapy and tracking of treatment progress. These apps assist different stakeholders including patients, clinicians, caregivers and family members in monitoring and improving sleep quality. Some general apps are designed for clinicians and health providers in areas of disease diagnosis, drug references, medical calculators, literature searches, clinical communication and medical training (Mosa et al. 2012). Psych Central is an example of a mobile app that summarizes behavioral health research findings for clinicians and health providers (Luxton et al. 2011).

With respect to screening and diagnosis of insomnia, sleep assessment generally depends on patients' information they provide on their sleep patterns and habits obtained from interviews and questionnaires. Therefore, the use of reliable and valid measures of sleep patterns obtained from smartphones with appropriate sensors could improve the quality of this information and result in better diagnosis and treatment. Collecting sleep data in a natural environment such as a home-setting is also one of the major advantages of the smartphone for well controlled data collection and disease assessment. Mobile apps, which are developed for insomnia diagnosis, include sleep questionnaires and standardized sleep scales, actigraphy, sleep diaries, audio records (for recording and analyzing the sound signals generated by the movements of sleeping subjects) and monitoring of sleep duration. These apps are categorized as follows:

Questionnaires and Sleep Scale. Questionnaires are commonly used as a first screening layer for insomnia and apps have been developed for their administration. ESS and ISI are examples of sleep scales that have been implemented in several mobile applications such as the Sleep Tester (Behar 2013).

Actigraphy and Apps Measuring Sleep Quality. Measuring the amount and quality of sleep is the starting point for dealing with a sleep issue. Mobile apps such as Proactive Sleep Alarm Clock calculate sleep quality by combining total sleep time and the amount of movement. Mobile apps with true actigraph capability provide an acceptable measure of sleep onset latency and sleep fragmentation. They use the accelerometer in the smartphone to detect movement and using an algorithm can estimate some standard sleep parameters such as sleep duration and sleep efficiency. Sleep Checker, Smart Alarm Clock, Sleep Cycle Alarm Clock, Relax Timer-Sleep Cycle are some examples of sleep apps with actigraph ability (Behar 2013).

Gartenberg and Parasuraman (2010) designed a mobile app in order to assess effect of sleep quality on vigilance task performance under naturalistic conditions and in home setting. Psychomotor vigilance task is a neurobehavioral test in studies of sleep and circadian rhythm research, and measures the speed with which subjects respond to a visual stimulus. Test results generally reflect the arousal and attentional state of the individual (Drummond et al. 2005) and are good indicators of the outcome of poor sleep quality.

Sleep Journaling Apps. These types of applications allow for the entry of sleep activities into an electronic diary and also can record diet, exercise and overall energy levels during the day as well as habits that affect sleep. An algorithm is applied and can provide correlation analysis showing how these various habits influence the sleep experience. Sleep Bot Tracker Log, Sleep Diary Lite and Sleep Tracker are among the apps that provide sleep activity and journaling capabilities. Proactive Sleep has ability to track behaviors that sleep clinicians use to assess sleep hygiene and can combine embedded capture of sleep data with manual retrospective data input (Gartenberg et al. 2013). Some apps such as eMood can be beneficial for other stakeholders such as family members, caregivers, or clinicians by generating and sending them reports (Luxton et al. 2011).

Sound Apps. White Noise, Relax Melodies, Sleep, Esleep Lite and Nature Sound Relax are a few examples of sleep apps that offer an ambient and relaxing sound with white noise, pink noise and natural sound. These apps help users by distracting them from stressful thoughts and also by masking outdoor sounds. Sleep Genius is a sleep application that uses a mixture of neurosensory algorithms, multi-band binaural beats, pink noise and psychoacoustic music to help the brain sleep. Neurosensory algorithms activate the vestibular system in the brain to simulate a rocking motion, and multi-band binaural beats synchronize brainwave patterns to certain sleep frequencies. Pink noise, which masks distracting noises, has a calming effect and slows heart rate and breathing patterns. Psychoacoustic music uses specific tempos, pitches and rhythms, and is reported to calm the brain and prepare it to go to and maintain sleep (sleepGenius 2013).

Relaxation and Hypnotic Apps. Several apps with relaxing, guided meditation are designed to provide a good night's rest for users. Deep sleep with Andrew Johnson, Insomnia Cure-Sleep Now with Max Kirsten, Yoga for Insomnia, Relax and Sleep Well, Relax and Rest Guided Meditations and Long Deep Breathing are examples of deep relaxation sleep sessions that are designed to improve sleep patterns and sleep quality.

Sleep Cycle Monitoring. Some mobile applications on the market monitor the sleep cycle and wake the user during the lightest phase of sleep. Although the most accurate way to determine sleep cycle is by measuring brain activity, most of the existing mobile applications use the subject's movement information to distinguish between light and deep sleep. Sleep Cycle Alarm Clock and Sleep as Android are examples of apps that use the accelerometer to measure the amount of movement. This movement is correlated with four phases: falling asleep/dreaming/waking up, light sleep, medium/deep sleep, and deep sleep, and are able to optimize time of waking to wake users from light sleep within a 30-minute window from their final alarm (Kelly et al. 2012). This feature is designed to reduce sleep inertia, which is a sensation of grogginess, confusion and decreased functioning, experienced immediately on waking (Tassi and Muzet 2000). Wake IQ is another example of an app which is designed to reduce sleep inertia (Devender and Shang 2013).

In addition to CBT- based apps for treatment of insomnia, some apps have virtual coaches that provide real-audio and visual instruction while a patient practices CBT at home. They also can record and date-stamp homework reviews that ensure the patient's willingness to follow the treatment (Luxton et al. 2011).

In terms of follow up, smartphones have the capability of providing real-time remote communication between patients and providers by synchronizing excellent-quality digital audio and video capabilities. Integration of calendar and phone contacts by smartphone apps can improve tracking of therapy appointments and can hold therapist contact information for convenient and immediate reference.

5.2 Physiological Sensor Level Advancements in Sleep Monitoring

Accurate sleep assessment requires information from multiple physiological sensors attached to the body including EEG, EOG, EMG, ECG, oxygen saturation and respiratory sensors. To date, such comprehensive tests have been completed in sleep labs with expensive and bulky equipment. However, with the need for repeated measurements, evaluation of temporal patterns and self-assessment of sleep, advancements in sensor technologies and home sleep monitoring with portable and wireless sleep systems are emerging. Although many home-based sleep monitoring systems have focused on sleep apnea diagnosis, there is also an opportunity for insomnia patients to benefit from advances in home-sleep-monitoring devices. We do not focus on devices used for detection or diagnosis of sleep apnea as they do not usually detect physiological signals such as EEG and usually focus on respiratory signals. In addition, mHealth application for sleep apnea is an extensive research area in itself and has been presented in different literature (Bsoul et al. 2011, Oliver and Flores-Mangas 2007, Alqassim 2012).

Although the diagnostic classifications of insomnia provided by the American Academy of Sleep Medicine (AASM) and by ICSD do not include any objective criteria, most investigators rely on polysomnographic data for insomnia diagnoses (Lee-Chiong 2008). Traditionally, PSG has been conducted in laboratory settings. Sleep patterns in the lab may not be representative of typical sleep because subjects must adapt to the unfamiliar environment. Sleep is also a dynamic process that varies from day to day, and it is very important to measure it for multiple nights in order to collect and analyze the data using longitudinal approaches. It also should be correlated with the timing of other activities such as exercise, naps, food, caffeine, alcohol, and stress. This approach is not possible with the existing technology because it is expensive and inconvenient to set up. (Kelly et al. 2012)

Based on AASM recommendations, a comprehensive PSG test requires three EEGs (frontal, central, occipital), two EOGs, three EMGs for the chin, one ECG plus oxygen saturation sensor, leg EMGs and breathing sensors. Commercial systems such as Somte PSG, Sapphire, Cadwell Easy III PSG and Trackit Sleep Walker Home PSG meet such a requirement. However, they are still expensive and bulky solutions with no direct connection with, and associated applications for, smartphones. In the current systems, the collected data are transferred to a computer through a proprietary or, in some cases, Bluetooth interface. The collected data is then be analyzed by software that runs on the computer. To the authors' knowledge, none of the existing PSG systems have the capability of sending the data to smartphones directly; however, it is expected that future generations of such systems will add this feature to the portable PSG to take advantage of the available resources on smartphones.

Zeo is a sleep-monitor system based on brain activity signals and has a wireless-enabled, sensor-equipped headband that picks up the electrical signals including EEG, EOG and EMG and feeds data to a smartphone. It has the ability to classify data into sleep stages in 30-second epochs and the capacity to monitor sleep for several months. Zeo does not meet AASM criteria for sleep assessment since it has only frontal EEG derivation and does not have other sensors such as breathing sensors and leg EMGs. In addition, Zeo cannot distinguish stage one and two of sleep (N1, N2) due to combining N1 and N2 into a single category. Excessive stage N1 sleep may indicate sleep fragmentation, whereas N2 represents the majority of sleep time in a normal individual (Kelly et al 2012) .

Commercial wireless brain computer interface systems such as Emotiv's EPOC system, OCZ technology's Neural Impulse Actuator, Advanced Brain Monitoring's B-Alert X series and Enobio have emerged over the last few years for entertainment, in smart living environments, and in cognitive neuroscience research. They use EEG, EOG, EMG and, in some cases, ECG physiological signals (Lee 2013). The technology is, in principle, similar to the wireless technology that is required for sleep monitoring; however, in their current format, most are not suitable for wireless sleep monitoring applications and need customization for sleep monitoring. Additional sensors such as pulse oximeter, breathing sensors and leg EMGs are also needed for accurate sleep assessment.

Sleep Monitors system such as EarlySense Mattress and Air Cushion (bed-based system), Fitbit and Lark (movement- based system) and Heally Recording System and Sleepimage (autonomic signals- based systems) are among products on the market that extract sleep information by measuring some physiological data, movement, body position, respiration rate and snoring. Sleeprate is a chest belt that measures sleep based on heart rate and can transfer data to smartphones (Kelly et al. 2012).

6 Challenges and Limitations of Existing mHealth Technology in Insomnia Diagnosis and Treatment

Wireless sensing and smartphone technologies for health care monitoring and especially sleep monitoring applications are in their infancy. Existing technologies in the areas of quantified-self applications and wearable sensor devices suffer from a number of challenges and limitations that need to be addressed:

Reliability of Detection and Reduction of False Alarms. Due to extensive use of medical applications, highly reliable of detection and a reduced number of false alarms are critical requirements. Even at 1% false alarms could lead to unnecessary consumption of health care resources (Behar 2013). More research is needed to evaluate the reliability requirements for portable sleep monitoring systems to determine how false alarms and failure to detect sleep disorders would impact health of individuals and use of health care resources.

Credibility, Quality and Accuracy. Currently no scientific evidence is available that validates the clinical effectiveness of any of the existing sleep monitoring applications, and recommends them as medical devices that meet FDA requirements (Behar 2013). There is a lack of published research in this area, and more data is needed to address the quality and efficacy of sleep monitoring and behavioral health associated apps that are being produced.

Data Security and Privacy. Using smartphone apps for medical applications poses a risk to patient information security and confidentiality. It is possible that some user information may be sent to the software developer and other companies without user consent (Thurm and Kane 2010). Clearly, protecting collected information and the associated risks regarding breaches of confidentiality are important considerations as the technology moves forward.

Regulatory Issues. Regulations in the context of mobile apps are evolving and becoming a complex problem. Definition and classification of medical apps as medical devices, their degree of risk and their regulation are among the regulatory issues need to be addressed (Behar 2013) .

Challenges for the Elderly/Disabled. Elderly people and those with disabilities may have trouble using and interpreting the information provided by the smartphone apps about sleep monitoring. Therefore, implementation of features such as memory aids, visual aids, haptic aids and safety features to simplify the use of app is very important (Boulos et al. 2011) . Some apps are designed to improve accessibility for patients with disabilities. The app TalkBack provides spoken feedback that can assist the visually impaired (Luxton et al 2011).

Usability and Acceptance Issues. Usability and acceptance of smartphones are different from person to person. Some may embrace it and some may prefer not to use it or cannot afford it. On the other hand, sometimes, older clinicians may be reluctant to use the new technologies in their medical practices. In these situations an alternative form of therapy is always recommended (Luxton et al.2011).

Patient Safety. Although with the help of mHealth technology health providers can stay in close contact with patients, clinicians need to assess risk on a case-by-case basis and determine whether a planned method is appropriate. Reliance on smartphone technology for high risk patients such as those with insomnia along with severe depression may not be recommended unless there is a backup plan in the case of technology failure (Luxton et al. 2011).

Ethical and Policy Related Issues. The expectation of immediate response to emails or texts, texting during off hours or outside of the therapeutic relationship, the consequences of patient self-diagnosis by using apps are examples of ethical or policy related issues (Luxton et al.2011). In some countries such as the U.S, the insurance system usually reimburses the face to face appointments thus there is a disincentive for clinicians and patients to take advantage of mHealth technology by teleconferencing, sending text and emailing (Ozdalga et al. 2012).

Disposability and Patient Commitment. Due to the availability of plenty of free apps that can be easily install and uninstall, patients may not be committed to use one app on regular basis or for a long time. They may be curious to try out new apps if the current one does not meet their expectations in short period of time (Dennison 2013).

Distraction and Interference with Sleep. Electronic media including smartphones have often been considered to have a negative impact on the sleep. Delayed bedtime and shorter total sleep time are the most consistent findings related to media use(Cain and Gradisar 2010). A recent study showed that mobile phone usage was positively associated with insomnia and negatively with morningness (Fossum et al. 2014).

Some other issues such as limited battery life, small screen size, potentially erroneous data input, computer viruses and potentially inefficient patient-physician interactions also need to be addressed as limitations of existing smartphone technology (Mosa et al. 2012). Over the coming years, advancements in several areas of these technologies in both wearable sensors and smartphone apps as well as their applications in sleep medicine are expected to emerge to address the existing limitations and challenges.

7 Future Directions in Research and Technology in Sleep Monitoring

In the current PSG systems, the information from sensors attached by wires to the different parts of the body goes into a central processing unit. The central unit collects all the information and sends it to a remote computer where the associated software stores the real time data for post processing and analysis. This method requires many wires going from different sensors to the central unit and may not be very convenient or comfortable for patients.

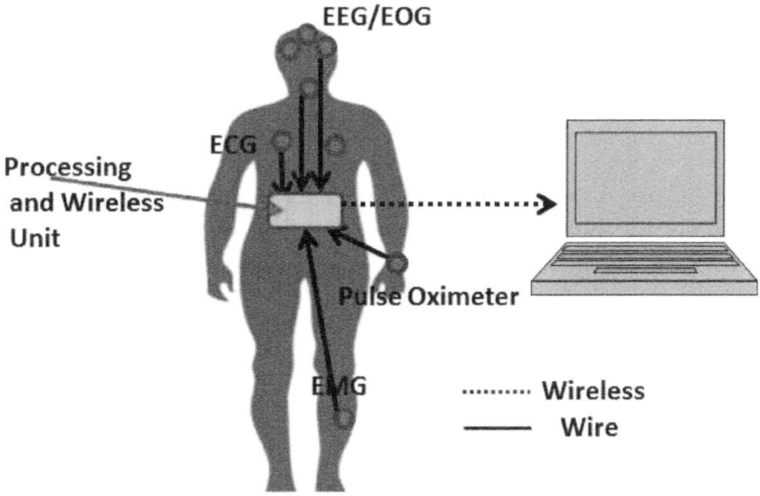

Fig. 1 Existing wireless PSG system technology for sleep monitoring

Moving towards the use of smartphone technology as a processing unit and as a gateway to the internet and cloud requires a number of new features in the sensor system and on the mobile phone. The wireless sensor systems need to support Bluetooth protocol to be able to transfer the data directly to the smartphone. Moreover, less expensive, more compact and energy efficient sensor technologies need to be developed to make the adoption of such a solution more widespread. From the perspective of patient comfort, the ideal scenario would be a technology where each individual sensor could communicate directly to the mobile phone. In this approach the wiring from each individual sensor to a central processing and transceiver unit would be removed. In the individual wireless sensor scenario, the power consumption of each individual node would be extremely low since, in this case, each sensor would have its own data acquisition and transceiver. In this scenario, minimal processing would be done on each node, and further advanced processing could be transferred to the mobile phone processor or other servers available in the cloud.

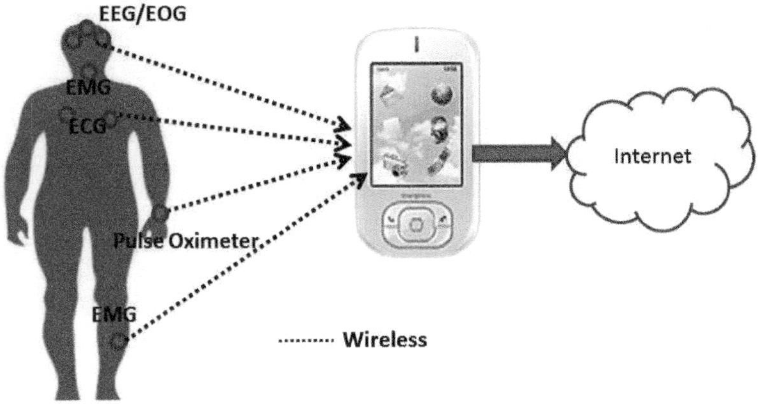

Fig. 2 mHealth vision of sleep monitoring technology

With a solution where a combination of ultra-low power physiological sensors and more advanced sleep analysis applications are used, more features are expected to emerge in mHealth sleep monitoring. For instance, if EXG signals (i.e., EEG, ECG, EOG, and EMG) are continuously monitored, insomnia treatment can be assessed precisely and information can be used for more efficient and effective treatment.

Another area of improvement is automatic sleep scoring and analysis tools. This feature, which works in combination with the wearable sensors and analyses the collected data by EXG (EEG, ECG, EOG, and EMG) sensors, is expected to be utilized by mobile apps using novel algorithms to analyze the EXG data and create an accurate sleep staging report based on physiological data.

Future mHealth technology could connect the physiological wireless sensors to the smartphone to offer more precise diagnosis, monitoring and treatment techniques for insomnia and other sleep disorders. This technology needs to be more evolved at both the sensor and smartphone levels in order to support the collection of high quality data from the physiological sensors with transfer directly and seamlessly to smartphones where automated algorithms would be used to distinguish among sleep stages and wakefulness.

References

Alqassim, S., et al.: Sleep Apnea Monitoring Using Mobile Phones. In: IEEE International Conference on E-Health Networking, Application and Services (HealthCom), Beijing, China (October 2012)

Behar, J., et al.: A review of current sleep screening applications for smartphones. Physiological Measurement 34(7), R29–R46 (2013)

Boulos, M.N.K., et al.: How smartphones are changing the face of mobile and participatory healthcare: an overview, with example from eCAALYX. Biomedical Engineering Online 10, 24 (2011)

Breslau, N., et al.: Sleep disturbance and psychiatric disorders: a longitudinal epidemiological study of young adults. Biological Psychiatry 39(6), 411–418 (1996)

Bsoul, M., Minn, H., Tamil, L.: Apnea MedAssist: real-time sleep apnea monitor using single-lead ECG. IEEE Transactions on Information Technology in Biomedicine: A Publication of the IEEE Engineering in Medicine and Biology Society 15(3), 416–427 (2011)

Cain, N., Gradisar, M.: Electronic media use and sleep in school-aged children and adolescents: A review. Sleep Medicine 11(8), 735–742 (2010)

Carskadon, M.A.: Patterns of sleep and sleepiness in adolescents. Pediatrician 17(1), 5–12 (1990)

Chesson, A.L., et al.: Practice parameters for the nonpharmacologic treatment of chronic insomnia. An American Academy of Sleep Medicine report. Standards of Practice Committee of the American Academy of Sleep Medicine. Sleep 22(8), 1128–1133 (1999)

Chevalier, H., et al.: Evaluation of severe insomnia in the general population: results of a European multinational survey. Journal of psychopharmacology (Oxford, England) 13(4 suppl. 1), S21–S24 (1999)

Chong, Y., Fryer, C.D., Gu, Q.: Prescription sleep aid use among adults: United States, 2005-2010. NCHS Data Brief (127), 1–8 (2013)

Colten, H., Altevogt, B.B.M., Colten, H.: Sleep Disorders and Sleep Deprivation: An Unmet Public Health Problem. In: Colten, H.R., Altevogt, B.M. (eds.) Committee on Sleep Medicine and Research, pp. 217–223 (2006)

Cover, H., Irwin, M.: Immunity and depression: insomnia, retardation, and reduction of natural killer cell activity. Journal of Behavioral Medicine 17(2), 217–223 (1994)

Dennison, L., et al.: Opportunities and challenges for smartphone applications in supporting health behavior change: qualitative study. Journal of Medical Internet Research 15(4), e86 (2013)

Drummond, S.P.A., et al.: The neural basis of the psychomotor vigilance task. Sleep 28(9), 1059–1068 (2005)

Fossum, I.N., Nordnes, L.T., Storemark, S.S., Bjorvatn, B., Pallesen, S.: The Association Between Use of Electronic Media in Bed Before Going to Sleep and Insomnia Symptoms, Daytime Sleepiness, Morningness, and Chronotype. Behavioral Sleep Medicine (2014), Gartenberg, D.: Understanding Brain Arousal and Sleep Quality Using a Neuroergonomic Smart Phone Application. Advances in Understanding, 200–210 (2010)

Gartenberg, D., et al.: Collecting health-related data on the smart phone: mental models, cost of collection, and perceived benefit of feedback. Personal and Ubiquitous Computing (2012)

Hillman, D.R., Murphy, A.S., Pezzullo, L.: The economic cost of sleep disorders. Sleep 29(3), 299–305 (2006)

Irwin, M., et al.: Partial night sleep deprivation reduces natural killer and cellular immune responses in humans. The FASEB Journal: Official Publication of the Federation of American Societies for Experimental Biology 10(5), 643–653 (1996)

Jacobs, G.D., et al.: Cognitive Behavior Therapy and Pharmacotherapy for Insomnia. Arch. Intern. Med. 164, 1888–1896 (2004)

Kelly, J.M., Strecker, R.E., Bianchi, M.T.: Recent Developments in Home Sleep-Monitoring Devices. ISRN Neurology, 1–10 (2012)

Kripke, D.F., et al.: Mortality associated with sleep duration and insomnia. Archives of General Psychiatry 59(2), 131–136 (2002)

Lee, S., et al.: Review of Wireless Brain-Computer Interface Systems. In: Fazel-Rezai, R. (ed.) Brain-Computer Interface Systems - Recent Progress and Future Prospects. InTech (2013) ISBN: 978-953-51-1134-4, doi:10.5772/56436

Lee, J.: Smart health: Concepts and status of ubiquitous health with smartphone. In: ICTC 2011 (2011)

Lee-Chiong, T.: Sleep Medicine: Essential and review, 1st edn., ch. 3, P. Oxford University Press (2008)

Littner, M., et al.: Practice parameters for using polysomnography to evaluate insomnia: an update. Sleep 26(6), 754–760 (2003)

Luxton, D.D., et al.: mHealth for mental health: Integrating smartphone technology in behavioral healthcare. Professional Psychology: Research and Practice 42(6), 505–512 (2011)

McCall, W.V.: Sleep in the Elderly: Burden, Diagnosis, and Treatment. Primary Care Companion to the Journal of Clinical Psychiatry 6(1), 9–20 (2004)

McNeil, J., Doucet, E., Chaput, J.-P.: Inadequate sleep as a contributor to obesity and type 2 diabetes. Canadian Journal of Diabetes 37(2), 103–108 (2013)

Miskelly, F.: Electronic tracking of patients with dementia and wandering using mobile phone technology. Age and Ageing 34(5), 497–499 (2005)

Morin, C.M., et al.: Behavioral and pharmacological therapies for late-life insomnia: a randomized controlled trial. JAMA: The Journal of the American Medical Association 281(11), 991–999 (1999)

Mosa, A.S.M., Yoo, I., Sheets, L.: A Systematic Review of Healthcare Applications for Smartphones. BMC Medical Informatics and Decision Making 12(1), 67 (2012)

Nakata, A., et al.: Sleep-related risk of occupational injuries in Japanese small and medium-scale enterprises. Industrial Health 43(1), 89–97 (2005)

Oliver, N., Flores-Mangas, F.: HealthGear: Automatic Sleep Apnea Detection and Monitoring with a Mobile Phone. Journal of Communications 2(2) (2007)

Ozdalga, E., Ozdalga, A., Ahuja, N.: The Smartphone in Medicine: A Review of Current and Potential Use Among Physicians and Students. Journal of Medical Internet Research 14(5), e128 (2012)

Pigeon, W.R.: Insomnia as a risk factor for disease. Insomnia: diagnosis and treatment. Informa Healthcare, New York (2010)

Roehrs, T., et al.: Neuropsychological function in obstructive sleep apnea syndrome (OSAS) compared to chronic obstructive pulmonary disease (COPD). Sleep 18(5), 382–388 (1995)

Roth, T.: Insomnia: definition, prevalence, etiology, and consequences. Journal of Clinical Sleep Medicine: JCSM: Official Publication of the American Academy of Sleep Medicine 3(5 suppl.), S7–S10 (2007)

Roth, T., Ancoli-Israel, S.: Daytime consequences and correlates of insomnia in the United States: results of the 1991 National Sleep Foundation Survey. II. Sleep 22(suppl. 2), S354–S358 (1999)

Schutte-Rodin, S., et al.: Clinical guideline for the evaluation and management of chronic insomnia in adults. Journal of Clinical Sleep Medicine: JCSM: Official Publication of the American Academy of Sleep Medicine, 487–504 (2008)

Shahly, V.: The Associations of Insomnia With Costly Workplace Accidents and Errors: Results From the America Insomnia Survey: Insomnia and Workplace Accidents and Errors. Archives of General Psychiatry 69(10), 1054 (2012)

Simon, G.E., VonKorff, M.: Prevalence, burden, and treatment of insomnia in primary care. The American Journal of Psychiatry 154(10), 1417–1423 (1997)

Sivertsen, B., et al.: Cognitive behavioral therapy vs zopiclone for treatment of chronic primary insomnia in older adults: a randomized controlled trial. JAMA: The Journal of the American Medical Association 295(24), 2851–2858 (2006)

Sleep Genius, `http://sleepgenius.com/` (January 14, 2014)

Smith, M.T., et al.: Comparative meta-analysis of pharmacotherapy and behavior therapy for persistent insomnia. The American Journal of Psychiatry 159(1), 5–11 (2002)

Drummond, S.P., et al.: The sleep of abstinent pure primary alcoholic patients: natural course and relationship to relapse. Alcoholism, Clinical and Experimental Research 22(8), 1796–1802 (1998)

Suka, M., Yoshida, K., Sugimori, H.: Persistent insomnia is a predictor of hypertension in Japanese male workers. Journal of Occupational Health 45(6), 344–350 (2003)

Tassi, P., Muzet, A.: Sleep inertia. Sleep Medicine Reviews 4(4), 341–353 (2000)

Taylor, D.J., Roane, B.M.: Treatment of insomnia in adults and children: a practice friendly review of research. Journal of Clinical Psychology 66(11), 1137–1147 (2010)

Thase, M.E.: Correlates and consequences of chronic insomnia. General Hospital Psychiatry 27(2), 100–112 (2005)

Thurm, S., Kane, Y.I.: What They Know: Your apps are watching you and sharing the personal data. Wall Street Journal Asia 35(79), 14–15 (2010)

Tsuno, N., Besset, A., Ritchie, K.: Sleep and depression. The Journal of Clinical Psychiatry 66(10), 1254–1269 (2005)

Van Devender, M., Shang, Y.: Wake IQ: Using a smartphone to reduce sleep inertia. In: 2013 IEEE 10th Consumer Communications and Networking Conference, CCNC 2013, p. 649 (2013)

Walsh, J.K., Engelhardt, C.L.: The direct economic costs of insomnia in the United States for 1995. Sleep (Rochester) 22, S386–393 (1999)

Wilson, S.J., et al.: British Association for Psychopharmacology consensus statement on evidence-based treatment of insomnia, parasomnias and circadian rhythm disorders. Journal of Psychopharmacology (Oxford, England), 1577–1601 (2010)

Chong, Y., Fryer, C.D., Gu, Q.: Prescription sleep aid use among adults: United States, 2005-2010. NCHS Data Brief (127), 1–8 (2013)

Zammit, G.K., et al.: Quality of life in people with insomnia. Sleep 22(suppl. 2), S379–S385 (1999)

Questions

Q1. What is the definition of insomnia based on International Classification of Sleep Disorder?

Q2. What are the main factors which have role in causing insomnia?

Q3. What is the first of line treatment approach for chronic insomnia? What are methods involved in this treatment approach?

Q4. Below are some examples of sleep graphs generated by Sleep Cycle application.

Part II

Remote Health Monitoring

Chapter 7: Accelerometer-Based Human Activity Recognition in Smartphones for Healthcare Services

Chapter 8: A Formal Approach for Dependability Assessment of a mHealth Monitoring Systems

Chapter 9: Wireless Monitoring System for Wheelchair Users with Severe Mobility Impairment

Chapter 10: Remote Health/Vital Sign Monitoring: mHealth and Remote Vital Sign Monitoring-Trends and Applications for ECG Analysis on Cell Phones

Chapter 11: Operation, Analysis and Optimization of Wireless Sensor Devices in Health Oriented Monitoring Systems

Chapter 12: mHealth Sensors and Applications for Personal Aid

Chapter 13: Location-Aware Services Using Android Mobile Operating Platform for Safety, Emergency and Health Applications

Chapter 14: mHealth Monitoring System for Hospitalised Older Adults – Current Issues and Challenges

Chapter 15: Ubiquitous Health Monitoring: Integration of Wearable Sensors, Novel Sensing Techniques, and Body Sensor Networks

7. Accelerometer-Based Human Activity Recognition in Smartphones for Healthcare Services

Cesar Torres-Huitzil and Andres Alvarez-Landero

LTI, CINVESTAV-Tamaulipas, Mexico
{ctorres,aalvarez}@tamps.cinvestav.mx

Abstract. The synergy of communication, computation and sensing capabilities in mobile systems-on-chip artifacts such as smartphones has made possible the development of wearable smart sensor systems for user activity monitoring and recognition. Assessing physical activity is useful to enhance people health as experts have evidenced a clear correlation between physical activity and overweight, obesity and metabolism-related syndromes. Due to user acceptability and the convenient nonintrusive manner for measuring data, smartphones have the advantage of taking proprioceptive motion measurements outside of a controlled environment for rather long periods of time using embedded sensors such as the accelerometer, however, activity recognition poses several challenges. Particularly, though work has been reported for accelerometer-based activity recognition using smartphones, the portability of the device to a single fixed tight position has been a major constraint to easy the interpretation of the collected data on resource-limited devices. In this chapter, a human activity hierarchical recognition system based on neural networks without the need of the smartphone to be constrained to a single fixed position is presented. Yet it is used as a representative example to show the challenges, the role and some of the main potential impacts that smartphones have and will have in mHealth. Experimental results on Android-capable smartphones on four on-body locations show that the recognition system achieves high classification rates, above 92%, for five activities including static, walking, running, and up-down stairs walking, which outperforms other proposals. The system is fully implemented in a smartphone running continuously in near real-time with reduced power consumption in a proof-of-concept client-server application for mHealth.

Keywords: human activity recognition, accelerometer, smartphone, mHealth, time-domain features, autoregressive model, neural networks.

1 Introduction

Recently, several studies have shown that knowledge of human activities and behaviors performed by a subject are good indicators of human health status [10]. Medical

S. Adibi (ed.), *Mobile Health*,
Springer Series in Bio-/Neuroinformatics 5, DOI: 10.1007/978-3-319-12817-7_7

experts have demonstrated a strong correlation between the lack of physical activity and overweight, obesity, and metabolic-related syndromes [26]. Thereby, automatic activity monitoring might provide clinicians and medical experts with the ability to monitor and diagnose patients using continuously generated data, rather than data from a time-bounded medical appointment [27]. Furthermore, activity recognition can be performed daily in order to detect deviations from a typical routine or deterioration of a patient current physical status. In a healthcare system, activity recognition can facilitate rehabilitation of patients undergoing physical therapy by monitoring their status in finer detail so as therapists are able to ensure patient quality of care with more accurate evaluations [11][6]. Therefore, human activity monitoring and recognition, in a mobile health (mHealth) system, can be used as evidence for medical diagnosis, treatment of diseases and mitigate unhealthy effects associated to sedentarism, within a general framework as conceptually shown in figure 1. The purpose of mHealth is to improve human health and well-being by continuously monitoring their status, rapidly diagnosing, recognizing behaviors and delivering just-in-time interventions in the mobile user environment [17]. Within this context, robust human activity recognition might be considered as a low-level enabler of mHealth systems to help people better understand their own unhealthy behavior, potentially decreasing the working load for the medical personnel, shortening the hospital stay for patients, and improving people quality of life.

Thanks to the advances in microelectromechanical (MEM) technologies, one of the current approaches employed to monitor human activity is the use of wearable inertial sensors directly attached to the human body or embedded into portable devices such as smartphones [5]. Due to their communication, computation and sensing capabilities, smartphones have become everyday devices that many people carry with them all day long, see chapter 1 for further details. As a consequence, smartphones may facilitate both remote acquisition and on-device processing of personal, social or environmental data, captured through embedded physical sensors in uncontrolled environments [13][18][16]. Regarding human activity recognition, it is known that the acceleration generated during human movement varies across the body and depends upon the activity being performed. However, motion signals captured by inertial sensors, such as accelerometers and gyroscopes, are highly dependent on both the performed human activity and the location where the sensor is placed. The acquired signals might be totally different when a subject performs the same activity with sensors placed on different locations. See figure 2 for an illustrative example of this situation where variation in accelerometer orientation and position affects the measured acceleration pattern. Moreover, a light displacement of the sensor from its original position might decrease the accuracy of the analysis.

For practical applications, under real mobile user scenarios, knowledge of the actual location and orientation of the sensing device is not available a priori, and the recognition system must somehow solve this problem or impose some placement constraints so as to reduce the problem complexity and easy signal interpretation [28]. Furthermore, from a pragmatic point of view, several other aspects should be considered for porting and deployment of an activity recognition solution to a mobile device in order to reduce the cost and guarantee a good performance. Factors

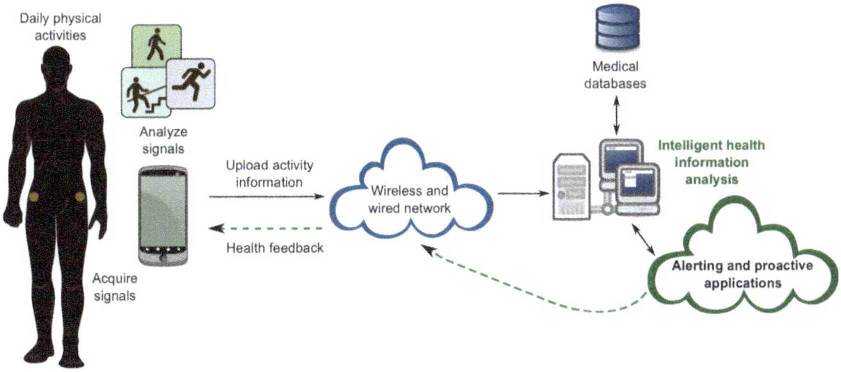

Fig. 1 A general overview of a smartphone-based activity recognition for health monitoring and assessment en a mHealth ecosystem

such as sensor sampling rate, processing time, memory size and power consumption, are particularly important for portable systems [21]. Thus, the collection of longterm user activity information with ubiquitous coverage still remains an open technological challenge. Although high-end smartphones are equipped with high performance microprocessors, any activity recognition approach should be light, i.e., incurring in a low computational overhead, as other applications in the software stack run simultaneously on top of the operating system and the whole processing power is not fully available at any time for just a single task.

In this chapter, a light-weight position-independent activity recognition system, implemented on a smartphone, based on a single embedded triaxial accelerometer, is presented and discussed. Recall that accelerometers are typically electromechanical devices that measure the applied acceleration acting along their sensitive axis. The measured acceleration can be static, like the constant force of gravity ($g = 9.8$ m/s^2), or dynamic, caused by moving or shaking the accelerometer. A triaxial accelerometer is intended for simultaneous measurement of acceleration in 3 perpendicular axes. The proposed activity recognition system is used as a representative example to show the challenges, the role and some of the main potential impacts that smartphones have and will have in the mHealth technology roadmap. Overall, the main goal was to design an activity recognition mobile system that works naturally and unobtrusively, at some extent, for the user and that can be used for continuous real-time monitoring of some human activities that provide useful and reliable information for mHealth. The stand-alone recognition system, fully implemented in an Android capable smartphone device, does not require any stringent a priori information about smartphone location. To be more precise, the smartphone device is allowed to be in some typical locations on the subject body and no restriction is imposed on the orientation of the device. With respect to on-body device positions, situations when the user carries the mobile device in the front and back trouser pockets are considered.

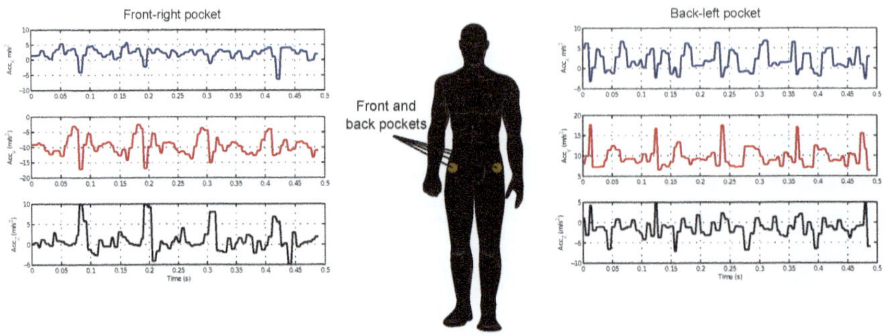

Fig. 2 Examples of on-body placement and accelerometer signal variability for the same activity, walking, when the smartphone is located in the front-right and the back-left pockets.

A set of fundamental activities associated to mobility (static, walking, running, upstairs and downstairs walking that are common in ambulatory daily routines) was chosen as their recognition has a direct translation into healthcare applications such as energy expenditure, activity level and ambulatory monitoring. Developing such applications requires to discriminate between different user physical activities contained within the accelerometer data ranging from coarser levels, such as moving or stationary modes, to finer levels of movement, such as running, walking, sitting, or standing, and even the transition patterns between them [13]. Current accelerometers embedded in smartphones are sufficient for the detection of such activities as they typically provide measurements with a configurable range of ± 2 o ± 8 g and an output rate of 100 Hz or 400 Hz that cover the generated acceleration range when performing physical activities. However, in practice, the full capabilities of embedded accelerometers are not achievable due to implementation constraints and restricted access to the full resolution and sampling frequencies of the majority of mobile software platforms [4][30].

The rest of this chapter is organized as follows. In section 2, some relevant works related to portable accelerometer-based activity recognition are briefly described. Section 3 presents in detail the proposed hierarchical neural recognition system and its main processing steps. Experimental results and analysis of the achieved classification precision are provided in section 4, as well as performance and power consumption of a proof-of-concept application ported to Android capable smartphones. Section 5 and 6 presents case studies and future directions, respectively. Finally, concluding remarks are presented.

2 Related Work

There are different approaches to activity recognition that vary depending on the underlying sensor technologies that are used to monitor activities, the alternative machine learning algorithms that are used to model the activities, and the complexity of the activities that are being modeled [5][24][1]. Regarding sensors technologies, activity recognition can be classified into two main categories: vison based and sensor-based recognition systems. The first one is based on visual sensing facilities to monitor user behavior and environmental changes. The generated data are video sequences that are analyzed using computer vision techniques. Different from video-based activity recognition systems leveraging on video cameras, in the wearable sensor paradigm, physical sensors are typically deployed to collect observations. Data from sensor-based monitoring systems are mainly time series. Herein, research works that commonly use portable inertial measurement units to recognize ambulatory movements and concretely those based on accelerometers are briefly reviewed. Interested readers in further details are referred to [5][24].

Several accelerometer-based activity recognition systems have been proposed and developed in the past for recognizing a wide set of daily physical activities. However, the majority of the advances has taken place only recently thanks to the advances in MEM and smartphone technologies. Table 1 provides a summary of some selected relevant works found in the literature. However, the revision is not exhaustive and a fair comparison among different approaches is not straightforward as the employed materials and methods are diverse. Most of these works use multiple accelerometers attached to different locations on a subject body [2][3], not the main concern in this work, and fewer proposals use a single accelerometer attached firmly to a specific location [10][8][15][19]. Under these highly constrained scenarios, those systems achieve good recognition results but perform offline processing, i.e., gather and then process data to recognize activities outside the mobile device, using the device just as a wearable sensor capable of sampling and storing sensor data. Regarding feature extraction, most works targeted to mobile devices use lightweight time domain features as the input to the classifier. However, some works have used frequency domain features employing the Fourier transform or the wavelet transform. These transforms usually mean a computational load for most resource-constrained mobile platforms but they provide a good precision. Additionally to the computational overhead, these frequency-based features are not suitable to detect short-time duration movements. On the other hand, several classification methods have been used for activity recognition ranging from heuristic classifiers to artificial neural networks (ANNs) combined with kernel discriminant analysis (KDA) and support vector machines (SVM). Such classification methods subject to real-time feedback must be adapted to the computational limitations, available memory, and the respective energy constraints of battery-powered mobile devices.

From the reviewed related work, it is important to highlight the proposals presented in [15] and [19] due to their obtained precision and closeness to our proposed approach. In these works, authors present an accelerometer-based approach for human-activity recognition based on a hierarchical scheme applied on statistical signal features and ANNs. The upper level recognition in this approach uses the autoregressive (AR) modeling of the acceleration component signals, incorporating the derived AR-coefficients along with the signal-magnitude area and tilt angle to form an augmented-feature vector. The resulting feature vector is further processed by the linear-discriminant analysis and ANNs to recognize a particular human activity. The proposed activity-recognition method recognizes fifteen activities with an

Table 1 Summary of selected accelerometer-based activity recognition systems. All works except [3] mostly use time-domain features. Acronyms: AR (Autoregressive), SVM (Support Vector Machine), ANN (Artificial Neural Network), KDA (Kernel Discriminant Analysis).

First author	No. Accel.	Position	Key features	Classifier	On-device
Bicocchi [3]	3	fixed (arm, waist, leg)	raw data	instance-based	no
Fuentes [8]	1	fixed (chest)	forward and vertical acceleration	SVM	yes Openmoko
Hache [10]	1	fixed (waist)	signal magnitude area inclination angle	decision tree	yes Blackberry
Khan [15]	1	fixed (chest)	AR model tilt angle	ANN	no
Lee [19]	1	fixed (chest)	AR model spectral entropy	ANN + KDA	no

average accuracy of 97% using only a single triaxial accelerometer attached to the user body. However, the accelerometer is fixed to a single position and their proposal is not validated in a smartphone device. Further in [14], authors extend their work following the same research direction so as to address position-independent activity recognition. Their results are encouraging but offline processing is performed without aiming at a real-time recognition system for continuous monitoring.

3 Activity Recognition System Design

In this section, the proposed activity recognition system using a single triaxial accelerometer in a smartphone is presented in detail. Recall that the main focus of this study is to not only classify ambulatory human activities, but also perform position-independent recognition, the device is not fixed to a single location, able to be integrated in a mobile application that aims at continuously monitoring and evaluating the activity level of a user. Figure 3 shows a general block diagram of the main processing steps that most accelerometer-based activity recognition systems typically perform. The particularities for each step in the proposed approach are also outlined in figure 3 and they are described in more detail in the following subsections.

In the initial sensing step, the accelerometer produces raw measurement data that are refined in the preprocessing step. Due to mobile device energy constraints, the configuration of the sensing stage, number of sensors, and sampling frequency, is a challenging and an important issue to be addressed [22][29]. Data redundancy and noise are minimized in the preprocessing stage and more computationally efficient representations of data, called features, are extracted using lightweight methods due to constraints of computation and memory resources. Feature selection for a classifier is often performed offline in a training-evaluation loop to asses classification performance. Finally, the selected features are used as inputs to the classification

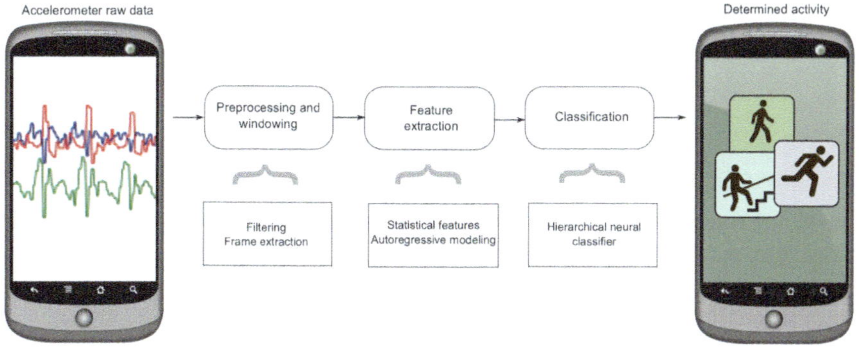

Fig. 3 General block diagram of the main processing steps for activity recognition from a single triaxial accelerometer embedded in a smartphone device.

technique, which determines the corresponding activity being performed. As stated in [13], mobile sensing system architectures require feature extraction control mechanisms well coupled with classifiers in order to better adapt to the resource constraints of mobile sensing platforms. The work presented in this chapter provides an important step in this direction as a full on-device implementation for continuous activity monitoring is pursued, but with a time resolution in the order of seconds.

3.1 Preprocessing and Windowing

The raw output of the embedded accelerometer was first preprocessed as it presents some artifacts and noise associated to the data acquisition process which are not really relevant for activity recognition. Recall that the acceleration as measured by the accelerometer can be static, due to the constant force of gravity, or dynamic, caused by moving or shaking the accelerometer. To remove gravity, a simple first order high pass filter was employed on each acceleration component. For noise reduction a three-point moving smoothing average filter was also applied [30][31].

A further issue that should be considered in the preprocessing stage is that the maximum claimed sampling frequency of accelerometers embedded in smartphones is in practice not achievable due to implementation constraints and restricted access by the software stack in mobile operating systems and frameworks such as Android. Moreover, according to experimental observations, the accelerometer was likely sampled irregularly because the restricted implementation of the sampling mechanism in the software stack of Android. Thus linear interpolation was applied in order to estimate missing samples according to the fixed sampling rates options.

With the information of a single triaxial accelerometer it is not possible to determine orientation of the device, thus the magnitude of the acceleration vector was computed so as to reduce the effect of device orientation. Formally, consider the acceleration be $\mathbf{a} = (a_x, a_y, a_z)$ sampled at a time t, where a_x, a_y, a_z are the three components of the acceleration projected on the smartphone reference frame. Thus, the magnitude of the acceleration for all samples is computed according to:

$$|\mathbf{a}| = \sqrt{a_x^2 + a_y^2 + a_z^2} \quad (1)$$

After the accelerometer samples have been filtered out and interpolated, a non-overlapping sliding window approach was used to segment the accelerometer magnitude into small time windows or frames. This segmentation consists in partitioning the accelerometer signal into consecutive windows of fixed length. The size of the window depends on the desired resolution for activity monitoring and the available sampling frequency in the device. It should be noted that the window size, usually in order of seconds, is an important parameter that affects both the computation and the power consumption of classification algorithms and must be taken into account when ported to a given portable device [12][25]. The windowing technique offers a simpler approach to learn the activity models during the training phase over the explicit segmentation approach and it further reduces the computational complexity. Then, time-domain features were extracted from accelerometer samples in each window that are fed to the classifier.

3.2 Feature Extraction

An important processing step in activity recognition is to select the relevant features with good discriminatory power. Recall that, a feature is a characteristic measurement, transform, or structural mapping extracted from the input data to represent important patterns of a desired phenomena with reduced dimension. Given a data set, selecting an accurate model that best approximates the distribution of the data is not an easy problem and is still an open issue. There are several features, time domain, frequency domain and heuristic features, that can be extracted from accelerometer data for human activity recognition. See [7] for a comprehensive review on this subject and [13] for a further feature classification based on their main context of application, namely user physical activity, social interactions, and environment.

In this work we mainly extract time-domain statistical features such as mean, standard deviation and percentiles from the acceleration magnitude as limitations in processing time, memory, and energy consumption in mobile platforms do exist. Mean and standard deviation are the most commonly used features for accelerometer signals to differentiate body postures and for estimating the intensity of an activity [13]. Figure 4 shows a typical distribution of class instances in the training dataset, see details in section 4, when these features are considered. Note the discriminative ability of standard deviation and median to differentiate static and running from other activities. Another empirical observation might be derived, features should be adapted to each activity itself, see subsection 3.3 for details. On the other hand, recall that a percentile is a statistical measure indicating the value below which a given percentage of observations in a group of observations falls. The extracted features considered for activity recognition are summarized in Table 2; they were selected based on their efficacy as demonstrated in the literature.

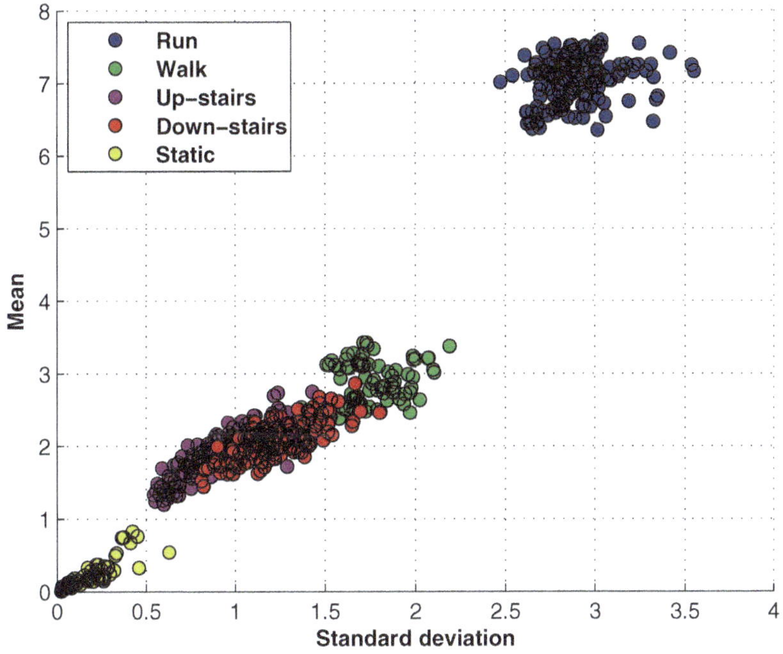

Fig. 4 Feature distribution for some instances in the training dataset

In addition to statistical features, an autoregressive model over the triaxial acceleration signal is performed since it has been proven to be useful for human activity classification as shown in [15]. In natural phenomena, the behavior of a variable evolving with time is largely influenced by the past values and to understand the generation mechanism, the mutual interaction among its samples is analyzed. Let the samples $x(n), x(n-1), x(n-2), \ldots$, represent the accelerometer signal when performing an activity. Hence, the current signal sample can be approximated by taking a linear combination of its past M samples:

$$\hat{x}(n) = \sum_{k=1}^{M} a(k)x(n-k) \tag{2}$$

where M is the order of the AR model for the signal $x(n)$, and $a(k)$, $k = 1, 2, \ldots, M$, are the AR coefficients, which can be obtained by solving the Yule-Walker equation using the Levinson algorithm [20][32].

The feature vector with the statistical measures and the AR-coefficients was used to recognize the following physical human activities, static, walking, running, upstairs and downstairs walking with a time resolution of five seconds. Principal components analysis (PCA) was applied to remove any correlation among features and

Table 2 Selected time-domain features used for accelerometer-based activity recognition

Feature	Formula
Mean	$\mu = \frac{1}{n}\sum_{i=1}^{n} x_i$
Standard deviation	$\sigma = \sqrt{\frac{1}{n}\sum_{i=1}^{n}(x_i - \overline{x})^2}$
Percentiles	$P_k = k(n+1)/100$
Coefficients of an AR model	$x(n) = \sum_{k=1}^{M} a_i x(n-k) + \varepsilon(n)$

to reduce further the feature vector dimension. An important advantage of PCA is that effectively the dimension of the feature space may be reduced without much loss of information, meaning less computational overhead on the mobile device and energy savings. The PCA method, which uses only multiplications and additions after the projection weights are obtained, is considered a computationally efficient method and it has been successfully implemented in a smartphone.

3.3 Hierarchical Neural Classifier

Several classification schemes have been used in previous works concerning activity recognition as outlined in section 2. Based on the proposals and results reported in [19] and [9], a hierarchical neural recognition scheme was used in this work. Artificial neural networks, which resemble the parallel computation in the brain, are a very popular technique that have been widely used in nonlinear pattern recognition problems thanks to their accuracy rates and generalization capabilities. Basically, neural networks use a connectionist approach to compute outputs through a feedforward network of inputs, hidden states, and possible outputs. The correct topology and numbers of hidden neurons is found experimentally by training and evaluating the performance of neural classifiers with different parameters on datasets.

A block diagram of the proposed hierarchical classifier is shown in figure 5. The hierarchical neural structure is based in the following observations. First the acceleration generated during human movement varies across the body and depends upon the activity being performed. As a consequence the activity recognition requires the ability to discriminate between different physical activities ranging from coarse levels, such as moving or stationary, to finer levels of motion, such as down-stairs and upstairs walking, probably using different features. Moreover, in physical activity recognition it has been shown that neural networks perform particularly well when a single activity needs to be detected instead of using a large network for the

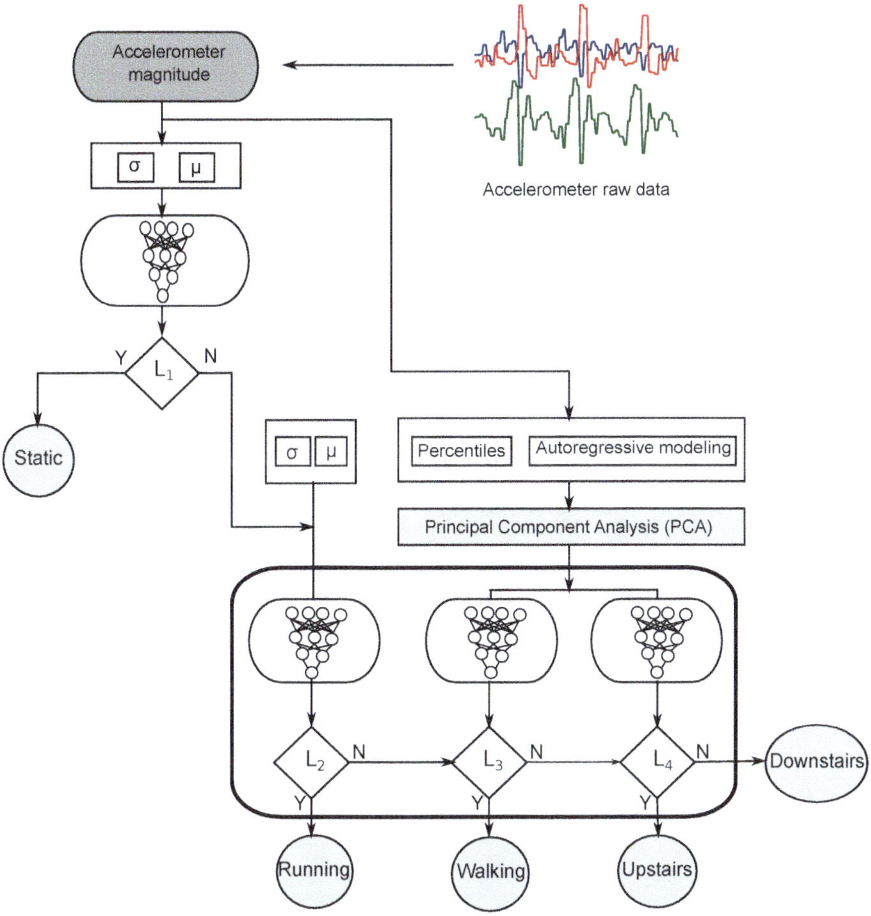

Fig. 5 Block diagram of the hierarchical neural classifier used for activity recognition. The processing overhead increases from top to bottom in the hierarchy, L_1 to L_4.

recognition of various physical activities [9]. In the proposed approach, the state recognition, static or dynamic, is performed at the first level in the hierarchy, L_1, using only statistical features using a simple neural network, the perceptron. For the L_2 level also a perceptron was used for classification so as to promote performance in the mobile device. At lower levels from L_3 to L_4, as the interclass correlation increases among the remaining activities, the number of features is augmented (autoregressive coefficients and percentiles with PCA) and more complex multilayer feedforward neural networks are employed.

Under this hierarchical approach, a specialized neural network is designed for each class per level in the hierarchy, and each network potentially receives different features of the total set available. This hierarchy model is very similar in principle to decision trees, but training on each level of the hierarchy to determine the threshold is performed in supervised mode. From an implementation point of view, note that a hierarchy of thresholds might be used as a simple and computationally efficient model for mobile platforms but with lower precision and generalization capabilities than neural networks. In the proposed recognition scheme, once a network in the hierarchy has asserted a given class, the rest of the networks are not evaluated. In this way, when the classifier is ported to a mobile device, performance and power consumption are improved as no additional computations are performed.

It is important to point out that training neural networks is usually considered computationally expensive and consequently it is performed offline and not in the mobile platform. Additionally note that PCA was also applied offline over the original features so as to reduce correlation. Once the projection matrix is found, an online transformation is applied to obtain a reduced number of transformed features compared to the original one. This introduces a little overhead on the mobile device but reduces the complexity and size of the required neural networks.

4 Experiments and Evaluation

Having described the proposed activity recognition scheme, in this section the experimental evaluation and results are presented and discussed.

4.1 Dataset and Experimental Setup

The dataset used in the experimentation was collected using a custom application running on a smartphone device, Samsung Galaxy Ace model GT-S5830M, with the Android operating system version 2.3.6. However, some other Android capable devices were used for testing the system. A smartphone application was designed as a proof-of-concept to measure and record the acceleration data provided by the embedded triaxial accelerometer. See figure 6 for some screenshots of the developed Android-based application which can operate in three different modalities: data acquisition, offline processing for debugging purposes, and real-time activity monitoring. As this work focuses on exploring the use of smartphones to perform activity recognition without interfering in the user's lifestyle, the application was designed to be continuously executed in a mobile device in background mode. Previous works have demonstrated that continuous activity recognition can rapidly drain the battery of mobile devices, thus it is important to reduce the energy overheads of continuous mobile sensing. Continuous background monitoring and processing with power consumption considerations was supported in Android by using services and partial wakelocks [23].

In Android capable smartphone devices four different sampling frequencies (fastest, game, normal and UI) can be selected for accelerometer data acquisition, however a constant sampling rate or a real-time response is not guaranteed as it depends on the computational load of the software stack and the computing capacities of the smartphone. The game sampling rate was selected for data acquisition, which is around 40 Hz. In such a way, a time window of five seconds contains around

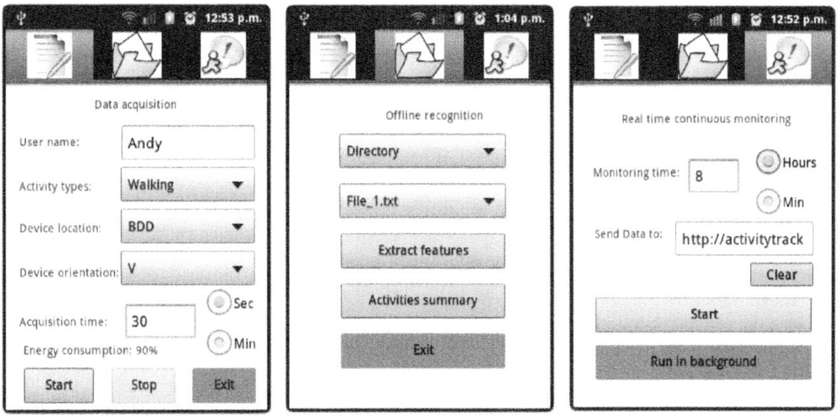

Fig. 6 Screenshots of the three main operation modes of the Android based application: data acquisition (left), offline processing and debugging (middle), and real-time activity monitoring (right).

200 samples. However, it is important to point out that the developed application is able to use any other of the discrete sampling rates and a different time window as well. Data was collected for three subjects performing five different activities wearing the smartphone device at four different locations during their performance, two front pockets and two back pockets without restriction in the device orientation. For each activity-location combination, each subject performed for 5 trials of one minute long. The accelerometer data and the corresponding labeled activity were stored in the internal memory of the smartphone.

4.2 Classification Performance Evaluation

Several experiments were conducted for parameter tuning and to verify the accuracy of the hierarchical activity recognition system evaluated with the collected dataset. Before going into details, it is important to point out the following key points. For all the experiments a time window of 5 seconds was established for the segmentation of the input data stream, i.e., around 200 samples per window. The features summarized in table 2 were extracted for each time window. For AR modeling of the acceleration magnitude, 50 coefficients were used as this number was enough to recover the original signal as shown by the performed experiments. In other words, it was found that a higher order AR model increases the computational complexity without increasing the accuracy of the classification. To reduce the size of the feature vector, PCA was applied and 25 coefficients were kept as these represent 97% of the variance in the analyzed data.

Table 3 Main characteristics of the neural networks used in the different levels of the hierarchical neural classifier for activity recognition.

Level	Network	Input features	Transfer function	Output
L_1	Perceptron	σ, μ	hard limit	Static
L_2	Perceptron	σ, μ	hard limit	Run
L_3	Feedforward (25-2-1)	$PCA_{25}(\sigma, AR_{Coeff})$	tangent sigmoid	Upstairs
L_4	Feedforward (25-2-1)	$PCA_{25}(P_{95}, AR_{Coeff})$	tangent sigmoid	Walk/Downstairs

As the proposed hierarchical neural classifier is based on supervised learning, the dataset for a single user was divided into training and testing sets, containing 70% and 30% of all samples, respectively. Additionally, a k-fold ($k = 10$) cross-validation approach was adopted so as to reduce the dependence of the results on the particular data used from the overall dataset in the training phase. Cross-validation simulates best the goal of measuring the performance on a randomly selected test set which is excluded from the training procedure. The details of the neural network and its main parameters used in each level in the proposed hierarchical approach are summarized in table 3. It is important to point out that the training of the networks was performed in MATLAB version 7.12.0 (R2011a) on a desktop workstation using the neural network toolbox. Then the trained networks of the recognition system were implemented in the smartphone, inheriting the trained parameter values, such as the weight matrices and bias vectors.

The activity recognition results, precision and recall, for the proposed classifier over the collected dataset, for a single user, are summarized in table 4. The proposed approach achieves an accuracy of 93.31% on the collected dataset. It is important to point out that this experiment was conducted by using data from the same user in the training and validation phase. Note that the activities that are more difficult to recognize are upstairs/downstairs walking as usually reported in the literature, even when a single location of the smartphone is considered. In this sense, the importance of the obtained performance results should be assessed considering the number of different location allowed for the smartphone, which is extremely important for naturalistic monitoring. According to preliminary experiments and results, user-specific trained classifier generalizes rather well since when it was tested on data from other three users, a precision over 86% was obtained, given evidence of user-independent recognition.

Some other standard classifiers were implemented for a rough comparison with the proposed approach. The obtained results for NB (Naïve Bayes), K-NN (k-nearest neighborhood), and SVM (Support Vector Machine) are summarized in table 5. The results points out that the proposed solution exhibits better average performance than other commonly used classifiers on the set of activities and dataset that herein were considered. Note once again that for most classifiers, the up-stairs and downstairs walking are the more difficult physical activities to recognize accurately.

Table 4 Representative confusion matrix for the proposed activity recognition system applied on the collected dataset using data from a single user

Class	Static	Running	Walking	Up-stairs	Down-stairs	Overall	Precision
Static	**150**	0	0	0	0	150	**100%**
Running	0	**152**	0	0	0	152	**100%**
Walking	0	0	**148**	3	7	158	**93.67%**
Up-stairs	0	0	6	**139**	18	163	**85.27%**
Down-stairs	0	0	7	11	**136**	154	**88.31%**
Total	150	152	161	153	161	**777**	
Recall	**100%**	**100%**	**91.92%**	**90.85%**	**84.47%**		**93.31%**

Table 5 Comparison with other classifiers. Acronyms: NB (Naïve Bayes), K-NN (k-nearest neighborhood), SVM (Support Vector Machine), P (Precision), R (Recall).

	NB		K-NN		SVM		Proposed	
Class	P	R	P	R	P	R	P	R
Static	100	99.36	100	99.36	100	99.36	**100**	**100**
Running	100	100	100	100	100	100	**100**	**100**
Walking	91.03	85.71	79.27	84.42	91.00	59.09	**93.67**	**91.92**
Up-stairs	80.98	88.69	78.07	86.90	57.30	93.45	**85.27**	**90.85**
Down-stairs	82.28	79.27	79.41	65.85	57.52	39.63	**88.31**	**84.47**
Average	90.85	90.60	87.35	87.30	81.16	78.30	**93.45**	**93.44**

Recall that the proposed approach uses a single accelerometer without posing stringent constraints on the location and orientation in which the smartphone is placed, being this the main advantage and the full on-device implementation compared with other approaches. In spite of the encouraging classifications results, the proposed system has some limitations that should be further addressed. The number of activities is still reduced and some other activities in a free-living monitoring environment despite the clear limits on using a single accelerometer. A larger dataset comprising more expressions of the same activities should be recorded so as to guarantee used-independent recognition. The problem of selecting the optimal length of the time window should be further addressed for a proper activity granularity recognition. If a very small interval is chosen, there is a possibility that it will not contain any relevant activity information for making any useful decision. If the time interval is too wide, then information pertaining to multiple activities can be embedded into it and the activity that dominates the time interval will have a greater influence in the classification decision.

Table 6 Execution time, in milliseconds, for different processing steps in the activity recognition system implemented in three Android devices. The feature extraction column provides the time required to extract all features and the time required for AR modeling is enclosed in parenthesis.

Smartphone device	Preprocessing	Feature extraction	Classifier
Samsung Galaxy Ace ARM 11, 832 MHz, 278 MB Android 2.3.6, 1350 mAh	185	553 (550)	573
Samsung Galaxy S Cortex-A8, 1 GHz, 512 MB Android 2.1, 1500 mAh	56	161 (160)	195
Samsung Galaxy Note II Cortex-A9, 1.6 GHz, 2 GB Android 4.1.1, 3100 mAh	35	146 (145)	156

4.3 Execution Time and Power Consumption Issues

The activity recognition system was implemented in Android capable smartphone devices so as to evaluate in-situ its performance for real time monitoring in terms both of the execution time and the power consumption. Three different smartphone devices were used as implementation platforms whose main characteristics and the activity recognition performance results are summarized in table 6. Note that one of the most power hungry stages in the proposed activity recognition is the AR modeling used for feature extraction. No optimizations were performed to fully exploit the multi-core processor in Galaxy Note II, however, better performance are obtained as Android is tailored to this platform. The increasing processing capabilities trend of smartphones will facilitate more complex computations on the device.

A typical power consumption profile of the activity recognition implemented on different smartphone devices is shown in figure 7. For evaluating power consumption, the battery of each smartphone was fully charged, ensuring an identical starting condition, and then the activity recognition application was allowed to run in background until the battery was depleted as the users went about performing their daily activities. The Android BatteryManager API was used to record the battery level, i.e., the estimation of power consumption of the activity recognition system. The power consumption was measured as a whole and includes all the power drawn by the active software and hardware components of the smartphone. It is interesting to note that the proposed system runs well on resource constrained device, ranging from 10 to 16 hours of autonomy in continuous monitoring modality. However, as expected, the smartphone application autonomy depends on multiple factors and these estimations are illustrative but not definitive.

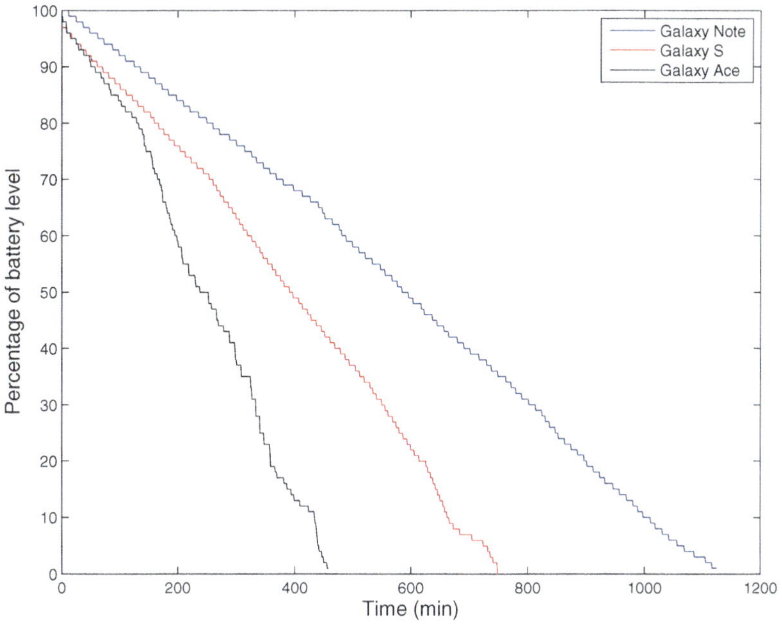

Fig. 7 Energy consumption profile of the activity recognition system on three different Android capable smartphone devices

4.4 Cloud-Based Proof-of-Concept Application

The smartphone-based activity recognition system has been further extended to develop an enhanced proof-of-concept application that allow to deliver sensed data and activity classification to a backend server and to store data in a remote database. The mobile application is able to consume information derived from an analysis of historical records. Data uploading to a server provides an opportunity for the exploitation of data aggregation from a large number of users and further perform health activity analysis considering a large time scale. The proof-of-concept application fits in the general interaction scheme as shown in figure 8 and it exploits push notifications to receive information in an opportunistic and timely manner. Recall that, push notification is the ability of a mobile device to receive messages that are pushed from a server and for the Android-capable devices this is enabled by the Google Cloud Messaging infrastructure.

The remote backend server receives recognized or labeled physical activities on the device, and it simply generates graphs summarizing how much time is spent on each activity in the course of a day. The server then automatically notifies or feedbacks this information to the user by using push notifications to inform him about potential sedentary habits. Adding other contextual information, for instance location or information from other embedded sensors, the backend server activity analysis application could recommend consistently and timely possible routines to

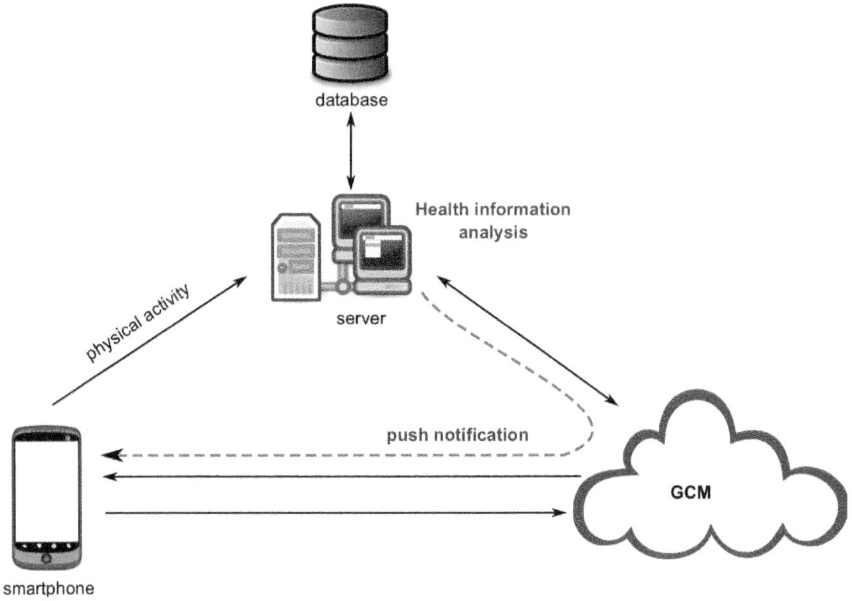

Fig. 8 Framework of proof of concept application based on Google Cloud Messaging (GCM) and push notifications

improve user health. Under this general framework it is possible to develop persuasive applications intentionally designed to change a person attitude or behavior in a predetermined way in order to motivate people to lead a healthier life style by mediating prevention and treatment.

5 Case Studies

5.1 Mobile Applications to Prevent Life-Style Diseases

There exist physical habits and other factors that can lead to chronic diseases such as diabetes, stroke, hypertension and cardiac failure, which have been rapidly increasing over the past decades. For instance, obesity, a complex physical condition resulting from the interaction of many factors, is considered as epidemic that has a great impact in health. In order to deal with lifestyle diseases, mobile applications able to measure the physical activity of a user, in a free-living monitoring environment, must be required so as to advise and guide them about how to deal with and prevent chronic medical conditions.

5.2 Activity Recognition for Rehabilitation Services

Another practical application of activity recognition is to assist in rehabilitation services, as a patient can perform specific exercises at home instead of into a specialized rehabilitation center. A smartphone-based activity recognition can be used to estimate training accuracy in daily basis manner so as therapist are able to ensure patient quality of care with more accurate evaluations. During these exercises the recognition systems capture the physical activity and classify according how accurately they perform exercises according to previously learned metrics established under supervision of a specialist. Under this scenario smartphone-based solutions can decrease healthcare services cost.

5.3 Personal Life-Log Mobile Middleware for mHealth

Telemonitoring involves remote monitoring of patients far away form health service providers. In order to telemonitoring be applied in a population-wise fashion, cost-effective and time-efficient mobile systems are required. Such systems must allow to collect rich information from physical health sensors and provide mechanisms to provide personalized support for several real-world applications. Thus, personal life mobile middleware are required to abstract common services such as data containing, sharing and data mining to speed up development cycles of applications.

6 Future Directions

Micro electromechanical systems (MEMS) sensors have already enable smartphones to provide services that deeply affect the user style of life. For instance, through embedded inertial sensors, smartphones are able to respond to hand gestures, support virtual reality applications, or provide information about physical activities that a user performs and infer knowledge about health status. Currently smartphones are able to hear, see, touch, feel and smell and truly act as sensor hubs and their derived services increasingly rely on how well they know about the user and user context. Such smartphone perceptual capabilities are possible thanks to the seamless integration of sensors, at hardware level, into the smartphone complex software stack. Moreover, smartphones are be able to seamlessly work with wireless sensors that the user worn or even with sensor deployed and distributed in the environment to collect data thanks to wireless communication facilities.

Since the complexity, quantity, and heterogeneity of information provided for a large set of embedded multimodal sensors increase, new ways to handle all these data will be required in next generation smartphone devices. Improvements in hardware, middleware and application software will be required so as to provide the computational capabilities to efficiently interpret data, i.e, more intelligence distributed across the mobile device. In spite of the continuous advance of mobile multicore processors, the resources demands and bandwidths of new and future sensors will become so significant that current mobile operating systems, such as Android, for sensor-streams handling will no be enough. Sensors with high bandwidth demands will require additional processing power that can only be provided by additional hardware working closely to the sensor chip. Therefore, intelligent or smart sensors will play a key role as it will be up to the sensor devices to process data and provide higher level information to the software stack in a smartphone. Meanwhile the operating system and the application software should be in charge of concurrently exchanging and processing information from different sensor devices so as to hide the complexity of sensor handling in such a way that the designer uses simple control points for information fusion. Thus, developing industry standards for sensor integration in smartphone platforms coupled with the almost unlimited computing capabilities from the cloud is required so as to met the sensor-processing demand of future cognitive smartphones. With the advance of sensing technology for gathering data about human health and the environment integrated in smartphones it is possible to envisage a great impact in mHealth for the next years.

7 Concluding Remarks

In this paper a robust human activity recognition approach based on a single triaxial accelerometer has been proposed, which does not impose a stringent constraint on the location of the smartphone. The proposed approach considers some typical location where users carry out the smartphone and the activity recognition system does not require a firm attachment of the smartphone to the subject. Experimental results show that the system performance compares favorably with previously reported works. A fully implementation on smartphone devices shows near real-time performance without a considerable power consumption thanks to the lightweight processing performed in the device. Despite these encouraging results, further work is still needed to improve the system portability across individuals and to better identify new classes at run time. Future work will be focused in including more physical activities to recognize and to assess the suitability of the proposed approach in mHealth applications in a more realistic proof-of-concept application.

Activity recognition as a case of study shows that research involved in developing mHealth applications lies at the intersection of several diverse disciplines, including sensing systems, signal processing, machine learning, data mining, data management, and large-scale distributed systems. Thus, further improvements in mHealth depend on the ability to automatically mine data from different types of sensor platforms in order to monitor, alert, learn from, and in some cases affect and control our surroundings. Moreover low-level improvements in supporting sensing capabilities throughout the software stack in smartphone platforms are needed for significantly reducing the energy overheads of continuous mobile sensing for mHealth.

Acknowledgements. Authors want to acknowledge the partial support received from Conacyt, Mexico, through the research grant No. 99912.

References

[1] Acampora, G., Cook, D., Rashidi, P., Vasilakos, A.: A survey on ambient intelligence in healthcare. Proceedings of the IEEE 101(12), 2470–2494 (2013), doi:10.1109/JPROC.2013.2262913

[2] Atallah, L., Lo, B., King, R., Yang, G.Z.: Sensor positioning for activity recognition using wearable accelerometers. IEEE Transactions on Biomedical Circuits and Systems 5(4), 320–329 (2011), doi:10.1109/TBCAS.2011.2160540

[3] Bicocchi, N., Mamei, M., Zambonelli, F.: Detecting activities from body-worn accelerometers via instance-based algorithms. Pervasive and Mobile Computing 6(4), 482–495 (2010), doi:http://dx.doi.org/10.1016/j.pmcj.2010.03.004

[4] Brezmes, T., Gorricho, J.L., Cotrina, J.: Activity recognition from accelerometer data on a mobile phone. In: Omatu, S., Rocha, M.P., Bravo, J., Fernández, F., Corchado, E., Bustillo, A., Corchado, J.M. (eds.) IWANN 2009, Part II. LNCS, vol. 5518, pp. 796–799. Springer, Heidelberg (2009)

[5] Chen, L., Hoey, J., Nugent, C., Cook, D., Yu, Z.: Sensor-based activity recognition. IEEE Transactions on Systems, Man, and Cybernetics, Part C: Applications and Reviews 42(6), 790–808 (2012), doi:10.1109/TSMCC.2012.2198883

[6] Cooper, R.A., Dicianno, B.E., Brewer, B., LoPresti, E., Ding, D., Simpson, R., Grindle, G., Wang, H.: A perspective on intelligent devices and environments in medical rehabilitation. Medical Engineering and Physics 30(10), 1387–1398 (2008), doi:http://dx.doi.org/10.1016/j.medengphy.2008.09.003

[7] Figo, D., Diniz, P., Ferreira, D., Cardoso, J.: Preprocessing techniques for context recognition from accelerometer data. Personal and Ubiquitous Computing 14(7), 645–662 (2010), http://dx.doi.org/10.1007/s00779-010-0293-9, doi:10.1007/s00779-010-0293-9

[8] Fuentes, D., Gonzalez-Abril, L., Angulo, C., Ortega, J.: Online motion recognition using an accelerometer in a mobile device. Expert Systems with Applications 39(3), 2461–2465 (2012), doi:http://dx.doi.org/10.1016/j.eswa.2011.08.098

[9] Györbíró, N., Fábián, A., Hományi, G.: An activity recognition system for mobile phones. Mob. Netw. Appl. 14(1), 82–91 (2009), doi:10.1007/s11036-008-0112-y

[10] Hache, G., Lemaire, E., Baddour, N.: Wearable mobility monitoring using a multimedia smartphone platform. IEEE Transactions on Instrumentation and Measurement 60(9), 3153–3161 (2011), doi:10.1109/TIM.2011.2122490

[11] Hadjidj, A., Souil, M., Bouabdallah, A., Challal, Y., Owen, H.: Wireless sensor networks for rehabilitation applications: Challenges and opportunities. Journal of Network and Computer Applications 36(1), 1–15 (2013), doi:http://dx.doi.org/10.1016/j.jnca.2012.10.002

[12] Himberg, J., Korpiaho, K., Mannila, H., Tikanmaki, J., Toivonen, H.: Time series segmentation for context recognition in mobile devices. In: Proceedings IEEE International Conference on Data Mining, ICDM 2001, pp. 203–210 (2001), doi:10.1109/ICDM.2001.989520

[13] Hoseini-Tabatabaei, S.A., Gluhak, A., Tafazolli, R.: A survey on smartphone-based systems for opportunistic user context recognition. ACM Comput. Surv. 45(3), 27:1–27:51 (2013), http://doi.acm.org/10.1145/2480741.2480744, doi:10.1145/2480741.2480744

[14] Khan, A., Lee, Y.K., Lee, S., Kim, T.S.: Accelerometers position independent physical activity recognition system for long-term activity monitoring in the elderly. Medical & Biological Engineering & Computing 48(12), 1271–1279 (2010), doi:10.1007/s11517-010-0701-3

[15] Khan, A., Lee, Y.K., Lee, S., Kim, T.S.: A triaxial accelerometer-based physical-activity recognition via augmented-signal features and a hierarchical recognizer. IEEE Transactions on Information Technology in Biomedicine 14(5), 1166–1172 (2010), doi:10.1109/TITB.2010.2051955

[16] Khan, W., Xiang, Y., Aalsalem, M., Arshad, Q.: Mobile phone sensing systems: A survey. IEEE Communications Surveys Tutorials 15(1), 402–427 (2013), doi:10.1109/SURV.2012.031412.00077

[17] Kumar, S., Nilsen, W., Pavel, M., Srivastava, M.: Mobile health: Revolutionizing healthcare through transdisciplinary research. Computer 46(1), 28–35 (2013), doi:10.1109/MC.2012.392

[18] Lane, N., Miluzzo, E., Lu, H., Peebles, D., Choudhury, T., Campbell, A.: A survey of mobile phone sensing. IEEE Communications Magazine 48(9), 140–150 (2010), doi:10.1109/MCOM.2010.5560598

[19] Lee, M.W., Khan, A., Kim, T.S.: A single tri-axial accelerometer-based real-time personal life log system capable of human activity recognition and exercise information generation. Personal and Ubiquitous Computing 15(8), 887–898 (2011), `http://dx.doi.org/10.1007/s00779-011-0403-3`, doi:10.1007/s00779-011-0403-3

[20] Makhoul, J.: Linear prediction: A tutorial review. Proceedings of the IEEE 63(4), 561–580 (1975), doi:10.1109/PROC.1975.9792

[21] Martin, H., Bernardos, A., Iglesias, J., Casar, J.: Activity logging using lightweight classification techniques in mobile devices. Personal and Ubiquitous Computing 17(4), 675–695 (2013), doi:10.1007/s00779-012-0515-4

[22] Misra, A., Lim, L.: Optimizing sensor data acquisition for energy-efficient smartphone-based continuous event processing. 2011 12th IEEE International Conference on Mobile Data Management (MDM) 1, 88–97 (2011), doi:10.1109/MDM.2011.76

[23] Pathak, A., Hu, Y.C., Zhang, M.: Where is the energy spent inside my app?: fine grained energy accounting on smartphones with eprof. In: Proceedings of the 7th ACM European Conference on Computer Systems, EuroSys 2012, pp. 29–42. ACM, New York (2012), doi:10.1145/2168836.2168841

[24] Rashidi, P., Cook, D.J., Holder, L.B., Schmitter-Edgecombe, M.: Discovering activities to recognize and track in a smart environment. IEEE Trans. on Knowl. and Data Eng. 23(4), 527–539 (2011), doi:10.1109/TKDE.2010.148

[25] Santos, A., Tarrataca, L., Cardoso, J., Ferreira, D., Diniz, P., Chainho, P.: Context inference for mobile applications in the upcase project. In: Bonnin, J.M., Giannelli, C., Magedanz, T. (eds.) Mobilware. LNICST, vol. 7, pp. 352–365. Springer, Heidelberg (2009)

[26] Sherwood, N.E., Senso, M.M., Fleming, C.K., Roeder, A.M.: Chapter 26 - behavioral risk factors for overweight and obesity: Diet and physical activity. In: Coulston, A.M., Boushey, C.J., Ferruzzi, M.G. (eds.) Nutrition in the Prevention and Treatment of Disease, 3rd edn., pp. 479–499. Academic Press (2013), doi:`http://dx.doi.org/10.1016/B978-0-12-391884-0.00026-3`

[27] Tentori, M., Favela, J.: Activity-aware computing for healthcare. IEEE Pervasive Computing 7(2), 51–57 (2008), doi:10.1109/MPRV.2008.24

[28] Xu, W., Zhang, M., Sawchuk, A., Sarrafzadeh, M.: Robust human activity and sensor location corecognition via sparse signal representation. IEEE Transactions on Biomedical Engineering 59(11), 3169–3176 (2012), doi:10.1109/TBME.2012.2211355
[29] Yan, Z., Subbaraju, V., Chakraborty, D., Misra, A., Aberer, K.: Energy-efficient continuous activity recognition on mobile phones: An activity-adaptive approach. In: 2012 16th International Symposium on Wearable Computers (ISWC), pp. 17–24 (2012), doi:10.1109/ISWC.2012.23
[30] Yang, J.: Toward physical activity diary: Motion recognition using simple acceleration features with mobile phones. In: Proceedings of the 1st International Workshop on Interactive Multimedia for Consumer Electronics, IMCE 2009, pp. 1–10. ACM, New York (2009), doi:10.1145/1631040.1631042
[31] Yang, J., Lu, H., Liu, Z., Boda, P.: Physical activity recognition with mobile phones: Challenges, methods, and applications. In: Shao, L., Shan, C., Luo, J., Etoh, M. (eds.) Multimedia Interaction and Intelligent User Interfaces, Advances in Pattern Recognition, pp. 185–213. Springer London (2010)
[32] Zhang, H.M., Duhamel, P.: On the methods for solving yule-walker equations. IEEE Transactions on Signal Processing 40(12), 2987–3000 (1992), doi:10.1109/78.175742

Questions

Q1. Which aspects should be considered for porting and deployment of an activity recognition solution to a mobile device?

Q2. Why is the triaxial accelerometer magnitude widely used for accelerometer-based activity recognition?

Q3. What techniques are commonly used for preprocessing raw accelerometer data?

Q4. Why are time-domain features preferred to frequency-based features for mobile human activity recognition?

Q5. What is the role of Principal Component Analysis in accelerometer.based activity recognition?

Q6. Mention two current technologies that enhance the smartphone capabilities for mHealth.

8. A Formal Approach for a Dependability Assessment of a mHealth Monitoring System

Alessandro Testa[1], Marcello Cinque[2],
Antonio Coronato[1], and Giuseppe De Pietro[1]

[1] Institute for High-Performance Computing and Networking,
via P. Castellino 111, Naples, Italy
{alessandro.testa,antonio.coronato,
giuseppe.depietro}@na.icar.cnr.it

[2] DIETI - University of Naples "Federico II", via Claudio, 21 - Naples, Italy
macinque@unina.it

Abstract. The implementation of mHealth monitoring systems is attracting increasing attention in academia and industry due to rising healthcare costs and the aging of the world population. However, the problem of failure detection and management in mHealth monitoring systems is becoming more and more critical and the use of wireless technologies and the adoption of commodity hardware/software platforms pose new challenges in terms of their correct functioning. Wireless channels can be affected by packet loss and cheap and wireless-enabled medical devices can exhibit wrong readings and temporary disconnections inducing medical staff to take wrong decisions.

In this chapter, we present an approach, relying on an event-based formalism, for the dependability assessment of mHealth monitoring systems. In particular, after a detailed failure modes and effects analysis, we identify some main critical events that could undermine the dependability of a Body Area Network (BAN) which is the most critical component of a mHealth monitoring system. Finally an application scenario concludes the chapter.

Keywords: Health Monitoring Systems, Wireless Sensor Networks, Dependability, Failure Detection, Formal Methods.

1 Introduction and Motivation

The design and realization of mHealth monitoring systems has attracted the interest of large communities both from industry and academia, particularly due to rising healthcare costs and the aging of the world population [10]. The monitoring of patient's vital signs is the target of an emerging business market that aims both to improve the quality of patients' lives and to reduce the costs of national healthcare services [6].

MHealth monitoring systems have been shown to be effective in helping to manage chronic disease and post-acute care, and in monitoring the safety of the older

S. Adibi (ed.), *Mobile Health*,
Springer Series in Bio-/Neuroinformatics 5, DOI: 10.1007/978-3-319-12817-7_8

adult population [7]. They can help older adults to slow the progression of chronic diseases and to experience a continued recovery after being discharged from an acute care setting.

Such systems, however, are particularly critical from the point of view of dependability. In mHealth monitoring systems the transmission of vital sign data and the alarms have to be dependable in order to ensure a prompt intervention and the time latency has to be limited and predictable. Moreover, the coverage and lifetime of the networks are important, since mHealth systems are based on *Wireless Body Area Networks* (WBANs) [17] in which sensor nodes are devoted to the monitoring of a physiological sign (such as in Electrocardiography or ECG, in the monitoring of oxygen saturation, etc.). To achieve this objective, cabled measurement equipment is already used to guarantee reliable and robust control of vital signs. However such systems complicate patient autonomy and mobility. Hence, wireless technologies and mobile devices are starting to be applied to build more comfortable and patient-friendly health monitoring systems [18].

Nevertheless, the use of wireless technologies and the adoption of commodity hardware/software platforms, such as smartphones, pose new challenges in terms of the correct functioning of health monitoring systems. Wireless channels can be affected by packet loss, due to shadowing and an absence of signal coverage. Smartphones can be subjected to unpredictable failures, which could affect the correct functioning of the system. Finally, cheap and wireless-enabled medical devices can exhibit wrong readings and temporary disconnections.

These issues may induce the medical staff to take wrong decisions, e.g., to administer wrong dosages of medicine, which can prove to be fatal for the patient.

For these reasons, the problem of dependability assessment in mHealth monitoring systems is starting to be addressed in the literature. It is particularly important to evaluate dependability within the WBAN where vital signs are sensed and transmitted to a mobile device. Thus the starting point in order to build a dependable mHealth monitoring system has to be the analysis of possible failure events that could occur in a WBAN.

However, several studies have been based on simplistic failure assumptions or on basic fault tolerance schemes (such as, sensor redundancy), which are not guaranteed to cover all possible failure scenarios. For instance, sensor replication is ineffective against smartphone failures.

To overcome the limitations of current solutions, in this chapter, considering a typical architecture of a mHealth monitoring system composed of a BAN with wired and wireless medical devices, a gateway device and a medical station, we present and discuss the steps of a process that a designer should perform in order to define specifications for an assessment of the dependability of a mHealth monitoring system using an event-based formal language (event calculus [12]). In particular we have placed particular attention on the definition and verification of formal specifications for the detection of critical events (identified by means of a preliminary *Failure Modes and Effects Analysis (FMEA)* [13]) that could occur in the BAN component of the system.

In fact, from an FMEA study we identify some of the most critical failure events; then we define two sets of formal specifications (one for the detection of failure events and one to define the configuration of a target BAN). Finally, exploiting an application scenario, we demonstrate the effectiveness of our methodology.

Nevertheless, the guidelines of the process are valid in order to define specifications for the detection of failures in the other components of the mHealth monitoring system (such as the gateway device, or medical center).

The rest of the paper is organized as follows. In Section 2 we present related work, while in Section 3 we illustrate a typical mHealth monitoring system. In Section 4 we generally introduce the proposed approach; the informal modeling phase is discussed in Section 5, general specification definition in Section 6 and a structural specification definition in Section 7; the verification phase is described in Section 8 and some suggested case studies are reported in Section 9. Future directions are discussed in Section 10 and finally in Section 11 we conclude the paper.

2 Related Work

Several new mHealth monitoring systems have been proposed in literature focusing on security and privacy issues but not on dependability issues within a WBAN [14, 25, 23].

In [14] the authors present a new kind of mHealth monitoring system that is based on cloud computing; the paper is focused on security and privacy concepts since the authors design a cloud-assisted privacy preserving mHealth monitoring system to protect the privacy of the involved parties and their data. However dependability requirements such as coverage or connection resiliency are not considered.

Additionally, in [25], a WBAN-based e-health monitoring system is implemented and evaluated which is sensitive to security and privacy issues; it is proposed a mixed localization algorithm based on received sensor data and a received signal strength indicator (RSSI) is proposed, which is exploited to perform the passive localization of monitored elderly people. However, the proposed system does not take into account failures like a packet loss or the isolation of part of a WBAN.

Triantafyllidis et al. [23] present the design and development of a pervasive health monitoring system which integrates patient monitoring, through mobile wearable multisensing devices, and status logging to capture various problems or symptoms encountered. The authors focus on the security issues of a gateway device disregarding any failure that could occur in a WBAN which is the source of the needed data.

There are also some mHealth monitoring systems based on reasoning techniques. Benlamri and Docksteader in [1] design the Mobile Ontology-based Reasoning and Feedback (MORF) health-monitoring system, which monitors a patient's health status using a mobile unit. Unlike previously cited papers, the system uses an ontology-based context model to process and determine a patient's health status; the system can process the incoming sensor data by means of ontologies and various reasoning methods. In respect of this feature, this work is similar to ours but also in this case no dependability assessment is performed.

In [20] Sneha and Varshney investigate an approach based on a mobile ad hoc network to address the challenge of enhancing communication dependability in the context of health monitoring; they propose power management protocols to overcome issues of low battery power management of patient monitoring devices increasing communication dependability. They assert that ensuring reliable end to end communication in a mHealth monitoring system is a critical requirement. However, their research is based on dependability issues that affect this kind of system, and therefore, in this respect this work is far more limited than ours since only power consumption is considered as a dependability parameter while we consider also coverage, connection resiliency and packet loss. Moreover they consider end-to-end dependability and not dependability within a WBAN.

Therefore, in summary, several mHealth monitoring systems have been proposed in literature but dependability assessment remains an ongoing challenge. We believe that a more performant way to evaluate the dependability degree of a mHealth monitoring system is the use of formal methods since they are widely adopted in the literature to verify the correctness of a system and thus its dependability.

3 A Mobile Health Monitoring System

To define a typical mHealth monitoring system, we analyze three of the most popular systems: the *MedApps System* [9], the *Nicolet Ambulatory Monitor System* [24] and a system used by the *Center for Technology and Aging* [15].

The *MedApps System* provides a healthcare connectivity platform that delivers a scalable and flexible remote distribution using cellular, wireless and wired technologies with cloud-based computing. This system can work with multiple internal and external devices. Patient data is collected, analyzed and forwarded, via a cell phone to servers, guaranteeing a more robust picture of the patients' health. The *Nicolet Ambulatory Monitor System* comprises a flexible, high quality diagnostic unit, ideal for patients of all ages. It is a flexible, robust system specifically realized to provide the requirements of long-term monitoring. This system diagnoses the patients' cerebral functions (premature neonates to older adults) monitoring continuously ill patients at risk of brain damage and secondary injury. Finally, the authors in [15] discuss two areas of opportunity for remote patient monitoring: i) Patient Safety and ii) Chronic Disease Management and Post-Acute Care Management. In alignment with the mission of the *Center for Technology and Aging*, they focus on technology-enabled innovations, such as wireless connectivity, mainly aimed at improving the health of older adults and promoting independent living in community-based, home, and long-term care settings.

Observing these architectures, we can assert that a mHealth monitoring system is usually composed of a number of wired or wireless sensors (medical devices), a gateway device (a handheld device) and a medical station; typical communication means are bluetooth and ZigBee (within the BAN - *Intra BAN communication*),

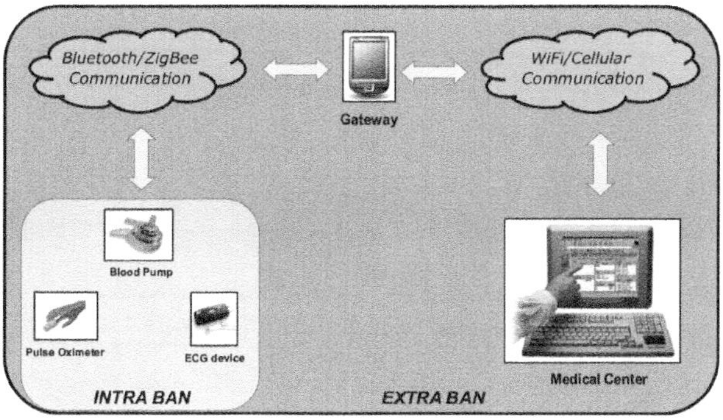

Fig. 1 mHealth monitoring system schema

WiFi and cellular (external to the BAN - *Extra BAN communication*). Vital signs are sensed by sensors (e.g. an oximeter, electrocardiogram - ECG, insulin pump, etc.) and transmitted to a mobile device over a bluetooth or ZigBee network. Afterwards, data are sent to a medical station deployed, for an example, in a hospital by means of either a WiFi or a cellular connection (the medical center location).

In this typical network, we can note that possible failures can occur in medical devices, in the bluetooth/ZigBee communication, in the mobile device, during the WiFi/cellular communication and finally in the local monitoring station of the caregiver. Figure 1 depicts the components of a generic mobile health monitoring system.

4 Approach

In this paragraph we present and discuss a process to define specifications for the dependability assessment of mHealth monitoring systems using *event calculus*[19] as an event-based formal language.

The process illustrated in figure 2 shows the steps that we followed in order to realize verifiable specifications. It is characterized by four sequential phases: *Informal Modeling*, *General Correctness Specification Modeling*, *Structural Correctness Specification Modeling* and *Verification/Testing*; the processes consist, in total, of 14 tasks.

5 Informal Modeling

In the *Informal Modeling* phase the application domain and correctness properties are analyzed.

In this Section we present the FMEA [3] study performed during the *Informal Modeling* phase. By taking advantage of our past experience and detailed field studies on the dependability of mobile devices, wireless communication technologies, such as Bluetooth, and wireless sensor networks (WSNs) [4], the FMEA allowed

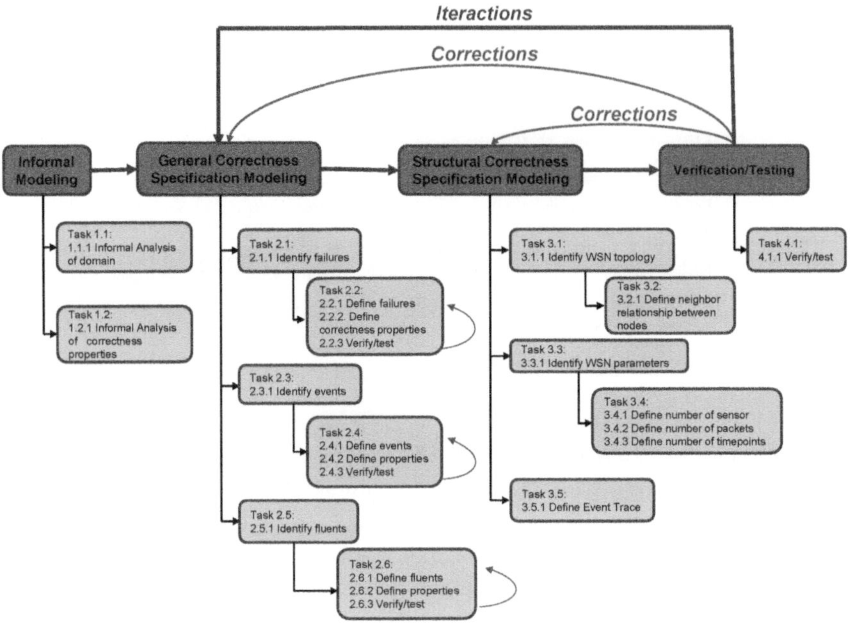

Fig. 2 The process for designing the specifications for WSN-based applications (4 phases - 14 tasks)

us to identify the failure modes of the main components of a mHealth monitoring system and thus its main criticalities.

An FMEA table, produced by analyzing a typical mHealth monitoring system, is presented as a set of results in which the main criticalities are detected.

5.1 FMEA Results

In this paragraph the results of the FMEA performed on generic monitoring systems are presented.

Let us consider a mHealth monitoring system described in Section 3. In this typical system, possible failures can occur in the medical devices, in the bluetooth/ZigBee communication, in the mobile device, during the WiFi/cellular communication and finally in the medical center. The most frequent failure occurrences have been identified from past experiences in real architectures and from the existing literature, trying to relate failure occurrences with potential causes (faults).

Considering the general architecture, the medical center location has been omitted, since it should be more reliable and under the direct control of the staff, who can immediately intervene in cases of failure (e.g., they can connect to the system using a different machine). Hence, the focus is on the components which have to be used by other users, who might not be technology experts and who need to rely on a monitoring system able to work even in case of accidental failures.

To perform the FMEA, four components/functions have been identified: the node (i.e., the sensor used to monitor a phenomenon like an ECG or an insulin pump), the Intra BAN communication, the Extra BAN communication, and the gateway (i.e., a smartphone).

In figure 3 an organization of components and subcomponents is shown.

Eight sub-components/subfunctions have been identified for the node component: the sensor board, the power supply unit, the CPU and the OS as the general components of a node, their analysis being based on our previous study on sensor networks [5]; the ECG device adhesive, ECG device electrolyte, patient cable and insulin pump as particular medical devices, the failures of such devices having been identified starting from existing studies, such as [21].

Three sub-components/sub-functions have been identified for the *Intra BAN communication* function: the transport/routing, Bluetooth stack and Bluetooth channel. These are based on our previous studies on sensor networks and on the Bluetooth protocol [2].

Finally, one component/function has been identified for the *Extra BAN communication*, starting from [5], and two sub-components/sub-functions have been identified for the *Gateway* component: the device (i.e., the smartphone), starting from our previous experiences on smartphone failures [4], and the Bluetooth application, which is responsible for gathering the measurements from Bluetooth medical devices (in fact, the majority of wireless medical devices use Bluetooth as the communication technology). In [2] we noticed that several failures may affect Bluetooth applications, due to problems in the underlying Bluetooth modules.

In table 1 the FMEA results are presented. Furthermore, for each failure mode the related severity and probability of occurrence are reported, such factors being represented by a value between 1 and 4. With the lower values we identify a weak severity/probability while instead with the higher values a strong severity/probability. For example, if a failure is classified with a severity of 4 and a probability of 4, this means that a failure is very dangerous and very probable. But if a failure is classified with a severity of 1 and a probability of 1, then there is almost no reason for concern.

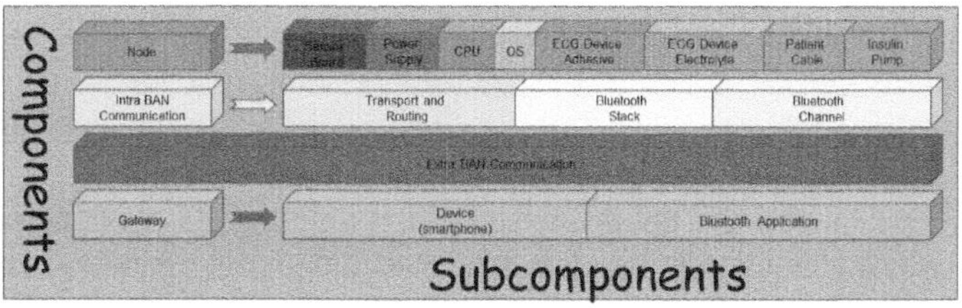

Fig. 3 Components and subcomponents considered in the FMEA

Table 1 Failure Mode and Effect Analysis of a mHealth monitoring system

Component	Sub-component	Potential Failure Mode	Potential Effects of Failure	Potential Causes of Failure	Sev.	Prob.
Node	**Sensor Board**	Stuck at Zero	The device is out-of-order; it does not deliver any output to inputs	Sensing hardware	4	2
		Null Reading	The device delivers null output values	Sensing hardware	4	4
		Out of Scale Reading	The device delivers no meaningful values	Sensing hardware	3	4
	Power supply	Stuck at Zero	The device is out-of-order; it does not deliver any output to inputs	Natural energy exhaustion	4	4
		Reset	The node resets itself to its initial conditions	Anomalous current request that cannot be supplied by batteries	3	1
	CPU	Stuck at Zero	The device is out-of-order; it does not deliver any output to inputs	Micro-controller	4	4
	OS	Software Hang	The device is powered on, but not able to deliver any output	Operating system's corrupted state	4	3
	ECG Device Adhesive	Incorrect reading	Wrong data values, irritation or rash of skin	Skin contact	2	3
	ECG Device Electrolyte	Incorrect reading	Wrong data values, irritation or rash of skin	Skin contact	2	3
	Patient Cable	Discontinuous readings	Noise, wrong data values	Defective wire	2	2
	Insulin Pump	Insulin overdose	Low blood sugar levels (hypoglycemia) which can be quite dangerous	Incorrect sugar level measured	3	4
		Insulin underdose	Patient at risk: sugar accumulates in the blood	Incorrect sugar level measured	1	3
		Power failure	Desired dose was not given	Natural energy exhaustion	2	2
		No delivery failure	Desired dose was not given	Corrupted rewinding mechanism	4	3
Intra BAN Communication	**Routing**	Packet Loss	The radio packet is not delivered	Packet corruption Buffer overrun	3	2
		Isolation	The node is not longer connected to the sink node	Failure of all forwarding nodes	4	2
	Bluetooth Stack	Bluetooth stack failure	A Bluetooth module (e.g. L2CAP, BNEP, etc.) fails	Bluetooth stack's corrupted state	3	1
	Bluetooth Channel	Header corruption	Header delivered with errors	Packet corruption	2	1
		Header length mismatch	Header length deviates from the specified one	Packet corruption	2	1
		Payload corruption	Payload delivered with errors	Packet corruption	3	1
Extra BAN Communication		Data Delivery Failures	The network is not able to deliver the required amount of measurements	The number of failed nodes is more than a given threshold	3	3
		Cellular/WiFi network unavailable	Monitoring stopped	Area without cellular/WiFi signal	4	3
Gateway	**Device (the smartphone)**	Freeze	The device's output becomes constant; the device does not respond to the user's input.	System's corrupted state	3	3
		Self-shutdown	The device shuts down itself; no service is delivered at the user interface.	Natural energy exhaustion or self-reboot due to corrupted state	4	2
		Unstable behavior	The device exhibits erratic behavior without any input inserted by the user	System/Application corrupted state	4	2
		Output failure	The device delivers an output sequence that deviates from the expected one	System/Application corrupted state	3	4
		Input failure	User inputs have no effect on device behavior	System/Application corrupted state; Natural energy exhaustion	1	1
	Bluetooth Application	Inquiry/Scan Failure	The scan procedure terminates abnormally	A Bluetooth module fails or device out of range	2	3
		Discovery Failure	The discover procedure terminates abnormally	A Bluetooth module fails or device out of range	2	3
		Connect Failure	The device is unable to establish a connection	A Bluetooth module fails or device out of range	4	3
		Packet Loss	Expected packets are not received	Packet corruption	3	1
		Data mismatch	Packets are delivered with errors in the payload	Memoryless channel with uncorrelated errors	3	1

The severity and probability values have been obtained from a consideration of our previous work [4, 2, 5] and our personal experience during past experimental campaigns of WSN dependability assessment.

Table 1 is structured in seven columns. Every row contains the description of a single failure mode. So, considering a possible failure mode that may occur in the monitoring system, we identify from left to right the relevant component (and subcomponent if it exists) interested, the possible effects of the failure, the possible causes of the failure and finally the severity and probability of the failure occurrences, in order to highlight the more dangerous and frequent failures. All of these analyzed failures cause abnormal data readings, even to the extent that a value is not received at the medical center location; in this case an inaccurate monitoring is provided, potentially resulting in a significant hazard to the monitored system. MHealth monitoring systems must be aware of all the possible failures, in order to react to them or, at least, to detect them. For instance, in the case of a failure detection, a possible action could be to call the caregiver or an emergency contact to immediately check the system and restore its normal operation. At the same time, a dependability assessment tool, to be used at design time, should emulate the main failure modes, such as node failures and packet losses, to evaluate the dependability level of the BAN against possible undesired events.

5.2 Critical Events in the BAN

Considering the results of the FMEA study conducted and discussed in the previous section, we concentrate on the following three main critical events: i) an ***isolation event*** (this occurs when a node is no longer able to reach the sink node, the main node of a BAN), ii) a ***packet loss event*** (this occurs when a packet is lost during the traversal of the network) and iii) a ***battery exhaustion event*** (this occurs when a node stops working since it has run out of battery).

However, the formal approach presented in this chapter allows the extension of the specification to assess the dependability of a mHealth monitoring system against the critical events of the other components.

5.3 Dependability Metrics

We define the following dependability metrics, already adopted in [8, 22], in order to assess the dependability of the BAN component of a mHealth monitoring system against previously described critical events occurring:

- the *Coverage*, being the time interval in which the BAN can operate, while preserving a given number of nodes connected to the sink node.

- the *Connection Resiliency*, representing the number of node failures and disconnection events that can be sustained while preserving a given number of nodes connected to the sink node.

- the *Data Delivery Resiliency*, being the number of node failures and disconnection events that can be sustained while preserving a given number of correct packets delivered to the sink node.

- the *Power consumption*, being the battery power consumed by each sensor, useful to estimate the expected lifetime of the BAN.

Overall these metrics allow to evaluate an evaluation of the expected dependability of the BAN component in terms of its robustness to failure events, its capacity to cover a given area, and its duration (lifetime). Clearly, other metrics can be defined for other purposes and other components.

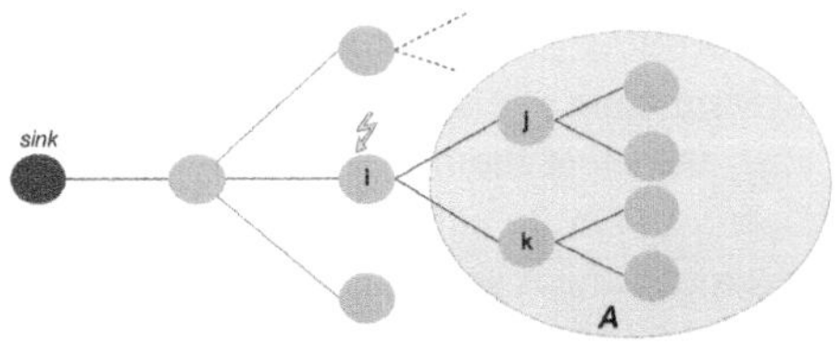

Fig. 4 Isolation of a WSN subnet

6 General Correctness Specification Modeling

In this phase failures, events and fluents are identified and defined using event calculus language; they are also properly refined after a verification and testing subprocess. Moreover, this phase includes the formal definition of the correctness properties, which are valid for every BAN, in order to detect previously chosen critical events.

The proposed verification framework is founded on the definition of a core formal specification in event calculus language. The main idea is to formalize the correctness properties allowing engineers to verify if a given BAN design, specified in terms of the number of nodes, position of nodes, channel quality, and initial battery level, is able to satisfy given design constraints. The metrics, defined in Section 5.3, are evaluated by analyzing the *narrative*, generated by a *Discrete Event Calculus Reasoner* (DECReasoner) [16], based on the specification of the target BAN.

The general correctness specification is described as follows: it specifies that a BAN performs correctly if none of the identified critical events (an *isolation event*, *packet loss event* or *battery exhaustion event*) happens.

The first two kinds of events can be caused by the stop of one or more nodes (e.g., due to a crash or battery exhaustion) or the temporary disconnection of a node from its neighbour(s) due to a transmission error. The last kind of event is generated by considering the initial battery charge and the energy requirement of the nodes due to packet sending and receiving activities.

6.1 Isolation Event

An isolation event happens when a node is no longer able to reach the sink node of the BAN, i.e., the gateway node where data are stored or processed.

For instance, let us consider figure 4 which represents part of the MEDiSN architecture topology [11] (as an example of a real-life deployed mHealth monitoring system); node i is the only one allowing the transmission of data between the sink node and the subnet A. We want to check when the subnet A is isolated from the rest of the network. We suppose that node i is connected with nodes j and k. If node i fails, nodes j and k (and all the nodes of the subnet A) are alive but isolated and so the whole subnet A is isolated. More generally, if a subnet depends on a node and this node becomes isolated then all of the nodes of the subnet are isolated.

Rules 6.1.1-4 show the correctness properties that represent the core of the specification for an isolation event.

We consider $\{S\} \subset \mathbb{N} - \{0\}$ as a set of *sensor* and *from_sensor* variables. $|S| = \{number\ of\ nodes\}$.

Rule 6.1.1: ***Isolation event detection***
$\forall$ *sensor, from_sensor* $\in$ *S, sensor*$\neq$*from_sensor:*
Neighbor(sensor, from_sensor) $\wedge$ *IsReachable(sensor)* $\wedge$ *IsAlive(sensor)* $\wedge$ ($\nexists$ *from_sensor2* $\in$ *S, from_sensor2*$\neq$*sensor: IsAlive(from_sensor2)* $\wedge$ *IsReachable(from_sensor2)* $\wedge$ *IsLinked(sensor, from_sensor2)* $\wedge$ *Neighbor(sensor, from_sensor2))*
$\Rightarrow$ ***Isolation(sensor)*** *event* ***happens***.

Rule 6.1.2: ***Join event detection***
$\forall$ *sensor, from_sensor* $\in$ *S, sensor*$\neq$*from_sensor:*
$\neg$*IsReachable(sensor)* $\wedge$ *IsAlive(sensor)* $\wedge$ *IsAlive(from_sensor)* $\wedge$ *IsReachable(from_sensor)* $\wedge$ *IsLinked(sensor, from_sensor)* $\wedge$ *Neighbor(sensor, from_sensor)*
$\Rightarrow$ ***Join(sensor)*** *event* ***happens***.

Rule 6.1.3: ***When Isolate event cannot occur***
$\forall$ *sensor, from_sensor* $\in$ *S, sensor*$\neq$*from_sensor:*
((IsAlive(from_sensor) $\wedge$ *IsReachable(from_sensor)* $\wedge$ *IsLinked(sensor, from_sensor))* $\vee$ $\neg$*IsAlive(sensor)* $\vee$ $\neg$*IsReachable(sensor))* $\wedge$ *Neighbor(sensor, from_sensor)*
$\Rightarrow$ ***Isolate(sensor)*** *event* ***cannot happen***.

Rule 6.1.4: ***When Join event cannot occur***
$\forall$ *sensor, from_sensor* $\in$ *S, sensor*$\neq$*from_sensor:*
(IsReachable(sensor) $\vee$ $\neg$*IsAlive(sensor)* $\vee$ $\neg$*IsLinked(sensor, from_sensor)* $\vee$ $\neg$*IsAlive(from_sensor)* $\vee$ $\neg$*IsReachable(from_sensor))* $\wedge$ *Neighbor(sensor, from_sensor)*
$\Rightarrow$ ***Join(sensor)*** *event* ***cannot happen***.

In the first rule we want to verify when a node becomes isolated. A sensor can be isolated if it was initially reachable and alive and, considering a link with another sensor, there is no sensor that is alive, reachable and connected with the sensor. Additionally, we report another rule 6.1.2 which allows the checking of a *Join* event. In particular the rule declares that if a sensor is not reachable, because it is isolated and alive, also its neighbor sensor is alive and reachable and there is a connection between them then the sensor can join the network and becomes reachable again.

Only for an isolation event, we report in Listing 1, as a proof of event calculus formalism, the rule 6.1.1 written also in our chosen formalism.

In table 2 we report the basic elements (sorts, events and fluents) used for the specification. The *generated events* are generated by the reasoner on the basis of the defined specification.

Listing 1 Rule 6.1.1 written in event calculus language

```
[sensor,from_sensor, time] Neighbor(from_sensor,sensor) & HoldsAt(
      IsReachable(sensor),time) & HoldsAt(IsAlive(sensor),time) &
      (!{from_sensor2} (HoldsAt(IsAlive(from_sensor2),time) &
      HoldsAt(IsReachable(from_sensor2),time) & HoldsAt(
      IsLinked(sensor,from_sensor2),time)) &
      Neighbor(from_sensor2,sensor)) ->
Happens(Isolate(sensor),time).
```

Table 2 Basic elements of the specification for the isolation event

Elements	Name	Description
Sorts	*sensor*	reference sensor for events and fluents
	from_sensor	sensor used in case of disconnection (i.e. a sensor disconnects **from** another sensor)
Generated Events	*Isolate(sensor)*	it occurs when a sensor is isolated from the network
	Join(sensor)	it occurs when there is at least a connection between a sensor and one or more sensors
Fluents	*IsAlive(sensor)*	true when a *Start* event occurs for a sensor
	IsLinked(sensor, to_sensor)	true when a *Connect* event occurs
	IsReachable(sensor)	true when a sensor is reachable from the sink node

6.2 Packet Loss Event

When there is a failure in a node or a link between a couple of nodes is disrupted, then all of the packets that are in delivery towards this node are lost. In turn, these packets are not delivered to the sink.

For instance, let us consider figure 5 which is a part of MEDiSN architecture topology. Node *B* sends a packet *pkt* to node *sink* by means of node *A*. If node A crashes it cannot receive the packet from node B and forward it to a node towards the sink; so the packet is lost.

Rules 6.2.1-5 show the correctness properties that represent the core of the specification for a packet loss event.

We consider $\{S\} \subset \mathbb{N}$ - $\{0\}$ as a set of *sensor*, *from_sensor*, *to_sensor* and *source* variables; $\{P\} \subset \mathbb{N}$ - $\{0\}$ as a set of *pkt* variables.

$|S|= \{number\ of\ nodes\}$; $|P|= \{number\ of\ packets\ sent\ by\ each\ node\}$.

Rule 6.2.1: ***First Forward event detection***
$\forall$ *pkt* $\in P \wedge \forall$ *sensor, to_sensor* $\in S$, *sensor*$\neq$*to_sensor:*
sensor sends pkt $\wedge$ *Neighbor(to_sensor, sensor)*
$\Rightarrow$ ***Forward(pkt, sensor, sensor, to_sensor)*** *event* ***happens***.

Rule 6.2.2: ***Catch event detection***

$\forall$ *pkt* $\in P \wedge \forall$ *sensor, from_sensor, source* $\in S$, *sensor*$\neq$*from_sensor*:
IsOnChannel(pkt, source, from_sensor, sensor) $\wedge$ ($\nexists$ *sensor2, from_sensor2, source2* $\in S$, *sensor2*$\neq$*from_sensor2*, $\wedge \nexists$ *pkt2* $\in P$: *IsOnChannel(pkt2, source2, from_sensor2, sensor2)* $\wedge$ *source*$\neq$*source2* $\wedge$ *sensor=sensor2)* $\wedge$ *IsAlive(sensor)* $\wedge$ *IsLinked(from_sensor, sensor)* $\wedge$ $\neg$*IsLost(pkt, source)*

$\vee$

$\forall$ *pkt, pkt2* $\in P$ $\wedge\forall$ *sensor, from_sensor, from_sensor2, source, source2* $\in S$, *sensor*$\neq$*from_sensor*$\neq$*from_sensor2, source*$\neq$*source2*:
IsOnChannel(pkt, source, from_sensor, sensor) $\wedge$ *IsOnChannel(pkt2, source2, from_sensor2, sensor)* $\wedge$ *source*<*source2* $\wedge$ *IsAlive(sensor)* $\wedge$ *IsLinked(from_sensor, sensor)* $\wedge$ $\neg$*IsLost(pkt, source)*

$\Rightarrow$ ***Catch(pkt, source, sensor, from_sensor)*** *event* ***happens***.

Rule 6.2.3: ***Next Forward event detection***

$\forall$ *sensor, to_sensor, from_sensor, source* $\in S \wedge \forall$ *pkt* $\in P$:
Catch(pkt, source, sensor, from_sensor) event happens $\wedge$ *Neighbor(sensor, from_sensor)* $\wedge$ *Neighbor(to_sensor, sensor)* $\wedge$ *IsAlive(sensor)* $\wedge$ $\neg$*IsLost(pkt, source)*

$\Rightarrow$ ***Forward(pkt, source, sensor, to_sensor)*** *event* ***happens***.

Rule 6.2.4: ***Receive event detection***

$\forall$ *from_sensor, source* $\in S \wedge \forall$ *pkt* $\in P$:
Catch(pkt, source, 1, from_sensor) event happens

$\Rightarrow$ ***Receive(pkt, source)*** *event* ***happens***.

Rule 6.2.5: ***PacketLoss event detection***

$\forall$ *sensor, to_sensor, source* $\in S \wedge \forall$ *pkt* $\in P$:
$\neg$*IsLost(pkt, source)* $\wedge$ *IsOnChannel(pkt, source, sensor, to_sensor)* $\wedge$ *Neighbor(to_sensor, sensor)* $\wedge$ ($\neg$*IsLinked(from_sensor, sensor)* $\vee$ *Stop(to_sensor) event happens* $\vee$ $\neg$*IsAlive(sensor))*

$\Rightarrow$ ***PacketLoss(pkt, source)*** *event* ***happens***.

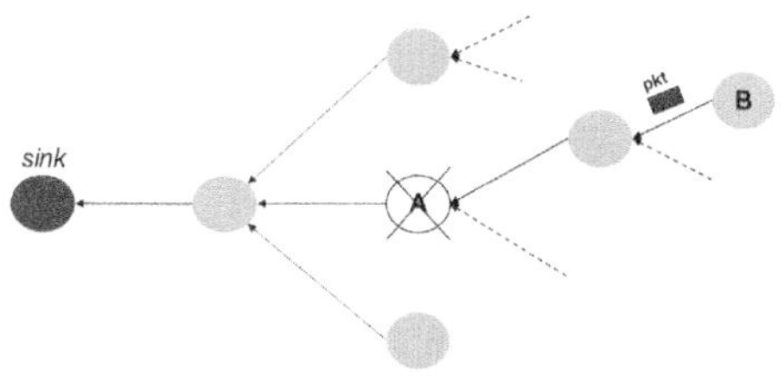

Fig. 5 Example of packet loss

Table 3 Basic elements of the specification for the packet loss event

Elements	Name	Description
Sorts	*pkt*	it is the packet id
	source	it is the sensor that sends the packet
Generated Events	*Receive(pkt, source)*	it occurs when sink node receives a packet
	Catch(pkt,source, sensor, from_sensor)	it occurs when a sensor catches a packet from one of its neighbor sensors
	Forward(pkt,source, sensor, to_sensor)	it occurs when a sensor, once caught a packet sent by a sensor, forwards it to its neighbor sensor
	PacketLoss(pkt, source)	it occurs when a packet is lost
Fluents	*IsOnChannel(pkt,source, sensor, to_sensor)*	true when a packet is in transmission on the channel between two nodes
	IsLost(pkt, source)	true when a packet is lost

By means of rule 6.2.1, we assert that when a sensor sends a packet, a *Forward* event is generated and the packet delivery starts.

In rule 6.2.2, we define the *Catch* event that occurs when, being the packet on the channel between a couple of nodes, the receiving sensor is alive, the packet is not lost and there is a connection between the nodes. Note that we have also included a concurrency control in order to manage packets that come from different nodes.

Also, we can check the end of the packet delivery (rule 6.2.4) when it is caught by the sink node (node 1) and so if it is received.

Finally, we report rule 6.2.5 that checks when a packet is lost. Considering a packet, that is not lost, in delivery towards the sink node and is on the channel between two nodes, when a disconnection event between the two nodes or the failure of a receiving node occurs then there is a packet loss event.

In table 3 we report the basic elements (sorts, events and fluents) used for the specification related to the packet loss event.

6.3 Battery Exhaustion Event

A battery exhaustion event happens when a node completely consumes its available energy. For instance, considering the MEDiSN architecture, we can observe that if a node sends a certain number of packets, its battery level continually decreases until it causes a crash of the node.

To achieve this aim, we adopt a sort in the specification for each sensor node, called a *level*, which represents the level of the battery of the node, and which is decreased each time the node sends, catches, or forwards a packet.

Rules 6.3.1-2 show the correctness properties that represent the core of the specification for a battery exhaustion event.

We consider $\{S\} \subset \mathbb{N}$ - $\{0\}$ as a set of *sensor*, *to_sensor* and *source* variables; $\{P\} \subset \mathbb{N}$ - $\{0\}$ as a set of *pkt* variables; $\{L\} \subset \mathbb{N}$ as a set of *level* and *levelnew* variables.

$|S|= \{number\ of\ nodes\}$; $|P|= \{number\ of\ packets\ sent\ by\ each\ node\}$; $|L|= \{number\ of\ charge\ levels\ of\ the\ battery\}$.

Rule 6.3.1: ***Battery level change event detection***
$\forall$ *sensor, to_sensor, source* $\in S \wedge \forall$ *pkt* $\in P \wedge \forall$ *level* $\in L$:
Forward(pkt, source, sensor, to_sensor) event happens $\wedge$ *BatteryLevel(level, sensor)* $\wedge$ *level>0*
$\Rightarrow$ ***Old_Consume(level, sensor)*** *event* ***happens*** $\wedge$ ($\exists$ *levelnew* $\in L$, $0 \leq levelnew \neq level$: ***levelnew=level-1*** $\wedge$ ***New_Consume(levelnew, sensor)*** *event* ***happens***)

Rule 6.3.2: ***Stop event detection due to battery exhaustion***
$\forall$ *sensor* $\in S$:
BatteryLevel(0, sensor) $\wedge$ *IsAlive(sensor)*
$\Rightarrow$ ***Stop(sensor)*** *event* ***happens***

When a *Forward* event occurs (due to the sending of a packet) and the battery level of the sensor is positive then we consider the new consumption level (rule 6.3.1).

In rule 6.3.2, we can see that when the battery level of a sensor, that is alive, is zero then a failure for the node occurs due to battery exhaustion.

In table 4 we report the basic elements (sorts, events and fluents) used for the specification related to the battery exhaustion event.

Table 4 Basic elements of the specification for the battery exhaustion event

Elements	Name	Description
Sorts	*level*	the current battery level of a sensor
	levelnew	the new battery level of a sensor after RX/TX operations related to a packet
Generated Events	*New_Consume(levelnew,sensor)*	it occurs when the battery level of a sensor decreases after the forwarding of a packet
	Old_Consume(level,sensor)	event created for specification reasons
Fluents	*BatteryLevel(level,sensor)*	true when the battery charge of a sensor is equal to a certain *level*

7 Structural Correctness Specification Modeling

This phase is mainly related to the topology of the BAN and is completed by parameters regarding the structure of the BAN (number of sensors, number of packets) and by the chosen observation time (number of timepoints). Differently from the previous step, this specification varies depending on the basis of the characteristics of the target BAN.

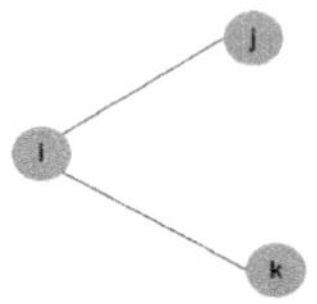

Fig. 6 Example of BAN topology

Listing 2 Example of Neighbor predicate Specification

```
[sensor1,sensor2] Neighbor(sensor1,sensor2) <-> (
(sensor1=i & sensor2=j) |
(sensor1=i & sensor2=k)
).
```

To specify the topology, we use the *Neighbor* predicate (already used in the previous specifications) to indicate how nodes are linked in the topology.

For instance, considering the topology in figure 6, let us suppose node *i* is connected with *j* and *k* and let us consider a tree graph where the sink node (root node) is the node *i* and the nodes *j* and *k* are the child nodes.

The resulting specification is reported in listing 2, where *sensor1* is the parent node (*i*) and *sensor2* are the child nodes (*j* and *k*). Clearly, this specification can be changed easily if the topology of the BAN changes. The role of the *Neighbor* predicate is very important in order to understand when an axiom can be applied. Let us examine the axiom related to a possible isolation (Rule 6.1.1) and let us apply it to figure 6. The described implication is true when, given a couple of nodes (*sensor*, *from_sensor*), the conditions about isolation are true and there is a link between the nodes (in this case, between nodes *j* and *i* or between nodes *k* and *i*). This, for instance, can never be true for the couple of nodes *j* and *k*, since there is no physical link between them.

Regarding the parameters, their values can be used to check the correctness properties of the BAN under different conditions, i.e., under different assumptions on the initial charge of the batteries (e.g., to verify a BAN in the middle of its life), or under different environmental conditions affecting the quality of the channels (impacting on the probability of having a disconnection event when checking the robustness of the BAN).

8 Verification/Testing

The last phase concludes the workflow. The aim is to formally verify or test the entire set of specifications by means of the DECReasoner. If some error is detected at this phase, possible corrections have to be performed in one or more of the previous steps. In this Section we are going to verify and test the defined sets of

Listing 3 Example of initial event trace

```
Happens(Stop(j),1).
Happens(Stop(k),3).
Happens(Start(k),4).
Happens(Disconnect(k,i),5).

completion Happens
```

specifications adopting two techniques of analysis: *what-if analysis*, to verify how the BAN behaves in response to a given sequence of events of interest for the designer, and *coverage robustness checking*, to verify the long term robustness of the BAN against random sequences of undesired events, useful to identify corner cases and dependability bottlenecks.

8.1 What-If Analysis

The goal of a *what-if* scenario analysis is to verify the behavior of the target BAN under precise circumstances. To achieve this aim, we need to indicate an initial event sequence (Event Trace). For example, as listing 3 shows, by means of *Happens* predicates, we can declare that at timepoint 1 sensor j stops, at timepoint 3 sensor k stops, etc. In this way, by means of the reasoner, we can observe the consequences of any initial sequence of events of interest for the designer, e.g., to test the robustness of the designed topology against the temporary unavailability (failure/recovery) of a given set of nodes, or to quantify to what extent a modification of the topology can be beneficial for the network.

8.2 Coverage Robustness Checking

By *coverage robustness checking* we mean a technique to statically verify the behavior of a BAN in the face of a number of failures that can occur during its operation. This test is aimed at verifying how many node failures the network can tolerate, while guaranteeing a given minimum level of coverage. For example if we consider a network composed of m nodes and a threshold coverage equal to 50%, we may want to understand what the sequences of failures causing more than $m/2$ nodes to be isolated are (i.e., coverage under the specified threshold) and how the resiliency level varies when varying the sequences of failures. This allows an evaluation of the maximum (and minimum) resiliency level reachable by a given topology and what the critical failure sequences are, i.e., the shortest ones causing a loss of coverage. These are particularly useful in order pinpoint weak points in the network.

8.3 Application Scenario

In this paragraph we want to verify the presented formal approach and the defined sets of specifications by means of an application scenario applying the what-if analysis technique; the aim is to describe how failures are detected in a BAN and to assess its dependability level (in terms of coverage and connection resiliency) after their effects.

We consider a BAN topology that has been realized in our laboratory using TMOTE sensors. The network (figure 7) is constructed by mounting nine sensor nodes. Node 1 is the sink of the network. Each node consists of a TMOTE SKY XM1000 (Tiny-OS and ContikiOS compatible) which follows the IEEE 802.15.4 standard.

We want to observe the behavior of this BAN in 10 timepoints, supposing that, after two timepoints, node 6 stops, a disconnection between node 7 and node 4 occurs at timepoint 4 and that node 2 stops at timepoint 7. In particular, we are interested in evaluating if the reasoning performed by the DECReasoner on our specifications and the expected dependability assessment are correct.

Setting a coverage threshold at 60%, we should observe a coverage interval equal to [0;4] (i.e., when node 7 disconnects from node 4 at timepoint 4, four nodes are not reachable, namely 6, 7, 8 and 9), and a connection resiliency equals 1 (i.e., only the first stop event is tolerated).

8.3.1 Structural Specification Generated by Event Calculus

Previously, we have discussed the application scenario we want to observe giving an occurring particular sequence of events. Now, we describe the structural specification file (Listing 4) generated using the event calculus syntax (in particular we focus on the formal definition of the BAN topology and the event trace to consider).

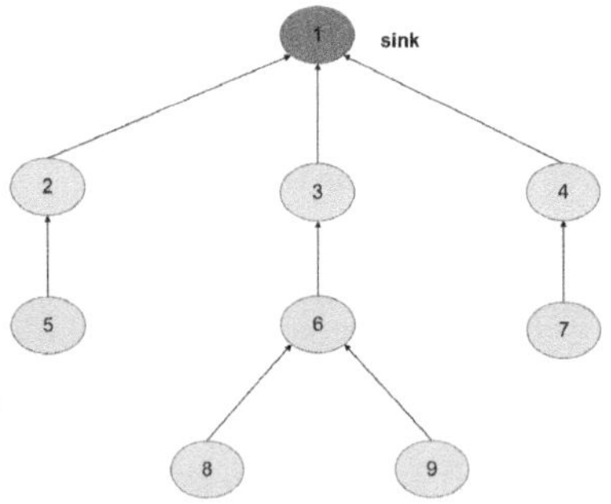

Fig. 7 The BAN topology of the application scenario

Listing 4 Structural specification written in event calculus language for BAN-based application scenario

```
1  load foundations/Root.e
2  load foundations/EC.e
3
4  load /home/alessandro-64/IsolationWorld.e
5
6  [sensor1,sensor2] Neighbor(sensor1,sensor2) <-> (
7  (sensor1=1 & sensor2=2) | (sensor1=1 & sensor2=3) | (sensor1=1 & sensor
       2=4) |
8  (sensor1=2 & sensor2=5) |
9  (sensor1=3 & sensor2=6) |
10 (sensor1=4 & sensor2=7) |
11 (sensor1=6 & sensor2=8) | (sensor1=6 & sensor2=9)
12 ).
13
14
15
16 [sensor] HoldsAt(IsAlive(sensor),0).
17 [sensor,from_sensor] Neighbor(from_sensor, sensor) <-> HoldsAt(IsLinked(
       sensor,from_sensor),0).
18
19 Happens(Stop(6),2).
20 Happens(Disconnect(7 , 4),4).
21 Happens(Stop(2),7).
22
23
24
25 completion Happens
26
27 range from_sensor 1 9
28 range sensor 1 9
29 range to_sensor 1 9
30 range time 0 10
31 range offset 1 1
32 option n_models 1
33 option modeldiff on
```

Formal definition of a topology

Analyzing the links of the BAN topology, we establish (lines 6-12 of the Listing 4) how the nodes are linked by means of the *Neighbor* predicate.

Event Trace

We insert an event trace (lines 19-21 of the Listing 4) that is composed of a list of *Happens* predicates that specify nodes and timepoints in which a given event occurs on the basis of the event sequence initially supposed for this application scenario. The *completion* statement specifies that a predicate symbol (i.e. *Happens*) should be subject to predicate completion.

Outcome and Metrics computation

Listing 5 reports the outcome produced by the DECReasoner. In the lines 11-36 the outcome shows the conditions of the BAN at timepoint 0: all of the nodes are alive, reachable and every node is linked with its neighbor on the basis of the considered topology. The event trace confirms our expectations. We can observe that at timepoint 2 node 6 stops (line 39), due to a failure, and it is no longer alive isolating nodes 8 and 9 (lines 42-43). Thus, these two nodes become not reachable (lines 45-46). Considering that, because of a disconnection from node 4 (line 47) at timepoint 4, node 7 is isolated (line 50) becoming not reachable (line 52), then this means that at timepoint 5 a total of 4 nodes are isolated. The coverage is computed as the timepoint of the last failure event causing such isolation, that is 4. Consequently, the connection resiliency is computed by counting the number of failures and disconnection events in the interval [0; 4], excluding the last event; hence, it is equal to 1. At the last stage, the failure of node 2 at timepoint 7 (line 54) and thus the isolation of node 5 (line 57) does not alter the results. The dependability metrics can be evaluated by analyzing the event flow (the *narrative*) generated by the DECReasoner based on the specification of the target BAN.

Listing 5 Outcome produced by the DECReasoner for BAN-based application scenario

```
1  loading /home/alessandro-64/Dottorato/EventCalculus/EC_Application/
       topology1803EC_20130318174304.e
2  loading foundations/Root.e
3  loading foundations/EC.e
4  loading /home/alessandro-64/IsolationWorld.e
5  15329 variables and 129112 clauses
6  relsat solver
7  1 model
8  ---
9  model 1:
10 0
11 IsAlive(1).
12 IsAlive(2).
13 IsAlive(3).
14 IsAlive(4).
15 IsAlive(5).
16 IsAlive(6).
17 IsAlive(7).
18 IsAlive(8).
19 IsAlive(9).
20 IsLinked(2, 1).
21 IsLinked(3, 1).
22 IsLinked(4, 1).
23 IsLinked(5, 2).
24 IsLinked(6, 3).
25 IsLinked(7, 4).
26 IsLinked(8, 6).
27 IsLinked(9, 6).
28 IsReachable(1).
29 IsReachable(2).
30 IsReachable(3).
31 IsReachable(4).
32 IsReachable(5).
33 IsReachable(6).
34 IsReachable(7).
35 IsReachable(8).
36 IsReachable(9).
37 1
38 2
39 Happens(Stop(6), 2).
40 3
41 -IsAlive(6).
42 Happens(Isolate(8), 3).
43 Happens(Isolate(9), 3).
44 4
```

```
-IsReachable(8).
-IsReachable(9).
Happens(Disconnect(7, 4), 4).
5
-IsLinked(7, 4).
Happens(Isolate(7), 5).
6
-IsReachable(7).
7
Happens(Stop(2), 7).
8
-IsAlive(2).
Happens(Isolate(5), 8).
9
-IsReachable(5).
10
P
Neighbor(1, 2).
Neighbor(1, 3).
Neighbor(1, 4).
Neighbor(2, 5).
Neighbor(3, 6).
Neighbor(4, 7).
Neighbor(6, 8).
Neighbor(6, 9).
!Happens(Connect(1, 1), 0).
!Happens(Connect(1, 1), 1).
!Happens(Connect(1, 1), 10).
```

9 Case Studies

1. Download and use the ADVISES tool[1] in order to test the BAN topology shown in figure 7 when an isolation event occurs (i.e. stop a node at timepoint x, stop a second node at timepoint y, etc).
2. Repeat case study n.1 in order to test the same BAN topology when a packet loss event occurs (i.e. a packet sent towards the sink is lost due to a node crash).

10 Future Directions

Presenting the proposed approach, we assert that it is important to detect and analyze the dependability issues of mHealth monitoring systems since their inception (at design time) in order to increase confidence about the robustness of the designed solutions before putting them into operation. Indeed, the chapter has highlighted, by means of a case study, that also in a simple deployment, the failure of a single node can be responsible for the failure of a piece of the network. In other cases, a node that is close to the sink (i.e., the gateway of the BAN that has the role of collecting all of the measures detected by the sensors) is more likely to fail due to the great stresses it is subjected to, and its failure would likely cause the isolation of a set of nodes from the sink.

To address these issues, we have defined the formal specifications used for the behavioral checking of mHealth monitoring systems (in particular we have focused

[1] http://sourceforge.net/projects/advises/

on BANs); the correctness specifications have been obtained using event calculus as the formal language since the behavior of a BAN can be characterized in terms of an event flow (e.g. a node turns on, a packet is sent, a node stops due to failure, etc.) and the event calculus formalism allows an easy specification of the system in terms of events. For this purpose, formal methods are a suitable solution for dependability assessment at design time.

The proposed modeling methodology, analysis and specifications can be adapted/extended to other systems considering the specific failure modes of the components used. In fact, it is possible to define new metrics of dependability not oriented only towards BANs but which consider the failure events that may occur in other components of a mHealth monitoring system (i.e. the gateway, Extra-BAN communication, etc). Additionally, another direction could be to adopt all of this approach (verification with formal methods) in other mHealth scenarios: for example adopting the approach for gesture and activity recognition for patients with cognitive disorders.

Moreover, we would like to propose a formal approach for the dependability assessment of mHealth monitoring systems also at runtime and we plan to use the specification also for mobile scenarios specifying further events that notify wireless sensor movements within clusters and from cluster to cluster.

11 Conclusions

In this chapter we have presented an approach, based on an event-based formalism, for the dependability assessment of a mHealth monitoring system. In particular, after a detailed failure modes and effects analysis (FMEA), we have identified some main critical events that could undermine the dependability of a BAN which is the most critical component within a mHealth monitoring system. Finally, by means of an application scenario, we have concluded our approach testing the defined formal specifications and at same time showing the benefits that we can obtain. We base our approach on sensors that are fixed (such as beacons) and with an established data routing reducing a topology like a spanning tree that is valid for every BAN (or more generally WSN).

Acknowledgements. This work has been partially supported by the "Se@me: Sustainable e-maritime @ssistance for Maritime Employees, Passengers and Yachtsmen" project, funded by Regione Campania, Italy under the "POR Campania 2007/2013" initiative.

References

1. Benlamri, R., Docksteader, L.: Morf: A mobile health-monitoring platform. IT Professional 12(3), 18–25 (2010), doi:10.1109/MITP.2010.3
2. Carrozza, G., Cinque, M.: Modeling and analyzing the dependability of short range wireless technologies via field failure data analysis. Journal of Software 4(7) (2009), `http://ojs.academypublisher.com/index.php/jsw/article/view/0407707716`

3. Cinque, M., Coronato, A., Testa, A.: Dependable services for mobile health monitoring systems. IJACI 4(1), 1–15 (2012)
4. Cinque, M., Cotroneo, D., Kalbarczyk, Z., Iyer, R.K.: How do mobile phones fail? a failure data analysis of symbian os smart phones. In: 37th Annual IEEE/IFIP International Conference on Dependable Systems and Networks, DSN 2007, pp. 585–594. IEEE (2007)
5. Cinque, M., Cotroneo, D., Martinio, C.D., Russo, S.: Modeling and assessing the dependability ofwireless sensor networks. IEEE Symposium on Reliable Distributed Systems, 33–44 (2007), http://doi.ieeecomputersociety.org/10.1109/SRDS.2007.28
6. Coronato, A.: Uranus: A middleware architecture for dependable aal and vital signs monitoring applications. Sensors 12(3), 3145–3161 (2012)
7. Coronato, A., Testa, A.: Long-term monitoring of vital signs for mobile patients. In: PECCS 2012, pp. 15–20 (2012)
8. Di Martino, C., Cinque, M., Cotroneo, D.: Automated generation of performance and dependability models for the assessment of wireless sensor networks. IEEE Trans. Comput. 61(6), 870–884 (2012), http://dx.doi.org/10.1109/TC.2011.96, doi:10.1109/TC.2011.96
9. Dicks, K.E.: Telemedicine 2.0 has arrived. Future Healthcare Magazine (2007)
10. Hao, Y., Foster, R.: Wireless body sensor networks for health-monitoring applications. Physiological Measurement 29(11), R27–R56 (2008)
11. Ko, J., Lim, J.H., Chen, Y., Musvaloiu-E, R., Terzis, A., Masson, G.M., Gao, T., Destler, W., Selavo, L., Dutton, R.P.: Medisn: Medical emergency detection in sensor networks. ACM Trans. Embed. Comput. Syst. 10(1), 11:1–11:29 (2010), http://doi.acm.org/10.1145/1814539.1814550, doi:10.1145/1814539.1814550
12. Kowalski, R., Sergot, M.: A logic-based calculus of events. New Gen. Comput. 4(1), 67–95 (1986), http://dx.doi.org/10.1007/BF03037383, doi:10.1007/BF03037383
13. Latino, R.J., Flood, A.: Optimizing fmea and rca efforts in health care. Journal of Healthcare Risk Management 24(3), 21–28 (2004)
14. Lin, H., Shao, J., Zhang, C., Fang, Y.: Cam: Cloud-assisted privacy preserving mobile health monitoring. IEEE Transactions on Information Forensics and Security 8(6), 985–997 (2013), doi:10.1109/TIFS.2013.2255593
15. Lindeman, D.: Technologies for remote patient monitoring in older adults. Public Health Institute
16. Mueller, E.: Decreasoner, http://decreasoner.sourceforge.net
17. O'Donovan, T., O'Donoghue, J., Sreenan, C., Sammon, D., O'Reilly, P., O'Connor, K.: A context aware wireless body area network (ban). In: 3rd International Conference on Pervasive Computing Technologies for Healthcare, PervasiveHealth 2009, pp. 1–8 (2009)
18. Paksuniemi, M., Sorvoja, H., Alasaarela, E., Myllyla, R.: Wireless sensor and data transmission needs and technologies for patient monitoring in the operating room and intensive care unit. In: 27th Annual International Conference of the Engineering in Medicine and Biology Society, IEEE-EMBS 2005, pp. 5182–5185 (2005)
19. Shanahan, M.: The Event Calculus Explained. In: Veloso, M.M., Wooldridge, M.J. (eds.) Artificial Intelligence Today. LNCS (LNAI), vol. 1600, pp. 409–430. Springer, Heidelberg (1999), http://citeseerx.ist.psu.edu/viewdoc/summary?doi=10.1.1.43.3267

20. Sneha, S., Varshney, U.: A framework for enabling patient monitoring via mobile ad hoc network. Decis. Support Syst. 55(1), 218–234 (2013), http://dx.doi.org/10.1016/j.dss.2013.01.024, doi:10.1016/j.dss.2013.01.024
21. Sommerville, I.: Software Engineering, 7th edn. Pearson Addison Wesley (2004)
22. Testa, A., Coronato, A., Cinque, M., Augusto, J.C.: Static verification of wireless sensor networks with formal methods. In: 2012 Eighth International Conference on Signal Image Technology and Internet Based Systems (SITIS), pp. 587–594. IEEE (2012)
23. Triantafyllidis, A., Koutkias, V., Chouvarda, I., Maglaveras, N.: A pervasive health system integrating patient monitoring, status logging and social sharing. IEEE Trans. Inf. Technol. Biomed. (2012)
24. Ximo, C.: Carefusion. Ject Press (2012), http://books.google.it/books?id=TOONtgAACAAJ
25. Yan, H., Huo, H., Xu, Y., Gidlund, M.: Wireless sensor network based e-health system?? implementation and experimental results. IEEE Transactions on Consumer Electronics 56(4), 2288–2295 (2010), doi:10.1109/TCE.2010.5681102

Questions

Q1. Describe the strengths and weaknesses of a formal-based dependability assessment approach of a mHealth monitoring system.

Q2. Considering a BAN with the topology shown in figure 7, and assuming that node 2 fails at timepoint 2, and node 3 at timepoint 4, what would be the narrative generated by the tool in 10 timepoints in accordance with our specifications?

Q3. What are the critical events that can be detected by means of our formal specification?

Q4. Explain the dependability metrics used for the WSN dependability assessment.

Q5. Describe the typical components of a mHealth monitoring system.

Q6. How have the critical events been chosen?

Q7. What is the difference between general and structural correctness specifications?

9. Wireless Monitoring System for Wheelchair Users with Severe Mobility Impairment

Diego E. Arias[1], Esteban J. Pino[1], Pablo Aqueveque[1], and Dorothy W. Curtis[2]

[1] Universidad de Concepción, Chile
{diegoarias,pablo.aqueveque}@udec.cl, epino@ieee.org
[2] Massachusetts Institute of Technology, MA, USA
dcurtis@csail.mit.edu

Abstract. According to the World Health Organization, around the world there are more than 785 million people with a disability. This phenomenon is produced to a great extent by the aging of the population and the increase in chronic diseases. These patients may be affected physically and/or cognitively, requiring nursing care and/or family assistance to perform daily activities. Bringing new tools to families and professional caregivers to improve the care of these patients is essential to increase the quality of life for both the disabled people and the caregivers. Recent advances in sensors, wireless communication systems and information technologies make possible the development of portable and wearable systems to monitor mobility impaired patients continuously during daily activities. Collecting vital signs, patient activity and ambient conditions allow the patient's health status to be assessed, providing an extra level of safety in cases of emergency. Also this information is useful for clinicians to manage treatment and rehabilitation therapies. However, the main challenge is to acquire this information unobtrusively, with a minimal impact on patients' daily life. To this end, new ways of collecting physiological information are needed.

Keywords: Pressure ulcers, vital signs, heat stress, powered wheelchairs, monitoring system.

1 Introduction

People with physical or mental disabilities are impaired in one or more major life activities. According to the World Health Organization, as of 2011, there were more than 785 million people with a disabling condition in the world [1]. As life expectancy increases, people are spending more years of their life with some kind of disability, usually due to a chronic disease [2]. Depending on the type of disability, patients may be affected in their physical and/or cognitive skills, requiring continual nursing care and assistance to perform daily activities. Usually, family members take the responsibility for supporting and caring for them at home. As an alternative, they can be put in a nursing home or assisted living facility under the charge of specialized caregivers. This global scenario challenges researchers to

S. Adibi (ed.), *Mobile Health*,
Springer Series in Bio-/Neuroinformatics 5, DOI: 10.1007/978-3-319-12817-7_9

develop new technological solutions designed for families and/or professional caregivers to improve the care of these patients, increasing their quality of life.

As technologies advances, new assistive devices can be designed to improve the care of people with disabilities. Recent advances in sensors, wireless communication systems and information technologies make possible the development of portable and wearable biomedical systems to monitor patients continuously during daily activities through mHealth. Collecting vital signs, patient activity and ambient conditions allow the patient's health status to be assessed, providing an extra level of safety in cases of emergency. Also this information is useful for clinicians because they can use it to improve treatment and rehabilitation therapies. However, the main challenge is to acquire this information unobtrusively, with a minimal impact on patients' daily life. For instance, conventional sensors such as skin electrodes can cause dermatitis during prolonged use, producing discomfort [3]. Also, respiration belts can be difficult to wear and uncomfortable for impaired patients. To this end, new ways of collecting physiological information are needed.

This chapter describes a monitoring system designed for wheelchair users who suffer severe motor disabilities. The system consists of several sensors and portable equipment deployed on a wheelchair, able to sense physiological data, patient activity level and ambient conditions in a non-invasive way. The selected sensors are able to capture respiration and heart activity from ballistocardiogram (BCG) signals, seat and back pressure changes, wheelchair tilt angle, ambient temperature and relative humidity [4,5]. This portable equipment consists of data acquisition hardware, a central processing unit and a communication system. The information obtained from sensors is sent to a patient monitoring center and is saved in a database. The communication can be done in two ways, depending on whether the patient is in a nursing home or not. In the first case, the good connectivity and easy implementation of a Wi-Fi network is preferred. For patients moving outside of the nursing home (moving around in a city), data transfer using GSM or GPRS technologies can be used. Electrical power for the portable system is supplied by the wheelchair batteries. The sensing procedure, data computing and data transfer are carefully designed to minimize the energy consumption.

A pilot study was conducted to evaluate the system in an actual clinical environment [6]. We analyze the results obtained during a two week evaluation to highlight the capability of such a monitoring system. Six Multiple Sclerosis (MS) patients, all of them permanent wheelchair users, were recruited from an assisted living facility to use the system during daily activities. Results show the feasibility of implementing alarms and relevant information capture to monitor wheelchair users. The implemented system is able to provide relevant information such as time on wheelchair, moving vs resting, environmental conditions, vital signs and pressure relief, which are useful for guiding or warning patients and caregivers.

Also in this chapter, we describe the several processing stages used for this application. For instance, multi-level wavelet decomposition is used to extract respiration and heart activity from BCG signals, peak detection algorithms are implemented to estimate heart and respiration rate, an algorithm is used to detect pressure relief behavior based on both pressure sensors and accelerometer data, patient activity level is measured during the day and a heat index based on temperature and humidity to avoid heat exposure is calculated.

1.1 Chief Problems for Mobility Impaired People

Pressure ulcers (PU) are injuries on the skin produced by excessive pressure during long periods of time. Excessive pressure, usually over bony prominences, reduces the blood flow producing tissue ischemia and death, resulting in an ulcer. Wheelchair users spend long hours sitting, making them particularly prone to developing PU. To avoid PU, a person has to change position regularly. This way, the excessive pressure is relieved and blood can flow into the previously compressed tissue. Clinicians recommend that patients perform push-ups, side-to-side leans or lean forward over the knees as methods to relieve pressure. However, people who suffer severe mobility impairment cannot relieve pressure by themselves. In severely impaired people, pressure relief motions need assistance, from either family members, specialized caregivers or powered systems. Powered wheelchairs equipped with seat functions such as a tilt-in-space are usually prescribed in these cases. The tilt-in-space system allows the seat angle orientation to be changed in relation to the ground while maintaining the seat to back angle [7]. This mechanism redistributes the seat pressure, transferring it from the seat to the back.

However, the use of a tilt-in-space system depends on its user as well as caregivers. It is a conscious, manual activity, whose importance must be clear to those in charge of practicing it. Usually, people need training and/or reminders to tilt the wheelchair. Furthermore, some patients such as those with Spinal Cord Injury (SCI) and MS can suffer loss of awareness, sensory impairment and cognitive loss which can result in forgetting to tilt their wheelchairs [8,9]. Also, the high workload in nursing homes can lead to caregivers forgetting to tilt the patient's wheelchair or reminding patients to tilt their chairs. The same can happen when a family member is responsible for the patient's care. In those cases, it would be valuable to be able to alert the patient and/or the caregiver when such action is necessary.

Under severe mobility impairment, people can develop comorbidities related to respiratory and cardiovascular function. For instance, people with MS can suffer respiratory muscle weakness, bulbar function impairment and complications in breathing control which produce difficulty in breathing [10]. Also, the most severe cases of SCI are unable to breathe due to paralysis of respiratory muscles requiring mechanical ventilation support. Furthermore, this population presents problems in the autonomic nervous system, in charge of cardiovascular functions regulating the heart and blood vessels. Autonomic dysfunction increases the probability of suffering from heart diseases [11,12]. These examples illustrate why respiratory and cardiovascular function variables should be monitored. Vital signs such as heart rate and respiratory rate provide information related to health status and they have been reported as predictors of cardiovascular and respiratory failure [13]. Also, measuring vital signs is an important part of the nursing assessment, therefore monitoring this information continuously will facilitate this task and help prevent further complications.

Some patients can also suffer thermoregulation problems related to the heat. Particularly, most MS patients are very sensitive to heat, producing an exacerbation of symptoms such as fatigue [14]. Also, SCI patients are affected by heat due to impaired innervation of sweat glands and cutaneous blood vessels [15] reducing their ability to dissipate heat. For this reason, caregivers need to supervise patients when they are outside. Long periods of time exposed to heat without caregiver's supervision can produce dehydration and sunburns. Deaths have been reported in nursing homes produced by heat exposure due to an impaired self-awareness and failure in supervision. Also, patients in their homes have been found dead due to heat exposure in situations like sunbathing [16] and hot baths [17] without supervision.

1.2 Relevant Physiological Variables

To assess the physiological state of a person with mobility impairments, the relevant variables are the ones related to their main problems. In this case, the main concerns are: PU, activity level, cardio-respiratory problems and heat stress.

1.2.1 Pressure Ulcers

To avoid PU, the variable to be measured is time between tilts and the tilt angle, to ensure a proper pressure relief event. Maximum recommended time between tilts varies [18,19], but should be somewhere between 15 minutes and 2 hours. There are different visions as to what constitutes a proper tilt. Depending on the source, the recommended minimum tilt angle can vary between 25° up to 65° from the horizontal (regular) sitting position. A safer approach is to combine information from tilt angle change and pressure distribution on the chair. A change in angle accompanied by a corresponding change in pressure distribution is a clear indicator that a proper pressure relieving tilt has been accomplished.

1.2.2 Activity Level

The activity level of a subject can be a good predictor of their general health. A sudden decrease in activity in consecutive days can be an indicator of a deteriorating condition, such as an increasing pain level, fatigue exacerbation or a bout of depression [20]. Activity level is also related to PU and cardio-respiratory issues.

1.2.3 Cardio-Respiratory Problems

The main variables to monitor cardio-respiratory condition are Heart Rate (HR) and Respiratory Rate (RR). Also, with good measurements, it is possible to detect dangerous events such as arrhythmias and apneas. To detect arrhythmias, the physiological variable measured must somehow contain information regarding the heart activity. Several non-invasive approaches exist, such as capacitive coupling

ECG, sensors embedded into clothing, accessories or through ambient sensors embedded in everyday objects [21]. Contactless ECG is prone to artifacts from movement and external electromagnetic interference. An alternative modality for heart activity is BCG, a mechanical signal produced by the ejection of blood from the heart [22]. While still sensitive to movement artifacts, it has more immunity from electromagnetic sources. The respiratory signal also can be acquired from multiple modalities, such as movement, air flow or blood gas measurements [23]. The least invasive is this case is pressure-based, detecting the thoracic movement to derive respiration rate. Even though this method does not provide information related to the actual gas exchange, it is sufficient to detect respiratory stress, such as tachypnea and apnea.

1.2.4 Heat Stress

There are several recommendations regarding the dangers of high temperature and humidity combinations. The usual indexes related to heat stress are Heat Index (HI) and Dew point. Ambient temperature (T_{AMB}) and relative humidity (RH) are used to calculate the dew point value and the HI. These indexes provide information of heat exposure and its possible effects on human body. They correspond to two methods to assess the same phenomenon. The dew point is a measure of the amount of moisture in the air. It is defined as the temperature to which air must be cooled in order to reach saturation. The dew point value has been described as a possible indicator of exacerbation of the symptoms produced by heat in MS patients [24]. High dew point values increase the probability of the symptoms worsening, because it becomes increasingly difficult to evaporate sweat and regulate body temperature. A dew point over 17 °C is reported to have severe effects on MS symptoms [25].

Heat index or apparent temperature is an index to quantify how hot the body feels when it is exposed to an environment with high temperature and humidity. As with the dew point, in high humidity conditions sweat evaporation is reduced, producing the sensation of being overheated. The HI value is calculated using the equation presented in [26], for RH greater than 40%. Table 1 summarizes the recommendations regarding the effect of HI in high risk group people.

Table 1 Heat Index effect on people in high risk group

Heat Index (°C)	**General effect**
Above > 54	HIGHLY LIKELY heat or sunstroke with continued exposure
41-54	LIKELY sunstroke, heat cramps or heat exhaustion. POSSIBLE heatstroke with prolonged exposure
32-40	POSSIBLE sunstroke, heat cramps or heat exhaustion with prolonged exposure and/or physical activity
26-31	POSSIBLE fatigue with prolonged exposure and/or physical activity

1.3 Monitoring Biomedical Variables

A monitoring system for wheelchair users should be based on unobtrusive sensors, combined with a wireless link to relay alerts and possibly data logging to a central station. Since the system's priority is to enhance patient's quality of life, both at home or in a nursing home, it is recommended that the sensors are installed into the wheelchair, not requiring skin contact with the user. Data is then collected and analyzed locally in a netbook computer or equivalent low-power device with wireless connectivity to be able to alert a central station. The whole system can be powered from the electric wheelchair batteries. A general system diagram is shown in Fig. 1. As mentioned, the relevant variables for mobility impaired people are time between tilts, activity level, physiological variables such HR and RR, and ambient temperature and humidity. Monitoring those variables requires specialized sensors.

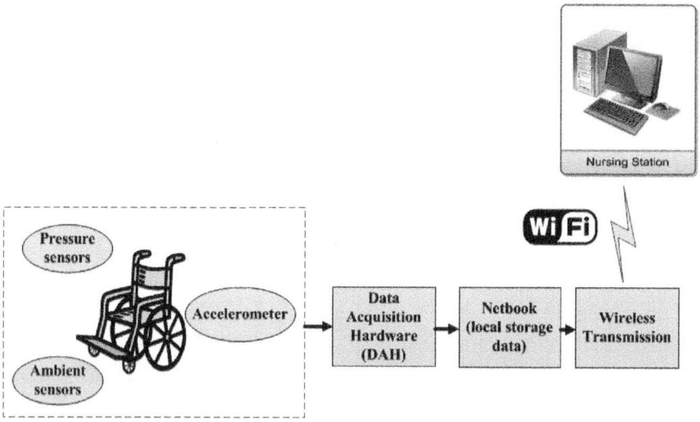

Fig. 1 General system diagram. Unobtrusive sensors on the wheelchair collect data and send them to a netbook for local processing and storage. Wireless capability allows relaying relevant information to a central station.

1.3.1 Time between Tilts

An accelerometer attached to the tilting structure of the wheelchair is able to measure angle to the horizontal plane by analyzing the g-force vector direction in static condition. Instead of measuring the actual tilt angle, to accommodate all wheelchair models and avoid complicated calibration procedures, it is preferable to monitor angle variations rather than absolute values.

1.3.2 Activity Level

The same accelerometer used for angle measurement serves as activity monitor. Activity is easily detected with the accelerometer. During wheelchair movement, the accelerometer shows forces in multiple directions and with varying amplitudes mainly due to floor imperfections.

1.3.3 Heart and Respiratory Rate

Heart and respiratory rates are measured with unobtrusive pressure sensors. Highly sensitive pressure sensors based on Electromechanical Films can detect BCG signals, usually mixed up with respiration signals. Less sensitive pressure sensors such as Force Sensing Resistors (force/resistance transducers) are also able to record respiration signals. In that case, BCG signals are too small to produce a noticeable variation. Both sensor types allow unobtrusive monitoring, by installing them in furniture that is used daily, such as beds or wheelchairs.

1.3.4 Temperature and Humidity

There are many commercially available ambient sensors that report temperature and humidity in both analog and digital format. As an ambient sensor, it can be installed in any convenient location.

2 Proposed System

Based on the general requirements for an assistive device for mobility impaired people, four prototypes were built on electric powered wheelchairs. The implemented system is shown in Fig. 2.

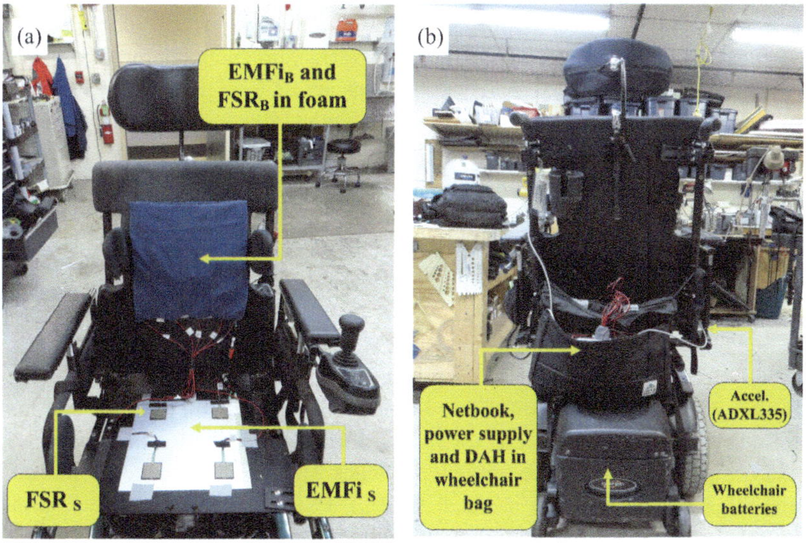

Fig. 2 Implemented system on electric powered wheelchair, (a) front and (b) back view. Unobtrusive sensors are installed on the wheelchair, under backrest foam and pressure relief cushion on the seat. Ambient sensor and accelerometer are attached to wheelchair structure, away from the user. The system is powered from the wheelchair batteries.

2.1 Sensor Selection and Acquisition Circuits

The prototypes carry an assortment of sensors deployed unobtrusively on the wheelchair to monitor vital signs, pressure relief behavior, patient activity and ambient conditions. To facilitate system acceptance, the key is to minimize impact in patient's daily life. Therefore, the selected sensors are able to capture all the information in a non-invasive way avoiding discomfort to the patient. Accordingly, the following sensors were selected: 3-axis accelerometer, EMFi, FSR, temperature and humidity sensor.

2.1.1 Accelerometer

The ADXL335 is a 3-axis accelerometer with ± 3g measurement range and analog output data. It is used to measure wheelchair tilt based on the static acceleration of gravity and detect wheelchair vibration, crashes and movement artifacts based on dynamic acceleration.

2.1.2 Electromechanical Films (EMFi)

An EMFi is a very sensitive sensor which can measure small pressure variations. EMFi consists of a film structured with flat voids separated by thin polyolefin layers. According to the sensor structure, EMFi sensors respond better to thickness change rather than strain. Thickness changes in the EMFi sensors generate a corresponding charge and hence, a voltage can be measured using a conditioning circuit.

Two EMFi sensors (model L-3030, 290x300 mm, manufactured by EMFIT Ltd.) were deployed on a wheelchair to measure BCG signals. Using this information, heart and respiratory rate can be calculated. One sensor was placed on the seat ($EMFi_S$), under a pressure relief cushion and another on the backrest, inside a thin foam ($EMFi_B$). Once the patient sits on the wheelchair, the system starts to collect respiration and heart activity without requiring any sensors on the skin, which is the main advantage over conventional ECG.

2.1.3 Force Sensing Resistor (FSR)

An FSR sensor is a polymer thick film which reduces its resistance when a force is applied in its surface. Nine small 44x38 mm FSR sensors (model FSR 406 manufactured by Interlink Inc.) are deployed on the wheelchair to measure prolonged pressure on the skin over the buttock and back area. Four sensors are put on the seat (FSR_S) and five on the backrest (FSR_B). FSR_B can also capture respiration activity based on thoraco-abdominal movement.

2.1.4 SHT15 Sensor

The SHT15 (manufactured by Sensirion) measures ambient conditions during outdoor and indoor activities. This information will be used to avoid dangerous heat exposure. This sensor integrates T_{AMB} and RH in one chip. Its digital output data are sent to the microcontroller unit (MCU) using a proprietary communication protocol.

2.2 Portable Processing Unit

A Data Acquisition Hardware (DAH) board captures the data from the different sensors and routes them to a netbook for storage and further processing. The DAH includes conditioning circuits for FSR and EMFi signals, an MCU to acquire, tag and pack the data and a serial transmission stage to the netbook.

The EMFi and FSR sensors require analog conditioning circuits for signal transduction, amplification and filtering before the analog to digital converter (ADC). The conditioning circuit for EMFi sensors was based on [27]. It consists of a charge amplifier followed by a second order Sallen-Key low pass filter with a cut off frequency of 30 Hz. The circuit has calibration potentiometers to modify offset, gain and sensor sensitivity. For FSR, the conditioning circuit is a current to voltage amplifier followed by another 30 Hz low pass filter. The amplifier for FSR_B is sensitive enough to capture respiration signals.

The information acquired by the sensors is sent into an ATxmega128A3 MCU running at 12 MHz. The main tasks of the MCU are sampling the signals from the ADC, establishing the SHT15 communication protocol and sending all the information to a netbook for further processing and storage. Each analog signal is sampled at 100 Hz using 11 bits of resolution. Ambient data from SHT15 sensor is read digitally using a two-wire serial protocol. T_{AMB} and RH are sampled at 0.1 Hz and their resolutions are 12 bits and 8 bits respectively.

Each sensor datum transferred to the netbook has 24 bits: 12 bit of data and 12 bit for a unique identifier used to unpack the data. There are 16 identifiers, one for each measured variable. The MCU sends the data to a netbook installed on a subject's wheelchair via serial communication at 57.6 Kbps. The devices are connected to each other by a RS-232 / USB adapter.

2.3 Power Supply

The system is fed from a 24V battery located under the wheelchair (Fig. 2). Normally, the 24V are composed of two 12V batteries with up to 70Ah. The battery voltage has a good regulation when the wheelchair is in use. DC Motors for operating the wheelchair are the main load and consume several amps without affecting the voltage when running. It is not necessary to install additional equipment (filters) to protect the electronic circuits against transients or voltage fluctuations on the batteries voltage. A schematic circuit is shown in Fig. 3.

The power consumption of the monitoring system must be low enough so as to not affect the wheelchair autonomy. The estimated power consumption of the whole system is shown in Table 2. In the prototype system, the highest power

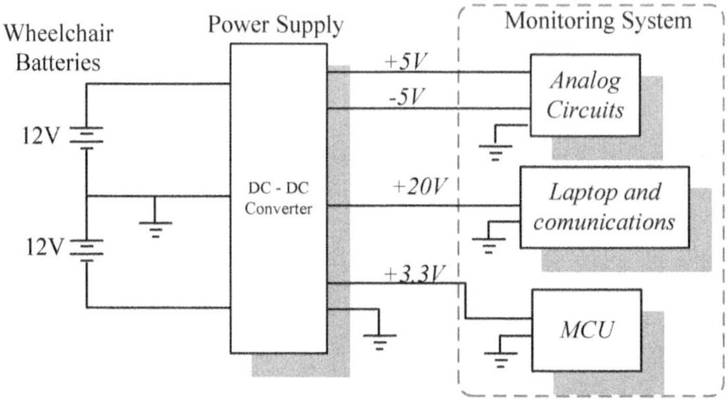

Fig. 3 Monitoring system power supply draws power from the electric wheelchair batteries. Multiple voltage outputs are necessary for different subsystems.

consumers are the netbook (24W) and the power supply (8W). A total consumption of almost 33W is obtained. In a new design, the netbook was replaced by a Raspberry PI® system that consumes only 3.5W with Wi-Fi communication. In this case, the power supply consumption is reduced to 1.4W to energize the whole system, yielding over 80% power reduction.

The power supply uses switching mode voltage regulators. These regulators have efficiencies between 70% and 90% when the voltage is regulated from 24 V to 3.3V and 5V. The low voltage required by the electronic circuits requires the use of Buck converters. Switching converters can affect communications systems because they generate electronic noise (EMI / RFI). For this reason, the power supply must be encapsulated in a Faraday cage.

2.4 Communications Systems, Data Transfer and Databases

The acquired data is stored and processed in the netbook mounted on the powered wheelchair. A Python script reads the raw data sent by the MCU through the serial port and writes it to a local PostgreSQL database [28] in the netbook. Another script unpacks the data according to the identifiers. The netbook will be used to generate alarms and patient recommendations in a stand-alone configuration and to send the data to a nursing station for caregiver's alarms in the final version.

The real time conditions envisioned for this project are 5:

1 Normal condition
2 Inadequate pressure relief
3 Excessive heat exposure
4 Abnormal HR (high or low)
5 Abnormal RR (high or low)

Table 2 Power system consumption

Device	Description	Voltage (V)	Power consumption (W)
Netbook and communication	A netbook with Wi-Fi communications. In the new design, this equipment is replaced by a Raspberry PI®	20V	24W
Power supply	Isolated and unisolated DC-DC converters. A switching mode power supply is recommended to increase the efficiency of the power system.	24V	8W
Analog circuits	Sensors and analog circuits for data acquisition and conditioning	±5V	1W
MCU	An ATxMEGA128 MCU is used. The consumption includes AD converters, serial communications and full 12MHz working clock.	3.3V	60mW
		Total	33W

Conditions 1 to 3 can be communicated directly to the wheelchair user, as voiced recommendations ("Please tilt your wheelchair") or alarms ("There is risk of heat stroke. Go back to a cooler location"). Depending on the situation, some or all of these alarms can be forwarded to a caregiver or family member via email notification, SMS messages or through pop-up windows in the nursing station.

Data from the netbooks is sent to the central station with maximum secrecy. No personal data is sent, only an arbitrary wheelchair identifier which can only be de-identified by authorized personnel (or relatives) at the server. Furthermore, all data communication is encrypted using ssh tunnels and public/private encryption keys, readily available in all modern operating systems.

At the central station, all data is logged into a main database. Here, secondary data processing occurs, which does not require real time performance. At this stage, long term trends are analyzed, such as activity level in the last 7 days, tilt compliance or dangerous exposure to heat. This also allows a stand-alone system to upload its information once a day if necessary. In the case of assisted living facilities or nursing homes with many residents, this also serves as a way to avoid overloading the wireless network during the day, when it may be in use for other purposes.

2.5 *Signal Processing and Algorithms*

After acquiring the raw data, several algorithms extract relevant information required to alarm the user, the caregivers or for off-line analysis and decision support.

2.5.1 Wavelet Decomposition

Raw EMFi signals are filtered to extract respiration and heart activity. To this end, a filter bank is implemented based on the Discrete Wavelet Transform (DWT).

The filter bank allows decomposing the signals in *n* levels using low-pass and high-pass filters designed with a specific mother wavelet. The original signal is decomposed in details (D_i) and approximations (A_i), where *i* is *i-th* decomposition level. D_i and A_i correspond to high and low frequencies, respectively. $EMFi_B$ and $EMFi_S$ are decomposed in 6 levels using a *symlet 5* and *daubechies 10* as mother wavelets, respectively. The mother wavelets were selected based on the similarity with the clean BCG waveforms from each sensor. Fig. 4 shows an example of wavelet processing. The raw $EMFI_B$ is decomposed into A_6 and D_1-D_6 levels. Fig. 5 shows BCG and respiration signals obtained from the selected D_i and A_i. Respiration signals are reconstructed using A_6 and BCG signals are reconstructed adding the details D_3, D_4 and D_5.

2.5.2 Respiratory Rate Algorithm

An algorithm for respiratory rate calculation was implemented on the respiration signal extracted from raw EMFi data. The algorithm is based on [29] and [30]. It detects inspiration and expiration cycles from respiratory waveform using a smoothed version of the original signal and extracting features such as area, width, maximum and minimum of each cycle. Once the features are calculated for both an inspiration and expiration cycle, the algorithm moves on to a decision stage to discriminate valid respirations from artifacts produced by movements. Two consecutive valid breaths are required to calculate RR.

2.5.3 Heart Rate Algorithm

An algorithm for heart rate calculation was implemented using BCG signals extracted from raw EMFi data. The algorithm consists of looking for possible beats in zones of BCG signal with high energy. Energy of BCG signals was calculated using a 30 ms sliding window and a peak detection algorithm was implemented to detect high energy peaks. When a peak is detected, a window of 700 ms is scanned in the BCG signal to detect potential beats. Maxima and minima of the BCG signal inside this 700 ms window are identified and the distance between consecutive peaks is calculated. The segment with the greater distance corresponds to a beat.

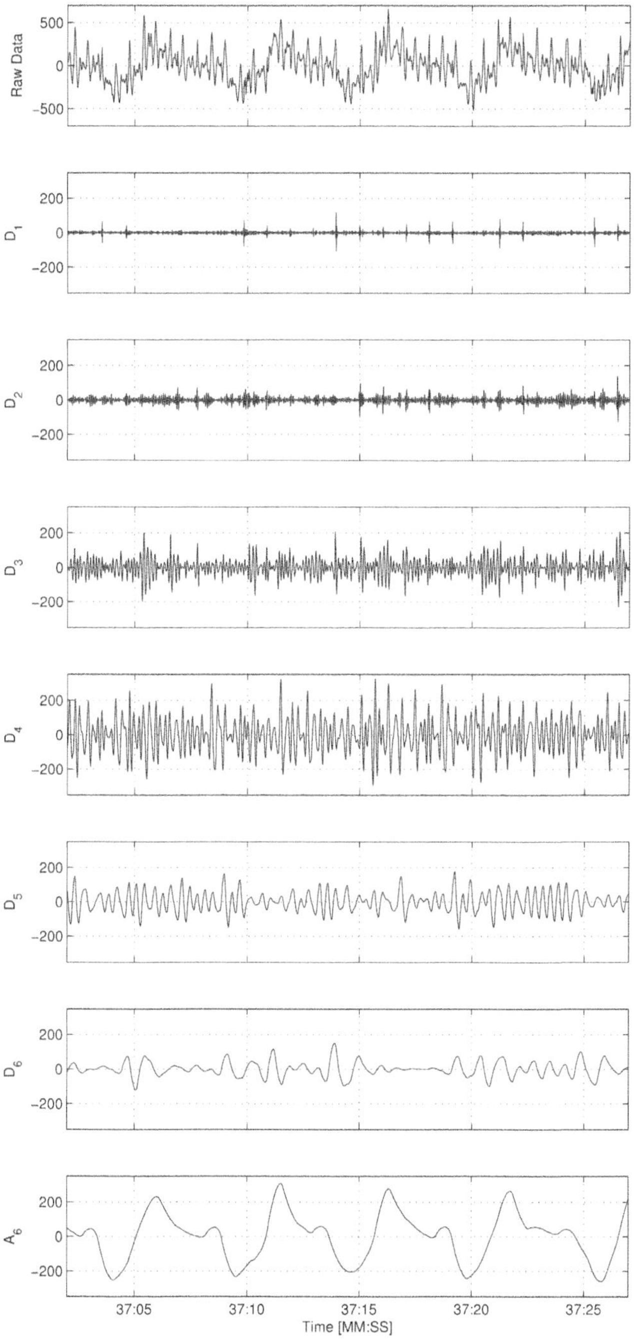

Fig. 4 Wavelet decomposition of raw data from $EMFi_B$ into A6 (approximation) and D1-D6 (details)

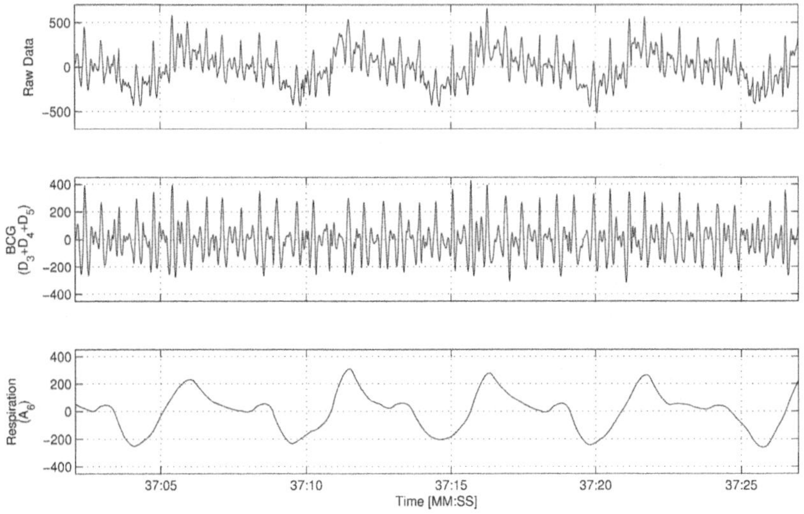

Fig. 5 Wavelet recomposition of BCG and respiration signal from selected components of the raw data

2.5.4 Pressure Relief Detection

A pressure relief by tilt (PRT) occurs when a wheelchair uses its tilt-in-space capability to produce pressure relief over the buttocks area. According to the study presented in [31], pressure relief is achieved if the skin is unloaded for at least 3:30 minutes, enough for the tissue to be oxygenated. For increased safety, in this project, a threshold of 5 minutes is used.

The PRT detection algorithm uses the accelerometer combined with the pressure sensors data. The reported angle is averaged using a 10 second sliding window to detect changes. When the tilt angle decreases by at least 5 degrees, which means that the wheelchair is tilted back, and the pressure sensors reduce their level sharply, the tilt is marked as possible PRT. To consider the tilt as an actual PRT, the subject must maintain this position or tilt the wheelchair even further for at least 5 minutes. Otherwise, the algorithm is restarted. When the subject returns to the initial position, the tilt is finished. In Fig. 6, a full day record shows 4 PRT detections. PRT are detected from the onset (▼) to their end (□). It is possible to observe how the pressure on the seat is transferred to the back during a PRT (Fig. 6 a, b). The pressure sensors are also used to determine wheelchair occupancy and for user activity estimation. In Fig. 6 the volunteer is on the wheelchair for 6:41 hours.

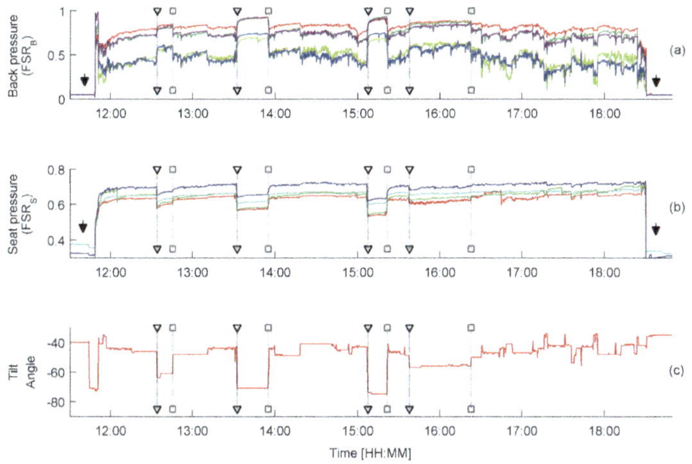

Fig. 6 PRT detection algorithm. (a) FSR_B, (b) FSR_S and (c) tilt angle obtained from the accelerometer. Time between ▼ and □ marks indicate a valid PRT. The beginning and end of wheelchair occupancy is also deduced from the pressure sensor data.

2.5.5 Activity Level Estimation

An Activity Index (AI) was introduced to quantify the activity level of each subject during the time spent on the wheelchair. AI is defined as the time showing high activity divided by the total wheelchair occupancy time. The AI estimate is based on the variations of the accelerometer data which reflect the subject's movement and the wheelchair vibration during driving. To extract this component, the standard deviation of the accelerometer data is calculated using a sliding window. High standard deviation values indicate high activity segments. A fixed threshold of 0.5 was used to distinguish between high and low activity, as shown in Fig. 7.

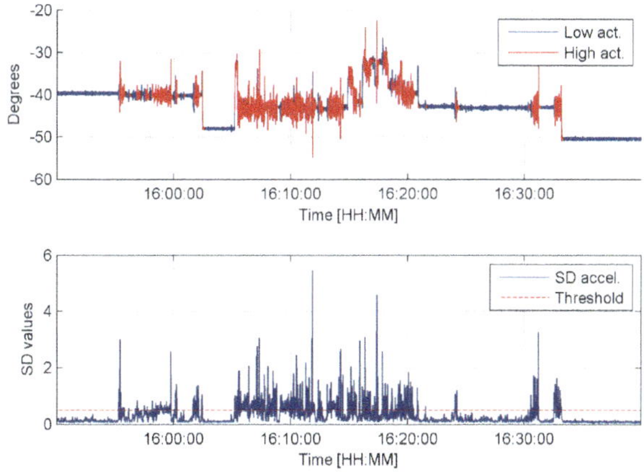

Fig. 7 The accelerometer data (top) is used to estimate the activity level for each subject. Highlighted segments show high activity periods (e.g. driving, tilting or moving). When the standard deviation of the accelerometer data (bottom) is low, there is no activity (e.g. resting, wheelchair stopped).

2.6 Common Issues

Raw EMFi signals are affected by artifacts due to a patient's movement and wheelchair vibration during driving. When this occurs, despite the filtering process, the artifacts hide the underlying signal completely, rendering the RR and HR algorithms useless. For this reason, a noise detection stage is implemented in each algorithm to suspend its execution when noisy signals are detected.

The noise detection stage uses the activity level estimation presented in 2.5.5. When high activity is detected, the algorithms are suspended until low activity is detected again. The noise detection stage is also complemented by analyzing raw EMFi data to discard saturated signals produced by sensor disconnections and movements that are not captured by the accelerometer. Signals are considered saturated if the maximum value found in a 5 second window is equal to 2^n, where n is the resolution of the AD converter used for signal acquisition (in this case, 11 bits). Fig. 8 shows how the noise detection stage works. During high activity or when the signal is saturated, the algorithm to detect RR and HR is not running. The FSR sensors are also used to stop the algorithms when the subjects are out of the wheelchair or when the subjects do not lean against the sensors placed on the wheelchair backrest.

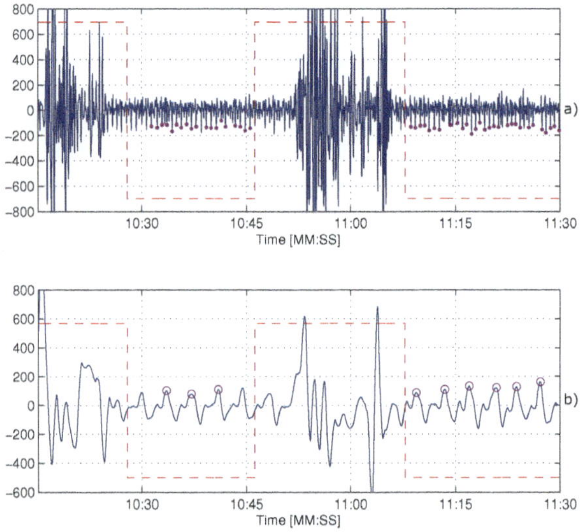

Fig. 8 When noisy segments are detected, RR and HR algorithms are suspended. Peak detection resumes once the signal returns to normal levels

A second issue detected is wireless network failure due to wandering out of coverage while on Wi-Fi or network downtime. In this case, local storage capability proves useful. Once the network is re-established, data forwarding can be resumed. It is also advisable that the system is able to work in a stand-alone mode, without network connectivity. Most of the powered wheelchair users are able to take actions to overcome a situation. It can be expected that tilting reminders and some high HI alarms are going to be the most common events. In those cases, a stand-alone system is sufficient to remind or request some action from the user. Other data can be saved locally for off-line analysis or forwarded to the central station at the earliest convenience.

3 Experience with MS Patients: A Pilot Study

Volunteers for a pilot study were recruited from The Boston Home (TBH), a specialized care residence for people with advanced MS and other neurological diseases. The inclusion criteria were full time wheelchairs users with severe mobility impairment, using electric-powered wheelchairs equipped with a tilt-in-space system and a pressure relief cushion.

3.1 Study Protocol

The prototypes are installed on the wheelchairs at the beginning of the study. Every day the prototypes are checked to detect and correct any technical issues before the volunteers start to use their wheelchairs. Prior to the data collection, the participants are asked if the sensors deployed on the wheelchair's backrest and/or seat produce discomfort. Then, the prototypes start to collect data continuously during the whole time the patient spends on the wheelchair. As the data collection is performed in a nursing home, researchers contact the nursing staff in case of any problems.

To validate the information obtained from the BCG signals, once a day the residents were requested to use a Nonin® finger clip pulse oximeter to acquire HR and SpO_2 level. This sensor was used during short periods of time to minimize discomfort to the residents. This protocol was approved by The MIT Committee on the Use of Humans as Experimental Subjects (COUHES).

3.2 Activity Reports

Table 3 shows the wheelchair occupancy and the activity level from all subjects. On average, the participants spent 6:23 ± 1:43 hours per day in their wheelchair. The maximum wheelchair occupancy registered was 10:36 hours and the minimum was 2:11 hours. The distribution of wheelchair occupancy for all subjects is shown in Fig. 9. A total of 39 days of data were collected. The subjects presented an Activity Index (AI) level under 10% in 8 days (31.8%). Most cases (29 days), the subjects presented an AI between 10% and 30%. Only on 2 days (2.2% of time) did the subjects present an AI greater than 30%. Fig. 10 shows the AI level and the wheelchair occupancy registered for S1 during 9 days. S1 occupied his/her wheelchair on average 7:28 ± 0:55 hours per day and the AI level was between 20% and 30% during the 88.9% of monitored days, except for day 3 where the AI level reached 48%. The increase in the AI is coincides with the day when S1 left the nursing home to do some errands.

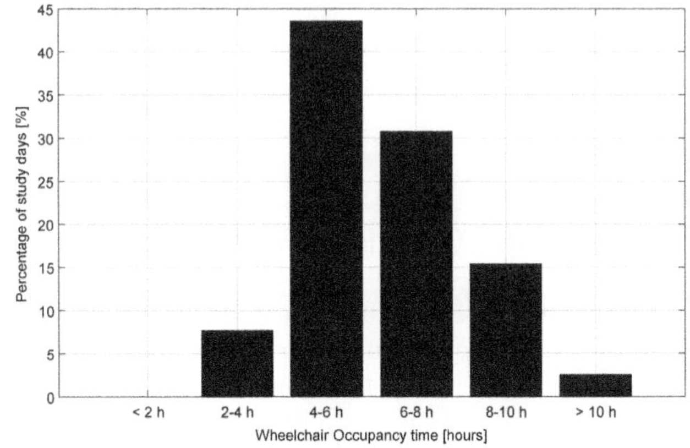

Fig. 9 Wheelchair occupancy distribution from all subjects during the study days

Table 3 Wheelchair occupancy and the activity level from all participants

Subject (S)	N° study days	Wheelchair occupancy [hh:mm]					Activity index [%]				
		min	*max*	*mean ± sd*			*min*	*max*	*mean ± sd*		
S1	9	6:36	9:28	7:28	±	0:55	20.4	48.5	26.8	±	8.7
S2	11	4:23	6:57	5:27	±	0:51	8.5	27.3	13.7	±	5.3
S3	6	2:10	5:06	4:11	±	1:08	14.7	29.9	23.2	±	5.8
S4	5	5:22	10:35	8:03	±	2:09	7.6	61.0	21.0	±	22.5
S5	4	3:55	4:36	4:19	±	0:17	8.0	24.0	15.3	±	8.5
S6	4	5:42	9:11	7:52	±	1:30	5.0	10.0	7.5	±	2.1

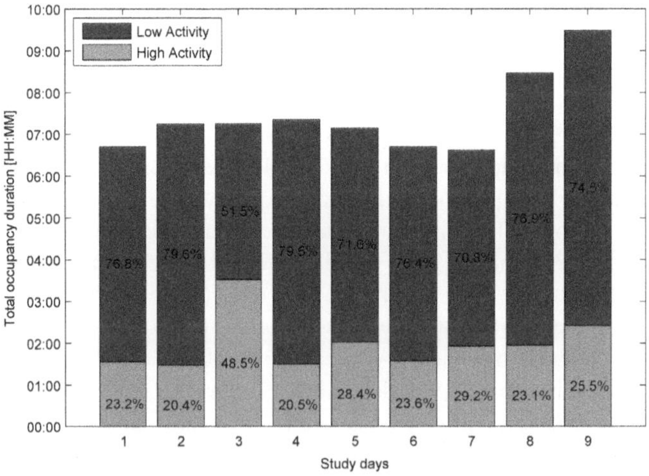

Fig. 10 Wheelchair occupancy and AI level registered by S1 during 9 days of study

3.3 Tilt Compliance

There are several clinical recommendations about the frequency pressure relieving maneuvers. Some guidelines for prevention of PU recommend performing a pressure relief every 15 minutes [18], other guidelines recommend longer intervals such as every 1 hour or every 2 hours [19]. Only 2.1% of PRT performed by the subjects meet the recommendation if the interval is every 15 minutes, 25.8% if the interval is every 1 hour and 57.7% if the interval is every 2 hours. Fig. 11 shows a histogram of the intervals between tilts for each subject. S1 and S2 logged the shorter intervals. 18.9% of the time intervals for S1 and 5.1% for S2 were under 1 hour. For S1, most of the intervals (41.7%) were between 1 and 2 hours. On the other hand, more than 50% of the intervals spent by the rest of the subjects (S2 to S6) were longer than 3 hours, against all recommendations. The maximum was registered by S6 reaching 93.9% of the wheelchair time in the same position (tilt angle). The average duration in a PRT position is 1:34 hours. However, those who perform frequent PRT spend a shorter time in a tilted position than the subjects who don't. On the other hand, S6 spent around 4:30 hours in a tilted position.

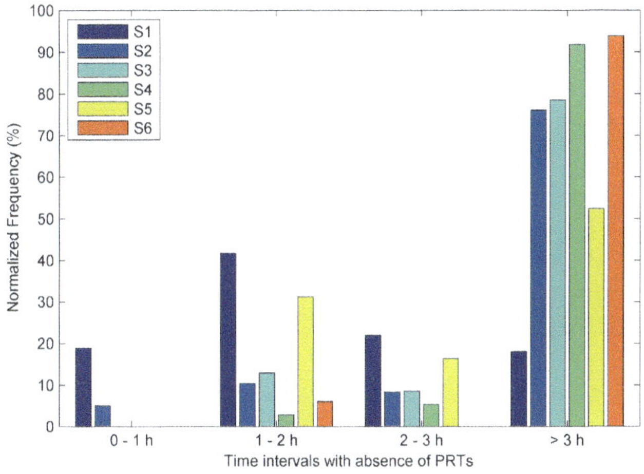

Fig. 11 Time intervals with absence of PRTs

3.4 Vital Signs Monitoring

Fig. 12 shows HR and RR trends obtained from one of the volunteers when his/her wheelchair was stopped. This information is useful to improve patient's care through assessing their vital signs continuously to quickly assist in case of emergency. Also, measuring vital signs allows knowing patients normal ranges which can be different from healthy people due to their impaired conditions.

During data collection, it is usual to observe the residents resting in their wheelchair or taking a nap. Fig. 13 shows an EMFi record of volunteer S3 while taking a nap, where it is possible to detect an 18 s apnea, condition that would be very difficult to detect otherwise.

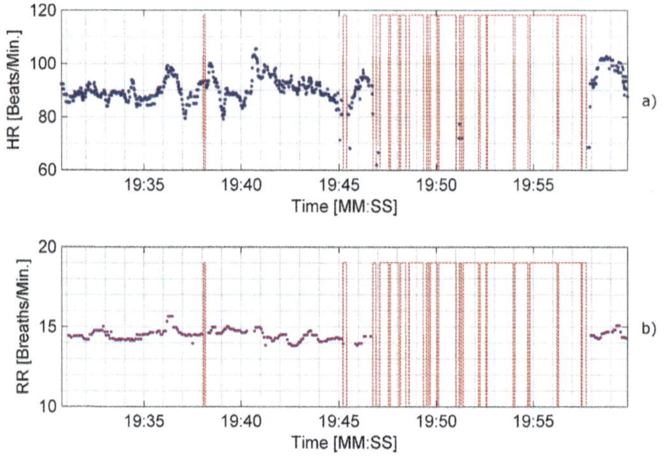

Fig. 12 HR (a) and RR (b) as detected by the algorithms. Red lines mark zones where it was not possible to estimate vital signs due to noise.

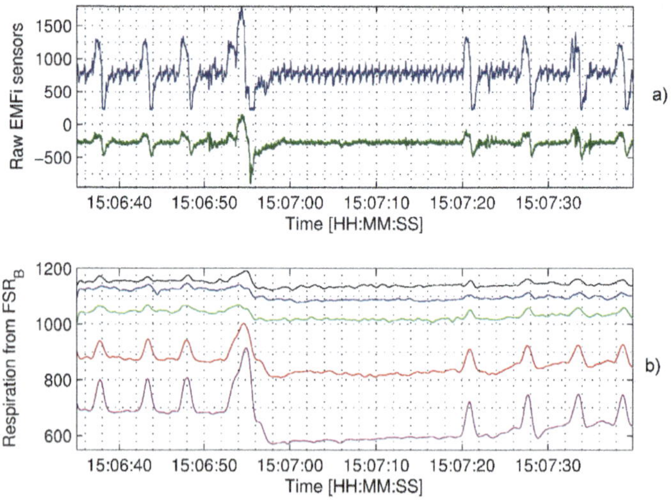

Fig. 13 Apnea captured by EMFi sensors. A clean BCG from the heart activity can be appreciated in (a) during the respiratory pause.

3.5 *Ambient Conditions*

People with a high level of disability, such as MS patients, are very sensitive to heat. For this reason, extra care is necessary when the patients spend time outside during the summer. The implemented system allows the continuous monitoring of ambient conditions, information that can easily be used for an alarm system.

Fig. 14 shows a wheelchair user who went outdoors for over 1 hour. In the middle of the summer, the temperature can rise above 40 °C. Depending on the RH, the HI and the Dew point can reach dangerous values.

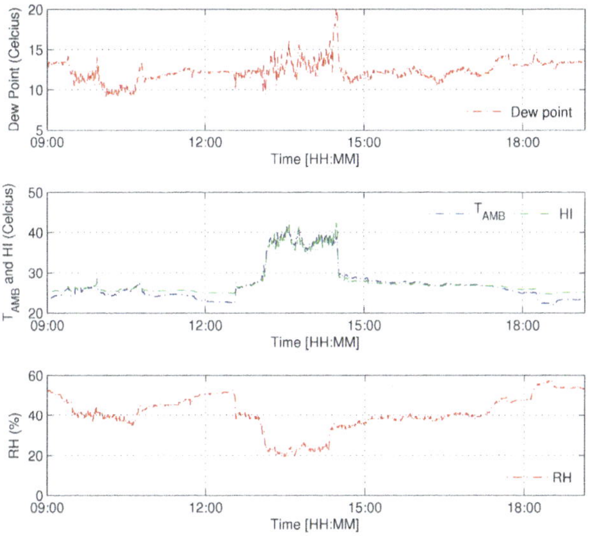

Fig. 14 HI and dew point values as a wheelchair user moves outdoors after lunch, in the summer

4 Future Directions

Current advances in sensors, networks and computer science are expanding the possible range of applications in mHealth that can be implemented. The monitoring system presented here for severely impaired people is just one possible application, but we believe it is headed in the right direction. We feel that the main goals in future mHealth developments should be: 1) unobtrusiveness, 2) pervasiveness and 3) robustness.

In order to provide supporting technologies for better quality of life, next generation systems should be 'invisible' and avoid any interference on daily living activities. Sensors should act behind the scenes, like the pressure sensors presented in this application. Ambient sensors, sensors embedded in everyday items such as beds, chairs, driving wheels, etc. and/or wearable sensors, coupled with advanced and powerful information processing systems will be able to provide just-in-time recommendations, warnings or alarms to prevent adverse events or to improve daily living in a variety of settings.

Pervasiveness of future mHealth applications is very desirable to provide constant monitoring and advice. The continuous drop in technology prices, or the increase in technology at constant price, allows us to envision an expansion of application settings, such as monitoring systems for independent patients with chronic diseases at home, in schools and gyms to detect at risk people on time, at work in adverse ambient conditions and in rural areas or in currently 'low tech' health centers to provide better diagnostic tools and decision support. This possible physical expansion of mHealth applications will provide basic health screening capabilities in non-standard settings, decreasing patient requirements for treatment in traditional health centers. Again, using the system presented as an example, pervasively monitoring physiological variables and general activity information would bring valuable insight to people with disabilities for deciding whether an appointment with a physician is due or an emergency visit to the hospital is required. Even during the visit, the pervasiveness of the system can prove useful by providing a large amount of previously recorded health data to assist the caregiver taking a prompt and correct action based on a correct diagnosis.

To support these advances, more research is necessary in algorithm robustness. Unobtrusive and pervasive systems mean obtaining data and making decisions based on low quality information. Clearly, as seen in our pilot study, there are certain conditions where the system is unable to properly monitor the physiological variables due to the noise level. Even after filtering out the intervals detected as noisy, the RR and HR series present errors. Better noise detection algorithms combined with better signal processing techniques will help making mobile monitoring systems more robust in the future.

5 Discussion and Conclusion

The proposed system allows the capture of physiological data without causing discomfort to the patient. The system is set up on a wheelchair by deploying unobtrusive sensors in a pillow or a piece of foam and under a pressure relief cushion to avoid direct contact. This way, patient stress is kept to a minimum, and the system can be used to capture relevant information during long-term monitoring. These features are key for patient acceptance.

However, the main problem for monitoring and assessing vital signs in daily life is noise. EMFi sensors which were used to measure BCG signals and then obtain HR and RR trends, are very sensitive. This sensitivity, which is required to detect small signals such as BCG, render the system useless in noisy conditions such as when the wheelchair is moving or the patient is changing positions. To achieve more accurate results in the RR cycle detections, a signal quality estimation algorithm is essential to decide when the algorithms will perform adequately. However, during periods of low activity, the algorithms are able to provide valid information regarding main physiological variables such as RR and HR.

The implemented system can also capture pressure changes and tilt usage in a non-invasive way, continuously during daily activities. Combining the data acquired from FSR sensors and the accelerometer, it is possible to detect when the patients relieve pressure by tilting. Analyzing the information collected by our system, useful parameters are obtained such as wheelchair occupancy, activity level, PRT per day, time spent in a tilted position, intervals without performing a PRT, etc., which can be useful for providing objective information to clinicians about pressure relieve behavior.

The results also show that the subjects did not relieve pressure as recommended using the tilt-in-space system. They spent long periods (over 3 hours) without performing a PRT. In addition, during the study, there were entire days when the subjects did not perform PRT. If these situations recur frequently and the subjects spend long hours on the wheelchair without relieving pressure, the probability of developing PU increases.

An interesting unintended finding was the apnea detection. Caregivers at the living facility also showed interest in evaluating this particular system with regard to sleep apnea, mainly because polysomnography studies are particularly complicated for this group of people. A reliable apnea screening system helps improve the quality of care delivered to their residents.

The pilot study also confirmed the need and the feasibility of implementing personalized alarms or reminders for different aspects of activities of daily living that may lead to danger or health complications. It is also clear that this system can be readily used in a variety of settings such as home care, nursing homes and/or assisted living facilities.

Acknowledgments. The authors would like to thank to the MIT International Science & Technology Initiatives and the Biomedical Engineering program at Universidad de Concepción. The authors would like to express our gratitude to Professor Seth Teller and the Robotics, Vision and Sensor Networks (RVSN) for his advice and support. We also would like to thank Don Fredette and The Boston Home for being so enthusiastic about this project.

References

1. World Health Organization. World report on disability, Geneva, p. 325 (2011)
2. Murray, C.J.L., Vos, T., Lozano, R., Naghavi, M., Flaxman, A.D., Michaud, C., et al.: Disability-adjusted life years (DALYs) for 291 diseases and injuries in 21 regions, 1990-2010: a systematic analysis for the Global Burden of Disease Study 2010. Lancet 15(380), 2197–2223 (2012)
3. Avenel-audran, M., Goossens, A., Zimerson, E., Bruze, M.: Contact dermatitis from electrocardiograph-monitoring electrodes: role of p-tert-butylphenol-formaldehyde resin. Contact Dermatitis. Munksgaard International Publishers 48, 108–111 (2003)
4. Pino, E.J., Arias, D.E., Aqueveque, P., Vilugron, L., Hermosilla, D., Curtis, D.W.: Monitoring technology for wheelchair users with advanced multiple sclerosis. In: 2013 Annu. Int. Conf. IEEE Eng. Med. Biol. Soc., pp. 961–964. IEEE (2013)

5. Pino, E., Arias, D., Aqueveque, P., Curtis, D.: Assistive Devices for HealthCare: Multiple Sclerosis. In: AMIA 2012 Annu. Symp. 2012, p. 1899 (2012)
6. Arias, D.E., Pino, E.J., Aqueveque, P., Curtis, D.W.: Data Collection Capabilities of a New Non-Invasive Monitoring System for Patients with Advanced Multiple Sclerosis. In: AMIA 2013 Annu. Symp. 2013, pp. 61–68 (2013)
7. Dicianno, B.E., Arva, J., Lieberman, J.M., Schmeler, M.R., Souza, A., Phillips, K., et al.: RESNA position on the application of tilt, recline, and elevating legrests for wheelchairs. Assist. Technol. Off. J. RESNA 21, 13–22 (2009)
8. Byrne, D.W., Salzberg, C.A.: Major risk factors for pressure ulcers in the spinal cord disabled: a literature review. Spinal Cord. Nature Publishing Group 34, 255–263 (1996)
9. Hagisawa, S., Ferguson-Pell, M., Herbert, J.: Reactive Hyperemia Response to Pressure-Induced Localized Ischemia in Patients with Multiple Sclerosis. Neurorehabil Neural Repair 8, 193–201 (1994)
10. Howard, R.S., Wiles, C.M., Hirsch, N.P., Loh, L., Spencer, G.T., Newsom-Davis, J.: Respiratory involment in multiple sclerosis. Brain, 479–494 (1992)
11. Jadidi, E., Mohammadi, M., Moradi, T.: High risk of cardiovascular diseases after diagnosis of multiple sclerosis. Mult. Scler. 19, 1336–1340 (2013)
12. Myers, J., Lee, M., Kiratli, J.: Cardiovascular disease in spinal cord injury: an overview of prevalence, risk, evaluation, and management. Am. J. Phys. Med. Rehabil. 86, 142–152 (2007)
13. Chester, J.G., Rudolph, J.L.: Vital signs in older patients: age-related changes. J. Am. Med. Dir. Assoc. Elsevier 12, 337–343 (2011)
14. Marino, F.E.: Heat reactions in multiple sclerosis: An overlooked paradigm in the study of comparative fatigue. Int. J. Hyperth. 25, 34–40 (2009)
15. Wilsmore, B.R., Cotter, J.D., Bashford, G.M., Taylor, N.A.S.: Ventilatory changes in heat-stressed humans with spinal-cord injury. Spinal Cord 44, 160–164 (2006)
16. Henke, A.F., Cohle, S.D., Cottingham, S.L.: Fatal hyperthermia secondary to sunbathing in a patient with multiple sclerosis. Am. J. Forensic Med. Pathol. 21, 204–206 (2000)
17. Kohlmeier, R.E., DiMaio, V.J., Kagan-Hallet, K.: Fatal hyperthermia in hot baths in individuals with multiple sclerosis. Am. J. Forensic Med. Pathol. 21, 201–203 (2000)
18. Consortium for Spinal Cord Medicine. Pressure Ulcer Prevention and Treatment Following Spinal Cord Injury: A Clinical Practice Guideline for Health-care Professionals. Paralyzed Veterans of America, p. 80 (2000)
19. Porter, R.S., Kaplan, J.L., Homeier, B.P.: The Merck Manual Home Health Handbook. Merck (2009)
20. Tawashy, A., Eng, J., Lin, K., Tang, P., Hung, C.: Physical activity is related to lower levels of pain, fatigue and depression in individuals with spinal-cord injury: a correlational study. Spinal Cord 47, 301–306 (2008)
21. Zhang, Y.-T., Zheng, Y.-L., Lin, W.-H., Zhang, H.-Y., Zhou, X.-L.: Challenges and opportunities in cardiovascular health informatics. IEEE Trans. Biomed. Eng. 60, 633–642 (2013)
22. Baek, H.J., Chung, G.S., Kim, K.K., Park, K.S.: A smart health monitoring chair for nonintrusive measurement of biological signals. IEEE Trans. Inf. Technol. Biomed. 16, 150–158 (2012)
23. Folke, M., Cernerud, L., Ekström, M., Hök, B.: Critical review of non-invasive respiratory monitoring in medical care. Med. Biol. Eng. Comput. 41, 377–383 (2003)
24. Johnson, R.: Warm, sultry weather and multiple sclerosis. Weather 57, 395–396 (2002)

25. Johnson, R.: The influence of dew-point temperature on multiple sclerosis symptoms. Weather 60, 170–170 (2005)
26. Rothfusz, L.P.: The heat index equation (or, more than you ever wanted to know about heat index). Technical Attachment, National Weather Service, 90–23 (1990)
27. Junnila, S., Akhbardeh, A., Värri, A.: An Electromechanical Film Sensor Based Wireless Ballistocardiographic Chair: Implementation and Performance. J. Signal Process. Syst. 57, 305–320 (2008)
28. The PostgreSQL Global Development Group. PostgreSQL database [Internet]., `http://www.postgresql.org` (cited August 5, 2013)
29. Chen, L., McKenna, T., Reisner, A., Reifman, J.: Algorithms to qualify respiratory data collected during the transport of trauma patients. Physiol. Meas. 27, 797 (2006)
30. Lu, W., Nystrom, M.M., Parikh, P.J., Fooshee, D.R., Hubenschmidt, J.P., Bradley, J.D., et al.: A semi-automatic method for peak and valley detection in free-breathing respiratory waveforms. Med. Phys. AAPM 33, 3634–3636 (2006)
31. Coggrave, M.J., Rose, L.S.: A specialist seating assessment clinic: changing pressure relief practice. Spinal Cord 41, 692–695 (2003)

Questions

Q1. This system needs to connect any sensor to the patient to collect physiological data?

Q2. The system can be used only inside a nursing home?

Q3. What is the difference between EMFi and FSR measurements in this system?

Q4. The power consumption of the monitoring system reduces the autonomy of the battery powered wheelchair?

Q5. Can the accelerometer be used to measure the movement of the body?

Q6. Does the system generate alarms for the patient, the nursing personnel or both?

Q7. If an alarm occurred, can the patient be located quickly?

10. Remote Health/Vital Sign Monitoring

mHealth and Remote Vital Sign Monitoring: Trends and Applications for ECG Analysis on Cell Phones

Oleg Starostenko[1], Vicente Alarcon-Aquino[1], Jorge Rodriguez-Asomoza[1], Oleg Sergiyenko[2], and Vera Tyrsa[3]

[1] Department of Computing, Electronics and Mechatronics,
Universidad de las Americas Puebla, Cholula, Pue., México
{oleg.starostenko,vicente.alarcon,jorge.rodriguez}@udlap.mx
[2] Engineering Institute, Autonomous University of Baja California, Mexicali, BC, Mexico
srgnk@iing.mxl.uabc.mx
[3] Universidad Politécnica de Baja California, Mexicali, BC, Mexico
vera_tyrsa@upbc.edu.mx

Abstract. This chapter presents analysis of emerging mHealth applications as well as the exploration of novel trends supporting healthcare intelligent environments assisted by mobile devices. The case of study is mHealth and remote vital sign monitoring. Particularly, we present a methodology for recollecting, processing and real-time monitoring heart activity with the main purpose to interpret electrocardiogram ECG signals, detect and manage situations of risk and provide the interaction between medical practitioner and patient into smart healthcare environment. The proposed architecture and approach provide continuous detection and interpretation of the patient's QRS complex. The challenge is to adapt some approaches for data gathering, processing, compression, storage, analysis, and visualization to capabilities of mobile devices. The designed system for monitoring vital signals has been tested using standard MIT-BIH Arrhythmia Database achieving satisfactory ECG interpretation accuracy with relative error in range from 4 % to 10 % for signal sampling frequency of 360 and 128 samples per second respectively. It is important to note that the proposed prototype does not substitute diagnosis by physician. Our intention is to propose methodologies that serve as guide for development of complex health assistance tool expanding coverage of medical services.

Keywords: mobile healthcare environments, remote vital signal monitoring, ECG signal processing, QRS complex recognition.

1 Mobile Health Applications: Reality or Fantasy

In recent years computing power and capabilities of mobile devices and particularly of cell phones have increased considerably with the addition of new processors, video cameras, sensors, and actuators. Development of mobile operating systems

S. Adibi (ed.), *Mobile Health*,
Springer Series in Bio-/Neuroinformatics 5, DOI: 10.1007/978-3-319-12817-7_10

such as Apple iOS, Google Android, Windows Phone 8 from Microsoft, and BlackBerry 10 makes cell phones capable of performing complex tasks previously restricted to these devices. Smart phones are now frequently used for remote control of electronic devices, high speed and reliable multimedia data exchange, information acquisition and dissemination from sensor networks, enhanced services in wireless environments, etc. (Nieto 2011; Starostenko 2013; Want 2009).

Mobile devices and cell phones, designed to accompany users anywhere and keep them in touch at all times, may provide users with access to health services and medical care applications. These systems and applications for mobile health (mHealth) are oriented to care of different aspects of monitoring and maintaining the good health of patients. The key applications and mobile services related to mHealth usually include remote data collection, remote monitoring, communication and data exchange with healthcare personnel, and diagnostic or treatment support. There are several platforms available for developing mobile applications; among the most popular are Binary Runtime Environment for Wireless (BREW) Software Development Kit (SDK), Adobe Air, BlackBerry JDE, DragonRAD of Seregon Solutions, IBM Worklight, Java Micro Edition (Java Me), previously J2ME, and more recently Android developed by Open Handset Alliance. iOS, Windows Phone, BlackBerry, and Android are complete development platforms that provide set of API libraries reused in multiple applications (Curtis et al. 2009; Muñoz-Ramos 2013; Pawar 2012).

Nowadays mHealth can be seen as the future of low-cost healthcare. Mobile devices used in mHealth can be categorized as portable by user or as data access tool. The portable or wearable by user devices operate usually in Wireless Body Area Networks (WBANs) or Wireless Personal Area Networks (WPANs), which are available communication technologies closely connected to users. Additionally, Bluetooth (BT) and Wireless Sensor Networks (WSN) are used for obtaining raw data from the human body using portable sensors. Data access tool provides information exchange with remote computers that may be located within a wireless local area network (mainly Wi-Fi) or on the Internet.

One important area of remote data collection and vital signal monitoring on mobile devices is the control of heart diseases because they have high incidence among population. Usually, personal mHealth monitoring systems support the rehabilitation process of patients after heart surgery or recovering from a heart attack (Gay 2009 and Tzu-Hao 2013), recollecting patient's electrocardiograms ECG, heart rate, and oxygen level in the blood (Kyriacou 2009 and Zhang 2013), remote control of cardio implants (Biotronik 2013), tracking of patients by GPS in case of emergency (Belgacem 2009), expanding coverage of medical services to rural communities (Kulkarni 2008), real time interaction and communication between physicians and patients (CardioNet 2013), providing teleconsulting, treatment compliance, mobile patient monitoring (Pawar 2012), and others for diagnosis assistance (AliveCor 2013; Handy 2013; IMEC 2013).

For vital sign monitoring specific data processing algorithms and methods are required. The widely used algorithms for analysis of heart activity are based on interpretation of ECG, detection of QRS complex, heart rate, and other vital signs

(Aghajan 2012; Muñoz-Ramos 2013; Zheng 2013). The well-known approaches include neural networks, genetic algorithms, wavelet transforms, heuristic methods, etc. (Alarcon-Aquino 2009; Zhang 2013, Yun-Hong 2011). However, the complexity of these approaches and significant amount of computation during ECG analysis do not permit direct utilization of cell phones supporting health care services. Therefore, instead of aforementioned data processing approaches simplified algorithms are suggested to apply. They use filters, analyze amplitude and slope of ECG signal, operate with thresholds and gradient to detect the QRS complex in systems with limited resources of cell phones (Muñoz-Ramos 2013 and Tzu-Hao 2013). For example, the most used approaches for detection of the QRS complex in real time are the Pan-Tompkins and Hamilton algorithms based on signal filtering and computing derivatives (Pan 1985; Hamilton 2002).

Table 1 Mobile healthcare systems and their characteristics

System, device	Data acquisition	Output parameters	Used Algorithms	Platform
Personal Health Monitoring (Gay 2009)	ECG via Bluetooth	heart rate, arrhythmia detection	Hamilton	Windows Mobile
Mobile Personal ECG Monitoring System (Belgacem 2009)	temperature, ECG, pressure via Bluetooth	ECG samples	retransmission to server	---
Wireless Children Monitoring (Kyriacou 2009)	ECG via Wireless	heart rate, arrhythmia detection	Hamilton	Windows Mobile
WE-CARE system (Zheng 2013)	ECG via Wireless	QRS complex	Multi-Resolution SVM	---
AliveECG (AliveCor 2013)	ECG via Wireless	heart rate	variable sampling	iOS, Android
Handy Sana (Handy 2013)	ECG, Embedded	heart rate, blood pressure, glucose	retransmission to server	---
Mobile Heart Monitoring (IMEC 2013)	ECG via Bluetooth	heart rate	retransmission to server	Android
Mobile Biotelemetry Syst. (CardioNet 2013)	ECG via Bluetooth	heart rate, arrhythmia detection	100-1000 sps*	Windows Mobile
HeartToGo (Kulkarni 2008)	ECG via Bluetooth	heart rate, arrhythmia detection	Pan Tompkins, 300 sps	Windows Mobile

*sps stands for samples per second

In order to determine the more efficient prototypes that will be used as a basis for development of new systems for real-time patient monitoring and ECG interpretation in mobile environments, the detailed analysis of well-known reports has been done. Particularly, we considered such important characteristics of prototypes as developing and programming platform, communication technology, designing methodology, number of ECG leads, format of input data and sampling frequency, performance and precision of used algorithms, speed and complexity of data processing, and visualization capabilities. Table 1 resumes several relevant commercial systems for heart activity monitoring.

2 Generalized Architecture of Healthcare Monitoring System

Currently, mHealth applications operate within intelligent environments that combine physical infrastructure (sensors, actuators, displays, smart computing devices, networks) with related software, services, interfaces, and humans that interact, sense, and communicate with each other (Aghajan 2012 and Pawar 2012). After performing the analysis of healthcare systems mentioned in Table 1, the structure of health-monitoring system on mobile devices may be introduced by three-level generalized architecture containing physical layer of data acquisition, mobile device-server communication layer, and application layer for data processing and visualization. The proposed architecture is shown in Fig. 1.

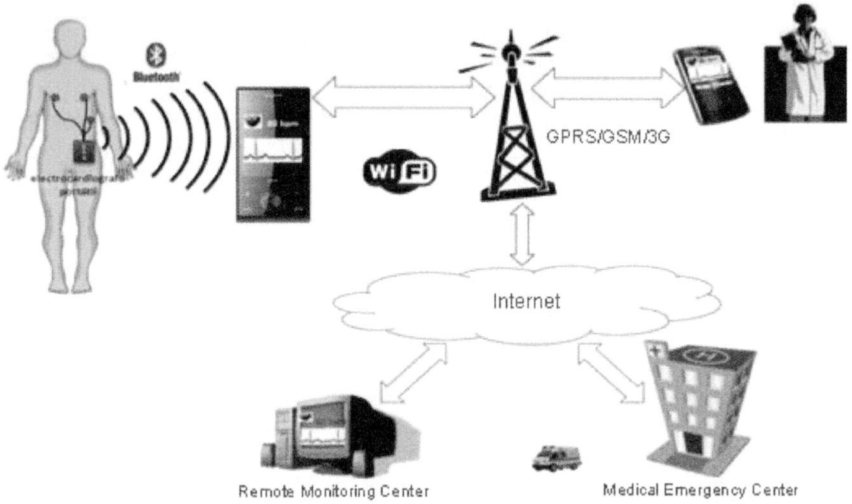

Fig. 1 Generalized architecture of healthcare system assisted by mobile devices

The input data acquired by sensors represent patient's vital and physiological signs for example, ECG, heart rate, blood pressure, temperature, volume of oxygen in the blood, etc. A mobile device operates as communication node for connected sensors, preprocessing recollected data before transmitting them to remote monitoring devices or to emergency centers. The analysis of data acquired by sensors has the purpose purpose of detecting conditions that may affect the health of patient such as an abnormal heart rate, falls, and failures in the sensors etc. Thus, if dangerous situations are detected, the mobile phone sends notifications and alarms to emergency services looking for help and providing medical care in the shortest time. The preprocessing stage is used for data sampling, analog-to-digital conversion, compression, storage, and visualization on mobile devices. Data processing is provided usually on server side because mobile devices, particularly cell phones, have restricted computing capabilities and storage. This implementation technique used for exploiting mobility and access to information and services is called server-thin-client approach. Additionally, a mobile device receives and visualizes patient´s clinical history, notifications, alerts, and recommendations as result of applying different healthcare services.

3 Communication Technologies for mHealth Applications

In order to design mHealth system for remote healthcare monitoring it is necessary to select appropriate technology to be used for data acquisition, processing, and visualization. Some results of analysis of wireless technology standards and their characteristics are introduced in this section. As it was mentioned early, Wireless Body Area Networks (WBANs) and Wireless Personal Area Networks (WPANs) are portable communication technologies wearable or closely connected to users. WBANs are used to interconnect devices and sensors that can be worn on clothes or implanted into the body of patient. WBANs are currently being standardized by the IEEE 802.15.6 Task Group. In IEEE 802.15.6 2013 standard the description of WBAN is presented in the following way: "low power devices operating on, in or around the human body (but not limited to humans) to serve a variety of applications including medical, consumer electronics or personal entertainment and others". BAN operational range is less than 10 meters with ultra-low power consumption (around 100μW) and low data rate (less than 256kbps).

For some areas including health care, telediagnostics and remote monitoring Wireless Sensor Networks (WSN) are used. WSN is a wireless network consisting of spatially distributed autonomous devices using sensors to monitor physical or environmental conditions. Some of available standards used in WSN include 2.4 GHz radios based on IEEE 802.15.4 (WPAN), IEEE 802.15.1 (Bluetooth) or IEEE 802.11 (Wi-Fi) standards or proprietary radios, which are usually 900 MHz. Bluetooth (BT) is a wireless telecommunication technology installed in more than 3000 million devices that has been designed to eliminate cables (BSIG 2013). BT is regulated under the standard IEEE 802.15.1. BT's most general characteristics are short operational range of less than 100 meters, star topology or peer-to-peer, transfer rate from 1 Mbps in BT version 1.2 up to 24 Mbps in version 3.0 +HS (High Speed) and version 4.0, frequency band from 2400 to 2483.5 MHz, and transmitting power from 1 mW to 100 mW. The complete core of BT v3.0+HS and v.4.0 can be downloaded from Bluetooth Special Interest Group Web page (BSIG 2013).

Today, WSN technology covers a wide variety of applications: e-health, military, domotic, agriculture, etc. A WSN is composed of one or more nodes (sensors) working together to carry out a specific task. These nodes with limited memory and capacity are characterized by their low power consumption. A valuable reason for choosing these networks to implement mHealth application is that, they employ reliable handshake protocol and use the Advanced Encryption Standard (AES) for secure data transfer. Therefore, Bluetooth and IEEE 802.15.4 networks provide excellent capabilities to retrieve patient´s vital information like temperature, heart rate, blood pressure, sugar and oxygen levels, and others.

Wi-Fi networks, in their five versions a/b/g/n/ac, are the most visible wireless access technologies in the consumer market. Most portable computers, smart phones, tablets, etc. support at least one of IEEE 802.11 network variations. Wi-Fi network can operate in two distinct modes, namely, ad hoc and infrastructure mode. In the infrastructured mode it operates via Wireless Access Point (WAP) device, which acts as central node that connects wireless terminals. The WAP usually enables access to the Internet, intranet, or other wireless network. Ad hoc networks can be formed "on the fly" without help of a WAP. In ad hoc networks Wi-Fi nodes may communicate directly with each other but in a local way. Therefore, ad hoc networks additionally require a device enabling access to the Internet. Hence, Wi-Fi networks in mHealth applications will mainly operate in the infrastructured mode. The role of WAP is to share data between mobile nodes and wired networks. Table 2 shows a comparison of the surveyed wireless technologies and their role in mHealth applications.

Table 2 Wireless technology standards and characteristics

Standard/Wireless Technology	Used by User as	Operating Band and Data Rate	Range (m)
IEEE 802.15.6 Body Area Network	Portable or Into	868/915MHz, 2.4GHz 76/101Kbps, 485Kbps	10
IEEE 802.15.4 Wireless Sensor Networks	Portable	868/915MHz, 2.4 GHz 20/40/250 Kbps	100
IEEE 802.15.4a ultra-wideband (UWB)	Portable	250–750 KHz/3.1–5 GHz/ 6–10.6 GHz, 675 Mbit/s	100
IEEE 802.15.1 Bluetooth v.4	Portable	2.4 GHz 24 Mbps	100
IEEE 802.11 a/b/g/n/ac Wi-Fi	Access	5 /2.4 /2.4 /2.4 or 5 GHz 54/11/54/600Mps/6.93Gbps	120/140/140/ 250

However, portable technologies (BANs and PANs), in which sensors will be embedded, still lack adequate interfaces for human interaction. As a result, the information they sense (get from the patient or environment) needs to be delivered to device providing adequate human interface and corresponding data processing. The use of mobile devices is a good solution because they offer appropriate display and audio output that can be used to generate graphics or alarms easily understandable by patients and health providers.

4 ECG: Interpretation of Electrical Activity of Heart

The electrocardiogram is a representation of heart activity in the Cartesian plane. Since its invention in 1901 by Willem Einthoven, electrocardiography is considered a principal medical diagnostic tool that provides valuable information about heart function and structure.

As illustrated in Fig. 2, ECG waves are denoted by the letters P, Q, R, S and T. These waves represent phases of the cardiac cycle, which is described below (Wagner 2009).

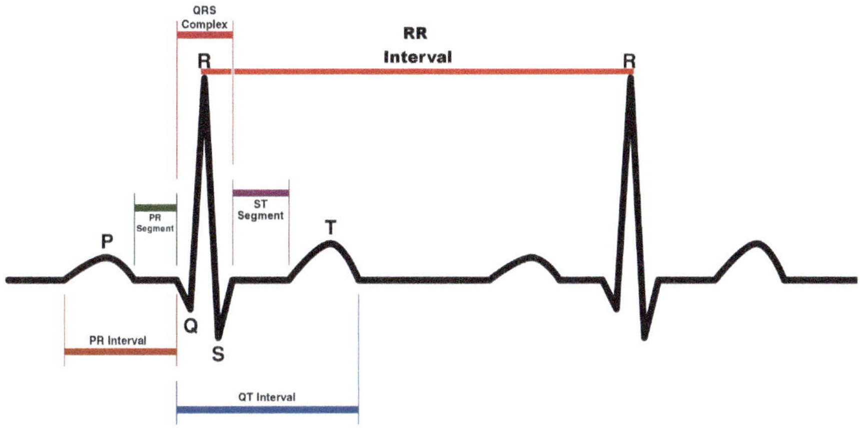

Fig. 2 ECG waves denoted by the letters P, Q, R, S and T

The depolarization and contraction of the atria mark the beginning of the cardiac cycle, an event that is recorded as P wave in the ECG. After P there is a pause that represents the time, when blood from the atria goes into the ventricles. This pause is recorded on the ECG as a horizontal line that appears after P wave and called as PR segment. The depolarization and contraction of the ventricles is recorded on the ECG as QRS complex. The QRS complex has particular importance because many of the existing algorithms are based on the detection of this component of ECG. After QRS the ST segment appears followed by T wave. The ST segment and T wave represent depolarization of the ventricles and recovery before the start of a new cardiac cycle. During ST segment the repolarization of ventricles is minimal and therefore, under normal conditions it appears as a horizontal line. In contrast, during T wave the ventricles are repolarized completely and return to their resting state waiting the next cardiac cycle.

Describe above ECG waves represent a cardiac cycle consisting of P wave, QRS complex, T wave, and the line segment, which continues until occurrence of the next P wave. Thus, the cardiac cycle comprises atrial contraction followed by ventricular contraction and the resting phase extending to beginning of the next cycle. For interpretation of ECG a special grid is used that allows easily determining the duration and amplitude of ECG waves. This grid shown in Fig. 3 has divisions every 1 and 5 mm, which correspond to 0.04 and 0.2 seconds respectively on the horizontal axis. In this way, 25 mm of the grid represent 1 second. Furthermore, on the vertical axis, which represents amplitude of waves in millivolts (mV), 0.1 mV corresponds to 1 millimeter, while 5 mm represent 0.5 mV. Using this grid, a physician can analyze the ECG and determine certain characteristics such as heart rate, rhythm, and others.

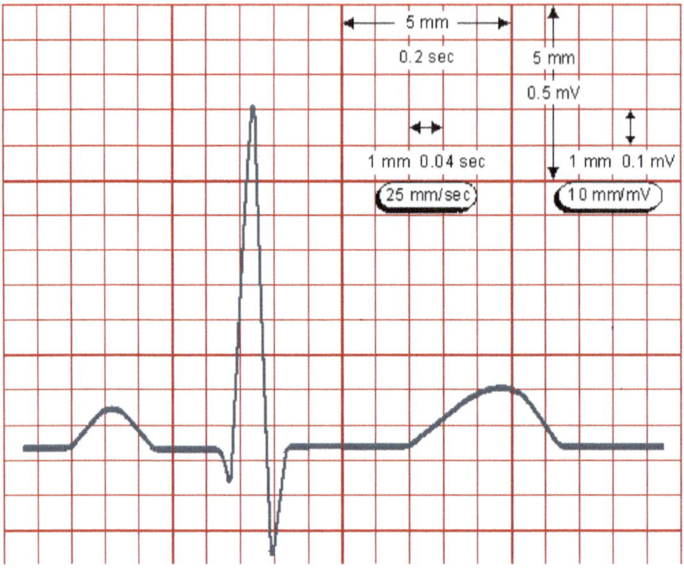

Fig. 3 Grid scale for the interpretation of ECG waves

5 ECG Monitoring Systems: mHealth Application Development

The block diagram of the proposed system for ECG wave monitoring is presented in Fig. 4. It has been designed with some modifications of typical used infrastructures for healthcare monitoring. All operations such as ECG acquisition, processing, and visualization are implemented on cell phone, which does not delegate any data processing task to server. This mode of data processing now is possible, in contrast to server-thin-client approach because modern cell phones have enough processing power and sufficient amount of memory. The interaction of cell phone with a server is provided for emergency messages or alerts including situations, when a medical personnel requests data from mobile device of monitored patient. In this way, a cell phone is converted to a true smart device, which may provide new services that it could not done before.

The proposed system consists of some modules and operates with remote *ECG Acquisition Device*. It includes *Sensors* that generate analog ECG signal to be digitized by *Analog-to Digit Converter* and then sent by *Bluetooth Transmitter* to cell phone. The *Bluetooth Connector* of mobile monitoring system handles address acquisition and user preferences for transmission and establishes communication with ECG acquisition device that may be defined by patient or physician.

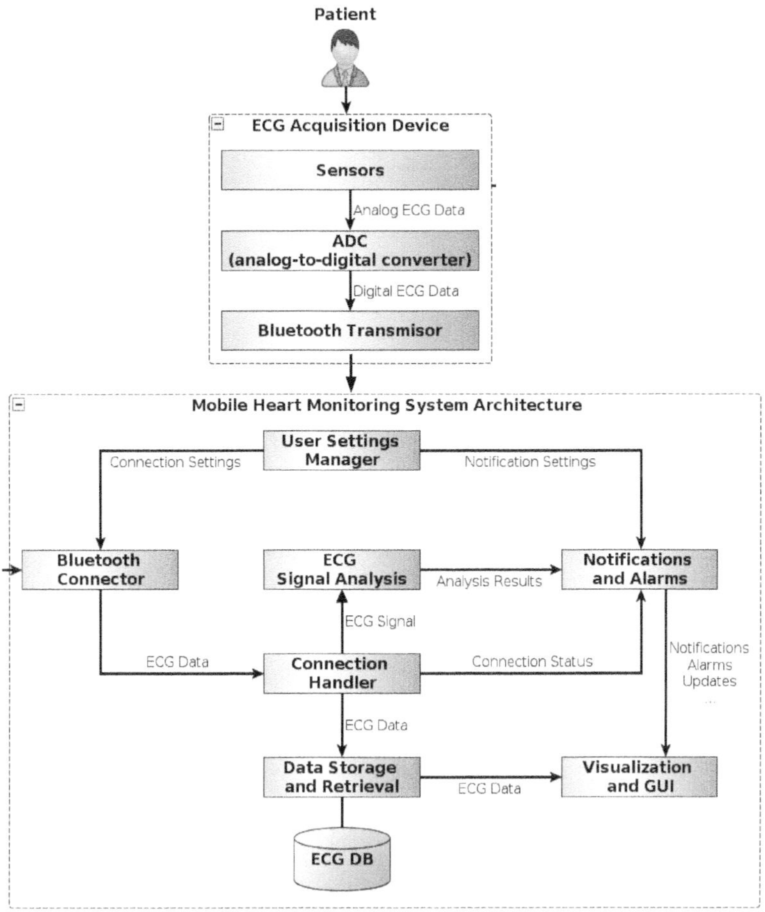

Fig. 4 Block diagram of system for ECG acquisition and preprocessing on mobile device

The *Connection Handler* is responsible for reception of data sent by ECG acquisition device, preparing them for analysis, storage, and monitoring changes in the state of connection. Additionally, it generates notifications or alerts for user. The *ECG Signal Analysis* module implements the proposed algorithm to detect QRS complex and patient's heart rate. The obtained results are processed by module of *Notification and Alarm*, which generates and distributes the corresponding messages about events that may occur during analysis of ECG or because of changes in connection status for example, loss of connection, interruption, changes in heart rate, and others. *Data Storage and Retrieval* module is responsible for persistent and secure storage of ECG data in *ECG DB* and ensures their availability for heart monitoring applications. The control of the proposed system is provided according to preferences defined by *User Settings Manager*. It is responsible for recording, storage, or sharing information and defines user identification, duration of analysis session, definition of default acquisition device, and others.

The *Visualization and Graphical User Interface* GUI module provides a set of applications that allow users to control the system visualizing ECG and its characteristics on cell phone display and show notifications and alarms. Using this architecture with modular structure a real-time system for monitoring and analyzing electrocardiogram in mobile environment may be designed.

5.1 The Proposed Algorithm for Detection of QRS Complex

The proposed algorithm for ECG analysis is based on detection of QRS complex, which is the most notable representation of cardiac cycle.

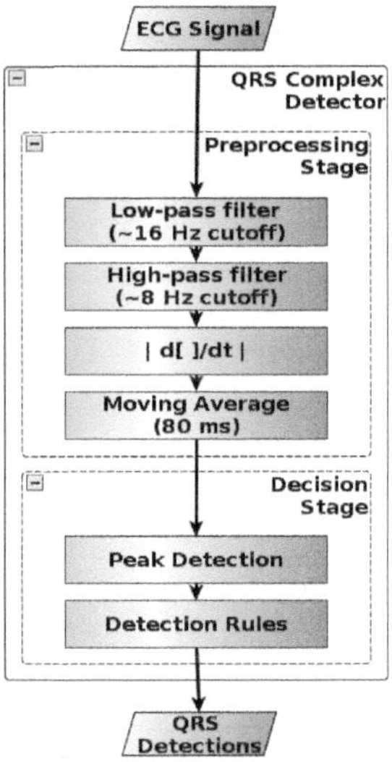

Fig. 5 QRS complex detector: flow chart of the suggested algorithm

The QRS detection algorithm has been implemented using as basis Pan-Tompkins QRS detector modified by Hamilton (Pan 1985; Hamilton 2002). Data flow chart of the suggested algorithm is shown in Fig. 5.

In the preprocessing stage, the ECG signal passes through series of filters. The low pass filter is used for reduction of signal noise. The high pass filter is applied for detection of signal gradient and identification of components with the greatest slope. To highlight the steepest parts, the absolute value of gradient is averaged over a moving window of 80 milliseconds considered as the minimum duration of QRS complex. In the decision phase a peak detector is used. The height of detected peaks is compared to a threshold determined as signal-to-noise ratio corresponding to particular ECG acquisition device. If the height of peak is higher than threshold, it is taken as a signal peak, otherwise it is interpreted as a noise peak. Finally, the detected signal peaks corresponding to QRS complex are selected and used for computing heart frequency. The results of applying filters on each step of the preprocessing and decision stages of algorithm are shown in Fig.6.

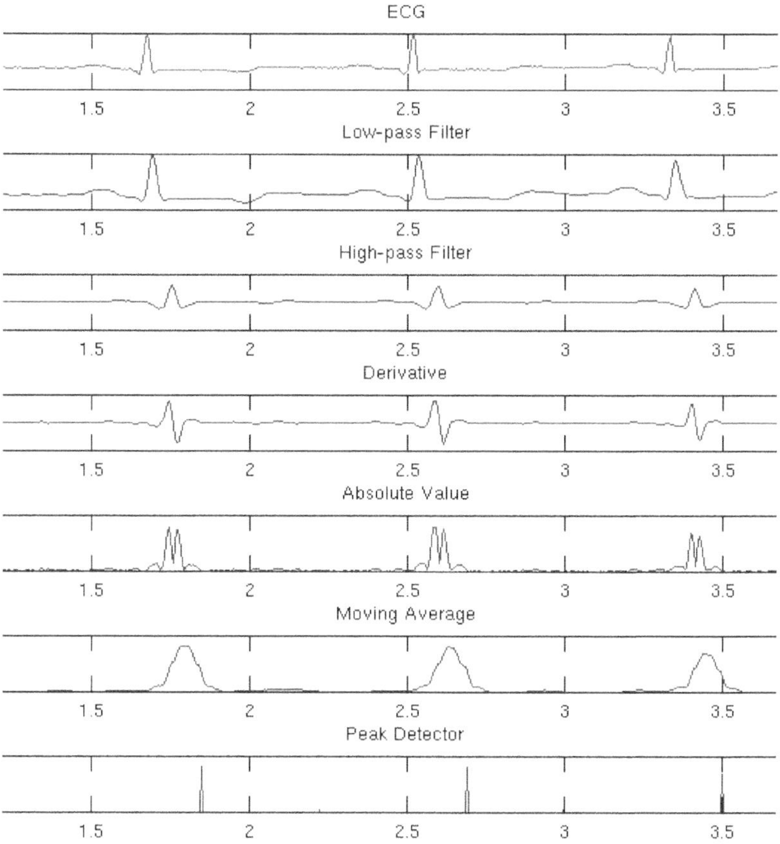

Fig. 6 Application of filters in the QRS complex detector

This algorithm is simple and fast for filtering ECG signal and detecting QRS complex. The algorithm is fast enough and may operate in real-time applications. Because the algorithm has been implemented using shift operations with integer numbers, the computational complexity is minimal, providing satisfactory results on cell phone with limited processing power and storage capability.

5.2 Framework for Real-Time ECG Monitoring on Smart Phone

In order to evaluate performance and efficiency of the proposed architecture and algorithm, a framework for ECG analysis on cell phone has been designed. The ECG acquisition, processing, and visualization consist of following steps, namely, ECG data recovery, computing heart activity characteristics, and ECG monitoring, management, and storage. The class diagram of the framework for heart activity monitoring is shown in Fig. 7.

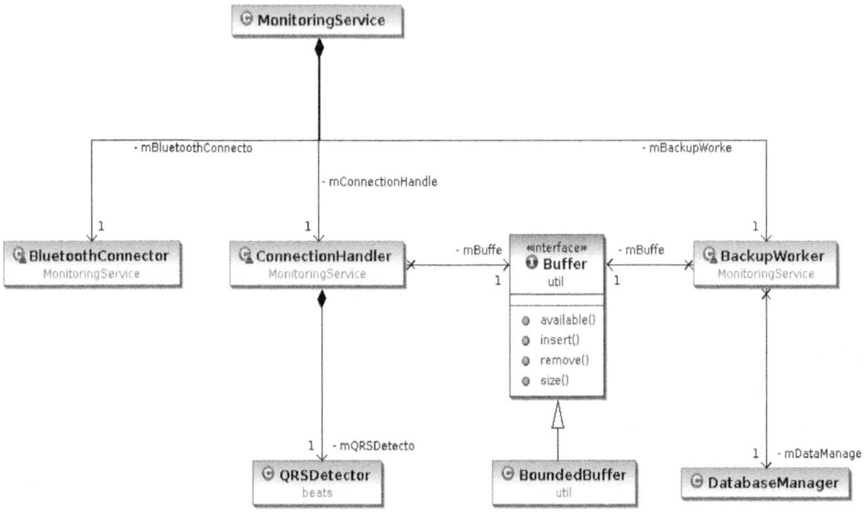

Fig. 7 Class diagram of the framework for heart activity monitoring

Electrocardiograms used by this framework are obtained from *MIT-BIH Arrhythmia Database* available in (Goldberger 2013). This database contains 48 records of 30 minutes two-channel ECGs with annotations of experts, indicating occurrence of the QRS complex. ECG acquisition module via *BluetoothConnector* supports communication between *Cardiograph* and cell phone by RFCOMM protocol, which allows emulating a serial port on both devices to send the connection request and desired recording time of ECG. Before ECG acquiring the *ConnectionHandler* defines signal features such as sample rate and calibration values of analog-to digital converter (*0* volts - *ADC zero* and gain factor - *ADC gain*). During data recovery ECG samples are converted to corresponding values in millivolts (*mV*) according to (Eq. 1) used then for ECG visualization on cell phone display.

$$\mathrm{mV} = \frac{sample\ \ value\ \ -\ \ ADC\ \ zero}{ADC\ \ gain} \tag{1}$$

ConnectionHandler includes *ConnectionListener* that defines a set of methods, which describe changes in state of the connection. It supports ECG data acquisition, processing and analysis, management of database, user interaction interface, and system testing. Additionally, it delegates responsibilities to associated classes and defines a set of methods to describe the connection status (on hold, online, offline, etc.), state of heart activity and irregular vital signs in order to generate alarms and send notifications to patient and medical personnel. *QRSDetector* determines the time of event occurrence and updates the value of heart rate using time interval between the last two QRS complexes. The *BackupWorker* receives data from *Buffer* and feeds *DatabaseManager*. *BoudedBuffer* supports a graphical interface that allows users to control the system, select data acquisition device, define user preferences, start or stop transmission, display ECG, and others.

The diagram of relationship between Java classes that implement presented in Fig. 6 algorithm for QRS complex detection is shown in Fig. 8.

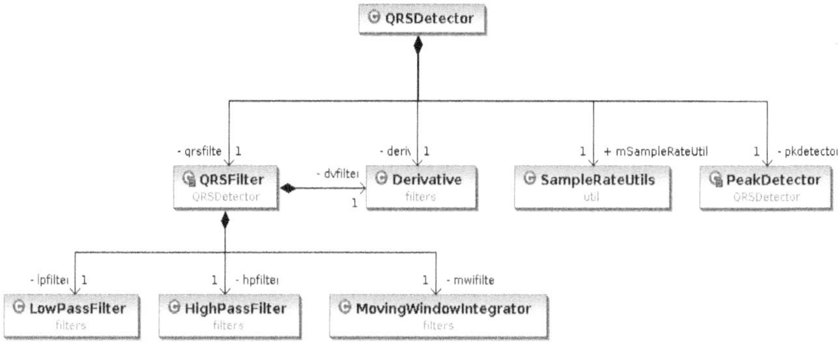

Fig. 8 Designed Java classes for QRS complex detector

Another important module is *MonitoringService*. Its class diagram is presented in Fig. 9. It includes a set of applications that provide access to information about patient's ECG via *DatabaseManager* class. This class consists of methods for querying, storage, and retrieval of ECG information. The reception of ECGs is controlled by service defined in class *MonitoringService*. This service is the most important element of the system. It is responsible for controlling the connection with ECG acquisition device, processing electrocardiograms, their storage in database, and notification about results of ECG analysis and detection of changes in the connection status for example, when system initiates data exchange with data acquisition device, or when connection is lost, or when transmission ends normally.

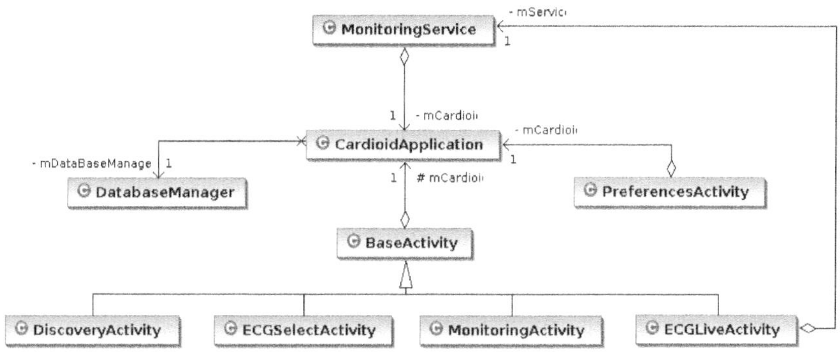

Fig. 9 Class diagram of monitoring services of the designed framework

Finally, one of the most important requirements is the ability of cell phone to store ECG records during long period of time. The storage capacity is limited mainly by ECG signal sampling rate, which is an indicator of the amount of data needed to represent one second of that signal. Representation of ECG record contains a significant volume of data therefore, we consider that the most appropriate way to store, manage, and retrieve electrocardiogram on cell phone is by using a database. In order to obtain efficient functionality of database, SQLite has been implemented because it is supported by several operating systems for mobile devices such as Symbian, iOS, BlackBerry, Android, etc. For each ECG the corresponding record is generated that includes parameters used by data acquisition device such as number of samples of signal, identification number, and the date of creation of record. The QRS complex is used as a reference point for determining the patient's heart rate. For this reason, the results obtained by QRS complex detector are stored including the identification number and time when QRS complex occurs. Several experiments with the proposed database show satisfactory storage of ECG with wide range of sampling frequencies taking into account restricted memory of cell phone.

The system prototype for heart activity monitoring has been implemented on Linux operating system particularly, on the Ubuntu version 10.10 for 64-bit processors, using Eclipse development environment 3.6 and Toolkit WFDB library functions (Goldberger 2013). The cell phone is Samsung Galaxy S, IG9000 with ARM Cortex A8 processor at 1 GHz, with 515 MB of RAM, 5 GB of internal memory and Bluetooth radio. Some examples of designed interfaces for continuous heart monitoring are shown in Fig. 10 and Fig. 11. Designed interfaces provide *Menu of Options* that allows user to select application (ECG View Stored Records, ECG Live View, Bluetooth Scan, or Preferences (Fig 10a), define ECG acquisition device (Fig. 10b, c), authenticate user with password and specify desired transmission time (Fig. 10a), analyze a list of records stored in database (Fig 11b), and visualize result of ECG analysis in real-time (Fig. 11c). The option of *Preferences* in main menu allows user to set personalized preferences for transmission of ECG signal, store Bluetooth address of the acquisition device, and specify the time duration of each monitoring session. The maximum duration that this application can support is 30 minutes according to standard test records used in *MIT-BIH Arrhythmia Database* (Goldberger 2013).

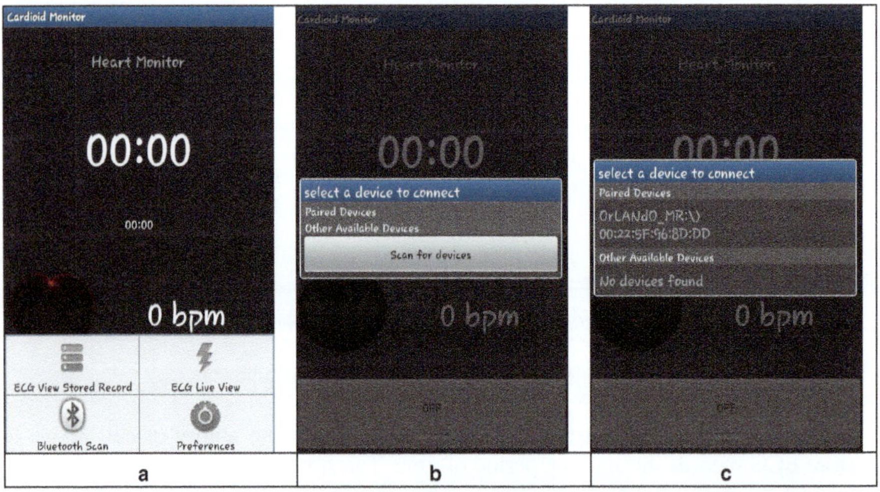

Fig. 10 Examples of designed interfaces: **a** principal menu of options; **b** and **c** selection of ECG acquisition device

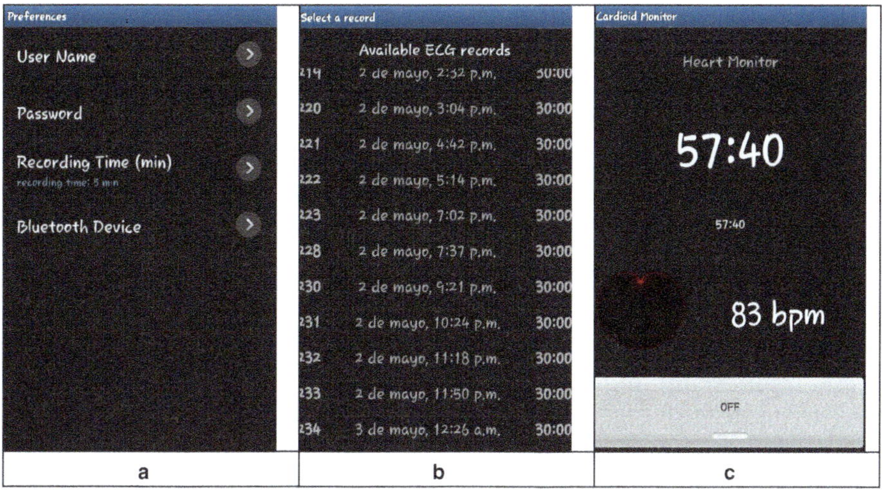

Fig. 11 Examples of designed interfaces: **a** user authentication interface; **b** view of available ECG records; **c** main GUI for continuous monitoring of heart rate.

6 Tests and Evaluation of ECG Analyzer on Cell Phone

To determine the feasibility of the designed prototype and evaluate performance of the proposed algorithm some tests have been conducted. They include analysis of battery behavior during continuous heart monitoring, computing precision of heart rate interpretation by the algorithm, and the evaluation of required time for transmission of ECG.

The power consumption is defined by the time of ECG recording without re-charging a battery. With the battery of Samsung Galaxy S IG9000 cell phone with 100% of capacity (1500 mAh) the operational time was 6 hours 14 minutes used for sequential transmission of 12 complete (30 minutes each) and one partial records, supporting monitoring until the battery is completely discharged. Even though the battery is low, the system processes input signals and detects emergency events, when the value of heart rate is less than 60 (bradycardia) or greater than 100 (tachycardia) beats per minute (bpm). Cell phone generates audible alarm, which is maintained until frequency heart returns to normal range of 60-100 bpm.

Some examples of these notifications and alarms are depicted in Fig. 12a and 12c. In Fig. 12b the normal heart rate is presented. In Fig. 12c the system detects and notifies occurrence of rapid heartbeats while the interface still presents the previous heartbeats (61 bpm).

For evaluating performance of the designed prototype with respect to ECG transmission, processing and storage time, the 30-min ECG records have been transmitted from *MIT-BIH Normal Sinus Rhythm Database* and *MIT-BIH Arrhythmia Database* using different sampling frequency of 128, 200 and 360 sps.

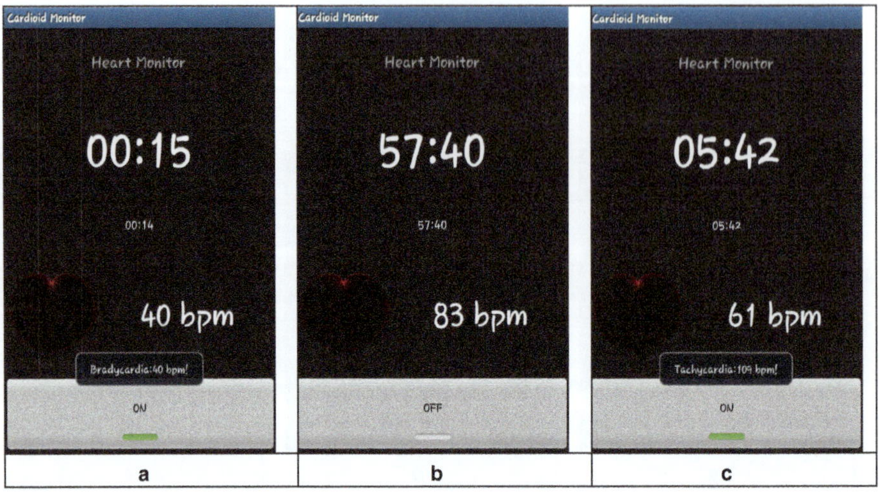

Fig. 12 Visualization of notifications and alarms for slow and rapid heartbeats: **a** slow heartbeat - bradycardia; **b** normal heart rhythm; **c** rapid heartbeats – tachycardia

For registration of record transmission and processing-storage time the prototype uses two timers shown on display. The big size timer shown in the middle of display measures the time of ECG transmission to cell phone. The second small size timer measures the time from start of transmission until ECG is stored to the database. As it is shown in display, these values are similar, confirming that the time for ECG processing and storage to database is less than one second.

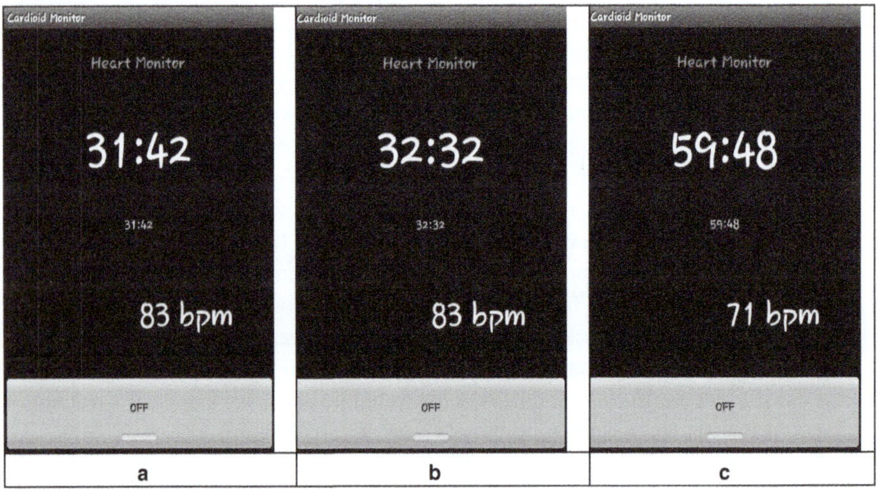

Fig. 13 Transmission and storage time measured in tests with sampling frequencies of **a** 128, **b** 200 and **c** 360 sps

When ECG signal is sampled with higher frequency, the amount of transmitted data to cell phone is increased. The results of test for measurement of transmission time are illustrated in Fig. 13. In the cases of 128 and 200 sps the transmission time approaches the duration of 30-minute ECG record achieving 31:42 and 32:32 min, respectively (Fig. 13a and b). However, in the case of 360 sps, the transmission time is increased almost in 50% (59:48 min for transmission of 30 minute ECG record due to increased amount of data to represent ECG signal with 360 sps) as it is shown Fig 13c. It is important to mention, that the values of small size timer, which represents a storage time, in all cases were similar to the transmission time shown by big size timer indicating that the time for processing and storage still takes less than one second with respect to duration of transmission.

In Fig. 13c the heart rate value (71 bpm) obtained with 360 sps is different comparing it with the values obtained in other cases (83 bps for 128 and 200 sps) for the same ECG record. This means that the computed result with 360 sps is more accurate due to higher quality of received signal.

After performing statistical analysis of multiple tests with different ECG records and sampling frequencies the relative error $\delta(\%)$ of computing heart rate obtained according to the following equation (Eq. 2)

$$\delta(\%) = \frac{|N_{\max/\min} - N_{real}|}{N_{real}} * 100\% \quad (2)$$

is around of *±10%* for 128 sps, *±8%* for 200 sps and *±4%* for 360 sps. $|N_{\max/\min} - N_{real}|$ is the absolute error of measuring real value N_{real} of heart rate with respect to maximum or minimum values $N_{\max/\min}$ computed and displayed on cell phone. However, because of significant increment (almost twice) of transmission time for high sampling rate, it is not recommended to apply sampling with more than 200 sps despite using powerful phone like Samsung Galaxy S IG9000 with ARM Cortex A8 processor. The increment of transmission time for 128 and 200 sps does not exceed *5%* and *8%* respectively. This is a good compromise between high speed and satisfactory accuracy, which does not fall below the precision of commercial systems for heart monitoring. For example, the Real-time ECG analysis application for wearable healthcare (Yun-Hong 2011) reports the average accuracy of ECG recognition about 92.46%. In the WE-CARE project based on MR-SVM algorithm (Zheng 2013) the achieved precision is about 93%. Other well-known systems such as the ECG intelligent monitoring system with MSP430 microcontroller (Zhang 2013) and the Portable real-time ECG recognition system based on smartphone (Tzu-Hao 2013) report average accuracy about 92.1% and 98% respectively.

Visualization of ECG in real time is very useful feature of the proposed heart monitoring framework. However, some concurrent operations such as ECG acquisition by cell phone, ECG retrieval from database, and plot of ECG on display have significant impact on system performance. For example, in Fig. 14a the ECG plot of 4:11 min is shown even though cell phone has completed 5:14 min of data reception.

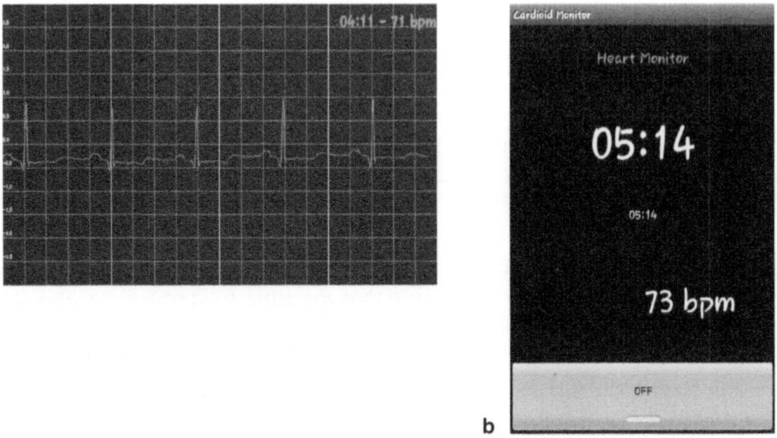

Fig. 14 a ECG plot during real-time monitoring on cell phone; **b** corresponding characteristics of heart activity.

Additionally, the computed heart rate is more precise after 5:14 min displaying 73 bpm in Fig. 14b than after 4:11 min that gives 71 bps in Fig. 14a. Even though the most critical operation for cell phone is ECG plotting, the achieved relative error of heart rate computing in the case of ECG signal encoded by 200 sps does not exceed ±3%. In order to increment performance of the proposed system, a cell phone with high speed and sufficient processing power must be used as well as the extension of bandwidth of data communication channel between phone and ECG transmission device is required.

7 Case Studies

1. What are the potential opportunities and benefits of Information and Communications Technologies (ICT) in healthcare support systems?

- Lewis D, Hodge N, Gamage D, Whittaker M (2011) Understanding the role of technology in health information systems, Knowledge Hub, Working Paper Series Number 17, June 2011, School of Population Health, University of Queensland. Pp 1-22. http://www.uq.edu.au/hishub/docs/WP_17.pdf
- Tan J (2011) New technologies for advancing healthcare and clinical practices, IGI Global, USA
- Nieto J I, Vazquez M, Cervantes H, Martinez M E, Sergiyenko O (2011) m-Health: software development and wireless technologies applications. In: Aquino Santos R, Rangel Licea V, Edwards Block A (eds) Emerging technologies in wireless ad hoc networks: applications and future development, IGI Global, USA, Ch. 5, pp 92–114
- Want R (2009) When cell phones become computers, IEEE Pervasive Computing J 8(2):2-5

2. Evaluate the effectiveness of mobile-health technologies to improve health care services

- Shrinath A (2012) The evolution of and trends in mobile health, Massachusetts Institute of Technology
- Free C, Phillips G, Galli L, Lambert F (2013) The effectiveness of mobile-health technologies to improve health care service delivery processes: a systematic review and meta-analysis, http://www.plosmedicine.org/article/info%3Adoi%2F10.1371%2Fjournal.pmed.1001363
- Sana H. (2013) The doctor in the pocket. http://handysana.com/
- Kulkarni S, Öztürk Y (2007) Requirements and design spaces of mobile medical care, ACM SIGMOBILE Mobile Computing and Communications Review J 11(3):12-30

3. Biomedical signal analysis: methods and applications.

- Chazal P, O'Dwyer M, Reilli R B (2004) IEEE, automatic classification of heartbeats using ECG morphology and heartbeat interval features, IEEE Trans on Biomed Eng. 51(7):1196-1206
- Yun-Chi Y, Wen-June W (2008) QRS complexes detection for ECG signal: the difference operation method, computer methods and programs in biomedicine. Elsevier J. 9(1):245–254
- Hamilton P (2002) Open source ECG analysis. J Computers in Cardiology 101-104
- Jinkwon K, Hang S, Kwangsoo S, Lee M (2009) Robust algorithm for arrhythmia classification in ECG using extreme learning machine, Biomedical Engineering pp 1-12. http://www.biomedcentral.com/content/pdf/1475-925X-8-31.pdf

8 Future Research Directions

mHealth is used to retrieve and transmit data in real-time between end-users and healthcare providers, helping considerably to improve patient quality of care. The possibility of transmitting vital information in real time can be crucial to saving the patient's life. Moreover, the use of mobile devices can help to eliminate common errors that may occur in paper-based communication. However, there are still many issues that need to be improved in order to provide effective services of *mHealth*.

Concerning development of mobile applications, the existence of several non-compatible platforms and languages represents a difficulty when working with different devices. Increased efforts need to be made in order to provide single development environment for all devices. For example, a common virtual machine could be useful allowing execution of application regardless of the platform for which it was implemented. This is still open research field that can help to motivate wider adoption of *mHealth* services.

From practical point of view, the modern communication technologies still require advances in design of high speed wireless data transmission tool linked to mobile computing. The *mHealth* architectural solutions for development of efficient healthcare environments, which include physical infrastructure, related applications and humans, remain still very challenging tasks. Finally, the developing *mHealth* applications and services requires new efforts for creation of new data and signal processing approaches and their migration to mobile devices with limited capabilities. New healthcare applications have the potential to offer a wide range of benefits to patients, medical personnel and society through continuous monitoring in ambulatory setting, early detection of abnormal conditions, supervised rehabilitation, among others.

9 Conclusions

Nowadays, the advances in development of computing and communication technologies serve as a platform for emerging mHealth applications that provide very useful and never before possible services supporting healthcare intelligent environments assisted by mobile devices. The review of several platforms available for development of mobile applications as well as popular wireless technologies have been analyzed with the main purpose to select appropriate infrastructure, data exchange environments, and methodologies for designing mHealth systems.

The proposed architecture, which provides continuous data transmission and processing on mobile device with limited computing resources, may be considered as good example of developing mHealth applications. This approach has sufficient merit to be used as a reference for design of system that may provide personal health monitoring, mobile learning assistant, remote detection, and control. Additionally, the proposed algorithm for ECG signal acquisition, processing, storage, and visualization provides efficient monitoring of heart activity. The relative error of computing heart rate is less than *±10%* in the worst case, when the sampling frequency of ECG signal is equal to 128 sps, and the relative error is less than *±4%* for 360 sps. In terms of practical contribution, the proposed approach integrates various emerging technologies. The implemented system with minimum computational resources has satisfactory performance, enough to be considered as precise heart monitoring tool on cell phone with the ability to process and visualize in real time ECG signals, detect and manage situations of risk, and provide the interaction between medical personnel and patient into smart healthcare environment.

It is important to note that prototype does not exclude diagnosis by a physician. Our intention is to propose approaches that serve as guide for development of more complex health assistance equipment expanding coverage of medical services.

Acknowledgments. This research is sponsored by two European Grants: S2EUNET - Security, Services, Networking and Performance of Next Generation IP-based Multimedia Wireless Networks, CASES - Cloud Computing-Based Advisory Services Towards Energy-Efficient Manufacturing Systems, and by Mexican National Council of Science and Technology, CONACyT, project #154438.

References

Alarcon-Aquino, V., Starostenko, O.: Detection of micro calcifications in digital mammograms using the dual-tree complex wavelet transform. J. of Eng. Intel. Sys. 17(1), 49–63 (2009)

Aghajan, H., Augusto, J.C.: Home-based health and wellness measurement and monitoring. J. Amb. Intel. and Smart Env. 4, 397 (2012)

AliveCor, AliveCor - mHealth for iHumans (2013), http://alivecor.com/ (accessed May 3, 2013)

Belgacem, N., Boumerdassi, S.: Mobile personal electrocardiogram monitoring system with patient location. In: Proceedings of the ACM Workshop on Medical-Grade Wireless Networks, Louisiana, pp. 69–72 (2009)

Biotronik, BIOTRONIK - excellence for life (2013), http://www.biotronik.com/es/lam/2246 (accessed December 10, 2013)

BSIG, Bluetooth Special Interest Group, Specification of BT v3.0+HS (2013), http://www.bluetooth.com/Pages/Bluetooth-Home.aspx (accessed January 6, 2014)

CardioNet, Mobile cardiac outpatient telemetry (MCOT) – cardiac telemetry cardionet event monitors (2013), http://www.cardionet.com/ (accessed May 10, 2013)

Curtis, D.V., Bailey, J., et al.: Using ambient intelligence for physiological monitoring. J. Amb. Intel and Smart Env. 1(2), 129–142 (2009)

Gay, V.: A mobile rehabilitation application for the remote monitoring of cardiac patients after a heart attack or coronary bypass surgery. In: Proceedings of the Conference on Pervsive Technologies, Greece, pp. 235–238 (2009)

Goldberger, A.L., et al.: PhysioBank, PhysioToolkit, and PhysioNet: components of a new research resource for complex physiologic signals. Circulation 101(23), e215–e220 (2013), http://www.physionet.org/physiobank/ (accessed April 6, 2013)

Hamilton, P.: Open source ECG analysis. J. Comput. in Cardiology, 101–104 (2002)

Handy, S.: The doctor in the pocket (2013), http://handysana.com/ (accessed April 8, 2013)

IEEE 802.15.4a, IEEE p802.15.4a/d7, Part 15.4: Wireless Medium Access Control and Physical Layer Specifications for low-rate wireless personal area networks, Technical report. IEEE, New York (2007)

IEEE 802.15.6, IEEE 802.15 WPAN™ Task Group 6 (TG6) Body Area Networks (2013), http://www.ieee802.org/15/pub/TG6.html (accessed January 10, 2014)

IMEC, Monitoring your health with your mobile phone (2013), http://www2.imec.be/be_en/press/imec- news/wirelesshealthnecklaceinterface.html (accessed April 8, 2013)

Kyriacou, E.: System for monitoring of children with arrhythmias. In: Proceedings of the Conference on Pervsive Technologies, Greece, p. 668 (2009)

Kulkarni, S.: Smartphone driven healthcare system for rural communities in developing countries. In: Proceedings of the Workshop on Systems and Networking Support for Health Care Environments, Colorado, pp. 1–3 (2008)

Muñoz-Ramos, O., Starostenko, O., Alarcon-Aquino, V., Cruz-Perez, C.: Real-time system for monitoring and analyzing personal electrocardiogram on cell phone. In: Khaled, E., Tarek, S. (eds.) Innovations and Advances in Computer, Information, Systems Sciences and Engineering. LNEE, vol. 152(28), pp. 327–338. Springer, Heidelberg (2013)

Nieto, J.I., Vazquez, M., Cervantes, H., Martinez, M.E., Sergiyenko, O.: m-Health: software development and wireless technologies applications. In: Aquino Santos, R., Rangel Licea, V., Edwards Block, A. (eds.) Emerging Technologies in Wireless Ad Hoc Networks: Applications and Future Development, vol. 5, pp. 92–114. IGI Global, USA (2011)

Pan, J., Tompkins, W.J.: A real-time QRS detection algorithm. IEEE Trans. on Biomed. Eng., 230–236 (1985)

Pawar, P., Jones, V., Beijnum, B.-J., Hermens, H.: A framework for the comparison of mobile patient monitoring systems. J. of Biom. Informatics 45(3), 544–556 (2012)

Tzu-Hao, Y.: A portable real-time ECG recognition system based on smartphone. In: 35th Annual International Conference of IEEE Engineering in Medicine and Biology Society, Osaka, pp. 7262–7265 (2013), `http://ieeexplore.ieee.org/xpl/articleDetails.jsp?tp=&arnumber=6611234` (accessed January 10, 2014)

Wagner, G.S.: Marriot Electrocardiografía Práctica, 11th edn. Lippincott Williams & Wilkins, Barcelona, Españ (2009)

Want, R.: When cell phones become computers. IEEE Pervasive Comp. J. 8(2), 2–5 (2009)

Yun-Hong, N.: Implementation of real-time abnormal ECG detection algorithm for wearable healthcare. In: 6th Conference on Computer Sciences and Convergence Information Technology, Seogwipo (2011), `http://ieeexplore.ieee.org/xpl/articleDetails.jsp?tp=&arnumber=6316585` (accessed January 10, 2014)

Zhang, Y., Tian, Y., Wang, Z.: An ECG intelligent monitoring system with MSP430 microcontroller. In: 8th Workshop on Systems, Signal Processing and their Applications, Algiers (2013), `http://ieeexplore.ieee.org/xpl/articleDetails.jsp?tp=&arnumber=6602364` (accessed January 15, 2014)

Zheng, Q., Chen, C., Li, Z.: A novel multi-resolution SVM (MR-SVM) algorithm to de-tect ECG signal anomaly in WE-CARE project. In: Conference on Biosignals and Biorobotics, Rio de Janerio (2013), `http://ieeexplore.ieee.org/xpl/articleDetails.jsp?tp=&arnumber=6487453` (accessed January 15, 2014)

Questions

Q1. Bluetooth devices have an increment of data transfer rate from *1.2 Mbps* in version BTv1.2 to *24 Mbps* in version 3.0+HS. What is the minimum required Nyquist bandwidth of the channel in both versions, if the signal levels are encoded by *4-bit* word?

Q2. The remote ECG acquisition system includes sensors that generate analog ECG signal, which to be digitized by Analog-to-Digit converter and then sent by Bluetooth transmitter to cell phone.

What is the channel capacity for ECG signal reception, if its bandwidth equals to *500Hz* and the signal-to-noise ratio is *3dB* (consider Shannon formula for theoretical maximum of data rate that can be achieved)?

Q3.Calculate the required processing time for each sample of ECG wave by system for heart monitoring on cell phone taking the worst case of tachycardia with *150* heartbeats per minute. Consider that:

- the ECG signal is sampled with frequency of *360* samples per second;
- the time for each sample transmission to cell phone and the time for sample storage to data base is about *0.3ms* each one;
- at least two highest peaks of the ECG wave (QRS complex) must be detected for computing the required number of heart beats

Q4. Fig.3 of the chapter shows the duration of ECG waves on the horizontal axis (*1mm* of grid corresponds to *0.04* seconds). On the vertical axis, which represents the amplitude of waves in millivolts (mV), *0.1mV* corresponds to *1mm*.
Calculate a correspondence of the standard ECG grid (on paper) and cell phone display (*1mm* of ECG wave on horizontal and vertical axes to number of pixels on display), if the area for ECG plot is *210x360* pixels.

Consider that at least *5* peaks of ECG wave must be shown on the plot in the case of bradycardia with *50* heartbeats per minute. The upper limit of amplitude of ECG signal is *3mV*.

Q5. Calculate the increment of precision of computing heart rate if ECG signal with the same characteristics is transmitted by *200* and then by *360* samples per second. Take into account the results from interfaces presented in this chapter in Fig. 13b and c. The result of heartbeat on *360* samples per sec is more accurate due to higher quality of received signal and its heart rate value may be used as N_{real} .

11. Operation, Analysis and Optimization of Wireless Sensor Devices in Health Oriented Monitoring Systems

Santiago Led, Leire Azpilicueta, Miguel Martínez-Espronceda, Luis Serrano, and Francisco Falcone

{santiago.led,leyre.azpilicueta,miguel.martinezdeespronceda, lserrano,francisco.falcone}@unavarra.es

Keywords: Biomedical Signals, Wearable Sensors, Remote Diagnostics, Personal Area Networks, Device Ergonomics.

1 Introduction

One of the main concerns from Government Bodies, Health Authorities and the General Public is to assure the sustainability of the Health Care Systems [1]. In this context, Ambient Assisted Living is becoming an alternative to traditional methods of assistance that requires the presence of specialized medical staff. In this context, wireless technologies are one of the resources which enable these services [2-3]. The high mobility, ergonomics, moderate cost and interconnectivity are the main factors in the successful use of wireless systems in Electronic Health (eHealth) and Mobile Health (mHealth) implementations [4].

The use of wireless technologies, however, implies to overcome certain limiting factors [5-6]. In order to give adequate service in terms of reliability, adaptability and comfort, several parameters such as the nature of the signals to transmit, the location of the wireless transceivers, the selected frequency band and the complexity of the environment in which the sensors must operate has to be considered. In addition, there are a variety of wireless systems which can be employed and all of the previous considerations will determine the election of one or more probably, a combination of these technologies in order to provide the necessary service [7-8].

Within mHealth applications, ubiquitous management of chronic cardiovascular disease is the most widespread healthcare service because these pathologies have high prevalence in the society [9-12]. Special attention is required to those ambulatory Electrocardiogram (ECG) monitoring applications aimed to following-up patients that suffer from no risk sporadic arrhythmias and syncopes; since these symptoms may occur occasionally, the ambulatory services require long-term monitoring periods in order to acquire the outstanding data and diagnose the patient's cardiovascular disease. With this aim, a new ambulatory ECG monitoring service is being developed to fulfill the whole healthcare staff's functional requirements and a suitable example of Wireless technologies applied to mHealth scenarios.

S. Adibi (ed.), *Mobile Health*,
Springer Series in Bio-/Neuroinformatics 5, DOI: 10.1007/978-3-319-12817-7_11

In this chapter, an overview of the requirements as well as the challenges and performance of different devices based on wireless technology applied in health monitoring and control will be presented. A holistic approach will be used, from the characteristics and impact of the physical radio channel to a final system overview, in which the interaction of gathered data with medical specialists is achieved. The aim of the chapter is to provide a clear view on the potential capabilities of mHealth systems as a function of the different wireless technologies available and their interaction. Adequate consideration of the environment in which the wireless mHealth systems will operate will minimize power consumption while increasing the overall performance.

2 Classification of Wireless Systems and Signals

The advances in microelectronics as well as in Radio Frequency (RF) and microwave devices have led to the implementation of a broad range of wireless communications systems, usually classified according to their scope of action or coverage area. In this way, Personal Area Networks and Body Area Networks provide communication between devices within the range of centimeters to several meters. Wireless Local Area Networks (WLAN), provided by WiFi systems based on IEEE 802.11 standard give access in the range from a few meters to several hundreds of meters. When mobility is required in large coverage areas in the range of several kilometers, Public Land Mobile Networks, spanning from 2G systems like GSM or IS-95, to 3G UMTS and incumbent 4G systems such as LTE are employed. A schematic view of this heterogeneous network architecture is depicted in Figure 1.

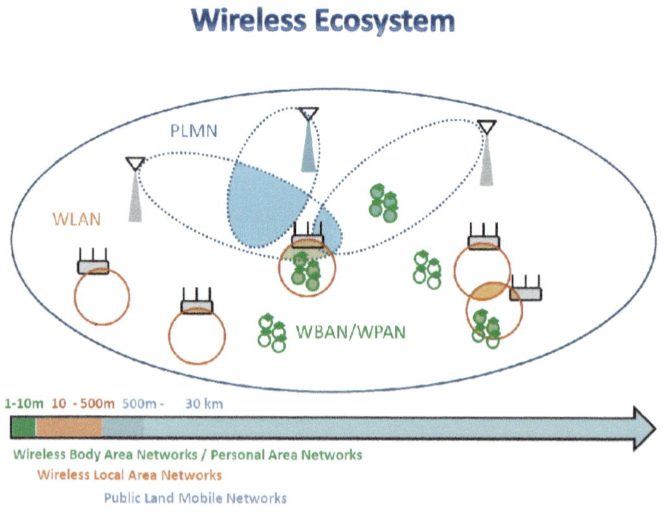

Fig. 1 Schematic view of co-existent wireless networks within the framework of mHealth systems

The resulting mHealth ecosystem will be usually composed by a combination of wireless systems, capable of capturing the required signals by means of acquisition sensors. A relevant feature of monitoring systems is the communication technologies used between acquisition sensors and gateway systems. Although wired technologies (RS-232, USB) are still implemented on monitoring devices, recently advances in transceivers and RF protocols have made possible to replace them by short range and medium/low power consumption wireless technologies: Bluetooth® Classic, Bluetooth® Smart, ZigBee®, Wi-Fi and NFC standard are some important examples.

The main acquisition sensors that are usually included in a monitoring system, the information transmitted, and the required communication features are shown in Table 1. There are systems focused on monitoring of vital signs (electrocardiogram, blood oxygen saturation, heart rate) that use some wireless standard previously mentioned, although most of them use Bluetooth® technology due to several features: robustness, low cost, reduced power consumption, and essentially high market penetration; all the personal mobile systems (Smartphone and Tablet PC) support Bluetooth® for short range exchange data applications. Nevertheless, technologies like ZigBee®, Bluetooth® Smart, and NFC will be getting incorporated on monitoring systems due to low power consumption and reduced bit rate features. It is possible to find these technologies implemented on numerous acquisition sensors: heart rate monitors, blood glucose meters, door security detectors, smart watches, blood pressure cuffs, etc.

Table 1 Description of Acquisition Sensors Employed in an m-Health environment

	Sensor	Parameter/Signal	Transmission Type	Communication Channel Type
Personal Health Sensors	ECG monitor	ECG signal Pulse rate	Real time	Streaming
	ECG recorder	ECG cardiac event Heart rate	Store&Forward	Realiable
	Impedance cardiograph	Heart rate Cardiac output Stroke volumen Thoracic fluid content	Real time	streaming
	Weigh scale	Body weight Body mass Index	Store&Forward	Realiable
	Blood pressure monitor	Systolic pressure Diastolic pressure Pulse rate	Store&Forward	Realiable
	Pulse oximeter	Blood oxygen percentage Pulse rate	Real time Store&Forward	Streaming Realiable
	Blood glucose meter	Blood glucose level	Store&Forward	Realiable
	Peak flow meter	Maximum exhale flow	Real time	Reliable

Table 1 *(continued.)*

Context Sensors	Room temperature sensor	Temperature	Real time	Streaming
	Humidity monitor	Humidity	Real time	Streaming
	Motion detector	User movements	Real time	Realiable
	Water consumption meter	Water consumption level	Store&Forward	Realiable
	Open/close door detector	Digital door state	Real time	Realiable
	Gas detector	CO gas level CO2 gas level	Real time	Realiable
	Flood detector	Digital flood state	Real time	Realiable
	On/off electrical appliance	Digital appliance state	Real time	Realiable

Gathering the required information in order to achieve effective remote health assistance is vital within a mHealth framework, in which sensors are key elements. In general, acquisition sensors can be put into following categories:

- pHealth sensors. These sensors are usually portable and wearable devices that acquire biomedical parameters and signals like weight, blood pressure, blood oxygen saturation, electrocardiogram, etc. Portable sensors are frequently placed in the same location and they are used by the user only when the biomedical data must be acquired. Wearable sensors are essential acquisition devices because they provide to the monitoring system several standout features such as ubiquitous and non-intrusive monitoring ability. It requires sensors to have the following characteristics: reduced size and weight, high operation autonomy, wireless communication, and intelligence. Moreover, both an ergonomic design and a power supply based on batteries are outstanding features. The development of wearable sensors has been possible thanks to micro- and nano-electronic advances, MEMS technology, and design of wireless communication technologies. Recent advances on textile manufacturing are allowing sensors to be integrated in clothes and accessories (shirt, shoes, rings, watch, among others). These devices are called e-textile sensors and provide high ergonomic and comfort levels. However, this type of acquisition sensors are not widespread used because the quality of acquisition signal is rather low given that movements and sweat can modify the acquisition process.
- Context sensors. Both early detection of chronic relapses and follow-up of degenerative diseases require the acquisition of information related to user's environment and his/her daily activities (context awareness). Thus, monitoring systems are integrating independent living and activity sensors in order to gather both user's activity and environment status along with

health status including room temperature, relative humidity, motion and presence detection, water consumption, flood detection, etc. The features and requirements of these context sensors are similar to wearable personal health devices: reduced size and weight, wireless communication, ultra-low power consumption, and an easy to use design. These features make system's deployment possible with minimal infrastructure.

As it has been previously mentioned, monitoring systems can include a great number of acquisition sensors. This allows the healthcare staff to have lots of information parameters and signals: blood pressure, weight, electrocardiogram, lung capacity, temperature, water consumption, on/off electrical appliances, motion detection, among others.

In addition to the own nature of data transmitted by sensors, there are several features related to the communication which are really important when wireless technology must be analysed. These features are following:

- Real time and store-and-forward transmission. Some monitoring parameters like electrocardiogram, impedance cardiograph, gas detection, and flood detection, among others, require the communication between sensor and gateway device to be instantaneous and in a continuous way in order to detect significant changes allowing the user and healthcare staff to act accordingly. This real time transmission imposes to wireless technology several requirements mainly in terms of bit rate. Other parameters such as weight, blood pressure, temperature, and humidity, among others, does not require a real time transmission. Delays between acquisition and transmission processes are not relevant and, further, the information can be temporarily stored by the sensor and transmitted to the gateway device later on. This store-and-forward communication imposes less restrictive requirements to wireless technology than real time transmission. In this case, bit rate establishes the time required to send all the information and consequently the total power consumption.
- Reliable and streaming communication. A reliable communication requires the wireless technology to guarantee gateway device the reception of acquired information because its value can be relevant in diagnostic and/or control processes. Some parameters and signals that require reliable communication are outstanding cardiac events, gas detection, flood detection, and blood pressure, among others. In a streaming communication the gateway device does not require the correct reception of all data frames. An example is the continuous ECG transmission in a real time monitoring process.

Thus, the type of information acquired by sensors all together with required communication and transmission modes are features related to wireless technology that must be considered and analyzed.

3 Characteristics and Impact of Wireless Channel Properties in Overall System Operation

The communication channel employed in mHealth solutions are electromagnetic carriers modulated with a variety of data schemes. The interaction between electromagnetic waves and the surrounding environment determines the level of losses that the incoming signal to a wireless transceiver faces. It is necessary to understand the underlying radio propagation mechanisms in order to assess if the proposed networks will operate with the necessary performance in terms of link availability, quality of service and transmission speed. In order to evaluate the performance and limitations of wireless systems, it is necessary to consider the following parameters:

- Carrier frequency of the employed wireless system, since an increase in the carrier frequency leads to a tradeoff between higher penetration losses, transmission capacity and usually smaller interference level.
- Coding and Modulation schemes, which give rise to a tradeoff between achievable spectral efficiency versus required Signal to Noise Ratio.
- Transmission Power, which implies a tradeoff between maximum distance between transceivers and required energy consumption of the devices.

In the case of mHealth systems, small devices with a moderate to low power consumption and high ergonomics are required. This leads to the use of compact antenna systems, usually embedded in the housing of sensors or mobile terminals, with low transmit power. In this context in which limited antenna gain and transmit power is available, the deployment of the required wireless systems should be analyzed with care, in order to account for radio propagation losses. In the case of complex indoor environments, losses are given by the following circumstances:

- Losses due to distance between transmitter and receiver elements, which increase as the operational frequency is increased.
- Losses due to absorption from elements within the radio propagation path (i.e., furniture, walls, human body).
- Losses due to diffraction from edges and scattering from non-homogeneous surfaces within the environment.

This leads to the consideration that indoor environments, as those on domestic households or in a medical environment, present complex radio propagation characteristics, which can change substantially in a relatively small area. It is true that healthcare mobile devices may be used anywhere. However, indoor environments present significant complexity in comparison with other surroundings because of the high intensity of clutter they could have. Electromagnetic phenomenon like reflection and refraction are relevant, as well as diffraction from the edges of the multiple obstacles and scattering from the rough surfaces of the walls. It is therefore necessary to consider the potential limitations that these systems can face. In order to gain insight in the behavior of wireless transceivers employed in m-Health systems, several scenarios have been simulated and analyzed with the aid of an in-house developed deterministic 3D ray launching code.

In the past, several indoor models have been developed based on semi-empirical approaches [13-15]. The advantage of these methods is that they require low computational cost but they have limited accuracy in comparison with high precision full wave techniques, which also exhibit very high computational cost [16-22]. A commitment between accuracy and computational time is acquired with deterministic methods, which are based on Geometrical Optics [23]. The Ray Tracing approaches combined with uniform theory of diffraction (UTD) is most commonly employed to radio coverage estimation [24-27].

The ray launching (RL) technique is based on recognizing a single point of the radiated wave with a ray that propagates along the space following a combination of optic and electromagnetic theories. RL techniques can be used in environments where the wavelength of interest is much smaller than the dimensions of the surroundings. A 3D in-house developed RL algorithm has been used for the characterization of wireless propagation of a conventional office, emulating a typical scenario where a healthcare mobile device could be used. The algorithm has been validated in the literature for different applications, like the analysis of wireless propagation in closed environments [28-31], interference analysis [32] or electromagnetic dosimetry evaluation in wireless systems [33]. It is based on Geometrical Optics (GO) and Geometrical Theory of Diffraction (GTD). GO approach consider only direct, reflected and refracted rays, bringing about abrupt areas, which correspond with the boundaries of the regions where these rays exist. Due to this constraint, the diffracted rays are introduced with the GTD and its uniform extension, the Uniform GTD (UTD).

Rays at a predetermined power level are launched in a predefined solid angle from a specific transmitter location and the resulting power levels along the path are calculated and stored. The whole scenario is divided into a uniform hexahedral mesh with cuboids of a given dimension. The rays will be reflected, transmitted and diffracted by the structure of the walls, the furniture of the office, the windows, the seats, etc. The material properties for all the elements within the scenario are considered, given the dielectric constant and permittivity at the frequency range of operation of the system under analysis. The dimensional maps of power levels can be obtained, as well as delay spread maps, power delay profiles and Signal to Noise ratio levels.

The considered scenario for the assessment of wireless propagation of a healthcare system is a conventional office location, in which furniture, appliances, walls and doors are present. The real scenario and the implemented simulation model are depicted in Figure 2.

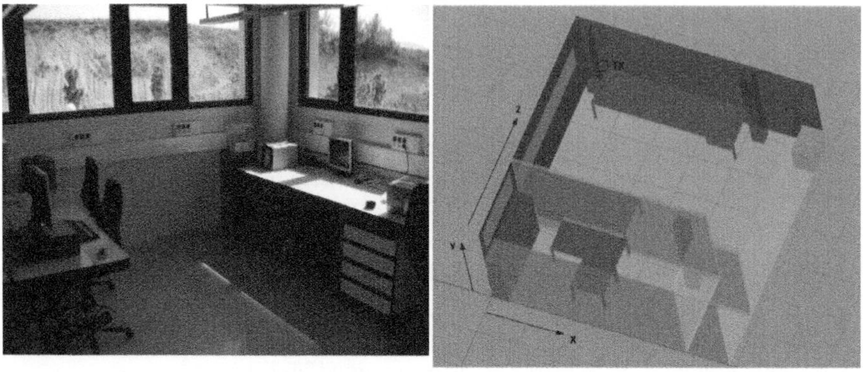

Fig. 2 View of the real and schematic representation of the considered scenario with the transmitter point depicted with a red circle

The simulation parameters are the following:

RAY LAUNCHING PARAMETERS

Frequency	868MHz/2.1GHz/2.4GH z
Transmitter power	0dBm
Antenna gain	5dBi
Horizontal plane angle resolution ($\Delta\Phi$)	1°
Vertical plane angle resolution ($\Delta\theta$)	1°
Reflections	6
Cuboids resolution	12cm x 12cm x

The results have been obtained for transceivers operating in the 868Mhz frequency range, a possible spectral allocation for ZigBee sensors. Figure 3 shows the distribution of received power as a function of location for two different cut plane heights (1m and 2m). As it can be seen, power distribution is not uniform, strongly depending on the elements within the vicinity of the propagation path. More detail can be observed in Figure 4, in which received power levels as a function of linear distance between transmitter and receiver is depicted for different heights. There strong variations in the received signal as a function of distance, with deviations in the order of 20 dB. This is given mainly by strong multipath components which appear in indoor environments. The influence of such multipath components, which are the main reason for such strong variability, can be observed by computing power delay profiles of the established radio links within the indoor scenarios. A schematic of the position in which the simulation power delay probes are located is shown in Figure 5, whereas the results for the power delay profiles is given in Figure 6 for each one of the considered probes.

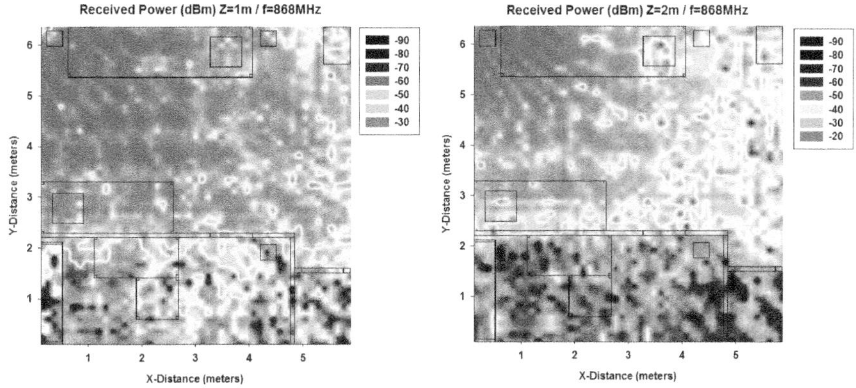

(a) (b)

Fig. 3 Bi-dimensional plane of received power (dBm) for 868MHz frequency (a) Z=1m (b) Z=2m

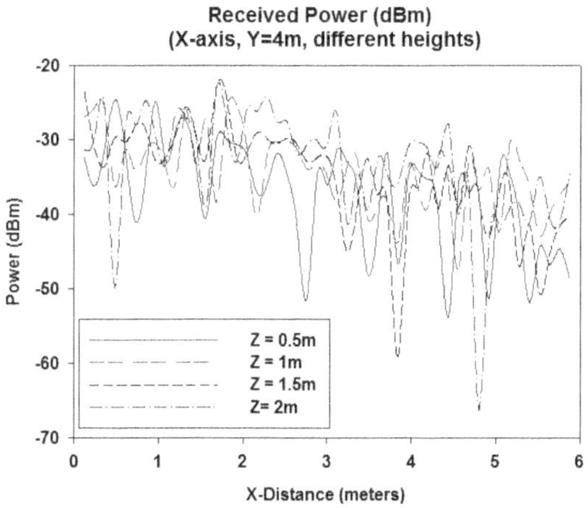

Fig. 4 Linear distribution of received power as a function of transmitter to receiver distance in the case of employing transceivers operating at 868MHz

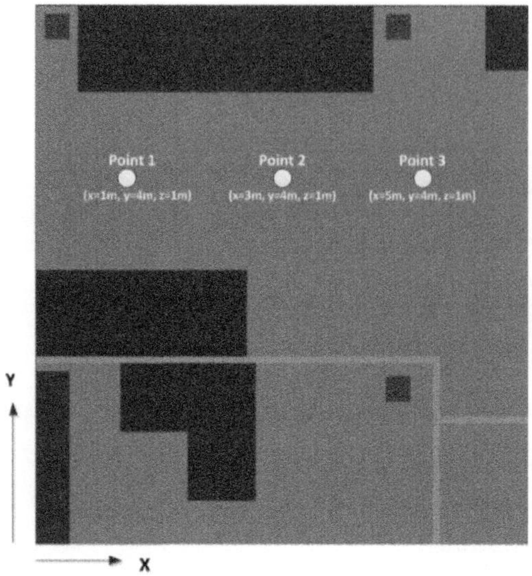

Fig. 5 Schematic view of the considered scenario with the considered points for the Power Delay Profiles

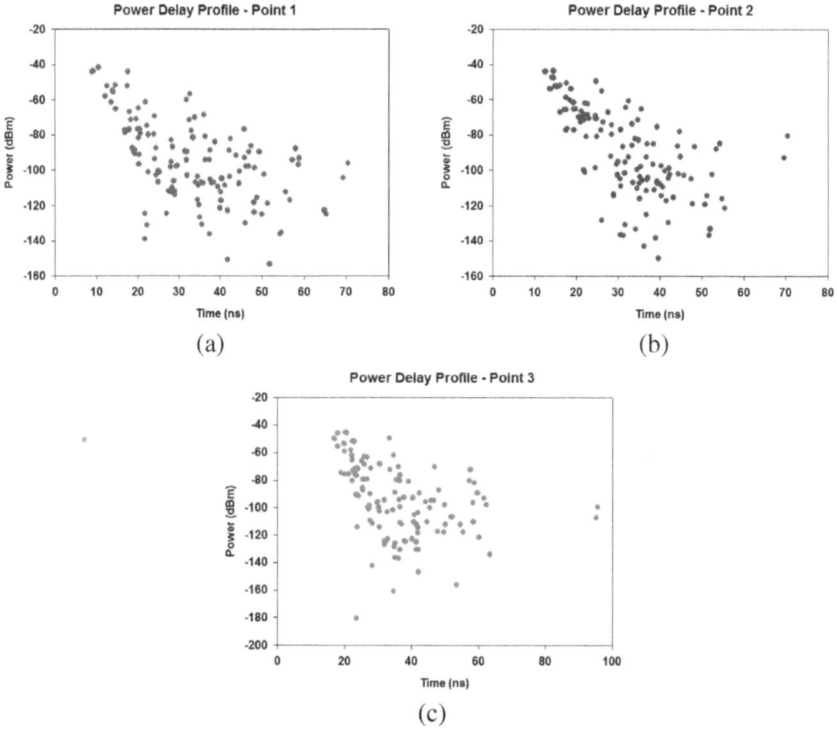

Fig. 6 Power Delay Profiles for different points in the considered scenario

As it can be seen, a large amount of components are propagated within the scenario, given the interactions of the transmitter with the surrounding environment, mainly due to reflection and diffraction. These variations are within the order of several tens of nanoseconds, as can be derived from delay spread computations, depicted in Figure 7 and that once again indicate the importance of multipath components within the scenario and their relation with the topology and the morphology of the indoor environment. A simple height variation of 1 meter can substantially modify the resulting response of the wireless channel, leading to larger losses and hence, degradation of system performance.

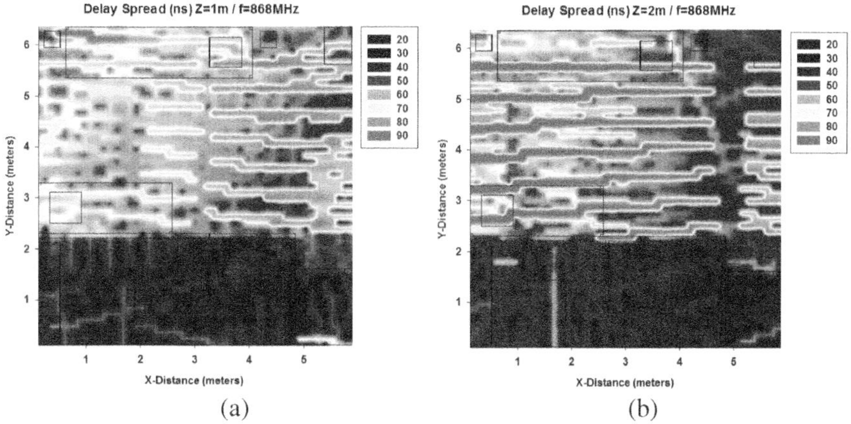

Fig. 7 Bi-dimensional plane of Delay Spread (ns) for 868MHz frequency (a) Z=1m (b) Z=2m

Another case has been considered, in which transceivers operating in the 2.41 GHz (i.e., Bluetooth or ZigBee sensors) are employed. The bidimensional power distribution planes have been obtained for two different heights, as depicted in Figure 8. As it can be seen, power levels are lower than in the previous case in which 868Mhz frequency was employed, a consequence in the use of a higher frequency band. Moreover, the 2.4GHz band is intensively used, with larger probability of interference. Figure 9 shows calculation of Signal to Noise ratio distributions within the scenario for two different cut planes, considering potential interference sources (i.e., adjacent non coordinated WLAN networks or WPAN devices). As it can be seen, the distribution once again strongly depends on the location considered.

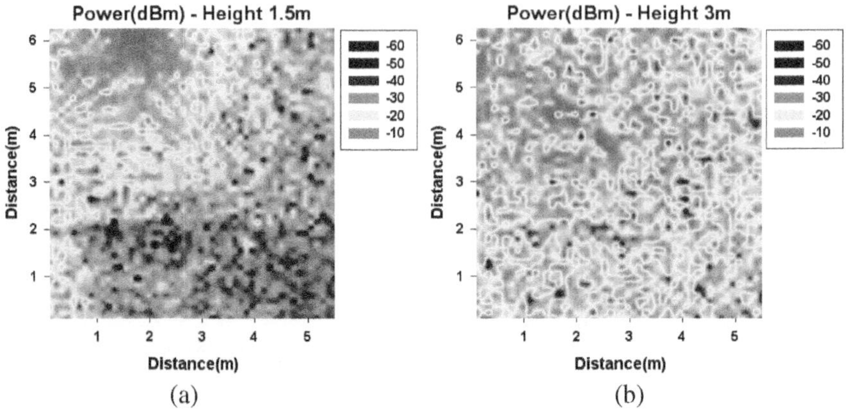

Fig. 8 Bi-dimensional plane of received power (dBm) for 2.41 GHz frequency (a) Z=1m (b) Z=2m

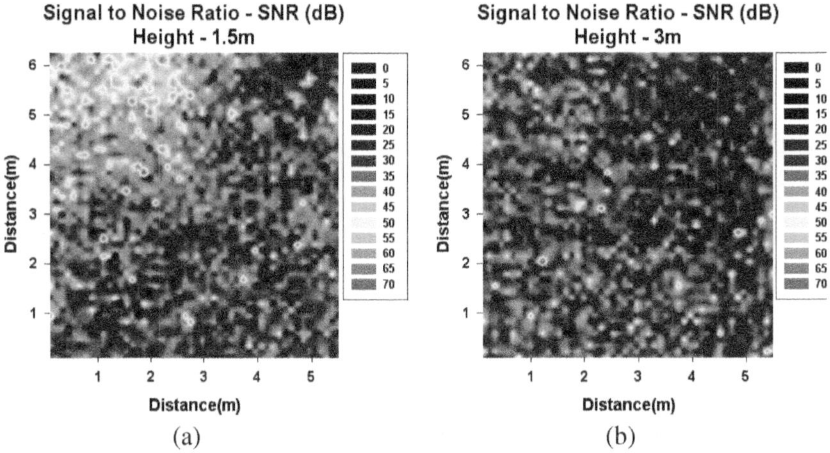

Fig. 9 Signal to Noise Ratio for 2.41GHz frequency (a) Z=1.5m (b) Z=3m

One last case is presented, corresponding to a mobile terminal operating in the 2.1GHz band, typical of 3G and future 4G systems. The bidimensional distribution of received power is depicted in Figure 10. In this case, received power levels are higher, given by the lower operating frequency. Once again, the location of the transceivers determines the received power levels and hence, the overall performance of the terminals.

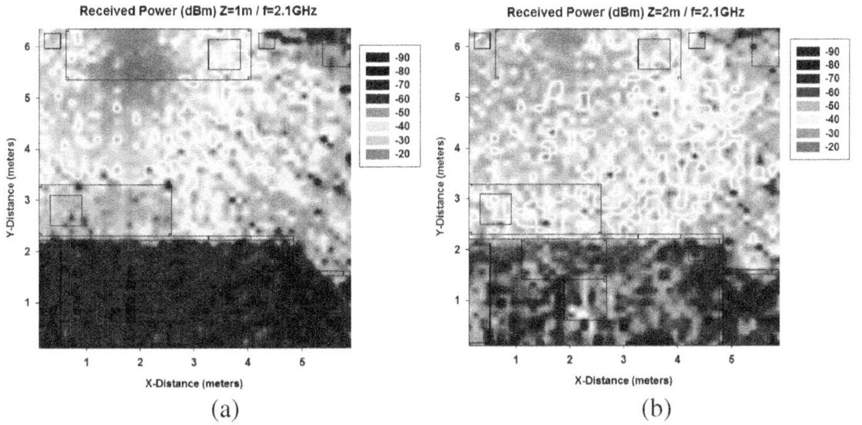

Fig. 10 Bi-dimensional plane of received power (dBm) for 2.1GHz frequency (a) Z=1m (b) Z=2m.

As it can be seen from the previous results, the election of the frequency of operation as well as the location of the devices, as a function of the scenario in which the sensors will be operating, are key parameters in order to gain insight in the expected overall performance. This must be carefully taken into account in order to guarantee that the received power levels will be above receiver sensitivity for a given data rate (which is usually low) to satisfy coverage/capacity requirements. Moreover, if adequate radioplanning choices are performed in the deployment of the wireless sensor network, received power levels can be maximized whilst reducing undesired interference and lowering overall power consumption.

4 mHealth Service Realization

HOLTIN service is been implemented as a wearable system for long-term and continuous cardiac activity monitoring. The system architecture consists of three elements: wearable ECG sensor, gateway device, and montoring server or back-end. The acquisition sensor is a smart device placed on the patient's chest designed with low form factor, very low power consumption, and ergonomic aspect. These features provide a noteworthy comfort levels to the patient. A real view of ECG sensor is shown in Figure 11. The device is able to monitor ECG signal and detect specific cardiac events suffered by the user. The outstanding cardiac information is temporarily stored in order to be transmitted later to a near gateway device through Bluetooth® technology.

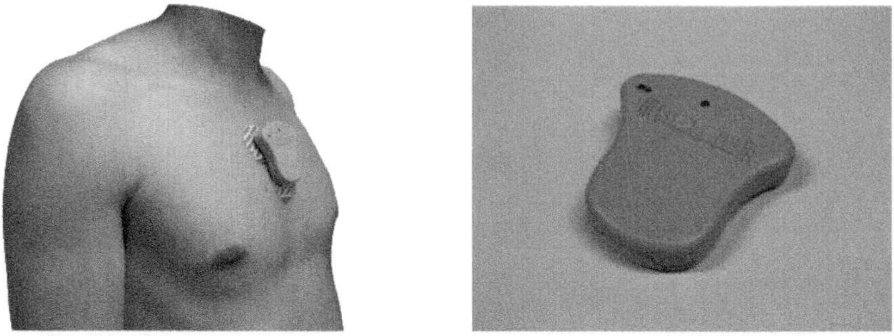

(a) (b)

Fig. 11 Image of the HOLTIN wearable ECG recorder (a) placed on a patient chest (b) detail of the device

The gateway device is implemented in a commercial smartphone with customized service software, and performs data transmission from wearable sensor to a management center through 3rd and 4th Generation (3G/4G) mobile phone network. This bridge device manages all the short/large range wireless communications, storage of patient's information, and visual/in voice alarm messages. With this aim, a customized software application for Android (Version 2.3 and above) Operating System (OS) is implemented in the Smartphone device. In this way, a minimal intervention of the patient is required thanks to the whole device functionality is achieved in autonomous way; that is, wireless communications are established with no requirement of patient confirmation. Smartphone device provides high usability and user friendly interfaces to the patient and HOLTIN service.

From a functional point of view, HOLTIN system includes all the requirements of healthcare staff. The main functional aspects are following:

- Cardiac event detection. Wearable device is able to detect the user's cardiac information in two ways: automatic detection and user notification. In any operation mode, the sensor always stores the outstanding ECG signal around the suffered cardiac event, allowing the healthcare specialist to diagnose user's disease.
- Data communication. Wearable device establishes Bluetooth® communication with Smartphone in order to transmit all the ECG information. In a similar way, the Smartphone device establishes 3G/4G communication channels with the back-end system. Moreover, wearable and gateway devices are able to store all the acquired information for later communication attempts. These functional features ensure the healthcare specialist to have the whole of user's information as soon as the environment and communication technologies make it possible.

- Alarm notifications. The monitoring system includes several notifications related to functional operation. Some alarm notifications are following: low battery state, low quality signal, user notification of cardiac events.

Once the user's data is available at the management center, the healthcare specialist can access to it using a client application. In this way, the application appropriately shows all the required data in order to diagnose the user cardiac disease. Figure 12 shows a screenshot of HOLTIN system's PC application.

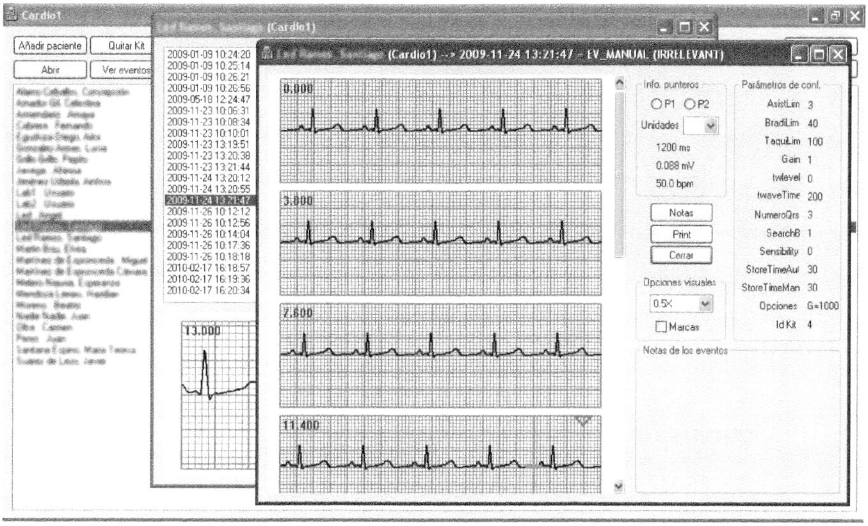

Fig. 12 Image from the PC-based data monitor within the described HOLTIN system

The automatic events are detected using configurable tachycardia and bradicardia thresholds applied to instant RR distance which, in turn, are calculated with consecutive QRS complexes. Asystolias are detected as absence of QRS complexes during a configurable span. The QRS complexes are detected using a Pan-Tompkins detector [33]. These algorithms are executed in a MSP430™ microcontroller [34]. The ECG events are stored in internal memory, including 64Mbit of flash memory and 16Mb of Random Access Memory (RAM) for permanent and temporal storage, respectively. Then, the events are wirelessly transmitted to the gateway over 2.4GHz Industrial, Scientific and Medical (ISM) band. The selected RF technology is Bluetooth® version 2.1, given its availability in Android-based smartphones. The Original Equipment Manufacturer (OEM) module used to manage Bluetooth communications is the BlueCore4™ [35]. The antenna is embedded in the PC-board design. The detected ECG events are transmitted in burst every five minutes, which limits the time that the wireless module is active and connected to the Smartphone and, therefore, the overall power consumption is restricted. The capacity of the batteries is 640mAh. With this configuration, the duration of the batteries is over one week. The device correctly communicates with the smartphone if both of them are within the same room. Communication between adjacent rooms was possible depending on wall materials, doors state, and distance between both devices.

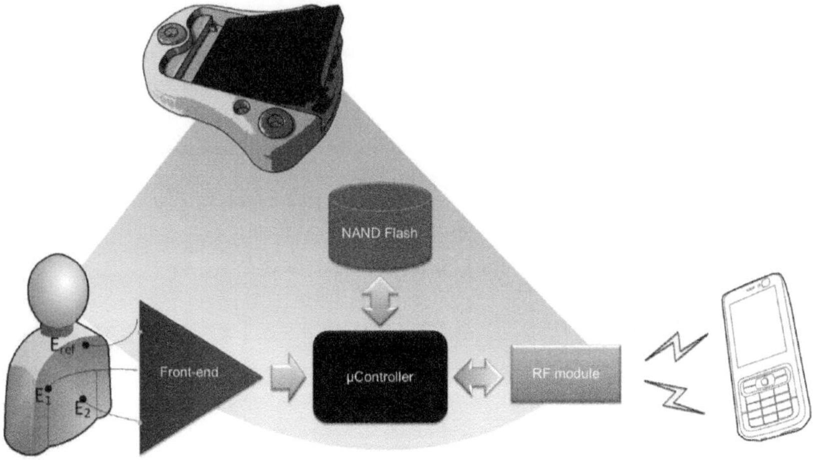

Fig. 13 Hardware blocks used to implement HOLTIN

5 Conclusions

The existence of Personal Area Network devices in combination with mobile terminals can offer an attractive solution in order to provide for m-Health services. It is necessary to consider the adequate acquisition sensors and there related signals in order to choose the right wireless technology to employ. The requirements for ergonomics, low power consumption and low cost lead to transceivers with moderate performance in which the impact of the environment in wireless channel behavior must be carefully addressed. In indoor environments, multipath propagation and material absorption are the main causes for highly variable signal fading. Therefore, detailed understanding of the performance of the wireless sensors as a function of location is compulsory in order to optimize wireless sensor network rollout. Overall, these type of solutions offer a flexible and low cost method to deploy m-Health systems.

6 Future Direction of Technology

The evolution in m-Health systems is strongly related to the advances in wireless transceivers in personal area network devices as well as in mobile networks. Smartphones are increasing the number of functionalities as well as the possibility of including multisystem transceivers. It will be possible to use mobile devices not only as information gateways but also as compact diagnostic devices, which are inherently connected, giving real time monitoring. The evolution in wireless sensors is heading towards ultra low power consumption transceivers, in which full energy harvesting techniques can be applied. Moreover, the advent of Internet of Things will lead to a fully connected environment, in which a myriad of sensors will be interconnected. In this sense m-Health systems will be integrated in such environment, in which Machine to Machine (M2M) communications and Cloud services will provide for global services.

References

[1] Luis, J., Peña, M., Gil, O.M.: Ehealth for patient empowerment in europe (2009), http://ec.europa.eu

[2] Teng, X.-F.: Wearable Medical Systems for pHealth. IEEE Reviews in Biomedical Engineering 1, 62–74 (2008), doi:10.1109/RBME.2008.2008248

[3] Bonato, P.: Wearable sensors/systems and their impact on biomedical engineering. IEEE Engineering in Medicine and Biology Magazine 22, 18–20 (2003), doi:10.1109/MEMB.2003.1213622

[4] Continua Health Alliance, http://www.continuaalliance.org (last visit: January 2014)

[5] Gatzoulis, L., Iakovidis, I.: Wearable and Portable eHealth Systems. IEEE Engineering in Medicine and Biology Magazine 26(5), 26–51 (2007),

[6] doi:10.1109/EMB.2007.901787

[7] Wireless technologies for telemedicine. Ng, H. S., y otros. 2, 2006, BT Technology Journal, Vol. 24, pp. 130 - 137. DOI: 10.1007/s10550 - 006 - 0050 - 9

[8] Lee, J.-S., Su, Y.-W., Shen, C.-C.: A Comparative Study of Wireless Protocols: Bluetooth. In: Annual Conference of the IEEE Industrial Electronics Society, pp. 46–51 (2007), doi:10.1109/IECON.2007.4460126

[9] Aragüés, A., Escayola, J., Martínez, I., Del Valle, P., Muñoz, P., Trigo, J.D., García, J.: Trends and challenges of the emerging technologies toward interoperability and standardization in e-health communications. IEEE Communications Magazine 49(11), art. no. 6069727, 182–188

[10] Led, S., Azpilicueta, L., Aguirre, E., Martinez-Espronceda, M., Serrano, L., Falcone, F.: Analysis and Description of HOLTIN Service Provision for AECG monitoring in Complex Indoor Environments. In: Sensors 2013, vol. 13(4), pp. 4947–4960 (2013), doi:10.3390/s130404947

[11] Led, S., Serrano, L., Galarraga, M.: Intelligent Holter: A new Wearable Device for ECG Monitoring using Bluetooth Technology. In: Proc. of International Federation for Medical and Biological Engineering (2005)

[12] Led, S., Martinez-Espronceda, M., Redondo, J., Baquero, A., Serrano, L., Cabezas, L., Niegowski, M.: Wearable electrocardiogram (ECG) recorder for a Mobile Point-of-Care based on recent interoperability standards. In: IEEE EMBS Special Topic Conference on Point-of-Care (POC) Healthcare Technologies, POCHT 2013, art. no. 6461341, pp. 287–290 (2013)

[13] Martínez-Espronceda, M., Martínez, I., Serrano, L., Led, S., Trigo, J., Marzo, A., Escayola, J., García, J.: Implementation methodology for interoperable personal health devices with low-voltage low-power constrains. IEEE Transactions on Information Technology in Biomedicine 15(3), 398–408 (2011)

[14] Hata, M.: Empirical formula for propagation loss in land mobile radio services. IEEE Trans. Antennas and Propag. 29(3), 317–325 (1980)

[15] Ikegami, F., Yoshida, S., Takeuchi, T., Umehira, M.: Propagation factors controlling mean field strength on urban streets. IEEE Trans. Antennas and Propag. 32(8), 822–829 (1984)

[16] Phaiboon, S., Phokharatkul, P.: Path loss prediction for low-rise buildings with image classification on 2-D aerial photographs. Progress in Electromagnetics Research 95, 135–152 (2009)

[17] Dimitriou, A.G., Sergiadis, G.D.: Architectural features and urban propagation. IEEE Trans. Antennas and Propag 54(3), 774–784 (2006)

[18] Franceschetti, M., Bruck, J., Schulman, L.J.: A random walk model of wave propagation. IEEE Trans. Antennas and Propag. 52(5), 1304–1317 (2004)

[19] Kanatas, A.G., Kountouris, I.D., Kostaras, G.B., Constantinou, P.: A UTD propagation model in urban microcellular environments. IEEE Trans. Veh. Tech. 46(1), 185–193 (1997)

[20] Lee, D.J.Y., Lee, W.C.Y.: Propagation prediction in and through buildings. IEEE Trans. Veh. Tech. 49(5), 1529–1533 (2000)

[21] Lee, S.H.: A Pyroelectric Infrared Sensor-based Indoor Location-Aware System. IEEE Transactions on Consumer Electronics 52(4), 1311–1317 (2006)

[22] Prieto, J.B., Toledo, R.L., Reguero, P.F., Abril, E., Martínez, A.B., Franco, S.M., Bullido, D.: A new metric to analyze propagation models. Progress in Electromagnetics Research 91 (2009)

[23] Tan, S.Y., Tan, H.S.: A microcellular communications propagation model based on the uniform theory of diffraction and multiple image theory. IEEE Trans. Antennas and Propag. 44(10), 1317–1326 (1996)

[24] Iskander, M.F., Yun, Z.: Propagation Prediction models for wireless communications systems. IEEE Trans. Microwave Theory Tech 50, 662–673 (2002)

[25] Gennarelli, G., Riccio, G.: A uapo-based model for propagation prediction in microcellular environments. Progress In Electromagnetics Research B 17, 101–116 (2009)

[26] Son, H.W., Myung, N.H.: A deterministic ray tube method for microcellular wave propagation prediction model. IEEE Transactions on Antennas and Propagation 47(8), 1344–1350 (1999)

[27] Song, H., Wang, H., Hong, K., Wang, L.: A novel source localization scheme based on unitary esprit and city electronic maps in urban environments. Progress in Electromagnetics Research 94, 243–262 (2009)

[28] Tayebi, A., Gómez, J., de Adana, F., Gutierrez, O.: The application of arrival and received signal strength in multipath indoor environments. Progress in Electromagnetics Research 91, 1–15 (2009)

[29] Azpilicueta, L., Falcone, F., Astráin, J.J., Villadangos, J., García Zuazola, I.J., Landaluce, H., Angulo, I., Perallos, A.: Measurement and modeling of a UHF-RFID system in a metallic closed vehicle. Microwave and Optical Technology Letters 54(9), 2126–2130 (2012)

[30] Moreno, A., Angulo, I., Perallos, A., Landaluce, H., Zuazola, I.J.G., Azpilicueta, L., Astráin, J.J., Falcone, F., Villadangos, J.: IVAN: Intelligent Van for the Distribution of Pharmaceutical Drugs. Sensors 12, 6587–6609 (2012)

[31] Nazábal, J.A., Iturri López, P., Azpilicueta, L., Falcone, F., Fernández-Valdivielso, C.: Performance Analysis of IEEE 802.15.4 Compliant Wireless Devices for Heterogeneous Indoor Home Automation Environments. International Journal of Antennas and Propagation (2012)
[32] Iturri, P.L., Nazábal, J.A., Azpilicueta, L., Rodriguez, P., Beruete, M., Fernández-Valdivielso, C., Falcone, F.: Impact of High Power Interference Sources in Planning and Deployment of Wireless Sensor Networks and Devices in the 2.4GHz frequency band in Heterogeneous Environments. Sensors 12(11), 15689–15708 (2012)
[33] Pan, J., Tompkins, W.J.: A real-time QRS detection algorithm. IEEE Transactions on Biomedical Engineering 32(3), 230–236 (1985)
[34] MSP430 microcontroller family, `http://www.ti.com` (last visit: January 2014)
[35] BlueCore 4, `http://www.csr.com`, (last visit: January 2014)

Questions

Q1. What are the possible wireless systems or combination can be employed to implement a low cost m-Health system?

Q2. What type of acquisition sensors are used in order to obtain biomedical data used within the mHealth system?

Q3. Does the surrounding environment have an impact on the performance of wireless transceivers employed for biomedical data acquisition and transfer?

Q4. What parameters must be taken into account in the design of the employed transceivers within an m-Health solution based on the combination of personal area network sensors and a mobile terminal gateway?

Q5. Indicate the relevant features of wireless communication in the case of transmitting data in an m-Health system.

12. mHealth Sensors and Applications for Personal Aid

P.S. Sousa, D. Sabugueiro, V. Felizardo, R. Couto, I. Pires, and N.M. Garcia

ALLab – Assisted Living Computing and Telecommunication Laboratory,
Instituto de Telecomunicações, University of Beira Interior, Covilhã, Portugal
{paulasousa,danieloliveira,virginie,
rgcouto,impires}@it.ubi.pt, ngarcia@di.ubi.pt

Abstract. The evolution of medical equipment and health care involves the miniaturization and autonomy of devices that are responsible for medical monitoring, screening and even therapeutic actions.

The latest generation of smartphones is increasingly being considered as handheld computers rather than as cell phones, due to their powerful on-board computing capability, capacious memories, large screens and open operating systems that encourage new application development. Recent medical applications for smartphones, such as the ones based on Android, Apple iOS, BlackBerry OS, Symbian, and Windows Phone, can be highlighted.

In this chapter, a review on this thematic is presented, and some of the most relevant projects in the monitoring and training lifestyles framework, such as a physical activity application, a platform for motivating behavior change, and an application to detect moments of stress using georeferencing.

The development of smartphone platforms and applications allows the creation of portable solutions for personal aid that are accessible anywhere and, most important, easily accepted by the user.

Keywords: Mobile health, Smartphones, Mobile communication, Mobile applications, Monitoring lifestyles, Training lifestyles, Behavior change, Emotional cartography.

1 Introduction

Given the lifestyle of today's society it has been increasingly important to adapt new solutions to technologies that have already been implemented. This is the case of mobile applications, to which there is already a high level of acceptance by the users.

Several applications related to fitness, health and medicine have emerged, aimed at making it easier for people to control and monitor several aspects of their daily life (*e.g.* nutrition) and daily activities (*e.g.* physical activity/exercise). Some of these applications may be used to alter lifestyles, motivating the user to be more active and/or change certain behaviors.

S. Adibi (ed.), *Mobile Health*,
Springer Series in Bio-/Neuroinformatics 5, DOI: 10.1007/978-3-319-12817-7_12

Some of this type of applications are discussed throughout the chapter, as well as the reasons that led to its development.

The remainder of this chapter is as follows: in Section 2, the relevance and challenges of the mobile solutions for Mobile Health are addressed; Section 3 covers some of the most to-date used communication architectures and technologies that are implemented in this kind of Mobile Health solutions; Section 4 presents applications developed in University of Beira Interior (Portugal) for monitoring and training lifestyles, namely physical activity, motivation for changing behavior and emotional states.

2 Relevance and Challenges

Smartphones have been a great success in the last decade. In a relatively short period of time, smart mobile technology has penetrated significantly into society, capturing an entire age spectrum of subscribers in western industrialized nations. Such progress has built upon a long history of the use of mobile communication devices and its rapid adoption in the latter part of the last century (Boulos *et al.*, 2011).

Smartphones have achieved a pervasive presence in society. Users use it to self-organize themselves, and many have adopted a culture of 'always connected' to their peer groups, communities of practice and information. Furthermore, mobile phones provide a portal into the entire world wide web of knowledge. Such continuous and pervasive social connectivity has important implications for society, and holds a lot of potential for use in education, healthcare and medicine, in particular (Boulos *et al.*, 2011).

In 2013, there were 6.8 billion mobile subscribers (Allied Health World, 2012). Analysts forecast over 7 billion subscriptions and more than 1 billion smartphones in 2014. The incredibly fast innovation of the past 10 years has put amazing new capabilities in the hands of consumers. Devices are expected to offer faster processors, bigger screens, and high quality cameras, while being always connected. But above all, they need to have energetic autonomy for at least a whole day (Cordova, 2010).

As smartphones are now part of our everyday life, smartphone applications have also become part of people's lives. There are now hundreds of smartphone apps focusing on wellness, fitness, and nutrition. Mobile health applications are on the rise. Currently, there are more than 500 mHealth projects worldwide and approximately 40,000 medical apps available for smartphones and tablets (Allied Health World, 2012).

Most health applications focus on prevention, and they are divided in two groups: those for individuals interested in mobile solutions to manage their health, and those intender for health professionals with the purpose of continuously training them, helping monitoring of remote patients, and managing health records. Many clinicians and allied health workers already adopted smartphones successfully in a diverse range of practices. Patients are also accessing health information, actively participating in their own care, and maintaining contact with their healthcare providers through smartphones (Boulos *et al.*, 2011).

By 2015, nearly 500 million users around the world will be using health applications. At this point, the health applications will probably have evolved to be perceived as "life facilitators" for patients, as well as for health organizations, enabling people to reduce the number of doctors making home visits, reduce hospital stays length and improve patient prescription tracking.

Although smartphones and smartphone health applications are on vogue nowadays, there are still many challenges that need to be considered and overcome (Boulos *et al.*, 2011):

- The usability design for this type of mobile platform must cater for older users' needs in two main areas, namely physical handling and maintenance of the smartphone, and the usage of the phone software itself;
- The reliability of the system and service, due to the possible negative sensation that the application may give to the user in case of malfunctioning, and due to the physical distance between technical maintenance teams and users;
- Loss or theft of devices that may impact upon the security of confidential digital health records or data held on mobile phones. The security of patient data held on mobile phones and the security risks of using mobile instant messaging in healthcare have been a concern;
- Patient attitudes and perceptions. When used as a method for monitoring the health status of remote patients, mobile phones should be applied only after due consideration to patient perceptions and feelings. Not all, mainly older, patients welcome the use of mobile technologies for remote monitoring. However, senior citizens of tomorrow will include the young and middle aged of today, who are more familiar with smartphones and the Internet, and are increasingly acquainted with the use of these technologies on a daily basis for study, work and leisure;
- Autonomy of the mobile platform in its operation, in order to provide a usable service to a target group that usually does not have any familiarity with technology and might even be unconscious during times of medical emergency and not able to manually operate any device or software; and, among others,
- Limitations of mobile phone applications, associated with the internal storage capacity, processing power and screen size of the mobile phone.

3 Communication Architectures and Technologies

Looking at the State of the Art, there are four major architectures at use on mobile health applications: the Wireless Body Area Network (WBAN), the Wireless Personal Area Network (WPAN) and the Wireless Local Area Network (WLAN) (Ngoc, 2008).

The WBAN network typically consists of one or more sensors, which measure data from the patient and send it to a Mobile Base Unit (MBU), represented by a smartphone, a Personal Digital Assistant (PDA) or even a laptop that is in contact with the Healthcare center. The data is sent to the Healthcare center by Global Packet Radio Service (GPRS), Telecommunications System (UMTS) or even over the Internet. Usually the WBAN uses ZigBee.

WPAN is a short-ranged network, with a wide compatibility, normally associated with low power consumption of technologies such as ZigBee, Bluetooth, RFID and Wi-Fi. Some of these technologies that are applied in the medical field will be discussed further in this section.

Being the first and oldest wireless technology used in the medical field, WLAN is a must for data transmission inside hospitals between departments and so on. It benefits from a simple installation, several levels of range (no need to use a network with a range above the needed), fairly low price point and its security make it a good solution to implement.

In the previous paragraphs some architectures were presented, but they are only useful when applied on technologies that require them. Further, some of these technologies and examples of their applications will be shown.

i. Bluetooth

Bluetooth is a wireless technology primarily designed to be a low-power short-range radio technology that could replace cables (Hewlett-Packard, 2002). It has now several years, but currently is still being used, and it comes bundled in almost every new smartphone, computer, tablet, etc. Bluetooth technology has a low cost and low power consumption since it can enter power saving sleep states, maximum range of one kilometer (not so short-ranged anymore), excellent resistance to WLANs interference since they use different frequencies, interoperability, best security in its class including immunity to "man-in-the-middle" attacks and, among others, the possibility of any device to be the master or slave in any communication (Hunn, 2009).

It can use two architectures known as *piconet* and *scatter net*. In *piconet* there is a master and up to seven slaves that only communicate with the master. A *scatter net* is nothing more than two or more linked *piconets*.

There are several examples of applications of this technology in the medical field. An example of application lies with a Blood Glucose Monitor for diabetic patients (Omre, 2010). Diabetes can evolve into some other complications if not managed correctly; one way to manage it is by measuring the blood glucose in order to control the normal levels. With Bluetooth built in a measuring device, the results from the measurements can be sent to the patient cell phone and from there to the physician's computer so he can analyze them. An early analysis by the physician can spot persistent abnormal cases and, in these situations, he can advise his patients accordingly by a simple phone call or short message service (SMS) much earlier than the typical quarterly review.

There are also products available in the market that use Bluetooth, such as the "*Accu-Check Combo*" which is a combination of a glucometer and an insulin pump. The Bluetooth technology is used by the combo for the communication between devices, enabling full remote control of the pump without the need to reach for it. Other example is bioPLUX (developed in Portugal by PLUX wireless biosignals S.A.), which captures several physiological signals (depending on the connected sensor) in real time and sends them to a computer via Bluetooth to further analysis (Figure 1). This product will be further discussed in section 4.C.

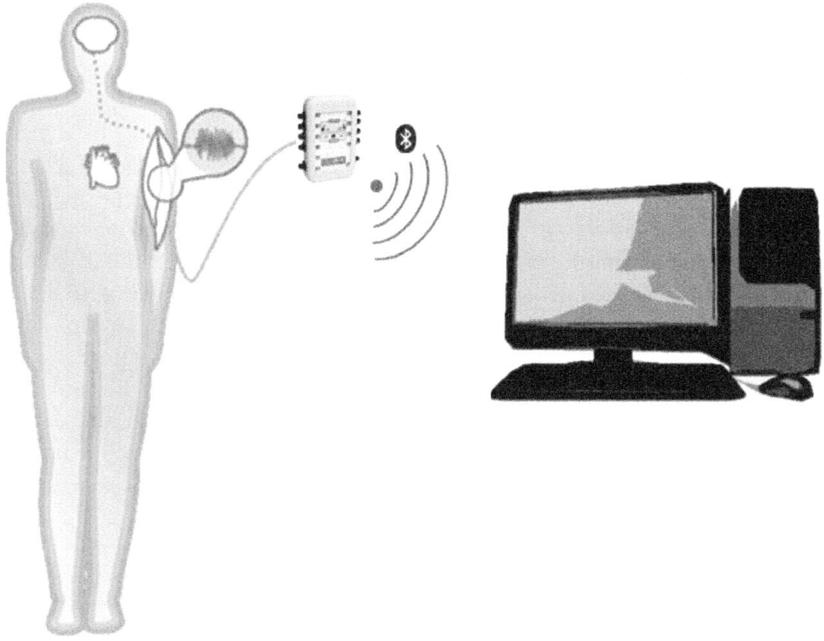

Fig. 1 Bioplux Bluetooth communication diagram (adapter from PLUX, 2011)

ii. Wi-Fi

One of the most used wireless technologies available to date is Wi-Fi, being implemented in devices such as cell phones, computers, video-game consoles, digital cameras, and printers, among others. It has undergone some changes and improvements up until the standard 802.11n, reaching a point that many industry analysts expect it to replace wired Ethernet as the network of choice (Aruba Networks Inc., 2012).

With this standard we can expect a jump on bandwidth, for example allowing more than two high definition videos streams at once, an increase of maximum number of users at one given time, a much higher communication speed (Cisco Systems Inc., 2009) and an increase in maximum indoor range (Barkay *et al.*, 2010).

In medical healthcare we can find applications that use Wi-Fi phones to allow the medical staff to be reached wherever they are working; wireless cardiac monitoring that allows the patients' hearts to be monitored all over the facility; real time location of medical staff and equipment allow for a simple monitoring, tracking and retrieval whenever it is necessary (Aruba Networks Inc., 2012).

iii. ZigBee

ZigBee is a wireless technology often used on home automation, since it can create personal area networks that use low power, don't demand high bandwidth or speed and is secure and low-cost. Hereupon, there are several topology models that can be used to implement a ZigBee network: Star, Peer-to-Peer and Cluster Tree.

Each one of the topology models has its own characteristics, although all have the same FFD (full-function device, used as the network coordinator(s)) and RFD (reduce-function device, used for simple applications) components (Jung *et al.*, 2007).

Besides home automation, ZigBee technology has been implemented in the healthcare field. According to ZigBee Alliance (Heile, 2012), there are several ZigBee Certified Products for Health and Fitness (ZigBee Alliance, 2013). Among those, there are the *Bed Sensor Agent/Controller*, that can be used with a bed or chair sensor to make an alarm call if a person that is sited on a chair or lied on a bed gets up and doesn't return within the preset time; the *Blood Glucose Reader* device that uses the ZigBee technology to wirelessly transmit the information collected from the One Touch Ultra 2 Blood Glucose meter, being an ideal device for remote patient monitoring; and the *Pill Dispenser* that keeps the medical staff alerted to the medication schedule compliance from patients in real-time and remotely, if need be.

iv. Radio-Frequency Identification

Radio-Frequency Identification (RFID) is a wireless non-contact short-range technology developed to be used to identify and track tags attached to objects. Commonly, tags are integrated circuits with an antenna or coupling element used for communication with the RFID reader, that don't have an on-board power source, the energy comes in the form of radio frequency signals. These readers use radio frequency channels to identify the tag, and to identify the object in which the tag is attached to (Weis, 2007).

One important aspect of RFID is the low price of the tags, although the readers have a fairly higher price point. Normally, tags used in the medical field have a passive power source and costs between $0.7 and $0.15, but it is expected a decrease to $0.5 if the production of these tags becomes massive every year. The readers are indeed more expensive, ranging from $500 up to $2000, depending on the required features, but for that high investment comes the very low price priced investment of the tags, since each reader can operate with a vast number of tags (operating with the same frequency as the reader).

In the medical field there have been several uses for RFID, such as patient tracking systems inside medical facilities (Motorola Solutions Inc., 2011). With this type of system, patients can be tracked at any moment; if a patient shouldn't leave a location an alarm can be triggered if he tries to do so, and it can even prevent mistakes like wrong medication or treatment.

RFID solutions can also be used for the working staff as well. A team of three researchers (Seth Freedman, Dave Mullinix and Dr. Andrew Fine) have developed a smart bracelet named IntelligentM (Owano, 2013). This bracelet comes with embedded RFID tags and an accelerometer, which allows monitoring how much time a healthcare provider has been without washing his hands, and it will even warn him in the case of not washing his hands properly. This solution is very important, since many infections are caught in hospitals due to the lack of hand hygiene (World Health Organization, 2007).

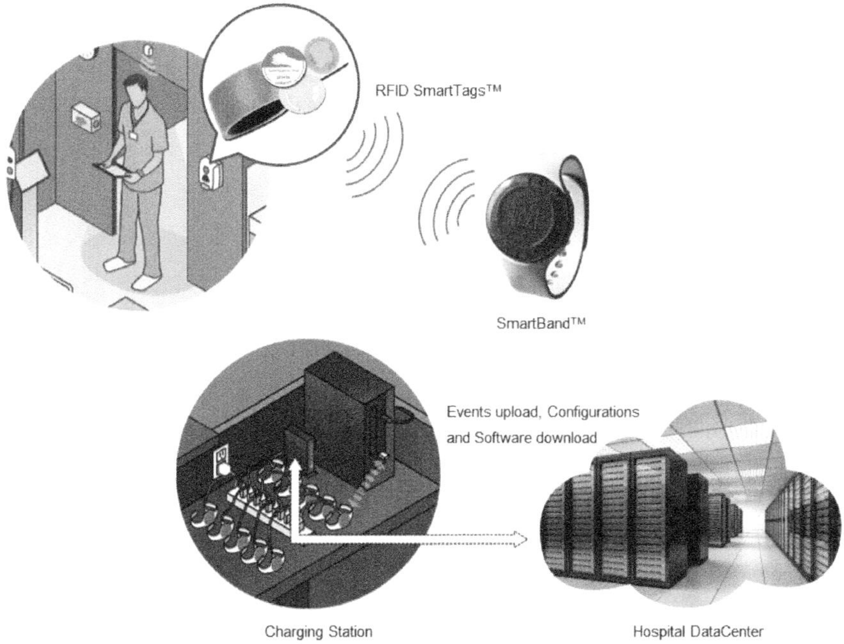

Fig. 2 Operating architecture of intelligentM (adapted from IntelligentM, 2013)

v. Long Term Evolution (4G LTE)

Long Term Evolution is the latest generation of mobile network technologies that offers higher speed and latency than previous generations, improved responsiveness allowing real-time communication and real-time applications, enhanced security and Quality of Service differentiation and besides all that it is backwards compatible and future-proof (Verizon Wireless, 2009; Place et al., 2012).

Since it's still a recent technology, there aren't many mobile health applications/devices that really need or use its full capacity, but for example, there is the InTouch RP-Xpress that allows remote monitoring of patients vitals using Wi-Fi or 4G LTE for its communication.

4 Monitoring and Training Lifestyles

This section presents three exploratory projects developed within the Assisted Living Computing and Telecommunications Laboratory (ALLab) in University of Beira Interior (Portugal):

- Physical activity – a mobile application developed to motivate physical activity that measures the jump flight time;
- Motivation for change in behavior – an application to motivate individual behavior change, more specifically, smokers;
- Emotional stress – an application aimed at creating an emotional rendered map.

4.1 Physical Activity

Regular physical activity is recommended for combating tension, with beneficial effects at several levels, for example, in the control of chronic diseases such as obesity or diabetes (Choi *et al.*, 2005).

Physical activity can be monitored using several sensors such as accelerometer, gravity sensors, location sensors or GPS (Global Positioning System), gyroscope, among others that allow motion detection (Lau and David, 2010).

Mobile devices embody multiple sensors that measure the physical activity, and allow performing the estimation of the travelled distance, number of steps, and energy expenditure, among others.

The estimation of energy expenditure (in calories) during physical activity has been discussed by several authors, which validated different methods for estimating it using the accelerometer collected data (Felizardo *et al.*, 2011). Depending on the effort made during physical activity, the accelerometry standards allow identifying its level of intensity (Hendrick *et al.*, 2009).

There are several online platforms related to physical activity, such as *Fatsecret.com* and *MyFitnessPal.com*, that allow to keep track on a diet and exercise plan, which estimate the amount of calories in ingested food and calories expended in activities such as walking, cycling, resting, and sleeping. The users can also access the Web platform to manage and monitor their data. Both applications are available for Android and iOS operating systems.

Other applications developed only for iOS include *Calories Burned*, which estimates the caloric expenditure during specific activities based on the data entered by the user; *Oracle Fitness*, which estimates the energy consumption and the real and basal metabolic rates using the coefficient of exercise based on the application's running time. For the Android operating system, there is the *Calories Burned Calculator*, which estimates the calories burned based on the metabolic rates.

In this scope, a project is being developed in University of Beira Interior (Portugal) with the aim of promoting the practice of physical activity, providing information about the energy expenditure, distance travelled, etc., during several small activities.

The main challenge of this project was to find a method for estimating the number of calories expended using the accelerometer and GPS. To overcome this challenge, an application was developed for calculating a vertical jump flight time using the accelerometer embedded in an Android Smartphone. A vertical jump consists of several phases (take-off, flying, and landing) that can be easily identified by the data collected by the accelerometer (Quagliarella *et al.*, 2010).

Thereby, and considering the project main objectives, the mobile application is constituted by three stages of research and development:

- Calculation of jump flight time (using accelerometry sensors);
- Estimation of travelled distance (using a GPS receiver); and
- Calculation of calories spent during physical activity (using accelerometry sensors).

For the jump flight time calculation, several patterns were identified (within the three jump phases) and used to define an algorithm for its calculation, using root mean square method for a 95% confidence level, taking in account the 550 tests performed. The caloric expenditure was calculated using the validation equation in Felizardo *et al.* (2011).

Besides the mobile application, the project also consists of a web platform. The web platform offers a way for the user to monitor daily physical activity, manage his data and define goals to improve his life quality, while establishing a network of people who care about their health and feel motivated to maintain a healthy lifestyle.

The mobile application for calculating the jump flight time is available for Android (IMSPDev, 2013) and iOS (Pires, 2013).

4.2 Motivation for Change in Behavior

Recent studies have reaffirmed that keeping up with constant motivation helps people to easily alter habits and behaviors.

There is a set of pathologies or behavioral dysfunctions that are primarily associated with specific lifestyles, such as chronic anxiety, smoking, alcoholism and in some cases obesity. This and other pathologies and behavioral dysfunctions may see their negative impact minored if patients get to be monitored and trained to correctly change their lifestyles.

An experiment conducted in the United Kingdom by the London School of Hygiene and Tropical Medicine (Free *et al.*, 2011) aimed to evaluate the effect of a smoking cessation program delivered through mobile phone automated text messages.

This experiment was performed using two groups of smokers. One group included subjects who tried to quit smoking using only their own motivation and perseverance, and the other group subjects who tried to quit smoking using an auxiliary motivational tool. The auxiliary motivational tool consisted in a set of five daily-programmed SMS (short message service) messages sent to the subjects of the second group during five weeks. The SMS messages were motivational messages *e.g.* "Today is the day you will quit smoking!" or "Keep up the good work".

The results showed that 10.7% of the volunteers quit smoking after completing the SMS messages experiment. This index represented more than double of the smokers of the group that tried to quit smoking but did not receive the motivational SMS messages, which had only a 4.9% success rate. In total, almost 6 thousand British smokers participated in this experiment, including the ones that received the SMS messages and the ones that didn't receive it.

The emergence and pervasive use of information and communication technologies brings new tools that can be used in this type of treatment program. A technological solution has been proposed by the University of Beira Interior (Portugal), which consists in a platform that interacts with its users employing Internet communication, SMS communication or Multimedia Message Service (MMS) communication. This solution combines privacy, pervasiveness, and an easy way to use mobile communication services like SMS and MMS with Internet rich content interfaces for exchanging, requesting and presenting data to the end user.

This solution, as shown in Figure 3, can be seen as a service that a person may enroll, by sending a SMS message with specific keywords, or by filling in an online form. This service will then evaluate the user's motivation by running him through a set of specific questionnaires. After these quick questionnaires, the user will enter in a treatment cycle, where he will receive a set of programmed motivational messages and interaction messages many times every day. The system may keep up his motivation level and at the same time his change in behavior is being monitored, and this way the system is able to personalize the interactions with his most recent treatment status. At the end of this cycle, the system will decide whether the user shall restart the cycle or not by matching the user's objective in the program with the current results.

This solution, designed for monitoring and modeling lifestyles, can be used, for example, to complement smoking cessation programs, obesity control, or even for anxiety control. The effectiveness of this type of solutions has been proved, as it can be stated in the experiment conducted by the London School of Hygiene and Tropical Medicine (Free *et al.*, 2011).

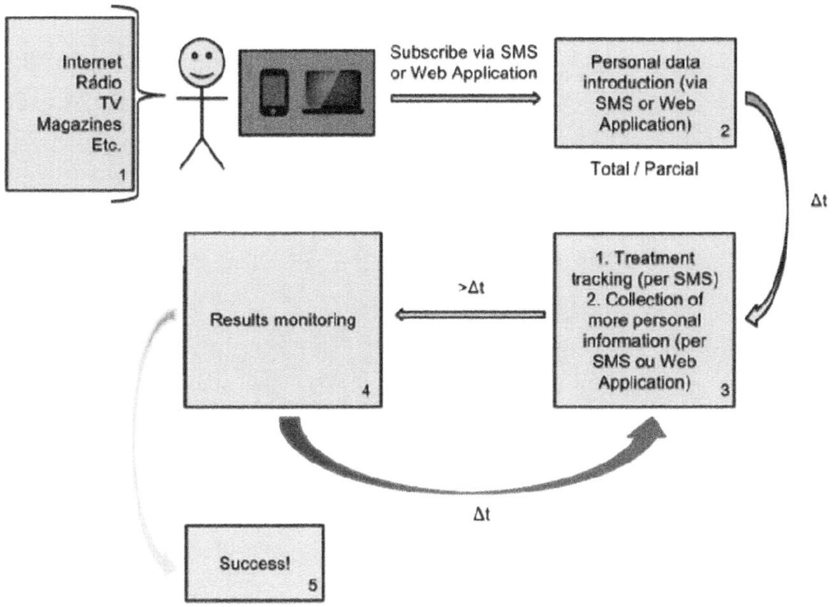

Fig. 3 Workflow for the proposed solution for monitoring change of behavior

4.3 Emotional States

This section provides a new approach based on physiological, emotional and geographical aspects to further the project on "emotional cartography", and introduces the concept of subjective or emotional element of space. Emotional cartography is the creation of a rendered map which represents several physiological parameters in its different points. Physiological measures that might be related to a person's emotional response to a given situation on a given environment have been the focus of research. This is the basis for the argument that emotional responses to space are highly relevant as an aid in navigating our environment (Gartner, 2012).

The utility of an emotional response to space is thought to enhance (positively or negatively) the ability to develop a cognitive map; emotional processing is presumed to be automatic and without conscious choice. The ability to link the emotion to a specific event may facilitate not only way finding and navigation during times when one's cognitive state is overloaded, but it may actually speed up the everyday process of cognitive mapping (Gartner, 2012).

Since today's society is increasingly governed by information, where control of all data is important, the data obtained from the daily activities may be relevant. By harnessing this information, it is possible to obtain data on the physical stress and the body reactions from the daily activities as well as to the surrounding environment. This information will be useful to outline health and exercise plans, and to evaluate the emotions in response to an object or a route in a specific location.

In this context, an application was developed with the aim of recording physiologic data along with the location of the user, combining the bioPlux (Plux, 2013) and a Global Positioning System (GPS).

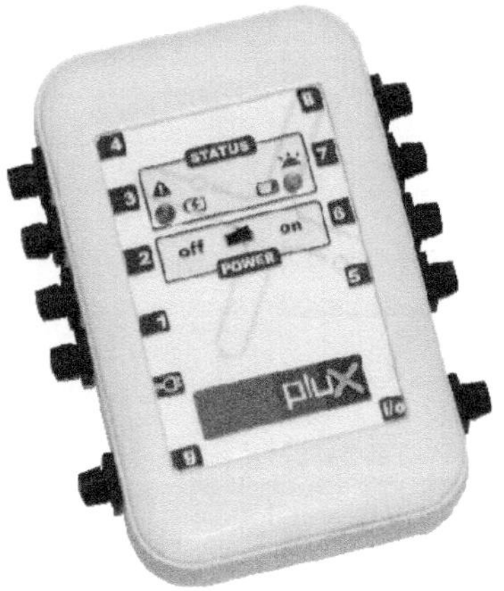

Fig. 4 bioPlux, biosignal acquisition system (Plux, 2013)

The bioPlux (Figure 4) is a device that collects and digitizes the signals from the sensors (electromyography, electrocardiogram, electrodermal activity, triaxial accelerometer, peripheral temperature, respiratory, electroencephalography), transmitting them via Bluetooth® (Connectivity - 10m range, with standard adapter; >100m with high range adapter) to a computer or smartphone, where they are viewed in real time. The 8 channels of the bioPlux have 12 bits, and its sampling frequency is 1 kHz. The bioPlux has also a digital port, a terminal for connecting the AC adapter and charge the internal battery (allows a lifespan of approximately 12 hours), and a channel to connect the reference electrode, which is essential for the proper monitoring of electromyography signal (Plux, 2013).

The GPS provides the following data: time, latitude and longitude. The date and time are provided by the computer. This information will allow synchronizing the values gathered from the bioPlux with the values collected by the GPS. Thus, in the sample of information it will be possible to identify the time *hh:mm* in which the user obtained x values in the coordinates (x,y).

This kind of solution allows obtaining a variety of data about the user in real time in a particular location. Changes in physiological parameters, such as the cardiac frequency and electrodermal activity, can provide important information about the emotional state (particularly stress) of the person. The geolocalization of physiological data allows the monitorization of daily activities, adding location information and its relation to the stress level.

5 Future Trends

User acceptance of eHealth or mHealth solutions is still pretty much an unsolved issue despite of the literally thousands of applications in the Google, Apple and other online mobile application stores, some of them sponsored by health product manufacturers, such as *e.g.* Durex. It seems that smartphones are viewed by all major players as a useful device when it comes to personal health care, as suggested by the effort the developer community is putting on producing mobile applications that run on these platforms. Recently, manufacturers have developed additional hardware to connect to smartphones, in order to allow them to perform tasks that were never intended to be deployed on a smartphone, such as, for example, to transform a smartphone into a microscope (iClarified, 2013). Other researchers focus on developing chips that can analyze body fluids and communicate with the smartphone to assess and transmit the results to a remote healthcare provider (Ryan, 2011). Other more resourceful users have used the integrated camera to achieve similar results (H., 2013).

If on one hand user acceptance is still low, on the other hand, smaller devices have been developed to serve as aid to user in fields related to health, such as body fitness or diagnosis. Recently, one of the Microsoft Imagine Cup contest winners developed a device that identifies blood type in an inexpensive and fast manner (Microsoft, 2013). In the group of manufacturers of independent monitoring and sensing devices that communicate with the user's smartphone, the major players are the manufacturers of sports equipment, such as Nike, Adidas, Polar, and so many others. The review and comparison of these devices would alone be a topic for a full book, although this would be almost useless because of the fast innovation rate and of the fierce commercial competition.

From the authors' perspective, in a short and medium timeframe period, there are two major trends on the area of Sensors and Applications for Personal Aid, in the context of mHealth.

The first trend is clearly related to the continued development of applications that run on commercial smartphones. This is an area where there is still a huge amount of work and research to be done, particularly on the field of validation and verification of results. As an example, a vast majority of the applications that claim to assess the caloric expenditure of a user, relying on the inbuilt sensors of a smartphone, lack statistical validation of its results. From another perspective, the set of sensors that usually are integrated into a smartphone is stable for quite some time now, and it will be interesting to see a new generation of smartphones with new sets of sensors, such as, for example, infrared cameras and stereoscopic vision, temperature, altitude, humidity and barometric pressure, or bio-signal sensors (*e.g.* electrocardiography and/or electromyography sensors). The addition of a new family of sensors, many of them present in other commodity devices such as for example some Casio, Tissot and Citizen watches, will allow the deployment of a new family of applications that will further enrich user's experience and contribute to the advancement of the mobile health area. The limitation for sensor integration is of course related to the limitation in space and energy consumption and it will take some amount of time, as foreseen by Moore's Law, to allow increased miniaturization of additional components.

The second trend is related to the development of devices that, being small, portable or even wearable, use the smartphone as a gateway for data processing and communications, in a context close to ubiquitous computing. Such devices, such as the microscope or the body fluid analysis chip referred previously, act as additional sensors to the smartphone, allowing it to perform tasks that are uncommon to the device, and possibly that do not have a wide market as to justify the integration of such sensors in the device itself. In many cases, the device will not need the smartphone as communications or data processing gateway, and will have the ability to perform its tasks in a standalone manner, relegating the integration of the collected data with the user health profile to a later point in time.

Without being experts at marketing, we propose a traditional approach. Considering the proposed five "P" factors of marketing and business success, that is Product, Price, Promotion, Place and People, we can reach interesting conclusions regarding why some applications are not user adopted while others are downloaded by the millions. The new distribution platforms for mobile applications on the Internet, known as mobile applications stores, eliminate the "People" and "Place" success factors for traditional marketing, as every potential customer is exposed to the same interaction with the platform and the product is available everywhere at the same cost. One can always argue that "People" is still a factor because of the need of post-sell service, but we believe that on the time of the first purchase, this factor weights almost zero on user decision. Moreover, known online applications stores fail to provide a human interface to the end user. "Price" is also an attenuated factor, not only because many of these mobile applications cost zero, but because the vast majority of the applications is sold at extremely low prices. So, following the traditional marketing rules, only two factors have impact on a mobile application success: "Product" and "Promotion".

Considering these two factors, the user adoption rate for a mobile health application is an issue for two major reasons: primarily, because of product inefficiencies; and secondly because of poor promotion. Product inefficiencies are, on its turn, due to two causes, the first one is related to poor application and interface design, and the second one, related to the user perceived value of the application.

In the first case, the user is not seduced by the application look and feel, rendering the application unusable and to some extent, un-promotable. Usability and accessibility issues are a well-studied area of science, and this limitation falls entirely in its scope, *i.e.*, the product must look good and feel good so the user feels comfortable using it. The second cause for user non-adoption of a solution is that the user does not perceive the value of using the solution. The tradeoff between "use-effort" and perceived "use-benefit" does not appeal to the user, and therefore, he/she chooses not to use the application. The design, usability and accessibility issues of applications, as mentioned before, falls within the scope of a particular area of engineering and design, both for interfaces and for software. In a competitive market such as the online applications, good usability and user-appeal plays the role of natural selection. The second factor may be attenuated with other approaches, namely, by improving user education and information, by adding application gamification, and of finally, by adding a social network component to the application. The gamification and social networking aspects of mobile applications have proved to be a sure asset to increase the user adoption ratios in cases such as the applications developed by Nike, Polar and Adidas, to use with their fitness equipment. Finally, user education and information must be implemented when the user really needs the mobile application to improve his/her health condition, and these tasks must be considered and planned while planning and developing the application.

Extending the concept of health to the wider concept of Ambient Assisted Living, other types of devices will be added to the list of equipment that will interact with the user and with the user's environment, and in a ubiquitous computing perspective, collect, analyze and share data in order to monitor, assess and control the user environment. In such a scenario, the smartphone will be used as a gateway to the monitoring platform of the environment, allowing the user to personalize environment parameters, to view results and take action on alarms.

References

Allied Health World, Smartphone = Smarter Healthcare? The rapid rise of mobile heath management tools (2012)

Aruba Networks Inc., Secure Wi-Fi For Healthcare Applications, Aruba White Paper (2012)

Barkay, O., Quimby, D., Vallejo, M.: Accelerating the Enterprise Network Using 802.11n Wireless, IT@Intel White Paper, Intel Corporation (2010)

Boulos, M.N.K., Wheeler, S., Tavares, C., et al.: How smartphones are changing the face of mobile and participatory healthcare: an overview, with example from eCAALYX. Biomed. Eng. OnLine 10(24) (2011)

Choi, J.H., Lee, J., Hwang, H.T., et al.: Estimation of activity energy expenditure: accelerometer approach. In: Conf. Proc. IEEE Eng. Med. Biol. Soc., vol. 4, pp. 3830–3833 (2005)

Cordova, S.: Harvesting the Future of Mobile, ST-Ericsson (2010)

Cisco Systems Inc., Key Performance Benefits of 802.11n, White Paper (2009)

Felizardo, V., Gaspar, P.D., Garcia, N.M., et al.: Acquisition of Multiple Physiological Parameters During Physical Exercise. International Journal of E-Health and Medical Communications 2(4), 37–49 (2011)

Free, C., Knight, R., Robertson, S., et al.: Smoking cessation support delivered via mobile phone text messaging (txt2stop): a single-blind, randomised trial. The Lancet 378(9785), 49–55 (2011)

Gartner, G.: Openemotionmap.org - Emotional response to space as an additional concept in cartography, International Archives of the Photogrammetry, Remote Sensing and Spatial Information Sciences, vol. XXXIX-B4 (2012)

H. M., Scientists use iPhone microscope to diagnose intestinal worms (2013), `http://www.phonearena.com/news/Scientists-use-iPhone-microscope-to-diagnose-intestinal-worms_id40939` (accessed September 2013)

Heile, B.: ZigBee Alliance Chairman, ZigBee Health Care (2012)
Hendrick, P., Bell, M.L., Bagge, P.J., et al.: Can accelerometry be used to discriminate levels of activity? Ergonomics 52, 1019–1025 (2009)
Hewlett-Packard, Understanding Bluetooth, White Paper (2002)
Hunn, N.: Bluetooth – The Wireless Ecosystem for Health, Fitness and Assisted Living, WiFore Consulting (May 2009)
iClarified, First Professional Microscope Released for iPhone and iPad (2013), http://www.iclarified.com/30227/first-professional-microscope-released-for-iphone-and-ipad (accessed September 2013)
IMSPDev, JumpTimeCalc, Google Play (2013), https://play.google.com/store/apps/details?id=com.imspdev.jumptimecalc
intelligentM, intelligentM website (2013), http://www.intelligentm.com (acessed December 2013)
Jung, S., Chang, A., Gerla, M.: Comparisons of ZigBee Personal Area Network (PAN) Interconnection Methods. In: IEEE 4th International Symposium on Wireless Communication Systems 2007, Trondheim, October 17-19, pp. 337–341 (2007)
Lau, S.L., David, K.: Movement recognition using the accelerometer in smartphones. In: Future Network and Mobile Summit 2010, Florence, June 16-18, pp. 1–9 (2010)
Microsoft, Portugal: World Citizenship Competition (2013), http://www.imaginecup.com/Finalists/Team?teamName=forabetterworld&year=2013#?fbid=yUSlhaT01UJ (accessed September 2013)
Motorola Solutions Inc., RFID Solutions for Healthcare for Healthcare Reducing Costs and Improving Operational Efficiency, White Paper (2011)
Ngoc, T.V.: Medical Applications of Wireless Networks (2008), http://www.cse.wustl.edu/~jain/cse574-08/ftp/medical/index.html (accessed August 2013)
Omre, A.H.: Bluetooth low energy: wireless connectivity for medical monitoring. Journal of Diabetes Science and Technol. 4(2), 457–463 (2010)
Owano, N.: Smart bracelet uses RFID to bleep dirty hands (2013), http://phys.org/news/2013-03-smart-bracelet-rfid-bleep-dirty.html (accessed August 2013)
Pires, I.: JumpTimeCalc, iTunes store (2013), https://itunes.apple.com/us/app/jumptimecalc/id654811255?mt=8
PLUX wireless biosignals S.A., biosignalsplux wearable body sensing platform (2013), http://www.biosignalsplux.com/
Quagliarella, L., Sasanelli, N., Belgiovine, G., et al.: Evaluation of standing vertical jump by ankles acceleration measurement. Journal of Strength and Condition Research 24, 1229–1236 (2010)
Ryan, S.: iPhone could be used to diagnose illnesses (2013), http://www.dailyedge.ie/iphone-could-be-used-to-diagnose-illnesses-292230-Nov2011/ (accessed September 2013)
Weis, S.A.: RFID - Radio Frequency Identification: Principles and Applications (2007)
ZigBee Alliance, ZigBee Certified Products – Health and Fitness, Website (2013), http://www.zigbee.org/Products/ByFunction/HealthFitness.aspx (accessed August 2013)

Questions

Q1. Why mHealth applications for personal aid are relevant?

Q2. What is the applicability of mobile solutions using GPS and physiological data acquisition system simultaneously?

Q3. What benefits can we achieve from the use of technologies like RFID, Bluetooth, among others applied in the medical field?

Q4. What is the aim of creating an application that measures the jump flight time and calories burned?

Q5. Why using SMS technology is preferred in order to communicate with systems for behavior change?

13. Location-Aware Services Using Android Mobile Operating Platform for Safety, Emergency and Health Applications

Prabhu Dorairaj[1,] Abbas Mohammed[2], and Nedelko Grbic[2]

[1] Blekinge Institute of Technology, Sweden
[2] ExAudio Limited, Sweden

Abstract. The introduction of smart phones has redefined the usage of mobile phones in the communications world. Smart phones are now equipped with various sophisticated features such as Wi-Fi, Global Positioning System (GPS) navigation capability, high-resolution camera, touch screen, and broadband access which helps the mobile phone users to stay connected and updated. Many of these features are primarily integrated with the mobile operating system and are out of reach of the public; that is the users can not manipulate these features. However, this situation has changed with the advent of an innovative, user friendly and versatile operating system known as "Android". It was developed by the Open Handset Alliance, an open source initiative piloted by Google. The open system architecture with customizable third party development and debugging environment features help the users to manipulate the features and create their own customizable applications.

In this chapter, we propose a location-aware remote system using Google's Andriod mobile platform for safety, emergency and health scenarios and develop an Andriod application for this purpose. Emergency is divided into three categories; heart abnormalities, security threats, and road accidents. The first scenario we investigated is targeted towards heart abnormalities monitoring. The heart rate monitoring device is integrated with the proposed application to measure the heart rate of a user driving a vehicle, and if abnormalities are detected the application performs a dual role. The application uses GPS to track the location information of the user and sends this information via SMS, email and posts it on the user's Facebook wall. Simultaneously, an emergency signal is sent to an Arduino microcontroller which will in turn trigger an alarm in the vehicle to notify the emergency situation. The second scenario is targeted towards road accidents. The proposed application is designed to detect that an accident has occurred using the sensors in the Andriod mobile phone and, consequently, an action is taken by the application to notify the emergency services that an accident has occurred and its location. Finally, the third scenario we investigated is security threats which can occur anywhere and the proposed application can also offer assistance in personal safety situations. In this case, the user interacts with the application by pressing a designated button and the application will automatically calculate the geographical information and send the location information via SMS and email to a pre-stored emergency contact, and also post the information on the user's Facebook wall.

S. Adibi (ed.), *Mobile Health*,
Springer Series in Bio-/Neuroinformatics 5, DOI: 10.1007/978-3-319-12817-7_13

1 Introduction

The use of mobile devices has given rise to a variety of innovative technologies, and the exchange of information globally has exploded and become a crucial part of everyday life. Smart phones have provided new and extended dimensions to the usage of conventional mobile phones. Apart from the basic functionalities such as calling, messaging and cameras, smart phones have laid the path to portray a personal computer. Not only the mobile phone looks newer, it is the operating systems and applications which are built to meet various features of the hardware that also made a considerable difference [2, 4].

The mobile phone has now become a major source of information and is an integral part in many people's lives. Mobile devices with sufficient computing capability are widely used to access networks via mobile communication networks nowadays. Different categories such as games, social networks, and healthcare applications are being developed to meet the user's needs [9, 10]. Each mobile user is unique, e.g., one wants to use the basic functionalities of the smart phone, the other wants to use the built-in applications, and the most advanced users want to manipulate the hardware to develop their own customized applications [1, 3, 5].

There are four major smart phone operating systems at present; namely, Symbian, Windows Mobile, iOS and Andriod [2]. Out of these operating systems, Andriod is the most dynamically developing mobile platform where almost half of the world's mobile users are using Andriod enabled devices.

The addition of GPS in mobile communication devices is regarded as one of the most useful applications that have been developed recently [6]. GPS is a satellite-based radio navigation system that provides reliable positioning, navigation and timing (PNT) information for devices equipped with appropriate GPS receivers. However, how can the location of a person be tracked and notified using Google's Andriod platform in case of an emergency [7, 8] and with no user interaction (e.g., in case of a heart attack)? In this chapter we present a GPS application for remote emergency tracking systems to provide location-based information of users in emergency situations using Google's Andriod mobile platform.

2 Integration between Hardware

2.1 Integration of Andriod Mobile Device with Arduino Microcontroller

In a mobile phone, every sent request has its own corresponding response. For instance, a phone call is alerted to the user by a ringtone, a text message received is displayed on the screen, a photo taken by the camera is saved on the phone's SD card and so on [5, 6]. These events are generated on the phone itself; however they can also be activated or triggered through a sensor (accelerometer) or on a microcontroller. In this situation Amarino is used [17, 18]. Amarino is a tool kit which helps in the integration of Andriod with Arduino. It consists of the Andriod application and libraries required for Arduino. Amarino helps to connect a mobile device enabled with Andriod and an Arduino microcontroller via Bluetooth.

2.2 Integration of Andriod Mobile Device with Heart Rate Monitoring Device

The heart is a main organ in the human body and the heart rate is an important factor to be considered for health monitoring of patients with heart complaints. The heart rate tells us how many times the heart beats in a minute and it is usually measured by checking the pulse on any area near an artery. Low or high heart rate can be dangerous to health, and so the heart rate should be properly monitored for people with heart complaints.

Heart rate monitoring is usually done by regular medical checks at hospitals or health centers. This is a normal scenario, however there are many situations where the heart rate is not monitored, e.g., while driving a vehicle or during exercise and so on. Thus, the use of a heart rate monitoring device is very essential for a patient who needs it. Heart rate monitoring is done at hospitals using electrocardiogram (ECG) tests. Even though ECG tests are very accurate they are also costly. In addition they require regular hospital visits, which will increase the already strained healthcare expenditure in many countries and also impact adversely on the carbon footprint. In addition, it is important for a person with a heart complaint to be able to monitor his/her condition continuously. To mitigate these problems, a heart rate monitoring device has to be purchased and used for regular personal healthcare check. Nowadays heart rate monitors are increasingly being used by ordinary people rather than only by health care professionals in a hospital or health clinic.

A heart rate monitor helps to detect the abnormalities in the heart and displays the result directly to the user. This feature has motivated us to use such a device in emergency situations especially when a person is driving a vehicle and is suddenly faced with a heart attack. To get assistance in this situation, the monitoring device can send an alert to the Andriod enabled mobile phone which is in turn halts the vehicle to avoid accidents. There are various heart rate monitoring devices available on the market such as the Zephyr Bluetooth (BT) heart rate (HR) monitor, polar Bluetooth heart rate monitor, Wahoo Fitness ANT plus Dongle, etc. [11].

The Zephyr HR Bluetooth heart rate monitor (*figure 1*) uses a fabric sensor to detect the data. This heart rate monitor is best suited for our proposed application for various reasons. Firstly, it is an open source and can be easily programmed. It is equipped with Bluetooth connectivity and so the transfer of data is done wirelessly which improves mobility and reduces hardware costs [12, 13]. In addition, speed and distance can also be displayed on this device which helps to show a simple pictorial representation of a person's heart rate. Finally, an important feature about this device is its tolerance to extreme body motion such as running, jumping, jogging and so on.

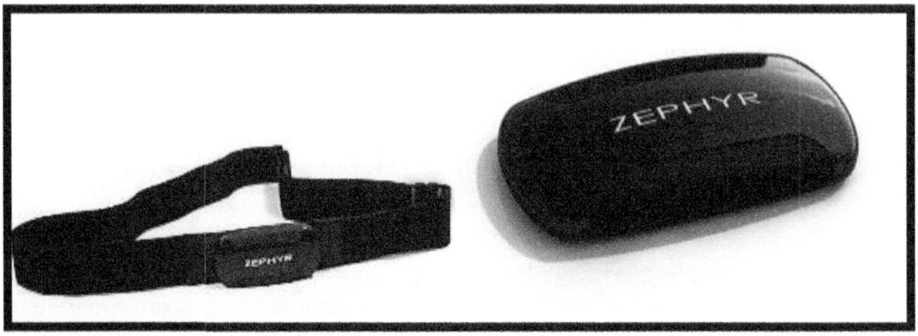

Fig. 1 Heart rate monitoring Device

3 Research Methodology

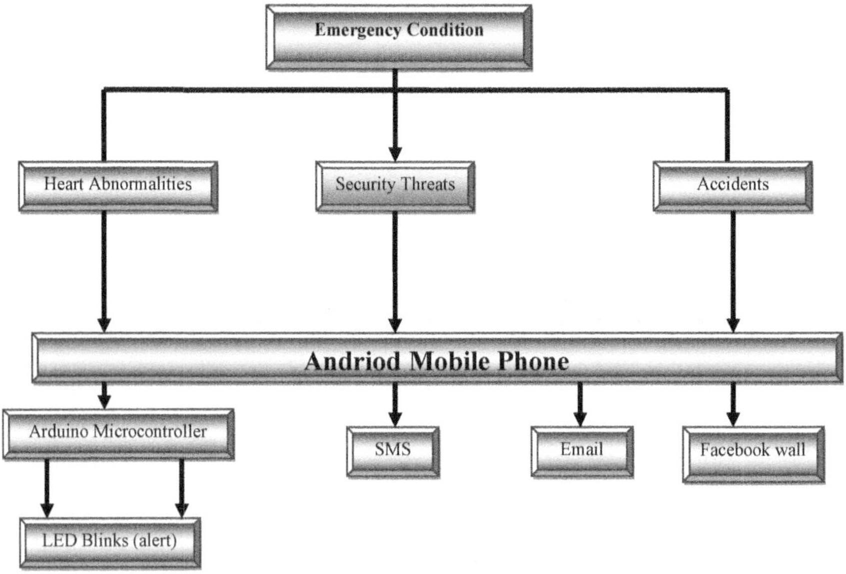

Fig. 2 Tree structure showing the overall workflow of the Emergency Tracking System (ETS) application

3.1 Main Research Questions

In this chapter we will try to answer the following important question:

1. How can the user's location be tracked and notified using Google's Andriod platform in case of an emergency?
2. How to monitor the heart rate and analyze the registered data to detect an emergency situation using an Andriod mobile platform?
3. How to integrate Arduino microcontroller with an Andriod mobile device?

3.2 Expected Outcomes

The expected outcomes of the proposed Emergency Tracking System (ETS) application are:

- The exact location of users will be tracked using GPS, and the location information is then sent to a pre-stored contact via SMS, email and posting of a message on the user's Facebook wall.
- The heart rate will be monitored and if abnormalities are detected a notification is sent to an Andriod mobile device. In addition, the location of the user will be tracked and sent to an emergency number, email and Facebook wall as shown in *figures 8.1, 8.2 and 8.3*, respectively.
- Under critical situations, the Andriod mobile sends a signal to a microcontroller and a LED in that the microcontroller blinks to indicate an alert signal as shown in *figure 2*.

4 Case Studies

4.1 Heart Rate Abnormalities Monitoring Implementation

The scenario investigated in this chapter deals with a user driving a vehicle. A typical scenario is, when a person driving on isolated roads, wearing the Zephyr HR monitoring device around the chest. The Zephyr HR monitors the heart rate of the driver and sends the information every second to an Andriod mobile phone via Bluetooth. The heart rate level is normal between 60 and 100 beats per minute (BPM); if the heart rate is less than 60 BPM it is known as *Bradycardia* and when more than 100 BPM it is known as *Tachycardia*. In most conditions the heart beat is lower when there is dehydration and decreased protein intake, and it is higher in uncontrolled hypertension [8].

4.1.1 Modeling of Heart Rate Abnormalities Monitoring Scenario

The proposed application is designed to sense the heart rate, and if there are any abnormalities in the heart rate (e.g., if it goes below 60 or above 100 BPM), then the ETS application will automatically send a signal to an Arudino microcontroller. The mobile phone and microcontroller are connected wirelessly via Bluetooth. The Arduino microcontroller in turn will trigger an alert signal which in our application is indicated by a blinking LED. Simultaneously, the application will track the location information of the user under an emergency situation and sends it to a remote pre-stored emergency contact number. This scenario is shown in *figure 3*.

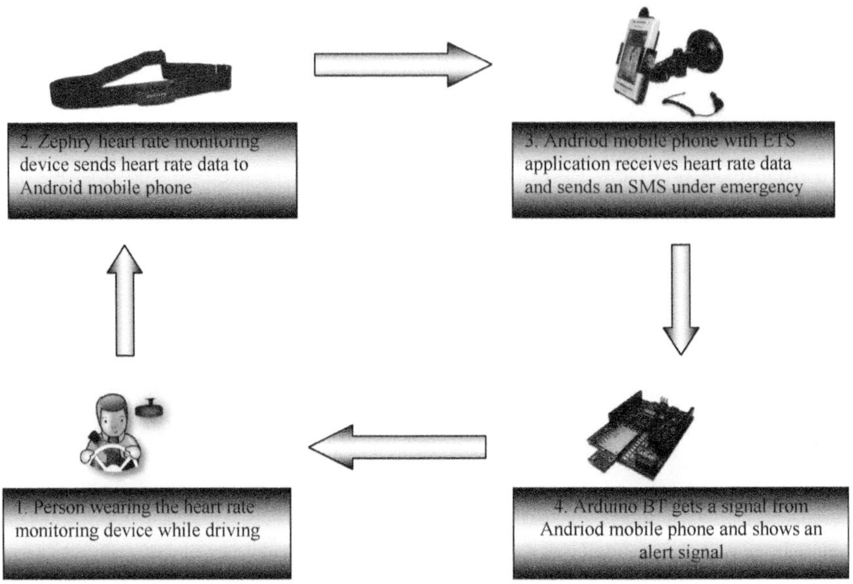

Fig. 3 Modeling of heart rate abnormalities monitoring scenario

4.1.2 Implementation of Heart Rate Abnormalities Monitoring Scenario

The implementation of the communications between the Zephyr HR monitor and Andriod mobile phone starts with designing an Application Programming Interface (API). *Figure 4* shows the Zephyr Bluetooth wirelessly communicating with an Andriod mobile phone over the Bluetooth link. The Zephyr Bluetooth HR device uses a Bluetooth SPP (Serial Port Profile) to communicate with the low level protocol which has the following specifications:

- 115,200 baud rate
- 8 data bits
- 1 stop parity bits
- No parity.

The application employs the API and enables the HR device to transmit different packet types such as heart rate, speed and distance. The following steps describe the most important aspects of the source code in the application used to enable the General Packet and display the data on the Andriod phone:

1. On clicking the Connect button, a Bluetooth adaptor type object is created and passed to an object of the *BT Client* class type. The *BT Client* object is essentially a thread that manages the overall Bluetooth connectivity of the phone with the HR device.

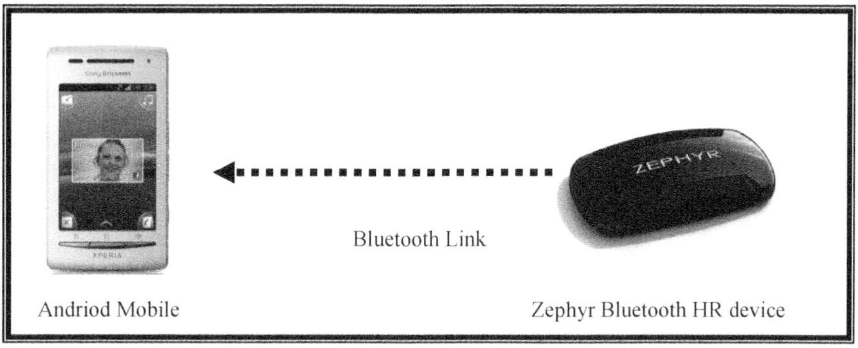

Fig. 4 Communications between Zephyr HR device and Andriod mobile phone

2. Next, an object of the *NewConnectedListener* class is created which eessentially implements the *ConnectedListener* interface, and one that extends the *ConnectedListener* class. This object is responsible for reacting differently to different kinds of messages. Then we override the parent class *connected* method and define our own method. Next, we create a *ZephyrProtocol* object and call its *addZephyrPacketEventListener* method. This method takes a *ZephyrPacketListener* argument in whose *ReceivedPacket* method we define what message we are interested in, and how we want the data to be displayed on the phone's screen.
3. This *ConnectedListenerImpl* object then needs to be connected to the *BTClient* object type via *addConnectedEventListener* function call to tie this object to respond to a received packet from the HR device.
4. Calling the *start* function of the *BTClient* thread kicks off the communications between the Application and HR device.

4.1.3 Validation of Heart Rate Abnormalities Monitoring Scenario

The HR monitoring device powers on automatically when worn. If there is insufficient skin conductivity, e.g., excessively dry skin and/or strap sensor pads,

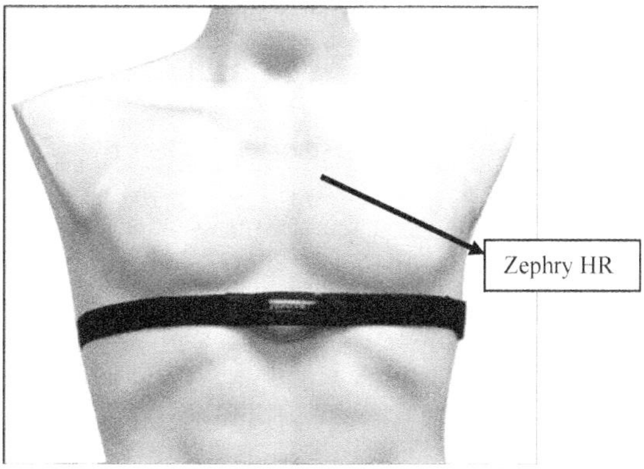

Fig. 5 Zephyr heart rate device worn

the wear-detect circuitry may not trip. Thus, the skin and sensor pads should be moistened with water before use. The heart rate device has to be worn around the chest region just near the sternum as shown in *figure 5*.

The open ETS application is installed on the Andriod mobile. As soon as the ETS is clicked, a home page with menus such as Pair HR Device, Pair Controller, Enter Details, Start and Disconnect is opened. In order to get the heart rate data the user needs to pair the HR device with the Andriod mobile phone. As soon as these devices are paired, the heart rate data starts to be displayed on the mobile phone screen. The entire workflow can be seen in *figure 6*.

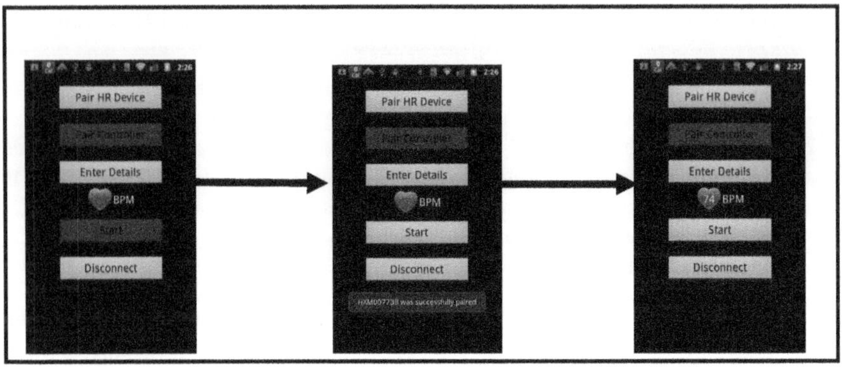

(1) **(2)** **(3)**

Fig. 6 The entire workflow of pairing HR device to Andriod phone

(1) Connect to the HR device by tapping on "Pair HR Device"

(2) A message is shown confirming the HR device is successfully paired

(3) Heart rate data is displayed on the screen "74 BPM (beats per minute)"

The emergency contact number, email address and Facebook account can be integrated in the "Enter Detail page" in the main menu of the ETS application as shown in *figure 7*.

When the HR device starts to transmit the heart rate data to the mobile phone, the ETS application performs a condition check with respect to $60<HR<100$ BPM; if the heart rate is less than 60 BPM or more than 100 BPM then the application decides that this condition is critical and starts to track the location of the user using GPS API and simultaneously sends a message to a pre-stored emergency contact number and email address. Finally a message containing "**This person is under emergency take necessary action**" followed by the geographical location of the person is posted on the user's Facebook wall. Simultaneously, a signal is sent to the Arduino microcontroller and a LED blinks to indicate an alert emergency signal. The entire workflow is presented in *figure 8*.

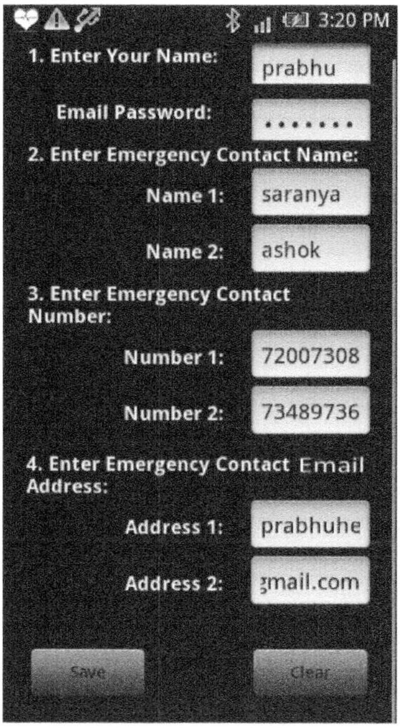

Fig. 7 Enter details page

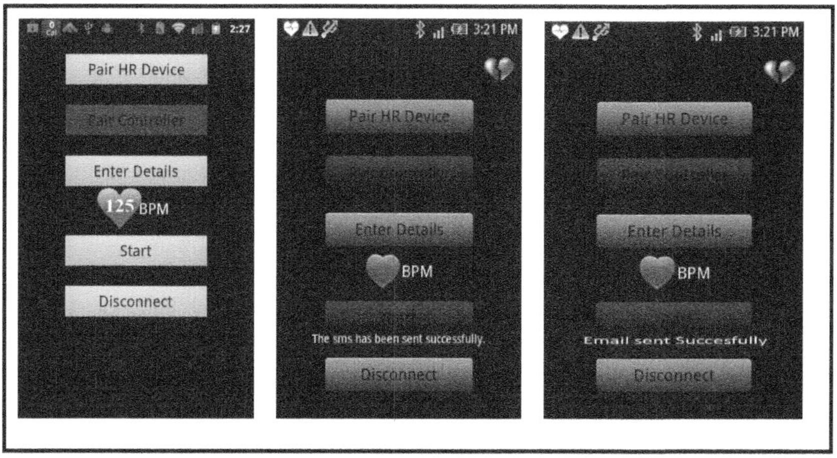

(1) (2) (3)

Fig. 8 Workflow of an abnormal heart rate scenario

(1) ETS reads an abnormal heart rate of 125 BPM

(2) SMS is sent to a pre-stored phone number

(3) Email is sent to a pre-stored email address

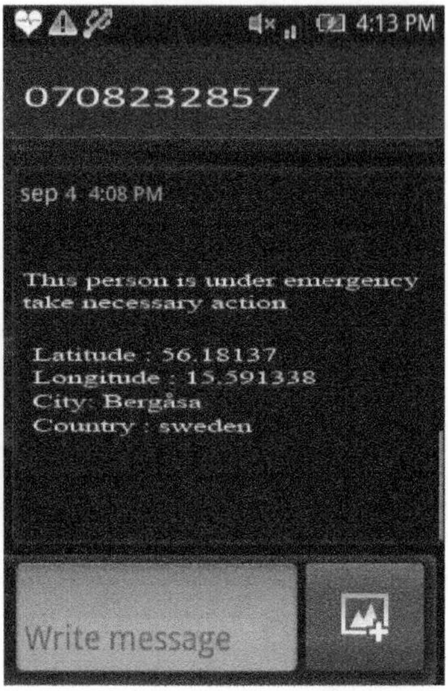

Fig. 9 An SMS message with the user's location is received

Fig. 10 An email with the user's location is received

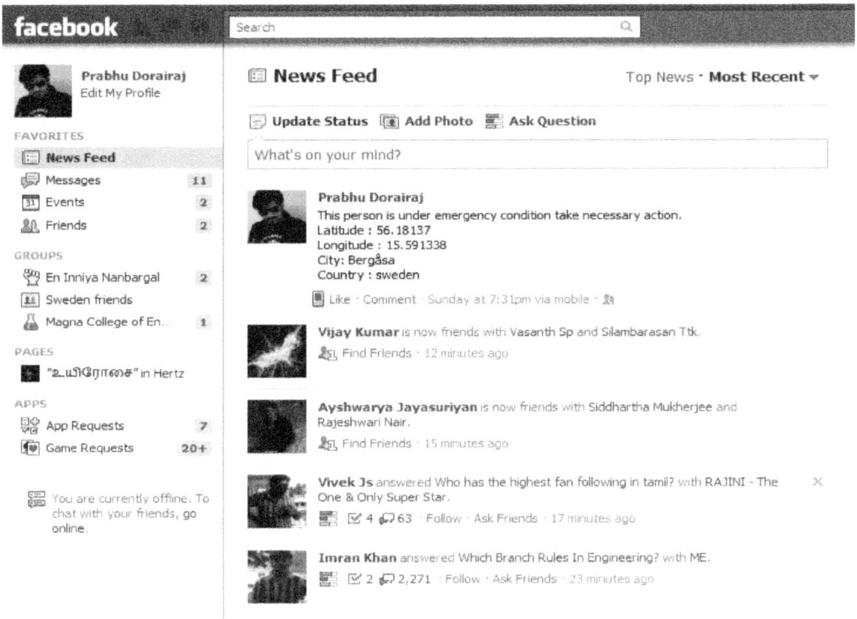

Fig. 11 An alert message is posted on the Facebook wall of the user experiencing emergency condition

4.2 Implementation and Validation of Security Threats Scenario

Security threats can occur anywhere and at any time. The scenario we investigated is designed to overcome a personal security threat under user request. If the user is in an emergency and feels that his/her location has to be sent to a remote user then the proposed application can be used to send the exact location information to a pre-stored number and email address under the user request as follows. When the heart symbol button is pressed on the ETS home page as shown in *figure 12*, the ETS application senses that the user is under a security risk and waits for the user to press the *send* button. In case the user is not interactive to press the button after pressing the heart symbol, the system automatically counts down for 14 seconds and then starts to get the location information (latitude and longitude coordinates) using a GPS API. Subsequently, the latitude and longitude coordinates are converted into city and country location information and sent via SMS and email along with the message "This person is under emergency take necessary action" which will also be posted on the Facebook wall of the registered user. *Figure 13* shows the workflow of the person safety scenario.

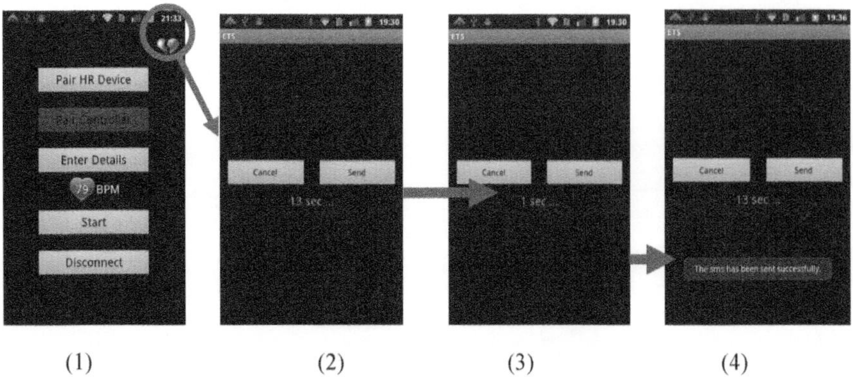

Fig. 12 Workflow of personal safety scenario

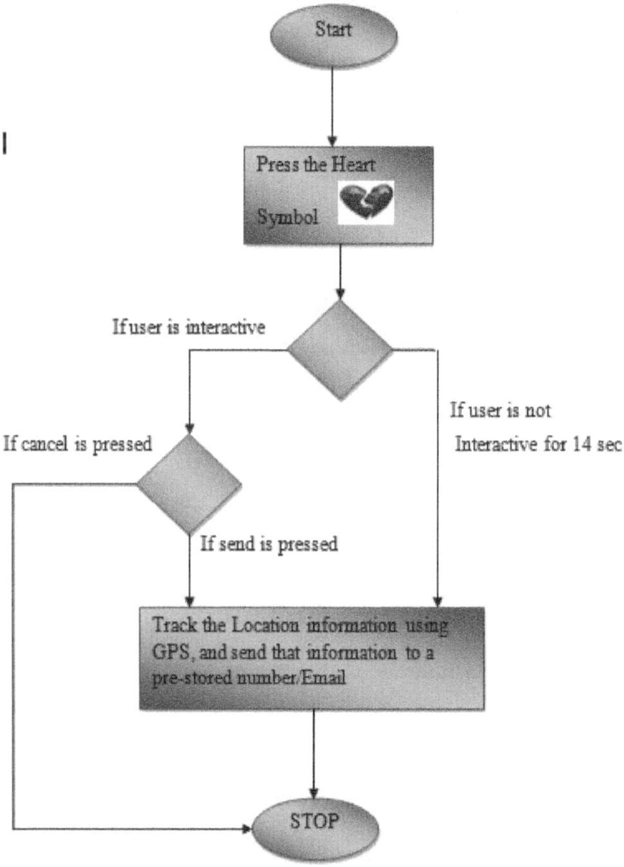

Fig. 13 Implementation of personal safety scenario

4.3 Implementation and Validation of an Accident Scenario

The design of this application is for the purpose of a person who is driving a vehicle with the mobile phone kept in the car in a secure fixed position. The mobile should not be kept in a pocket or held in the hand. Any abnormal movement (like tilt between landscape and portrait) of the mobile device would be detected by an inbuilt sensor (Accelerometer), and an action is taken by ETS to notify the emergency services that an accident has occurred and its location. *Figure 14* shows how the accident scenario is handled by the application.

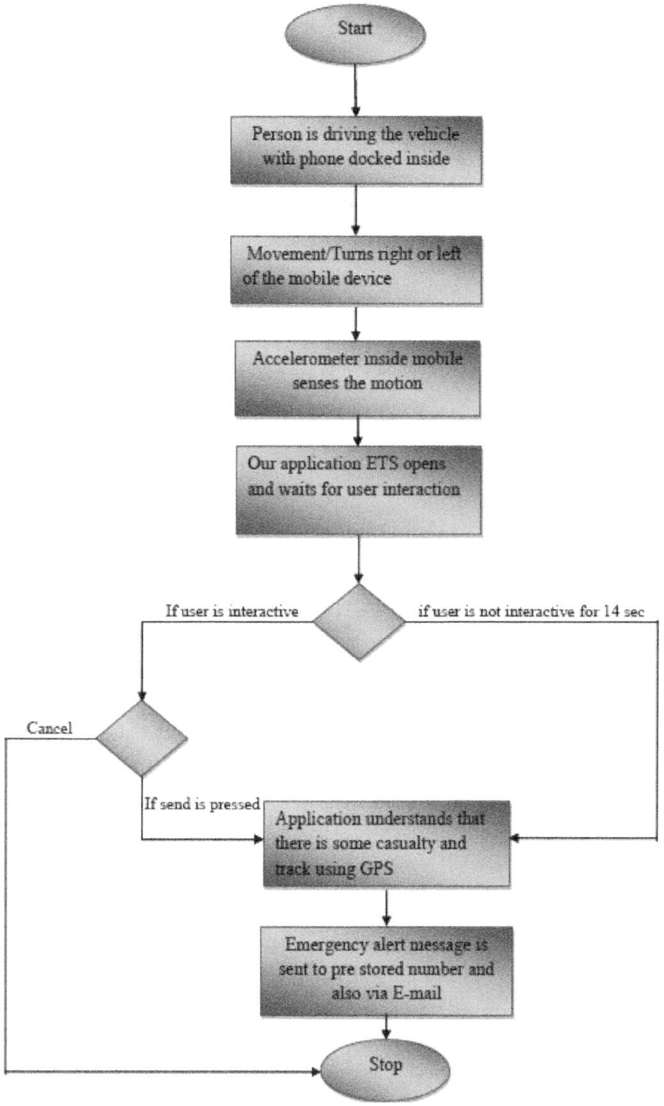

Fig. 14 Modeling of accident scenario

5 Overview of the Complete ETS under Different Scenarios

The complete ETS using Google's Andriod mobile platform and the required actions under different scenarios is shown in *figure 15*.

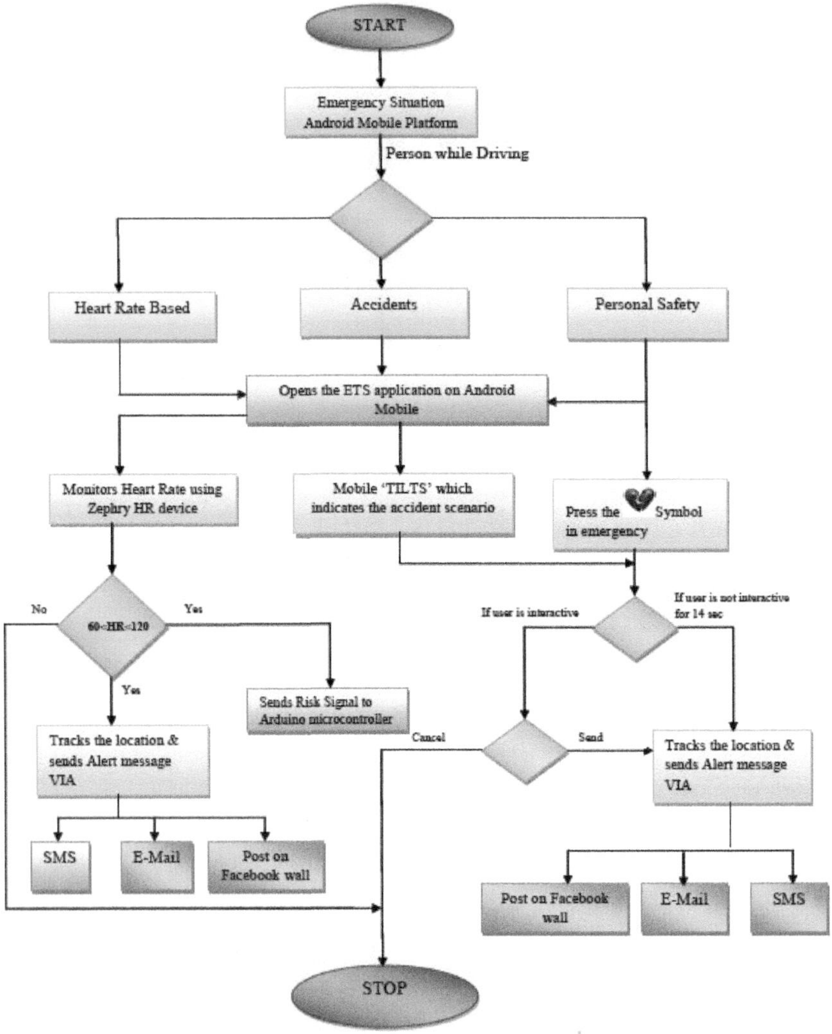

Fig. 15 Complete workflow of the ETS under different scenarios

6 Future Directions

In future work, it would be interesting to investigate the ETS system in a real vehicle, so that when an emergency signal is sent from an Andriod mobile phone, the vehicle will be brought to a halt. One of the issues in using GPS is the signal loss in some situations; in this case a terrestrial navigation system or an aerial platform can be incorporated with this system to improve the performance. This system can also be deployed like a black box in the vehicle, so that when there is an emergency a call can be triggered out automatically where the user can speak out or record a short message explaining the situation.

7 Conclusions

In this chapter we proposed an Emergency Tracking System using Google's Andriod Mobile platform and a related application that can used in many emergency situations and take necessary actions correspondingly. Emergency situations were successfully handled by the ETS on the Andriod mobile with regards to the heart rate, accidents and personal safety, respectively. Each individual emergency scenario was investigated, designed, and implemented. This research was entirely targeted towards an emergency situation for users driving a vehicle. The user's heart rate was monitored using a specialized HR device which sends the heart rate data to the mobile phone, which in turn makes a decision with regards to the abnormal heart rate and sends an alert signal to the Arduino microcontroller. Simultaneously, a message is sent to a pre-stored phone number, email address and posted on the user's Facebook wall. The accident scenario was also managed successfully by the proposed intelligent application on the Andriod mobile using an accelerometer, and an alert message containing the GPS location information is sent to the emergency services and the user's contact information. Finally, personal safety was also investigated successfully by clicking an interactive button which tracks the location information and sending it via SMS, email and post on the user's Facebook wall.

References

[1] Javier, G.: First Google's Andriod Phone Launched. IEEE Vehicular Technology Magazine 3 (September 2008)

[2] `http://www.xcubelabs.com/evolution-of-mobile-operating-systems.php` (accessed on June 10, 2011)

[3] Conti, J.P.: The Andriods Are Coming. IEEE Engineering and Technology 3(9), 72–77 (2008)

[4] Open Handset Alliance, Open Handset Announces 14 New Members

[5] http://www.openhandsetalliance.com/Andriod_overview.html (accessed on June 10, 2011)

[6] Shu, X., Du, Z., Chen, R.: Research on Mobile Location Service Design Based on Andriod. In: 5th International Conference on Wireless Communication Networking and Mobile Computing, Beijing, pp. 1–4 (2009)

[7] Whipple, J., Arensman, W., Boler, M.S.: A Public Safety Application of GPS-Enabled Smart Phones and the Andriod Operating System. In: IEEE International Conference on System, Man and Cybernetics, San Antonio, pp. 2059–2061 (2009)

[8] Yang, C.T., Chu, Y.Y., Tsaur, S.C.: Implementation of a Medical Information Service on Andriod Mobile Device. In: 4th International Conference on News Trends in Information Science and Service Science, Gyeongju, pp. 72–77 (2010)

[9] Hou, Q.: Research and Implementation of Remote Heart Rate Monitoring System Based on GSM and MCU. In: 2nd International Conference on Information Science and Engineering, Hangzhou, p. 2293 (2010)

[10] `http://developer.Andriod.com/reference` (accessed on December 21, 2013)

[11] Dai, J., Teng, J., Bai, X., Shen, Z., Xuan, D.: Mobile Phone Based Drunk Driving Detection. In: 4th International Conference on Pervasive Computing Technologies for Healthcare, pp. 1–8 (2010)

[12] Cheng, Y.H., Kuo, W.K., Su, S.L.: An Andriod System Design and implementation for Telematics Services. In: 5th International Conference on Intelligent Computing and Intelligent Systems, p. 206 (2010)

[13] Yang, S.-Y., Lee, D.-L., Chen, K.-Y.: A New Ubiquitous Information Agent System for Cloud Computing - Example on GPS and Bluetooth Techniques in Google Andriod Platform. In: International Conference on Electric Information and Control Engineering, pp. 1929–1932 (2011)

[14] `http://developer.Andriod.com/guide` (accessed on December 18, 2013)

[15] Wu, Z., He, J., Zeng, X.: Key Technology Research on Adaptation Layer of Andriod Platform. In: International Conference on Consumer Electronics, Communications and Networks, April 16-18, pp. 4154–4156 (2011)

[16] Kato, Y.: Splish: A Visual Programming Environment for Arduino to Accelerate Physical Computing Experiences. In: 8th International Conference on Creating Connecting and Collaborating through Computing, January 25-28, pp. 3–10 (2010)

14. mHealth Monitoring System for Hospitalised Older Adults – Current Issues and Challenges

Mirza Mansoor Baig[1,*], Hamid Gholamhosseini[1], and Martin J. Connolly[2]

[1] Department of Electrical and Electronics Engineering, School of Engineering
Auckland University of Technology
Private Bag 92006, Auckland 1142, New Zealand
{mirza.baig,hgholamh}@aut.ac.nz

[2] Freemasons' Professor of Geriatric Medicine, University of Auckland
North Shore Hospital, Takapuna, Auckland, New Zealand
martin.connolly@waitematadhb.govt.nz

Abstract. The mobile healthcare (mHealth) applications are becoming increasingly important in monitoring and delivery of healthcare interventions. They are often considered as pocket computers, due to their advanced computing features and diverse capabilities. Their sophisticated sensors and advanced software applications make mHealth based applications more feasible and innovative. Advanced engineering, communication and information technologies combined with medical and clinical knowledge enable the possibility of remote, wireless, continuous monitoring of physiological parameters. These technologies facilitate the implementation of mHealth based patient monitoring and diagnostic systems virtually anywhere: home, hospital and outdoors (on the move). The proposed mHealth vital sign monitoring system in this chapter is aimed to help clinicians by illustrating the trace of critical physiological parameters, generating early warning/alerts and indicating any significant changes to the data. The system was validated with different set of collected data from 20 hospitalised older adults and achieved an accuracy of 95.83%, sensitivity of 100%, specificity of 93.15%, and predictability of 90.38% in compare with a clinician assessment for tachycardia, hypertension, hypotension, hypoxemia and hypothermia. Another important aspect of this chapter is to investigate challenges and critical issues related to the use of such applications in healthcare including reliability, efficiency, mobile phone platform variability, cost effectiveness, energy usage, user interface, quality of medical data, and security and privacy.

Keywords: Vital signs monitoring, mHealth monitoring, decision support system, fuzzy diagnosis system, early warning, mHealth challenges and issues.

---

* Corresponding author.

S. Adibi (ed.), *Mobile Health*,
Springer Series in Bio-/Neuroinformatics 5, DOI: 10.1007/978-3-319-12817-7_14

1 Introduction

This decade is witnessing one of the greatest technological shifts in every area of life, especially in daily life activities and healthcare delivery. Today, mHealth monitoring systems are already in use for basic electronic health services. Email reminders, short message service (SMS), electronic health records, medical history and e-prescriptions are among these services and are cost effective and reliable. Vital signs or physiological parameters include heart rate (HR), blood pressure (BP), respiratory rate (RR), oxygen saturation (SpO2) and body temperature (T) [1]. The use of information and communications combined with medical and engineering technologies enable healthcare researchers to enhance patient monitoring at home, hospital or outdoors, using 2G, 3G, WiFi or online World-Wide-Web. Today, clinicians, nurses and family members can receive instant alert/messages about health information of their concerned person/patients on a smartphone, laptop, and personal digital assistant (PDA) or personal computer (PC). The most commonly performed monitoring is the collection of patient's vital signs using state-of-the-art medical devices, sensors and wearable textile for collecting and transmitting data to a remote server or processing unit for further analysis, storing and generating alerts [2].

Current healthcare focus is changing from the traditional/original medical treatment to the prediction, prevention and/or early detection and warning [3]. A number of reviews, surveys and interviews have been conducted to evaluate the benefits of mHealth systems. The use of internet based social and professional services offered across desktop computers, laptops, tablets and/or smartphones potentially enable mHealth applications to revolutionize and enhance the delivery of healthcare. Patient monitoring systems in particular are playing a critical role in decision support, early diagnosis and knowledge-based systems to the healthcare professionals [2, 4, 5].

The remainder of this chapter is structured as follows. The next section presents a background and reviews related mHealth monitoring systems. Section 3, describes the protocols we used in this research for data collection, medical devices used to collect data at the hospital with their communication and wireless connectivity and defines the system modelling and design. Section 4 discusses current issues and challenges related to the current mHealth applications.

2 Current mHealth Applications

The use of mHealth based, expert systems [6], clinical decision support systems [7] and rule-based systems [8] have proved to be useful in medical diagnosis. These techniques have been established as expanding areas of research using their features and capabilities to turn 'data' into 'useful information'. It is reported that an ideal vital signs monitoring system should be able to: (i) collect high quality data via medical devices/sensors; (ii) interpret and present the collected data in a meaningful and valuable manner; (iii) facilitate decision support via expert knowledge into the real health situation and; (iv) perform appropriate actions with clinicians' feedback to the patient on the basis of collected data [9]. mHealth applications/devices not only serves as key computing and communication of choice, but they also come with a rich set of embedded and advanced sensors, such as an accelerometer, digital compass, gyroscope, GPS, microphone, and camera. Collectively, these sensors are enabling new applications across a wide variety of domains. Tackling diabetes is likely one of the major concerns for global public health community where smartphone can play an effective role. Figure 1 shows variety of mHealth based blood glucose detection applications which are currently in use and/or available in market today and developed by; (a) Farmer et al. [10], (b) Tatara et al. [11, 12], (c) Cho et al. [13] and (d) AgaMatrix [14].

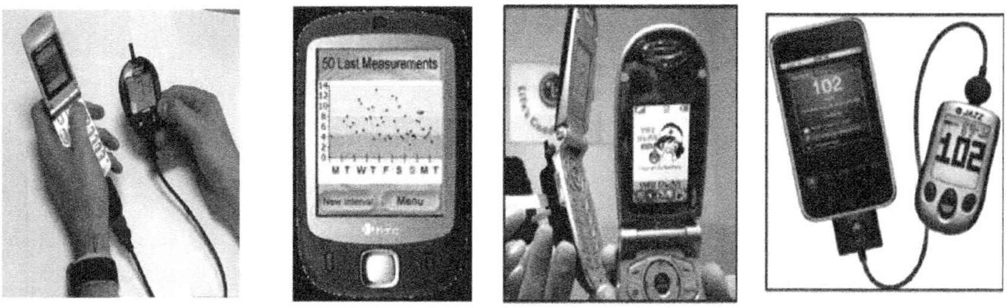

a b c d

Fig. 1 Shows the variety of smartphone based blood glucose detection systems developed by; (a) Farmer et al. [10], (b) Tatara et al. [11, 12], (c) Cho et al. [13] and (d) AgaMatrix [14].

Usually these systems can be connected wired or via wireless BT (Accu-Chek compact plus blood glucose meter) and often record the readings as a text and present data with time and history information on the user screen. Each reading is then converted into a text format which can be sent manually or automatically as a SMS or over the internet to doctors, nurses or family members. For blood glucose meter, user has to buy a user's phone compatible device, a mobile application and a mobile phone service subscription. However, main features of this technology, such as; low cost, easy to use, simple to adopt are often neglected or ignored in such cases.

3 A Novel System Design and Modelling for mHealth Applications

The generic system architecture is illustrated in Figure 2. The set-top-box is capable and responsible for vital data collection from all wireless Bluetooth (BT) connected devices. An optional feature of audio/video recordings may be added for transmitting the data in real-time to the clinician's computer directly or via central processing system where data analysis and fuzzy logic diagnosis is performed. High importance has been given to the security and privacy of the system due to the wireless and remote connectivity of the system. Security and privacy of patient information in remote monitoring systems is one of the biggest concerns and considered as a barrier to the adoption of this healthcare technology worldwide [15, 16]. The set-top-box has a secure gateway which allows data connectivity to certain devices which are registered and linked to specific patient profile using devices' details (serial number and MAC address). Data received at the clinicians' end can only be accessed by entering a unique username and password. We purposely avoid any public network connectivity due to the possibility of third-party access to the sensitive medical information and therefore mobile 3G data has been used via secure router and firewall settings for connecting to the set-top-box and laptop (wired as well as wireless). The system can also work on other networks such as WiFi, mobile data (GPRS, 2G or 3G), LAN or UMTS.

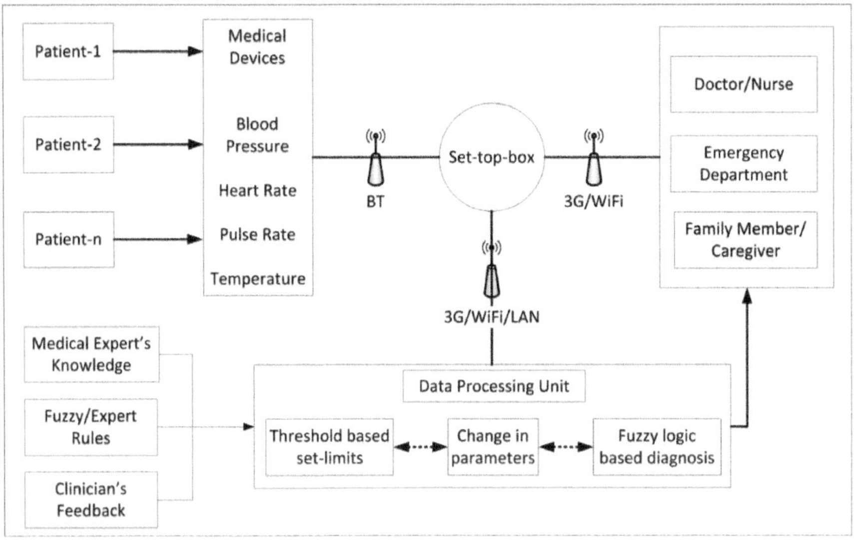

Fig. 2 The generic system architecture for the proposed monitoring and diagnosis system, three bold lines connecting the set-top-box show the wireless connectivity to other devices

3.1 Selection of Medical Devices

Figure 3 shows the medical devices and set-top-box selected for this design. We use market-available clinically proven medical devices incorporated into one system. The system (set-top-box) is able to collect multiple data simultaneously from multiple patients and transmit the data in real-time to the laptop/PC for processing and then to a clinician's device.

1. Set-Top-Box: This runs the software application which receives patients' physiological data from different medical devices and transmits it in real-time over the secure internet connection to personal PC or laptop.
2. Blood Pressure Monitor: Boso-medicus prestige blood pressure monitor [17] is a wireless (Bluetooth) device, which provides measurements at user defined time intervals It measures blood pressure (systolic and diastolic) and pulse. Records also show regular/irregular heartbeat.
3. Pulse Oximeter: Nonin's Onyx II finger clip oximeter [18] is a wireless Bluetooth device which records oxygen saturation and heart rate continuously. In the first 10 seconds, the device checks for stability in the pulse signal, and once stable readings are received (indicated by green light) the device transfers data to the set-top-box.
4. Blood Glucose Meter: Accu-Chek Compact plus blood glucose meter [19] is wireless infrared connected device which records the blood glucose level.
5. Ear Temperature: Omron's instant ear thermometer [20] is an accurate and fast tympanic (ear) thermometer.

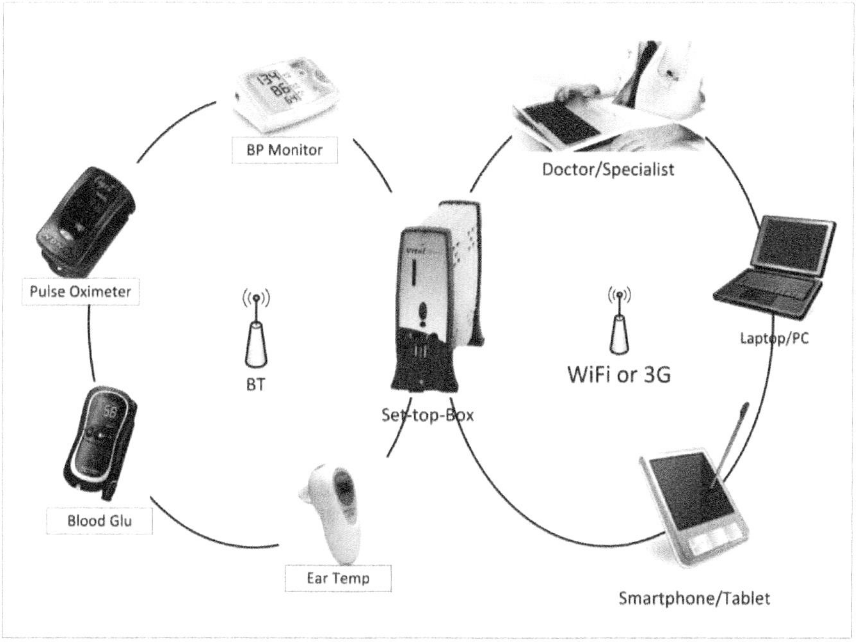

Fig. 3 The proposed monitoring system with a set-top-box (centre) connected wirelessly to various medical devices (left circle) and transmits the collected data to various devices (right circle) in real-time, BT is class 2 Bluetooth connectivity and Blood Glu is blood glucose monitor

3.2 Data Collection

For evaluating the design and measuring the performance of the proposed system 20 set of patient data were collated with informed consent from hospitalised patients at the North Shore Hospital and Waitakere Hospital, Auckland, New Zealand. The patients were hospitalised at older adults' wards and the data collection was performed after seeking approval from the respective local and national ethical committees (NTX/12/EXP/073, Waitemata District Health Board 0980712176, and AUTEC 12/117). Table 1 describes the type of information collected from the patients.

Table 1 Statistical information of data collected for the diagnosis system

Parameter	Mean Value	Range/Standard Deviation
Number of Patients	20	-
Age	82Y 4M	20Y 8M/6.52
Sex (M/F) %	60/40	-
BP (Systolic/Diastolic)	123/71	(127/66)/(21.55/12.02)
Heart rate	79 BPM	77/16.06
Oxygen Saturation	96%	20/3.88
Blood Glucose	142.41mg/dl (7.9mmol/l)	[170/49.70 mg/dl] or [9.43/2.75 mmol/l]
Tympanic (ear) Temperature	36.5 °C	2.2/0.41

We implemented a three-way cross validation data collection method: firstly, vital signs were collected by wireless medical devices described below and transmitted in real-time to the base machine (laptop); secondly, the registered nurse performed blinded manual readings of the same parameters, using standard ward techniques; every measurement made by the medical devices was recorded manually by the researcher to check any wireless transmission data loss, inaccurate data transmission, and transmission delay time.

3.3 Vital Signs Diagnosis Conditions

A fuzzy logic model may be employed to assess and interpret multiple physiological parameters using collected vital signs: HR, BP, PR, SpO2 and T as indicated in Table-2. H, N and L represent high, normal and low values of these parameters respectively. It is important to mention that the normal range (Table-2) represents the range of normality across the whole group of older people, but vital signs of each individual will have different normal and abnormal levels. Thus, fuzzy logic plays an important role in the diagnosis of individualized vital sign monitoring. The system learns the normal parameters for each particular patient using adaptive-neuro fuzzy inference system modelling (ANFIS). The relation between physical signs and vital signs was established by consulting with physicians and medical assessment (Table 2).

Table 2 The relation of vital signs with physical signs

Physical Signs/Parameters	Heart Rate (HR)	Blood Pressure (BP)	Oxygen Saturation (SpO2)	Temp. (T)
Bradycardia	L	N/A	N/A	N/A
Tachycardia	H	N/A	N or L	N/A
Hypotension	N/A	L	N/A	N/A
Hypertension	N/A	H	N/A	N/A
Hypoxaemia	N/A	N/A	Often L	N/A
Fever	H or N	N/A	N/A	H
Hypothermia	L	N or L	N/A	L
Normal Range	60-90 bpm	100-140/60-80 mm/Hg	94%-99%	36.5-37.5 ^{0}C

Normal ranges are adopted from literature [21] as well as medical expert consultation, which may be different for some people (when clustered into groups: age, and/or sex) but they are acceptable over the data set. H = high, L = low, N = normal and N/A = not applicable

3.4 *Priority Allocation and Fuzzy Set Modelling*

The aim of the proposed model is to provide multiple combinations of extracted parameters in order to assist clinicians with detection and estimation of health conditions and/or early diagnosis. Physical signs are classified as, priority-2-warning and priority-1-alarm. Priority 2 warnings are generated when the vital sign(s) changes above the set standard deviation (SD) limit (individualised optimised limits). The vital sign(s) may be reversed to normal state but if significant changes are detected then priority-1 type alarm will be generated. The proposed system is designed to reduce false alarms and to achieve high clinical reliability. Due to the nature of physiological parameters, which are variable and changing throughout monitoring, fuzzy logic is one of the best approaches in such type of monitoring and diagnosis [22].

Vital signs have been categorised into two outcome priorities for a reliable and robust interpretive system with high clinical accuracy. Each vital sign has been given several levels of importance in relation to different health events, for example: blood pressure has more weight/importance when considering Hypotension and Hypertension. The mapping and linking of multiple vital signs for detecting a single event using fuzzy model gives high accuracy with reliable indication of health events. For example, Hypotension is defined as a 'low blood pressure', irrespective of the other parameters. Whether the Hypotension is of any clinical relevance will depend on (to some degree) the other parameters measured. When the proposed fuzzy model classifies a health event as Hypotension, it is due to giving high importance as well as high weightage to blood pressure. Instead of considering only blood pressure as one of the possibility of Hypotension we also consider 'Low pulse volume' as other possible clinical relevance of Hypotension from the clinician's point of view. This criterion is applied to all other possible physical signs which can provide a robust indication of possible Hypotension (Table 2).

Instead of setting crisp numeric limits, we employed linguistic variables, such as, fuzzy sets, membership functions MFs and rules [23], to describe the degree of occurrence of a certain medical event. Fuzzy sets are defined as:

- Low BP (Lbp), High HR (Hhr) and High Pulse volume (Hp) for priority-2 Hypotension,
- Very Low BP (VLbp), Very High HR (VHhr) and Very High pulse volume (VHLp) for priority-1 Hypotension.

Patient physical signs (events) (E1-E7) and their priorities (P1-P2) are assigned as;

E1: Bradycardia.
E2: Tachycardia.
E3: Hypotension.
E4: Hypertension.
E5: Hypoxemia.
E6: Fever.
E7: Hypothermia.

Bradycardia: Let us consider the first physical sign E1 to show how the proposed model is working. Figure 4 shows the initial patient status as normal (N in green) then, when there is a low heart rate (denoted as 'Lhr') the status changes to E1P2, i.e. possible Bradycardia with priority-2 warning. Further, if the heart rate goes very low (denoted as 'VLhr') then the current status changes to possible Bradycardia with priority-1 alert and returns to normal state when the heart rate is back to the normal range for that particular data pattern (patient).

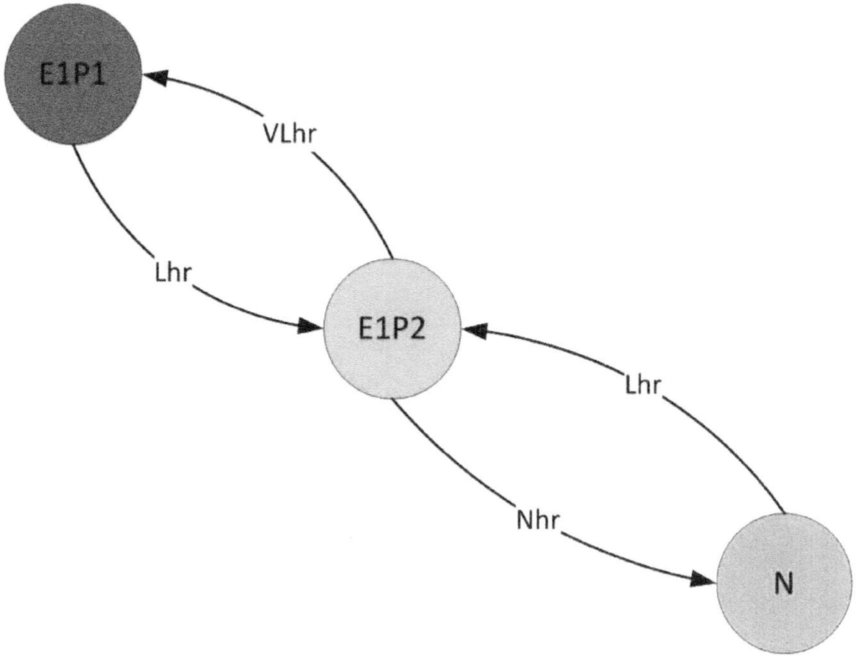

Fig. 4 Possible Bradycardia using two categories of heart rates: low and very low with two levels of priority

Now, considering E1-E7 physical signs, depicted in Figure 5, using the combination of five vital signs: HR, BP, SpO2 and T with two levels of priority (P1 and P2). In this case, we considered all possible scenarios where the physical sign extracted from any physiological measurement include additional and finer fuzzy sets, with the inclusion of *Very Low (VL), Low (L), High (H) and Very High (VH)*. For example, priority-1 Hypotension (E3P1) is possible when the combination of vital signs indicates Very Low BP (VLbp), as shown in Figure 5. In some other cases such as Hypoxaemia, it is necessary to consider Low SpO2, but also, HR and BP, which can be high or normal or low. In such case, the proposed model gives more weight to SpO2 for 'Low' and considers HR and BP with normal weight value (Table 2). Centre N (green) is the initial status and E1-E7-with P2 (orange) is the priority 2 classified physical signs and P1 (red) are the priority-1 health outcomes as an alert. Prioritisation of outcomes gives clinicians a high degree of control over the system's outcome because of two levels of priorities. This structure potentially gives high reliability and stability with only priority 1 generating alarms.

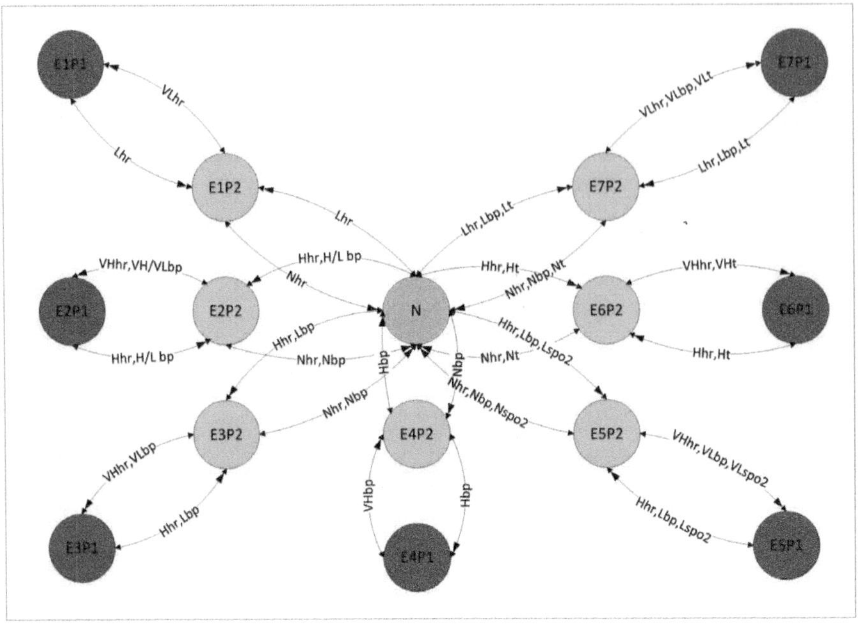

Fig. 5 System working model with all possible seven physical signs using four vital signs and two levels of priority and a centre green normal state as initial patient's condition

3.5 Diagnostic Results and Validation

Kappa analysis [24] was used to measure the level of agreement/disagreement between the proposed system and a medical expert. Kappa is used as the measure of how accurately the system can mimic humans' performance. We have visited each patient three times (on average) during the trial and collected four sets of vital signs during each visit. Each dataset contains blood pressure measurement – standing and sitting with the time difference of five minutes on average, heart rate, oxygen saturation, ear temperature and blood glucose readings. The same data was again collected at time difference of four hours during the same visit. On average, we collected 12 datasets from each patient. Therefore, with 20 patients we collected a total of 240 datasets. The diagnostic performance of the system was verified through a series of (retrospective) tests using 240 datasets from 20 hospitalized patients. The system was trained with data from 10 patients (120 datasets) and tested with data from the remaining 10 patients (120 datasets). The training and testing datasets were selected randomly from the whole 20 patients' data. Table 3 shows the total number of alarms generated by the proposed system and classified into the related physical signs as well as expert diagnosis of the same event.

Table 3 Classification of physical signs by the proposed system as well as expert diagnosis

Vital signs	Physical Sign-No of alarms	Expert Diagnosis
Blood Pressure (systolic/diastolic)	Hypotension-8 Hypertension-11	Hypotension-8 Hypertension-9
Heart rate	Tachycardia-12	Tachycardia-12
Oxygen Saturation	Hypoxemia-15	Hypoxemia-15
Temperature	Hypothermia-6	Hypothermia-3
Total Alerts	52	47

Table 4 Kappa values

System/Expert	Expert (+ve)	Expert(-ve)	Total
System (+ve)	47 (TP)	5 (FP)	52
system (-ve)	0 (FN)	68 (TN)	68
Total	47	73	120

Based on values from Table 4, Kappa analysis is carried out. Agreements between the two diagnoses may be affected by chance. Kappa (k) is a measurement of agreement between the expert and system which has been corrected for error by chance. Kappa (k) is calculated by subtracting the proportion of readings that are expected to agree by chance (Pe) from the overall agreement (Po) and dividing the remainder by the number of cases on which agreement is not expected to occur by chance. Table-5 summarises the Kappa analysis results for the system's performance. P_o, P_{pos}, and P_{neg} are overall, positive, and negative agreements, respectively. SE represents the standard error and $CI_{95\%}$ is 95% confidence intervals for Kappa.

Table 5 Kappa analysis of patients' data

Overall Agreement	Positive Agreement	Negative Agreement	Agreement by Chance	Standard Error	95% Confidence Intervals for K
Po	Ppos	Pneg	Pe	SE	CI95%
0.95	0.94	0.97	0.51	0.04	0.98 and 0.83

For the offline-tests (retrospective analysis) performed, the Kappa based statistical analysis showed substantial level of agreement (k = 0.91 or 91%) between the experts' and system's diagnoses. Table-5 shows that proposed system has achieved the overall agreement of 95% with K value of 91%. Now, we calculate the quantitative categories such as accuracy, sensitivity, specificity and predictability using the standard equations (Table 4 for TP, TN, FP and FN values). The proposed system has achieved the accuracy of 95.83%, sensitivity of 100%, specificity of 93.15% and predictability of 90.38%.

3.6 *Discussion*

There are five disagreements (FP) from the medical expert (Table 4, false positive) whereon three occasions 'possible hypothermia' was detected due to detection of below normal value of ear temperature for that particular patient. The system generated a possible alert but the medical expert disagreed as it was a borderline value and he would have wait to see next few values before actually considering it as a possible alert. On two other occasions system detected 'possible tachycardia and hypertension' for two particular patients from their individual measured vital sings values. Blood pressure was slightly on higher side and heart rate (pulse) was very high, which indicated possible significant changes in the patient's vital sign values when compared with previous values and overall normalised data trend. The medical expert agreed with the alert for 'tachycardia' but disagreed with the 'hypertension' alert, and explained that blood pressure values were borderline and it is not justifiable to say 'possible hypertension' at this stage, and would go for possible tachycardia instead.

4 Current Issues and Challenges

Among a number of critical issues in mHealth applications, duration of monitoring, frequency of data collection and transmission, amount of data transmitted and nature of monitoring in terms of alert, periodic or continuous are considered important. Some of these issues and future challenges of mHealth monitoring systems as well as their diversity and complexity will be addressed in the following sections.

4.1 *Reliability, Efficiency and Acceptability*

With the ever growing mHealth applications, end-user acceptability is becoming an important aspect in the design of such systems. There is still an open research question to be addressed. More specifically, a particular question of "Do mHealth applications make a difference to the patient's (end-users) well-being?" needs to be answered. In that aspect many researchers have included patients as well as medical professionals' considerations at every stage of the design and development [25]. We believe the acceptance of any system in the healthcare industry depends on the user awareness and acceptability. Adaptation of a device within the clinical field is stuck when they are negatively perceived. User-centred design is essential in order to incorporate these perceptions into the product, especially at the earlier stages of the project development. When analysing user's needs, contextual inquiry, user's profiling the designer should consider a number of factors such as task analysis, surveys, interviews and focus groups to address the user acceptability [26]. We support the proposal of Steele et al. [27] that future studies should document any attitudes, perceptions and concerns of users. It is known that highly sophisticated technology and data analysis techniques become irrelevant if users do not wear the sensor systems for the allocated periods of time [28].

The reliability of monitoring systems is an important and open research question which often considered as a critical parameter for the acceptability of the system. In the context of reliability and efficiency, main purpose is to connect mHealth monitoring system to the user within their activity area (Bluetooth range) and model the regular activities. An alert is not triggered when the person is outside coverage area or specific range by more than a predefined threshold range. Several methods are proposed for determining when an alert/alarm should be triggered [29]. Bergmann and McGregor [30] carefully considered several systems and recommended the best and most efficient body worn sensor design.

4.2 Platform Variability and Cost Effectiveness

Software platform is becoming a drawback towards the development and implementation of mHealth applications due to its multiple/different operating systems. The development environments for handsets cover a wide range of operating systems including, Microsoft Windows Mobile, Symbian, Blackberry, Palm OS, Mobile Linux, J2ME, Apple's iOS and the Android platform by Google. Another issue with use of mobile platforms is variability and compatibility between the programming language and application environment. At present, efforts to make mHealth systems fully functional in all available (common) platforms are slowly approaching [31]. The involvement of healthcare professionals in the development of such systems and their participation in the policy discussions would be important in order to achieve full potential of mHealth applications [32]. Another barrier is mobility of data as most of the mHealth applications transfer patients' vital data and/or key physiological parameters via mobile communication links, that are; GPRS, 2G, 3G, 4G and 5G (under development) networks. Costly mobile phone contracts and expensive termination fees could create a barrier to access specific services such as medical data processing [33].

4.3 User Interface and Quality of Patient's Medical Data

From user point of view, graphical user interface (GUI) feature of a mHealth applications one of the most appealing functions that should be easy to use, simple and functional. Such feature can be difficult to develop for different types of users including the elderly and handicapped. Today's market offer simple and easy process to download an application (by looking at 3-4 screen shots shown by the manufacturer). Thus, there is a possibility that if the user dislikes the application or finds it complicated, then the application will be removed as easy as it was downloaded. Therefore it is essential to have a user-friendly and simple to use mHealth application for all common mobile platforms.

Various techniques were applied to collect high quality and up to the medical standard data for reliable communication in mHealth applications [34]. However, in some studies short time data measurement was worrisome because, either the data quality was substandard or high false positive alarm rates were reported [35]. It is important to capture and transfer high data quality for vital signs monitoring systems and set a theoretical framework to manage physiological parameters at early stage of application development. At each stage of the development, the feedback of medical professionals should also be considered and discussed in relation with the end user (patient).

4.4 Security and Privacy

Security and privacy are the most important issues of any healthcare systems which often get less attention during the development of mHealth based healthcare system. Usually mHealth applications deal with personal information and should be safe and secure. As transmission of data in mHealth based applications is wireless, it may result in various security threats with potential bridging privacy and confidentiality matters. Security issues in wireless sensor networks are a major area of research in recent years and many researchers have specifically addressed security issues with respect to healthcare applications [36]. In the following section some of the security and privacy issues related to mHealth based system will be discussed along with some recommendation for improvement.

A number of security and privacy frameworks have been developed, designed and tested for their reliability in mHealth applications, 22 free web based personal health record privacy and security policies have been analysed by Carrion et al. [37] and reported a high level of user's security is in place with applications such as Google Health, ZeabraHealth, Keas and Microsoft Health Vault. Al Ameen et al. [15] divided security and privacy into two aspects; firstly, system security, which includes administrative, physical and technical level security and secondly, information security, which includes data encryption, data integration, authentication and freshness protection.

A robust security framework using encryption and decryption [38] was designed to secure mobile wireless-networked sensors with a middle ware component to transmit and receive patient information securely using a two tier architecture. The system was built based on three main functions: data protection on the device, secure authentication and data encryption [16]. Design consideration for secure long term monitoring is presented in [15], with some of the recent development carried out by the researchers in addressing the most common and vital security and privacy issues. If any of the above security frameworks is adopted for a mHealth application then the system would have basic security standards. The advancement in mHealth technology can improve the acceptance of these systems among general public and facilitate development of more secure and reliable applications in the global market [39].

5 Case Studies

5.1 *Case Study 1*

Environmental Multimodal for Tele-vigilance Medical (EMUTEM) for home monitoring has been developed [40]. It consists of three subsystems: the Anason subsystem includes a set of microphones for remote monitoring of the acoustical environment of the older adult; the RFpat, a wearable device that can measure physiological data, such as heart rate, motion activity, posture and a fall; infrared sensors called Gardien which detect the presence of the person. EMUTEM fuzzy inference engine was developed using two groups of fuzzy IF-THEN rules: output variable localization and the output variable alarm according to all inputs. As a result the system has achieved 95% of accuracy for alarm generation and 97% accuracy for localisation.

5.2 *Case Study 2*

mHealth applications are beginning to emerge as a useful technology for healthcare delivery. Using a basic cell phone calling service or SMS, people with type 1 diabetes mellitus could be supported through self-management systems by text message communication. This method has introduced favourable changes to diabetes self-efficacy and adherence to treatment and their behavioural changes [41]. There is also an effective and positive response from smokers, which is one of the world's current biggest problems. This is done through mobile phone based projects such as 'Text2Quit' and 'txt2Stop' [42].

5.3 *Case Study 3*

A mHealth system called 'Blue Box' [43] has been developed as a novel hand-held device capable of collecting and wirelessly transmitting key cardiac parameters ECG, Photoplethysmography (PPG) and bio-impedance. It also measures RR interval and QRS duration, HR, systolic time intervals and assesses their values in correlation with cardiac output measured by an echo-doppler. In ECG measurement a 30-60 seconds time delay has been reported. Other common issue reported in literature is, simulation or testing of the system using small sample size and healthy subjects which often gives low accuracy results when tested in real-time [43].

6 Future Direction of the Technology

The mHealth technology is likely to be expanded in near future due to the advancement of embedded sensors in mobile devices (e.g., the accelerometer and gyroscope). Smart phones and mobile devices will soon be programmed to support new disruptive sensing applications wirelessly such as sharing user's real-time activity with friends on social networks such as Facebook, keeping track of a person's carbon footprint, or monitoring a user's well-being. Another aspect is the increasing awareness of chronic diseases, growing market of mobile devices, advanced communications, expansion of 3G and 4G networks, and a promise of cost-effective healthcare. The application of mHealth is likely to be expanded in near future due to the following reasons. Finally, the programmability feature of smartphones which will lead to developing a variety of healthcare applications with high acceptance among users. This will further enhance the ease of mHealth adoption. The combination of these advances opens the door for innovative research and will lead to the development of mHealth applications that are likely to revolutionize a large number of existing business sectors and ultimately significantly impact on healthcare delivery as well as our everyday lives.

7 Conclusion

The proposed system achieved an overall positive agreement (P_o) and accuracy 95% and Kappa value 91%, i.e. the system and the medical expert agreed 91% of the identified health events. This chapter also discusses current state of the art and open challenges in emerging field of mHealth applications. One of the mHealth's biggest fears and technical barriers is the security and privacy of the personal information. Once these technical barriers are resolved, mHealth technology will emerge as a reliable and trustworthy technology across many health domains as predicted by majority of clinicians, patients (users) and manufacturers. However, the current study highlights the fact that there are still challenges and issues that need to be resolved for such systems to become more applicable to real-life situations and to be accepted by patients and healthcare professionals as a reliable, multifunctional, easy-to-use, and cost effective technology that can enhance the quality of living.

References

[1] Nangalia, V., Prytherch, D.R., Smith, G.B.: Health technology assessment review: remote monitoring of vital signs–current status and future challenges. Crit Care 14(5), 233 (2010)

[2] Klingeberg, T., Schilling, M.: Mobile wearable device for long term monitoring of vital signs. Computer Methods and Programs in Biomedicine 106(2), 89–96 (2012)

[3] Yu, V.L., Madoff, L.C.: ProMED-mail: an early warning system for emerging diseases. Clinical Infectious Diseases 39(2), 227 (2004)

[4] Lin, C.-F.: Mobile Telemedicine: A Survey Study. Journal of Medical Systems 36(2), 511–520 (2012)

[5] Klasnja, P., Pratt, W.: Healthcare in the pocket: Mapping the space of mobile-phone health interventions. Journal of Biomedical Informatics 45(1), 184–198 (2012)

[6] Garibaldi, J.: Fuzzy Expert Systems. In: Gabrys, B., et al. (eds.) Do Smart Adaptive Systems Exist? STUDFUZZ, vol. 173, pp. 105–132. Springer, Heidelberg (2005)

[7] Basilakis, J., Lovell, N.H., Redmond, S.J., Celler, B.G.: Design of a Decision-Support Architecture for Management of Remotely Monitored Patients. IEEE Transactions on Information Technology in Biomedicine 14(5), 1216–1226 (2010)
[8] Kundu, M., Nasipuri, M., Kumar Basu, D.: Knowledge-based ECG interpretation: a critical review. Pattern Recognition 33(3), 351–373 (2000)
[9] Kyriacos, U., Jelsma, J., Jordan, S.: Monitoring vital signs using early warning scoring systems: a review of the literature. Journal of Nursing Management 19(3), 311–330 (2011)
[10] Farmer, A., Gibson, O., Hayton, P., Bryden, K., Dudley, C., Neil, A., Tarassenko, L.: A real-time, mobile phone-based telemedicine system to support young adults with type 1 diabetes. Informatics in Primary Care 13(3), 171–178 (2005)
[11] Tatara, N., Årsand, E., Hartvigsen, G.: Patient-user involvement for designing a self-help tool for Type 2 diabetes. Therapeutic Strategies A Challenge for User Involvement in Design 336(7639), 53 (2010)
[12] Tatara, N.: Designing mobile patient-centric self-help terminals for people with diabetes. In: Proceedings of the 11th International Conference on Human-Computer Interaction with Mobile Devices and Services Bonn, MobileHCI 2009, Germany (2009)
[13] Cho, J.H., Lee, H.C., Lim, D.J., Kwon, H.S., Yoon, K.H.: Mobile communication using a mobile phone with a glucometer for glucose control in Type 2 patients with diabetes: as effective as an Internet-based glucose monitoring system. Journal of Telemedicine and Telecare 15(2), 77–82 (2009)
[14] AgaMatrix (2012), JAZZ-Blood Glucose Monitoring System, `http://www.wavesense.info/wavesense-jazz`
[15] Al Ameen, M., Liu, J., Kwak, K.: Security and Privacy Issues in Wireless Sensor Networks for Healthcare Applications. Journal of Medical Systems 36(1), 93–101 (2012)
[16] Weerasinghe, D., Rajarajan, M., Rakocevic, V.: Device Data Protection in Mobile Healthcare Applications. In: Weerasinghe, D. (ed.) Electronic Healthcare. LNICST, vol. 1, pp. 82–89. Springer, Heidelberg (2009)
[17] CorScience. Boso-Medicus Prestige + BT: blood pressure meter, http://www.corscience.de/en/medical-engineering/products/blood-pressure/boso-medicus-prestige-bt-device.html
[18] Nonin. Onyx II, Model 9560- fingertip pulse oximeter, `http://www.nonin.com/PulseOximetry/Fingertip/Onyx9560`
[19] Accu-Chek. Accu-Chek Compact Plus Blood Glucose meter system, `https://www.accu-chek.com/us/glucose-meters/compact-plus.html`
[20] Omron. Omron MC-510 Instant Ear Thermometer, `http://www.omronhealthcare.com.au/Omron-Instant-Ear-Thermometer.html`
[21] McPhee, S.J., Papadakis, M.A., Rabow, M.W.: Current Medical Diagnosis & Treatment 2012. McGraw-Hill Medical (2010)
[22] Sadegh-Zadeh, K.: Fuzzy health, illness, and disease. Journal of Medicine and Philosophy 25(5), 605–638 (2000)
[23] Innocent, P.R., John, R.I.: Computer aided fuzzy medical diagnosis. Information Sciences 162(2), 81–104 (2004)
[24] Kundel, H.L., Polansky, M.: Measurement of observer agreement. Radiology 228, 303–308 (2003)

[25] Lee, H.J., Lee, S.H., Ha, K.-S., Jang, H.C., Chung, W.-Y., Kim, J.Y., Chang, Y.-S., Yoo, D.H.: Ubiquitous healthcare service using Zigbee and mobile phone for elderly patients. International Journal of Medical Informatics 78(3), 193–198 (2009)
[26] Schleyer, T., Mattsson, U., Ni Riordain, R., Brailo, V., Glick, M., Zain, R., Jontell, M.: Advancing oral medicine through informatics and information technology: a proposed framework and strategy. Oral Diseases 17(suppl. S1), 85–94 (2011)
[27] Steele, R., Lo, A., Secombe, C., Wong, Y.K.: Elderly persons' perception and acceptance of using wireless sensor networks to assist healthcare. International Journal of Medical Informatics 78(12), 788–801 (2009)
[28] Dilmaghani, R.S., Bobarshad, H., Ghavami, M., Choobkar, S., Wolfe, C.: Wireless Sensor Networks for Monitoring Physiological Signals of Multiple Patients. IEEE Transactions on Biomedical Circuits and Systems 5(4), 347–356 (2011)
[29] Imhoff, M.: Medical Device Alarms–The Clinician, pp. 17–18 (2009)
[30] Bergmann, J., McGregor, A.: Body-Worn Sensor Design: What Do Patients and Clinicians Want? Annals of Biomedical Engineering 39(9), 2299–2312 (2011)
[31] Macropinch, Personal Heart Rate Meter (2012), `http://macropinch.com/cardiograph`
[32] Lorenz, A., Oppermann, R.: Mobile health monitoring for the elderly: Designing for diversity. Pervasive and Mobile Computing 5(5), 478–495 (2009)
[33] Patrick, K., Griswold, W.G., Raab, F., Intille, S.S.: Health and the mobile phone. American Journal of Preventive Medicine 35(2), 177 (2008)
[34] Lim, Y., Hong, K., Kim, K., Shin, J., Lee, S., Chung, G., Baek, H., Jeong, D.-U., Park, K.: Monitoring physiological signals using nonintrusive sensors installed in daily life equipment. Biomedical Engineering Letters 1(1), 11–20 (2011)
[35] Apiletti, D., Baralis, E., Bruno, G., Cerquitelli, T.: Real-Time Analysis of Physiological Data to Support Medical Applications. IEEE Transactions on Information Technology in Biomedicine 13(3), 313–321 (2009)
[36] Hung, M.-C., Jen, W.-Y.: The Adoption of Mobile Health Management Services: An Empirical Study. Journal of Medical Systems 36(3), 1381–1388 (2012)
[37] Carrión, I., Fernández-Alemán, J.L., Toval, A.: Usable Privacy and Security in Personal Health Records. In: Campos, P., Graham, N., Jorge, J., Nunes, N., Palanque, P., Winckler, M. (eds.) INTERACT 2011, Part IV. LNCS, vol. 6949, pp. 36–43. Springer, Heidelberg (2011)
[38] Yonglin, R., Pazzi, R.W.N., Boukerche, A.: Monitoring patients via a secure and mobile healthcare system. IEEE Wireless Communications 17(1), 59–65 (2010)
[39] Wang, Y., Lin, J., Annavaram, M., Jacobson, Q.A., Hong, J., Krishnamachari, B., Sadeh, N.: A framework of energy efficient mobile sensing for automatic user state recognition, pp. 179–192 (2009)
[40] Medjahed, H., Istrate, D., Boudy, J., Dorizzi, B.: A fuzzy logic system for home elderly people monitoring (EMUTEM). Fuzzy Systems (2009)
[41] Fjeldsoe, B.S., Marshall, A.L., Miller, Y.D.: Behavior change interventions delivered by mobile telephone short-message service. American Journal of Preventive Medicine 36(2), 165–173 (2009)
[42] Abroms, L.C., Ahuja, M., Kodl, Y., Thaweethai, L., Sims, J., Winickoff, J.P., Windsor, R.A.: Text2Quit: Results From a Pilot Test of a Personalized, Interactive Mobile Health Smoking Cessation Program. Journal of Health Communication 17(suppl.1), 44–53 (2012)

[43] Pollonini, L., Rajan, N., Xu, S., Madala, S., Dacso, C.: A Novel Handheld Device for Use in Remote Patient Monitoring of Heart Failure Patients—Design and Preliminary Validation on Healthy Subjects. Journal of Medical Systems 36(2), 653–659 (2012)

Questions

Q1. What are the enabling factors for adopting mHealth applications?

Q2. What are the major barriers preventing worldwide mHealth adoption?

Q3. What are the major challenges currently facing by mHealth applications?

Q4. Which mHealth application has the fastest development among patient monitoring applications?

Q5. What are the critical factors for a successful mHealth application?

15. Ubiquitous Health Monitoring: Integration of Wearable Sensors, Novel Sensing Techniques, and Body Sensor Networks

Kevin Hung, C.C. Lee, and Sheung-On Choy

School of Science and Technology, The Open University of Hong Kong
khung@ouhk.edu.hk

Abstract. Emergence of the Internet and widespread use of mobile computing have brought traditional eHealth beyond the boundary of the clinical setting, evolving to mHealth which is patient-centered and ubiquitous. Faced with the world's rapidly ageing population and its burden on the healthcare system, one of the intense areas of development in mHealth is continuous patient monitoring. It requires careful integration of wearable sensors and wireless body sensor networks. Unlike traditional ambulatory monitors, sensors for mHealth may appear in various forms, such as watches, jewelry, eyewear, and even smart garments. Recent work have focused on design for wearability, alternative sensing techniques, and mHealth-specific network topologies.

Keywords: ambulatory monitoring, wearable sensors, pupil size variability, smart fabric, fiber optic sensor, body sensor network (BSN), body area network (BAN), intra-body communication (IBC).

1 Introduction

In response to the world's ageing population and prevalence of chronic diseases, there is intense interest in providing chronic and geriatric care with mHealth [1]. Specifically, recent developments have focused on patient-centered and ubiquitous health monitoring services, which require seamless integration of wearable sensors and wireless body sensor networks (WBSN). Since comfort, unobtrusiveness and low-cost are key features of mHealth, some of the convention physiological sensing techniques, which were originally designed for monitoring in the hospital setting, may no longer be the best approaches.

There are new ways to measure biosignals and health-related parameters for mHealth. For example, instead of using special-purpose accelerometers, we can use those readily embedded in a mobile phone to measure body movements and postures for activity monitoring [2-5]. By placing a fingertip on a smartphone's camera to measure photoplethysmogram (PPG), parameters such as heart rate

S. Adibi (ed.), *Mobile Health*,
Springer Series in Bio-/Neuroinformatics 5, DOI: 10.1007/978-3-319-12817-7_15

(HR), respiratory rate and blood oxygen saturation level (SaO2) can be estimated with reasonable accuracy. The underlying algorithms are expandable to detect atrial fibrillation and blood loss [6]. Some researchers bring it to the next level by estimating HR solely based on facial videos captured by the phone's camera [7]. It works by measuring slight color variations of the face that correspond to cardiac pulses. Meanwhile, others have focused on contactless sensing and wearable sensors [8][9].

This chapter will first give an overview of traditional health monitoring sensors, followed by descriptions of a range of wearable sensors used in common mHealth applications. It will then discuss the integration of these sensors into Wireless Body Sensor Network (WBSN). The conclusion will outline the future trends of wearable sensors for mHealth.

2 Review of Wearable Physiological Sensors

2.1 Traditional Wearable Sensors

Many traditional sensors for patient monitoring are only intended for the clinical setting, where the patients are bedbound. Examples of are electrocardiogram (ECG) leads and electrodes, finger probe of desktop pulse oximeter, acoustic respiration sensor attached on the neck, and electroencephalogram (EEG) hat for brain activity measurement. Although these sensors appear "wearable", they need to be hardwired to bedside equipment. Holter monitor, ambulatory blood pressure (BP) monitor, and fingertip pulse oximeter can offer better mobility, but are still too bulky for long-term wearing and require setting up by health providers. Moreover, data recorded by these devices can only be retrieved and analyzed offline (i.e. after each monitoring session), so timely detection and prediction of abnormality cannot be effectively achieved [10]. The much smaller implantable devices like pacemakers and glucose sensors possess monitoring capability, but still lack the communication links for real-time and remote monitoring.

2.2 Emergence of mHealth-Specific Sensors

Noting the widespread use of mobile phones and high penetration of mobile data services, numerous medical device companies have added communication capabilities into their existing products. Examples are BP meters, blood glucose meters, HR monitors, thermometers, spirometers and digital peak flow meters with USB, Bluetooth and / or Wi-Fi connectivity that were launched in the past decade. In a typical usage scenario, the mobile phone or access point would act as a patient-side hub, which relays the stored readings from the medical device to a remote center for store-and-forward monitoring, analysis and ultimately disease management [11-15]. Along with these variants of devices, a number of mHealth-specific wearable sensors emerged.

3 Overview of mHealth-Specific Sensors Applications

3.1 *Mission-Critical and Sports Training Applications*

A number of early mHealth developments focused on mission-critical applications. A common characteristic of these sensors is their rugged design for use in extreme environments (e.g. extreme temperature, humidity, vibration and pressure). In military applications, wearable sensors are mainly used to monitor soldiers' vital signs. Measurement of sudden acceleration, presence of wounds, projectile impact / penetration, and state of consciousness have been proposed for integration into solders' vests [16][17]. In space exploration, detectors of radiation and biochemical agents are used to monitor the astronauts' surrounding in addition to the vital signs [18][19].

Another area which has a comparatively long history with wearable sensors is sports training, specifically for assessing athletes' performance and recovery. While postures and movements can be captured with high speed cameras during training, wearable sensors are often simultaneously used to deepen coaches' understanding of athletes' conditions. It is common to monitor breathing pattern for respiratory training; body kinematics and posture for reinforcing body awareness and movement efficiency; foot plantar pressure for gait analyzes; electromyogram (EMG) for detecting muscle fatigue; muscle oxygenation for evaluating aerobic efficiency, as well as amount of sweat for prescribing rehydration strategies [20-22]. According to a recent market research, mHealth is in the commercialization phase with the first batch of users who are mainly interested in sports and fitness. The second generation of mHealth, however, will shift the focus to chronic disease management [23].

3.2 *Chronic Care*

The number of persons aged 60 years or older is projected to be almost 2 billion by 2050, and it will be the first time in human history that the number of elderly will surpass that of children [24]. Population statistics show that ageing of the world's population is pervasive, profound, and enduring. As the prevalence of chronic conditions among the elderly is higher than among younger persons, the demand for chronic care is expected to escalate tremendously. Chronic diseases are now the leading causes of mortality, accounting for 63% of all deaths around the world [25]. If not properly managed, they will become the most expensive problems faced by society [26]. There is thus an urgent need to explore new healthcare provision models and technologies.

As symptoms in chronic patients tend to be subtle and vague, long term and continuous health monitoring is needed for timely and early detection of abnormalities. In wearable mHealth systems for chronic care, the commonly monitored signals or parameters are BP, blood pressure variability (BPV), ECG, HR, heart rate variability (HRV), respiration activity, blood glucose level, blood oxygen saturation (SpO2), body temperature, medicine adherence, and physical activity.

4 Wearable Sensors for mHealth

This section will give an overview of the common and emerging wearable sensors, as well as sensing techniques used in mHealth applications.

4.1 Biopotential Monitoring

ECG, EMG and EEG recordings offer non-invasive means to monitor the electrical activities of the heart, muscle and brain, respectively. The most common method to acquire these signals in the clinical setting is to use self-adhesive, wet silver-silver chloride (Ag/AgCl) electrodes connected to a high-input-impedance preamplifier stage. These electrodes contain conductive gel for reliable electrical contact with the patient's skin. Although widely used and readily available, Ag/AgCl electrodes have several disadvantages when applied in mHealth. The electrodes are not reusable and costly; and the gel may cause skin irritation and dry out after long-term contact with the skin [27][28].

A more appropriate alternative for long-term monitoring is gel-free dry electrodes. There are mainly three types: i.) metal plates; ii.) flexible rubber electrodes, which are made by mixing conductive filler materials such as nickel-coated graphite with an elastomer; and iii.) fabric electrodes woven from conductive yarns [29]. These electrodes are reusable, washable, and can be easily embedded into wearable harnesses for long-term use. However, the main drawbacks of biopotential measurement with these electrodes are high sensitivity to different skin conditions, such as dry skin and presence of hair, and high susceptibility to motion artifacts. One way to tackle this problem is to use active electrodes, which are characterized by minimal distance between the electrode-skin interface and preamplifier, very high input impedance, and localized shielding [30]. Circuits and electrodes arranged in this configuration have been integrated into bandage-like patches, some even with wireless charging and communication features [31]. A number of such ECG monitoring patches, either in one-lead or three-lead configurations, have been launched into the market in the past few years.

Recently, non-contact capacitive sensing of biopotentials have received massive attention in mHealth. Because this approach does not require an ohmic connection between the dry electrodes and the skin, biopotentials can be measured through clothing and hair. The electrodes can operate up to a few millimeters' distance from the skin, so they can be easily integrated into garments and furniture for truly ubiquitous monitoring. The major technical challenge in capacitive sensing is the large and varying skin-electrode coupling impedance / capacitance, which causes low frequency noise. Substantial effort has been spent on designing low-noise mixed-signal Application Specific Integrated Circuits (ASIC) that comprise ultra-high input impedance preamplification stage, multi-stage bandpass filters and analog-to-digital convertor [32]. Such chips have been fabricated and successfully tested for contactless measurement of EEG, ECG and EMG with dry and textile electrodes [33][34].

4.2 Cardiac Activity and SpO2 Monitoring

Heart activities can be reflected in multiple biosignals and parameters. Among these, HR and HRV are found to be reliable indicators of the cardiovascular system's health [35]. They can be extracted from ECG, as well as from photoplethysogram (PPG) and phonocardiogram (PCG). PCG, which is the recording of heart sounds with a sensitive microphone, is traditionally monitored along with other cardiac signals in the clinical setting. Studies have shown that diagnosis based on combined ECG and PCG information can lead to higher-performance heart diagnostics [36]. PCG monitors have now been developed into wearable and wireless systems [37]. PPG has been extensively used in mHealth because of its advantages as a noninvasive, inexpensive and convenient diagnostic tool.

A PPG sensor assembly consists of a Light Emitting Diode (LED), which emits light on a selected site of the body and a photodiode (or phototransistor), which measures the transmitted or reflected light from the site. For the estimation of blood oxygen saturation level (SpO2) based on the Beer-Lambert Law, PPG is acquired at two optical wavelengths (red and infrared). The measured optoelectronic signal would fluctuate with the pulsation of the blood flow underneath the skin; its AC component corresponds to the arterial oxygen saturation, while its DC component represents light absorption by other tissues [38].

PPG and SpO2 are conventionally measured from the fingertip, earlobe and toe, but recent mHealth projects have explored sensing from other locations, such as forehead, chest and limbs, using wearable sensors operating in reflective mode. Different from conventional pulse oximeter probes, these sensor assemblies appear in the form of adhesive patch, finger ring, earring, earphone and headband [39][40]. As an example, the finger-ring sensor developed by Asada *et al.* measured PPG signals at the base of the finer to extract SpO2 and heart rate information, and wirelessly sent them to a host computer. To minimize motion artifact, they used a double-ring design, which lowered the influence of external force, acceleration, and ambient light. Further motion artifact reduction was achieved with noise-reference signal and mechanical design [41]. Other groups have proposed different sensing techniques for the finger-ring, such as adaptive filter and PID control, to improve the PPG signal quality sensed by the finger ring at the digital arteries [42][43]. Figure 1 shows the block diagram of the system developed in [43]. The finger sensing assembly contained arrays of infrared LEDs and photodiodes. It worked by switching on individual LEDs in the array with a sequential pattern, sensing along the finger surface until the optimal location for sensing PPG was found.

Some groups have developed the finger-ring sensor as part of a multi-parameter monitoring system. In one prototype, heart-rate signal detected from the ring was transmitted to a wristwatch for calculation and display. Further enhancements included usage in the estimation of SpO2 and BP, and development of a HR measurement ASIC, which used current-steering technique and very low LED pulsing duty cycle to achieve very low cutoff frequency and low power consumption [44].

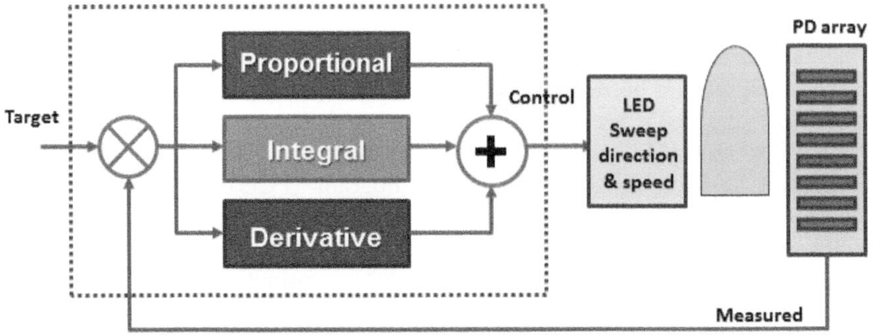

Fig. 1 Intelligent control of a finger-ring sensor for PPG monitoring

4.3 Blood Pressure Monitoring

Clinical studies have revealed that not only office BP is a useful parameter; but BP variability (BPV) also contains important and independent prognostic information for chronic patient monitoring [45]. BPV has been found to be a reliable predictor of cardiovascular events, morbidity and mortality. There is thus an advantage to have wearable blood pressure meter (BPM) to achieve continuous BP monitoring.

A number of wearable BPM based on the traditional oscillometric method have been developed. They have eventually evolved into watch-like devices which can measure BP with an inflation cuff over the radial artery at the wrist instead of on the brachial artery at upper arm. A few recently emerged wrist-worn devices can estimate BP based on arterial tonometry, which uses an array of pressure sensors to capture radial pulse waves [46]. Although no cuff is needed, the contact pressure may still cause skin irritation and discomfort after long-term wearing. The measurements are not continuous but sporadic, and accuracy is location sensitive.

A popular category of BP measurement techniques is based on the pulse-transit-time (PTT), which is a parameter derived from the time delay between the PPG and ECG waves arriving at a measurement site on the body. Because this technique does not require any cuff or pressure sensor, and allows beat-by-beat measurement, the BPM can be feasibly made into a truly wearable device for continuous patient monitoring. PTT-based BPM has been integrated into various forms, including mobile phone, watch, bed and chair. The first PPT-based watch-type BPM was produced by Casio back in the early 1990's. Starting in the 2000's various groups have further investigated in the effects of several factors, including finger temperature, finger contacting force, and pre-ejection period, on BP estimation accuracy [47-50]. A major challenge is to derive an accurate estimation algorithm and a convenient calibration procedure.

4.4 Respiration Monitoring

Traditional respiratory inductance plethysmography (RIP) sensor, magnetometer and strain gauge are commonly used for evaluation of respiratory patterns. Their working principles are similar -- sensors are wrapped around the torso at the ribcage and abdomen to measure the breathing-induced displacements. Setup by the healthcare professional is often required, and the sensors have to be connected to specialized equipment.

More convenient ways to monitor respiratory activities for mHealth are available. One approach is to combine conductive yarn, elastomer, polyester fibers, and smart textile to fabricate piezo-resistive sensors, which can be interwoven into garments [51]. Fiber optic sensor has also been considered, partly due to its immunity to electromagnetic interference (EMI). Among the numerous types of sensing methods with fiber optics [52-55], Fiber Bragg Grating (FBG) sensor has the advantage of easy implementation in terms of its size and stability. For example, FBG sensor has been used to measure respiration during Magnetic Resonance Imaging (MRI) scan [56-58]. In this scenario, the optical fiber (sensor) is interwoven into the patient's garment, while the bench-top light source and photodiode detector are placed outside the MRI scanner. Changes in the reflected wavelength would correspond to the respiratory movement. Multiple sensors can be made on one single fiber by using Wavelength Division Multiplexing (WDM) and / or Time Division Multiplexing (TDM) schemes. Some groups have developed wearable FBG sensors for ambulatory monitoring, by using small-sized optoelectronic components [59]. A small distributed feedback (DFB) laser source is connected to the FBG respiration sensor. The transmitted power of the DFB laser after passing through the FBG sensor is detected by the circuit with a small photodetector. As the FBG sensor is pulled by the respiratory movements, the center wavelength of the FBG is shifted, resulting in changes in the transmitted power through the FBG. A calibration curve is used to relate the received optical power and respiratory motions. As a DFB laser source is used instead of LED light source with board spectrum, the optical filter is not required for the transformation of changes in wavelength to changes in power.

4.5 Body Temperature Monitoring

Body temperature reveals important information about the physiological state. The specific body sites on the body must be selected carefully, so that the readings can reflect the body core temperature even under environmental changes. Currently, a number of mHealth applications measure temperature at the ear, armpit, chest, and abdomen with wearable sensors. There are four major categories of temperature sensors: thermocouples, thermistors, thermopile and fiber-optic sensor. Fiber optics temperature sensors have the technical advantages of flexibility, small dimensions and immunity from EMI [60]. During measurement, one end of the fiber terminal is unclad and attached to the site. Refractive index of the site would decrease as temperature increases, and intensity of light remained in the fiber core would then vary accordingly. By measuring the reflected light in the core, one can derive the behaviour of the light intensity as a function of temperature. Temperature can also be estimated indirectly with wearable sensors. Instead of measuring temperature directly, some groups have proposed monitoring the wearer's shiver motion with accelerometers in order to classify ranges of core body temperature [61].

4.6 Eye Monitoring

It has been suggested that pupil size variability (PSV) and eye dynamics contain information of prognostic and diagnostic value, and can be acquired in a non-invasive and non-contact fashion such as with a camera-mounted eyewear [62]. PSV was first only found to contain respiratory rhythm [63][64], but recent researches have reported the presence of HRV and BPV frequency components [65][66]. This phenomenon is likely due to the common influence of the autonomic nervous system, which regulates glands, cardiac muscle, and smooth muscles throughout the body. The iris' sphincter and dilator muscles, the heart, muscles around blood vessels, and various other organs are under the control of parasympathetic and sympathetic nervous systems of the ANS. This fascinating relationship between PSV and biosignals suggest that it may be worthwhile to monitor PSV for prognosis.

Based on this idea, a group has proposed an eye-based mHealth system. They first performed experiments on human subjects using a wearable eye tracker, verifying the presence of frequency components of respiration activity and systolic blood pressure in the frequency spectrum of PSV, as shown in Figures 2 and 3.

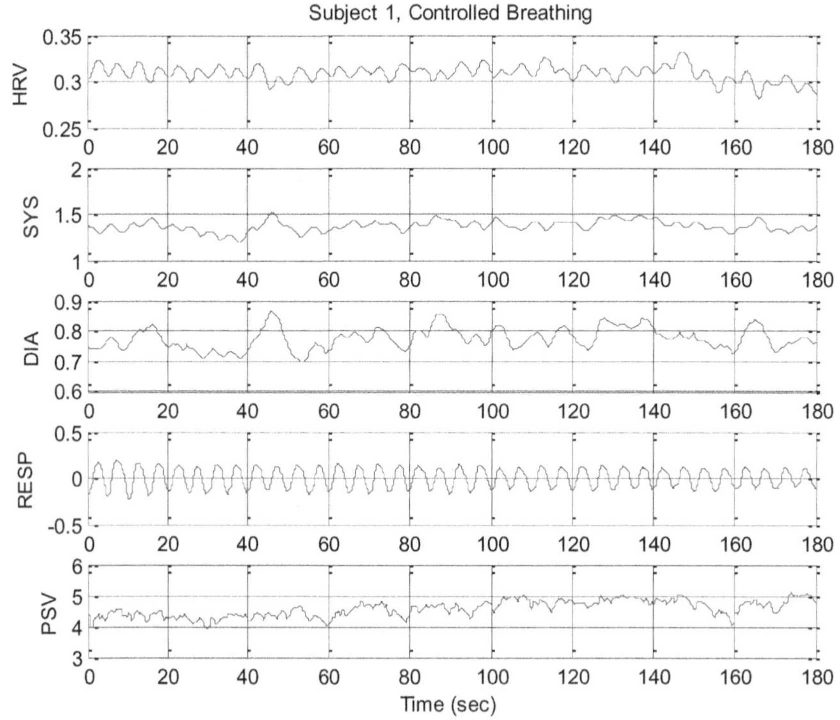

Fig. 2 PSV, HRV, blood pressure waveforms and respiration activity measured from a subject

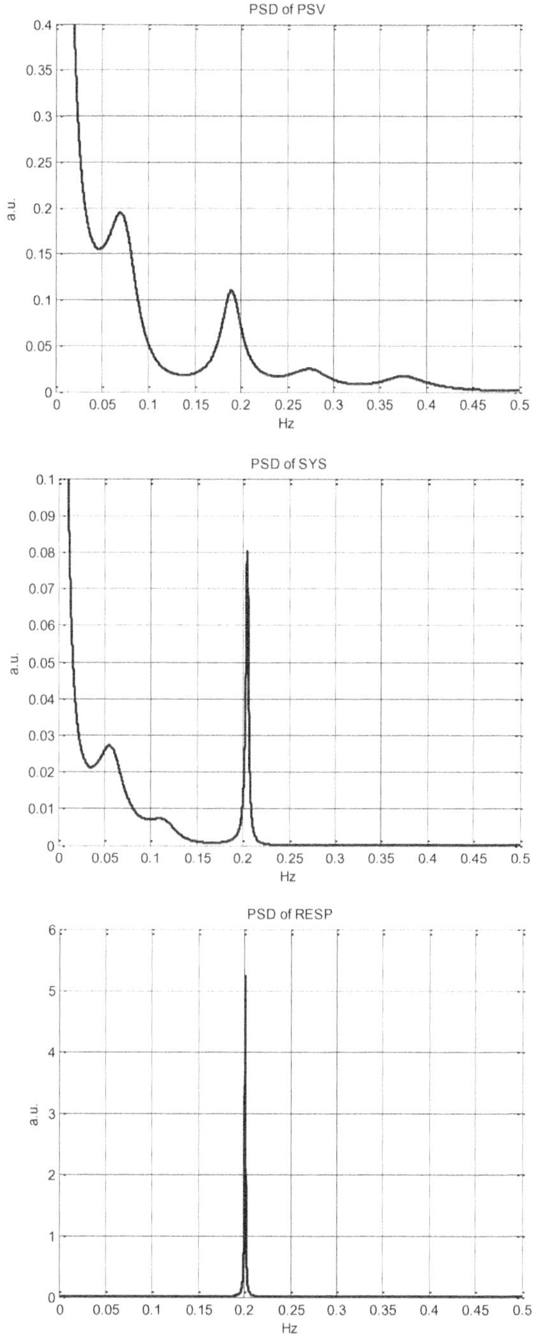

Fig. 3 Frequency spectra of PSV (left), systolic blood pressure (middle) and respiration activity (right) measured from a subject

Their experimental results also show that PSV amplitudes were maximal for mid-range pupil size, and minimal when pupil size was at its extremities, suggesting that subject-centered calibration is needed if PSV is considered for evaluating the LF/HF ratio as an autonomic balance indicator (just like in HRV) [67]. The group also proposed integrating iris biometrics for user authentication and data encryption. The proposed system architecture is shown in Figure 4. Using the same image sensor on the eyewear for capturing PSV and iris image, there is minimal user intervention for authentication and encryption during usage. It also eliminates the system's vulnerability to attacks with "dummy" irises, which do not exhibit PSV or pupillary reflex.

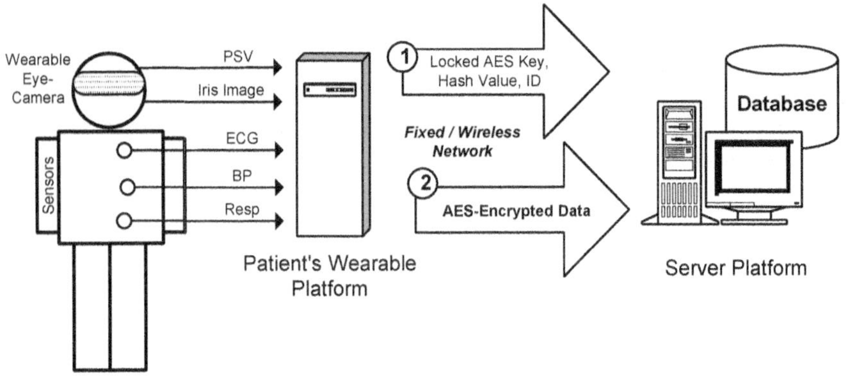

Fig. 4 Eye-based mHealth system

4.7 Fall Detection

As falling is the leading cause of fatal and non-fatal injuries among the elderly population [68], some mHealth systems focus on physical activity monitoring to detect human's falls. Most fall detection techniques are either image-based with computer vision or sensor-based [69][70].

The sensor based methods typically use wearable tri-axial accelerometers to derive the body's acceleration and posture information. The challenge is to reduce false positives and false negatives, distinguishing real falls from fall-like events such as sitting down quickly or jumping. To improve fall detection, some have suggested using at least two tri-axial accelerometers at separate body locations and additional gyroscope to measure body orientation before and after fall-like events. It has been shown in experiments that detection accuracy of over 80% can be achieved with different combinations of methods [69-73]. In one study which used Hidden Markov Models (HMM), fall could be predicted a few hundred milliseconds before the person collided with the floor.

Off-the-shelf fall detectors are either chest-worn or waist-worn devices. There were attempts in making wrist-worn monitors, which is much preferred from a user's perspective; however, the technical challenge is the large number of false positives induced by arm movements. Another interesting on-going development is the design of collision avoidance system – i.e. a wearable "airbag" which is automatically actuated upon fall detection [74].

4.8 Blood Glucose Level Monitoring

More than 150 million people have diabetes worldwide, and this figure is expected to double by 2025. Blood glucose level is traditionally measured with the invasive (fingerstick) method. A number of wearable solutions have been suggested for mHealth. One early example is the wrist-worn GlucoWatch, which uses reverse iontophoresis [75]. It works by applying a small electric current between two skin-contacting electrodes on the underside of the watch, while small amount of body fluid is drawn into the device. The watch then measures the glucose concentration of the fluid with an embedded electrochemical sensor. The reading is shown on the device's LCD screen. Although appears convenient, the watch actually requires daily calibration with standard test strips. Another interesting wearable product is the FDA-approved MiniMed Continuous Glucose Monitoring System, which uses a needle-like glucose sensor to measure blood glucose concentrations subdermally [76]. Its sensor is wired to the embedded processor with an LCD to show the blood glucose concentration. Meanwhile, some other groups are developing wearable patches that measure blood glucose with photoacoustic spectroscopy [77]. Google X Lab recently announced that they have started developing a smart contact lens that will measure glucose levels in tears using a small wireless chip and miniaturized sensor [78].

4.9 Monitoring with Multifunction Devices

While efforts have been put into developing individual sensors for mHealth, there is also a trend to package these sensors into a multifunction and wearable system. An example is the AMON project sponsored by the EU FP5 IST program [79]. The AMON was a wrist-worn monitoring / alert device for high-risk cardiac and respiratory patients. The prototype was integrated with sensors for SpO2, ECG, BP, movement, and temperature monitoring. During their trials, medical personnel at a telemedical center remotely monitored the patients via a GSM link. The system was tested and successfully validated on 33 subjects. Another example is the CodeBlue from Harvard University, which was integrated with pulse oximeter and sensors for ECG and EMG monitoring [80]. The system provided wireless protocols for device discovery and multihop routing.

4.10 Monitoring with "Smart Shirts"

There are increasing mHealth systems and services that utilize sensing technologies in the form of garment. For example, the LifeShirt from VivoMetrics is embedded with RIP sensors, accelerometer, and ECG electrodes [81]. Data were recorded in a

PDA where they were stored or sent to a monitoring center via cellular link. Another group integrated sensors in the fabric of a shirt and developed their prototype which contains dry ECG electrodes, shock / fall sensor, respiration sensor, temperature sensor, GPS receiver and wireless module [82]. The Wearable Health Care System (WEALTHY), developed by Paradiso *et al.*, was based on wearable sensors, electrodes, and connections in fabric forms [83]. Piezoresistive fabric sensors were realized using lycra fabric coated with carbon-loaded rubber and commercial electroconductive yarn, while contact electrodes were fabricated with stain-less-steel-based yarns. The system could simultaneously acquire ECG, respiration, activity, and temperature. Another group developed a shirt which was integrated with an intelligent finger ring sensor, textile ECG electrodes, eye monitoring sensor, FBG sensors and GPS [84][85].

Although PPT-based BP monitoring has been implemented on various mobile platforms, users still need to perform measurement by placing touching the sensing electrodes. The h-Shirt, developed by Zhang *et al.* tackled this problem by using conductive fabrics (or e-textile) as ECG electrodes and medium for signal transmission [86]. ECG was captured by the sleeves from the two wrists, with reference to an electrode placed on the forearm to avoid noise induced by respiration, while a photo-reflective sensors on a fingertip captured PPG. Using fasten snap buttons as connection terminals, the signals were fed into a wristwatch for continuous signal processing and parameter display. On-going development include real-time biofeedback mechanisms for therapeutic treatment.

5 Sensor Integration in Body Sensor Networks

A typical mHealth system comprises the following subsystems: sensing technologies, patient terminal, local station, central server, physician terminal, and communication technologies. Transmission of sensor data to the patient terminal can be via wires or wireless links. Monitoring sessions in some existing mHealth systems are patient-activated; however, automatic and semi-automatic configurations are expected to dominate. A wireless body sensor network (WBSN) is formed by physiological parameter sensors and wearable computing devices placed in, on or around the human body. A WBSN, which is sometimes referred to as a body area network (BAN) or a wireless body area network (WBAN), is a kind of wireless sensor networks (WSNs). In general, a WBSN consists of sensor nodes for collecting physical and physiological data of a human body. The data are then fed to a sink node (or a body area aggregator) which has signal processing, data fusion, feature extraction, and wireless communication capabilities. Usually, the sink node will transmit data to a nearby base station which is connected to the Internet. Figure 5 depicts the general architecture of a WBSN and its environment. A WBSN might interact with existing systems, such as healthcare servers in hospitals or medical centers.

WBSNs are an emerging research area in recent years as their applications are in great demand in the sectors of healthcare, medical treatment and social welfare. There are three main areas of research on WBSNs: sensor technology, data fusion, and network communications.

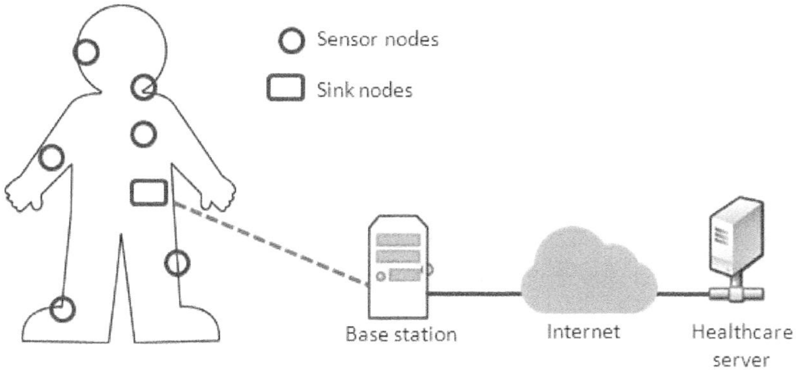

Fig. 5 General Architecture of a Wireless Body Sensor Network (WBSN)

5.1 Design Issues of WBSNs

There are multiple challenges in the design and implementation of WBSNs [87][88]. First, the nodes of a WBSN are deployed in, on or around the human body, so the total number of nodes cannot be large. Therefore, the accuracy and reliability of nodes is an important requirement as it is not practical to use redundant nodes to solve the problems. Second, the size of sensor nodes must be small and have high wearability and biocompatibility. Moreover, the shape and material used should not cause irritation, allergy or hamper the efficiency of individual. Third, in terms of power sources of WBSNs, nodes cannot be equipped with high-energy battery, and they may need to be powered by kinetic energy or heat energy. So, low power consumption is also an important requirement in WBSNs. Due to the energy constraints, the sensors may have less computational power and cannot perform large bit computations. Fourth, considering data communication and networking, since the data types and channel assignments will be different among WBSN nodes, the transfer rates will vary among different types of nodes. Finally, the users' safety and privacy is also an important issue in WBSNs. So, Quality of Service (QoS), confidentiality, authenticity, and other security issues in data communication must be considered in the implementation of WBSNs.

5.2 Data Communications and Networking in WBSNs

To satisfy the specific needs of WBSN communications, many works and studies have been carried out in the aspects of radio technology, communication protocol, network topology, energy control, and data security. For example, the IEEE 802.15.6 working group was established to standardize certain aspects of WBSNs [89]. The IEEE 802.15.6 standard defines both physical and media access control (MAC) layers that can satisfy various needs of WBSNs.

5.3 Frequency Bands and Channel Selection

In general, WBSN frequency bands can be divided into Human Body Communication (HBC) band, Narrowband (NB), and Ultra wideband (UWB). The selection of each band depends on the sensor nodes as well as the application. The HBC is used for Intra-body Communication, taking the human body as the transmission medium, and the primary technology is electromagnetic coupling such as capacitive coupling and galvanic coupling [90][91]. Some researchers such as [92] proposed using human body as the electromagnetic waveguide in WBSN communication.

In the range of NB, it includes the following well-known frequency bands suitable for WBSN communication, namely MICS (Medical Implant Communications Service), WMTS (Wireless Medical Telemetry Service), and ISM (Industrial, Scientific and Medical) bands. MICS specifies using a frequency band between 402 and 405 MHz in communication and the maximum transmit power is very low (25 uW) so that the risk of interfering with other users of the same band can be reduced. The maximum bandwidth is 300 kHz, which makes it only suitable for low data rate networks. WMTS is a wireless service specifically defined for transmission of data related to a patient's health and is mainly used in wireless medical telemetry in hospitals. The WMTS is in the 608 – 614, 1395 – 1400, and 1427 – 1432 MHz range [93]. The ISM bands are frequency bands reserved internationally for industrial, scientific and medical purposes other than telecommunications. The ISM band commonly used in WBSNs operates at 2.4 GHZ. Medical equipment operating in this band must tolerate any interference generated by other ISM equipment. Compared to HBC band and NB, UWB is a high frequency band that can provide high data rate and is promising in WBSN applications [94].

5.4 MAC Protocols

MAC protocols for WBSNs should require minimum power consumption. WBSNs that are operated in unlicensed bands may suffer from interference and may result in performance degradation. These are the issues that need to be considered in selecting MAC protocols suitable for WBSNs. MAC protocols can be divided into three main types: contention protocols, schedule protocols, and contention-schedule protocols [95]. The contention protocols such as Carrier Sense Multiple Access/Collision Avoidance (CSMA/CA) protocols may encounter serious collision problems for high traffic nodes and may not suitable for WBSN applications [96]. On the other hand, existing schedule protocols such as Time Division Multiple Access (TDMA) protocols require frequent time synchronization that may be difficult to achieve for WBSNs. In [97], researchers proposed a MAC protocol that is particularly suitable for medical applications of WBSNs. The protocol is an enhanced Distributed Queuing Medium Access Protocol (DQ MAC) and has high network energy efficiency and scales well for very dense WBSNs. In [98], researchers proposed using heartbeat rhythm information to perform time synchronization in MAC. The proposed MAC protocol is based on TDMA and can achieve high energy efficiency. Other TDMA-based MAC protocols, such as WhMAC protocol [99] and BodyMAC protocol [100], has been proposed to achieve flexible bandwidth allocation and data transfer and improve node energy efficiency. In [101], researchers proposed a contention-schedule MAC protocol called ATLAS that can provide better energy efficiency, higher capacity-utilization, and less delay in MAC.

5.5 Security and Privacy Issues of WBSNs

BSNs are extremely limited in terms of power, computation, and communication. Current research trends focus more on reliable communication, energy-efficient protocols, and network mobility. However, WBSNs are often deployed in accessible areas and the data collected has a close relationship with individual information. Therefore, deploying WBSNs for healthcare applications needs to consider security and privacy issues.

A survey in [102] discusses the security and privacy issues in healthcare application using WBSNs. The paper highlights some popular wireless healthcare projects and discusses their security threats and privacy issues. There are a number of security threats that would be harmful for the wireless healthcare success. First, monitoring and eavesdropping on patient vital signs are common threats to the patient privacy [103]. Second, there exist some threats to data integrity when WBSN nodes are sending information (i.e. in transit). In transit attacks can modify the actual data and can then trigger a false alarm or hide the true patient conditions. Third, there are various threats relating to unguarded rely nodes in wireless networks. An attacker can use an illegal rely node to act as a real node to the network and impose masquerade and replay threats to wireless healthcare systems. Fourth, some healthcare applications using WBSNs have the capabilities of location tracking and activity tracking of users. These users' detailed information, if not be protected properly, could directly infringe a user's privacy.

5.6 Standardization in WBSNs and Healthcare Systems

The IEEE 802.15 Task Group 6 (BAN) is developing a communication standard optimized for low power devices operated on, in or around the human body [104]. The IEEE 802.15.6 has put forward the fundamental content about physical layer channels and their characterization, and has defined many WBSN band standards. ISO/IEEE 11073 Personal Health Data standards are a group of standards addressing the interoperability of personal health devices such as weighing scales, blood pressure monitors, blood glucose monitors and the like. These standards facilitate exchange of health information between network devices as well as hosting devices.

The Continua Health Alliance [105] is an alliance of technology, medical device and healthcare companies aiming to make personal telehealth a reality through developing guidelines and standards to enable interoperability between systems and devices. The American Health Insurance Portability and Accountability Act of 1996 (HIPAA) [106] address the security and privacy of health data. The standards are meant to improve the efficiency and effectiveness of healthcare systems by encouraging the widespread use of electronic data interchange in healthcare systems. The Health Information Technology for Economic and Clinical Health Act (HITECH) [107] includes provisions to enlarge the use of information technology to store, capture, transmit, properly share and use health data. The HITECH Act introduces the new Act that states that those who manage patient health information should notify the affected individuals if there is any breach that discloses their information.

6 Future Trends of Wearable Sensors for mHealth

Wearability of sensors is an important issue in mHealth which involve patient monitoring. More researchers are putting efforts in improving sensor wearability and reducing the failure rate of devices. Those efforts include reducing size and weight, enhancing the wear resistance, improving softness and biocompatibility (for avoiding skin inflammation and allergy) of sensors.

The power consumption issue of WBSNs will also be a hot research trend. The researches on power consumption problems are not limited to how to reduce energy consumption of sensors, but also how to increase the battery capacity, and how to harness energy from the surroundings. More work will be done in the aspects of circuit design, data fusion algorithms, and network communication protocols towards the goal of reducing the power consumption of the WBSNs as a whole.

Large-scale and community mHealth applications involve management of a large number of disparate WBSNs. A huge amount of data streams from WBSNs needs to be processed and analyzed both online and offline. In this regard, a powerful and scalable storage and computing infrastructure will be another intesnse field of research in mHealth. Cloud computing technology is likely to be applied in the management and monitoring of WBSNs.

References

[1] mHealth: New Horizons for Health through Mobile Technologies: Based on the findings of the second Global Survey on eHealth, World Health Organization (WHO), Geneva (November 2011), `http://www.who.int/goe/publications/goe_mhealth_web.pdf`

[2] Lan, N.D., Miluzzo, E., Lu, H., Peebles, D., Choudhury, T., Campbell, A.T.: A Survey of Mobile Phone Sensing. IEEE Communications Magazine 48(9), 140–150 (2010)

[3] Viet, V.Q., Lee, G., Choi, D.: Fall Detection based on Movement and Smart Phone Technology. In: Proc. IEEE International Conference on Computing and Communication Technologies, Research, Innovation, and Vision for the Future, Ho Chi Minh City, Vietnam, February 27-March 1, pp. 1–4 (2012)

[4] Kaenampornpan, M., Anuchad, T., Supaluck, P.: Fall Detection Prototype for Thai Elderly in Mobile Computing Era. In: Proc. 9th International Conference on Telecommunications and Information Technology, May 17-19, pp. 446–449 (2011)

[5] Kwon, S., Lee, J., Chung, G.S., Park, K.S.: Validation of heart rate extraction through an iPhone accelerometer. In: Proc. 33rd Annual International Conference of the IEEE EMBS, Boston, USA, August 30-September 3, pp. 5260–5263 (2011)

[6] Scully, C.G., Lee, J., Meyer, J., Gorbach, A.M., Granquist-Fraser, D., Mendelson, Y., Chon, K.H.: Physiological Parameter Monitoring from Optical Recordings With a Mobile Phone. IEEE Transactions on Biomedical Engineering 59(2), 303–306 (2012)

[7] Kwon, S., Kim, H., Park, K.S.: Validation of heart rate extraction using video imaging on a built-in camera system of a smartphone. In: Proc. 34th Annual International Conference of the IEEE EMBS, San Diego, USA, August 28-September 1, pp. 2174–(2177)

[8] Mikhelson, I.V., Lee, P., Bakhtiari, S., Elmer II, T.W., Katsaggelos, A.K., Sahakian, A.V.: Noncontact Millimeter-Wave Real-Time Detection and tracking of Heart Rate on an Ambulatory Subject. IEEE Transactions on Information Technology in Biomedicine 16(5), 927–934 (2012)

[9] Teng, X.F., Zhang, Y.T., Poon, C.C.Y., Bonato, P.: Wearable Medical Systems for p-Health. IEEE Review in Biomedical Engineering 1, 62–74 (2008)

[10] Raviele, A., Giada, F., Bergfeldt, L., et al.: Management of patients with palpitations: a position paper from the European Heart Rhythm Association. Europace 13(7), 920–934 (2011)

[11] Lee, R.-G., Chen, K.-C., Hsiao, C.-C., Tseng, C.-L.: A Mobile Care System With Alert Mechanism. IEEE Transactions on Information Technology in Biomedicine 11(5), 507–517 (2007)

[12] Song, W., Yu, H., Liang, C., Wang, Q., Shi, Y.: Body Monitoring System Design Based on Android Smartphone. In: Proc. 2012 World Congress on Information and Communication Technologies, Trivandrum, India, October 30-November 2, pp. 1147–1151 (2012)

[13] Lim, J.-H., Park, C., Park, S.-J.: Home Healthcare Settop-box for Senior Chronic Care using ISO/IEEE 11073 PHD Standard. In: Proc. 32nd Annual International Conference of the IEEE EMBS, Buenos Aires, Argentina, August 31-September 4, pp. 216–219 (2010)

[14] Cao, Z., Zhu, R., Que, R.-Y.: A Wireless Portable System With Microsensors for Monitoring Respiratory Diseases. IEEE Transactions on Biomedical Engineering 59(11), 3110–3116 (2012)

[15] Dal Negro, R.W., Micheletto, C., Tognella, S., Turati, C., Bisato, R., Guerriero, M., Sandri, M., Turco, P.: PIKO-1, and effective, handy device dor the patient's personal PEFR and FEV1 electronic long-term monitoring. Monaldi Archives for Chest Disease 67(2), 84–89 (2007)

[16] Lim, H.B., Ma, D., Wang, B., Kalbarczyk, Z., Iyer, R.K., Watkin, K.L.: A Soldier Health Monitoring System for Military Applications. In: Proc. 2010 International Conference on Body Sensor Networks, Singapore, June 7-9, pp. 246–249 (2010)

[17] Park, S., Jayaraman, S.: Adaptive and Responsive Textile Structures (ARTS). In: Tao, X. (ed.) Smart Fibres, Fabrics and Clothing, ch. 13, pp. 226–245. CRC Press, Boca Raton (2001)

[18] Montgomery, K., Mundt, C., Thonier, G., et al.: Lifeguard – A Personal Physiological Monitor for Extreme Environment. In: Proc. 26th Annual International Conference of the IEEE EMBS, San Francisco, USA, September 1-5, pp. 2192–2195 (2004)

[19] Canina, M., Newman, D.J., Trotti, G.L.: Preliminary Considerations for Wearable Sensors for Astronauts in Exploration Scenarios. In: Proc. 3rd IEEE-EMBS International Summer School and Symposium on Medical Devices and Biosensors, Boston, USA, September 4-6, pp. 16–19 (2006)

[20] Morris, D., Schazmann, B., Wu, Y., et al.: Wearable Sensors for Monitoring Sports Performance and Training. In: Proc. 5th International Summer School and Symposium on Medical Devices and Biosensors, Hong Kong, June 1-3, pp. 121–124 (2008)

[21] Coyle, S., Morris, D., Lau, K.-T., Diamond, D., Moyna, N.: Textile-based wearable sensors for assisting sports performance. In: Proc. 6th International Workshop on Wearable and Implantable Body Sensor Networks, Berkeley, USA, June 3-5, pp. 307–311 (2009)

[22] Ermes, M., Parkka, J., Mantyjarvi, J., Korhonen, I.: Detection of Daily Activities and Sports with Wearable Sensors in Controlled and Uncontrolled Conditions. IEEE Transaction in Information Technology in Biomedicine 12(1), 20–26 (2008)

[23] Mobile Health Market Trends and Figures 2013-2017 – The Commercialization of mHealth Applications, research2guidance, vol. 3, Berlin (2013)

[24] World Population Ageing: 1950 - 2050, Population Division, Department of Economic and Social Affairs, United Nations, New York (2002), http://www.un.org/esa/population/publications/worldageing19502050/

[25] Global status report on noncommunicable diseases 2010: description of the global burden of NCDs, their risk factors and determinants, World Health Organization (WHO), Geneva (April 2011), http://www.who.int/nmh/publications/ncd_report2010/en/

[26] Chronic Disease and Palliative Care, Global Business Group on Health (December 2012), http://www.businessgrouphealth.org/global/issues/chronic.cfm

[27] Chi, Y.M., Jung, T.-P., Cauwenberghs, G.: Dry-Contact and Noncontact Biopotential Electrodes: Methodological Review. IEEE Reviews in Biomedical Engineering 3, 106–119 (2010)

[28] Neuman, M.R.: Biopotential Electrodes. In: Webster, J.G. (ed.) Medical Instrumentation: Application and Design, 4th edn., ch. 5, pp. 189–240. Wiley & Sons, New Jersey (2010)

[29] Ryu, C.Y., Nam, S.H., Kim, S.: Conductive Rubber Electrode for Wearable Health Monitoring. In: Proc. 27th Annual International Conference of the IEEE EMBS, Shanghai, China, January 17-18, pp. 3479–3481 (2006)

[30] Lee, S.H., Jung, S.M., Lee, C.K., Jeong, K.S., Cho, G., Yoo, S.K.: Wearable ECG monitoring system using conductive fabrics and active electrodes. In: Jacko, J.A. (ed.) HCI International 2009, Part III. LNCS, vol. 5612, pp. 778–783. Springer, Heidelberg (2009)

[31] Yoo, J., Yan, L., Lee, S., Kim, Y., Yoo, H.-J.: A 5.2mW self-configured Wearable Body Sensor Network Controller and a 12(W Wirelessly Powered Sensor for a Continuous Health Monitoring system. IEEE Journal of Solid-Sate Circuits 45(1), 178–188 (2010)

[32] Valchinov, E., Rutkovskis, A., Pallikarakis, N.: Wireless Dry-Contact Biopotential Electrode. In: Proc. 3rd International Conference on E-Health and Bioengineering, Iasi, Romania, November 24-26, pp. 1–4 (2011)
[33] Sullivan, T.J., Deiss, S.R., Cauwenberghs, G.: A Low-Noise, Non-Contact EEG / ECG Sensor. In: Proc. IEEE Biomedical Circuits and System Conference, Montreal, November 27-30, pp. 154–157 (2007)
[34] Taelman, J., Adriaensen, T., van der Horst, C., Linz, T., Spaepen, A.: Textile Integrated Contactless EMG Sensing for Stress Analysis. In: Proc. 29th Annual International Conference of the IEEE Engineering in Medicine and Biology Society, Lyon, August 22-26, pp. 3966–3969 (2007)
[35] Breiser, K.H., Kluttig, A., Schumann, B., Kors, J.A., Swenne, C.A., Kuss, O., Werdan, K., Haerting, J.: Cardiovascular Disease, Risk Factor and Heart Rate Variability in the Elderly General Population: Design and Objectives of the CARdiovascular Disease, Living and Ageing in Halle (CARLA) Study. BMC Cardiovascular Disorder 5(33) (November 2005)
[36] Phanphaisarn, W., Roeksabutr, A., Wardkein, P., Koseeyaporn, J., Yapin, P.P.: Heart Detection and Diagnosis Based on ECG and EPCG Relationships. Medical Devices: Evidence and Research 4, 133–144 (2011)
[37] Abdel-Motaleb, I.M., Sambaraju, V.K.: Development of a Wireless Cardiogram System for Acute and Long-term Healthcare Monitoring. In: Proc. 2012 IEEE International Conference on Electro/Information Technology, Indianapolis, USA, May 6-8, pp. 1–6 (2012)
[38] Webster, J.G.: Measurement of Flow and Volume of Blood. In: Webster, J.G. (ed.) Medical Instrumentation: Application and Design, 4th edn., vol. 5, pp. 338–376. Wiley & Sons, New Jersey (2010)
[39] Poh, M.-Z., Kim, K., Goessling, A.D., Swenson, N.C., Picard, R.W.: Heartphones: Sensor Earphones and Mobile Application for Non-obtrusive Health Monitoring. In: Proc. International Symposium on Wearable Computers, Linz, Austria, September 4-7, pp. 153–154 (2009)
[40] Kim, S.H., Ryoo, D.W., Bae, C.: Adaptive Noise Cancellation Using Accelerometers for the PPG Signal from Forehead. In: Proc. 30th Annual International Conference of the IEEE EMBS, Lyon, France, August 22-26, pp. 2564–2567 (2007)
[41] Asada, H.H., Shaltis, P., Reisner, A., Rhee, S., Hutchinson, R.C.: Mobile Monitoring with Wearable Photoplethysmographic Biosensors. IEEE Engineering in Medicine and Biology Magazine 22(3), 28–40 (2003)
[42] Han, H., Kim, M.-J., Kim, J.: Development of real-time motion artifact reduction algorithm for a wearable phototplethysmogrphy. In: Proc. 29th Annual International Conference of the IEEE EMBS, Lyon, pp. 1538–1541 (2007)
[43] Hung, K., Chan, W.M., Lee, C.C., Choy, S.-O., Kwok, P.: Preliminary Design of an Intelligent Finger-Ring Sensor for Physiologic Monitoring in Mobile Health. Presented in 2nd AMA-IEEE Medical Technology Conference, Boston, October 16-18 (2011)
[44] Wong, A., Pun, K.P., Zhang, Y.T., Hung, K.: A Near-infrared heart rate measurement IC with very low cutoff frequency using current steering technique. IEEE Trans. Circuits Syst. I, Fundam. Theory Appl. 52(12), 2642–2647 (2005)
[45] Parati, G., Ochoa, J.E., Salvi, P.: Prognostics Value of Blood Pressure Variability and Average Blood Pressure Levels in Patients With Hypertension and Diabetes. Diabetes Care 36(suppl. 2), 312–324 (2013)

[46] Tanaka, S., Matsumoto, Y., Wakimoto, K.: Unconstrained an non-invasive measurement of heart-beat and respiration periods using a phonocardiographic sensor. Med. Biol. Eng. Comput. 40(2), 246–252 (2002)

[47] Poon, C.C.Y., Wong, Y.M., Zhang, Y.T.: M-Health: The Development of cuff-less and Wearable Blood Pressure Meters for Body Sensor Networks. In: Proc. IEEE / NLM Life Sci. Syste. Appl. Workshop, Bethesada, MD, July 13-14 (2006)

[48] Zhang, X.Y., Zhang, Y.T.: The Effects of Local Mild Cold Exposure on Pulse Transit Time. Physiological Measurement 27(7), 649–660 (2006)

[49] Teng, X.F., Zhang, Y.T.: Theoretical Study on the Effects of Sensor Contacting Force on Pulse Transit Time. IEEE Transactions on Biomedical Engineering 54(8), 1490–1498 (2007)

[50] McCombie, D.B., Reisner, A.T., Asada, H.H.: Motion Based Adaptive Calibration of Pulse Transit Time Measurements to Arterial Blood Pressure for an Autonomous, Wearable Blood Pressure Monitor. In: Proc. 30th Annual International Conference of the IEEE EMBS, Vancouver, Canada, August 20-24, pp. 989–992 (2008)

[51] Guo, L., Berglin, L., Li, Y.J., Mattila, H., Mehrjerdi, A.K., Skrifvars, M.: Disappearing Sensor – Textile Based Sensor for Monitoring. In: Proc. 2011 International Conference on Control, Automation and Systems Engineering, Singapore, July 30-31, pp. 1–4 (2011)

[52] Pavlath, G.A.: Fiber-optic gyroscopes. In: Proc. IEEE Lasers and Electro-Optics Society (LEOS) Annual Meeting, Boston, October 31-November 3, vol. 2, pp. 237–238 (1994)

[53] Danisch, L., Englehart, K., Trivett, A.: Spatially continuous six degree of freedom position and orientation sensor. Sens. Rev. 19(2), 106–112 (1999)

[54] Donno, M., Palange, E., Di. Nicola, F., Bucci, G., Ciancetta, F.: A new flexible optical fiber goniometer for dynamic angular measurements: Application to human joint movement monitoring. IEEE Trans. Instrum. Meas. 57(8), 1614–1620 (2008)

[55] Poeggel, S., Leen, G., Bremer, K., Lewis, E.: Miniature Optical fiber combined pressure- and temperature sensor for medical applications. In: IEEE Sensors 2012, pp. 1–4 (2012)

[56] D'Angelo, L.T., Weber, S., Honda, Y., Thiel, T., Narbonneau, F., Lüth, T.C.: A system for respiratory motion detection using optical fibers embedded into textiles. In: Proc. 30th Annual International Conference of the IEEE EMBS, Vancouver, August 20-25, pp. 3694–3697

[57] De Jonckheere, J., Narbonneau, F., D'Angelo, L.T., Witt, J., Paquet, B., Kinet, D., Kreber, K., Logier, R.: FBG-based smart textiles for continuous monitoring of respiratory movements for healthcare applications. In: Proc. 2010 12th IEEE International Conference on e-Health Networking Applications and Services, Lyon, France, July 1-3, pp. 277–282 (2010)

[58] Witt, J., Narbonneau, F., Schukar, M., Krebber, K., et al.: Medical Textiles With Embedded Fiber Optic Sensors for Monitoring of Respiratory Movement. IEEE Sensors Journal 12(1), 246–254

[59] Lee, C.C., Hung, K., Chan, W.-M., Wu, Y.K., Choy, S.-O., Kwok, P.: FBG sensor for physiologic monitoring in M-health application. In: Proc. Asia Communications and Photonics Conference and Exhibition 2011, Shanghai, November 13-16, pp. 1–14 (2011)

[60] Scheggi, A.M., Brenci, M., Conforti, G., Falciai, R.: Optical-fibre thermometer for medical use. IEE Proceedings H Microwaves, Optics and Antennas 131(4), 270–272 (1984)
[61] Sung, M., DeVaul, R., Jimenez, S., Gips, J., Pentland, A.: Shiver Motion and Core Body Temperature Classification for Wearable Soldier Health Monitoring Systems. In: Proc. 8th International Symposium on Wearable Computers, Arlington, USA, October 31-November 3, vol. 1, pp. 192–193 (2004)
[62] Hung, K., Zhang, Y.T.: Preliminary investigation of pupil size variability: toward non-contact assessment of cardiovascular variability. In: Proc. 3rd. IEEE EMBS International Summer School in Medical Devices & Biosensors, Cambridge, pp. 137–140 (2006)
[63] Ohtsuka, K., Asakura, K., Kawasaki, H., Sawa, M.: Respiratory fluctuations of the human pupil. Exp. Brain Res. 71(1), 215–217 (1988)
[64] Borgdorff, P.: Respiratory fluctuations in pupil size. Am. J. Physiol. 228(4), 1094–1102 (1975)
[65] Calcagnini, G., Giovannelli, P., Censi, F., Bartolini, P., Barbaro, V.: Baroreceptor-sensitive fluctuations of heart rate and pupil diameter. In: Proc. 23rd Annual International Conference of the IEEE EMBS, Istanbul, Turkey, October 25-28, vol. 1, pp. 600–603 (2001)
[66] Calcagnini, G., Censi, F., Lino, S., Cerutti, S.: Pupil Diameter Variability in Humans. In: Proc. 22nd Annual International Conference of the IEEE EMBS, Chicago, USA, July 23-28, vol. 3, pp. 2298–2301 (2000)
[67] Hung, K.: Pupillary Dynamic Monitoring for Mobile Health. In: Proc. 39th International Conference of the IEEE IES, Vienna, Austria, November 10-13 (2013)
[68] Falls Among Older Adults: An Overview, Centers for Disease Control and Prevention, USA (September 2013), `http://www.cdc.gov/homeandrecreationalsafety/falls/adultfalls.html`
[69] Purwar, A., Jeong, D.U., Chung, W.Y.: Activity Monitoring from Real-Time Triaxial Accelerometer data using Sensor Network. In: Proc. International Conference on Control, Automation and Systems, Seoul, Korea, October 17-20, pp. 2402–2406 (2007)
[70] Fernandez-Caballero, A., Sokolova, M.V., Serrano-Cuerda, J., Castillo, J.C., Moreno, V., Castineira, R., Redondo, L.: HOLDS: Efficient Fall Detection Through Accelerometers and Computer Vision. In: Proc. 2012 8th International Conference on Intelligent Environments, Guanajuato, June 26-29, pp. 367–370 (2012)
[71] Sorvala, A., Alasaarela, E., Sorvoja, H., Myllyla, R.: Activity Classification Using a State Transition Diagram and Activity Levels. In: Proc. 2012 6th International Symposium on Medical Information and Communication Technology, La Jolla, March 25-29, pp. 1–4 (2012)
[72] Tong, L., Song, Q., Ge, Y., Liu, M.: HMM-Based Human Fall Detection and Prediction Method Using Tri-Axial Accelerometer. IEEE Sensors Journal 13(5), 1849–1856 (2013)
[73] Li, Q., Stankovic, J.A., Hanson, M.A., Barth, A.T., Lach, J., Zhou, G.: Accurate, Fast Fall Detection Using Gyroscopes and Accelerometer-Derived Posture Information. In: Proc. 6th International Workshop on Wearable and Implantable Body Sensor Networks, Berkeley, USA, June 3-5, pp. 138–143 (2009)

[74] Tamura, T., Yoshimura, T., Sekine, M., Uchida, M., Tanaka, O.: A Wearable Airbag to Prevent Fall Injuries. IEEE Transactions on Information Technology in Biomedicine 13(6), 910–914 (2009)

[75] Garg, S.K., Potts, R.O., Ackerman, N.R., Fermi, S.J., Tamada, J.A., Chase, H.P.: Correlation of Fingerstick Blood Glucose Measurements with GlucoWatch Biographer Glucose Results in Young Subjects with Type 1 diabetes. Diabetes Care 22(10), 1708–1714 (1999)

[76] Chee, F., Fernando, T., Savkin, A.V., van Heerden, P.V.: The Use of MiniMed CGMS in Real-Time Glucose Monitoring. In: Proc. 7th Australian and New Zealand 2001 Intelligent Information Systems Conference, November 18-21, pp. 159–164 (2001)

[77] MacKenzie, H.A., Ashton, H.S., Spiers, S., Shen, Y., Freeborn, S.S., Hannigan, J., Lindberg, J., Rae, P.: Advances in Photoacoustic Noninvasive Glucose Testing. Clinical Chemistry 45(9), 1587–1595 (1999)

[78] Tweed, K.: Google Working on Smart Contact Lens to Monitor Diabetes. IEEE Spectrum (January 17, 2014), `http://spectrum.ieee.org/tech-talk/biomedical/devices/google-working-on-smart-contact-lens-to-monitor-diabetes`

[79] Anliker, U., Ward, J.A., Lukowicz, P., et al.: AMON: A Wearable Multiparameter Medical Monitoring and Alert System. IEEE Transactions on Information Technology in Biomedicine 8(4), 1–11 (2004)

[80] Shnayder, V., Chen, B.-R., Lorincz, K., Fulford-Jones, T.R.F., Welsh, M.: Sensor Networks for Medical Care. In: Proc. 3rd International Conference on Embedded Networked Sensor Systems, San Diego, USA, November 2-4, p. 314 (2005)

[81] Grossman, P.: The LifeShirt: a Multi-function Ambulatory System Monitoring Health, Disease, and Medical Intervention in the Real World. Studies in Health Technology and Informatics 108, 133–141 (2004)

[82] Noury, N., Dittmar, D., Corroy, C., et al.: A Smart Cloth for Ambulatory Telemonitoring of Physiological Parameters and Activity: the VTAMN project. In: Proc. 6th International Workshop on Enterprise Networking and Computer in Healthcare Industry, Grenoble, June 28-29, pp. 155–160 (2004)

[83] Linz, T., Kallmayer, C., Aschenbrenner, R., Reichl, H.: Fully Integrated EKG Shirt Based on Embroidered Electrical Interconnections with Conductive Yearn and Miniaturized Flexible Electronics. In: Proc. International Workshop on Wearable and Implantable Body Sensor Networks, Boston, USA, April 3-5, pp. 23–26 (2006)

[84] Zhang, Y.T., Poon, C.C.Y., Chan, C.H., Tsang, M.W.W., Wu, K.F.: A Health-shirt Using e-Textile Materials for the Continuous and Cuffless Monitoring of Arterial Blood Pressure. In: Proc. 3rd IEEE-EMBS International Summer School and Symposium on Medical Devices and Biosensors, Boston, USA, September 4-6, pp. 86–89 (2006)

[85] Hung, K., Lee, C.C., Chan, W.M., Choy, S.-O., Kwok, P.: Development of a wearable system integrated with novel biomedical sensors for ubiquitous healthcare. In: Proc. 2012 Annual International Conference of the IEEE EMBS, San Diego, August 28-September 1, vol. 1, pp. 5802–5805 (2012)

[86] Hung, K., Lee, C.C., Chan, W.M., Choy, S.-O., Kwok, P.: Development of Novel Wearable Sensors for Mobile Health. In: Proc. 2012 IEEE-EMBS International Conference on Biomedical and Health Informatics, Hong Kong, January 5-7, pp. 745–747 (2012)

[87] Hanson, M.A., Powell, H.C., Barth, A.T., et al.: Body Area Sensor Networks: Challenges and Opportunities. IEEE Computer 42(1), 58–65

[88] Shah, S.A., Raazi, S.M.K., Khan, R.A.: Wireless Sensor Networks Health Monitoring: Trends and Challenges. Journal of Emerging Trends in Computing and Information Sciences 3(3), 319–324 (2012)

[89] IEEE 802.15 Working Group Documents, `https://mentor.ieee.org/802.15/dcn/08/15-08-0792-00-0006-channel-model-updates-for-802-15-6.ppt` (accessed on September 7, 2013)

[90] Wegmueller, M.S., Kuhn, A., Froehlic, J., Oberle, M., Felber, N., Kuster, N., Fichter, W.: An attempt to model the human body as a communication channel. IEEE Transaction on Biomedical Engineering 54, 1851–1857 (2007)

[91] Wang, W.C., Nie, Z.D., Guan, F., Leng, T.F., Wang, L.: Experimental Studies on Human Body Channel Communication Characteristics Based upon Capacitive Coupling. In: Proc. of 2011 International Conference on Body Sensor Networks, Dalas, USA, May 24-25, pp. 180–185 (2011)

[92] Hachisuka, K., Nakata, A., Takeda, T., et al.: Development and Performance Analysis of an Intra-Body Communication Device. In: Proc. 12th International Conference on Transducers, Solid-State Sensors, Actuators and Microsystems, Boston, USA, June 8-12, pp. 1722–1725 (2003)

[93] FCC Encyclopedia: Wireless Medical Telemetry Service (WMTS), `http://www.fcc.gov/encyclopedia/wireless-medical-telemetry-service-wmts` (accessed on September 7, 2013)

[94] Ben, A., Tony, B., Katja, S., Ernesto, Z., et al.: Ultra Wideband: Applications, Technology and Future Perspectives. In: Proc. International Workshop on Convergent Technologies, Oulu, Finland, June 6-10, pp. 1–6 (2005)

[95] Ullah, S., Shen, B., Riazul, I.S., Khan, P., Saleem, S., Kwak, K.S.: A study of MAC protocols for WBANs. Sensors 10, 128–145 (2010)

[96] Ullah, S., Kwak, D., Lee, C., Lee, H., Kwak, K.S.: Numerical Analysis of CSMA/CA for Pattern-Based WBAN System. In: Proc. Biomedical Engineering and Informatics, Tianjin, China, October 17-19, pp. 1–3 (2009)

[97] Otal, B., Alonso, L., Verikoukis, C.: Energy Efficiency Analysis of a Distributed Queuing Medium Access Control Protocol for Biomedical Wireless Sensor Networks in Saturation Conditions. Sensors 11, 1277–1296 (2011)

[98] Li, H.M., Tan, J.D.: Heartbeat-driven Medium-access Control for Body Sensor Networks. IEEE Transactions on Information Technology in Biomedicine 14, 44–51 (2010)

[99] Marinkovic, S., Popovici, E.: Ultra low power signal oriented approach for wireless health monitoring. Sensors 12, 7917–7937 (2012)

[100] Fang, G., Dutkiewicz, E.: BodyMAC: Energy Efficient TDMA-Zbased MAC Protocol for Wireless Body Area Networks. In: Proc. 9th International Symposium on Communications and Information Technology, Incheon, Korea, September 28-30, pp. 1455–1459 (2009)

[101] Rahman, M.O., Hong, C.S., Lee, S., Ban, Y.C.: ATLAS: A traffic load aware sensor MAC design for collaborative body area sensor networks. Sensors 11, 11560–11580 (2011)

[102] Kumar, P., Lee, H.-J.: Security Issues in Healthcare Applications Using Wireless Medical Sensor Networks: A Survey. Sensors (Basel) 12(1), 55–91 (2012)

[103] Dimitriou, T., Loannis, K.: Security Issues in Biomedical Wireless Sensor Networks. In: Proc. 1st International Symposium on Applied Sciences on Biomedical and Communication Technologies, Aalborg, Denmark, October 25-28 (2008)

[104] IEEE 802.15 WPAN™ Task Group 6 (TG6) Body Area Networks, `http://www.ieee802.org/15/pub/TG6.html`

[105] Continua Health Alliance, `http://www.continuaalliance.org/`

[106] Office for Civil Rights, United State Department of Health and Human Services Medical Privacy. National Standards of Protect the Privacy of Personal-Health-Information, `http://www.hhs.gov/ocr/privacy/hipaa/administrative/privacyrule/` (accessed on September 7, 2013)

[107] Health Information Technology for Economic and Clinical Health Act (HITECH), `http://waysandmeans.house.gov/media/pdf/110/hit2.pdf` (accessed on September 7, 2013)

Questions

Q1. List three types of biosignals which can be used to extract heart rate information.

Q2. Which physiological parameters can be measured with an FBG sensor?

Q3. Describe, in brief, the three main areas of research in wireless body sensor networks.

Q4. There are a number of security threats in wireless body sensor networks. Give any three of them.

Part III

Interoperability

16. Interoperability and mHealth – Precondition for Successful eCare*

Mate Beštek[1] and Andrej Brodnik[2]

[1] University of Ljubljana, Faculty of Computer and Information Science (PhD Student), Ljubljana, Slovenia
mate.bestek@gmail.com
[2] University of Ljubljana, Faculty of Computer and Information Science, Ljubljana, Slovenia and University of Primorska,
Department of Information Science and Technology, Koper, Slovenia
andrej.brodnik@fri.uni-lj.si

Abstract. *Background*: Lately, the world's biggest Professional Services company has recognized that mHealth is enabling and accelerating three major global trends in healthcare: *regulatory reform driven by demographic changes*, *industrialization of the healthcare sector* and *personalized medicine*. They have identified interoperability as a key enabler of scalable mHealth. Continua Health Alliance is a non-profit organization that globally certifies mobile health solutions. They publish interoperability guidelines for connecting health related devices. Unfortunately, the adoption of these guidelines has not been consistent.

Problem: We want to define a platform and a framework to support the development of different e&m-health interventions for various domains. The solution has to support sharing of information between interventions within the system and also with external systems in a standardized way.

Methodology: In the chapter we describe an approach towards interoperability that was achieved in several different aspects. First, the enterprise interoperability is supported by using IHE profiles.. Second, organizational interoperability is supported by use of BPMN2; and third, the semantic interoperability is supported by using OpenEHR, MLHIM and standard vocabularies.

Results: We built the platform that permits iterative incremental development process of new interventions. The developed interventions are fully interoperable within e&m-health environments. On the platform we deployed five different interventions that were also clinically tested. In this chapter we describe eDiabetes intervention in greater detail. Besides, we give guidelines on how to develop and deploy a new intervention.

Keywords: Interoperability Stack, Shared ontologies, OpenEHR, MLHIM, Care Model, Architecture Homogeneity, Intervention Design, eCare, SOA, Health Monitoring.

---

* Supported by the Public Research Agency of the Republic of Slovenia and Ministry of Health of the Republic of Slovenia.

S. Adibi (ed.), *Mobile Health*,
Springer Series in Bio-/Neuroinformatics 5, DOI: 10.1007/978-3-319-12817-7_16

1 Introduction

Healthcare is going through a transformation where the main focus has shifted from acute care to managing chronic diseases. This process, together with the extended life expectancy, is the major source of sustainability problems that healthcare systems face today. Thus, collaboration and coordination between different healthcare providers need to be established. This endeavor is a complex process and is ongoing in many developed and developing countries.

Chronic no communicable diseases represent an important and growing threat to sustainability of health care systems. Key components for efficient treatment are self-empowerment of patients and development of patient's adherence to the prescribed treatment plan. It is generally accepted that information-communication technology (ICT) can be used to support these key components, and in particular the mobile technologies. For mHealth to become successful, considerable innovation in fundamental science and engineering is needed as well as integration of the technology with healthcare systems [1].

Traditionally, mHealth has been perceived as part of eHealth but has gained traction recently as new opportunities have been identified, both in the industry and research [2]. The world biggest Professional Services company have recognized that mHealth is enabling and accelerating three major global trends in healthcare: *regulatory reform driven by demographic changes*, industrialization *of the healthcare sector* and *personalized medicine* [3]. They have identified **interoperability as a key enabler of scalable mHealth**. According to [4], sustainability relies on interoperability in today globally networked environments and, conversely, interoperability relies on all three pillars of sustainability. "Triple Bottom Line (TBL) approach to business sustainability"[4] is based on only on economic performance, but also environmental and social accomplishment [4].

Continua Health Alliance is a non-profit organization that globally certifies mobile health solutions. They publish interoperability guidelines for connecting health related devices. In spite of, the adoption of these guidelines has not been consistent [3]. The major reason for this can be found in connected health classical business model. Vendors offer closed solutions in order to increase market share. This creates information silos that are not easily connected together. Service Oriented Architecture (SOA) can solve the problem of connecting these silos, but additional solutions are needed in order to solve interoperability problems. Structure and meaning of data should be shared. Data acquired in different mobile devices should be transferable to a backend or cloud service. Thereon, data, should be transferable to eHealth services without integration. The same data should be viewable by doctors, nurses and other healthcare personnel, and last but not least, it should be viewable by patients.

All this data should be modeled outside of traditional approaches to developing ICT solutions. Data should be semantically described by means of all the important terminologies used in healthcare today (e.g. SNOMED, LOINC, ICD-10,...). These data elements together with the mappings to different terminologies should become standardized. There is great work being made towards achieving this in the field of eHealth with standards such as Health Level 7 (HL7) [5] and future standards such as OpenEHR [6, 7] and recently the Clinical Information Modeling Initiative (CIMI) [8] that is based on OpenEHR. Therefore, the joint effort in achieving interoperability between eHealth and mHealth solutions is unavoidable. In [9] authors argue that the approach to interoperability in healthcare that is defined by ISO 14258 should base on predefined shared ontologies. This involves management of ontologies that describe concepts and relations between concepts on different layers, and the harmonization of these. Focusing on the pathway through a set of ontologies starting from top-level ontologies all the way to the application level ontology or ICT implementation ontology is the enabler of semantically interoperable, sustainable and trustworthy eHealth systems [10] and also mHealth systems.

Boulos et al. [11] provide a brief state-of-the-art overview of health and healthcare smartphone applications on the market today. In [12] authors propose a system for monitoring patients with diabetes. The system included wearable sensors, an Android smart phone, smart home technologies and Google Health for the control of the disease. Pioggia [13] describes a platform supporting the monitoring of patients with chronic diseases, notably CHF patients. The platform includes multiple wearable biosensors connected to an Android smartphone. Doukas et al. [14] present "the implementation of a mobile system that enables electronic healthcare data storage, update and retrieval using Cloud Computing. The mobile application is developed using Google's Android operating system and provides management of patient health records and medical images". Most of the systems that exist today are not interoperability driven which is the key to sustainable future solutions.

In addition to the theoretical background, this chapter also presents the experience with a prototype implementation of the eCare [15] system for the Slovenian health system that includes interventions for the treatment of asthma, diabetes, schizophrenia, and depression, screening and prevention activity in obesity controlling, and development in individual's sport activity[1].

Our eCare project solution consists of the following key components:

- the mobile subsystem,
- the middleware connecting all components and storing and validating the data,
- the subsystem that improves the outcome of the patients treatment based on collected personal data and personal intervention plan,
- standardized protocols and tools used in all components.

The chapter was structured as follows: in 2. Methods we introduce interoperability and standardization both in theory and practice, and also present interventions which are applications in different specialized sub-domains that were supported on the eCare platform. In 3. Results we present the main achievements and explain in detail

---

1 See also `http://www.clinicaltrials.gov/ct2/show/NCT01461642`, `http://clinicaltrials.gov/ct2/show/NCT01566981`, and `http://www.clinicaltrials.gov/ct2/show/NCT01716325`.

the most important outcomes such as the architecture that is common to both server side system and mobile system. We end this chapter with 4. Discussion and Conclusion.

2 Methods

We developed a platform that allows intervention design and development using Service Oriented Architecture (SOA) based architecture, ontologies and agile methodologies. The platform was developed using BPMN2 modeling, openEHR and Multi Level Health Information Modeling (MLHIM) [16] methodologies, open source process engine, XML database and established java enterprise edition (JEE) environment to which all the components of the platform were deployed.

OpenEHR methodology enabled separation of clinical knowledge from the rest of ICT development. Medical professionals adopted the role of a business analyst in BPMN2 setting, developing key clinical concepts and care management processes, which were further used in ICT development using metadata models definition for deployment. A synchronization of models is needed between the models that Business Analysts create and the implementation models to close the "business-IT gap"[17]. We have managed to find a solution [18] that supports this "no business-IT gap" approach to process modeling and more importantly have successfully used it in the eCare project.

Based on agile methodologies used, iterative process was applied in all modeling and development phases.

Our approach towards achieving interoperability between eHealth and mHealth is based on the OpenEHR methodology. One of our concerns in supporting general OpenEHR based architecture on different devices is the amount of resources that are needed to make such architecture feasible. OpenEHR models and the engine that manages these models consume a lot of resources since all the models are being transformed into object models in runtime. This is a great drawback of using OpenEHR models on mobile devices such as smartphones. The logical approach towards solving this problem is making the engine or the specification lighter. Something like this was made in the MLHIM method which is based on XML schema for which we have very efficient parsers, and validator thus allowing the usage of MLHIM models on devices that have limited resources. OpenEHR and MLHIM are related in terms of a common reference model (the basic building block). In fact, MLHIM reference model is a subset of the OpenEHR reference model. This means an MLHIM model can be managed by the OpenEHR engine. Obviously the other direction is not possible but this is not a problem since the server solution is based on OpenEHR.

The basic artifacts of OpenEHR are archetypes and templates. They present a way to express the underlying clinical process ontology. Archetypes are used to model the concepts, whereas templates are used to create new structures that are use case specific. Similarly, MLHIM is also used to model clinical concepts, but with a different philosophy. MLHIM is more general and lightweight than OpenEHR. This makes it interesting for use with mHealth.

We developed interventions in these sub-domains: asthma, diabetes, obesity, schizophrenia, depression, and sports activity. The interventions reused different parts of the underlying platform. All clinical knowledge was modeled by means of OpenEHR artifacts. For mHealth, the modeling was done by using MLHIM. Data that was collected from the users was first validated against the artifacts and then stored in a separate OpenEHR storage server. This solution allows queries over structured health related data about a patient or population. All the interventions reused a layer of BPMN2 models. These models described processes of care which were defined by healthcare professionals. They were stored in a repository of BPMN2 models and were loaded and executed in runtime. A process model execution coordinated different roles and internal and external systems towards performing a care process. The main point is that all the clinical data coming from external systems were also modeled by means of OpenEHR and MLHIM artifacts.

Further on, different sensors and mobile devices were used as parts of our interventions. With these in place, users did not have to enter their data manually since the data was transferred to our cloud based solution automatically. The data coming from the sensors was modeled using OpenEHR archetypes and templates and MLHIM models, as well. It was then easily added into an existing OpenEHR/MLHIM storage server. With this, even more data was available for being queried or transferred. Since all the data is fully structured, it is easily converted to standards such as HL7 Clinical Document Architecture (CDA) which are used in most national eHealth initiatives and EHR systems worldwide.

2.1 Interoperability and Standardization

The main focus of achieving interoperability in healthcare is shown in terms of an Interoperability stack in Figure 1 where three methods together with different layers and technologies used in the USA are shown. It is interesting to note that the idea of semantic interoperability is something we hear a lot about in the field of healthcare. The idea itself is not new and is defined by ISO 16100 only as interoperability. The systems that are integrated and can use the shared data meaningfully are automatically also semantically interoperable. The problem is then in the architecture that should provide answers on scalability of such a connection which depends heavily on current business models. These are based on locking data in and do not promote interoperability. Issues arise with supporting numerous existing legacy systems, which act as information silos. We need to achieve consensus on how to bring all the systems to a common denominator, or better said how to make all the systems interoperable. This includes not only legacy systems in one domain, but also on systems that exist in different domains (healthcare, insurance, public health, research etc.). The prerequisites for enabling cross-domain interoperability are: **architecture homogeneity** and **shared ontology systems** [19]. Since each domain focuses on its own set of standards for achieving interoperability, a problem arises when trying to talk about cross-domain interoperability. This is the problem of over specifying domain specific interoperability standards. A framework like the one described in [19] can be useful for bridging the problem.

An example of healthcare domain standards on all the layers is given in Figure 1. We can see here the typical layers of healthcare specific software solutions that are used in the USA at the national level. These systems allow different types of integrations between different healthcare service providers. Most efforts in healthcare ICT address the top two layers. These define vocabularies used to add semantics to clinical concepts. This represents one of the foundations for achieving semantic interoperability. International Health Terminology Standards Development Organization (IHTSDO) is a global organization that focuses on different healthcare terminologies and standards. It is a mayor player in achieving a truly interoperable global healthcare system.

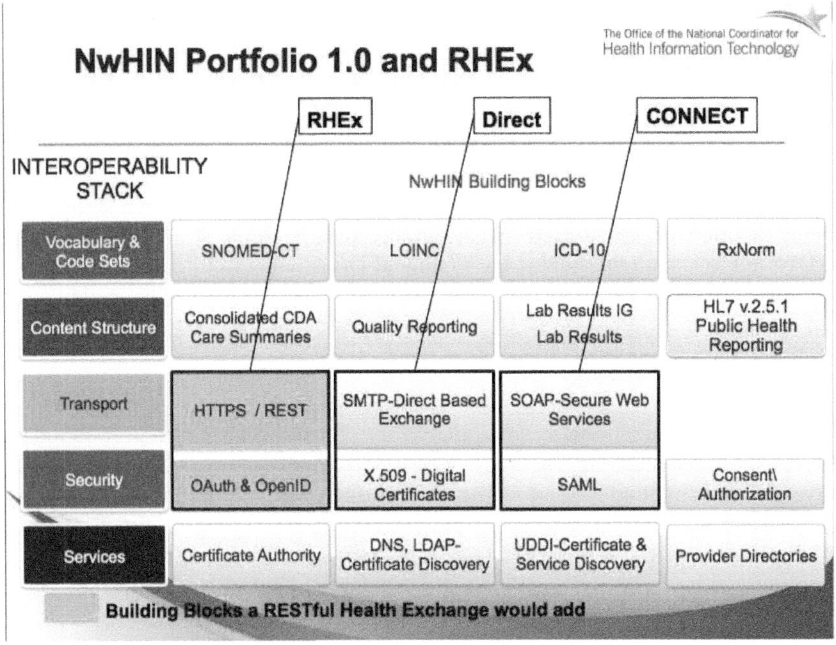

Fig. 1 Interoperability Stack, Source: [20]; Direct is based on the federative architecture, while Connect is based on a centralized architecture; RHEx is focused towards using Representational State Transfer (REST) for information exchange

Interoperability as a concept is not always used with the same meaning. Besides semantic interoperability, we can talk about many other types of interoperability that we want to define further.

We should use a systematized body of knowledge like the one presented in [21, 22] where the basic foundations of studying interoperability include **frameworks, theories and models**. Frameworks provide a meta-theoretical language. It is used for development of interoperability theories. Frameworks provide the most general list of variables used in studying interoperability problems. Theories are used to define which elements of the frameworks are important for a particular kind of problems and also for defining general assumptions about these elements. Models make precise assumptions about a limited set of parameters and variables [23, 21]. A generic interoperability framework proposed a structure for studying interoperability in different layers [24] consisting of **technical**, **semantic**, **organizational** and **enterprise** interoperability as shown in Table 1.

Table 1 General Interoperability Framework

Interoperability Facet	Contents
Enterprise Interoperability	Legislation and business rules representation and alignment, collaborative business models, collaborative enterprise goal setting, cross enterprise and international supply chain collaboration.
Organizational Interoperability	Process representation and alignment, collaborative process design, business process re-engineering for interoperability, collaborative organizational models.
Semantic Interoperability	Data and representation schemas, XML standards (generic and sectoral), general ontologies, metadata sets, unified code lists, core components, data elements, glossaries.
Technical Interoperability	Technical standards and agreements on system interconnection, web services collaboration, information representation formats, middleware, platforms, network connectivity, accessibility and security.

Our set of shared vocabularies includes the following standards on different interoperability levels as shown in Table 1:

- **Enterprise Interoperability:** Integrating the Healthcare Enterprise (IHE)[2] is used for the purpose of cross-domain interoperability. The focus is on the IHE profiles that serve the purpose of cross-domain or cross-enterprise information exchange, mainly: XCPD (Cross-Community Patient Discovery), XCA (Cross-Community Access), and XDR (Cross-Enterprise Document Reliable Interchange).
- **Organizational Interoperability:** BPMN2 for specifying business processes – both orchestration and choreography are supported. Most of the IHE profiles are dedicated to solve problems at this level (e.g. XDS …).
- **Semantic Interoperability:** the basics for achieving this type of interoperability in our system are the usage of OpenEHR and MLHIM clinical models. These define the clinical concepts in terms of data structures that are mapped to terminologies like ICD-10, LOINC, SNOMED-CT, national terminologies, etc. These models are used to verify all the data that travels through the system. Prior to this being enabled, the process of mapping has to be performed. An example of such mapping would be between a data set that comes from a sensor

---

[2] http://www.ihe.net

device to clinical models that define the concepts for which the sensor is designed to retrieve data for. This means we need a mapping between the format of the data that comes from the sensor and the format of the clinical models. This process of harmonization is greatly optimized if sensors conform to the latest Continua Alliance profiles. Otherwise, custom mappings are needed for each sensor or more often the hub (hub-and-spoke design pattern [19, 25]) that collects data from the sensor and transfers it to the server side. Greater details on connecting medical devices and achieving interoperability can be found in [25]. Since all the clinical data is stored in the form of validated clinical models, there is only a matter of transformation from this structure to any other format like HL7 CDA.

- **Technical Interoperability:** at this level we are using standard methods of connecting different components of the system by means of well-established standardized protocols (e.g. IEEE 802.15 etc.).

Explicit theories on interoperability do not exist. Rather, there are theories that are somewhat related to interoperability problems [21]. One of such theories is also the theory of complexity we can use for grounding theories that explain interoperability problems and challenges. These include [24]:

- intelligent reconfiguration of components in evolutionary networked systems (theories should explain the learning process that supports adaptation of network nodes and keeping the interoperable network – complex integrated system),
- conformance testing and checking in complex systems (how to add new nodes, update or remove them),
- harmonization of ontological structures to support dynamic ecosystems.

In our system, we are touching all of the presented layers. Most of our work is focused on technical and semantic interoperability but at the same time we are taking into account also the organizational and enterprise interoperability problems.

Use of standards as the approach towards technical and semantic interoperability taken in the field of healthcare. The used standards can be seen on all the layers of the interoperability block in Figure 1. Our method adds a separate layer between the top two layers on the same figure (*Vocabulary and Code Sets*, and *Content Structure*). This layer follows from the OpenEHR/MLHIM methodology. Clinical concepts are modeled separately by means of basic structures called archetypes and templates. The concepts are then mapped to different vocabularies and code sets (e.g. LOINC, SNOMED-CT,…). After defining such models, we can easily transform them into the needed content structure that is then used to transport data between different parties. Most used standard for defining content structure is HL7 CDA standard which is XML based.

Interoperability is of utmost importance in healthcare. We do not gain, as a society, if we support the development of new information silos, which is, unfortunately, happening already. Therefore, all models of clinical concepts should be publicly available. This is enabled by using the OpenEHR and MLHIM methodologies. In [26] authors present four generations of classifications of complexity. Healthcare concepts can only be unambiguously modeled if the fourth generation of classification is used. This means healthcare is complex enough that concepts should be modeled by means of ontologies (For comparison: the third generation gave methods such as Object Oriented Programming and Relational Databases).

2.2 *Interventions*

Our approach to developing new interventions is similar to those that are based on decomposition. The basic idea noted in [27] states that "Identifying which core components of treatment are necessary for everyone, as well as determining which additional modules are needed by only some users, can help provide treatments that are more efficient, effective, and relevant."

Examples of such deconstruction studies are [28–30]. Consequently, simple interventions could be a reasonable approach of creating interventions since it was found that many interventions are not used as intended. Authors of [31] stressed that in order to design and develop meaningful eHealth/mHealth interventions, we need to take into consideration human characteristics, socioeconomic and social environments, and technology since all of these is considered to be closely related. Therefor our intervention development is based on interdisciplinary collaboration that results in simple interventions.

From features of interventions that have the potential to influence adherence of interventions, as described in [32], we have focused on the following:

- tailoring [33],
- credibility (National Medical Ethics Committee of the Republic of Slovenia),
- regular "push factors" with human support, and
- periodic prompts (email, sms, and phone).

The features were chosen primarily on the basis of our past experience on creating an intervention for patients with depression, named eDepression [34] that was shortly described in the context of [35]. We want to see if similar results can be expected in other types of interventions.

Furthermore, we added new components that were not available in the eDepression intervention, but were shown as important in the literature [32].

We designed interventions for following domains: diabetes, asthma, obesity, schizophrenia, and sports activity. Interventions target different groups of people. This includes **healthy people**, **people with identified chronic conditions**, and **people with identified risk factors**. Diabetes, asthma and schizophrenia are chronic diseases thus the interventions are oriented to chronic disease management. Intervention for obesity targets people with greater risk for developing a disease. Intervention for sports activity targets different groups of people as a special module that can be used in prevention purposes.

3 Results

The main results of our project are the following:

- Loosely coupled future proof component based platform.
- Dual modeling approach – separating clinical knowledge modeling from other ICT modeling.
- Repositories of models (clinical models and care process models),
- Validation of clinical data against clinical knowledge models – OpenEHR on the server and MLHIM on mobiles.
- Convergence of mHealth and eHealth at the level of processes.
- Common architecture of the server side system and the mobile subsystem.

Component based platform was built upon ten essential components which create a holistic development environment as can be seen on Figure 5, namely **Access Module, Integration Module, Cross Domain Integration Module, Process Execution Engine, Clinical Models Execution Engine, Clinical Model Repository, EMR Repository, PHR Repository, Process Repository** and **Persistence**. The presented ten components create a framework with components either reusable or adjustable for additional applications development. The defined architecture shows how these components behave. We have also found that this architecture is similar for both the server side system and the mobile subsystem.

All four interoperability layers shown in Table 1 have been addressed. Enterprise interoperability is supported by means of IHE profiles that focus on cross-domain exchange of data. Organization interoperability is focused in business processes we have modeled separately by using BPMN2. Semantic interoperability and most importantly, clinical knowledge developed in each intervention is integrated in existing repositories, leveraging the scope of clinical knowledge upon which further interventions are created. As a result, all components with clinical and process knowledge are reusable and adaptive in further development. Also, platform supports scalability both in the number of users as in the number of applications running. The development environment was used to build five different clinical interventions: asthma, diabetes, obesity, schizophrenia and sports activities. All interventions were used in clinical trials showing positive clinical outcomes.

3.1 Intervention Development Process

The following activities comprise our Intervention Development Process:

- modelling of care processes and clinical concepts,
- interweaving care processes and clinical models,
- development of logic that handled runtime execution of the care processes,
- development of unit tests and integration tests,

- deployment of the interweaved process and clinical models,
- user interface integration with the process models,
- delivery and validation of the application to the doctors and nurses,
- Inclusion of patients.

Modelling of Care Processes and Clinical Concepts

In the intervention design phase, we used knowledge from medical, technology, social and behaviour sciences. This knowledge has been constructed to some degree in a previous project [34]. New medical knowledge was obtained by conducting interviews with healthcare personnel including specialist doctors, primary care doctors, care managers and nurses, and other experts from the fields of prevention and healthy lifestyle. In these interviews, we gathered information on care processes that are special for each of the interventions. Medical professionals were asked to prepare written descriptions of the intervention as they envisioned it.

Identified information items were organised into a few main categories common for each intervention. They include:

- Patient medical condition,
    - Patient master data - data collected during the medical investigation,
    - Medical history,
    - Diagnosis,
    - Laboratory findings,
    - Medication Therapy,
    - Patient data from self-monitoring.
- Data and information about the care processes supported by the eCare system,
- Time intervals and reminders,
- Rules for data analysis and interpretation.

We used this information to design a process model of intervention. To describe processes we used the BPMN2 modelling language from business process modelling community, since it supports service orchestration and choreography. By using BPMN2, we designed the care processes of intervention. It also enabled collaborative work with the doctors and care managers in order to capture knowledge and rules that are part of care processes. OpenEHR methodology was used for modelling clinical knowledge. Each task that the user received to his/her web user interface was modelled by an OpenEHR template. A task was a primary unit of a care process. Each OpenEHR template defines additional constraints to one or more OpenEHR archetypes. We used existing OpenEHR archetypes and also developed new ones. Overall, 36% of archetypes we used were obtained from the OpenEHR global repository of Table 2. We developed the rest of archetypes. 36% of these newly developed archetypes were reused at the common platform level – by all interventions. The rest of archetypes were specific to the intervention. On average, each intervention required 5,6% of custom archetypes.

Table 2 Archetypes reusability ratio

Source of archetypes	Ratio
OpenEHR.org public repository	36%
Custom archetypes reused on all interventions	36%
Intervention specific archetypes (all interventions together)	28% (5.6% on average for each intervention)

Interweaving Care Processes and Clinical Models
For the purpose of a deployment to an execution environment, we had to combine both types of models (BPMN2 and OpenEHR models). This is how our universal language was backed up by these models and then also implemented in an execution environment. This interweaving was done with additional metadata and Java programming language which enriched the BPMN2 models. The execution environment was a process engine [18] that natively uses BPMN2. This means it can read the models and the metadata an execute them. Parts of the metadata were also the names of Java classes. The process engine is a self-standing software components that can be deployed as a part of an existing solution or as a self-standing solution. The processes are thus deployed separately from other logic. They represent a separate layer of the SOA architecture strategy. This means they can also be redeployed separately from e.g. user interface application.

Development of Java Classes That Managed the Execution of the Care Processes
Java objects that were created in the runtime from these classes, were called by the Process engine. Each step of a BPMN2 model was handled by a Java object (many already built into the Process engine). The logic inside these Java objects used the OpenEHR object model to verify all the input data coming from users against the constraints in the templates and the underlying archetypes. After successful validation, all the data was serialized and stored in XML format. The XML was the OpenEHR template that contained the data in its value elements. Such implementations also became reusable libraries. Instances of these classes can be seen as services. Many of them were reusable and became part of the common platform.

Development of Unit and Integration Tests
For each Java class there was a test developed that tested the correctness of the Implementation. When all the tasks were supported by Java classes, integration tests were developed that tested entire processes and communication between different processes. When such tests gave positive results, we were ready to go to the next step of the development. This is generally acceptable way of assuring quality in software intensive projects.

Deployment of the Interweaved Process and Clinical Models
Such processes could then be deployed to the Java based application server. The process engine we used already supported external integration with the processes by means of REST web services which used JavaScript Object Notation (JSON) as the data protocol. We had a unique development part - the back end process developer that worked mostly on supporting different processes. Another

development part was the web developer who had the responsibility of developing a web user interface for all the three roles of the system (patient, care manager, doctor). The user interface was a separate web application and can be deployed on different application server. There is a clear separation of the common platform and the user interface applications that were provided for the end users. Clinical models and care process models can be changed and redeployed separately from the user interface. Each deployment receives a new version number which is important for the purpose of supporting tasks that were part of the old processes and were still active. Just when all tasks of a set process version were in the finished state, we could remove the processes from the execution environment.

User Interface Integration with the Process Models

The user interface is a separate application developed separately from the common platform. It interacts with the process engine by means of JSON protocol. It receives new tasks and data for a user and posts back new data that was input by the web application user. The process running in the process engine waits until a particular user task is completed. Then it can continue to process next steps of the care process. This means that an external web application had to provide timely information and data back to the process engine. Otherwise, the process engine executed exception handling mechanisms that were part of the care processes. Web user interface functionalities were partly defined by the care processes, partly by experience from our previous project [34] and partly by the requests made by the future users of the system (doctors and care manager).

Delivery and Validation of the Application to the Doctors and Nurses

When the development was complete, we presented the application to doctors and nurses. This usually involved additional requests mostly of little complexity. The system was available on our test environment. Everyone, except the patient had access so that they could perform testing. A particular time was appointed for the purpose of testing and education of the new users. In the case of updates to the application, all parties were informed of updates and were invited to test them.

Inclusion of Patients

After confirming that the ICT environment was finalized, the inclusion of the patients could start. Patients were invited by the doctors and care managers. The web application supported an administration action for adding a new patient. This involved a thorough information form that enabled entering most of the usual data that a regular Electronic Medical Record (EMR) system supports. Care managers and doctors also had the "New Visit" action which is also a core functionality of EMR systems. Initial examination data that was the result of a patient visit was entered into the system. Since clinical trials were performed for all the interventions, each patient was randomly assigned to the control or experimental group.

3.2 Interventions

An intervention is an application on top of the developed platform. The platform has a process layer that includes an extendable abstract care process model. We can say that this abstract process model is actually a configurable process model that can be adapted for different usages. A new intervention is defined in an interdisciplinary team including sub-domain specialists. Components of the abstract care process model are reused to support the functional specification of the intervention. The specification is built by defining:

1. User interface functionalities that are unique to the current sub-domain (by default, users, get a user experience that is shared across interventions); this includes web user interface and mobile user interface.
2. Specialized care processes are created based on the abstract care process model by injecting a new intervention specific context.
3. Devices and sensors that are used in intervention are defined and included in the care processes.

There is further work on the platform side if the sensors are not supported yet, or if the abstract process model needs to be extended. Similarly, clinical models sometimes might require additional extensions.

As an example, we give a description of intervention for the diabetes subdomain.

The Diabetes Intervention

The diabetes intervention (DI) includes three different user roles: patient, nurse, and a doctor. Figure 2 presents the functional description of the user interface that a nurse or a doctor uses. It shows eight basic functionalities that are common to all interventions. These functionalities are meant to be shown on a web based user interface. On a smartphone, a different set is used to bring a different user experience. A web user interface interacts with the process layer to get the assignments for the user. In fact,web interface is just a part of the orchestration defined with BPMN2 models. So only the functionalities of the user interface that are part of an orchestration, communicate with the process layer. Other functionalities can be added as needed.

Generally, we have three groups of user interface functionalities:

- Functionalities that participate in care process orchestrations:
  - **My Assignments**
  - **Administration**
- Functionalities that perform queries against the clinical data:
  - **Health Status**
  - **Reporting**
  - **Status Bar**
- Communication and Education functionalities:
  - **Questions and answers**
  - **Live Chat**
  - **Content**

We should stress that, for a new intervention, customizations are required only in the **Health Status, Reporting**, and **Administration**. The first two are focused on regular parameters that are crucial to determining the intervention specific health status of the patient. The last one needs to support the management of patient data that is related to the domain of a particular intervention since we do not capture all the data about a patient but only data specific to the intervention.

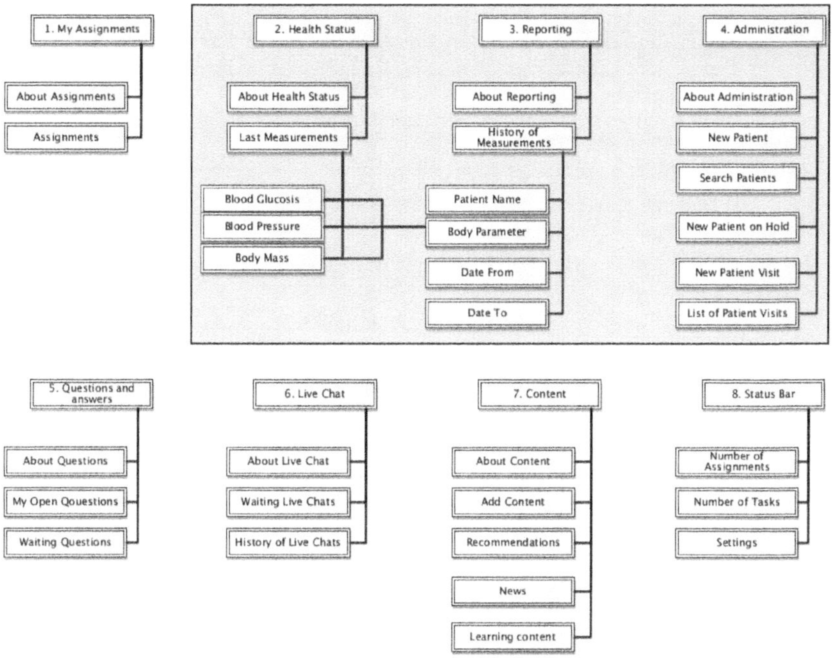

Fig. 2 Structure diagram depicting functionalities for nurses and doctors. Health Status, Reporting and Administration are the three components that need to be customized in various interventions.

Let us look at an example of the Administration module functionality **New Patient**. It supports inclusion of a new patient into the system and the (clinical) trial. Again, we divide the data into two sets. The first set consists of data common to all interventions. The second set includes intervention specific data. If we take into consideration that all the data is modeled by means of OpenEHR and MLHIM models and that most data occur on all interventions, we see that for each intervention we need to setup only the intervention specific clinical model for data.

Using OpenEHR we defined a template for the purpose of inclusion of a new patient. It includes several archetypes: **admission details, family history, alcohol consumption, smoking, healthy lifestyle, body mass, body height, body mass index, blood pressure, description of the rest of the history, WONCA questionnaire data, diagnosis, and therapy**.

An example of the WONCA questionnaire archetype as seen in the Archetype Editor tool is shown in Figure 3.

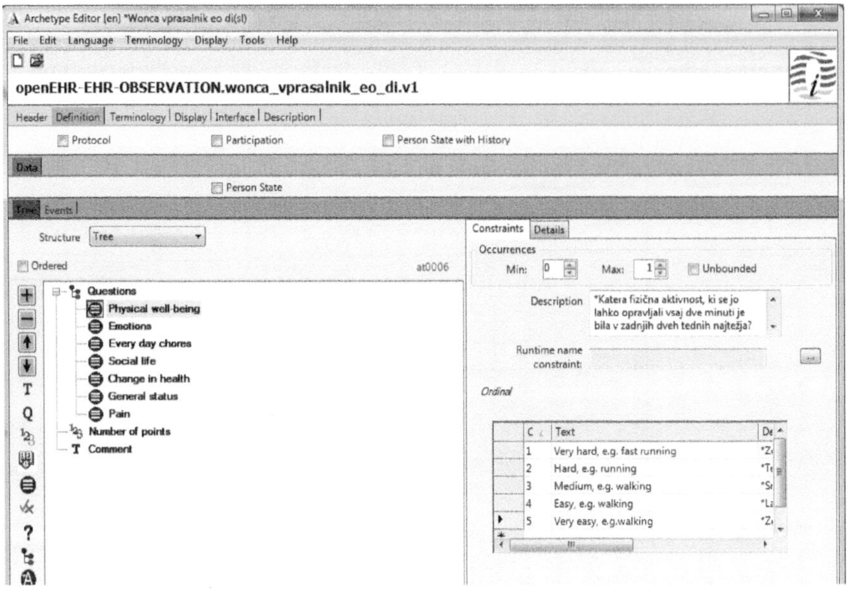

Fig. 3 WONCA questionnaire modeled as an archetype

Interventions can have a specialized user interface on the web and mobile devices. It needs custom care processes, and it also needs clinical OpenEHR model for intervention specific data. The model is used to verify the acquired data. The data is collectable through the web forms or directly from sensors connected to a mobile device. More details on the overall architecture and the process models are given in the following subsections 3.2 and 3.4, while the subsection 3.3 introduces the mobile subsystem.

3.3 Architecture of the eCare System

The eCare platform architecture is service oriented. It provides the necessary abstraction at the business rule level and maintains standard vendor independent lower-level technologies (e.g. web services) [19]. We focused on the development of a platform that is patient centered and supports integrated care, in which care processes are designed around patients. Further, the platform is designed to support different interventions. When designing we paid special attention to mayor prerequisites for cost-effectiveness which is the possibility to reach a high number of users [36] as also noted in [32]. Our platform was designed to be future proof.

In Figure 4, the basic layers of our SOA based architecture are shown. For the purpose of the eCare project, we used only three main roles: patient, doctor and care manager. New roles can be added upon identified need.

The three roles are involved in care processes that were modeled based on clinical guidelines and some other organization specific procedures. These processes have been modeled by using BPMN2. We tried different BPMN2 modeling environments but finally decided to use the academic licenced Signavio Modeler. During the development, doctors and nurses were in the role of business analysts. They were actively participating in the process of modeling processes.

At the same time, they were also providing the development team with the clinical concepts that were to be modeled by means of OpenEHR and MLHIM. One of very important strengths of OpenEHR/MLHIM methodologies is the fact that many models already exist and can be reused. This enabled faster development since only a small number of archetypes had to be developed from start. These content models include not only basic data structures, but also mappings to terminologies like LOINC and ICD-10 and other meta-data that give further meaning to data.

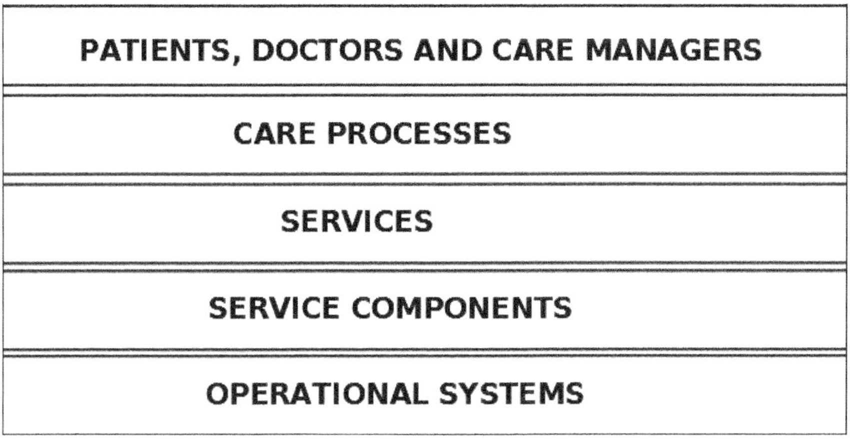

Fig. 4 eCare SOA stack

The care processes and services layers include two important components: repository of care processes and repository of content models. The BPMN2 models that describe care processes were orchestration of tasks that different roles had to perform and services that were used to support the care processes. The bottom two layers in Figure 4 represent the actual java based implementation of the services logic.

In Figure 5, the architecture of the whole system is shown. We see that a central part of the system is given to the Process Execution Engine (PEE) that executes the orchestrations modeled as BPMN2 models. Since interoperability is achieved by means of shared ontologies [10], we immediately see all the points where we need to consider how to achieve interoperability (the connections from the Integration Module that lead to different external systems, and also relationships within the eCare system that lead from the PEE system).

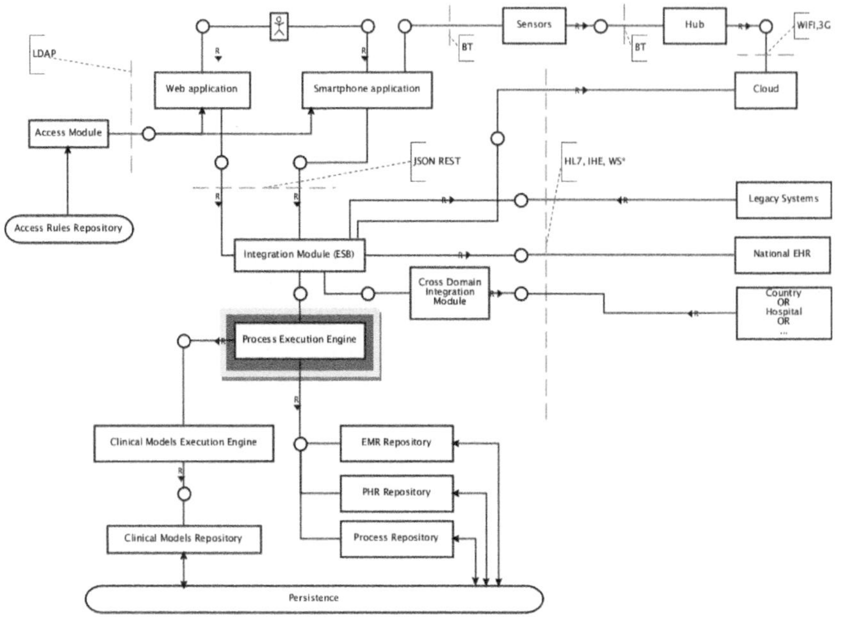

Fig. 5 Architecture of the system as a FMC (Fundamental Modeling Concepts) [37] diagram

If we look at our mobile subsystem, (Smartphone application component from the architecture of the system shown in Figure 5) we can see that its architecture is similar to the architecture of the overall system. The mobile subsystem uses a set of process models, a set of MLHIM content models which are stored in local repositories, has an integration module etc. We can say that we can look at such architecture as description of an agent that can run on many different physical devices or networks of devices depending on the level of distribution of the main components (e.g. the repository of clinical content can be deployed on the same device or can be deployed on a different device). The topology of all the components is not really important. What is important is the fact that these agents are able to communicate between each other by exchanging messages which are created by using shared ontologies, standards and terminologies that are determined in advance. This type approach to interoperability is categorized as the unified approach (the other two are full integration and federation)[9]. This architecture homogeneity and shared ontology systems [19] are the basic steps towards cross-domain interoperability. Cross-domain interoperability needs to be achieved if we want to achieve the overall goal of global virtual health record.

3.4 Mobile Subsystem

The architecture of the mobile system was introduced in the previous section and in this section we focus on the sensors and other sources of data that the mobile subsystem aggregates, validates against MLHIM models, and eventually synchronizes with the server side.

The sensor system presented was developed to 24/7 monitor vital parameters of patients with chronic diseases and older persons. It consists of an Android smartphone, and wearable sensors connected to the phone via a Bluetooth connection. The system permits triggering of alerts in case the vital parameters exceed predefined thresholds, live monitoring of the essential parameters either on the smartphone or remotely (e.g. by a doctor,care coordinator, or automatically by the system itself) using another Android device or via a web browser on the computer. The parameters such as ECG, heart rate, breathing rate, GPS location, accelerations, and skin temperature are monitored. Among others, the system is used in the Artemis JU Chiron[3] project, that includes a nine month long observational study with patients suffering from congestive heart failure.

Since the architecture of the mobile subsystem enables standards harmonization, we can also look at our mobile subsystem as being a hub in the hub-and-spoke design pattern. The mobile subsystem runs the same BPMN2 models as the server side runs. The lower two levels of Figure 4 – the services implementation are different on the server side and the mobile side. Separately modeling business processes is a great step forward in ensuring organization and enterprise interoperability. Since BPMN2 processes play such an important role, we are giving a deeper explanation in 3.5 Process control subsystem.

3.5 Process Control Subsystem

We identified concepts such as steps, tasks, roles, and schedules of tasks. For each intervention, we needed to identify what is specific for a particular disease or prevention plan including also specifics of the institutions with which we collaborated. Typically a patient first visits a healthcare service provider who performs an initial examination. Then he is given a treatment plan and scheduled next visit. This is the primary model of care. We expanded the model by including knowledge from the following domains:

- depression chronic disease,
- asthma chronic disease,
- diabetes chronic disease,
- obesity preventive program,
- schizophrenia chronic disease,
- and sports activity program.

After considering all domains, we produced an abstract care process model that consists of a set of abstract steps and tasks that are common to all domains. Tasks are performed by different roles that define the context. An abstract care process model receives a context in the execution environment (as an input variable). Context is created of all the details regarding the tasks that should be performed. This includes content of messages, medical rules, and parameters that the patient needs to input. Steps contained in the abstract care process model are the following:

[3] http://www.chiron-project.eu/

- **Assessing the Current State of Patient Health -** data about the patients clinical, social and psychological state is assessed.
- **Further investigation** is performed if needed (laboratory testing etc.).
- **Definition of the Treatment Plan -** for the purpose of a clinical trial, the experimental group was assigned to use the web and mobile based system. This intervention was not personalized in the sense of the protocol but was personalized at the level of feedback messages and tasks to be performed.
- **Monitoring -** patients received tasks that they needed to perform. These tasks were questionnaires, input forms, articles, etc. This is mostly what the part of the home-based interventions is.
- **Follow up** according to the health status of the patient, additional visits to the doctor were planned. If the data about the patient showed a managed condition, then additional visits were not needed. Besides, patients and doctors/care managers were able to engage in live chats and forum-like activity. In case a patient was not compliant with the intervention (part of the treatment plan), the care manager was notified. According to the maximum waiting time limit, the care manager reached the patient over the telephone line and personally encouraged him to follow the treatment plan. In case the patient did not want to continue being involved in the intervention, the care manager could exclude him.

We need to stress that this abstract care process model takes into account that the patient is in home care (out of hospital). Since our system collects different data, it can propose visits to the doctor office by tracking the change in the data that the patient inputs during home care. In case a critical pattern is observed, the patient, care manager and the doctor are notified. A phone call is made to the patient for urgent follow-up and also in the case when a patient does not perform the required tasks. This holds for our eCare project. In general, there are no limitations to the type of action that can be supported by the platform. If we look at the Process Layers Architecture diagram in Figure 6 we can see that our basic care processes are described as agents, which can also be orchestrated for achieving higher level functionalities.

In Figure 6, we see the basic three layers of processes defined at the process level of our platform. The three core elements reused in different settings are the Reminder, Notifier and Data Entry.

The connections between different elements at different levels represent the inheritance operator at the BPMN2 level. If we take the Notifier-Doctor element on the right side of the bottom layer as an example, we can see that connections to this element are made from the elements in the second layer called Notifier-Caremanager-Doctor and Notifier-Doctor-Ext respectively. This means that both the elements at the second layer inherit functionality that is defined for the Notifier-Doctor element and include some additional functionality.

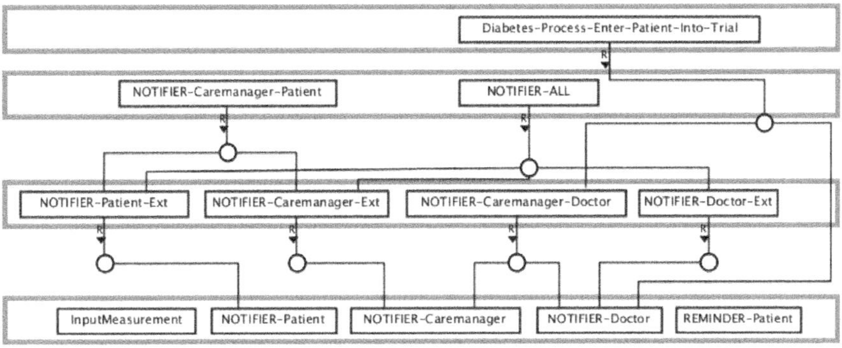

Fig. 6 Process Layers Architecture including a Diabetes specific process at the top 4th layer. The links describe extensions of processes.

Reminders and Notifiers

A reminder is a notification about an ongoing activity where patient participation is needed. Notifications are a type of reminders about the outcome of an ended action, which is used for new decisions, and to notify one or many users. Data Entry elements are used for simple acquisition and persistence of new data that collected from the users. We explicitly defined a set combination of component-role types to cover all the usage scenarios of the interventions. This was easily achieved using the BPMN2.0, which permits simple definition of new processes from existing ones. By using the inheritance operation we defined the following first level processes:

- **Reminder-Patient:** We used reminders only for patients. A reminder is a short message service message in our interventions. The patient is usually reminded of a waiting task that he/she has to perform. Otherwise, a care manager calls him/her and in case of non-cooperation, a patient is excluded from the intervention.
- **Notifier-Patient:** the patient is notified of e.g. results of the latest measurements.
- **Notifier-Caremanager:** a care manager is notified about e.g. deteriorating of a patient health state.
- **Notifier-Doctor:** a doctor is notified in case of emergencies - if a health state parameter becomes critical.

The second layer in Figure 6 consists of the second level processes which reuse those at the first layer. They are mostly defined as extensions of the first layer processes. This means they contain the functionality of the first layer processes and add new functionality. This is off course a known principal in object-oriented software development methods called inheritance. The difference here is that we are defining inheritance between processes. The following second level processes have been defined and used in our interventions:

- **Notifier-Patient-Ext:** created by extending the Notifier-Patient process.
- **Notifier-Caremanager-Ext:** created by extending the Notifier-Caremanager process.
- **Notifier-Caremanager-Doctor:** created by Notifier-Caremanager and Notifier-Doctor processes. Multiple inheritances are thus possible.
- **Notifier-Doctor-Ext:** created by extending the Notifier-Doctor process.

At the third layer of our process hierarchy, we have another layer of processes that extend processes in lower layers:

- **Notifier-Caremanager-Patient:** created by extending the Notifier-Patient-Extended and Notifier-Caremanager-Ext processes.
- **Notifier-All:** formed by extending three processes from the second layer: Notifier-Patient-Ext, Notifier-Caremanager-Ext and Notifier-Doctor-Ext.

These three layers of our process hierarchy represent all the processes that are used in all our interventions. Furthermore, there are also higher level processes, which include more context information.

The idea of layering processes is to make the architecture scalable. The architecture supports the addition of new layers and new basic processes without the need to stop the execution environment.

The names of processes usually include the name of the domain (intervention domain e.g. diabetes) followed by the use case name. As shown on Figure 6, high level process called **Diabetes-Process-Enter-Patient-Into-Trial** extends one process from the second layer (Notifier-Caremanager-Doctor) and one process from the first layer (Notifier-Doctor).

As an example, our diabetes intervention uses the following 4th layer processes:

- **Diabetes-Process-Periodic-Notify-Regular-Visit:** A process that sends notifications to patients that should visit the doctor.
- **Diabetes-Process-Periodic-DSCA:** a notification is sent to the patient that he has to enter a questionnaire.
- **Diabetes-Process-Periodic-DataEntry:** a notification is sent to all the patients that have a task waiting - to enter a new measurement.
- **Diabetes-Process-Update-Data:** this process is executed when doctors or care managers update or enter data.
- **Diabetes-Process-Include-New-Patient-In-Trial:** this process is used only once when the patient is first registered into the system.

The above 5 processes represent the core functionalities of the diabetes intervention. All of the interventions can be described by such a set of executable process models. The common subset of all these sets is the abstract care process model.

We have decomposed interventions in a way that permits us to add a new layer of the system. This approach to modeling interventions thus supports reusability and scalability. By working as an interdisciplinary team, using a universal language that is backed by a common model, we can create new interventions faster, cheaper and more reliably. The first intervention was designed and developed in 6 months with 3 people working. The last intervention was developed in 2 months with 2 people working. Additional information about the platform including the documentation and the source code is available on the eCare project site [15].

Additionally, we have to stress the connection between eHealth and mHealth at the process level. The described process layers architecture shows processes that are deployed to a process execution engine. This process execution engine is lightweight which means, it can run on the server side or a smartphone. This means we can create interventions by means of applications which are able to run locally on the smartphone. At some point, only the data transfer needs to occur so that all the data is safely stored on the server side.

4 Future Directions

We have based our work on architecture homogeneity and shared ontologies as the basic steps towards achieving a global virtual electronic health record. All developed and many developing countries are working on national projects with the aim of moving all the patient data to electronic form so that the data can be accessed and updated at the point of care. The USA and EU have also done a lot of work in exchanging clinical information between states and countries. Great effort is currently being put in the cross-ocean exchange between USA and EU. This shows us that the global trend is in interconnected healthcare systems that enable ubiquitous healthcare. But there are many obstacles on this path. We speak different languages, we use different standard and non-standard terminologies and technologies, our organizations have different goals and business processes, etc. We need to focus on interoperability on different layers. We should be able to describe data together with the surrounding context at the capture time. This enables artificial systems to use the data later without the need of human intervention. This is one of main focuses of research in the field of interoperability.

The technology we present in the chapter is going to be used in new projects that deal with remote patient monitoring and care. This means our abstract care process model is going to become stronger in the sense of capturing even more information that is common to all care processes. We are going to add another component that will use artificial intelligence and machine learning algorithms to find new knowledge in the gathered data. We will have the ability to predict future relapses of patients, or visits to the hospital etc. All this will enable us to help our healthcare system to become more optimized and thus sustainable.

All the artifacts are going to be released as open source technology. This is a normal step towards if we want to help healthcare systems towards a future that does not include data lock in as the primary business model. All the models and all the code has to be free. But this does not mean that the data will be accessible to everyone. Security and access control is a separate question but of utmost importance. Our future work also includes testing of different access and security mechanisms for protecting the data at one side and enabling patients to decide what to do with the data on the other side. The use of elliptic curves is of interest here to us, as also other well accepted mechanisms.

5 Discussion and Conclusions

Healthcare systems are becoming highly distributed, specialized and cooperative structures with goals of increased quality and safety of care as well as efficiency and efficacy of delivery processes [10]. Advanced ICT solutions are needed that support the communication and collaboration of different parties involved. This leads to advanced models of care that are based on personalization of care that includes prevention, home care, elderly and lifestyle services, and can be called personal health or pHealth [10] systems.

Most important requirements for interoperable information systems that allow this advanced care paradigm defined in [10] are openness, scalability, flexibility, portability, standards based, service-orientation, user-friendliness, lawfulness, and trustworthiness.

The ICT solutions support communication and cooperation of various entities. As noted in [10], the main interoperability challenge here can be divided into a computation-independent level, including business objectives, policies and different organizational and social limitations like culture, environmental conditions, care setting workflow, education, specialized methodologies and knowledge expression resources, and the computation-level. This holds for all eHealth/mHealth/pHealth systems. The main questions here are how to explain different aspects and how to harmonize them. Ontologies can be used for this. They help to solve problems with concept and knowledge representation, mapping different representational artifacts between disciplines by means of ontology management and harmonization. As noted in [9], the approach to Interoperability that fits healthcare best is described as the unified approach. The unified approach assumes that the shared ontology is negotiated in advance in order to achieve semantic interoperability whereas the other two methods, full integration and federalization, are less appropriate. In general, interoperability refers to the coexistence, autonomy and federated environments, and integration refers to coordination and coherency [17].

We based our work on the use of the currently only ontology based (ontology of clinical processes expressed as Archetypes) approach to knowledge representation in the field of healthcare namely OpenEHR [38]. For the same purpose, we later presented the latest approach that eliminates the EHR context from the OpenEHR, namely MLHIM. MLHIM was partly created due to the size of the OpenEHR method in terms of the processing power needed to manage archetypes and templates in runtime. This becomes an issue when developing systems on devices with limited resources. We introduced MLHIM for the mobile subsystem. The knowledge needs to be available on the mobile devices as well as on the server. We don't see our mobile system as a mere gateway for data transport between patient devices and the server. Our mobile subsystem has the power to verify data on the device itself. This was possible by using MLHIM knowledge models. Lately, CIMI has the potential of defining a general format for all clinical models independent of the logical and implementation perspective, thus focusing on the conceptual aspect [39, 8]. This way a common format for all clinical models will be produced. These conceptual, clinical models will be then represented in different lower level modeling languages, primarily by using Archetype Description Language (ADL) that was developed within the OpenEHR initiative, and Unified Modeling Language (UML). The support for CIMI modeling approach is something that we plan for future work.

Since our interventions are primarily a set of process models interweaved with the clinical models, our overall architecture is based on business process modeling to explain cross domain aspects of the system and also to model medical treatment processes. These are distinguished from the other type of processes, namely organizational processes that focus on e.g. medical order entry or result reporting [40]. Business processes mostly explain parts of care (advanced care paradigm). We were able to treat each process merely as a component. Composition of components allows a higher complexity of the systems structure and behavior [10]. The architecture of a system is believed to be a semantically interoperable approach if it formally describes enterprise and information models [41]. We have used BPMN2 to describe business processes and rules, and OpenEHR/MLHIM to describe information semantics and therefore semantically interoperable components and services development is supported [41]. Since the architecture of our server side system and the mobile subsystem are the same, we have successfully performed both first steps towards cross-domain or cross-enterprise interoperability as an example of the unified approach to interoperability in healthcare.

Healthcare systems are modeled as complex adaptive systems (CAS)[42]. Historically healthcare systems scaled up but were never designed and planned properly. The growth and development were natural. Therefore, it is practically impossible to create a platform that would support all the processes in a typical healthcare system. Moreover, the systems need to be also personalized. A typical CAS system is modeled as a set of independent agents that can exchange information by means of orchestration or choreography. We approached the modeling of care processes from this stand point and thus decided to use the BPMN2 specification that supports both mechanisms of communication. Our platform thus permits modeling of ad hoc personalized processes which use rich clinical models for data validation and decision support. This makes our platform future proof since we do not need to know all the processes in advance but rather just have tools to extend for new add-ons. Our abstract process models can be characterized as configurable process models that define the rules of how a reference process model can be adapted [43]. This means that, in our system, the configurable process models can show the originating process models whereas the reference models do not enable this [43]. We have thus created configurable process models for different interventions. Authors of [44] provide a method for developing a reference process model which can be then used to create configurable process models which define a family of process models. Author of [43] gives more in depth information on this method along with the similarity measures, tools etc. that are needed to execute the mentioned method efficiently. In our eCare project, such measures have not been used since the interventions were mostly simple and allowed a relatively easy recognition of related processes that occurred in different interventions. In future work, we plan to use such similarity measures in order to support the development of a reference process models for more complex interventions.

All the processes that were fully implemented can be deployed on the server side or the mobile devices. We thus say that the layer containing process models is shared across different environments (server, mobiles…). This permits loose coupling of mobile mHealth applications that at the same time support semantic interoperability. Our mHealth and eHealth applications are similar in the sense of a common process layer that uses clinical models for validation of data, and in the sense of the overall architecture of both systems. What is left to differentiate applications is the user interface where user friendliness is the main requirement.

In recent reviews of the United States Agency for Healthcare Research and Quality (AHRQ) [45] (also Volumes 4, 5 and 7) no conclusive results have been found regarding different interventions for managing chronic diseases, coordinated care and prevention which are the focus of our work. Similar results were shown in [31]. The results of the reviews can be seen as a rationale for an approach similar to ours. We developed a platform that supports a stepped approach by means of simple interventions that span the eHealth and mHealth spaces, which are then clinically tested and finally brought to clinical practice if outcomes are shown to be satisfying. This way we can design and develop simple personalized interventions. This allows the concept of dosage of interventions. Additionally, we can create new and more complex interventions by orchestrating these simple ones.

Our future work includes development of new interventions that will give us valuable feedback on the overall strategy to mHealth and eHealth interventions design and development. We also plan to give more attention on the adoption of interventions in clinical practice and other environments depending on the type of environment.

A simple and standardized ICT platform permits faster and more cost efficient development of interventions than a traditional approach. These can be iteratively tested and adjusted for final clinical use. Furthermore, since clinical and process knowledge developed in iterations is reused, this also supports faster development and deployment of applications which was at the crux of our study. Nonetheless, the study presents methods and approaches to build a sustainable ICT environment for interoperable mHealth/eHealth intervention development.

6 Case Studies

Case Study 1: eHealth Interreg

A project called eHealth, a cross border exchange of data, started in 2010. It is a project that is trying to establish cross border exchange of clinical data across borders between countries Italy and Slovenia. This means that the project participants need to address all layers of the general interoperability framework which includes also law differences between the countries. There are 12 hospitals taking part in the project, 6 from Slovenia and 6 from Italy. University of Primorska is acting as the technology provider and main coordinator. The project is focusing on cross-domain interoperability. IHE profiles are being used for defining all the requirements for the solution that will enable the exchange. Two national contact points will be used, one in Italy and one in Slovenia. All the exchange will always go through these contact points.

One of the project's work packages includes a scenario in which patients are using different sensors and measuring devices in order to track their health status. This data will be automatically transferred to the server side. In case of relapses, doctors and nurses will be notified.

It has been identified that great problems have to be solved in order for this kind of scenario to work across borders, mainly, the law differences that are found regarding patient rights in both countries. Due to these problems, the project has still not been able to implement all the scenarios envisioned. Another set of problems arose due to not using the OpenEHR methodology for modeling clinical content. Data mapping schemas are planned to be used, which is not an optimal solution, and is definitely not scalable or future proof.

Case Study 2: epSOS

In 2008 the European Commission has started ~~a project~~ the European Patients Smart Open Services (EPSOS) project. There are 45 project beneficiaries consisting of national ministries of health, national/regional competence centers, a consortium of industry and the Project Management team. There are 25 countries that are involved in the project, twenty-two of which ~~22~~ are European Union (EU) member countries and 3 are non-EU member countries. All the countries form a circle of trust. All the data exchange is done through national contact points (NCP). An NCP is both an organizational concept as well as a technological one. A solution is used in each country that performs different services towards other NCPs on one side, and services towards countries national eHealth systems (to ne bo v redu, ampak bi rabila slovensko verzijo, ker ni 100 % razumljivo, kaj bi moralo biti pravilno) (e.g. National Electronic Health Record). The NCP solution is based on IHE profiles, mostly the ones that are focused on cross-domain interoperability. The NCP solution used by many countries is open sourced. It is called OpenNCP and can be downloaded and used.

In the piloting phase, patients will be able to go to see a doctor in any one of the 25 countries. The doctor will be able to access the patient's clinical data from the patient's country of origin. Such services will be common in the near future and give a new sense of freedom of mobility to the European citizens.

The project did not use OpenEHR for modeling clinical concepts and has identified great problems due to this decision. Mostly, problems occur because different countries use different coding standards, and also problems occur with the automated translations between different languages. This kind of problems are well addressed with the OpenEHR approach and are one of the main reasons why the methodology was developed.

References

1. Kumar, S., Nilsen, W., Pavel, M., Srivastava, M.: Mobile Health, Revolutionizing Healthcare Through Transdisciplinary Research. Computer (Long Beach Calif), 28–35 (January 2013)
2. Fiordelli, M., Diviani, N., Schulz, P.J.: Mapping mHealth research: a decade of evolution. J Med Internet Res 15:e95 (2013)
3. Current and future state of mhealth: PwC, http://www.pwc.com/gx/en/healthcare/mhealth/index.jhtml#&panel1-1 (accessed April 13, 2013)
4. Dassisti, M., Goncalves, R., Molina, A., Noran, O., Panetto, H., Zdravković, M.: Sustainability and Interoperability: Two Facets of the Same Gold Medal. In: 8th Int. Work. Enterp. Integr. Interoperability Netw., pp. 250–261 (2013)
5. About Health Level Seven International, http://www.hl7.org/about/index.cfm?ref=common (accessed April 13, 2013)
6. Browne, E.: Archetypes for HL7 CDA Documents (2008), http://www.openehr.org/wiki/download/attachments/3440870/Archetypes_in_CDA_4.pdf
7. Beale, T., Heard, S.: OpenEHR Architecture Overview (2006), https://github.com/openEHR/specifications/raw/Release-1.0.1/publishing/architecture/overview.pdf
8. CIMI Reference Model Report, Draft V 0.4 (May 2012), http://informatics.mayo.edu/CIMI/images/3/34/CIMI_Reference_Model_Takforce_Report_v0.5_May_2012.docx
9. Noran, O., Panetto, H.: Modelling a sustainable cooperative healthcare: An interoperability-driven approach. In: Demey, Y.T., Panetto, H. (eds.) OTM 2013 Workshops 2013. LNCS, vol. 8186, pp. 238–249. Springer, Heidelberg (2013)
10. Blobel, B., Kalra, D., Koehn, M., Lunn, K., Pharow, P.: The Role of Ontologies for Sustainable. Semantically Interoperable and Trustworthy EHR Solutions, 953–957 (2009)
11. Boulos, M.N.K., Wheeler, S., Tavares, C., Jones, R.: How smartphones are changing the face of mobile and participatory healthcare: an overview, with example from eCAALYX. Biomed Eng Online 10(24) (2011)
12. Zhou, F., Yang, H.-I., Álamo, J.M.R., Wong, J.S., Chang, C.K.: Mobile personal health care system for patients with diabetes. In: Lee, Y., Bien, Z.Z., Mokhtari, M., Kim, J.T., Park, M., Kim, J., Lee, H., Khalil, I. (eds.) ICOST 2010. LNCS, vol. 6159, pp. 94–101. Springer, Heidelberg (2010)
13. Pioggia, G.: Unobtrusive wearable multi-sensorial platform for remote monitoring of patients. In: 10th Int. Conf. Smart homes Heal. Telemat, June 12-15 (2012)
14. Doukas, C., Pliakas, T., Maglogiannis, I.: Mobile healthcare information management utilizing Cloud Computing and Android OS. In: Conf. Proc. IEEE Eng Med. Biol. Soc., pp. 1037–1040 (2010)
15. e-Oskrba - E-podpora procesa zdravstvene oskrbe, https://eoskrba.iam.upr.si/ (accessed December 20, 2013)
16. Hissam, S.A., Russo, B., Mendonça Neto, M.G., Kon, F. (eds.): Open Source Systems: Grounding Research (2011), doi:10.1007/978-3-642-24418-6
17. Panetto, H., Cecil, J.: Information systems for enterprise integration, interoperability and networking: theory and applications. Enterp. Inf. Syst., 7:1–7:6 (2013)

18. Activiti BPM platform, http://www.activiti.org/
19. Dehmoobad, A., Sartipi, K.: Minimized Domain Knowledge for SOA-Based Interoperability. 2008 IEEE Asia-Pacific Serv Comput Conf., 500–506 (2008)
20. Standards & Interoperability (S&I) Framework - NwHIN Portfolio and RHEx.png - Detail, http://wiki.siframework.org/file/detail/NwHIN+Portfolio+and+RHEx.png/342461254 (accessed August 5, 2013)
21. Grilo, A.: Interoperability Framework, Theories and Models (2005), http://www.fines-cluster.eu/jm/Publications/Download-document/54-Interoperability-Frameworks-Theories-and-Models-Grilo.html
22. Jardim-Goncalves, R., Grilo, A., Agostinho, C., Lampathaki, F., Charalabidis, Y.: Systematisation of Interoperability Body of Knowledge: the foundation for Enterprise Interoperability as a science. Enterp Inf. Syst. 7, 7–32 (2013)
23. Ostrom, E.: Understanding Institutional Diversity. Princeton University Press (2005)
24. Popplewell, K.: Towards a Scientific Foundation for Interoperability (2009), ftp://ftp.cordis.europa.eu/pub/fp7/ict/docs/enet/20090603-presentacion-charalabidis-goncalves-liapis-popplewell_en.pdf (accessed January 14, 2014)
25. Murray, J.F., Pascoe, G.A.: Medical device interoperability (2012), http://www.aami.org/interoperability/Materials/MDI_1203.pdf
26. Grütter, R.: Knowledge Media in Healthcare: Opportunities and Challenges, p. 296. IGI Global (2001)
27. Ritterband, L.M., Thorndike, F.P., Cox, D.J., Kovatchev, B.P., Gonder-Frederick, L.A.: A behavior change model for internet interventions. Ann Behav Med 38, 18–27 (2009)
28. Tate, D.F., Jackvony, E.H., Wing, R.R.: A randomized trial comparing human e-mail counseling, computer-automated tailored counseling, and no counseling in an Internet weight loss program. Arch Intern Med 166, 1620–1625 (2006)
29. Greist, J.H., Marks, I.M., Baer, L., Kobak, K.A., Wenzel, K.W., Hirsch, M.J., Mantle, J.M., Clary, C.M.: Behavior therapy for obsessive-compulsive disorder guided by a computer or by a clinician compared with relaxation as a control. J Clin Psychiatry 63, 138–145 (2002)
30. Wright, J.H., Wright, A.S., Albano, A.M., Basco, M.R., Goldsmith, L.J., Raffield, T., Otto, M.W.: Computer-assisted cognitive therapy for depression: maintaining efficacy while reducing therapist time. Am. J. Psychiatry 162, 1158–1164 (2005)
31. Van Gemert-Pijnen, J.E.W.C., Nijland, N., Van Limburg, A.H.M., Ossebaard, H.C., Kelders, S.M., Eysenbach, G., Seydel, E.R.: A holistic framework to improve the uptake and impact of eHealth technologies. J. Med. Internet. Res. 13, e111 (2011)
32. Murray, E.: Web-Based Interventions for Behavior Change and Self-Management: Potential. Pitfalls, and Progress. Med. 20, 1:e3 (2012)
33. Hawkins, R.P., Kreuter, M., Resnicow, K., Fishbein, M., Dijkstra, A.: Understanding tailoring in communicating about health. Health Educ. Res. 23, 454–466 (2008)
34. Meglic, M., Furlan, M., Kuzmanic, M., et al.: Feasibility of an eHealth Service to Support Collaborative Depression Care: Results of a Pilot Study. J. Med. Internet. Res. 12, e63 (2010)
35. Riper, H., Andersson, G., Christensen, H., Cuijpers, P., Lange, A., Eysenbach, G.: Theme issue on e-mental health: a growing field in internet research. J. Med. Internet. Res. 12, e74 (2010)

36. Tate, D.F., Finkelstein, E.A., Khavjou, O., Gustafson, A.: a, Khavjou O, Gustafson A Cost effectiveness of internet interventions: review and recommendations. Ann. Behav. Med. 38, 40–45 (2009)
37. Tabeling, P., Apfelbacher, R., Wappler, S.: Fundamental Modeling Concepts Metamodel (2003), http://www.fmc-modeling.org/download/metamodel/FMC-Metamodel_Explained.pdf
38. Beale, T., Garde, S.: Openehr Knowledge governance (2010), http://www.mz.gov.si/fileadmin/mz.gov.si/pageuploads/eZdravje/Novice/gradiva_predstavitve_dogodkov/Open_EHR/3.dan/4_Governance.pdf
39. Ahn, S., Huff, S.M., Kim, Y., Kalra, D.: Quality metrics for detailed clinical models. Int. J. Med. Inform. 82, 408–417 (2013)
40. Lenz, R., Reichert, M.: IT support for healthcare processes – premises, challenges, perspectives. Data Knowl. Eng. 61, 39–58 (2007)
41. Lopez, D.M., Blobel, B.G.M.E.: Blobel BGME Semantic interoperability between clinical and public health information systems for improving public health services. Stud Health Technol Inform 127, 256–267 (2007)
42. Plsek, P.E., Greenhalgh, T.: Complexity science: The challenge of complexity in health care. BMJ 323, 625–628 (2001)
43. Dalinghaus, W.R.: An Introduction to Reference Process Modeling in Healthcare. Master Thesis,Eindhoven University of Technology (2013)
44. Pajk, D., Indihar-Stemberger, M., Kovacic, A.: Reference model design: An approach and its application. In: Proc. ITI 2012 34th Int Conf. Inf. Technol. Interfaces (2012), doi: 10.2498/iti.2012.0419
45. Williams, J., Jackson, G., Powers, B., et al.: The Patient-Centered Medical Home. Closing the Quality Gap: Revisiting the State of the Science (2012), http://www.effectivehealthcare.ahrq.gov/reports/final.cfm

Questions

Q1. What transformation process is currently ongoing in most healthcare systems?

Q2. What are the major global trends in healthcare?

Q3. What has been recognized as the major enabler of scalable mHealth?

Q4. What are the foundations of studying interoperability?

Q5. What are the main components of the General interoperability Framework?

Q6. What are the main contents of semantic interoperability? Give examples.

Q7. Where do we place the OpenEHR and MLHIM methodologies in the interoperability stack?

Q8. What do we understand under the term abstract care process model in the eCare project?

Q9. What are the basic steps towards a global virtual electronic health record?

Q10. Why is the OpenEHR approach a better approach towards semantic interoperability?

17. Recent Advances in mHealth: An Update to Personal Health Device Interoperability Based on ISO/IEEE11073

Hector Gilberto Barrón-González*, Miguel Martínez-Espronceda, J.D. Trigo, Santiago Led, and Luis Serrano

Department of Electronic and Communications, Universidad Autónoma de Nuevo León, San Nicolas De Los Garza, México
gilbertobarron@ieee.org

Abstract. In recent years important advances have been made in the development of health care systems in specific areas owing to different uses and applications. One of the foremost advancements is the so-called mobile Health (mHealth), due to its potential social and economic impact. Within this scope, several technologies – such as smart phones and ultra-slim tablet-PCs, as well as low-voltage and low-power wireless sensor networks – provide telemonitoring services at home and ubiquitous support when they are strongly integrated. Therefore, the integration of international interoperability standards is, nevertheless, required for a regulated transmission of health information in mHealth scenarios. However, such implementation tends to increase the processing load and battery size in order to ensure its autonomy. To tackle this problem, different strategies are being tested in this set of mobile technologies which are aimed at providing a solution to such critical aspects. Another important aspect that should be addressed is the management of communication between different parts that constitutes the system. The new communication capabilities of smart phones enable the implementation of new techniques that improve data transmission from the mobile system to the telemonitoring center, facilitating the integration of Personal Health Devices (PHDs) with telemonitoring systems. This makes the system more accessible to the user.

This work highlights some new features of X73PHD and the benefits of using them. In addition, current needs to enhance the operational capacity of the different specializations of X73PHD are also shown, which include the definition of new remote command and control functionalities. These new management functions are being developed in order to allow any X73PHD-compliant manager to modify the operation of any X73PHD-compliant agent. This functionality could expand the technological boundaries of interoperable mHealth to healthcare services and improve Quality of Service (QoS) by providing remote configuration of medical devices to health professionals. The definition of this a new package extension for X73PHD involves the analysis of previous studies and related use cases, work which is currently being carried out within the PHD Working Group

* Corresponding author.

S. Adibi (ed.), *Mobile Health*,
Springer Series in Bio-/Neuroinformatics 5, DOI: 10.1007/978-3-319-12817-7_17

(PHD-WG), in which the authors are actively collaborating. Future lines include the proposal and adoption of the new extension remote command and control package, along with its application to other specializations.

1 Introduction

In recent years, two areas of eHealth have evolved together, namely mHealth – developed to facilitate user mobility – and home telemonitoring – designed to enable the follow-up of patients at home [1]. As such paradigms have been gaining momentum, international standards, protocols and norms have emerged to regulate the information that is being sent, both syntactically and semantically [2, 3]. In this context, the ISO/IEEE11073 (X73) set of norms guarantees interoperability between different medical devices – also referred to as agents – and concentrator devices or managers, which administer the health data, regardless of the transport technology [4,5].

The new branch of X73 has been designed for Personal Health Devices (PHDs) (X73PHD). X73PHD leverages the foundations of a previous version of X73 – also referred to as Point-of-Care (X73PoC). While X73PoC provided a complex – albeit robust – way to handle medical information at the point of care, X73PHD proposes a lighter version to enable interoperability between agents and managers in Body Area Networks (BAN) and Personal Area Networks (PAN). Today, X73PHD has more than 10 specializations and the PHD-WG is working on the development of new specializations.

In traditional home telemonitoring scenarios, concentrators – such as set top boxes – are used to collect the measurements of the different PHDs, and subsequently retransmit such data using fixed technologies, e.g. the Public Switched Telephone Networks (PSTN) or, more recently, Digital Subscriber Lines (DSL). In mHealth solutions, however, smart phone devices can be used as concentrators, thereby gathering X73PHD data from the different sensors used in BAN / PAN and subsequently retransmitting such data via existing mobile phone networks, such as Global System for Mobile communications (GSM). It is noteworthy that GSM-like networks coverage reaches 90% of the world's population, according to a study published by GSM Association at 2010 [6]. The benefits that could be derived from the integration between X73PHD and mHealth using GSM technologies are diverse: the improvement of the response of health professionals in monitoring the disease of a patient, the reduction of healthcare costs by using existing GSM infrastructure, compatibility between manufacturers, scalability, and a marked improvement in quality of life.

Besides, the PHD Working Group (PHD-WG) is constantly developing new use cases (i.e. specializations) that would enhance the integration of X73PHD in mHealth scenarios. A number of use cases have been selected to illustrate the potential of standard-based mHealth systems. These include the basic Electrocardiograph, the Sleep Apnea Breathing Therapy Equipment (SABTE), and the medication monitor. Furthermore, new features are constantly being discussed within the PHD-WG. Such new capabilities would enable a full range of novel possibilities of X73PHD in a mHealth scenario, such as, for example, remote control of medical devices.

This chapter is organized as follows: a general review of the state of art regarding X73 is provided in Section 2. Then, recent command and control use cases, which justify the need for remote command and control, will be exposed and analyzed in Section 3. These use cases will be used as a starting point to determine the requirements of a new command and control service within X73PHD. Research benefits of PHD interoperability are analyzed in Section 4. Within this section, the advantages and disadvantages of interoperability standards will be discussed against real application scenarios. In Section 5, requirements that were found in the use cases are outlined, including details such as main configuration parameters and their data types. Using these requirements, several implementation alternatives to the command and control services are proposed and analyzed in Section 6. The first approach is an extensive service which is able to meet the requirements of existing specializations and new use cases. The second approach is previously proposed to meet basic requirements of existing specializations. Future lines are highlighted in Section 7 and, finally, conclusions are drawn in Section 8.

2 The ISO/IEEE 11073 Standard: Evolution and Implementations

X73 implements the 7-layer Open Standard Interconnection (OSI) to enable intercommunication between the agent and the manager [7]. This characteristic results in the device being independent of any other transport technology, allowing the interoperability between agents and managers who normally use other types of transport technologies such as ZigBee, Bluetooth, Wifi and GSM. The X73 family is divided into two versions: the X73PoC, also known as the "classic version", and the X73PHD for PHDs. There are several implementation experiences for each one. Thus, to analyze the evolution and the implementation of this standard, a thorough analysis of both versions has been conducted.

2.1 ISO/IEEE 11073 – Point-of-Care (X73POC)

This family of standards was developed for Plug-and-Play interoperability in clinical environments. It was originally created from other standards such as the American Medical Information Bus (MIB) and the European VITAL (ENV13734) and INTERMED (ENV13735) standards [8,9,10]. Its core documents were developed and approved by 2004. These documents are named as ISO/IEEE-11073-xyyzz (or X73-xyyzz) [11,12]. The X73-1yyzz documents which are also known as Medical Device Data Language (MDDL) refer to device modeling. Within this part, the main documents are the X73-10101 (nomenclature and terms), the X73-10201 –(Domain Information Model (DIM) and services offered) and the X73-103zz (device specializations). The X73-2yyzz documents define the application profile which determines the exact realization to access the services and data in the device model. There are two main application profiles, namely a polling mode profile (X73-20201) and a baseline profile (X73-20202). The X73-3yyzz documents define the transport and physical profiles.

In its mainstream, X73 defines a peer to peer, single-link protocol between the agent and the manager. The background philosophy in X73 is to identify the information within a device, and model this information as a set of objects with attributes. Then changes in the values of these attributes are transmitted to the other peer using announcement services. In addition, the other peer can execute remote procedure calls to initiate predefined operations. These operations are limited to collecting data from the other peer in the basic profiles, but it is possible to extend its possibilities by implementing the optional remote control package (X73-20301)[13].

Several classic X73-compliant proof-of-concept have been implemented to date. Taboada et al. (1997) presented an implementation for Intensive Care Unit (ICU) using X73 [14]. Kousoris et al. (2005) developed a similar system to gather ECG [15]. Clarke and Jones (2005) presented a telemonitoring platform based on the principles of X73POC gathering several health measurements (weigh, blood pressure, temperature, SpO2, ECG among others) [16]. They incorporated remote agent control and configuration without the need to go to a patient's home. Yao and Warren (2005) implemented X73POC over Bluetooth [17]. Garguillo and S. Martinez (2007) published a tool to test X73 compliance [18]. In addition, I. Martinez et al. (2008) implemented and End-to-End platform based on X73POC and ISO/EN13606 [19]. Later on, Stylianides et al. (2011) presented a home-based remote control based on X73POC [20].

2.2 ISO/IEEE 11073 – Personal Health Devices (X73PHD)

This new branch of the X73 standard was originally launched by Continua Health Alliance [21]. In this way, Continua defined the use cases which are needed to further the requirements of this new protocol and its later elaboration. There are two objectives to having this new family: first, to incorporate the latest wireless technologies, and second, to adapt the classic version to low-voltage, low-power, low-cost hardware which was being used in consumer specific products. By then, the X73POC required too much processing capacity so that implementing X73POC required high featured hardware.

Therefore, the new branch is a simplified version of the X73POC standard. It maintains the basic modeling concepts but simplifies the layer stack and the services provided. X73PHD defined the X73-20601 application profile, also known as optimized exchange protocol, which is a self-contained document. Moreover, the X73PHD defines specializations of the X73-20601 document, numbered as X73-104zz [22,23]. At the moment, more than 10 specializations have already been published.

Goga et al. (2008) formally analyzed X73PHD both manually and automatically [24]. They applied model-checking techniques, the Promela language, and the Spin tool to automatize their analysis. Their technique could be used to probe future extensions to the standard. Martinez et al. (2010) implemented an end-to-end standards-based solution integrating X73PHD and ISO/EN 13606 [25]. The platform used smart phones to adapt vital signs obtained from non-wireless commercial devices into X73PHD-compliant MDs. The data gathered by these devices was released to an ISO/EN 13606-compliant EHR server. Trigo et al. (2010) conducted a review of the wide heterogeneity on existing digital ECG formats and standards for interoperable exchange of digital ECG signals and integration of health information systems (HIS) of the hospitals. The results show that a large number of particular solutions hinder the integration of HIS and exposes the need for a single standardized ECG format [26]. Carot-Nemesio et al. (2010) built a Free/Libre Open Source Software (FLOSS), interoperable platform based on an Android platform for the MORFEO OpenHealth project. The results of this project were integrated into Bluez, the open source Bluetooth-compliant implementation in the GNU/Linux world [27]. Martínez-Espronceda (2011) presented lessons learned and several guidelines to implement X73PHD in low-cost hardware [28]. Ji et al. (2012) presented a personalized middleware for mHealth services for use in the Ubiquitous Consumer Wireless World (UCWW) [29]. Norgall and Wichert presented the universal project that uses X73PHD due to its benefits to both users and the development itself [30]. Lasierra et al. (2012) presented an SNMP-based solution to enable remote ISO/IEEE 11073 technical management [31]. Clarke et al. (2013) presented a Continua-compliant X73PHD implementation using ZigBee Health Care [32]. Egner et al. (2013) presented an enhanced version of X73PHD that includes a security protocol [24].

3 New Remote Command and Control Use Cases for Interoperable X73-Based on PHDs

Among the most recent developments in the ongoing work of standardization for PHDs is an important requirement related to the command and control service that has attracted attention lately. This requirement is exposed by three use cases specializations, i.e. X73-10472 (medication monitor), X73-10406 (basic ECG) and X73-10424 (SABTE). These new use cases were able to group configurable parameters for this kind of services into two parts: the operational parameters that refer to the characteristics that may modify the applications of a therapy and data acquisition parameters for monitoring devices, and the operational modes, which include parameters that allow changes in the operational mode of PHDs – i.e. application of therapy, monitoring mode, test mode or another present. It is also important to mention that these use cases grouped together all of the similar features that are present in other specializations of the standard and future specializations.

3.1 Holtin

The new use case HOLTIN is an advanced version of the basic 1 lead ECG device that can be found within the context of the X73-10406 specialization [33]. This device performs the functions of an ECG 1 lead, which, as part of its functions, is that it has the ability to record data acquired when developing a critical event. This event can be generated automatically when an algorithm detects a cardiovascular event, using a set of limits and ranges associated with the functioning of the heart. Similarly, the user can also activate the recording of the data manually. The recording would include the duration of the event, plus an amount of data acquired before and after it. It is worth mentioning that all the parameters of the use case HOLTIN which were exposed below require a remote service that allows them to be changed by means of a manager.

3.1.1 Automatic Detection Algorithm

The use case HOLTIN proposes the implementation of an automatic detection algorithm, which is designed to detect cardiovascular events. These are generated by comparing the data obtained in real time, with the limits and ranges established within the detection algorithm.

A list of parameters for the implementation of an automatic detection algorithm is set out in the first column in Table 1. The second column shows the parameters which are necessary for the implementation of automatic detection algorithm in the new use cases, which are contemplated change through a remote control service.

Table 1 Automatic detection algorithm parameters

Parameter Reference	Description
Asist-Lim	Asystole threshold
Brady-Lim	Bradycardia threshold
Taqui-Lim	Tachycardia threshold
QRS-Number	RR Intervals threshold

The following describes the usefulness of each of these parameters within the automatic detection algorithm.

<u>Brady-Lim:</u> Limit bradycardia in beats per minute (bpm). For recording a bradycardia event the heart rate should be less than this threshold for at least one beat.

<u>Taqui-Lim and Qrs-Number:</u> Limit tachycardia in beats per minute and the number of consecutive QRS complexes with a heart rate greater than Taqui-Lim respectively. For the recording of an event type, an automatic tachycardia heart rate must exceed the value Taqui-Lim beats which is a number greater than or equal to the value of an Qrs-Number.

Asist-Lim: The time interval (in seconds) without heart rate detected in the patient, once exceeded, determines the existence of an arrhythmic episode of asystole. The typical value of this parameter is 2-4 seconds.

3.1.2 Storage Parameters

The storage parameters set the time duration of the ECG signal records that are stored on the device as part of the information associated with cardiac events detected / reported. The first column of the table 2 shows whether there exist parameters that can be modified to increase or decrease the recording time of the ECG signal in the X73-10406 specialization. In the second column, the parameters associated with the duration of the recording of the events detected in form and automatic and manual – which would be modified using a remote control service – are exposed.

Table 2 Storage parameters for recording time

Parameter Reference	Supported by current specs	Description
ECG-Recorded-Span	Read only	Recording time events detected automatically
Automatic-Detection-Time	No	Recording time events detected automatically
Manual-Detection-Time	No	Recording time of events reported manually

The following describes the usefulness of each of these storage parameters:

Automatic-Detection-Time: The time interval (in seconds) of ECG signal, around the detection instant, at which information is stored as associated with arrhythmic episode. The value of this time may be located within the range of 30 seconds to 120 seconds.

Manual-Detection-Time: Time interval (in seconds) of ECG signal, around the time of notification, at which information is stored as associated with cardiac episode. The value of this time is rather longer than the time of detected events automatically, and may be within the range of 300 seconds to 600 seconds, or even higher.

3.1.3 Detection Parameters

The wearable front-end has several operating parameters to optimize the sensitivity and specificity of the detection of heart rate, based on the morphological characteristics of the patient's ECG signal. Configurable parameters intrinsic to the detection algorithm used by the device are shown in Table 3.

Table 3 Detection Parameters of Heart Rate

Parameter Reference	Description
SearchB	Blanking window sensitivity
T-Wave-Time	Blanking window span

The following describes the usefulness of each of these parameters associated with the automatic detection algorithm.

SearchB (Sensitivity): These are parameters whose setting ensures detection of all QRS complexes and prevents sub-detecting them.

T-Wave-Time: Time interval (in milliseconds) for disqualification of the algorithm once QRS complex in the ECG signal is detected during which the algorithm suspends the process of searching for new waves R. The setting of this time interval avoids situations on - as produced when detecting the T wave amplitude and has morphology similar to the R-wave, that is, to increase the specificity of detection algorithm heartbeat.

3.1.4 Operational Mode

The use case HOLTIN proposes two additional operating modes to the X73-10406 specialization. The current operation mode only allows data to be sent continuously from the agent to the manager. The two additional operating modes for this specialization are shown in Table 4.

Table 4 New modes of operation proposed

Operational mode	Span	Connected to manager	Real Time	Supported by current specs
Real time	Whole ECG	Always	Yes	Yes
Event driven	Cardiac events	When needed	No	No
Classic holter	Whole ECG	Periodically (by configuration)	No	No

The first case that is observed in the table is already in the current specialization. The second automatic-event uses an events detection algorithm that enables the transmission of a data set stored in the agent. The first part is made up of data stored for a period prior to the event and the second part consists of a period of data acquired from the detection of the event.

The following operating mode (Continuous off-line) sends a piece of data stored on the device for every certain period of time.

3.2 Medication Monitor

This use case is derived from specialization [34]. This is an advanced PHD specialization, consisting of a device with a carousel loaded with medication. It is programmed to turn at specific times to make the medication available to the patient. At these times it flashes and/or beeps to remind them to take medication. On tilting to remove the medication, it generates an event. If the medication is not taken after a predefined interval, it also generates an event.

3.2.1 Medication Dosage Parameters

For this specialization, this use case proposes to modify the dosage to the user, using the remote control service.

Table 5 Medication dosage parameters

Parameter Reference	Supported by current specs	Description
Event-Dosage-Number	Read Only	Dosage Number
Time-Dosage-Available	No	Start and end time service
Alarm-Duration	No	Start and end alarm

The parameters within the use case to be used for remote configuration by remote control service are shown in Table 5.

Event-Dosage-Number: Sets the number of doses available to the user per day.

Time-Dosage-Available: These alert the patient about the disposition of the drug (Start and end times).

Alarm-Duration: This parameter sets the duration of the alarm to indicate the start of any event.

It is important to mention that there may be several alarm types to indicate the time in which the drug is available, as well as the fact that there may also be different numbers of doses for different drugs and different alarms. Therefore, this PHD only requires one mode of operation, since no parameters that would enable a different mode of operation are present.

3.3 Sleep Apnea Breathing Therapy Equipment (SABTE)

The SABTE is a new use of case which is used for the treatment of various types of sleep-related diseases by the application of continuous positive airway pressure to the patient's respiratory tract during the night. This device has the ability to apply different types of therapy – e.g. Continuous Positive Airway Pressure (CPAP), Auto-CPAP and Bi-Level Positive Airway Pressure (Bi-Level PAP).

3.3.1 Therapy Parameter

This use case proposes the implementation of different modes of operation, and within the application mode of therapy, we propose the application of four therapies.

The proposed therapies and the parameters necessary for the implementation of the same are shown in Table 6.

Table 6 Therapys Selection

Parameter Reference	**Supported by current specs**	**Description**
Therapy-Mode	W/R	Therapy-Mode
CPAP / hPa, PAP Waveform / %.	W/R	CPAP
Min APAP/hPa, Max APAP/hPa, PAP Waveform.	W/R	Auto-CPAP
IPAP / hPa EPAP / hPa Inspiration Trigger Sensitivity / % or Self-Adjusting Expiration Trigger Sensitivity / % or Self-Adjusting Inspiration Rise Time / msec or Self-Adjusting I:E Ratio / % or Self-Adjusting Minimum Respiratory Frequency / bpm or Self-Adjusting	W/R	Bi-Level-PAP
Min IPAP / hPa , Max IPAP / hPa, Min EPAP / hPa , Max EPAP /hPa	W/R	Auto-Bi-Level-PAP

CPAP: This is defined as the application of continuous positive air pressure, and it is the original type of breathing therapy which delivers air at a constant pressure, whether you are inhaling or exhaling.

Bi Level PAP: It delivers a parameter preset to inspiratory positive airway pressure (IPAP) and expiratory positive airway pressure (EPAP).

Auto Bi Level PAP: At this mode of operation, the agent features air that is delivered automatically and adjusts depending on the user's breathing patterns.

3.3.2 Operational Mode

This use case proposes the implementation of different modes of operation, and within the application mode of therapy, we propose the application of four therapies. The proposed therapies and the parameters necessary for the implementation of the same are shown in Table 7.

Table 7 Operational Modes

Operational mode	Connected to manager	Supported by current specs
Standby	When needed	Yes
Therapy	Always	Yes
Drying	Periodically (by configuration)	Yes
Mask Test	Periodically (by configuration)	Yes

These PHDs have four modes of operation which are detailed below:
Standby: This mode of operation places the device in standby.
Therapy: This enables the therapeutic mode in the agent.
Drying: This enables the drying mode in the agent.
Mask Test: This configures the agent to operate in a test mode.

4 Research Benefits of PHD Interoperability

By implementing the standards defined in the group X73PHD, we can be sure that significant benefits to consumers and manufacturers will emerge without the need of a complex analysis. The integration of different experiences from several manufacturers will facilitate the development of new devices in less time, with a significant decrease in the production cost. In addition, those integrations will lower the product purchase price for consumers. More benefits like this could be obtained immediately by adopting this set of rules. In the following sections, a description of the advantages, benefits and impact obtained from this standard will be exposed.

4.1 Benefits and Drawbacks (One Way Specializations)

The main purpose of telemonitoring and home hospitalization is to integrate a network with PHDs, which allow for monitoring of the evolution of a disease, the health or the welfare of a user. This has been achieved largely through the development of the standard X73PHD. But what was the problem that required the need to develop a standard for interoperability for personal health environments? The most representative features that led to the development of this standard are shown in Table 8.

Table 8 Benefits and drawbacks in home telemonitoring

	Non Standard	Standard
Solution	Ad-Hoc	Shared
Costs	Expensive home telemonitoring (High number of devices)	Lower cost (integration of different PHDs manufacturers)
Scalability	Hard	Easy
Plug&Play	Limited	Granted
	Higher requirements, largest number of devices	Selection of devices according to user needs
Competency	Close	Open

4.1.1 Proprietary Solutions Compared to the Interoperability Standard - X73PHD

Before the development of a standard, proprietary solutions provided usually just services, which integrated agents and managers to monitor specific physiological variables. Consequently, if a new vital sign was required to be monitored, the purchase of new devices was also a requirement. This often led to other problems, such as low compatibility with Plug & Play existing technologies in PAN, when ad-hoc solutions were used. Therefore, the user could not reuse previous devices with devices from other manufacturers, thus losing the ability to scale their home telemonitoring system at low cost and making the implementation of this new type of mHealth service expensive.

The X73PHD standard for interoperability implements plug and play technology, allowing the user to select PHDs based on their needs, costs and future requirements, regardless of whether they were made by different manufacturers. It is noteworthy that the standard also facilitates the use of different transport technologies, such as Wi-Fi, Bluetooth, IrDA, or GPRS.

4.1.2 Advantages for Manufacturers and Users

The benefits obtained for the user after the implementation of the standard are diverse. For example, the wide variety of specializations are defined within the standard. This enables the following up of a considerably high number of diseases, given that such specializations cover most of the physiological variables used to monitor different diseases. This makes possible the efficient extension of health care services to the home.

Another of the most significant benefits for the user is the opportunity to continue with medical treatment from their own home independently. Thus, the health care professionals can monitor the evolution of the patient's condition remotely.

An important additional benefit for users is that they can use their personal devices – such as smart phones or PDAs – as manager devices. This empowers the user to manage the information stored at the agents, in order to gather feedback about the evolution of performance or some other statistics. Integration is a straightforward process for PHDs in BANs, to the extent that the patient makes use of devices which they handle in their daily lives.

Regarding manufacturers, integrating all experiences and synergies in developing a standard for PHDs produces a reduced time-to-market and, moreover, lower costs in the development thereof. In the same way that standardization accelerates the development of new devices, it also encourages competition among manufacturers, reducing monopolies and encouraging the development of interoperable devices with third party tools.

4.2 *Benefits and Drawbacks (Remote Control Use Cases)*

In the previous section, the benefits of implementing the new features of the new use cases in existing specializations were covered. But the three new use cases presented above outline the functional characteristics that could benefit a programming service and remote control from the manager to the agent. The advantages of implementing a service of programming and remote control alongside the X73-20601 for the aforementioned new use cases – HOLTIN, SABTE and medication monitor – is presented as follows.

4.2.1 HOLTIN

This use case proposes to use a service programming and remote control to set the values used for the implementation of automatic detection algorithm (limit of asystole, bradycardia, tachycardia and QRS number) and other parameters that adjust the sensitivity of the algorithm (sensitivity and disabling time detection algorithm).

The parameters associated with the storage time of the ECG signal for the detection of events manual and automatic will also benefit from this service. Having the capability of changing these parameters will allow health professionals to adjust the parameters of the PHD adequately to the current evolution of the disease, and will also allow monitoring of a person's physical performance remotely or even allow users to send a file of their physical performance feedback from the agent to a Smartphone which is operated as a manager.

4.2.2 SABTE

This use case has the highest amount of benefits, if a service of programming and remote control was implemented. The two main areas of specialization of this proposal, which contain the most important functional characteristics with respect to the medical service provided to the user and device operational capabilities, are as follows:

- Operational mode: This feature allows the operational status of the PHD to be changed. This agent has four operational modes: standby, therapy, drying and mask test. The Standby operational mode changes the status of the device to standby. Therapy is used to treat a patient's condition, Drying and Mask-test active a test mode to verify proper operation of the device.

- Therapy Mode. There are four modes of therapy that the device can apply: C-PAP, Auto-CPAP, Bi Level PAP and Automatic Bi Level PAP. The professional health care can modify the implemented therapy from the evolution in the treatment of apnea which the patient is following.

4.2.3 Medication Monitor

The medication monitor is used to provide the necessary medication to treat a condition. It is expected to use the service configuration and remote control to set the number of doses that will be supplied to a patient as well as the times in which the drug will be available to the user (dosage number, start and end time service, and start and end alarm). Then, from the analysis of the supply of the drug, the treatment implemented could be modified, along with the number of doses, intervals in which were supplied and the times in which the drug will be available to the patient.

5 Discussion

By analyzing new uses, proposed in the PHDWG, it was observed that monitoring devices require specific parameters which can set limits in the data acquisition. Similarly, the devices used in applied therapy require a set of parameters to establish limits in their implementation. Moreover, in the case of devices that have more than one way of operation, it is important to consider parameters that allow the selection of those modes.. Afterwards, it was observed that current specializations share requirements posed by new cases. For this reason, the parameters required for monitoring and therapeutic devices can be grouped in different sets representing the operation parameters necessary for the PHDs operation, while the parameters required to select different modes of operation can be integrated in a single set of parameters. This is very noteworthy because it allows the development of a configuration and remote control to provide a service to existing specializations and develop them in the future with the X73PHD.

In the following subsections, the requirements of each use case are set out. The main objective is to analyze whether specializations meet the requirements necessary to implement these new functional characteristics, or whether it is necessary to implement new attributes that support these requirements.

5.1 Requirements for New Use Cases

The required set of parameters for each use case is obtained by analyzing the parameters required by the agent to perform its primary functions. Regardless of whether the PHDs are intended for monitoring or therapeutic purposes, certain parameters are required that allow them to set operational limits to work. HOLTIN and SABTE use cases exhibit different ways of operation but, in both cases, additional parameters are required to allow the selection of the desired mode of operation. In the next section, a brief description of the required parameters for each use case to perform all the implemented functions is given. They are exposed for each use case according to the required data, the function for which they are intended, the operating parameter with which they are associated and, finally, an example of its full application.

5.1.1 HOLTIN

For this use case to be implemented, a number of parameters related to the automatic detection algorithm and manage storage capacity are required. The necessary parameters suggested to be included in ISO/IEEE 11073-20601 in order to support the new features proposed by the use cases are shown in Table 9.

Table 9 Requirements for use case HOLTIN

Type	Category	Associated to	Examples
Numeric range with accuracy	Automatic Detection Algorithm Parameters	Asist-Lim	1.0-3.0 sec
		Brady-Lim	30-100 bmp
		Taqui-Lim	100-200 bmp
		QRS-Number	1-5 complexes
	Storage Parameters	Manual-Event-Store-Time	1-65.535 sec
		Automatic-Event-Store-Time	
		ECG-Recorded-Span	
Numeric	Sensitivity	SearchB	0 - 100%
Set Value			
Select item (from list)	Operational Mode Parameters	Operational Mode	Online-Continuous Offline-Continuous Automatic-Events
RealTime-Sample Array	T-Wave-Time		
Relative Time			

5.1.2 Medication Monitor

Although this use case requires a smaller number of parameters, some new parameters are still required. These include parameters for the number of doses, time intervals for drug delivery and alarms that indicate the disposition thereof. The necessary parameters within the ISO / IEEE 11073-20601, to implement the features of the new use case within the specialization 11073-10472 Medication Monitor are shown on table 10.

Table 10 Medication monitor parameters

Type	Associated to	Examples
Numeric Set Range	Time-Dosage-Available Alarm-Duration	10 minutes 1 minute
Numeric Set Value	Event-Dosage-Number	24 dairy

5.1.3 SABTE

The following use case contains a rich amount of data that can benefit from the service configuration and remote control. Therefore the following table sets forth the most important parameters that benefit the user in terms of medical treatment.

The parameters necessary for implementing different therapies are shown in Table 11.

Table 11 SABTE parameters

Type	Category	Associated to	Examples
Select item (from list)	Therapy Selection	Therapy-Selection	CPAP Auto-CPAP Bi-Level-PAP Auto-Bi-Level-PAP
	Operational Mode	Operational-mode	Standby Therapy Drying Mask-Test
Numeric Set Range	CPAP Therapy Parameter	CPAP / hPa, PAP Waveform / %	4cm to 20cm H20
	Auto-CPAP Therapy	Min APAP/hPa, Max APAP/hPa, PAP Waveform / %	4cm to 25 cm H20
	Bi-Level-PAP Therapy	IPAP / hPa, EPAP / hPa, Inspiration Trigger Sensitivity, Expiration Trigger Sensitivity, Inspiration Rise Time msec, Minimum Respiratory Frequency, % or Self-Adjusting	4cm to 25cm H2O
	Auto-Bi-Level-PAP	Min IPAP / hPa , Max IPAP / hPa, Min EPAP / hPa , Max EPAP /hPa	

6 Command and Control Approaches

The current state of the X73PHD allows the exchange of preset data between agents and managers, as it is defined within the Optimized Exchange Protocol (11073-20601). However, the constant development of PHDs encourages the need for new functionalities, as discussed in relation to the use cases described above. It has been explained that there is a lack of a remote control service that would allow modify different parameters in the agents from the manager. In order to develop this service, within PHD-WG two alternatives have been proposed. First, a specialization that includes remote service is proposed. The second approach has been proposed reusing some features extracted from 11073-10201 to integrate them into the Optimized Exchange Protocol 11073-20601. In the following sections, the most significant features of these two proposals are presented.

6.1 11073 – 20301 Optional Package – Remote Control

The PHD-WG first approach to implement remote control in ISO/IEEE11073 was to draft part 11073-20301 (Optional Package Remote Control). This proposal basically consists of using the Control Package Object within 11073-10201, as a new specialization. It was a feasible alternative at that time, because the requirements of the specializations were minor. However, the requirements of the new use cases pose more functionality that cannot be covered only with this package. Therefore, in this paper it is proposed to add the alert package and the set service, and include all of them in an optional package within ISO/IEEE 11073-20601.

6.2 11073 – 10201 Control, Alert Package Object and Set Service

The Control Package Model was originally designed for managing the remote control capabilities between medical devices at the point of care. The information model of the control package contains a number of specialized objects for several operations (see Figure 1).

The Alert Package Model is used to represent status information about the condition of the patient or the technical condition of the device caused by a measurement or device functioning. The model of the alert packet is shown on Figure 2, which contains an alarm-related object identified by the term alert. The alert object represents the state of the condition of a simple alarm which may come from the measurement of a variable physiological or technical condition of a medical device related. The alert status object represents the final state of a process alarm, which takes into account all the conditions present within the scope and type of alarm and also allows it to be determined whether the alarm is real or false. Finally the Alert Monitor object is used to represent the output of a device or system alarm processor and it provides a list of all alarm conditions of the system in terms of scope.

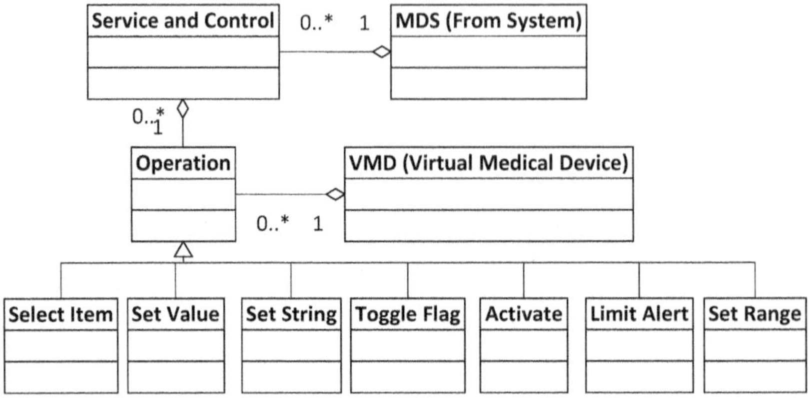

Fig. 1 Control package model
Source: 11073-10201 Point of Care, rev 2004, Objects in the Control Package.

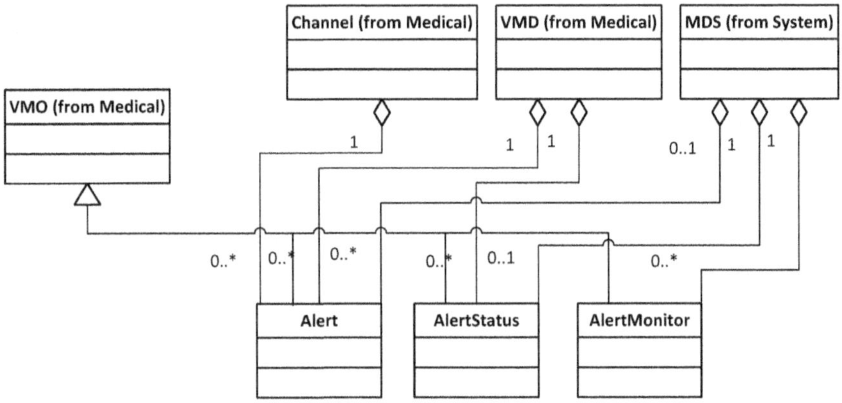

Fig. 2 Alert package models.
Source: 11073-10201 Point of Care, rev 2004, Object model of Alert package.

The Set Service allows modification of data attributes within object instances. This has two modes of operation: confirmed or unconfirmed. For the confirmed mode, a Set service call requires a response. When the manager invokes this service to modify one or several attributes of a selected object instance in an agent, the response must contain the attribute ID, a modify operator, and the attribute value (see Figure 3).

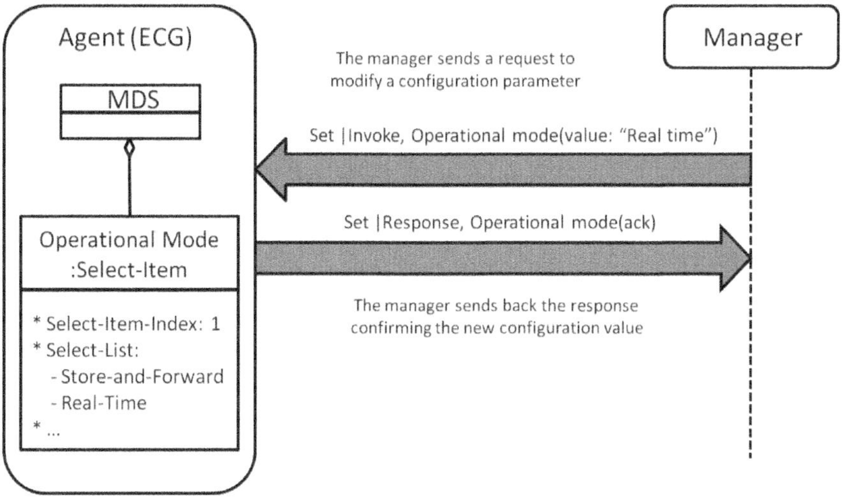

Fig. 3 Set Service

One alternative to implementing a service configuration and remote control within X73-20601 is to use the Control and Alert Package Object, which belongs to the standard X73-10201 PoC medical device communication, and the Set Service, which belongs to the standard X73-20601. The purpose of reusing these objects and services is to use tools that have been successfully implemented within their respective standards. These comply within the requirements proposed by the case studies analyzed. The Control Package Object can be used to modify the attributes of the instances of the objects implemented in the agents of the three proposed use cases. The Alert Package Object can be used to set or modify any type of alarm implemented by agents, such as those required in the SABTE and medication monitor specialization. Finally, the Set service is necessary to set the values of attributes of the agent's object using the manager device.

7 Future Lines

X73PHD and their constituent specializations are focused on three domains: disease management, health and fitness and independent living. These domains constantly integrate new transport technologies to provide the user with state-of-the-art devices. For the first case, disease management, the technologies used should comply with the requirements for using PHDs, whose operation can be managed remotely from a medical center. This must be accomplished using existing communication networks from the patient environment to the medical center. In the second domain, health and fitness, and the third domain, independent living, the standard should allow the user to use their own devices, like smart phones or PDAs. Therefore, potential new lines of research include the examination of the behavior of managers developed within the user's smart phones as well as the wireless technologies that they use.

Additionally, new proposals for specialization within X73PHD – such as the insulin pump, the Continuous Glucose Monitor (CGM), the insulin pump and the urine analyzer – can be further analyzed for possible integration in mHealth systems.

8 Conclusion

The X73PHD standard provides a feasible solution for integrating interoperability standards in mHealth solutions. The existing X73PHD specializations, along with the hardware capabilities and energetic performance of the smart phones, facilitate the design, development and deployment of state-of-the-art mobile healthcare systems in compliance with international norms.

New use cases – i.e. X73-10472 (medication monitor), X73-10406 (basic ECG) and X73-10424 (SABTE) – can be used to illustrate how forthcoming specializations – i.e. X73-20301 (remote command and control) – shall be designed and which new features shall be included. These new features may be grouped into two groups: the first group refers to characteristics that can modify (a) the therapy applied by the devices, which is to vary the medication, as in the case of the medication monitor, or (b) the way the devices apply a physical therapy to the patient, as in the case of equipment for treating apnea. The second group includes the modification of parameters that allow changing the operational mode, i.e. the ability to apply therapies, the monitoring mode, or methods of testing devices. As a result of the analysis presented here, it is proven that mHealth solutions can benefit from standardization, but further specializations and features – such as remote command and control – are required for fully exploiting the inclusion of interoperability standards in mHealth.

References

[1] Poon, C., Gu, W., Zhang, Y.: Health informatics for low-cost and high-quality health care. In: 2010 Annual International Conference of the IEEE Engineering in Medicine and Biology Society, EMBC 2010, pp. 1755–1758 (2010)

[2] Commission, E.: Mandate M/403: Standardization mandate addressed to CEN, CENELEC and ATSI in the field of information and communication technologies (2007) (last visit: September 2013)

[3] Kun, L.: Interoperability: The cure for what ails us (government affairs). IEEE Engineering in Medicine and Biology Magazine 26(1), 87–90 (2007)

[4] International Organization for Standardization, http://www.iso.org (last visit: September 2013)

[5] Institute of Electrical and Electronics Engineers, http://standards.ieee.org/ (last visit September 2012)

[6] GSM Association, Universal Access - How Mobile can Bring Communications to All, http://www.gsma.com/publicpolicy/wp-content/uploads/2012/03/universalaccessfullreport.pdf (last visit: September 2013)
[7] Russell, A.L.: OSI: The Internet That Wasn't. IEEE Spectrum (2013), http://spectrum.ieee.org/computing/networks/osi-the-internet-that-wasnt
[8] Gardner, R., Hawley, W., East, T., Oniki, T., Young, H.-F.: Real time data acquisition: Recommendations for the Medical Information Bus (MIB). International Journal of Clinical Monitoring and Computing 8(4), 251–258 (1991)
[9] ENV13734 - Health informatics. Vital signs information representation (VITAL), European Committee for Standardization, European Standard (1999)
[10] ENV13735 - Interoperability of patient-connected medical devices (INTERMED), European Committee for Standardization, European Standard (1999)
[11] ISO/IEEE health informatics - point-of-care medical device communication – part 10101: Nomenclature, ISO/IEEE 11073-10101:2004(E) (2004)
[12] ISO/IEEE health informatics - point-of-care medical device communication – part 10201: Domain information model, ISO/IEEE 11073-10201:2004(E) (2004)
[13] ISO/IEEE health informatics - point-of-care medical device communication – part 20301: Application profile - optional package, Remote Control ISO/IEEE 11073-10201:2004(E) (2004)
[14] Taboada, J., Arcay, B., Arias, J.: Real time monitoring and analysis via the medical information bus, Part I. Medical and Biological Engineering and Computing 35(5), 528–534 (1997)
[15] Anagnostaki, A., Pavlopoulos, S., Kyriakou, E., Koutsouris, D.: A novel codification scheme based on the VITAL and DICOM standards for telemedicine applications. IEEE Transactions on Biomedical Engineering 49(12), 1399–1411 (2002)
[16] Clarke, M.: A reference architecture for telemonitoring. Studies in Health Technology and Informatics 103, 381–384 (2004)
[17] Yao, J., Warren, S.: Applying the ISO/IEEE 11073 standards to wearable home health monitoring systems. Journal of Clinical Monitoring and Computing 19(6), 427–436 (2005)
[18] Garguilo, J., Martinez, S., Rivello, R., Cherkaoui, M.: Moving toward semantic interoperability of medical devices. In: Proceedings - 2007 Joint Workshop on High Confidence Medical Devices, Software, and Systems and Medical Device Plug-and-Play Interoperability, HCMDSS/MDPnP (2007)
[19] Martinez, I., Fernandez, J., Galarraga, M., Serrano, L., de Toledo, P., Jimenez-Fernandez, S., Led, S., Martinez-Espronceda, M., Garcia, J.: Implementation of an end-to-end standard-based patient monitoring solution. IET Communications 2(2), 181–191 (2008)
[20] Stylianides, N., Dikaiakos, M., Gjermundrd, H., Panayi, G., Kyprianou, T.: Intensive care window: Real-time monitoring and analysis in the intensive care environment. IEEE Transactions on Information Technology in Biomedicine 15(1), 26–32 (2011)
[21] Carroll, R., Cnossen, R., Schnell, M., Simons, D.: Continua: An interoperable personal healthcare ecosystem. IEEE Pervasive Computing 6(4), 90–94 (2007)
[22] ISO/IEC/IEEE Health informatics - Personal health device communication - Part 20601: Application profile - Optimized exchange protocol, ISO/IEEE 11073-20601:2010(E), pp. 1–208 (January 2010)

[23] ISO/IEEE Health informatics - Personal health device communication – Device specializations, ISO/IEEE 11073-10400
[24] Egner, A., Moldoveanu, F., Goga, N., Moldoveanu, A., Asavei, V.: Enhanced communication protocol for ISO/IEEE 11073-20601. UPB Scientific Bulletin, Series C: Electrical Engineering 75(2), 3–16 (2013)
[25] Martinez, I., Escayola, J., Martinez-Espronceda, M., Muñoz, P., Trigo, J., Muñoz, A., Led, S., Serrano, L., Garcia, J.: Seamless integration of ISO/IEEE 11073 personal health devices and ISO/EN13606 electronic health records into an end-to-end interoperable solution. Telemedicine Journal and E-Health: The Social Journal of the American Telemedicine Association 16(10), 993–1004 (2010)
[26] Trigo, J., Chiarugi, F., Alesanco, A., Martinez-Espronceda, M., Serrano, L., Chronaki, C., Escayola, J., Martinez, I., Garcia, J.: Interoperability in digital electrocardiography: Harmonization of ISO/IEEE X73-PHD and SCP-ECG. IEEE Transactions on Information Technology in Biomedicine 14(6), 1303–1317 (2010)
[27] Carot-Nemesio, S., Santos-Cadenas, J., De-Las-Heras-Quirs, P., Bustos, J.: Openhealth the openhealth OSS implementation of the ISO/IEEE 11073-20601 standard. In: Proceedings of the HEALTHINF 2010 - 3rd International Conference on Health Informatics, pp. 505–511 (2010)
[28] Martinez-Espronceda, M., Martinez, I., Serrano, L., Led, S., Trigo, J., Marzo, A., Escayola, J., Garcia, J.: Implementation methodology for interoperable personal health devices with low-voltage low-power constraints. IEEE Transactions on Information Technology in Biomedicine 15(3), 398–408 (2011)
[29] Ji, Z., Zhang, X., Ganchev, I., O'Droma, M.: A personalized middleware for ubiquitous mHealth services. In: 2012 IEEE 14th International Conference on E-Health Networking, Applications and Services, Healthcom 2012, art. no. 6379465, pp. 474–476 (2012)
[30] Norgall, T., Wichert, R.: Towards interoperability and integration of Personal Health and AAL ecosystems. Studies in Health Technology and Informatics 177, 272–282 (2012)
[31] Lasierra, N., Alesanco, Á., García, J.: An SNMP-based solution to enable remote ISO/IEEE 11073 technical management (2012)
[32] Clarke, M., De Folter, J., Palmer, C., Verma, V.: Building point of care health technologies on the IEEE 11073 health device standards. In: IEEE EMBS Special Topic Conference on Point-of-Care (POC) Healthcare Technologies: Synergy Towards Better Global Healthcare, PHT 2013, art. no. 6461298, pp. 117–119 (2013)
[33] ISO/IEEE health informatics - point-of-care medical device communication – part 10406: Basic Electrocardiograph," ISO/IEEE 11073-10406:2011(E) (2011)
[34] ISO/IEEE health informatics - point-of-care medical device communication – part 10472: Medication Monitor. ISO/IEEE 11073-10472:2010(E) (2010)

Questions

Q1. What is X73PHD?

Q2. What are the use cases and how are they obtained?

Q3. How works the communication between Agent and Manager using X73PHD?

Q4. What is the application environment where the standard is used?

Q5. Is there another area of application for PHDs different to medical area?

Q6. Does it ensure the validity of the standard to the constant technological changes?

Q7. What is command and control? Is it possible to add remote control functions to standard X73PHD?

Q8. What is the objective of this work?

Part IV

Framework, Architecture, Software, and Hardware Systems

Chapter 18: Design Guidelines for Wireless Sensor Network Architectures in mHealth Mobile Patient Monitoring Scenarios

Chapter 19: Evaluation of Health Care System Model Based on Collaborative Algorithms

Chapter 20: Software Architecture for Emergency Remote Pre-hospital Assistance Systems

Chapter 21: mHealth Portable Systems and Platforms

Chapter 22: A Product Line Architecture for Mobile Patient Monitoring System

Chapter 23: The Design of Integrated Circuit for Biomedical and mHealth Application

18. Design Guidelines for Wireless Sensor Network Architectures in mHealth Mobile Patient Monitoring Scenarios

Manuel Casillas, Salvador Villarreal-Reyes*, Ana Lilia González, Edwin Martinez, and Aldo Perez-Ramos

CICESE Research Center, Electronics and Telecommunications,
Km. 107 Carretera Tijuana-Ensenada, 22860, Ensenada, México
{mcasilla,svillar,anagonza,emartine,aramos}@cicese.edu.mx

Abstract. Mobile patient monitoring has been widely recognized as a viable solution to improve the quality of patient care and decrease health care costs. Depending on the mobile patient monitoring scenario and the patient pathology, it may be necessary to monitor different bio-physiological variables by means of specialized sensors. The information collected by these sensors is sent to the healthcare provider for its analysis, processing, and storage. Thus, the collected data can be used for remote diagnosis, ambulatory patient monitoring and to trigger emergency services. In mobile patient monitoring systems it is common to use wireless network technologies to transmit the data generated by the sensors, thus defining a wireless sensor network (WSN). The implementation of the WSN is not a trivial task, as different application scenarios have different requirements in terms of the number of variables to monitor, the number of supported users, mobility, data loss tolerance, power consumption, and coverage range, among others. All these requirements must be considered when designing the WSN architecture to be used in a particular mobile patient monitoring system.

There are several network architectures and wireless technologies that can be used for the implementation of WSNs in mobile patient monitoring scenarios. Thus, the main wireless technologies that can be used for this purpose are reviewed within this chapter. Then, the advantages and disadvantages offered by homogeneous and heterogeneous WSN architectures are discussed and analyzed by comparing performance metrics for a particular application scenario. Lastly, several guidelines for the proper design of WSN architectures in mHealth mobile patient monitoring scenarios are provided.

Keywords: Mobile patient monitoring, WBAN, WPAN, WLAN, WSN, homogeneous WSN architecture, heterogeneous WSN architecture, network guidelines.

---

* Corresponding author.

S. Adibi (ed.), *Mobile Health*,
Springer Series in Bio-/Neuroinformatics 5, DOI: 10.1007/978-3-319-12817-7_18

1 Introduction to mHealth Mobile Patient Monitoring Architectures

Remote patient monitoring (or telemonitoring), enabled by information and communication technologies (ICTs), is widely recognized as an effective tool to address current and future challenges related to the provision of healthcare services (Chen et al., 2010) (Sneha & Varshney, 2013). For example, home patient monitoring may help to increase the healthcare systems capacity by reducing the hospitals bed occupancy rate, (Goldberg & Wickramasinghe, 2003). As well, it can be used as an effective tool for the assessment of the health condition and wellbeing of elderly people (Chen et al., 2012) (Teng et al., 2008). Because of its potential impact in the provision of healthcare services by means of new paradigms, remote patient monitoring has attracted considerable attention from the industrial and academic communities. If wireless network technologies are used in remote monitoring, such that patients can freely move within a predefined coverage area, then remote patient monitoring is commonly referred to as mobile patient monitoring.

Depending on the application and patient pathology, it may be necessary to monitor different kinds of variables by means of specialized sensors placed in different parts of the human body (Kailas & Ingram, 2009). Once the data is collected, it must be sent to the healthcare provider for its analysis, processing, and storage. Based on the application scenario, the collected data can be used for different purposes such as in-hospital monitoring, remote diagnosis, ambulatory patient monitoring or to trigger emergency services. Thus, depending on the application scenario, retransmitting the data through one or several networks by means of different wired or wireless technologies may be required.

The current trend in mHealth is to use wireless transceivers to interconnect the sensors that monitor the bio-physiological variables (Egbogah & Fapojuwo, 2011) (Ko et al., 2010), thus forming a wireless sensor network (WSN). Depending on the specific transmission range and the technology used, the sensors will form a wireless body area network (WBAN), wireless personal area network (WPAN) or even a wireless local area network (WLAN). As the collected data must be sent to the healthcare provider, a device capable of retransmitting the WBAN/WPAN information through a wide area network (WAN), a metropolitan area network (MAN) or a local area network (LAN – e.g. a hospital or home internal network connected to the IP cloud) is usually needed. This device may also be required to process and store the collected data, and will be referred to as the "patient information sink" within this chapter. The patient information sink may be a smartphone, a computer, a laptop, a tablet with 3G or 4G connectivity, an access point, a gateway or another similar device.

Fig. 1 shows the generic mobile patient monitoring architecture model assumed throughout this chapter. The wireless communication between the patient's sensors and the information sink is not restricted to be performed in a single wireless hop. In fact, depending on the wireless transceivers range, the use of more than one wireless hop may be required to reach the information sink. Once the collected data reaches the patient information sink, it must be sent to the healthcare provider for further processing and analysis. Note that the healthcare provider may be located in the same building or campus as the patient (e.g. mobile patient monitoring within a hospital or a nursing home), in a remote location (e.g. telehomecare scenarios) or both.

Retransmissions of the collected data within the same building or campus may be performed using wired or wireless LANs (e.g. Ethernet or WiFi). On the other hand, communication with remote locations normally requires the use of WANs or MANs which provide direct (e.g. WiMAX) or IP cloud (e.g. ADSL, cellular) connectivity to the healthcare provider facilities.

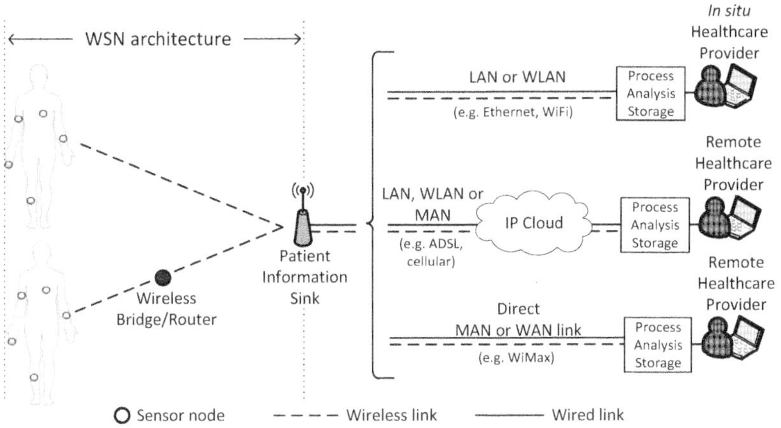

Fig. 1 Generic mobile patient monitoring architecture used in this chapter

As previously mentioned, a current trend in mHealth is to use WSNs to transmit the patient's data towards the information sink. Thus, this chapter provides an overview of the different wireless network technologies that can be used to implement such WSNs. Technical specifications of technologies such as Bluetooth, IEEE 802.15.4/ZigBee, Bluetooth Low Energy, IEEE 802.15.6 and some proprietary technologies (e.g., ANT/ANT+, Sensium, Z-Wave, RuBee, Zarlink) are discussed within this chapter, this in order to evaluate its suitability to enable the sensors wireless connectivity. Additionally, a comparative study of homogeneous and heterogeneous WSN architectures based on analytical and simulated results is presented. Finally, guidelines for the design of WSN architectures in mobile patient monitoring scenarios are provided.

2 Wireless Sensor Network Technologies and Architectures in Mobile Patient Monitoring Scenarios

The main requirements for the deployment of WSNs in mobile patient monitoring scenarios are: low complexity, low power consumption, robustness, and medium or low data rates. Technologies able to cover such requirements are classified within the WPAN and WBAN categories (because of their transmission ranges). It has become a common practice to use WPAN solutions to form WBANs by reducing the transmission range of the wireless transceivers. However, there are several specific challenges that must be solved for the effective deployment of WBANs. Some of such challenges are: very low power consumption (required for the sensors to operate for a long time), miniaturization (to facilitate user adoption), security, and low costs

(Penders et al., 2008). It is noteworthy that WLANs solutions like IEEE 802.11/WiFi have recently gained popularity for the deployment of WSNs. The reasons for this growing acceptance lie in the advances achieved in the reduction of power consumption on WiFi transceivers (Li et al., 2011). However, WPAN technologies such as Bluetooth and ZigBee still offer lower power consumptions than WiFi.

To properly design a WSN solution for mobile patient monitoring systems, it is necessary to define the application scenario based on aspects such as: number of supported users, variables to be monitored, data loss tolerance, security, power consumption, and coverage range. These aspects must be considered when defining the WSN architecture to be used and the wireless network technology that will support it. Thus, it is necessary to know the capabilities offered by different wireless technologies in order to make an adequate selection for the WSN implementation. With this purpose, Section 2.1 introduces different WBAN and WPAN technologies that can be used for the implementation of WSN, while the study of different WSN architectures is addressed in Section 2.2.

2.1 Overview of the Main WBAN and WPAN Technologies

The main wireless technologies used in low power wireless sensors for healthcare applications can be grouped in two principal categories: open-standard technologies supported by scientific communities or alliances, and proprietary wireless technologies. Standards such as IEEE 802.15.4/Zigbee, Bluetooth, Bluetooth Low Energy, and IEEE 802.15.6 belong to the first category; while technologies such as ANT/ANT+, Sensium, RuBee, Zarlink, and Z-Wave belong to the second group.

Most of the open-standard and proprietary technologies transmit within the industrial, scientific and medical (ISM) radio bands, (Chen et al., 2011). The main reason for using these bands is because they are internationally available and most of them allow unlicensed operations. The downside of using the ISM bands is that unlicensed devices operating in them must tolerate any interference generated by other devices. Another problem is that some bands are becoming (or are already) crowded, e.g. the 2.4 GHz band. Thus, it is important to carefully choose the ISM band to use when designing a mobile patient monitoring system. For example, if the 2.4 GHz band is already too crowded it would be wise to use a technology transmitting in other ISM bands like the 915 MHz or the 5.8 GHz bands.

Before providing an overview of the main open-standard and proprietary WBAN/WPAN technologies, it is important to introduce the concept of network topology, as the supported network topology (or topologies) may decide the use of a particular technology within a specific WSN deployment.

2.1.1 Network Topologies

One important characteristic on every network is the topology, which determines how the different network nodes (e.g. the sensors) are arranged, how they communicate with each other, and the way the data flows within the network.

The selection of the network topology is an important step in the design of the WSN architecture to be used in a particular mobile patient monitoring solution.

Several factors such as: energy consumption, cost, complexity, mobility, and robustness must be considered when performing the selection. Table 1 presents the most common topologies used in WBANs and WPANs implementations altogether with their main characteristics, advantages, and disadvantages.

Table 1 Common topologies for WBAN/WPAN networks

Topology	Characteristics	Advantage	Disadvantage
Point-to-Point	Consists of a pair of endpoints directly linked with each other.	Simplicity	Consists of only two nodes
Star	Each node is connected to a unique central controller. All traffic passes through the central controller.	Simplicity Centralized system	Dedicated central node Single point of failure Poor scalability Asymmetric power consumption
Mesh	Any node can communicate with others within its range.	Distributed processing Fault tolerant Scalable Energy can be balanced	Complexity of routing Nodes require processing capacities
Tree	Based on a hierarchy of nodes. The first node becomes the root. Additional nodes can join in as branches.	Low power consumption on leafs Many nodes possible High complexity	Medium scalability Low reliability Asymmetric power consumption

It is noteworthy that although each topology presents its own advantages and disadvantages, some characteristics like energy consumption, cost, and robustness will also depend on the particular technology used to deploy the network.

Once the network topology concept has been introduced, the next sections present a brief overview of the main open-standard and proprietary WBAN/WPAN technologies commonly used in mHealth applications.

2.1.2 Overview of Open-Standard WBAN/WPAN Technologies

As previously mentioned, the main open standard WBAN/WPAN technologies used in mHealth applications are IEEE 802.15.4/Zigbee, Bluetooth, Bluetooth Low Energy, and IEEE 802.15.6. Table 2 summarizes the main technical characteristics of these technologies, while a brief description of each technology is provided next.

IEEE 802.15.4/Zigbee

The IEEE 802.15.4 standard, first introduced in 2003, provides a specification for low rate WPANs, (IEEE Computer Society, 2006). It defines physical (PHY) and medium access control (MAC) layers for low data rate wireless connectivity. Because of its characteristics such as low power consumption, low complexity, and low cost, IEEE 802.15.4 technology has become an attractive alternative for WBAN/WPAN deployments. The supported network topologies can extend the coverage area of an

IEEE 802.15.4 network by means of multi-hop communications. Furthermore, linking IEEE 802.15.4 with other wireless technologies can be a powerful solution in mHealth mobile patient monitoring deployments (Vaidehi et al., 2013) (Boquete et al., 2012).

Table 2 Main technical characteristics of open-standard wireless sensors link technologies

Standard name	IEEE 802.15.4 / ZigBee	Bluetooth	Bluetooth Low Energy	IEEE 802.15.6
Frequency bands	868/915 MHz 2.4 GHz	2.4 GHz	2.4 GHz	402–958 MHz, 2360–2483 MHz 3.1 GHz – 10.6 GHz
Peak data rate	20,40,250 kbps	1,3,24 Mbps	1 Mbps	971.4 kbps (NB) 15.6 Mbps (UWB)
Modulation	BPSK,O-QPSK	GFSK, π/4DQPSK, 8DPSK	GFSK	GFSK, π/2DBPSK, π/4DQPSK
Application throughput (max)	151 kbps	723kbps (1.1) 2.1Mbps (2.0)	236 kbps	674.7 kbps (NB)
Number of nodes	65000	7+1	App limited	64
Coverage	10-100 m	1-100 m	1-10 m	< 3 m out-body
Medium access	CSMA-CA, GTS	TDMA	TDMA	CSMA-CA, slotted Aloha, EAP
Network Topology	Star, cluster-tree, mesh	Star, scatternet	Star	Star, multi-hop star
Security	128 bit AES-CCM	Shared key and AES-CCM	128 bit AES-CCM	128 bit AES-CCM
Peak Current	20 mA	40 mA	15 mA	3 mA (NB)
mHealth applications	Homecare (Tung, 2013) Multiparameter (Chen, 2012)	Patient monitoring (Yu., 2010) ECG monitor (Zhang, 2009)	Consumer electronics ECG monitor (Park, 2013)	In-body implants communications Patient monitoring (Jung, 2012)

ZigBee is a higher layer specification based on the PHY and MAC layers of IEEE 802.15.4. This specification defines several features such as: protocols and profiles for end-to-end compatibility; routing capabilities; more complex network topologies (e.g., mesh); and additional security services. It is important to note that ZigBee includes a health care profile ratified as a transport protocol for ISO/IEEE 11073 personal health data (PHD) devices. Furthermore, ZigBee has been widely used in mHealth remote monitoring applications (Tung et al., 2013) (Fraile et al., 2010) (Chen et al., 2012).

Bluetooth

Bluetooth is a wireless communications standard enabling information exchange between devices over WPANs (Bluetooth SIG, 2010). The protocol stack provides a complete end-to-end solution, from the PHY to the application layers. Similar to ZigBee, Bluetooth offers a health device profile (HDP) ratified as a transport protocol for ISO/IEEE 11073 PHD devices. Bluetooth is based on a master-slave structure with a typical star topology. Each slave is synchronized with the master clock with a maximum number of 7 slaves forming a piconet.

Although classic Bluetooth has been used for patient monitoring systems (Sharma et al., 2011) (Yu et al., 2010) (Zhang & Xiao, 2009), it is not considered to be a strong candidate for the development of mHealth applications because of its relatively high transmission power, high duty cycles and the limitation of having a maximum of 8 nodes per network.

Bluetooth Low Energy

Bluetooth Low Energy (BLE) is a set of protocols defined in the Bluetooth standard, and it is mainly focused on applications requiring low power consumption, low data rates, and short ranges. The main objective of BLE is to enable wireless transmission of information in scenarios such as healthcare, sports, security, and home entertainment (Bluetooth SIG, 2010). The low power consumption is mainly achieved by making use of very small duty cycles and smaller packet sizes than those used in classic Bluetooth. Different to classic Bluetooth, there is not a limit in the number of slaves supported by the master. However, the number of active slaves connected is limited by the transmission requirements of the network.

In the healthcare context BLE has standardized profiles and products for monitoring blood pressure, heart rate, body temperature, and cycling (Bluetooth SIG, 2013). Additionally, BLE has been used for transmitting other bio-physiological signals, such as electrocardiogram (ECG) (Park & Cho, 2013) (Yu et al., 2012). Because of its characteristics, BLE provides a strong solution for mHealth mobile patient monitoring deployments where multi-hop communication across the WSN is not required.

IEEE 802.15.6

The IEEE 802.15.6 standard provides a solution for monitoring systems that operate over, inside, or around the human body (IEEE Computer Society, 2012). This is the only standard designed for WBANs that specifically defines a human body communication (HBC) layer for medical implants. In order to obtain a better system performance by reducing mutual interference, IEEE 802.15.6 allows transmission over a wide range of frequency bands. (Martelli et al., 2011) (Ullah & Kwak, 2011). In IEEE 802.15.6, nodes are organized in WBANs sets with one-hop communications using a star network topology formed by only one hub (or coordinator) and a specific number of sensor nodes. The standard also defines seven user priorities for quality of service (QoS), which include different priority levels for medical and emergency data.

Because of IEEE 802.15.6 was specifically designed for WBANs, this standard has shown performance metrics adequate for the deployment of mobile patient monitoring applications (Jung et al., 2012). Remarkable characteristics that make IEEE 802.15.6 a strong solution for the development of mHealth applications are: low vulnerability to mutual interference, multiple frequency bands, user priorities for QoS, low error rates, and ultra-low power consumption.

2.1.3 Overview of Proprietary WBAN/WPAN Technologies

Although there are numerous low power wireless proprietary technologies, most of them were conceived as projects or are only used in research and development laboratories. This subsection covers some of the most suitable commercially-available technologies for healthcare monitoring: ANT/ANT+, Sensium, RuBee, Zarlink, and Z-Wave. Table 3 summarizes the main technical characteristics of these technologies, while a brief description of each technology is provided next.

Table 3 Main technical characteristics of proprietary wireless sensors link technologies

Standard name	ANT / ANT+	Sensium	RuBee	Zarlink	Z-wave
Frequency bands	2400- 2524 MHz	862-870, 902-940 MHz	131 kHz	402-405 MHz, 433-434 MHz	908.4, 868.4, 919.8 MHz
Peak data rate	1 Mbps	50 kbps	9.6 kbps	800 kbps	9.8, 40 kbps
Modulation	GFSK, π/2DBPSK, π/4DQPSK	GFSK, FSK	ASK, BPSK, BMC	2FSK, 4FSK	BFSK, GFSK
Application throughput (max)	20 kbps	Not available	~ 8 kbps	Not available	~ 30 kbps
Number of nodes	64000+1	8+1	2	2	232
Coverage	< 10 m	< 10 m	< 30 m	2 m (in-body)	30 – 100 m
Medium access	TDMA	TDMA	N/A	N/A	CSMA
Network Topology	Peer-to-peer, star, mesh, cluster tree	Star	Peer-to-peer	Peer-to-peer	Mesh
Security	128 bit AES-CCM	None	Public private key and AES	None	None
Peak Current	17 mA	3 mA	3 mA	5 mA	23 mA
mHealth applica-tions	Consumer electronics (Mahmood, 2011)	Multiparamete r monitor (Toumaz, 2010)	Patient record Homecare (Yu, 2011)	Med. implants Defibrillators (Chen, 2010)	Multiparemete r monitor (Csernath, 2008)

ANT/ANT+

ANT is a proprietary WSN protocol specifically designed to target ultra-low power networking applications (Dynastream, 2013) (Texas Instruments, 2011). The characteristics of the ANT logical connections (named "ANT channels") make possible the formation of complex topologies such as star, mesh, tree or shared cluster (Texas Instruments, 2011). Because of the small amount of information transmitted on each packet and the periodicity of the transmissions, ANT networks are mainly focused on low data rate applications.

ANT+ defines a set of profiles typically tied to a specific application scenario. These set of profiles are built on the top of the ANT protocol and ensure standardized communications among ANT+ branded devices. Because of the ultra-low power consumption and low data rate, ANT+ has been used in mHealth solutions that do not require the transmission of large amounts of information (Mahmood et al., 2011).

Additionally, there are several ANT+ profiles and products for health care, such as heart rate bands, watches, distance and speed monitors, weight scales, etc.

Sensium

Sensium is an on-body ultra-low power proprietary radio technology (Wong et al., 2009). It is mostly used in healthcare and medical areas to monitor variables such as temperature, ECG, heart rate, oxygen levels, and blood sugar levels. Similar to Bluetooth, this technology works in a star topology with a maximum of 8 slaves and 1 master. In this topology, all the sensor nodes are capable of transmitting multiple vital signs to a personal server which forwards the information to a health professional (Movassaghi et al., 2012). As a proprietary ultra-low power solution, Sensiumvitals (Toumaz, 2010) allows healthcare providers to monitor patients continuously, wirelessly, and at low cost.

RuBee

RuBee (IEEE 1902.1) is a low frequency inductive (magnetic) wireless protocol that can handle a two-way active communication (IEEE Communications Society, 2009). The main application is known as RuBee tags, which has lower power needs than other high frequency communications, making a battery last from 5 to 10 years (RuBee, 2009). The major disadvantage of RuBee is its low data rate, which makes it a slow technology compared to other WBAN/WPAN technologies. Nevertheless, RuBee can be used in mHealth applications related to electronic patient record, emergency telemedicine and home monitoring (Yu et al., 2011).

Zarlink

Zarlink is a company that has produced an ultra-low power wireless RF chip (ZL70101) that can be used as an implantable medical device (IMD) such as pacemakers, implantable cardioverter-defibrillators (ICDs), etc. Zarlink has been used in the world first swallowable camera capsule, which transmits two movie-quality images per second, allowing a more thorough and non-invasive examination of the gastrointestinal tract (Chen et al., 2010).

Z-Wave

Z-Wave is another proprietary low power sensor-based technology used in home automation and recently in the healthcare industry. Similar to Zigbee, this technology works with a mesh network topology. The cost associated with Z-Wave is typically lower compared to ZigBee or similar WBAN/WPAN technologies. Although Z-Wave has been used in remote monitoring systems (Csernath et al., 2008), its power consumption and low data rates make it less attractive for the development of health-related applications.

2.1.4 Comparison of WBAN/WPAN Technologies

From the previous overview of the main open-standard and proprietary WBAN/WPAN technologies and the information provided in Tables 2 and 3, it can be noticed that most of the proprietary standards offer a good trade-off between data rates and energy consumption. This is partly because they were specifically designed for health monitoring applications. Nevertheless, it is believed that the future of mHealth WSN technologies will be based on non–proprietary and open standard technologies (Adibi, 2012). For example, IEEE 802.15.4/ZigBee features are suitable for the deployment of health monitoring WBAN/WPANs, and the use of this technology in mobile patient monitoring scenarios has already been demonstrated (Benocci et al., 2009). It is noteworthy the emergence of power optimized standards like BLE and IEEE 802.15.6, whose aim is to provide better power consumption and data rates features for the deployment of mobile patient monitoring WBAN/WPANs. In fact, BLE is becoming a major player in the area of health monitoring, fitness and sports; partly because of the penetration of previous Bluetooth versions and the support from companies like Microsoft, Apple, Intel, and Ericsson.

The IEEE 802.15.6 standard released on 2012 was specifically developed for the deployment WBANs. This standard offers several advantages for the implementation of body area WSNs compared to previous WBAN/WPAN technologies. However, since it is a recent standard, there are still amendments to be made (based on the input from the scientific community) and the adoption from manufacturers is still ongoing. There are some non-technical factors for the success of WBAN technologies in the mass market such as: affordability, regulatory and ethical issues, comfort, and user friendliness (Chen et al., 2011). Also, in order to become a viable long-term technology, WBANs still need the widespread acceptance of key players in the mHealth ecosystem such as the medical-electronics industry, policy makers, patients, caregivers, and patient advocacy groups.

2.2 *Homogeneous and Heterogeneous Wireless Network Architectures*

When deploying wireless networks to provide mHealth services, the specific application requirements directly affect aspects like network topology, throughput, energy consumption, and the radio transceiver range. Thus, these requirements must be considered in the selection of the network technology or technologies that will be used in the WSN architecture. When a single network technology is used for the WSN deployment, then the network architecture will be referred to as "homogeneous wireless network architecture". Similarly, when more than one network technology is used to implement the WSN, then the architecture will be referred to as "heterogeneous wireless network architecture".

Several studies related to remote patient monitoring consider the use of homogeneous WSN architectures, since they use the same technology to transmit data from the sensors towards the information sink (González-Valenzuela et al., 2011) (Benocci et al., 2009) (Koh & Kong, 2006) (Sharma et al., 2011) (Zhang & Xiao,

2009). However, there are some proposals that consider the use of bridge devices to interconnect different wireless network technologies in remote patient monitoring scenarios, (Misic & Misic, 2010) (Ullah & Kwak, 2011), thus defining a heterogeneous WSN architecture. This section discusses the main characteristics of homogeneous and heterogeneous WSN architectures.

2.2.1 Homogeneous WSN Architectures

In homogeneous WSN architectures, each of the sensors and network nodes use the same technology to transmit the collected data towards the information sink. Three scenarios can be identified depending on the distance between the patient and the information sink: a) the patient and the information sink are in close proximity (e.g. the information sink is a mobile phone); b) the patient and the information sink are within reach separated by just a few meters; and c) the information sink is separated by several meters from the patient.

In the first scenario, the star topology is the most commonly implemented. As it is assumed that the information sink and the patient are in close proximity, it is convenient to adjust the transmit power of the transceivers to form a WBAN, or alternatively use a specifically designed WBAN technology, e.g. IEEE 802.15.6. In this case, patient mobility is achieved by moving the information sink with him/her as in the case of mobile phones (see Fig. 2a). An advantage is that if the information sink provides wireless WAN (WWAN) or wireless MAN (WMAN) connectivity, then the patient can freely move within the WWAN or WMAN coverage area. Nevertheless, note that this usually implies that the WSN and the information sink will provide connectivity for just one patient. Therefore, if more than one patient has to be monitored, then a WSN and information sink per patient must be used. Another advantage is that if a WBAN solution is implemented, then mutual interference from sensors worn by different patients should be minimal, unless patients are located within close proximity of each other. An issue that must be considered when implementing this kind of WSNs is the operation frequency band and the possible presence of transmissions from other wireless network technologies.

In the second scenario, a WSN with star topology is usually deployed using WPAN technologies (see Fig. 2b). In this scenario, patient mobility is limited by the transmission range of the transceivers. As in the previous scenario, it is necessary to properly plan frequency band usage such that the WPAN performance is not degraded by transmissions from other wireless technologies. The number of users that can be served by a particular WPAN technology depend on its PHY and MAC layer characteristics, as well as the number of sensors, nodes and data rates (defined by the requirements of the bio-physiological variables) that each user needs. For example, if a particular mobile patient monitoring scenario requires the use of 5 wireless sensors, then a Bluetooth piconet (see Section 2.1.3) will only be able to serve a single user, since the maximum number of nodes per Bluetooth piconet is 8 (1 master and 7 slaves). If the data rate generated by such nodes is low, then a single IEEE 802.15.4/ZigBee deployment may be able to serve more users. In contrast, if only a single variable will be monitored requiring transmission data rates above 64 kbps, then a single IEEE 802.15.4/ZigBee deployment working in

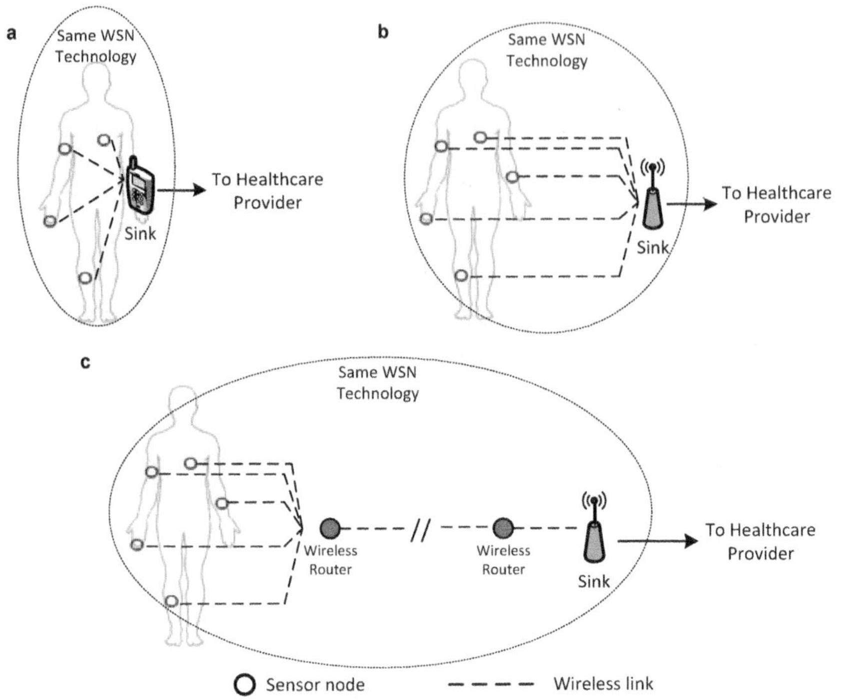

Fig. 2 Generic architecture for homogeneous WSN in mHealth scenarios. **a** - WBAN Star topology. **b** - WPAN Star topology. **c** - Mesh or cluster tree topology.

the 2.4 GHz ISM band will only be able to support one user, whereas a Bluetooth 2.0 piconet should be able to support more than one user.

In the third scenario, network topologies such as star, mesh or tree can be used for the WSN deployment depending on the separation between the patient and the information sink (Fig. 2c). A drawback of the star topology is that the transmit power of the sensor nodes may have to be increased in order to reach the information sink, increasing the node power consumption as consequence. On the other hand, with the mesh and tree topologies the WSN coverage area can be expanded without increasing the transmission range of each single transceiver. This is achieved by re-transmission of the data packets by some nodes, such that the collected data can reach the information sink through one or more wireless "hops". Although all these WSN topologies provide a good performance in low data rate scenarios, as the number of users and sensors increases the overall network performance may be degraded. For example, in an IEEE 802.15.4/ZigBee homogeneous WSN with tree topology, each one of the sensors sends the collected data towards the information sink (which may act as the IEEE 802.15.4/ZigBee network coordinator) through one or several wireless hops, e.g. (González-Valenzuela et al., 2011). If the sensors of all patients transmit over the same channel, then the contention for the channel will consist of both: transmissions from sensors collecting data, and transmissions from nodes forwarding data (towards the information sink) previously transmitted by other wireless nodes. As shown further in the chapter, this contention quickly degrades the network performance as the number of users is increased. Therefore this issue must be considered when designing a homogeneous WSN architecture for this kind of scenarios.

2.2.2 Heterogeneous WSN Architectures

As previously explained, in homogeneous WSN architectures the network performance may be degraded as the number of users is increased. This issue poses a scalability problem that could be solved with the use of heterogeneous WSN architectures. Implementation of these architectures requires an effective segmentation of the WSN and the use of bridge devices to interconnect different wireless network technologies. In heterogeneous WSN architectures, sensors transmit the collected data towards a bridge device located over or within close proximity of the patient's body. Thus the first WSN segment consists of a WBAN. This implies that unless patients are located within close proximity of each other, the mutual interference from sensors worn by different patients should be minimal. The bridge transmits the information from all the sensors worn by the patient towards the information sink by using a different WPAN or WLAN technology (which by definition has larger transmissions ranges than a WBAN), such that there is minimal or no interference between WBAN and WPAN/WLAN transmissions (e.g. the WBAN and WPAN/WLAN work in different and non-overlapping frequency bands). When using a robust WPAN/WLAN technology for the bridge-to-sink WSN segment, a good mobility range for patients and network scalability can be provided. Fig. 3 shows a generic architecture for heterogeneous WSN.

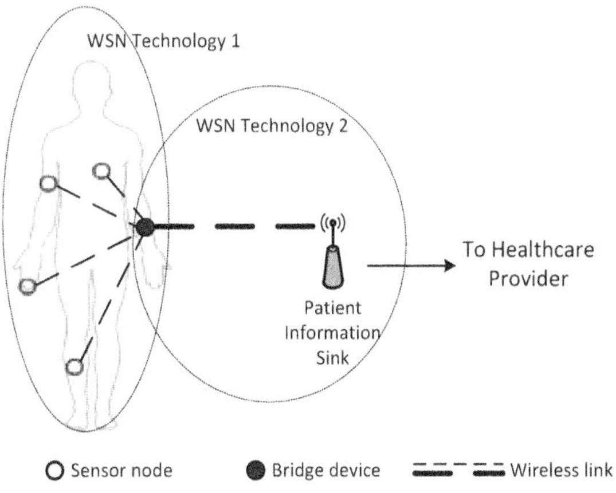

Fig. 3 Generic architecture for heterogeneous WSN in mHealth scenarios

An example of a heterogeneous WSN implementation is as follows. The WBAN segment uses IEEE 802.15.4 transceivers whose transmit power has been accordingly adjusted. Then the coordinator of the IEEE 802.15.4 WBAN can act as the bridge device. The bridge must have the capability of retransmitting the information collected from the sensors towards the information sink using another WPAN/WLAN technology. For this purpose it would be convenient to use a network technology offering larger transmission ranges and higher data rates than IEEE 802.15.4 (e.g., IEEE 802.11/WiFi). Because of the reduced radio range, in this architecture, the IEEE 802.15.4 sensors of different users will not interfere with each other as much as in the homogeneous IEEE 802.15.4 WSN architecture with multiple wireless hops. Thus, the number of supported users will be mainly limited by the capabilities offered by the technology used in the second WSN segment. A drawback of this approach is the high power consumption that the technology used for the second segment may have.

3 Comparative Analysis of Different WSN Architectures

Since each mobile patient monitoring system has specific requirements regarding patient mobility, number of bio-physiological variables to monitor, network reliability, and scalability (i.e., number of patients to be served); the wireless network technologies and WSN architecture (heterogeneous or homogeneous) to be used in the system must match these requirements. Thus, in order to properly design the WSN it is convenient to review the transmission requirements for bio-physiological variables commonly found in mobile patient monitoring applications. These requirements define a patient monitoring traffic profile that will be introduced in the following sub-section. Depending on the bio-physiological variables to be monitored, a specific traffic profile for specific mobile patient monitoring implementations is generated. One of such specific traffic profiles will be used in Section 3.2 to provide a performance comparison between homogeneous and heterogeneous WSN architectures.

3.1 Mobile Patient Monitoring Traffic Profile

In a mobile patient monitoring system the kind and number of bio-physiological variables to monitor will vary depending on the application scenario and the clinical condition of the patient. For example, for heart related problems, ECG and cardiac output monitoring may be required; whereas in diabetes, glucose, and blood pressure might be more relevant to monitor.

There are plenty of bio-physiological variables that can be considered for different health monitoring applications. This section presents the bio-physiological variables most commonly found in mobile patient monitoring scenarios (Pantelopoulos & Bourbakis, 2010) (Vouyioukas et al., 2007) (Gallego et al., 2005) (Skorin-Kapov & Matijasevic, 2010) (Adibi, 2012) (Misic & Misic, 2010) (Latré et al., 2011). Table 4 presents such variables altogether with typical transmission parameters such as: sampling frequency, delay sensitivity, and data rate. These parameters were adapted from (Misic & Misic, 2010) (Skorin-Kapov &

Matijasevic, 2010) and (Adibi, 2012). Monitoring these variables is important in elderly patient care or in the prevention and treatment of diseases such as: hypertension, diabetes, heart related problems, epilepsy, and sleep disorders among others. However, it is noteworthy that more bio-physiological variables can be added or discarded depending on the particular application scenario.

Most of the bio-physiological variables presented in Table 4 are not considered to be delay sensitive (Malindi, 2011). The exceptions are the electroencephalography (EEG) and ECG signals, which have a delay sensitivity of around one second (Skorin-Kapov & Matijasevic, 2010). In general, for mobile patient monitoring purposes the latency must be limited to be within a range of a few seconds up to one minute (Adibi, 2012). Additionally, the supported packet loss rate tolerance for remote patient monitoring applications reported in literature varies from zero to 5% (Supriyanto et al., 2009).

Table 4 Typical transmission requirements for bio-physiological variables in remote monitoring scenarios

Bio-physiological variable	Sampling frequency (Hz)	Bits per sample	Delay sensitivity	Data rate (bps)
Electroencephalography (EEG)	256 (x24 Ch.)	16	~ 1 s	98,304
Electrocardiography (ECG)	200 (x3 Ch.)	12 – 16	~ 1 s	7,200 – 9,600
Motion detection	50 (x9 Ch.)	16	NO	7,200
Glucose monitoring	40 – 200	16	NO	640 – 3,200
Blood pressure	120	16	NO	1,920
Blood oxygen (SpO2)	60 (x2 Ch.)	16	NO	1,920
Cardiac output	40	16	NO	640
Respiration	50	6	NO	300
Body temperature	0.2	12	NO	2.4

From the information presented in Table 4 it can be seen that, with the exception of the EEG, the transmission requirements of bio-physiological variables can be covered by the WBAN/WPAN technologies presented in Section 2. In fact, the peak data rates of most of these technologies (see Tables 2 and 3) seem to be relatively high compared to the data rates required for the transmission of bio-physiological variables (see Table 4). Thus, it would be tempting to design the WSN using the peak data rate as the main constraint that limits the maximum number of users and variables supported by the mobile patient monitoring system. However, other aspects such as the MAC mechanism and the data frame structure (e.g., headers, addressing, error check, etc.) will limit the effective data rate and the number of supported users within a WSN deployment. Thus, it is important to carefully review the performance characteristics and limitations of the specific technology and architecture to be used.

3.2 Traffic Capacity Analysis

In this section a performance comparison of different homogeneous and heterogeneous WSN architectures is presented. The performance metrics were obtained by means of the ns-2 network simulator, which is widely used by the academic community for the analysis and evaluation of wired and wireless network architectures (ns2, 2013).

For the homogeneous network architectures, it was assumed that each node in the WSN uses IEEE 802.15.4 transceivers. This WPAN technology was chosen because it has been widely used for the implementation of different mobile patient monitoring applications (Corchado et al., 2010) (Benocci et al., 2009).

For the heterogeneous network architecture, the use of IEEE 802.15.4 devices to form the WBAN was considered, while the bridge-to-sink communication was assumed to be enabled by an IEEE 802.11 link. Therefore the bridge has an IEEE 802.15.4 and an IEEE 802.11 transceiver, thus defining an IEEE 802.15.4 – IEEE 802.11 heterogeneous WSN architecture.

As previously mentioned, the number of sensors and patients supported by a given technology will be bounded by its peak data rate and MAC mechanism. Both standards, IEEE 802.15.4 and IEEE 802.11, implement a carrier sense multiple access with collision avoidance (CSMA-CA) mechanism at the MAC layer. In CSMA-CA, each node has to sense if the channel is busy before transmitting. If the channel is busy, the node has to wait for a random back-off interval before sensing again. Nevertheless, it is worth to mention that there are some differences between the CSMA-CA mechanisms implemented in each standard, see (IEEE Society, 2006) (IEEE Society, 2012). Both CSMA-CA back-off mechanisms can be modeled by using Markov chains and probability theory, (Zhai et al., 2004) (Kim et al., 2008) (Pollin et al., 2008) (Misic & Misic, 2009), however only simulations results are presented within this section for sake of conciseness.

In order to fairly evaluate the heterogeneous and homogeneous architectures, a representative set of bio-physiological variables for a particular mobile patient monitoring scenario must be defined. Thus, to obtain the performance metrics presented in this section, it was assumed that each patient transmits information collected from three bio-physiological variables: ECG, blood oxygen, and blood pressure. This means that the total data rate generated by each patient is equivalent to 13,440 bps. Additionally it is important to note that for these variables each sample consist of 2 bytes (see Table 4).

Two important parameters to consider when evaluating the network architectures are the data payload length (i.e., how many samples are transmitted within each packet) and the overhead per packet (i.e., preamble, headers, address, control, error check, security). An overhead percentage can be calculated dividing the overhead per packet by the total packet length (i.e., the data payload length plus the overhead per packet). Notice that when considering the transmission of only one bio-physiological sample (2 bytes for the evaluation scenario) per packet, the overhead percentage increases significantly. For instance, the typical overhead for short addressing in IEEE 802.15.4 is 13 bytes. This means that to transmit one bio-physiological sample of 2 bytes another 13 additional bytes are required, resulting in an overhead percentage of 86%. On the other hand, considering that IEEE 802.15.4 defines a maximum packet length of 127 bytes, a total of 57 samples (equivalent to 114 bytes of data payload) per packet can be transmitted. This results in an overhead percentage of 10%. Thus, in order to show the overhead percentage effects over the network performance, the use of two different packet sizes for the IEEE 802.15.4 transceivers was considered: 60 bytes (medium) and 127 bytes (maximum).

The homogeneous and heterogeneous WSN architectures considered for the performance evaluation presented in this section are:

- A homogeneous IEEE 802.15.4 architecture where each sensor transmits directly (one hop) towards the information sink (see Fig. 2b).
- A homogeneous IEEE 802.15.4 architecture where the data reach the information sink by means of two hops (see Fig 2c).
- A heterogeneous IEEE 802.15.4 – IEEE 802.11 architecture, where a bridge device retransmits the data towards the information sink (Fig. 3).

For the three WSN architectures, the packet loss rate and the number of concurrent sensors and users supported were obtained considering that each patient transmits the three variables previously defined (ECG, blood pressure, and blood oxygen). A packet loss rate of 2.5% is considered to be acceptable for comparison purposes.

3.2.1 Homogeneous IEEE 802.15.4 WSN Architecture

In this architecture each of the nodes in the WSN collects and transmits the information directly towards the ward room information sink, which consists of an IEEE 802.15.4 access point. In order to evaluate the worst case scenario for this homogeneous WSN architecture, it was considered that all the transceivers were within the transmission range of each other.

As previously described, it was assumed that each patient has three sensor nodes transmitting: ECG, blood pressure, and blood oxygen variables. For evaluation purposes, non-beacon enabled operation of the IEEE 802.15.4 transceivers was assumed. Additionally, no retransmissions and no acknowledgements were considered at the MAC layer. Fig. 4 presents the packet loss rate obtained for this architecture in function of the number of patients. In this figure it can be readily seen that as the number of users increases, the packet loss rate increases significantly. Additionally, it can be observed that in terms of the packet loss rate, the use of 127 bytes packet lengths presents a better performance than the use of 60 bytes packet lengths. Note that this behavior comes from the consideration that the total data rate per patient is fixed; thus, reducing the packet size implies shorter inter packet generation times. This leads to higher channel occupancy because the nodes will try to transmit packets more frequently. Thus, even though the over-the-air transmission time for 60 bytes packets is shorter, the overall channel occupancy is larger. Therefore, when transmitting using packets of 60 bytes the packet loss rate is higher than that obtained when transmitting using packets of 127 bytes.

Lastly, note in Fig. 4 that if a maximum packet loss rate tolerance of 2.5% is assumed, then only one patient per IEEE 802.15.4 channel can be served when using packets of 60 bytes. On the other hand, when using packets of 127 bytes up to 2 users per channel can be supported. Considering that the IEEE 802.15.4 standard defines 16 channels in the 2.4 GHz band, a maximum of 32 users could be supported with this architecture by using packets of 127 bytes. However, this number will be reduced if the 2.4 GHz frequency band is being used by other technologies such as WiFi.

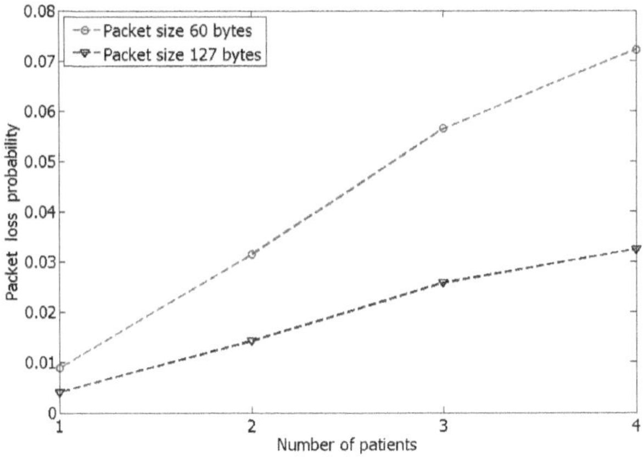

Fig. 4 Packet loss probability of the IEEE 802.15.4 homogeneous network with one hop

3.2.2 Two-Tier Homogeneous IEEE 802.15.4 WSN Architecture

The two-tier homogeneous network architecture considered in this section assumes that the patient's IEEE 802.15.4 sensors form a WBAN. Each of the WBAN nodes transmits the collected information towards a WBAN coordinator in the first hop. The WBAN coordinator then retransmits the information towards an information sink using a different IEEE 802.15.4 channel in the second wireless hop. Thus, the information sinks in this homogeneous architecture consist of IEEE 802.15.4 access points (APs) located in the ward rooms. This architecture is similar to the one introduced in (González-Valenzuela et al., 2011).

The utilization of different frequency channels seems to be a good solution to prevent mutual interference between the first and second wireless hops. Nevertheless, the coordinator will have to use two IEEE 802.15.4 transceivers in order to be able to transmit over the two different channels simultaneously. It is also worth noticing that a particular WBAN may experience interference or the hidden node problem caused by coordinators and APs transmitting within the same IEEE 802.15.4 frequency channel. In order to avoid this, a specific number of channels should be assigned for second-tier coordinator-to-AP communications and the rest should be reserved for first-tier WBANs transmissions (e.g., 8 channels for the WBANs and 8 channels for the coordinator-to-AP hop in the 2.4 GHz band).

In order to find the overall packet loss rate achieved when using this architecture, it was assumed that WBANs from different users do not interfere with each other. Thus, for the first-tier WBAN communications the packet loss rate equals the obtained in Fig. 4 for a single patient. Then, the packet loss rate in coordinator-to-AP communications must be obtained to calculate the overall packet loss rate. In order to determine the number of users supported by this architecture, it was assumed that more than one patient can use the same channel for the coordinator-to-AP hop. Fig. 5 shows the overall packet loss rate obtained for a single coordinator-to-AP channel as a function of the number of patients served.

From Fig. 5 it can be seen that if a 2.5% packet loss rate tolerance is assumed, then at most 2 concurrent patients per coordinator-to-AP channel can be served for both packet sizes. Considering this, if only one IEEE 802.15.4 channel in the 2.4 GHz band is reserved for WBANs transmissions and the remaining 15 channels are used for the coordinator-to-AP transmissions, then a maximum of 30 patients within the 2.4 GHz band can be attended.

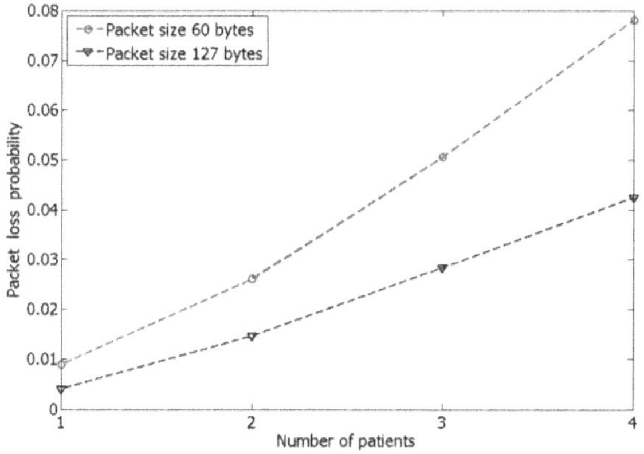

Fig. 5 Packet loss probability of the two-hop IEEE 802.15.4 homogeneous network

It is noteworthy that despite the number of users that can be served, the two-tier homogeneous architecture has the advantage of offering wider coverage areas than the single tier homogeneous architecture previously presented.

3.2.3 Heterogeneous IEEE 802.15.4 – IEEE 802.11g WSN Architecture

In the heterogeneous WSN architecture it is assumed that each of the sensors transmits the collected information towards an IEEE 802.15.4 - IEEE 802.11g bridge device worn by the user. The bridge function is to retransmit the information towards a patient information sink (e.g. an AP) by means of an IEEE 802.11g link. This architecture is similar to the one presented in (Misic & Misic, 2010).

The transmit power of the IEEE 802.15.4 transceivers is limited to form a WBAN, this in order to prevent mutual interference between WBANs worn by different patients. The bridge device acts as the coordinator of the IEEE 802.15.4 network and simultaneously as a station in the IEEE 802.11g network. The packets generated by the WBAN are seamlessly transmitted in the IEEE 802.11g network. This means that the information payload length is not modified for the transmission in the IEEE 802.11g segment. The typical PHY and MAC overhead of IEEE 802.11g is 62 bytes. Therefore, while the packets in the IEEE 802.15.4 WBAN have total lengths of 60 and 127 bytes, in the IEEE 802.11g network they will have total lengths of 109 and 176 bytes respectively. Since both technologies work on the 2.4 GHz band, it is important to adequately choose the transmission channel for each technology in order to avoid overlapping. For instance, if one particular IEEE 802.11g channel is used, then one of the 12 non-overlapping IEEE 802.15.4 channels must be used to deploy the WBAN. For the evaluation of this architecture, the peak data rate of the IEEE 802.11g segment was set to 6 Mbps. Note that this data rate provides longer transmission ranges and is more robust in presence of interference.

For evaluation purposes, default operation parameters of IEEE 802.15.4 and IEEE 802.11g were considered. It is assumed that the mutual interference between WBANs of different patients is minimal. Therefore, the packet loss rate for the sensor-to-bridge communications can be obtained from the metrics reported in Fig. 4 for single patient transmissions. These metrics do not change even when the number of patients is increased because of the previous assumption. In the IEEE 802.11g network tier, the packet loss due to collisions remains negligible until the saturation state of the network is reached. At this point, packets arrive to the network faster than they can be served and the network performance cannot be guaranteed. Fig. 6 presents the overall packet loss rate for the heterogeneous IEEE 802.15.4 – IEEE 802.11g architecture as a function of the number of patients being served. Note that, as expected, the 176 bytes packets offer a lower packet loss probability than the 109 bytes packet.

From Fig. 6 it can be seen that if a 2.5% packet loss rate tolerance is assumed, with this network architecture up to 80 patients could be (theoretically) served when using 109 bytes packets. Furthermore, when using a packet size of 176 bytes, up to 140 users could be (theoretically) attended. Therefore, this heterogeneous architecture significantly increases the number of patients that could be served within a single coverage area, compared with both IEEE 802.15.4 homogeneous architectures previously analyzed. Nevertheless, it is important to mention that before adopting a specific network architecture, it is necessary to consider other aspects such as power consumption, transceivers size and cost, etc.

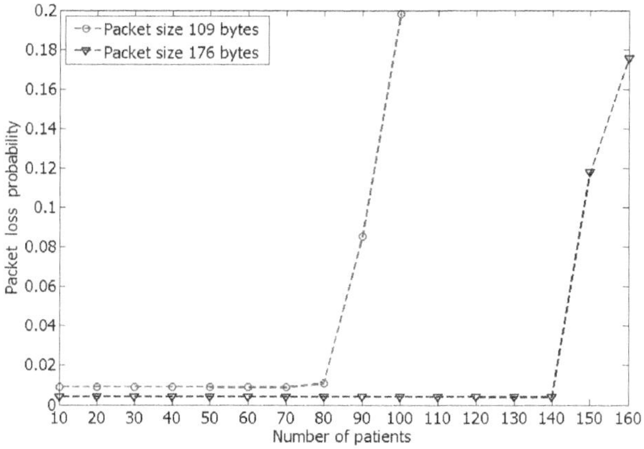

Fig. 6 Packet loss probability due collisions of the IEEE 802.15.4 – IEEE 802.11g heterogeneous network

4 Design Guidelines for WSN Architectures in mHealth Mobile Patient Monitoring

In order to properly design the WSN architecture for a particular mobile patient monitoring application, an adequate characterization of the different constraints found in the application scenario is required (i.e. user mobility, number of bio-physiological variables transmission requirements, etc.). Nevertheless, this is not a trivial issue. Therefore in order to facilitate this process the following design guidelines are provided as a reference:

- The first step in the design of the WSN is to verify the transmission requirements of the bio-physiological variables to monitor, this in order to define a specific traffic profile (e.g. see Table 4).
- To adequately select the wireless technology or technologies to use in the WSN, different network characteristics such as peak data rate, number of sensors, network topology, required throughput and energy consumption must be considered. It is worth mentioning that monitoring high data rate variables like the EEG poses different challenges than those presented when monitoring low data rate variables (e.g., blood oxygen, blood pressure or body temperature). For instance, if a specific application requires EEG and ECG monitoring, then IEEE 802.15.4 may not offer the required throughput to even attend a single patient.
- The application scenario and location-specific constraints (i.e., home, hospital, ambulance, anywhere) must be considered when designing the WSN network architecture. Particularly, the following aspects should be addressed:
  - For low patient mobility and low data rate scenarios (see Fig. 4b-c) or when each patient carries his/her own information sink (see Fig. 4a), a

homogeneous WSN architecture is preferred. In these scenarios there is no need for a bridge device as the patient's sensors and information sink are always within coverage range. Furthermore, there will be a relatively low density of sensors and patients accessing the same information sink. This could help to lower energy consumption and reduce the WSN complexity.
  - For high patient densities, a heterogeneous WSN architecture may be required. In order to mitigate interference between patients, the transmission power of the sensors must be adjusted to WBAN boundaries. Additionally, the technology used for the bridge-to-sink communication must provide higher transmission data rates than the WBAN technology, this in order to increase the number of users that can be served.

- For fixed data rates, when using technologies based on the CSMA-CA MAC mechanism the number of samples transmitted per packet can set boundaries for network performance in terms of packet loss rates and number of supported users. Although longer packet lengths represent a little increase in packet delays, these delays may become tolerable when considering the offered benefits in terms of lower packet loss rates and higher number of supported users. For this reason the number of samples transmitted per packet must be chosen considering a trade-off between packet delay, packet loss rates and number of supported users.
- Adequate selection of the operating frequency is really important to guarantee the performance of the WSN. Thus, operating in the same frequencies as other wireless networks deployed within the same coverage area must be avoided. When using multi-tier homogeneous or heterogeneous WSN architectures, it is important to select non-overlapping frequency channels for the patients WBANs and the bridge-to-sink link to avoid mutual interference.

A reference framework is suggested in Fig. 7 in order to simplify the process of designing a WSN for mHealth mobile patient monitoring.

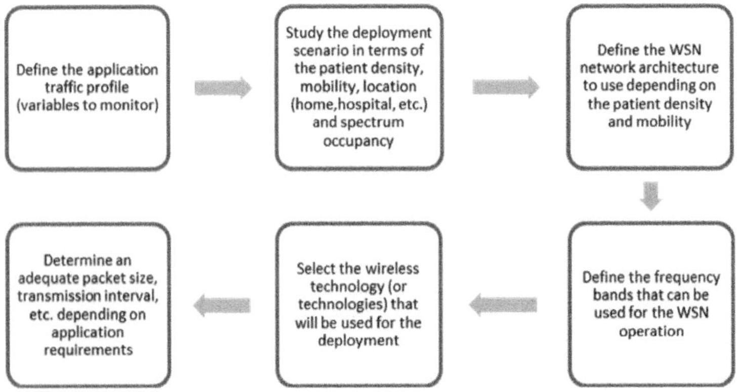

Fig. 7 Reference framework for the design of mHealth mobile patient monitoring WSNs

5 Future Direction and Open Research Issues

Future direction of mHealth mobile patient monitoring suggests anywhere, anytime scenarios for prevention and tracking of a patient's health. Moreover, for in-hospital disease treatment and emergency response, the overall hospital network architecture should be robust enough to effectively provide robust solutions for the different healthcare services. For instance, in emergency situations or intensive care units (ICU) the transmissions requirements of bio-physiological signals are harsher. Thus, a minimum network performance must be guaranteed for the WSN to be useful in life-threatening situations. Additionally, interoperability between sensors, medical devices and display equipment will be indispensable for the deployment of comprehensive mHealth solutions. In this context, network equipment should be compatible with higher layer standards like HL7 for the exchange, integration, sharing, and retrieval of electronic health information.

There are still some open research issues regarding the full implementation of mobile patient monitoring systems. One of the most important is the end-to-end security and privacy of the transmitted information. Although there are some levels of security within different network infrastructures, there are still weak points across the network where the decryption-encryption processes reside. Another important research issue is the effective QoS mapping between different networks. Since different network technologies may offer different QoS classifications, an effective mapping between networks, exclusive for mHealth variables, should be considered. From the bio-physiological signal processing point of view, there are still several open research issues. For instance, development of application-specific codecs for bio-physiological signals is a relevant research area. Additionally, current trends suggest that future sensors in mHealth scenarios will be based on energy scavenging systems with ultra-low power consumption, such that they will be able to operate unattended over their complete lifespan.

6 Conclusions

The effective implementation of WSNs for mobile patient monitoring systems requires the consideration and evaluation of different constraints specific to the application scenario. In this chapter, several guidelines for the design of WSN architectures in mHealth mobile patient monitoring scenarios were presented. The most significant requirements to consider when designing WSNs for mHealth applications are: patient mobility, transmission requirements of the bio-physiological variables to monitor, reliability of information, network scalability (i.e., number of concurrent users being served), required security, and the distance between the patient(s) and the information sink.

It is believed that the future of WSNs for patient monitoring will be dominated by open-standard technologies, since these technologies provide a better platform for interoperability between different implementations and manufacturers. From the technologies analyzed, it seems that the recent IEEE 802.15.6 standard will offer several characteristics that will lead to its wide adoption for WBANs deployments.

Some of those characteristics are: low interference, multiple frequency bands and medium access mechanisms, user priorities for medical QoS, ranging capabilities, low error rates, and ultra-low power consumption.

Although designing WSNs architectures that can support a high density of users may seem trivial, it is important to remember that aspects such as the MAC mechanism and the supported network topologies may adversely affect the network performance in terms of the packet loss rates and throughput. Therefore the use of tools such as network simulators is recommended to properly assess the capabilities offered by a given WSN architecture and wireless network technology.

Homogeneous WSN architectures are best suited for low patient mobility and low data rate applications or when each patient carries his/her own information sink. In this sense, by adequately implementing the homogeneous architecture energy consumptions savings can be achieved, since there is no need for a bridge device to retransmit the information through another wireless technology. On the other hand, homogeneous WSN architectures can be seriously limited when there is a high density of users connected to the patient information sink.

Heterogeneous WSN architectures can be used when the density of patients to be served within the same coverage area is a critical design parameter. Interference between sensors from different patients can be mitigated by decreasing its transmission power to WBAN levels. Compared to homogeneous architectures, by selecting a bridge-to-sink communication technology that provides higher data rates than the one used in the WBAN tier, a significant increment on the number of supported users can be achieved. Nevertheless, the overall energy consumption in this architecture may be higher because of the bridge device.

References

1. Adibi, S.: Link Technologies and BlackBerry Mobile Health (mHealth) Solutions: A Review. IEEE Transactions on Information Technology in Biomedicine 16(4), 586–597 (2012), doi:10.1109/TITB.2012.2191295.
2. Benocci, M., Farella, E., Benini, L., Vanzago, L.: Optimizing ZigBee for data streaming in body-area bio-feedback applications. In: 3rd International Workshop on Advances in sensors and Interfaces, IWASI, pp. 150–155 (June 2009), doi:10.1109/IWASI.2009.5184786
3. Bluetooth SIG, Bluetooth Core Especification V.4.0 (2010), `https://www.bluetooth.org/docman/handlers/downloaddoc.ashx?doc_id=229737` (accessed September 2013)
4. Bluetooth SIG, Bluetooth Smart Devices List (2013), `http://www.bluetooth.com/Pages/Bluetooth-Smart-Devices-List.aspx` (accessed September 2013)
5. Boquete, L., et al.: A portable wireless biometric multi-channel system. Measurement 45(6), 1587–1598 (2012), ISSN 0263-2241
6. Chen, M., et al.: Body Area Networks: A Survey. Mobile Networks and Applications 16(2), 171–193 (2011), doi:10.1007/s11036-010-0260-8

7. Chen, S.-K., et al.: A Reliable Transmission Protocol for ZigBee-Based Wireless Patient Monitoring. IEEE Transactions on Information Technology in Biomedicine 16(1), 6–16 (2012), doi:10.1109/TITB.2011.2171704
8. Chen, B., et al.: Patient vital signs monitoring using Wireless Body Area Networks. In: Proceedings of the 2010 IEEE 36th Annual Northeast, Bioengineering Conference, pp. 26–28 (March 2010), doi:10.1109/NEBC.2010.5458139
9. Corchado, J.M., Bajo, J., Tapia, D.I., Abraham, A.: Using Heterogeneous Wireless Sensor Networks in a Telemonitoring System for Healthcare. IEEE Transactions on Information Technology in Biomedicine 14(2), 234–240 (2010)
10. Csernath, G., et al.: A novel ECG telemetry and monitoring system based on Z-Wave communication. Engineering in Medicine and Biology Society. In: 30th Annual International Conference of the IEEE, EMBS 2008, pp. 2361–2364 (2008), doi:10.1109/IEMBS.2008.4649673
11. Dynastream, I.: ANT Message Protocol and Usage V. 5.0 (2013), `http://www.thisisant.com/resources/ant-message-protocol-and-usage/` (accessed August 2013)
12. Egbogah, E.E., Fapojuwo, A.O.: A Survey of System Architecture Requirements for Health Care-Based Wireless Sensor Networks. Sensors 11(5), 4875–4898 (2011), doi:10.3390/s110504875
13. Fraile, J.A., Bajo, J., Corchado, J.M., Abraham, A.: Applying wearable solu-tions in dependent environments. IEEE Transactions on Information Technology in Biomedicine 14(6), 1459–1467 (2010), doi:10.1109/TITB.2010.2053849
14. Gallego, J.R., et al.: Performance analysis of multiplexed medical data transmission for mobile emergency care over the UMTS channel. IEEE Transactions on Information Technology in Biomedicine 9(1), 13–22 (2005), doi:10.1109/TITB.2004.838362
15. Goldberg, S., Wickramasinghe, N.: 21st century healthcare - the wireless panacea. In: Proceedings of the 36th Annual Hawaii International Conference on System Sciences, pp. 6–9 (January 2003)
16. González-Valenzuela, S., Chen, M., Leung, V.C.M.: Mobility Support for Health Monitoring at Home Using Wearable Sensors. IEEE Transactions on Information Technology in Biomedicine 15(4), 539–549 (2011), doi:10.1109/TITB.2010.2104326
17. IEEE Communications Society, IEEE Standard for Long Wavelength Wireless Network Protocol, p. 25 (2009), doi:10.1109/IEEESTD.2009.4810102
18. IEEE Computer Society, Part 15.4: Wireless Medium Access Control (MAC) and Physical Layer (PHY) Specifications for Low-Rate Wireless Personal Area Networks (WPANs), p. 323 (2006)
19. IEEE Standard for Local and metropolitan area networks - Part 15.6: Wireless Body Area Networks, p. 271 (2012)
20. IEEE Computer Society, Part 11: Wireless LAN Medium Access Control (MAC) and Physical Layer (PHY) Specifications. IEEE Standard for Information technology, p. 2793 (2012)
21. Jung, B.H., Akbar, R.U., Sung, D.K.: Throughput, energy consumption, and energy efficiency of IEEE 802.15.6 body area network (BAN) MAC protocol. In: IEEE 23rd International Symposium on Personal Indoor and Mobile Radio Communications (PIMRC), pp. 584–589 (September 2012), doi:10.1109/PIMRC.2012.6362852
22. Kailas, A., Ingram, M.A.: Wireless Aspects of Telehealth. Springer Wireless Personal Communications 51(4), 673–686 (2009)

23. Kim, T.O., et al.: Performance analysis of IEEE 802.15.4 non-beacon mode with the unslotted CSMA/CA. IEEE Communications Letters 12(4), 238–240 (2008), doi:10.1109/LCOMM.2008.071870
24. Koh, B.K., Kong, P.Y.: Performance Study on ZigBee-Based Wireless Personal Area Networks for Real-Time Health Monitoring. ETRI Journal 28(4), 537–540 (2006)
25. Ko, J., et al.: Wireless Sensor Networks for Healthcare. Proceedings of the IEEE 98(11), 1947–1960 (2010), doi:10.1109/JPROC.2010.2065210.
26. Latré, B., et al.: A survey on wireless body area networks. Wireless Networks 17(1), 1–18 (2011), doi:10.1007/s11276-010-0252-4.
27. Li, L., Xiaoguang, H., Ke, C., Ketai, H.: The applications of WiFi-based Wireless Sensor Network in Internet of Things and Smart Grid. In: 2011 6th IEEE Conference on Industrial Electronics and Applications (ICIEA), pp. 21–23 (June 2011), doi:10.1109/ICIEA.2011.5975693
28. Mahmood, N.H., et al.: LED indicator for heart rate monitoring system in sport application. In: 2011 IEEE 7th International Colloquium on Signal Processing and its Applications (CSPA), pp. 64–66 (2011), doi:10.1109/CSPA.2011.5759843
29. Malindi, P.: QoS in Telemedicine. In: Graschew, G. (ed.) Telemedicine Techniques and Applications, pp. 119–138 (2011), doi:10.5772/724
30. Martelli, F., Buratti, C., Verdone, R.: On the performance of an IEEE 802.15.6 Wireless Body Area Network. In: 11th European Wireless Conference - Sustainable Wireless Technologies (European Wireless), pp. 1–6 (April 2011) ISBN: 978-3-8007-3343-9.
31. Misic, J., Misic, V.B.: Bridging between ieee 802.15.4 and IEEE 802.11b networks for multiparameter healthcare sensing. IEEE Journal on Selected Areas in Communications 27(4), 435–449 (2009), doi:10.1109/JSAC.2009.090508
32. Misic, J., Misic, V.B.: Bridge performance in a multitier wireless network for healthcare monitoring. IEEE Wireless Communications 17(1), 90–95 (2010), doi:10.1109/MWC.2010.5416355
33. Movassaghi, S., Arab, P., Abolhasan, M.: Wireless technologies for Body Area Networks: Characteristics and challenges. In: International Symposium on Communications and Information Technologies (ISCIT), pp. 42–47 (October 2012)
34. ns2, The Network Simulator - ns-2 (2013), `http://www.isi.edu/nsnam/ns/` (accessed September 2013)
35. Pantelopoulos, A., Bourbakis, N.G.: A Survey on Wearable Sensor-Based Systems for Health Monitoring and Prognosis. In: IEEE Transactions on Systems, Man, and Cybernetics, Part C: Applications and Reviews, vol. 40(1), pp. 1–12 (2010), doi:10.1109/TSMCC.2009.2032660
36. Park, Y., Cho, H.: Transmission of ECG data with the patch-type ECG sensor system using Bluetooth Low Energy. In: 2013 International Conference on ICT Convergence (ICTC), pp. 289–294 (2013), doi:10.1109/ICTC.2013.6675359
37. Penders, J., et al.: Human++: Emerging Technology for Body Area Networks. In: De Micheli, G., Mir, S., Reis, R. (eds.) VLSI-SoC: Research Trends in VLSI and Systems on Chip. IFIP, vol. 249, pp. 377–397. Springer, Boston (2008)
38. Pollin, S., et al.: Performance Analysis of Slotted Carrier Sense IEEE 802.15.4 Medium Access Layer. IEEE Transactions on Wireless Communications 7(9), 3359–3371 (2008), doi:10.1109/TWC.2008.060057
39. RuBee, RuBee and RFID Differences (2009), `http://www.rubee.com/White-SEC/RuBee-SellingPoints-190208.pdf` (accessed August 2013)

40. Sharma, S., et al.: Wireless Body Area Network for health monitoring. In: 4th International Conference on Biomedical Engineering and Informatics (BMEI), vol. 4, pp. 2183–2186 (2011), doi:10.1109/BMEI.2011.6098710
41. Skorin-Kapov, L., Matijasevic, M.: Analysis of QoS Requirements for e-Health Services and Mapping to Evolved Packet System QoS Classes. International Journal of Telemedicine and Applications, 1–18 (2010), doi:10.1155/2010/628086
42. Sneha, S., Varshney, U.: A framework for enabling patient monitoring via mobile ad hoc network. Decision Support Systems 55(1), 218–234 (2013) ISSN 0167-9236.
43. Stevens, J., Weich, C., GilChrist, R.: RuBee (IEEE 1902.1) – The Physics Behind, Real-Time, High Security Wireless Asset Visibility Networks in Harsh Environments (2010), `http://www.rubee.com/White-SEC/RuBee-Security-080610.pdf` (accessed August 2013)
44. Supriyanto, E., Satria, H., Mulyadi, H., Putra, E.H.: A novel low cost telemedicine system using wireless MESH network. In: South East Asian Technical Universities Consortium(SEATUC) - 3rd SEATUC Symposium Proceeding, pp. 382–388 (February 2009)
45. Teng, X., et al.: Wearable Medical Systems for p-Health. Biomedical Engineering. In: IEEE Reviews, vol. 1, pp. 62–74 (2008), doi:10.1109/RBME.2008.2008248
46. Texas Intrsuments, 1- and 8-Channel ANT™ RF Network Processors (2011), `http://www.ti.com/lit/ds/symlink/cc2570.pdf` (accessed August 2013)
47. Toumaz, Towards the improvement of patients' safety with continuous wireless monitoring: Pilot at Saint John's Health Center (2010), `http://www.toumaz.com/sites/default/files/White%20paper%20-%20v1%206%20NL.pdf` (accessed August 2013)
48. Tung, H.Y., et al.: The design of dual radio ZigBee homecare gateway for re-mote patient monitoring. IEEE Transactions on Consumer Electronics 59(4), 756–764 (2013), doi:10.1109/TCE.2013.6689686
49. Ullah, S., Kwak, K.-S.: Throughput and delay limits of IEEE 802.15.6. Wireless Communications and Networking Conference (WCNC), pp. 174–178. IEEE (March 2011), doi:10.1109/WCNC.2011.5779126
50. Vaidehi, V., et al.: Agent Based Health Monitoring of Elderly People in Indoor Environments Using Wireless Sensor Networks. Procedia Computer Science 19, 64–71 (2013), ISSN 1877-0509
51. Vouyioukas, D., Maglogiannis, I., Komnakos, D.: Emergency m-Health Services through High-Speed 3G Systems: Simulation and Performance Evaluation. SIMULATION The Society for Modeling and Simulation International 83(4), 329–345 (2007), doi:10.1177/0037549707083113
52. Wong, A., et al.: Sensium: an ultra-low-power wireless body sensor network platform: Design & application challenges. In: Annual International Conference of the IEEE Engineering in Medicine and Biology Society, EMBC 2009, pp. 6576–6579 (2009), doi:10.1109/IEMBS.2009.5334001
53. Yu, B., Xu, L., Li, Y.: Bluetooth Low Energy (BLE) based mobile electro-cardiogram monitoring system. In: International Conference on Information and Automation (ICIA), pp. 763–767 (2012), doi:10.1109/ICInfA.2012.6246921
54. Yu, B., Yang, L., Chong, C.C.: ECG Monitoring over Bluetooth: Data Compression and Transmission. In: IEEE Wireless Communications and Networking Conference (WCNC), pp. 1–5 (2010), doi:10.1109/WCNC.2010.5506641

55. Yu, X., Xia, X., Chen, X.: Design and Application of RuBee-Based Tele-medicine Data Acquisition System. In: 2011 IEEE/ACIS 10th International Conference on Computer and Information Science (ICIS), pp. 365–370 (2011), doi:10.1109/ICIS.2011.64
56. Zhai, H., Kwon, Y., Fang, Y.: Performance analysis of the IEEE 802.11 MAC protocols in wireless LANs. In: Wiley Wireless Communications and Mobile Computing, vol. 4, pp. 917–931 (2004), doi:10.1002/wcm.263
57. Zhang, Y., Xiao, H.: Bluetooth-Based Sensor Networks for Remotely Monitoring the Physiological Signals of a Patient. IEEE Transactions on Information Technology in Biomedicine 13(6), 1040–1048 (2009), doi:10.1109/TITB.2009.2028883

Questions

Q1. Which are the main challenges and factors for the effective deployment of WBANs?

Q2. What are the advantages and disadvantages of heterogeneous and homogeneous WSN architectures in mobile patient monitoring applications?

Q3. How does packet size affects the performance metrics of heterogeneous and homogeneous WSN architectures?

Q4. Which are the more significant application requirements to consider when developing mHealth networks?

Q5. According to the technologies discussed in this chapter, which types of standards are expected to prevail for the mobile patient monitoring systems?

Case Studies

Consider a mHealth mobile patient monitoring scenario to serve a low density number of users locally within a nursing home. Also consider that low data rate bio-physiological variables will be monitored and that the patient information sink is in close proximity for all the patients (few meters away). Following the guidelines presented in Section 4, design a homogeneous WSN architecture capable of providing access to a fixed number of patients with the less power consumption possible. Define the wireless technology, network topology and an estimated number of patients to be served.

Consider a mHealth mobile patient monitoring scenario to serve the maximum number of users within a hospital with the capability of transmitting the information towards a remote healthcare provider. Also consider that high data rate bio-physiological variables will be monitored and that there are different patient information sinks distributed within the hospital. Following the guidelines presented in Section 4, design a heterogeneous WSN architecture capable of providing access to the maximum number of patients. Define the wireless technologies to be used, network topology and an estimated number of patients to be served.

19. Evaluation of Health Care System Model Based on Collaborative Algorithms

Vladimir Trajkovik*, Saso Koceski, Elena Vlahu-Gjorgievska, and Igor Kulev

Faculty of Computer Science and Engineering, University "Ss Cyril and Methodious", "Rugjer Boshkovikj" 16, P.O. Box 393 1000 Skopje Macedonia
{trvlado,igor.kulev}@finki.ukim.mk,
saso.koceski@ugd.edu.mk, elena.vlahu@uklo.edu.mk

Keywords: Personal healthcare, recommendation algorithm, fuzzy-discretization, classification, data preprocessing, distance metric, mathematical model, evaluation.

Introduction. The rapid development and use of information and communication technologies in the last two decades has influenced a dramatic transformation of public health and health care, changing the roles of the health care support systems and services. Recent trends in health care support systems are focused on developing patient-centric pervasive environments and the use of mobile devices and technologies in medical monitoring and health care systems [1].

Nowadays, myriad of different sensors are used to monitor human's vital parameters such as: heart rate, blood pressure, blood sugar level. These sensors can be connected to a smartphone. The patient can use the software installed on the smartphone to read the sensor data and the software can present this data directly to the patient or can send it to the doctor which can further analyze it. This is the simplest scenario in mHealth. Smarter system can make conclusions about the current health condition of the patient by using the sensor data and integrated expert knowledge. The system could inform the patient or the doctor when some of the patient's health parameters are not in the normal range.

The big challenge in mHealth is to develop a system which could give advices to the patients in order to improve their health condition. Patients could be advised to eat some particular food more often, to perform specific activities, to take some medicine or to visit a doctor for physical examination. If the process of making advices is done entirely by a doctor which has remote access to the mobile application used by the patient, it could be expensive and slow. Better approach could be to automate the process of making advices and recommendations. For this purpose the system will need all data that could be useful: the current health condition, the history of measurements, the patient's diagnosis and an expert's knowledge. Additionally, the types of recommendations and the algorithm which will be used to generate them should also be defined. Although the expert's knowledge will be used, the elimination of the direct human factor from the process of making advices means that the generated advices will be with smaller

---

* Corresponding author.

S. Adibi (ed.), *Mobile Health*,
Springer Series in Bio-/Neuroinformatics 5, DOI: 10.1007/978-3-319-12817-7_19

relevance – additional detailed physical examination could reveal more useful information. It is advised to use the system in a domain where the generated recommendations could potentially improve the health condition of the patient, but they could not worsen it.

In this context, introduction of a novel patient-centric collaborative health care system model COHESY [1], with an integrated intelligent recommendation algorithm, could give a new dimension to preventive and curative medicine. The proposed model helps its users to actively participate in their health care and prevention, thereby providing an active life in accordance with their daily responsibilities at work, family and friends.

COHESY uses mobile, web and broadband technologies, so the citizens have ubiquity of support services where ever they may be, rather than becoming bound to their homes or health centers as pointed out by different authors [2]. Broadband mobile technology provides movements of electronic care environment easily between locations and internet-based storage of data allows moving location of support. Integrated social network in COHESY, allows communication between users with same or similar condition and exchange of their experiences. The social network provides data from all users which can be used by the recommendation algorithm.

The recommendation algorithm in COHESY gives recommendations about physical activities that should be performed by the patients in order to improve their health condition. The algorithm is based on the dependence between the values of the health parameters and the users' physical activities (e.g. walking, running, biking). The basic idea is to find out which physical activities cause change (improvement) of the value of health parameters. This dependence continues to be used by the algorithm to recognize the same or similar health conditions found in another user with similar characteristics. To achieve this, classification and filtering algorithms are applied in the proposed recommendation algorithm in order to group users with similar characteristics. The usage of classified data provides relevant recommendations based on prior knowledge of users with similar health conditions and reference parameters. In this way, COHESY bridges the gap between users, clinical staff and medical facilities, strengthening the trust between them and providing relevant data from a larger group of users, grouped on the basis of various indicators.

1 COHESY Overview

COHESY is deployed over three basic usage layers (Fig.1). The first layer consists of the bionetwork (implemented from various body sensors) and a mobile application that reads sensor data and allows the users to specify the activities they perform. The application collects users' bio data during various physical activities (e.g. walking, running, cycling). Additionally, the application collects other sensor data (e.g. GPS coordinates) that can be used to describe the performed activities (e.g. speed, duration).

The second layer is presented by the social network implemented as a web portal which enables different collaboration within the end user community. The social network keeps the data obtained from the mobile application. It consists of various web services that give a restricted access to this data to different end users. The third layer enables interoperability with the primary/secondary health care information systems which can be implemented in the clinical centers and different policy maker

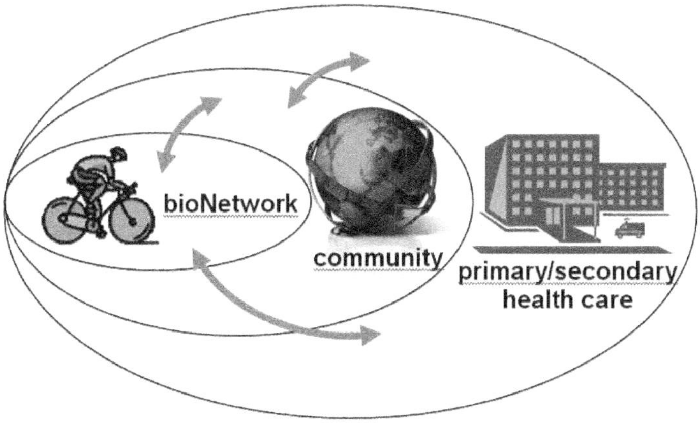

Fig. 1 Collaborative healthcare system model (COHESY)

institutions. The communication between the first and the second layer is defined by the users' access to the social network where users can store their own data (e.g. personal records, healthcare records, bionetwork records, readings on physical activities). Users can also receive average results from all patients that share the same health condition. Some of these results can be the average levels of certain bio data calculated for certain geographical region, age or sex. Additionally, the data from the social network is used by the recommendation algorithm to recommend physical activities that could be useful to the user.

The communication between the first and the third layer is determined by the communication between the patient and the health care centers. The patient has 24 hour access to medical personnel and a possibility of sending an emergency call. The medical personnel remotely monitors the patient's medical condition, reviews the medical data (e.g. blood pressure, blood-sugar level, heart rate) and responds to the patient by suggesting the most suitable therapy (if different from the one that is encoded in the mobile application) as well as sending him/her various notifications (e.g. tips and suggestions) regarding her/his health condition. The second and the third layer can exchange data and information regarding a larger group of patients grouped by any significant indicator (region, time period, sex, type of the activities) which can be later used for research, policy recommendations and medical campaign suggestions.

COHESY creates an opportunity to increase users' health care within their homes by 24 hour monitoring on one hand, and to increase the medical capacity of health care institutions on the other hand. This results in reducing the overall costs for users and hospitals and improves the user's quality of life. It provides a better health care allowing suggestions and recommendations based on knowledge from other users, cases and experiences. This makes COHESY different from other health care systems. MobiCare [3] and Personal Care Connect [4] are systems that facilitate the remote monitoring of the patients. Both systems consist of a body sensor network, communication infrastructure and servers on which the data is stored. They allow medical personnel and other applications to use the information gathered from sensors. The need for quality of service support in wireless e-health and e-emergency services is discussed by Gama, Carvalho, Alfonso and Mendes in [5]. They emphasize that the network must prioritize the transmission of critical data when sudden change occurs in the patient medical condition, and because of this it is important to distinguish all collected information. Most similar to our system model is Jog Falls system [6]. Jog Falls system is an end to end system to manage diabetes that blends activity and energy expenditure monitoring, diet-logging, and analysis of health data for patients and physicians. This is an integrated system for diabetes management providing the patients with continuous awareness of their diet and exercise, automatic capture of physical activity and energy expenditure, simple interface for food logging, ability to set and monitor goals and reflects on longer term trends.

These examples include research of systems that cover only parts of the presented model. The main difference between the aforementioned examples and COHESY is the social network, in which the recommendation algorithm that we present in this chapter is implemented.

2 Recommendation Algorithm in COHESY

The problem of recommending physical activities cannot be solved using the existing algorithms from collaborative and content-based filtering. The existing algorithms need to have information about the ratings given to items by users. In our case, the items are analogous to physical activities. In other words, we need to know how the physical activities cause change of the value of health parameters and whether that change improves or worsens the health condition of the user. This is another problem that needs to be solved if we want to use the existing recommendation algorithms. Additionally, people that have some diagnosis, for example people with heart problems, should not be recommended some types of activities, for example fast running, because these activities can worsen their health condition. The system should be carefully designed in order to use efficiently all the available information. We propose recommendation algorithm that takes into account all the available information and the complexity of the domain.

For every person and at every moment there is a set of useful activities that can potentially improve her/his health condition. Our recommendation algorithm discovers and recommends these useful activities to the user. The algorithms for activity recommendation must be based on few main principles: (1) Except the physical activities, there are other factors that cause change of the people's health condition (medicines, food and psychic condition) and the deficiency of this additional information brings to bigger inaccuracy in the recommendations; (2) People can be grouped according to their characteristics (diagnosis, place of living) and for each of these groups activities have specific effect on the parameters which is similar for people in the same group, and is different for people in different groups; (3) For every person activities do not influence on his/her health with the same intensity and in the same way.

We will separate the health parameters into two groups: descriptive and control parameters. Descriptive parameters express the factors that are important for the people's health (for example age or place of living) and that cannot be affected. Control parameters express the characteristics of the people's health (for example blood pressure or blood sugar level) that can be affected. The goal of the proposed algorithm is to find activities that could change the control parameters towards preferable state which indicates normal health condition.

2.1 Data Representation

The main data which will be needed by the recommendation algorithm are:

- The measurements of the health parameters. Each measurement is represented by a vector $Measurement(user, parameter, time, value)$
- The executions of the physical activities. Each execution is represented by a vector $Activity(user, type, time, duration, difficulty)$

For example, if user Alexander runs 5 km in 30 minutes on 29th July, 2013 at 11:00, the $Activity$ vector will look like this:

$$Activity(Alexander, running, 30\ minutes,$$
$$29th\ July\ 2013\ 11{:}00, 30\ minutes, 5\ km)$$

Next we give few examples of the Measurement vector:

$$Measurement(Alexander, weight, 28th\ July\ 2013\ 15{:}00, 78\ kg)$$
$$Measurement(Alexander, weight, 30th\ July\ 2013\ 14{:}00, 76\ kg)$$

The difficulty of the activity is a quantitative representation of the given effort to complete that activity. When there is more than one type of activity, the measure of difficulty should be chosen so that if two activities have the same difficulty, the given effort to complete both activities is the same, independently of the types of activity. The measure of difficulty can be analogous to the calories spent to complete the activity.

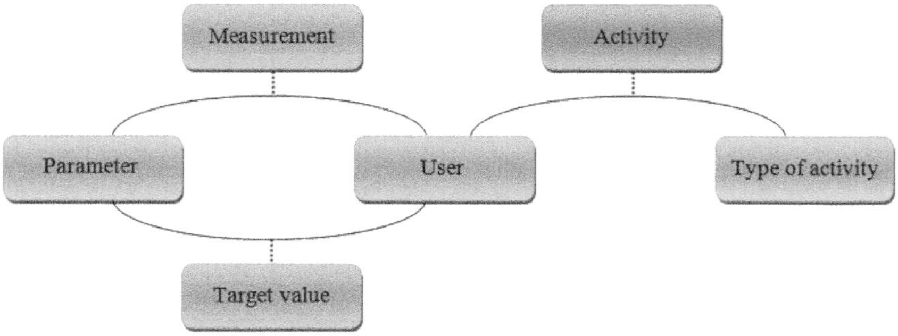

Fig. 2 Connections between the different types of data

We will need one more type of data: the users' diagnoses (e.g. heart problems, diabetes). This information will be needed to make the first filtering of users. We can assume that we are given D different types of diagnoses. The vector that describes the diagnoses of some particular user is given by:

$$Restriction(user, diagnosis_1, diagnosis_2 \ldots diagnosis_D)$$

Each $diagnosis_i$ represents a discrete variable that can take one of two values: zero (the user does not have the diagnosis) or one (the user has the diagnosis), or a continuous variable defined over the range [0, 1].

2.2 *Data Preprocessing*

Some of the health parameters have big variance i.e. the measurements performed in close moments differ a lot. These kinds of measurements can be obtained when measuring the blood sugar levels in people that are insulin deficient. The deficiency is caused by the reduced production of insulin in the pancreas [7]. The insulin levels depend on three actions: diet, exercises and insulin injections. The blood sugar levels changes a lot during the day in people that are insulin deficient, so if we are given only two measurements we cannot know for sure whether the health parameter is improved or worsened.

We need to make a transformation of the raw sensor data that will give a more accurate reflection of the health condition of the user at the time of measurement. For that purpose, we can use a function that will transform the readings. This function should be determined by an expert. One function that can be used to find the transformed measurement is:

$$transformedMeasurement(x) = \int_{x-C}^{x} \frac{C-(x-t)}{C} \cdot measurement(t) \cdot dt \tag{1}$$

where C is the time period before the moment x in which the measurements have an effect on the transformed measurement of the parameter at the moment x. In this function we assume that the older measurements have weaker influence on the final value of the parameter. This formula refers to the ideal case when we are given a continuous function $measurement(t)$ that can be defined mathematically. For more practical application we can use a formula in which the integral is replaced by a weighted sum of all measurements made in the time period C before the moment x multiplied by a factor $\frac{C-(x-t)}{C}$ where t is the moment in which the particular measurement is made. We should note that the result of the transformation of the raw sensor data should represent the parameter that we will try to control through recommendation of physical activities.

Diabetes Data Set [7] contains measurements of the blood sugar level of 70 patients that are diagnosed with diabetes. We chose data from one patient and we applied two transformations on the raw sensor data: weighted average and average from the last 24 hours. We can notice that the transformed measurements are more stable and they have smaller variance.

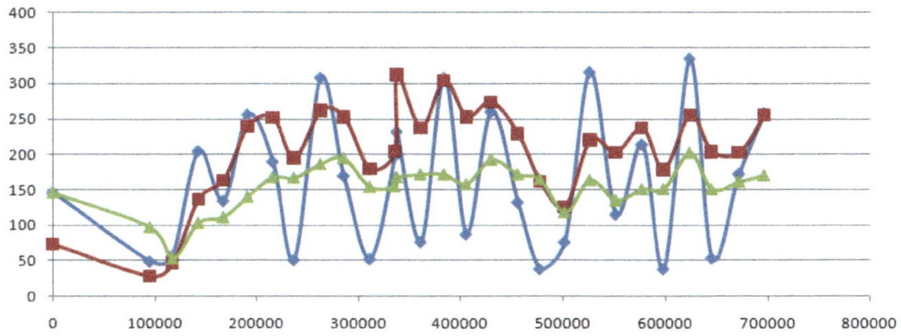

Fig. 3 Raw sensor data (blue), weighted average from the last 24 hours (red), average from the last 24 hours (green). The time after the first measurement is represented on the x-axis and the blood sugar level is represented on the y-axis (mg/dl).

2.3 Data Fuzzification

In reality, health parameters do not have to be one-dimensional, for example when measuring the blood pressure we obtain two values: systolic and diastolic pressure. In our algorithm each dimension should be treated as a separate health parameter. Another possibility is to map the more-dimensional health parameter into one-dimensional health parameter. Each health parameter has its own characteristics and its value changes on a particular way. We also define a range for each health parameter. Some of the values indicate normal health condition, and some of them indicate worsened health condition. The information about which values are desirable, and which are not, is very important for our recommendation algorithm. That is why we need to map the measurements (optionally transformed) into new classes that will give them semantic meaning. For example, the classes can be: "under normal", "normal" or "above normal", or if we are considering the age: "young", "adult" or "old". This mapping is made by an expert. On Figure 4 we can see the membership functions for BMI.

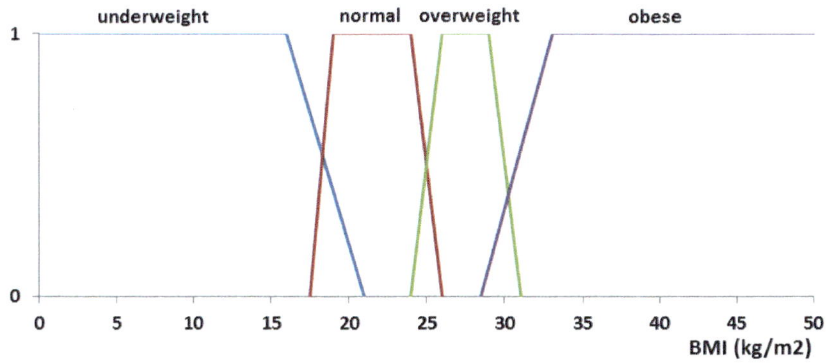

Fig. 4 Membership functions for BMI

If we separate the range of the parameter into smaller disjunctive subintervals where each subinterval represents a special class, then we reduce our mapping problem into discretization problem. In this case, if we consider heart rate as a health parameter with normal range [60, 100] heart beats in one minute, two people that have 61 and 59 heart beats in one minute should be placed in different classes. This would not be very useful because the small difference could be because of the noise in the sensor data. Additionally, the two values are on the boundary between two classes and we cannot be fully sure that the first value can be considered as a "normal", and the second value can be considered as "under normal". Each measurement should be assigned a membership to each class. For that purpose we should use fuzzy logic and data fuzzification. This process is also called fuzzy discretization and should be done by an expert.

Let's define the set of membership classes for the parameter p with L_p. For each measurement that refers to user u we should define the membership $MV_{u,p,l}$ to each parameter class $l \in L_p$. We will denote the membership of the last measurement with $MV_{u,p,l}$, and we will denote the membership of the last measurement before the moment t with $MV_{u,p,l,t}$. One measurement belongs to all classes, possibly with different degrees of membership. This means that the mapping of the measurement made at moment t can be represented as a vector $FuzzifiedMeasurement$ that consists of all $MV_{u,p,l,t}$ where $l \in L_p$.

2.4 Health Profiles

For each user and for each moment there is a health profile that can be defined as a combination of the values of her/his health parameters at that moment. The current health profile is represented by the combination of the most recent parameter readings. Some of the health profiles can be considered as "good" health profiles because the values of the health parameters belong in the normal ranges. Our algorithm tries to arouse transition of the health condition of the user from "bad" health profile to "good" health profile through the physical activities it recommends. A health profile can be represented as a vector:

$$Profile(user, time, FuzzifiedMeasurement_1, \dots, FuzzifiedMeasurement_P) \quad (2)$$

Health profiles can be generated at regular time intervals (in ideal case after each new measurement). However, we don't need to keep all the profiles in the database, but only the profiles which are considerably different from all health profiles that belong to the user saved so far. In a case where we have new profile $hp_{u,t}$ that belongs to the user u at time t, first we need to need to calculate the Euclidean distance between the two profiles $ProfilesDistance$ and if the distance between all existing profiles is larger than a given threshold T, the new profile is added to the database. The distance metric is defined as:

$$ProfilesDistance(hp_{u,t_1}, hp_{v,t_2}) = \sum_{p \in P} t_p \cdot MeasurementsDistance_p(hp_{u,t_1}, hp_{v,t_2}) \quad (3)$$

$$ProfilesDistance(hp_u, hp_v) = \sum_{p \in P} t_p \cdot MeasurementsDistance_p(hp_{u,t_1}, hp_{v,t_2}) \quad (4)$$

where

$$MeasurementsDistance_p(hp_{u,t_1}, hp_{v,t_2}) = \sqrt{\sum_{l \in L_p} \left(MV_{u_i,p,l,t_1} - MV_{u_j,p,l,t_2}\right)^2} \quad (5)$$

$$MeasurementsDistance_p(hp_u, hp_v) = \sqrt{\sum_{l \in L_p} \left(MV_{u_i,p,l} - MV_{u_j,p,l}\right)^2} \quad (6)$$

In the second formula we use the most recent measurements. We define the significance of the parameter p by t_p. Larger value of t_p means that the measurements of the parameter p will have bigger impact in the calculated distance between two profiles. If we are interested only in improving the state of only one health parameter, then the significance of that parameter should be positive, and the significance of all other parameters should be equal to zero.

2.5 *Recommendation Algorithm*

The algorithm for recommendation of physical activities consists of four main phases:

- Categorization of the users according to their diagnosis and filtering of all users that do not belong to the same category with the active user
- Selection of the users most similar to the active user according to the history of the health profiles by using collaborative filtering
- Calculating the usefulness of the activities to the active user and his similar users by using their health history and history of performed activities
- Generation of recommendations by using the calculated usefulness of the activities

2.5.1 Categorization According to Diagnosis

Some physical activities can worsen the health condition of people that have certain diagnosis and that is why they should not be recommended to those people. If the conclusions for the usefulness of the activities are based only on the data from the users belonging to the same group as the active user, they would be more accurate because the activities that have a negative effect to the group of people with a certain diagnosis would get much lower relevance so they would not be recommended. Categorization of the users is made to that users having the same diagnosis and the same set of permissible activities are grouped together. In this phase we need expert knowledge to define the categories and the conditions for membership in these categories.

We need to automate this process because the expert might not be always available. There are two ways to perform categorization: through implementation of expert rules or through building a model using classification algorithms. The first way is more expensive in a case when there are many categories, many different diagnoses or complicated conditions for membership in the categories. That is why the best choice is to use classification algorithms that can build a model using a relatively small set of labeled samples. In this phase of the algorithm we will use the vector $Restriction$ which contains information about the users' diagnoses. First of all, medical expert should define the categories and afterwards she/he should manually classify part of the set of vectors which will be used as a training set. All users should be classified using the classification model. If U is the set of all users and u^* is the active user for which we want to generate recommendations, we should design a function $FilterAccordingToDiagnosis$ that will return the users that belong to the same category as the active user:

$$\begin{gathered} U' = FilterAccordingToDiagnosis(U, u_*) \\ U' \subseteq U \end{gathered} \tag{7}$$

If all diagnoses are considered as independent features, we can use Naïve Bayes classifier. We can assume Gaussian distribution for all features and after we determine the parameters of the distributions we will assign the most probable category to each unclassified user.

Decision tree is one kind of inductive learning algorithms that offers an efficient and practical method for generalizing classification rules from previous concrete cases that already solved by domain experts [8]. These kinds of algorithms are useful for many real life applications because the rules are easy to understand. Typically automatically generated diagnostic rules slightly outperformed the diagnostic accuracy of physicians specialists [9]. In our recommendation algorithm we will use decision trees. The most popular decision tree algorithms are IDE, C4.5 and CART [10]. Using these algorithms we can obtain very accurate rules that reflect the expert knowledge used to classify the training set.

The users from the same category have similar health problems. For each category there could be a set of activities that shouldn't be recommended to the users that belong to that category because those activities could worsen their health condition. An expert should determine the set of undesirable activities for all categories. After the user is assigned a category, we should filter out all undesirable activities that refer to that category. In this way we will obtain a set of activities that are potential candidates for recommendation. We define function $FilterAccordingToCategory$ that filters out all activities that could be harmful for the category in which u_* belongs:

$$A' = FilterAccordingToCategory(A, u_*)$$
$$A' \subseteq A \tag{8}$$

In the subsequent phases of the algorithm we should find the usefulness of each type of activity from A' and recommend the most useful activities.

2.5.2 Selection of the Most Similar Users

The second phase of the algorithm is used to select the users that belong to the same class as the active user and that are very similar to the active user regarding their health history. In this way we want to obtain better data that would result with better recommendations. Here we will use the set of health profiles. The main assumption is that if two users had the same combination of parameter values in the past, there is bigger probability that similar latent factors affect their health condition. The current health profile of the active user is analyzed and is compared with the saved health profiles of all other users from the set U'. If some user has at least one health profile similar enough (according to some measure such as Euclidean distance) to the current health profile of the active user, then we declare this user as similar to the active user and his data are used in the next phase of the algorithm. For selection we define the function:

$$U'' = FilterAccordingToProfile(U', u_*)$$
$$U'' \subseteq U' \tag{9}$$

so that for every $i \in U''$, there is at least one $hp_{u,t}$ for which:

$$ProfilesDistance(hp_{u^*}, hp_{u,t}) < T \tag{10}$$

Data that belongs to users from the set U'' (health history and history of performed activities) is used to find the connection between the change of the active user's parameter value and the performed activities. If there are many users that are similar enough to the active user, then we should select k most similar users (for example the first 100). In a case when there are not enough most similar users, the threshold should be increased (in this way there is bigger probability that the generated recommendations will be less relevant).

This type of filtering selects data which could be more useful. If some user in the past had worsened health condition which is the same with the current health condition of the active user and succeeded to improve it on better, her/his data can improve the recommendations for the active user because it potentially contains information about which activities contributed towards improving the health condition. This is on the assumption that in the health history of the similar user there is some improved health condition some period after the moment of storing the profile which is declared as similar to the current health profile of the active user.

2.5.3 Calculating the Usefulness of the Activities

The main part of the recommendation algorithm is the phase where we calculate the usefulness of the activities. We are given the current health condition of the user (the combination of the values of her/his health parameters) and we want to find activities that would improve the values of the parameters that are into worsened state.

In the previous phase we used both types of health parameters (descriptive and control) to find the most similar users, but in this phase we will use only the set of control parameters P'. We examine the history of measurements and performed activities for each user from the set of similar users U''. For each performed activity we check its influence on the change of the parameter value. Our main assumption is that the activities that happened in the interval between two measurements influenced the parameter change with intensity that depends on the moment of occurrence and that is proportional to the effort given to complete the activity.

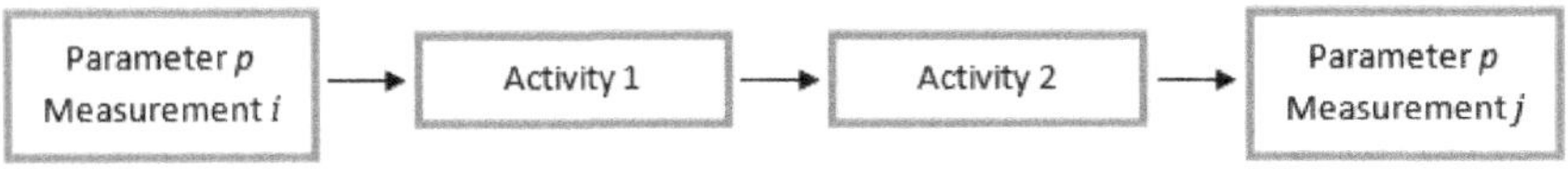

Fig. 5 Activities that happened in the interval between two measurements i and j $(i < j)$ contributed to the change of the parameter value

We cannot assume that immediately after the completion of some activity there will be a change on the parameter value because some time should pass until the effect of the activity could be visible. On the other side, we cannot assume that the results of a single activity will be visible after a long time period (few months). We create a model that will represent the effect of the activity on the parameter value after its completion. By using this model we find out which measure made after the activity best reflects the effect of that activity. We say that this measurement has the biggest validity. We also need to find out the closest measurement before the activity is started, and its validity can be calculated using the model shown on Figure 6.

The function that will define the model should assign bigger validity to measurements that were made shortly before the activity was performed. We need to find the measurement that has the biggest validity and that happened before the activity is executed. On the basis of our requirements for the shape of this validity function, we chose a function that has the form of cumulative normal distribution. Our choice for a function additionally decreases the computational complexity of our algorithm. We also need to find a measurement after the activity is completed that will best reflect the effect of the activity. The validity after the activity should slowly increase, then it should reach a maximum and afterwards it should slowly decrease. We chose Gamma distribution to model the validity of the measurements after the activity. This distribution has two parameters and its

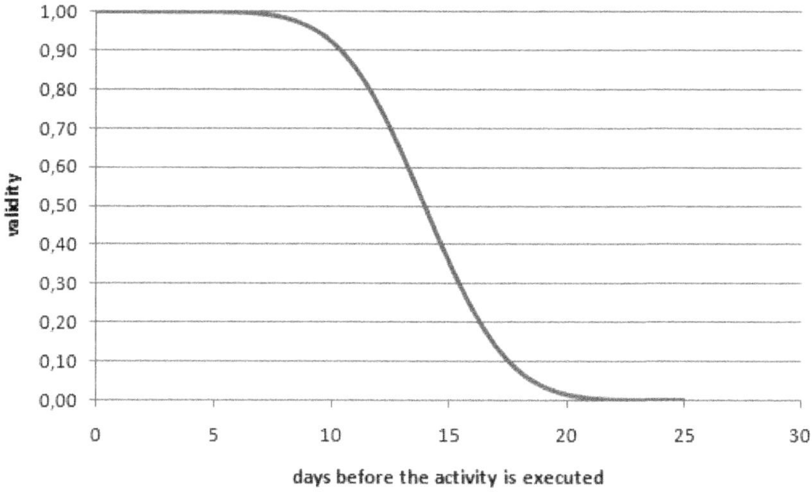

Fig. 6 Validity of a measurement before the activity is executed

values are chosen to that the distribution has the form shown on Figure 7. One of the most important characteristics of this validity function is the moment when the maximum is reached. Before and after that moment the validity decreases. In our implementation we choose this moment to be 7 days after the activity is completed. In this way we assume that the maximal influence of the activity is after 7 days. This value can be changed if we are presented with more accurate information.

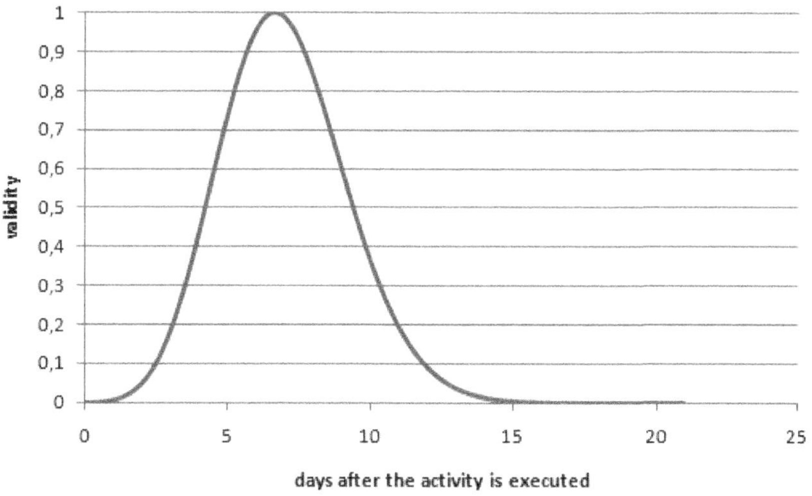

Fig. 7 Validity of a measurement after the activity is executed

Let's define a function $dir(u_*, p)$ that as a result returns 1, -1 or 0 depending on whether the value of the parameter p for the active user u_* should be increased, decreased or it shouldn't be changed. The desirable parameter values indicate good health condition and depend on what the active user wants to achieve. By using the validity models (before and after the activity) we define functions $prev_p(activity)$ and $next_p(activity)$ that as a result return the measurements with the biggest validity before and after the activity. We use them in the formula for calculating the usefulness of each type of physical activity to the parameter p for each user u from the set of similar users U'':

$$V_{u,a,p} = \frac{imp_p \cdot \sum_{a_u} \left(\frac{next_p(a_u) - prev_p(a_u)}{timeSpan(duration(a_u))} \right) \cdot val(a_u) \cdot dir(u_*, p) \cdot difficulty(a_u)}{num(a_u)} \tag{11}$$

$$val(a_u) = validityPrev(a_u) \cdot validityNext(a_u)$$

$$\forall u \in U'', \forall a \in A', \forall p \in P'$$

The variables and the functions that are used in the formula are:

- U'' is the set of similar users for the active user u_*
- A' is the set of different types of activities
- P' is the set of control health parameters
- a_u is a variable that alters all activities of type a performed by user u
- $next_p(a_u)$ is a function that returns the value of the measurement made after activity a_u that has the biggest validity
- $prev_p(a_u)$ is a function that returns the value of the measurement made before the activity a_u that has the biggest validity
- $duration(a_u)$ denotes the duration of the activity a_u
- $difficulty(a_u)$ denotes the difficulty of the activity a_u
- $num(a_u)$ is the number of activities of type a performed by user u
- $val(a_u)$ is a function that returns the validity of the measurements that are used in the analysis of a_u
- imp_p denotes the importance of the parameter in improving the health condition of the active user. The parameters which are into worsened state should have bigger importance. We need to assign even bigger importance to the parameters that are into worsened state and for which it's harder to cause more significant change on their values.
- $timeSpan(x)$ is a function of x where x represents the duration of the activity. This function is an important factor in calculating the change of the parameter value per unit of time. Longer execution of the activity might mean that the user has put smaller effort in completing the activity. The simplest model of this function is $timeSpan(x) = x$. However, in reality we cannot expect that the effect of the activity is disproportionate to the duration of the activity. That is why we suggest this function to have logarithmic form i.e. $timeSpan(x) = \log x$.

In the next step of this algorithm we calculate how useful each type of activity is to user u:

$$V_{u,a} = \sum_{p \in P} V_{u,a,p} \tag{12}$$

We insert these values into a matrix of useful types of activities:

$$M = \begin{matrix} & a_1 & \dots & a_r \\ u_1 \\ \vdots \\ u_m \end{matrix} \begin{bmatrix} V_{1,1} & \cdots & V_{1,r} \\ \vdots & \ddots & \vdots \\ V_{m,1} & \cdots & V_{m,r} \end{bmatrix} \tag{13}$$

2.5.4 Generation of Recommendations

There are few different ways to utilize the results produced in the previous phase in order to make recommendations. The simplest method is to calculate the usefulness V_a for each type of activity according to the formula:

$$V_a = \sum_{u \in U''} V_{u,a} \tag{14}$$

and to recommend the activity with the maximal V_a. This is not always practical because some type of activity could have significantly bigger usefulness to some user (comparing with the other users) and that value will dominate in the sum. In this way a type of activity that is very useful only to a small subset of the set of similar users would be recommended to the active user. Another way is to recommend the activity which is considered as the most useful to the largest number of similar users. First, for each user from the set of similar users we should create a vector of useful activities:

$$\begin{gathered} \forall u \in U'', \quad UA_u = k \quad where \quad V_{i,k} = \max_{j \in A'} V_{i,j} \\ UA = [UA_1 \ldots UA_m] \end{gathered} \tag{15}$$

$$\begin{gathered} \forall k \in A' \quad count(k) = \textstyle\sum_{i=1}^{m} r_i \quad \text{where} \quad r_i = \begin{cases} 1, k = UA_i \\ 0, k \neq UA_i \end{cases} \\ RA = u \quad where \quad count(u) = \max_{k \in A'}(count(k)) \end{gathered} \tag{16}$$

Activity type RA is recommended to the active user u_*.

3 Evaluation by Simulation

Our recommendation algorithm tries to find the usefulness of each type of activity on the bio-medical parameter change. Activity is considered useful if it changes the global parameter value towards the desired one. The change of a parameter value might be influenced by many factors. It is impossible to make a mathematical model that takes into account all these factors, so we tried to make a model for the parameter change, under the influence of the activities performed, that is simple and as closer to the reality as possible. We assume that each performed activity has some influence on the parameter change and that the parameter change is influenced only by the effect of the activities (pharmacological influence is neglected).

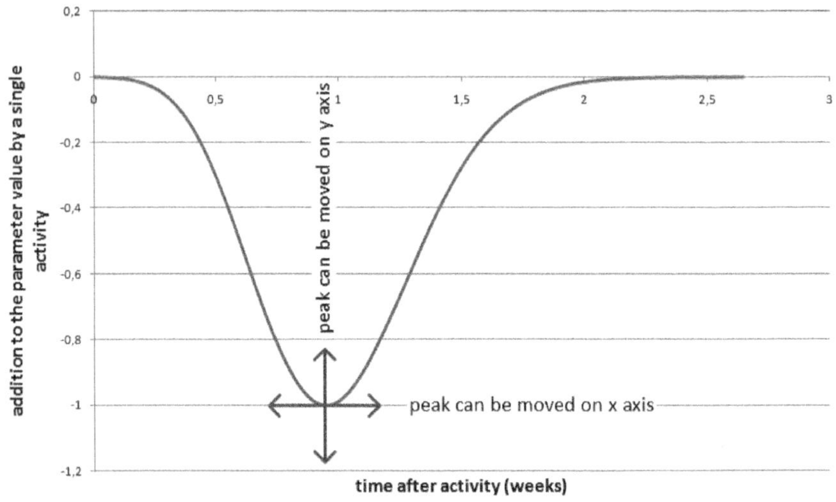

Fig. 8 Influence on the global parameter change by a single activity

We model a single activity influence to the global parameter change by a function whose shape is similar to a Poisson probability mass function (Fig. 8). In our model, it depends on the type of activity. The peak of the model function can have either positive or negative amplitude depending on whether it has a stimulatory or inhibitory effect correspondingly. Gaussian noise is added to the model function. When multiple activities from different types are performed, each of them influences the parameter value. So, the parameter value at each moment of time can be defined by the expression:

$$f(t) = \sum_{i} h(type_i, timeOfStart_i, t) \tag{17}$$

Gaussian noise is also added to $f(t)$. On Figure 9 and Figure 10 we can observe two global parameter functions, the first one without noise and the second one with noise.

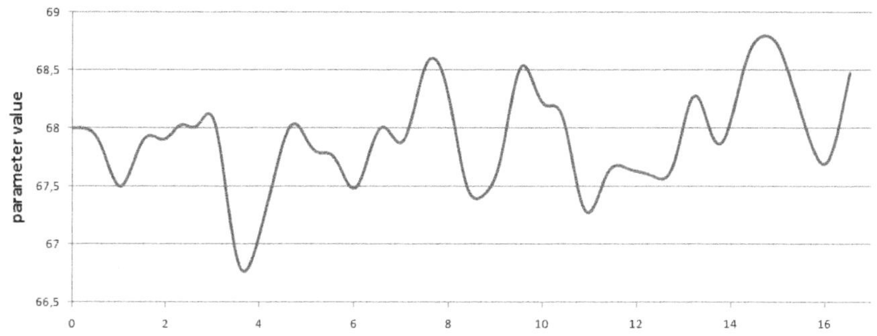

Fig. 9 Global parameter function without noise

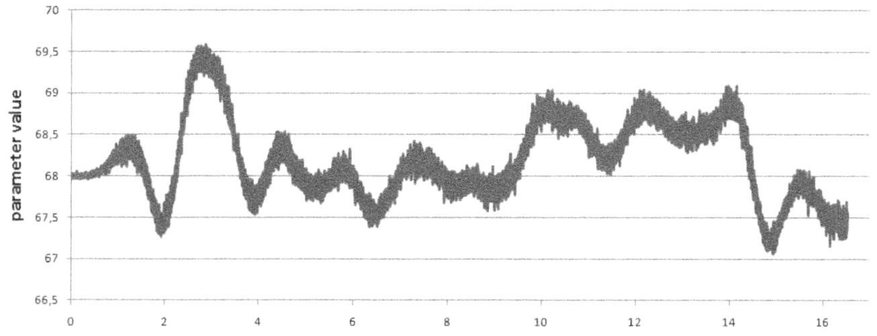

Fig. 10 Global parameter function with noise

The simplest case to test the correctness of our algorithm is to use two types of activities which are symmetrical. The first one has positive peak amplitude and the second one has negative peak amplitude. Our algorithm should "guess" which activity increases the parameter value and which doesn't. If we flipped a coin we could guess with 50% accuracy which activity has more usefulness (in our experiment we assume that activity is useful if it increases the global parameter value). In our simulator we use these parameters:

- Duration of the simulation
- Average time between consecutive activities
- Average time between consecutive measurements
- X coordinate of the peak (relative to the start of the activity and same for both activities)
- Y coordinate of the peak (both activities have peaks symmetric about the y-axis)
- Standard deviation of the peak (x coordinate)

- Standard deviation of the peak (y coordinate)
- Standard deviation of the Gaussian noise (single influence)
- Standard deviation of the Gaussian noise (global parameter function)

We expected that by increasing the standard deviation we would get lower accuracy, but by increasing the duration of the simulation and the average time between consecutive activities we would get higher accuracy. Longer duration of the simulation means more activities and more data and longer average time between consecutive activities means that it could be easier to distinguish between consecutive activities by observing the global parameter function. On the other side, longer average time between consecutive measurements means that we will have less data for our recommendation algorithm and less accurate recommendations. We applied our recommendation on generated data and we obtained correct results when we didn't have standard deviation of the peak and of the noise. We wanted to know how the algorithm behaves when we change the parameters of the simulation. We were especially interested how the accuracy of the recommendation algorithm changes in the border cases (Figure 11 and Figure 12).

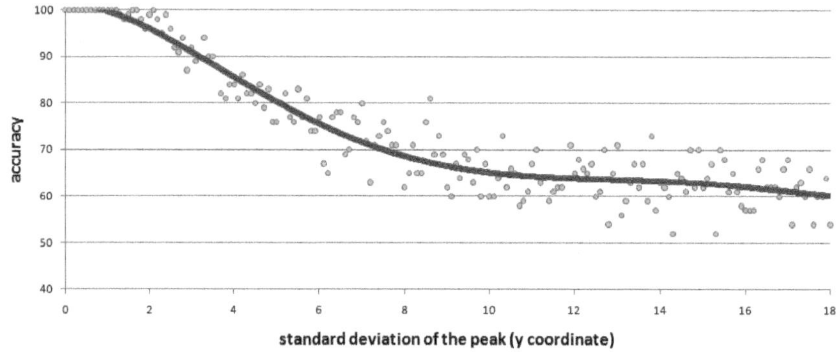

Fig. 11 Accuracy as a function of the standard deviation of the peak (y coordinate)

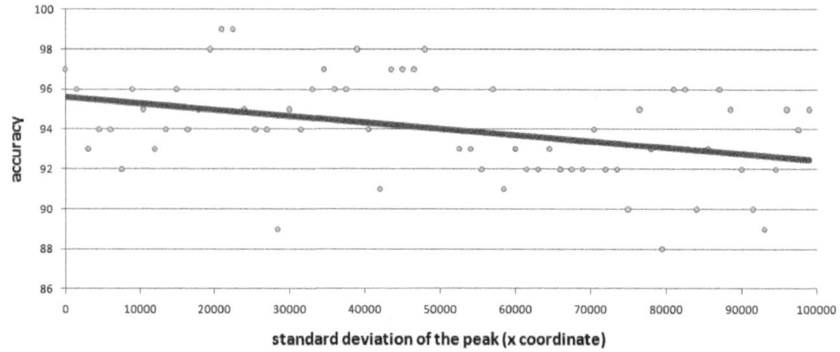

Fig. 12 Accuracy as a function of the standard deviation of the peak (x coordinate)

We chose a set of values for the parameters of the simulator in order to see more clearly the way the accuracy changes. We tried to define a curve that fits the results we obtained. In the first case, when we observed the accuracy as a function of the standard deviation of the peak (y coordinate) we noticed that there is some threshold after which the algorithm gives bad results, however, this threshold is relatively big comparing to the amplitude of the peak. Surprisingly, the increase of the standard deviation of the peak (x coordinate) didn't cause significant worsening of the results.

The increase of standard deviation of the noise caused exponential decline of the accuracy. This is shown on Fig. 13 and Fig. 14.

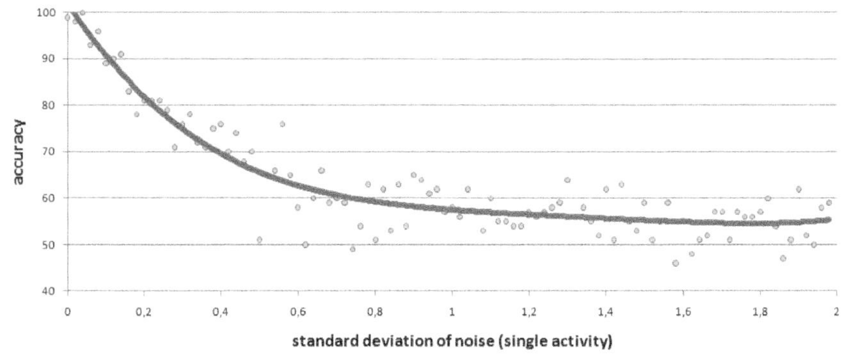

Fig. 13 Accuracy as a function of the standard deviation of the noise (single activity)

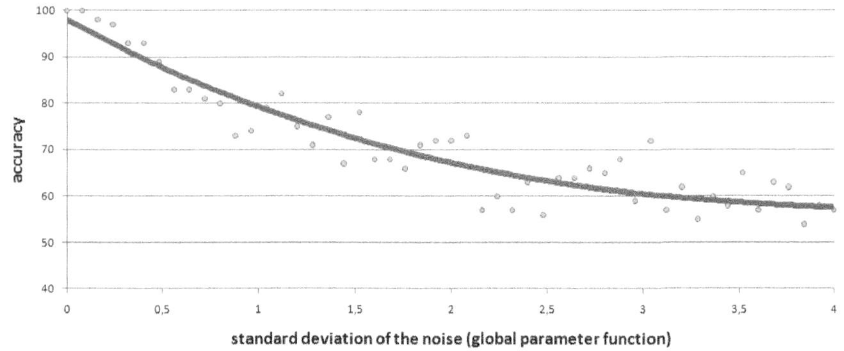

Fig. 14 Accuracy as a function of the standard deviation of the noise (global parameter function)

Case Studies

Case Study 1

The user (with heart problems) switches on the application on his phone. The application connects to Bluetooth devices which measure her/his heart rate, blood pressure and blood sugar level. The application uses its knowledge base to determine whether her/his health parameters are in the normal ranges. The application searches for users that have the same diagnosis and it concludes that walking for one hour could improve her/his health condition. The user tells the application that she/he will start her/his activity (walking). The application tracks the GPS coordinates at every moment during the activity. In the end of the activity the user tells the application that she/he has finished the activity.

Case Study 2

Some person A decides to practice more cycling. She/he uses the mobile application to record data each day about her/his weight, blood pressure, blood-sugar level and heart rate. Every two days she/he cycles for about 1 hour. She/he starts the application before she/he starts with the activity. The application measures the distance passed and the velocity. After the activity, the application decides what the level of difficulty of the activity was. All these data are saved in the social network. After 2 months the user notices that she/he has lost weight (3 kilograms). After that some other user B installs the application on her/his mobile phone and tries to get recommendations for losing weight. First she/he records data about her/his health parameters and she/he tells the application that she/he wants to lose weight. The application finds other users which are similar to the user B (gender, weight, heart rate, diagnosis). Among them is the user A. Some of these users succeeded to lose weight in the past. One of them is user A. The application concludes that cycling is the best activity to lose weight and this activity is recommended to the user A.

Future Directions

The future healthcare systems will have sense making component in order to increase the understanding of patient's personal condition, clinical or living environment and, in that way, to increase the healthcare system efficiency by increasing its availability. It is very important to discuss the ways in which information systems that support healthcare can make that kind of sense making. COHESY is a health care system that gathers a lot of information from the patient's bionetwork, his medical record, the environment, his interaction with the mobile application, and this information is used to make conclusions about the health condition of the patient and the ways in which it can be improved. Its recommendation algorithm gives the people bigger control over their own health and supports their actions. We expect that besides the presented algorithm, other decision support algorithms will appear that will utilize the data provided by

COHESY. They could focus on making different types of conclusions. The increase of the number of users will bring more data and this means that the generated conclusions and recommendations could be more accurate. The patterns that are hidden behind the connections between patients can lead to better understanding of their current health condition. That is why we believe that collaborative algorithms could be very beneficial for health care systems.

There are few other reasons why we can expect that collaborative health care systems will be intensively used in the future. The first reason is that the technology is advancing rapidly and this means that the sensors would become more accurate and would be easier to use. We can also expect that new sensors would emerge that could measure different health parameters. This leads to bigger and better data available to the collaborative algorithms. The second reason is that the recommendation systems and collaborative systems could be a promising approach for preventive and curative healthcare solutions. The development of these algorithms along with additional information which can be relevant in health care can lead to more accurate recommendations.

Modern medical technology helped people a lot, and there will be even deeper integration of the technology in the health care systems. These systems will be more aware of the health condition of the patients and the types of conclusions will be even closer to the conclusions made by doctors. Our proposed system is a step towards smarter health care. The recommendations help people to be more aware of their actions and the consequences that they cause.

Conclusion

The recommendation algorithm presented in this chapter is the main component of the collaborative health care system model – COHESY. The purpose of the algorithm is to find the dependency of the users' health condition and the physical activities she/he performs. It gives recommendations for performing specific activities that would improve users' health. To achieve this we consider datasets from the history of measurements and the history of performed activities of users. Data is obtained by the user through its interaction with the mobile application and by sensors connected to the smartphone.

The algorithm consists of four phases: categorization of the users according to their diagnosis, selection of the users most similar to the active user, calculating the usefulness of the activities and generation of recommendations. First two steps are not necessary, but they contribute to improving the quality of the recommendations by selecting the most relevant data. We assume that if two users had the same combination of parameter values in the past, there is bigger probability that similar latent factors affect their health condition. The selection of the similar users and the usage of their data represent the collaborative component of our algorithm.

The proposed algorithm is different from the standard recommendation algorithms because of the complexity of the problem it solves. The time dimension and the uncertainty of the feedback make the solving of the problem difficult. We want to find the influence of the physical activities on the change of the health condition, but there are many other factors that affect on this change, for example: medications, food, and emotional state. That is why we cannot assume that the generated recommendations will be always correct. We try to overcome this downside by increasing the amount of data used for calculating the usefulness of the activities.

The fuzziness which is included in the processing of the sensor data allows us to obtain their semantic meaning. In this way the algorithm receives information about which parameter values indicate normal state, and which indicate worsened state. Also, this representation allows the algorithm to have equal look at the health parameters regardless of their characteristics and the different ranges in which their values belong.

The experiments conducted with generic data show that the accuracy of the algorithm increases with the duration of the observation i.e. with the number of activities observed and the amount of data obtained. The behavior of the algorithm on the increased data uncertainty is also evaluated. The results show that the algorithm is very robust and promising. It can be implemented efficiently and it offers a lot of possibilities for adaptation. The generated recommendations allow the user to adapt and align her/his physical activities in order to improve her/his health condition with bigger confidence. In this way, users have bigger control on their own health.

There are few reasons why we can expect that collaborative health care systems will be intensively used in the future. The first reason is that the technology is advancing rapidly and this means that the sensors would become more accurate and would be easier to use. We can also expect that new sensors would emerge that could measure different health parameters. The second reason is that the recommendation systems and collaborative systems could be a promising approach for preventive and curative healthcare solutions. The development of these algorithms along with additional information which can be relevant in health care can lead to more accurate recommendations.

Acknowledgement. The authors would also like to acknowledge the contribution of the COST Action IC1303 - AAPELE, Architectures, Algorithms and Platforms for Enhanced Living Environments.

References

1. Vlahu-Gjorgievska, E., Trajkovik, V.: Towards collaborative health care system model – COHESY. In: Proceedings of the 2011 IEEE International Symposium on a World of Wireless, Mobile and Multimedia Networks (WoWMoM), pp. 1–6. IEEE (2011)
2. Taylor, C., Dajani, L.: The future of homecare systems in the context of the ubiquitous web and its related mobile technologies. In: Proceedings of the 1st International conference on Pervasive Technologies Related to Assistive Environments, p. 43. ACM (2008)
3. Chakravorty, R.: A programmable service architecture for mobile medical care. In: Proceedings of UbiCare 2006: The first workshop on Ubiquitous and Pervasive Healthcare, p. 5. IEEE (2006)
4. Blount, M., Batra, V.M., Capella, A.N., Ebling, M.R., Jerome, W.F., Martin, S.M., Nidd, M., Niemi, M.R., Wright, S.P.: Remote health-care monitoring using personal care connect. IBM Systems Journal 46(1), 95–113 (2007)

5. Gama, O., Carvalho, P., Afonso, J.A., Mendes, P.: Quality of service support in wireless sensor networks for emergency healthcare services. In: 30th Annual International Conference of the IEEE Engineering in Medicine and Biology Society, EMBS 2008, pp. 1296–1299. IEEE (2008)
6. Nachman, L., Baxi, A., Bhattacharya, S., Darera, V., Deshpande, P., Kodalapura, N., Mageshkumar, V., Rath, S., Shahabdeen, J., Acharya, R.: Jog falls: A pervasive healthcare platform for diabetes management. In: Floréen, P., Krüger, A., Spasojevic, M. (eds.) Pervasive 2010. LNCS, vol. 6030, pp. 94–111. Springer, Heidelberg (2010)
7. Kahn, M.: Diabetes Data Set (1994), `http://archive.ics.uci.edu/ml/datasets/Diabetes`
8. Chao, S., Wong, F.: An incremental decision tree learning methodology regarding attributes in medical data mining. In: 2009 International Conference on Machine Learning and Cybernetics, vol. 3, pp. 1694–1699. IEEE (2009)
9. Kononenko, I.: Inductive and Bayesian learning in medical diagnosis. Applied Artificial Intelligence: An International Journal 7(4), 317–337 (1993)
10. Lavanya, D., Rani, D.K.: Performance evaluation of decision tree classiffers on medical datasets. International Journal of Computer Applications 26(4), 1–4 (2011)

Questions

Q1. How many usage levels does COHESY have? Please describe each of them.

Q2. How do we separate the health parameters?

Q3. What types of recommendations are generated by the recommendation algorithm in COHESY?

Q4. What types of data are used by the recommendation algorithm?

Q5. What is a health profile?

Q6. Which are the four main phases of the recommendation algorithm in COHESY?

Q7. Which phases can be omitted by the recommendation algorithm, but are important to generate more accurate recommendations?

Q8. How do we choose which measurements made before and after a particular activity are most relevant to describe the effect of that activity on the health parameter?

20. Software Architecture for Emergency Remote Pre-hospital Assistance Systems

Juan C. Lavariega, Alfonso Avila, and Lorena G. Gómez-Martínez

Tecnológico de Monterrey, Av. E. Garza Sada 2501, Monterrey, NL, Mexico
{lavariega,aavila,lgomez}@itesm.mx

Abstract. Vehicle accidents resulting in trauma occupy the fourth place as a cause of death, in middle and low-income countries and the tenth cause of death worldwide. Paramedics, physicians, and emergency room staff coincide that in order to decrease the possibility of a patient suffering an irreversible damage, qualified medical attention must be given within the first hour of the accident (the so called golden hour). However, first-response emergency workers in low and middle income countries are mostly volunteers with minimal medical training; additionally, there is a shortage of these first aid responders worldwide. In this chapter we show that, by using pervasive and ubiquitous information technology currently available, it is possible to alleviate part of the problems stated above. In particular we include the use of mobile devices, wireless communication and remote body health sensors to build a system that allows the delivery of professional medical assistance in emergency situations. In the chapter we present and discuss the *software architecture* for assisting paramedics while attending victims of traffic accidents. Our discussion is based on a project that we have developed called ERPHA (Emergency Remote Pre-Hospital Assistance).

Keywords: mHealth, Software Architecture, wireless sensor networks, Emergency Systems, vital signs monitoring, video and data transmit.

1 Introduction

In its fact sheet No. 310 [2] the World Health Organization indicates that the top ten causes of death in decreasing order are: ischemic heart disease; stroke and other cerebrovascular diseases; lower respiratory infections; chronic obstructive pulmonary disease; diarrhoeal diseases; HIV/AIDS; trachea, bronchus and lung cancers; tuberculosis; diabetes mellitus; and road traffic accidents. Ischemic heart disease was the cause of 7.25 million deaths and road traffic accidents caused 1.21 million deaths in 2011 according to the same source (equivalent to the 2.1 % of total deaths). In a more recent study [1, 2], worldwide the total number of road traffic deaths remains unacceptably high at 1.24 million per year.

Depending in the income level (low, medium or high) of the country included in the WHO report [1], the leading cause of death slightly vary, however traffic accidents move to the top four causes of death in low and middle income countries

S. Adibi (ed.), *Mobile Health*,
Springer Series in Bio-/Neuroinformatics 5, DOI: 10.1007/978-3-319-12817-7_20

where most of the world population resides. In Mexico, for example, 4.3% of total deaths are due to traffic accidents (that is more that the double worldwide) [3,5].

In Mexico, over 24000 people mostly minors and adults under the age of 35 died due to traffic accidents. A similar number (about 20,000) result with permanent severity injures. It is estimated that the cost of injuries and disabilities caused by the traffic accidents is around 9,200 million dollars, enough to feed 3 million children for a year according to UNICEF [19, 20].

Paramedics, physicians, and emergency room staff coincide that in order to decrease the possibility of a patient suffering an irreversible damage, qualified medical attention must be given immediately [1,4]. For this reason, the first hour after an accident occurs is called the "golden hour." Paramedics and emergency technicians (EMT) give pre-hospital care to the extent of their knowledge, but most of the time paramedics perform their duties under no medical supervision. Thus, the importance of having trained paramedics becomes extremely relevant; however the reality in many countries is that paramedics are under-prepared and under-trained. As an example, the Department of Public Health in Mexico reported in 2006 [3,4] that around 75% of paramedics do not meet the requirements dictated by international standards, but they cannot be replaced due to the insufficiency of personnel. This could lead to a patient not receiving the proper care before reaching the hospital during the "golden hour."

More recently IT and mobile technology have been applied to medicine and health system trying to alleviate the lack of specialized health personnel problems, such as monitoring specific illness groups (i.e., diabetes patients, elderly people) [8,9,10,12] or assisting professionals with a more expedite access to information (i.e., doctors, nurses and staff at hospitals) [9,10,11]. Our development is focused in emergency-remote assistance, and it is called "Emergency Remote Pre-Hospital Assistance" or just ERPHA.

Efforts in applying IT to health systems have derived in the development of tools for improving communication [12,]; allowing better and faster access to medical data [11,12]; increasing the understanding of pathologies, allowing distant patient monitoring and assisting with decision making. Several of these approaches are found in the accompanying chapters of this book.

Information technology and particularly, mobile IT, can reduce the rate of errors in three mainly ways: by preventing errors and adverse events, by facilitating a more rapid response when an accident has occurred, and by tracking and providing feedback about adverse events. Our approach strives for accomplishing the second strategy: facilitating a more rapid response and in some degree, we also support the last one: providing feedback about an adverse event.

ERPHA combines wireless technologies, hardware sensors and software architecture to provide continuous monitoring of a patient's condition during the pre-hospital period, enhancing the care by supporting the early intervention of hospital staff with key data such as the patient's vital sings, video, and geo-position of the incident and/or ambulance. The patient's information is collected via a body sensor network and the camera of the mobile device, such information is transmitted through a wireless network to the medical center where medical doctors can assist remotely to paramedics at the place of the incident. The information from patients is stored in a database; therefore a historical record of the patient is created that later it can be used to identify future treatment patterns.

In this chapter we describe how the architecture of the project enables the flow of information from accident victim all the way to the hospital professionals. The chapter also discusses on the type of functionality that these systems must provide, the tradeoffs in quality attributes that need to be kept in mind, and the description of the architecture and prototype details. We conclude by addressing future improvements regarding the project and the trends we envision in mHealth technology.

The remainder or the chapter is organized as follows. Section 2 presents an overview of other systems and emphasizes differences with respect to ERPHA. Section 3 describes the functionality and other requirements for ERPHA. Section 4 discusses the software architecture of our project. Section 5 presents details of the prototype implementation. Finally, Section 6 discusses future improvements and presents our conclusions regarding the project and the impact of technology in the health care sector

2 Health Systems, Mobile Technology and Wireless Communication

The introductory chapter of this book presents the concepts and definitions that surround the application of mobile information technology to health systems. Here, we briefly reinforce those concepts.

As Bates [6] addresses, health care continues growing extremely complex and demands the use of information technologies for keeping the pace with processing needs. One of the key opportunities for IT in health care is to improve safety by reducing errors and improving communication. Bates mentions that failures in communication remain among the most common factors contributing to the occurrence of adverse events and even fatalities.

A new generation of technology that includes computerized coverage systems for signing out, hand-held personal digital assistants, and wireless access to electronic medical records may improve the exchange of medical information. Also, the use of mobile devices has established a new trend in the development of e-health systems. Firstly, mobile devices are easy carried from place to place; secondly they provide enough computing capacity to perform several tasks; and thirdly, there is a wide availability of such devices for the general public.

Strategies for preventing medication errors and adverse events in primary attention include tools that can improve communication, make knowledge more readily accessible, require key pieces of information (such as the maximum dose of a drug), and assist with calculations. Other tools perform checks in real time, assist with monitoring, and provide decision support. ERPHA deals mainly with improving communication, accessibility to information, and monitoring. Other chapters in this book present excellent proposals and discussions on diverse technological challenges for mHealth, such as preventive and curative medicine; or mobile data collection.

Some other examples of mHealth systems, besides the ones presented in this book, include:

A system for decision support in treatment of patients with type I diabetes mellitus [15]. This system provides a risk assessment for long-term complications. Data exchange between a unit for the patient and a central manager occurs; data is also stored in a database server. By combining database technologies, simulation algorithms and data mining, the system provides an advanced processing of the stored data to support the decision making for the physician. Here mobility is restricted to input data to the patient unit and transmitting it to the database server

Another system called MoTeCh [16] takes advantage of the proliferation of mobile devices and helps to keep track of patients' schedules. The systems notifies via mobile phones to patients and community workers of incoming medical revisions. The system automates the delivery of information for routine reports and integrates with existing software applications for mobile data collection.

Using satellite communications instead of cellular networks a South American system [17] called TeleConsult establishes a network, among medical centers, for sharing images from ultrasound examinations, electrocardiogram and blood imaging. TeleConsult pretends to cover disciplines such as cardiology, gynecology pediatrics and infections from the region. In this case the coverture area is wider than in mobile networks but it is only available between fixed geographical points.

ERPHA is different from other proposals that focus mainly on supporting monitoring of patients with a known diagnosis (Diabetes, Alzheimer, or after some surgery) [7,8], and in preventing diseases by monitoring population at risk. In ERPHA, the relationship patient-doctor happens in real time and under unanticipated circumstances. Also in more conventional remote monitoring systems users (patients) are responsible of establish the communication by subscribing to a service through a mobile device[8]. Information collected by the device or input by the user is then transmitted to his/her personal doctor who can suggest an action to take. In ERPHA monitoring is not for one specific previously known diagnosis. ERPHA monitors several vital sings continuously in addition to the attention that the paramedic may give to the patient.

In the following sections we share our experience at building a system for remote monitoring victims in traffic accident situations.

3 Requirements in Health Systems for Remote Assistance

The first questions we need to answer while building a software system are: what is the objective that we are pursuing? And in which context the system is going to be used?. Then we can determine its functionality and architecture properties. The simplest way to contextualize the use and objective of our work is through visualizing scenarios, just as the following:

A car accident happens in a remote rural route, there is a person with severe injuries which put on risk the life of the individual, time is critical. Fortunately, the closest town provides primary medical services; an ambulance with paramedics arrives at the scene within minutes. One of the paramedics starts reviewing the condition of the patient, due to the severity of the injuries and the limited training by the paramedic; she decides to start her ERPHA system with her mobile device. ERPHA starts collecting vital signs from the patient, recording video from the scene and establishing connection with the central servers of ERPHA at the main hospital at a larger city. Now at one end of the systems lays the paramedic close to the patient, at the other end a professional certified traumatologist watches for the vital signs sent by ERPHA, additionally the video and geo-localization information sent by ERPHA give to the medic a better context of the situation; now paramedic can provide a better care for the patient.

The above scenario helps us to clarify our objective; therefore we can say that ERPHA´s objective is to provide a high-technology technical solution that will enable the monitoring of a patient's condition by a specialist physician during the pre-hospital period. This means that from the moment paramedics arrive at the accident scene, a specialist can monitor the patient's status by analyzing audio, video and data from his vital signs, in order to be able to treat the patient before he/she arrives to the hospital.

With the scenario in mind, we came to a better picture of the components required to build our system. Hence, ERPHA consists of a mobile device which gets readings from vital sign sensors, gets geo-localization information, and takes video; transmit all these data to a remote application that displays the video and data, records audio feedback from the medical expert at the medical center and sends it back to the mobile device. Vital sign sensors are enabled with Bluetooth technology. Therefore the main functionality in ERPHA is given by the interaction of the different sensors, the capture of video and the transmission and reception of data from the accident site to the central servers and vice versa.

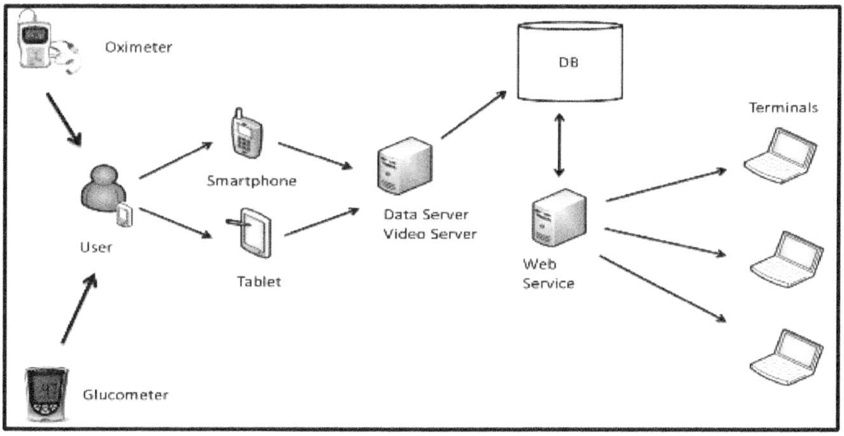

Fig. 1 . ERPHA overall architecture, physical view

The overall architecture of ERPHA is shown in Figure 1, where we can see that our system is a link between patient and paramedic, and a link between paramedic and doctor. The doctor is located at the central servers' site. Also in Figure 1, we can see the topological distribution of the major hardware components: body sensors (represented in Figure 1 only by an oximeter and glucometer); the mobile device acting as communication unit between the sensors and the main servers; the main servers (data, video and web) are located at a medical center. Computer terminals, located at the medical center too, access information in the servers and allow physicians to interact with paramedics at the accident site.

Once a clear picture of ERPHA has been formed we need to define and refine the requirements of the system. Requirements for a system include elements for functional requirements, quality attribute requirements, and constraints [13]. It is important to mention that quality attributes and constraints often referred as non-functional requirements. Following, we describe these elements for ERPHA.

Functional Requirements. These requirements state what the systems must do and how the system must behave to external stimuli. Other alternative for functional requirements is to refer to them as services. In ERPHA, our main functional requirements are:

At the mobile device:

1. To. collect bio-data from body sensors attached to the patient, including: blood pressure, breathing rhythm, pulse, temperature, and eventually any other vital sign if a sensor reader is available.
2. To activate camera and collect video.
3. Send sensor data, video and geo-localization coordinates to a data, video and web servers
4. If available, transmit patient´s electronic health record (ERH) identifier

At the medical center:

1. Allow medical experts select incident from list of incidents available at the data servers.
2. From the selected incident
    a. Display information from sensor in separated graphs, depending of the reading (blood pressure, breathing rhythm, pulse, temperature, etc.)
    b. Interpret the geolocalization data, and display in a map the localization of the unit
    c. Display the video transmitted from the accident site.
    d. If patient´s EHR id is available search and retrieve the corresponding patient medical history.

Quality Attributes Requirements. These are qualifications of the functional requirements such as how fast the response time must be. Quality attributes are also qualifications of the overall product such as the time to deploy the system, or how fault-resistant is the system [13]. Following we describe the main quality attributes for ERPHA, preceding by a brief definition for each quality attribute.

Interoperability is about the degree in which two or more systems can usefully exchange information via interfaces in a particular context. In ERPHA, we require the system to be interoperable by communicating externally with vital sings sensors and if available with electronic health records. Internally, we also have some sort of interoperability because a communication mechanism must exist between the mobile part (at the accident site) and the centralized component where the doctors are.

Modifiability is the property of a system of been prepared to accept changes and to minimize the cost of them. In ERPHA, we have to be prepared to incorporate new vital sings sensors or to change one of them for an updated model. Also we have to prepare the system for new requirements an future development such as perform data analytics in the incidents data base or in the video server.

Usability is the system´s ability to facilitate users to accomplish a desired task and the user support the system provides. The people whom ERPHA is intended to serve are doctors, paramedics, and the patients derived from an accident. Another possible audience may be hospital personnel and database administrators. The most relevant of them are paramedics and doctors. ERPHA has two front ends: the mobile unit interacting with the paramedic, and the web interface at the desktop or laptop computer connected to the central server interacting with the doctor. At the mobile device the user interface has to be very simple and intuitive so the paramedic can start the system to read and transmit data as soon as possible. The web interface must allow the doctor to select an accident event easily from a list of events; and must present in just one view all the information related to that incident.

Performance is the system´s ability to respond to events (interrupts, messages, request from users, etc.) in an acceptable time frame. For ERPHA, performance becomes relevant at the mobile device, where it is critical to read continuous information from the body sensors and to retransmit this information plus the captured video to the central components at the medical center. Also the response time must be within acceptable limits to provide an accurate diagnostic and valuable feedback to the paramedic.

Availability is the software property that determine the software is there and ready to carry out its task when the user needs it. Availability also implies how resilient to failures the system is. ERPHA is required to be always available to be used. We don't have control of when and where accidents happen. Also, once the system starts working, data transmitted between the mobile device and the centralized component should not be lost or interrupted.

Constraints are design decisions with zero degrees of freedom, those are design decisions already been made, such as the use of some standard for documentation; or the selection of a programming language; or the restriction to use a specific hardware platform. Constraints also set the scope of the system due limitations on technology, use environment, or legal regulations. ERPHA is limited to work in geographic areas where 3G or Wi-Fi service is available. Also ERPHA is only available for functioning with devices equipped with Windows Mobile, or Android operating systems. The mobile device must also support Bluetooth connectivity and must be equipped with a camera. At the centralized component the system is restricted to work only in Windows operating system (currently WindowsXP and Windows7). The feedback channel from the doctor to the paramedic is not part of the system at this moment, but in such a case we suppose doctors and paramedics can use a civil band radio communication.

4 Software Architecture for Health Systems for Remote Assistance

A software architecture is the blueprint that captures the relevant decisions throughout the accomplishment of functionality and quality attributes of a system including hardware and software components. In this section we describe such hardware and software elements and the relationships among them for ERPHA.

We follow the common practice for describing software architectures using the Krutchen 4+1 view model approach [14,13]. The 4+1 view model uses four complementary descriptions of a software product: the physical view, the logical view, the development view, and the process view; plus one integral view based on use cases and scenarios.

4.1 Software Architecture Description

The physical view depicts the topology of the system with its software components on the physical layer, as well as the physical connections between these components. The ERPHA physical view was already illustrated in Figure 1 above. In Figure 1, we can see that ERPHA architecture consists of six major components: the Body Sensor Network (BSN), a mobile device, two servers, a database and terminals or workstations with Internet access.

- The BSN is a set of sensors that provides readings of vital signs from an accident victim. The BSN connects, via Bluetooth, each sensor to a mobile device which acts as communication unit. The vital signs read in ERPHA are blood pressure, heart rate, respiration rate, and temperature. These vital sing are the most widely used to assess the basic body functions [18]. ERPHA is prepared to add other health metrics such as oxygenation of the blood or the heart's electrical activity with additional sensors.The BSN is constructed using commercially available sensors that use the Bluetooth communication protocol.
- The mobile device reads the sensed data from the BSN and sends it to the processing information server which processes data and video. If the mobile device has camera capabilities, it will also send the video of the patient as she/he is being transported to the hospital. The communication with the server is done through any of the following wireless technologies: 3G, Wi Max, or Wi-Fi. When a paramedic activates ERPHA in the mobile device, an incident session-patient id is created.
- The information processing server for data and video receives the information from the mobile device and saves it into the database using the patient session id. Therefore, it is possible to receive data simultaneously from different incidents and patients.
- The database records information of each session for each patient that is being assisted at the accident site and during his/her transportation to the hospital.
- The second server is the web server that is accessed by doctors at the hospital site. The web server provides content to terminals or workstations and displays the patients' status. Each patient can be monitored individually.

- The terminals or workstations are the means to present information to hospital doctors. Terminals only require a web browser supporting basic HTML and JavaScript features to view the patient vital signs and video.

The logical view is concerned with the functionality that the system provides to end-users and the information the system process. A key elements of the logical view in ERPHA is its simple data model, that allow us to keep track of the patient vital sings and monitor his/her evolution while reaches the hospital. Figure 2, shows ERPHA data model. The data model is implemented in the database.

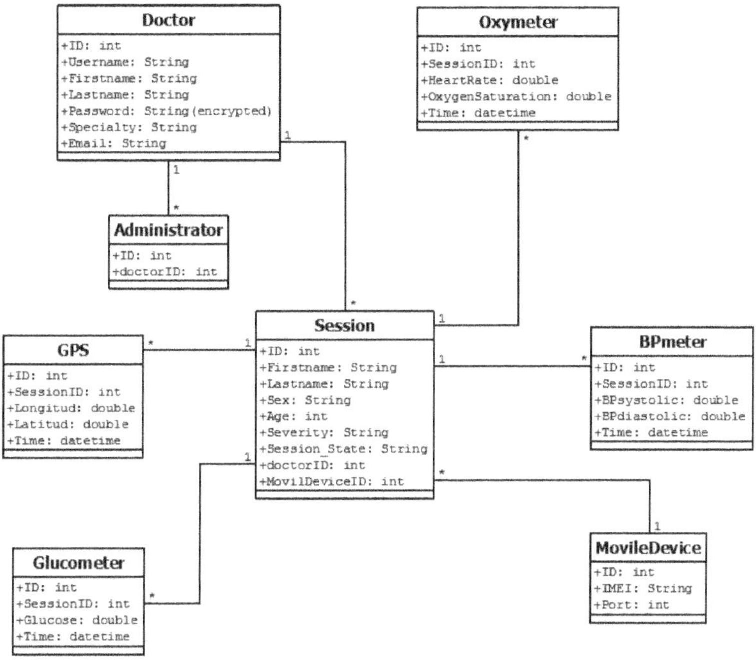

Fig. 2 ERPHA data model, architecture logical view

The main entity in our data model, shown in Figure 2, is the *Session* which keeps record of the patient being monitored and relate him/her to his/her vital sings, in addition to her/his location and the doctor at the hospital that is reviewing the patient´s evolution. Generals for patients are stored here like name, approximate age and sex. This information can be omitted in benefit of starting monitoring vital sings as soon as possible. Unique value of the Session entity is its ID created at the moment the paramedic arrived at the accident site and started ERPHA in the mobile device. The MobileDevice entity keeps information about the device sending the vital sings and video. With the device information we can identify the camera and link the video stream to the corresponding active session.

The development view illustrates a system from a programmer's perspective and is concerned with software management. It describes the software modules, packages, subsystems and/or software libraries. Figure 3 shows the ERPHA architecture in its development view, where the software environments are assigned to the main components. In particular Figure 3, shows the development view for our Android OS architecture. In fact the difference between the Windows Mobile OS and the Android OS is on the mobile devices. The servers for data video and web as well as the web interfaces are almost identical for both versions, except minor differences in the user interface.

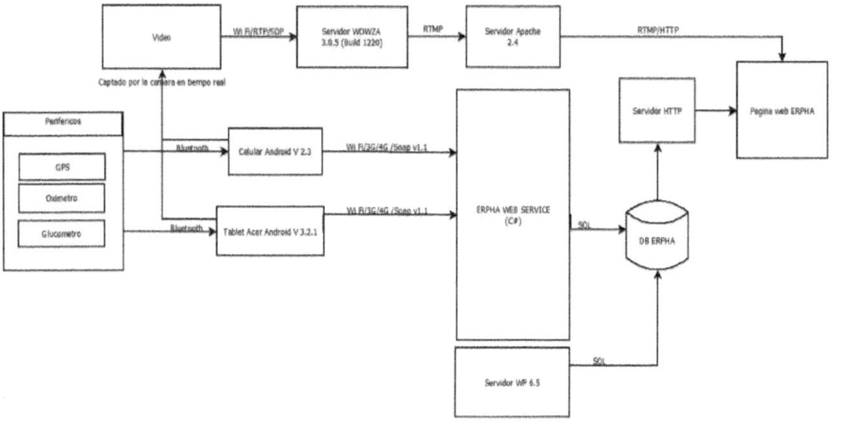

Fig. 3 ERPHA architecture, deployment view

The process view deals with the dynamic aspects of the system, explains the system processes and how they communicate, and focuses on the runtime behavior of the system. Figure 4 shows ERPHA's activity diagram for describing the interaction and flow of events between the main components and actors. In Figure 4 we identify, using the swim lane style of activity diagrams, the actions a paramedic and a doctor perform or initiate; and the two ERPHA front end components: the mobile device (identified as App in Figure 4) and the web interface at the terminals (identified as Site in Figure 4).

Lastly the use case view (also known as the additional view or fifth view) illustrates a system architecture using a small set of use cases, or scenarios. The scenarios describe sequences of interactions between objects, and between processes. Part of this description was already presented with the motivational scenario at the beginning of Section III. Scenarios and formal use cases complement each other to form the integrated view that put a system in the context of its use. The main use case diagram for ERPHA is illustrated in Figure 5, where three type of actors interact with ERPHA: Administrators which set the system and create other user accounts; Doctors which are localized at the hospital and select from a list of active incidents one to monitor; and Paramedics which initiate and terminate the vital sing transmission from the mobile device.

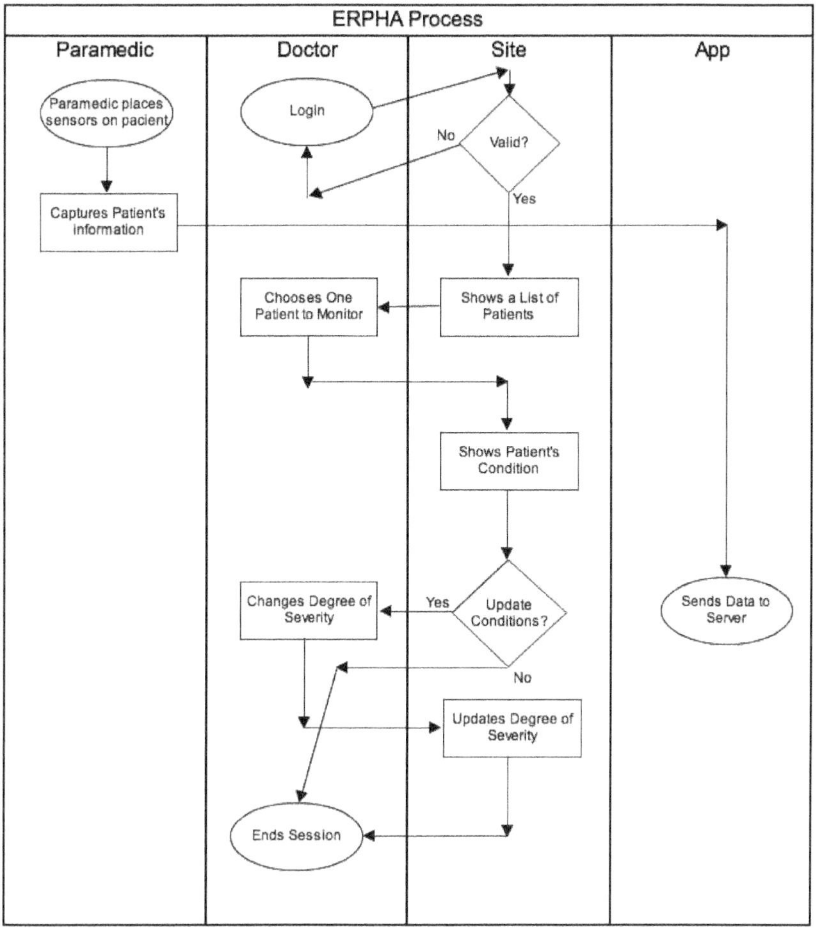

Fig. 4 ERPHA activity diagram, process view

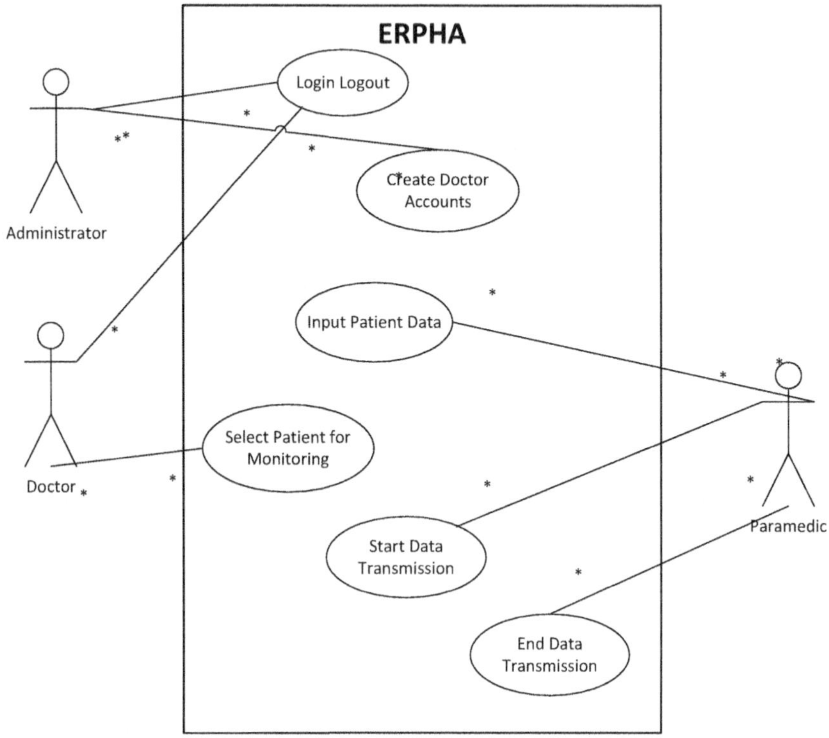

Fig. 5 ERPHA main use case diagram, use case view

4.2 Addressing Quality Attributes in the Architecture

In Section 2, we define the ERPHA requirements; there we included functionality, quality attributes and constraints. Here we discuss how the quality attributes are reflected in the architecture we presented and the tactics we use to include them.

For the interoperability attribute in order to communicate with sensors and other systems we make use of standard protocols for data exchange and communication. In the case of sensors communication is via Bluetooth protocol. If information about EHR of a patient is available we are defining a considering the use of the HL7 standard communication protocol.

With respect to the modifiability attribute, we need to be prepared for changes in the sensors boxes or for adding new sensors easily. We accomplish modifiability by making use module decomposition and change localization to the mobile device only. Also internally for the front end applications (web and mobile) we use the Model View Controller (MVC) architectural style that give us flexibility to add new services with minimum effect in other components such as the presentation layer.

In the case of the usability requirement, again the use of MVC allows to focus user interface characteristic based on the type of mobile device without affecting other components or layers. Besides the motivational scenario included in this chapter, we define several scenarios to understand the flow of events and the type of users we will have in ERPHA.

For performance the main concern was to reduce the computational overhead that may occur by reading each sensor, sending its data to the server and also sending video to the central site. We decided to use just one channel for joint transmission of video and data in the mobile, at the central site with much computational power decomposition of the various signals take place. Details of how we accomplish joint transmission of video and vital sign data using wireless facilities such as 3G is out of the scope of this chapter[1]. Also use case scenarios and interviews with medicine experts help us to define a scheduling policy for sampling the sensors (i.e., there are vital sings that do not change very frequently like temperature).

Lastly, the availability requirement is the reason behind why we have implemented two prototypes with different operating systems in the mobile device (Windows Mobile and Android). Also the ability to use different communication protocols such as 3G, WiMax and Wi-Fi.

5 Prototype Implementation Details

Two ERPHA prototypes have been built: one using the Windows Mobile 6.5 operating system in the mobile device and other uses Android4.1 operating systems in another mobile device. Except for the Android mobile version the application was developed using the C# SDK provided for Visual Studio 2008. This same tool was used to manage the data and web servers, which connect to a SQL Server 2008 database, and the results are displayed on the hospital computers using ASP.NET with C#. Figure 6, next, indicates the functions currently implemented in the components.

The graphical interfaces for the mobile application are illustrated in Figure 7. The first and second screens are used to capture a patient's information, such as name, age, gender, and severity. The severity is assigned with the paramedics' judgment, and it can have three levels: stable, unstable, and critical. All these values can be left blank, which is the default value, in order to start the session as quickly as possible. This information is used to start the session with the server which is a unique integer used as ID. Once the mobile device application pairs with the Bluetooth sensors, it starts displaying video, the vital sings and the latitude and longitude.

---

[1] A detailed description of our approach for transmitting video and data from sensors has been submitted to a specialized journal and should be available by December 2013.

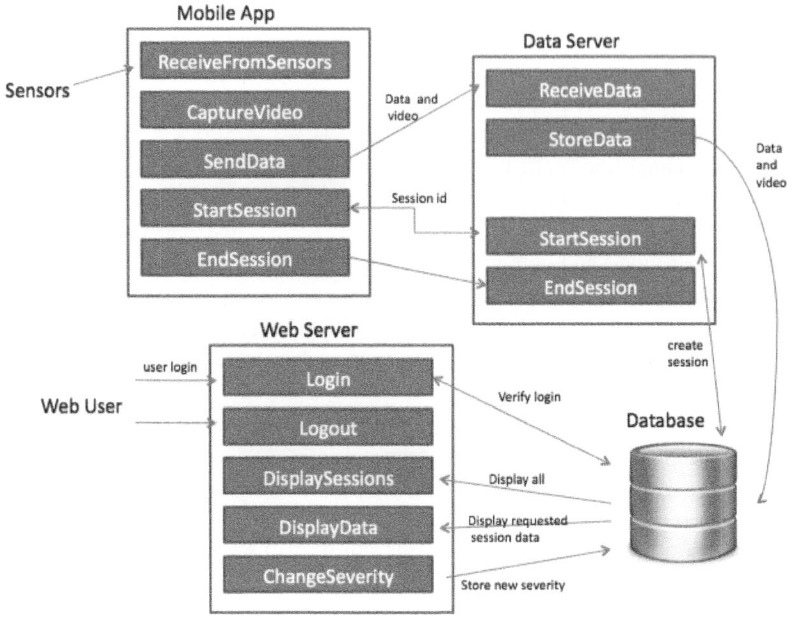

Fig. 6 ERPHA implementation modules and functionality

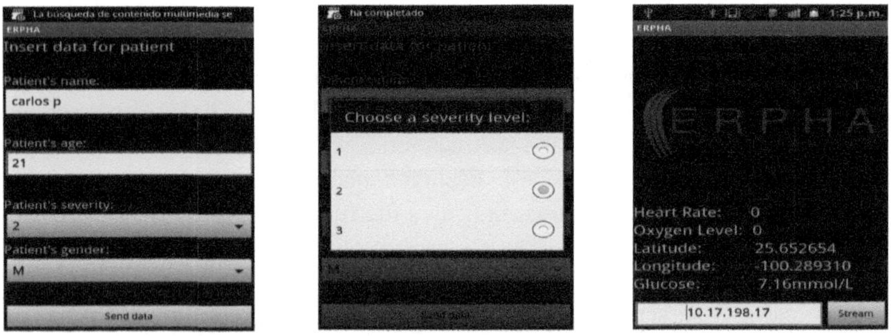

Fig. 7 ERPHA mobile aplicación

When the patient arrives to the hospital, or there is no need to keep monitoring the patient with ERPHA, the paramedic stops ERPHA at the mobile device.

The Data Server's functionality is split into two, handling the sessions and the data reception. When the server receives a request to create a session from the mobile application, it creates a unique patient ID in the database with the basic information gathered from the device and sends it back. Whenever it receives a session termination, it records that the session is no longer active and should not receive anything further from it. The server is constantly listening for data from two ports, one for the video and the other for the patient's vital signs. Whenever it receives data, it verifies if it is a video record or a vital signs record, reads the session number from the data structure, and writes it to the database using the session number as its identifier.

Finally, the Web Server accesses the most recent information about a patient from the database by sorting the received information with its timestamp. The design decision to use the database as a middle-man instead of having the mobile device connect directly to a session in the terminal was taken since it was more maintainable and easier to scale, although the tradeoff would be to have a better performance.

Using the database as a middle-man allows the storage of historical records and it lets us have independent layers that can be modified without altering the rest of the project, which proved useful when dividing the tasks between the developers.

The hospital staff has to Login and Logout, to protect the patient information. Once inside, they can review all patients sorted by severity, appearing on top the critical patients. If hospital staff wants to assist a certain patient, they can view the vital signs, video, and location of the ambulance. If the hospital staff believes the patient's severity is not correct or it has changed after certain treatment, they can change the severity from the website so it will be sorted accordingly whenever the rest of the staff is monitoring the patients.

The data visualization in the browser (Figure 8) was done using ASP.NET with C# and JavaScript. Charts with the vital signs were created, as well as showing the

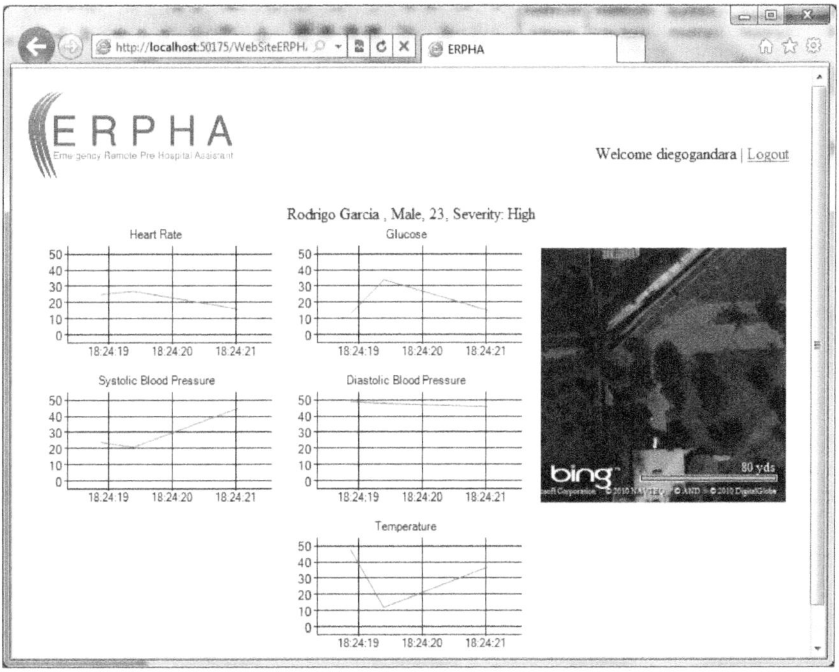

Fig. 8 ERPHA. Web interface.

video in the browser and the location of the ambulance in Bing maps. Physicians and hospital staff can click on assist a patient session, so others will see the sessions that have not been assisted. As well, sessions are sorted by severity, so they are assisted first. Physicians can change the severity of a patient whenever they feel the patient has become more stable.

6 Conclusions and Future Work

In this chapter, we have described ERPHA as a system that aids in the continuous monitoring of patients involved in traffic accidents; and anticipates his/her arrival at hospitals preparing hospital nurses and medics for a better patient treatment. ERPHA deals mainly with improving communication, accessibility to information, and monitoring of patients. As it, we consider ERPHA a great improvement in pre-hospital treatment; however, we consider there are several areas of opportunity.

At the moment, a smartphone is used as the device to which everything connects, this means that the paramedics must take care of what the video is displaying and make sure there is a connection with the Bluetooth to the BSN. A group of invited collaborators of ERPHA are working in the development of special hardware for integrating the mobile device's camera and wireless capabilities with the BSN. Under this approach it would be possible to control the camera remotely by the physician; therefore the paramedic only would need to place the device closer to the patient and he/she would have more action freedom.

In terms of software improvement, the new Windows Phone 7 SDK and Windows 8 contains more multimedia APIs which can improve the performance and maintainability of the video and also improve the streaming algorithm.

The patient information is saved in the database as historical records, but there is no access to it by the paramedic from the mobile device. Accessing health historical data will allow paramedics to decide what's best for the patient in the accident area. Also a great improvement would be to have a computer system for diagnosis health support by including Big Data and Data Analytics technology.

At this time, we are testing the project in a wider scale. As the project progress in the field, we expect to have quantifiable benefits in terms of saved lives or reduced hospitalization costs. We are also expecting to collect information with respect to problems of scalability, robustness and usability-

As mentioned before, in developing countries there is a problem of paramedics undertraining. The videos recorded by ERPHA could be used as a training tool, showing what other paramedics are doing, as well as preparing on-the-go by receiving feedback as they are performing their duties Our project ERPHA deals mainly with improving communication, accessibility to information, and monitoring of patients involved in traffic accidents.

Future Trends (Vision)

We consider that the last word in mHealth is very distant. Current trends in making smaller and smaller devices allows to have more personal and less intrusive own health care. But not only technology development shall pave the roadway of health systems. Important legal aspects taking place in all economies of the world to have health records available for better practice of the discipline.

If we focus only in the technological side, currently we found technologies such as robotics that start to taking care of small duties from the doctor or the nurse. In the same sense, we foresee more common the use of remote medical care assisted by robots in distant areas. The use of medical devices for patient monitoring shall be less intrusive and more ubiquitous and wearable in the future. Technology as ERPHA using out of the shelf components will become common, smaller with multiple connection between the components and mobile devices.

Communication means shall have a major impact for presetting services in distant areas. The whole communication spectrum shall be revised to adapt to new needs and probably will have to dedicate a special part of the band for medical remote assistance.

One key factor for the adoption of these technologies is the empowering of nurses and patients. Empowering nurses by providing more information and better tools to keep track of patients. Patients will empowered by having better control of their illness, by the use of cell pones and or any personal, mobile device that reminds him/her of medical dates and also store relevant medical information.

Internet of things is a new research area from which several data providers, usually from sensors that surround our environment are constantly producing relevant info. Sensors such as video cameras, radios, fridge, air conditioning and others are generate information about our daily habits that could produce and accurate profiles for better treatment. In this case data processing may be based on Technologies such as Big Data.

References

[1] World Health Organization, Global Status Report on Word Safety (March 2013)

[2] World Health Organization, The top 10 causes of death, Fact sheet No. 310 (June 2011)

[3] Department of Public Health (Secretaría de Salud), Segunda Evaluación de Personal, México (Available only in Spanish) (2006)

[4] Van Der Veldenl, M.W.A., Ringburg, A.N., Bergs, E.A., Steyerberg, E.W., Patka, P., Schipper, I.B.: Prehospital interventions: time wasted or time saved? An observational cohort study of management in initial trauma care. Emergency Medicine Journal 25, 444–449 (2008)

[5] National Institute for Statistics and Geography (Instituto Nacional de Estadística y Geografía), Mujeres y Hombres en Mexico (Available only in Spanish) (2010) ISBN 978-607-494-097-8

[6] Bates, D.W., Gawade, A.A.: Improving safety with information technology. New England Journal of Medicine 348, 2526–2534 (2003)

[7] Chen, W., Wei, D., Zhu, X., Uchida, M., Ding, S.: Mobile Phone-based Wearable Vital Signs Monitoring System. In: Proceedings of the 2005 The Fifth International Conference on Computer and Information Technology (2005)

[8] Chung, B., Fawzy, A., Jan, M., Muraskin, J.: Accessible Home Vital Signs Monitoring System, Technical Report, Columbia University (2009)

[9] Thatcher, M.: IT Governance in Acute Healthcare: A Critical Review of Current Literature. In: George, C., et al. (eds.) Ehealth: Legal, Ethical and Governance Challenges, pp. 349–370. Springer, Heidelberg (2013)
[10] Gunnar, M.L., Annika, B.: European Journal of Clinical Pharmacology. 69(1), 65–72 (2013)
[11] Carspecken, W.C., Sharek, M., et al.: Clinical case of electronic health record drug alert fatigue: consequences for patient outcome. Pediatrics 131, e1970–e1973 (2013)
[12] DeMello, J., Deshpande, S.P.: Factors Impacting Use of Information Technology by Physicians in Private Practice. International Journal of Healthcare Information Systems and Informatics 7(2), 17–28 (2012)
[13] Bass, L., Clements, P., Kazman, R.: Software Architecture in Practice ISBN: 978-0321815736
[14] Krutchen, P.: Architectural Blueprints The "4+1" View Model of Software Architec-ture. IEEE Software 12(6), 42–50 (1995)
[15] Macleod, B., Phillips, J., Stone, A., Walji, A., Awoonor-Williams, J.: The Architecture of a Software System for Supporting Community-based Primary Health Care with Mobile Technology: The Mobile Technology for Community Health (MoTeCH) Ini-tiative in Ghana. Online Journal of Public Health Informatics 4(1) (2012), doi:10.5210/ojphi.v4i1.3910
[16] Mougiakakou, S.G., Bartsocas, C.S., Bozas, E., Chaniotakis, N., Iliopoulou, D., Kouris, I., Nikita, K.S.: SMARTDIAB: A communication and information technology approach for the intelligent monitoring, management and follow-up of type 1 diabetes patients. IEEE Transactions on Information Technology in Biomedicine 14(3), 622–633 (2010)
[17] Sachpazidis, I., Rizou, D., Menary, W.: Broadband Health Care Network in Brazil and Peru. 1-5. x, Scopus database (2008)
[18] Simel, D.L.: Approach to the patient: history and physical examination. In: Goldman, L., Schafer, A.I. (eds.) 24th Goldman's Cecil Medicine, Saunders Elsevier, Philadelphia (2011)
[19] Cruz-Martinez, A.: Mexico among the top 10 countries with most deaths caused by traffic accidents. La Jornada (March 23, 2013)
[20] National Institute for Statistics and Geography (Instituto Nacional de Estadística y Geografía) (2014), `http://www.inegi.org.mx` (last accessed date: January 12, 2014)

Questions

Q1. What is the golden hour? Perform further investigation about the safe time limits for start delivering medical care in different types of accidents.

Q2. Consider there exists limits in the communication or that communication is restricted to low bandwidth. Propose an alternate strategy to ensure high data quality under these condition on the communication.

Q3. Propose an alternate architecture for ERPHA that favors high definition video. Which quality attributes should be sacrificed?

Q4. Which are the quality requirements for ERPHA?

Q5. Propose at least one different/new requirement in functionality, quality and constraints for ERPHA.

Q6. In the chapter, we discussed how architecture supports different quality attributes. Which quality attributes were defined for ERPHA and what a good metric is for each of them.

Q7. What is the proportion of deaths due to car accidents compared to the total number of deaths?

Q8. What are the two types of networks found in ERPHA?

Q9. What are the aspects that enable ERPHA reduce mortality and morbidity?

Case Studies

1. Tokyo, San Francisco and Mexico City are among the most prone cites in the world for suffering a earthquake of great proportions. It is possible that ERPHA finds to be helpful for an earthquake. Yes/no? Why?

2. What other advances in technology could be applied to ERPHA to make a more reliable product?

21. mHealth Portable Systems and Platforms

Saifal Zahir and Radwa Hammad

Image Processing, Graphics and Multimedia Lab, Department of Computer Science
University of N. British Columbia, PG, BC, V2N 4Z9, Canada
{zahirs,hammad}@unbc.ca

Abstract. In this chapter, we introduce the state-of-the-art of medical oriented portable devices and smartphones that are capable of communicating with sensors and healthcare providers effectively. In this chapter, we propose a typical model for the infrastructure of smartphone-based mHealth service. This infrastructure includes hardware, software, wireless protocols, and collaborative applications. The current smartphone operating systems included in this chapter are: Android (Google), iOS (Apple), Symbian (Nokia), Windows Mobile (Microsoft), and QNX (BlackBerry). A dedicated section is provided on the amalgamation of wireless equipment with smartphone to provide specific mHealth service. Examples on such devices are the Glucophone which is an integrated cell phone-glucometer system and the Cellscope which takes a standard cell phone and transforms it into a compact, high-resolution, handheld microscope. The proposed infrastructure is efficient and cost effective.

Keywords: Smartphones, mHealth, mobile health, smartphones medical apps, medical sensors, operating systems, wireless communications, personal health records, electronic health records.

1 Introduction

A smartphone is a cellular portable phone with computer-like processors, operating systems, Internet connectivity, as well as built-in and downloadable features and applications [1]. Current smartphones have high resolution touch screens, high pixel value digital cameras, and are capable of sending and receiving text messaging, playing MP3 as well as viewing videos. Also, most smartphones have Bluetooth and/or Wi-Fi capabilities, which allow two way information transmissions wirelessly over short distances [1]. In addition, smartphones come with downloadable global positioning systems (GPS), which can be used to map the location of services in their vicinities including but not limited to medical resources such as providers' clinics, hospitals, pharmacies, and emergency departments as well as the patient location. Such features are critical for clinicians and patients in a healthcare setting as it makes smartphones easier and more intuitive to use [1].

Fundamentally, mHealth includes basic health services, remote diagnosis, remote patient monitoring and chronic disease management. Such services rely on several

S. Adibi (ed.), *Mobile Health*,
Springer Series in Bio-/Neuroinformatics 5, DOI: 10.1007/978-3-319-12817-7_21

hardware, software, and collaborative programs and systems. These programs and systems are becoming ubiquitous and sophisticated.

The amalgamation of smartphones and mHealth applications and devices is fueling a rapid expansion in healthcare services and delivery. Given the pace of rapid changes and advances in mobile health apps, smartphone-based mHealth is expected to grow more rapidly than mHealth services delivered through conventional medical devices.

The number of smartphones in use globally crossed the one billion mark for the first time in the third quarter of 2013 and will double by 2015 [2]. The number reached 1.038 billion in the three-month period, a forty seven percent 47% increase from a year earlier as reported by the Boston-based industry researcher in a statement earlier this month, July 2013. That translates into one in every seven people worldwide owning a handheld device that works like a computer, according to the statement.

The growth of the smartphone market is estimated by Bloomberg Industries to be about $219 billion in 2012. Such numbers helped Apple and Samsung Electronics Co. rack up record earnings as they take sales from rivals including Espoo, Finland-based Nokia. In one month, Cupertino, California-based Apple sold more than 5 million iPhone-5s in the device's debut weekend, shattering a record set by the previous model. Samsung, the Suwon, South Korea-based maker of Galaxy smartphones, ended Nokia's 14-year reign as the world's biggest maker of mobile phones this year, 2013, according to shipment estimates by industry researchers including Strategy Analytics.

In this chapter, we introduce smartphones as crucial mHealth devices that can play a major role between patients and healthcare providers. In section 2 we present a model for the infrastructure for a complete two way mHealth communication network including the hardware implementation, software implementation, collaborative software, and the wireless protocols. In section 3, we introduce the various smartphone operating systems, while section 4 presents the state-of-the-art in the new merger of the two technologies: smartphones and healthcare delivery. Finally in section 5, we provide a look ahead in the mhealth sector and present some anticipated additional smartphone applications and services to mHealth on a large scale.

2 The Proposed mHealth Infrastructure Model

A typical mHealth infrastructure model contains four main parts: (i) hardware implementation; (ii) software implementation; (iii) wireless protocols; and (iv) mHealth collaborative software. Technically, a mobile health system is used to add mobility to the quasi static public healthcare services so that patients need not be confined to a specific place or to their residence. In addition, it enables them to participate in their usual day-to-day activities. The mHealth system depicted in Figure 1 consists of a patient segment and a healthcare providers segment. The patient segment includes: the user(s) with a number of implanted or wearable wireless devices; one or more microcontroller(s); and one or more smartphone(s) with Internet connectivity to a remote server [3]. The healthcare providers segment includes one or more smartphone(s) with Internet connectivity, one or more server(s), and one or more PC that can communicate the content or messages to healthcare providers. In the next section, we discuss these segments in some details.

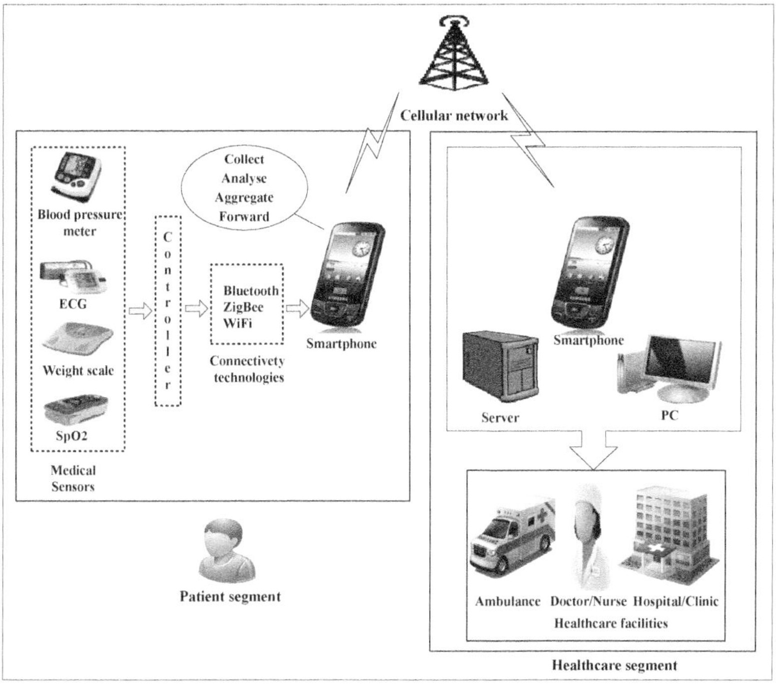

Fig. 1 The Proposed mHealth Infrastructure Model

2.1 *Hardware Implementation*

The hardware implementation for a mHealth system comprises three main components namely: (1) a set of wearable and/or wireless sensors-based medical equipment that are connected to microcontrollers; (2) smartphones; and (3) computers and servers.

(1) The first component of the hardware implementation is a set of wireless medical sensors that measure human medical conditions such as blood pressure, heart rate, oxygen saturation, temperature and the like. These sensors can be either implanted or attached to the patient's body. For example, an electrocardiography (ECG or EKG) sensor is used to measure the rate and regularity of heartbeats and the size and position of the chambers, the presence of any damage to the heart, and the effects of drugs or devices used to regulate the heartbeats. Another example of such medical sensors is a blood oxygen saturation sensor (SpO2) which measures

hemoglobin saturated with oxygen at the time of the reading. A third and final example is a blood pressure sensor that can constantly and automatically measures blood pressure and records/transmit it to the microcontroller which in turn send it to the healthcare providers via a smartphone [3].

(2) The second component of the hardware implementation is the smartphone. In this configuration, smartphones act as the control node in the network as it resides on both segments of the infrastructure. In the patient segment, a smartphone acts as a sender and in the healthcare segment, the smartphone acts as a receiver [3].

(3) Finally, the third component is the computer(s) and server(s), which may be in a doctor's office, clinic/hospital or a remote monitoring service. Usually a microcontroller monitors the wearable sensors' readings and compares it to acceptable (normal) ranges and sends only abnormal events and alarming readings to the server via a smartphone. In such cases, the clinic/hospital collects the data from the server for further action [3].

2.2 Software Implementation

Software applications or, apps, represent a significant feature of smartphones. Each smartphone operating system has an online store where apps can be purchased and downloaded to increase its functionality. Some of these apps monitor exercise trackers, note taking, and GPS navigation and tracking. With the recent significant growth of smartphone applications and the increase in smartphone performance, a number of health monitoring and wellness applications have been introduced. These apps can be classified into three main categories: (i) hospital workflow management; (ii) medical reference; and (iii) health and wellness management [4].

(1) Hospital Workflow Management: Hospital workflow management applications assist medical professionals in their everyday activities such as remotely accessing patients' historical health records, their current vitals, or pharmaceutical calculations. These applications can be categorized into six main services [4].

   a. Disease Diagnosis Applications: Disease diagnosis applications were designed to access diagnosis and treatment information on infectious diseases, pathogens, diagnosis, treatment, medications, and differential diagnosis using a smartphone. There are many applications including Johns Hopkins Antibiotic Guide (JHABx), 5-Minute Clinical Consult (5MCC), 5-Minute Infectious Diseases Consult (5MIDC), Sanford Guide to Antimicrobial Therapy (SG), and ePocrates ID just to name a few [5].

   b. Medical Calculator Applications: A medical calculator is a software program for calculating various clinical scores and indices such as body mass index (BMI), body surface area (BSA), coronary heart disease risk, individual drug dosing, etc. [5].

c. Clinical Communication Applications: Smartphones support several means of communication including voice calling, video calling, text messaging, email messaging, multimedia (text, image, and video) messaging, and conferencing through the cellular phone service provider. The use of smartphone communication is critical as it relays important information quickly, reducing the risk of medical errors. Nurses may need to go through a elaborate process to find the responsible physician for a patient using traditional paging [5].
d. Hospital Information Systems Client Applications: Client applications for hospital information systems (HIS), such as electronic health records (EHR), electronic medical record (EMR), and picture archiving and communication systems (PACS), provide the flexibility of accessing patient information securely from anywhere at any time. OsiriX Mobile is the client application for OsiriX PACS, which processes and displays images using the DICOM standard for digital image storage [5].
e. Medical Training Applications: Smartphones are also used for medical training and continuing medical education (CME). CME provides training in the most current evidence-based medical practice. iCPR and iResus are two medical training applications are designed for the iOS platform and available for free. This application measures the chest compression rate and gives audiovisual feedback, improving the performance of chest compression by helping the user to achieve the correct chest compression rate [5].
f. General Healthcare Applications: There are other general applications on smartphones such as: HCSIT, Borboleta, LIFE- reader, Multimedia Paging Based Clinical Alarm, Out - breaks Near Me, H1N1 Swine Flu Update, and WISER. A real- time clinical alarm monitoring system was developed to monitor intensive care unit patients by smartphone, displaying the alarms for an entire unit or one bed, with the vital signs at the moment of the alarm color-coded by severity [5].

(2) Medical reference applications: Medical reference applications help medical professionals and other users in finding information related to a broad spectrum of medical topics, such as anesthesiology, cardiology, and dermatology. Typical example of a medical reference application is Netter's Atlas of Human Anatomy on the Android platform. This application provides detailed information and graphical illustrations of the human anatomy [4].

a. Drug Reference Applications: There are several drug reference applications available for smartphones. These include, Skyscape's RxDrugs, Epocrates, Medscape, SafeMed Pocket, FDA drugs, and DrugDoses.net. A drug reference application generally includes the names of drugs, their indications, dosages, pharmacology, drug-to-drug interactions, contraindications, cost, and identifying characteristics. Epocrates was cited as the most commonly accessed drug reference application. Skyscape's RxDrugs, is another drug reference that has seen widely used in the U.S. which is capable of checking multiple drug-to-drug interaction at the same time. Smartphone-based drug reference applications can be useful evidence-based resources at the point of care [5].

b. Literature Search Applications: these applications are used by healthcare professionals to facilitate searching biomedical literature databases such as PubMed/MEDLINE, Essie, and the like to find and display medical reference information [5].

(3) Health and Wellness Management: Most of health related smartphone applications are dedicated to health and wellness management. Such applications include prenatal & infant care, cardio fitness, diet, medication adherence, women's health, strength training, stress, hearing check, smoking cessation, sleep, mental health, and chronic disease management, cardio mobile, pulmonary rehabilitation, PAL Calculator, and Fall Detector. Other applications such as "Wellness diary" are designed to support users in learning about their behavior, and both making and maintaining behavior changes [4].

2.3 *mHealth Collaborative Software*

Patients can use smartphones to access their personal health records, PHR, which include their medical information from different sources, such as insurance claims, pharmacy data, and electronic health records, EHRs, [6]. Customary, EHRs contain the whole health history of the patient that can be shared among healthcare providers (e.g. physicians). Evidently, PHR systems are not only static store for patient information; they combine data, software tools which allow patients to become active participants in their own care [7]. Therefore, by accessing the PHR from smartphones the physicians and patients can share the information. Many smartphone applications allow physicians to review patient files, write prescriptions, and provide quicker response times when they are away from their offices. Examples on such applications are: Care360 Mobile, Allscripts Remote, Atlas Mobile, IQMax Mobile, iMedDoc EMR, and Acumen Mobile [8]. Several companies, nowadays, offer applications or interfaces with EHRs to save the time of typing by allowing physicians to convert their speech into readable text and add it to the patient health record [9].

2.4 *Wireless Network Protocols*

Smartphones use a variety of communication techniques and protocols to communicate with other smartphones, equipment (e.g. medical sensors), and computers. In this chapter, we focus on three main communications technologies.

(1) Zigbee: This protocol is a suite of high level communication protocols that is highly used in smartphones. It is a low cost, low power consumption, and short distance wireless communication technology, which is developed for wireless personal area network, WPAN, [10]. Zigbee is a synonym of IEEE 802.15.4 protocol [4], [11]. It is widely used in building automatic control, industry autoimmunization, monitoring and control of hospital and home to name a few [11].
(2) Wi-Fi: This protocol refers to the wireless Ethernet IEEE 802.11b standards for wireless local area network, WLAN, that allow wearable devices to exchange data and/or connect wirelessly to the Internet [12]. Wi-Fi can cover up to several square miles.
(3) Bluetooth: This technology is the most commonly used wireless communication standard in personal area network, PAN, and uses IEEE 802.15.1 standard [4]. Bluetooth can exchange data over short distances from fixed and mobile devices, with high levels of security. Bluetooth operates in the international 2.4 GHz industrial, scientific, and medical, ISM, band with a data rate of 1M bit / second [13].

3 Smartphones as a Wearable Device in mHealth System

3.1 Smartphones Operating Systems

Smartphone operating systems vary in specifications and capabilities. The most prominent operating systems in the market today are iOS (Apple), Android (Google), Symbian (Nokia), QNX (Blackberry), and Windows (Microsoft). All these operating systems are capable of delivering and receiving content wirelessly that makes them excellent candidates for mHealth services. Table 1 shows the worldwide shipment volume for smartphones using these operating systems Q1 – Q4, 2012 and Q1 – Q2, 2013. In this section we provided the fundamentals of each of these systems.

Table 1 Smartphone's worldwide share (2012 Q1 – 2013 Q2)

	Operating system market shares					
Quarter-year	**Android**	**iOS**	**Windows Phone**	**Blackberry**	**Symbian**	**Others**
1Q-2012 [19]	59%	23%	2.20%	6.40%	6.80%	2.60%
2Q-2012 [18]	68.10%	16.90%	3.50%	4.80%	4.40%	2.30%
3Q-2012 [17]	75%	14.90%	2%	4.30%	2.30%	1.50%
4Q-2012 [16]	70.10%	21%	2.60%	3.20%	0.70%	2.40%
1Q-2013 [15]	75%	17%	3%	3%	1%	1%
2Q-2013 [14]	79.20%	13.20%	3.70%	2.90%	0.20%	0.80%

3.1.1 Android Operating System

Android operating system is developed by Android Inc. in 2003 and it is based on the Linux operating system [20]. At a later date, the giant search engine 'Google' acquired Android in 2005. Android uses the Java programming language. Android is open source and Google releases the code under the Apache License [21], [22], [23]. Individuals and companies can develop applications (apps) for this OS, and any company can release a phone using it. This open source code and permissive licensing allows the software to be freely modified and distributed by device manufacturers, wireless carriers and enthusiast developers also. Android platform can provide solutions for a wide range of industries which are highly portable as well as cost effective. Also it makes no distinction between applications whether static core applications or dynamic third-party applications, are treated identically and have equal access to the device's functionality [23]. Table 1 shows that Android share market is about 79% in the second quarter of 2013 [14]. This represents a significant portion of worldwide smartphones users.

3.1.2 Symbian Operating System

Symbian operating system is primarily available on Nokia phones. The first Symbian based smartphone was released in the year 2000 [20]. Symbian is also an open source operating system where anyone can use it without penalty or having to pay. This system is widely used, but it is not a full-featured mobile phone operating system like Android. For instance, most Symbian based smartphones are low-end devices [20]. Table 1 shows that the worldwide share for this OS is shrinking [14] - [19].

3.1.3 Apple Operating System

The iOS is the mobile OS of Apple's iPhones. This operating system is a UNIX-based system and it is the same technologies used by Mac OS X. This operating system is entirely closed-source system. Applications are primarily developed using Objective-C, a superset of ANSI C, extended with syntactical and semantic features (derived from Smalltalk) to support object-oriented programming. Cocoa Touch is the application environment to build applications for iOS and consists of a suite of object-oriented libraries, a runtime, and an integrated development environment (e.g., XCode or Interface Builder). Two core frameworks are essential to application development for this system: the Foundation and the UIKit [21]. Table 1 shows that the worldwide share of iOS is still far distance second after Android operating system [14] - [19].

3.1.4 BlackBerry Operating System

BlackBerry operating system, QNX, is one of the most well-known operating systems for mobile phones [20]. It was developed by Research in Motion, RIM, in 2003 [24]. BlackBerry smartphones are available on all communication service providers. The BlackBerry OS is tailor-made for business, with functionality as a target rather than personality or appearance. This system is focused on messaging,

BBM, SMS, and email as well as other communication services. BlackBerry smartphones are particularly appealing to individuals and corporates with security concerns as it supports full device encryption and encrypted communication [22]. Blackberry operating system represents nearly 3% of the worldwide shipment volume as can be seen in Table 1.

3.1.5 Windows Mobile Operating System

Windows Mobile operating system, also known as Windows Phone, is the mobile phone version of Microsoft's. Such privilege made many Windows programs compatible to this operating system especially the well-known Microsoft Office. This helped make it popular choice for businesspeople. Windows Mobile was originally designed for Microsoft's line of Pocket PCs before being adapted for use in phones [20]. This operating system has nearly 3.7% of the worldwide market, as shown in Table 1.

Figure 2 shows the most recent versions of the five operating system devices discussed above. The Figure shows that all five systems (companies) offer a large touch screen smartphones. In Table 2, we present the different features of the popular operating systems (iOS, Android, Windows Phone, Blackberry and Symbian). This Table shows that each operating system has its own characteristics, capabilities and limitations. Although there are some technical differences between these systems but they all have similar capabilities in terms of transmitting and receiving signals wirelessly and also capable of processing different types and formats of content. In addition, these systems have a large volume of applications for a variety of sectors including healthcare intervention and monitoring. One important aspect of these smartphones systems is their highly secure communication as many of them use encryption as an additional layer of smartphones are transportable to different countries and can use GPS and produce maps internationally at any point.

Blackberry Z10 (QNX)	HTC (Android)	iPhone5 (iOS)	HTC Windows - 8X	Nokia603 (Symbian)

Fig. 2 Different types of recent smartphones [25]

Table 2 General features of the different types of smartphone operating systems [26], [27]

Item/Feature	Operating system				
	iOS	Android	Windows Mobile	BB - QNX	Symbian
Make	Apple, Inc	Google	Microsoft	BlackBerry	Symbian
Market share	13.2%	79.2%	3.7%	2.9%	0.2%
Current version	6.1.4	4.3	8	10 OS	10.1
OS Family	Darwin	Linux	Windows CE 7 / Windows NT 8+	Mobile OS	Mobile OS
Remote locking	Yes	Yes	Yes	Yes	Yes
On-device encryption	Yes	Yes	Yes	Yes	Yes
Ext. storage encryption	External storage not supported	No	No	Yes	Yes
Internet connectivity	Yes	Yes	Yes	Yes	Yes
Application store	App Store	Google Play	Windows Phone Store	App World	Symbian Horizon, Nokia store
Email sync protocols	POP3, IMAP, MAPI	POP3, IMAP, MAPI	POP3, IMAP, MAPI	BES, BIS, Push e-mail	POP3, IMAP, Exchange
Voice recognition	Yes	Yes	Yes	Yes	Yes
Offline Voice Recognition	No	4.1+	No	No	No
USB On-The-Go	No	Yes	No	Yes	Yes
Turn-by-turn navigation	Yes	Yes	3rd party software or free Nokia Drive	3rd party software	free global Nokia Maps and 3rd party software
Offline maps	3rd party software	Yes	3rd party software	3rd party software	free Nokia Maps and 3rd party software
Multitasking	Yes	Yes	Yes	Yes	Yes
Bluetooth keyboard	Yes	Yes	No	Yes	Yes
File transfer: Wi-Fi	No	Yes	No	No	No
File transfer over Bluetooth	3rd party SW on jailbroken	Yes	Yes	Yes	Yes
Direct file transfer over NFC	No	Yes	Yes	Yes	No
VPN	Yes	Yes	No	BB Enterprise Server	Yes
GPU Accelerated GUI	Yes	Yes	Yes	Yes	Yes
Official SDK platform(s)	Mac OS X using iOS SDK	Windows Mac OS X, Linux,	Windows	Windows, Mac OS X	Symbian SDK, Linux, OS X, Nokia Qt SDK
Relocation to other countries	Yes	Yes	Need hard reset and new live id	Yes	Yes

4 Portable mHealth-Based Service

4.1 Smartphones mHealth Service Capabilities

Smartphones offers numerous advanced features that are especially needed for remote and standard healthcare services. In this section, we discuss some of these features and its usage in the healthcare sector.

(1) Text Messaging

Text messaging or short message service, SMS, is a service that enables 160 character messages to be sent from one smartphone to virtually any other smartphone worldwide. As it is supported by all smartphones, SMS is widely used not only in the first world but in the developing countries as well, where mobile phone penetration has reached nearly 70%. This widespread use makes text messaging the most universally accessible technology for phone-based healthcare delivery and interventions. Beyond its accessibility, text messaging has seen broadly adopted as a health technology for two reasons: (i) text messages are a push technology, allowing intervention messages to be delivered without any effort on the part of the recipient; (ii) text messages can be sent and received by both phones and computers, text messaging provides a way for users to log their health-related activities and physiological parameters (e.g., exercise, peak flow in asthma management), and to receive customized feedback based on these data. For example smoking cessation intervention for young adults uses precisely this kind of SMS information loop. The diversity of these interactions provides evidence for just how flexible text messaging can be as a system for delivering health interventions [10].

(2) Cameras

In recent years, cameras have become a standard feature on all smartphones. The quality of phone cameras is still below the quality of dedicated digital cameras, but their constant availability makes phone cameras a useful tool for collecting health-related data throughout the day. So far, health interventions have used cameras in three primary ways: (i) as an alternative way to journal health-related behaviors, such as food consumption, (ii) as a way to provide healthcare providers with additional information about a condition, such as the appearance of psoriasis lesions, and (iii) as a way to document circumstances relevant to the self-management process, such as the contextual factors that might influence diabetic patients' ability to effectively manage their blood glucose levels. Improved smartphone camera technology allows users to record and document events and instances in high definition mode and exploits new imaging applications. For instance, the healthcare sector can use high-resolution imaging for skin and eye scans for remote diagnosis and prevention [10].

(3) Automated Sensing

Nearly all contemporary smartphones can connect to sensing devices over Bluetooth, Wi-Fi, Zigbee, or another form of personal area networking. In this way, smartphones can connect to digital scales, blood pressure monitors, glucose meters, portable electrocardiograms (ECG), pedometers and gym equipment. A smartphone that is connected to such devices can act as transmitter/receiver and stores data for a variety of health-intervention data. In addition, over the last 2 years, an increasing number of smartphones have begun to ship with built in sensors such as accelerometers and GPS, enabling them to detect users' behaviors even without the use of an external device. For example, the iPhone application RunKeeper uses built-in GPS to automatically track how far users run or cycle, to create maps of their exercise routes, and to calculate how many calories were burned during these workouts [10].

(4) Internet Access

Smartphones always-on connectivity with the Internet means that users' data, such as blood glucose levels or peak flow readings, can be uploaded to providers' servers as soon as they are captured, thus enabling early detection of critical events. The data can also be uploaded to websites where users can easily view, chart, and edit their information. The use of online resources makes it easier to keep the content of an intervention up to date without requiring users to repeatedly install updated versions of the application [10].

4.2 Hybrid Smartphones – Medical Devices Service

Recently, the health industry has introduced a number of medical devices that can be attached to smartphones to perform specific health tasks and functions. In the next section, we discuss some examples to show how smartphones have been merged with such medical devices.

(1) Cellscope

Cellscope takes a standard cell phone and transforms it into a compact, high-resolution, handheld microscope with the capability of on-site disease diagnosis and wireless transmission of patient data to clinical centers for remote diagnosis and treatment as shown in Figure 3(a) [28].

(2) Microskia

Microskia is a microscope peripheral created out of off-the-shelf components that is intended for developing nations. The device has been tested in Africa 2 years ago and proved to be viable. This device is shown in Figure 3(b) [28].

(3) Agamatrix's IBGStar

IBGStar is a blood glucose meter. It is a plug-in for the iPhone from Sanofi, with the plugin module created by Agamatrix. Agamatrix already offers such an application for iPhone users under its Wavesense brand. This device is shown in Figure 3(c) [28].

(4) Glucophone

Glucophone is an integrated cell phone-glucometer system introduced in 2006 that uses custom software on certain LG mobile phones along with a Glucopack fitted onto the device's back. Users place the test strip in a small slot on the side of the GlucoPak, with the results uploaded seconds later to the HealthPia website, and securely stored there. The results can be automatically sent to your physician, guardian, or family member in real-time. Finally, in August 2007, HealthPia announced that they have received FDA approval for the device and were beginning to ship the product for certain phones. This device is shown in Figure 3(d) [28].

(5) Near-Eye Tool for Refractive Assessment

Near-Eye Tool for Refractive Assessment, NETRA, is a new system for prescribing eyeglasses. This invention is created by a team of researchers at the

MIT Media Lab that uses a smartphone and a plastic lens attachment. A user/patient looks through the lens attached to the smartphone and runs an application that shows them two parallel lines, one is red and the other is green. The user uses arrow keys on the smartphone to adjust the image until the two lines overlap. At this point, the smartphone takes the reading and produces an eye prescription. This device is shown in Figure 3(e) [28].

(6) Smartphone as Wearable Cameras

A smartphone wearable camera/sensor platform is able to gather data from all on-board sensors continually during the day. A secondary aim is to automatically identify specified daily activities using the passively captured image and sensor data. The smartphone includes sensors that can capture a rich life experience archive, life-logging software was developed that captured photos at a configurable rate along with data gathering from all the available on-board sensors as shown in Figure 3(f) [29].

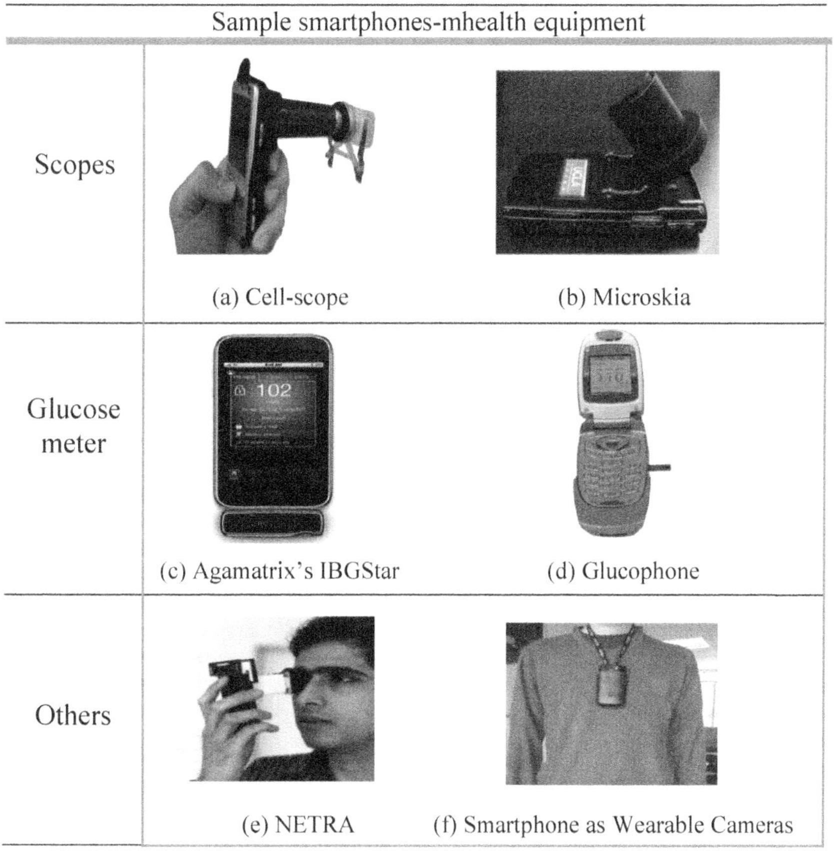

Fig. 3 Smartphones as an integral part of medical equipment [28], [29]

5 Conclusions and Future Trends

The number of medical apps for healthcare is rapidly increasing by the day, therefore it is difficult for physicians and other healthcare professionals to gain familiarity with such apps and recommend them for their patients at the same rate they are introduced. As healthcare providers cannot cope with such overwhelming number of apps, it is recommended that a specific app, for a given disease or a practice, may be recommended for patients taking into consideration their level of education and technical abilities [1]. It is also recommended that prior to using or recommending any app, the following points should be considered: (i) is the app evidence-based or is it based on recognized guidelines?; (ii) what is the patient's level of understanding of both smartphones and apps?; (iii) can the patient use an app with or without training? [1]. In addition It is recommended not to place any sensitive information including photos or personal health information on personal devices in a public health or other healthcare setting prior to consulting with the telecommunication carrier as well as your healthcare provider [1].

Smartphones role in mHealth is expanding and covering a wide range of medical aspects. Each smartphone generation comes with additional capabilities. It is anticipated that we see smartphones with a 3-D display and a 3-D web browser in a short period of time. So instead of just viewing web pages on your smartphone, one will be able to go into environments (or stores or showrooms) and maneuver around in them, just as we do on devices like the Xbox [30]. In five years, smartphones will be two-piece systems: a "CPU," maybe imbedded in something you already wear such as your shoes, a belt, watch or clipped to your clothes. You'd view data on a variety of screens that wirelessly connect to this smartphone CPU, whether it's smart glasses, like Google Glass, or a heads-up display in your car. Maybe you'll even opt for an ocular implant that lets you see your "screen" floating out in front of you [31]. Moreover, you won't need to locate a facility map to find out You Are Here. Instead, you'll use your phone to not only find where you are, but to get directions to where you need to go within a building [31]. By 2018, you'll be able to leave your house, travel to an airport, get on a plane and land in a foreign city and never lose your Wi-Fi connection. In fact, your cellular connection (4G, LTE Advanced or whatever) will end up being your backup data network, since pass-point Wi-Fi will be close to ubiquitous worldwide [31]. Soon the wireless charging feature will become quite common in future smartphones, negating the need to connect your device to a charger with cables. Users will be able to wirelessly charge their smartphone anywhere with ease [32]. In addition, smartphones will interface with smart surfaces. We are already seeing the beginning of using touch and voice-operated intelligent screens as tabletop computers that can access the internet. Simply by placing the smartphone on these surfaces, the two will link together. Additionally, your ultra intelligent agent will flow from your smartphone to the screen. [30], [33]. Finally, smartphone users will not have to remember numerous passwords instead they can access data and sites on smartphones using multiple biometric authentications such as fingerprints, voice, and the user face [33].

References

[1] Bedno, S.A., Vicsik, D.M.: The smartphone revolution in healthcare. Medscape (2013), http://www.medscape.org/viewarticle/778212

[2] Yang, J.: Smartphones in Use Surpass 1 Billion, Will Double by 2015. Bloomberg Technology (October 2012), http://www.bloomberg.com/news/2012-10-17/smartphones-in-use-surpass-1-billion-will-double-by-2015.html

[3] Crilly, P., Muthukkumarasamy, V.: Using smartphones and body sensors to deliver pervasive mobile personal healthcare. In: 2010 Sixth International Conference Intelligent sensors, Sensor networks and Information Processing (ISSNIP), Brisbane, QLD, pp. 291–296 (2010)

[4] Milosevic, M., Shrove, M.T., Jovanov, E.: Applications of Smartphones for Ubiquitous Health Monitoring and Wellbeing Management. Journal of Information Technology and Applications 1(1), 7–15 (2011)

[5] Mosa, A., Yoo, I., Sheets, L.: A systematic review of healthcare applications for smartphones. BMC Medical Informatics and Decision Making 12, 12–67 (2012)

[6] Detmer, D., Bloomrosen, M., Raymond, B., Tang, P.: Integrated personal health records: transformative tools for consumer-centric care," BioMed Central. BMC Medical Informatics and Decision Making 8(45) (October 2008)

[7] Tang, P., Ash, J., Bates, D., Overhage, J., Sands, D.: Personal health records: definitions, benefits, and strategies for overcoming barriers to adoption. Journal of the American Medical Informatics Association 13(2), 121–126 (2006)

[8] Dolan, B.: 12 EMRs with remote access iPhone apps. Mobi Health News (2010), http://mobihealthnews.com/6093/12-emrs-with-remote-access-iphone-apps/

[9] Chaudhari, R., Brady, W.: What other technology can enhance my EHR. American Academy of Dermatology (2013), http://www.aad.org/members/practice-management-resources/hit-kit/ehr-implementation/what-other-technology-can-enhance-my-ehr

[10] Klasnja, P., Pratt, W.: Healthcare in the pocket: Mapping the space of mobile-phone health interventions. Journal of Biomedical Informatics 45(1), 184–198 (2012)

[11] Hu, S., Fu, Y., She, C., Yao, H.: Enabling Zigbee communications in Android devices. In: 2nd International Conference on Computer and Information Application (ICCIA 2012), pp. 659–662 (2012)

[12] Al-Alawi, A.: WiFi technology: Future market challenges and opportunities. Journal of Computer Science 2(1), 13–18 (2006)

[13] Haartsen, J., Naghshineh, M., Inouye, J., Joeressen, O.J., Allen, W.: Bluetooth: vision, goals, and architecture. Mobile Computing and Communications Review 2(4), 38–45 (1998)

[14] International Data Corporation (IDC). Apple cedes market share in smartphone operating system market as android surges and windows phone gains (August 2013), http://www.idc.com/getdoc.jsp?containerId=prUS24257413

[15] International Data Corporation (IDC). Android and iOS combine for 92.3% of all smartphone operating system shipments in the first quarter while windows phone leapfrogs blackberry (May 2013), http://www.idc.com/getdoc.jsp?containerId=prUS24108913

[16] International Data Corporation (IDC). Android and iOS combine for 91.1% of the worldwide smartphone OS market in 4Q12 and 87.6% for the year (February 2013), http://www.idc.com/getdoc.jsp?containerId=prUS23946013
[17] International Data Corporation (IDC). Android marks fourth anniversary since launch with 75.0% market share in third quarter (November 2012), http://www.idc.com/getdoc.jsp?containerId=prUS23771812
[18] International Data Corporation (IDC). Android and iOS surge to new smartphone OS record in second quarter (August 2012), http://www.idc.com/getdoc.jsp?containerId=prUS23638712
[19] International Data Corporation (IDC). Android- and iOS-powered smartphones expand their share of the market in the first quarter (May 2012), http://www.idc.com/getdoc.jsp?containerId=prUS23503312
[20] Drede, J.L.: What are the different types of mobile phone operating systems? Conjecture Corporation (June 2013), http://www.wisegeek.org/what-are-the-different-types-of-mobile-phone-operating-systems.htm
[21] Lettner, M., Tschernuth, M., Mayrhofer, R.: Mobile Platform Architecture Review: Android, iPhone, Qt. In: Moreno-Díaz, R., Pichler, F., Quesada-Arencibia, A. (eds.) EUROCAST 2011, Part II. LNCS, vol. 6928, pp. 544–551. Springer, Heidelberg (2012)
[22] Cromar, S.: Smartphones in the U.S.: Market report. Creative commons attribution 3.0 United States. US (2010), https://www.ideals.illinois.edu
[23] Woodbridge, J., Nahapetian, A., Noshadi, H., Sarrafzadeh, M., Kaiser, W.: Wireless health and the Smartphone conundrum, http://www.cs.ucla.edu/~ani/publications/smartphoneconundrum.pdf
[24] Blackberry. A short history of the Blackberry (August 2013), http://www.bbscnw.com/a-short-history-of-the-blackberry.php
[25] Latest smartphones (2013), http://www.google.ca/search?q=latest+smartphones+2013
[26] Wikipedia.Comparison of mobile operating systems (August 2013), http://en.wikipedia.org/wiki/Comparison_of_mobile_operating_systems
[27] Kawasumi, Y., Ernst, P., Abrahamowicz, M., Tamblyn, R.: Smartphones in Clinical Practice, Medical Education, and Research. Arch Intern Med, American Medical Association 171(14), 1294–1296 (2011)
[28] Dolan, B.: 7 medical phone peripherals you should know. mobi health news (July 2011), http://mobihealthnews.com
[29] Gurrin, C., Qiu, Z., Hughes, M., Caprani, N., Doherty, A.R., Hodges, S.E., Smeaton, A.F.: The smartphone as a platform for wearable cameras in health research. American Journal of Preventive Medicine 44(3), 308–313 (2013)
[30] Burrus, D.: From 3D graphics to biometric scans: How your smartphone will get smarter. CNN Tech (February 2012) http://www.cnn.com/2013/02/21/tech/mobile/smartphone-developments-mobile-strategy
[31] Wolpin, S.: Your smartphone in 2018: 15 futuristic features (June 2013) http://news.yahoo.com/smartphone-2018-15-futuristic-features-222906401.html
[32] Passary, A.: 5 features to expect in future smartphones (June 2013), http://www.itechpost.com/articles/9866/20130615/5-features-expect-future-smartphones.htm
[33] Burrus, D.: Your smart phone is getting smarter (May 2013), http://bigthink.com/flash-foresight/your-smart-phone-is-getting-smarter

22. A Product Line Architecture for Mobile Patient Monitoring System

Gaurav Paliwal[1,*] and Arvind W Kiwelekar[2]

[1] Department of Computer Engineering
R C Patel Institute of Technology (RCPIT),
Shirpur 425405 Dist Dhule (MS)
gvpaliwal@gmail.com
[2] Department of Computer Engineering
Dr. B. A. Technological University,
Lonere 402103 Dist Raigad (MS)
awk@dbatu.ac.in

1 Introduction

The term mobile health (mHealth) means to adopt Information and Communication Technology (ICT) especially mobile hand held devices to provide health-care and delivery services [1]. The ICT is mainly adopted with an intention to provide cost-effective, efficient and easily accessible health care services. Patient monitoring is one of the health care activities that can effectively adopt mobile devices and allied techniques for chronic disease management and continuous patient monitoring. The Mobile Patient Monitoring (MPM) systems have promise and potential to provide as an effective alternative for traditional patient monitoring techniques.

A number of MPM systems have been developed earlier that offer a varied range of functional and non-functional features. In this chapter, we provide a product-line architecture that essentially captures the similarities and variabilities of these MPM systems. Product Line (PL) architecture is a software design approach especially useful for evolving domains to achieve a higher degree of reuse and customization within the cost and time constraints. The basis of a *PL* is a collection of features, called the scope and a collection of reusable components called the *core assets* [2]. Feature-Oriented Reuse Method (FORM) is an approach for PL architecture development [34] [50] [21] that use feature models to capture both commonalities and variabilities of PL. The output of the feature model analysis can be further used to explore reusable components for product line development.

Keywords: Mobile Patient Monitoring, Product Line Architecture, Feature Oriented Domain Analysis, Domain modeling, context analysis, feature analysis.

---

* Corresponding author.

S. Adibi (ed.), *Mobile Health*,
Springer Series in Bio-/Neuroinformatics 5, DOI: 10.1007/978-3-319-12817-7_22

2 Generic Architecture of MPM Systems

The generic architecture for MPM[28] includes system development processes (e.g., planning and designing) and various MPM product variants. From the product perspective, the generic architecture for MPM systems has two components, namely, *Body Area Network* (BAN) and Back *End System* (BESys). The BAN is defined as a network of communicating devices worn on or around the body which is used to acquire health related data to provide mobile health services to the patient. As shown in Figure 1, the BAN consists of a *Mobile Base Unit* (MBU) and a set of BAN devices such as sensors, actuators, or other wearable devices used for medical purpose. The sensors may directly transmit the bio-signal data to the MBU or do it via the *Sensor Front-End* (SFE). The BESys comprises of the back-end server which can be of two types: *back-end server* to which the MBU transmits bio-signal data and clinical back-end server which may host custom health-care applications.

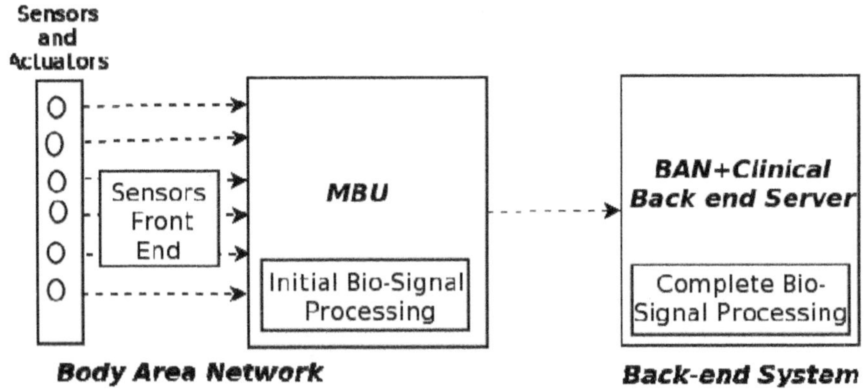

Fig. 1 A Generic Architecture for Mobile Patient Monitoring System [28]

From the process perspective, we can represent the MPM architecture into five steps as: (i) Data acquisition (ii) Data Transmission (iii) Bio-signal Interpretation and Processing (iv) Notification (v) Intervention. BAN acquires the physiological data actively or passively by interacting with the patient. Data can comprise of vital signs, blood glucose levels, responses to specific health conditions and general health questions as well as patient location. Data is then packaged, transferred and delivered to the caregivers and third parties via different modes of communication. Algorithms and/or healthcare workers interpret and review the data and determine if there are any areas for concern. If needed clinicians, family, caregivers, third parties or the patients themselves are notified of a potential problem through either the device itself or an intermediary health-care worker. They can then take an action when intervention is needed. This separation between product and process perspective help system modelers to individually focus on structural and functional elements of MPMS facilitating systematic development of MPMS.

3 Survey of Existing Mobile Patient Monitoring Applications

During last few years, mobile patient monitoring has evolved as the most promising and active application area [3] in the domain of m-Health. As a result a number of mobile patient monitoring systems have been proposed by various researchers. These systems are either designed for a specific community or for people suffering from a specific disease such as cardiovascular disease [4], depressive illness [5], dementia [6], hypertension [7], and diabetes [8.]. Few of these systems are dedicated to the older age group patients [9] and mass casualty victims [26]. The basic functionality provided by these systems include continuous monitoring and delivery of bio-signals [11] [12] and real time signal detection [13]. These systems vary in terms of non-functional features supported by them. For example, few systems provide *interoperability* between the various heterogeneous medical information servers by adopting Service-*Oriented Architecture (SOA)* [14] [15] for application deployment and adopting HL7 standards [16] [17] for data encoding. Some of the additional functionalities realized by these systems included to predict current health status [18] through various data mining techniques, to provide secured access to the monitored data [20] through password authentication, to interpret medical data and to make decisions [22].

Table 1 shows a comparison of MPM systems, in which various systems have been compared against the following functionalities: to process bio-signals (FR1), to deliver bio-signals(FR2), to raise emergency alarm (FR3), to interpret bio-signals (FR4), to perform differentiation of bio-signal (FR5), to acquire data (FR6), to communicate bio-signals (FR7), and medicine infusion (FR8). Table 1 also compares these systems against a set of non-functional parameters such as Genericity (NFR1), Security (NFR2), Unique Patient Identification (NFR3), Interoperability (NFR4), Privacy (NFR5), Intelligence (NFR6), Availability (NFR7), Response Time (NFR8), Easy Wear-ability (NFR9), Graphical interface (NFR10), Accuracy (NFR11), Data loses (NFR12) and adoption of standards (NFR13).

Table 1 Comparison of mobile patient monitoring systems

	FR1	FR2	FR3	FR4	FR5	FR6	FR7	FR8	NFR 1	NFR 2	NFR 3	NFR 4	NFR 5	NFR 6	NFR 7	NFR 8	NFR 9	NFR 10	NFR 11	NFR 12	NFR 13
IMHMS	✓	×	✓	✓	✓	✓	✓	×	✓	✓	✓	✓	×	✓	×	✓	✓	✓	×	×	×
MHCS	×	✓	×	✓	×	✓	✓	×	×	×	×	×	×	✓	✓	✓	✓	✓	×	×	×
MTDDS	×	✓	×	✓	×	✓	✓	×	×	×	×	✓	×	×	×	✓	✓	✓	×	×	×
WISS	×	×	×	✓	✓	✓	✓	×	×	×	×	×	×	×	×	✓	✓	✓	✓	×	×
MCWA	✓	✓	×	✓	✓	✓	✓	×	×	×	×	×	×	✓	×	✓	✓	✓	×	×	×
PHM	✓	✓	✓	✓	✓	✓	✓	×	×	×	×	×	×	✓	×	✓	✓	✓	✓	×	×
MHS	×	✓	×	✓	×	✓	✓	×	✓	×	×	×	×	×	×	✓	✓	✓	×	×	×
THC	✓	✓	✓	✓	×	✓	✓	×	×	×	×	×	×	✓	×	✓	✓	✓	×	×	×
UMHMSE	✓	✓	✓	✓	×	✓	✓	×	×	×	×	×	×	✓	×	✓	✓	✓	×	×	×
PBEHS	×	✓	×	✓	×	✓	✓	×	×	×	×	×	×	✓	×	✓	✓	✓	×	×	×
CS	✓	✓	✓	✓	✓	✓	✓	×	×	×	×	×	×	✓	×	✓	✓	✓	×	×	×
AID-N	✓	✓	✓	✓	×	✓	✓	×	×	×	✓	×	×	✓	×	✓	✓	✓	×	×	×

This comparison reveals that numerous MPMS exist but there is an absence of a *reference architecture* against which these existing systems can be compared. One reason behind the absence of reference architecture is that, most of the time MPMS products are *developed* and they are rarely *engineered* by following systematic development process. A systematic approach to product development typically includes adoption of engineering processes of requirements, analysis and architecture design. While comparing these MPMS, it has also been observed that few approaches exist that perform requirements analysis for MPMS [23] and architecture design for MPMS [24] [25] [26] [6]. However, they overlook the aspect of capturing *variability* and *commonality* among numerous MPMS products. One of the consequences of adopting adhoc approach to MPMS development is that the design information and knowledge about the MPMS is distributed and scattered all over the existing implementations.

4 Feature Oriented Domain Analysis (FODA)

4.1 Background

The FODA is a method of requirements analysis which defines both the products and the processes of domain analysis. In FODA, a feature is a prominent or distinctive user-visible aspect, a quality, or a characteristic which can be identified as mandatory, optional or alternative property for a system. In FODA, domain analysis is defined as a process of collecting, identifying, representing and organizing the relevant information in a domain based on the study of existing systems and their expansion histories, knowledge taken from domain experts, emerging technology, and underlying theory within an application domain [31].

The main objective behind performing FODA is to identify similarities and differences that are present in multiple realizations of a software system. Such an analysis helps to build a set of derivative products efficiently on a common platform. FODA provides a generic description of the requirements of systems and a set of approaches for their implementation [31].

To identify similarities in multiple software systems, the FODA can exploit reuse characteristics of a system platform. For example, in case of MPMS the followings are hardware and software reuse characteristics.

- *Hardware:* The reuse of sensors for monitoring the bio-signals as well as reuse of electronic interfacing control units with hardware modules and for implementing the software.
- *Software*: The reuse of software subsystems for interpretation, evaluation, transmission, and display of sensor data to aid advanced level applications.

The variations exhibited through adoption of multiple technological can be used to identify variabilities among multiple software systems. In case of MPMS, variabilities arise mainly because of different sensor technologies, different numbers of sensors, and different types of measurements, different bio-signal transmission media, and evaluation. Domain Analysis is frequently used to capture these type of variabilities and commonalities of the systems in a domain [30]. The output of FODA would count the following benefits:

- To reduce the development cost and improves the quality due to reuse of well tested components in various systems.
- To lower hardware costs.
- To achieve faster integration and development due to focused application specific development.

The FODA as a method of requirements analysis also develops a common understanding of an application domain in terms of the features, common requirements and scope of the domain.

4.2 A Feature Oriented Domain Analysis for Mobile Patient Monitoring Systems

This section describes the results of performing FODA for the domain of mobile patient monitoring with an intention to capture similarities in terms of functionality, implementation, hardware specification, software specifications, requirements and components. The analysis presented in this section is based on the methodology provided by Kyo. C. Kang et al [31]. The Feature Oriented Domain Analysis is typically carried out in two phases [32] namely, context analysis and feature analysis.

4.2.1 Context Analysis

The purpose of context analysis is to define the boundaries of a domain by using the domain data and constraints. This phase establishes a proper scope inside which the domain analysis is performed. In context analysis, relationships of the candidate domain with peripheral elements are analyzed, and variabilities are evaluated. The output of context analysis is described through structure and context diagrams.

4.2.1.1 Structure Diagram

The structure diagram in FODA decomposes the target domain as higher, lower, and peer-level domains. A structure diagram for mobile patient monitoring system is shown in Figure 2. The diagram covers complete application domain and it consists of subsystems necessary for realizing a full functional mobile patient monitoring system. The layered architecture style is used for structuring the functionalities of MPMS. The various functionalities of MPMS are grouped into three layers.

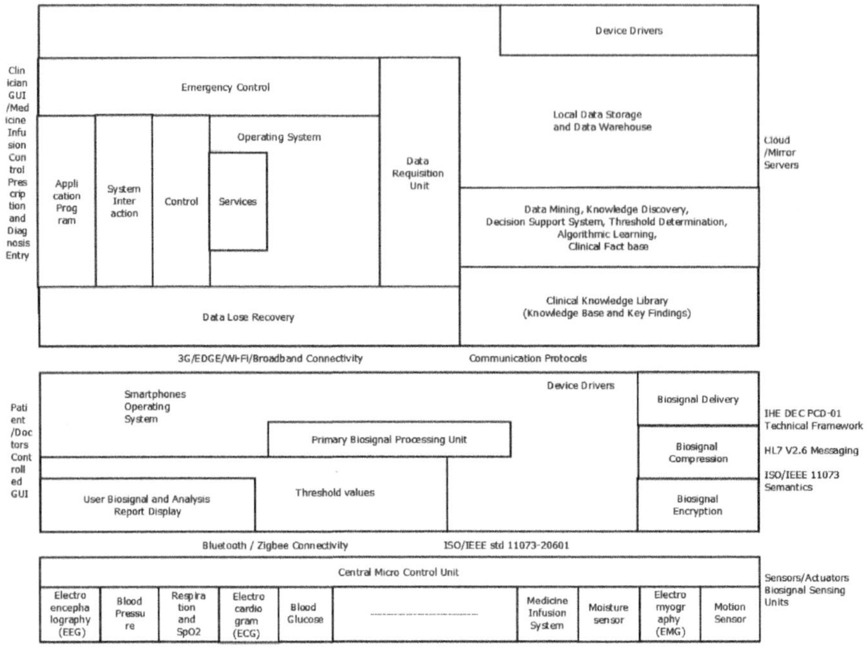

Fig. 2 Structure Diagram

The lowermost layer is the Body Area Network (BAN) which contains the wearable bio-sensing units and central micro-controller unit. The layer falls in the application domain of electronic micro control unit's design where body worn bio-sensors and actuators are developed. In this subsystem, we have included some of the sensing units that are not included in earlier mobile patient monitoring systems previously with an intention to provide additional sensing of physical parameters.

The system also interacts with some of the entities external to it through the defined application interface for the layer. These entities mainly include the standards and semantics defined for medical domain and communication network standards and protocols. Some of the standards defined by IEEE, ISO, HL7, IDE, IHE and DEC are included because these are commonly used in health informatics. The standards are ISO/IEEE std 11073-20601 [35], ISO/IEEE std 11073-104zz [36], IHE DEC PCD Technical Framework [33] and HL7 V2.X [37] and V3 [38] Messaging. These standards mainly deal with optimized exchange protocol for personal health device communication and device specialization for a specific agent. They describe the device capabilities and its implementation details appropriate for sharing of medical information. These standards facilitate optimal patient care and packaging of health-care data in the form of messages.

These standards are primarily intended to be used in clinical environment but it is well suited for wearable sensors [40]. The standards altogether makes the monitoring of data (Electronic Health Record (EHR) and Personal Health Record (PHR)) vendor independent [40] and hence provides a complete interoperability solution for mobile patient monitoring systems.

4.2.1.2 Context Diagram

The Context Diagram in FODA is a data-flow diagram showing data flows between a generalized application within the domain and the other entities and abstractions from system environment with which it communicates [32]. A context diagram showing comprehensive data flows for the mobile patient monitoring system is shown in Figure 3. This diagram captures the major data flows in the form information requisition, information provision, Biosignal transmission and request and response calls. In the diagram, the closed square boxes represent an automated request and response systems and they act as a source and a sink for the information whereas the open messages represent the user interaction. The arrows represent the direction of information flow.

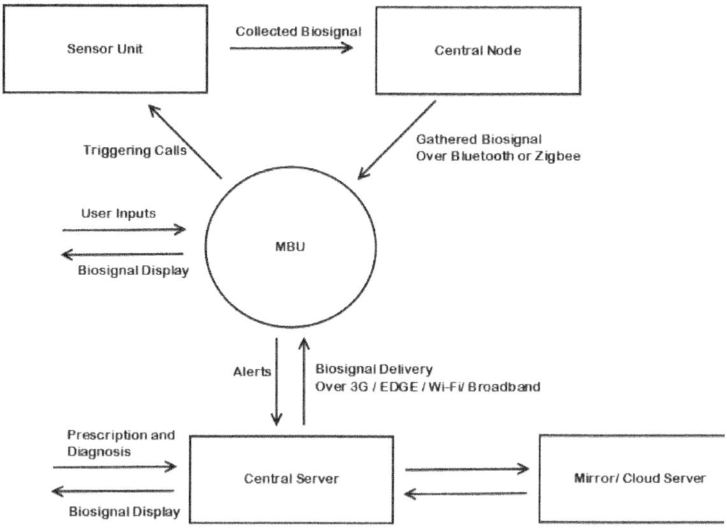

Fig. 3 Context Diagram

The MBU is represented in the middle of the diagram and it acts as a core element of the system which handles the major data flow whereas the central node and central server act as a major source and sink respectively. The data originated from the sensors is gathered by the central node and passed to the MBU where primary bio-signal analysis is done and data is forwarded to central node for further bio-signal processing. Wireless medias like Bluetooth, Zigbee, Wi-Fi, 3G are the most convenient and preferable choice for the data transfer. Smart-phones and mobile networks are the most convenient, powerful [41] [42], wide spread [43] and cost effective [44] choice for the mobile patient monitoring system MBU and data transfer respectively.

4.2.2 Domain Modeling

As defined in Section 4.1.1, the purpose of domain modeling is to define scope for the application to be developed. In our case, the application is mobile patient monitoring system. It provides a better understanding of the systems by systematically describing functions, objects, data, and relationships. The domain modeling phase mainly consists of three phases, namely, feature modeling, entity relationship modeling and operational modeling. The output produced by each phase describes what a system is, what the functions to be performed by an application are and how they perform it. Domain modeling provides activities, features, vocabulary, entities, control flow, data flow and requirements of the system.

4.2.2.1 Feature Analysis

The feature model captures the capabilities of a system from end users point of view. The analysis is mainly focused on the functionalities of the system, performance parameters, and operating environment in which the system application run. We have defined features as the attributes of a system that directly or indirectly affects the end-users by affecting the system performance, quality, security and services. The features shown in Figure 4 are identified on the basis of deliverable functionalities, capabilities, technologies, operating environment and functional and non-functional [29] requirements. These features can be broadly divided into three types as functional, operational and presentation. The functional feature represents the services provided by the applications, operational features represent the actions or tasks performed, and presentation features deals with the user interaction and information representation.

The feature diagram shown in Figure 4 encompasses the features of mobile patient monitoring systems in three levels. The first level of the diagram mainly includes the functional features of MPMS and the components present in the system. The level 2 of the diagram contains the features expected to be delivered by the respective components and some of these features has been further extended in the level 3. The features containing a white circle on the end of a connector represent an optional feature, whereas, the features containing a black circle on the end of a connector is a mandatory feature and the arc on a connector represents an alternative feature.

Bio-signal delivery and bio-signal processing are the only features, which are fundamentally necessary for a mobile patient monitoring system but these features are not sufficient for practically usability. The features we have included in the feature diagram contain many features that are not essential but we have treated them as necessary features to improve the user responses towards the system. Some of these features are discussed below:

- *Genericity*: The System should be modified according to the patient monitoring needs. It should not be specific to certain disease, group or community of people. In mobile patient monitoring systems genericity can be achieved with help of generic architectures [27] [28]. This architecture allows the number of sensors and system functionalities to be modified according to the disease monitoring needs.
- *Intelligence*: System should be capable of taking decisions on the basis of past and current records. It should assist the physician in interpreting the medical data, decision making and automating the patient monitoring process through

artificial intelligence [45] and a data management system [46] which can extract, interpret and effectively encapsulate real world context aware information ensuring that physicians get the correct data every time. The data server should react differently depending on the medical data and real time readings of the patients in different condition. To support these type of systems some techniques like data mining for predicting current health status [18] of a patient, applying clustering algorithm to both real time and historic data have been proposed.

- *Unique patient identification*: Patient should be provided with a patient ID by which he/she can be globally uniquely identified. 2G RFID [47] tags are the most promising choice for the feature; it not only offers unique identity to the patient but also provide solutions for security and patient tracking issues.
- *Availability and Speed:* The system should be available 24*7 and should be able to process gigabytes of data in real time. Cloud servers [9] can ensure the availability and in time processing of data due to its theoretically unlimited processing power.
- *Data delivery*: In Bio-signal delivery, the quality of received Biosignal should be good enough so that the clinical decision making should not be affected. To fulfill the Goodput requirements in consideration with the data loss ratio and round trip time, [48] 3G connections are the most viable option for data transfer. Data compression is also one of the possible solutions for the problem since it reduces the amount of data transferred but introduces significant delay which is intolerable for patient monitoring systems.

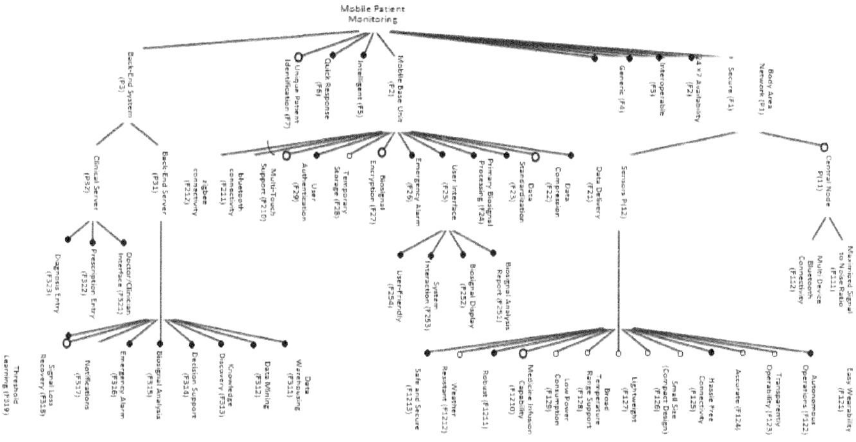

Fig. 4 Feature Diagram

- *Multi-touch support*: For clinical decision making medical images needs to be interpreted in more than one dimension called as multimodal analysis. For multi-modal analysis the device should have multi touch support.
- *Bio-sensors*: biosensors are normally deployed on the patient body to acquire Biosignals. The biosensors deployed in mHealth systems should be easy wearable, accurate, small in size and power efficient.

- *Medicine infusion capability*: The capability is very useful for diabetic, cardiac and asthmatic patients. It can improve the quality of scheduled medicines intake and emergency response time by automatically triggered infusion of medicine into the patient's body. Shinji Ichimori et al. [49] have shown a successful implementation of highly responsive needle-type glucose sensor for diabetic patient. The sensor has been successfully tested independently and can be potentially incorporated into the mobile patient monitoring systems.

4.2.2.2 Entity Relationship Model

The ER model captures and defines the domain knowledge essential for implementing applications in the system domain. The ER model mainly consists of entity classes which are related through *is-a* relationship and *consist-of* relationship. The main aim behind developing ER model is to deliver the system knowledge in terms of entities and relationships for the functional analysis.

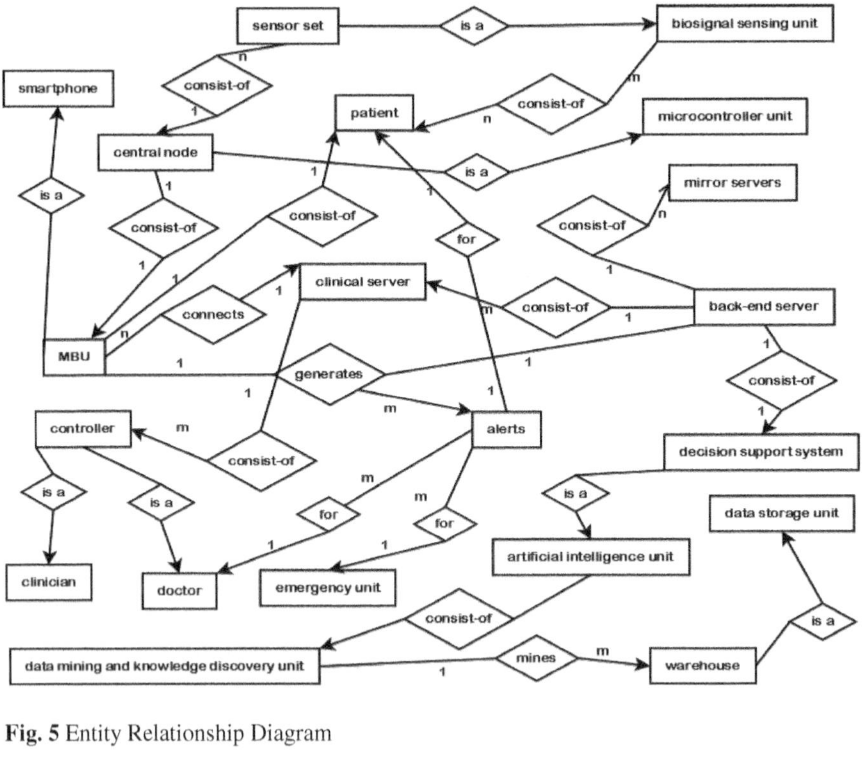

Fig. 5 Entity Relationship Diagram

The entities and relationships shown in Figure 5 are mainly taken from some existing systems whereas few entities like mirror servers and medicine infusion sensors are taken from possible systems. In our ER model, we have replaced some of the *consists-of* relationships with active nouns to describe the connection between two entities.

4.2.2.3 Operational Model

The operational model provides an immediate understanding of the system to the user by describing that how the system works. The model captures functional and behavioral relationships from feature model and entity relationship model. The operational model mainly provides the data and control flow of the system.

The model can be represented by a single operational diagram which identifies the complete system functionalities or with the help of functional model consist of separate state chart and activity diagrams. In operational diagram, a transition causes transfer from one state to the next in response to an event controlled by the conditions and events.

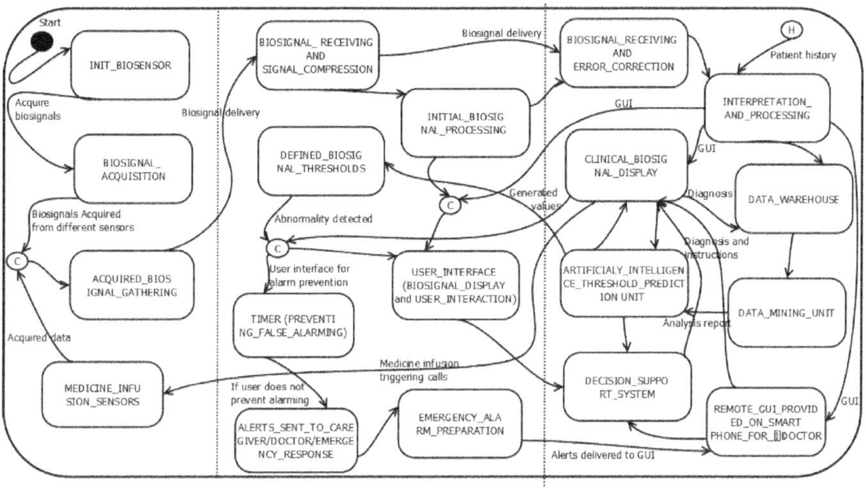

Fig. 6 Operational Diagram

The operational diagram shown in the Figure 6 shows three parallel sub-systems of mobile patient monitoring operating simultaneously and continuously. The first sub-system, BAN offers major functionalities related to the bio-signal acquisition and bio-signal delivery. MBU offers bio-signal processing, emergency alarm and bio-signal delivery. The bio-signal delivery in BAN and MBU is more or less the same except the delivering media. BAN delivers bio-signal to MBU over the Bluetooth or Zigbee link whereas the MBU delivers bio-signals to *BESys* over internet. The major functionalities implemented by the *BESys* consist of bio-signal interpretation, decision support, threshold generation, data mining, data storage, emergency alarm and bio-signal processing. The signal processing in MBU and BESys mainly differs in data processing. In bio-signal processing MBU processes only the current data available for abnormalities whereas on BESys current data is processed with historic data to find any type of abnormalities in the Biosignals. Thresholds generated by *BESys* on the basis of complete data processing are delivered to MBU for primary data processing. The threshold values should normally be defined over a group of patients because it varies with age, sex, weight and many other factors. While generating these values the diagnosis feed by the doctor manually in the system should also be considered. The system also provides the interpreted bio-signal display on various devices to various user mainly differs in the details and functionalities provided to the user. We have introduced some medicine infusion sensors in the MPM system that are successfully implemented in some systems but not integrated with an existing mobile patient monitoring system. These sensors can play an important role in emergency conditions to save patient's life but it also raises certain security issues with patient safety.

5 Deriving the Product Lines from Feature Model

Product Line is an approach to software development that considers development concerns observed in all phases of the production life-cycle of a system. Product lines are developed with an intention to describe experience of system builders in the areas of requirements, data management, algorithm, human-computer interfaces, design processes, testing and validation of a product. Purpose of product lines is to provide systematic guidelines for product development.

In this section, we describe product lines for the mobile patient monitoring system using FODA grammar. Here we are representing product variability observed in MPMS using Extended Modal Transition Systems [51] [52].

PL 1. $\rightsquigarrow$ *MPM; (*$\rightsquigarrow$*P1;* ∧ *F*1; ✓ ∧ F2; ✓ ∧ F3; ✓ ∧ F4; ✓ ∧ $\rightsquigarrow$ P2; ∧ F5; ✓ ∧ F6; ✓ ∧ F7; ✓ ∧ $\rightsquigarrow$P3;)

PL 2. $\rightsquigarrow$P1; ($\overline{P11}$; ∧ P12;)

PL 3. $\rightsquigarrow$P11; (F111; ✓ ∧ $\overline{F112}$; ✓)

PL 4. $\rightsquigarrow$ *P12; (* *F*121; ✓ ∧ $\overline{F122}$; ✓ ∧ $\overline{F123}$; ✓ ∧ F124; ✓ ∧ $\overline{F125}$; ✓ ∧ $\overline{F126}$; ✓ ∧ $\overline{F127}$; ✓ ∧ $\overline{F128}$; ✓ ∧ $\overline{F129}$; ✓ ∧ $\overline{F1210}$; ✓ ∧ F1211; ✓ ∧ $\overline{F1212}$; ✓ ∧ F1213; ✓)

PL 5. $\rightsquigarrow$ P2; (F21; ✓ ∧ $\overline{F22}$; ✓ ∧ F23; ✓ ∧ F24; ✓ ∧ $\rightsquigarrow$ F25; ✓ ∧ F26; ✓ ∧ $\overline{F27}$; ✓ ∧ $\overline{F28}$; ✓ ∧ F29; ✓ ∧ $\overline{F210}$; ✓ ∧ F211; ✓ □ F212; ✓)

PL 6. $\rightsquigarrow$ F25; (F251; ✓ ∧ F252 ✓ ∧ F253; ✓ ∧ F254; ✓)

PL 7. $\rightsquigarrow$ P3; (P31; ∧ P32;)

PL 8. $\rightsquigarrow$ P31; (F311; ✓ ∧ F312; ✓ ∧ F313; ✓ ∧ F314; ✓ ∧ F315; ✓ ∧ F316; ✓ ∧ $\overline{F317}$; ✓ ∧ $\overline{F318}$; ✓ ∧ F319;✓)

PL 9. $\rightsquigarrow$ P32; (F321; ✓ ∧ F322; ✓ ∧ F323; ✓)

In the expressions from PL1 to PL9, feature identifiers are used to identify a feature and A;✓and $\overline{A}$;✓ represents the mandatory and optional feature respectively, whereas, ∧ and □ represent the conjunction and choose-one operations respectively. The variability represented through these expressions can play an important role in designing the different variants of MPMS.

6 Product Line Architecture

Product Line (PL) Architecture core assets required for development are identified with an objective to support effective product variability management. During development, core assets can be either shared or reused among the members of the product line with an explicit treatment of variability. The typical objects to be included as core asset are architecture, reusable components, requirements statements, domain models, specifications and documentation, performance models, budgets, schedules, test cases, test plans, work plans, and process descriptions [53]. The elements from feature models are the most effective way to illustrate commonality and variability in products by showing the variation points explicitly [54]. In this section, we provide mapping rules from requirements to features and mapping rules from features to product line [55].

Mapping rules from requirements to features

1. $\exists F_i \in PF, \exists R_j \in PR$, specify $(R_j, F_i) \wedge SC_r(R_j) =$ mandatory $\Rightarrow SC_f(F_i) =$ mandatory
2. $\exists F_i \in PF, \exists R_j \in PR$, specify $(R_j, F_i) \wedge ((SC_r(R_j) =$ alternative$) \vee (SC_r(R_j) =$ mutual exclusion $\vee (SC_r(R_j) =$ *multiple alternatives*$) \vee (SC_r(R_j) =$ optional$))$ $\Rightarrow SC_f(F_i) =$ optional
3. $\exists PF_k \subseteq \exists PF, \exists R_j \in PR, \forall F_i \in PF_k$ specify $(R_j, F_i) \wedge SC_r(R_j) =$ mandatory $\Rightarrow \forall F_i \in PF_k, SC_f(F_i) =$ mandatory
4. $\exists PF_k \subseteq PF, \exists R_j \in PR, \forall F_i \in PF_k$, specify $(R_j, F_i) \wedge ((SC_r(R_j) =$ alternative$) \vee SC_r(R_j) =$ mutual exclusive$) \vee (SC_r(R_j) =$ multiple alternatives$) \vee (SC_r(R_j) =$ optional$)) \Rightarrow SC_f(F_i) =$ optional
5. $\exists R_i, R_j \in PR, \exists F_i, F_j \in PF, SC_r(R_i) = SC_r(R_j) =$ mutually exclusive $\wedge$ specify $(R_i, F_i) \wedge$ specify $(R_j, F_j) \Rightarrow (i \neq j) \wedge SC_f(F_i) = SC_f(F_j) =$ mutually exclusive
6. $\exists F_i, \in PF, \exists R_j \in PR$, specify $(R_j, F_i) \wedge \exists R_K \in PR$, required$(R_k, R_j) \wedge SC_r(R_j) =$ optional $\wedge SC_r(R_k) =$ mandatory $\Rightarrow SC_f(F_i) =$ mandatory
7. $\exists F_i, \in PF, \exists PR_k \subseteq PR$, specify $(PR_k, F_i) \wedge ((\forall R_l \in PR, SC_r(R_l) =$ optional $\Rightarrow R_l \in PR_k) \vee (\exists R_l \in PR_k, SC_r(R_l) =$ mandatory $)) \Rightarrow SC_f(F_i) =$ mandatory
8. $\exists F_i \in PF, \exists PR_k \subseteq PR$, specify $(PR_k, F_i) \wedge (\forall R_l \in PR_k, SC_r(R_l) =$ optional$)$ $\Rightarrow SC_f(F_i) =$ optional

Mapping rules from features to SPL

9. $\exists A_i \in PA, \exists PF_k \subseteq PF$, specify $(PF_k, A_i) \wedge (\exists F_j \in PF_k, SC_f(F_j) =$ mandatory$)$ $\Rightarrow SC_a(A_i) =$ mandatory
10. $\exists A_i \in PA, \exists F_j \in PF$, specify $(F_j, A_i) \wedge \exists F_k \in PF$, required $(F_k, F_j) \wedge SC_f(F_j) =$ optional $\wedge SC_f(F_k) =$ mandatory $\Rightarrow SC_a(A_i) =$ mandatory
11. $\exists A_i \in PA, \exists PF_k \subseteq PF$, specify $(PF_k, A_i) \wedge (\forall F_l \in PF_k, SC_f(F_l) =$ optional$) \Rightarrow$ $SC_a(A_i) =$ optional
12. $\exists F_i, F_j, \in PF, \exists A_i, A_j, \in PA$, excluded $(F_i, F_j) \wedge$ specify $(F_i, A_i) \wedge$ specify $(F_j, A_j) \Rightarrow (i \neq j)$

Based on these mapping rules [55] [56] [57] [10] [19] elements from an under constrained architecture [39] are depicted in the Table 2 given below mainly consist of core assets of product line architecture.

Table 2 Product Line Architecture Core Assets

Sr. No.	Requirement	Identifier	Type	Map Rule	Feature	Identifier	Type	Map Rule	Identifier	Asset	Type
1	Data loses	NFR12	M	1	Maximized Signal to Noise Ratio	F111	M	10	A1	Noise Ratio	O
2	Easy Wear-ability	NFR9	MA	7	Easy Wearability	F121	M	9	A2	Wearability	M
3	Privacy	NFR5	MA	2	Transparent and Autonomous Operability	F123, F122	O	11	A3	Transparent and Autonomous Operability	O
4	Accuracy	NFR11	MA	7	Accurate	F124	M	9	A4	Accurate	M
5	Biosignals Delivery	FR2	M	2	Hassle Free Connectivity	F125	O	11	A5	Hassle Free Connectivity	O
6	Easy Wear-ability	NFR9	MA	2	Small Size (Compact Design)	F126	O	11	A6	Compact Design	O
7	Easy Wear-ability	NFR9	MA	2	Lightweight	F127	O	11	A7	Lightweight	O
8	Accuracy	NFR11	MA	4	Broad Temperature Range Support	F128	O	11	A8	Broad Temperature Range Support	O
9	Availability	NFR7	M	2	Low Power Consumption	F129	O	11	A9	Low Power Consumption	O
10	Medicine Infusion	FR8	O	8	Medicine Infusion Capability	F1210	O	11	A10	Medicine Infusion Capability	O
11	Response Time	NFR8	M	1	Robust	F1211	M	9	A11	Robust	M
12	Accuracy	NFR11	MA	4	Weather Resistant	F1212	O		A12	Weather Resistant	O
13	Security	NFR2	MA	7	Safe and Secure	F1213, F1	M	9	A13	Safe and Secure	M
14	Availability	NFR7	M	1	24 ×7 Availability	F2	M	9	A14	Availability	M
15	Interoperability	NFR4	M	1	Interoperable	F3	M	9	A15	Interoperable	M
16	Genericity	NFR1	M	1	Generic	F4	M	9	A16	Generic	M
17	Biosignals Delivery	FR2	M	1	Data Delivery	F21	M	9	A17	Data Delivery	M
18	Biosignals Delivery	FR2	M	2	Data Compression	F22	O	11	A18	Data Compression	O
19	Standards	NFR13	M	1	Data Standardization	F23	M	9	A19	Data Standardization	M
20	Biosignal Processing	FR1	M	1	Primary Biosignal Processing	F24	M	10	A20	Primary Biosignal Processing	O

Table 2. (*continued*)

21	Biosignals Interpretation	FR4	M	1	Biosignal Analysis and Display	F251, F252	M	9	A21	Biosignal Analysis and Display	M
22	Graphical interface	NFR10	M	1	User-Friendly	F254	M	9	A22	User-Friendly	M
23	Security	NFR2	MA	2	Biosignal Encryption	F27	O	11	A23	Biosignal Encryption	O
24	Biosignals Delivery	FR2	M	2	Temporary Storage	F28	O	11	A24	Temporary Storage	O
25	Privacy	NFR5	MA	7	User Authentication	F29	M	9	A25	User Authentication	M
26	Biosignals Interpretation	FR4	M	2	Multi-Touch Support	F210	O	11	A26	Multi-Touch Support	O
27	Biosignals Delivery	FR2	ME	5	Bluetooth connectivity	F211, F112	ME	12	A27	Bluetooth connectivity	ME
28	Biosignals Delivery	FR2	ME	5	Zigbee connectivity	F212	ME	12	A28	Zigbee connectivity	ME
29	Response Time	NFR8	M	1	Quick Response	F6	M	9	A29	Quick Response	M
30	Unique Patient Identification	NFR3	M	1	Unique Patient Identification	F7	M	9	A30	Unique Patient Identification	M
31	Intelligence	NFR6	M	3	Data Warehousing	F311	M	9	A31	Data Warehousing	M
32	Intelligence	NFR6	M	3	Data Mining	F312	M	9	A32	Data Mining	M
33	Intelligence	NFR6	M	3	Knowledge Discovery	F313	M	9	A33	Knowledge Discovery	M
34	Intelligence	NFR6	M	3	Decision Support	F314, F5	M	9	A34	Decision Support	M
35	Biosignal Differentiation	FR5	M	1	Biosignal Analysis	F315	M	9	A35	Biosignal Analysis	M
36	Raise Emergency Alarm	FR3	M	1	Emergency Alarm	F316, F26	M	9	A36	Emergency Alarm	M
37	Communication	FR7	O	8	Notifications	F317	O	11	A37	Notifications	O
38	Accuracy	NFR11	MA	2	Signal Loss Recovery	F318	O	11	A38	Signal Loss Recovery	O
39	Intelligence	NFR6	M	3	Threshold Learning	F319	M	9	A39	Threshold Learning	M
40	Data Requisition, Communication	FR6, FR7	M	1	User / Doctor / Clinician Interface	F321, F25, F253	M	9	A40	User / Doctor /Clinician Interface	M
41	Graphical interface	NFR10	M	1	Diagnosis and Prescription Entry	F322, f323	M	9	A41	Diagnosis and Prescription Entry	M

7 Future Directions

The systematic development and deployment of MPMS systems will benefit millions of chronically ill patients by providing better healthcare services at reduced cost. The MPMs have both, promise and potential to change the face of traditional patient care but some issues like system security and patient privacy still needs the attention. User friendliness is another area of MPMS systems which should be improved to achieve better response from the end user. In this chapter we have considered almost every technical issue but non-technical issues such as legal, economical, and managerial that we have not included should also be taken into account for end user deployment of these systems.

8 Conclusion

Patient monitoring is one of the health-care activities for which techniques developed in the field of mobile communications can be effectively used. However, deployment of these techniques poses the challenge of understanding varied requirements. To overcome this challenge, we suggest adopting the *product line approach* for the development of mobile patient monitoring systems. The rationale behind adopting product-line approach is that numerous MPMS surveyed in this chapter exhibit few similarities and variations. By using the product-line approach, system builders develop a common understanding of MPM and they gain the benefits of reusing *core assets* across multiple variants of MPMS. To facilitate the adoption product-line approach, a feature-oriented domain modeling for the domain of mobile patient monitoring is presented in this chapter. A product line architecture which represents variations among features across multiple variants of MPMS and which describes core assets that have presence across multiple variants of MPMS is also presented. The various models and mapping presented in this chapter provides the greater insights about the functional and non-functional requirements of mobile patient monitoring.

9 Case Study

9.1 Mobile Cardiac Wellness Application (MCWA) [60]

The MCWA is a mobile patient-centric, self-monitoring, symptom recognition and self-intervention system that supports chronic cardiac disease management. The systems consists of back-end data repository, data mining, knowledge discovery, knowledge evolution and knowledge processing system, providing clinical data collection, procedural collection, intervention planning, medical situational assessment and health status feedback for users. It utilizes patient information and evidence based nursing knowledge to offer real-time guidance. The systems architecture has been presented from three different viewpoints as; an informational view (utilize multiple sources of information to construct patient specific health assessments and wellness), an operational perspective (data collection, patient assessments, patient evaluation, intervention planning and execution) and an architectural design view. For the feature analysis we have divided the system into three major parts according to the standard architecture presented in the Figure 1.

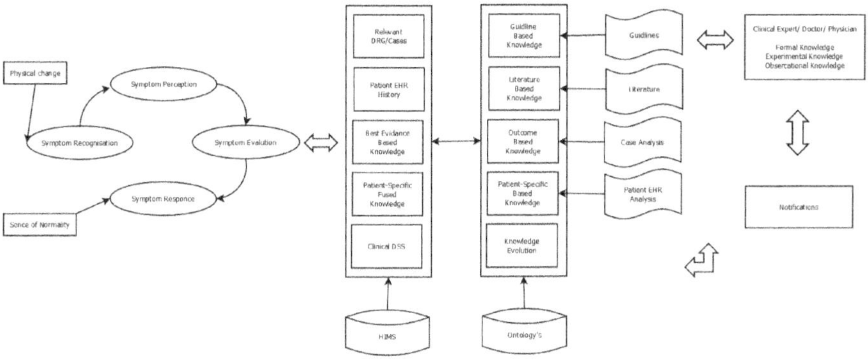

Fig. 7 Mobile Cardiac Wellness Application Conceptual Architecture

Feature Oriented Domain Analysis for MCWA

- **Structural Analysis**
  From the structural perspective the MCWA very clearly defined HIMS, various ontologies, User-System interactions as internal entities to the system in which the system will works but the external entities are not identified in the system. The MCWA also does not provide any type of standards or protocols that are used for data standardization or interoperability.
- **Context Analysis**
  MCWA very clearly recognize every single bit generated in the system. System provides the facility of data collection through patient front end, nurse mobile monitoring devices and patient wearable sensors to improve the clinician's ability to diagnose subtle changes in symptoms which act as a Procedural collection, clinical data collection, instructional feedback and medical situational assessment platform. This information is collected and fused into Systems knowledge base using four forms of case equivalence (structural equivalence, functional equivalence, conceptual equivalence and temporal equivalence) forming new clinical use cases mined for new clinical knowledge.
- **Domain Analysis**
  MCWA provides all the major functional requirements except emergency alarm and medicine infusion but from the non-functional point of view System provide only Intelligence, Response Time, Easy Wear-ability and Graphical interface. The main non-functional parameters such as Genericity, Security, Unique Patient Identification, Interoperability, Privacy, Availability, Accuracy, Data loses and adoption of standards are overlooked in the implementation of the system. Knowledge discovery and Decision support system provides the core strength to MCWA but when it comes to other features the System looks very weak.

9.2 *Intelligent Mobile Health Monitoring System (IMHMS)* [59]

The main objective of IMHMS system is to intelligently predict patient's health status and to provide them feedback through their mobile devices. The IMHMS system is a pervasive mobile health monitoring system that uses Wearable Wireless Body or Personal Area Network to collect data from patients. The system also stores examination and treatment results in the central database. The key strength of the system is its learning capabilities implemented using state-of-the-art data mining techniques. In particular IMHMS provides a flexible and simple MPMS with user-friendly interface however the security mechanisms for the system still require improvements.

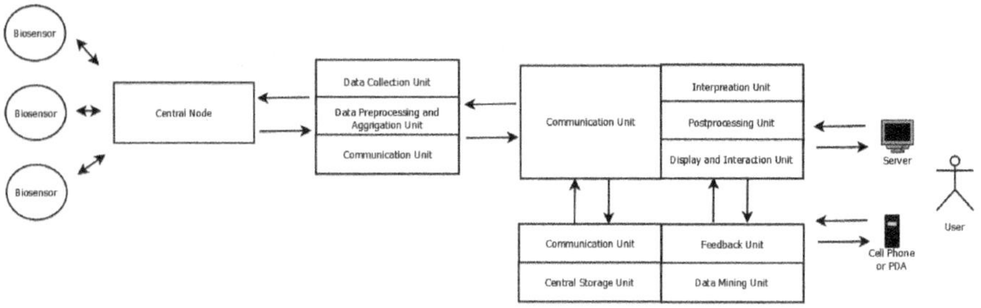

Fig. 8 Intelligent Mobile Health Monitoring System Architecture

Feature Oriented Domain Analysis for IMHMS

- **Structural Analysis**
  From the structural perspective the IMHMS very clearly defined Wearable Body Sensor Network [WBSN], Patients Personal Home Server [PPHS], Intelligent Medical Server [IMS] as internal entities to the system in which the system will works but the external entities are not identified in the system. The IMHMS included RFID tags for unique patient identification. IMHMS does not provide any type of standards that are used for data standardization or interoperability.
- **Context Analysis**
  IMHMS collects patient's physiological data through the bio-sensors. The data is aggregated in the sensor network and a summary of the collected data is transmitted to a patient's personal computer or cell phone/PDA. These devices forward data to the medical server for analysis. After the data is analyzed, the medical server provides feedback to the patient's personal computer or cell phone/PDA. The patients can take necessary actions depending on the feedback.
- **Domain Analysis**
  IMHMS not only provides all the major functional requirements except emergency alarm and medicine infusion but from the non-functional point of view also System provides Intelligence, Response Time, Easy Wear-ability and Graphical interface, Genericity, Security, Unique Patient Identification,

Interoperability. The non-functional parameters such as Privacy, Availability, Accuracy, Data loses and adoption of standards are still overlooked in the implementation.

The characteristics features of IMHMS are described below:

Simplicity: The system architecture of IMHMS is a simple one with no complex system or communication architecture.

Cost-Effective: IMHMS is cost effective. WBSN setup consists of some low cost bio-sensors. The communication from WBSN is also cheap due to the use of low cost Bluetooth or ZigBee adapters. PPHS setup is cost effective due to use of personal computer of normal configuration or low cost cell phones.

Secure: Security is a major issue in IMHMS. The IMS maintains patients profile information with the RFID in the central database. So malicious attacks can be blocked using this information because a patient can be easily tracked using RFID. Moreover large volume of data to is transmitted between the three components of the IMHMS. So the transmitted data will be in encrypted form between the components to protect from security vulnerability.

Flexible communication protocol: The communication protocols of IMHMS are flexible. The WBSN central controller can communicate with the PPHS using any of the three protocols: Bluetooth, WLAN or ZigBee.

The implemented prototype of IMHMS is very user friendly; people with little or no technical knowledge can use it without any difficulties. The communication architecture of IMHMS is very simple and flexible as so there is no complexity in communication between the components of IMHMS. The data encryption is provided with Advanced Encryption Standard (AES) which the system effectively slower.

References

1. Robyn, W.: Issues in mHealth: findings from key informant interviews. Journal of Medical Internet Research 14(5) (2012)
2. Swarup, M., Ramesh, S., Millo, J.-V., Krishna, S.N., Narwane, G.K.: Tracing SPLs precisely and efficiently. In: Proceedings of the 16th International Software Product Line Conference, vol. 1, pp. 186–195. ACM (2012)
3. Organization, W.H.: New horizons for health through mobile technologies: second global survey on ehealth. Global Observatory for eHealth series 3 (2011)
4. Paul, F., Viall, B.: Development of a mobile cardiac wellness application and integrated wearable sensor suite. In: The Fifth International Conference on Sensor Technologies and Applications, SENSORCOMM 2011, pp. 301–306 (2011)
5. Dickerson, R.F., Gorlin, E.I., Stankovic, J.A.: Empath: a continuous remote emotional health monitoring system for depressive illness. In: Proceedings of the 2nd Conference on Wireless Health, p. 5. ACM (2011)

6. Phyo, W.A.A., Fook, F.S., Jayachandran, M., Song, Z., Biswas, J., Nugent, C., Mulvenna, M., Lee, J.-E., Yap, L.: Smart wireless continence management system for elderly with dementia. In: 10th International Conference on E-health Networking, Applications and Services, HealthCom 2008., pp. 33–34. IEEE (2008)
7. Miao, F., Miao, X., Shangguan, W., Li, Y.: Mobihealthcare system: Body sensor network based m-health system for healthcare application. E-Health Telecommunication Systems and Networks 1(1), 12–18 (2012)
8. Logan, A.G., McIsaac, W.J., Tisler, A., Irvine, M.J., Saunders, A., Dunai, A., Rizo, C.A., et al.: Mobile phone–based remote patient monitoring system for management of hypertension in diabeticpatients. American Journal of Hypertension 20(9), 942–948 (2007)
9. Abderrahim, B., Feham, M., Bouchachia, A.: Ubiquitous Mobile Health Monitoring System for elderly (UMHMSE). arXiv preprint arXiv:1107.3695 (2011)
10. Abd, H.S., Norhayati, D., Jawawi, A., Deris, S.: Variability Integration in Software Product Line Core Assets. In: The 3rd Software Engineering Postgraduate Workshop (SEPoW), pp. 63–68 (2011)
11. Tahat, A.: Mobile messaging services-based personal electrocardiogram monitoring system. International Journal of Telemedicine and Applications 2009 4 (2009)
12. Hariton, A.N.A.M.A.R.I.A., Creu, M., Nia, L., Slcianu, M.: Database security on remote patient monitoring system. International Journal of Telemedicine and Applications 9 (2011)
13. Lee, D.H., Rabbi, A., Choi, J., Fazel-Rezai, R.: Development of a mobile phone based e-health monitoring application. Development 3(3) (2012)
14. Abousharkh, M., Mouftah, H.: SOA-driven Sensor-based Patient Monitoring System with XMPP based Event Notification
15. Gonçalo, P.R., Carlos, R.: Promoting ubiquity and interoperability among health information systems using an soa based architecture (2012)
16. Lars, S., Falck, T., Wartena, F., Simons, D.: Towards plug-and-play interoperability for wireless personal telehealth systems. In: 4th International Workshop on Wearable and Implantable Body Sensor Networks (BSN 2007), pp. 257–263. Springer, Heidelberg (2007)
17. De Toledo, P., Lalinde, W., Del Pozo, F., Thurber, D., Jimenez-Fernandez, S.: Interoperability of a mobile health care solution with electronic healthcare record systems. In: 28th Annual International Conference of the IEEE Engineering in Medicine and Biology Society, EMBS 2006, pp. 5214–5217. IEEE (2006)
18. Dipti, P., Vijay, W.M., Gund, M., Biyani, R., Andhalkar, S., Agrawal, B.: An Adaptive parameter free data mining approach for healthcare application. arXiv preprint arXiv:1211.1788 (2012)
19. Jaejoon, L., Muthig, D., Naab, M.: A feature-oriented approach for developing reusable product line assets of service-based systems. Journal of Systems and Software 83(7), 1123–1136 (2010)
20. Johannes, B., Karahan, H., Meyer, U.: Security and privacy for mobile electronic health monitoring and recording systems. In: 2010 IEEE International Symposium on World of Wireless Mobile and Multimedia Networks (WoWMoM), pp. 1–6. IEEE (2010)

21. Kyo, K.C., Lee, J., Donohoe, P.: Feature-oriented product line engineering. IEEE Software 19(4), 58–65 (2002)
22. Gonçalo, P.R., Cunha, C.R., Morais, E.P.: A SOA based architecture to promote ubiquity and interoperability among health information systems (2011)
23. Prajakta, K., Öztürk, Y.: Requirements and design spaces of mobile medical care. ACM SIGMOBILE Mobile Computing and Communications Review 11(3), 12–30 (2007)
24. van Halteren, A.T., Bults, R.G.A., Wac Konstantas, K.E., Widya, D.I.A., Dokovski, N.T., Koprinkov, G.T., Jones, V.M., Herzog, R.: Mobile patient monitoring: The mobihealth system (2004)
25. Valérie, G., Leijdekkers, P.: A health monitoring system using smart phones and wearable sensors. International Journal of ARM 8(2), 29–35 (2007)
26. Tia, G., Massey, T., Selavo, L., Crawford, D., Chen, B.-R., Lorincz, K., Shnayder, V., et al.: The advanced health and disaster aid network: A light-weight wireless medical system for triage. IEEE Transactions on Biomedical Circuits and Systems 1(3), 203–216 (2007)
27. Val, J., Gay, V., Leijdekkers, P.: Body sensor networks for mobile health monitoring: Experience in europe and australia. In: Fourth International Conference on Digital Society, ICDS 2010, pp. 204–209. IEEE (2010)
28. Pravin, P., Bert-Jan, V.J., Van Beijnum, F., Hermens, H.: A framework for the comparison of mobile patient monitoring systems. Journal of Biomedical Informatics 45(3), 544–556 (2012)
29. Paliwal, G., Kiwelekar, A.W.: A comparison of mobile patient monitoring systems. In: Huang, G., Liu, X., He, J., Klawonn, F., Yao, G. (eds.) HIS 2013. LNCS, vol. 7798, pp. 198–209. Springer, Heidelberg (2013)
30. Fei, C., Bryant, B.R., Burt, C.C., Huang, Z., Raje, R.R., Olson, A.M., Auguston, M.: Automating feature-oriented domain analysis. Alabama Univ in Birmingham (2003)
31. Kyo, K.C., Cohen, S.G., Hess, J.A., Novak, W.E., Peterson, A.S.: Feature-oriented domain analysis (FODA) feasibility study. No. Cmu/Sei-90-Tr-21. Carnegie-Mellon Univ Pittsburgh pa Software Engineering Inst. (1990)
32. Robert, K., Zalman, N.: Domain Analysis Workshop Report for the Automated Prompt and Response System Domain. No. CMU/SEI-96-SR-001. Carnegie-Mellon Univ Pittsburgh pa Software Engineering Inst (1996)
33. ACC, HIMSS, and RSNA. IHE Patient Care Device Technical Framework In Volume 1Integration Profiles, Year 1 (2006-2007)
34. Kyo, K.C., Kim, S., Lee, J., Kim, K., Shin, E., Huh, M.: FORM: A feature-; oriented reuse method with domain-; specific reference architectures. Annals of Software Engineering 5(1), 143–168 (1998)
35. International Standard, ISO/IEEE 11073-20601, Health informatics - Personal health device communication Part 20601: Application profile Optimized exchange protocol. ISO/IEEE 11073-20601:2010(E)
36. 11073-10471, Health informatics - Personal health device communication, Part 10471: Device specialization Independent living activity hub. IEEE Engineering in Medicine and Biology Society, IEEE Std 11073-10471™-2008

37. Health Level Seven, Version 2.x, Message Profiling Specification, Version 2.2 (November 2000)
38. Beeler, G.W., Huff, S., Rishel, W., Shakir, A.-M., Walker, M., Mead, C., Schadow, G.: Message Development Framework, hl7 modeling & methodology committee, HL7 Version 3, Version 3.3 (December 1999)
39. Perry, D.E.: Generic architecture descriptions for product lines. In: van der Linden, F.J. (ed.) Development and Evolution of Software Architectures for Product Families. LNCS, vol. 1429, pp. 51–56. Springer, Heidelberg (1998)
40. Maohua Catherine, Y., Chronaki, E., Lupkes, C., Thiel, A., Plosnig, M., Hinterbuchner, L., Arbelo, E., et al.: Guideline-driven telemonitoring and follow-up of cardiovascular implantable electronic devices using IEEE 11073, HL7 & IHE profiles. In: 2011 Annual International Conference of the IEEE Engineering in Medicine and Biology Society, EMBC, pp. 3192–3196. IEEE (2011)
41. Maged, B.N.K., Wheeler, S., Tavares, C., Jones, R.: How smartphones are changing the face of mobile and participatory healthcare: an overview, with example from eCAALYX. Biomedical Engineering 10(1), 24 (2011)
42. Hao, W., Liu, J.: Mobile phone based health care technology. Recent Patents on Biomedical Engineering 2(1), 15–21 (2009)
43. Darrell, W.: How mobile devices are transforming healthcare. Issues in Technology Innovation 18, 1–14 (2012)
44. Dermot, R., Price, D., Musgrave, S.D., Malhotra, S., Lee, A.J., Ayansina, D., Sheikh, A., Tarassenko, L., Pagliari, C., Pinnock, H.: Clinical and cost effectiveness of mobile phone supported self-monitoring of asthma: multicentre randomised controlled trial. BMJ: British Medical Journal 344 (2012)
45. Dimitris, F.I., Likas, A., Protopappas, V.: Intelligent patient monitoring. Wiley Encyclopedia of Biomedical Engineering (2006)
46. John, O., Herbert, J.: Data management system: A context aware architecture for pervasive patient monitoring. In: Proceedings of the 3rd International Conference on Smart Homes and Health Telematic (ICOST 2005), pp. 159–166 (2005)
47. Min, C., Gonzalez, S., Leung, V., Zhang, Q., Li, M.: A 2G-RFID-based e-healthcare system. IEEE Wireless Communications 17(1), 37–43 (2010)
48. Pravin, P.: Context-aware vertical handover mechanisms for mobile patient monitoring (2011)
49. Shinji, I., Ichinose, K.: Development of a highly responsive needle-type glucose sensor using polyimide for a wearable artificial endocrine pancreas. Journal of Artificial Organs 9(2), 105–113 (2006)
50. Lei, T., Lin, Y., Ye, H.: Quality-Oriented Software Product Line Architecture Design. Journal of Software Engineering and Applications 5(7), 472–476 (2012)
51. Carlos, C.G.: Software product lines using FODA: a formal approach. PhD diss (2012)
52. Dario, F., Uchitel, S., Braberman, V.: A foundation for behavioural conformance in software product line architectures. In: Proceedings of the ISSTA 2006 workshop on Role of software architecture for testing and analysis, pp. 39–48. ACM (2006)
53. Abd, H.S., Jawawi, D.N., Ibrahim, N., Deris, S.: An Approach for Representing Domain Requirements and Domain Architecture in Software Product Line. Software Product Line–Advanced Topic 23 (2012)
54. Faheem, A., Capretz, L.F.: The software product line architecture: An empirical investigation of key process activities. Information and Software Technology 50(11), 1098–1113 (2008)

55. Chongxiang, Z., Lee, Y., Zhao, W., Zhang, J.: A Feature Oriented Approach to Mapping from Domain Requirements to Product Line Architecture. Software Engineering Research and Practice, 219–225 (2006)
56. Critina, G., Anastasopoules, M.: Implementing product line variabilities. ACM Sigsoft Software Engineering Notes 26(3), 109–117 (2001)
57. Benjamin, K., Krogmann, K.: Model-Driven Product Consolidation into Software Product Lines. In: Workshop on Model-Driven and Model-Based Software Modernzation (MMSM 2012), Bamberg (2012)
58. Northrop, L.M., Clements, P.C.: A Framework for Software Product Line Practice, Version 5.0. Software Engineering Institute, Carnegie Mellon
59. Shahriyar, R., Bari, M., Kundu, G., Ahamed, S., Akbar, M.: Intelligent mobilehealth monitoring system (imhms). Electronic Healthcare, 5–12 (2010)
60. Fortier, P., Viall, B.: Development of a mobile cardiac wellness application and integrated wearable sensor suite. In: The Fifth InternationalConference on Sensor Technologies and Applications, SENSORCOMM 2011, pp. 301–306 (2011)

Questions

Q1. Define the following terms with reference to Mobile Patient Monitoring Systems.

- Body Area network
- Mobile Base Unit
- Back end systems

Q2. Explain functions performed by various system elements included in the generic architecture of Mobile Patient Monitoring System (MPMs). Why the elements included in the generic architecture of MPMs are separated between product and process perspective?

Q3. What is Feature Oriented Domain Analysis? Explain various steps involved in FODA with output.

Q4. Give the various pros and cons of Feature Oriented Domain Analysis.

Q5. Define fodaA. How the FODA diagrams are converted into the fodaA grammer.

Q6. Defines Software Product Lines and core assets of the software product lines. What are the Benefits offered by the software product lines?

23. The Design of Integrated Circuit for Biomedical and mHealth Application

Yayu Cheng and Ye Li*

Shenzhen Institute of Advanced Technology, Chinese Academy of Sciences
1068 Xueyuan Avenue, Shenzhen University Town, Nanshan District
Shenzhen, China
{yy.cheng,ye.li}@siat.ac.cn

Abstract. Portable, wearable and networked medical devices for the diagnosis, monitoring, and treatment of diseases are vital and desired for future mHealth applications. For the pursuit of convenience, low power and low cost, the system of mHealth instrumentations should be developed as small as possible and the power consumption should be the lower the better, therefore; integrated circuits (ICs) designed for biomedical fields are necessary in these mHealth applications.

The key challenge in designing biomedical ICs is the pursuit of low frequency, low noise, and low power, and thus, there is a tradeoff between noise and power, and tradeoff between area, cost and bandwidth, because: 1) biomedical signals are usually low frequency, week and noisy, and 2) the relevant biomedical devices should be wearable and portable in the practical applications.

In this chapter, the recent low power, low noise and low frequency IC design technologies which have been used in multi-parameters, human-body signal monitoring and processing are introduced. These technologies are suitable for biomedical and mHealth applications including the design of analog circuits such as amplifiers and filters, digital circuits, and radio-frequency (RF) circuits which are widely used in physiological signal acquisition, signal processing and in the signal transmission respectively. In the last, some synthetic systems are introduced.

Keywords: Integrated Circuit, mHealth, Low Power, Low Noise, Low Frequenc.

1 Introduction

The growing aging population, skyrocketing healthcare costs, and prevalence of chronic diseases are the main driving forces to propel the fundamental transformation of the current healthcare systems and look for more cost-effective and responsive ways to deliver healthcare services. The conventional hospital-centered healthcare system, which focuses on diagnosis and treatment, is shifting its focus on an individual-centered healthcare system with emphasis on early detection, early diagnosis and long-term monitoring [1].

---

* Corresponding author.

S. Adibi (ed.), *Mobile Health*,
Springer Series in Bio-/Neuroinformatics 5, DOI: 10.1007/978-3-319-12817-7_23

Portable, wearable and networked medical devices for the early detection, early diagnosis, monitoring, and treatment of diseases are vital and desired for the individual-centered healthcare system in mHealth applications [2]. Some recent advances in IC design technology have led to these devices that can supplement, or even replace some complex biomedical operations or instruments. Indeed, the processing capabilities of the design of circuits are unlimited, but in the future, in biomedical application they are highly limited. For example, in the application of biomedical body sensor device, it should be designed as small as possible for convenience. Then, the area of the circuits are much limited, but for the characteristic of biomedical signal, the circuits should be operated in the low frequency domain, since large areas are needed for implementing very large capacities or very large resistors. Similarly, because many bioelectric signals are in the microvolt range, noise form electrodes, amplifiers, and the environment is significant compared to the signal level, and the power consumption should be lower for the longer lifetimes of biomedical devices, so the development of integrated circuit (IC) for biomedical signal monitoring and processing is needed for further advances in these micro mHealth devices and sensor nodes.

As mentioned above, because 1) biomedical signals are usually low frequency, week and noisy, and 2) the devices are wearable and portable in application situations, when designed IC, there has a tradeoff between noise and power, and between area, cost and bandwidth. So, the key challenge in designing circuits for biomedical application is the pursuit of low noise, and low frequency. Certainly, low power is always the pursuit in circuit design anytime.

In this chapter, the recent low power, low noise and low frequency IC design technologies which have been used in multi-parameters, human-body signal monitoring and processing are introduced. These technologies especially are suitable for biomedical and mHealth applications including the design of analog circuits such as amplifier and filter, are widely used in physiological signal acquisition, digital circuits usually applied in signal processing and radio-frequency (RF) circuits in signal transmission. In the last, some synthetic systems are introduced.

The remainder of the chapter is organized as follows. In Section 2, we present some recent and classical circuits applied in biomedical signal acquisition, signal processing, signal transmission respectively. Then two classical synthetic circuits are introduced. For the limitation of chapter length, some details don't be discussed. In Section 3, we put forward some future research and development direction of the IC design for biomedical application.

2 Research Progress of Biomedical Integrated Circuits

2.1 Biomedical Signal Acquisition

In this section, we introduce some integrated circuits published in recent years, which are used for biomedical signal acquisition. They are mostly suitable for multi-parameter biomedical signals, such as electroencephalogram (EEG), electromyography (EMG), and electrocardiography (ECG), etc.

2.1.1 Amplifier Design

Due to electrochemical effects on the electrode-tissue interface, DC offsets are common across differential recording electrodes. So some bio-amplifier designs use off-chip capacitors in the nano-farad range to obtain a low frequency cutoff that rejecting large dc offsets. However, this approach is not feasible for large numbers of implanted electrodes. A novel neural amplifier developed by Dr. Reid R. Harrison in 2003 uses a MOS-bipolar pseudo-resistor element to amplify low frequency signals down to the milli- Hertz range. The rejecting large dc offset is presented in Figure 1. The mid-band gain A is set by C_1/C_2 for the case with C_1, $C_L \geq C_2$, and the bandwidth is approximately $g_m/(A \times C_L)$, where gm is the transconductance of the operational transconductance amplifier (OTA).

Transistors M_a-M_d shown in Figure 1 (a) are MOS-bipolar devices acting as pseudo-resistors. For small voltages across this device, its incremental resistance r_{psr} is extremely high. This circuit uses two MOS-bipolar devices in series to reduce distortion for large output signals. The low frequency cutoff ω is given by $1/(2\ r_{psr}\ C_2)$. Figure 1 (b) shows a schematic of the current-mirror OTA used in the bio-amplifier. Although the circuit topology is a standard design suitable for driving capacitive loads, the sizing of the transistors is critical for achieving low noise at low current levels.

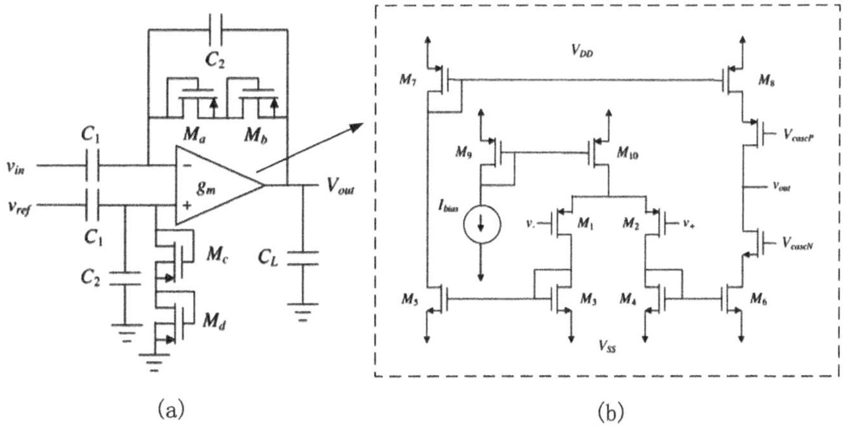

Fig. 1 Schematic of (a) neural amplifier, (b) OTA used in neural amplifier [3]

A fully integrated, low noise tunable bandwidth front-end amplifier (TB-FEA) is developed by the team of Prof. Yong Liang in 2009, which is shown in Figure 2. In this design, a novel balanced tunable pseudo-resistor structure is introduced for the adjustment of high-pass corner frequency. This is an improvement to the structure of circuit in Figure 1. The proposed TB-FEA acts as a low amplifier as well as a reconfigurable band-pass filter. The mid-band gain of this TB-FEA is determined by the capacitance ratio C_1/C_2. The low-pass corner frequency is adjusted by changing the bandwidth of the OTA, and the high-pass corner frequency is adjusted by the balanced tunable pseudo-resistors.

The existing pseudo-resistor, as shown in Figure 1, exhibits asymmetric and nonlinear resistances when the voltage across the pseudo-resistor varies. The asymmetric resistances lead to unavoidable signal-dependent output shift as well as early clippings at one of the power rail as the signal level increases. The proposed balanced tunable pseudo-resistor technology is used to solve this problem. Each transistor of M_1 and M_2 works as an active resistor for only half of each cycle. The control voltages $V_{ctrl(b)1}$ and $V_{ctrl(b)2}$ are formed by (V_{out}-V_1) and (V_{ref}-V_2), whose values are defined by the gate to source voltages of M_3 and M_4, respectively, which are in direct association with the drain currents of M_3 and M_4 and independent of V_{out} and V_{ref}. In this way, the resistance of pseudo-resistor is not affected by the change of V_{out}.

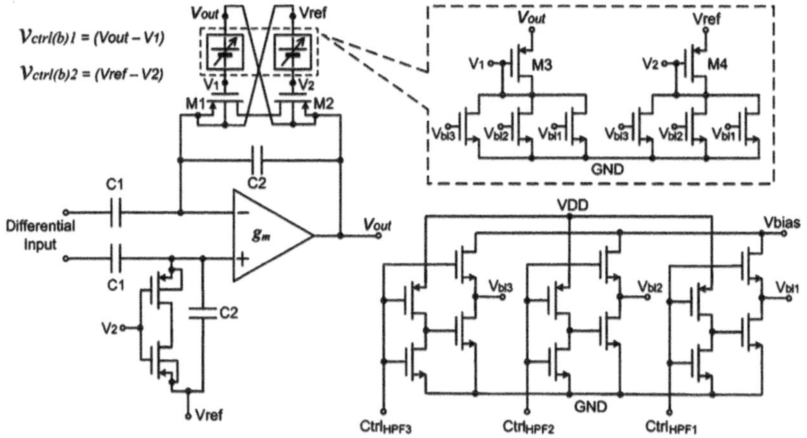

Fig. 2 Schematic of tunable bandwidth front-end amplifier developed by the team of Prof. Yong Liang [4]

A new high performance rail-to-rail instrumentation amplifier (IA), developed by the team of Prof. P. K. Chan in 2005, is shown in Figure 3. This IA provides high and balanced input impedance and supports a rail-to-rail input common mode range in a reduced supply environment. In this realization, two chopper-stabilized differential difference amplifiers (DDAs) are arranged in parallel configuration. Dependent on the input common-mode range, only one DDA is active at any time. The selection of either DDA is made by continuously monitoring the dc level of the negative input terminal which is extracted through a simple RC low-pass filter with respect to 0V. A hysteresis of ±80mV is built into the comparator to prevent noise from chatter effect when the input common voltage is close to 0V. For a fixed gain in this design, the output of the IA easily saturates to either supply rail because of the dc offset voltage, which is resulted from the difference between the input electrode-skin interfaces. So, filtering circuits are added to the basic chopper-stabilized DDA for suppressing the input dc offset voltage.

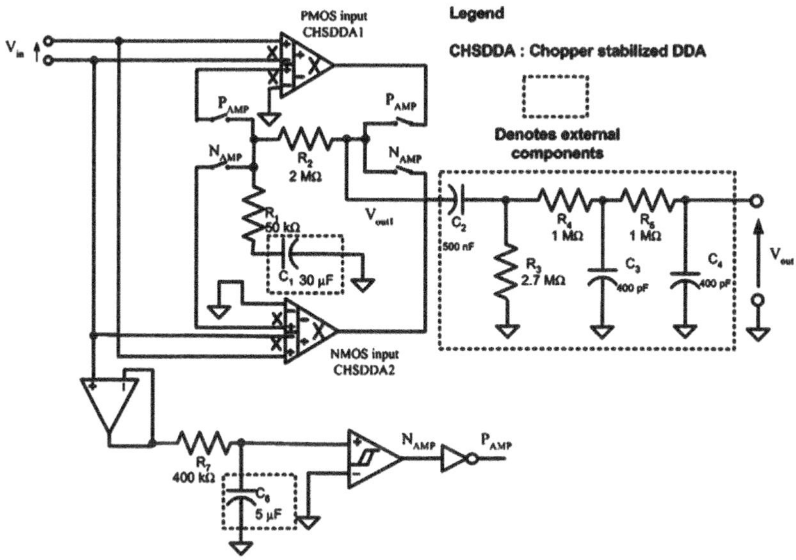

Fig. 3 Schematic of high performance rail-to-rail instrumentation amplifier (IA)[5]

The topology of DDA and its circuit schematic of PMOS input architecture are shown in Figure 4. It requires only one active amplifier plus two resistors to set the instrumentation gain, and the CMRR performance is related only to the mismatch of the input ports, where the mismatch between resistors $R_1$and R_2 only affects the gain factor. It does not degrade the CMRR of the amplifier. Incorporating the chopper stabilization technique into the DDA, it attains high CMRR and low input offset and 1/f noise simultaneously, and more importantly, the component matching issues are relaxed. The DDA here is based on the chopper-stabilized DDA described in [6].

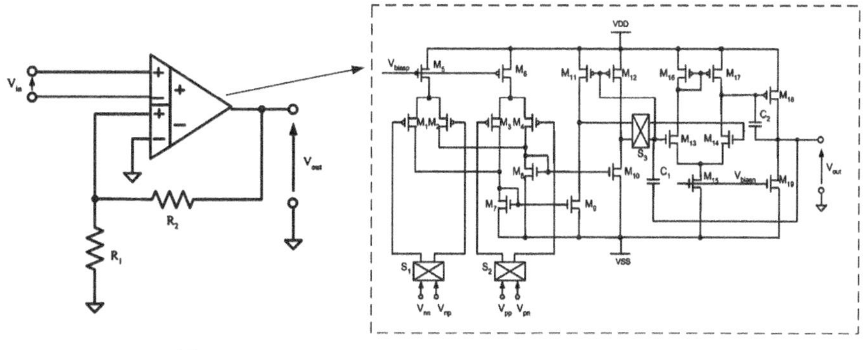

Fig. 4 (a) topology of DDA, (b) circuit schematic of PMOS input DDA

Figure 5 shows a low power and low noise readout front-end for the acquisition of EEG, ECG, and EMG signals. The readout front-end consists of an AC-coupled chopped instrumentation amplifier (ACCIA), a chopping spike filter (CSF) stage, a digitally programmable gain stage and an output buffer. The most important in power consuming building block of the front-end is the ACCIA, which uses a low power current feedback instrumentation amplifier (IA). It defines the noise performance and the CMRR of the front-end while chopping filters the 1/f noise of CMOS transistors and filters the differential electrode offset. A digitally programmable gain stage with selectable gain and bandwidth is used to adjust the gain and bandwidth of the readout front-end for different biomedical signals. The variable gain stage of this circuit operates in continuous-time mode and gain is defined by the ratio of two capacitors.

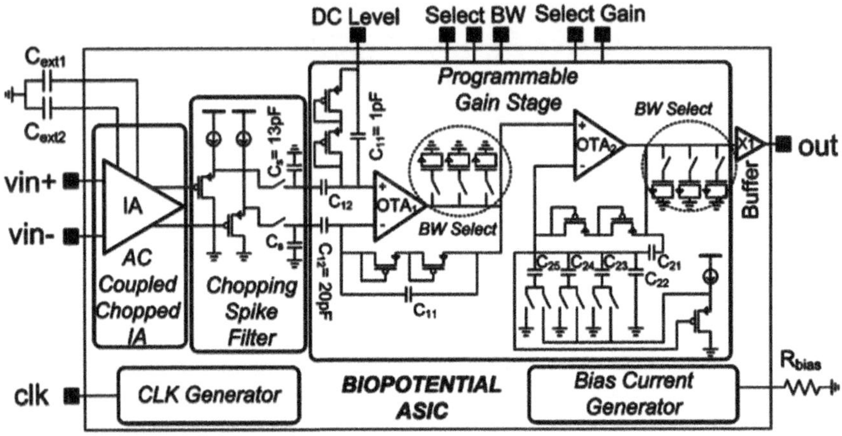

Fig. 5 Schematic of bio-potential readout front-end developed by Refet Forat Uazocioglu [7]

Figure 6 shows the implementation of the ACCIA module that can eliminate the 1/f noise, while the offset voltage of electrodes (V_{offl}) and IA are filtered. The offset voltage of electrode, is modulated by the input chopper and copied to the terminals of R1. This voltage creates a current through R_1, which is copied to R_2 and defines the output voltage after demodulation by the output chopper. A transconductance module, which is implemented by the OTA_2-C_{ext2} filter and g_{m2} module, has an equivalent transconductance of $A_v g_{m2}$ and a low-pass cutoff frequency f_p, where A_v is the voltage gain of OTA_2. The transconductance module filters the DC component of the output voltage and converts it into current. Then, a chopper modulates this current since the current of electrodes' offset voltage through R_l is modulated by the input chopper. At steady-state, the current supplied by transconductance module equals V_{off1}/R_1. Therefore, no current supplied by the IA and the current passing through R_2 is zero, and thus the output is zero. At the same time, another transconductance module which is implemented by the OTA_1-C_{ext1} filter and g_{m1} module is designed to cancel the offset voltage of IA.

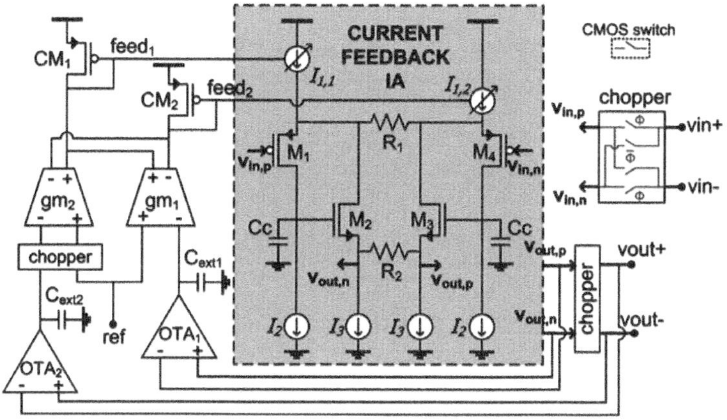

Fig. 6 Schematic of ACCIA implementation

A novel logarithmic transimpedance amplifier for oxygen saturation of blood with inherent contrast sensitivity, distributed amplification and automatic loop gain control is proposed by the team of Prof. Rahul Sarpeshkar in 2010, as shown in Figure 7 (b). The value of oxygen saturation is obtained by $(i_{ac\text{-}red}/I_{DC\text{-}red})/(i_{ac\text{-}infra}/I_{DC\text{-}infra})$, where $i_{ac\text{-}red}$, $I_{DC\text{-}red}$, $i_{ac\text{-}infra}$, $I_{DC\text{-}infra}$ are the currents measured by red and infrared LED light respectively through a patient's finger.

The key incentive using this logarithmic sensing is that it is inherently sensitive to the contrast (ac/DC) of the input photocurrent signal since the derivative of a log(U) is equal to the derivative of U over U itself when the feedback transistor Q1 operates in its subthreshold regime and exhibits an exponential I-V characteristic. The circuit schematic of a well-known logarithmic transimpedance photoreceptor [8] is shown in Figure 7 (a), where the diode D1 and transistor Q1 form a logarithmic photoreceptor, transistors Q2-Q4 implement a single-stage inverting common-source amplifier, the elements C1, C2, and R1 form a high-pass characteristic for the photoreceptor and boost the gain for ac signal. The output voltage of the photoreceptor for ac signals is equal to

$$v_{ac,out} = \left(\frac{U_T}{\kappa_1}\right)\left(\frac{C1+C2}{C1}\right)\frac{i_{ac,D1}}{I_{DC,D1}} \tag{1}$$

Where κ_1 is the sub-threshold exponential coefficient of Q1, and $U_T=kT/q$ is the thermal voltage.

Figure 7 (b) shows a novel logarithmic transimpedance amplifier that proposed by the team of Prof. Rahul Sarpeshkar in 2010. This photoreceptor improves on the circuit of Figure 7 (a) in three ways. The first improvement is to distribute the gain of the amplifier over three stages to increase its gain-bandwidth (GBW). The second improvement is to incorporate an adaptive loop gain mechanism to automatically adjust its loop gain based on the light intensity. Finally, a common-gate stage Q_4 is used in the transimpedance amplifier to sense the input photocurrent and convert it to a logarithmic voltage, remove the unwanted feed-through coupling due to the gate-to-source capacitance C_{par} in Figure 7 (a).

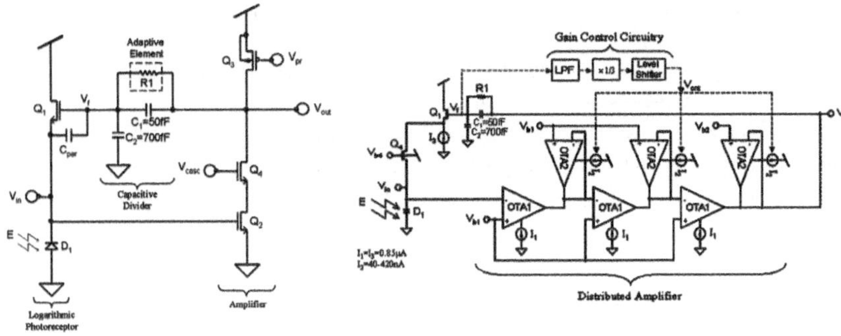

(a) (b)

Fig. 7 (a) circuit schematic of a logarithmic transimpedance photoreceptor with a single-stage amplifier, (b) circuit schematic of logarithmic transimpedance amplifier proposed by the team of Prof. Rahul Sarpeshkar[9]

Figure 8 shows a low noise front-end amplifier with configurable gain proposed by Dr. Robert Rieger in 2011, which is targeted at the recording of small biomedical signals. The circuit consists of an input operational transconductance stage (G_{mi}) whose output currents charge feedback capacitors C_f connected in a negative configuration around a second voltage amplifier. A fully differential architecture is used to reduce the effect of interference and switching noise. Reset switches connect all feedback capacitors C_f to the negative supply voltage (V_{ss}) when a digital control signal clr is high. When the switches are opened, the common-mode (CM) feedback circuit connected to the input transconductor senses the CM voltage at the input side of the capacitors and regulates it toward the reference voltage (V_{ref}). The V_{ref} is chosen slightly below the expected input CM voltage to ensure that bipolar transistors Q1 and Q2 are always saturated. The input terminal is realized by a BJT differential pair because MOS transistors in the input stage tend to dominate the noise performance at the low frequencies of interest duo to their intrinsic flicker noise. Clearly, there is a tradeoff between the noise performance and the power consumption of the device. For the function of Figure 8 (a), the overall voltage gain

$$\frac{V_o}{V_{in}} \approx \frac{G_{mi} \bullet t}{C_f} \tag{2}$$

Where G_{mi} is the transconductance of the input operational stage, C_f is the feedback capacitance, and t is the charging time of the capacitors. The output voltage is sampled after it is defined by a period t and the gain is tuned by variation of t.

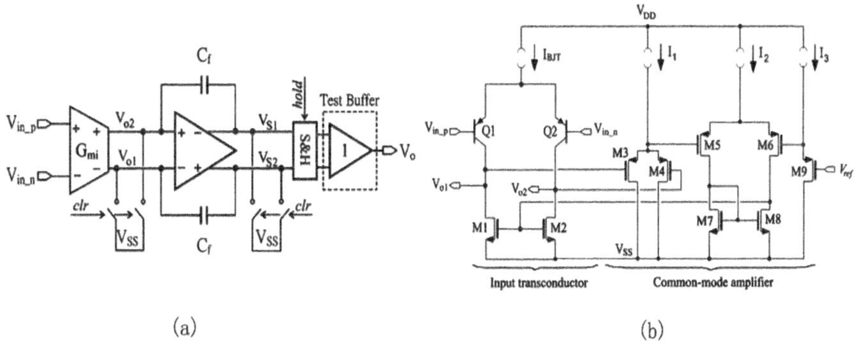

Fig. 8 (a) schematic of a low noise front-end amplifier with configurable gain proposed by Dr. Robert Rieger, (b) circuit schematic of the input transconductance stage G_{mi}[10]

2.1.2 Design of Filter

A fully integrated Gm-C, 0.5-7Hz band-pass filter proposed by Dr. Robert Rieger in 2004 is presented in Figure.9 (a). It has two cascaded stages with gain: a pre-stage formed by G_{m1}, G_{m2}, G_{m3}, C_1, C_2, and a gain stage formed by G_{m4}, G_{m5}, G_{m6}, C_3, C_4. Each stage has a low-pass structure given by $G_{m1(4)}$, $G_{m2(5)}$, $C_{2(4)}$, and a DC cancellation loop ($G_{m3(6)}$, $C_{1(3)}$). The overall function

$$V_{out2} = \frac{C_1 G_{m1} s}{G_{m1} G_{m3} + G_{m2} C_1 s + C_1 C_2 s^2} \bullet \frac{C_3 G_{m4} s}{G_{m4} G_{m6} + G_{m5} C_3 s + C_3 C_4 s^2} V_{in} \tag{3}$$

Figure 9 (b) shows the circuit schematic of the transconductors using series-parallel current division technology. The value of transcondutances of the Gm can be designed as low as 30pS to 100nS. It is a symmetrical OTA with a linear input pair and a current division factor N^2, the number of parallel/stacked transistors in limited by leakage currents only. The effective transconductance is calculated as

$$G_{mx} = g_m / N^2 \tag{4}$$

Here g_m is the transconductance of the input transistor.

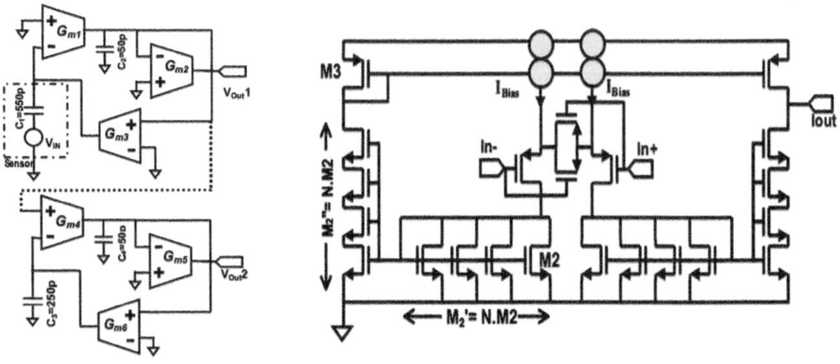

(a) (b)

Fig. 9 (a) schematic of a fully integrated 0.5-7Hz band-pass filter proposed by Dr. Robert Rieger[11], (b) circuit schematic of the transconductors using series-parallel current division[12]

A new architecture of Gm-C filter was proposed by the team of Prof. Y. T. Zhang in 2006, as shown in Figure 10. It is implemented by three Gm blocks whose circuit schematics are show in Figure 10 (b). Its cutoff frequency can be designed to only a few hertz when the capacitors ranging from 8pF to 0.01pF were used. The overall function

$$V_{\text{out}} = \frac{1}{(1+\frac{C_2}{G_m}s)(1+\frac{C_1}{G_m}s)} V_{in} \tag{5}$$

Here G_m is the transconductance of three Gm blocks.

The Gm block is implemented using sub-threshold and current reduction technique [13], the transistor biased in triode region, MR, is split into two transistors and its source-gate voltage is controlled by MC1, and the value of transconductance can be tuned by VB2 and VB3. Transistors MM, M1 and MN are operating as source followers. The differential input voltage (V_1-V_2) is converted to current through transistor MR, most of them flows to ground through MM because the $g_{mMM} \geq g_{mM1}, g_{mMN}$. Current reduction is designed by the partial cancellation of drain currents of M1 and MN. The small signal transconductance is

$$G_m = \frac{i_o}{V_1 - V_2} = \frac{N-1}{M+N+1} g_{MR} \tag{6}$$

Here g_{MR} is the small signal drain-souce conductance of transistor MR.

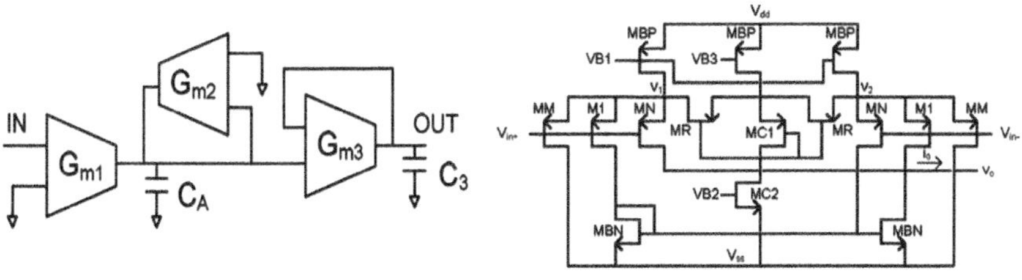

(a) (b)

Fig. 10 (a) schematic of a fully integrated 2nd Gm-C low-pass filter proposed by the team of Prof. Y. T. Zhang, (b) circuit schematic of the transconductors in (a)[14]

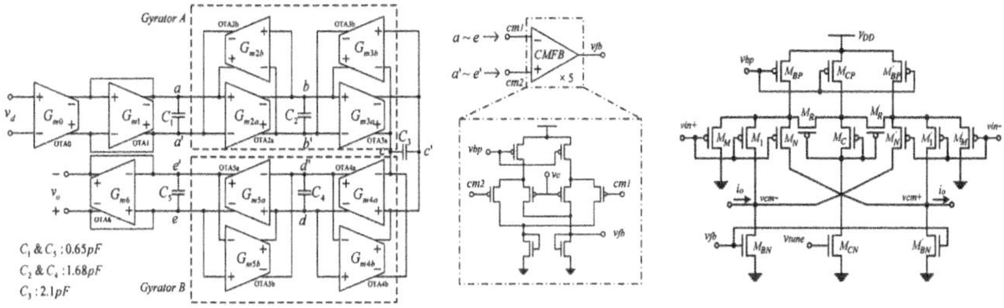

(a) (b)

Fig. 11 (a) schematic of a fifth-order Gm-C filter with common-mode feedback circuits proposed by Prof. S. Y. Lee in 2009, (b) circuit schematic of the transconductors in (a)[15]

A fifth-order Gm-C filter with common-mode feedback circuits was realized by Prof. S. Y. Lee in 2009, with the using of the sub-threshold and current reduction technique, which is shown in Figure 11. The two gyrators A and B are used to implement equivalently the inductors respectively. The Gm block that use the sub-threshold and current reduction technique has two output terminals, which are different from that in Figure 10, as shown in Figure 11 (b).

A switched capacitor low-pass filter was proposed by Dr. R. F. Yazicioglu in 2011, as shown in Figure 12. The input signal is demodulated with a frequency offset (Δ), where (Δ) is defined as the center of the band-pass filter. The width of the filter is defined by the switched capacitor (SC) low-pass filter (LPF) after the demodulators. In order to reduce the noise folding, the SC LPF is operated at 4 times the chopper demodulation frequency, $4\times(f+\Delta)$. The completer LPF consist of three stages, the cutoff frequency of a single stage is set by

$$f_p = \frac{1}{2\pi RC} = \frac{4\times(f+\Delta)\times C_R}{2\pi C_L} \tag{7}$$

Here C_R is the equivalent SC resistor of the capacitance implementation and C_L is the total load capacitance. C_R can be changed from C to 2C with four equivalent steps, and cascading three stages results in three different poles which can be set to change bandwidth configurability. With a few pF of capacitance, the cutoff frequency can be implemented less than 1Hz.

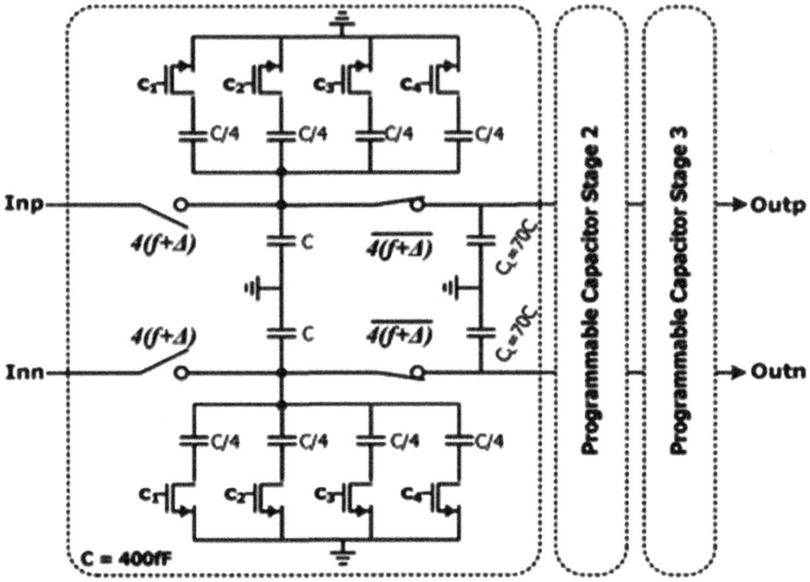

Fig. 12 Schematic of a switched capacitor low-pass filter proposed by Dr. R. F. Yazicioglu in 2011[16]

2.2 *Biomedical Signal Processing Part*

2.2.1 Image Compression

A new low complexity near-lossless image compression method and its ASIC design for the wireless endoscopy were proposed by the team of Prof. Z. H. Wang in 2005. Its architecture is shown in Figure 13. The Bayer color filter array (CFA) raw data from the image sensor is firstly filtered and saved in SRAM, and then the data will be compressed by JPEG-LS. Finally the compressed data is stored back to the SRAM so that the communication scheme can be used to assure the high quality image communication. The synchronous signals from the CMOS image sensor include vertical and horizontal synchronous outputs and pixel clock output, and they are used to count the location of current pixel, the column and row number. According to the synchronous signal, control unit of JPEG-LS controls which pixels should be filtered or not to realize lossless compression.

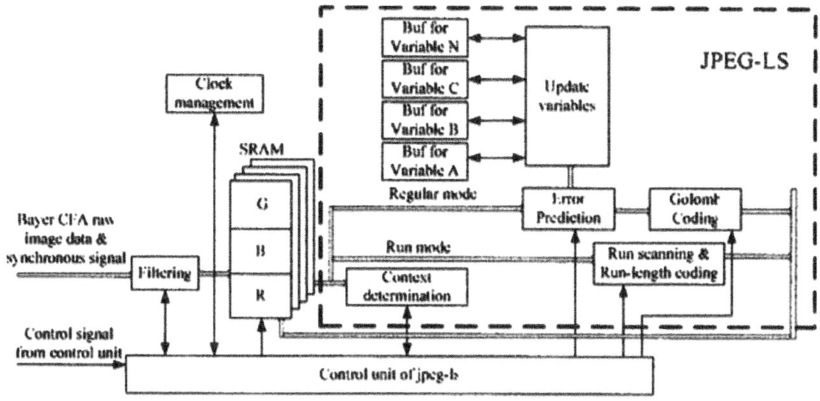

Fig. 13 The VLSI architecture of image compression encoder proposed by the team of Prof. Z. H. Wang in 2005 [17]

A mixed-signal VLSI CMOS image compression sensor was proposed by Alireza Nilchi in 2009, as shown in Figure 14 (a). It computes block-matrix and convolutional image transforms of each video frame with programmable digital kernels of up to 8x8 pixels, directly on the focal plane. It employs algorithmic multiplying analog-to-digital converters (MADCs) that perform spatial computation during pixel array readout, concurrently with quantization. The signal from pixel is multiplied by the bits of the kernel coefficient, C, sequentially from MSB to LSB. The outputs of the multiplexer are analog and binary-weighted, with their weight decreasing by a factor of two in each cycle, in order to match the weight of the corresponding residue signal. Figure 14 (b) depicts the circuit implementation of the column signal path from a pixel to the accumulator. The pixel has two frame memories to allow for multiple sampling of the same pixel. A gain stage following the pixel combines four functionalities in a single amplifier. Accumulation over V adjacent pixels is performed on the programmable feedback capacitor bank. The ouput dynamic range is adjusted by programming the gain to match the size of the block matrix.

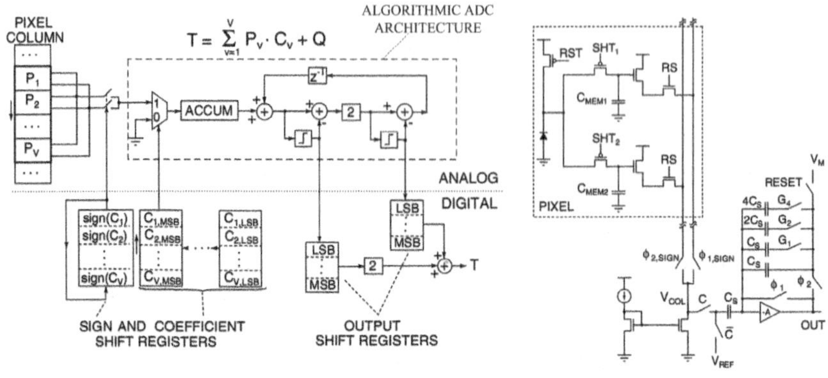

(a) (b)

Fig. 14 (a) architecture of one column of the image compression sensor proposed by Alireza Nilchi in 2009, (b) the pixel and difference/multiplier/accumulator circuit in (a)[18]

A 32×32 imaging integrated circuit that captures and compress gray scale images on the focal plane of the image sensor is shown in Figure 15, which was proposed by the team of Prof. Antonio Petraglia in 2013.

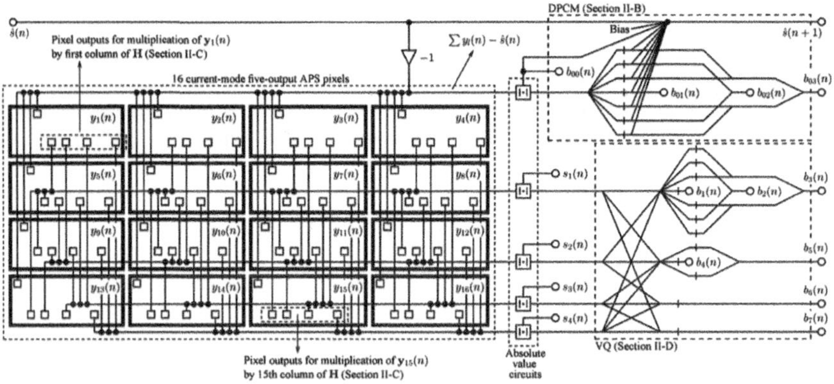

Fig. 15 The block diagram of the compression algorithm by the team of Prof. Antonio Petraglia in 2013[19]

In this design, every 4×4 pixel block, differential pulse-code modulation (DPCM), linear transform, and vector quantization (VQ) are implemented by using analog circuits. Image is first divided into 4×4 pixel blocks. The compression algorithm which combines DPCM, a linear transform, and VQ is thus applied to 16-component vectors denoted as y(n), where n=1,..., N is the block index in an image that has N blocks. The details of the compression algorithm are described in [20]. The summation of the 16 values is a representation of the block average, that representation is subtracted from $\widehat{s}(n)$, which is a block-summation estimate generated by the previous block. The sign of the difference is signed by $b_{00}(n)$ and its absolute value is also computed. The absolute value is encoded by three bits: $b_{01}(n)$, b02(n), $b_{03}(n)$. Then four DPCM bits are obtained from which the block average is reconstructed at the decoder. The four highest-energy components computed linearly from y(n) also going through absolute value circuits, and thus it can yield the four bits of the ac component signs, $s_m(n)$. the vector x(n) containing the absolute values of $s_m(n)$ is encoded into the bits $b_1(n)$ to $b_7(n)$, which compose the VQ output.

2.2.2 Signal Data Compression

A highly area and power efficient VLSI circuit that implements multichannel, multilevel lifting–based discrete wavelet transform (DWT) used for data compression without compromising the information content, which was proposed by Awais M. Kamboh in 2007. DWT is a very effective method for compressing data[21], with quantized filter coefficients and integer computation. This integer lifting DWT maintains a high signal to noise ratio, and the architecture is described in Figure 16. It composes of a customized computation core (CC), i.e., a digital controller, and five memory modules for incoming data, filter coefficients, intermediate CC products and intermediate values necessary to sequentially process multiple levels and channels.

The CC can be implemented by using two multipliers using Booth multiplication algorithm and a three-term adder formed by cascading two-term adders. The input buffer is constructed by using FIFO shift registers for the 32-channel implementation. In this module, the hold cycle is used o compute all results for higher levels of decomposition. A separate coefficient memory block is designed to store the eight 5-bit filter coefficients. This block is most efficiently implemented using hardwired connections because this data is static. The CC memory block is used to load all four bytes in parallel and move data to the appropriate CC input as its cycles through the filter functions in DWT algorithm. It was implemented by using flip-flops for parallel load/store and byte-wise shift operations. And the pairing memory is designed to store one 10-bit byte which is generated by all computation cycles except those of the highest level. The main functions of the controller are to manage overall system operation and route data between DWT circuit modules at the proper time.

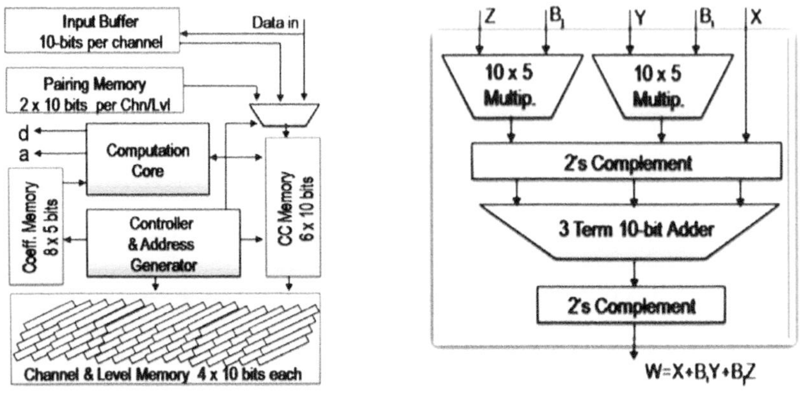

(a) (b)

Fig. 16 (a) system diagram of sequential calculation of DWT proposed by Awais M. Kamboh in 2007, (b) computation core architecture for integer lifting DWT in (a)[22]

Figure 17 shows a highly integrated VLSI implementation of a mixed bio-signal lossless data compressor to reduce storage and data communication bandwidth requirements proposed by the team of Prof. W. C. Fang in 2011. The proposed lossless data compression algorithm is largely based on a basic discrete pulse code modulation (DPCM) predictor followed by Golomb-Rice entropy coding, and its block diagram and algorithm flow is shown in Figure 17 (a).

The hardware architecture of the proposed compressor unit is shown in Figure 17 (b), which comprises four pipeline stages including a priority data selector (PDS) stage, a prediction and parameter estimation stage, a Golomb-Rice entropy coding stage, and an output packaging stage. The PDS stage serves as a multiplexer, serially sending each sample to the main compression unit and indicating the biomedical signal type CHSEL and sample precision SP. The prediction stage receives the current sample, determines its context and loads the context statistics from memory. Then K parameter is calculated; the prediction error remapped, and the context variables updated and written back to memory. The Golomb-Rice stage implements the Goomb-Rice coding algorithm for various values of K. It calculates the quotient and remainder based on the estimated K parameter and input prediction error, output the result to the next stage within a clock cycle. The final packing stage maintains four separate output buffers for the different biomedical signals, which are filled up as samples are encoded into bit streams.

A ready-acknowledge handshaking mechanism is employed throughout the pipeline in every stage to prevent pipeline overflow. The bias cancellation mechanism was forfeited due to restriction in chip real estate, resulting in around 7% performance in compression ratio.

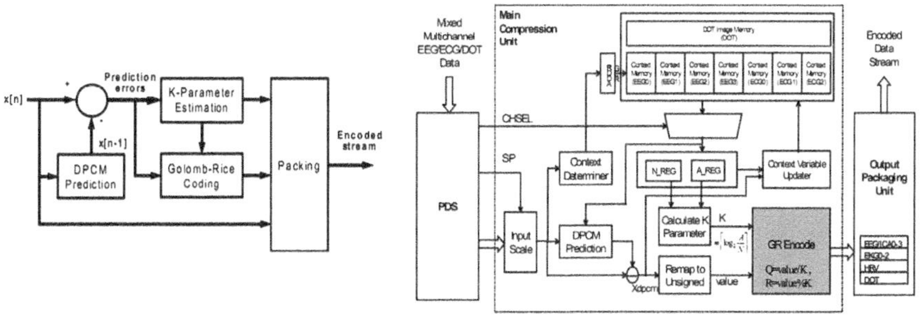

(a) (b)

Fig. 17 (a) block diagram of lossless data compression algorithm proposed by the team of Prof. W. C. Fang in 2011 (b) architecture of mixed bio-signal lossless compressor[23]

2.3 *Biomedical Signal Transmission Part*

A low power low noise, 2.4GHz RF transceiver which consists of a receiver and a transmitter is proposed by Dr. J. P. Carmo in 2007.

The front-end schematic of the receiver is shown in Figure 18 (a). It uses direct demodulation by means of envelope detection. The low noise amplifier (LNA) is the first gain stage where the signal must be amplifier as much as possible to achieve low noise figure as small as possible. Its structure is an inductively degenerated common source amplifier. This structure allows the input impedance at 2.4GHz to be tuned to 50Ω, for matching with antenna. The inductance L_s is implemented with the bonding connection to the external PCB. Inductance $L_{s,d}$ is used to reduce the effects of inductance generated by the wires connected the die to the external PCB. A minimum RF level which defines the receiver's sensitivity at the envelope detector is achieved by means of further amplification at the LNA output.

The schematic of the transmitter is shown in Figure 18 (b) by using ASK modulation. The modulated signal is generated by mean of a switched power amplifier which has a cascade of five inverters to drive the ASK output signal to the input of the power amplifier. The L_1C_1 is tuned to the carrier frequency, while the L_2C_2 reduces the emissions outside of the 2.4 GHz band.

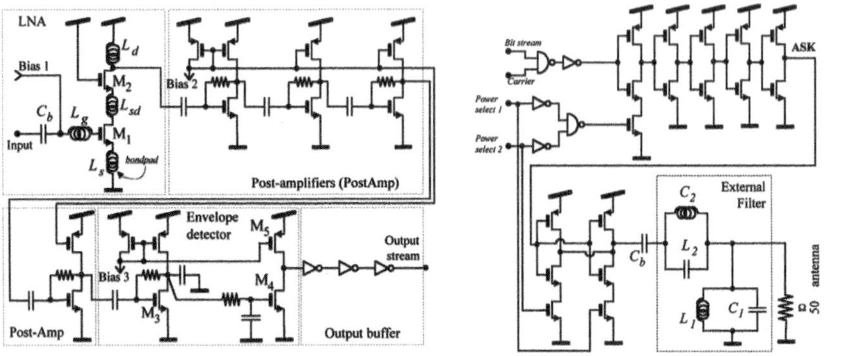

(a) (b)

Fig. 18 (a) the receiver's front-end schematic of a 2.4GHz RF transceiver proposed by Dr. J. P. Carmo in 2007 (b) the transmitter's schematic of the 2.4GHz RF transceiver [24]

A low power fully-integrated CMOS RF front-end circuit for a passive 13.56 MHz biomedical implant is proposed by Prof. S. Y. Lee in 2013, as shown in Figure 19. The front-end circuit is composed of a full-wave bridge rectifier, a linear regulator, a binary phase shift keying (BPSK) signal demodulator, and a clock/data recovery (CDR). The full-wave bridge rectifier converts the waveform with the BPSK signal to an unregulated DC voltage. Then the DC voltage is stabilized by the linear regulator to 1.8V that serves as the DC source for the implant. The BPSK demodulator translates the incoming signal from the internal coil to the CDR which can successfully recover the clock and data for the system controller.

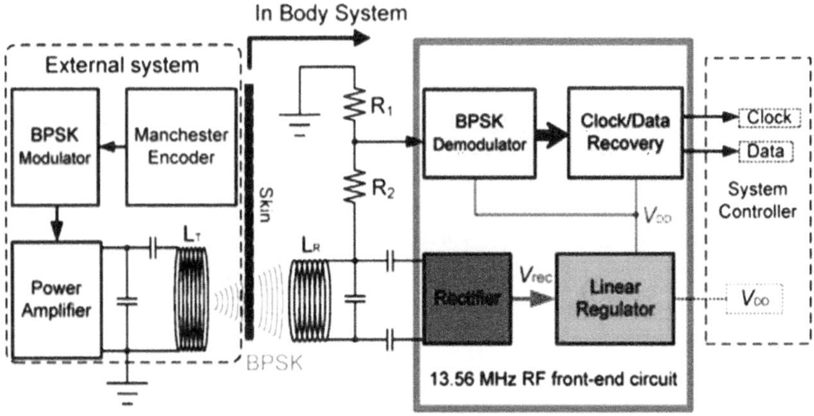

Fig. 19 Block diagram of a low power 13.56MHz RF front-end proposed by Prof. S. Y. Lee in 2013[25]

For the implanted biomedical applications, a fully CMOS integrated full-wave bridge rectifier is used with an off-chip capacitor as its output for ripple rejection [26]. For small area and low output ripples, a linear regulator without an inductor is preferred, and the design without an output capacitor based on unconditional stability is better [27]. A novel BPSK demodulator based on the squaring loop architecture is shown in Figure 20 (a). It is composed of a CMOS adjustable Schmitt trigger S_3, a phase detector, two delay cells with different timing delays, an exclusive NOR gate as a mixer, and a D flip-flop as an LPF. The circuit diagram of CDR is shown in Figure.20 (b), as called the Manchester decoder. As the most blocks are the same as the proposed BPSK demodulator. The circuit can be directly revised from the structure of Figure. 20 (a), and the operation is similar to that of the BPSK demodulator. The non-ideal effects, such as the jitter and decoding latency, can be also overcome by the delay cells.

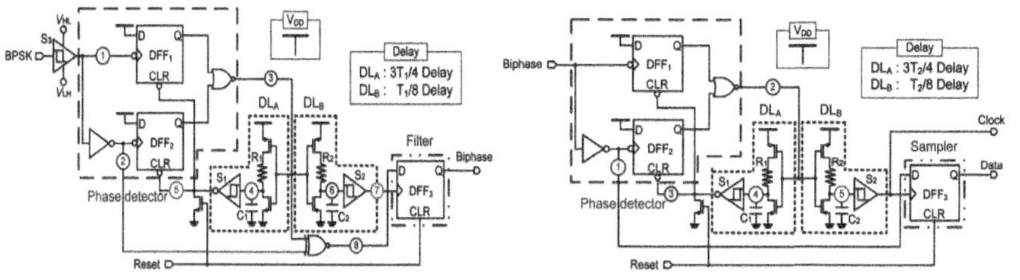

(a) (b)

Fig. 20 (a) circuit diagram of BPSK demodulator (b) circuit diagram of clock and data recovery

2.4 *Synthetic Application*

A programmable processor for bionic ears and speech-recognition front ends is developed by Prof. Rahul Sarpeshkar in 2005. This processor can transform frequency patterns in sound into corresponding spatial electrode-stimulation patterns for the auditory nerve. The signal processing blocks of the processor are shown in Figure. 21. An audio front end (AFE) converts sound signal over a 77dB range into an electrical signal. Then the automatic gain control (AGC) circuit compresses this electrical signal to a narrower 57dB internal dynamic range at which each of the 16 channels of the processor operate. Each channel is made up of a programmable Gm-C band-pass filter, an envelope detector, a logarithmic dual-slope A/D converter. The output bits of the A/D converters are sequentially scanned via a shift register which is clocked by a clock.

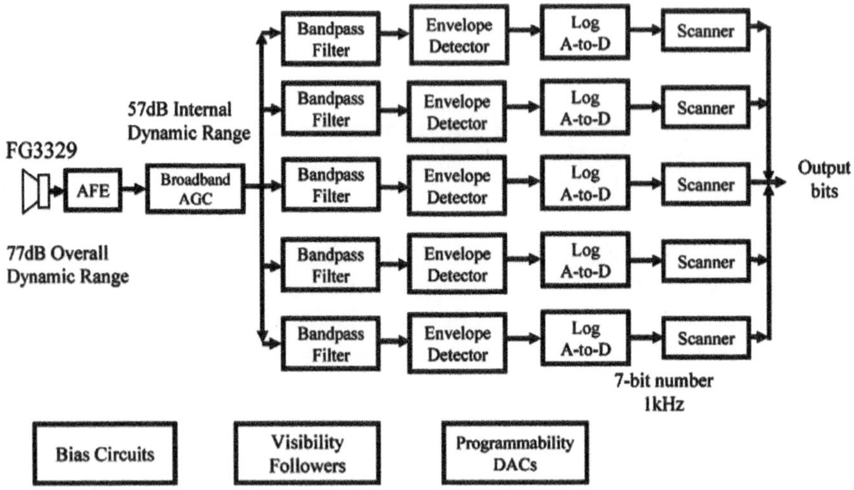

Fig. 21 A programmable analog bionic ear processor developed by Prof. Rahul Sarpeshkar in 2005[28]

The audio front end (AFE) is shown in Figure 22 (a). The circuit of microphone is shown inside the dotted box. This AFE regulate the drain output of the microphone to a reference voltage via negative feedback, and sense and amplify the output current in a transimpedance or sense-amplifier topology. The low frequency negative feedback loop formed by the Gm transconductor M1, and the capacitor subtracts the bias current of the microphone before it is amplified. This feedback circuit is operational only at low frequencies such that ac microphone currents are normally amplified. The input-referred noise of the operational amplifier and the noise due to M1 must be kept low through appropriate device sizing and biasing. The band-pass filter shown in Figure 22 (b) is a fourth-order Gm-C filter and consists of two stages of second-order filter [29]. In each stage C_1 and AC_1 implement capacitive-attenuation to widen linear range in the transconductors, G_1 and attenuating capacitors implement the high-pass function, G_2 and its output capacitors implement the low-pass function, G_3 provides low frequency adaptation to set the floating node of G_2's capacitive divider.

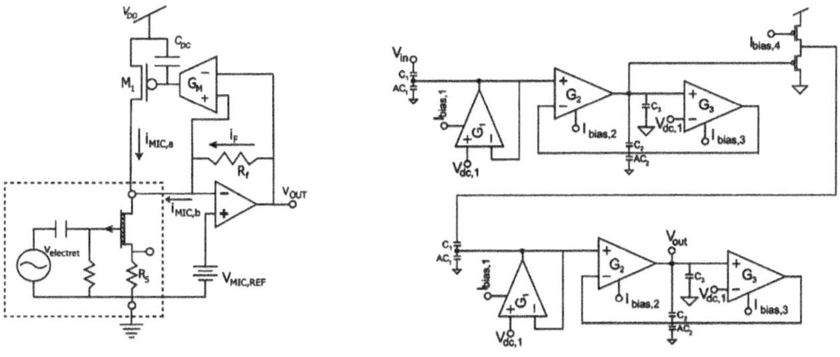

(a) (b)

Fig. 22 (a) circuit diagram of the audio front end (b) circuit diagram of the fourth-order Gm-C filter

Figure 23 (a) shows the circuit of envelope detector. It consists of a rectifier followed by a current-mode peak detector that estimates the dc component of the rectified signal. The rectifier is implemented by inserting a class AB current conveyor inside a Gm-C filter in a negative-feedback configuration, and the peak detector is a current-mode low-pass filter, modified to have asymmetric attack and release time constants. This detector quickly follows the rising portion of the full-wave-rectified input and holds this value with little decay in between cycles during a slow release. Figure 23 (b) shows the schematic of the logarithmic A/D with offset and temperature compensation. The WLR circuit is a wide linear-range transconductor with gate inputs, bump-linearization and source-degeneration [30], and the comparator is a stand operational transconductance amplifier. The output current from the envelop detector is converted to a logarithmic voltage on a diode. The WLR circuit linearly reconverts the diode voltage to a current. The ratio of this current to the bias current of the transconductor is then digitized in a dual-slope analog-to-digital conversion.

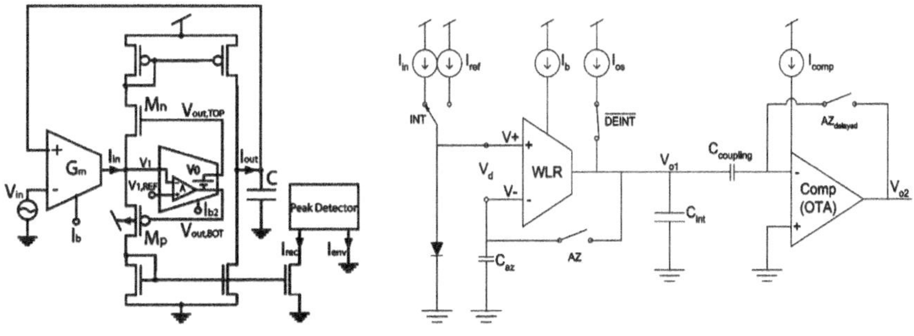

(a) (b)

Fig. 23 (a) circuit diagram of the envelope detector (b) circuit diagram of the logarithmic A/D

A 128 channel neural recording integrated circuit with on-the-fly spike feature extraction and wireless telemetry is proposed by Prof. W. T. Liu in 2009. This circuit consists of eight 16 channel front-end recording modules, spike detection and feature extraction processor, ultra wideband transmitter.

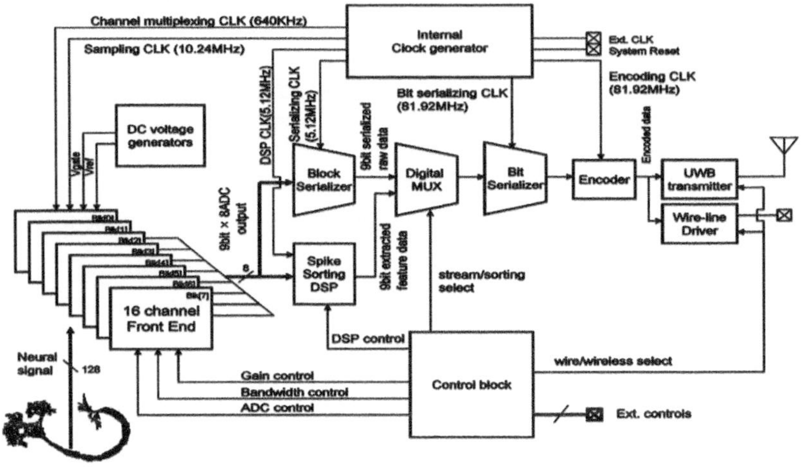

Fig. 24 Circuit diagram of the integrated neural recording system proposed by Prof. W. T. Liu in 2009[31]

The 16 channel front-end modules are shown in Figure 25 (a). First, weak and noisy neural signals from electrodes are amplified by fully differential preamplifiers, and time-multiplexed by an analog multiplexer. A second amplifier provides additional gain, then a successive approximation register (SAR) ADC is used to digitize the 16-1 time multiplexed analog signal. The amplifiers are designed to have programmable gain and bandwidth to meet the requirements of various biological experiments. As shown in Figure 25 (b), the preamplifier uses ac-coupling technology to reject the large dc offset at the electrode-tissue interface. As mentioned in chapter 2.1.1, the gain is set by the ratio of C_1/C_2, the high-pass cutoff frequency is tunable by changing the gate voltage V_B of the NMOS used as bias resistors, and the low-pass cutoff frequency is configurable by varying the load capacitance C_L. A high common mode rejection ratio (CMRR) is preferred to suppress the power interference in the design, which makes the circuit saturated and disables at the later stages.

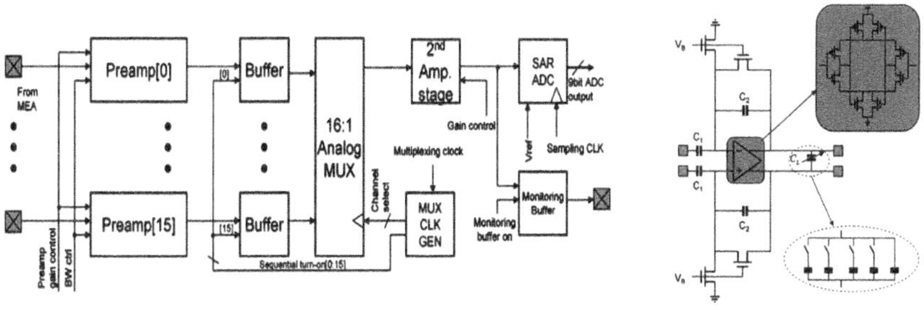

(a) (b)

Fig. 25 (a) circuit diagram of the 16 channel front-end modules (b) circuit diagram of the preamplifier uses ac-coupling technology

3 Further Research and Development

Some recent technologies for the design of low power, low noise and low frequency which have been used in multi-parameter, human-body signal monitoring and processing are presented. They are introduced in biomedical signal acquisition, signal processing, and signal transmission respectively. Most of them have good design of transistor size and capacitance area for the tradeoff of low power and low noise, low frequency and small area. But there are still some problems existing and need more research; such as how to ensure the accuracy and reliability of the acquired signal because some biomedical signals have the similar or same frequency domain, and how to remove the affection of people's motion to biomedical signals effectively in the circuit level, etc. After all, the design of IC for biomedical and mHealth applications needs more advanced combining circuit design skills and the biological sciences together in the state of the art.

References

[1] Zhang, Y.T., Poon, C.C.Y.: Editorial Note on Biomedical and Health Informatics. IEEE Transactions on Information Technology in Biomedicine 15, 175–177 (2011)
[2] Teng, X.F., Zhang, Y.T., Poon, C.C., Bonato, P.: Wearable medical systems for p-Health. IEEE Rev. Biomed. Eng. 1, 62–74 (2008)
[3] Harrison, R.R., Charles, C.: A low-power low-noise CMOS amplifier for neural recording applications. IEEE Journal of Solid-State Circuits 38, 958–965 (2003)
[4] Zou, X.D., Xu, X.Y., Yao, L.B., Lian, Y.: A 1-V 450-nW Fully Integrated Programmable Biomedical Sensor Interface Chip. IEEE Journal of Solid-State Circuits 44, 1067–1077 (2009)
[5] Ng, K.A., Chan, P.K.: A CMOS analog front-end IC for portable EEG/ECG monitoring applications. IEEE Transactions on Circuits and Systems I-Regular Papers 52, 2335–2347 (2005)

[6] Chan, P., Ng, K., Zhang, X.: A CMOS chopper-stabilized differential difference amplifier for biomedical integrated circuits. In: The 2004 47th Midwest Symposium on Circuits and Systems, MWSCAS 2004, vol. 3, pp. iii-33-6 (2004)
[7] Yazicioglu, R.F., Merken, P., Puers, R., Van Hoof, C.: A 60 mu W 60 nV/root Hz readout front-end for portable biopotential acquisition systems. IEEE Journal of Solid-State Circuits 42, 1100–1110 (2007)
[8] Delbruck, T., Mead, C.: Adaptive photoreceptor with wide dynamic range. In: 1994 IEEE International Symposium on Circuits and Systems, ISCAS 1994, pp. 339–342 (1994)
[9] Tavakoli, M., Turicchia, L., Sarpeshkar, R.: An ultra-low-power pulse oximeter implemented with an energy-efficient transimpedance amplifier. IEEE Trans Biomed Circuits Syst 4, 27–38 (2010)
[10] Rieger, R.: Variable-Gain, Low-Noise Amplification for Sampling Front Ends. IEEE Transactions on Biomedical Circuits and Systems 5, 253–261 (2011)
[11] Arnaud, A., Galup-Montoro, C.: A fully integrated 0.5-7 Hz CMOS bandpass amplifier. In: Proceedings of the 2004 IEEE International Symposium on Circuits and Systems, vol. 1, pp. 445–448 (2004)
[12] Arnaud, A., Galup-Montoro, C.: Pico-A/V range CMOS transconductors using series-parallel current division. Electronics Letters 39, 1295–1296 (2003)
[13] Solís-Bustos, S., Silva-Martínez, J., Maloberti, F., Sánchez-Sinencio, E.: A 60-dB dynamic-range CMOS sixth-order 2.4-Hz low-pass filter for medical applications. IEEE Transactions on Circuits and Systems II: Analog and Digital Signal Processing 47, 1391–1398 (2000)
[14] Wong, A., Pun, K.P., Zhang, Y.T., Choy, C.S.: An ECG measurement IC using driven-right-leg circuit. In: Proceedings of the 2006 IEEE International Symposium on Circuits and Systems, vol. 1-11, pp. 345–348 (2006)
[15] Lee, S.Y., Cheng, C.J.: Systematic Design and Modeling of a OTA-C Filter for Portable ECG Detection. IEEE Transactions on Biomedical Circuits and Systems 3, 53–64 (2009)
[16] Yazicioglu, R.F., Kim, S., Torfs, T., Kim, H., Van Hoof, C.: A 30 mu W Analog Signal Processor ASIC for Portable Biopotential Signal Monitoring. IEEE Journal of Solid-State Circuits 46, 209–223 (2011)
[17] Xie, X., Li, G.L., Chen, X.K., Zhang, C., Wang, Z.H.: A low complexity near-lossless image compression method and its ASIC design for wireless endoscopy system. In: 2005 6th International Conference on ASIC Proceedings, Books 1 and 2, pp. 118–121 (2005)
[18] Nilchi, A., Aziz, J., Genov, R.: CMOS Image Compression Sensor with Algorithmically-Multiplying ADCs. In: Iscas: 2009 IEEE International Symposium on Circuits and Systems, vol. 1-5, pp. 1497–1500 (2009)
[19] Oliveira, F.D.V.R., Haas, H.L., Gomes, J.G.R.C., Petraglia, A.: CMOS Imager With Focal-Plane Analog Image Compression Combining DPCM and VQ. IEEE Transactions on Circuits and Systems I-Regular Papers 60, 1331–1344 (2013)
[20] Oliveira, F.D., Haas, H.L., Gomes, J.G.R., Petraglia, A.: A CMOS imaging system featuring focal-plane compression based on DPCM and VQ. In: 2012 IEEE Third Latin American Symposium on Circuits and Systems (LASCAS), pp. 1–4 (2012)
[21] Mallat, S.: A wavelet tour of signal processing: Access Online via Elsevier (1999)

[22] Kamboh, A.M., Raetz, M., Oweiss, K.G., Mason, A.: Area-Power Efficient VLSI Implementation of Multichannel DWT for Data Compression in Implantable Neuroprosthetics. IEEE Transactions on Biomedical Circuits and Systems 1, 128–135 (2007)

[23] Chua, E., Fang, W.C.: Mixed Bio-Signal Lossless Data Compressor for Portable Brain-Heart Monitoring Systems. IEEE Transactions on Consumer Electronics 57, 267–273 (2011)

[24] Carmo, J.P., Dias, N.S., Silva, H.R., Mendes, P.M., Couto, C., Correia, J.H.: A 2.4-GHz low-power/low-voltage wireless plug-and-play module for EEG applications. IEEE Sensors Journal 7, 1524–1531 (2007)

[25] Lee, S.Y., Hong, J.H., Hsieh, C.H., Liang, M.C., Kung, J.Y.: A Low-Power 13.56 MHz RF Front-End Circuit for Implantable Biomedical Devices. IEEE Transactions on Biomedical Circuits and Systems 7, 256–265 (2013)

[26] Ghovanloo, M., Najafi, K.: Fully integrated wideband high-current rectifiers for inductively powered devices. IEEE Journal of Solid-State Circuits 39, 1976–1984 (2004)

[27] Chen, D., He, L., Yan, X.: A low-dropout regulator with unconditional stability and low quiescent current. In: 2006 International Conference on Communications, Circuits and Systems Proceedings, pp. 2215–2218 (2006)

[28] Sarpeshkar, R., Salthouse, C., Sit, J.J., Baker, M.W., Zhak, S.M., Lu, T.K., et al.: An ultra-low-power programmable analog bionic ear processor. IEEE Trans. Biomed. Eng. 52, 711–727 (2005)

[29] Baker, M.W., Sarpeshkar, R.: A low-power high-PSRR current-mode microphone preamplifier. IEEE Journal of Solid-State Circuits 38, 1671–1678 (2003)

[30] Sarpeshkar, R., Lyon, R.F., Mead, C.: A low-power wide-linear-range transconductance amplifier. In: Neuromorphic Systems Engineering, pp. 267–313. Springer (1998)

[31] Chae, M.S., Yang, Z., Yuce, M.R., Hoang, L., Liu, W.T.: A 128-Channel 6 mW Wireless Neural Recording IC With Spike Feature Extraction and UWB Transmitter. IEEE Transactions on Neural Systems and Rehabilitation Engineering 17, 312–321 (2009)

Questions

Q1. What topic is this chapter talking about?

Q2. Why the design of ICs is needed and necessary in mHealth application?

Q3. The system of mHealth instrumentations, which part are these IC design technologies mainly used?

Q4. Are there some challenges in the design of biomedical ICs?

Q5. Why are there these challenges?

Part V

Cloud Applications

Chapter 24: A Secured Hybrid Cloud Architecture for mHealth Care

Chapter 25: Analysis of mHealth Systems with Multi-cloud Computing Offloading

Chapter 26: An Overview of MHealth Medical Video Communication Systems

Chapter 27: Mobile Health Improves Healthcare Delivery

Chapter 28: Terahertz Technology in the Future of Health and Medical Applications

Chapter 29: Developing Multi-agent Systems for mHealth Drug Delivery

24. A Secured Hybrid Cloud Architecture for mHealth Care

Khaled Ahmed Nagaty

Faculty of Informatics and Computer Science,
Ain Shams University,
British University in Egypt,
Cairo, Egypt
khaled.nagaty@bue.edu.eg

Abstract. This chapter presents a secure mHealth application which is based on hybrid cloud architecture combined with cryptographic techniques to protect data privacy, integrity and security of patients and health care givers and with role based access control to authenticate and authorize cloud users. Hybrid cloud platform combines the advantages of both the private cloud which guarantees privacy and safety of data and the public cloud which provides a platform for reduced services costs. Integrating cryptography and role based access control with hybrid cloud computing ensures the safety of patients' medical records and enables user authentication and authorization for access control. This integrated technology can provide mHealth care the required safety and privacy to flourish.

Keywords: mHealth, Mobile computing, Cloud computing, Cryptography, Hybrid cloud, Public cloud, Private cloud, Data privacy, Data security, Data confidentiality, Users Authentication, Users Authorization, Electronic health care.

1 Introduction

In cloud computing, storage and processing resources are all located remotely where users share these resources in a controlled and efficient manner with the illusion of unlimited compute network and storage capacity. Cloud computing resources can be used to run users applications and can be released when they are no longer needed. Users can acquire storage and processing resources as demanded which helps organizations with limited budget to lower their IT costs. Although cloud computing is becoming more popular among users but the number of cloud subscribers are beyond expectations as indicated by a survey conducted by the International Data Corporation [1]. Most IT Executives and CEOs are not interested in adopting cloud services due to the risks associated with security and privacy because cloud datacenters are managed by cloud providers and not owned or managed by cloud users which raises many concerns about security, data privacy, confidentiality, accidental or malicious attempts. Huge efforts have been put to build secure cloud computing environments and infrastructures in order to guarantee secure communication and more specifically authentication, integrity

S. Adibi (ed.), *Mobile Health*,
Springer Series in Bio-/Neuroinformatics 5, DOI: 10.1007/978-3-319-12817-7_24

and confidentiality between users and processes in the cloud. Itani W. et al. in [2], present privacy as a service (PaaS) where security protocols which enforce user data privacy, legal compliance and user trust need to be considered at every phase of cloud computing architectures. PaaS allows for secure storage and processing of users' confidential data by leveraging the tamper-proof capabilities of cryptographic coprocessors. The fundamental operations for security in cloud computing are encryption and role based access control (RBAC). Encryption transforms data from something understandable to an intruder into something not understandable. Encryption helps users to check if their data had been tampered with or not which means that encryption provides the cloud with data integrity. RBAC supports integrity which means that data and processes must be modified only in authorized ways and by authorized users. In this manner, integrating cloud computing with cryptography and RBAC can help increasing privacy, confidentiality and protecting data in the cloud which encourages more security and privacy concerned organizations to adopt cloud computing environment. Secure communication is the first level of security and can be implemented using encryption and signature schemes. Secured cloud storage is the second level of security. Security in cloud storage should be provided to cloud users as a service with minimal additional cost. Providing secured cloud storage is considered the major obstacle that prevents organizations and enterprises from adopting cloud computing. Some cloud providers are providing secure storage services but offering different levels of protection. Efficient methods are required to verify on-demand data correctness in the cloud storage on behalf of the data owners [3]. Cryptography can help cloud computing to provide secured storage. This chapter is organized as follows: cryptography primitives will be discussed in section 2. RBAC will be discussed in section 3. Secure communication will be discussed in section 4. Secure cloud storage will be discussed in details in section 5. A model for a secured mHealth system which integrates cryptography, RBAC and hybrid cloud is discussed in section 6. Section 7 is dedicated for case studies. Section 8 for future directions and section 9 for conclusions.

2 Cryptographic Primitives

Cryptographic primitives are the building blocks for all crypto systems. They include: encryption schemes, cryptographic hash functions and digital signatures.

2.1 Encryption Schemes

2.1.1 Symmetric Cryptography

It is also known as secret key cryptography. It is used to maintain the privacy and authenticity of communicated data in which knowledge of the encryption key is explicitly or implicitly equivalent to knowing the decryption key. In other words, the same key is used to encrypt and decrypt a message:

$$P = D_K(E_K(P)) \tag{1}$$

Where P is the plain text, K is the encryption & decryption key, E_K is the encryption process using key K while D_K is the decryption process using the same key. Symmetric key encryption can use either stream ciphers or block ciphers. In stream ciphers, the message is encrypted one bit at a time while in block ciphers the message is encrypted one block at a time where a block size of 64-bits is commonly used. Text padding is used so that the plain text is a multiple of the block size. Although, this technique is simple but unfortunately it is less secure as the sender and receiver must initially communicate to exchange their secret keys before sending any encrypted messages.

2.1.2 Asymmetric Cryptography

It is also known as public-key cryptography. It is used to overcome the drawback of symmetric cryptography which uses the same key for both encryption and decryption. Digital signature uses asymmetric cryptography to demonstrate authenticity. In this scheme the encryption key is effectively public knowledge, without giving any useful information about the decryption key which is kept private. In other words, the keys for encryption and decryption are different, but together form a unique pair that is mathematically related. There is a separate key K_E for encryption and one for decryption K_D such that:

$$P = D_{K_D}(E_{K_E}(P)) \tag{2}$$

Where K_E is the public key, K_D is the private key, E_{K_E} is the encryption process using the public key K_E and D_{K_D} is the decryption process using the private key K_D.

2.1.3 Threshold Cryptography

This scheme is useful to organizations which own valuable information such as certificate authorities, governments and military organizations. These organizations need to distribute their valuable information among multiple servers in a way that no group of cooperated servers less than a specified threshold can decrypt this encrypted information. A threshold cryptography scheme allows secret sharing between n parties where each party is assigned a secret share. To recover the secret a pre-specified number of parties must collaborate in order to obtain the secret. A threshold scheme of (n, r) means that a secret is divided among n parties however any r parties out of the n parties can jointly obtain the secret [4]. The private key k of the service is divided into n shares $(s1, s2, \ldots, sn)$, assigning one share to each server. The n servers share the ability to generate the crypto-service. As an example, suppose we want to generate a crypto- service using (3,2) threshold cryptographic scheme where 3 is the number of servers that share the private key and 2 is the number of servers that can collaborate to generate the crypto-service using their shares. Fig.1 shows how servers generate a crypto service $(3, 2)$ threshold cryptographic scheme where 3 is the number of servers, K is the public key and k is the private key which is divided among the 3 servers, where each server gets a share of the private key

$(s1, s2, s3)$. Suppose that the required crypto-service is to encrypt message m, so server i performs a partial crypto-service $PS(m, si)$ using its share si of the private key k. In this example, servers 1 & 3 can submit their $PS(m, s1)$ and $PS(m, s3)$ respectively to the combiner c in order to generate the required service, although server 2 did not submit its share $PS(m, s2)$ to the combiner c.

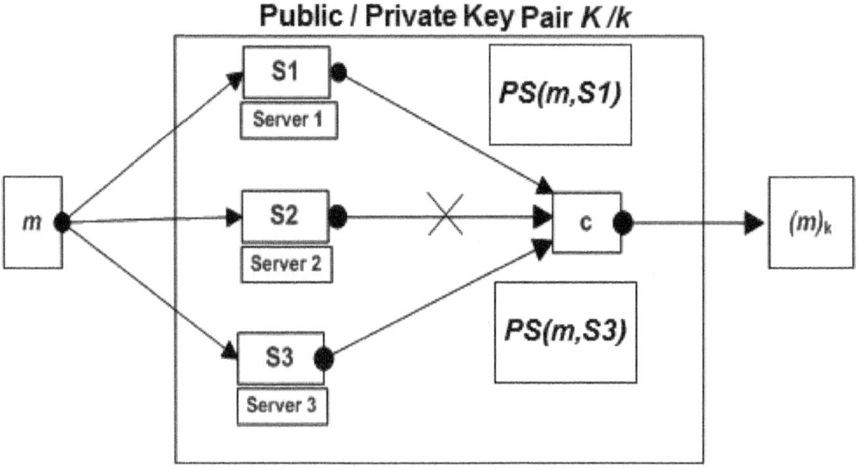

Fig. 1 Threshold Cryptographic Scheme (3,2)

2.1.4 Identity Based Encryption (IBE)

It is a primitive of identity based cryptography. It is used to solve the difficulties in using asymmetric cryptography. These difficulties include public keys distribution because senders want to send encrypted messages must first find the recipient's public key which may be not an easy task for users outside an organization. Identity based cryptography is considered a promising solution to this difficulty as it can be initiated by the sender. A public known string which acts as a unique identifier of the recipient such as some information representing the identity *ID* of a user or text value of the name in the email address can be used to generate a public key that can be used for encrypting information. A trusted third party, called the private key generator (PKG) publishes a master public key to generate public keys using the identity *ID* and retains the corresponding master private key to generate the corresponding private keys. By using the master public key any user can generate a public key that corresponds to the identity *ID* by combining the master public key with the identity value. To generate the corresponding private key, the user authorized to use the identity *ID* contacts the PKG, which uses the master private key to generate the private key for identity *ID* [5]. Dan Boneh et al. in [6] defined a set of four algorithms that form a complete IBE cryptosystem:

Setup: This algorithm runs by the PKG system one time for creating the whole IBE environment. The master private key is kept secret and used to derive users' private keys, while the system parameters are made public. It accepts k a security parameter (i.e. binary length of key material) and outputs:

- A set $P$of system parameters, including the message space and cipher text space M and C respectively.
- A master private key K_m.

Extract: This algorithm runs by the PKG when a user requests his private key. It takes as input P, K_m and identifier $ID \in \{0,1\}^*$ and returns the private key d for user ID.

Encrypt: This algorithm takes, a message $m \in M$ and $ID \in \{0,1\}^*$ and outputs the encryption $c \in C$.

Decrypt: This algorithm accepts, P and $c \in C$ and returns $m \in M$.

Example
Assume that Alice wants to store confidential information in the cloud storage:

1. Alice encrypts and signs her identity *ID* and then sends it to PKG to retrieve her public key.
2. The PKG uses the master public key and Alice identity *ID* to generate Alice public key.
3. The PKG then sends to Alice her public key.
4. Alice encrypts her secret information using her public key and uploads it on the Cloud.

Fig.2 shows the generation of public key and sending encrypted information to the cloud computing storage using IBE.

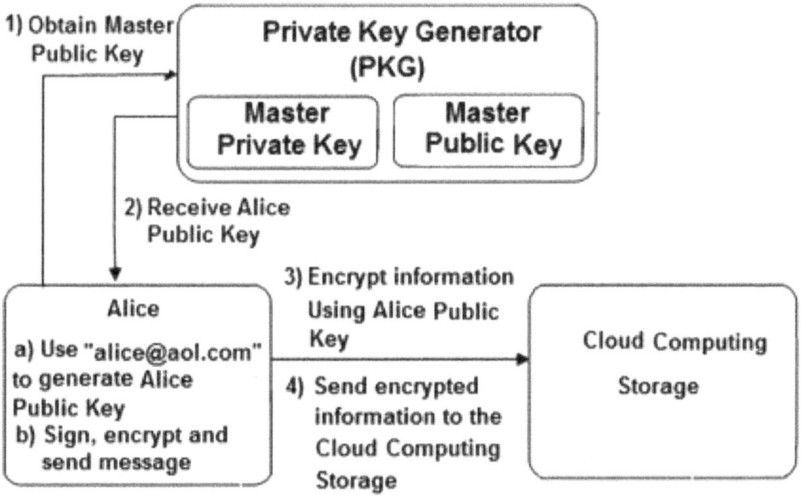

Fig. 2 Generate public key and send encrypted information to the cloud computing storage using IBE

Now assume that Alice wants to retrieve her encrypted information from the cloud storage:

1. Alice encrypts and signs her identity *ID* and then sends it to PKG to retrieve her private key.
2. The PKG uses the master private key and Alice identity *ID* to generate Alice private key.
3. The PKG then sends to Alice her private key.
4. Alice decrypts her secret information using her private key.

Fig.3 shows the generation of private key and decrypting the received encrypted information from the cloud computing storage using IBE.

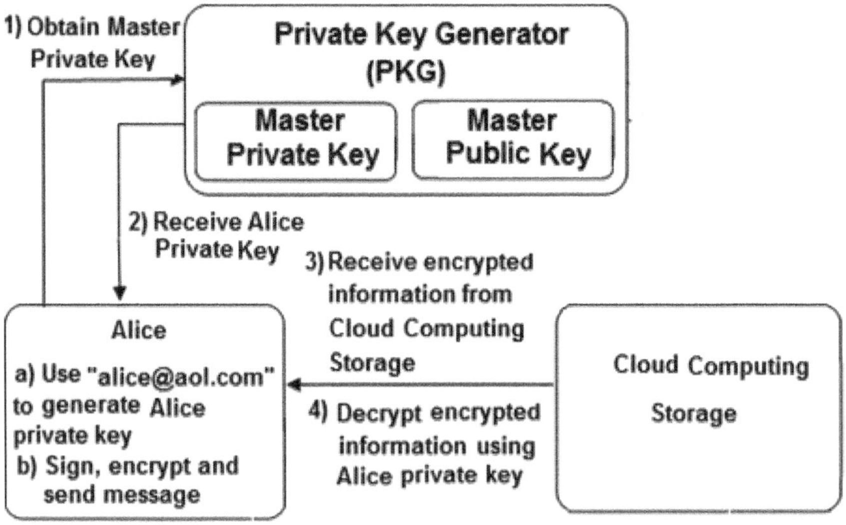

Fig. 3 Generate private key and decrypt the received encrypted information from the cloud computing storage using IBE

2.1.5 Identity Based Signature

This scheme is used to allow different parties to securely communicate and verify each other's signatures without exchanging private or public keys and without keeping key directories. The framework of identity based signature scheme consists of algorithms described below [5]:

Setup: The private key generator (PKG) provides the security parameter as the input to this algorithm, generates the systems parameters and the master private key.

Extract: The user provides his identity ID to the PKG which runs this algorithm with identity ID, parameters and master private key as the input and obtain the private key d which is sent to user through a secure channel.

Sign: For generating a signature on a message m, the user provides his identity ID, his private key d, parameters and the message m as input. This algorithm generates a valid signature on message m by the user.

Verify: This algorithm on input a signature on message m by the user with Identity ID, parameters, checks whether signature is valid on message m by ID.

2.2 Linear Search Algorithm

This algorithm is used to perform secure searches over encrypted data without leaking any information to the host server. In the linear search algorithm, a symmetric encryption algorithm is used to encrypt the plain text. For the cipher text of each keyword under symmetric encryption scheme a pseudo-random sequence is generated with a length less than that of the cipher text. Meanwhile, a check sequence is generated with a length less than that of the cipher text and a check sequence is generated based on the pseudo random sequence and the cipher text. The sum of the lengths of the pseudo random sequence and the check sequence equals to the length of the cipher text. Finally, the pseudo random sequence and the check sequence equals to the length of the cipher text again by modulo 2 addition. When searching, a user submits the cipher text sequence under symmetric encryption schemes. On the server side modulo 2 additions with each sequence is performed to each sequence. If the result satisfies the checking, the sequence is the encryption of the cipher text; otherwise, the sequence is not encryption of the cipher text [5].

2.3 Homomorphic Encryption

This scheme is used to guarantee better data security and to keep cloud users private information confidential. Homomorphic encryption allows executing operations on encrypted data without knowing the private key to decrypt it which helps protecting data confidentiality[7]. When decrypting the obtained results they are the same as if the operations were carried out on the raw data. In other words, an encryption is Homomorphic, if: from Enc(a) and Enc(b) it is possible to compute Enc(f (a, b)), where f can be: +, ×, $\oplus$ and without using the private key.

2.4 Public Key Encryption with Keyword Search

This scheme is used for searching data that is encrypted using a public key system. It allows a recipient to search the encrypted message for specific keyword without the need to decrypt the entire message and without learning anything else. A public key encryption with keyword search (PEKS) scheme consists of four polynomial time randomized algorithms [10]:

KeyGen(*s*): Takes as input a security parameter "*s*" and generate a public/private key pair $(\boldsymbol{pk}, \boldsymbol{sk})$.

PEKS $(\boldsymbol{pk}, \boldsymbol{W})$: Takes as input the receiver public key $\boldsymbol{pk}$, word $\boldsymbol{W}$ and produces a searchable encryption of $\boldsymbol{W}$, *i.e.* $\boldsymbol{S} = \boldsymbol{PEKS}(\boldsymbol{pk}, \boldsymbol{W})$

Trapdoor $(\boldsymbol{sk}, \boldsymbol{W})$: Takes as input the receiver private key $\boldsymbol{sk}$, a word $\boldsymbol{W}$ and produces a trapdoor $\boldsymbol{Tw}$.

Test $(\boldsymbol{pk}, \boldsymbol{C}, \boldsymbol{Tw})$: Takes as input the receiver public key $\boldsymbol{pk}$, a searchable encryption

$\boldsymbol{C} = \boldsymbol{PEKS}(\boldsymbol{pk}, \boldsymbol{W'})$, a trapdoor $\boldsymbol{Tw}$ and produces "**yes**" if $\boldsymbol{W} = \boldsymbol{W'}$ and "**no**" otherwise.

2.5 *Attribute Based Encryption*

This scheme is useful for sharing data between users where standard cryptography is inefficient since the data must be re-encrypted using every user's public key. Also, this scheme is used to avoid collusion which is a crucial security feature. In an attribute-based encryption scheme, we label each cipher text with a set of descriptive attributes such as the country in which the user lives or the kind of subscription the user has, then we provide each user with a (Key, Attributes) pair. The decryption of a cipher text is possible only if the set of attributes of the user key matches the attributes of the cipher text.

2.5.1 Searchable Encryption

Searchable encryption is a broad concept that is useful for searches in encrypted data where we are able to conditionally retrieve or query data without having to decrypt all the data [8]. The first searchable encryption scheme is the symmetric searchable encryption (SSE): in SSE schemes the party who searches over the data is also the one who encrypted it. The typical usage scenario for SSE is single writer/single reader (SWSR) which makes SSE more efficient for the party doing the encryption because SSE schemes are based on symmetric encryption and for the party performing the search because data are stored in efficient data structures[9]. Security is guaranteed as follows:

- Without any token, which is small bits of the data, the server learns nothing about the data except its length.
- Given a token of keyword w, the server learns which encrypted documents contain w without learning w.

Another approach is the asymmetric searchable encryption (ASE): In ASE schemes the party searches over the data is different from the party who encrypted it. The typical usage scenario for ASE is many writer/single reader (MWSR). With Public-Key Encryption with Keyword Search (PEKS) scheme, proposed by Boneh et al. [10] it is possible for multiple persons to encrypt data using public cryptography but only the owner of the private key will be able to search for a keyword.

2.6 Key Management

In a public key infrastructure, each party has a public/private key pair. Public keys can be known to all other nodes, while each node keeps its own private key confidential. Key management is used in the generation, exchange, storage, use, and replacement of cryptographic keys. Key management is implemented by a trusted entity called Certification Authority (CA) [11]. The CA has a public/private key pair, with its public key is known to every node. With this public key the CA signs certificates which bind public keys to nodes. If the party owning a public key is no longer trusted or left the network then his public key should be revoked, also a party should refresh his key pair on regular basis in order to reduce the possibility of a successful brute-force attack on its private key.

2.6.1 Key Recovery

Key recovery is used in case of emergency or losing the decryption key (typically a secret key or private key) to access encrypted data. One recovery method is known as recovery agent who recovers cryptographic keys using a master key. Other recovery methods are key escrow and key encapsulation. With key escrow a copy of the decryption key is stored in a safe place to be recovered later. The decryption key can be split into one or several parts and distribute these parts to escrow parties or ``trustees". In an emergency situation or a third party needs to decrypt data as defined by law such as in case of court order, these trustees can use their ``shares" of the keys either to reconstruct the missing key or simply to decrypt encrypted communications directly. Key encapsulation encrypts data in a communication using a ``session key" which varies from communication to communication. The session key is encrypted with a trustee public key and sent with encrypted communication. So, the trustee is able to decrypt the communication when necessary. A variant of this method, in which the session key is split into several pieces, each encrypted with a different trustee public key.

2.6.2 Cryptographic Hash Functions

Cryptographic hash functions are used to achieve many security goals which include data authentication, data integrity, digital signatures and digital time stamping. A cryptographic hash takes a message of arbitrary length and returns a fixed length bit string and the cryptographic hash value such that any change in the message either accidentally or deliberately will change with a very high probability the hash value [13]. The hash value sometimes called the message digest. The main important property of a hash function is that it is a one way hash function which means that given the output of a hash function it is impossible to recover the original message using only this output. Another important property is that it is difficult to find two distinct messages give the same digest. Examples of practical hash functions are: MD4, MD5, SHA-1, SHA-256.

2.6.3 Digital Signatures

A digital signature is a mathematical scheme which employs asymmetric cryptography to prove that a digital message or document is originated from a known sender, a process known as authenticity [14]. Digital signatures are used in e-commerce to provide non-repudiation where a user cannot deny having done a transaction or signed a document. Also, digital signatures are used when it is important to detect forgery or to demonstrate that a message or document has not been tampered with during transmission, a process known as data integrity. To sign a document, the sender's software use one of the common cryptographic hash functions to crunch down the data into just a few lines by a process known as a hashing. These few lines are called a message digest. The message digest is encrypted with the sender's private key. The output is the digital signature. Finally, the digital signature is appended to the document. At the receiver, the receiver's software decrypts the signature using the sender's public key thus recovering the message digest. If this process is successful, then this shows that the sender actually signed the document, because only the sender has his own private key. The receiver's software hashes the received document into a message digest. If the obtained message digest matches with the message digest restored from the decrypted signature, then the receiver knows that the signed document has not been changed.

2.6.4 Cryptographic Coprocessors

They are secure crypto coprocessors carrying out cryptographic operations using keys that are protected within the device from physical attacks and network attacks. Secure crypto-processors are used for tamper resistance to enforce data-usage policies. Smart cards holding decryption keys are one of the low-cost uses of embedded cryptographic processors [15]. Electronic payment systems use secure crypto-processors to securely store the payment card authentication keys and secure communications between banks and clients. Trusted computing (TC) aims at making TC machines trusted to certify both programs and platform on which they are running. The key idea is to embed a special crypto-processor module in current computing platforms like desktop PCs and mobile devices.

3 Role Based Access Control (RBAC)

Users and service authentication is extremely important for increasing security and enhancing the trust of cloud computing users. RBAC is a scheme used by many cloud computing applications to allow administrators create users, groups, and assign roles and policies to all users. RBAC goes through five stages [16]:

3.1 Provisioning and De-provisioning

Every user of the system is given an identity. A user is authorized to access the system based on the role. Cloud management uses techniques such as identity management as a service (IDaaS).

3.2 Authentication and Authorization

Authentication and authorization applications are required to fulfill the business goals.

3.3 Self-service

Self service in the identity management can help users to reset their information like password and uphold their data from any location.

3.4 Password Management

Single sign on (SSO) support system is to access cloud-base services. Password management comprises how the password will be stored in the cloud database.

3.5 Compliance and Audit

Access to the system will be tracked in order to monitor the security breaches in the system. It assists auditing the fulfillment to diverse access control policies, periodic auditing and reporting.

3.6 Formal Description of RBAC

A simple formal description of role based access control in terms of sets and relations [17]:

1. For each subject, the active role (AR) is the one that the subject is currently using:

$AR(s{:}\, subject) \;=\; \{the\ active\ role\ for\ subject\ s\}.$

2. Each subject may be authorized to perform one or more roles:

$RA(s{:}\, subject) \;=\; \{authorized\ roles\ for\ subject\ s\}.$

3. Each role may be authorized to perform one or more transactions:

$TA(r{:}\, role) \;=\; \{transactions\ authorized\ for\ role\ r\}.$

4. Subjects may execute transactions. The predicate $exec(s, t)$ is true if subject s can execute transaction t at the current time, otherwise it is false:

$exec(s{:}\, subject, t{:}\, tran) \;=\; true\ iff\ subject\ s\ can\ execute\ transaction\ t.$

Three basic rules are required:

1. Role assignment: A subject can execute a transaction only if the subject has selected or been assigned a role:
   $\forall s : subject, t : tran, (exec(s,t) \Rightarrow AR(s) \neq \phi)$. The identification and authentication process (e.g. login) is not considered a transaction. All other user activities on the system are conducted through transactions. Thus all active users are required to have some active role.

2. Role authorization: A subject active role must be authorized for the subject: $\forall s : subject, (AR(s) \subseteq RA(s))$.
   With (1) above, this rule ensures that users can take on only roles for which they are authorized.

3. Transaction authorization: A subject can execute a transaction only if the transaction is authorized for the subject active role: $\forall s : subject, t : tran, (exec(s,t) \Rightarrow t \in TA(AR(s)))$.
   Using the above rules a doctor could be assigned the privilege of read/write access to a patient medical file, while the hospital pharmacist might have only read access.

4 Secure Communication

A secure mHealth application involves a secure communication between cloud servers and the mobile devices. A secure communication channel between mobile devices and the cloud server is guaranteed by using secure routing protocols [12]. Routing protocols are subject to different types of attacks that can affect the selection of correct paths or allow for the denial-of-service attacks. A secure routing protocol must ensure that route signals cannot be spoofed or altered during transition and fabricated routing messages cannot be injected in the network. It must also ensure that routing loops cannot be formed through malicious actions and routes cannot be redirected from the shortest path due to malicious actions. In the management of open environment, unauthorized nodes should be excluded from route computation and discovery. Cryptographic certificates are helpful to offer routing security. They are used with authenticated routing for ad hoc networks protocol (ARAN). ARAN uses certification to guarantee end-to-end authentication. A trusted certificate server T has its public key K_{T+} is well known to all valid nodes. At priori the public keys of all valid nodes are generated and exchanged with the certificate server. Before entering the ad hoc network, a valid node A must securely authenticate itself to the certificate server T. If it is authenticated then node A will receive only one certificate $cert_A$ which includes node A IP address IP_A, public key K_{A+}, time stamps of certificate creation t and expiration e, all are concatenated and signed by the private key of the certificate server:

$$K_{T-} \text{ [12]: } T \rightarrow A: cert_A = [IP_A, K_{A+}, t, e]K_{T-} \quad (3)$$

5 Secure Cloud Storage

Cloud storage is a pool of computer data storage where user's data is stored by the cloud provider. With secured cloud storage end users can securely access the required information while revoked users cannot. Cloud storage techniques fall into two major classes: Class A which refers to cloud storage that is designed using cryptographic techniques not within the framework of cryptography theory. Class "B" refers to cloud storages that are designed using cryptographic techniques that are within the framework of cryptography theory [18]. Cloud storages in class "A" are based on non-standard cryptographic techniques. Kamara et al. [19] designed a cloud storage technique for both consumer and enterprise scenarios based on non-standard cryptographic techniques. For the consumer scenario a client application must be downloaded which consists of:

1. Data processor (DP): processes data before being sent to the cloud.
2. Data verifier (DV): checks data integrity.
3. Token generator (TG): generates tokens which are small bits of data that enable the cloud storage provider to retrieve segments of customer data.
4. A credential generator implements an access control policy by issuing credentials to the various parties in the system. These credentials will enable the parties to decrypt encrypted files according to the policy.

First, the client's application generates a unique cryptographic key which is called the master key and it is locally stored in the client's system and kept secret from the cloud storage provider.

- To upload medical data to the cloud storage the patient attaches some metadata such as the time and size to the medical data. The DP first indexes the medical data and the metadata and then encrypts this data with the unique cryptographic key. Encryption is implemented with a symmetric encryption scheme such as the advanced encryption standard (AES) using the unique key. The DP encrypts the index using a searchable encryption scheme and encrypts the unique key with an attributed-based encryption scheme under appropriate policy. To check the integrity of the medical records the DV uses the patient's key to interact with the cloud storage provider.
- To retrieve data the token generator is used to create a token which is sent to the cloud storage provider to retrieve the required encrypted medical file and return it to the patient. Then the patient uses the decryption key to decrypt the encrypted medical file. If the patient wants to share the medical file with the doctor the client application invokes the token generator to generate a token and the credential generator generates a credential for the doctor. The token and credential are sent to the doctor who sends the token to the cloud storage provider which in turn uses the token to retrieve and return the encrypted file to the doctor. The doctor uses his credential to decrypt the patient encrypted file. Fig. 4 shows the interaction between Alice, Bob and cloud storage provider [18].

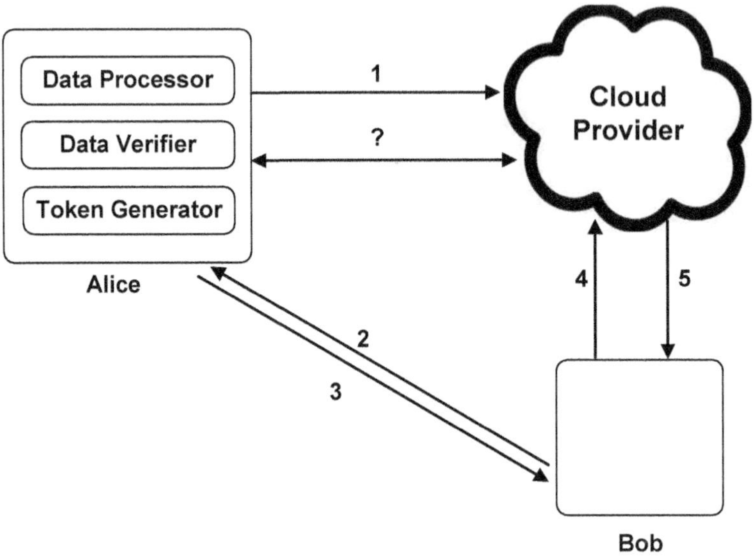

Fig. 4 The interaction between Alice, Bob and cloud storage provider where [As adapted from 18]:

1. Alice's data processor prepares the data before sending it to the cloud;
2. Bob asks the patient for permission to retrieve the medical file;
3. Bob's token and credential generators send a token for the requested file and a credential back to Bob;
4. Bob sends the token to the cloud;
5. The cloud uses the token to find the appropriate encrypted file and returns them to Bob.
6. (?) At any point in time, patient data verifier can verify the integrity of the data.

The other scenario is the enterprise architecture which consists of enterprise MegaCorp that stores its data in the cloud, a business partner PartnerCorp that wants to share data and a cloud storage provider that store's MegaCorp's data. In the case that MegaCorp is a hospital while its partner are doctors, to use that service the MegaCorp deploys dedicated machines that run various core components. These components must be kept protected and the master secret key must be kept secret from the PartnerCorp and the cloud storage provider. Fig. 5 illustrates the enterprise architecture and the interaction between MegaCorp and PartnerCorp.

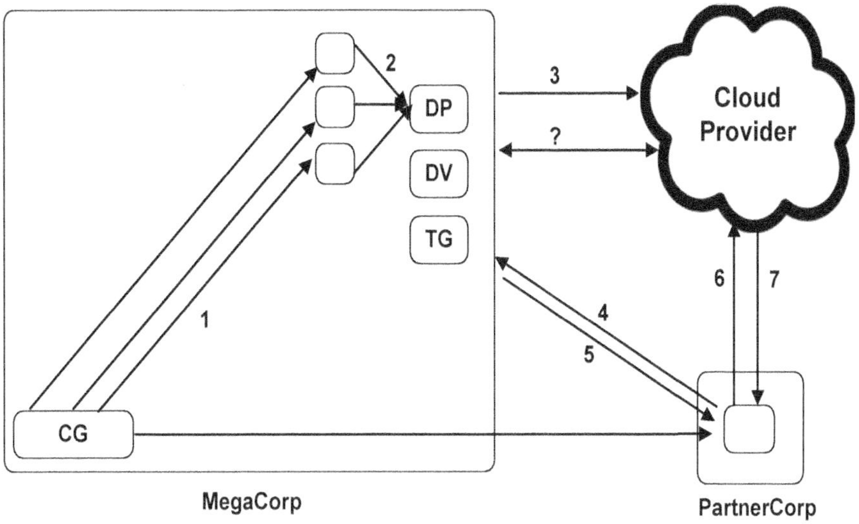

Fig. 5 Enterprise architecture and the interaction between MegaCorp and PartnerCorp [As adapted from 18]

6 mHealth in Secured Hybrid Cloud Architecture

mHealth based on cloud computing is revolutionizing the way healthcare is provided to patients. Applications of mHealth installed on mobile devices such as mobile phones, tablet computers and PDAs help patients to continuously receive healthcare and medical treatment anytime, anywhere and anyhow and help practitioners to real-time monitoring their patients and direct provision of healthcare. Usually, mHealth applications store health information in an electronic health records system (EHR). Due to the limited resources of mobile devices such as low CPU, low memory, low storage capacity and battery powered environment, mHealth clients should use the unlimited resources of the cloud providers such as unlimited storage and high processing power of the cloud servers. EHR systems can increase their storage capacity by uploading the medical information to the cloud storage where healthcare practitioners in different hospitals and medical institutions can share the medical records of the patients in order to make correct diagnosis and give them the proper medication they need. EHR systems and mHealth applications should be characterized by scalability, security, non-repudiation and availability in a ubiquitous interoperable manner in order to provide patients with quality on demand healthcare services [20]. Scalability refers to the ability of the mHealth systems to handle the growing demand on accessing the EHR system. Users of mHealth applications are interested in storing their medical and personal information on cloud servers unexposed. Security challenges that may be caused by intruders which must be targeted in order to provide a secured mHealth environment include confidentiality, integrity and authentication. Confidentiality refers to the process that ensures that information is accessible only to those authorized to have access to it and never disclosed to unauthorized entities. Integrity refers to ensure that patient medical information during its transfer is not modified or corrupted by unauthorized accesses to ensure patient safety. A security framework should provide privacy and security features to all mHealth users [21] and must be highly scalable to be adaptively handled without degradation in performance. Security also involves auditing, where an independent third party systematically examines medical records activities, unauthorized accesses, modifications and performance [22]. Non-repudiation is an important feature to ensure that a message sender cannot deny having sent this message. Availability of a service is defined as the property of being accessible and useable upon demand by an authorized entity [25]. Availability of health information is critical to effective healthcare delivery system. Sharing medical information with practitioners across many administrative boundaries can violate patient information confidentiality, integrity and privacy [23, 24]. So, authentication enables a party to ensure that he/she is communicating with the correct party on the other side. Without authentication, unauthorized access to resources and confidential information could be gained by an unauthorized party. Role Based Access Control (RBAC) is a scheme that restricts unauthorized users to access a system. It offers a satisfactory level of safety & security for authorized mHealth users who accessing EHR system resources & medical records through a set of rules & policies put into effect for patients, doctors or any other personnel in the form of login & password [26]. In [27] Yu-Yi Chen et al. proposed an EHR sharing and integration system in healthcare clouds and analyze the arising security and privacy issues in access and management of EHRs. Securing mobile health applications running on a mobile device is therefore an important area, if we want the applications to be trustworthy and reliable [28]. A mHealth cloud computing system consists of three main components which are [1]:

6.1 Mobile Client

Clients of mHealth applications utilize the storage services and the processing power provided by the cloud service provider. Applications of mHealth are embedded in the mobile devices of clients to interact with EHR system on the cloud server using mobile networks. Mobile networks are composed of three parts [29]:

- Base stations: send and receive transmissions to and from clients.
- Mobile telephone switching offices: transfer calls between national or global phone networks and base stations.
- Subscribers (clients): connect to base stations by using communication devices.

Mobile devices should register to a carrier service or provider which is licensed to offer services within a certain geographic areas. If a mobile device is out of the geographic area of its provider then roaming occurs where service connectivity can be extended in locations far from the base station where the service was first registered. Two technologies are used to improve digital communications: Time Division Multiple Access (TDMA) which divides each channel into six time slots one for transmission and one for reception. Code Division Multiple Access (CDMA) which transmits multiple encoded messages over a wide frequency and decode them at the receiver. CDMA is very useful in mHealth systems [30].

6.2 Cloud Service Provider

The cloud service provider efficiently manages, operates, maintains, allocates cloud resources and provides mHealth users security services such as intrusion detection, key management, encryption, authentication and authorization. The healthcare cloud system is consisted of public and private cloud providers where hospitals can construct besides the public cloud environment their own private cloud environment. Medical information for patients can be stored in both private cloud server and public cloud server [27]. Auditing can be implemented using a cloud service monitor which records clients' activities such as accessing or modifying medical records chronologically.

6.3 Trusted Third Party

A trusted third party is needed for configuring and installing tamperproof crypto-coprocessors on the cloud. Multiple registered mobile clients are associated with each crypto-coprocessor. The crypto-coprocessor distributes secret key (SK) with associated mobile clients and generates a message authentication code on behalf of mobile clients.

Fig. 6 shows the architecture of mHealth hybrid cloud computing environment.

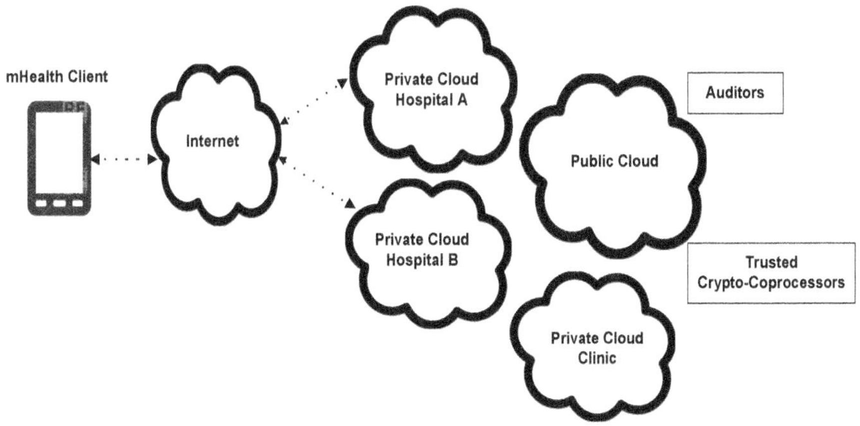

Fig. 6 The architecture of mHealth hybrid cloud computing environment

A hybrid cloud EHR scheme is composed of the following phases [27]:

1. New Electronic Record Creation.
2. Electronic Medical Record Access.
3. Emergency Access for Electronic Medical Record.

Our mHealth system that depends mainly on using mobile devices to create and access medical records on a secured hybrid cloud has the above mentioned phases.

Phase 1: mHealth Medical Record Creation

The hospital owns the medical information of patient (i) creates a new medical record for this patient. The medical record consists of: *Patient ID* (ID_i), *Medical record file* $(file_i)$, *Patient Secret Key* (pSK). The patient secret key (pSK) is generated using the patient ID_i and identity based encryption (IBE). The medical file corresponding license ACL is composed of: *Patient ID* (ID_i), *Patient Right* $(pRight)$, *Patient Secret Key* (pSK).

The patient medical file $file_i$ is encrypted by the patient secret key (pSK) to generate the encrypted medical file $Mfile_{ci}$. Identity based cryptography uses the patient ID_i to generate the patient public key (e_i) that can be used for encrypting information. The trusted third party, called the private key generator (PKG), generates the corresponding patient private key (d_i). For the purpose of data privacy and confidentiality the generated corresponding license ACL_i is double encrypted firstly by the patient public key (e_i) to generate $eACL_{ci}$ and secondly by the hospital public key (h_{public}) to generate $hACL_{ci}$ as follows:

$$Mfile_{ci} = E_{pSK}(file_i) \tag{4}$$

$$Hfile_{ci} = E_{h_{public}}(Mfile_{ci}) \tag{5}$$

$$\begin{aligned} eACL_{ci} &= E_{e_i}(ACL_i) = E_{e_i}\left(ID_i||pRight||pSK\right) \\ &= \left(ID_i||pRight||pSK\right)^{e_i} \ mod\ N \end{aligned} \tag{6}$$

The $eACL_{ci}$ is then encrypted by the hospital public key (h_{public}) as follows:

$$hACL_{ci} = E_{h_{public}}(eACL_{ci}) \tag{7}$$

The hospital signature $sACL_{ci}$ is generated as follows:

$$sACL_{ci} = E_{h_{public}}\left(H(eACL_{ci})\right) \tag{8}$$

Where: H is cryptographic hashing function such as *MD5, SHA1* or *SHA2* and h_{public} is the patient's hospital public key.

The hospital's private cloud generates a corresponding signature as follows:

$$ScID = E_{h_{public}}(ID_i, time) \tag{9}$$

The $Hfile_{ci}$, $hACL_{ci}$, $sACL_{ci}$ and $ScID$ are all transmitted on a secure communication channel from the hospital's private cloud server to be stored on the public cloud server. The signature $ScID$ is stored in the public cloud for auditing purposes.

Algorithm 1: mHealth Medical File Creation

After finishing examination of a patient the doctor either updates the medical file of this patient or creates a new file using the patient's mobile device. Using this algorithm the patient encrypts the created or updated medical file and its corresponding license using his mobile device. First of all, the patient's mobile device must be authenticated by its associated trusted crypto-coprocessor in order to access the hospital's private cloud. If authentication is succeeded the RBAC method validates the identity of the patient and checks his authorized role. If both validation and authorization of the patient are succeeded the doctor can update the medical file or create a new one for the patient using the patient's mobile device. After the doctor finishing updates the patient uses his secret key (pSK) to encrypt the created or updated medical file in order to generate $Mfile_{ci}$ and uses his public key (e_i) to encrypt the corresponding license in order to generate $eACL_{ci}$. The mobile device sends patient's ID_i , $Mfile_{ci}$ and $eACL_{ci}$ to the trusted crypto-coprocessor associated with it in order to send them to the patient's hospital private cloud. The hospital private cloud re-encrypts both the patient's encrypted medical file $Mfile_{ci}$ and the encrypted corresponding license $eACL_{ci}$ with the hospital's public key h_{public} in order to generate a double encrypted patient's medical file $Hfile_{ci}$ and a double encrypted corresponding license $hACL_{ci}$ respectively. In the mean time, the hospital's private cloud uses a cryptographic hash function to generate a message digest $H(eACL_{ci})$ for the encrypted license $eACL_{ci}$ and encrypts this message digest using the hospital's public key h_{public} in order to generate the hospital signature $sACL_{ci}$. The hospital signature $sACL_{ci}$ is used to verify that the encrypted license $eACL_{ci}$ was not modified or tampered with by any intruder while it was being stored in the public cloud. The hospital's private cloud generates a corresponding signature $ScID$ using the patient's ID_i and system's time. $Hfile_{ci}$, $hACL_{ci}$, $sACL_{ci}$ and $ScID$ are all transmitted on a secured communication channel to be stored in the public cloud. The $ScID$ is stored in the public cloud for auditing purposes. Fig.7 shows the flowchart for using a patient's mobile device for creating a medical file in the hybrid cloud.

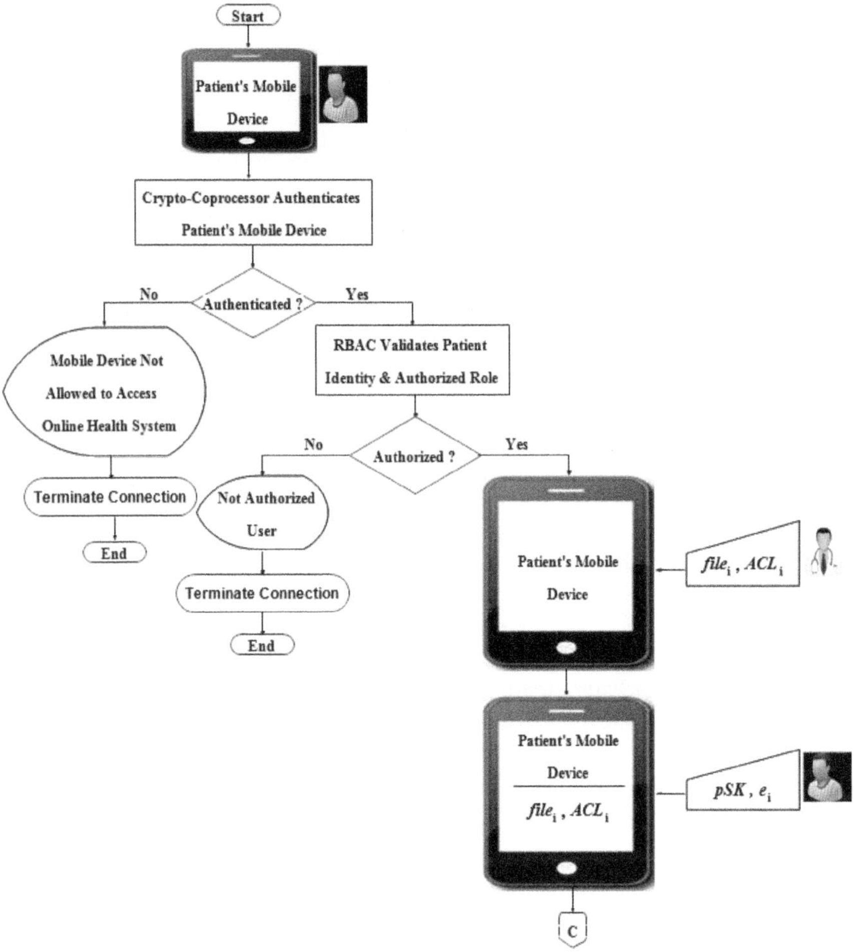

Fig. 7 Creation of a patient's medical file in the hybrid cloud using patient's mobile device

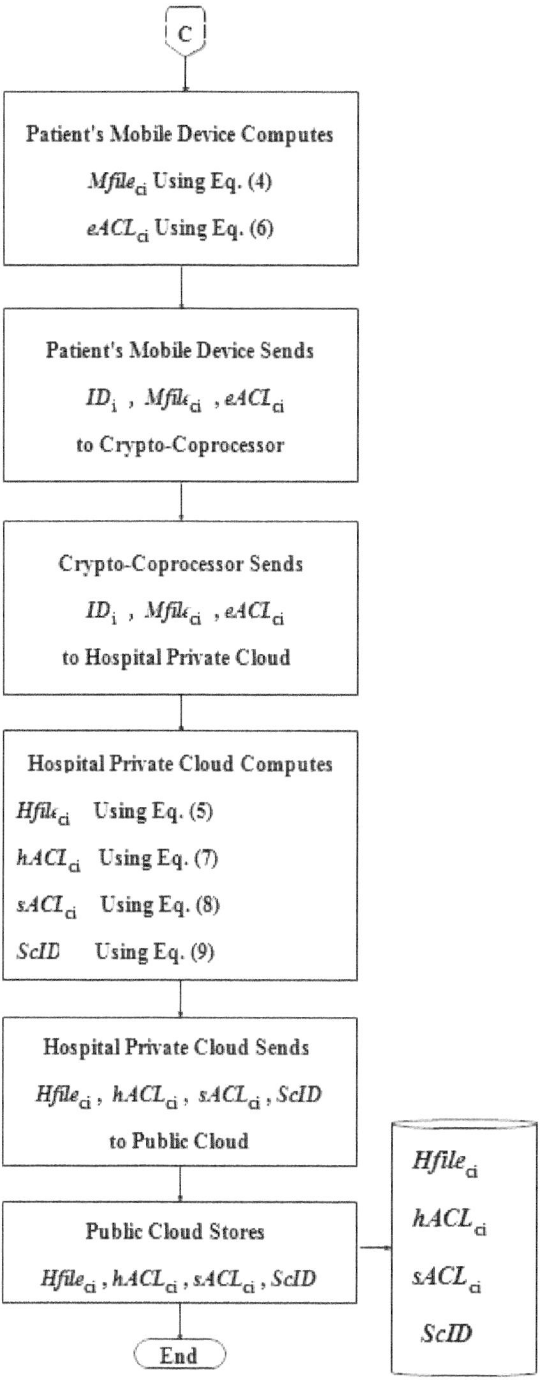

Fig. 7 (*continued*)

Fig. 8 shows a block diagram for the interaction between a patient's mobile device, its trusted crypto-coprocessor, public and private cloud servers for creating an electronic medical file for that patient.

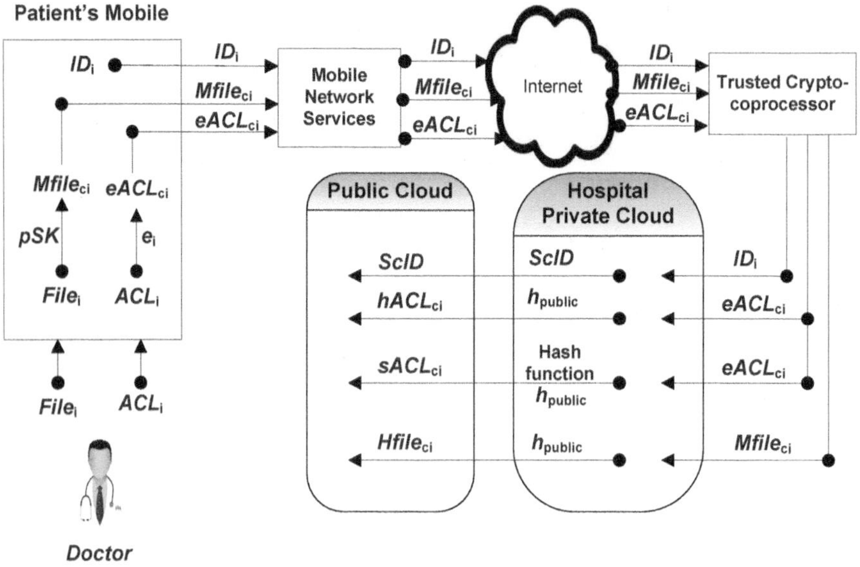

Fig. 8 Interaction between the patient's mobile device, its trusted crypto-coprocessor, public and private cloud servers for mHealth medical file creation

Phase 2: mHealth Medical File Access

We have three access situations:

1) Same hospital access situation: a doctor in a hospital wants to access a patient electronic health file $file_i$ that was created by the same hospital using the patient's mobile device.
2) Cross hospital access situation: a doctor from another hospital wants to access the medical file of a patient that is owned by another hospital. This can be done after being granted permission from the owner hospital.
3) Emergency access situation: the requesting emergency center can bypass the patient's permission to decrypt the medical file in emergency cases.

Algorithm 2: Same Hospital Access Situation

This algorithm is used when a patient from a hospital wants to access his medical file that was created and owned by the same hospital in order to be investigated by the doctor. First of all, the patient's mobile device must be authenticated by its associated trusted crypto-coprocessor in order to access the hospital's private cloud. If authentication succeeded the RBAC method validates the identity of the patient and checks his authorized role. If validation and authorization are both succeeded the mHealth application installed on the patient's mobile device invokes the token generator to generate a "*token*". The "*token*" and patient's ID (ID_i) are both sent to the trusted crypto-coprocessor which sends them to the hospital's private cloud. The hospital's private cloud generates a signature $ScID$ and sends the "*token*" and $ScID$ to the public cloud. $ScID$ is stored in the public cloud for auditing purposes while the "*token*" is used to retrieve the double encrypted patient's medical file $Hfile_{ci}$, double encrypted corresponding license $hACL_{ci}$ and the hospital signature $sACL_{ci}$ from the public cloud. The public cloud sends $Hfile_{ci}$, $hACL_{ci}$ and $sACL_{ci}$ to the hospital's private cloud which uses the

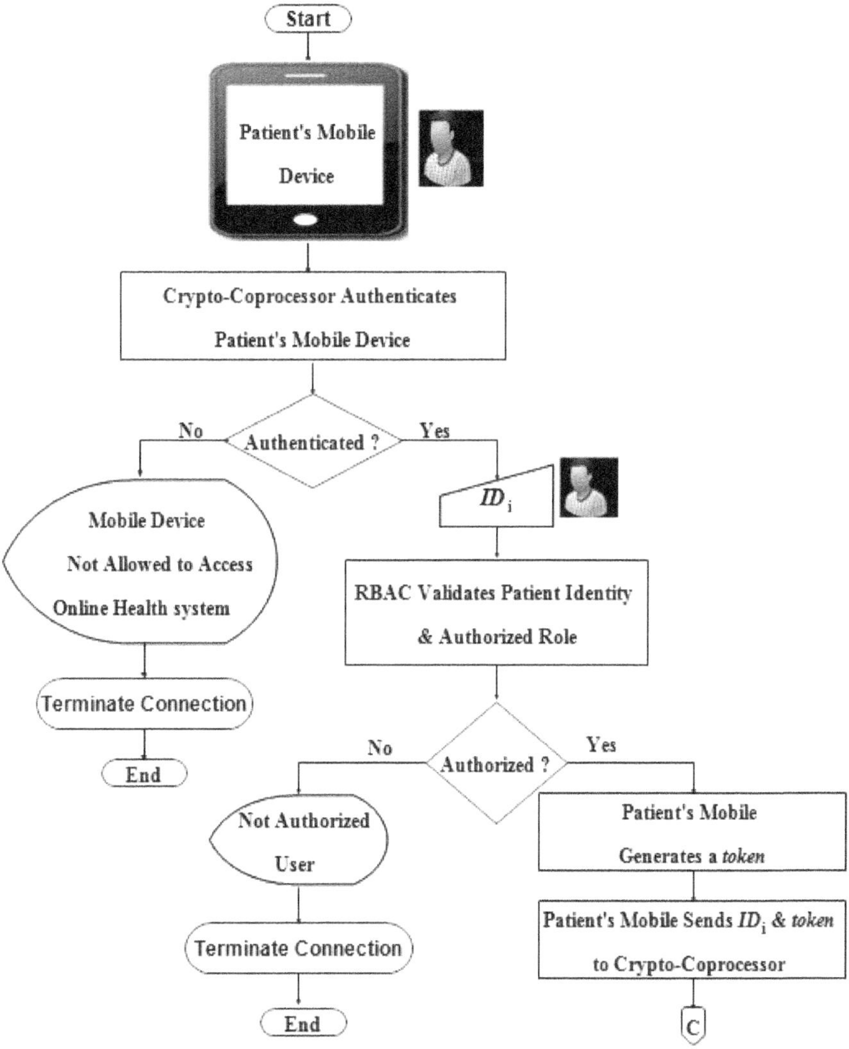

Fig. 9 Retrieval of a medical file which was created and owned by the same hospital

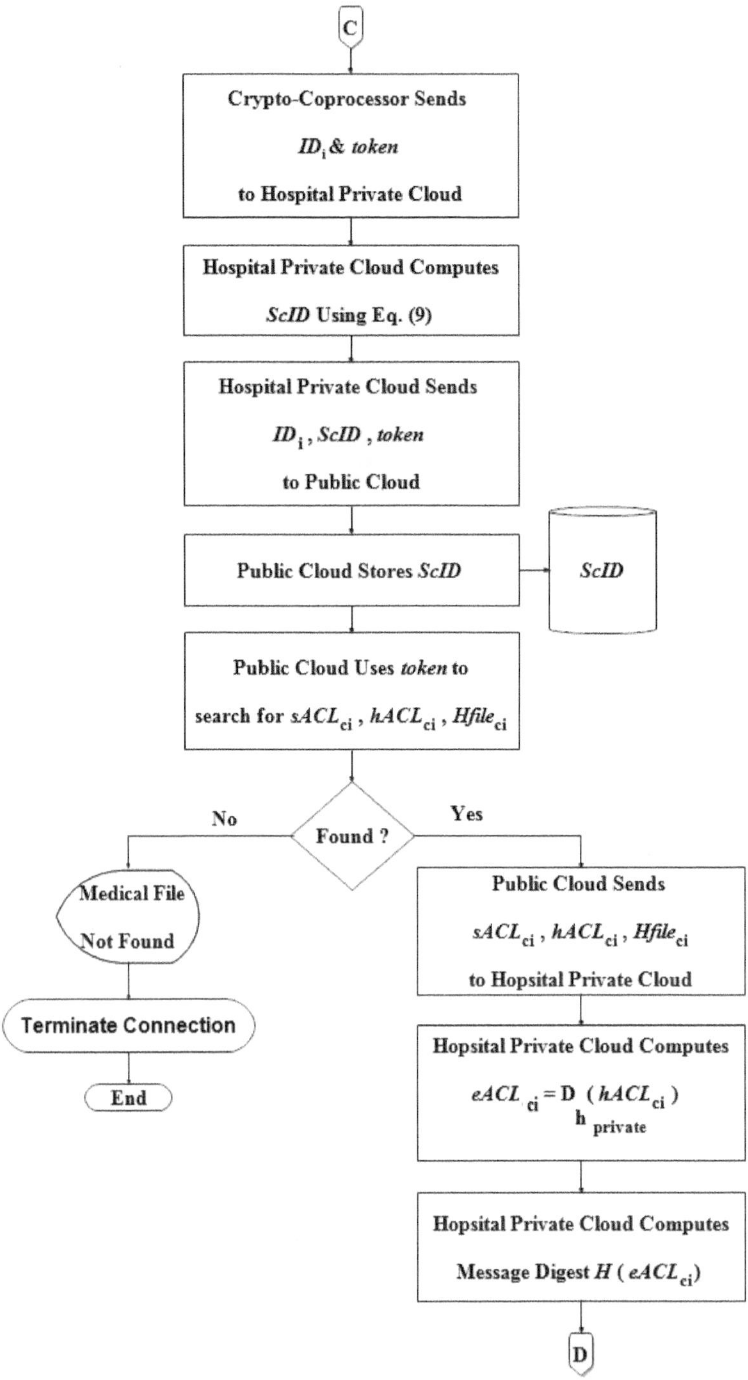

Fig. 9 (*continued*)

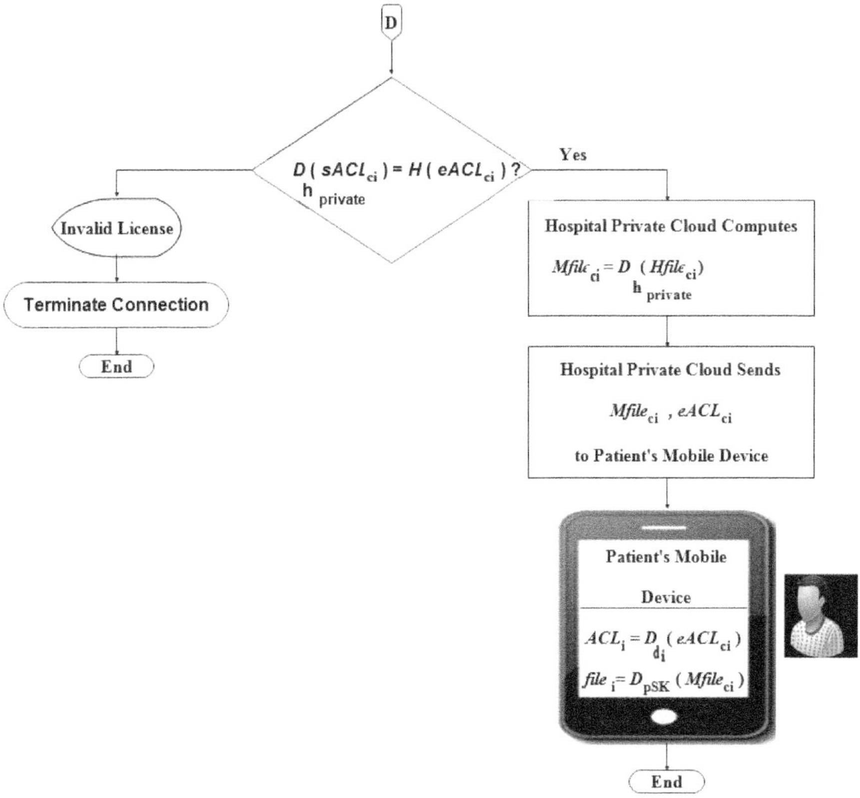

Fig. 9 (*continued*)

hospital's private key h_{private} to restore both the encrypted medical file $Mfile_{\text{ci}}$ and the corresponding encrypted license $eACL_{\text{ci}}$. The hospital's private cloud verifies the signature $sACL_{\text{ci}}$ which if verified both $eACL_{\text{ci}}$ and $Mfile_{\text{ci}}$ are sent to the patient's mobile device where the patient uses his private key (d_{i}) to decrypt $eACL_{\text{ci}}$ and restores the corresponding license ACL_{i} and uses his secret key (pSK) to restore his medical file $file_{\text{i}}$. The doctor can read the restored patient's medical file $file_{\text{i}}$ and the corresponding license ACL_{i} on the patient's mobile device. Fig. 9 shows the flowchart for retrieving a patient's medical file which was created and owned by the same hospital.

Fig. 10 shows the interaction between the mobile device, trusted crypto-coprocessor, public and private cloud servers to retrieve a medical file created by the hospital.

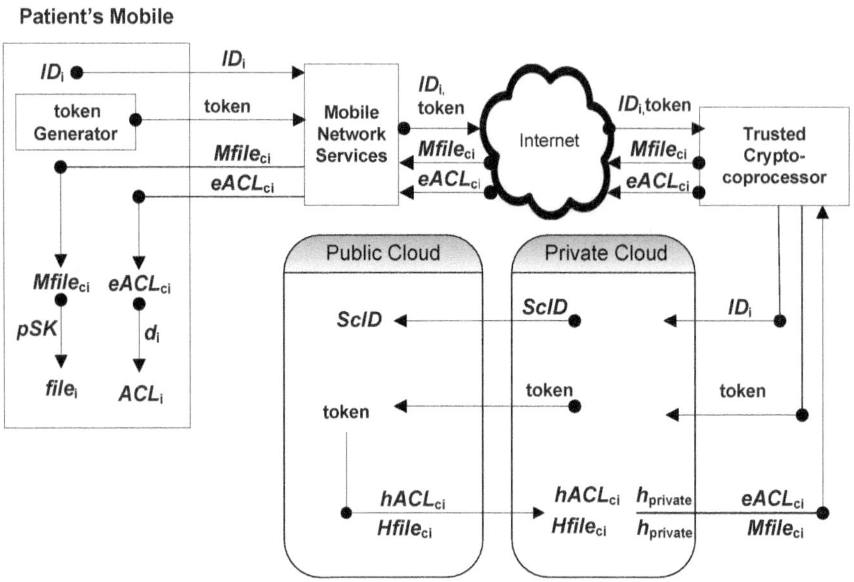

Fig. 10 Interaction between the mobile device, trusted crypto-coprocessor and the cloud server in order to retrieve the medical file of a patient

Algorithm 3: Cross Hospital Access Situation

This algorithm is used when a doctor from hospital "A" wants to access using his mobile device a patient's medical file that was created and owned by hospital "B". First of all, the mobile device at hospital "A" authenticates itself with its associated crypto-coprocessor "A". If authenticated the RBAC of hospital's "A" private cloud validates the identity of the doctor and checks his authorized role. If both validation and authorization are succeeded then the doctor from hospital "A" signs the patient ID_i to generate a doctor's signature $SdOC$ and generates a message Msg to be sent to the required patient. A doctor's signature is generated using equation (10), where the doctor uses his private key (d_{doc}) to encrypt the message digest $H(ID_i)$. The doctor's mobile device at hospital "A" sends patient ID_i , Msg, $SdOC$ and the doctor's public key (e_{doc}) to its associated trusted crypto-coprocessor "A" which in turn sends them to hospital's "A" private cloud where the patient ID_i is used to generate $ScID$ signature. The hospital's "A" private cloud sends $ScID$, ID_i , Msg , $SdOC$, hospital's "A" public key $h^A{}_{public}$ and e_{doc} to the public cloud. The public cloud searches for the hospital that owns the double encrypted medical file and double encrypted corresponding license of the required patient. We assume it to be hospital "B". The public cloud stores $ScID$ for auditing purposes and sends $h^A{}_{public}$, e_{doc} , $SdOC$, Msg and ID_i to hospital's "B" private cloud. The private cloud of hospital "B" verifies doctor's signature using ID_i and $SdOC$. To verify a doctor's signature, the private cloud of hospital "B" uses the doctor's public key e_{doc} to restore $H(ID_i)$ using $SdOC$ and uses the same cryptographic hash function H to compute the message digest of the received ID_i. If both messages digests matched then the doctor's signature is verified because we know that the doctor at hospital "A" is the only one who owns the private key d_{doc}.

The RBAC of hospital's "B" private cloud validates the patient's identity and checks his authorized role, which if validated and authorized the private cloud of hospital "B" sends both e_{doc} and Msg to the patient's mobile device. If the required patient from hospital "B" agrees to share his medical file with the doctor from hospital "A" then the mHealth application embedded in the patient's mobile device invokes the token generator to generate a token "*token*". This "*token*" is sent to hospital's "B" private cloud which in turn sends it to the public cloud in order to retrieve the double encrypted patient's medical file $Hfile_{ci}$ and the corresponding double encrypted license $hACL_{ci}$. Hospital's "B" private cloud verifies the signature $sACL_{ci}$ which if verified it uses the hospital "B" private key $h^{B}{}_{private}$ to decrypt both $Hfile_{ci}$ and $hACL_{ci}$ in order to restore $Mfile_{ci}$ and $eACL_{ci}$. Those files are both sent to the patient's mobile device at hospital "B" where the patient uses his secret key (pSK) to decrypt $Mfile_{ci}$ in order to restore his medical file $file_i$ and uses his private key (d_i) to decrypt $eACL_{ci}$ in order to restore the corresponding license ACL_i. Finally, the patient uses the doctor's public key (e_{doc}) to encrypt both the medical file $file_i$ and the corresponding license ACL_i to generate $efile_{di}$ and $eACL_{di}$ respectively.

The following set of equations is used:

$$SdOC = E_{d_{doc}}(H(ID_i)) \tag{10}$$

$$Mfile_{ci} = D_{h^{B}{}_{private}}(Hfile_{ci}) \tag{11}$$

$$eACL_{ci} = D_{h^{B}{}_{private}}(hACL_{ci}) \tag{12}$$

$$file_i = D_{pSK}(Mfile_{ci}) \tag{13}$$

$$ACL_i = D_{d_i}(eACL_{ci}) \tag{14}$$

$$efile_{di} = E_{e_{doc}}(file_i) \tag{15}$$

$$eACL_{di} = E_{e_{doc}}(ACL_i) \tag{16}$$

The patient at hospital B sends both $efile_{di}$ and $eACL_{di}$ to hospital's "B" private cloud using the embedded mHealth application in the mobile device. Hospital's "B" private cloud double encrypt $efile_{di}$ and $eACL_{di}$ using hospital's "A" public key $h^{A}{}_{public}$ in order to generate $Hfile_{di}$ and $hACL_{di}$ respectively.

The following set of equations is used:

$$Hfile_{di} = E_{h^{A}{}_{public}}(efile_{di}) \tag{17}$$

$$hACL_{di} = E_{h^{A}{}_{public}}(eACL_{di}) \tag{18}$$

Hospital's "B" private cloud sends both $Hfile_{di}$ and $hACL_{di}$ to the public cloud which in turn sends them to hospital's "A" private cloud. Hospital's "A" private cloud uses its private key $h^{A}{}_{private}$ to decrypt both $Hfile_{di}$ and $hACL_{di}$ in order to restore $efile_{di}$ and $eACL_{di}$ respectively.

Then hospital's "A" private cloud sends $efile_{\mathrm{di}}$ and $eACL_{\mathrm{di}}$ to the doctor's mobile device at hospital "A". The doctor uses his private key (d_{doc}) to decrypt $efile_{\mathrm{di}}$ and $eACL_{\mathrm{di}}$ and restore the patient's medical file $file_{\mathrm{i}}$ and the corresponding license ACL_{i}.

The following set of equations is used:

$$efile_{\mathrm{di}} = D_{\mathrm{h}^{\mathrm{A}}_{\mathrm{private}}}(Hfile_{\mathrm{di}}) \tag{19}$$

$$eACL_{\mathrm{di}} = D_{\mathrm{h}^{\mathrm{A}}_{\mathrm{private}}}(hACL_{\mathrm{di}}) \tag{20}$$

$$file_{\mathrm{i}} = D_{\mathrm{d_{doc}}}(efile_{\mathrm{di}}) \tag{21}$$

$$ACL_{\mathrm{i}} = D_{\mathrm{d_{doc}}}(eACL_{\mathrm{di}}) \tag{22}$$

Fig. 11 shows the flowchart for a request issued by a doctor at hospital "A" to restore the medical file of a patient that was created and owned by hospital "B".

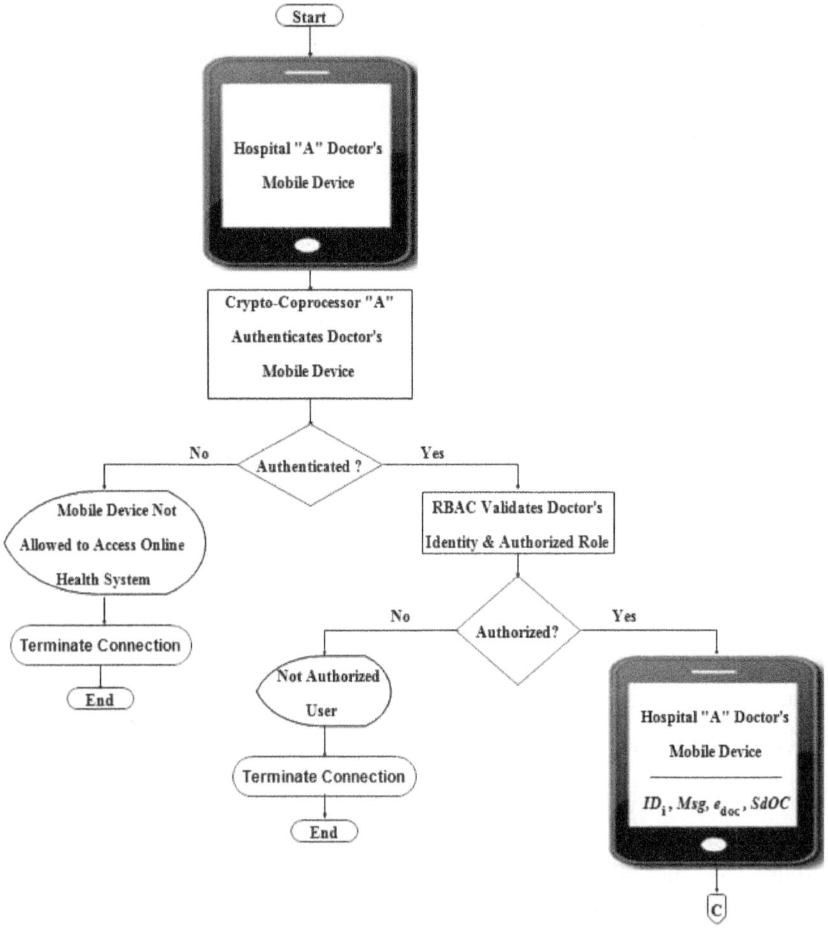

Fig. 11 Retrieval of a medical file that was created and owned by hospital "B" by a doctor from hospital "A"

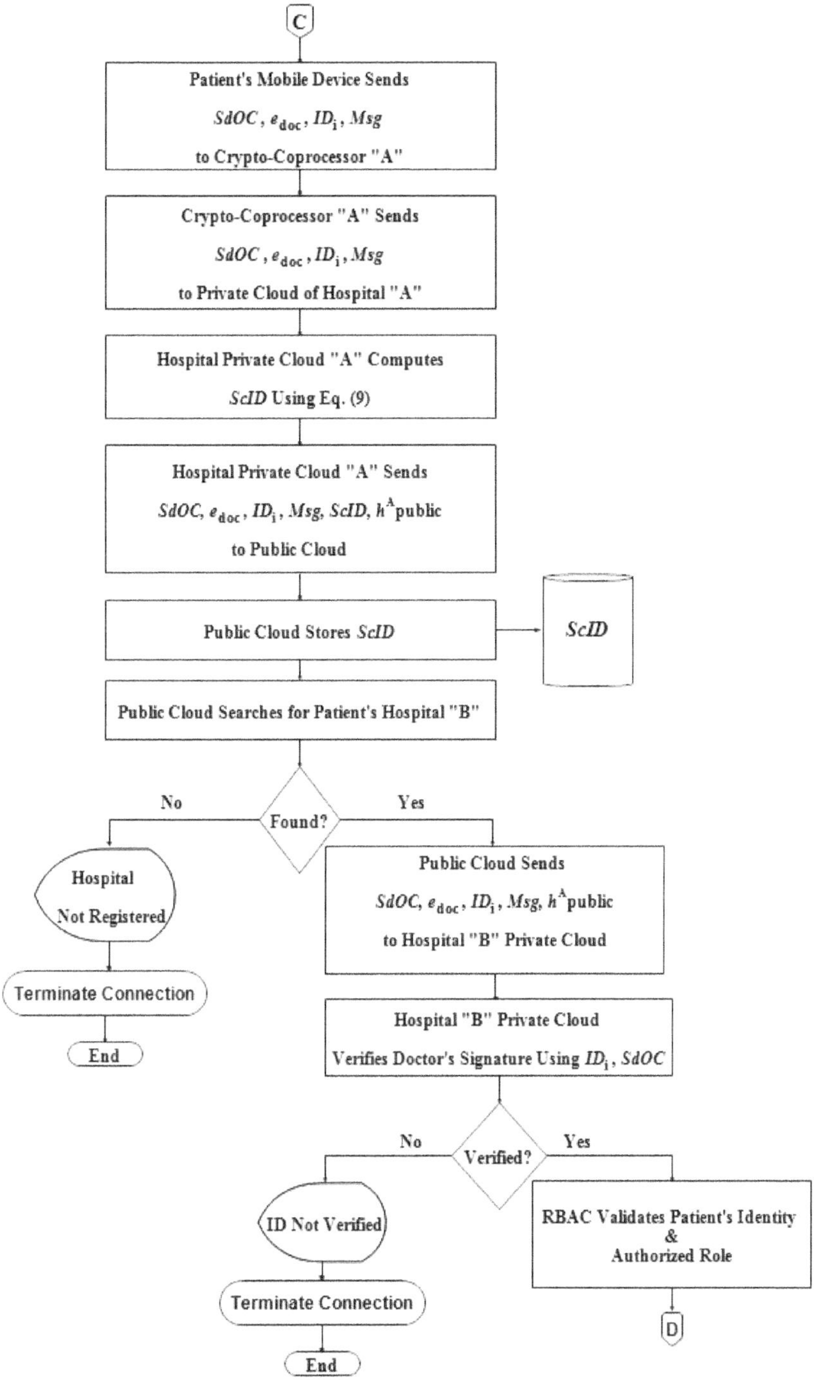

Fig. 11 (*continued*)

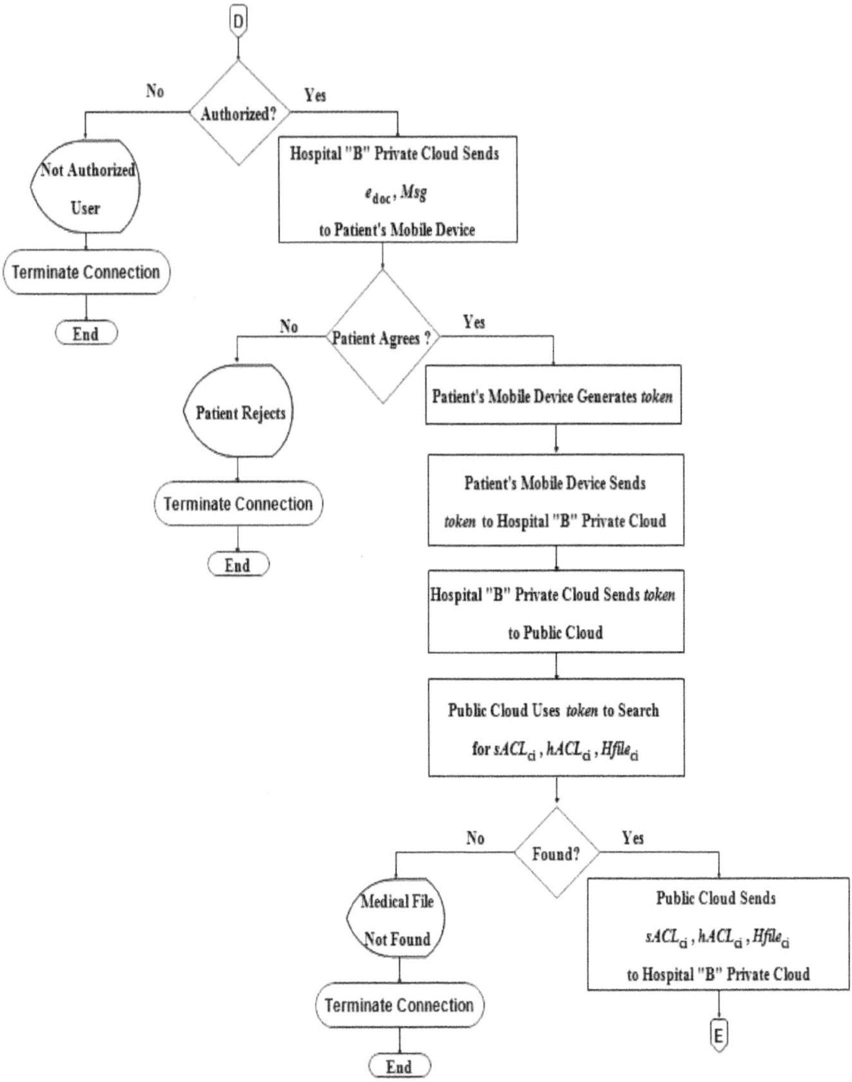

Fig. 11 (*continued*)

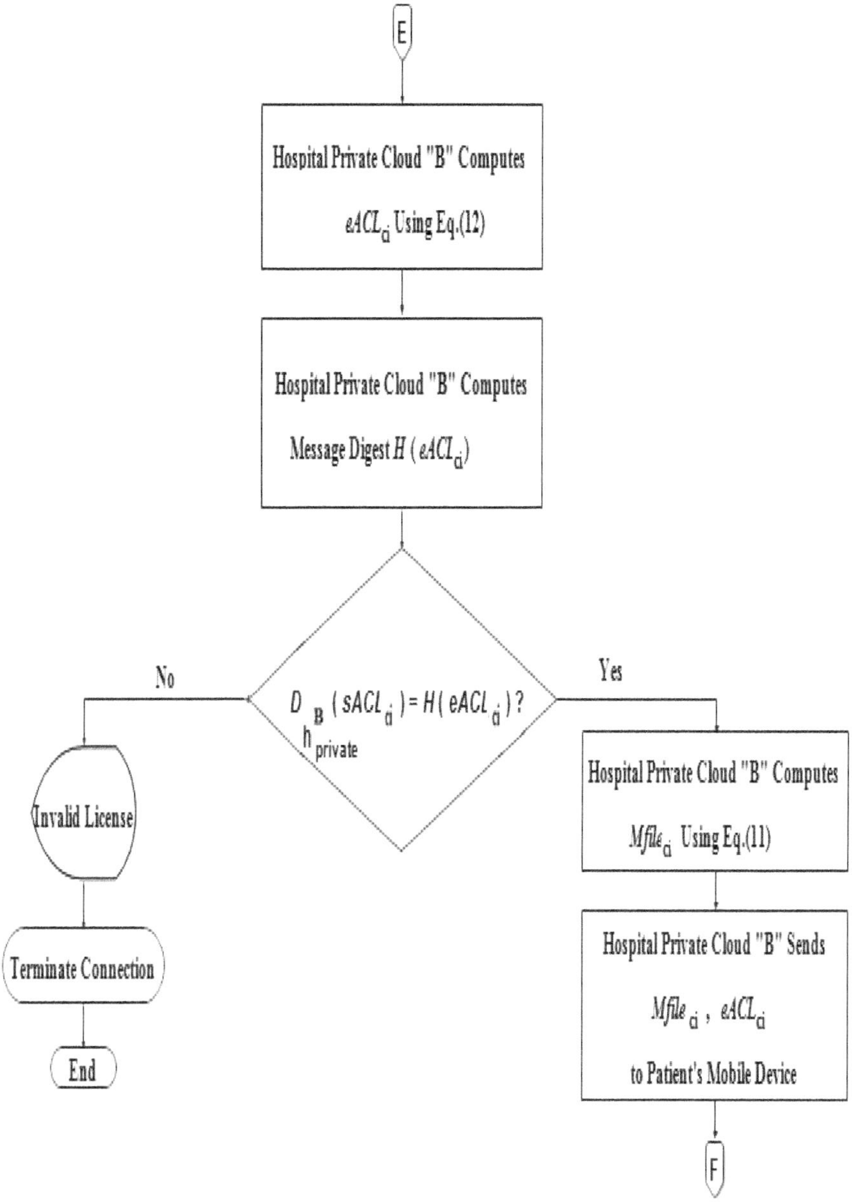

Fig. 11 (*continued*)

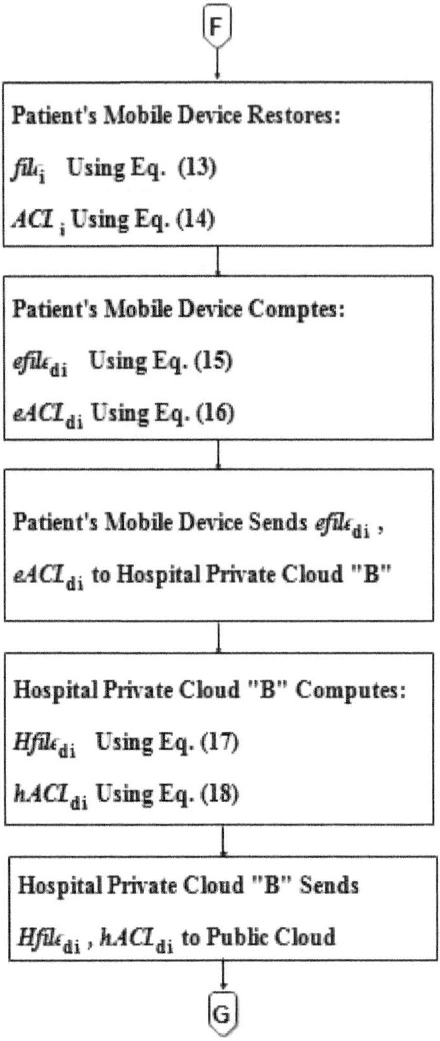

Fig. 11 (*continued*)

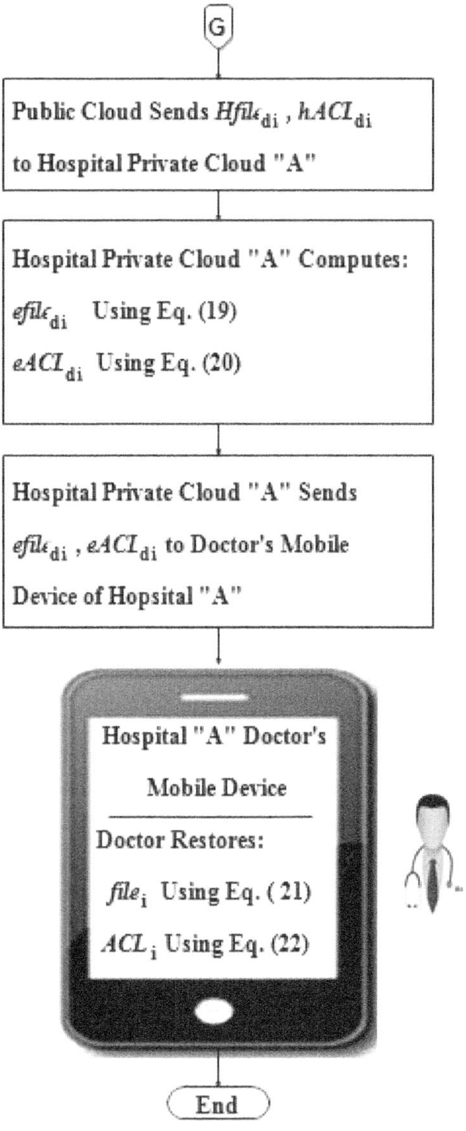

Fig. 11 (*continued*)

Fig. 12 shows the interaction between the mobile device, trusted crypto-coprocessor and the cloud server to retrieve a medical record for a patient that is owned by another hospital

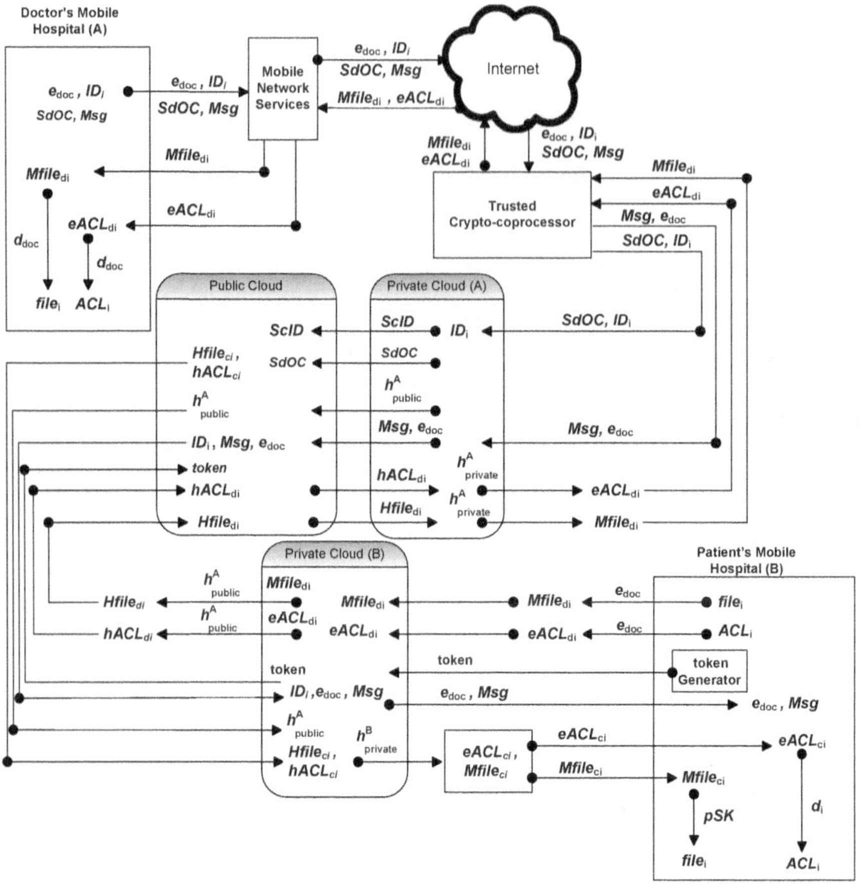

Fig. 12 Interaction between the mobile device, trusted crypto-coprocessor and the cloud server to retrieve a medical file that was created and owned by another hospital

Algorithm 4: Emergency Hospital Access Situation

If a patient in the emergency center can use his mobile device algorithm 2 can be used to retrieve the patient's medical file from the hospital owns this file for investigation by the emergency doctor. If the patient could not use his mobile device the doctor in the emergency center can use this algorithm to access the medical file of the patient which was created and kept by a hospital. First of all, the doctor's mobile device at the emergency center authenticates itself with its associated crypto-coprocessor. If mobile device is authenticated the RBAC at the private cloud of emergency center validates the identity of the emergency doctor and checks his authorized role. If both validation and authorization are succeeded the emergency doctor signs the patient ID_{i} to generate doctor's signature $SdOC$. The mobile device at the emergency center sends the emergency center public key em_{public} , $SdOC$ and ID_{i} to the PKG. The PKG validates doctor's signature which if validated the PKG uses its master private key to generate (d_{i}) which is the private key for the patient's ID_i. PKG encrypts the generated patient's private key (d_i) with em_{public} as follows:

$$d_{emi} = E_{em_{public}}(d_i) \tag{23}$$

PKG sends d_{emi} to the emergency center private cloud which decrypts the d_{emi} using the emergency center private key $em_{private}$ in order to restore the patient's private key (d_i) and sends it to the emergency center as follows:

$$d_i = D_{em_{private}}(d_{emi}) \tag{24}$$

The private cloud of the emergency center sends a query that contains its public key em_{public} and patient ID_i to the public cloud in order to restore $file_i$ and ACL_i for the patient in emergency. Firstly, the public cloud searches for the patient's hospital which if found the public cloud sends em_{public} and patient ID_i to the patient's hospital private cloud. The patient's hospital private cloud RBAC method validates the identity ID_i of the patient which if validated the hospital's private cloud sends ID_i and em_{public} to the patient's hospital which invokes the token generator to generate *token*. The patient's hospital sends the *token* to the private cloud which consequently sends it to the public cloud. The public cloud uses the *token* to find $HFile_{ci}$ and $hACL_{ci}$ for the patient and sends those files to the patient's hospital private cloud. The patient's hospital private cloud uses its private key $h_{private}$ to decrypt $HFile_{ci}$ and $hACL_{ci}$ as follows:

$$Mfile_{ci} = D_{h_{private}}(Hfile_{ci}) \tag{25}$$

$$eACL_{ci} = D_{h_{private}}(hACL_{ci}) \tag{26}$$

The patient's hospital private cloud sends $Mfile_{ci}$ and $eACL_{ci}$ to the patient's hospital which encrypts them using the emergency center public key em_{public} as follows:

$$Efile_{em} = E_{em_{public}}(Mfile_{ci}) \tag{27}$$

$$EACL_{em} = E_{em_{public}}(eACL_{ci}) \tag{28}$$

The patient's hospital private cloud sends $Efile_{em}$ and $EACL_{em}$ to the public cloud which sends them to the emergency center's private cloud. The emergency center's private cloud uses its private key $em_{private}$ to decrypt $Efile_{em}$ and $EACL_{em}$ as follows:

$$Mfile_{ci} = D_{em_{private}}(Efile_{em}) \tag{29}$$

$$eACL_{ci} = D_{em_{private}}(EACL_{em}) \tag{30}$$

The emergency center private cloud sends $Mfile_{ci}$ and $eACL_{ci}$ to the emergency center which uses the recovered patient's private key (d_i) to decrypt the $eACL_{ci}$ as follows:

$$ACL_i = D_{d_i}(eACL_{ci}) \tag{31}$$

The emergency center extracts the patient's secret key (pSK) from the decrypted corresponding license ACL_{i} in order to decrypt the encrypted medical record $MFile_{\mathrm{ci}}$ as follows:

$$file_{\mathrm{i}} = D_{\mathrm{pSK}}\,(MFile_{\mathrm{ci}}) \tag{32}$$

Fig.13 shows the flowchart for a doctor at an emergency center retrieving the medical file that was created and owned by a hospital.

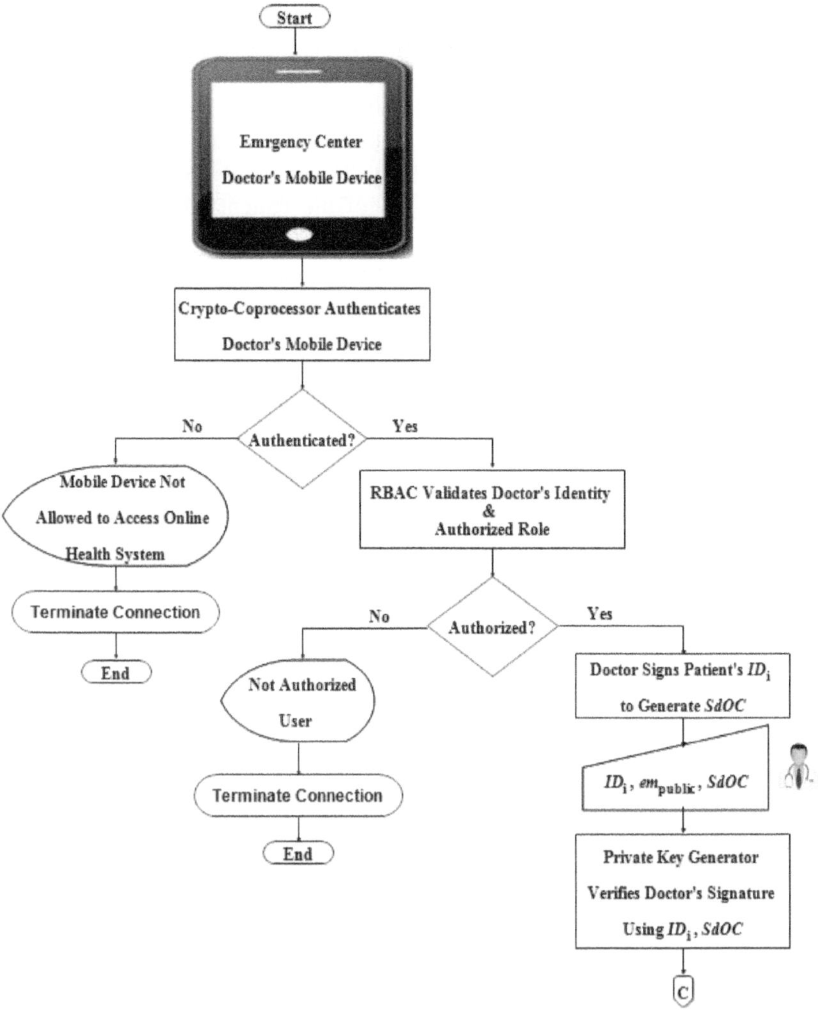

Fig. 13 Retrieval of a medical file that was created and owned by a hospital by a doctor from an emergency center

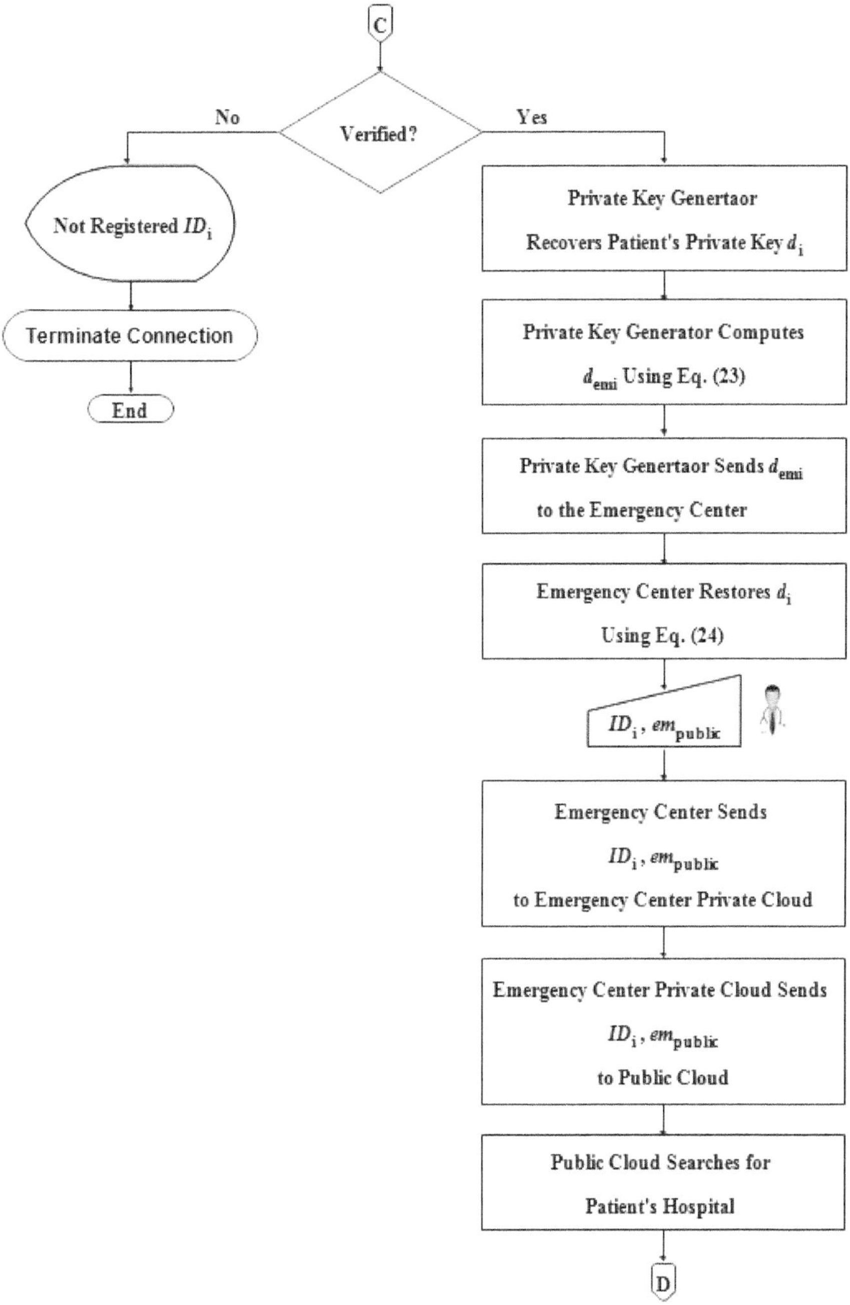

Fig. 13 (*continued*)

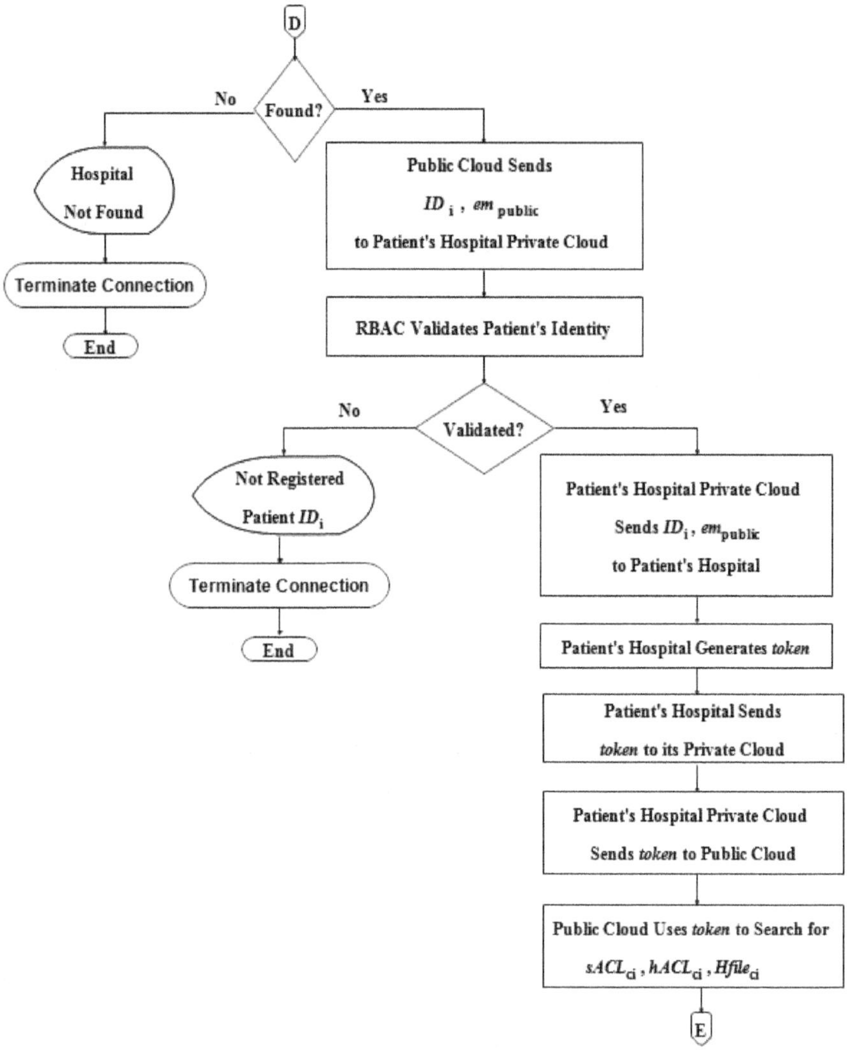

Fig. 13 (*continued*)

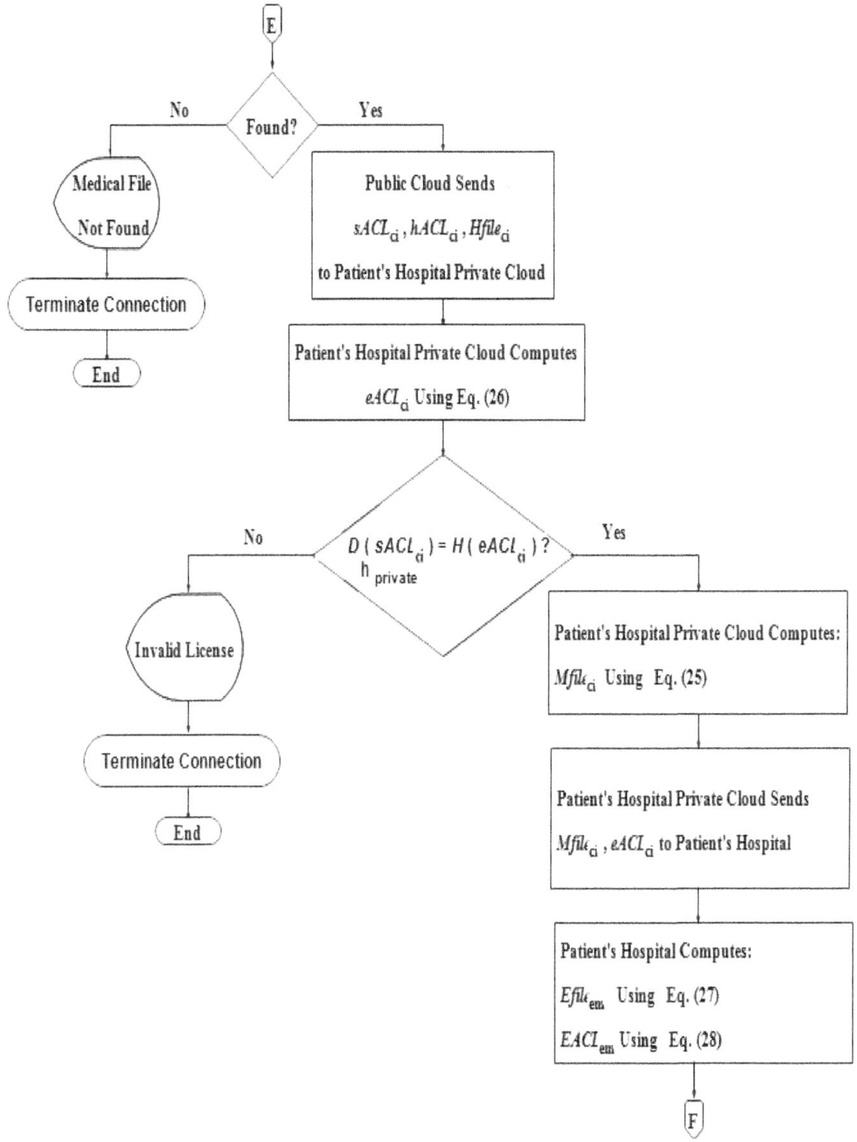

Fig. 13 (*continued*)

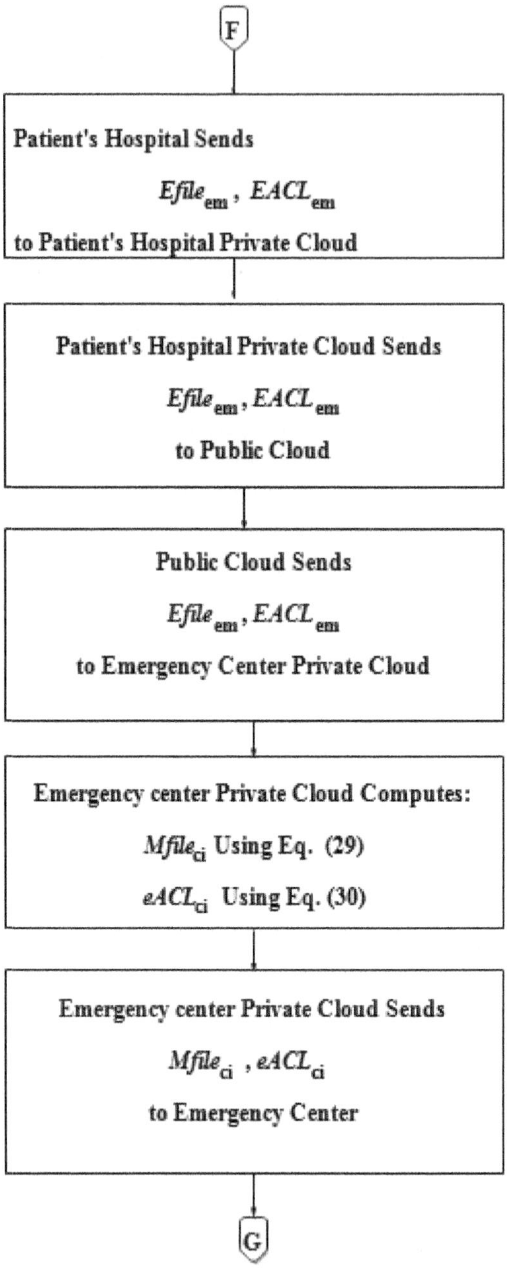

Fig. 13 (*continued*)

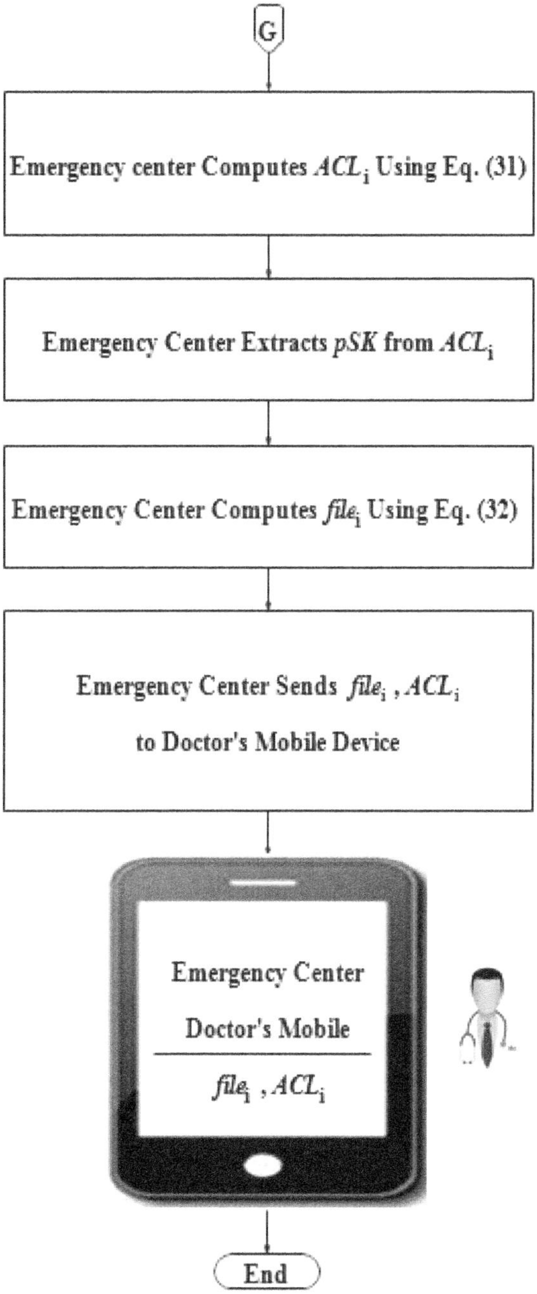

Fig. 13 (*continued*)

Fig. 14 shows the interaction between the emergency center, emergency center private cloud, owner hospital private cloud.

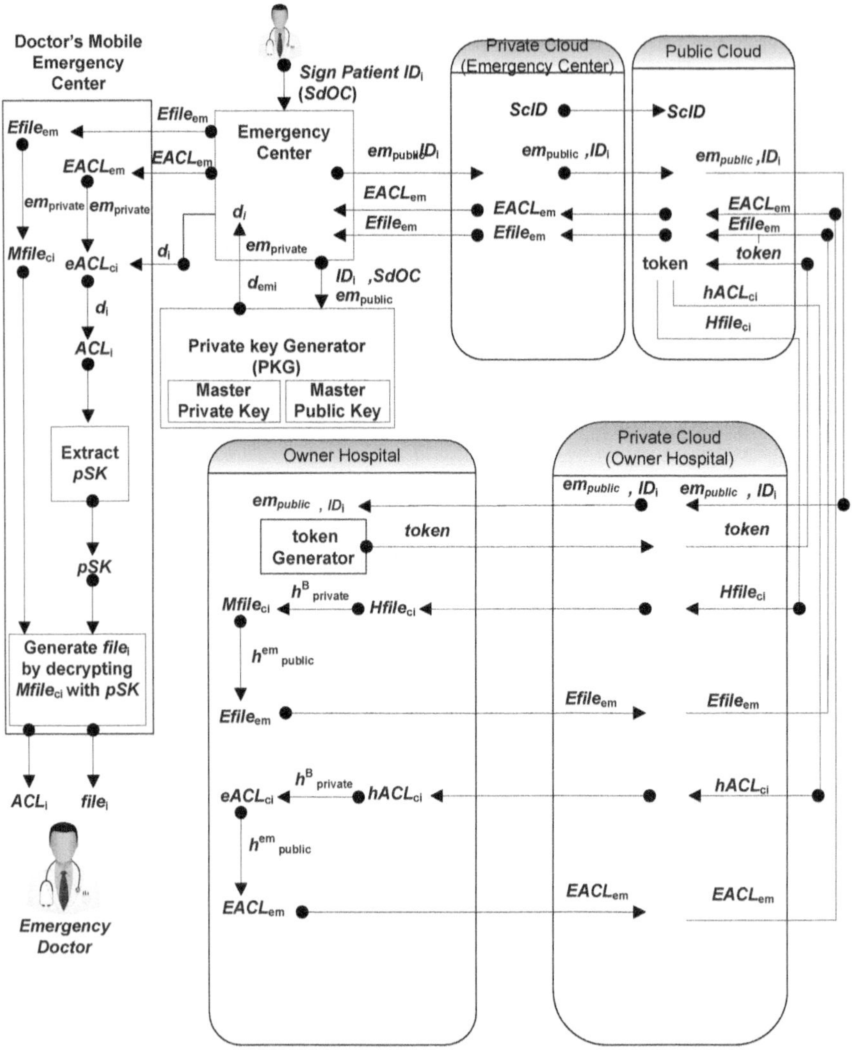

Fig. 14 The interaction between the emergency center, emergency center private cloud, owner hospital private cloud

7 Case Studies

7.1 Case Study: 1

PROTECTING THE SECURITY AND PRIVACY OF MOBILE DEVICES
Many health and medical applications use smart phones to provide their services to their clients. However, a growing number of mobile devices are dedicated to provide personal health and fitness. Many specialized devices provide clients with biometric data such as heart rate, blood pressure or blood glucose level. Keeping this valuable information protected provide people with big picture of their health status. Cairo a very crowded city and home to 15 million people is also the home for Hossein age 52. Hossein is a very busy man that he has neglected to take care of himself and adopted some unhealthy habits, he rarely find the time to exercise and always making unhealthy food choices. Upon his annual check Hossein's doctor warned him that he is on the verge of congestive heart failure and he is also diagnosed of type 2 diabetes and sleep apnea which are two chronic diseases. To monitor Hossein's health status the doctor suggested to use mobile device to monitor his weight scale and another mobile device to measure his blood glucose. The doctor advised Hossein to download a mobile health app to track his exercise, weight, calories and biometrics. These mobile devices provide the doctor with Hossein's health score and the doctor can provide Hossein with the recommended action items on how to improve his health score.

Answer the following questions:

1) What type of information can be stolen from Hossein's mobile devices?
2) What types of security threats Hossein's mobile device are vulnerable to?
3) How can these devices automatically track Hossein's location?

7.2 Case Study: 2

MOBILE COMPUTING: A KEY PLAYER FOR FUTURE HEALTH CARE
Faced with strong competition by health organizations, health care givers are looking for new methods to provide better medical services to their patients with lower operational costs. They found mobile technology is the key to speed up, improve medical services, keep monitoring of their patients' health score and provide medical advices anytime and anywhere. Examples of mobile devices for monitoring biometric measures are: A&D Medical wireless weight scale, MyGlucoHealth wireless blood glucose meter and Macaw^{TM} mobile health app. developed by U.S. Preventive Medicine in collaboration with Qualcomm life to track exercise, weight, calories and biometrics. Elderly people can wear 3G-enabled Lifecomm device which is mobile personal emergency response device that tracks old people whereabouts and their activities. In case of emergency this device allows old people to call for emergency personnel. mHealth devices are smart enough to alert the patients against critical situations such as environments

that may trigger allergy attack and can give real-time advice to them about treatments via mobile health applications.

Answer the following questions:

1) Where valuable data obtained from mobile devices should be stored?
2) How information obtained from different mHealth devices is used to give a big picture about the patient health status?
3) How can we prevent loss of valuable data?

7.3 Case Study: 3

ELECTRONIC HEALTH RECORD STANDARDS

Systems keeping medical information of their patients on paper records have difficulties in getting comprehensible data on a band of patients from these paper records. Compiling data from a large number of paper records is cumbersome and susceptible to significant errors. Also, sharing medical data on paper records between hospitals is time-consuming and expensive to copy these records and send them by post mail, fax or personally which may result in vital information not be shared or late arrived. The administration of G.W. Bush has laid the ground for US use of electronic health systems and by the time of Obama administration a Healthcare IT Standards Panel had issued a set of technical standards to ensure that EHR systems are strong, secure and can communicate with each other. If two systems use common standards they can communicate with each other and can share information with no transformation. On the other hand, healthcare systems using no standards, different standards, partial standards or even their own standards will face communication problems for sharing medical information. Technical standards also include encryption and decryption standards in order to guarantee data confidentiality, integrity and privacy when medical records are sent over the Internet to other hospitals or EHR systems. Joining the network of EHR systems require an EHR system to be certified before being used which requires testing through an authorized testing body such National Institute of Standards and Technology (NIST). Testing and certifying a complete vendor-supplied EHR system will cost $33,000 which is considered an extra cost that will be passed to hospitals and providers who seek to adopt the new system.

Source: Health Policy Briefs: http://www.healthaffairs.org/healthpolicybriefs/brief.php?brief_id=26, visited on 11/07/2013.

Answer the following questions:

1) What standards and regulations do mHealth systems satisfy?
2) What benefits the US government can gain if standards-compliant EHR systems are in widespread use?
3) How do you expect standards for EHR systems will affect mHealth systems market to grow? Explain.

8 Future Directions

The technology of cloud computing will grow in the future as more companies and businesses grow and their needs for a bigger storage to store, processing and management their valuable information with a lower cost. Scalability of cloud computing motivates businesses and organizations migration to cloud computing which will continue in the future. Users can access the requested information and manipulate it anytime and from anywhere. The trust in cloud computing will also be improving as stricter techniques in information security, user identification and access control are expected to be improved. As a result more organizations will be moving in to cloud computing in running their businesses and transactions. Mobile applications that use cloud computing will grow exponentially in the future as developers are expecting a more active participation from various industries. Community clouds for every industry are also expected. Mobile apps will increasingly live in the cloud as more mobile applications will employ the computing resources and storage capacity of the cloud. Using of mobile devices such as smart phones, portable PCs, cellular phones and personal digital assistants (PDAs) will be doubled in the next decade which will make the role of mHealth to improve patient safety and the quality of healthcare critical. In the future, high-speed communication networks will be closer to reality in remote rural areas in developing countries which will help patients in these areas to use mHealth apps that rely on high speed communication capabilities to deliver high-quality, timely voice, video and data in a multitude of formats. However, we expect that compatibility difficulties between telecommunication services and devices will continue to be challenging for mHealth applications. mHealth will continue gaining popularity in the future regardless of the many concerns of healthcare CIOs that see the combination of mobile computing and cloud computing is an environment with many problems and vulnerabilities. Attempts will continue in the future to enhance the security of medical information in cloud storage and the security and reliability of delivered information from a variety of resources. Cloud providers and healthcare givers must formally designate human resource security to work with a team of health information technology experts to continuously evaluate the security and privacy measures applied, identify security weaknesses and threats and assign a risk or likelihood of security concerns in the organization and address them. Creating useful mHealth record systems will require the healthcare expertise, information management and cloud technology professionals, ethicists, administrative personnel, and patients. The collaboration of healthcare expertise, medical information, cloud and mobile computing will improve the quality of mHealth care. Future mHealth applications must focus on medical records standardization to facilitate the integration between different healthcare databases in a seamless way. This enables the storage, updating, search and retrieval of useful information. The full collaboration of mHealth applications requires a secured and reliable collaboration between wireless technologies such as Bluetooth, WiFi, GSM and satellite communications, wired communications and the Internet. This will make the health service delivery to be much more decentralized and much more widely distributed. In the future, hospitals will expand the mHealth system to have all patients monitored 24/7. Benjamin Kanter, Palomar Health Chief Medical Information Officer said: "We absolutely, fundamentally believe that every single hospital in the United States is going to have this kind of technology in place over the next 10 years. I don't think you can afford to be a patient in the hospital and not have monitoring."

9 Conclusions

In this chapter, we discussed the cryptographic primitives that can be used to protect data privacy, confidentiality and integrity in the cloud storage. We discussed the role based access control method (RBAC) that is used to authenticate and authorize cloud users. We also discussed the technology of the secured cloud storage which allows end users to securely access cloud storage. Finally, we presented a secured mHealth system that is based on hybrid cloud architecture and uses some of the above mentioned techniques. The proposed system uses crypto-processors to authenticate mobile devices which want to access the cloud servers and uses the identity based encryption to generate the patient's secret key, public key and private key. The secured storage technology is used to enable end users to securely access the requested information from cloud servers using generated tokens. Three modes for using the mHealth system were discussed: the first mode is the same hospital access which when a doctor from the same hospital that owns the patient's medical records wants to access the patient's data. The second mode is the cross hospital access which when a doctor from a hospital wants to access the medical record of a patient that is owned by another hospital. The third mode is in case of emergency which when a patient is entered to an emergency center and the doctor wants to access the patient's medical record that is owned by a hospital.

References

1. Khana, A.N., Mat Kiah, M.L., Khanb, S., Madanic, S.: Towards secure mobile cloud computing: A survey. Journal Future Generation Computer Systems 29, 1278–1299 (2013)
2. Itani, W., Kayssi, A., Chehab, A.: Privacy as a Service: Privacy-Aware Data Storage and Processing on cloud Computing Architectures. In: Eighth IEEE International Conference on Dependable, Autonomic and Secure Computing, pp. 711–716. IEEE Press, Chengdu (2009)
3. Wang, C., Ren, K., Lou, W., Li, J.: Toward Publically Auditable Secure Cloud Data Storage Services. IEEE Network 24(4), 19–24 (2010)
4. Al-Muhtadi, J., Hill, R., Al-Rwais, S.: Access control using threshold cryptography for ubiquitous computing environments. Journal of King Saud University – Computer and Information Sciences 23, 71–78 (2011)

5. Kaur, S.: Cryptography and Encryption on cloud Computing. VSRD-International Journal of Computer Science & Information Technology 2(3), 242–249 (2012)
6. Boneh, D., Franklin, M.: Identity-Based Encryption from the Weil Pairing. SIAM J. of Computing 32(3), 586–615 (2003)
7. Tebaa, M., El Hajji, S., El Ghazi, A.: Homomorphic Encryption Applied to the Cloud Computing Security. In: World Congress on Engineering, London, U.K, vol. I (2012)
8. Agudo, I., Nuñez, D., Giammatteo, G., Rizomiliotis, P., Lambrinoudakis, C.: Cryptography goes to the cloud. In: Lee, C., Seigneur, J.-M., Park, J.J., Wagner, R.R. (eds.) STA 2011 Workshops. CCIS, vol. 187, pp. 190–197. Springer, Heidelberg (2011)
9. Song, D., Wagner, D., Perrig, A.: Practical techniques for searching on encrypted data. In: IEEE Symposium on Research in Security and Privacy, pp. 44–55. IEEE Computer Society, Berkeley (2000)
10. Boneh, D., Di Crescenzo, G., Ostrovsky, R., Persiano, G.: Public key encryption with keyword search. In: Cachin, C., Camenisch, J.L. (eds.) EUROCRYPT 2004. LNCS, vol. 3027, pp. 506–522. Springer, Heidelberg (2004)
11. Gasser, M., Goldstein, A., Kaufman, C., Lampson, B.: The digital distributed systems security architecture. In: 12th National Computer Security Conference, pp. 305–319. National Institute of Standards and Technology (NIST), National Computer Security Center (NCSC), Baltimore (1989)
12. Sanzgiri, K., Dahill, B., Levine, B.N., Shields, C., Belding-Royer, E.B.: Secure Routing for AdHoc Networks. In: 10th IEEE International Conference on Network Protocols, pp. 78–87. IEEE Press, Paris (2002)
13. Charles, D., Lauter, K.: Cryptographic Hash Functions from Expander Graphs. J. Cryptol. 22, 93–113 (2009)
14. Rivest, R., Shamir, A., Adleman, L.: A Method for Obtaining Digital Signatures and Public-Key Cryptosystems. Communications of the ACM 21(2), 120–126 (1978)
15. Anderson, R., Bond, M., Clulow, J., Skorobogatov, S.: Cryptographic processors-a survey. Proceedings of the IEEE 94, 357–369 (2006)
16. Mather, T., Kumarasuwamy, S., Latif, S.: Cloud Security and Privacy. O'Rielly, California (2009)
17. Ferraiolo, D., Richard, D.: Role-Based Access Control. In: 15th National Computer Security Conference, pp. 554–563. Interaction Design Foundation, Baltimore (1992)
18. Yong, P., Wei, Z., Feng, X., Zhong-hua, D., Yang, G., Dong-qing, C.: Secure Cloud Storage Based on Cryptographic Techniques. The Journal of China Universities of Posts and Telecommunications 19(2), 182–189 (2012)
19. Kamara, S., Lauter, K.: Cryptographic cloud storage. In: Sion, R., Curtmola, R., Dietrich, S., Kiayias, A., Miret, J.M., Sako, K., Sebé, F. (eds.) RLCPS, WECSR, and WLC 2010. LNCS, vol. 6054, pp. 136–149. Springer, Heidelberg (2010)
20. Germanakos, P., Mourlas, C., Samaras, G.: A Mobile Agent Approach for Ubiquitous and Personalized eHealth Information Systems. In: Proceedings of the Workshop on Personalization for e-Health 10th International Conference on User Modeling (UM 2005), pp. 67–70. Edinburgh (2005)
21. Yeratziotis, A., Van Greunen, D., Pottas, D.: A Framework for Evaluating Usable Security: The Case of Online Health Social Networks. In: 6th International Symposium on Human Aspects of Information Security & Assurance, pp. 97–107. Interaction Design Foundation, Greece (2012)

22. Fernández-Alemán, J.L., Señor, I.C., Lozoya, P.Á., Toval, A.: Security and privacy in electronic health records: A systematic literature review. Journal of Biomedical Informatics 46(3), 541–562 (2013)
23. Zhang, R., Liu, L.: Security models and requirements for healthcare application clouds. In: 3rd IEEE International Conference on Cloud Computing, pp. 268–275. IEEE Press, Florida (2010)
24. Linden, H., Kalra, D., Hasman, A., Talmon, J.: Interorganization future proof HER systems-A review of the security and privacy related issues. Int. J. Med. Inform. 78, 141–160 (2009)
25. ISO/EN 13606, `http://www.iso.org/iso/home.htm`
26. Bacon, J., Evans, D., Eyers, D., Migliavacca, M., Pietzuch, P., Shand, B.: Enforcing End-to-End Application Security in the Cloud. In: Gupta, I., Mascolo, C. (eds.) Middleware 2010. LNCS, vol. 6452, pp. 293–312. Springer, Heidelberg (2010)
27. Chen, Y., Lu, J., Jan, J.: A Secure EHR System Based on Hybrid Clouds. Journal of Medical System 36(5), 3375–3384 (2012)
28. Nkosi, M.T., Mekuria, F.: Cloud Computing for Enhanced Mobile Health Applications. In: 2nd IEEE International Conference on Cloud Computing, pp. 629–633. IEEE Press, Indianapolis (2010)
29. Bidgoli, H.: Management Information Systems. Cengage Learning, Connecticut (2013)
30. Falconer, D., Adachi, F., Gudmundson, B.: Time division multiple access methods for wireless personal communications. IEEE Communications Magazine 33(1), 50–57 (1995)

Questions

Q1. What security and privacy features do mHealth systems should have?

Q2. Discuss the psychological effects of using secured mHealth systems?

Q3. What are limitations do a secured mHealth system have?

Q4. What can we expect from mobile health (mHealth) services as consumers of e-health services, either as patients or patients' family?

Q5. Many experts believe that mHealth systems can reduce the cost of healthcare as they provide accurate information on timely basis and on demand which improves the efficiency of medical care delivery. However, keeping privacy and confidentiality of personal information becomes important if these systems to flourish.

a) Discus three advantages of mHealh systems.
b) How privacy and confidentiality of personal information improved using these systems?

25. Analysis of mHealth Systems with Multi-cloud Computing Offloading

Huaming Wu

Freie Universität Berlin
Takustr.9, 14195 Berlin, Germany
huaming.wu@fu-berlin.de

Keywords: mHealth; cloud computing; mobile device; sensor network; offloading; serve topology; graph.

1 Introduction

Owing to the latest technologies in wireless communication and the development of mobile devices, related issues in mobile computing are becoming more and more concerned [1-2]. However, it is challenging to run very complex applications on the mobile devices because of the strict constraints on their resources such as memory capacity, network bandwidth, CPU speed and battery power [3]. This is not just a temporary limitation of current mobile hardware technology, but is intrinsic to mobility [4]. Cloud computing is becoming increasingly popular these days due to its features like elasticity, scalability and inexpensive. Along with the maturity of cloud computing [5], offloading the data or program from mobile devices to remote clouds is one of the attractive ways to overcome the above problems. Offloading brings many potential benefits, such as energy saving, performance improvement, reliability improvement, ease for the software developers, better exploitation of contextual information and so on.

Mobile healthcare (mHealth) applications, which are supported by mobile devices for delivering medical and healthcare services, can benefit from offloading the computationally intensive operation onto the cloud. Recently, the mobile healthcare market is showing a significant growth. According to a new report from *Research and Markets* [6], the market for mobile health applications and associated devices will grow at a compound annual growth rate of 61% to reach $26 billion in revenue by 2017 due to sales of mobile monitoring devices and integration with mainstream medicine. About 11% of cell phone users and 19% of smartphone users had mHealth apps on their devices in 2012, according to a Pew Internet survey [7]. By 2015, more than 500,000 people are expected to be using healthcare smartphone applications.

mHealth systems frequently use body sensors to collect healthcare data on patients and the data should be transmitted to a nearby or a remote mobile device.

S. Adibi (ed.), *Mobile Health*,
Springer Series in Bio-/Neuroinformatics 5, DOI: 10.1007/978-3-319-12817-7_25

Thus, minimizing the energy consumption needs to be considered and energy-efficient transmission schemes must be deployed. Fortunately, if we allow individual sensor nodes to cooperate with each other, a cooperative MIMO system can be constructed for the data transfer [8].

The mHealth systems are characterized by low coupling and powerful parallel computing capabilities, therefore offloading can be beneficial in these systems. Since there are a number of multimedia sensor signals to be processed on different servers, a traditional offloading scheme that offloads the program from a single mobile device to a single server is not sufficient. Therefore, finding some suitable offloading schemes is very important for the mHealth cloud systems. One multi-cloud offloading scheme that offloads the program from a single mobile device to a different server at each time is described, but the communication cost for such scheme is very huge since the network bandwidth between the mobile device and the cloud is small. However, the bandwidth between different clouds is much larger, and thus a multi-cloud offloading scheme where the data is shifted among the clouds is presented in this chapter.

The fast growth of cloud computing has attracted more and more companies to migrate their in-house IT applications into cloud and it also occurs in the medical field. A mHealth system with cloud offloading is considered in this chapter and it can be divided into two stages: sensor network and cloud offloading. In the first stage, information collected by body sensors should be transmitted to a remote mobile device. In order to save energy, an energy-efficient transmission scheme called cooperative multi-input multi-output (MIMO) is constructed for data transfer when allowing individual sensor nodes to cooperate with each other. In the second stage, two offloading schemes called self-reliant multi-cloud offloading system and multi-cloud offloading system are proposed and further analyzed based on service topology and optimal graph partition. The former provides stability but with high communication cost, while the latter reduces communication cost but is less stable. Both schemes can be effectively applied to the mHealth systems in which we would like to perform offloading on multiple cloud servers.

This chapter is an extended version of the conference paper [9] presented at IEEE MDM 2013 and it involves the area of sensor network, mobile offloading and mHealth cloud, and the purposes of this chapter are as follows:

1) To make a comparative study of the existing mHealth cloud systems.
2) To construct an energy-efficient transmission scheme in the body sensor network.
3) To find the most suitable architecture of multi-cloud offloading for the mHealth applications.

2 Overview of Mobile Cloud Healthcare Systems

2.1 mHealth

2.1.1 Overview of mHealth

mHealth is defined as the practice of medical and public health supported by mobile devices such as mobile phones, tablet computers and PDAs for delivering medical and healthcare services and information [10]. It is currently being heavily developed to keep pace with the continuously rising demand for personalized healthcare. With the development of wearable medical devices and wireless communication technology, mHealth is getting increasingly popular.

In order to fully utilize wireless technology between the wearable medical devices, we use body sensor network (BSN), which is a kind of wireless sensor network (WSN) around human body. Base on this, the mHealth systems that provide a personalized healthcare based on BSN are further developed. A typical mHealth system includes various body sensors to collect physiological signals specifically for different requirements, a mobile device such as smartphone to facilitate the joint processing of spatially and temporally collected medical data from different parts of the body for resource optimization and systematic health monitoring, and a server cluster with great data storage capacity, powerful analysis capabilities to provide data storage, data mining and visualization [11].

2.1.2 Limitations of Current mHealth Systems

We will explore the limitations of current mHealth systems:

1) Heavy algorithms cannot be run and processed on mobile devices since the mobile devices have limited computational capacity and run on small batteries. Therefore, the performance of mHealth applications is limited.
2) Lack of protection of the healthcare data capture and communication, more attention should be paid to the privacy and security for patients' data.
3) Lack of the design and development standards for mHealth clouds.

2.2 Cloud Computing

2.2.1 Overview of Cloud Computing

The emergence of cloud computing is promising to solve some of the concerns facing mobile computing platforms. Cloud computing refers to an on-demand, self-service internet infrastructure that enables the user to access computing resources such as processing, memory and storage anytime and anywhere in the world [12].

mHealth service is one of the applications that can benefit from the combination of mobile computing and cloud computing technologies. Some healthcare providers have found an opportunity to shift the burden of managing and maintaining complex health information to the cloud or more appropriately to the cloud service providers. For example, Rolim et al proposed a cloud-based system to automate the process of collecting patients' vital data via a network of sensors connected to legacy medical devices, and to deliver the data to a medical center's cloud for storage, processing, and distribution [13].

The special type of cloud computing used for improving patient care is called mHealth cloud and it provides opportunities to solve some of the current limitations facing mHealth systems. In addition to providing independent per-healthcare provider solutions, the mHealth cloud also has the potential to support collaborative work among different healthcare sectors through connecting healthcare applications and integrating their high volume of dynamic and diverse sources of information. By doing this, dispersed healthcare professionals and hospitals will be able to establish networks to coordinate and exchange information more efficiently.

2.2.2 mHealth Cloud Applications

Cloud computing can be one of the most suitable technologies for healthcare infrastructures. However, several serious issues concerning security, data protection and ownership, quality of services, and mobility need to be resolved before cloud computing is widely applied. The strengths and weaknesses of mHealth cloud applications are listed as follows, separately.

- **Strengths**:
  - Provide real-time data collection.
  - Relocate data through embedded database storage.
  - Data pre-processing and intelligent monitored data analysis.
  - User easy access.
  - Reduce spending on technology infrastructure.
- **Weaknesses**:
  - Lack of a strong security mechanism.
  - Weak data privacy protection.
  - Data can only be accessed when the network connection is available.

2.3 Offloading Approaches

Cloud offloading is an effective solution for solving the limitation on mobile devices. When a mobile device needs to run a heavy application, it offloads the heavy application to a cloud server. The result is sent back to the mobile device after executing the particular heavy application in the cloud. Classes for offloading can be divided into weak and strong classes. The weak class requires low security and computation, while the strong class requires complex multimedia processing and security. Therefore, the weak class will be entirely performed in the mobile device while the strong class prefers to be performed in the cloud under varied conditions. Through the approach of offloading, the mobile device is relieved from executing those heavy applications.

Normally, the limitation of mobile cloud computing is the battery life of mobile devices is very low when compared with that of other desktop devices. The main memory available is not enough to run complex applications. When the data storage is outsourced to the cloud servers, it is hard to ensure that the data is absolutely safe and will remain confidential in the cloud. Therefore, lack of trust in data security and privacy is at the heart of the resistance that many customers have to offload the healthcare data to the cloud [14].

Since mobile devices have very limited computational capacity and small battery life, they are unable to run heavy multimedia and security algorithms. A cloud computing framework is proposed in [15] to relieve the mobile devices from executing heavier multimedia and security algorithms in delivering mobile health services. They suggest that we offload the security algorithm to another cloud separately from the cloud where the institute's data is stored. Whereas, the user's mobile device only sends data and receives the result from the cloud through the network. In this way, the power consumption of the mobile device can be reduced, and on the other hand, it increases the database security that is stored in the cloud. In other words, multimedia and security operations can be performed in the cloud, allowing mobile health service providers to subscribe and extend the capabilities of their mobile health applications beyond the existing limitations of mobile.

3 mHealth Systems with Cloud Computing

3.1 Architecture

The architecture of mHealth systems with cloud offloading is shown in Fig.1. It is comprised of three main components: body sensors for collecting physiological signals, mobile devices for joint processing medical information and delivering healthcare services via mobile technology, cloud servers including the database server, data mining server and graphic server for signal processing [11].

This architecture has many advantages: it not only provides round the clock and real-time data collecting from patients, but also eliminates manual data collecting work and the mistakes made during manual data collecting. Medical staff can always access patients' information data through the cloud storage, and thanks to the wireless network connection on these devices, setting up of extra cable is not a must.

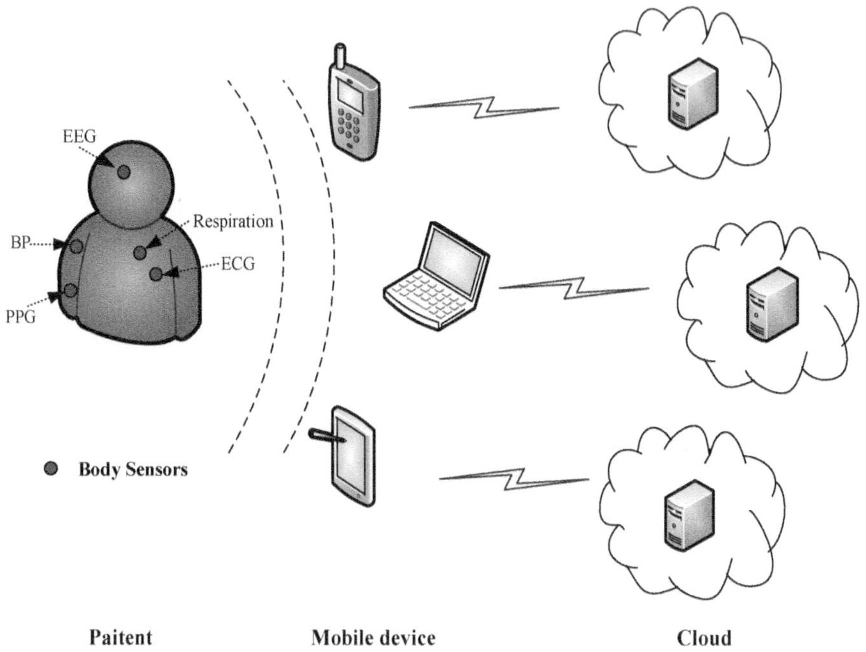

Fig. 1 Architecture of mHealth systems with mobile cloud offloading

Body Sensors: The process of a patient's data collection required a lot of manual work. A solution was proposed to attach sensors to medical equipment to automate the processes of the patient's vital data collection and make the patient's information becomes available in the cloud environment. In a typical system, each sensor node collects various physiological signals in order to monitor the patient's health status no matter their location and then instantly transmit all information in real time to the medical or the doctors. In this way, sensors act as data sensing where medical data will be shifted into the mobile devices.

Many body sensors such as electrocardiograph (ECG), photoplethysmograph (PPG) and blood pressure (BP) and so on, can be used for collecting vital signals for further analysis [11]. For example, body temperature can be detected from the patient through the mobile agent and this data is then processed through this intelligent model to produce useful data.

mHealth systems are designed to meet the requirements of different users. For example, the patients with heart diseases who need a long-term monitoring after recovery to prevent its relapse, the hypertension patients who are under the process of medicine adjustment and sub-healthy people who want to have a good knowledge of and follow up their health conditions to prevent some kinds of chronic diseases.

Finally, physiological signals collected by the sensors will be transferred to the mobile device via a sensor network.

Mobile Devices: Mobile devices such as mobile phone, laptop and PDA, aim to jointly process medical information and deliver healthcare services. Some special software is installed in the mobile device. After collecting data through Bluetooth sent from body sensors, we could get preliminary analysis results such as heart rate, abnormalities of a single test, and etc.

With the modern cloud technology, mobile devices can be considered to be a platform for delivering health information. However, there are some challenges to run heavy applications on a mobile device. No matter how powerful a smartphone device can be, it will not be as powerful as a computer. Because of the size constraint of a mobile device for portability purpose, a mobile device has a very limited battery capacity. This becomes the main reason why mobile devices may have the computation and memory limitation, and it will be an issue when a mobile device has to run a heavy application such as delivering health information.

Therefore, heavy multimedia and signal processing are unable to run on the mobile devices. In order to get further insights, the processing is offloaded to the server in the cloud [15].

The mobile devices mediate interaction between a wireless sensor network and a back-end intelligent cloud platform.

Clouds: Recently, there are a large number of cloud platforms appearing in the "sky" with different cost patterns and conditions, such as Amazon's EC2 [16], Apple's iCloud [17], Microsoft's Azure [18], Google's App Engine and so on for data storage and processing. These systems use a proprietary cloud platform to provide a personalized service. The cloud data center specifically designed for healthcare service can provide a platform for large data storage and parallel computing capabilities for data mining. It can support tens of thousands of people login and upload data simultaneously with response time of less than 1 second.

Healthcare data that is collected by the sensors will be transferred to the cloud middleware through the mobile device. At this stage, the data are analyzed, processed in the cloud middleware. Three different functions of clouds can be used for mHealth systems, *Security Cloud* that consists of security algorithm to encrypt the data, *Data Storage Cloud* for storing the patient data, and the *Application Cloud* to run the calculation and services provided by the applications.

3.2 Stages of mHealth Systems

The whole mHealth system can be divided into two stages.

First of all, we need to transmit the information collected by the body sensors to a mobile device. In some cases, the mobile device may be close to patients, where the Bluetooth can be used, but in most situations, the distance may be a challenge. For example, mobile devices used by the hospital are far from patients at home. In such cases, other transmission methods should be used to save energy consumption and reduce transmission time.

In fact, the body sensors can be seen as sensor networks, and the data collected by multiple body sensors need to be transmitted to a remote mobile device. If the sensors do not cooperate with each other, this is actually the case of multiple SISO (single input single output) transmission from an individual sensor to the mobile device. The energy cost would be huge due to the long distance between the local nodes and the remote mobile device. Therefore, a new strategy is required to minimize the total energy consumption of the entire nodes and transmission instead of reducing the energy cost of the individual node.

3.2.1 Cloud Offloading Stage

In this stage, programs in the mobile devices should be offloaded to the cloud servers for further processing.

The traditional scheme for cloud offloading is constrained by offloading data or computation from a single mobile device to a single server in cloud. However, it can't be applied to full range of scenarios in which we would like to perform offloading. For example, it is not suitable in the above healthcare system since different signal processing methods such as ECG, PPG, BP and so on, could not be completed only in one server and thus multiple offloading servers should be considered. Therefore, new offloading schemes should be further explored to overcome such complex signal processing.

The above two stages will be analyzed in detail in the following sections.

4 Analysis of Sensor Network Stage

In this section, we will study the sensor network stage of mobile healthcare systems as illustrated in Fig.1, i.e., transferring the collected information from local sensors to the remote mobile devices.

4.1 System Model

In the sensor network stage, a sensor network can be abstracted as a mathematical model, and it is depicted in Fig.2.

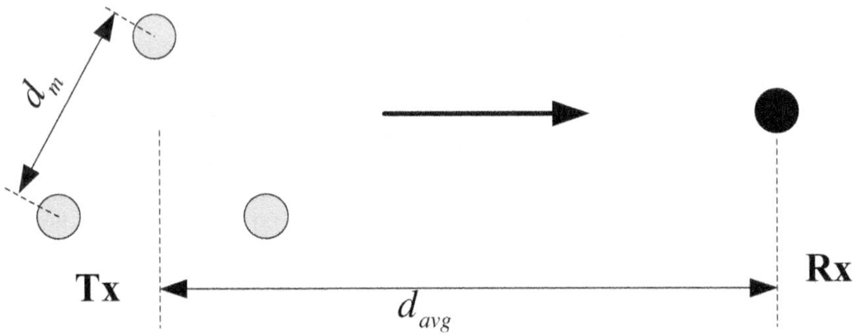

Fig. 2 Architecture of a sensor network

From Fig.2, we assume that there are three body sensors at Tx side, which are abstracted as three nodes, and the mobile device at Rx side is also abstracted as a node. The local distance between each node at Tx side is d_m, and the average distance from the sensors to the mobile device is d_{avg}. Since the mobile device is far away, we usually have $d_m \ll d_{avg}$ that the local distance is negligible when compared to the transmission distance..

4.1.1 Non-cooperative Approach

For the non-cooperative approach, the model in Fig.2 can be treated as M_t mutually independent SISO schemes. Each node at Tx side just transmits information to the remote node at Rx side on its own, and the local nodes do not cooperate with each other. The total energy consumption per bit in transmission and circuitry for a single SISO scheme with uncoded MQAM modulation is easily found to be [19]:

$$E_{SISO} = E_{tr} \frac{\gamma_0 d_{avg}^2}{bB} + E_c \frac{2}{bB} \tag{1}$$

where γ_0 denotes the SNR required in SISO scheme, $b = \log_2 M$ is bits per symbol in modulation and M is the constellation size, B is the bandwidth for a fixed transmission, E_{tr} is a constant factor for transmission, and E_c is a constant factor for circuitry.

We assume that there are M_t nodes at Tx side and each has L_i bits to transmit, where $i = 1, \cdots, M_t$. As a result, the total energy consumption for the non-cooperative approach is given by:

$$J_{Non-coop} = \sum_{i=1}^{M_t} L_i E_{SISO} \tag{2}$$

4.1.2 MISO Approach

As far as we know, MIMO (including MISO and SIMO) can save energy in fading channels [8]. Thus, if the multiple local nodes work together in transmission to the destination node, they can be treated as multiple antennas and an equivalent MIMO system can be constructed. As shown in Fig.2, only one receiver is given, therefore, we construct an equivalent MISO for example, and the other equivalent SIMO and MIMO can be constructed accordingly.

In order to make the cooperative transmission possible, the M_t nodes at Tx side will cooperate before the long transmission. Information on each node is broadcasted to all the other local nodes in different time slots. After each node receives all the information from other nodes, they encode the transmission sequence according to the STBC [20].

As for the MISO strategy, according to the reference [21], the total transmission energy and circuitry energy consumption per bit is:

$$E_{MISO} = E_{tr}\frac{d_{avg}^2}{bB}P_e^{-\frac{1}{M_t}} + E_c\frac{M_t + M_r}{bB} \tag{3}$$

where P_e is bits error rate (BER).

When the system in Fig.2 is treated as a cooperative MISO, in addition to the transmission and circuitry energy cost in Equation (3), the energy consumption at the Tx side due to the cooperation overhead needs to be considered. We denote the energy cost per bit for local information flow at the Tx side as E_i, which can be expressed as:

$$E_i = E_{tr}\frac{\gamma_0 d_m^{\ 2}}{b_i B} + E_c\frac{M_t}{b_i B} \tag{4}$$

where $b_i = \log_2 M_i$ is bits per symbol in modulation and M_i is the constellation size during the local transmission at the Tx side.

Therefore, the total energy consumption for the cooperative MISO approach is calculated as

$$J_{MISO} = \sum_{i=1}^{M_t} L_i(E_i + E_{MISO}) \tag{5}$$

4.2 Numerical Results

To give numerical examples, we assume that M_t=3, M_r=1, and the distance between each sensor is d_m=1m. The information is transmitted from the sensors to the mobile device by using QPSK modulation, and thus we have M=4. We also set the fixed bandwidth and the BER as B=10KHz and $P_e=10^{-3}$, respectively. Besides, the energy constants E_c=40uJ and E_t=10nJ.

The data to transmit in each node is L_1=500Kb, L_2=1Mb, L_3=2Mb. In the broadcast process of each node, we use 4QAM, 16QAM and 64QAM modulations, respectively. From the BER for SISO scheme $P_e = \frac{1}{2}\cdot\left[1-\sqrt{\gamma_0/(1+\gamma_0)}\right]$ [22], we can calculate the γ_0 for a certain P_e.

In order to compare the energy consumption in Equation (2) and (5), we obtain the numerical results as shown in Fig. 3.

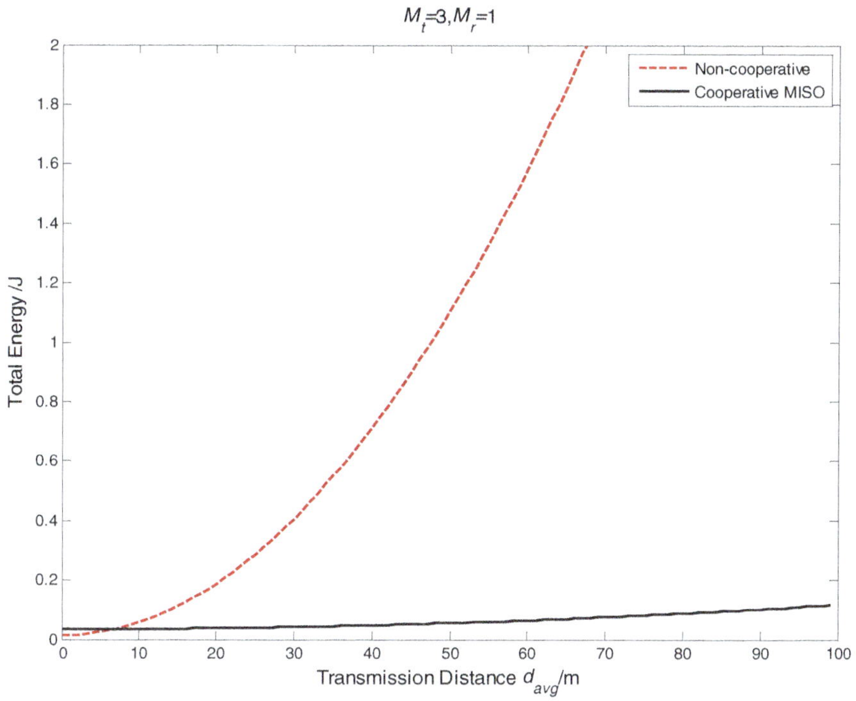

Fig. 3 Total energy consumption over d_{avg}

As shown in Fig.3, the total energy cost of the non-cooperative approach and the MISO scheme are plotted over the average transmission distance d_{avg}. It can be seen obviously that when d_{avg} is very small (e.g., $d_{avg} < 8$m), the non-cooperative transmission can still be more energy efficient than the scheme with MISO. However, when d_{avg} becomes larger, the transmission energy dominates the energy consumption, and the MISO approach becomes much more energy efficient than the scheme without cooperation.

5 Analysis of Cloud Offloading Stage

In this section, we will investigate the second stage of the mHealth systems as illustrated in Fig.1, i.e., offloading the program from the mobile devices to remote clouds for further execution.

The traditional scheme for cloud offloading is constrained by offloading computation from a single mobile device to a single server in the cloud. However, it cannot be applied to full range of scenarios in which we would like to perform offloading. For example, it is not suitable in the above healthcare system since different signal processing options such as ECG, PPG and BP could not be finished only in one server and it requires distribute across multiple offloading servers.

Therefore, a new scheme called multi-cloud offloading for mobile healthcare systems is proposed, which offloads portions of the program to multiple remote clouds.

5.1 *Self-reliant Multi-cloud Offloading System*

A self-reliant multi-cloud offloading system is described as Fig.4. Actually, it is equal to multiple offloading from single mobile device to a single server one by one.

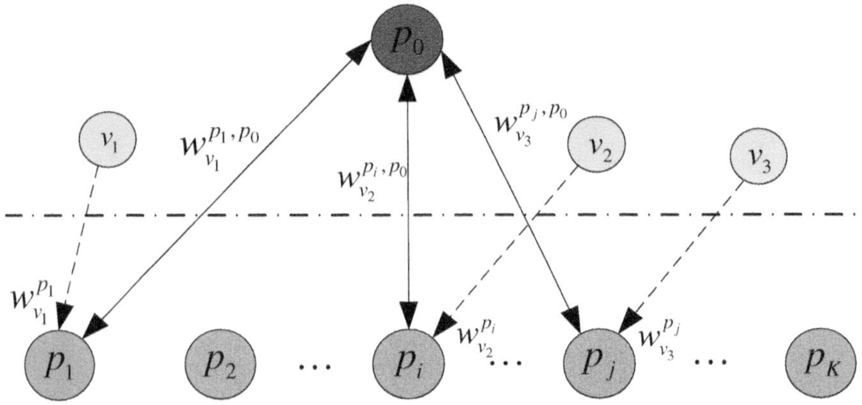

Fig. 4 Self-reliant multi-cloud offloading system

We assume that the number of available remote servers is k, $V=\{v_1,v_2,v_3\}$ means that there are three tasks named v_1, v_2, v_3 in a mobile device that need to be offloaded to cloud servers, $P=\{p_0,p_1,\cdots,p_k\}$ where p_0 represents the mobile device, and $p_1,\cdots,p_k$ represent the offloading servers. The dotted-line arrow denotes the allocation of tasks, and with a computation cost on it, while the solid-line arrow denotes the communication between each element of P. Besides, $w_v^{p_i}$ is the computation cost when v is assigned to server p_i and $w_v^{p_i,p_0}$ is the communication cost between p_i and p_0 when task v is assigned to p_i.

Here, the tasks of v_1, v_2 and v_3 are allocated to p_1, p_i and p_j, respectively, the mobile device offloads the computation to each server separately, and after executing the program, the server then sends the result back to the mobile device.

The disadvantage of the self-reliant multi-cloud offloading system is that the programs can only be offloaded to different servers one by one, separately and the approach cannot support more complex offloads, e.g., parallel offload of different application parts to different servers.

5.2 *Multi-cloud Offloading System*

5.2.1 Architecture

In order to overcome the limitations of self-reliant multi-cloud offloading system, another multi-cloud offloading system is depicted in Fig.5, where $w_{v_m,v_n}^{p_i,p_j}$ is the communication cost between p_i and p_j when v_m is assigned to p_i and meanwhile v_n is assigned to p_j.

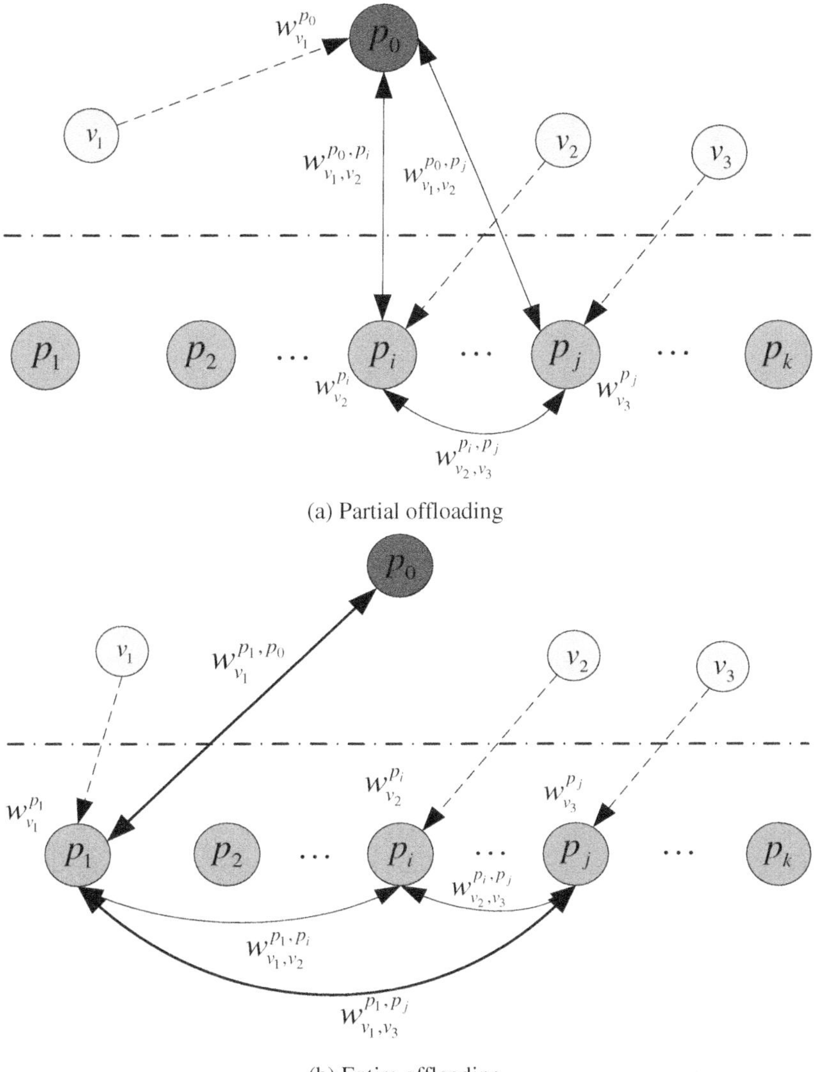

(a) Partial offloading

(b) Entire offloading

Fig. 5 Multi-cloud offloading systems

The system depicted in Fig.5a expresses a partial offloading scheme that partial programs are offloaded to servers while partial programs are executed locally by mobile device. It can be seen that the task of v_1 is allocated to the mobile device p_0 while the tasks of v_2 and v_3 are allocated to p_i and p_j. We compare the self-reliant multi-cloud offloading system in Fig.4 with the scheme in Fig.5a, it is found that the clouds with the tasks can communicate with each other and also communicate with the mobile device in Fig.5a while the clouds only communicate with the mobile device in Fig.4.

The system shown in Fig5b indicates an entire offloading scheme where all the tasks are offloaded to different servers. It can be seen that the tasks of v_1, v_2 and v_3 are allocated to p_1, p_i and p_j, respectively. The difference between Fig.4 and Fig.5b is that the clouds with the tasks communicate with each other in Fig.5b while the clouds only communicate with the mobile device in Fig.4.

Since the communication between two cloud-resident servers such as p_i and p_j may be very fast, while the communication between the mobile and the cloud may be much slower [23], the servers communicates with each other when executing the allocated tasks.

5.2.2 Case Study

An example of multi-cloud offloading is depicted in Fig.6. It can be seen that in addition to the mobile device, three different kinds of clouds named *data storage cloud*, *security cloud* and *application cloud* are combined together to overcome the limitation of mobile device: small memory, low security and small battery, respectively [23].

Data storage cloud: it is used for storing the patients' healthcare data.

Security cloud: or private cloud, such a cloud should be protected and can only be visited by the user. The security cloud uses security algorithm to encrypt the healthcare data before it is sent to the data storage cloud.

Application cloud: it contains main algorithms to run the calculation and services provided by the applications.

The workflow is such that, first, the healthcare data are put into the mobile device through the embedded sensor from the patient. And then the mobile device sends the data to the security cloud to encrypt them before they are stored in the data storage cloud. When the patient requires a service, the application cloud is invoked to acquire necessary data from the data storage cloud and then it decrypts the data for further processing. The result of the process is sent to the patient's mobile device in encrypted form, and it needs to be decrypted by the user.

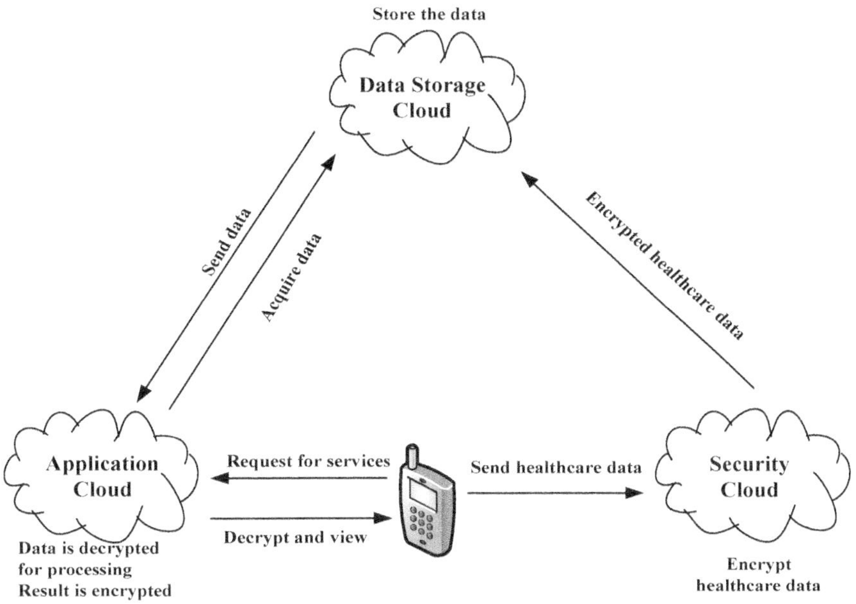

Fig. 6 Example of multi-cloud offloading

This architecture provides several benefits:

1) **Better performance**: Through offloading, the mobile device is prevented from running complex and massive algorithms that require a lot of memory and CPU power. The mobile device only receives the healthcare data from the patient, sends them to the security cloud, invokes the application cloud and decrypts the result. This could solve the limitations of small battery and memory of mobile device, and meanwhile provide better performance.
2) **Better security**: It is very crucial to protect the patient's healthcare data. In order to keep the healthcare data confidential and secure in the cloud, all the data are transferred in the encrypted form except the sending of raw data from the mobile device to the security cloud. It is difficult for the intruder to get the right decryption key to the right set of data since the encryption key in the security cloud is specific to the user. Therefore, it can provide better security.
3) **High dependence**: It provides loose coupling among the components. The mobile device and each cloud have their own responsibility in the architecture. If any failure occurs, it only affects one of the components, the other components are still workable and would not be greatly affected.

5.3 Service Topology

In order to compare the difference between the self-reliant multi-cloud offloading system and multi-cloud offloading system much more vividly, service topologies of multi-cloud offloading systems are depicted as Fig.7.

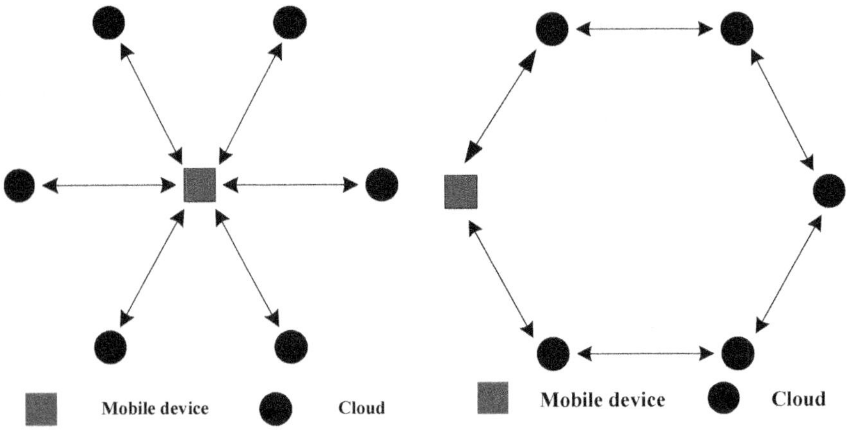

(a) Self-reliant multi-cloud offloading system (b) Multi-cloud offloading system

Fig. 7 Serve topology of multi-cloud offloading systems

It can be found in Fig.7a that the service topology for the self-reliant multi-cloud offloading system is star. The mobile device needs to communicate with different servers one by one. Thus, a lot of time and energy are spent on communication especially when the network condition is bad. However, it is very stable due to its separate offloading.

From Fig.7b, it can be seen that the service topology for the multi-cloud offloading system is ring. The mobile device only needs to communicate with two servers. Since the communication among clouds usually is very fast, while the data transfer between the mobile and the cloud may be much slower, this topology woks much fast than the star topology. However, its stability is much worse since it depends on every server, and if any failure happens in one of the middle servers, the program couldn't be executed successfully.

5.4 Partition Problem

We can formulate the multi-cloud offloading as a graph partitioning problem with $G=(V, E)$, where V is the set of vertices and the set of edges $E \in V \times V$. Each cloud has different computational and storage capacities, and hence it requires different weights of nodes. Besides, it requires different edge weights for communication since the network bandwidths are different [24].

We can cast this partition problem as an optimization problem, and the optimization goal can be to minimize the battery consumption, minimize the local storage needs while keeping communication low, or minimize the computation time.

The optimization problem for given tasks V and servers P [25] is shown as follows:

$$C = \underset{V,P}{\text{minimize}} \left[\sum_{v \in V} \sum_{p_i \in P} w_v^{p_i} \cdot m_v^{p_i} + \sum_{v_m, v_n \in V} \sum_{p_i, p_j \in P} w_{v_m, v_n}^{p_i, p_j} \cdot m_{v_m, v_n}^{p_i, p_j} \right] \tag{6}$$

where $\sum_{v \in V} \sum_{p_i \in P} w_v^{p_i} \cdot m_v^{p_i}$ is the total cost of computation and $\sum_{v_m, v_n \in V} \sum_{p_i, p_j \in P} w_{v_m, v_n}^{p_i, p_j} \cdot m_{v_m, v_n}^{p_i, p_j}$ denotes the total cost of communication, $m_v^{p_i} = \begin{cases} 1 & \text{if } v \text{ is assigned to } p_i \\ 0 & \text{otherwise} \end{cases}$ and $m_{v_m, v_n}^{p_i, p_j} = \begin{cases} 1 & \text{if } v_m \text{ is assigned to } p_i \text{ and } v_n \text{ to } p_j \\ 0 & \text{otherwise} \end{cases}$. $\forall v \in V$, we further have $\sum_{p_i \in P} m_v^{p_i} = 1$, which enforces that each vertex is assigned to exactly one partition.

Each node and each edge are assigned a different cost depending on the partition of the application graph where it finally ends up in.

The example given in Fig.5b can be expressed as a cost matrix:

$$\begin{array}{c} \\ p_0 \\ p_1 \\ \vdots \\ p_i \\ \vdots \\ p_j \\ \vdots \\ p_k \end{array}
\begin{array}{c} \begin{array}{ccccccccc} p_0 & p_1 & \cdots & p_i & \cdots & p_j & \cdots & p_k \end{array} \\
\begin{bmatrix}
0 & w_{v_1}^{p_1,p_0} & \cdots & w_{v_1,v_2}^{p_1,p_i} & \cdots & w_{v_1,v_3}^{p_1,p_j} & \cdots & \infty \\
w_{v_1}^{p_1,p_0} & 0 & \cdots & \infty & \cdots & \infty & \cdots & \infty \\
\vdots & \vdots & \ddots & \vdots & \ddots & \vdots & \ddots & \vdots \\
\infty & w_{v_1,v_2}^{p_1,p_i} & \cdots & 0 & \cdots & w_{v_2,v_3}^{p_i,p_j} & \cdots & \infty \\
\vdots & \vdots & \ddots & \vdots & \ddots & \vdots & \ddots & \vdots \\
\infty & w_{v_1,v_3}^{p_1,p_j} & \cdots & w_{v_2,v_3}^{p_i,p_j} & \cdots & 0 & \cdots & \infty \\
\vdots & \vdots & \ddots & \vdots & \ddots & \vdots & \ddots & \vdots \\
\infty & \infty & \cdots & \infty & \cdots & \infty & \cdots & 0
\end{bmatrix} \end{array}$$

where the graph partition is assumed as undirected and the edge weights that do not exist are set as ∞.

6 Conclusion

A typical mHealth system with cloud offloading is investigated and it can be divided into two stages of sensor network and cloud offloading.

Sensor nodes are attached to patients' medical equipment to transmit and collect data through wireless network communication. In order to save energy consumption on the entire sensor nodes, an energy-efficient transmission scheme is constructed for the data transfer when allowing individual nodes to cooperate with each other. The cooperative MISO approach seems to be much more energy efficient than the scheme that without cooperation when the mobile device is far away from the mobile device.

The proposed two new schemes of multi-cloud offloading for mobile healthcare infrastructure are analyzed and compared based on the graph partitioning and service topologies. There is a tradeoff between the stability and communication cost. Both schemes can be applied to other range of scenarios in which we would like to perform offloading on multiple servers. An example of multi-cloud offloading architecture that combines three different kinds of clouds are proposed, which can solve the limitations of computation, security and memory for the mHealth systems.

Through these systems, people can have knowledge of their own health information and even the risk factor of some chronic diseases in the future.

References

[1] Gu, X., Messer, A., Greenberg, I., Milojicic, D., Nahrstedt, K.: Adaptive Offloading for Pervasive Computing. IEEE Pervasive ComputingMagazine 3(3) (2004)

[2] Wolski, R., Gurun, S., Krintz, C., Nurmi, D.: Using Bandwidth Data to Make Computation Offloading Decisions. In: Proceedings of theInternational Parallel and Distributed Processing Symposium (IPDPS 2008), High-Performance Grid Computing Workshop (April 2008)

[3] Wu, H., Wang, Q., Wolter, K.: Tradeoff between Performance Improvement and Energy Saving in Mobile Cloud Offloading Systems. In: Proceedings of IEEE International Conference on Communications 2013: IEEE ICC 2013-1st International Workshop on Mobile Cloud Networking and Services (MCN), pp. 738–742 (2013)

[4] Satyanarayanan, M., Bahl, P., Caceres, R., Davies, N.: The case for vm-based cloudlets in mobile computing. IEEE Pervasive Computing 8(4), 14–23 (2009)

[5] Armbrust, M., Fox, A., Griffith, R., Joseph, A., Katz, R., Konwinski, A., Lee, G., Patterson, D., Rabkin, A., Stoica, I., Zaharia, M.: Above theClouds: ABerkeley View of Cloud Computing, Technical Report No. UCB/EECS-2009-28, University of California at Berkley, USA (February 10, 2009)

[6] Mobile Health Trends and Figures 2013-2017 (July 2013), http://www.researchandmarkets.com/research/nhc8j7/mobile_health

[7] Terry, K.: Mobile Health Market to Reach $26B by 2017 (July 30, 2013), http://www.informationweek.com/healthcare/mobile-wireless/mobile-health-market-to-reach-26b-by-201/240159145
[8] Cui, S., Goldsmith, A.J., Bahai, A.: Energy-Efficiency of MIMO and Cooperative MIMO Techniques in Sensor Networks. IEEE Journal on Selected Areas in Communications 22(6), 1089–1098 (2004)
[9] Wu, H., Wang, Q., Wolter, K.: Mobile Healthcare Systems with Multi-Cloud Offloading. In: Proceedings of the 2013 IEEE 14th International Conference on Mobile Data Management (MDM), vol. 2 (June 2013)
[10] Cipresso, P., Serino, S., Villani, D., Repetto, C., Selitti, L., Albani, G., Mauro, A., Gaggioli, A., Riva, G.: Is Your Phone So Smart to Affect Your States? An Exploratory Study based on Psychophysiological Measures. Neurocomputing 84, 23–30 (2011)
[11] Miao, F., Miao, X., Shangguan, W., Li, Y.: MobiHealthcare System: Body Sensor Network based M-Health System for Healthcare Application. E-Health Telecommunication Systems and Networks 1(1), 12–18 (2012)
[12] Mell, P., Grance, T.: The NIST Definition of Cloud Computing. Commun ACM 53(6) (2010)
[13] Rolim, C., Koch, F., Westphall, C., Werner, J., Fracalossi, A., Salvador, G.: A Cloud Computing Solution for Patient's Data Collection in Health Care Institutions. In: Proceedings of the 2nd International Conference on eHealth, Telemedicine, and Social Medicine, February 10-16 (2010)
[14] Everett, C.: Cloud Computing: AQuestion of Trust. Computer Fraud and Security 6, 5–7 (2009)
[15] Nkosi, M.T., Mekuria, F.: Cloud Computing for Enhanced Mobile Health Applications. In: Proceedings of the 2010 IEEE Second International Conference on Cloud Computing Technology and Science (CloudCom 2010), pp. 629–633 (2010)
[16] http://aws.amazon.com/ec2/
[17] http://www.apple.com/icloud/
[18] https://windows.azure.com/
[19] Liu, W., Li, X., Chen, M.: Energy Efficiency of MIMO Transmissions in Wireless Sensor Networks with Diversity and Multiplexing Gains. In: Proceedings of the 2005 IEEE International Conference on Acoustics, Speech, and Signal Processing (ICASSP 2005), vol. 4, pp. 897–900 (2005)
[20] Sandhu, S., Paulraj, A.: Space-Time Block Codes: ACapacity Perspective. IEEE Communication Letters 12(12), 384–386 (2000)
[21] Cui, S., Goldsmith, A.J., Bahai, A.: Modulation Optimization under Energy Constraints. In: Proceedings of the 2003 IEEE International Conference on Communications (ICC 2003), vol. 4, pp. 2805–2811 (May 2003)
[22] Wu, H., Wang, Q., Wolter, K.: Methods of Cloud-Path Selection for Offloading in Mobile Cloud Computing Systems. In: Proceedings of the 2012 IEEE 4th International Conference on Cloud Computing Technology and Science (CloudCom 2012), pp. 443–448 (December 2012)

[23] Yuan, J.N.Z., Ping, W.W., Wen, Y.H., Husain, W.: Healthcare Applications on Mobile Cloud Computing. In: Proceedings of the Third International Conference on Digital Information Processing and Communications (ICDIPC2013), the Society of Digital Information and WirelessCommunication, pp. 514–522 (2013)
[24] Sinha, K., Kulkarni, M.: Techniques for Fine-Grained, Multi-Site Computation Offloading. In: Proceedings of the 2011 11th IEEE/ACM International Symposium on Cluster, Cloud and Grid Computing (CCGrid 2011), pp. 184–194 (May 2011)
[25] Chun, B.G., Maniatis, P.: Dynamically Partitioning Applications between Weak Devices and Clouds. In: Proceedings of the 1st ACM Workshop on Mobile Cloud Computing and Services: Social Networks and Beyond (MCS), vol. 7 (June 2010)

26. An Overview of mHealth Medical Video Communication Systems

Andreas S. Panayides[1], Zinonas C. Antoniou[2], and Anthony G. Constantinides[1]

[1] Communication and Signal Processing Group,
Department of Electrical and Electronic Engineering, Imperial College, London, UK
{a.panagidis,a.constantinides}@imperial.ac.uk
[2] Electronic Health (eHealth) Lab., Department of Computer Science,
University of Cyprus, Cyprus
antoniou.zinonas@cs.ucy.ac.cy

Abstract. Significant technological advances over the past decade have led mHealth systems and services to a remarkable growth. It is anticipated that such systems and services will soon be established in standard clinical practice. MHealth medical video communication systems progression has been primarily driven by associated advances in video coding and wireless networks technologies. Responsive, reliable, and of high-diagnostic quality systems are now feasible. Such systems build on compression ratios and error resilience tools found in current state-of-the-art video coding standards, linked with low-delay and high-bandwidth communications facilitated by new wireless systems. To achieve this however, these systems need to be diagnostically driven. In other words, both encoding and transmission need to adapt to the underlying medical video modality's properties, for maximizing the communicated video's clinical capacity. Moreover, the proper mechanisms should be developed that will guarantee the quality of the transmitted clinical content. Current video quality assessment (VQA) algorithms are unsuccessful to replicating clinical evaluation performed by the relevant medical experts. Clearly, there is a demand for new clinical VQA (c-VQA) metrics.

This chapter reviews medical video communication systems. It highlights past approaches and focuses on current design trends and future challenges. It provides an insight to the most prevailing diagnostically driven concepts and the challenges associated with each system component, including pre-processing, encoding, wireless transmission, and quality assessment. It discusses how exploiting high efficiency video coding (HEVC) standard, together with the emergence of 4G and beyond wireless networks, is expected to deliver the mHealth medical video communication systems that will rival in-hospital examinations. The latter, linked with c-VQA that will correlate with clinical ratings is expected to aid the adoption of such systems and services in daily clinical practice.

S. Adibi (ed.), *Mobile Health*,
Springer Series in Bio-/Neuroinformatics 5, DOI: 10.1007/978-3-319-12817-7_26

1 Introduction

Driven by greater socioeconomic aspects and materialized by hardware and software technology advances, mHealth systems and services continue to grow in unparalleled numbers [1]-[4]. Foreseeing the benefit in global healthcare delivery provided by information and communication technologies, the World Health Organization (WHO) founded the global observatory for eHealth in 2005 [5]. The global observatory is responsible for providing an insight to the prevalence of eHealth solutions to the relevant stakeholders (such as governments, private sector, academia), as well as identify, suggest, and trigger eHealth paradigms that can contribute to advancing both regional and global healthcare delivery [6], [7]. Technology advances push healthcare frontiers in new directions, transforming traditional medical practice (e.g., in remote diagnosis and preventive care, see Figure 1). Wider adoption of mHealth systems and services in standard clinical practice is capable of removing geographical, social, and economical barriers, in the provision of specialized healthcare. While still debatable -in the absence of a sufficient sample of large scale studies-, projected significant financial savings are expected to bring further growth and accelerate the deployment of mHealth solutions [8], [9].

Figure 1 depicts a subset of typical application scenarios where the use of mHealth systems and services is currently being investigated and foreseen to be established in standard clinical care. This comprises of wireless medical video communication systems, the use of body area networks (BANs) for personal health (pHealth) and remote monitoring of individuals, the combination of the latter two approaches in disaster crisis management, as well as electronic health records, mHealth cloud based services and smartphone applications in general.

Medical video communication is a key bandwidth-demanding component of mHealth applications ranging from emergency incidents response, to home monitoring, and medical education [10]-[14]. In-ambulance video (trauma and ultrasound) communication for remote diagnosis and care can provide significant time savings which can prove decisive for the patient's survival. Similarly, emergency scenery video can assist in better triage and preparatory hospital processes. Remote diagnosis allows access to specialized care for people residing in remote areas, but also for the elderly, and people with chronic diseases and mobility problems. Moreover, it can support mass population screening and second opinion provision, especially in developing countries. Medical education also benefits from real-time surgery video transmission as well as ultrasound examinations.

Body area networks can be used for (home) monitoring of frail individuals as well as manpower under particularly stressed conditions (such as first responders, firefighters, policemen, soldiers, etc.). The idea is that wearable physiological and kinetic sensor devices are carried by each individual, which communicate the sensed data to a monitor/ gateway device (i.e. smartphone). Such data include electrocardiogram, respiratory rate, oxygen rate, etc. Gateway devices can be equipped with appropriate medical decision software which can monitor a person's health and issue alerts if needed. The data can be stored or communicated for further processing and real-time monitoring.

Disaster crisis management can benefit from real-time emergency scenery video transmission for the assessment of the emergency incident. Bracelet sensors can be used for assessing an individual's health status for treatment, monitored via the on-site management center, increasing responsiveness and enabling more efficient triage.

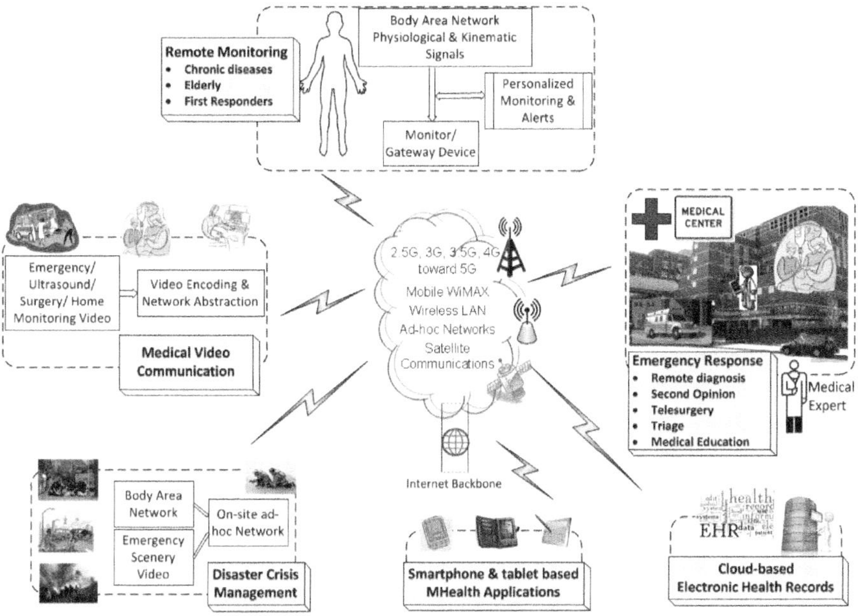

Fig. 1 Selected mHealth systems and services range from medical video communications and remote monitoring, to emergency response and disaster crisis management, and electronic health records and mHealth cloud, smartphone & tablet based applications [14]

Electronic health records (EHR) can improve the level of healthcare provision in multifarious ways. EHRs reduce storage requirements and costs over non-EHR systems, minimize time delays, errors, and losses, while also providing support for remote access and processing of data. MHealth cloud can include m-pharmacies, which will be connected to electronic health records, as well as insurance companies and relevant stakeholders for minimizing the volume of paperwork through automation. Toward this direction, familiarity of patients with technology capabilities achieved via the wide acceptance of smartphone devices and associated applications already set the foundations for adoption of such approaches.

Overall, wireless medical video communication poses significant challenges that stem from limited bandwidths over noisy channels. In terms of both bandwidth and processing requirements, medical videos dominate over other biomedical signals. Clearly, the evolution of future mHealth systems will depend and also benefit from the development of effective medical video communication systems.

The rest of the chapter is organized as follows: Section 3 provides an insight to video coding standards and wireless technologies that are the key enabling technologies in mHealth medical video communication systems. Section 4 discusses the current trends and design considerations of such systems, while Section 5 provides an overview of the most representative studies of the past decade. Finally, the last section provides some concluding remarks and discusses future directions.

2 Enabling Technologies

MHealth medical video communication systems' impressive growth over the past decade is primarily attribute to associated advances in video compression and wireless network technologies. The former allow real-time, robust and efficient encoding, while recent video coding standards also encompass a network abstraction layer for higher flexibility. The latter, facilitate constantly increasing data transfer rates, extended coverage, reduced latencies and transmission reliability. In what follows, we highlight video coding standards and wireless transmission technologies evolution timeframe.

2.1 Video Coding Standards

In terms of video coding standards, efficient video compression systems were introduced over the last decades. In early 1990's the international telecommunication union-telecommunication sector (ITU-T) developed the first, H.261 [15] video coding standard, which was originally designed for video-telephony and videoconferencing applications. H.261 supported the quarter common intermediate format (QCIF-176x144) and the common intermediate format (CIF-352x288) video resolutions. Subsequently, H.262 [16] released in 1995 facilitated interlaced video provision and increased video resolutions support to 4CIF (720x576) and 16CIF (1408x1152). Its successor, termed H.263 [17] introduced in 1996 provided for improved quality at lower bit rates and also allowed lower, sub-QCIF (128x96) video resolution encoding. The highly successful H.264/AVC [18] standard was released by ITU-T in 2003 and accounted for bit rate demands reductions of up to 50% for equivalent perceptual quality compared to its predecessor [19]. The current state-of-the-art video coding standard is the High Efficiency Video Coding (HEVC) standard [20], standardized in 2013. HEVC supports video resolutions ranging from 128x96 to 8192x4320 and provides 50% bit rate gains for comparable visual quality compared to H.264/AVC [21].

2.1.1 H.264/AVC

H.264/AVC was jointly developed by the ISO/IEC motion pictures experts group (MPEG) and ITU-T video quality experts group (VCEG) who formed the joint video team (JVT). H.264/AVC met the growing demand for multimedia and video services by providing enhanced compression efficiency significantly outperforming all prior standards. H.264/AVC comprises of numerous advances in standard video coding technology, coding efficiency improvement, error robustness enhancement, and flexibility for effective use over a variety of network types and application domains [19]. H.264/AVC defines a video coding layer (VCL) and a network abstraction layer (NAL) to enable transportation over heterogeneous networks. VLC is responsible for video coding, maintaining the block-based motion-compensated hybrid video coding concept. Compared to prior standards, VCL provides enhanced entropy coding methods, uses a small block-size exact-match transform, facilitates adaptive in-loop deblocking filter and enhances motion prediction capability. On the other hand, NAL is a novel concept aiming at a network-friendly adaptation of VCL content to candidate heterogeneous networks and/or storage devices (or cloud).

H.264/AVC defines different profiles and levels. Each profile and level specify restrictions on bitstreams, hence limits on the capabilities needed to decode a bitstream [18]. Baseline, main, extended and high profiles assume different processing devices tailored for different applications and provide incremental level capabilities (and therefore complexity). Figure 2 demonstrates unique features of each profile.

A level is a specified set of constraints that indicate a degree of required decoder performance for a profile. For example, a level of support within a profile specifies the maximum picture resolution, frame rate, and bit rate that a decoder may use. A decoder that conforms to a given level must be able to decode all bitstreams encoded for that level and all lower levels. Lower levels attribute to lower resolutions, lower bitrates and lower memory to store reference frames. A level is primarily used to designate device compatibility. For example, the iPhone 5S supports H.264/AVC high profile level 4.2, which means that a video's peak bitrate can't exceed 50 Mbps and the maximum supported video resolution and frame rate is up to 1080p at 60 frames per second.

H.264/AVC facilitates a broad range of error resilience techniques for a wide variety of applications. Toward this end, one of H. 264/AVC key error resilience features is flexible macroblock ordering (FMO) which defines macroblock transmission order to allow for easier recovery. Using this feature, each frame may be partitioned in up to seven different slices. This feature also allows region-of-interest (ROI) coding and recovery (a scheme extensively used in mHealth medical video communication systems, see Section IV) or arbitrary spatial placement of blocks for better concealment during interpolation. Another important feature is redundant slices (RS) where a slice is redundantly inserted in the communicated bitsream to maximize the video's error resilience.

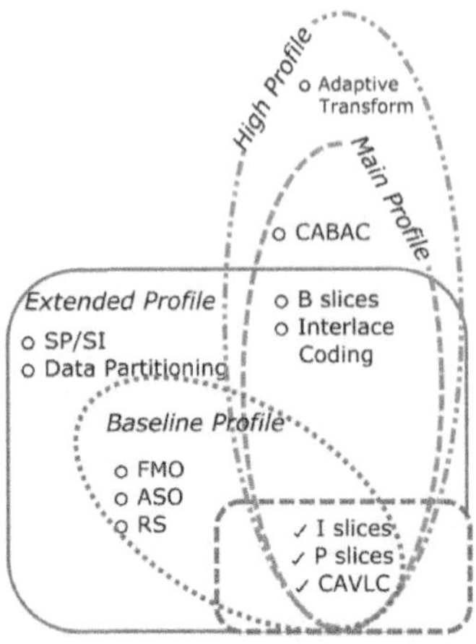

Fig. 2 H.264/AVC baseline, main, extended, and high profiles features (taken from [12]). Context-adaptive variable-length coding (CAVLC), flexible macroblock ordering (FMO), arbitrary slice ordering (ASO), redundant slice (RS), switching-predictives/switch (intra SP/SI), context-adaptive binary arithmetic coding (CABAC).

2.1.2 High Efficiency Video Coding (HEVC)

On February 2011 a total of 27 proposals were submitted to the joint collaborative team on video coding (JCT-VC) following a call for proposals for the next video coding standard in 2010. On April 2013 HEVC was officially approved as an ITU standard and on June 2013 [20] was formally published on the ITU-T website. HEVC is also known as H.265 and MPEG-H Part 2.

To meet the ever increasing requirements for cost effective video encoding process HEVC optimizes video quality, compression efficiency, spatial and temporal resolution, and finally computational complexity. HEVC introduces new coding tools as well as significant improvements of components already known from H.264/AVC. New tools in HEVC include variable size block partitioning using quadtrees for the purpose of prediction and transformation and an additional in-loop filter, namely sample adaptive offset (SAO). Improvements include additional intra-prediction angles, advanced motion vector prediction (AMVP), a new block merging mode that enables neighboring blocks to share the same motion information, larger transform sizes and a more efficient transform coefficient coding [22]. HEVC incorporates only one entropy coder, context adaptive binary arithmetic coding (CABAC) borrowed from H.264/AVC.

One of the key differences of HEVC compared to H.264/AVC is the frame coding structure. In HEVC each frame is partitioned into coding tree blocks (CTBs) [21]. The luma CTB and the two chroma CTBs, together with the associated syntax, form a coding tree unit (CTU). The CTU is the basic processing unit of the standard to specify the decoding process and replaces the macroblock structure found in all prior video coding standards. A CTB may contain a single coding unit (CU) or can be split to form multiple CUs [22]. CUs can be further split into prediction units (PUs) used for intra- and inter-prediction and transform units (TUs) defined for transform and quantization. As the picture resolution of videos increases from standard definition to high definition (HD) and beyond, the chances are that the picture will contain larger smooth regions, which can be encoded more effectively when larger block sizes are used. Based on the afore-mentioned assumption, HEVC supports up to 64x64 pixels encoding blocks compared to H.264/AVC that only supports up to 16x16 pixels blocks.

Similar to H.264/AVC, HEVC uses block-based intra-prediction to take advantage of the spatial correlation within a picture. HEVC follows the basic idea of H.264/AVC intra prediction but makes it far more flexible. HEVC has thirty five luma intra prediction modes compared to nine in H.264/AVC. HEVC also includes a planar intra prediction mode, which is useful for predicting smooth picture regions. In planar mode, the prediction is generated from the average of two linear interpolations (horizontal and vertical). Also, a DC prediction can also be used at surface with a value matching the mean value of the boundary samples [22].

HEVC is the first video coding standard to incorporate features that provide for parallel processing [22]. The new tiles tool allows the partitioning of a picture into independently decoded rectangular segments of approximately equal CTUs. Tiles increase flexibility compared to normal slices in H.264/AVC and incorporate considerably lower complexity than FMO. Wavefront parallel processing (WPP) splits slices into rows of CTUs, which are then processed in parallel, provided a certain time window has advanced since the processing of the immediately prior row to allow time for decisions relating to entropy coding. Dependent slices allow a slice using tiles or WPP coding tools to be fragmented and associated with different NAL packets. Dependent slices are associated with lower encoding time but a slice may be decoded provided part of the decoding process of the dependable slice has been performed.

2.2 Wireless Transmission Technologies

The global system for mobile communications (GSM) signified the transition from analog 1st generation (1G) to digital 2nd generation (2G) technology of mobile cellular networks. In the past two decades, mobile telecommunication networks are continuously evolving. Milestone advances range from 2.5G (general packet radio service (GPRS) and enhanced data rates for GSM evolution (EDGE)) and 3G (universal mobile telecommunications system (UMTS)) wireless networks, to 3.5G (high speed downlink packet access (HSDPA), high speed uplink packet access (HSUPA), high speed packet access (HSPA), and HSPA+), mobile WiMAX networks, and long term evolution (LTE) systems. The afore-described wireless networks facilitate incremental data transfer rates while minimizing end-to-end delay. In other words, features in wireless communications follow those of wired infrastructure with a reasonable time gap of few years. The latter enables the development of responsive mHealth systems suitable for emergency telemedicine

[11]. Evolving wireless communications networks' theoretical upload data rates range from 50 kbps - 86 Mbps. In practice, typical upload data rates are significantly lower. More specifically typical upload data rates range from (i) GPRS: 30-50 kbps, (ii) EDGE: 80-160 kbps, (iii) evolved EDGE: 150-300 kbps, (iv) UMTS: 200-300 kbps, (v) HSPA: 500 kbps - 2 Mbps, (vi) HSPA+: 1-4 Mbps, and (vii) LTE: 6-13 Mbps [26].

2.2.1 4G Networks Conforming to IMT-Advanced Requirements

Mobile WiMAX and LTE are today's state-of-the-art deployed networks, incorporating a plethora of sophisticated technologies. However, while they do meet some of the international mobile telecommunications-advanced (IMT-advanced) requirements specified by the international telecommunication union-radio communication sector (ITU-R) [27], they both fail to uniquely address all listed requirements. This led to the development of even more efficient techniques and concepts, in order to facilitate conformance with aforementioned specifications. Resulting WirelessMAN-Advanced and LTE-advanced technologies based on IEEE 802.16m and 3GPP Release 10 specifications respectively, participated in IMT-advanced evaluation process, following ITU-R call for candidate technologies. Evaluation processes concluded that both candidate technologies met IMT-advanced requirements and are now officially considered as 4G technologies.

Being backwards compatible, this family of technologies targets improved uplink and downlink rates of 100 Mbps and 1 Gbps respectively, increased coverage and throughput, enhanced mobility support, reduced latencies, enhanced quality of service (QoS) provision, efficient spectrum usability and bandwidth scalability, and security, with simple architectures, in favor of the end user.

2.2.2 Worldwide Interoperability for Microwave Access (WiMAX)

Worldwide Interoperability for Microwave Access (WiMAX) was firstly standardized for fixed wireless applications in 2004 by the IEEE 802.16-2004d and then for mobile applications in 2005 by the IEEE 802.16e standards. After an initial hype of the WiMAX, lately there has been skepticism as to its successful wide deployment, in favor of the LTE. Current standardization 802.16m [28], also termed as IEEE WirelessMAN-Advanced, aims to achieve higher market penetration.

WiMAX frequency today lies typically in the range of 2.3, 2.5-2.7, 3.5, and 5.8 GHz, while 4G frequency bands will facilitate deployment between 450-3600 MHz [28]. Channel bandwidth allows great flexibility in the sense that it allows WiMAX operators to consider channel bandwidths between 1.25, 2.5, 5, 10, and 20 MHz (802.16e). In 802.16m scalable bandwidth between 5-40 MHz for a single RF carrier is considered, extended to 100 MHz with carrier aggregation to meet IMT-advanced requirements. WiMAX employs a set of high and low level technologies to provide for robust performance in both line-of-sight (LOS) and non-line-of-site (NLOS) conditions. A thorough overview of WiMAX standardization process, evolving concepts, technologies and performance evaluation of IEEE 802.16m appears in [28]-[30]. Key features of physical (PHY) and medium access control (MAC) layers are discussed next.

2.2.2.1 Physical Layer Features

As already mentioned above, WiMAX standards define the air interface and more specifically MAC and PHY layers. PHY layer's central features include adaptive modulation and coding, hybrid automatic repeat request (HARQ), and fast channel feedback. Key technology in the success of WiMAX systems in general is the orthogonal frequency division multiplexing (OFDM) scheme employed in the PHY layer. OFDM, and more specifically scalable orthogonal frequency division multiple access (SOFDMA), allows dividing transmission bandwidth into multiple subcarriers. The number of subcarriers starts from 128 for 1.25 MHz channel bandwidth and extends up to 2048 for 20 MHz channels. In this manner, dynamic QoS tailored to individual application's requirements can be succeeded. In addition, orthogonality among subcarriers allows overlapping leading to flat fading. In other words, multipath interference is addressed by employing OFDM while available bandwidth can be split and assigned to several requested parallel applications for improved system's efficiency. The latter is true for both downlink (DL) and uplink (UL). Multiple input multiple output (MIMO) antenna system allows transmitting and receiving multiple signals over the same frequency. Two types of gain are possible, namely spatial diversity and spatial multiplexing. For spatial diversity, unique configuration enables enhanced link quality by combining independent faded signals resulting from simultaneously transmitted duplications of the same information. For spatial multiplexing, increased throughput is achieved via the parallel spatial channels transmission of multiple streams.

2.2.2.2 Medium Access Control Layer Features

In the MAC layer, the most important supported features can be summarized in QoS provision through different prioritization classes, direct scheduling for DL and UL, efficient mobility management, as well as security. The 5 QoS categories facilitated by WiMAX networks are discussed in [28]. According to individual application requirements the appropriate QoS class is considered and the corresponding UL burst is scheduled and data rate assigned. For real-time video streaming, as in the case of emergency telemedicine scenarios, real-time Polling Service (rtPS) QoS class best suits the applications requirements. Mobility management is also well addressed in 802.16e and current 802.16m standards, which was an issue in 802.16d primary standard for fixed connections. With a theoretical support of serving users at 120 km/h in 802.16e, established connections provide adequate performance for vehicles moving with speeds between 50-100 km/h. In 802.16m, mobility support is extended for mobile speeds up to 350 km/h as depicted in the evaluation of IMT-advanced requirements. Enhanced security, especially when compared to competing technologies (like WLANs) is one of the key features in WiMAX networks, shielding the end-user from a variety of threats. Improved security is based on extensible authentication protocol and advanced encryption.

2.2.3 Long Term Evolution (LTE)

Long term evolution (LTE) mobile communication networks have been standardized through the 3rd generation partnership project (3GPP) Release 8. LTE facilitates significant improvements with respect to 3G and HSPA systems. It provides increased data rates, improved spectral efficiency and bandwidth flexibility ranging between 1.4-20 MHz, reduced latency and seamless handover. While being backwards compatible enabling seamless deployment on existing infrastructure, LTE shares a set of new cutting edge technologies and simple architecture.

In the physical layer, multiple-carrier multiplexing OFDMA is adopted for the downlink, while single carrier frequency division multiple access (SC-FDMA) is the access scheme used in the uplink. SC-FDMA utilizes single carrier modulation, orthogonal frequency multiplexing, frequency domain equalization, and has similar performance to OFDMA. Frequency division duplex (FDD) and time division duplex (TDD) are jointly supported in a single radio carrier. LTE allows multi-antenna applications for single and multi users through MIMO technology (up to 4-layers in the downlink and 2-layers in the uplink), and supports different modulation and coding schemes. Automatic repeat request (ARQ) and hybrid-ARQ are implemented for increased robustness in data transmission. Enhanced mobility support, efficient multimedia broadcast multicast service, QoS provision, security, and cell capacity up to 200 active users summarize the key features provided by LTE systems.

2.2.3.1 LTE-Advanced

LTE-advanced bridges the gap between LTE Release 8 and IMT-advanced requirements. LTE-advanced is standardized in 3GPP Release 10. Novel technologies found in LTE Release 8 and enhancements in Release 9 are incorporated in the design of LTE-advanced systems. Compared to LTE, further improvements in peak data rate, spectrum efficiency, throughput and coverage, as well as latency reductions are facilitated.

Data rates in the order of 1 Gbps for low mobility and 100 Mbps for high mobility are achieved via the adoption of new technologies such as carrier aggregation. Carrier aggregation enables wider bandwidth transmission up to 100 MHz utilizing a combination of frequency blocks, thus increasing system's peak data rates. An example utilization scenario of 100 MHz system bandwidth utilizes 5 LTE 20 MHz blocks, for uplink and downlink, at 3.5 GHz and FDD. Enhanced MIMO techniques in LTE-advanced systems include 8-layer transmission in the downlink and 4-layer transmission in the uplink. Coordinated multipoint transmission and reception (CoMP) is another novel technique defined in the standard which provides for increased throughput on the cell edge. The key idea is that multiple e-Node Bs cooperate to coordinate transmission relevant aspects that provide for reduced interference and increased throughput for user equipment located near the cell edge. Relaying techniques for efficient cell deployment and coverage, and parallel processing for even greater reductions in latency are also exploited.

A comprehensive review of LTE and LTE-advanced technologies, comparative analysis, utilization scenarios combining different technology components to demonstrate conformance to IMT-advanced requirements, and also IMT-advanced evaluation results can be found in [31]-[32].

3 MHealth Medical Video Communication Systems: Design Approaches

In this section we review mHealth medical video communication systems' approaches spanning over the last decade (see Table I). We focus on recent, diagnostically relevant approaches and discuss state-of-the-art methods. We highlight the impact of latest advances in video compression and wireless networks technologies in enhancing the clinical capacity of the transmitted medical video. We further comment on the imperative need of developing new, diagnostically driven video quality assessment (VQA) metrics.

3.1 Diagnostically Driven MHealth Systems

Diagnostically driven systems exploit the properties of the underlying medical video modality aiming to maximize the clinical capacity of the communicated medical video. They range from diagnostically relevant and resilient encoding, to reliable wireless communications based on the communicated video's clinical significance, and clinical video quality assessment methods.

3.2 Diagnostic Region(s)-of-Interest

Each medical video modality is characterized by unique properties which are assessed by the relevant medical expert during an examination. A diagnosis is provided based on a clinically established protocol that considers different clinical criteria. These criteria often relate to specific video regions, as for example in routine ultrasound screening examinations. In other words, specific video regions carry the clinical information required by the medical expert to assess the patient's status during a visit. The latter observation has been the driving force of diagnostically relevant approaches using diagnostic regions-of-interest (d-ROI).

Diagnostic ROI outline the clinically important video regions. Essentially, there exist two ways of defining d-ROI: a) using modality aware segmentation algorithms, and b) denoted by the medical expert. Both approaches can be established in standard clinical practice, and the selected mode is application specific and depends on resources availability. MHealth medical video communication systems using d-ROI dominate over other diagnostically driven approaches [14]. They have become the standard method to use in a very short time since they were first proposed. Besides the obvious efficiency that such systems introduce, the latter is also attributed to the plethora of different approaches and remarkable flexibility that these schemes enjoy.

3.3 Diagnostically Relevant Encoding

Variable quality slice encoding based on the video region's diagnostic significance is a popular diagnostically driven method that builds on d-ROI. This scheme assigns quality levels according to the clinical significance of each d-ROI. Higher quality (less compression, more bits) is assigned to the most sensitive –clinically wise- regions while lower quality (higher compression, less bits) to the background, non-diagnostically important regions. The effectiveness of this approach has been demonstrated in [33]-[39] where significant bitrate demands reductions were documented without compromising perceptual quality. In earlier studies based on MPEG-2 and MPEG-4 video coding standards [33], [34], smoothing filters have been applied on the non-diagnostically important regions. On the other hand, more recent studies based on H.264/AVC make use of a customized version of the FMO error resilient technique [35]-[38], or use a similar concept for macroblock-level variable quality slice encoding [39]. Diagnostically relevant encoding approaches based on echocardiogram's mode properties which also lower bitrate requirements but without considering d-ROI is presented in [40], [41].

3.4 Diagnostically Resilient Encoding

Diagnostically resilient encoding/decoding adapts error resilience/concealment techniques to the d-ROI. Error resilience techniques such as intra updating (frame level) can be tailored to match the periodicity of the cardiac cycle. Error-free cardiac cycles assist in communicating diagnostically lossless medical video. Intra updating (MB level prior to HEVC, CTU onwards) can match the d-ROI so that only the clinically sensitive regions are intra updated at specified intervals. Moreover, redundant slices of only the diagnostically important regions can be inserted in the communicated bitstream for maximizing error resilience [36]-[38]. Similarly, sophisticated error concealment or post processing techniques can be applied on the d-ROI at the receiver's end [42]. Afore-described schemes emphasize on enhancing/preserving the diagnostic capacity of the communicated medical video, while efficiently tackling the available resources.

3.5 Reliable Wireless Communication

Reliable wireless communication is of paramount importance in mHealth systems. The quality of the transmitted medical video cannot be compromised by the error-prone nature and varying state of wireless channels. Therefore, the appropriate mechanisms should be installed so that adaptation to the current wireless network state is performed. The latter involves both *a priori* adaptation models, as well as real-time adaptation methods. Diagnostically relevant procedures include unequal error protection (UEP) schemes for the clinically sensitive regions such as forward error protection (FEC) codes [35], [41]. Where applicable, retransmissions of data packets conveying d-ROI information can be also adopted. This is the case for 3.5G and beyond wireless networks provided low end-to-end delay for ARQ and HARQ schemes, yet this scheme's efficiency remains to be investigated. Toward this end, MAC layer service prioritization is now a standard feature in 3.5G systems, enabling key video streaming applications -such mHealth systems- to receive traffic prioritization, in addition to guaranteed bitrate and low latencies. This service allows for *a priori* adaptation to the wireless channel's specification and has become standard in medical video communication [35], [38], [42], [43].

Cross layer approaches are widely used in the literature for adapting to the wireless channel's varying state [41]-[45]. Real-time monitoring of wireless network's QoS parameters allows a decision algorithm to trigger a switch to a pre-constructed (encoder) state that preserves the desired QoS threshold values. These threshold values that correspond to medical video of acceptable diagnostic quality are determined a priori. A switch criterion is usually based on a weighted cost function that includes a fusion of source encoding and network parameters, objective VQA measurements (e.g., Peal Signal to Noise Ratio (PSNR)), and c-VQA (mean opinion score (MOS) of clinical ratings provided by the relevant medical experts).

3.6 Clinical Video Quality Assessment

Clinical VQA validates medical video communication systems objective of delivering medical video of adequate diagnostic quality over the wireless medium to a remote location. It largely differs from conventional perceptual quality evaluation, termed subjective quality. While low rate transmission errors may be generally tolerable in video streaming applications, in medical video communication, these errors cannot be allowed to compromise the diagnostic capacity of the transmitted medical video. At the remote physician's site the reconstructed medical video has to reproduce the quality of the in-hospital examination.

Central to the success of mHealth medical video communication systems is to *a priori* establish the range of diagnostically acceptable source encoding parameters. This is an essential pre-processing step which was primarily attributed to the inability of wireless networks to support medical video communication at the clinically acquired setup. The latter is expected to fade in the near future once 4G systems are widely deployed. Still, the minimum acceptable compression levels have to be clinically established by the relevant medical expert, given the variations between different medical video modalities. This procedure is a major component of the clinical evaluation procedure. Acceptable video resolutions that do not compromise the geometry characteristics, frame rates that preserve clinical motion, and compression ratios (both in terms of measured PSNR and quantization parameters) that maintain diagnostically acceptable quality levels are shown for different medical video modalities in [33], [36], [37], [40], [45], [46].

Presently, objective VQA algorithms do not correlate with medical experts mean opinion scores. As a result, there is a need to design new, diagnostically driven c-VQA metrics that will be eventually used to predict clinical scores. Clinical VQA is based on clinically established protocols that assess different clinical criteria. As already described, these clinical criteria often relate to specific d-ROI. This property can be exploited in future c-VQA metrics that will provide a weighted output function based on the region's clinical significance. The latter is briefly highlighted in [36], where VQA measurements performed over the primary d-ROI correlated better to the clinical ratings. The most detailed protocol for echocardiogram evaluation using c-VQA principles appears in [47]. Physiological properties -such as the periodicity of the cardiac cycle- are key features that relate to diagnostic capacity as highlighted earlier. Consecutive error-free cardiac cycles allow the medical expert to reach a confident diagnosis. This is a challenging research subject that has not been adequately addressed in the current literature.

Table 1 Selected MHealth Medical Video Communication Systems

	Author	**Year**	**Resolution, Frame Rate, BitRate**	**Encoding Standard**	**Wireless Network**	**Medical Video Modality**
Non-Diagnostically Driven Systems	Chu *et al.* [44] [2]	04	{320x240 and 160x120} <5fps 50-80 Kbps	M-JPEG	3G-CDMA	Trauma video
	Garawi *et al.* [45] [2,3]	06	176x144 @ 5fps 18.5-60 Kbps	H.263	3G-UMTS	Cardiac Ultrasound
	Alinejad *et al.* [43] [1, 2]	12	{176x144, 352x288} @ 10/20 fps {220, 430} Kbps, 1.3 Mbps	Windows Media Video (WMV)	3.5G - Mobile WiMAX, HSDPA	Cardiac Ultrasound
	Istepanian *et al.* [46] [2,3]	09	176x144 @ 8-10fps 50-130 Kbps	H.264/ AVC	3.5G –HSDPA	Abdomen Ultrasound
	Pedersen et al. [53] [2,3]	09	320x240 @ 10fps 349 Kbps	H.264/ AVC (Scalable)	WLAN and 3.5G - HSPA	Echocardiogram
	Hewage et al. [54] [1,3]	11	NA, 50fps, NA	H.264/AVC (MVC)	LTE	Endoscopy Procedures
	Panayides et al. [52] [2,3]	13	{176x144, 352x288, 560x416} @ 15fps, 64 – 768 Kbps	H.264/AVC	3.5G - HSPA	Carotid Artery Ultrasound
	Panayides *et al.* [49] [1,3]	13	560x416 @ 50fps, up to 1.7 Mbps	H.264/AVC and HEVC	3.5G and beyond	Carotid Artery Ultrasound

Table 1 (*continued*)

Diagnostically Driven Systems	Rao *et al.* [33][1,3,4]	09	360x240 @ 30 fps 500 Kbps	MPEG-2	3G and beyond	Pediatric respiratory distress related videos
	Tsapatsoulis et al. [34][1,3,4]	07	352x288 @ 10fps 10 videos average: 507 Kbps	MPEG-2 and MPEG-4	3G and beyond	Carotid Artery Ultrasound
	Martini *et al.* [35][1,4]	10	480x256@15fps 300 Kbps	H.264/AVC	3.5G - Mobile WiMAX	Cardiac Ultrasound
	Panayides *et al.* [36][1,3,4]	11	352x288 @ 15fps 197-421 Kbps	H.264/AVC	3G and beyond	Carotid Artery Ultrasound
	Panayides *et al.* [37][1,3,4]	12	{176x144, 352x288, 640x480} @15/30fps 200 Kbps -1.1 Mbps	H.264/AVC	3G and beyond	Abdominal Aortic Aneurysm
	Debono *et al.* [42][1,4]	12	640x480 @ 25fps	H.264/AVC	3.5G - Mobile WiMAX	Cardiac Ultrasound
	Khire *et al.* [39][1,3,4]	12	720x480 @ 30fps, 125 – 200 Kbps	H.264/AVC	3G and beyond	Maxillofacial Surgery Clips
	Panayides *et al.* [38][1,3,4]	13	704x576@15fps 768 Kbps -1.5 Mbps	H.264/AVC	3.5G - Mobile WiMAX	Carotid Artery Ultrasound
	Cavero *et al.* [40][1,3]	13	720x576 @ 25fps 40 Kbps (M-mode), 200 Kbps (B-mode)	SPIHT	3G and beyond	Cardiac Ultrasound
	Cavero *et al.* [41][1,3]	12	720x576 @ 25fps, 200 Kbps	SPIHT	3.5G - HSUPA, mobile WiMAX	Cardiac Ultrasound

[1]Simulation, [2]Real-Time, [3]Clinical Evaluation, [4]d-ROI.

4 Selected MHealth Medical Video Communication Studies

A synopsis of mHealth medical video communication systems developed over the previous decade which encompasses both diagnostically and non-diagnostically driven methodologies is depicted in Table I. A first interpretation of the table shows that mHealth systems are consistent with technology advances. Earlier studies employ earlier video coding standards, such as motion-JPEG, MPEG-2, and H.263 standards, while recent studies rely on the H.264/AVC standard. Undeniably, H.264/AVC has been one of the most successful video coding standards, and the profound choice in medical video communication systems over the previous decade. However, its successor HEVC, is already gaining ground in the literature [48], [49]. Similarly, progression in wireless networks infrastructure is imprinted via the utilization of 3G-based systems in the past decade, gradually replaced by early 3.5G wireless networks (e.g. HSDPA), before reaching currently dominant mobile WiMAX and HSPA(+) networks. Despite the wide presence in the research literature, WiMAX market penetration is rather poor compared to HSPA(+) and LTE -current and scheduled- deployments. As depicted in Table I, video resolution and frame rates are constantly increasing building on new codecs and wireless transmission media. It is important to note here that clinical quality is highly correlated to the specific source encoding parameters. Higher upload data rates accommodate higher video resolutions and frame rates that can now rival the clinically acquired setups. Communicating medical video of sufficient diagnostic quality at the clinically acquired setup is a milestone that will aid in the adoption of mHealth medical video communication systems in standard clinical practice.

4.1 Diagnostic-Driven Medical Video Communication Systems

In [33], the authors vary quality levels in the primary, d-ROI, and the background. Six different encoder states are considered based on different bit allocations for the two ROIs that result in different bitrate requirements. Based on bandwidth availability, a switch is triggered to the most relevant state. A switch is considered on a group of pictures intervals. The examined medical video modality is pediatric respiratory distress videos and the system is validated for diagnostically lossless performance based on a set of pre-determined clinical criteria.

A study associated with the transmission of ultrasound video of the common carotid artery (CCA) appears in [34]. The d-ROI is identified using an automated atherosclerotic plaque segmentation algorithm. Moreover, d-ROI specification is also based on a visual attention model of what human readers would find interesting. The remaining, non-diagnostically important regions are then blurred using a smoothing filter. Smoothed regions occupy significantly less bandwidth, given the high-compression ratios achieved when encoding smoothed areas using existing video coding standards. Medical experts verify that the proposed approach does not compromise the ultrasound videos' clinical capacity.

The authors in [35] employ variable quality slice encoding for cardiac ultrasonography video. The primary d-ROI is denoted by the clinician, followed by the fan-shaped sector produced by the ultrasound scanner, and the background region, comprising of electrocardiogram (ECG) and patient data. Different quantization parameters are assigned to different d-ROI based on diagnostic significance. Unequal error protection is used to protect more strongly the clinically sensitive regions, and preserve the ultrasound video's clinical capacity when transmitting over error-prone wireless channels.

In [36], [38] d-ROIs are identified using automated segmentation algorithms for atherosclerotic plaque ultrasound video [50]. The proposed diagnostically relevant encoding setup, which assigns quality levels as a function of the ultrasound video region's diagnostic significance, provides for substantial bitrate demands reductions for scalable resolutions (QCIF-4CIF). Diagnostically resilient encoding using methods available in the H.264/AVC standard shows that error resilient tools such as FMO, intra-updating, and redundant slices can deliver acceptable diagnostic quality even at severe packet loss rates. In-depth examination of the correlation between clinical ratings and VQA scores reveals that existing VQA methods correlate poorly to the medical experts' ratings. At the same time, the need for diagnostically relevant VQA algorithms is highlighted, as correlation was considerably improved when measuring VQA metrics over the primary d-ROI [36], rather than the whole video.

Efficient mobile WiMAX network parameter selection for maximizing the communicated video's diagnostic capacity is considered in [38]. Typical scenarios for emergency telemedicine evaluating different MAC and PHY layer characteristics, including service prioritization classes, modulation and coding schemes, fading channel's conditions, and mobility are investigated. Experimentation demonstrates that high-resolution low-delay ultrasound video communication is possible using existing mobile WiMAX networks. Recommendations for network parameter fine tuning are deducted based on c-VQA.

The afore-described diagnostically relevant (variable quality slice encoding) and resilient (redundant slices) encoding is also validated for abdominal aortic aneurysm (AAA) medical video transmission in [37]. Scalable video resolutions and frame rates for video communication in noisy environments (incremental packet loss rates) are investigated. Recommendations for diagnostically acceptable encoding setups are provided based on clinical evaluation using AAA clinical assessment protocol which validates different clinical criteria.

The authors in [39] propose an ROI-based system for the communication of emergency surgical videos, targeting tele-mentoring in disaster relief episodes and military combat fields. Real-time and low-bandwidth medical video communications compose the strict requirements for enabling expert physician feedback to be exploited during the surgical intervention. The first requirement is met by the open-source x264 codec, the most efficient H.264/AVC software implementation to date, which allows for real-time encoding of adequate quality video. The second criterion is satisfied by employing variable quality slice encoding by slightly modifying the x264 codec. The latter allows all H.264/AVC compliant decoders to render the transmitted video. An interesting feature of the proposed system is that the video camera transmitting the surgical video is mounted on the surgeon's head. Moreover, the d-ROI is fixed to the capturing device's screen center for simplicity purposes. This property allows for accommodating different emergency surgical videos communication.

A diagnostically resilient method applied at the receiver's side is proposed in [42]. The technique utilizes a post processing error-concealment algorithm, applied only on the d-ROI outlined by the relevant physician. The diagnostically resilient method is evaluated over a simulated mobile WiMAX topology, also employing the cross-layer approach introduced in [46] to adapt to the variable wireless channel's state. Experimental evaluation shows that the proposed approach achieves higher PSNR scores at the expense of slightly higher computational complexity.

A diagnostically driven technique for echocardiogram compression is proposed in [40]. Building on different echocardiogram modes properties, namely sweep and 2D modes, diagnostically relevant encoding is performed that minimizes bitrate requirements for equivalent clinical quality. More specifically, when in sweep modes, only the updated slice region is compressed at each frame, that corresponds to the new clinical information. When in 2D modes, a hybrid encoding process is adopted, where greyscale information is compressed using the 3D set partitioning in hierarchical trees (SPIHT) codec, whereas the color information using run length encoding (RLE). The afore-described method outperforms all approaches in the literature that neither distinguish between encoding modes nor take into account individual mode properties. Recommendations for source encoding parameters and associated bitrate demands that guarantee the integrity of the communicated ultrasound video are provided based on clinical evaluation. Clinical evaluation uses a customized version of the clinical distortion index introduced in [47]. The clinical distortion index is modified accordingly to reduce assessment time without compromising the efficacy of the evaluation process.

An application layer protocol for reliable wireless medical video communication, termed reliable clinical video transmission protocol (RCVTP) is proposed in [41]. RCVTP uses a combination of packet retransmissions and read-solomon FEC codes to enhance the clinical capacity of the communicated video. Essentially two modes of operation exist: a) retransmission mode, provided low transmission error rates, and b) a combination of both retransmissions and FEC codes, provided packet loss rates exceed a certain threshold. The latter is modeled using a markov state model and allows adaptation to the varying channel conditions. Experimental evaluation demonstrates the efficacy of the proposed approach when compared to conventional retransmission and FEC schemes employed separately.

4.2 Non-diagnostically Driven Systems Medical Video Communication Systems

The use of d-ROI provides for very efficient systems that target specific diseases. While the d-ROI provide for a direct way to address clinical criteria, defining ROIs can be a very difficult task and may not be always possible. Similarly, in the absence of diagnostically driven techniques that adapt to the underlying medical video modality, mHealth systems still need to be reliable to provide a dependable diagnosis. In this case, alternative approaches need to be explored.

One of the earlier studies in mHealth medical video communication systems appears in [44]. A portable tele-trauma system is proposed that features simultaneous transmission of trauma video, medical images, and ECG signals. The proposed system uses media transformation, data prioritization, and application-level congestion control to adapt to the wireless network's conditions during transmission. Data transformation uses JPEG and M-JPEG standards, for image and video compression, respectively. In order to preserve bandwidth in low-bitrate communication scenarios, the system choses to communicate only the d-ROI, denoted by the medical expert at the receiver's side upon the first received frame. Data prioritization assigns different priority levels to each of the transmitted media formats. More specifically, a high priority is selected for ECG transmission, followed by medium priority image transmission, while video communications are classified at the lowest priority scale. While the afore-described prioritization scheme may seem awkward as video is usually assigned a higher priority, limited data transfer rates of early 3G channels prohibited video communication at adequate bitrates. Moreover, video signals would dominate over other biomedical signals and often result in network congestion. During simultaneous transmission of the aforementioned priority data classes, when congestion occurs, the incorporated algorithm reduces the trauma video's frame rate to alleviate congestion effects leading to packet losses. System's performance evaluation investigates the maximum video resolution and frame rate that is possible for simultaneous streams transmission.

A performance analysis of a real-time end-to-end mobile tele-echography using an ultra-light robot (OTELO) system is presented in [45]. Low bitrate echography ultrasound medical video transmission over 3G channels is investigated and functional clinical bounds are provided for pre-diagnosis purposes. Ultrasound video quality (measured in terms of PSNR), video resolution, frame rate, and maximum delay that provide for acceptable diagnostic performance are the key source encoding and QoS parameters that are investigated. Acceptable values are based on clinical evaluation performed by the expert physician, who is also responsible for remotely monitoring patients using the tele-operated robot.

Alinejad et al. [43] describe a cross-layer approach for real-time adaptation to the wireless channel's varying state. The cross-layer optimizer uses the reinforcement learning approach described in [46]. The objective is to minimize a cost function which comprises of source and channel encoding parameters, as well as objective VQA measurements, for maximizing the clinical capacity of the communicated video. More specifically, medical video quality is measured in terms of video resolution and frame rate, in addition to the PSNR scores. Channel parameters are abstracted to end-to-end delay. The efficacy of the cross-layer optimizer approach is validated over commercially available HSUPA mobile cellular network and mobile WiMAX wireless network (in a controlled lab environment using own equipment). Mobile WiMAX achieves better objective ratings than HSUPA, while it provides for higher resolution medical video communication.

Based on the concept of reinforcement learning, the authors in [46] propose an ultrasound video streaming rate control algorithm based on QoS. The proposed algorithm varies the ultrasound video's frame rate based on the channel's state. At the same time, a pre-defined medical QoS criterion prohibits the algorithm to reduce the frame rate and hence quality indexes in a way that clinical motion will be compromised. The medical QoS criterion and clinically acceptable bounds are described in [45]. Robotic tele-ultrasonography system OTELO is also used in this study for abdominal ultrasound medical video transmission. The efficiency of the proposed method is validated with respect to the default rate–control algorithm of the JM H.264/AVC reference software. Both simulations and real time setups (3.5G HSDPA wireless network) are performed and the achieved results are verified by the relevant medical expert.

The OTELO system is also used in [51], where cross layer control is used to fine tune source and channel encoding parameters for maximizing the clinical quality of the transmitted ultrasound video. Different encoder states are constructed by varying key source encoding parameters like compression ratio, frame rate, and intra-update intervals, as well as channel encoding parameters such as average code-rate channel protection. A switch is triggered to the most relevant state based on medical video's quality index, underlying wireless channel's signal-to-noise ratio, and other QoS network parameters (packet loss rates, delay, and delay jitter) on a per second basis. Experimentation using wireless local area networks shows that the proposed cross layer adaptation scheme significantly enhances communicated video's quality (measured in PSNR).

An open-source platform for reliable medical video communication is presented in [52]. The software package is based on widely used open-source software tools (such as x264, ffmpeg, videolan) and VQA algorithms. Experimental evaluation using different scenarios including a plethora of different source encoding parameters shows that diagnostically acceptable medical video communication is possible using existing wireless infrastructure and open-source video processing software. Medical videos are transmitted over commercially available HSPA wireless network in Cyprus. The clinical evaluation shows that while clinical motion is affected by temporal freezes introduced by real-time communications, error-free cardiac cycles allows the physician to proceed to a confident diagnosis. Toward this end, when variable frame delay (VFD) algorithm -which removes temporal mismatch between compared frames- is used as a pre-processing step to VQA, it enhances correlation between objective VQA measurements and clinical scores. Implementing an mHealth medical video communication system using low cost means (open-source software) is of paramount importance in establishing tele-healthcare systems in standard clinical practice, especially in developing countries.

Scalable video coding for generating different ultrasound video instances for communication over different wireless networks is depicted in [53]. The effective spatiotemporal setting supported by the underlying wireless channel is investigated using QoS parameters and medical expert's feedback. More specifically, data transfer rates, packet loss rates, end-to-end delay and delay jitter are measured for different scalability layers streamed over wireless local area networks and early 3.5G channels (HSDPA). Based on the examined parameters and in conjunction with the MOS provided by medical experts, recommendations regarding the higher SVC layer supported by each wireless channel are provided. Clinical evaluation assesses different clinical criteria.

A very challenging task due to the relatively limited usage of 3D capturing devices is 3D medical video communication over wireless networks. Challenges stem from additional bandwidth requirements, real-time video processing requirements, investigation of the effect of channel errors on perceived clinical quality, and so forth. Preliminary results in [54] suggest that 3D endoscopy clips transmission over LTE wireless networks is possible using multi-view coding (H.264/MVC). Naturally, the introduction of HEVC and higher data rate channels will trigger further studies to investigate the clinical benefits of 3D wireless medical video communication.

5 Concluding Remarks and Future Work

The objective of mHealth medical video communication systems is to establish how diagnostic decisions can be safely made based on the communicated medical video. The ultimate goal is to establish such systems and services in standard clinical practice. Despite the numerous approaches in the literature over the past decades, there has been limited adoption in daily clinical practice and/or routine examinations. Recently released HEVC video coding standard and already deployed 4G (and fast approaching 5G) wireless networks, are expected to trigger a new era in mHealth video systems. The latter will emerge from supporting medical video communication with the low delay and quality requirements (video resolution, frame rate, and compression ratio) for reliable and dependable remote diagnosis. Beyond high definition coding capabilities of the new HEVC standard linked with wired comparable upload data transfer rates and near negligible latencies provided by 4G channels, are expected to materialize the afore-described predictions. Medical video communication that can match the quality of in-hospital examinations is expected to play a decisive role toward wider adoption.

The challenge is developing the appropriate mechanisms that will guarantee that the communicated video is diagnostically acceptable. In other words, the remote medical expert should have a strong indication while the medical video is being rendered on the remote screen that the particular video quality is of adequate clinical capacity and has not been compromised in any way. Likewise, in case quality has been compromised, the appropriate signaling should inform the physician that the diagnostic capacity of the displayed medical video does not match the original's one. The absence of efficient VQA metrics that correlate with clinical scores strongly highlights the need of developing new methods that will reproduce the ratings provided by the relevant medical experts.

Such metrics need to be diagnostically driven. Unique properties of the communicated medical video modality ought to be considered in the design of such metrics, and therefore the resulting metrics should be dynamically adjustable. To cater for different modality characteristics, weighted VQA scores based on d-ROI is of paramount importance. Most VQA algorithms today do not distinguish between video regions, and as a result do not qualify for use in c-VQA. Clinically sensitive regions should be assigned individual VQA scores, which will then be weighted accordingly. The weight factor should reflect the associated contribution to the evaluation of individual clinical criteria. In this manner, the VQA metric will, in a sense, mimic the clinical evaluation performed by the medical expert, bridging the gap between objective measurements and clinical scores. Moreover, physiological properties such as the number of consecutive diagnostically lossless cardiac cycles that are needed to provide a confident diagnosis for each modality should be determined. The latter is a key component in the development of efficient, diagnostically driven video quality metrics. It allows the VQA metric to quantify the qualitative clinical information displayed. Toward this end, source encoding parameters -such as resolution and frame rate- that do not compromise the diagnostic capacity of the communicated video should be assigned the proper weight in a diagnostically relevant VQA metric. Incremental (scalable) video resolution and frame rate correspond to incremental clinical capacity. *A priori* constructed look up tables for each medical video modality will aid in the design of reliable VQA metrics. Ultimately, the goal is to *predict clinical quality using automated techniques*, facilitated by computer aided diagnosis systems in standard clinical practice.

References

[1] Pattichis, C.S., Kyriacou, E., Voskarides, S., Pattichis, M.S., Istepanian, R.: Wireless telemedicine Systems: An Overview. IEEE Antennas and Propagation Magazine 44(2), 143–153 (2002)

[2] Istepanian, R.H., Laxminarayan, S., Pattichis, C.S.: M-Health: Emerging Mobile Health Systems. Springer, New York (2006)

[3] Kyriacou, E., Pattichis, M.S., Pattichis, C.S., Panayides, A., Pitsillides, A.: M-Health e-Emergency Systems: Current Status and Future Directions. IEEE Antennas and Propagation Magazine 49(1), 216–231 (2007)

[4] American Institute for Medical and Biological Engineering, C. Medical and Biological Engineering in the Next 20 Years: The Promise and the Challenges. IEEE Transactions on Biomedical Engineering. 60(7), 1767–1775 (2013)

[5] Resolution WHA58.28. eHealth. In: Fifty-eighth World Health Assembly, Geneva. Annex. Resolutions and decisions. Geneva: World Health Organization (WHA58/2005/REC/1), May 16–25 (2005), `http://apps.who.int/gb/or/e/e_wha58r1.html` (accessed September 2013)

[6] World Health Organization. Telemedicine: Opportunities and Developments in Member States. Global Observatory for eHealth Series, vol. 2 (2011)

[7] World Health Organization. mHealth: New Horizons for Health Through Mobile Technologies. Global Observatory for eHealth Series, vol. 3 (2011)

[8] Mistry, H.: Systematic review of studies of the cost-effectiveness of telemedicine and telecare. Changes in the economic evidence over twenty years. Journal of Telemedicine and Telecare 18(1), 1–6 (2012), doi:10.1258/jtt.2011.110505

[9] West, D.: How mobile devices are transforming healthcare. Issues in Technology Innovation 18, 1–14 (2012)

[10] Panayides, A., Pattichis, M.S., Pattichis, C.S., Schizas, C.N., Spanias, A., Kyriacou, E.C.: An Overview of Recent End-to-End Wireless Medical Video Telemedicine Systems using 3G. In: Proc. of IEEE EMBC 2010, Buenos Aires, Argentina, August 31-September 4 (2010)
[11] Panayides, A., Pattichis, M.S., Pattichis, C.S., Pitsillides, A.: A Tutorial for Emerging Wireless Medical Video Transmission Systems [Wireless Corner]. IEEE Antennas & Propagation Magazine 53(2), 202–213 (2011)
[12] Panayides, A.S.: Diagnostically Resilient Encoding, Wireless Transmission, and Quality Assessment of Medical Video, Ph.D Dissertation, Department of Computer Science, University of Cyprus, Nicosia (2011)
[13] Panayides, A., Pattichis, M.S., Constantinides, A.G., Pattichis, C.S.: M-Health Medical Video Communication Systems: An Overview of Design Approaches and Recent Advances. In: Proc. of IEEE EMBC 2013, Osaka, Japan, July 3 – 7, pp. 7253–7256 (2013)
[14] Panayides, A.S., Pattichis, M.S., Pattichis, C.S.: Mobile-Health Systems Use Diagnostically Driven Medical Video Technologies [Life Sciences]. IEEE Signal Processing Magazine 30(6), 163–172 (2013)
[15] ITU-T, Video codec for audiovisual services ar px64 kbit/s, ITU-T Recommendation H.261 (November 1990)
[16] ITU-T, Information Technology – Generic Coding of moving pictures and associated audio information: video, ITU-T Recommendation H.262 (July 1995)
[17] ITU-T, Video coding for low bitrate communication, ITU-T Recommendation H.263 (November 1995)
[18] ITU-T, Advanced video coding for generic audiovisual services, ITU-T and ISO/IEC 14496-10 Recommendation H.264 (MPEG4-AVC) (May 2003)
[19] Wiegand, T., Sullivan, G.J., Bjontegaard, G., Luthra, A.: Overview of the H.264/AVC video coding standard. IEEE Transactions on Circuits and Systems for Video Technology 13, 560–576 (2003)
[20] ITU-T, H.265: High efficiency video coding, ITU-T Recommendation H.265 (June 2013)
[21] Ohm, J.-R., Sullivan, G.J., Schwarz, H., Tan, T.K., Wiegand, T.: Comparison of the Coding Efficiency of Video Coding Standards – Including High Efficiency Video Coding (HEVC). IEEE Trans. Circuits and Systems for Video Tech. 22(12), 1669–1684 (2012)
[22] Sullivan, G.J., Ohm, J.-R., Han, W.-J., Wiegand, T.: Overview of the High Efficiency Video Coding (HEVC) Standard. IEEE Trans. Circuits and Systems for Video Tech. 22(12), 1649–1668 (2012)
[23] Marpe, D., Schwarz, H., Wiegand, T., Bobe, S., Bross, B., Lakshman, H., Nguyer, T., Oudin, S., Siekmann, M., Karsten, S., Winken, M.: Improved video compression technology and the emerging high efficiency video coding standard (Berlin). IEEE (2011)
[24] Tikekar, M.M.D.: Circuit implementations for high-efficiency video coding tools. PhD thesis, Massachusetts Institute of Technology (2012)
[25] Pourazad, M.T., Doutre, C., Azimi, M., Nasiopoulos, P.: HEVC: the new gold standard for video compression. IEEE Consumer Electronics Magazine (July 2012)
[26] Rysavy Research, L.L.C.: Mobile Broadband Explosion: The 3GPP Wireless Evolution (August 2013), http://www.4gamericas.org/
[27] ITU-R Rep. M.2134, Requirements Related to Technical Performance for IMT-Advanced Radio Interface(s) (November 2008)

[28] The Draft IEEE 802.16m System Description Document (SDD). IEEE 802.16 Broadband Wireless Access Working Group (July 2008)
[29] Ahmadi, S.: An overview of next-generation mobile WiMAX technology. IEEE Communications Magazine 47(6), 84–98 (2009)
[30] Papapanagiotou, I., Toumpakaris, D., Jungwon, L., Devetsikiotis, M.: A survey on next generation mobile WiMAX networks: objectives, features and technical challenges. IEEE Communications Surveys & Tutorials 11(4), 3–18 (2009)
[31] Ghosh, A., Ratasuk, R., Mondal, B., Mangalvedhe, N., Thomas, T.: LTE-advanced: next-generation wireless broadband technology (Invited Paper). IEEE Wireless Communications 17(3), 10–22 (2010)
[32] Parkvall, S., Furuskär, A., Dahlman, E.: Evolution of LTE toward IMT-advanced. IEEE Communications Magazine 49(2), 84–91 (2011)
[33] Rao, S.P., Jayant, N.S., Stachura, M.E., Astapova, E., Pearson-Shaver, A.: Delivering Diagnostic Quality Video over Mobile Wireless Networks for Telemedicine. International Journal of Telemedicine and Applications Article ID 406753, 9 (2009)
[34] Tsapatsoulis, N., Loizou, C., Pattichis, C.: Region of Interest Video Coding for Low bit-rate Transmission of Carotid Ultrasound Videos over 3G Wireless Networks. In: Proc. of 29th Annual Conference of the IEEE Engineering in Medicine and Biology Society, IEEE EMBC 2007, Lyon, France, August 22-26, pp. 3717–3720 (2007)
[35] Martini, M.G., Hewage, C.T.E.R.: Flexible Macroblock Ordering for Context-Aware Ultrasound Video Transmission over Mobile WiMAX. International Journal of Telemedicine and Applications 2010, Article ID 127519, 14 (2010)
[36] Panayides, A., Pattichis, M.S., Pattichis, C.S., Loizou, C.P., Pantziaris, M., Pitsillides, A.: Atherosclerotic Plaque Ultrasound Video Encoding, Wireless Transmission, and Quality Assessment Using H.264. IEEE Transactions on Information Technology in Biomedicine 15(3), 387–397 (2011)
[37] Panayides, A., Antoniou, Z., Barberis, V., Pattichis, M., Pattichis, C., Kyriacou, E.: Abdominal aortic aneurysm medical video transmission. In: Proc. IEEE-EMBS Int. Conf. Biomed. Health Inf., Hong Kong, China, January 2-7, pp. 679–682 (2012)
[38] Panayides, A., Antoniou, Z., Mylonas, Y., Pattichis, M.S., Pitsillides, A., Pattichis, C.S.: High-Resolution, Low-delay, and Error-resilient Medical Ultrasound Video Communication Using H.264/AVC Over Mobile WiMAX Networks. IEEE Journal of Biomedical and Health Informatics 17(3), 619–628 (2013), 10.1109/TITB.2012.2232675
[39] Khire, S., Robertson, S., Jayant, N., Wood, E.A., Stachura, M.A., Goksel, T.: Region-of-interest video coding for enabling surgical telementoring in low-bandwidth scenarios. In: Proc. of MILCOM 2012, October 29 - November 1, pp. 1–6 (2012)
[40] Cavero, E., Alesanco, A., Castro, L., Montoya, J., Lacambra, I., Garcia, J.: SPIHT-Based Echocardiogram Compression: Clinical Evaluation and Recommendations of Use. IEEE Journal of Biomedical and Health Informatics 17(1), 103–112 (2013)
[41] Cavero, E., Alesanco, A., Garcia, J.: Enhanced protocol for real time transmission of echocardiograms over wireless channels. IEEE Transactions onBiomedical Engineering 59(11), 3212–3220 (2012)
[42] Debono, C., Micallef, B., Philip, N., Alinejad, A., Istepanian, R., Amso, N.: Cross Layer Design for Optimised Region of Interest of Ultrasound Video Data over Mobile WiMAX. IEEE Transactions on Information Technology in Biomedicine 16(6), 1007–1014 (2012)

[43] Alinejad, A., Philip, N., Istepanian, R.: Cross Layer Ultrasound Video Streaming over Mobile WiMAX and HSUPA Networks. IEEE Transactions on Information Technology in Biomedicine 16(1), 31–39 (2012)

[44] Chu, Y., Ganz, A.: A Mobile Teletrauma System using 3G Networks. IEEE Transactions onInformation Technology in Biomedicine 8(4), 456–462 (2004)

[45] Garawi, S.A., Istepanian, R.S.H., Abu-Rgheff, M.A.: 3G Wireless Communication for Mobile Robotic Tele-ultrasonography Systems. IEEE Communications Magazine 44(4), 91–96 (2006)

[46] Istepanian, R.S.H., Philip, N.Y., Martini, M.G.: Medical QoS Provision based on Reinforcement Learning in Ultrasound Streaming over 3.5G Wireless Systems. IEEE Journal on Selected Areas in Communications 27(4), 566–574 (2009)

[47] Alesanco, A., Hernandez, C., Portoles, A., Ramos, L., Aured, C., Garcıa, M., Serrano, P., Garcıa, J.: A clinical distortion index for compressed echocardiogram evaluation: recommendations for Xvid codec. Physiological Measurement 30(5), 429–440 (2009)

[48] Panayides, A., Antoniou, Z., Pattichis, M.S., Pattichis, C.S., Constantinides, A.G.: High Efficiency Video Coding for Ultrasound Video Communication in M-Health Systems. In: Proc. of IEEE EMBC 2012, California, USA, August 28 – September 1, pp. 2170–2173 (2012)

[49] Panayides, A.S., Pattichis, M.S., Loizou, C.P., Pantziaris, M., Constantinides, A.G., Pattichis, C.S.: An Effective Ultrasound Video Communication System Using Despeckle Filtering and HEVC. IEEE Journal of Biomedical and Health Informatics (in press)

[50] Loizou, C.P., Pattichis, C.S., Pantziaris, M., Nicolaides, A.: An integrated system for the segmentation of atherosclerotic carotid plaque. IEEE Trans. Inf. Technol. Biomed. 11(5), 661–667 (2007)

[51] Martini, M.G., Istepanian, R.S.H., Mazzotti, M., Philip, N.: Robust multi-layer control for enhanced wireless tele-medical video streaming. IEEE Trans. on Mobile Computing 9(1), 5–16 (2010)

[52] Panayides, A., Eleftheriou, I., Pantziaris, M.: Open-Source Telemedicine Platform for Wireless Medical Video Communication. International Journal of Telemedicine and Applications 2013, Article ID 457491, 12 (2013), doi:10.1155/2013/457491

[53] Pedersen, P.C., Dickson, B.W., Chakareski, J.: Telemedicine applications of mobile ultrasound. In: Proc. of MMSP 2009, October 5-7, pp. 1–6 (2009)

[54] Hewage, C., Martini, M.G., Khan, N.: 3D medical video transmission over 4G networks. In: Proc. 4th Int. Symp. on Appl. Sci. in Biomedical and Commun. Technologies, Barcelona, Spain (October 2011)

27. Mobile Health Improves Healthcare Delivery

Israel R. Kabashiki[*]

IZ New Consulting
consulting@izglobalexpertise.com

Abstract. The healthcare industry suffers from a high degree of fragmentation which results in inefficiencies. It is widely recognized that the use of Information and Communication Technology (ICT) systems in health care has the potential to improve healthcare systems' efficiencies. The use of ICT in health care is referred as electronic health (eHealth). Four key steps are deemed essential for eHealth to improve healthcare delivery.

First, the healthcare industry should adopt ICT systems that can allow effective health information management. Health ICT systems, such as Electronic Health Record (EHR), Electronic Medical Record (EMR), and Personal Health Record (PHR) can considerably improve the delivery of medical services.

Second, with today's increase of health consumerism, the integration of PHR and EHR systems is needed to facilitate physician-patient partnership and support patients' self-management.

Third, for ICT systems to improve healthcare delivery, they should provide healthcare providers and patients with uninterrupted services and timely access to patients' records. Distance-irrelevant and mobile connectivity systems are required to make relevant health information readily available to healthcare stakeholders irrespective of geography and time.

Fourth, to further improve healthcare delivery, eHealth should be supported by mobile technologies. This practice is referred to as mobile health.

Today, nearly six billion individuals across the world use mobile devices. The commercial wireless signals coverage is increasing. Thus, mobile health can significantly improve healthcare delivery – in terms of health costs containment and patients' access to care. For this to happen, several obstacles to the mobile adoption must be overcome.

1 Introduction

Several views expressed in literature suggest that electronic health (eHealth) is an essential tool for enhancing healthcare delivery and patient care. Even though the healthcare industry is information-intensive, it suffers from a high level of

[*] Corresponding author.

S. Adibi (ed.), *Mobile Health*,
Springer Series in Bio-/Neuroinformatics 5, DOI: 10.1007/978-3-319-12817-7_27

fragmentation that results in inefficiencies (Powner 2005). In this chapter, we examine four steps deemed critical to fixing such inefficiencies.

Information and Communication Technology (ICT) has empowered banks and airlines' customers to have access to a wide collection of services irrespective of time and location. ICT has not had the same effect on the healthcare industry. Even though it is widely recognized that effective use of ICT systems has the potential to better the quality of healthcare delivery, the adoption of modern ICT in the healthcare industry has been sluggish for several reasons. When used effectively, ICT systems can provide the healthcare industry with various benefits. These benefits include among others: (a) enhanced health information management, (b) timely access to patients' health records, (c) better collaboration among healthcare providers, and (d) improved communication between healthcare providers and their patients.

With the increase of health consumerism, individuals prefer to participate in, and influence decisions about their own care. Moreover, as healthcare providers' preference is to view patients as partners in healthcare processes, improved communication between the two is critical to establishing physician-patient partnership. The integration of electronic health record systems and personal health record systems has the potential to facilitate such partnership.

Also, the healthcare industry experiences various challenges, like the shortage of healthcare providers, especially in high-demand specialties. Furthermore, today's patient care requires an interdisciplinary teamwork approach involving healthcare professionals operating either locally or across national boundaries. Working over long distance however, poses challenges for effective collaboration and information sharing.

The quality of healthcare delivery depends on the degree of ICT systems adoption in health care. Not only can this adoption allow effective teamwork, but it can also connect scattered healthcare stakeholders together. For this reason, there is a need for ICT systems that have the potential to overcome time and distance barriers. Distant-irrelevant systems can provide patients with timely access to medical expertise from healthcare providers across the globe.

For ICT to truly create efficiencies and improve healthcare delivery, it is essential that all healthcare stakeholders have timely access to relevant health information. Therefore, to further improve healthcare delivery, eHealth should be supported by the use of mobile technologies. The need for timely and around-the-clock access to patients' health records calls for a seamless and secure cross platform. This cross platform will be necessary to support a broad range of hand-held devices used by healthcare providers and healthcare consumers.

Today, nearly six billion people across the world use wireless devices; the commercial wireless signals coverage represents over 85 percent of the world's population (World Health Organization, 2011). Thus, mobile health can significantly improve healthcare delivery – in terms of health costs containment and patients' access to care by reaching populations in underserved regions of our planet. For this to happen, several obstacles to the adoption of mobile technologies in health care must be overcome.

2 Electronic Health and Healthcare Delivery

Today, healthcare providers are required to operate in a team-oriented work environment to provide high quality care for patients. As patient care requires an interdisciplinary teamwork approach involving healthcare professionals operating either locally or across the globe, information management plays a considerable role in the provision of health care. The delivery of health care in a team-oriented setting necessitates effective management of not only the patient and expenditure flows, but also the health information flows.

To meet the increasing health needs of healthcare consumers and the expectations of the healthcare providers, it is crucial that healthcare practice be supported by electronic processes and communication. This is referred to as electronic health (eHealth). The goal of eHealth is to achieve the delivery of safer and more efficient care to healthcare consumers. Electronic health services examined in this chapter include: (a) personal health records or personal health portals, (b) electronic medical records, and (c) electronic health records.

When used effectively, these health ICT systems can revolutionize the healthcare industry and improve the quality of health care. This is possible for several reasons. They allows for: patients' participation in their care, better health information sharing, and improved relationship between healthcare consumer and healthcare provider. Also, ICT is seen as a vehicle to achieve two goals in health care: performance and transparency of the system.

The problem is that the healthcare industry experiences a high level of fragmentation that results in significant inefficiencies (Powner 2005). The majority of health ICT systems have been implemented in silos. Furthermore, the balkanization of the ICT infrastructure in the healthcare industry intensifies the fragmentation (Cebul *et al.* 2008). The use of various health ICT systems that cannot communicate with each other has been a major problem in health care, especially when the care of a patient is shared between different healthcare organizations. Therefore, a key step in fixing the healthcare delivery system is thought to be the adoption of common electronic and linguistic communication standards (Swartz 2006).

3 Health Information System in Support of Physician-Patient Partnership and Self-management

In this section, we provide a brief discussion about three characteristics of the twenty-first century health care: physician-patient partnership, health consumerism, and uninterrupted healing relationship. Herein, we discuss the need for integrating EHR with PHR systems to facilitate physician-patient partnership. We also discuss the use of ICT to promote health consumerism and engage individuals in their care.

3.1 Physician-Patient Partnership

Physician-patient partnership is critical to bettering healthcare delivery. Involving healthcare consumers in their care processes improves the quality of care they receive. To drive quality, healthcare consumers should become active participants in health care. Their responsibility should consist of working collaboratively with healthcare providers to achieve their treatment goals. Building working partnerships in the provision of care to patients is thought to be a useful approach to improving the quality of healthcare delivery (Hunter and Cameron 2006).

PHR systems can play a key role in building working partnerships between healthcare providers and patients. This is possible because PHR systems facilitate health information sharing between healthcare providers and patients. When used effectively, PHR systems provide patients and healthcare providers with means to develop effective collaboration. Effective use of PHR systems has the potential to transform the relationship between healthcare providers and patients and, to meet two requirements of the present century: effective physician-patient partnership and uninterrupted healing relationship.

Personal health record systems benefit healthcare consumers and healthcare providers. The use of PHR systems to support patients' self-management extends the reach of healthcare providers. This is possible because PHR systems allow for effective sharing of health information between healthcare consumers and healthcare providers (Solomon 2008). Furthermore, as healthcare providers prefer to view the patient as a valuable partner in the healthcare processes (Car and Sheikh 2004), improved communication between the two is essential for effective partnerships.

Effken and Abbott (2009) believe that integrating PHR systems with EHR systems can facilitate partnership between healthcare providers and their patients. An effective integration of these two systems has the potential to profoundly transform the relationship between healthcare providers and patients and meet three main requirements of the twenty-first-century health care. Those requirements include: effective physician-patient partnership, patient self-management, and uninterrupted healing relationship.

Tang and Lansky (2005) opined that new approaches to better health care should allow for an uninterrupted healing relationship between healthcare providers and healthcare consumers. Integrated health ICT systems in support of around-the-clock and mobile connectivity can play an essential role in this regard.

3.2 EHR-PHR Integration

In this information technology driven world, healthcare consumers should be able to interact with their healthcare providers without the travel hardship. The integration of these two health ICT systems provides the possibilities of delivering integrated and coordinated health care, as well as improved quality of care. According to Susilo and Win (2007), advances in ICT systems have led to collaborative research work and storage of health data in a single research database. The use of centralized research databases results in improved patient-oriented care (Katic *et al.* 2007).

The integration of EHR systems and PHR systems allows healthcare consumers to have access to some medical services at distance. It also provides healthcare providers with more time to spend with patients who require face-to-face consultations. Integrating these two systems empowers healthcare consumers as it allows them to collaborate with their healthcare providers as equal partners. Establishing a secure link between PHR systems and EHR systems provides individuals with full access to their medical charts. Such access provides them with several benefits.

First, it allows healthcare consumers to review the content of their charts. In other words, they will be able to review their prescription histories, medical notes, laboratory test results, and diagnostic imaging results. Second, it provides them with the ability to control their health records. In other words, healthcare consumers can learn who had accessed their medical records. Third, web-based PHR systems enable individuals to request and consume some health services available online from anywhere and anytime. For instance, medical services, like the booking of medical appointments or prescription refills, can be consumed online.

To reap the above benefits, fragmentations experienced in the healthcare industry must be addressed in an effective way. The healthcare landscape must be flattened. Four main requirements must be met to flatten the healthcare landscape. First, it is imperative to integrate all stand-alone EMR systems and EHR systems. Second, the adoption of common electronic and linguistic communication standards is essential. Third, it is necessary to use a seamless and secure cross platform to support EHR systems and a broad range of handheld devices used by healthcare providers and healthcare consumers. Fourth, integrated health systems require either the use of a shared database or effective exchange of health data housed in the databases of the systems involved.

3.3 Health Consumerism and Health Outcomes

Today, the healthcare industry is under pressure from increasing demands on the quality of medical services, the competency of healthcare professionals, and the efficiency of healthcare organizations. As consumerism drives more transparency in costs and quality of health care, the healthcare industry ought to demonstrate superior health outcomes and value (Lefton 2008). With the increase of health consumerism, healthcare consumers want to be active participant in their care. They like to influence decisions affecting their healthcare system (Effken and Abbott 2009).

As they have the potential to improve communication between healthcare consumers and healthcare providers, ICT systems, like PHR, increase patient's participation in health care. Grossman et al. (2009) believe in the potential of health ICT systems to engage patients in their health care. The implementation of web-based PHR systems allows healthcare consumers to be active participant in their care irrespective of time and location. In other words, ICT empowers healthcare consumers to play a critical role in the betterment of their safety by becoming actively and proactively involved in their health care.

Gonzalez (2007) argues that providing individuals with tools or resources that empower them to become better healthcare consumers is crucial to promoting consumerism. Health ICT empowers individuals to play a crucial role in the improvement of their safety. This is possible because health ICT allows people to become actively involved in their health care (Davis et al. 2007, Longtin et al 2010). With PHR systems acting much like automated teller machine cards, individuals have access to their own personal health information and several other health services regardless of time and location (Ball *et al.* 2008). In the Danish health care system, for instance, ICT has empowered individuals to improve self-care by enabling them to review their health information such as, laboratory results and prescriptions (Protti et al. 2008a).

4 Health Information Systems in Support of Uninterrupted Connectivity

In this section, we discuss the concept of uninterrupted connectivity and its impact on healthcare delivery. Our discussion covers: sensor and wireless technologies, distance-irrelevant systems, and cloud computing. Also, we touch on the impact of these technologies on health care.

4.1 Sensor and Wireless Technologies

The use of wireless, sensor, and portable device technologies can enable individuals to provide continual services to their organizations. Wireless access points, as well as temperature and location sensors points, can be embedded in the environment to automatically connect individuals entering the environment (Zheng and Lun Lee 2005). By allowing personal computing devices to be adapted to the context, sensors and wireless access make critical information readily available to individuals to deal with the tasks at hand (Zheng and Lun Lee 2005). With effective security protocols, wireless access points can be as secure as wired networks.

A crucial step toward the realization of this universal environment is timely distribution of relevant information to information seekers (Zheng and Lun Lee 2005). Two approaches are usually used in the dissemination of data to mobile clients. The first approach, known as on-demand access, uses point-to-point connection, in which a mobile client sends a request to the server. The server processes the query and returns the results directly to the mobile client. The second approach is known as periodic broadcast or data broadcast. In this approach, the server determines data to broadcast and their schedule on the wireless channel. It is the mobile client's responsibility to listen to the channel and choose the data to retrieve.

On-demand access provides fast service in a system with relatively small number of queries from clients. The performance of on-demand access deteriorates with the increase of the system workload because of the competition for bandwidth among mobile clients. In general, data broadcast or periodic broadcast is an appealing alternative to on-demand because of its potential to broadcast data concurrently to a large number of mobile clients.

The benefit of having a pervasive environment in the healthcare industry is the omnipresent access to health information by healthcare stakeholders. Timely and relevant health data are made available to healthcare providers, healthcare consumers, policy makers, and decision makers regardless of time and location. Moreover, ICT systems in support of distributed health teams must allow for uninterrupted medical services. By using portable devices, wireless, and sensor technologies to implant computing technologies transparently in their environments, healthcare systems can provide healthcare providers with uninterrupted health services.

4.2 Distance-Irrelevance and Mobile Connectivity

As the globalization of work has transformed today's workplace into a transparent landscape with less or no reliance on physical location, effective distribution of knowledge requires ICT systems that can span distance. "Wireless networking and portable digital devices provide people with unrestricted mobility" (Zheng and Lun Lee 2005 p. 109). ICT in support of distributed teams must have two critical attributes: distance-irrelevance and unrestricted connectivity. As distance-irrelevant ICT systems overcome time and space boundaries, organizations can benefit from the knowledge and expertise of their dispersed knowledge workers (Johnson *et al.* 2001). Cost and time savings are additional benefits that can be derived from the use of distance-irrelevant ICT systems. This is possible because these systems reduce the need for face-to-face meetings.

ICT systems in support of geographically dispersed health teams must allow for uninterrupted services and mobile connectivity. The advent of portable computing devices, wireless, and sensor technologies can enable healthcare systems to embed computing technologies transparently in their environment to provide uninterrupted health services. This way, healthcare providers and healthcare consumers entering that environment will automatically connect to it. In other words, sensor, wireless, and portable device technologies can allow personal computing devices to adapt to this context and to make relevant information readily available for effective decision making. Such pervasive environment provides healthcare providers, healthcare consumers, policy makers, and decision makers with better access to relevant health information to provide or consume medical services.

4.3 Cloud Computing and Health Care Delivery

Cloud computing is more than a technology. It is a business and service model in which ICT resources, such as networks, servers, data storage, software applications, are treated as a consumption-based service. Cloud computing allows organizations to use a web of remote servers hosted on the Internet, rather than local servers. Also, in this business and service model, organizations use hardware and software maintenance as a utility. By using ICT resources and related services as a utility, organizations can focus on core competencies.

There have been talks about integrating health ICT systems with the cloud. These talks have some merit as cloud computing can provide healthcare systems around the world with numerous benefits both at the micro and macro levels. At the macro level for instance, cloud computing can help deploy EHR systems in a cost-effective way. Cloud computing pay-as-you-grow billing model can address two big factors that hinder the implementation of EHR systems in this recovering economy. These factors include increasing costs of health care and insufficient financing.

At the micro level, the use of cloud-based electronic medical record systems can be greatly cost-effective. The majority of private medical practices lack the technical expertise required to support new technologies. Cloud computing lessens the burden of hiring internal ICT expertise to maintain and service in-house infrastructure for EMR systems and other applications used in healthcare delivery. Furthermore, small healthcare facilities and private medical practices do not have sufficient funds to purchase state-of-art health ICT systems. In the cloud, the pay-as-you-grow billing model will provide these healthcare facilities and medical practices with great opportunities to use modern health ICT systems without purchasing them. In this way, the cloud computing model has the potential to help medical practices maximize infrastructure usage, reduce marginal operating costs, and improve medical services delivery in term of quality and timeliness.

Cloud computing can help implement a shared database needed for public health research, health services research, and clinical research. The integration of EHR systems with the cloud provides healthcare systems stakeholders with access to health information stored on a remote secure server via either web browsers or mobile devices, such as smart phones or tablets. One of the benefits derived from the use of cloud in health care is the limitless amount of space available to store health information. For this reason, interests in using cloud computing in health care are sprouting.

However, perceived challenges relating to security and ownership of health data stand in the way of achieving the above-mentioned benefits that cloud computing promises. According to a survey conducted by PhoneFactor in 2010 in the United States of America, security was the primary barrier to cloud computing adoption, followed by compliance, portability, and data ownership. Effective security measures are therefore required to protect health information housed in the cloud. Unfortunately, security measures for cloud and EHR's integration are thought to be in their infancy (Blazek 2012). As a result, the integration of EHR systems with the cloud has been slow.

In the United States of America, the use of cloud computing in health care is considered as a way to help comply with some security concerns laid out in the Health Insurance Portability and Accountability Act regarding the mobile health world (Blazek 2012). In Canada, for instance, the concern about compliance with the Canadian Freedom of Information and Protection of Privacy Act regulations remains one of the most significant barriers to cloud computing adoption in the

Canadian healthcare system. The idea of migrating most or all of healthcare providers' ICT resources, including health data storage, to cloud providers remains a cause for concern in the Canadian healthcare system. Furthermore, one of the biggest concerns is the U.S. Patriot Act. Canadians are of the belief that the U.S. Patriot Act can violate the privacy of their health information stored by U.S. cloud providers.

5 Mobile Connectivity

In this section, we provide a brief discussion about mobile connectivity. We touch on two concepts: virtual teams and trust-building. We briefly discuss the impact of mobile connectivity on health care.

5.1 Virtual Team and Mobile Connectivity

Today, the world economy has given rise to virtual and mobile teams. Increasingly, virtual and mobile teams are becoming a standard and effective way of working. To carry out their duty in an effective way, virtual and mobile team members require ICT systems that provide mobile connectivity. According to Zheng and Lun Lee (2005), wireless networking and portable digital devices can provide individuals with unrestricted mobility. Mobile computing devices include mobile phones, personal digital assistants, patient monitoring devices, and other wireless devices. As the information age call has been to develop virtual organizations (Landry *et al.* 2005), distance-irrelevant ICT systems and mobile technologies play a crucial role in organizational performance.

Mobile connectivity has several impacts on health care. As distance-irrelevant ICT systems overcome time and geographic boundaries, healthcare organizations can benefit from the knowledge and expertise of healthcare professionals dispersed across the globe. Moreover, medical functions can be performed anywhere at any time thanks to the use of web-based applications. Furthermore, ICT systems in support of virtual teams in health care have the potential to provide the healthcare industry with cost and time savings by reducing time to travel for consultations or for meetings.

5.2 Real-Time Systems and Trust Building in Health Care

Working over long distance present several challenges. Collaboration, effective information sharing, and trust building are some of those challenges. As communication helps resolve problems and avoid misunderstandings, ICT systems that allow synchronous and effective interactions can help build relationships in virtual team environment. Being able to ask questions and receive answers in real-time helps build trust among virtual team members. Thus, real-time systems are required to support today's health care which is expected to be delivered in a team-oriented work environment.

As mentioned earlier, the use of mobile technologies in the delivery of health care is referred to as mobile health (mHealth). Mobile phone's simple utilities, such as voice and short messaging services, can be used in the provision of health care. Mobile health can also include the use of more advanced functionalities and applications, such as space-based satellite navigation systems, smart phones, and Bluetooth technologies. Tablets and other handheld computing devices can be used in health care. To improve healthcare delivery in terms of availability of health information, containment of health costs, and reaching populations in underserved regions of the world, electronic health should use mobile technologies.

6 Mobile Technologies Revolution in the Healthcare Industry

In this section, we discuss the current state of the healthcare landscape around the world. For the purpose of this chapter, we briefly examine the healthcare landscapes of the following countries: the United States of America, Australia, and Canada. Our discussion covers the potential of mobile health to flatten national and global healthcare landscapes.

6.1 Healthcare Fragmentation and the Role of ICT

A review of literature indicates that health care in most countries is fragmented and experiences significant disparities. In the United States of America, for instance, the healthcare landscape has been characterized by some racial and ethnic disparities (Jha *et al.* 2009). According to Sehgal (2009), the U.S. healthcare system is characterized by the presence of disparities based on gender, race, type of health insurance, socioeconomic status, and some other relevant categorizations. For instance, a study conducted in 2012 in New York City revealed that older people and individuals with lower education levels were significantly less likely to use the Internet (De Leon *et al.* 2012). The researchers argue that disparities in Internet access have the potential to lead to uneven access to medical services.

The above views are similar to that of Effken and Abbott (2009) who found that the U.S. healthcare landscape was characterized by geographic disparities. Effken and Abbott (2009) believe that there is a geographic disparity in healthcare delivery as rural communities experience a significant shortage of healthcare providers. Although 20 to 25 percent of the U.S. residents live in rural regions (Bahensky *et al.* 2008), only 9 percent of healthcare providers practice there (Effken and Abbott 2009). With respect to racial disparities, about 22 percent of primary care providers in the United States of America provide care to over 80 percent of aged African American citizens (Jha *et al.* 2009). These disparities, and especially the shortage of healthcare providers in outer urban communities, have bearing on healthcare delivery in the United States of America.

Like the U.S. healthcare system, the Australian healthcare system experiences geographic disparities in the provision of health care. Literature suggests that in comparison with metropolitan regions, access to medical services and health outcomes are in most cases poorer in remote and rural regions. The problem is that nearly 33 percent of Australian residents live in remote and rural regions of Australia (Wakemman *et al.* 2008).

In response to geographic disparities, those who lead the Australian healthcare system have looked to ICT as a means to overcome distance and improve healthcare delivery (Soar and Young 2007). From a review of literature, patients living in remote, regional, and outer metropolitan areas have been given the opportunity to tele-health. Australians are now allowed to have video consultations with specialist healthcare providers. The goal of this initiative is to provide individuals living in rural regions with easier access to specialized care without the hardship of travelling to major cities (Australian Government n.d).

A study conducted in 2005 in the United States of America led to the understanding that health ICT systems can facilitate the advancement of modern medicine into remote and rural regions (Landry *et al.* 2005). In 2009, another study conducted in the United States of America indicated that EHR systems have the potential to connect the U.S. healthcare landscape (Effken and Abbott 2009). These two studies led to believe that EHR systems have the potential to improve the quality of care in both urban and rural communities. From these studies, it can be concluded that effective use of EHR systems can address disparities experienced in healthcare systems.

It is worth mentioning that the success of EHR systems depends in part on the adoption of EMR systems. The problem is that the adoption of EMR systems has been slow in several countries. Also, the majority of healthcare organizations, that have adopted EMR systems, are working in silos. Furthermore, a review of literature suggests a slow adoption of EMR systems in rural regions of some countries.

In the United States of America, for instance, there is a slow adoption of EMR systems in rural areas, where 41 percent of community health facilities are located (Bahensky *et al.* 2008). Recommendations have been made for U.S. federal and state governments to work collaboratively with the private sector in order to accelerate the adoption of EMR and EHR systems in the rural regions of the United States of America (Bahensky *et al.* 2008). It is believed that ensuring equal access to electronic health records is an essential part of reducing disparities experienced in the U.S. healthcare landscape (Jha *et al.* 2009). For this reason, George W. Bush signed an execution order, on April 27, 2004, to establish within the Department of Health and Human Services the position of National Health Information Technology Coordinator. One of the key responsibilities of the coordinator is to ensure the utilization of EHR for every single individual in the United States of America by 2014 (Legal Information Institute n.d).

Recent surveys and data from the U.S. federal government indicate an increasing adoption of EHR systems. A study conducted in the summer of 2012 by MedScape, a research firm, suggested that more than 80 percent of physicians were using EHR systems (Desjardins 2012). Findings from the same study suggested that 82 percent of participants were either using EHR systems, or were in the process of implementing them. According to the same study, the adoption rate of EHR systems has more than doubled as a similar study conducted in 2009 indicated that only 38 percent of healthcare providers were using EHR systems. Findings from a recent survey administered to 1,291 healthcare professionals in the United States of America revealed that 50 percent of respondents claimed that their EHR systems were mobile-accessible (ZirMed 2013).

Like in the United States of America, the Canadian healthcare system is fragmented. There are disparities among provinces in the quality of health delivery. There are also inter-regional disparities within provinces or territories. Findings from a study conducted by Ravenscroft (2007) suggested that poor information management and ineffective communication among healthcare providers affect the continuity of patients' care. In response, the Canadian health care system has promoted the development of interoperable EHR systems. These systems have the potential to allow seamless health information sharing among healthcare providers. As the future of the Canadian healthcare system is expected to be influenced by EHR systems (Collier 2010), all levels of Canadian governments have invested significantly to achieve the goal of providing 50 percent of Canadians with adequate EHR systems by 2010 (Reid and Wagner 2008).

Although progresses are being made in this area, the attainment of this ambitious goal was constrained by several factors (Reid and Wagner 2008). These factors include among others: insufficient funding, complexity of requirements relating to data portability, slow rates of adoption, and concerns relating to privacy. With respect to slow adoption rates and privacy concerns, nearly 53 percent of Canadian patients would not want their health information in the government electronic health record systems (Atherley 2007). Furthermore, about 52 percent of primary care providers would not upload health information to the government's EHR systems without patients' specific consent (Atherley 2007).

To address security issues that hinder a widespread adoption of electronic health record systems in Canada, it has been suggested that the current focus be shift from the promotion of electronic health record systems to the protection of health information that these systems are intended to manage (Atherley 2007). This shift is expected to boost the adoption of EHR systems across Canada. Two more solutions have been put forth to promote the use of electronic records in the Canadian healthcare system. First, healthcare providers should be provided with the right incentives to adopt health ICT systems. Second, under the single-payer system, the use of EMR and EHR systems can be enforced. In some Canadian provinces, healthcare providers are required to submit their request for payment electronically.

6.2 Mobile Health and Global Health Care Landscape

The use of mobile technologies in health care has evolved into a new field, known as mobile health. Mobile health has the potential to address the disparities experienced in the provision of health care either within the same country or across the world. Mobile computing devices, such as mobile phones and smart phones, present numerous significant advantages over other forms of digital technologies for public health advance. The use of mobile phones and smart phones can facilitate the delivery of medical services to populations in underserved areas (Blazek 2012).

This is possible for three reasons. First, EHR and PHR systems can be downloaded onto smart phones. Second, nearly six billion individuals across the world use either mobile phones or smart phones (World Health Organization 2011). Third, the commercial wireless signals coverage represents over 85 percent of the world's population (International Telecommunication Union 2011, World Health Organization 2011). Therefore, mobile health has the potential to significantly improve healthcare delivery around the world. Health costs containment, timely access to care, and enlarged reach of healthcare providers are some of the benefits that can be derived from the use of mobile technologies in health care.

Today, the use of mobile phone networks is increasing in numerous developing countries. Because of their increasing reach and potential, mobile networks have become the platform of choice for functions once limited to work computers. In health care, the evolving sophistication of mobile applications and health databases can improve the way health information is accessed, delivered, and managed.

6.3 Mobile Health and Health Care Efficiency

According to Siau and Shen (2006), wireless and mobile technologies provide a new frontier in delivering medical services to patients. The use of wireless and mobile technologies in healthcare industry provides healthcare consumers and healthcare providers with timely access to critical health information. In other words, health information can be delivered to healthcare consumers and healthcare providers in real-time. When integrated into patient care, wireless and mobile technologies allow healthcare providers to monitor patient's health conditions from anywhere. The use of mobile devices in health care improves the efficiency of health information sharing. Furthermore, mobile health provides healthcare consumers and healthcare providers with on-the-run learning opportunity. Also, during natural disasters, mobile health provides evident and substantial advantages over other means of delivering health care.

The use of mobile devices, such as tablet computers and personal digital assistants, allows healthcare providers to access health information at the point of care. The availability of health information at the point of care improves healthcare delivery. For instance, being able to have access to test results allows healthcare providers to make informed and timely decisions regarding the treatment. Also, providing healthcare providers with effective tools to order tests and medications at the point of care can expedite healthcare delivery. For this reason, healthcare providers want timely access to health information in order to deliver better care to patients. In New Zealand for instance, the advent of online enquiries and *real-time messaging* has increased the need of healthcare providers for instantaneous access to health information (Protti et al. 2008b). Literature suggests that more than 80 percent of New Zealander primary healthcare providers use web-services technology and the first of a new generation of *real-time messaging* services (Protti *et al.* 2008b).

The use of mobile devices, such as tablet computers and personal digital assistants, provides healthcare providers with the opportunity to collect health information electronically at the point of care. The collection of health information is one thing. But integrating health information collected through portable computing devices into EHR systems is necessary. The argument for this integration is that health information stored in EHR systems is available to healthcare providers across several medical facilities. In other words, this integration results in improved patient care. It is however worth noting that the flow of health information between mobile technologies and EHR systems is possible if both systems have the ability to share data. This calls for a seamless and secure cross platform to support the integration of EHR systems and a wide range of handheld devices used by healthcare providers and their patients.

The use of mobile technologies and decision support systems in health care allows individuals to manage some health problems more effectively on their own. Not only do these technologies improve self-care, but they also provide healthcare stakeholders with on-the-fly access to critical health information. Mobile technologies can help people take more initiative in managing their overall health. Individuals can proactively communicate with their healthcare providers for either preventive care or specific intervention when certain criteria thresholds are met.

Mobile health ICT systems, like PHR systems, can connect to Cloud-based EHR systems and provide individuals with significant benefits. They allow healthcare consumers to manage their health at home or away from home. In case of emergency for instance, mobile PHR systems offer clear-cut advantages over other ICT systems used in health care. They provide first responders with patient's information needed to provide lifesaving care.

6.4 Emergence of New Health Care Model

Mobile health can bring a profound transformation in healthcare delivery to meet the requirements of the twenty-first century's healthcare delivery. In the United States of America, for instance, mobile health technologies have revolutionized the way healthcare providers interact with healthcare consumers, and the whole delivery of health care (Blazek *et al.* 2012). When used effectively, mobile health can transform the traditional healthcare model. Instead of seeing their healthcare providers only when they need medical treatment, individuals are expected to become more active and proactive partners in their care. In other words, mobile health has given rise to a new healthcare model, in which healthcare consumers have become partners pursuing the common and ultimate goal of improving the healthcare system and staying healthy.

Mobile health is considered as a catalyst for switching to a new healthcare model that can meet today's expectations of all healthcare stakeholders. These expectations include among others (a) patient's self-care, (b) effective health information sharing, (c) elimination of disparities experienced in healthcare delivery, (d) growth of health consumerism, and (e) containment of health costs and, simultaneously, improvement of the healthcare delivery. Mobile health has the potential to connect diverse healthcare areas such as, healthcare providers and patients' education, communication among healthcare stakeholders, and patient care, within a continuum of care. Mobile health enables healthcare providers to have a continuous interaction with healthcare consumers. Such interaction or dialogue results in an uninterrupted healing relationship between healthcare consumers and their healthcare providers, which is much needed today.

7 Challenges to Mobile Health

In this section, we discuss some challenges that stand in the way of a widespread adoption of mobile technologies in health care. Perceived risks, Bluetooth attacks, and regional disparities in internet bandwidth usage are briefly examined. To reap all benefits that mobile technologies promise in health care, these challenges must be addressed in an effective way.

7.1 Human Factor and Perceived Risks

Although mobile health promises enormous technological capabilities, the 'patient' component remains unquestionably critical to the success of any mobile health enterprise. Even though more and more individuals feel comfortable using mobile devices for their banking needs, they remain reluctant to use mobile devices in health care. And this is due to perceived security concerns. These concerns include among others potential risks and vulnerabilities associated with remote access to health data and offsite use of patient information.

Undoubtedly, mobile health provides healthcare providers with the opportunity to use smart phones and other wireless handheld devices to interact with their colleagues and patients, or to access patient health information. Even though these devices provide a convenient way to access, process, and share health information, they have become attractive to hackers because of their advanced Internet features, like Web browsing (Froemelt 2008). Mobile communication devices used by healthcare providers and patients, such as tablet computers, Blackberry, and other smart phones, can put healthcare organizations' networks at risk and compromise health information privacy. It is critical to mention that most of these perceived threats come from either the ignorance of users or improper security implementation.

To some degree, the success of mobile health undertaking will depend on the way real and perceived security concerns are addressed by all healthcare stakeholders. Literature suggests that due to security concerns posed by smart phones, banks were forced to spend more money to address security threats (Collins 2009). Thus, the development of mobile technologies and their use in health care must take into consideration real and perceived security concerns.

Cloud-based mobile health is thought to be the solution to most security issues relating to the use of mobile technologies in the provision of health care. The argument is simple. With cloud-based mobile computing, little data relating to patient health will be stored on mobile devices (Blazek *et al.* 2012). In other words, little health information will be at risk for theft because portable computing devices will be used mostly as a presentation layer of an n-tier architecture. Still, sufficient end-to-end data encryption is necessary to protect health information during the transmission between the server and the mobile client.

7.2 Bluetooth Attacks

Bluetooth is a technology that enables two wireless devices to establish ad hoc connection. For this reason, it has become a famous technology since its launch. It has found its way into every mobile device such as, smart phones, tablets, and laptops. Dunning (2010) asserts that this advance has made Bluetooth a primal target for hackers. Like every technology, Bluetooth has its weaknesses. All of the hacking problems that are related to computers are valid for smart phones and Bluetooth. Loo (2009) argues that, if used properly, risks of using Bluetooth and smart phones are relatively low compared with those of other technologies.

Bluetooth attacks are often directed toward mobile devices that have weak or no security features. One of main characteristics of Bluetooth attacks is their limited scope. They affect only mobile computing devices in the vicinity of the attacker (Dunning 2010). This makes it difficult to assess the real damage that hackers abusing Bluetooth technology inflict on the use of mobile technologies in health care.

7.3 Internet Bandwidth Disparities

Several developing countries are witnessing a proliferation of mobile computing devices whose software applications can advance healthcare delivery. The problem is that their citizens often use mobile-broadband networks as a primary method to access the World Wide Web (W3). There are, however, significant disparities between regions in terms of available Internet bandwidth per W3's user. For instance, European Internet consumers use on average 90,000 bits per second of bandwidth, whereas the average is 2,000 bits per second per consumer in Africa (International Telecommunication Union 2011). Unquestionably, this disparity put developing countries at a disadvantage.

8 Convergence of Challenges and Risks in Healthcare Perspective

In this section, we discuss how challenges and risks identified in this chapter converge in the healthcare perspective. Undeniably, the use of ICT to support healthcare delivery has introduced new challenges, such as privacy and security concerns. Furthermore, challenges, such as low level of individuals' proficiency with ICT systems, have bearing on the adoption of ICT in the provision and consummation of health care. Also, disparities in access to health ICT systems and to the Internet technology have prevented currently-implemented ICT systems from improving healthcare delivery. A brief discussion about the effects of these challenges on the delivery of health care is provided below.

8.1 Impact of ICT-Related Risks on Healthcare Delivery

If risks relating to the use of ICT are not addressed in an effective way, they will have a negative impact on the delivery of healthcare. The use of ICT in health care raises concerns about the protection of health data. For instance, the protection of centralized health databases such as, health research databases and clinical health databases, is critical because they contain longitudinal health information about health status of individuals and their identifiable information. Furthermore, the use of web-based health ICT systems has introduced security and privacy risks that must be addressed.

Furthermore, the implementation of personal health portals to allow the use of PHR systems regardless of time and location introduces new security attacks such as, cross-site scripting and cross-site request forgery. Also, as web-based PHR systems user interface accepts sensitive information such as usernames and password, security measures, like secure sockets layer (SSL) protocol, are required. SSL will provide the following features: (a) encryption of health data using the public key infrastructure, (b) protection of health data that are transmitted between EHR systems and PHR systems, and (c) prevention of unauthorized access to health data during transmission.

Also, the need for around-the-clock and mobile connectivity in health care amplifies security and privacy concerns. Because of their dynamic topology and lack of a trusted infrastructure (Liu et al. 2013), the use of wireless sensor networks in health care introduces new security challenges. For instance, every mobile endpoint used in the provision of health care will require some degree of information technology (IT) management. Effective IT management should achieve two main goals. First, it should help to insure that both ends of the connection are in fact who they say they are. Second, it should help to insure that health data are not compromised during their transmission. As long as healthcare providers and healthcare consumers are not sure that these two goals are successfully achieved, they will remain reluctant to use portable computing devices in the provision and consumption of medical services.

To promote the adoption of mobile devices in health care, effective security measures are required. Implementing a strong firewall to authenticate mobile devices users is one of the examples of the security measures. Strong firewalls should be used to filter network traffic at several different layers of the health ICT. Effective authentication and encryption technologies should be used to insure that health data remains secure.

8.2 Impact of Internet Bandwidth on Healthcare Delivery

The introduction of mobile technologies into new regions of the world is increasing. Despite the proliferation of mobile technologies across the globe (Singh 2012), the existence of significant disparities of the available internet bandwidth among countries has a negative impact of healthcare delivery. These disparities result in unequal access to medical services.

Literature suggests that people living in formerly disconnected regions of developing countries use primarily mobile-broadband networks to connect to the Internet. The problem is that there is a competition for limited bandwidth among the mobile devices' users. This situation has a negative impact on individuals' access to health services delivered online. Undoubtedly, the use of mobile-broadband networks and their bandwidth capacity have bearing on healthcare delivery. As a result, individuals who should benefit the most from mobile health, may be the least able to reap several benefits that mobile technologies promise in health care.

8.3 Impact of Bluetooth on Healthcare Delivery

As mentioned earlier, Bluetooth allows peer-to-peer ad hoc wireless connectivity. This makes Bluetooth an easy-to-use and vulnerable technology at the same time (Kaur 2013). To communicate, two mobile computing devices located in the same physical area need only a Bluetooth radio and Bluetooth software. Bluetooth's security vulnerabilities lie in the lack of centralized administration and the absence of security administration infrastructure.

Today, there is an urgent need to secure this open standard technology to promote the adoption of mobile technologies in health care. In other words, the degree of mobile computing devices' adoption in health care depends, to some extent, on the way Bluetooth attacks are addressed. To see a widespread adoption of mobile technologies in health care, real and perceived risks relating to the security and privacy of personal health information must be addressed in an effective way.

8.4 Impact of Human Factor on Healthcare Delivery

As a recent survey, sponsored by ZirMed (2013), found that 50 percent of respondents claimed that their EHR systems were mobile-accessible, healthcare stakeholders are expected to play a crucial role in addressing security risks. Concerns about the security and privacy of health information are greater in mobile health because of the use of portable computing devices. This is because portable computing devices, such as smart phones and tablets, have the reputation to be vulnerable to hackers' attacks. Also, they can be stolen easily. There is a necessity to ensure the confidentiality of health information collected or sent through mobile technologies in order to boost their adoption in health care.

The protection of personal health information should be at the heart of any attempt to develop health ICT systems. For this reason, health ICT architects or manufacturers should build secure systems. Privacy and security should be embedded in all phases of health ICT systems' development. Designing a system architecture that provides mobile computing devices with a secure connection to EHR systems would minimize risks of privacy breach. Moreover, strong end-to-end encryption is needed to protect health data during their transmission. Furthermore, to improve patient privacy, manufacturers should design health ICT systems with effective authentication technologies, such as smart health cards, tokens, one-time password, or biometrics.

As practitioners, healthcare providers must understand the necessity to integrate privacy into their clinical processes and practices. Smart phones or tablets used in their practice must always be protected to prevent their loss and compromise of patient health information. For this reason, mobile devices used in the provision of health care should not be left accidently in a public location. Also, healthcare providers should learn to turn off Bluetooth on their mobile computing devices whenever these devices are not used to provide medical services. This best practice can minimize hackers' access to patient information.

Healthcare consumers are responsible for managing, sharing, and controlling their personal health information. Due to the prevalence of portable computing devices' theft, healthcare consumers are expected to protect their smart phones or tablets used in the consumption of health care. Such protection is required to prevent the loss of mobile devices and ultimately privacy breach. To prevent the use of their tablets or smart phones by unauthorized individuals, individuals must create strong passwords. This calls for public education.

9 Future Directions

Convergence is the buzzword that can best define the future direction of information and communication technologies in health care. It is economically and efficiently justifiable to use cloud computing to support healthcare delivery. That is why interest in the use of cloud computing in health care is increasing. Healthcare organizations will increasingly embrace cloud-based electronic health record (EHR) systems. To have ubiquitous access to their health information, people will prefer cloud-based personal health record (PHR) systems. Thus, the healthcare industry will witness an increasing rate of cloud-based EHR and PHR systems adoption in the near future. For that to happen however, perceived security and privacy risks associated with the use of cloud computing in health care must be addressed in an effective and convincing way.

To meet the mobility requirements of the global economy, electronic health must be supported by mobile technologies and wireless networks. Undeniably, mobile technologies have the potential to revolutionize healthcare delivery in terms of health cost containment, health information management, and access to health care. Thus, mobile health will be the preferred healthcare delivery method in the near future. This is possible because of the expansion of wireless scope, the increasing use of mobile devices across the world, and the clear-cut advantage mobile health has over traditional healthcare delivery methods during natural calamities. For mobile health to become the healthcare delivery method of choice, there should be a widespread adoption of mobile technologies in health care. To witness a desired widespread adoption of mobile technologies in health care, security risks and challenges associated with the use of mobile technologies must be addressed convincingly.

It is critical to be aware of another challenge that has emerged from the widespread use of mobile and handheld computing devices across the world: high volume of data traffic. Because of its unique characteristics such as, rapid elasticity, broad network access, and on-demand self-service, cloud computing is indisputably the right option to manage this challenge. Due to the high demand of services, both from the healthcare provider's side and the healthcare consumer's side, the healthcare industry will soon realize the necessity to adopt mobile cloud computing (MCC). MCC refers to the convergence of mobile technologies, wireless networks, and cloud computing. The adoption of MCC platform in health care is critical to connect mobile computing devices and cloud computing.

In the MCC model, the heavy lifting of processing and storing large amounts of health data will be performed in the cloud. In the MCC infrastructure, handheld computing devices used by healthcare providers and healthcare consumers will serve mostly as a presentation layer. As little health information will be stored on mobile devices, concerns relating to the security and privacy of health information in the mobile health world are therefore addressed. In the near future, mobile cloud computing will be the infrastructure or the platform of choice in the healthcare industry.

10 Conclusion

Undoubtedly, the use of ICT in health care can significantly improve healthcare delivery in terms of cost containment, access to care, and health information sharing. The use of web-based applications in health care has the potential to make health information available to healthcare stakeholders anywhere at any time. To meet today's health consumerism and patients' needs for active participation in their care, it is crucial that new approaches to health system design allow for an uninterrupted healing relationship between healthcare consumers and healthcare providers. To achieve this desired uninterrupted healing relationship, inefficiencies experienced in health care must be addressed by taking four steps. First, as electronic health is critical to improving healthcare delivery, the healthcare industry should implement health ICT systems that can improve health information management. Second, electronic health record systems should be integrated with patient health record systems to support effective partnership between healthcare providers and healthcare consumers. Third, there is a need for safe, uninterrupted, distance-irrelevant, and around-the-clock health ICT systems built on a seamless and secure cross platform. This is crucial to support portable computing devices used in health care. Fourth, to meet the mobility requirements of the twenty-first century, mobile technologies should support electronic health. Mobile health provides healthcare providers and healthcare consumers with the opportunity to use smart phones and other wireless handheld devices in the provision and consumption of health care. Mobile health technologies have the potential to revolutionize healthcare delivery in terms of health cost containment, information management, and access to care. There are, however, several obstacles to overcome before their

widespread adoption in health care. Security, public education, common standards, and Internet bandwidth disparities are some of the key issues.
Case Studies

Case 1

Topic:
Access to patients' health and insurance information

Situation:
On July 18, 2013, an individual was found unconscious on Catalina Avenue, Redondo Beach, in California. This tourist held a New Zealander passport. His name was Brian Palmer. His driver's license and medical smart card were found in his wallet.

Challenges Facing First-Responders:
The challenges facing first-response's team were in:

- Determining what happened to Mr. Brian Palmer
- Accessing the patient's background information from the medical smart card (first-responders' smart card reader could not read Mr. Palmer's medical smart)
- Determining whether Mr. Brian Palmer has a chronic condition, such as high blood pressure or seizure
- Stabilizing the patient's condition

Challenges Facing ER Healthcare Providers:
The challenges facing Emergency Room (ER) healthcare providers were in:

- Accessing Mr. Palmer's health information if his medical smart card was password-protected.
- Collecting enough information relating to Mr. Palmer's current condition to determine a treatment plan
- Understanding why first responders' smart cards reader was unable to read Mr. Palmer's medical smart card.

Actions Taken by First-Responders:
The actions first-responders took to address their challenges were:

- Checked the patient's vital signs and found that his blood pressure was extremely high
- Administered him medication to lower the patient's blood pressure
- Drove Mr. Brian Palmer to a nearby hospital in Los Angeles

Actions Taken by ER Healthcare Providers (Hospital):
The actions the Emergency Room (ER) healthcare providers took to address their challenges were:

- They accessed the patient critical health information as his medical smart card was not password-protected
- They discovered that the patient had two chronic diseases: diabetes and hypertension
- They checked the patient's vital signs and found that his blood pressure was still high
- Based on this information, they administered medication to stabilize his condition and monitored the effects
- ER healthcare providers implemented a safe and effective treatment plan for hypetension
- Mr. Palmer used his medical smart card to pay medical expenses associated with the intervention

Results:
The use of medical smart cards resulted in:

- Providing valuable information regarding the patient's health background and health insurance
- Allowing a safe and effective treatment plan
- Facilitating timely payment of medical expenses

Questions:

- Why first responders were unable to access the information they needed from Mr. Palmer's medical smart card? (What actions must be taken to address this issue?)
- What would have happened if Mr. Palmer's medical smart card was password-protected?

Case 2

Topic:
Making treatment decisions at distance

Situation:
Mr. Larry Phillips, a U.S. citizen, was vacationing in the Democratic Republic of the Congo, in Africa. On August 10, 2013, he was in Kongolo Monji, a small village located 500 miles away from the closest hospital. All the sudden, he felt dizzy and extremely tired. He became worried and decided to contact Dr. John Zook, his family physician practicing in Denver, Colorado.

- Mr. Phillips had a smart phone
- Mr. Phillips had a wrist Wi-Fi blood pressure monitor (Bluetooth enabled)

Challenges:
The challenges facing Mr. Phillips and Dr. Zook were in:

- Determining what was happening
- Sharing enough health information between them
- Determining an effective treatment plan at distance

Actions:
The actions they took to address these challenges were:

- Dr. Zook advised Mr. Phillips to use his smart phone to access the online PHR
- Mr. Phillips was able to share vital signs information with his family physician in the U.S. as the doctor's office in Denver had successfully completed PHR-EHR integration
- Dr. Zook was able to diagnose Mr. Phillips at distance
- Dr. Zook was able to prescribe medication
- Mr. Phillips enabled Bluetooth on the blood pressure monitor and the smart phone to allow ad hoc data sharing between the two portable computing devices
- Dr. Zook was therefor able to monitor Mr. Phillips' condition any time thru his tablet and smart phone as the EHR system in the doctor's office was mobile-accessible

Results:
The use of smart phone, wrist Wi-Fi blood pressure monitor, tablets, and Bluetooth resulted in:

- Mr. Phillips being able to share real-time vital signs information with his family physician in the U.S. as the doctor's office in Denver had successfully completed PHR-EHR integration
- Dr. Zook being able to diagnose Mr. Phillips at distance
- Dr. Zook being able to treat Mr. Phillips at distance
- Dr. Zook being able to monitor Mr. Phillips' condition any time thru his tablet and smart phone as the EHR system in the doctor's office was mobile-accessible

Questions:

- What would have happened if Mr. Phillips did not have a smart phone?
- How was Mr. Phillips able to access his PHR from his mobile phone?
- What would have happened if Dr. Zook's office did not integrate EHR-PHR systems?

- What would have happened if one of Mr. Phillips' mobile computing devices was not Bluetooth enabled?
- What would have happened if the EHR system in Zook's office was not mobile-accessible?

References

Atherley, G.: Electronic health records strategy needs a second look. Fraser Forum, pp. 18–20 (2007)

Australian Government, Specialist video consultations under Medicare, `http://www.mbsonline.gov.au/telehealth` (accessed June 15, 2013)

Collier, R.: Health care had a decade of abundance. CIHI says. CMAJ: Canadian Medical Association Journal 182, E45-E45 (2010)

Collins, L.: Bank Technology Stress Test, Massachusetts Banker (4th Quarter), pp. 16–23 (2009)

Desjardins, D.: EHR adoption rates accelerate. Medicine on the Net 18(12), 3 (2012)

Dunning, J.P.: Taming the Blue Beast: A Survey of Bluetooth Based Threats. IEEE Security & Privacy 8(2), 20–27 (2010), doi:10.1109/MSP.2010.3

Froemelt, M.: PDA & Smart Phone Business Security Impact (2008), `http://www.schooleymitchell.com/english/pdf/Publicity/PDA%20Security%20Article.pdf` (accessed May 27, 2013)

International Telecommunication Union, The World in 2011: ICT Facts and Figures, `http://www.itu.int/ITU-D/ict/facts/2011` (accessed June 15, 2013)

Gonzalez, G.: Computer maker launches online tools to promote health care consumerism. Business Insurance 41(12), 19 (2007)

Kaur, S.: How to secure our bluetooth insecure world! IETE Technical Review 30(2), 95–101 (2013), `http://dx.doi.org/10.4103/0256-4602.110547`

Lefton, R.: Reducing variation in healthcare delivery: the lack of clinical and administrative standardization across the provider and payer spectrum creates variation and inefficiency. Health care Financial Management 62(7), 42 (2008)

Legal Information Institute, Office of the National Coordinator for Health Information Technology (June 11, 2013), `http://www.law.cornell.edu/uscode/text/42/300jj-11` (accessed)

Loo, A.: Technical Opinion: Security Threats of Smart Phones and Bluetooth. Communications of the ACM 52(3), 150–152 (2009)

PhoneFactor, Cloud Computing Survey: Security Concerns Hinder Adoption of Cloud Services (2010), `http://www.phonefactor.com/../PhoneFactor-2010CloudSecuritySurvey` (accessed June 10, 2013)

Ravenscroft, E.: A patient perspective on health care system navigation in the context of multi-morbidity: an interpretive description. CANNT 2007, CANNT Journal 17(3), 23 (2007)

Powner, D.: Health and Human Services' Estimate of Health Care Cost Savings Resulting from the Use of Information Technology: GAO-05-309R. GAO Reports p1, 11p. (2005)

Roman, L.: Combined EMR, EHR and PHR manage data for better health. Drug Store News 31(9), 40–78 (2009)

Sehgal, A.R.: Universal health care as a health disparity intervention. Annals of Internal Medicine (Ann. Intern. Med.) 150(8), 561–562 (2009)

Simborg, D.W.: The limits of free speech: the PHR problem. Journal of the American Medical Informatics Association 6(3), 282–283 (2009)
Singh, R.D.: Wireless for Communities: Empowering Communities through Wireless Connectivity. IEEE Internet Computing 16(3), 77–79 (2012), doi:10.1109/MIC.2012.58.
Solomon, M.R.: Information technology to support self-management in chronic care. Disease Management and Health Outcomes 16(6), 391–401 (2008)
Swartz, K.: Electronic medical records - Federal standards needed. Inquiry - Excel-lus Health Plan, Rochester 43, 307–309 (2007)
Webster, P.C.: National standards for electronic health records remain remote. CMAJ: Canadian Medical Association Journal 182(9), 888–889 (2010)
ZirMed. 2013 Technology Survey Results (2013), `http://www.physicianspractice.com/technology-survey/2013-technology-survey-results` (accessed July 20, 2013)
Car, J., Sheikh, A.: Email consultations in health care: 1–scope and effectiveness. BMJ (Clinical Research) 329(7463), 435–438 (2004)
Effken, J., Abbott, P.: Health IT-enabled care for underserved rural populations: The role of nursing. Journal of the American Medical Informatics Association: JAMIA (J Am Med In-form Assoc) 16(4), 439–445 (2009)
Hunter, R., Cameron, R.: Patient involvement in health care will improve quality. BMJ: British Medical Journal 333(7559), 147–148 (2006)
Lafky, D.B., Horan, T.A.: Personal health records. Health Informatics Journal, 2011 17(1), 63–71 (2011)
Reid, R., Wagner, E.: Strengthening primary care with better transfer of information. CMAJ: Canadian Medical Association Journal = Journal De L'association Medicale Canadienne (CMAJ) 179(10), 987–988 (2008)
Siau, K., Shen, Z.: Mobile health care informatics. Medical Informatics and the Internet in Medicine 31(2), 89–99 (2006)
Soar, J., Young, S.: Health and aged care enabled by information technology. Annals of the New York Academy of Sciences (Ann N Y Acad Sci) 1114, 154–161 (2007)
Susilo, W., Win, K.T.: Security and access of health research data. Journal of Medical Systems (J Med Syst) 31(2), 103–107 (2007)
Tang, P.C., Lansky, D.: The missing link: bridging the patient- Provider health information gap. Health Affairs (Project Hope) (Health Aff (Millwood)) 24(5), 1290–1295 (2005)
World Health Organization, mHealth New Horizons for Health through Mobile Technologies. Global Observatory for eHealth series, vol. 3 (2011) (accessed May 4, 2013)
Zheng, B., Lun Lee, D.: Information dissemination via wireless broadcast. Communications of the ACM 48(5), 105–110 (2005)
Archer, N., Fevrier-Thomas, U., Lokker, C., McKibbon, K.A., Strauss, S.E.: Personal health re-cords: a scoping review. Journal of the American Medical Informatics Association 18(4), 515–522 (2011)
Bahensky, J.A., Jaana, M., Ward, M.M.: Health care information technology in rural America: electronic medical record adoption status in meeting the national agenda. The Journal of Ru-ral Health: Official Journal of The American Rural Health Association and The National Rural Health Care Association (J Rural Health) 24(2), 101–105 (2008)
Ball, M.J., Costin, M.Y., Lehmann, C.: The personal health record: consumers banking on their health. Studies in Health Technology and Informatics (Stud Health Technol In-form.) 134, 35–46 (2008)

Blazek, N., Buddle, R.B., Kluge, E., Siegal, G.: Mobile health technologies improve communi-cation with patients. Orthopedics Today 32(12), 1–12 (2012)
Botts, N.E., Horan, T.A., Thoms, B.P.: Health ATM: personal health cyber infrastructure for underserved populations. American Journal of Preventive Medicine 40(5S2), S115–S122 (2011)
Cebul, R.D., Rebitzer, J.B., Taylor, L.J., Votruba, M.E.: Symposium: Health care: Organizational fragmentation and care quality in the U.S. Healthcare system. Journal of Economic Perspectives 22(4), 93–113 (2008)
Davis, R.E., Jacklin, R., Sevdalis, N., Vincent, C.A.: Patient involvement in patient safety: what factors influence patient participation and engagement? Health Expectations 10(3), 259–267 (2007)
De Leon, S.F., Silfen, S.L., Wang, J.J., Kamara, T.S., Wu, W.Y., Shih, S.C.: Patient experiences at primary care practices using electronic health records. The Journal of Medical Prac-tice Management: MPM 28(3), 169–176 (2012)
Grossman, J., Zayas-Cabán, T., Kemper, N.: Information gap: Can health insurer personal health records meet patients' and physicians' needs? Health Affairs 28(2), 377–389 (2009)
Hourcade, J.P., Chrischilles, E.A., Gryzlak, B.M., Hanson, B.M., Dunbar, D.E., Eichmann, D.A., Lorentzen, R.R.: Design lessons for older adult personal health records software from older adults. In: Stephanidis, C. (ed.) Universal Access in HCI, Part II, HCII 2011. LNCS, vol. 6766, pp. 176–185. Springer, Heidelberg (2011)
Hsieh, S., Hou, I., Cheng, P., Tan, C., Shen, P., Hsu, K., et al.: Design and Implementation of Web-Based Mobile Electronic Medication Administration Record. Journal of Medical Systems 34(5), 947–958 (2010), doi: http://dx.doi.org/10.1007/s10916-009-9310-9
Jha, A.K., Bates, D.W., Jenter, C., Orav, E.J., Zheng, J., Cleary, P., et al.: Electronic health records: use, barriers and satisfaction among physicians who care for black and Hispanic patients. Journal of Evaluation in Clinical Practice 15(1), 158–163 (2009), doi:10.1111/j.1365-2753.2008.00975.x.
Johnson, P., Heimann, V., O'Neill, K.: The wonderland of virtual teams. Journal of Work-Place Learning 13(2), 24–30 (2001)
Katić, M., Soldo, D., Ozvacić, Z., Blazeković-Milaković, S., Vrcić-Keglević, M., Bergman-Marković, B., et al.: Information systems and the electronic health record in primary health care. Informatics in Primary Care 15(3), 187–192 (2007)
Landry, B.J., Mahesh, S., Hartman, S.: The impact of the pervasive information age on health-care organizations. Journal of Health & Human Services Administration 27(4), 444–464 (2005)
Liu, A., Kim, M., Oliveira, L.B., Tan, H.: Wireless Sensor Network Security. International Journal of Distributed Sensor Networks (2013), http://dx.doi.org/10.1155/2013/362385
Longtin, Y., Sax, H., Leape, L.L., Sheridan, S.E., Donaldson, L., Pittet, D.: Current Knowledge and Applicability to Patient Safety. Mayo Clinic (Mayo Clin Proc) 85(1), 53–62 (2010), doi:10.4065/mcp.2009.0248
Protti, D., Bowden, T., Johansen, I.: Adoption of information technology in primary care physician offices in New Zealand and Denmark, part 3: medical record environment com-parisons. Informatics in Primary Care 16(4), 285–290 (2008a)
Protti, D., Bowden, T., Johansen, I.: Adoption of information technology in primary care physician offices in New Zealand and Denmark, part 4: benefits comparisons. Informatics in Primary Care 16(4), 291–296 (2008b)

Tang, P.C., Ash, J.S., Bates, D.W., Overhage, J.M., Sands, D.Z.: Personal health records: definitions, benefits, and strategies for overcoming barriers to adoption. Journal of the American Medical Informatics Association 13(2), 121–126 (2006)
Wakerman, J., Humphreys, J.S., Wells, R., Kuipers, P., Entwistle, P.: Jones JPrimary health care deliv-ery models in rural and remote Australia– a systematic review. BMC Health Services Research 8, 276 (2008), doi:10.1186/1472-6963-8-276.
Yamin, C.K., Emani, S., Williams, D.H., Lipsitz, S.R., Karson, A.S., Wald, J.S., et al.: The digital di-vide in adoption and use of a personal health record. Archives of Internal Medicine 171(6), 568–574 (2011)

Questions

Q1. What is electronic health?

Q2. What is mobile health?

Q3. Why does electronic health need to be supported by mobile and wireless technologies

Q4. What is an electronic health record?

Q5. What are two main factors that hinder a widespread adoption of electronic records in health care?

Q6. What are three main factors that hinder a widespread adoption of mobile technologies in health care?

28. Terahertz Technology in the Future of Health and Medical Applications

Abdorreza Heidari

PluraPix Inc.,
Waterloo, Ontario, Canada
reza@plurapix.com

1 Introduction

Terahertz refers to the electromagnetic waves with the frequency range between millimeter-wave and infrared, approximately from 300 GHz up to 10 THz. The THz spectrum, also known as the "Terahertz gap" is the last portion of the electromagnetic spectrum which has not been fully explored and exploited. Terahertz technology [1,2] is a fast-growing field with applications in biology and Medicine [3], medical imaging [4], material spectroscopy and sensing [5], security, monitoring and spectroscopy in pharmaceutical industry, and high-data-rate communications.

The non-ionizing nature of Terahertz makes it appealing for health applications, and is expected to replace some of the current imaging modalities. Terahertz imaging offers a non-invasive non-ionising alternative to x-rays and also provides higher contrast in clinical diagnostics. Terahertz imaging has already been shown to work for skin and other types of cancer. Terahertz wound and burn inspections, and terahertz dental imaging, are other immediate candidates with some promising early researches. THz imaging and spectroscopy has been employed in several areas of health and medical applications, such as diagnosis of cancer tumors [7], blood analysis [6], breath analysis, and air quality monitoring.

However, the low-power sources are one of the current challenges. Also Terahertz water absorption limits the penetration depth in the body especially at higher frequencies. Advanced imaging systems are being developed that can solve at least part of these problem. Also tremendous engineering efforts are at work to improve the portability and mobility of Terahertz systems. Mobile processing seems to be another challenge, where the solution lies at the mercy of the Moore's law and other advances in this area.

THz radiation can penetrate most of the dielectric materials, such as plastic, paper, wood, and clothing. This "See-through" property of THz imaging results in very useful and also cool applications, which before only seen in science fiction and superhero movies! Recent advances in THz components and systems, especially CMOS-based sources and detectors, provide a real possibility for the availability of such THz applications in the mass market. CMOS-based solutions can decrease the costs drastically, and make miniaturization of such system possible, even to the extent to be integrated into existing mobile devices. Early proof of

S. Adibi (ed.), *Mobile Health*,
Springer Series in Bio-/Neuroinformatics 5, DOI: 10.1007/978-3-319-12817-7_28

concept shows this as a real possibility in the near future, paving the way for a wide range of Terahertz mHealth applications.

2 Terahertz Medical Imaging

2.1 Cancer Detection

Epithelial tissues are the source of more than 80% of all adult cancers and include common cancers of the skin, lung, breast, colon, bladder, liver and prostate. Several studies have shown that THz radiation can distinguish between a cancerous and a healthy tissue. Although the exact reason is unknown, it is widely believed that the different water content of these tissues is the main cause [10, 7, 11, and references therein]. Due to the heavy THz absorption in aqueous media, deep body penetration is still challenging. Therefore, the deep in-vivo cases are still limited, although several research studies are underway worldwide. Given the current state of the art in THz imaging systems, the ex-vivo applications, as well as skin-deep examinations, are possible.

2.2 Skin Cancer

THz radiation has been successfully used for imaging the human skin [14-20], where a few millimeters of penetration depth is sufficient. Therefore, the early detection and diagnosis of skin cancers has been the subject of several studies [14-17]. This technology is envisioned to reduce or probably eliminate the need for the ex-vivo biopsies of anomalous skin tissues. Also the early stage recognition drastically reduces the risk of these cancers as they can still be treated effectively, as well as the treatment costs. Some of skin cancers are visible to the trained eye, however some that are small and flat are not easily recognizable. THz imaging, because of its ability to differentiate the spectral fingerprints of surface proteins that are markers for certain cancers, may be able to identify these cancers at early stages.

2.3 Wound and Burn Inspection

THz imaging has also been used as a non-invasive tool to inspect wounds and burns [18-20], mostly by using a THz tomography setup. THz tomography provides an axial image of a multilayered object. The point of interest is shined with a THz pulse, and the reflected THz pulse train is measured and analyzed. Each layer reflects a deformed pulse, which carry information about the depth and optical properties of that layer. When THz tomography used for noncontact inspection of skin burn injuries, the shape and size of each reflected pulse depends on the depth and the properties of the burn damage. A plastic over-layer (e.g., Polystyrene) may be applied on the surface to provide a reference for the signal analysis. Burned and unburned areas show a distinct difference in THz reflectivity [18]. Such imaging is conveniently possible even under layers of medical cotton (gauze) and bandage.

2.4 Dental Imaging

The three-dimensional (3D) THz imaging can be used to help the dentists in their various diagnoses, including the detection of tooth decays and enamel erosion [12, 13]. It has been observed that THz radiation has higher attenuation in decayed enamel relative to healthy enamel, providing a distinctive contrast in the THz image. THz imaging exploits the change in the refractive index between dental enamel and dentine tissues. X-rays are currently the state of the art modality for dental imaging. However, X-rays do not provide an accurate tool for detecting tooth decay in early stages when tissue damage can be effectively remedied.

Caries are a result of mineral loss from enamel, causing a change in the refractive index within the enamel. Detection of the changes in the refractive index profile of a tooth means that small lesions can be recognized. THz pulse imaging is a suitable candidate for in vivo dental imaging, which is able to examine surface features from the outer layer of enamel as well. In vitro experiments show that THz imaging can detect small induced changes on the surface of the enamel of the teeth. Imaging of enamel thickness with THz imaging would be considered a significant breakthrough in this area.

Article [13] describes a TPI system that can be used to generate images of dental tissue in 3-D. It presents data from a sample of 12 human tooth incisors, with a 91% success rate in detecting the enamel dentine junction. It also images artificially altered enamel thickness to within 10 um of the expected results. We have thus demonstrated that we can accurately and reliably make direct measurements of enamel thickness, a technology that is necessary to monitor enamel erosion, a common dental disorder. The article reports difficulty in imaging very thick samples and in measuring angular surfaces. Also total internal reflection in the tooth is another problem which needs to be addressed, for example, with the development of a probe that can be used in vivo.

2.5 THz Endoscopy (Access inside the Body)

In endoscopy, a usually flexible probe set is used to provide a visual from inside the body, for diagnostic or interactive surgical operations. Other than visible-range images, endoscopes are being extended to ultrasound and magnetic resonance imaging (MRI) as well. The development and use of THz endoscopes [20-21] is another step in this direction, providing a powerful tool for medical practitioners. In particular, THz endoscopy opens up a new medical field for diagnosis and treatment of several types of cancers residing inside the body.

As reported in article [20], the researchers have designed, fabricated, and characterized a miniaturized optical fiber-coupled terahertz (THz) endoscope system. In reflection mode, the endoscope provides a high signal-to-noise ratio and a wide frequency spectrum similar to the conventional THz time-domain spectroscopic system. The cross section of the endoscope including generator and detector head is (2 × 4 mm) × 6 mm, which is small enough to be inserted into a human body.

3 Terahertz Components

Lack of mature THz components has been one of the main reasons holding the THz technology back from getting into the mass market. In the following, the available technologies for THz source and detectors in briefly reviewed. Then, the availability of the semiconductor source and detectors for the THz region is discussed. This is because of the importance of the newly introduced field for the development of THz portable and mobile devices, including for mHealth applications.

3.1 Terahertz Sources

Several types of THz sources are available today. These sources fall into the following general categories:

- THz Lasers
- THz Electronic sources
- THz Electro-Optical sources

Laser sources are convenient, small in size, and provide a high level of spectral and temporal control. A variety of lasers are in use or under development, a few of those described below.

Optically Pumped THz Lasers (OPTL) are continuous wave (CW) THz sources, which require a carbon-monoxide pump laser.

Quantum cascade lasers (QCLs) are a type of semiconductor laser that can generate coherent radiation in the THz range with powers of the order of milliwatts or more. Recent QCLs can operate at the ambient temperature, however their power efficiency increases under cryogenic conditions. Narrow line-width and wavelength tunability makes QCLs appealing THz sources.

Free-Electron lasers (FELs) are very bulky and expensive, but can generate very high THz powers, reportedly up to 1 kW.

Electronic sources do not require any laser devices to generate THz. The cost, size and easier control are the appealing aspects of these sources. A few are named in the following.

Backward-wave oscillators (BWOs) generally emit at the lower end of the THz range, up to around 1.5 THz, with output powers up to 12 mW. These sources are mostly used for scientific research.

Direct multiplied THz sources take millimeter-wave sources and directly multiply their output up to THz frequencies. DM sources with frequencies up to about 2 THz are possible, but performance improves at lower frequencies. These sources with a few tens of milli-watt of lower THz power are available in the market, in compact sizes.

Electro-optical sources require one or more other laser sources to generate THz. The two main types are explained in the following.

Photoconductive Antennas (PCAs) is one of the oldest methods of producing THz. Ultra-short laser pulses excite carriers into the conduction band of a semiconductor, where an applied bias voltage produces THz radiation. PCA THz sources are commercially available and can generate up to several tens of microwatts.

Photomixing is a technique that uses two lasers to generate THz radiation, by mixing the lasers beams in a photomixer, to generate a THz difference frequency. An integrated antenna, designed to operate at THz frequencies, radiates the generated THz beam.

3.2 Terahertz Detectors

A variety of detectors are also in use or under development for THz detection. These detectors either detect the THz radiation directly, or in a heterodyne fashion.

PCA detectors, similar to their source counterparts, use PC antennas to detect THz coherently.

Bolometers are the most sensitive type detector for the measurement of THz radiation over a wide rage of THz frequencies. Their problem, however, is that they usually need to be cooled down. A *hot electron bolometer (HEB)* is particularly very sensitive and low noise over 1 THz.

Golay cells are very stable and reliable detectors that are widely used to measure the intensity of THz radiation.

Pyroelectric detectors are small, portable, and also cost-effective compared to bolometers and Golay cells. They operate at the room temperature too. However, their performance needs to be improved for the THz region.

Heterodyne receivers mix the incoming signal with a "local oscillator". The mixing produces a signal at the difference frequency, which is lower compared to the original frequency for easier processing. Several mixing technologies are available for heterodyne reception at THz frequencies, such as Schottky Mixers, Superconductor-Isolator-Superconductor (SIS) Mixers, and Hot Electron Bolometer Mixers.

3.3 THz Semiconductor Components

Moving towards high-volume deployment of compact and mobile THz systems, one of the main requirements is to have low-cost and easily integrate-able components. THz semiconductor components are an obvious solution. To the surprise of the researchers, CMOS-based THz sources and detectors have been shown to be feasible [8, 9, and the references therein]. Linking THz photonics to conventional CMOS process technologies are widely employed in the visible and near-infrared frequencies, for example, in digital cameras. Several design efforts are currently underway to solve the technical challenges and improve the performance of these components for the THz region.

4 What Are the Challenges?

4.1 Water Absorption

As opposed to infrared and visible light, THz radiation can penetrate fog, clouds and light rain. However, water strongly absorbs THz waves. Similar problem exists in the neighboring frequencies, including infrared and microwaves.

The Earth's atmosphere, especially in humid conditions, heavily attenuates THz radiation heavily. THz signal is also strongly distorted while propagating through air, which poses another challenge for standoff and long-range applications. Advanced signal processing techniques can enhance the performance of THz system in short propagating ranges.

4.2 Resolution versus Penetration

For conventional imaging systems, (spatial) resolution is limited to the wavelength of the radiated wave. This well-known limitation is known as the Diffraction limit. The quality of the optics, including the hardware calibration, is another factor. On the other hand, Signal to Noise ratio (SNR) of the image is proportional to image acquisition time (or the detector integration time), selectivity of the received waveform, and also may be bound by the sensitivity of the detector. The maximum viable penetration depth is determined by the available SNR.

To improve the resolution in conventional imaging system, one should move to higher frequencies. At higher THz frequencies, the source power is usually lower. Also the attenuations are more severe. Therefore, this move usually results in a lower SNR, and so a shorter penetration depth.

4.3 Portability and Mobile Processing

THz systems used to be built in the form of bulky setups. The current trend is towards miniaturization of such systems to improve their portability and ease of use. The improvement of semiconductor components, which are integrate-able to conventional electronics, can accelerate these efforts, even towards THz mobile solutions. Such integrations will also provide access to on-chip processing power for the development of advanced THz systems.

4.4 Is THz Safe?

THz photons have meV energies, which is significantly lower than X-ray photons with KeV energies. Unlike ionizing forms of radiation such as X-rays, nonionizing radiation lacks sufficient energy to remove tightly bound electrons. Therefore, THz radiation is unlikely to damage body tissues and DNA, and so it is considered to be safe to human bodies. It is also environmentally friendly. Hence, THz radiation is a safer alternative for medical and other applications.

5 Conclusion

THz radiation is safe, provides rich spectroscopic signatures for many molecule structures, and also penetrates most of dielectric materials, such as paper, wood, plastics, and clothing. THz imaging can be used for a wide range of applications, such as detection of hidden artifacts for security (currently used at the airports) and inspection, medical imaging diagnostics, and non-destructive tests.

The mass-market deployment of THz imaging has been struggling by shortage of enabling components, complicated and expensive hardware setups, and proper integration of hardware and software to achieve high performance imaging operation. In particular, the penetration depths of existing imaging setups are limited by the power of THz sources and the sensitivity of the detectors, let alone the heavy water absorptions. Also, the spatial resolution is constrained by the diffraction limit.

The feasibility of semiconductor (especially CMOS-based) THz sources and detectors, as well as development of novel THz systems can solve most of the aforementioned problems. In the next few years, more advanced THz systems are expected to be developed to address medical needs, including mHealth applicatons.

References

[1] Tonouchi, M.: Cutting-edge terahertz technology. Nature Photonics 1(2), 97–105 (2007)

[2] Siegel, P.: Terahertz technology. IEEE Transactions on Microwave Theory and Techniques 50(3), 910–928 (2002)

[3] Siegel, P.: Terahertz technology in biology and medicine. IEEE Transactions on Microwave Theory and Techniques 52(10), 2438–2447 (2004)

[4] Mittleman, D., Gupta, M., Neelamani, R., Baraniuk, R., Rudd, J., Koch, M.: Recent advances in terahertz imaging. Applied Physics B: Lasers and Optics 68(6), 1085–1094 (1999)

[5] Macpherson, E., Gallerano, G.P., Park, G.S., Hintzche, H., Wilmink, G.J.: Terahertz Imaging and Spectroscopy for Biology and Biomedicine. IEEE Trans. on Terahertz Science and Technology 3(4), 354–356 (2013)

[6] Reid, C.B., Reese, G., Gibson, A.P., Wallace, V.P.: Terahertz Time-Domain Spectroscopy of Human Blood. IEEE Trans. on Terahertz Science and Technology 3(4), 363–367 (2013)

[7] Peter, B.S., Yngevesson, S., Siqueira, P., Kelly, P., Khan, A., Glick, S., Karellas, A.: Development and Testing of a Single Frequency Terahertz Imaging System for Breast Cancer Detection. IEEE Trans. on Terahertz Science and Technology 3(4), 374–386 (2013)

[8] Lisauskasa, A., Boppela, S., Seliutab, D., Minkevičiusb, L., Kašalynasb, I., Valušisb, G., Krozera, V., Roskos, H.G.: Terahertz Detection and Coherent Imaging from 0.2 to 4.3 THz with Silicon CMOS Field-Effect Transistors. IEEE MTT-S International Microwave Symposium Digest, MTT (June 2012)

[9] Friederich, F., von Spiegel, W., Bauer, M., Meng, F., Thomson, M.D., Boppel, S., Lisauskas, A., Hils, B., Krozer, V., Keil, A., Löffler, T., Henneberger, R., Huhn, A.K., Spickermann, G., Bolívar, P.H., Roskos, H.G.: THz Active Imaging Systems With Real-Time Capabilities. IEEE Trans. on Terahertz Science and Technology 1(1) (2011)

[10] Application Note, Study to distinguish between healthy tissue and basal cell carcinoma, Teraview Ltd, Cambridge, UK

[11] Sim, Y.C., Ahn, K.M., Park, J.Y., Park, C.S., Son, J.H.: Temperature-dependent Terahertz Imaging of Excised Oral Malignant Melanoma. IEEE Trans. on Terahertz Science and Technology 3(4), 368–373 (2013)

[12] Arnone, D., Ciesla, C., Pepper, M.: Terahertz imaging comes into view. Physics World 4, 35–40 (2000)

[13] Crawley, D., Longbottom, C., Wallace, V.P., Cole, B., Arnone, D., Pepper, M.: Three-dimensional terahertz pulse imaging of dental tissue. Journal of Biomedical Optics 8(2), 303–307 (2003)

[14] Woodward, R.M., Cole, B.E., Wallace, V.P., Pye, R.J., Arnone, D.D., Linfield, E.H., Pepper, M.: Terahertz pulse imaging in reflection geometry of human skin cancer and skin tissue. Physics in Medicine and Biology 47(21), 3853 (2002)

[15] Woodward, R.M., Wallace, V.P., Arnone, D.D., Linfield, E.H., Pepper, M.: Terahertz pulsed imaging of skin cancer in the time and frequency domain. Journal of Biological Physics 29(2-3), 257–259 (2003)

[16] Wallace, V.P., Fitzgerald, A.J., Shankar, S., Flanagan, N., Pye, R., Cluff, J., Arnone, D.D.: Terahertz pulsed imaging of basal cell carcinoma ex vivo and in vivo. British Journal of Dermatology 151(2), 424–432 (2004)

[17] Pickwell, E., Cole, B.E., Fitzgerald, A.J., Pepper, M., Wallace, V.P.: In vivo study of human skin using pulsed terahertz radiation. Physics in Medicine and Biology 49(9), 1595 (2004)

[18] Taylor, Z.D., Singh, R.S., Culjat, M.O., Suen, J.Y., Grundfest, W.S., Lee, H., Brown, E.R.: Reflective terahertz imaging of porcine skin burns. Optics Letters 33(11), 1258–1260 (2008)

[19] Huang, S.Y., Macpherson, E., Zhang, Y.T.: A feasibility study of burn wound depth assessment using terahertz pulsed imaging. In: IEEE/EMBS International Summer School and Symposium on Medical Devices and Biosensors, ISSS-MDBS 2007, pp. 132–135. IEEE (2007)

[20] Dougherty, J.P., Jubic, G.D., Kiser Jr., W.L.: Terahertz imaging of burned tissue. In: Integrated Optoelectronic Devices 2007, p. 64720N. International Society for Optics and Photonics (2007)

[21] Ji, Y.B., Lee, E.S., Kim, S.-H., Son, J.-H., Jeon, T.-I.: A miniaturized fiber-coupled terahertz endoscope system. Optics Express 17(19), 17082–17087 (2009)

[22] Lu, J.-Y., Kuo, C.-C., Chiu, C.-M., Chen, H.-W., Pan, C.-L., Sun, C.-K.: THz interferometric imaging using subwavelength plastic fiber based THz endoscopes. In: Conference on Lasers and Electro-Optics. Optical Society of America (2008)

29. Developing Multi-agent Systems for mHealth Drug Delivery

Michael Mutingi[1,*] and Charles Mbohwa[2]

[1] Faculty of Engineering and the Built Environment, University of Johannesburg, Bunting Road Campus, P.O. Box 524, Auckland Park 2006, Johannesburg, South Africa
mmutingi@gmail.com

[2] Quality and Operations Management Department, University of Johannesburg, Bunting Road Campus, P.O. Box 524, Auckland Park 2006, Johannesburg, South Africa

Abstract. Recent research has explored how mobile networks are revolutionizing multiple aspects of healthcare in both developing and developed countries. As an area of innovation with the potential to make a huge difference, mHealth involves the utilization of mobile communication technologies to deliver healthcare services, such as SMS alerts that remind patients to take their prescription drugs at the appropriate time, remote diagnosis and even treatment for patients who do not have easy access to physicians, remote health monitoring devices that track and report patients' conditions, and scheduling drug delivery. Such services could have a major impact on healthcare outcomes and costs. Decision making in mHealth drug delivery systems is complex. Decisions involve satisficing multiple goals regarding customer service quality, service cost, and healthcare worker satisfaction. With the increasing world-wide need for effective home healthcare, the increasing elderly population, and the increasing pressure from governments and other stakeholders, developing effective approaches for mHealth drug delivery decisions is imperative. In this paper, we present a multi-agent architecture that facilitates decision making in mHealth drug delivery system. The approach integrates the capabilities of a multi-agent system and Web services so as to facilitate effective decisions for home healthcare services. The aim is to provide a multi-agent system, where decisions are based on intelligent agents that provide quick and intelligent alternative decisions in a dynamic environment.

Keywords: mHealth Systems, eHealth, mHealth Drug Delivery, Drug Distribution, Home Care, Home-based Care, Homecare Logistics, Multi-agent Systems, Software Agents, Decision Support.

1 Introduction

Mobile networks are increasingly revolutionizing multiple aspects of healthcare services all over the world (Ascari et al., 2010; Varnfield et al., 2012; Rossi et al.,

* Corresponding author.

S. Adibi (ed.), *Mobile Health*,
Springer Series in Bio-/Neuroinformatics 5, DOI: 10.1007/978-3-319-12817-7_29

2012). As an area of innovation with immense potential to make a huge difference in the healthcare status of every society, mHealth utilizes mobile communication technologies to coordinate and deliver healthcare services, for instance, SMS alerts to remind patients to take their prescription drugs at the appropriate time, remote diagnosis and even treatment for patients who do not have easy access to physicians, remote health monitoring devices that track and report patients' conditions, and mHealth drug delivery and distribution (Dolan, 2010; Krohn and Metcalf, 2012). Thus, in the last decade, the rapid progress of portable and mobile sensors has enhanced significant growth towards the use of related sensors in mHealth technologies. This has enabled remote evaluation of various bio-physiological signals such as blood-pressure, heart-pulse-rate, and respiration, and the ensuing decision making on operations such as drug delivery. In particular, the evolution over time of such signals may give the doctors (or caregivers) very relevant information about the evolution of the pathology itself. Over time, a number of advantages are inevitable: (i) the realization of substantial cost savings in terms of use of specialized human resources, (ii) the patient's comfort is enhanced, (iii) the quality of the social life of the patient is improved, and most importantly, and (iv) decision making on the part of healthcare operations managers and planners is improved, particularly in a home healthcare environment where dynamic decisions are made concerning drug delivery and distribution.

It is quite an advantage to allow patients and elderly people with varying disabilities and chronic diseases to receive medical and paramedical services while in their homes since long-term stay in nursing homes is often much more costly than home-based care (Eveborn et al., 2009). Consequently, home healthcare services can provide a more cost-effective and flexible mechanism. Overall, these mHealth applications could have even more desirable impact on healthcare outcomes and costs. In particular, drug distribution and delivery is one of the most important operational activities that highly lend themselves to the application of mHealth technologies, as well as multi-agent decision systems.

Drug delivery has increasingly been an operational activity of much concern in a number of homecare organizations (Woodward et al., 2004; Bachouch et al., 2009; Matta et al., 2012). Researchers have pointed out that transportation of perishable and personal drugs (e.g., anti-cancer drugs) to patients is a critical activity that requires reliable decision support systems (Matta et al., 2012; Fraile et al., 2009). What are the decisions to be considered when building decision support systems? How may these decisions be incorporated into decision support systems? Can a multi-agent system or framework be helpful in building dynamic mHealth drug delivery systems?

The purpose of this chapter is to propose a framework for developing multi-agent systems for mHealth drug delivery systems in a dynamic environment. We achieve this by exploring the common operational decisions in empirical case studies in the literature. Analysis of empirical studies reveals the important operational decisions that should be taken into account, together with their complexities, specifically in a dynamic environment. Therefore, the objectives of this research are as follows:

1. To explore the common operational decisions associated with mHealth drug delivery systems in a dynamic environment;
2. To propose a multi-agent system framework, highlighting agent functionalities, and the overall system services; and,
3. To apply the framework to dynamic task allocation problem in a dynamic mHealth drug delivery system.

The rest of this chapter is structured as follows. The next section presents an exploration of the key decisions in drug delivery in a home healthcare environment. This is followed by a general description of the drug delivery problem in home healthcare services in the third section. In the fourth section, a framework is proposed for developing a multi-agent system for drug delivery services in a home healthcare setting. The fifth section presents an application of the framework to homecare task allocation in a dynamic multi-objective environment. Conclusions and further work prospects are provided in the sixth section.

2 Decision Making for Drug Delivery in Homecare Services

Decision making in mHealth drug delivery systems in a home care setting is complex but crucial (Bachouch et al., 2009; Anthony, 2012). The mHealth drug delivery systems are necessitated by the ever-increasing need for care and support from people in the community, particularly those with physical disabilities, learning disabilities, chronic diseases, and the elderly people who need assistance and medical treatment in their own homes (Akjiratikarl et al., 2007; Fraile et al., 2009a; Fraile et al., 2009b; Maeda, 2012). Decisions involve satisficing multiple goals regarding customer service quality, service cost, and healthcare worker satisfaction (Matta et al., 2012. With the increasing world-wide need for effective home healthcare, the increasing elderly population, and the increasing pressure from governments and other stakeholders, developing effective approaches for mHealth drug delivery decisions is imperative (Bahdori et al., 2003; Anthony, 2012). In this paper, we present a multi-agent architecture that facilitates decision making in mHealth drug delivery system. The proposed framework is an event-based formalism, a mHealth monitoring system that integrates the capabilities of a multi-agent system and Web services so as to facilitate effective decisions for home healthcare services. The aim is to provide a multi-agent system, where decisions are based on a system of coordinated intelligent software agents that provide quick and intelligent alternative decisions in a dynamic environment.

Mobile networks are revolutionizing multiple healthcare systems in developing and developed countries. mHealth in general, and mHealth drug delivery in particular, seek to utilize mobile communication technologies to deliver quality healthcare services. Implementing real-time decision systems that are able to provide informed real-time decisions are essential for decision makers in home health care systems. As such, mHealth drug delivery is an important area of innovation that has a huge potential to make a significant difference in healthcare systems all over the globe.

The real issue in a mHealth drug delivery system is that of efficient and effective coordination of the players in the system (Fraile et al., 2009; Rossi et al., 2012). Multi-agent systems have been recognized as one of the technologies that would meet decision making needs in an agile and fast-changing environment (Monostori and Kadar, 1999; Zhang et al., 2006; Nyanga et al., 2012). An agent is a software object that mimics the role of a competent personal assistant to perform a specific task on behalf of the user, intelligently or not, independently of with little guidance (Zhang et al., 2006). In other words, an agent is a computational system that is situated in a dynamic environment and is capable of exhibiting autonomous and intelligent behavior.

3 The mHealth Drug Delivery Problem

Home healthcare service providers are faced with dynamically changing task allocation schedules (e.g., drug delivery and distribution tasks) due to evolving problem situations. Staff preferences and patient expectations change over time. In addition, management targets or goals are often adjusted in response to internal and external changes. Oftentimes, preferences and management goals are expressed without measurable precision, but rather, in a linguistic, ill-structured way, for instances, expressions such as "about 8 working hours", "preferably 80% service rate", and "preferably morning shifts" and other related forms, are difficult to quantify precisely (Topaloglu and Selim, 2010). Moreover, the interaction of the three players, the management, staff, and patients, is complex and difficult to model using closed form methods. Managers need intelligent interactive decision support tools in such complex, dynamic and multi-objective homecare environments. Decisions need to be made as regards to scheduling of tasks for nursing staff and assignment of homecare tasks that are to be performed at preferred time windows specified by the patients (Corchado et al., 2008). The use of intelligent agents for decision making is the most viable option (Tapia et al., 2006).

3.1 The Nature of mHealth Drug Delivery Systems

The home healthcare environment is often concerned with the construction of work schedules and the dynamic assignment of tasks to care givers or nursing staff over the course of day, such as drug delivery, patient monitoring, and other homecare related services (Corchado et al., 2008; Varnfiled et al., 2012). Care workers need to visit patients are their homes at specific time windows, defined by earliest start time and latest start time. In the process, the staff, patients, and management communicate through mobile and other network channels so as to update information and to revise staff schedules as needed. Meanwhile, a number of objectives have to be considered simultaneously when updating information on the staff schedules and the task assignments. The information update and the ensuing decision processes are difficult to manage in a dynamic and uncertain environment.

3.2 Complexities in mHealth Drug Delivery Systems

Due to the presence of multiple conflicting objectives in homecare decisions, multiple goals are considered when developing suitable solution to homecare staff scheduling and task assignment. Three goals are considered simultaneously: (i) minimizing the schedule cost, (ii) maximizing client satisfaction, and (iii) maximizing worker satisfaction. This essentially involves finding a judicious trade-off between the goals, considering preferences and choices of the decision maker. In a typical homecare environment, the decision maker needs to consider a number of objectives, including the following:

- *Schedule cost*. This is a management objective concerned with minimizing the cost associated with the trips followed by healthcare givers. The cost of each trip is estimated in terms of the total distance from the care giver's point of origin to the first client, the distances between successive clients, and from the last client back to the care giver's point of origin.
- *Patient satisfaction*. This objective concerns the maximization of client satisfaction or service quality which can be expressed as a function of the violations of time windows preferred by the clients. A penalty pseudo-cost is often imposed when a care giver reaches a patient's home too early (earliness) or too late (lateness) in comparison to the preferred time window of the patient.
- *Worker satisfaction*. In practice, it is essential to consider healthcare worker satisfaction, which entails meeting the worker preferences to the highest degree possible. Each worker indicates specific preferences in terms of individual working hours, work starting time, work finishing time, among others. One of the most common specifications is the desired total individual working hours. In essence, the overall worker satisfaction should be maximized satisfactorily.

In practice, the decision maker should consider a trade-off between client satisfaction, worker satisfaction and cost minimization. Such complex decisions can be handled conveniently by developing a suitable multi-agent system, consisting of software agents that are capable of independent and joint analysis of complex situations (Nyanga et al., 2102; Monostori and Kadar, 1999).

4 A Multi-agent System Framework

The proposed multi-agent system framework consists of different kinds of agents, each with specific roles, capabilities and characteristics. In this section, we briefly describe the overall system structure, its goals, and the basic individual agent functionalities.

4.1 The Overall System Structure and Goals

The suggested system framework incorporates intelligent agents, internet services, wireless networks, and mobile devices. The framework consists of a number of functional agents that are supervised by a supervisor agent, and managed by a manager agent through available communication channels. In particular, the system consists of Manager Agent, Supervisor Agent, Resource Agent, Scheduler Agent, Nurse Agent, and Patient Agent. The overall goal of the multi-agent system is to provide intelligent and robust decisions in a homecare environment whose characteristics evolve over time. Quick and robust decisions are especially needed in regards to homecare staff scheduling and task allocation, where task loading is dynamic over the course of day, staff capability and availability change over time due to unforeseen circumstances, and patients' requests are continually updated over the course of day.

Fig 1 shows the proposed framework for developing a multi-agent system for coordinated drug delivery in a home healthcare setting. We identify six agents for the multi-agent system. Brief descriptions of the basic functionalities of the agents are provided.

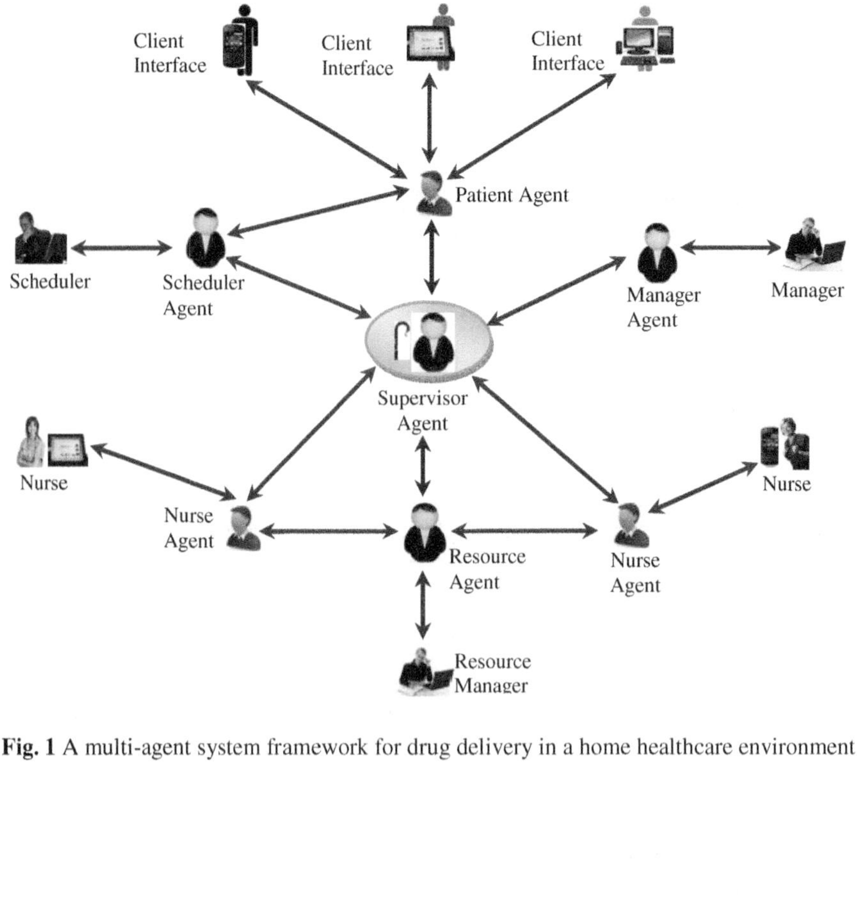

Fig. 1 A multi-agent system framework for drug delivery in a home healthcare environment

4.2 Agents and Their Functionalities

The agent platform forms the heart of the multi-agent system. These agents communicate and collaborate together through reliable communication channels to achieve a common goal. Each agent should have specific functionalities, capabilities and characteristics. This enhances system flexibility to add in new agents, as necessary. The major agents and their basic functionalities are described as follows:

4.2.1 Manager Agent

The manager agent has the overall responsibility of managing and resolving conflicts between functional agents, as communicated by the supervisor agent. In this regard, it holds and keeps track of the quality of service, the nurse preferences, and the management requirements or goals. This information is necessary for task scheduling and assignment decisions. In this respect, it sends requests for information updates from the supervisor agents. In summary, the manager agent has the following responsibilities:

- setting up and updating the overall goals or requirements of the management;
- regularizing the ever-changing preferences of the patients and the nursing staff; and,
- finalizing staff planning and scheduling decisions originally prepared by the supervisor agent.

In its operation, the manager agent closely interacts with the supervisor agent for relevant periodic information updates.

4.2.2 Supervisor Agent

The supervisor agent ensures that all the agents function correctly in the system. The agent periodically sends diagnostic messages to verify the status of all the agents. In the case that an agent malfunctions, another instance of the agent should be created. The supervisor agent is also responsible for conflict resolutions among other agents, as directed by the manager agent. In summary, the supervisor agent is responsible for the following tasks;

- ensuring, by periodic diagnosis, that all other agents function correctly in the system;
- ensuring resolution of conflicts that may arise between other agents in the system;
- obtaining, recording and keeping track of the status of other agents in the system; and,
- updating information on management goals, staff preferences, and patient preferences.

The supervisor agent closely interacts with the resource agent in order to obtain the relevant information updates concerning resource availabilities and capabilities.

4.2.3 Resource Agent

The resource agent keeps track of the information on capabilities and availabilities of staff resources in the system; compares the capabilities of staff with the workload, as advised by the supervisor agent and the nurse agents. It informs the supervisor agent on the times the nurse staff are available according to their current schedules. If there are any tasks that can be completed by the available staff, the information is relayed to the scheduler agent through the supervisor agent. Therefore, the resource agent is responsible for,

- maintaining updates on the overall staff resource availabilities and their capabilities;
- recording, updating, and keeping track of task requirements and task loading; and,
- comparing current loading and resource availabilities, and updates the supervisor agent.

The resource agent closely interacts with the nurse agents, obtaining resource information updates periodically.

4.2.4 Nurse Agent

The nurse agent schedules the nurse's working day and manages profiles, tasks, available time, and the resources. The agent must generate the plan considering the nurse's preferences such that all that all the patients assigned to the nurse receive care according to their time preferences. In addition the schedule must ensure that the nurse's working hours do not exceed the present maximum hours. Thus, the every nurse agent generates personalized plans based on the nurse's profile and working habits, as well as patient preferences. In summary, the nurse agent is responsible for the following tasks:

- managing the individual nursing staff profiles, tasks, availability, and resources;
- recording, updating and keeping track of the nurse preferences and their working habits;
- generating primary schedule plan, considering nurse preference, patient preferences and management goals.

4.2.5 Patient Agent

The patient agent manages patients' personal data and behavior, that is, monitoring location, daily tasks, and reporting anomalies, if any. The patient agent is the most sensitive agent and it must frequently update information regarding the beliefs, preferences, and expectations of patients. Requests from the patients are recorded and considered for the task scheduling decisions. Precisely, the patient agent is responsible for,

- updating information on the behavior patients and their healthcare preferences;
- updating information regarding nursing and drug delivery task requirements; and,
- reporting any patient anomalies, emergencies or any requests that may arise.

Thus, the patient agent frequently updates the supervisor agent as well as the scheduler agent concerning the patients' requested tasks so as to maintain or improve the quality of service.

4.2.6 Scheduler Agent

The scheduler agent integrates management target, resource, and task information from the manager agent, the resource agent, and patient agent, respectively. The aim is to generate the ultimate schedule, subject to management goals, staff preferences, patients' preferences, and other constraints. Thus, in the presence of imprecise or uncertain goals and preferences, the scheduler agent generates an overall satisficing schedule that seeks to suffice and satisfy the overall expectations of the three players in the actual system: the management, the staff, and the patients. In summary, the scheduler agent is responsible for the following tasks,

- integrating the evolving target, resource and task information from other agents;
- updating the changes in the choices and preferences of the scheduler to assist in the scheduling decision process; and,
- generating a satisficing solution to the drug delivery task scheduling for the care workers;

Therefore, the scheduler provides a satisficing solution, considering management targets, nurse preferences, and patient preferences that often evolve with time. This enables the healthcare system to provide improved quality of service across all the players in the home healthcare system.

5 An Application to Dynamic Task Allocation in a Homecare Environment

Drug distribution and delivery in a home healthcare set up involves complex issues. For example, each patient may specify a specific preferred time window for drug delivery, care workers may have their own preferences in regards to workloads, working times or schedules, and other tasks, and management often have their goals in regards to quality of service and patient satisfaction. Over the course of day, patient may raise new individual care requests, care workers or managers may also request changes to schedules as need arises. Therefore any changes in the drug distribution process may occur, raising the need to re-schedule the task allocation plans in a dynamic fashion. The drug delivery and distribution, in this instance, becomes dynamic and multi-objective, inducing more complexities in the problem.

It is noted in this study that a dynamic multi-objective homecare environment is characterised by (i) goals, expectations, and preferences that may change or evolve over time, (ii) multiple conflicting objectives, and (iii) imprecise or uncertain information. In the presence of fuzzy and dynamic data, coupled with conflicting objectives, the most appropriate decision is the one that seeks to satisfice the players in the system. Therefore, we propose a satisficing procedure that judiciously selects from a generated solution space at a given point in time, and then updates the solution space in accordance with the evolving information. In practice, for instance, a scheduler agent seeks to generate a task schedule that satisfies the care worker, the patient, and the management goals. The procedure is built into the agent in the form of a heuristic.

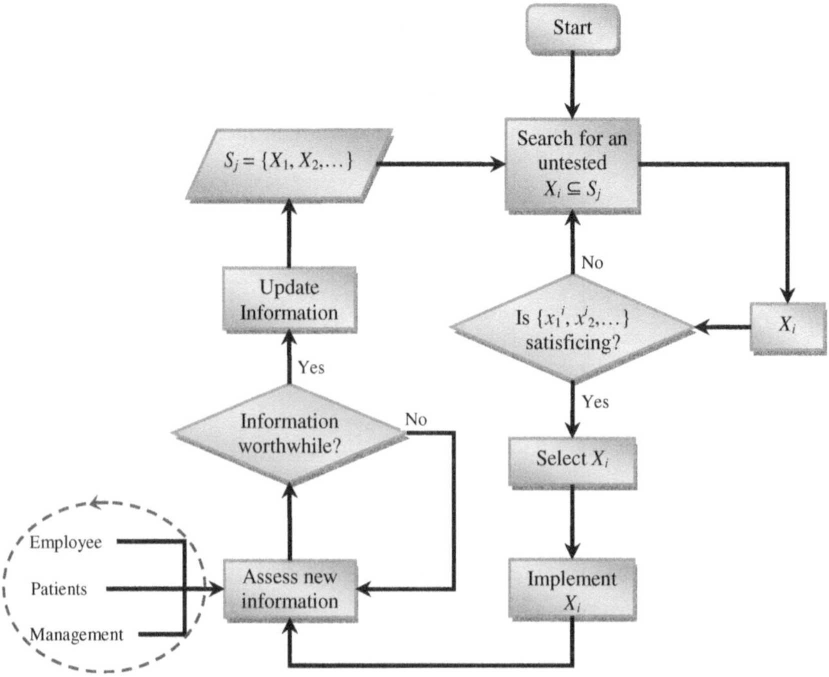

Fig. 2 A satisficing heuristic for drug delivery task allocation in home healthcare

Fig 2 illustrates the proposed satisficing heuristic for interactive homecare staff scheduling decisions. Formally, the heuristic begins by searching from the solution space S_j, an untested solution $X_i \subseteq S_j$, and testing whether or not it is satisficing. If X_i is not satisficing, the search process continues, otherwise the solution is selected and implemented. After the implementation of the schedule, the perceptions of the three players – employee, client, management – will change over time, such that the decision maker needs to acquire and assess new information on their satisfaction or preferences and goals. If the new preferences reach the pre-specified thresholds, then the solution space S_j should be updated accordingly. The decision maker updates his preferences according to the new information obtained. The search for a satisficing solution starts again with the update information. The heuristic makes the approach more applicable and adaptable to the real life human decision process. However, the information pertaining to the choices, expectations or preferences or the patients, the decision maker, and the management is usually imprecise or uncertain. Moreover, the information tends to change of time.

6 Concluding Remarks

Due to the ever-growing need to provide satisfactory care and support to patients at their homes and the widespread drive to offer such care services, novel and effective solutions to homecare decisions are essential. Developing robust decision making systems for homecare staff scheduling and task assignment, such as drug delivery is urgently necessary. This chapter proposed a framework for developing multi-agent systems for homecare drug delivery task assignment decisions when the management goals, the staff preferences, and the patients' preferences are dynamic and conflicting. Considering that the information so provided is often imprecise in practice, we proposed a satisficing algorithm based on a dynamic solution space generated at a specific time, subject to evolving preferences of the staff, the patients and the management. The proposed framework consists of a number of collaborating agents that communicate and are coordinated through efficient communication channels so as to achieve a satisficing solution at any point in time. It is anticipated that the framework forms a platform for efficiently coordinated decision making on drug delivery in homecare environments where information is imprecise and dynamic.

References

Akjiratikarl, C., Yenradee, P., Drake, P.R.: PSO-based algorithm for home care worker scheduling in the UK. Computers & Industrial Engineering 53, 559–583 (2007)

Anthony, J.M.: Tablet computers for mhealth: opportunities for personal healthcare. In: Proceedings of the IASTED International Conference, Health Informatics (AfricaHI 2012), Botswana, pp. 355–359 (2012)

Ascari, A., Bakshi, A., Grijpink, F.: mHealth: A new vision for healthcare. McKinsey & Company, Inc., and GSMA (2010)

Bachouch, R.B., Guinet, A., Hajri-Gabouj, S.: A model for scheduling drug deliveries in a french homecare. In: International Conference on Industrial Engineering and Sytems Management, IESM 2009, Montreal, Canada (2009)

Hahadori, S., Cesta, A., Grisetti, G., Iocchi, L., Leonel, R., Nardi, D., Oddi, A., Pecora, F., Rasconi, R.: RoboCare: Pervasive Intelligence for the Domestic Care of the Elderly. AI*IA Magazine Special Issue (January 2003)

Corchado, J.M., Bajo, J., Abraham, A.: GERAmI: Improving the delivery of health care. IEEE Intelligent Systems. Special Issue on Ambient Intelligence (March/April 2008)

Dolan, B.: Philips suggests dedicated mHealth spectrum (2010), `http://mobihealthnews.com/7178/philips-suggests-dedicated-mhealth-spectrum/`

Eveborn, P., Rönnqvist, M., Einarsdóttir, H., Eklund, M., Lidén, K., Almroth, M.: Operations Research Improves Quality and Efficiency in Home Care. Interfaces 39, 18–34 (2009)

Fraile, J.A., Tapia, D.I., Rodríguez, S., Corchado, J.M.: Agents in Home Care: A Case Study. In: Corchado, E., Wu, X., Oja, E., Herrero, Á., Baruque, B. (eds.) HAIS 2009. LNCS, vol. 5572, pp. 1–8. Springer, Heidelberg (2009a)

Fraile, J.A., Bajo, J., Abraham, J., Corchado, J.M.: HoCaMA: Home care hybrid multiagent architecture. In: Hassanien, A.E., et al. (eds.) Pervasive Computing: Innovations in Intelligent Multimedia and Applications, Computer Communications and Networks, pp. 259–285. Springer-Verlag London Limited (2009b)

Krohn, R., Metcalf, D.: mHealth: From Smartphones to Smart Systems (2012), `http://www.mhimss.org/sites/default/files/resource-edia/pdf/CHAPTER%202_mHealth.pdf`

Maeder, A.J.: Tablet computers for mhealth: opportunities for personal healthcare. In: Proceedings of the IASTED International Conference, Health Informatics (AfricaHI 2012), Botswana, pp. 355–359 (2012)

Matta, A., Chahed, S., Sahin, E., Dallery, Y.: Modelling home care organisations from an operations management perspective. Journal of Flexible Service Manufacturing (2012), doi:10.1007/s10696-012-9157-0

Monostori, L., Kadar, B.: Agent-based control of manufacturing systems. In: Proceedings of Second International Conference on Intelligent Processing and on Intelligent Processing and Manufacturing of Materials (IPMM 1999), Hawaii, USA, pp. 131–137 (1999)

Nyanga, L., Van Der Merwe, A.F., Matope, S., Tlale, N.: E-Manufacturing: A framework for increasing manufacturing resource utilization. In: Proceedings of the 42nd Computers & Industrial Engineering, Cape Town, South Africa, pp. 1–12 (2012)

de Rossi, D., Lorussi, F., Tognetti, A.: Wearable systems for e-health: telemonitoring and telerehabilitation. In: Proceedings of the IASTED International Conference, Health Informatics (AfricaHI 2012), Botswana, pp. 335–338 (2012)

Tapia, D.I., Bajo, J., De Paz, F., Corchado, J.M.: Hybrid Multi-agent System for Alzheimer Health Care. In: Rezende, S.O., da Silva Filho, A.C.R. (eds.) Proceedings of HAIS 2006, Ribeirao Preto, Brasil (2006)

Topaloglu, S., Selim, H.: Nurse scheduling using fuzzy modeling approach. Fuzzy Sets and Systems 161, 1543–1563 (2010)

Varnfield, M., Ding, H., Karunanithi, M.: Use of mobile phone based health applications in home care delivery of cardiac rehabilitation. In: Proceedings of the IASTED International Conference, Health Informatics (AfricaHI 2012), Botswana, pp. 360–365 (2012)

Zhang, D.Z., Anosike, A.I., Lim, M.K., Akanle, O.M.: An agent-based approach for e-manufacturing and supply chain integration. Computers & Industria Engineering 51, 343–360 (2006)

Questions

Q1. Briefly explain the complicating features encountered when making decisions regarding drug delivery and distribution in a typical home healthcare organization.

Q2. What is the essence of mHealth technology in the dynamic scheduling of drug delivery schedules?

Q3. What are the advantages of using multi-agent systems in mHealth drug delivery?

Q4. Suggest any disadvantages or limitations that may be associated with the use of multi-agent systems in mHealth drug delivery and distribution.

Part VI

Radio Technologies and Applications

30. Wireless Networks in Mobile Healthcare

Pantea Keikhosrokiani[1], Nasriah Zakaria[1,2], Norlia Mustaffa[1], Tat-Chee Wan[1,3], Muhammad Imran Sarwar[3], and Keyvan Azimi[4]

[1] School of Computer Sciences, Universiti Sains Malaysia, 11800 USM, Penang, Malaysia
[2] Medical Education Department, College of Medicine, King Saud University, Riyadh
[3] National Advance IPv6 Centre (NAv6), Universiti Sains Malaysia, Penang, Malaysia
[4] School of Electrical Engineering, Universiti Sains Malaysia, Penang, Malaysia
pantea.kia@ieee.org, {tcwan,norlia,nasriah}@cs.usm.my, imran@nav6.usm.my, keyvan.azimi.64@gmail.com

Keywords: Mobile healthcare, Wireless technologies, Cellular network, Satellite, Diagnosis, Emergency, Remote monitoring, Location-based services.

1 Introduction

With the advancement of telemedicine systems, health information can now be transferred as interactive video over wireless networks. The key features of a mobile wireless network based system are portability, long battery life, ease of use, full duplex support, and optional encrypted communication (Pattichis, 2002). Rapid advances in wireless and mobile communication technologies have led to new categories of services to support healthcare. Telemedicine can be used to deliver healthcare and medical information remotely using wired or wireless telecommunication technologies. POTS (plain old telephone system) and ISDN (integrated services digital network) are two types of wired communication technologies; GSM (global system for mobile communication), GPRS (general packet radio service), UMTS (universal mobile telecommunication systems) and satellite communication are examples of modern wireless telecommunication technologies (Pattichis, 2002).

Mobile wireless communication systems have improved reliable delivery of medical multimedia information services that can be applied to emergency ambulance service, mobile hospital (m-hospital), general health care, early warning systems for health care, and more. Patient information, including videos, images, electrocardiogram (ECG) can be transmitted from the hospital emergency room to other places via 4G cellular network, mobile WiMAX wireless networks and mobile satellite communication systems (MSC). These new options are beneficial for doctors, medical care staff in ambulances, and medical students who are learning and observing medical treatment long distance (Lin et al., 2010). Rapid advancements in network technologies and mobile wireless network infrastructure, along with

S. Adibi (ed.), *Mobile Health*,
Springer Series in Bio-/Neuroinformatics 5, DOI: 10.1007/978-3-319-12817-7_30

improvements in multimedia-based applications, and opening new horizons for those services in the health care management sector and medical informatics.

Mobile telemedicine is one of these new 21st century state of the art technologies. It provides an architecture for transferring biomedical patient information over a wireless communication system anytime, from anywhere with great speed and reliability. Telemedicine encompasses regular healthcare, mobile hospitals, emergencies, maritime, remote monitoring and more. By merging biomedical signals and information technology through a robust wireless communication system which can enable remote medical treatment or health monitoring medical service at any location (Lin and Lee, 2009). Using mobile healthcare systems enables healthcare personnel to improve the efficiency and quality of service (QoS) of medical care. Such healthcare systems can also decrease the number of patient visits to the hospital as well as length of patient stay, since adequate follow-up care can be delivered after the patient has been discharged (Lin et al., 2010).

The objective of this paper is to present a review of the different communication technology components used by healthcare systems as well as various types of healthcare systems that use wireless communication networks as a means of efficient and effective communication between remote patients and healthcare providers. Following this, we highlight the most used commonly used wireless technologies in healthcare systems today.

The structure of this chapter is as follows. Section 2 discusses wireless and cellular communication technologies in use in the healthcare followed by section 3 that highlights medical data transmission in healthcare systems. Section 4 offers a classification of healthcare systems based on publications. Section 5 contains discussion of findings, followed by concluding remarks in Section 6.

2 Communication Technologies in Healthcare

In this section, we review different communication technologies used in healthcare systems and highlight their pros and cons. Some communication technologies are suitable for healthcare systems while others are not. Selecting the right communication technology depends on the specific healthcare situation as well as on the user population.

Healthcare systems are increasingly delivering biomedical signals and other healthcare information over long distances using communication technologies. These same services can be delivered anytime from anywhere using mobile communication systems with high speed and high reliability. Such mobile communication technologies can be applied to a mobile hospital (M- hospital), emergency ambulance services, early warning systems for diseases, illness rehabilitation, general healthcare and so on. As part of pre-hospital medical care, patient information including videos, images, electrocardiogram (ECG) data and pulse biomedical signals can be transmitted via Global System for Mobile communication (GSM), General Packet Radio Service (GPRS), third generation and forth generation (3G and 4G) mobile cellular systems, WiMAX, and mobile satellite communication (MSC) systems (Pattichis et al., 2002, Rashvand et al., 2008). In this section we summarize the main wireless technologies that are used in healthcare systems today, such as GSM, GPRS, 3G, 4G, LTE, Wi-Fi, WiMAX, satellite, WAP, wireless IP, wireless LAN, Wireless Personal Area Networks (WPAN), Wireless Sensor Networks, and Wireless Body Area Network (WBAN). These technologies, along with their frequency band and data transmission rates are illustrated in Table 1.

Table 1 Mobile telephony and satellite communication networking technologies and standards (Kyriacou et al., 2007, Zain et al., 2012)

Type	Sub-Type	Frequency Band	Data Transfer Rate
3G	FDD,WCDMA	1920-1980 MHz UL, 2110-2170 MHz DL	144 hbps- 2Mbps. Typically between 220-384 kbps
	TDD, DTD, CDMA	1900-1920 MHz UL, 2010-2025 MHz DL	As above
	HSDPA	As above	Up to 14.4 Mbps. Typically between 400 kbps- 2 Mbps
Satellite	Satellite	1980-2010 MHz UL, 2170-2200 MHz DL	
	ICO	C, S band	2.4 kbps
	Globalstar	L, S , C band	Up to 56 kbps
	Iridium	L, Ka band	2.4-10 kbps
	Inmarsat	L, C band	4-492 kbps
	Thuraya	L, C band	9.6-144 kbps
	IntelSat	C, Ku band	64 kbps-45 Mbps
	MSAT	L band	1.2-4.7 kbps
Wireless LAN	IEEE 802.11	2.4 GHz	2 Mbps
	IIEEE 802.11a	5 GHz	20 Mbps
	IEEE 802.1 lb	2.4 GHz	11 Mbps
	IEEE 802.1 lg	2.4 GHz	22 - 54 Mbps
	Hiperlanl	5 GHz	20 Mbps
	Hiperlan2	5 GHz	54 Mbps
	HiS WANa	5 GHz	54 Mbps
WiMAX	TDD	3.5 GHz	45 Mbps/Channel
	FDD	3.5 GHz	45 Mbps/Channel
	TDD	5.8 GHz	45 Mbps/Channel
BAN, PAN	Bluetooth	2.4 GHz	2 Mbps
	Zigbee	10 channels at 902- 928 MHz, and 1 channel at 868	20-250 kbps
LTE			50 Mbps to 86.4 Mbps on the uplink, 100 Mbps to 326.4 Mbps on the downlink
4G		700 MHz, 1700/2100 MHz	Min 50-100 Mbps , average 200 Mbps

2.1 GSM

Global System for Mobile communications (GSM) is a second generation mobile communication network. Although GSM was designed for voice telephony, it can carry data as well (Kyriacou et al., 2007, 2012a). Standard GSM supports voice and data-transfer speeds of up to 9.6 kbps together with transmission of SMS (Short Message Service). Enhanced GSM, which is known as High-Speed Circuit-Switched Data (HSCSD), supports data transmission of up to 43.3kbps (Kyriacou et al., 2007,

Pattichis et al., 2002, Kumar et al., Kumar et al., 2010). Many frequencies are available for GSM according to the various GSM sub-types, but the most common are 900, 1800, and 1900 MHz (Hillebrand, 2002, Pattichis et al., 2002).

2.2 GPRS

General Packet Radio Service (GPRS) is a packet-based mobile data service. It is designed to work on 2G and 3G cellular communication systems that are used for voice communications, such as GSM and PDC (Personal Digital Cellular). Maximum downlink data rate for GPRS is around 171 kbps (Kumar et al., Kyriacou et al., 2007). GPRS enables users to access advanced online services on the move, including e-mail, multimedia messages, social networking, ftp, HTTP-based web services, and location-based services from almost anywhere (Kumar et al., 2010).

Most telemedicine applications use GSM/GPRS networks to support prehospital treatments. These applications are divided into two main groups: 1) those transmitting bio-signals such as ECG, oxygen saturation, blood pressure, etc.; and 2) those transmitting medical images (Kyriacou et al., 2007, Kumar et al., 2010). Some applications combine both types of information (Pavlopoulos et al., 1998, Istepanian et al., 2001). Some applications transfer bio-signals in order to allow healthcare providers to monitor chronic heart disease patients (Salvador et al., 2005) while other applications are limited to transferring only images (Schächinger et al., 1999, Voskarides et al., 2003). GSM/GPRS can also be used to access electronic patient records (Istepanian et al., 2006b).

2.3 Third-Generation (3G) Wireless Networks

Mobile telemedicine becomes faster and more effective as it moves from second generation (2G) to 2.5 G and 3G (W-CDMA, CDMA2000, TD-CDMA) systems as these systems provide a faster data transfer rate (Kyriacou et al., 2007). With 3G enabled systems, telemedicine systems can employ a data-transfer speed of 14.4 Mbps (typically, around 1 Mbps) which was formerly only feasible on wired communication networks (Kyriacou et al., 2007, 2012a). Because of this high data-transfer rate of 3G, many healthcare systems prefer it. The first application of 3G to medical service was by (Chu and Ganz, 2004) who presented a cost-effective portable teletrauma system to offer prehospital trauma care. The proposed system would transmit a patient's medical images, video, and electrocardiogram signals over 3G wireless cellular data service to the responsible healthcare center. (Navarro et al., 2006) proposed a multi-collaborative mobile healthcare system using 3G mobile networks.

The next application (Lim et al., 2007) was a system for delivering medical reports from a medical imaging department to the referring doctor via 3G GSM mobile phones. Another relatively recent application (Garawi et al., 2006) investigated transmission of real-time ultrasound video by presenting end-to-end mobile tele-echography using an ultra-light robot (OTELO) over the 3G mobile communications network. Y. Chu and A. Ganz (2004) proposed a mobile tele-trauma system that would provide pre-hospital care for patients residing at remote locations using 3G cellular networks. The system could be installed on mobile tablet PCs in ambulances and patient information such as video, image, and electrocardiogram signals could be transmitted through cellular networks (Chu and Ganz, 2004). Due to the low data rate of 3G networks, compression technology was used to make the transmission more efficient. A similar system employing Universal Mobile Telecommunication System (UMTS) was used for communication between ambulance and hospital (Gallego et al., 2005). Trauma care through the transmission of patient video, medical images, and ECG along with OTELO was extensively investigated and implemented as applications over 3G cellular networks by (Chu and Ganz, 2004, Garawi et al., 2006). The HYGEIAnet project encompassed the design, development and deployment of advanced e-health and m-health services at various levels of the healthcare system, including primary care, pre-hospital emergency management, and so on. The system accommodates 600,000 people across 7 hospitals and 16 primary health centers using. 3G cellular network for communication between different hospitals and centers (Tsiknakis et al., 2002). The deployment of HYGEIAnet project services has shown significant economic, clinical and access care benefits with 65% of pre-hospital care managed by paramedics, and won a 2005 eEurope award.

2.4 Fourth-Generation (4G)

Fourth Generation (4G) mobile phone communication is a more advanced version of 3G which provides mobile ultra-broadband internet access. In the next few years, 4G wireless networks will become more widespread and users will be able to access multiple networks without changing from network to network manually (Varshney, 2007). 4G systems have higher transmission speeds than 3G (min 50-100 Mbps, average 200 Mbps) and they are supported by internet protocols (IPv6). They can provide services from anywhere at any time because 4G networks are IP-based heterogeneous networks. In addition, users can access and run a broad range of applications from multiple networks. All of these factors demonstrate the high usability of 4G. Moreover, multimedia services provided by 4G systems have lower per-bit transmission costs with high data transfer rate and good reliability. The services can be more personalized in order to meet diverse users' demands and users are able to take advantage of multiple services from different service providers at the same time. Consequently, future healthcare networks will offer "personalized predictive healthcares" and "Virtual Mobile Hospitals and Specialised M-Health Centers." 4G technologies will allow the healthcare sector to develop effective medical care delivery that will benefit both patients and physicians (Istepanian et al., 2006c).

Istepanaian and Zhang (2012) introduced the new and novel concept of 4G health which they argue represents the long-term evolution of m-health, and discussed the long-term evolution, challenges, and future implementation issues of 4G m-health (Istepanaian and Zhang, 2012).

2.5 Long Term Evolution (LTE)

The recent growth of mobile applications motivated the 3rd Generation Partnership Project (3GPP) to work on a new project which they dubbed Long-Term Evolution (LTE). LTE is a standard for wireless data communications technology and an evolution of the GSM/UMTS standards. The main goal of LTE is to provide a unique and efficient mobile system for public use. In the beginning, the focus was on broadband service using GSM framework. LTE has been used in most parts of the world with advanced mobile communication technology.

Several features of 3G needed to be modified in order to improve mobile communication networks and their service sectors. Therefore, the next project of LTE was the achievement of 4G which is advanced personal mobile communication technology with "all packet-based system" and focused on IPv6 (Adibi et al., 2010). Technology remains a concern for wireless networks in mobile healthcare applications. Managing healthcare systems, especially in emergency situations, is crucial; network support is thus an important issue in healthcare, especially in location tracking and routing protocols. The healthcare system must be reliable in order to provide continuous monitoring of patients in various environments. To meet healthcare systems' various needs and requirements, Fourth generation (4G) networks are proposed as a way to tackle these limitations and enable high speed broadband communication within the service providers' networks. LTE could be the best candidate for supporting robust healthcare applications and making them available anytime and anywhere to a wide range of potential users (Zain et al., 2012).

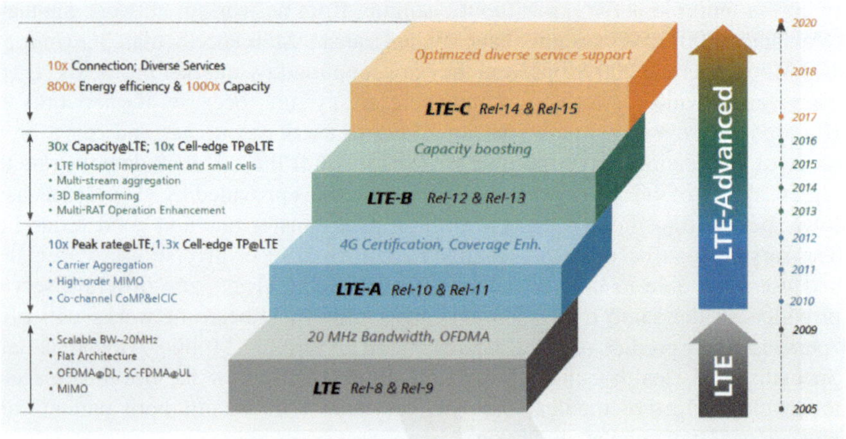

Fig. 1 The Evolution of LTE for 2005 to 2020

The cellular industry has, in the past several years, undergone a profound paradigm shift. The shift to mobile broadband has resulted in not just an increased appetite for mobile data services but also a philosophical shift in end-user habits, from relying on personal computers to relying on mobile smartphones. As a consequence of this socio-technological trend, the amount of data carried on cellular networks is growing explosively. Anticipating this, 3GPP initiated an investigation into E-UTRA (Evolved Universal Terrestrial Radio Access) in 2004; E-UTRA specified the first release of LTE (Rel-8) in March 2009, with later releases of LTE-Advanced to meet ITU-R IMTAdvanced (also called as 4G) requirements; while LTE-Advanced continues to evolve, introducing new features and supporting higher performance and broader services, including multiple market phases; LTE-A, and LTE-B, may be followed by LTE-C and so on in the future, as shown in the LTE roadmap in Figure 1.

2.5.1 LTE-A

Release 10 of the 3GPP specifications, released in 2011, is the initial phase of LTE-Advanced. LTE-A is one a framework for further advancements in LTE in order to meet 4G requirements and has been approved by the International Telecommunications Union- Radiocommunication sector (ITU-R). Attaining the data rates set by the International Mobile Telecommunications Advanced (IMT-Advanced) standard was one of the goals of LTE, and the LTE-A accomplishes this. LTE-A extends LTE to "the coverage of high data rates, group mobility, temporary network deployment, the cell-edge throughput or to provide coverage in new areas" (Lee and Vasilakos). LTE-A is expected to support instantaneous peak data rates of 1Gbps and 500Mbps for the downlink and uplink respectively. Moreover, the standard is also expected to provide wider bandwidth, lower latency, and higher peak and average efficiency (Saleh et al., 2009, Stencel et al., 2010). Other performance enhancements considered in LTE-A include coordinated multipoint transmission and reception (CoMP), relaying, support for heterogeneous networks, LTE self-optimizing network (SON) enhancements, home enhanced-node-B (HeNB) mobility enhancements, and fixed wireless customer premises equipment (CPE) RF requirements.

2.5.2 LTE-B

LTE-B is the second phase of LTE-Advanced and was developed for capacity boosting. LTE-B will reduce the cost and energy consumption per bit of transmission as well as offering excellent support for social networking, video sharing and so on. LTE-B provides stable access with more throughput at cell edge in comparisone with the LTE Rel-8 system. LTE-B is designed to improve the standard in several categories including: further multi-antenna enhancements with 3D-beamforming, LTE hotspot improvements along with small cell spectrum efficiency enhancements, small cell efficient operation with traffic adaption and interference coordination, heterogeneous networks (HetNet) mobility enhancement and multiple stream aggregation, HetNet frequency-division duplexing(FDD)/ time-division duplexing (TDD) joint operation, LTE interworking enhancement, WiFi interworking, minimization of drive test (MDT). It will also offering new services such as machine type communications (MTC), proximity based services (ProSe), and group communication service for LTE (GCSE_LTE).

LTE-B combines multiple technologies such as HetNet deployment with Macro-assistant coordinated small cells, 3D MIMO technology, LTE-centric multi-RAT operations, and more. This combination of technologies allows LTE-B to be a highly efficient, cost effective and flexible system more capable of fulfilling end user expectations. In particular the enabling technologies offer improved the capacity, more efficient network operation, flexible multi-radio access technology (RAT), all of which will enable it to support diverse applications and new services that are expected in the future. It will form the backbone of a new world of high definition mobile social video with ubiquitous broadband access and edge-free user experience.

2.5.3 LTE-C

LTE-C (Rel14/Rel15) the third phase of LTE-Advanced is currently being developed. Its goals are to further improve energy efficiency, capacity, connection and diverse services.

2.6 IEEE 802.16/WiMAX Systems

Worldwide interoperability for Microwave Access (WiMAX) is a wireless communication standard defined by IEEE 802.16. WiMAX provides broadband wireless access up to 50 km for fixed stations and 5-15 km for mobile stations. WiMAX thus has advantages over WLAN and is intended for wireless "metropolitan-area networks" (WMANs). WiMAX works between 2 GHz and 66 GHz with a maximum throughput of 70Mbps (Niyato et al., 2007, 2012b). WiMAX is considered a more suitable technology for healthcare applications than WLAN for several reasons. WiMAX can provide broadband access to both fixed mobile networks. It can use high bandwidth to transfer high-quality medical images and provide good video-conferencing between patient and healthcare center. It incorporates MAC layer security features (Niyato et al., 2007). Li, et al. (2012) designed a WiMAX-based emergency system in order to provide immediate online emergency treatment instructions via video and audio transmission between healthcare center, hospitals and emergency ambulances (Li et al., 2011). Lam, et al. (2009) proposed a WiMAX-based telemedicine system for emergency medical assistance in the ambulance; the system will shorten the emergency period and organize medical treatment properly to the arrival at the hospital (Lam et al., 2009).

2.7 Ad-Hoc and Sensor Networks in Healthcare

Most wireless systems require a built infrastructure and wired backbone network; ad-hoc and sensor networks, however, do not need any wired infrastructure. In Mobile Ad-hoc Networks (MANETs), mobile nodes are geographically distributed and interact with one another while on the move over a wireless medium. Thus, there is no communication with base stations. In terms of emergency systems where there are time limitations and deficiencies of wired infrastructure these kinds of networks are really helpful (Karl and Willig, 2005). A sensor network consists of many sensor nodes inside, or close to, the object or area of interest. Each sensor has an on-board processor and is able to collect and rout data. In the case of healthcare systems, these sensor nodes can be positioned to monitor patients (Akyildiz et al., 2002). Varshney (2008) designed a framework for patient monitoring that would support emergency messages by using a wireless ad hoc network. The system consists of patient monitoring devices, routing protocols and information presentation for vital signs and parameters. The performance results demonstrate the reliability of message delivery and low monitoring delays that can be accomplished through multicast or broadcast-based routing schemes (Varshney, 2008).

2.8 Satellite in Healthcare

Satellite has some advantages over other technologies. It has various data transferring rates (from 2.4 kbps to more than 2 x 64 kbps) and satellite links can function all around the world. However, satellite needs higher power as well as line-of-sight for transmission (Reinaldo, 1999, Pattichis et al., 2002). Thus, satellite has been used in many healthcare systems with different functions such as remote monitoring, emergency services, and telecardiology, especially in aircraft and ship settings (Pattichis et al., 2002). Adequate emergency healthcare services are essential for those traveling by aircraft or ship, as well those living in remote locations (e.g., mountains, islands forests). Mobile satellite communication (MSC) methods are a good solution for these kinds of situations, since they can provide interactive medical video conferencing. Numerous researchers (Murakami et al., 1994, Anogianakis et al., 1998, McDermott et al., 1999, Seon-Cheol and Myoung-Ho, 2000, Pierucci and Del Re, 2000) have explored the idea of video conferencing using a satellite communication system due to its advantages of wide-area coverage and ability to provide broadcast and multicast connections.

(Murakami et al., 1994) proposed a telemedicine technique using satellite communication for aircraft and ship. The study measured various characteristics such as channel capacity, real-time operation, size of the system, electromagnetic interference, and reliability of vital sign transmission along with overall effectiveness of the system. The proposed system would send color images, audio, ECG (3channels) data and blood pressure reading from the mobile device to the ground station, and transmit audio and error control signals from the ground station to the mobile device. Experiments with medical data transmission were conducted while participating ships and aircraft were moving. Therefore, the feasibility of this technique has been verified.

Satellite technology is particularly useful for remote healthcare delivery in a maritime environment. A number of projects have been proposed, including the EU's MERMAID project as well as others by Anogianikis, McDermott, Seon-Cheol & Myoung-Ho, and more. Refer to section 4.3.1 for a more detailed discussion of this research area (Anogianakis et al., 1998). McDermott et al. (1999) also proposed a maritime telemedicine system for high-speed medical data transmission using satellite networks (McDermott et al., 1999).

Seon-Cheol and Myoung-Ho, (2000) developed the telePACS maritime telemedicine system that can access a hospital`s picture archiving and communication system (PACS) servers using the Internet. They used Java to implement their system for browsing medical data and images via browsers such as Netscape. Finally, they proposed a fast and cost-effective PACS using web technology not only for inter-hospital but also for public use (Seon-Cheol and Myoung-Ho, 2000). (Pierucci and Del Re, 2000)developed a telemedicine application called Medical Environment for Diagnostic Images (MEDI) to provide remote medical care services. This system uses the broadcast and multicast ability of satellite as well as wide-band capacity. The system uses a Java client-server architecture based on TCP/IP. They evaluated the performance of the proposed application based on satellite connection. Li and Takahashi et al. (2009) proposed a Wireless Body Area Network (WBAN) system using satellite communication to sustain remote medical treatment and healthcare especially in emergency cases in rural or isolated areas. The system sends equivalent biosignals to the hospital so that doctors can assist patients remotely. The delay of system data delivery is measured that demonstrate that satellite link capacity has strong effect on transmission delay (Li et al., 2009).

2.9 WAP-Based Healthcare System

A combination of Internet and mobile communication along with improved handheld devices, wireless infrastructures, application programming languages, and protocols has made it possible to offer mobile Internet access. Wireless Application Protocol (WAP), which is a technical standard for accessing information over a mobile wireless network, has become a hot trend in telecommunication. WAP provides a communication protocol and application environment for the purpose of advanced telephony services and Internet access from mobile devices. E-banking, e-shopping, email, and games are typical examples of WAP application. WAP overcomes the limitation of interactive data services offered by service providers. Thus, WAP can be valuable in the healthcare field (Kumar et al., Hung and Zhang, 2003). Hung and Zhang (2003) developed a WAP-based telemedicine system for patient monitoring which utilized WAP devices as mobile access terminals for general inquiry in store-and-forward mode (Hung and Zhang, 2003). On the other hand, Paro (2001) pointed out that WAP has some limitations such as lower bandwidth, connection stability, and predictable availability; push operation is not specified; and a lack of cookies for session management. These disadvantages negatively affect the efficiency of WAP (Paro, 2001).

2.10 Wireless IP for Healthcare

Internet Protocol (IP) telephony is defined by the International Telecommunication Union (ITU) as "the conveyance of voice, fax and related services partially or wholly over packet-switched IP-based networks". It consists of Internet telephony and Voice-over-Internet-Protocol (VoIP); Internet telephony utilizes the public Internet, while VoIP employs privately managed IP-based networks, in addition to the public Internet (Zhao et al., 2002). Nowadays, VoIP is attracting more attention due to its ability to deliver real-time, two-way, synchronous voice and data traffic over packet-switched IP-based networks (Davidson and Peters, 2000). IP developed into a unifying platform for all methods of communication after deployment of 3G systems, and may be useful in the future development of public switched telephone network (PSTN), mobile network, fixed wireless and the Internet in the communication industry. IP telephony can be used for the transmission of voice as well as a combination of voice with other information because it uses network capacity more efficiently, reduces cost, manages enhanced services better, and offers more flexibility (Zhao et al., 2001).

Currently the use of wireless and IP based applications in healthcare systems is still in the development phase. These kinds of applications will improve healthcare systems by offering capabilities that are impossible through the usual telephony. Zhao, et al (2001) proposed a prototype wireless IP telephony system for telemedicine that would transmit voice and data in a wireless mobile environment (Zhao et al., 2001). Pandian et al. (2007) proposed a prototype wireless IP based store-and-forward telemedicine system that applied Very Small Aperture Terminal (VSAT), satellite communication and WLAN. Physiological data and medical images with videoconferencing were supported by this system (Pandian et al., 2007). (Husni et al., 2006) designed a mobile telemedicine system that would transfer medical healthcare data in emergency situations using a combination of Mobile Internet Protocol version 6 (MIPv6), and MANET, and IEEE 802.11 WLAN technology.

2.11 Wireless LAN in Healthcare

A Wireless Local Area Network (WLAN) is a wireless communication method that provides a connection to the wider internet via access points (cells). A WLAN data communication system is an extension of wired LAN but with higher flexibility. WLANs become popular in the healthcare environment after the introduction of the 802.11 standards; around 50% of hospitals have 802.11 local-area networks (LANs) installed (Baker and Hoglund, 2008). WLAN also became popular in retail, manufacturing, warehousing, and academia. It offered a data transmission rate of tens Mbps and limited user mobility. WLAN coverage is limited to about 100 meters per cell, making it sufficient for covering the area of a hospital (Kyriacou et al., 2007, Pattichis et al., 2002). Lin et al. (2010) proposed a WLAN transmission scheme based on the IEEE 802.11n standard to develop an effective wireless telemedicine application. This system is able to transmit clinical information with a high quality of service (Lin et al., 2010). In the case of a disaster situation where many people might be injured in a small areas, a WLAN-based monitoring application was proposed in order to monitor the condition of the injured (Palmer et al., 2005, Lenert et al., 2005). These applications could transmit CT (computed tomography) and MRI (magnetic resonance imaging) images collected by the bio-signal and sensor networks (Kim et al., 2005).

2.11.1 Wi-Fi (IEEE 802.11)

Wi-Fi can be defined as any WLAN that is based on the IEEE 802.11 standard. The original IEEE 802.11 was finalized in 1997. It specified a Medium Accesses Control (MAC) sub-layer combined with three physical layers: Frequency Hopping Spread Spectrum (FHSS), Direct-Sequence Spread Spectrum (DSSS) and Infra-Red (IR). These three physical layers ran at 1 and 2 Mbps. Two of these three layers FHSS and DSSS, were radio based with 2.4 GHz carrier. Two years later in 1999, IEEE 802.11a was ratified. It introduced a physical layer with orthogonal frequency division multiplexing (OFDM) operating in the UNII band (5GHz) and 54 Mbps. In mid 1999, IEEE 802.11b was proposed; it had a similar physical layer to the original IEEE but specified a higher net bit rate of 11 Mbps using complementary-code keying (CCK) (Paul and Ogunfunmi, 2008). In early 2003 IEEE 802.11g was released, which was a merging of IEEE 802.11a and IEEE 802.11b. 802.11g specifies four physical layers with a data rate of 2.4GHz and bit rate of 54Mbps. The four physical layers were Extended Rate Physicals (ERP)-DSSS/CCK, ERP-DSSS/PBCC, ERP-OFDM and DSSS-OFDM. The first two layers were the same physical layers used by IEEE 802.11b while the last two were new layers introduced with IEEE 802.11g. This new standard offers short preamble support, interoperability and CTS-to-self mechanism(Vassis et al., 2005).

(A) IEEE 802.11n: Although simulation testing proved the good performance of IEEE 802.11g, especially in the case of improving channel capacity, some issues remain. For example, a demand for simultaneous high-quality video-streaming by more than one user brought a motivation for providing higher throughput. IEEE 802.11n was the first attempt to fulfill this goal. It was published after 11 drafts in 2009. It is designed to provide a variety of services like higher data rates, better Quality of Service (QoS), more reliability, and network security. IEEE 802.11n improved both the physical layer and MAC layer to support throughputs beyond 100 Mbps, surpassing previous 802.11 standards. It maintains backwards compatibility by being compatible with previous IEEE 802.11 standards (Perahia and Stacey, 2008). The physical layer consist of: (1) Enhanced OFDM, which produces a high data rate of 65Mbps and achieves higher transmission speed; (2) Multiple-Input Multiple-Output/ spatial division multiplexing (MIMO/SDM) technology, which uses multiple antennas in both transmitter and receiver to send more than one data stream simultaneously. This is the most significant technology in the physical layer. Its pivotal feature is to increase data rate with SDM to 135Mbps in the real world. Link robustneed is improved by using MIMO along with optional techniques such as space-time block coding (STBC), spatial expansion, receive diversity, transmitter beamforming and low-density parity check (LDPC). (3) Channel bonding, which provides a 40MHz channel in both the 2.4GHz and 5GHz band by bonding two adjacent 20MHz channels. The combination of channel bonding with OFDM maximizes performance. Coupling MIMO with wider bandwidth can increase the data transfer rate up to 600Mbps (Perahia and Stacey, 2008).

The MAC layer consists of two components: (1) Frame aggregation, used in IEEE 802.11n to improve transmission efficiency. It aggregates multiple data frames coming from the upper layer into a single frame in order to reduce overhead. There are two types of aggregation: Aggregate MAC Service Data Unit (A-MSDU) and Aggregate MAC Protocol Data Unit (A-MPDU). (2) Block ACK (BA) mechanism used in IEEE 802.11n with a frame aggregation mechanism because a larger data frame causes an increased frame error rate. If an error is found in a frame, a BA will be generated to the sender stating that an error has occurred. The sender has to retransmit only those parts of the frame which are incorrect. BA can be applied on A-MPDU but not on A-MSDU (Perahia, 2008).

(B) IEEE 802.11ac: The rapid growth of mobile data traffic worldwide requires a high data rate to support video, live gaming and voice. Consequently, a standard is needed to be able to deal with next-generation mobile networks. IEEE 802.11ac builds upon IEEE 802.11n to provide throughputs of 500 Mbps for a single user and at least 1 Gigabit per second (Gbps) for multiple user on the 5 GHz band (Eng Hwee et al., 2011). IEEE 802.11ac takes advantage of existing IEEE802.11n features as well as adding some new features to afford Very High Throughput (VHT). New features specified in the physical layer including the following : (1) Multi-user MIMO technology, an evolution of MIMO, to transmit data to multiple clients simultaneously from a single access point; (2) Channelization for 20, 40, 80 and 160 MHz, to support wider channel bandwidth -- this is a low cost alternative to achieve higher data rates (Eng Hwee et al., 2011). New features in the MAC layer include explicit channel feedback to regularly provide information about channel state to the access point in order to achieve reliable communication in multi-antenna systems (Redieteab et al., 2012).

(C) IEEE 802.11ad: Throughput intensive applications like wireless I/O and uncompressed high-definition multimedia transmissions, which demand high data retests, motivated the industry to use 60MHz band . IEEE802.11ad is able to support multi-gigabit data transmission up to 7 Gbps (Perahia and Gong, 2011, Xiaoyi et al., 2011). Features proposed in the physical layer include: (1) Two operating modes in 2.16GHz channel bandwidth: OFTM for high performance application and Single Carrier (SC) for low power transceivers with low complexity; (2) a common preamble used by CS and OFDM for a simpler implementation and better coexistence (Cordeiro et al., 2010). Features proposed in the MAC layer include: (1) A combination of contention-based CSMA/CA along with a contention free TDMA, to support all kind of applications like web browsing with lower average latency or video transmission that demands a high QoS; three types of ACK mechanism offered are No- ACK, Block ACK and immediate ACK (Xiaoyi et al., 2011). (2) Personal Basic Service Set (PBSS), a new architecture, which enables two devices to have a direct communication with each other without any special devices like AP; this architecture makes possible new applications such as fast audio-video transmission to a projector or fast device synchronization; (3) Fast Session Transfer (FST), a pivotal major innovation mechanisms, which provides fast transition between different channel/ bands and enables devices to switch from a high frequency channel like 60 MHz to a lower frequency channel (Cordeiro et al., 2010).

Wi-Fi is widely used by hospitals today as it offers reliable, security-protected transmission of data and messaging. Wi-Fi can assist hospitals in track assets and patients inside its walls. As a consequence of the needs in hospitals -- from PC networking to the proliferation of Wi-Fi devices such as Smartphones, tablets, and patient monitoring devices -- Wi-Fi systems in hospitals are growing significantly. Wi-Fi networks offer hospitals a proven, interoperable and security-protected system for handling a broad range of devices and data transfers. However, in order to maximize the benefits of Wi-Fi for hospitals several points specific to the hospital environment must be considered. Among these are security requirements for patient records, the mobility needs of devices, "always on" network uptime requirements to accommodate patient monitors and applications, real-time access to records and data transfers, and continuously increasing bandwidth demands.

2.12 IEEE 802.15

2.12.1 Wireless Personal Area Networks (WPAN)

A wireless personal area network (WPAN) is a short-range wireless network defined by IEEE standard 802.15. It can cover only few meters and it can be used only for personal devices such as cell phones or personal digital assistants (PDAs), printers, and computers without any wired connection. The most relevant WPAN technologies in the healthcare field are Bluetooth (Blueooth, 2012) and ZigBee (Kumar et al., 2010). Bluetooth is a wireless technology that works at the 2.4 GHz frequency to allow communication electrical devices. The data transmission rate of Bluetooth is 723 kbps (Barnes, 2002, Rasid and Woodward, 2005). Since Bluetooth is a low-cost, low-size and low-power wireless technology, it is appropriate for short-range personal area network (PAN) applications. Data transmission range is 10-100 meters and obstacles such as walls do not have any effect (Kumar et al.). Rasid and Woodward (2005) discussed a processor that would transmit biomedical signals to a mobile phone using Bluetooth (Rasid and Woodward, 2005). ZigBee (IEEE 802.15.4 standard) has a low data rate of 20-250 kbps. and uses 16 channels at 2.4 GHz, ten channels at 902- 928 MHz, and one channel at 868-870 MHz. It has very long battery life and very low complexity. ZigBee can also be used in short-range wireless telemedicine applications (Kumar et al., Lin, 2010, Kumar et al., 2010).

2.12.2 Wireless Body Area Network (WBAN)

Body Area Networks (BANs) play an important role in healthcare systems today. A BAN consists of different nodes capable of sensing, sampling, processing, and communicating physiological signals (e.g., body temperature, heart rate, blood pressure, blood oxygen saturation, blood glucose level), physical activity (e.g., body posture, type and level of activity), and environmental parameters (e.g., location, temperature, humidity, light, atmospheric pressure). These nodes can be located directly on the human body or attached to clothing. If they communicate wirelessly, the system is referred to as a Wireless Body Area Network (WBAN) (Jovanov and Milenkovic, 2011). WBAN developed as a means of increasing the use of wireless networks in healthcare and minimizing the electrical devices. In a WBAN, these wearable sensors have ability of establishing wireless communication network. The wearable sensors are used to monitor patients continuously and send real time feedback to medical personnel (Arrobo and Gitlin, 2011, Crosby et al., 2012, Beno et al., 2011, Latre´ et al., 2010).

(W)BANs consist of three types of components, which may or not be wireless sensors, actuators, and (Wireless) personal device (PD). Sensors measure certain parameters of the human body internally or externally as shown in Figure 2, including heartbeat, body temperature or even recording a prolonged electrocardiogram (ECG). It comprises several components: sensor hardware, a power unit, a processor, memory and a transmitter or transceiver. Actuators perform specific actions consistent with the data they receive from the sensors or through interaction with the user -- for instance, an actuator equipped with a built-in reservoir and pump that administers the correct dose of insulin to a diabetics patients based on glucose level measurements. An actuator consists hardware (e.g. a reservoir to hold insulin), a power unit, a processor, memory and a receiver or transceiver. The personal device (PD) is responsible for collecting all the information obtained by the sensors and actuators and gives the user (patient, doctor, nurse, etc) the necessary information by displaying it on the device. PD components are a power unit, a (large) processor, memory and a transceiver (Latre´ et al., 2010).

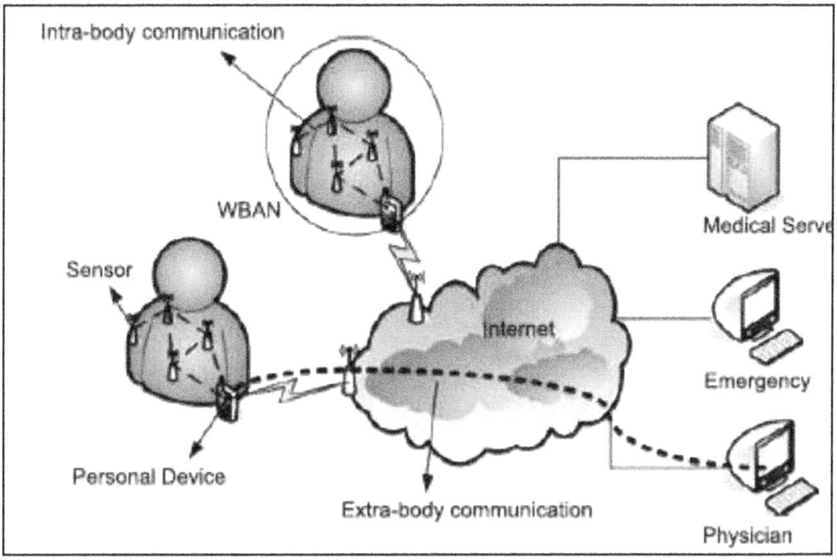

Fig. 2 Intra-body and Extra-body Communication in a WBAN

Data rates will be different depending on the application (simple data at a few kbs to video at several Mbps). Table 2 illustrates the data rates for a range of different applications. In addition, it shows the desired accuracy of the measurements as well as required bandwidth. As depicted in the table, the essential data rates for a WBAN are not high. However, the data rate can reach a few Mbps by adding devices such as motion sensors, ECG, EMG, glucose monitoring, etc. (Penzel et al., 2001, Arnon et al., 2003, Beno et al.). WBAN applications can be categorized according to usage of that application. WBAN is usually used for medical treatment and diagnosis in healthcare networks. There are WBAN applications for the diagnosis and treatment of many diseases, such as Cardiovascular Disease (CVD), Paraplegia, Cancer, Alzheimer`s, depression, Hypertension, Diabetes, Asthma, Defective Tooth positions, epileptic Seizures, Visual impairments, High Blood pressure, Parkinson's disease, renal failure, and Post-operative monitoring (Barakah and Ammad-uddin).

Table 2 Examples of Medical WBAN Applications (Beno et al., 2011, Theogarajan et al., 2006, Penzel et al., 2001, Arnon et al., 2003, Gyselinckx et al., 2007, Latre´ et al., 2010)

Application	Data Rate	Bandwidth (Hz)	Accuracy (bits)
ECG(12 leads)	288 kbps	100–1000	12
ECG (6 leads)	71 kbps	100–500	12
EMG	320 kbps	0–10,000	16
EEG (12 leads)	43.2 kbps	0–150	12
Blood saturation	16 bps	0–1	8
Glucose monitoring	1600 bps	0–50	16
Temperature	120 bps	0–1	8
Motion sensor	35 kbps	0–500	12
Cochlear implant	100 kbps	–	–
Artificial retina	50-700 kbps	–	–
Audio	1 Mbps	–	–
Voice	50-100 kbps	–	–

3 Medical Data Transmission in Healthcare

Transmitting various kinds of data and using different types of communication technologies makes for a wide diversity among remote healthcare systems. Data that is transmitted by healthcare systems can be bio-signals such as electrocardiogram (ECG), electroencephalogram (EEG), surface electromyogram (SEMG); bioacoustic signals (lung, heart, and bowel sounds) (Hadjileontiadis, 2006); vital statistics such as blood pressure, oxygen saturation, body temperature, pulse, saturation of haemoglobin (SpO2), blood glucose, etc. Remote healthcare systems may also want or need to transmit image or video data (Kyriacou et al., 2007, Pattichis et al., 2007, Lin, 2010, Hadjileontiadis, 2006). Remote access to electronic patient record (EPR) data will be mandatory for the healthcare system in order to improve the quality of care delivered in emergency situations (Hall et al., 2003). The underlying communication technology that sends the required patient data to the hospital, clinic, medical laboratory, or other facility can be GSM, GPRS, 3G, CDMA, satellite, wireless LAN, 4G, WiMAX etc. As discussed earlier, each of these communication technologies has its own frequency band, data transfer rate and coverage rate. The figure shows the abstract architecture for all the telemedicine systems that are discussed and covered in our paper. Telemedicine includes healthcare, mobile hospital, emergency situation, maritime telemedicine, remote monitoring system, etc.

4 Classification of Mobile Healthcare Systems

This section provides an overview of different healthcare services. We have categorized them into four main groups: (1) diagnostic healthcare systems (e.g., tele-cardiology and tele-radiology), (2) emergency healthcare systems (e.g., consultation), (3) remote monitoring systems (e.g., maritime healthcare; healthcare systems for diabetics), and (4) location-based healthcare systems. These are summarized in Figure 3 for further elaboration. Furthermore, we discussed about the wireless technologies used in these systems, type of data transmitted by these systems, and their system framework and features.

Fig. 3 Classification of Healthcare Systems

4.1 Diagnostic Healthcare Systems

Diagnostic healthcare systems refer to those systems that determine or identify the possible disease or disorder. Mobile health (mHealth) systems would allow physicians to diagnose, monitor, and treat patients remotely without compromising standards of care. The following sections discuss tele-cardiology and along with tele-radiology. Table 3 provides some examples of diagnostic healthcare systems along with the transferred biomedical data, the technology that they use, and their system features.

4.1.1 Tele-Cardiology

(Ernest et al., 2011) proposed a ZigBee-based Tele-cardiology system for remote healthcare service delivery that minimizes the communication gap between doctor and patient by giving doctors the ability to monitor and diagnose their patients remotely via the Internet. They proposed ZigBee communication technology be used to transmit data from sensors on the patient`s body to the hospital. (Chin-Teng et al., 2010) presented a novel and intelligent telecardiology system that was cost-effective and provided features like wireless communication, ambulatory service, real-time and auto-alarm intelligent telecardiology system. In their system, a small, light-weight, power-saving wireless device collects ECG signals and transmits them to a nearby mobile or ubiquitous device via Bluetooth, where they are then processed and sent to the remote database server that can be accessed via an Internet browser. Their system also has the capability to notify EMS if the sensors detect an emergency. In order to transmit ECG signals and images , (Sufi et al., 2009) proposed an ECG compressed algorithm, real-time and auto-alarm intelligent system that would use lightweight power saving wireless ECG mobile and ubiquitous devices with automatic expert wireless system that has the ability to activate EMS. Furthermore, (Iftikhar et al., 2010) proposed a novel approach for providing cost-effective cardiac services in remote rural areas; it would evaluate patients using a hybrid approach combining frequent tele-consultation with in-frequent in-person meetings, as well as offering an educational component.

4.1.2 Tele-Radiology

Teleradiology is a rapidly evolving specialty in telemedicine. Mobile based teleradiology was proposed and discussed by (Ninos et al., 2010). In their system, a PDA acts as a terminal with a secure connection for thyroid nodule diagnostics, enabling the hospital's picture archiving and communication through DICOM protocol using Wi-Fi communication technology. They evaluate their proposed system in terms of features such as mobility, usability, stability, and performance with their prime focus on ease of use for the medical practitioners who would use their system. (Ruotsalainen, 2011) proposed a privacy and security model for teleradiology in order to gain the trust of both patients and medical practitioners. Their proposal explains how teleradiology can be integrated into an eHealth telemedicine system using sms and email. Their security features include dynamic and context-aware security and privacy policies. (Abdullah et al., 1999) explained the brief history of teleradiology and its role in healthcare telemedicine with options using ISDN, ATM and satellite communication.

Table 3 Examples of Diagnostic Healthcare Systems

Author	Framework	Biomedical data	Technology	System features
Tele-cardiology				
(Ernest et al., 2011)	Remote healthcare service delivery	ECG serial stream data,	ZigBee, Internet	• Diagnosis and continuous monitoring remotely
(Chin-Teng et al., 2010)	System using wearable and wireless ECG to detect atrial fibrillation	ECG wireless signals	Bluetooth, Internet	• Wireless, ambulatory, real-time and autoalarm intelligent system • Lightweight and power saving wireless ECG device • Automatic expert wireless system • Ability to activate EMS
(Sufi et al., 2009)	ECG compressed algorithm	Wireless signals	SMS, MMS, HTTP, custom socket	• ECG compressed algorithm for bandwidth constrained wireless link • Algorithms for cardiovascular abnormality detection • Mobile phone based cardiovascular monitoring systems 6.75 times faster than existing
(Iftikhar et al., 2010)	Telecardiology for e-Diagnosis and e-Learning in rural area	Lab data, clinical findings, ECG, X-rays	Emai,l 3G, Wi-Fi, UMS	• A novel approach for Cardiac Services in rural medicine • Effective cardiac care for variable distance • Specialize services for the cardiovascular patients I remote location • Evaluate patients by hybrid approach (frequent teleconsultation and in-frequent real meeting) • Less healthcare cost and education system
Tele-radiology				
(Ninos et al., 2010)	PDA-based teleradiology	Image	Wi-Fi	• PDA serving as teleradiology terminal with secure connection for thyroid nodule diagnostics • Enabling picture archiving and communication through DICOM protocol • Mobility, usability, stability, performance • Easy to use
(Ruotsalainen, 2011)	Privacy and security in teleradiology	Image, X-rays, CT	SMS, Email, Internet	• Encourages teleradiology trust • Integrates teleradiology services with eHealth • Dynamic & context aware security and privacy features
(Abdullah et al., 1999)	Impact of teleradiology in clinical practice	Image, data	ISDN, ATM, Satellite	• Brief history of teleradiology • Role of teleradiology in health care • Telecommunication technology options for teleradiology

4.2 Emergency Healthcare Systems

Li et al. (2012) designed a WiMAX-based emergency system capable of transmitting physiological data and providing immediate online instructions on emergency treatment via video and audio. This system linked healthcare centers, hospitals and emergency ambulances (Li et al., 2011). An Android-based emergency alarm and healthcare management system has been proposed by Yuanyuan et al (2011). Using GPS and GSM technologies, the system can track the location of a patient in an emergency and send a message about the patient's location and situation to the doctor and the patient's family. Moreover, the system handles the health records of users so that it can send information regarding a patient`s physical condition to the appropriate doctor (Yuanyuan et al., 2011). Bergrath et al (2011) evaluated a multifunctional telemedicine system for emergency medical service (EMS) from the user's perspective. The system would transmit a patient`s vital signs and ECG data as well as pictures from an ambulance to the appropriate healthcare center. In addition, the system would allow voice communication and video transmission from the ambulance to the teleconsultation center. WLAN, LAN, and GSM have been used as communication technologies (Bergrath et al., 2011). Ren-Guey et al. (2007) proposed and implemented an intelligent mobile care system with alarm mechanism for chronic care environment (Ren-Guey et al., 2007). Patients, nurses, physicians and health providers are equipped with a mobile device that communicates with the servers in the health care center. Moreover, they designed physiological signal recognition algorithms that can be built into a chronic patient's Bluetooth-capable mobile device. Doctor can observe physiological parameters such as blood pressure, pulse, and saturation of haemoglobin (SpO2), and collect electrocardiogram (ECG) data. In addition to observation, important and/or abnormal physiological information can be uploaded to the healthcare center. An alert management mechanism in the healthcare center could handle emergency cases by sounding the alert if it receives an emergency message from a patient`s mobile device (Ren-Guey et al., 2007).

Takeda et al. (2010) proposed a mobile health monitoring system for emergency and regular situations based on 3G and Internet. The system could transmit a patient's vital medical data to a specialist while the patient is located in an ambulance, home or rural health care center, or in remote areas or any other outdoor environment. Doctors can access the patient's collected data in a real time mode or store-and-forward mode, and can initiate real-time video conferencing with the patient or other doctors if required (Takeda et al., 2010). El-Masri and Saddik (2012) designed an emergency system to automate ambulance calls in accident cases. The proposed system would first locate the nearest ambulance and the nearest hospital; once the patient is on board the ambulance it would access the patient's online health record. The mobile device would need access to the Internet and would be equipped with a GPS receiver. All of the ambulances would have high-speed wireless Internet access through General Packet Radio Service (GPRS) or Worldwide Interoperability for Microwave Access (WiMAX) (El-Masri and Saddik, 2012). Table 4 summarizes a few proposed projects in the emergency healthcare system category, showing their framework, biomedical data transmitted, technology, and system features.

4.2.1 Consultation in Healthcare Systems

Traditional consultation, whereby the patient must make an appointment with their doctor to visit a medical care center or hospital, is no longer the only option. Today it is possible to have remote monitoring and video consultation by means of mobile devices or computers using the latest advanced communication technologies like 4G and LTE. The patient can subscribe to a medical healthcare centre that will diagnose any disease and suggest medical consultation based on the patient`s current health conditions. Through the continuous efforts of researchers and medical advisors, it is becoming more convenient for remote patients to have video consultation sessions with medical doctors regardless of geographic distance. This is especially beneficial for geriatric patients, patients with a physical disability, or patients living in a remote location. Remote consultation not only benefits patients but also doctors, for example by enabling a doctor to solicit and receive valuable advice during a difficult surgery, or provide consultation to his hospital while traveling. Moreover, live or recorded video of medical activities and consultation can be broadcast to universities for distance learning/distance education purposes for medical students. An advance

Table 4 Examples of Emergency Healthcare Systems

Author	Framework	Biomedical Data	Technology	System features
(Yuanyuan et al., 2011)	Emergency Management System	Message, Audio	GSM, GPS	• Call • SMS • Online mode
(Li et al., 2011)	Emergency Care System (in hospitals and ambulances)	ECG, blood pressure,pulse, heart beat, body temperature, Video, Audio	WiMAX	• Physiological data transmission • Online instructions on emergency treatment via video and audio transmission
(Bergrath et al., 2011)	Multifunctional Telemedicine System in an Emergency Medical Service	vital signs data, ECG, Image, Audio, Video	WLAN, LAN, GSM	• Vital signs data • 12-lead-ECG data transmission • Sending pictures • Voice communication and video transmission from an ambulance
(Lam et al., 2009)	A telemedicine System for emergency medical assistance in the ambulance	Multimedia	WiMAX	• Simple diagnosis and resource arrangement within "the platinum ten minutes after receiving an emergency call • Shortening of the emergency period • Videoconferencing between doctor and emergency officer • Receipt of instructions from doctor, arrange medical treatment for patient • Can find the shortest path for ambulance with least congestion
(Ren-Guey et al., 2007)	A Mobile Care System With Alert Mechanism	Blood pressure, Pulse, Saturation of haemoglobin (SpO2), Electrocardiogram (ECG) data	Bluetooth, GSM/GPRS	• Physiological signal recognition algorithms • Transmission of s physiological parameters • Monitoring of important or abnormal physiological information • Alert management mechanism
(Takeda et al., 2010)	Mobile Health Monitoring System	Vital medical data, Multimedia	3G, Internet	• Transmission of vital medical data • Access to patient's data in a real-time or store-and-forward mode • Real time video conferencing
(El-Masri and Saddik, 2012)	Emergency System for Ambulance	Medical data	GPS, GPRS, WiMAX	• Locate nearest ambulance and hospital • Access the patient's online health record • Book a bed or cubicle for the incoming patient

tele-echography robotic system has been developed that allows a medical ultrasound expert to perform procedures on a distant patient. The ultrasound images are monitored and controlled at the expert station, while the patient station is equipped with a robotic preholder system, an ultrasound device and video conferencing link. The two sites are connected via ISDN, satellite and 3G communications link were (Vieyres et al., 2006).

4.3 Remote Monitoring Systems

Takeda et al. (2010) proposed a mobile health monitoring system for emergency and regular situations based on 3G and Internet. The system transmits a patient's vital medical data to the specialist while the patient is located in an ambulance, home or rural health care center, or in a remote area or other outdoor environment. The doctors can access the patient's collected data in a real-time mode or store-and-forward mode. The doctor can initiate real-time video conferencing with the patient or with other doctors if required (Takeda et al., 2010). We have subdivided remote monitoring mHealth systems into maritime healthcare system and healthcare systems for diabetics. The different remote monitoring systems are summarized in Table 5.

4.3.1 Maritime Healthcare System

Emergency healthcare services are essential for those who are traveling by aircraft or ship as well as those living in remote areas (mountains, islands, forests). Mobile satellite communication (MSC) methods are a good solution for these situations, as they provide for interactive medical video conferencing. Numerous research articles (Murakami et al., 1994, Anogianakis et al., 1998, McDermott et al., 1999, Hwang and Lee, 2000, Pierucci and Del Re, 2000) have discussed video conferencing using a satellite communication system, as satellites have the advantages of wide-area coverage and the ability to provide both broadcast and multicast connections.

(Murakami et al., 1994) proposed a telemedicine technique using satellite communication to support clinical diagnosis and emergency care services on a moving vehicle, aircraft or ship. Their study measured several characteristics such as channel capacity, real-time operation, size of the system, electromagnetic interference, and reliability of vital sign transmission along with overall effectiveness of the system. The proposed system can send color images, audio signals, ECG (3 channels) and blood pressure from a mobile device to the ground station, and transmit audio and error control signals from the ground station to the mobile device. Experiments were conducted while ship and aircraft were moving, verifying the feasibility of this technique.

MERMAID is a European Union (EU) project initiated in 1998 to provide maritime telemedicine services via Integrated Services Digital Network (ISDN) video conferencing. This system not only supports medical services for the crew but also for travelers aboard ship. MERMAID supports telemedicine conferencing as well as multimedia communication and file and image transfer using low, medium, and high-bandwidth requirements. MERMAID utilizes coaxial cable, optical fiber, satellites, WLAN, cellular radio, Bluetooth, and ultra-wideband (UWB) as communication media and employs the Internet, ISDN, and asynchronous transfer mode (ATM) as network technologies (Anogianakis et al., 1998). McDermott et al. (1999) proposed a maritime telemedicine system for high speed medical data transmission using satellite networks (McDermott et al., 1999). Seon-Cheol and Myoung-Ho (2000) developed the telePACS maritime telemedicine system that can access a hospital`s picture archiving and communication system (PACS) server using the Internet. They used Java to implement their system for accessing medical data and images via browsers such as Netscape. Finally, they proposed a fast and cost-effective PACS using web technology not only for inter-hospital but also for public use (Hwang and Lee, 2000). (Pierucci and Del Re, 2000) discussed a telemedicine application called Medical Environment for Diagnostic Images (MEDI) designed to provide remote medical care services by exploiting the broadcast and multicast ability of satellite as well as its wide-band capacity. Java was used for the client-server architecture based on TCP/IP and performance of the proposed application based on satellite connection was evaluated.

Table 5 Examples of Remote Monitoring Systems

Author	Framework	Biomedical Data	Technology	System features
Remote Monitoring				
(Takeda et al., 2010)	Mobile Health Monitoring System	Vital medical data, Multimedia	3G, Internet	• Transmission of vital medical data • Access the collected patient's data in a real-time mode or store-and-forward mode • Real time video conferencing
(Ernest et al., 2011)	Remote Healthcare Service Delivery	ECG Serial stream data,	ZigBee, Internet	• Diagnosis and continuous monitoring remotely
Maritime				
(Murakami et al., 1994)	Telemedicine using satellite communication by moving vehicle	Image, Audio signal,	3G, Satellite	• Supports clinical diagnosis and emergency care services on moving vehicle (aircraft ship) • Channel capacity, real-time operation, and size of the system, electromagnetic interference • Reliability of vital sign transmission • Data send from mobile to base station
(Anogianakis et al., 1998)	MERMAID	File, Image, Video	ISDN, Satellite, Wi-Fi, Bluetooth, UwB, ATM, Internet	• Telemedicine services via Integrated Services Digital Network (ISDN) • Supports video conferencing • Interactive and effective
(McDermott et al., 1999)	Satellite based maritime telemedicine system	Medical Images, Data	ATM, Ethernet, Satellite	• High speed medical data transmission using satellite networks • Detailed reports
(Hwang and Lee, 2000)	telePACS maritime telemedicine system	Medical Data, Image	Internet	• Picture archiving and communication system (PACS) server • Java-based • Access to medical data and images via browsers such as Netscape • Fast and cost-effective using web technology
(Pierucci and Del Re, 2000)	Medical Environment for Diagnostic Images (MEDI)	Medical Data, Image, Video	Satellite	• Remote medical care services • Broadcast and multicast ability of satellite as well as wide-band capacity • Client-server architecture based on TCP/IP • Effective performance
Diabetes				
(Mougiakakou et al., 2010)	SMARTDIAB platform	Medical Data, Image	Wi-Fi, 3G, GPRS, Web Services	• Support monitoring, management, treatment • Databases approach • Consist of patient unit, patient management unit • System can be used with PC/mobile
(JansÃ et al., 2006)	Structured Therapeutic Education Program	Medical Data, Image and Video	Bluetooth	• Assessment of telecare system • Evaluation of existing technology • Cost analysis • GlucoBeep Patient device
S. Franc, et al (Franc et al., 2009)	Flexible intensive insulin therapy (FIT)	Medical Data	GPRS	• Real-life application and validation • Algorithm to adjust prandial insulin • PDA-based electronic diary • Individualized parameters that permit fast and accurate adjustment

4.3.2 Healthcare Systems for Diabetics

Diabetes is one of the most common diseases in any part of the world, in both developed or developing nations. Millions of people around the globe die from diabetes every year due to lack of proper treatment and monitoring of their health conditions. SMARTDIAB was proposed by (Mougiakakou et al., 2010) to support the monitoring, management and treatment of the diabetic patient. Their system is very big and easily scalable, making it suitable for big hospitals. It consists of a patient unit and a patient management unit, accessible via PC and mobile device, and uses Wi-Fi to get medical data from the patient, 3G/GPRS to transmit the data from the patient to the hospital. The Structured Therapeutic Education Program (JansÃ et al., 2006) was proposed as a tool for the assessment of telecare systems and to evaluate the existing telemedicine systems and the communication technologies they are using. The researchers conducted a cost analysis of entire system and tried to make it affordable. In their system, patients use a GlucoBeep device that checks the patient's diabetic information and sends it to the server using Bluetooth technology. Flexible Intensive Insulin Therapy (Franc et al., 2009) was designed to use real-life application and validation and employs an algorithm to adjust the prandial insulin of the patient. It also uses a PDA-based electronic diary to reduce the amount of manual work conducted and make it more efficient. The system focuses on individualized parameters to permit fast and accurate adjustments.

4.4 Location-Based Healthcare Systems

A very recent trend in the field of telemedicine relates to applications in which the location of the patient is continuously monitored; these are referred to as Location-based Telemedicine systems. An Android-based emergency alarm and healthcare management system has been proposed by Yuanyuan, et al (2011). Using GPS and GSM technologies, the system can track the location of a patient in an emergency and send a message containing the patient's location and situation to the doctor and patient's family. Moreover, the system handles the health records of users and so can send information regarding the patient`s physical condition to the appropriate doctor (Yuanyuan et al., 2011). El-Masri and Saddik (2012) designed an emergency system for dispatching of ambulances in accident cases. The proposed system would first locate the nearest ambulance and the nearest hospital; once the patient is on board the ambulance it would access the patient's online health record. The mobile device would need access to the Internet and would be equipped with a GPS receiver. All of the ambulances would have high-speed wireless Internet access through General Packet Radio Service (GPRS) or Worldwide Interoperability for Microwave Access (WiMAX) (El-Masri and Saddik, 2012). (Yuanyuan et al., 2011) proposed a location-based telemedicine system that continuously monitors the location of the patient through GPS technology and sends location details as well as audio using GSM/GPRS. (El-Masri and Saddik, 2012) proposed a location-based emergency system for hospital ambulances. Their system would identify the quickest route in order to help the ambulance locate the emergency patient as well as the location of the nearest hospital using GPS, GPRS and WiMAX communication technologies. Their system would allow the medical care unit to access the patient`s health record history and even book a private room or bed in the surgical unit for the incoming patient before the ambulance reaches the hospital. EmerLoc proposed by (Maglogiannis and Hadjiefthymiades, 2007) was based on sensor technology attached on the body of the patients that send the signals to the hospital by the help of GSM, GPRS, GPS and indoor positioning system. Their proposed system was technically feasible and the transmission was based on DICOM protocol. Their proposed system was technically feasible. In order to improve their system and gain information about training the medical staff, they administered questionnaires to doctors and medical staff. ZUPS (ZigBee and ultrasound positioning system) (Marco et al., 2008) was designed to pinpoint the indoor location of the patient; its focus was especially on elderly and disabled people. Apart from continuous monitoring and navigation, it also had an alarm in case of any emergency.

Some location-based healthcare systems along with their framework, transferred biomedical data, their technology communication technology and system features are summarized in Table 6.

Table 6 Examples of Location-Based Healthcare Systems

Author	Framework	Biomedical Data	Technology	System Feature
(Yuanyuan et al., 2011)	Location-Based	Message, Audio	GSM, GPS	• Call • SMS • Online mode
(El-Masri and Saddik, 2012)	Emergency System for Ambulance	Medical data	GPS, GPRS, WiMAX	• Locates nearest ambulance and hospital • Access to patient's online health record • Book a bed or cubicle for incoming patient
(Maglogiannis and Hadjiefthymiades, 2007)	EmerLoc	Medical Data, Image	GSM, GPRS, GPS, indoor UCLA	• System based sensor technology • Technically feasible • Uses DICOM protocol • Questionnaire to physicians experiments
(Marco et al., 2008)	ZUPS (ZigBee and ultrasound positioning system)	Medical Data, Text	ZigBee	• Pinpoints indoor location • Sends location to innovative services for elderly and disabled people • Alarm and monitoring
(Chew et al., 2006)	Hybrid mobile-based patient location tracking (PLT)	Medical Data, Image, Text	GPS, GPRS, 3G	• Hybrid mobile-based location technique • GPS and cellular mobile network • Good accuracy for positioning tracking

5 Discussion

It evident that 3G performance is not sufficient to meet the needs of future high-performance applications, such full-motion video, wireless videoconferencing etc. Research shows that the current UMTS standard has fundamental capacity limitations for high user loads. Also, when the number of active users increases beyond a certain point, the aggregate system capacity starts to decline. On top of that, present systems can't support seamless handover and mobility among heterogeneous IP networks including cellular networks and wireless local area networks (WLANs), which are the driving factors for 4G telecommunications networks and systems. One of the terms used to describe 4G is MAGIC—Mobile multimedia, Anytime anywhere, Global mobility support, Integrated wireless solution, and Customized personal service. The ITU definition of a 4G system, known as IMT-Advanced, requires a target peak data rate of 100Mbps for high mobility and 1Gbps for low mobility applications.

Healthcare solutions comprises as variety of patient diagnostic methods used by hospital consulting teams to research the particular cause of the disease from which a patient is suffering. Once the patient`s disease has been diagnosed, remote monitoring can be conducted continuously in order to check and analyze the patient`s health conditions based on the diagnosis done earlier. In case of emergency, EMS will be activated and proper measures will be taken to treat the patient and, following discharge, provide the best health care services to maintain the good health of patients. Constant consultation is required along with continuous guidance and monitoring.

Location based services in healthcare not only allow remote monitoring of the patient but also link in with the emergency management system. In remote monitoring of the patient, the consultation team can analyze patient data that is collected and transmitted to them via a Bluetooth mobile device. The mobile device should always be near the patient since Bluetooth has limited data transmission coverage. Thus, in the personal area network (PAN), the strictest requirement should be data coverage range. Once the signal received from the Bluetooth device, the location of the patient will be requested using GPS and A-GPS technology. GPS is used if the patient is outside, while indoor location coordinates are calculated in a different way. Hence, the requirement for location retrieval will be a mobile device with GPS and A-GPS technology. Text-based health information does not have a hard QoS requirement and can be transmitted with even low signal coverage, while monitoring, consultation and emergency systems may require pictures or multimedia video in order to properly evaluate the patient`s condition. Multimedia audio or video rhas strict QoS requirements such as minimizing packet loss, jitter, and out-of-order-delivery and (most importantly) having adequate bandwidth. Weak wireless signals will directly impact QoS parameters resulting in disconnections, missing video or incomplete video transmission.

The QoS parameters are as follows:

Low Throughput: A wireless network is an open network wherein resources are shared among its users. When the threshold of the resource utilization is maximized, due to limited bandwidth users will compete to fulfill their application requirements. Similarly, if the patient is located in an area where the wireless throughput is low or limited then the data transmission will be affected. The most adversely affected system will be remote consultation or continuous remote monitoring that requires the transmission of video and audio. It is a responsibility of both patient and medical provides to have reliable wireless service from their Internet Service Provider (ISP) so the throughput issue will not arise.

Packet Loss: Due to limited resources in the wireless network, congestion can occur which might result in data corruption or packet drop on the router due to buffer overflow. Packet loss is the most crucial problem that can affect all the categories of healthcare systems discussed above. For instance, important patient information might be lost during remote monitoring and tele-consultation services. Similar to the solution of low throughput, both patient and healthcare provides should have a good wireless service carrier that will provide adequate data resources.

Error: Packets might become corrupt due to interference of line of sight (for example, being indoors or behind walls). The information can be transmitted in case of packet loss, but sometimes errors may not be detected. Imagine the danger if a patient's data gets corrupted and sent to the healthcare unit. With the wrong data transmitted, the healthcare unit will not have accurate information about the patient's situation, a potentially dangerous situation. It is important that there should not be any interference between the patient's Bluetooth device and the mobile device, in order to reduce the chances of data becoming corrupted.

Latency: Sometimes congestion in the network can cause a packet to reach its destination late. Latency is very important when audio or video transmission is taking place. In mild cases the patient information (audio/video) will simply be delivered a bit late to the other end, but if the lag is severe it may render the audio or video unintelligible that obviously is dangerous in emergency cases.

Jitter: Like latency, jitter causes packets to travel from source to destination with different delays. This unpredictable behavior varies with the packet`s position in the queue of the router or shortest path etc. This variation in delay might seriously affect the quality of video and audio as much of the information gets lost. Imagine a scenario where video arrives before the audio or vice versa. This is potentially dangerous for healthcare services that require audio and video transmission.

6 Case Studies

Case Study 1: Mobi-Dev: Mobile Devices for Healthcare Applications (Istepanian et al., 2006a)

Mobile Devices for Healthcare Applications (Mobi-Dev) is a project involving healthcare institutions, universities and IT companies from Italy, Spain, Greece, Belgium and the United Kingdom. This project started in January 2001 and finished after 30 months. In order to quality of healthcare services, this project aimed to offer a next generation mobile communication tool for healthcare professionals. Mobi-Dev concentrates on three main objectives: (1) Mobility: ability to access to the clinical information systems from inside and outside of the workplace anytime. (2) Ease of Use: enable to input data onto the Hospital Information System (HIS) by voice, (3) Security and confidentiality: protection of data transmission, management and authentication via smart card. This project was provided for hospital medical and nursing staffs who must be connected all the time with HIS in the hospitals. Figure 4 illustrates connectivity in Mobi-Dev using wireless technologies. Mobi-Dev user interface was designed in such a way the users could perform several daily tasks such as: (1) edit medical/nursing history, (2) handle pharmaceutical prescriptions by voice, (3) benefit from a "red flag" control system, (4) checking against the input of wrong or out-of-range values - e.g. for vital signs or lab results, (5) receive and give orders related to medical/nursing care, receive “early warnings”. Beside those functionalities, Mobi-Dev enable to store and retrieve medical information files and pharmaceutical formularies, install useful medical programs, exchange e-mails and browsing the Internet.

The user requirement analysis was conducted at the beginning of the project using an internal panel of qualified end users (hospital institutions and university polyclinics). The analysis aimed to evaluate to different technical solutions, respectively called (1) **Web Mobi-Dev** and (2) **Synch Mobi-Dev**. Web Mobi-Dev was arranged to meet some requirements such as: Natural Language Understanding (NLU), Wireless web access to the Hospital Information System (HIS), Digital signature based on smart cards MOBI-DEV, Portability. Synch Mobi-Dev aimed to provide Remote wireless synchronisation with the HIS, Digital signature based on smart cards.

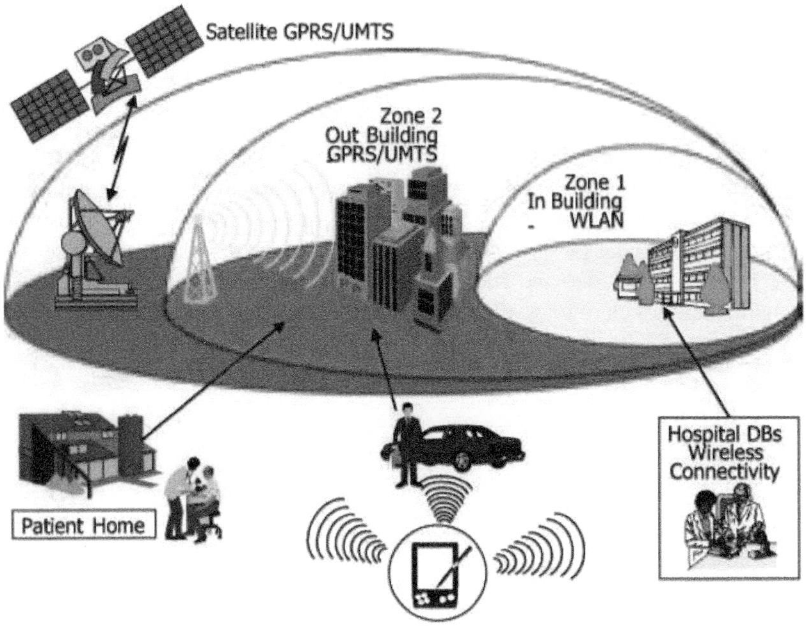

Fig. 4 Mobi-Dev's connectivity capabilities: healthcare practitioners can use the Mobi-Dev palm device to connect to the hospital DB, by exploiting WLAN connectivity if in the hospital or GPRS/UMTS mobile network in all other situations

Web Mobi-Dev: is based on client server architecture as shown in Figure 5. It provides wireless connectivity for handheld devices in order to access securely to data stored into a generic relational database (DB). For this purpose, a middleware software and an ad-hoc web interface is used. Specific application must be installed on the palm device in order to authenticate users by comparing user credentials with encrypted data stored on a smart card, perform data input by voice, initiate a wireless SSL session for data transmission and safely record data onto a remote DB.

Synch Mobi-Dev: Synch Mobi-Dev is different from Web Mobi-Dev as because of two reasons. First, there are no speech recognition functionalities in Synch Mobi-Dev because of the lack of solution available on the market yet for capturing and managing voice data in Greek. Nothing impedes to implement this functionality on the system at a further step, if required. Second, it uses remote synchronisation providing WWAN independency. Based on this method, data are selectively moving from/to the target database, to/from the distributed personal and/or PDA database systems. Figure 6 depicts high-level architecture of the Synch Mobi-Dev system.

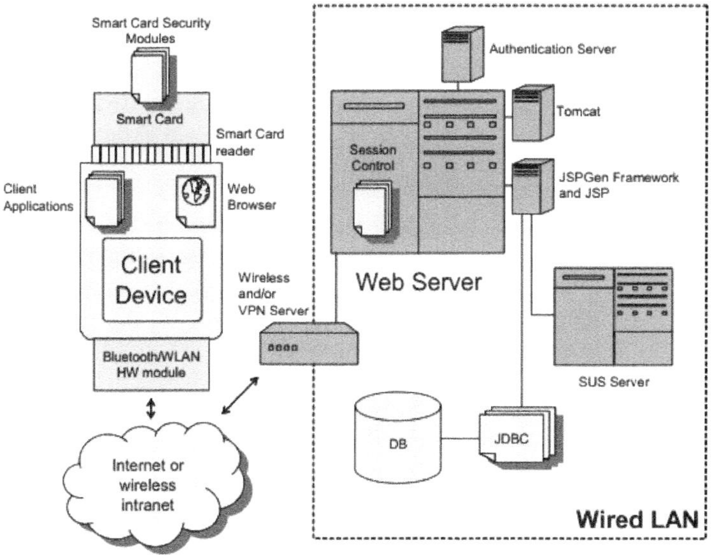

Fig. 5 Web Mobi-Dev general architecture

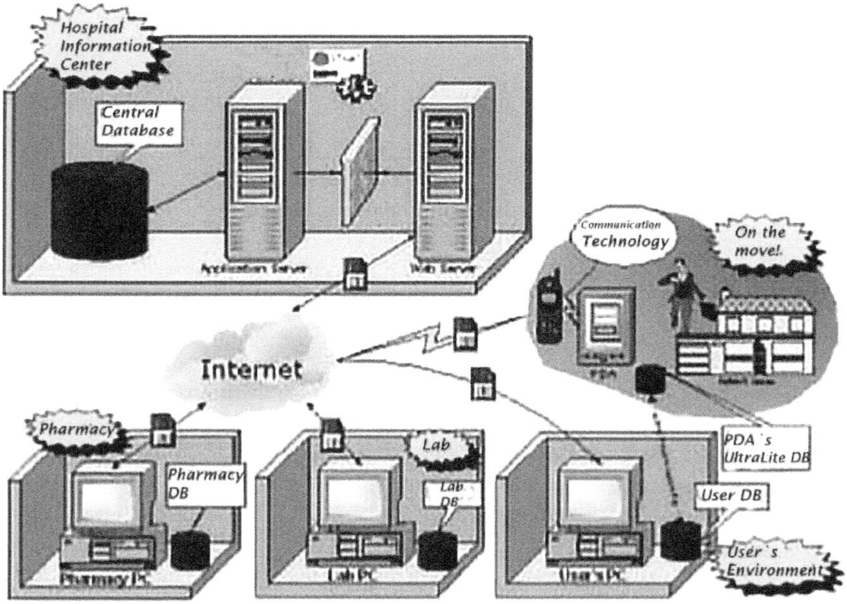

Fig. 6 High level architecture of the Synch Mobi-Dev system

Validation Methodology for Mobi-Dev: the most important assessment criteria for validation of this project consist of relevance, effectiveness, process evaluation, efficiency evaluation, impact evaluation. The validation plan covers all aspects of the Mobi-Dev system such as hardware, software and network components. Another important issue in evaluation was security. All of these issues and queries have been addressed in the validation and the first results. In order to yield information on either non-quantifiable characteristics of the system or intangible aspects of the Mobi-Dev service, such as those related to performance or costs, a standard approach has been followed, based on questionnaires. All users involved in the validation in the four pilot sites have been requested to respond to a series of pre-established questionnaires. An additional group of health professionals, who do not participate directly in the pilot phase, will also be contacted in order to get their opinion on the Mobi-Dev service.

Open Problems and Remaining Barriers: The main obstacles to the penetration of mobile solutions based on handheld devices and wireless technology includes limitation of current technology resources, the reluctance of healthcare professionals, the need to fulfill very restrictive regulations, ensure decision traceability in each step of care delivery and guarantee the patients against privacy violations.

Conclusion: finally, they conclude that Mobi-Dev is able to shorten the time needed to achieve the goal. They believed healthcare systems will continue technological path. Older physicians will be retired soon, and the young doctors who knows how to use modern healthcare systems will be taking over.

Case Study 2: A Tele-Operated Robotic System For Mobile Tele-Echography: The OTELO Project (Istepanian et al., 2006d)

OTELO project aimed to develop an advanced tele-echography robotic system. The system was able to perform echography along with a reliable echographic diagnosis on a distant patient located at the isolated site by medical ultrasound expert, located at the closest university hospital. The system consists of three main parts as illustrated in Figure 7: (1) the expert station where the experts receives ultrasound images of the patient's organ on the control monitor. (2) Patient station that is equipped with the robotic probe holder system an ultrasound device and a videoconference link. (3) Communication line between two stations that can be terrestrial when ISDN infrastructure is available in particular for homecare market or satellite (mobile or fixed), when there is coverage, in order to reach isolated populations such as islands or Amazonian regions.

Test was conducted in a clinical environment at the UMPS Unit (University of Tours, France) for the overall chain evaluation. The expert was located at the control desk in expert station while patient was placed in a room far from the expert station within the same hospital. Two internal ISDN lines, giving a total bandwidth of 128 kbps, connected the two stations. The experts from the ambient images monitored via videoconferencing supervised the positioning of the robot.

Echographic examinations were performed on 20 patients to obtain longitudinal and transverse views of four groups of organs. Using ISDN, satellite, and 3G communication links, the examination was successful. In addition, using satellite and three ISDN connections the system was exploited between the Nicosia General Hospital and the Kyperounta Medical Centre in Cyprus, with the University of Tours and the Barcelona Hospital Clinic.

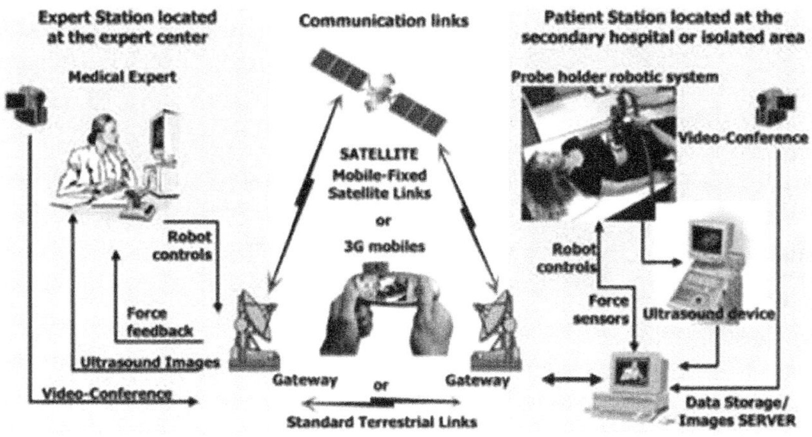

Fig. 7 The overall OTELO tele-echography chain

The OTELO project was conducted by nine partners, includes several disciplines such as mechanical design of the probe holder, robot control laws, graphic user interface, image coding and communication links. OTELO will provide better access to healthcare in remote areas, allow the maintenance of local infrastructures, such as small maternity units supported by large expert centers. It can offer an alternative for medical centers while there is lack of experts in order to assist isolated patients. Moreover, it can reduce the cost, as there is no need for patient transportation. OTELO can be used to train the future clinical ultrasound specialists.

7 Future Direction

Evolution of wireless technological advancement and rapid improvement of application solutions for mobile healthcare facilitates the remote patient. The coverage of wireless and cellular communication becomes vastly available even in the remote location that compliments the healthcare solutions for the patients anywhere anytime. Although in some cases, the stringent quality of service requirements for healthcare system can be concern in case of limited bandwidth availability and weak wireless architecture. Researchers and Industry experts are working on the improvement of wireless infrastructure, algorithm, and architecture suitable for healthcare solutions in even unstable wireless networks. Due to the efforts of researchers that it is possible that the current wireless communication technologies and infrastructure makes it possible to have reliable and efficient healthcare solutions like emergency systems, Tele-cardiology, tele-radiology, remote monitoring, consultation and location-based healthcare system with quality audio and visual medium to facilitate patients as well as hospital consulting team.

Evolution of these technologies will evolve further with advancement of application solutions that will require more requirements to facilitate the end user. With advancement of technologies, some technologies such as GSM, GPRS WiMAX, and 3G are already obsolete in healthcare. Recently progressed communication standard of LTE-A, B, C have more bandwidth and data coverage that will complements the healthcare sector. LTE-based cellular communication technology will not only offer excellent data transmission parameters but also lower the energy consumption per bit transmission as well as performance improvement for audio and video transmission that is highly required by the Telemedicine and healthcare applications for remote patients. Current issues such as network congestion, packet loss, jitter and error transmission can be minimized using the upcoming advance wireless and cellular communication transmission technologies. One of the key factor of LTE-based technology is its integration with existing wireless communication medium especially IEEE 802.11 WiFi networks that is commonly deployed in vast majority of urban and rural areas. This enabling technologies, will be highly efficient and cost effective with increase of flexibility and integration that promises to enhance the level, features and experience of future mobile healthcare systems. Fast and accurate location-based healthcare systems can be a new horizon in the future of healthcare systems in order to track the patient indoor and outdoor with high accuracy.

8 Concluding Remarks

The evolution of wireless technology and the rapid improvement of application solutions for mobile healthcare have the potential to vastly improve service to the remote patient. The coverage of wireless and cellular communication widely available even in remote locations, which complements the healthcare solutions for patients anywhere, at anytime, although the stringent QoS requirements for healthcare systems can be a concern in cases of limited bandwidth availability and weak wireless architecture. Researchers and industry experts are continually working on the improvement of wireless infrastructure, algorithms, and architecture suitable for healthcare solutions even in unstable wireless networks. Due to the efforts of researchers, it is possible that current wireless communication technologies and infrastructure make it possible to have reliable and efficient healthcare solutions including emergency systems, tele-cardiology, tele-radiology, remote monitoring and consultation and location-based healthcare systems with quality audio and visual. This will benefit patients as well as doctors and hospital consulting teams.

These technologies will evolve further in tandem with new application solutions that will require yet more robust standards to support them. Some technologies such as GSM, GPRS WiMAX, and 3G are already obsolete in healthcare. The progressively more powerful communication standards of LTE-A, B, C have more bandwidth and data coverage, which will help meet the needs of the healthcare sector. LTE-based cellular communication technology not only offers excellent data transmission parameters but also lower the energy consumption per bit transmission as well as improving performance for audio and video transmission, a necessity for telemedicine and healthcare applications for remote patients. Current issues such as network congestion, packet loss, jitter and error transmission can be minimized by using the upcoming advanced wireless and cellular communication transmission technologies. One of the key factors of LTE-based technology is its integration with existing wireless communication media especially IEEE 802.11 WiFi networks that are commonly deployed in the vast majority of urban and rural areas. This enables new technologies to be highly efficient and cost effective with an increased flexibility and integration that promises to enhance the level, features, and experience of future mobile healthcare systems. Fast and accurate location-based healthcare systems that accurately track the patient, whether indoors or out represent a new horizon in the future of healthcare.

References

GSM [Online]. GSAMA Company, `http://www.gsmworld.com` (accessed)

WiMax.com (2012b), `http://www.wimax.com/` (accessed)

Abdullah, B.J., Ng, K.H., Pathmanathan, R.: The impact of teleradiology in clinical practice–a Malaysian perspective. The Medical Journal of Malaysia 54, 169–174 (1999)

Adibi, S., Mobasher, A., Tofighbakhsh, M.: Fourth-Generation Wireless Networks: Applications and Innovations. IGI Global (2010)

Akyildiz, I.F., Weilian, S., Sankarasubramaniam, Y., Cayirci, E.: A survey on sensor networks. IEEE Communications Magazine 40, 102–114 (2002)

Anogianakis, G., Maglavera, S., Pomportsis, A.: Relief for maritime medical emergencies through telematics. IEEE Transactions on Information Technology in Biomedicine 2, 254–260 (1998)

Arnon, S., Bhastekar, D., Kedar, D., Tauber, A.: A comparative study of wireless communication network configurations for medical applications. IEEE Wireless Communications 10, 56–61 (2003)

Arrobo, G.E., Gitlin, R.D.: Improving the reliability of wireless body area networks. In: 2011 Annual International Conference of the IEEE Engineering in Medicine and Biology Society, EMBC, August 30-September 3, pp. 2192–2195 (2011)

Baker, S.D., Hoglund, D.H.: Medical-Grade, Mission-Critical Wireless Networks [Designing an Enterprise Mobility Solution in the Healthcare Environment]. IEEE Engineering in Medicine and Biology Magazine 27, 86–95 (2008)

Barakah, D.M., Ammad-Uddin, M.: A Survey of Challenges and Applications of Wireless Body Area Network (WBAN) and Role of a Virtual Doctor Server in Existing Architecture. In: 2012 Third International Conference on Intelligent Systems, Modelling and Simulation (ISMS), February 8-10, pp. 214–219 (2012)

Barnes, S.J.: Under the skin: short-range embedded wireless technology. International Journal of Information Management 22, 165–179 (2002)

Beno, Latr, T., Braem, B., Moerman, I., Blondia, C., Demeester, P.: A survey on wireless body area networks. Wirel. Netw., 17, 1–18

Latré, B., Braem, B., Moerman, I., Blondia, C., Demeester, P.: A survey on wireless body area networks. Wirel. Netw. 17, 1–18 (2011)

Bergrath, S., Rörtgen, D., Rossaint, R., Beckers, S.K., Fischermann, H., Brokmann, J.R.C., Czaplik, M., Felzen, M., Schneiders, M.-T.R.S., Skorning, M.: Technical and organisational feasibility of a multifunctional telemedicine system in an emergency medical service - an observational study. Journal of Telemedicine and Telecare 17, 371–377 (2011)

Blueooth. Bluetooth SIG (2012) `http://www.bluetooth.com` (accessed)

Chew, S.H., Chong, P.A., Gunawan, E., Goh, K.W., Kim, Y., Soh, C.B.: A Hybrid Mobile-based Patient Location Tracking System for Personal Healthcare Applications. In: 28th Annual International Conference of the IEEE Engineering in Medicine and Biology Society, EMBS 2006, August 30-September 3, pp. 5188–5191 (2006)

Chin-Teng, L., Kuan-Cheng, C., Chun-Ling, L., Chia-Cheng, C., Shao-Wei, L., Shih-Sheng, C., Bor-Shyh, L., Hsin-Yueh, L., Ray-Jade, C., Yuan-Teh, L., Li-Wei, K.: An Intelligent Telecardiology System Using a Wearable and Wireless ECG to Detect Atrial Fibrillation. IEEE Transactions on Information Technology in Biomedicine 14, 726–733 (2010)

Chu, Y., Ganz, A.: A mobile teletrauma system using 3G networks. IEEE Transactions on Information Technology in Biomedicine 8, 456–462 (2004)

Cordeiro, C., Akhmetov, D., Park, M.: Ieee 802.11 ad: introduction and performance evaluation of the first multi-gbps wifi technology. In: Proceedings of the 2010 ACM International Workshop on mmWave Communications: from Circuits to Networks, pp. 3–8. ACM (2010)

Crosby, G.V., Ghosh, T., Murimi, R., Chin, C.A.: Wireless Body Area Networks for Healthcare: A Survey. International Journal of Ad Hoc, Sensor & Ubiquitous Computing, IJASUC 3 (2012)

Davidson, J., Peters, J.: Voice over IP fundamentals. Cisco Press (2000)

El-Masri, S., Saddik, B.: An Emergency System to Improve Ambulance Dispatching, Ambulance Diversion and Clinical Handover Communicationâ-A Proposed Model. Journal of Medical Systems 36, 3917–3923 (2012)

Eng Hwee, O., Kneckt, J., Alanen, O., Zheng, C., Huovinen, T., Nihtila, T.: IEEE 802.11ac: Enhancements for very high throughput WLANs. In: 2011 IEEE 22nd International Symposium on Personal Indoor and Mobile Radio Communications (PIMRC), September 11-14, pp. 849–853 (2011)

Ernest, K., Lamei, C., Mohamed, S., Shakshuk, M., Badreldin, I., Elbabli, I.: A ZigBee-based telecardiology system for remote healthcare service delivery. In: 2011 1st Middle East Conference on Biomedical Engineering (MECBME), February 21-24, pp. 442–445 (2011)

Franc, S., Dardari, D., Boucherie, B., Riveline, J.P., Biedzinski, M., Petit, C., Requeda, E., Leurent, P., Varroud-Vial, M., Hochberg, G., Charpentier, G.: Real-life application and validation of flexible intensive insulin-therapy algorithms in type 1 diabetes patients. Diabetes & Metabolism 35, 463–468 (2009)

Gallego, J.R., Hernandez-Solana, A., Canales, M., Lafuente, J., Valdovinos, A., Fernandez-Navajas, J.: Performance analysis of multiplexed medical data transmission for mobile emergency care over the UMTS channel. IEEE Transactions on Information Technology in Biomedicine 9, 13–22 (2005)

Garawi, S., Istepanian, R.S.H., Abu-Rgheff, M.A.: 3G wireless communications for mobile robotic tele-ultrasonography systems. IEEE Communications Magazine 44, 91–96 (2006)

Gyselinckx, B., Penders, J., Vullers, R.: Potential and challenges of body area networks for cardiac monitoring. Journal of Electrocardiology 40, S165–S168 (2007)

Hadjileontiadis, L.J.: Biosignals and Compression Standards. In: Istepanian, R.S.H., Laxminarayan, S., Pattichis, C. (eds.) M-Health – Emerging Mobile Health Systems. Springer (2006)

Hall, E.S., Vawdrey, D.K., Knutson, C.D., Archibald, J.K.: Enabling remote access to personal electronic medical records. IEEE Engineering in Medicine and Biology Magazine 22, 133–139 (2003)

Hillebrand, F.: The Status and Development of The GSM Specifications. In: Zvonar, Z., Jung, P., Kammerlander, K. (eds.) GSM: Evolution Towards 3rd Generation Systems. Kluwer Academic Publisher (2002)

Hung, K., Zhang, Y.-T.: Implementation of a WAP-Based Telemedicine System for Patient Monitoring. IEEE Transactions on Information Technology In Biomedicine 7 (2003)

Husni, E.M., Heryadi, Y., Woon, W.T.H., Arifianto, M.S., Viswacheda, D.V., Barukang, L.: Mobile ad hoc network and mobile IP for future mobile telemedicine system. In: 2006 IFIP International Conference on Wireless and Optical Communications Networks, p. 5 (2006)

Hwang, S.-C., Lee, M.-H.: A Web-based TelePACS using an asymmetric satellite system. Trans. Info. Tech. Biomed. 4, 212–215 (2000)

Iftikhar, M., Singh, H.M.P., Arifianto, M.S.: Telecardiology for e-Diagnosis and e-Learning in Rural Area of Sabah, Malaysia: A Novel Approach for Cardiac Services in Rural Medicine. In: 2010 Third International Conference on Advances in Human-Oriented and Personalized Mechanisms, Technologies and Services (CENTRIC), August 22-27, pp. 83–88 (2010)

Istepanaian, R.S.H., Zhang, Y.T.: Guest Editorial Introduction to the Special Section: 4G Health, The Long-Term Evolution of m-Health. IEEE Transactions on Information Technology in Biomedicine 16, 1–5 (2012)

Istepanian, R.S.H., Laxminarayan, S., Pattichis, C., Altieri, R., Incardona, F., Kirkilis, H., Ricci, R.: Mobi-Dev: Mobile Devices for Healthcare Applications. In: M-Health. Springer US (2006a)

Istepanian, R.S.H., Laxminarayan, S., Pattichis, C., Kontaxakis, G., Sakas, G., Walter, S.: Mobile Tele-Echography Systems – Teleinvivo: A Case Study. In: M-Health. Springer US (2006b)

Istepanian, R.S.H., Laxminarayan, S., Pattichis, C.: Ubiquitous M-Health Systems and the Convergence Towards 4G Mobile Technologies. In: M-Health. Springer US (2006c)

Istepanian, R.S.H., Laxminarayan, S., Pattichis, C., Vieyres, P., Poisson, G.R., Courrèges, F., Smith-Guerin, N., Novales, C., Arbeille, P.: A Tele-Operated Robotic System for Mobile Tele-Echography: The Otelo Project. In: M-Health. Springer US (2006d)

Istepanian, R.S.H., Hadjileontiadis, L.J., Panas, S.M.: ECG data compression using wavelets and higher order statistics methods. Trans. Info. Tech. Biomed. 5, 108–115 (2001)

Jansã, M., Vidal, M., Viaplana, J., Levy, I., Conget, I., Gomis, R., Esmatjes, E.: Telecare in a structured therapeutic education programme addressed to patients with type 1 diabetes and poor metabolic control. Diabetes Research and Clinical Practice 74, 26–32 (2006)

Jovanov, E., Milenkovic, A.: Body Area Networks for Ubiquitous Healthcare Applications: Opportunities and Challenges. Journal of Medical Systems 35, 1245–1254 (2011)

Karl, H., Willig, A.: Protocols and Architectures for Wireless Sensor Networks. John Wiley & Sons, New York (2005)

Kim, D.-K., Yoo, S.K., Kim, S.H.: Instant wireless transmission of radiological images using a personal digital assistant phone for emergency teleconsultation. Journal of Telemedicine and Telecare 11, 58–61 (2005)

Kumar, B., Singh, S.P., Mohan, A.: Emerging mobile communication technologies for health. In: 2010 International Conference on Computer and Communication Technology (ICCCT), September 17-19, pp. 828–832 (2010)

Kumar, B., Singh, S.P., Mohan, A.: Emerging mobile communication technologies for health. In: 2010 International Conference on Computer and Communication Technology (ICCCT), September 17-19, pp. 828–832 (2010)

Kyriacou, E., Pattichis, M.S., Pattichis, C.S., Panayides, A., Pitsillides, A.: m-Health e-Emergency Systems: Current Status and Future Directions [Wireless corner]. IEEE Antennas and Propagation Magazine 49, 216–231 (2007)

Lam, K.L., Tung, H.Y., Tsang, K.F., Ko, K.T.: WiMAX telemedicine system for emergence medical service. In: 7th International Conference on Information, Communications and Signal Processing, ICICS 2009, December 8-10, pp. 1–5 (2009)

Latré, B.T., Braem, B., Moerman, I., Blondia, C., Demeester, P.: A survey on wireless body area networks. Wirel. Netw. (2010)

Lee, K.-D., Vasilakos, A.: Access stratum resource management for reliable u-healthcare service in LTE networks. Wireless Networks 17, 1667–1678

Lenert, L.A., Palmer, D.A., Chan, T.C., Rao, R.: An Intelligent 802.11 Triage Tag For Medical Response to Disasters. In: American Medical Informatics Association (AMIA), Annual Symposium Proceedings (2005)

Li, H.-B., Takahashi, T., Toyoda, M., Mori, Y., Kohno, R.: Wireless Body Area Network Combined with Satellite Communication for Remote Medical and Healthcare Applications. Wireless Personal Communications 51, 697–709 (2009)

Li, S.-H., Cheng, K.-A., Lu, W.-H., Lin, T.-C.: Developing an Active Emergency Medical Service System Based on WiMAX Technology. Journal of Medical Systems 36, 3177–3193 (2011)

Lim, E.Y., Chiang, L., Weidong, C., Dagan, F., Fulham, M.: Development of an Electronic Medical Report Delivery System to 3G GSM Mobile (Cellular) Phones for a Medical Imaging Department. In: 29th Annual International Conference of the IEEE Engineering in Medicine and Biology Society, EMBS 2007, August 22-26, pp. 6726–6729 (2007)

Lin, C.-F.: Mobile Telemedicine: A Survey Study. Journal of Medical Systems, 1–10 (2010)

Lin, C.-F., Lee, H.-W.: Wireless multimedia communication toward mobile telemedicine. In: Proceedings of the 9th WSEAS International Conference on Applied Informatics and Communications. World Scientific and Engineering Academy and Society (WSEAS), Moscow (2009)

Lin, C.F., Hung, S.I., Chiang, I.H.: An 802.11 n wireless local area network transmission scheme for wireless telemedicine applications. In, Proceedings of the Institution of Mechanical Engineers, Part H: Journal of Engineering in Medicine (2010)

Maglogiannis, I., Hadjiefthymiades, S.: EmerLoc: Location-based services for emergency medical incidents. International Journal of Medical Informatics 76, 747–759 (2007)

Marco, A., Casas, R., Falco, J., Gracia, H., Artigas, J.I., Roy, A.: Location-based services for elderly and disabled people. Computer Communications 31, 1055–1066 (2008)

Mcdermott, W.R., Tri, J.L., Mitchell, M.P., Levens, S.P., Wondrow, M.A., Huie, L.M., Khandheria, B.K., Gilbert, B.K.: Optimization of wide-area ATM and local-Area Ethernet/FDDI network configurations for high-speed telemedicine communications employing NASA's ACTS. IEEE Network 13, 30–38 (1999)

Mougiakakou, S.G., Bartsocas, C.S., Bozas, E., Chaniotakis, N., Iliopoulou, D., Kouris, I., Pavlopoulos, S., Prountzou, A., Skevofilakas, M., Tsoukalis, A., Varotsis, K., Vazeou, A., Zarkogianni, K., Nikita, K.S.: SMARTDIAB: A Communication and Information Technology Approach for the Intelligent Monitoring, Management and Follow-up of Type 1 Diabetes Patients. IEEE Transactions on Information Technology in Biomedicine 14, 622–633 (2010)

Murakami, H., Shimizu, K., Yamamoto, K., Mikami, T., Hoshimiya, N., Kondo, K.: Telemedicine using mobile satellite communication. IEEE Transactions on Biomedical Engineering 41, 488–497 (1994)

Navarro, E.A.V., Mas, J.R., Navajas, J.F., Alcega, C.P.: Performance of a 3g-based mobile telemedicine system. In: 3rd IEEE Consumer Communications and Networking Conference, CCNC 2006, January 8-10, pp. 1023–1027 (2006)

Ninos, K., Spiros, K., Glotsos, D., Georgiadis, P., Sidiropoulos, K., Dimitropoulos, N., Kalatzis, I., Cavouras, D.: Development and evaluation of a PDA-based teleradiology terminal in thyroid nodule diagnosis. Journal of Telemedicine and Telecare 16, 232–236 (2010)

Niyato, D., Hossain, E., Diamond, J.: IEEE 802.16/WiMAX-based broadband wireless access and its application for telemedicine/e-health services. IEEE Wireless Communications 14, 72–83 (2007)

Palmer, D.A., Rao, R., Lenert, L.A.: An 802.11 Wireless Blood Pulse-Oximetry System for Medical Response to Disasters. In: American Medical Informatics Association (AMIA), Annual Symposium Proceedings (2005)

Pandian, P.S., Safeer, K.P., Shakunthala, D.T., Gopal, P., Padaki, V.C.: Internet Protocol Based Store and Forward Wireless Telemedicine System for VSAT and Wireless Local Area Network. In: International Conference on Signal Processing, Communications and Networking, ICSCN 2007, February 22-24, pp. 54–58 (2007)

Paro, D.D.: Wireless Application Protocol (WAP) "What is it all about....How does it work". SANS Institute (2001)

Pattichis, C.S., Kyriacou, E., Voskarides, S., Pattichis, M.S., Istepanian, R., Schizas, C.N.: Wireless telemedicine systems: an overview. IEEE Antennas and Propagation Magazine 44, 143–153 (2002)

Pattichis, C.S., Kyriacou, E.C., Pattichis, M.S., Panayides, A., Mougiakakou, S., Pitsillides, A., Schizas, C.N.: A brief overview of m-Health e-Emergency Systems. In: 6th International Special Topic Conference on Information Technology Applications in Biomedicine, ITAB 2007, November 8-11, pp. 53–57 (2007)

Pattichis, C.S., Kyriacou, E., Voskarides, S., Pattichis, M.S., Istepanian, R., Schizas, C.N.: Wireless telemedicine systems: an overview. IEEE Antennas and Propagation Magazine 44, 143–153 (2002)

Paul, T., Ogunfunmi, T.: Wireless LAN Comes of Age: Understanding the IEEE 802.11n Amendment. IEEE Circuits and Systems Magazine 8, 28–54 (2008)

Pavlopoulos, S., Kyriacou, E., Berler, A., Dembeyiotis, S., Koutsouris, D.: A novel emergency telemedicine system based on wireless communication technology-AMBULANCE. IEEE Transactions on Information Technology in Biomedicine 2, 261–267 (1998)

Penzel, T., Kemp, B., Klosch, G., Schlogl, A., Hasan, J., Varri, A., Korhonen, I.: Acquisition of biomedical signals databases. IEEE Engineering in Medicine and Biology Magazine 20, 25–32 (2001)

Perahia, E.: IEEE 802.11n Development: History, Process, and Technology. IEEE Communications Magazine 46, 48–55 (2008)

Perahia, E., Gong, M.X.: Gigabit wireless LANs: an overview of IEEE 802.11 ac and 802.11 ad. ACM SIGMOBILE Mobile Computing and Communications Review 15, 23–33 (2011)

Perahia, E., Stacey, R.: Next Generation Wireless LANs: Throughput, Robustness, and Reliability in 802.11n. Cambridge University Press (2008)

Pierucci, L., Del Re, E.: An interactive multimedia satellite telemedicine service. IEEE Multimedia 7, 76–83 (2000)

Rashvand, H.F., Salcedo, V.T., Sanchez, E.M., Iliescu, D.: Ubiquitous wireless telemedicine. IET Communications 2, 237–254 (2008)

Rasid, M.F.A., Woodward, B.: Bluetooth telemedicine Processor for multichannel biomedical signal transmission via mobile cellular networks. IEEE Transactions on Information Technology in Biomedicine 9, 35–43 (2005)

Redieteab, G., Cariou, L., Christin, P., Helard, J.F.: PHY+MAC channel sounding interval analysis for IEEE 802.11ac MU-MIMO. In: 2012 International Symposium on Wireless Communication Systems (ISWCS), August 28-31, pp. 1054–1058 (2012)

Reinaldo, P.: Chapter 8 Space environment and interference. Wireless Communications Design Handbook. Academic Press (1999)

Ren-Guey, L., Kuei-Chien, C., Chun-Chieh, H., Chwan-Lu, T.: A Mobile Care System With Alert Mechanism. IEEE Transactions on Information Technology in Biomedicine 11, 507–517 (2007)

Ruotsalainen, P.: Privacy and security in teleradiology. European Journal of Radiology 73, 31–35 (2011)

Saleh, A.B., Redana, S., Raaf, B., Hamalainen, J.: Comparison of Relay and Pico eNB Deployments in LTE-Advanced. In: 2009 IEEE 70th Vehicular Technology Conference Fall (VTC 2009-Fall), September 20-23, pp. 1–5 (2009)

Salvador, C.H., Carrasco, M.P., De Mingo, M.A.G., Carrero, A.M., Montes, J.M., Martin, L.S., Cavero, M.A., Lozano, I.F., Monteagudo, J.L.: Airmed-cardio: a GSM and Internet services-based system for out-of-hospital follow-up of cardiac patients. IEEE Transactions on Information Technology in Biomedicine 9, 73–85 (2005)

Schächinger, U., Kretschmer, R., Neumann, C., Nerlich, M.: NOAH A Mobile Emergency Care System. Studies in Health Technology and Informatics 64, 85–92 (1999)

Seon-Cheol, H., Myoung-Ho, L.: A Web-based TelePACS using an asymmetric satellite system. IEEE Transactions on Information Technology in Biomedicine 4, 212–215 (2000)

Stencel, V., Muller, A., Frank, P.: LTE Advanced - A further evolutionary step for Next Generation Mobile Networks. In: 2010 20th International Conference on Radioelektronika (RADIOELEKTRONIKA), April 19-21, pp. 1–5 (2010)

Sufi, F., Qiang, F., Khalil, I., Mahmoud, S.S.: Novel methods of faster cardiovascular diagnosis in wireless telecardiology. IEEE Journal on Selected Areas in Communications 27, 537–552 (2009)

Sorwar, G., Ali, A.: Advanced Telemedicine System Using 3G Cellular Networks and Agent Technology. In: Takeda, H. (ed.) E-Health 2010. IFIP AICT, vol. 335, pp. 187–197. Springer, Heidelberg (2010)

Theogarajan, L., Wyatt, J., Rizzo, J., Drohan, B., Markova, M., Kelly, S., Swider, G., Raj, M., Shire, D., Gingerich, M., Lowenstein, J., Yomtov, B.: Minimally Invasive Retinal Prosthesis. In: IEEE International Solid-State Circuits Conference, ISSCC 2006. Digest of Technical Papers, February 6-9, pp. 99–108 (2006)

Tsiknakis, M., Katehakis, D.G., Orphanoudakis, S.C.: An Open, Component-Based Information Infrastructure for Integrated Health Information Networks. International Journal of Medical Informatics (2002)

Varshney, U.: Pervasive healthcare and wireless health monitoring. Mob. Netw. Appl. 12, 113–127 (2007)

Varshney, U.: A framework for supporting emergency messages in wireless patient monitoring. Decision Support Systems 45, 981–996 (2008)

Vassis, D., Kormentzas, G., Rouskas, A., Maglogiannis, I.: The IEEE 802.11g standard for high data rate WLANs. IEEE Network 19, 21–26 (2005)

Vieyres, P., Poisson, G., Courrèges, F., Smith-Guerin, N., Novales, C., Arbeille, P.: A Tele-Operated Robotic System for Mobile Tele-Echography: The Otelo Project. In: Istepanian, R.H., Laxminarayan, S., Pattichis, C. (eds.) M-Health. Springer US (2006)

Voskarides, S.C., Pattichis, C.S., Istepanian, R., Michaelides, C., Schizas, C.N.: Practical evaluation of GPRS use in a telemedicine system in Cyprus. In: 4th International IEEE EMBS Special Topic Conference on Information Technology Applications in Biomedicine, April 24-26, pp. 39–42 (2003)

Xiaoyi, Z., Doufexi, A., Kocak, T.: Throughput and Coverage Performance for IEEE 802.11ad Millimeter-Wave WPANs. In: 2011 IEEE 73rd Vehicular Technology Conference (VTC Spring), May 15-18, pp. 1–5 (2011)

Yuanyuan, D., Yu, C., Dan, W., Jinzhao, L., Yongqiang, L.: An android-based emergency alarm and healthcare management system. In: 2011 International Symposium on IT in Medicine and Education (ITME), December 9-11, pp. 375–379 (2011)

Zain, A.S.M., Yahya, A., Malek, M.F.A., Omar, N.: 3GPP Long Term Evolution and its application for healthcare services. In: 2012 International Conference on Computer and Communication Engineering (ICCCE), July 3-5, pp. 239–243 (2012)

Zhao, Y., Yagi, Y., Juzoji, H., Nakajima, I.: A Study of Wireless IP for Telemedicine. In: Fourth International Symposium on Multimedia Communications (WPMC 2001), Aalborg, Denmark, pp. 1597–1600 (2001)

Zhao, Y., Yagi, Y., Nakajima, I., Juzoji, H.: IP telephony - New horizon for telemedicine and e-health. Journal of Medical Systems 26, 277–283 (2002)

Questions

Q1. What Are Long Term Evolution Technology and How Many Standards it Currently Has?

Q2. How Technology Plays in Improving Mobile Healthcare? Give Examples.

Q3. What Are the Important Classification of Mobile Healthcare Systems?

Q4. What Is Quality of Service (QoS) and Its Parameters? How QoS Can Impact on Mobile Healthcare?

Q5. How Author Make Discuss about the Current Mobile Healthcare System and Evolution of Communication Technology?

Q6. What Are the Future Prospects of Mobile Healthcare?

31. Cross-Layer Adaptation Technology to Enable Just-Enough Services for mHealth in OFDM-Based Mobile Networks

Anpeng Huang

mHealth Lab., Peking University, Beijing, China, 100871
hapku@pku.edu.cn

Abstract. Link adaptation plays a unique role to handle uncertainties and variations while delivering emerging mHealth services in mobile networks. In terms of typical link adaptation strategies, AMC (Adaptive Modulation and Coding) is designed to maximize spectrum efficiency in the physical layer, and HARQ (Hybrid Automatic Repeat reQuest) is conceived for enhancing transmission reliability in the combination of physical and upper layers. In reality, the fundamental goal of a link adaptation technique should utilize just-enough radio resources to deliver QoS (Quality of Service)-benchmarked services, since the radio resources are always limited in any mobile networks, and QoS requirements are the baseline of any kinds of mobile services. Motivated by this fundamental goal, we propose a novel cross-layer link adaptation to integrate AMC and HARQ mechanisms, in which they are sharing link feedback information by a CQI-Judgement support mechanism. To be specific, this proposal is a cross-layer combination of transmission reliability inspection and transmission pattern recognition, in which AMC is a coarse self-adaptive mechanism and HARQ is applied for the fine self-adaptation. To facilitate this proposal, we conceive an implementation scheme to realize the CQI-Judgement support mechanism, namely, BLER-oriented WMS (Weighted Mean Scheme) judgement, and the advanced version by using NLMS (Normalized Least Mean Square) model with occasional channel measurement. System-level simulation experiments demonstrate that the proposal can finely tailor just-enough radio resources to accommodate QoS-benchmarked mHealth services in LTE-Advanced mobile networks. Results show that the proposal can obtain gains 8.4% in terms of system throughput. Obviously, this proposal is open to integrate any potential advances in the future, for example, ARQ in the RLC (Radio Link Control) layer, power control in the physical layer, etc.

Keywords: Cross-Layer Link Adaptation, mHealth Services, CQI Judgement; AMC, HARQ.

S. Adibi (ed.), *Mobile Health*,
Springer Series in Bio-/Neuroinformatics 5, DOI: 10.1007/978-3-319-12817-7_31

1 Introduction

According to the reports from WHO (World Health Organization) [1], [2], healthcare cost is becoming the unaffordable socio-economic problem in the world. Let us take cardiovascular disease (CVD) as an example. Only cardiovascular disease has contributed counting up to 29% of the total global deaths [2]. Official statistics in [3] show that 230 million people in China - 1/5 of Chinese adults -suffer from cardiovascular diseases. On average, one patient dies from CVD every 10 seconds in China. In turn, if the mortality rate from CVD could be reduced by 1% in next three decades, the reduction in total social cost would be about 10.7 trillion US dollars (68% of GDP in the 2010 Chinese fiscal year). The cost of CVD has attracted attention from academic and industry communities in order to develop an early warning system for CVD monitoring [4], [5], [6]. The main causes of fatal cardiovascular disease include serious myocardial ischemia (acute myocardial infarction), heart failure, and malignant arrhythmia. As shown in [7], most of these symptoms can be predicted by observing certain specific manifestations of electrocardiogram (ECG) signals. If a system can detect such manifestations at an early phase, it can save valuable time for taking precautions against the cardiovascular disease [8], [9], [10]. To satisfy the requirement above, health conditions of CVD affected people must be collected and delivered to a professional healthcare center online without unexpected disruption and distortion. Recent advances in wireless mobile networking technologies have provided an opportunity to accomplish this objective; this concept is known as mobile health (mHealth) [4], [5], [6], which is changing health-care delivery today and is at the core of responsive health systems.

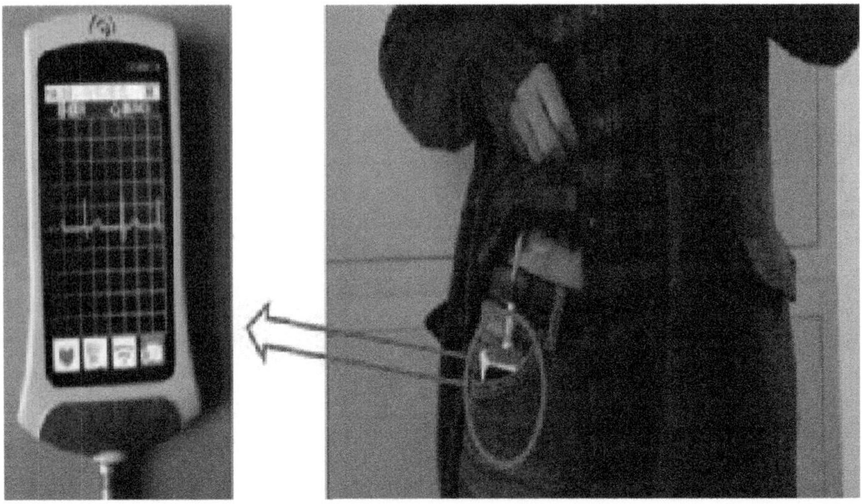

Fig. 1 WE-CARE system in clinical trials

Driven by this successful future of mHealth, we developed and tested WE-CARE (Wearable Efficient teleCARdiology systEm)[1], which generates real-time disease risk alerts without sacrificing user mobility or losing important clinical information, please see the usage style in Fig. 1 [9]. In the WE-CARE system, it is self-evidence that wireless mobile technologies are preferred for online health applications because they do offer real-time access services for mHealth users without time and location limitations [8], [9]. On the other hand, our tests also revealed that wireless radio channels are badly suffering from interference, fading, path loss, shadow and other many negative effects. Currently, the QoS (Quality of Service) of mHealth applications are guaranteed in the cost of radio resources by using all available wireless techniques. Since radio resources in any mobile networks are strictly limited, it is significant to effectively and efficiently utilize wireless bandwidths while providing QoS-benchmarked services for mHealth applications. Without any doubt, this tradeoff between QoS-benchmarked requirement and resource utility is the most critical challenge to wireless mobile networks. To deal with this challenge, link adaptive techniques may play a special role. In this chapter, we focus on this topic in the newly OFDM-based mobile networks [10].

2 Background and Related Works

A) Background for Link Adaption Technologies

To deliver mHealth services, the newly standardized LTE (Long-Term Evolution) Advanced mobile system should be agile to handle uncertainties and variations without lowering down service quality requirements. As observed in the WE-CARE project, an mHealth user needs health information to be 'accurate' and 'real-time.' To offer such a strict service for a user, its service quality needs to be guaranteed at the cost of radio resources. Unfortunately, the radio resources are extremely limited in any mobile networks forever. Furthermore, the overconsumption of radio resources could not be avoided in real scenarios because a radio channel is always volatile. To facilitate these mHealth applications, a link adaptation technology plays a special role to improve transmission reliability and enhance spectrum efficiency in mobile networks. In fact, a link adaptation strategy could adaptively adjust system transmission parameters based on current wireless channel conditions, so as to adapt to the dynamic channels and effectively increase user transmission rate. In general, a link adaptation technology is comprised two components: the first component is to obtain channel information and transmission state, which is commonly indicated by some indices, e.g. CQI (Channel Quality Indicator); the second component is to adjust transmission parameters, including MCS (Modulation and Coding Scheme), redundant information, transmit power, time-frequency resources, etc.

---

[1] In the WE-CARE system, several key enabling technologies are developed and tested upon requests emerging in our system experiments and clinical trials, e.g., de-noising technology for clean signal capture, power-saving wearable mobile user device, ECG signal frame design and data compression over TCP/IP in mobile networks for reliable transmission, privacy-sensitive data center architecture, and clinical decision-support system for data mining.

In terms of self-adaptive objects, conventional link adaptation technologies can be classified into four types as follows: (1) AMC (Adaptive Modulation and Coding) adjusts system modulation mode and coding rate for transmission according to wireless channel conditions. In other words, it can gradually upgrade system modulation and coding levels when noise and interference are decreasing, vice versa. (2) HARQ (Hybrid Automatic Repeat reQuest) adjusts redundant information during data transmission, in order to obtain retransmission/merging gain at a receiver, and realize dynamic adaptation to channel states for guaranteeing successful packet delivery. (3) Power Control is to adjust system transmit power based on channel states, while minimizing interference to co-existing users. (4) Channel Selective Scheduling arranges efficient RBs (Resource Blocks) for each UE (User Element) based on channel measurement. Among them, AMC with power control can address inter-cell interference, and HARQ can promote transmission reliability for UEs when suffering poor channel conditions. Generally, these link adaptation technologies can increase transmission rate for UEs with well channel conditions for maximizing system throughput.

However, these link adaptation technologies face common challenges. Firstly, they are all based on various network feedback mechanisms. As the public knowledge, feedback delay leads the greatest challenge to realize synchronization between link adaptation and channel state updates. Consequently, a mobile network has to tolerate high network overhead to get prompt feedback catching up with channel updates. Otherwise, link adaptation gain might be degraded. Thus, adaptive sensitivity is significant for any kinds of adaptation technologies if they are implemented based on a feedback mechanism. Secondly, in a mobile system, feedback mechanism is related to channel condition measurement. In LTE and LTE-A networks, channel measurement is implemented at the reference signals. This means the measured channels are totally different from the traffic-carrying channels. Thus, the measurement authenticity is really a problem if considering frequency-selective fading in real deployment scenarios (about the channel measurement problem, we will further discuss in the rest of this chapter). Therefore, feedback content might be not effective and real. Thirdly, a mobile system suffers from high complexity when various link adaptation technologies are deployed. So far, existing link adaptation strategies are all channel-dependent and feedback–dependent. In fact, any link adaptation strategy is functioned for either radio spectrum efficiency or packet delivery. In nature, there is an internal relationship between these two adaptation functions. Therefore, it is impossible to make the best use of the self-adaptive potentials, when each link adaptation function works independently. For a telecommunication system, its ultimate goal is to utilize the minimized radio resources to carry QoS-benchmarked services. This ultimate goal is the most critical to mHealth users in a mobile network because radio resources are limited, and radio channels are unstable. As a result, it is worthwhile studying a novel link adaptation, which can combine both adaptation functions for achieving this ultimate goal. Most significantly, if a novel link adaptation can be devised, not only can the channel-dependent and feedback-dependent response mechanisms be improved, and also minimize system overhead and operation complexity greatly.

B) Related Works for Link Adaptation Technology

As well known, the HARQ is the successful example to combine FEC (Forward Error Correction) at the physical layer and ARQ at the MAC (Media Access Control) layer seamlessly. On the other hand, FEC is the immanent function in the AMC mechanism. This phenomenon drives us to think about whether there is an option to further combine AMC with HARQ. Such a combination can further avoid or reduce negative effects aforementioned. In fact, either AMC or HARQ has some limitations to handle link uncertainties and variants if they are working separately. Actually, there are extensively studies about how to design a cross-layer solution by combining AMC and ARQ together. In particular, [11] showed that the ARQ combined with AMC could achieve high throughput in time-division multiple-access and time-division duplex systems. In [12], the effect of ARQ with the adaptive modulation systems was examined as well. More recently, the throughput performance of the ARQ with adaptive modulation over correlated fading channels was analyzed in [13], [14]. In [15], the authors studied the joint design of AMC together with a truncated ARQ, which was later extended to multiple-input-multiple-output systems in [16]. In [17], a chase combining based ARQ protocol with AMC was considered, in which the AMC rate was selected based on an analytical throughput criterion. In [18], authors studied a CQI delay compensation schemes based on Jakes' model in ITU (International Telecommunication Union) regulated scenarios. However, most of these cross-layer designs simply impose AMC into ARQ procedure and increase the redundancy of uplink feedback information, which affects the sensitivity of system adaptation. Furthermore, most of them assumed that the perfect CQI feedback is available and that the feedback channel has zero delay and is free of errors, which are not true in real scenarios. In addition, they did not consider the factor of BLER (Block Error Rate), which is the critical factor to a retransmission decision making.

In all existing solutions, the remaining greatest challenge is raised from the legacy adaptation implementation. Let us take AMC as an example. AMC function is still relying on the channel state feedback. According to 3GPP [19], there are two types of channel-state reports in LTE and LTE-A networks, aperiodic and periodic. Aperiodic reports are requested by the network on a per-need basis through PUSCH (Physical Uplink Shared CHannel) while periodic reports are configured by the network with a certain periodicity normally through PUCCH (Physical Uplink Control CHannel), possibly as often as once every 2 ms, which means the overhead from these reports is a considerable problem. Worsely, the limited payload supported on the PUCCH implies the information of a periodic report may not be able to transmit in a single sub-frame. In terms of CQI, there are two types as well, namely wideband CQI of the entire cell bandwidth and sub-band CQI of certain number of RBs (Resource Blocks) configured by the network, dependent of the bandwidth[2]. Here's the contradiction: In the scenario of continuous transmission

---

[2] In LTE-advanced mobile networks, the basic unit of radio resources is a Resource Element (RE) that consists of one 15 KHz subcarrier and an OFDM symbol, in which a RE is also an intact modulation symbol. One Resource Block (RB) consists of seven OFDM symbols (= 0.5ms, namely, one slot) and 12 subcarriers (= 180 KHz), which contain 84 REs. One sub-frame (two slots) with 12 subcarriers is formed into a RB pair, and 10 sub-frames form one radio frame.

from an eNB to a UE, in order to maximize the throughput of the network or the spectrum efficiency, timely periodic feedback of sub-band CQIs are mostly desirable, which may occupy more resources on PUCCH.

Another limitation of channel measurement is the delay from a UE to an eNB. The UE needs no less than 3 ms to estimate the channel, while upon receiving the channel-state report, the eNB also needs no less than 3 ms to decode the information. According to 3GPP specifications, channel measurement involves at least 10 ms. In case of high mobility, the delay may cause serious mismatch between CQI feedback and the actually channel quality.

In terms of HARQ, ACK/NACK feedback is to guarantee a packet delivery without damages. As soon as any damage is detected in a packet during the decoding process at a receiver, a NACK feedback event is triggered for a retransmission request. If comparing two feedback mechanisms between AMC and HARQ, it is self-evidence that the ACK/NACK feedback is indispensable since it is for the ultimate goal of the successful delivery. Essentially, the CQI feedback in AMC is critical as well. But, compared with ACK/NACK in HARQ, it is desired to find a feasible substitution for CQI feedback rather than ACK/NACK feedback, if necessary for designing a full link adaptation solution below.

3 A Novel Proposal: A Cross-Layer Link Adaptation

Below, we explore a novel link adaptation solution by combining AMC and HARQ together. For the combination of transmission reliability inspection and transmission pattern recognition, the proposed solution can tailor just-enough radio resources to carry mobile services at a required QoS level, named Just-Enough Services (JES). In this proposal, a CQI-Judgment support mechanism is necessary for performing the cross-layer link adaptation. Correspondingly, we conceive three CQI judgment schemes, namely, profiling-based judgment, BLER-oriented judgment and BLER-oriented WMS judgment. We also develop a method to verify the CQI-Judgment by using NLMS (Normalized Least Mean Square) model with rare channel measurement. In the cross-layer link adaptation, AMC is for the coarse adjustment and HARQ is for the fine self-adaptive adjustment.

A) Principles of Cross-Layer Link Adaptation

In HARQ process, the *correct decoding of the received data is the premise of traffic transmission*. In fact, ACK/NACK feedback mechanism in HARQ process is to inform the upper layer whether packets have been correctly received and decoded. Otherwise, a retransmission process is triggered to resend damage or wrong packets for the necessary reliability. Intuitively, there are some native relationship between ACK/NACK content and channel conditions. To be specific, *ACK/NACK frequency in a given period of a HARQ process could reflect channel quality as below*.

- When the number of ACKs in HARQ increases within a given period, it means the channel condition is getting better and MCS levels could be upgraded. Thus transmission rate and system throughput are both increased.

- When the number of NACKs in HARQ increases within the given period, it indicates the channel condition is getting worse. Correspondingly, it has to scale down MCS levels. For a lower MCS level, packet delivery is becoming more robust because more redundancy is applied to encoding, and lower transmission rate is adopted for the concern of reliability.

In a mobile network, AMC is at the core of link adaptation technologies to handle physical variations and uncertainties. As a consequence, AMC must be integrated into our expected full link adaptation. However, the AMC is the CQI-dependent link adaptation technology. As mentioned before, CQI measurement has serious limitations in deployment scenarios. If the relationship above is true, is it possible to get a link adaptation technique that is free of CQI measurement? Based on this motivation, a cross-layer link adaptation is proposed in Fig. 2.

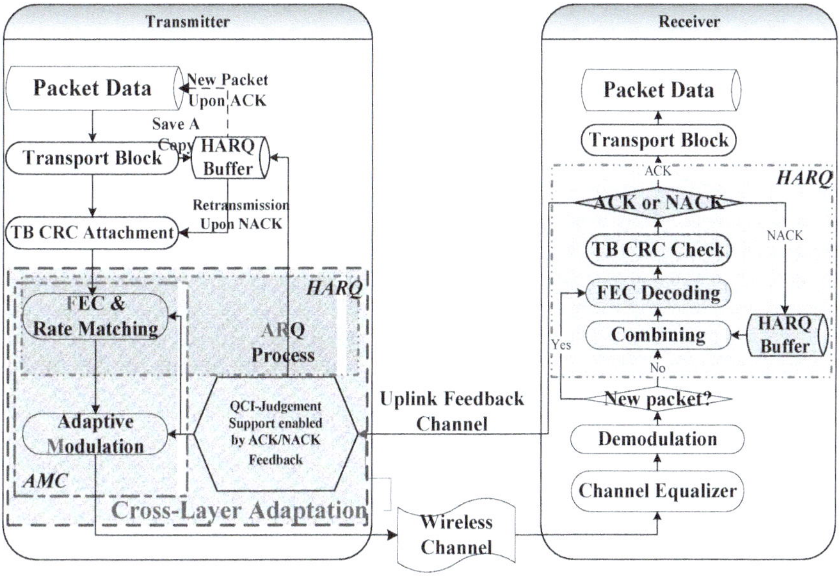

Fig. 2 On Design of Cross-Layer Link Adaptation

In the proposed solution, AMC can share the ACK/NACK feedback information from HARQ. Specifically, when a new packet arrives at a transmitter, it is segmented and formed into transport blocks. And then CRC attachment is added to each transport block. Meanwhile, a copy of the original transport block will be stored in the HARQ buffer at the transmitter in case of a retransmission needed. At the transmitter, AMC can make a decision to choose an appropriate MCS level according to the CQI-Judgment support mechanism. At the counterpart receiver, HARQ is to correct any error bits first if BER (Bit Error Rate) is under the constraint of a predefined threshold. Otherwise, the HARQ starts a retransmission by a NACK feedback if BER is beyond the threshold (that means the whole transport block is error now, not only a few number of bits are at error). At the receiver, the received transport block can be correctly demodulated and decoded if a new packet is properly received. If the transport block is error, a NACK message is sent to the transmitter to fetch its copy for retransmissions. As for retransmission, new transport blocks at the receiver are combined with these copies stored in the HARQ buffer at the receiver for any error recovery. In reality, an outdated ACK may be turned to a NACK report as well.

In the proposal, CQI *feedback mechanism is not necessary for AMC anymore.* In the cross-layer link adaptation, AMC shares the ACK/NACK content with HARQ, to save signaling overheads and improve performance. As aforementioned, FEC content in the AMC is used to indicate whether to trigger an ARQ in HARQ. Here, ACK/NACK feedback in HARQ is shared by AMC, which shows which states the active channel is at, instead of CQI measurement and report. In this proposal, AMC and HARQ are seamlessly integrated each other, in which the CQI judgment support mechanism is the vital, and is an enabling technology. Below, the CQI-Judgment support is highlighted.

B) Formulation of CQI-Judgment Optimal Problem

In a mobile network, BLER (Block Error Rate) is the real representation of a radio channel condition. Naturally, a BLER of a radio channel indicates channel condition states. On the other aspect, there is a native logic between BLER (Block Error Rate) and HARQ ACK and NACK events. Thus, it is reasonable to use ACK/NACK to capture channel states, instead of CQI measurement and feedback. To evaluate whether the CQI-Judgment from ACK/NACK content can substitute CQI measurement, the optimal problem can be formulated as below.

We denote A_t for CQI obtained at time *t* by channel measurement, A_t' for the CQI judgment result at time t, and $c_t = 1$ indicates an *ACK* transmission while $c_t = -1$ indicates an *NACK* transmission at time t, respectively. And we define $C(t) = [c_{t-1}, ..., c_{t-M}]$, $A(t) = [A_{t-1}', A_{t-2}', ..., A_{t-M}']$, in which *M* presents a time window. The optimal problem objective could be modeled as:

$$\min_{f(.)} | f(C(t), A(t)) - A_t |, \tag{1}$$

where f(.) expresses a function how to use historical CQI assignments and their corresponding BLERs to arrange a suitable CQI A_t' at time *t*, in case that channel measurement result A_t isn't available.

C) Implementation Scheme of CQI-Judgment Support

In this subsection, three implementations are conceived to provide an effective and reliable CQI judgment support mechanism for the above mentioned cross-layer link adaptation.

Suppose the judged CQI A_t' at time t is calculated by last M (CQI, c_t) pairs, which namely presents a given CQI leads to whether an ACK or an NACK.

Using Weighted Mean Scheme (WMS) judgment scheme, the A_t' can be calculated as:

$$A_t' = f(C(t), A(t)) = \sum_{i=1}^{M} g(c_{t-i}) * A_{t-i}' \quad (4)$$

$$g(c_{t-i}) = \varphi_t * w_i * \exp(-c_{t-i} * \lambda_t) \quad (5)$$

and
$$\lambda_t = \sum_{i=1}^{M} \frac{c_{t-i}}{M} - \overline{BLER} \quad (6)$$

where $g(c_{t-i})$ indicates the weight of A_{t-i}' which means the CQI at time t-I, and λ_t is used for measuring the gap between average BLER during window M and system $\overline{BLER}$.

The weight $g(c_{t-i})$ is dominated by two parameters: period to time t, namely w_i, and how to ensure system $\overline{BLER}$, namely, $\exp(-c_{t-i} * \lambda_t)$. When *BLER* at time t is high than $\overline{BLER}$ for window *M*, namely $\lambda_t > 0$, more *NACK* packets are needed, thus $c_{t-i} = NACK$ should be encouraged and $c_{t-i} = ACK$ should be restrained, vise versa. Additionally, φ_t is used for the coefficient normalization at time t.

D) The Advanced Version of CQI-Judgment Support

In the conceived CQI-Judgment support, no CQI measurement is applied. To enhance accuracy of the CQI-Judgment, CQI measurement may be used occasionally. It means that the CQI measurement is only conducted to adjust our CQI-Judgment to match with real situations if necessary. Below, Normalized Least Mean Square (NLMS) method is adopted to perform the calibration, in which particular CQI judgment can be logically selected to compare with occasionally measured CQI, and dynamically promote the CQI-Judgment accuracy.

Assume that **G(t)** and **A(t)** denotes the filter tap weight vector and filter input vector at time t, respectively. They can be expressed as:

$$G(t) = [g(c_{t-1}), g(c_{t-2}), \ldots, g(c_{t-M})] \quad (7)$$

$$A(t) = [A_{t-1}', A_{t-2}', \ldots, A_{t-M}'] \quad (8)$$

The judged CQI at time t is defined as A_t', which is calculated by:

$$A_t' = G(t) * A(t)^T \quad (9)$$

The judgment error is given as:

$$e(t) = A_t - A_t' \quad (10)$$

Note that A_t is the measured CQI and A_t' is the judged one via last M slots data transmission, and get their ACKs and NACKs as well as their assigned CQI levels.

NLMS weight vector adaptation equation is modeled as

$$G(t+1)=G(t)+v*\frac{A(t)e(t)}{A(t)A^T(t)}, \tag{11}$$

where control parameter $v\in[0,2]$.

Thus, the future CQI at time t can be denoted as,

$$A_{t+1}^{'}=A(t+1)*G^T(t+1), \tag{12}$$

where $A(t+1)=[A_t', A_{t-1}',..., A_{t-M+1}']$. Then the AMC mechanism should use $A_{t+1}^{'}$ as the CQI value at time t+1.

4 Performance Evaluation

In this section we will use our system-level simulator to evaluate this proposal. We first analyze negative effects from channel measurement, and then evaluate system performance with the designed full link adaptation strategy. System-level evaluations verify that the proposal is intelligently agile to radio resource assignment to satisfy users' QoS requirements, while maintaining resource efficiency and user fairness in multiuser OFDM-based mobile networks.

A) System-Level Simulator

The following experiments are all conducted on our own system-level LTE-Advanced simulation platform, which deploys the wrap-around technique to simulate the interference from surrounding cells of each of the 7 or 19 cells in the urban macro-cell scenario, please see parameters in Tables 1 and 2. In each cell, users from Table 2 are randomly dropped with a mixed profile [20], [21]. Throughput, spectrum efficiency, BL loss rate, BLER of each UE and the entire system are considered as the performance metric here.

Table 1 System-Level Simulator Configuration

Parameters	**Configuration**
Transmission mode	1 transmitter and 1 receiver
Topology	7 cells 21 sectors, 10UEs in each sector
Uplink and downlink sub-frame configuration	mode 1 (DSUUS DSUUD in a sub-frame), 7 HARQ processes for each UE
Bandwidth	5MHz, which contains 25RBs
Scheduling	PF (Proportional Fairness)
Traffic type	Full Buffer
Time	5000 ms

Table 2 Standardized QCIS for mhealth in LTE-A Networks [20]

QCI Index	Priority Level	Packet Delay	Error Loss	Typical Examples from mHealth Applications
1	2	100 ms	10^{-2}	Emergence VoIP Call
2	4	150 ms	10^{-3}	Consultation Video call
3	5	300 ms	10^{-6}	Daily Health Monitoring
4	3	50 ms	10^{-3}	Online Consultation Video Meeting
5	1	100 ms	10^{-6}	Real-time Operation Video
6	7	100 ms	10^{-3}	Medical and Healthcare Education
7	6	300 ms	10^{-6}	Medical Data Transmission with TCP
8	8	300 ms	10^{-6}	Daily Health Conditions Notices
9	9	300 ms	10^{-6}	Medical Image Download, etc.

In Fig. 3, the physical–layer SNR-to-BLER-to-CQI mapping is obtained from our system-level tests, which is calibrated by the China mobile industry platform in our joint national projects as well. This mapping is used as a look-up table that is the basis for system-level simulation experiments. But, we will analyze the major negative effects from the CQI measurement according to the results contained in this mapping first.

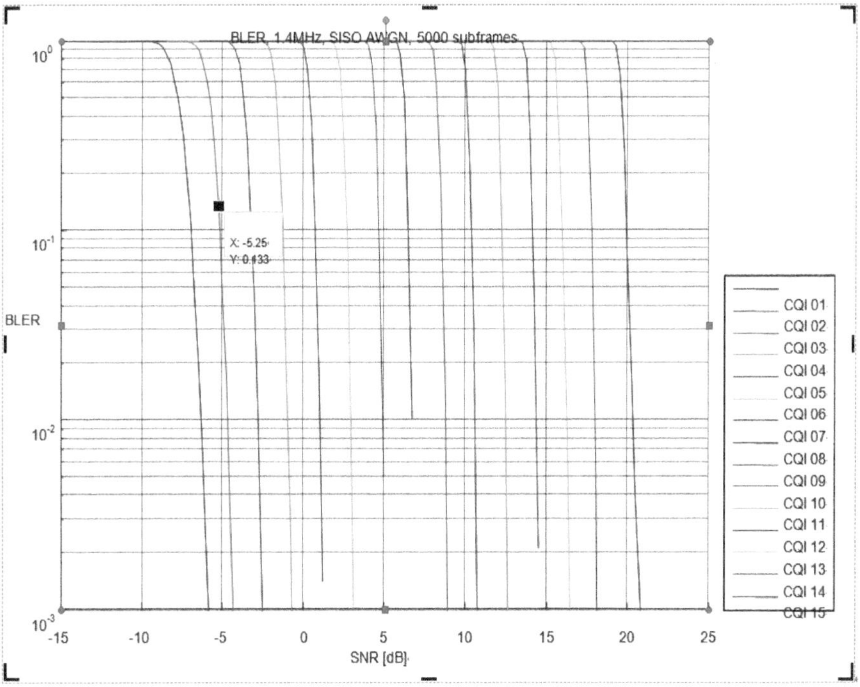

Fig. 3 SNR-BLER-CQI mapping

B) Channel Measurement Analysis

In the results in Fig. 3, we can observe several major phenomena. *Firstly, CQI levels for a UE is totally dynamically, which is depending on the UE real-time channel conditions. But the dynamic range is really large to be applied.* In our tests, we found there are some UEs experience CQI changes badly. For example, if the system CQI is sharply decreased from level 10 to level 2, it causes huge packets loss if MCS adjustment could not response promptly. *Secondly, fast fading has direct impact on CQI measurement and report delay.* Our results demonstrate that more than 65% UEs got CQI feedback delay due to the fast fading, which lead to low resource utilization. More interestingly, we found that *CQI varies often with ACK changes*, especially for channels with lower CQI levels. As shown in Fig. 4, there is an inverse proportion relationship between CQI levels and NACK events, when a MCS level is fixed. It means, when a fixed MCS level is applied, the higher probability loss may happen when a radio channel is suffering from the poor CQI; the more radio resources are wasted when the channel is turned to the higher CQI.

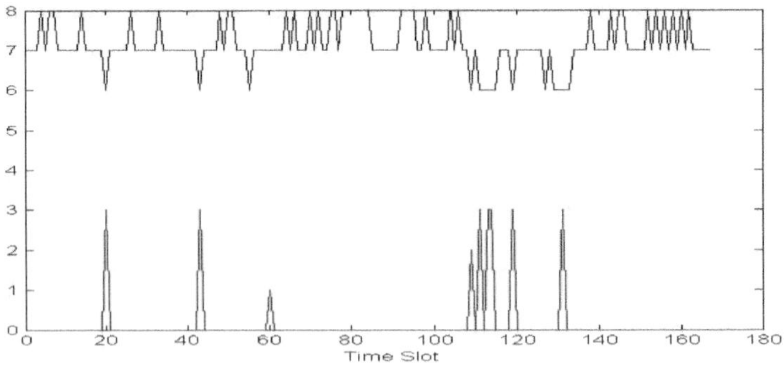

Fig. 4 Relationship between NACKs and CQI levels (in the figure, the CQI levels indicated by the upper lines in blue; and the numbers of NACK events shown in the lower lines in the red, in which the vertical axis is the numeric values)

C) System Performance with the Cross-Layer Link Adaptation

To compare with the alternates (Profile-based, and BLER-based solutions) [9], results in Figs. 5 and 6 show the CDF curves of the spectrum efficiency and BLER of each UE, respectively, in which the BLER-Oriented WMS is the conceived solution in this study.

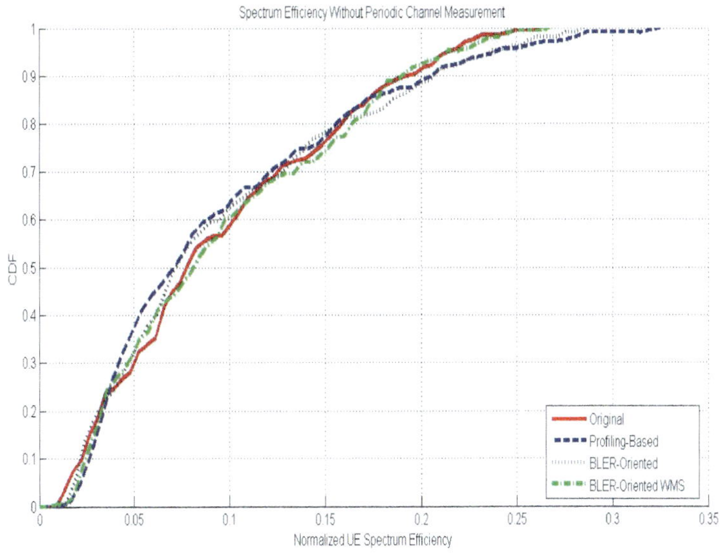

Fig. 5 Spectrum Efficiency for CQI-Judgment schemes

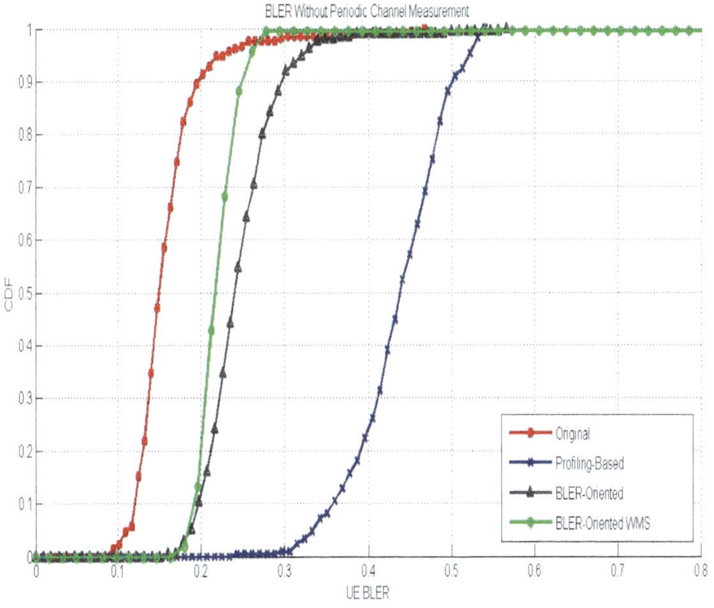

Fig. 6 BLER for CQI-Judgment schemes

In Fig. 5, we can see that our proposed CQI-Judgment support is ideally close to the CQI measurement (labeled as the original curve), compared with alternates. In real applications, the proposed CQI-Judgment support is useful when a reasonable adjustment is adopted. In Fig. 6, we can see that the proposed CQI-Judgment is close to the original CQI-measurement results.

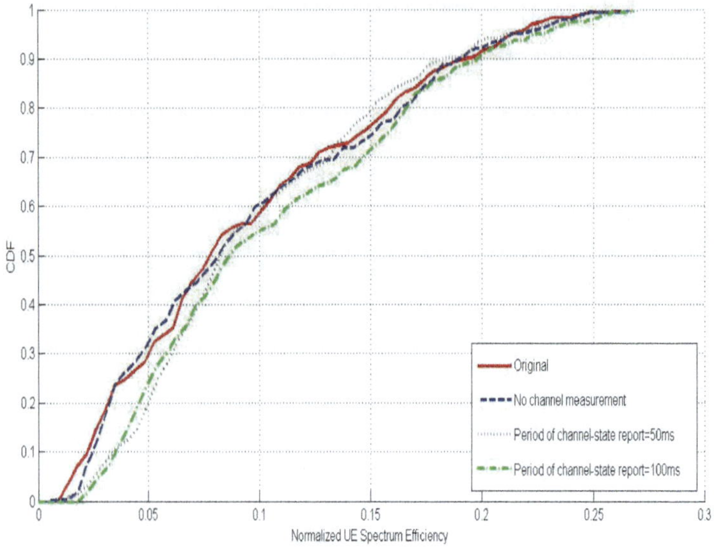

Fig. 7 Spectrum Efficiency for different periods of channel measurement

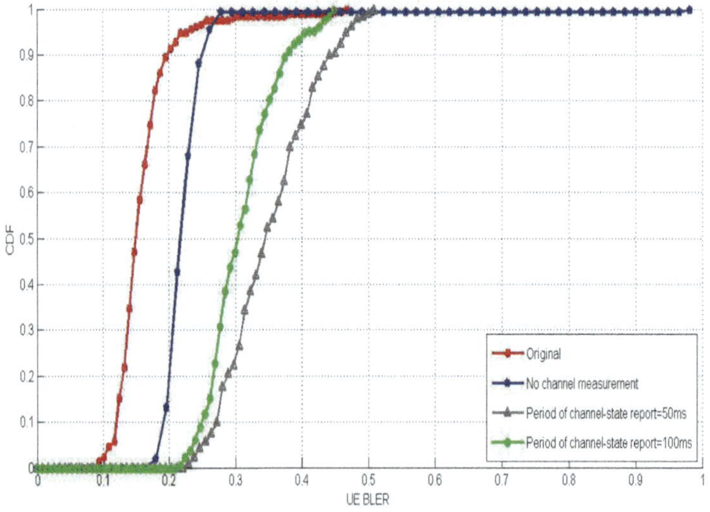

Fig. 8 BLER for different periods of channel measurement

In Figs. 7 and 8, test results demonstrate that the proposed full link adaptation can achieve about 8.4% throughput gain contrasted with alternates.

Obviously, it is significant to study how to offer the QoS-benchmarked services for mHealth and other emerging mobile internet applications, without jeopardizing resource efficiency in mobile networks. Our tests verifies that the proposed link adaptive strategy can be agile to handle variations and uncertainties over volatile radio channels in relation to the service quality requirements.

5 Conclusion

In this chapter, we focus on the most challenge, how to conceive a cross-layer link adaptation, which is dedicated to tailor just-enough radio resources to offer QoS-benchmarked services for mHealth users. To be specific, we use a CQI-Judgment support to combine AMC and HARQ seamlessly, in which AMC is a coarse self-adaptive mechanism and HARQ is applied for the fine self-adaptation. To perform the CQI-Judgment support mechanism, BLER-oriented WMS (Weighted Mean Scheme) judgment and advanced version are conceived for the integration of transmission reliability inspection and transmission pattern recognition. To test this proposal, we conduct system-level simulation experiments. Test results confirmed that the proposal can obtain 8.4% gains in terms of system throughput. Of course, there are some open problems left for future study, for example, how to incorporate a scheduling strategy into the proposal.

References

[1] World Health Organization, Noncommunicable diseases country profiles 2011, WHO Report (September 2011)

[2] Van Lerberghe, W.: The world health report 2008: Primary Health Care: Now more than ever. World Health Organization (2008)

[3] Central Government Report, Report on Cardiovascular Disease in China 2011, Ministry of Chinese Public Health (2012)

[4] Global survey report, mHealth: New horizons for health through mobile technologies. World Health Organization (2011), http://www.who.int/goe/publications/goemhealthweb.pdf

[5] Estrin, D., Sim, I.: Open mHealth Architecture: An Engine for HealthCare Innovation. Science 330(6005), 759–760 (2010)

[6] Collins, F.: The Real Promise of Mobile Health Apps: Mobile devices have the potential to become powerful medical tools. Scientific american (July 2012), http://www.scientificamerican.com/article.cfm?id=real-promise-mobile-health-apps

[7] De Bacquer, D., De Backer, G., Kornitzer, M., Blackburn, H.: Prognostic Value of Ecg Findings for Total, Cardiovascular Disease, and Coronary Heart Disease Death in Men and Women. Heart 80(6), 570–577 (1998)

[8] Zhang, Y., Huang, A., Wang, D., Duan, X., et al.: A 3R dataflow engine for restoring electrophysiological signals in telemedicine cloudplatforms. In: Prodeedings of 2012 IEEE 14th International Conference on e-Health Networking, Applications and Services (Healthcom 2012) (the best student paper award), Beijing, China, pp. 486–489 (October 2012)
[9] Huang, A., Chen, C., Bian, K., Duan, X., et al.: WE-CARE: An intelligent mobile telecardiology system to enable mHealth applications. IEEE Journal of Biomedical and Health Informatics (to be published)
[10] Huang, A., Xu, W., Li., Z., Sarrafzadeh, M., et al.: System light-loading technology for mHealth: Manifold-learning-based medical data cleansing and clinical trials in WE-CARE project. IEEE Journal of Biomedical and Health Informatics (to be published)
[11] Naijoh, M., Sampei, S., Morinaga, N., Kamio, Y.: ARQ schemes with adaptive modulation/TDMA/TDD systems for wireless multimedia communication services. In: Proc. IEEE Int. Symp. PIMRC, Helsinki, Finland, vol. 2, pp. 709–713 (September 1997)
[12] Falahati, S., Svensson, A.: Hybrid type-II ARQ schemes with adaptive modulation systems for wireless channels. In: Proc. IEEE VTC-Fall, Amsterdam, The Netherlands, vol. 5, pp. 2691–2695 (September 1999)
[13] Yun, J., Jeong, W., Kavehrad, M.: Throughput analysis of selective repeat ARQ combined with adaptive modulation for fading channels. In: Proc. MILCOM, October 7-10, vol. 1, pp. 710–714 (2002)
[14] Yun, J.: Markov error structure for throughput analysis of adaptive modulation systems combined with ARQ over correlated fading channels. IEEE Trans. Veh. Technol. 54(1), 235–245 (2005)
[15] Liu, Q., Zhou, S., Giannakis, G.B.: Cross-Layer combining of adaptive Modulation and coding with truncated ARQ over wireless links. IEEE Trans. Wireless Commun. 3(5), 1746–1755 (2004)
[16] Maaref, A., Aissa, S.: Combined adaptive modulation and truncated ARQ for packet data transmission in MIMO systems. In: Proc. IEEE GLOBECOM Conf., Dallas, TX, vol. 6, pp. 3818–3822 (December 2004)
[17] Peng, X., Song, M., Song, J.: Cross-layer Design for Adaptive Modulation and Coding with Hybrid ARQ. In: International Symposium on Microwave, Antenna, Propagation and EMC Technologies for Wireless Communications, pp. 138–141 (August 2007)
[18] Dai, H., Wang, Y., Shi, C., Zhang, W.: The Evaluation of CQI Delay Compensation Schemes Based on Jakes' Model and ITU Scenarios. In: Proc. IEEE VTC-Fall (2012)
[19] 3GPP, Evolved universal terrestrial radio access (E-UTRA); physical layer procedures, 3GPP, TS 36213 v11.1.0 (2013)
[20] Mehlfhrer, C., Wrulich, M., Ikuno, J., Bosanska, D., Rupp, M.: Simulating the long term evolution physical layer. In: 17th European Signal Processing Conference, Glasgow, Scotland (2009)
[21] Kovacs, I.Z., Pedersen, K.I., Kolding, T.E., Pokhariyal, A., Kuusela, M.: Effects of Non-Ideal Channel Feedback on Dual-Stream MIMO-OFDMA System Performance. In: Proc. IEEE VTC-Fall, pp. 1852–1856 (2007)

32. Telemedicine Services over Rural Broadband Wireless Access Technologies: IEEE 802.22/WRAN and IEEE 802.16 WiMAX

Roberto Magana-Rodriguez, Salvador Villarreal-Reyes*, Alejandro Galaviz-Mosqueda, Raul Rivera-Rodriguez, and Roberto Conte-Galvan

CICESE Research Center, Electronics and Telecommunications, Ensenada, México
{rmagana,svillar,agalaviz,rrivera,conte}@cicese.edu.mx

Abstract. Telemedicine services represent restrictive applications mostly related to patient's consultation, diagnosis and monitoring through their biomedical signals by health specialists. Since wireless telemedicine services depend on technological infrastructure, rural area coverage has been considered an important challenge, and the Institute of Electrical and Electronics Engineers (IEEE) 802.16/WiMAX standard has been used to provide broadband wireless access (BWA) because of its MAC and PHY characteristics. However, the switch-off of the analog terrestrial network presents the opportunity of delivering high data rates over large coverage areas by means of TV white spaces (TVWS) technology using cognitive radio capabilities. Hence, at the end of 2011 the IEEE Working Group for Wireless Regional Area Networks (WRAN) released the first TVWS standard named 802.22. Within this standard bandwidth availability depends on the geographical location of the base station (BS) and the customer premise equipment (CPE).Therefore, a model to evaluate the suitability of IEEE 802.22 and WiMAX for the deployment of rural telemedicine networks is proposed. The model considers specific traffic profiles based on the telemedicine services that will be offered over the rural wireless telemedicine network. The evaluation presented in this chapter is performed by calculating the number of telemedicine services that IEEE 802.22/WRAN and IEEE 802.16/WiMAX networks can support considering the available bandwidth and the telemedicine traffic profiles requirements. Frame preambles and a MAC/PHY overhead factor per active connection are considered within the analysis. A case study from the State of Chiapas, Mexico, is presented for the deployment of wireless rural telemedicine networks based on the IEEE 802.16/WiMAX and IEEE 802.22/WRAN standards.

Keywords: WRAN, WiMAX, telemedicine, TVWS, BWA, rural wireless networks.

* Corresponding author.

S. Adibi (ed.), *Mobile Health*,
Springer Series in Bio-/Neuroinformatics 5, DOI: 10.1007/978-3-319-12817-7_32

1 Introduction

Traditionally, medical practice services are based on patient consultation, diagnosis and follow-up scheme. These services involve several methods and specific medical instruments that require the patient's physical presence to treat multiple health conditions. In this context, recent technological advances have not only provided health professionals with new methods and instruments to assist them in the evaluation and decision making process, but also with new communication methods that allow patients to be located in remote sites. This method of reaching and interacting with patients through communication technologies is known as *Telemedicine*.

Telemedicine involves the use of information and communications technologies (ICT) to deliver health related medical information from one location to another [1]. Thus, telemedicine implies the distribution of medical services, traditionally offered through face-to-face consultation, by means of telecommunications technologies [2]. The use of ICT for clinical activities has increased in recent years, as a consequence of the development of diverse technological solutions (e.g. mobile telephony, voice over IP (VoIP), video streaming, digital imaging, etc) that enable the delivery of health services. The medical services offered by telemedicine can be classified as: teleconsulting, telemonitoring, telediagnosis, teleeducation and medical database access [3]. Furthermore, according to [4], telemedicine applications include: initial patient evaluations, triage decisions, surgical follow-up, routine consultation or second opinion (primary care encounters), medical imaging, public health, preventive medicine and patient education.

There are three main factors to consider when planning a telemedicine deployment: the medical application scenario, the medical services requirements, and the ICT infrastructure. Indeed, telemedicine deployment efforts consider the use of already available telecommunications infrastructure; the kind of telemedicine services that can be offered is limited by the capacity of the network (or networks) connecting the sites. Consequently, the variety and number of telemedicine services offered in urban areas are larger than those offered in rural areas.

Currently there exists an unequal distribution of medical services between ur-ban and rural areas, [5]. Providing traditional medical services in rural regions faces several challenges such as: mobility issues, poor technological infrastructure, large travelling times and natural geographical barriers among others, [6]. Thus, the use of telemedicine (which includes mHealth) represents an opportunity to provide rural regions with health services that otherwise would not be available. Rural telemedicine implies the use and/or deployment of telecommunication networks to deliver medical services in underserved, isolated and sparsely populated areas. In this matter, long range broadband wireless access (BWA) technologies present a feasible solution for the implementation of telemedicine and mHealth services in rural areas.

1.1 *Metropolitan and Rural Area Broadband Wireless Access Technologies*

The exponential growth of wireless networks has been a key factor in the development of mHealth technologies. Wireless networks can be categorized according to their maximum coverage area into: wireless local area networks (WLANs), wireless wide area networks (WWANs), wireless metropolitan area networks (WMANs) and wireless regional area networks (WRANs). WWANs and WLANs are the most widely adopted technologies and are able to deliver peak data rates in excess of 100 Mbps (e.g. LTE and IEEE 802.11n), [7]. However, the coverage area of WWANs and WLANs may restrict their use as a standalone solution for the implementation of rural telemedicine and mHealth services. Thus, the use of WMANs and WRANs seems to be more suitable for the deployment of rural telemedicine and mHealth solutions, as further discussed below throughout the chapter. Next, two BWA technologies, suitable for rural scenarios, are introduced.

The Institute of Electrical and Electronics Engineers (IEEE) 802.16 task group has been developing a family of standards for WMANs since 2001, which resulted in the IEEE 802.16 standards family, [8, p. 16]. Similar to the WiFi Alliance with the IEEE 802.11 family of standards, the Worldwide Interoperability for Microwave Access (WiMAX) forum released a certification based on IEEE 802.16 standards for WMANs. The certification aim is to guarantee technological interoperability between different IEEE 802.16 equipment vendors. With the release of the IEEE 802.16m standard in 2011 (known as mobile WiMAX release 2), the IEEE 802.16 family of standards included both fixed and high-speed mobile wireless services. The IEEE 802.16/WiMAX standard provides coverage ranges of up to 50 kilometers for line-of-sight (LOS) and up to 24 kilometers for non-line-of-sight (NLOS) transmissions. Therefore, it is considered a suitable technology to cover rural areas and a feasible solution to serve as a backbone network over a backhaul infrastructure. In contrast to fixed WiMAX, the IEEE 802.16m standard supports mobility and is fully capable of serving dense urban areas with a coverage ranging from 5 to 15 kilometers for NLOS transmissions. This version of the standard provides peak data rates of up to 70 Mbps, [9]. Through the adoption of quality of service (QoS) and scheduling mechanisms, WiMAX is able to support time constrained transmissions such as: multimedia streaming, real time surveillance, VoIP, and multimedia conferencing, [10].

As previously mentioned, the WRANs represent another alternative to enable BWA in rural areas. A recent trend in the implementation of WRANs is the use of cognitive radio networks (CRN) technology. The aim of a cognitive radio network is to be able to transmit over unused frequency bands commonly assigned to other technologies or services on a non-interference basis. This is achieved by means of active spectrum sensing, using frequency bands that are found to be free to transmit information. The active sensing must be kept on while exchanging information within the CRN, so that the frequency band can be released when transmissions from primary users are detected. The analog switch-off towards the migration from analog to digital television broadcasts, presents an opportunity for the development of WRAN solutions operating over frequency bands previously used by analog TV transmissions (the frequency bands are commonly referred to as TV white spaces (TVWS)). Answering the notice of proposed rule making (NPRM), issued by the United States Federal Communications Commission (FCC), regarding unlicensed operation in TV broadcast bands (i.e. TVWS) the IEEE formed the 802.22 working group to develop the first cognitive radio WRAN standard. As a result, the IEEE 802.22 standard was released in 2011, [11, p. 22]. The IEEE 802.22 standard supports fixed and mobile stations considering both LOS and NLOS transmissions with the aim of providing BWA to suburban and rural areas. This standard provides a maximum transmission range of up to 100 kilometers delivering peak data rates of up to 22.69 Mbps per channel. This way, the IEEE 802.22 standard doubles the coverage range of similar BWA technologies such as IEEE 802.16/WiMAX. A major contribution of the IEEE 802.22 standard is the inclusion of cognitive radio capabilities, which enable IEEE 802.22 networks to operate as secondary users in TVWS frequency bands. Thus, according to the FCC rules, the IEEE 802.22 transceivers must actively perform spectrum sensing to avoid or stop using frequency bands found to be occupied by transmissions from primary users, particularly digital/analog TV transmissions and Part 74 broadcast auxiliary services (i.e. wireless microphones).

Both technologies, IEEE 802.16/WiMAX and IEEE 802.22/WRAN, aim to offer wireless data services for fixed and mobile users comparable to those offered by wireline networks. Thus, both standards can be considered as emerging long range BWA technologies, [6].

1.2 Telemedicine Services over Metropolitan and Rural Area Broadband Wireless Access Technologies

Basic telemedicine deployments generally involve data acquisition from patients, transmission, processing and, depending on the offered services, remote storage. To be effective, any telemedicine solution must consider the QoS and bandwidth requirements for the different telemedicine services included within the deployment. Thus, in order to be a suitable alternative for the implementation of rural telemedicine networks, a particular BWA technology must be able to satisfy the QoS, bandwidth, and transmission range requirements of the network. In this chapter the feasibility of using the IEEE 802.22/WRAN and IEEE 802.16/WiMAX standards to deliver rural telemedicine services is explored.

It has been shown that the IEEE 802.16/WiMAX standard presents a viable solution for the deployment of telemedicine and mHealth services, [12]–[29]. The proposed mHealth applications cover different scenarios and can be categorized as: accident, clinical care, and home care, [12], [13]. In [14] and [15] the authors implemented a telecommunications backbone from a local clinic to a hospital using WiMAX. The network aim was to support a medical information system (MIS). The MIS provided remote locations with telemedicine services such as: expertise sharing between physicians; remote consultations; access to laboratory analysis and tests results for physicians; access to a medical scientific website; and an appointment scheduling application for patients. An integrated system using a heterogeneous network architecture with WiMAX and WLANs was proposed in [16]. The aim of the system was to deliver mHealth services such as patient monitoring, medical data access, emergency and pre-hospital care. In a different perspective, [17] introduces a mapping from multiple mHealth services parameters to mobile QoS variables in WiMAX. In contrast, [18] proposes a particular WiMAX dynamic subframe allocation for the support of mHealth services. Both proposals aim to enable telemedicine services under clinical care, home care, and emergency scenarios.

The development of telediagnosis services for ambulances has received particular attention because of its potential for improving the survival rate of patients in emergency situations. Typically, an emergency scenario requires providing real-time communications capabilities in order to support the transmission of biomedical signals, voice and/ or video, all of them in real time. Providing these capabilities will enable the provisioning of services such as pre-hospital care, remote assistance, and second opinion to paramedics. Different telemedicine and mHealth emergency systems such as WiMAX extensions for remote and isolated research data networks (WEIRD), [19], [20], Emergency Wi-Medicine (EWM), [21], and Focused Assessment with Sonography for Trauma using Tele-echography (FASTele), [22], have been developed as integral systems based on WiMAX technologies. The WEIRD project includes applications for remote diagnosis, monitoring and follow-up in general telemedicine scenarios. EWM tries to improve hospital admission times through enhanced videoconferencing services. This is achieved by modifying the maximum video bit rates over a WiMAX platform, [21]. The FASTele project uses WiMAX technology to provide a wireless link between an ambulance and a hospital. This link is used by a physician located in the hospital to remotely control a portable and attachable tele-echography robot system, [22]. Other proposals focus on modifications to the IEEE 802.16/WiMAX standard in order to transmit high quality video images from ultrasound devices in emergency situations, [23]–[25]. In [23], a cross-layer approach based on region of interest (ROI) algorithms and resource allocation mechanisms are used to guarantee QoS parameters in the transmission of ultrasound video data. Other approaches consider the use of H.264/advanced video coding (AVC), [24, p. 264], or high-efficiency video coding (HEVC) standards, [25], to enable the transmission of high resolution ultrasound video, resulting in quality comparable to that of in-hospital examination standards.

Another telemedicine and mHealth area that has drawn much attention is the development of telemonitoring systems for home care scenarios. These systems transmit the patient's biomedical signals, enabling health specialists to evaluate their health status. In this context, WiMAX and IEEE 802.22 can fulfill the requirements specified in [26] to enable extra-BAN communication for mHealth applications. For example, in [27] a telemonitoring system framework based on WiMAX technology for chronic hypertension patients is proposed.

Although all the previously mentioned projects show that WiMAX is viable technology to implement telemedicine and mHealth services, its coverage range below 50 kilometers can represent a drawback for its deployment in certain rural scenarios. In fact, deployment of telemedicine services in scarcely populated areas with urban centers separated by long distances may require the use of broadband satellite communications (BSC), leaving the use of WiMAX as a wireless last mile solution, [28] and furthermore, a drawback of using BSC solution is its cost of operational. Therefore, an alternative solution could be the deployment of wireless telemedicine networks based on the IEEE 802.22 standard. Theoretically, this standard could double the coverage range offered by WiMAX. The IEEE 802.22

standard defines a CRN technology that uses an opportunistic spectrum access scheme. Consequently, QoS provisioning as required by telemedicine and mHealth networks may present a major challenge. Nevertheless, in [29] it was shown that by designing resource allocation methods both the urgent and periodic traffic profiles, the CRNs standard is a suitable technology for wireless telemedicine implementations.

A performance evaluation of both, the IEEE 802.22/WRAN and the IEEE 802.16/WiMAX standards, is carried out in this section in order to assess their suitability for providing rural telemedicine services. Section 2 provides an overview of both standards. Section 3 introduces a wireless telemedicine model for rural areas. Performance evaluation results obtained using this telemedicine model are provided in Section 4. Conclusions are presented in Section 5.

2 IEEE 802.22/WRAN and IEEE 802.16/ WiMAX Standards Overview

An overview of the physical (PHY) and medium access control (MAC) layers of the IEEE 802.22/WRAN and IEEE 802.16/WiMAX standards is provided in this section. The section ends with a summary of the main parameters that must be considered for the deployment of rural telemedicine networks using these technologies.

2.1 IEEE 802.22/WRAN

The IEEE 802.22 air interface includes 50 channels ranging from 54 to 698 MHz in the US and Canada. A total of 60 channels with 6 MHz bandwidth, from 47 to 862 MHz, are considered for Western Europe. Additionally, for countries in Africa, Asia, and the Pacific, bandwidths of 7 MHz or 8 MHz per channel are contemplated. A single air interface based on 2048 carriers orthogonal frequency-division multiple access (OFDMA) is used to provide a reliable link for NLOS operation in a single time-domain duplex (TDD) mode. Because different channel delays must be supported, four different lengths of cyclic prefix defined as 1/4, 1/8, 1/16, and 1/32 are considered. The standard includes an adaptive modulation and coding (AMC) scheme defined by the PHY layer description. The AMC considers the quality variations in the wireless link (caused by fading, interference, etc.) to provide enhanced data rates between a base station (BS) and a customer premise equipment (CPE). Therefore, the use of quaternary phase shift keying (QPSK), 16-quadrature amplitude modulation (16-QAM), and 64-quadrature amplitude modulation (64-QAM), with four coding rates (1/2, 2/3, 3/4, 5/6) for each OFDMA subcarrier are considered in IEEE 802.22 to provide robustness over time variable wireless links.

The MAC layer includes cognitive capabilities for a reliable protection of primary users (PUs) in TVWS frequency bands and for enabling self-coexistence between IEEE 802.22 networks located nearby. The frequency bands used by an IEEE 802.22 cell will depend on the TV channels' availability, with is inherent to the geolocation of both the BS and CPE. Additionally, the use of frequency bands by the BS and the CPE will be restricted by Part 74 device broadcast transmissions (i.e. narrowband primary users -NPU- like wireless microphones). Therefore, the cognitive capabilities of IEEE 802.22 networks rely on their capacity to detect the primary users (PUs – TV transmissions and NPUs) over all the possible operational channels. Thus, the CPE and the BS will be forced to stop any transmission in channels where PUs are detected. In case a CPE detects a PU it will notify the BS via an urgent coexistence situation (UGS) message. Then, an incumbent detection recovery protocol (IDRP) will be performed such that the BS will reallocate the CPE transmissions to a backup channel. In case the BS detects a PU, it will run the IDRP to reallocate all the transmissions from the CPEs that were using the channel where the PU was found.

The IEEE 802.22 standard includes a normal operational mode, where one WRAN cell transmits over an entire channel using all the available frames merged into a superframe. Additionally, a self-coexistence operational mode is considered in the standard. This operational mode aims to coordinate transmissions from multiple WRAN cells by assigning one or several frames to one particular WRAN. The self-coexistence mode uses a coexistence beacon protocol (CBP) operated by BSs detecting concurrent transmissions from other WRANs. The CBP protocol requires broadcasting a packet containing BS identifiers through a dedicated self-coexistence window (SCW). This facilitates network discovery, coordination and spectrum sharing.

The IEEE 802.22 systems use a connection-oriented protocol to exchange information between the BS and the CPEs. This protocol allows the CPEs to request particular bandwidth allocations from the BS in a dynamic fashion. However, as granting all the bandwidth allocations requests from different CPEs may consume all the available system resources, the BSs must decide how to assign resources to a particular CPE. There is not a pre-defined way to perform this, leaving the resource allocation scheduling as an open issue. Commonly, the scheduling algorithms look to provide a certain degree of fairness regarding resource allocation to CPEs. However, in the case of telemedicine networks an "unfair" scheduling algorithm that prioritizes the traffic generated by telemedicine and mHealth applications may be required.

In addition to cognitive radio capabilities, the IEEE 802.22 MAC layer includes geolocation operations, channel database access and spectrum sensing initialization. These functionalities are needed for synchronization, ranging, capacity negotiation, registration, and connection setup authorization processes. The BSs and the CPEs use satellite-based geolocation technology to perform two main tasks: synchronization with other WRAN networks through a global time source, and verification of unused TV channels on the region by referring to an up-to-date database. Nevertheless, regardless of the channel status (used or unused) registered in the database, the spectrum sensing function must be performed by both the CPEs and BSs.

The IEEE 802.22 standard provides three types of QoS schemes: unsolicited grant service (UGS), polling service (PS), and best-effort service (BE). These schemes support constant bit rate (CBR) and variable bit rate (VBR) traffic to enable real and non-real time transmissions. The UGS service is designed to carry

CBR traffic where fixed-size data packets are transmitted periodically. Thus, the BS will reserve the required resources (if available) to support CBR traffic transmissions from the CPE with the aim of reducing the negative effects of delay and jitter. The PS service offers two service subtypes supporting two kinds of VBR traffic: real-time and no real-time. These service subtypes are called real-time polling service (rtPS) and non-realtime polling service (nrtPS). The rtPS service supports the transmission of real-time data streams by dynamically assigning system resources based on the QoS requirements. Contrary to rtPS, the nrtPS service just assigns a minimum bandwidth that can be used for the transmission of no real-time data streams. Lastly, the BE service does not guarantee any QoS, since its bandwidth allocation depends on the policies used for the other service types.

Because of the QoS schemes included in the IEEE 802.22/WRAN standard, it seems to be a viable technology for the deployment of wireless telemedicine networks. Nevertheless, an open research issue for the implementation of IEEE 802.22/WRAN telemedicine network concerns how to deal with possible throughput drops caused by sporadic PU transmissions (detected by either the BS or the CPEs).

2.2 IEEE 802.16/WiMAX

The air interface of IEEE 802.16/WiMAX operates at unlicensed spectrum between the 2-11 GHz frequency range or licensed spectrum between the 10-66 GHz frequency range. For NLOS scenarios, IEEE 802.16/WiMAX supports two different air interfaces at the PHY layer: OFDM and OFDMA, [30], [31]. Furthermore, the IEEE 802.16e includes a scalable OFDMA (S-OFDMA) air interface. The standard supports channel bandwidths ranging from 1.5 MHz to 20 MHz. Additionally, it considers the use of adaptive modulation and coding (AMC). Thus, the modulation and coding schemes employed in each OFDM subcarrier are set based on a link quality indicator. The standard includes a connection oriented MAC layer definition to support different types of QoS. Hence, the MAC protocol considers an unidirectional connection for all transmissions between BS and mobile station (MS), [32]. The standard defines two medium access schemes: time division duplex (TDD) and frequency division duplex (FDD). In both cases, bandwidth for QoS connections can be requested.

As the BS manage the network, it assigns the bandwidth reservation and transmission opportunities for each MS. However, each MS can dynamically request bandwidth from the BS by using a bandwidth-request (BWR) protocol data unit (PDU) in a contention or contention-free mode (e.g. polling). Additionally, IEEE 802.16e defines different BWR protocols such as: unsolicited request, poll-me bit, piggybacking, bandwidth stealing, codeword over quality indicator channel (CQICH) and code division multiple access (CDMA) code-based BWR.

Similar to the IEEE 802.22/WRAN standard, there are different scheduling service types in IEEE 802.16/WiMAX, namely: unsolicited grant service (UGS), extended real-time polling service (ertPS), real-time polling service, non real-time polling service, and best-effort service. Each of these service types has its own QoS parameters like minimum throughput requirement and delay/jitter constraints.

2.3 *IEEE 802.22/WRAN and IEEE 802.16/WiMAX Main Characteristics Comparison*

A comparison between the main characteristics of the IEEE 802.22/WRAN and IEEE 802.16/WiMAX standards is presented in Table 1. Although both standards show some similarities at the PHY and MAC layers (e.g. the use of OFDMA, ACM, QoS types of services), the bandwidth allocations, transmission ranges, and MAC frame sizes are different.

As previously mentioned, IEEE 802.22/WRAN networks operate as secondary users in TVWS frequency bands. Operation in these bands within the United States is unlicensed as far as there is not interference to PUs. The rules to use TVWS channels are still under review worldwide. However, Mexico, the European Community, South Africa, and China among other countries are considering following the United States TVWS regulation model. Therefore, it is fair to assume that IEEE 802.22/WRAN networks potentially have 60 unlicensed channels to operate worldwide. In contrast, most of the operating frequencies considered in the IEEE 802.16/WiMAX standard are licensed, except by the 2.4 GHz and 5.8 GHz frequency bands, which are already used by other wireless technologies (e.g. WiFi, ZigBee, Bluetooth).

Because of IEEE 802.16/WiMAX operates at frequencies above 2 GHz, propagation characteristics limit its maximum transmission range to 50 kilometers for LOS and to approximately 24 kilometers for NLOS. In comparison, the IEEE 802.22/WRAN standard operating bands are allocated below 900 MHz. Therefore, the maximum transmission range of IEEE 802.22/WRAN doubles the range offered by IEEE 802.16/WiMAX, reaching up to 100 kilometers for LOS and up to 33 kilometers for NLOS transmissions.

Table 1 Comparison between IEEE 802.16 and IEEE 802.22

System parameters	IEEE 802.16/WiMAX	IEEE 802.22/WRAN
Frequency band	2-10 GHz	54-862 MHz
Channel bandwidth	$\geq$ 5 MHz	6, 7, 8 MHz
Transmission rate	10-40 Mbps	23-31 Mbps
Cell radius	Up to 50 kms.	Up to 100 kms.
Multiple access	TDMA or OFDMA	OFDMA
MAC frame size	5 ms	10 ms

Another difference between the IEEE 802.16/WiMAX and IEEE 802.22/WRAN standards relies on their maximum data rates. The IEEE 802.16e standard is fully capable of achieving a maximum data rate of up to 40 Mbps. In comparison the IEEE 802.22 standard supports up to 31 Mbps using a 64-QAM modulation scheme, [11, p. 22], [33]. However, IEEE 802.22 achieves this data rate using a bandwidth channel of only 8 MHz instead of the 20 MHz required by the IEEE 802.16e standard.

Maximum data rate is only achieved when the distance between the BS and the user (MS in WiMAX and CPE in IEEE 802.22/WRAN) is relatively short, such that 64-QAM can be used without incurring in an unacceptable bit error rate (BER). In order to compare the capabilities offered by the IEEE 802.22/WRAN and IEEE 802.16/WiMAX standards, Fig. 1 shows a graphical representation of the modulation scheme used for a given coverage area in a typical deployment scenario. Since the coverage range of IEEE 802.22 is larger than that of IEEE 802.16, the overlapping between IEEE 802.22 QPSK and IEEE 802.16 is minimal. In contrast, it can be seen that the areas covered by IEEE 802.22 16-QAM and IEEE 802.16 QPSK are similar (they are almost completely overlapped). Furthermore, the area corresponding to IEEE 802.22 64-QAM completely covers the area corresponding to IEEE 802.16 64-QAM and a significant portion of the area corresponding to IEEE 802.16 16-QAM. This means that the range achieved with IEEE 802.22 for QPSK, 16-QAM, and 64-QAM is larger than that achieved with WiMAX.

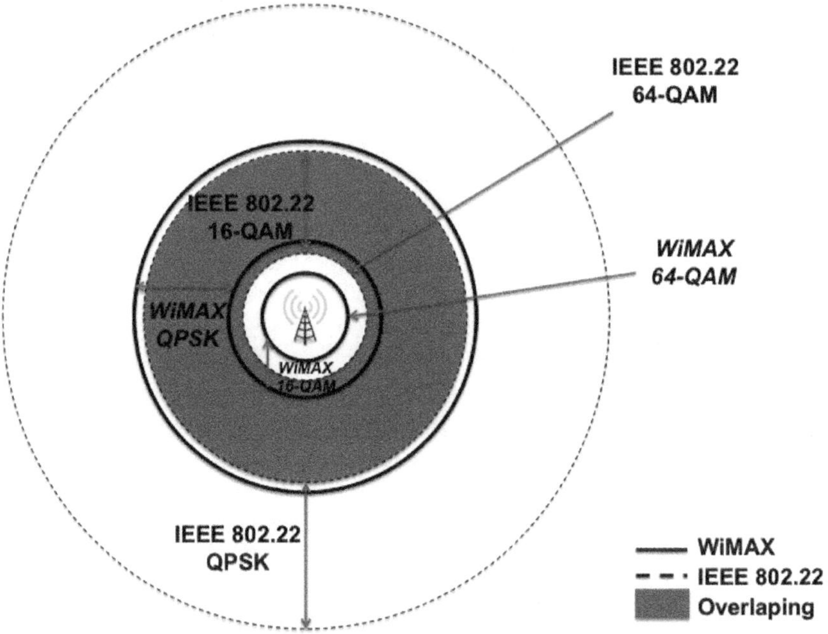

Fig. 1 Modulation schemes used for a given coverage range for typical IEEE 802.22/WRAN and IEEE 802.16/WiMAX deployments

Before reaching any conclusions based on the previous section regarding the suitability of IEEE 802.16/WiMAX and IEEE 802.22/WRAN for the deployment of rural telemedicine networks, it is first necessary to define a wireless telemedicine model for rural regions. In the following section, such model is proposed in order to assess the capabilities offered by each technology to allocate rural telemedicine services.

3 Wireless Telemedicine Model

One of the main challenges in rural telemedicine is mobility, [34]. Since health specialists are often concentrated in main urban centers, rural patients may be exposed to long traveling times in order to be attended. This condition may represent a major effort for post-operative, re-convalescing and elderly patients, [35]. Wireless rural telemedicine presents an alternative solution not only to reduce travel time issues, but also to extend health services by offering health care specialist consultation in situ.

A rural telemedicine model must take into account the health issues prevalent among rural regions and their particular needs. Thus, the telemedicine services offered should focus in attending critical health issues prevalent in rural regions, which may be different to those found in large urban centers. Another aspect that must be considered is the technological infrastructure found in the place where the rural telemedicine solution will be implemented. For example, [36] presents a study of the economic impact of rural hospitals in the United States, where it was assumed that the telemedicine sites had wired connectivity. Such assumption may not be used to perform a similar study for developing countries, as most rural clinics may not have wired connectivity. Indeed, there are several examples in the literature [34]–[50], that propose the deployment of wireless rural telemedicine solutions in order to offer medical services to rural populations in developing countries. Deploying a rural wireless telemedicine network based on IEEE 802.16/WiMAX and IEEE 802.22/WRAN may offer advantages where there are not wired solutions available. Thus, the rural telemedicine model proposed in this chapter for the evaluation of IEEE 802.16/WiMAX and IEEE 802.22/WRAN will be based in previous efforts for the deployment of wireless rural telemedicine services.

3.1 Telemedicine Services

As mentioned before, from a general perspective telemedicine services comprise teleconsulting, telediagnosis, telemonitoring, medical database access (MDBA) and teleducation, [3]. The inclusion of one or all services in a telemedicine model depends on the application scenario. For instance, because of the large technological and health infrastructure available in countries like Germany and the United States, rural telemedicine services focusing on home monitoring have been proposed, e.g. [37], [38]. In contrast with developing countries, (e.g. Malaysia and Bangladesh), rural telemedicine is mainly focused on interconnecting rural clinics with main hospitals. This, in order to assist local physicians remotely by extending health specialist services through telediagnosis and teleconsulting, [39], [40].

Table 2 shows examples of different rural telemedicine efforts around the world. The differences and similarities between telemedicine services offered by different projects can be observed.

Table 2 Rural telemedicine services offered by different projects around the world

Ref.	Country	Telediagnosis	Teleconsulting	Teleeducation	Telemonitoring	MDBA
[41]–[43]	China	Local physician collects data and communicates with health specialist for diagnosis purposes	The patient consulting is based on a health specialist decision performing diagnosis via a meeting to review the collected data	Academic and training meetings are scheduled abetween local and remote locations	Patient's monitoring is performed by local physicians based on scheduled appointments for follow-up and data collections purposes	The storage of collected data, from the hospital database, is available from local and remote locations
[34], [35], [44]–[46]	India	Health specialist based their diagnosis on collected data and described their recommendations to local physicians	The consulting is performed between local physician and health specialist only to discuss the patient file information	Available to postgraduate students, district hospital doctors, paramedical workers and hospital administrators to improve patient care skills	The local physician is in charge of performing the monitoring based on scheduled appointments to update the patient's file	The storage of collected data, from the hospital database, is available from local and remote locations
[47], [48]	Africa	Local physician collects data and communicates with health specialist for diagnosis purposes	The local physician can choose to perform a consultation with a local hospital specialist or a tertiary hospital specialist	Training of local physicians to improve their consulting medical skills	Patient's monitoring is performed by local physicians for follow-up monitoring locally, sending basic information to remote hospital	The access to a medical database available for local physicians allows them to identify and prevent particular diseases
[49]	Indonesia	Local physician collects data and update the patient record file for diagnosis purposes	The consultation awaits the update of patient record profile with health specialist recommendations		Patient's monitoring is performed by local physicians for follow-up monitoring updating the information to remote hospital	The storage of collected data, from the hospital database, is available from local and remote locations
[40], [50]	Latin America		The consultation update of patient record profile is performed between physician and health specialist		Patient's monitoring is performed by local physicians for follow-up monitoring updating the information to remote hospital	The storage of collected data, from the hospital database, is available from local and remote locations

3.2 Telemedicine Traffic Profile

The definition of a traffic profile for a telemedicine network depends on the services to be offered and the technological infrastructure available. For instance, the International MedioNet of China (IMNC) network is based on Internet access through plain old telephone service (POTS) to enable rural telemedicine consultation services, [41].

In the matter of technological infrastructure, connection availability towards the deployment of a telemedicine network may limit the medical services that can be offered. In other words, a telemedicine model designed to offer diagnosis services using high definition video (with a minimal bandwidth requirement of 400 Kbps connection) may require QoS provisioning that a modem connection over a POTS network (with a bandwidth of 56 Kbps) cannot offer.

The definition of a traffic profile for rural telemedicine should not be based only on its applications in telemedicine, but also on the particular health requirements of rural populations. Therefore, it is difficult to define a general telemedicine traffic profile for rural areas. A few examples of the traffic profiles used in different rural telemedicine projects are presented in Table 3. Note that most projects include VoIP services even though the used link offered different connection bandwidths (e.g. 1 Mbps ADSL in [42] vs. 128 kbps ISDN in [51]).

Table 3 Rural telemedicine traffic profiles in different projects around the world

Country	Ref.	Applications	Traffic Profiles	Available Bandwidth
China	[42]	VoIP Conferencing	Real Time Streaming Constant Bit Rate	ADSL Connection up to 1 Mbps.
		HTTP FTP E-mail	Non Real Time Streaming Variable Bit Rate	
	[43]	ECG Signal up. 5-12 lead	Store and Forward Non-Real Time Streaming Constant or Variable Bit Rate	Bluetooth up to 1 Mbps 3G Mobile Technology
India	[35]	VoIP Conferencing	Real Time Streaming Constant Bit Rate	ADSL Connection up to 512 Kbps.
		Web based Medical Application System	Non Real Time Streaming Variable Bit Rate	
	[44]	HTTP	Non Real Time Streaming Variable Bit Rate	Bluetooth up to 1 Mbps 3G Mobile Technology

Table 3 (*continued*)

	ECG Blood Pressure Heart Rate Breathing Rate Body Temperature Blood Glucose Cholesterol Oximeter	Store and Forward Non-Real Time Streaming Constant or Variable Bit Rate	
[34]	VoIP Conferencing	Real Time Streaming Constant Bit Rate	Connection up to 512 Kbps.
	Web based Medical Application System	Non Real Time Streaming Variable Bit Rate	
[45]	VoIP Conferencing	Real Time Streaming Constant Bit Rate	CDMA 2000 Mobile Technology
	Mobile X-Ray Portable Ultrasound Scanner ECG Ophthalmoscope Retinoscope Digital Microscope	Store and Forward Non-Real Time Streaming Constant or Variable Bit Rate	Link Aggregation up to 220 Kbps.
[46], [51]	VoIP Conferencing	Real Time Streaming Constant Bit Rate	
	High Definition Images via E-mail ECG	Store and Forward Non-Real Time Streaming Constant or Variable Bit Rate	ISDN up to 128 Kbps

3.3 Rural Telemedicine Model

As mentioned before, the particular telemedicine services offered in a rural telemedicine network depend on the rural population requirements. Based on the examples introduced in the previous sections, the rural telemedicine model proposed in this chapter includes the following services:

1) *Teleconsulting* has two cases: physician-specialist and patient-specialist. Physician-specialist teleconsulting is performed between a local physician and a remote specialist. Its aim is to discuss the patient's electronic data in order to obtain recommendations. In patient-specialist teleconsulting, a videoconference is set up to perform the patient's physical check required by the remote specialist for follow up purposes.
2) *Telediagnosis* has two cases: ambulance-clinic and clinic-hospital diagnosis. In ambulance-clinic telediagnosis, a local physician obtains

patient-collected data and emits recommendations to paramedics, while setting up pre-hospital treatment. In clinic-hospital telediagnosis, the local clinic physician collects the patient's data and sends the information to the health specialist's located in a hospital. The local physician can receive the specialist recommendations and instructions in real time or near real time.

3) In *Telemonitoring* a local nurse or physician collects medical data from the patient (e.g. blood pressure, ECG, EEG, etc.) and sends it to a remote database for follow-up and patient record update purposes.
4) *Teleeducation* is based on scheduled meetings between clinic physicians and the hospital specialist medical board for training purposes. In this service, access to remote hospital medical databases is provided to local clinic physicians in order to perform consultations of the patient's medical records.

The traffic profile for the offered telemedicine services described in the proposed model is shown in Table 4. In order to define the data rates, the information reported in [12], [24], [52] was considered. The traffic profile was defined considering average values, as the required uplink (UL) and downlink (DL) data rates differ from one medical specialty to another.

Table 4 Rural telemedicine model traffic profile, applications and services

Type of data	Description	Minimum Data Rate (Kbps)	Telemedicine Services									
			Consulting				Diagnosis				Monitoring	
			Physician-Specialist		Patient-Specialist		Ambulance-Clinic		Clinic-Hospital			
			UL	DL	UL	DL	UL	DL	UL	DL	UL	DL
Biomedical Signals	ECG (12 leads)	288					✓		✓		✓	
	Heart Rate	2					✓		✓		✓	
	Blood pressure	2					✓		✓		✓	
Videoconference	Normal quality videoconferene	425			✓	✓	✓	✓	✓	✓		
	Low quality videoconference	128	✓	✓								
Diagnosis Audio and Video	Diagnostic Sound	32					✓		✓			
	High quality diagnostic video	768					✓					
		Average data rate (Kbps)	128	128	425	425	1517	425	749	425	292	0

The number of fixed and mobile wireless connections needed in a rural wireless telemedicine deployment varies with the proximity of the population to the local clinics. Thus, when there is a relatively large number of rural settlements in a given area, the number of required services may also be large. Hence, in order to evaluate the suitability of IEEE 802.16/WiMAX and IEEE 802.22/WRAN of deploying rural telemedicine networks in this kind of scenario, a case study based on the State of Chiapas, Mexico will be presented.

In Mexico more than 37% of states have over 30% of rural population. The most drastic case is the State of Chiapas, with a rural population of 51%, [53] (which, incidentally, has the highest poverty rate in Mexico with over 74% of the population living in poverty, [54]). Therefore, Chiapas is a representative example where people need primary and extended medical services, including consultation by health specialists. At least 30 municipalities are located within a radius of 50 km from Tuxtla Gutierrez, Chiapas's capital (including the tourist City of San Cristobal de la Casas), and 15 of them can be classified as rural. Based on these conditions, the generic rural wireless telemedicine network architecture that will be used for the evaluation of IEEE 802.16/WiMAX and IEEE 802.22/WRAN capabilities is shown in Fig. 2. The network includes a 50 km backhaul link between points A and B. Additionally, other link distances are: 22 km (A to C), 17 km (A to D), 18 km (A to E), 24 km (A to F), 14 km (A to G), and 60 km (A to H).

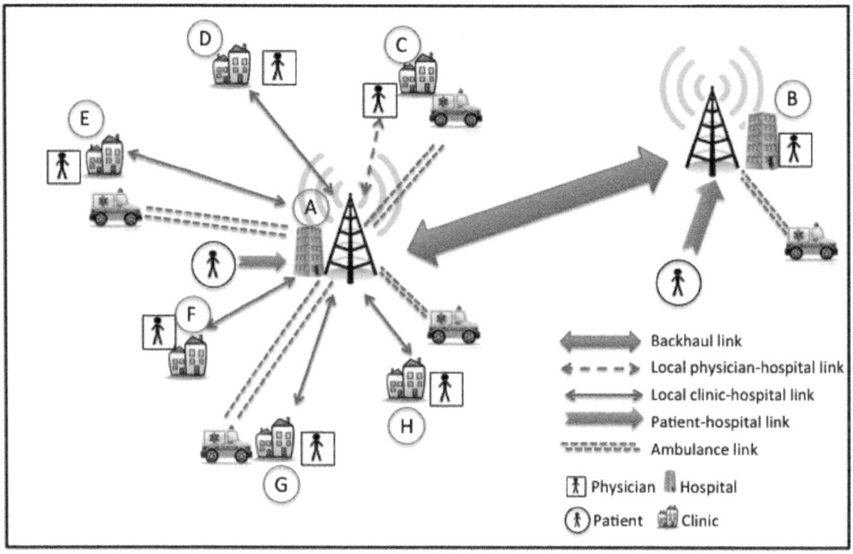

Fig. 2 Generic rural wireless telemedicine network architecture

By using the rural telemedicine model and the generic rural wireless telemedicine network architecture, the number of active services that can be supported by the IEEE 802.16/WiMAX and IEEE 802.22/WRAN standards will be estimated. The number of services will be limited by bandwidth and QoS requirements. Therefore, the system resources for the telemedicine services must be reserved by means of resource scheduling algorithms. In IEEE 802.16/WiMAX, the use of scheduling algorithms towards QoS provisioning has been widely studied in [9], [12], [55]. Nevertheless, from the bandwidth availability point of view, an evaluation using a throughput metric can be performed. Thus, in the following section the system capacity in terms of the number of telemedicine services supported will be estimated by using the maximum throughput.

4 System Performance Evaluation

As mentioned in Section 1.3, the IEEE 802.16/WiMAX standard has shown its suitability to enable rural telemedicine networks around the world. However, the IEEE 802.22/WRAN standard, as an emerging technology, represents an alternative for the deployment of rural telemedicine networks because of its capacity to operate through TVWS channels with ranges of up to 100 kilometers. Therefore, in order to establish the bandwidth availability per channel towards the deployment of rural wireless telemedicine networks, an evaluation based on maximum throughput is presented in this section for the IEEE 802.16/WiMAX and IEEE 802.22/WRAN standards.

4.1 Throughput Performance Calculation

The performance evaluation of a system based on throughput requires finding the maximum data rate achieved considering parameters such as modulation, frame overhead, and system reserved resources, among others. The calculation presented in this section was performed with the system parameters shown in Table 5.

Table 5 System parameters for the evaluation of IEEE 802.16/WiMAX and IEEE 802.22/WRAN

System parameters	IEEE 802.16/WiMAX	IEEE 802.22/WRAN
Bandwidth channel	5 MHz	6 MHz
Sampling frequency	5.6 MHz	6.856 MHz
NFFT	512	2048
Frame duration (ms)	5	10
Cyclic prefix	1/8	1/8
Modulation	QPSK, 16-QAM, 64-QAM	QPSK, 16-QAM, 64-QAM
Coding rate	1/2, 2/3, 3/4, 5/6	1/2, 2/3, 3/4, 5/6
Symbol time (ms)	102.9	336.05
TTG/RTG (ms)	105.71 / 60	209.88 / 305.86
Symbols per frame	48	30

The IEEE 802.16e standard calculation is based on the Partially Used SubCarrier (PUSC) scheme in order to obtain the maximum data rate, [56]. Different to 802.16e, the IEEE 802.22 standard defines a fixed number of subchannels available for any link according to a bandwidth request protocol.

4.2 Data Bandwidth Allocation

Data bandwidth allocation in the IEEE 802.16e and IEEE 802.22 standards depends on such system parameters as: channel bandwidth, coding and modulation schemes, number of OFDM subcarriers, etc. Four mandatory overhead symbols corresponding to downlink/uplink (DL/UL) MAPs messages, frame preamble and TTG/RTG are considered in both cases for throughput calculation. Using these considerations and the parameters shown in Table 5, the peak throughput results for both technologies are reported in Table 6. It is important to note that these results are consistent with the values presented in the standards definitions, [8], [11].

The system performance evaluation presented in this section uses the rural telemedicine model introduced in Section 3.3. In this context, IEEE 802.16/WiMAX technology offers connectivity up to 50 km for LOS and up to 24 km for NLOS.

Under similar conditions, IEEE 802.22/WRAN is able to offer a maximum transmission range of 100 km in LOS and 65 km in NLOS. However, these transmission ranges are only reached under ideal conditions. Therefore, in order to better assess the suitability of both technologies for the deployment of rural telemedicine networks, the propagation model introduced in [57] is used to adjust the maximum transmission ranges as follows:

- For IEEE 802.16/WiMAX, it is assumed that a range of: 24 km is achieved with QPSK-3/4; 11 km is reached with 16-QAM-3/4; and 6 km is achieved with 64-QAM-3/4.
- For IEEE 802.22/WRAN, it is assumed that a range of: 65 km is achieved with QPSK-3/4; 32 km is reached with 16-QAM-2/3; and 9 km is achieved with 64-QAM-3/4.

Regarding the relationship between modulation-coding and maximum transmission ranges from Fig. 2, it can be observed that point F is located at the boundary where IEEE 802.16/WiMAX can offer QPSK-3/4. Nevertheless, in the same point, IEEE 802.22/WRAN is able to offer 16-QAM-2/3. Furthermore, IEEE 802.22/WRAN is able to offer QPSK-3/4 coverage for point H, which is not reached by WiMAX.

Table 6 Peak throughput comparisons between IEEE 802.22 and IEEE 802.16e

Modulation scheme	Coding rate	IEEE 802.16/WiMAX (Mbps)	IEEE 802.22/WRAN (Mbps)
QPSK	1/2	5.56	4.54
	2/3	-	4.99
	3/4	8.34	5.62
	5/6	-	6.24
16-QAM	1/2	11.12	7.49
	2/3	-	9.98
	3/4	16.68	11.23
	5/6	-	12.48
64-QAM	1/2	16.68	11.23
	2/3	22.25	14.98
	3/4	25.03	16.85
	5/6	27.81	18.72

Because both technologies can operate in TDD mode, flexible UL and DL data bandwidth allocations can be used as long as at least 10% of the available data bandwidth is reserved for both the DL and the UL. Therefore, in order to evaluate the system performance for asymmetric DL/UL data bandwidth allocations, increments/decrements of 10% for the DL/UL will be used.

A total of 15 municipalities from the case study introduced in Section 3.3 are considered for the evaluation of the capacities offered by IEEE 802.16/WiMAX and IEEE 802.22/WRAN for rural telemedicine deployments. Thus, a maximum of 15 rural locations are used to calculate the number of active telemedicine services that can be offered for each DL/UL data bandwidth allocation. Additionally, it will be assumed that the users are located within the major overlapping area shown in Fig. 1. Hence, QPSK-3/4 transmissions will be available for WiMAX, and 16-QAM-2/3 transmissions will be available for IEEE 802.22/WRAN.

The data throughput is affected by the MAC and PHY layers overhead, which is a function of the number of active users in the network. No overhead reduction methods are considered for the analysis. Therefore, the MAC/PHY overhead will be comprised by the DL/UL MAP, the information elements (IE) formats, and ranging and bandwidth requests. Because of the similarities in MAC and PHY layers between IEEE 802.22/WRAN and IEEE 802.16e, the overhead percentage calculation method proposed in [58] will be used for both technologies. The calculated overhead percentage factors are shown in Table 7.

Table 7 MAC and PHY layers overhead percentage for IEEE 802.22/WRAN and IEEE 802.16e

Number of active users	IEEE 802.16e (%)	IEEE 802.22 (%)
1	32.91	30.24
2	34.97	33.60
3	37.03	36.96
4	39.09	40.32
5	41.14	43.68
6	43.20	47.04
7	45.26	50.40
8	47.31	53.76
9	49.37	57.12
10	51.43	60.48
11	53.49	63.84
12	55.54	67.20
13	57.60	70.56
14	59.66	73.92
15	61.71	77.28

As previously mentioned, data throughput will be affected by the overhead percentage factor. For example, although IEEE 802.16e QPSK-3/4 peak data rate is 8.34 Mbps and IEEE 802.22/WRAN 16-QAM-2/3 peak data rate is 9.98 Mbps (see Table 6), when two users are active within the network the throughputs are 5.42 Mbps and 6.63 Mbps respectively. This reduction in the data rates is due to the 34.97% and 33.60% overhead percentage factors of IEEE 802.16e and IEEE 802.22/WRAN, respectively (see Table 7). Therefore, if an ambulance service with telediagnosis requires 1,517 Kbps for the UL and 425 Kbps for the DL, then a maximum of two ambulance services can be enabled for this modulation scheme.

Following the procedure described before, it is possible to calculate the number of telemedicine services that can be enabled for a given DL/UL data bandwidth allocation. Thus, by using the model introduced in Section 3.3 and the information provided in Tables 4, 6, and 7, the number of telemedicine services that can be enabled for a given DL/UL data bandwidth allocation is shown in Fig. 3.

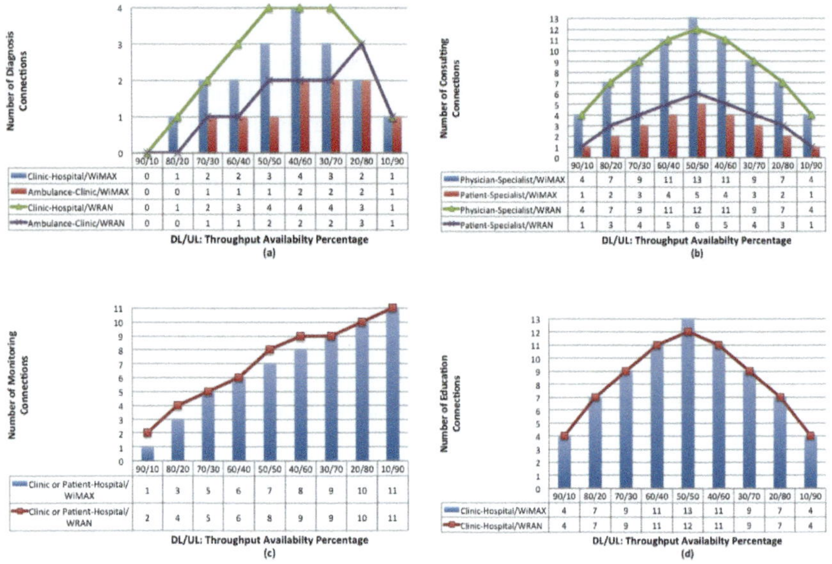

Fig. 3 Data Bandwidth allocation comparative between the IEEE 802.16e and the IEEE 802.22 standards for different telemedicine services. a) Telediagnosis, b) Teleconsulting, c) Telemonitoring, d) Teleeducation.

Fig. 3a shows that a maximum of 2 active ambulances transmitting telediagnosis services can be supported by IEEE 802.16/WiMAX. Similarly, a maximum of 3 active ambulances can be connected when using IEEE 802.22/WRAN. Both technologies achieved a maximum of 4 active clinic-hospital connections transmitting telediagnosis services. Fig. 3b shows that the IEEE 802.16/WiMAX standard can support up to 13 active physician-specialist connections transmitting teleconsulting services. In contrast, the maximum number of active physician-specialist connections achieved with IEEE 802.22/WRAN is 12. With respect to patient-specialist teleconsulting services, the IEEE 802.22/WRAN standard is able to allocate up to 6 active connections, but the IEEE 802.16/WiMAX standard supports a maximum of 5.

Similar to telediagnosis, an asymmetric DL/UL data bandwidth allocation was considered for telemonitoring services. Nonetheless, telemonitoring only sends traffic through the upstream link. Therefore, the number of allocated telemonitoring services rises when the UL available data rate is increased as shonw in Fig. 3c. Both technologies show a similar behavior in the number of supported telemonitoring services, with a maximum of 11 connections.

Note that when the maximum number of active services supported is relatively low, IEEE 802.22/WRAN can support up to one additional connection compared to IEEE 802.16/WiMAX. However, when the maximum number of active services supported is large, this behavior is inverted. The reason for this is that the overhead percentage factor of IEEE 802.22/WRAN is below that offered by IEEE 802.16/WiMAX when the number of active users is less than 4 (see Table 7).

As mentioned before, IEEE 802.16/WiMAX has been widely used to deploy telemedicine networks in several countries. For this reason, it is not surprising that Fig. 3 confirms the capability of WiMAX to support different telemedicine services. Furthermore, from Fig. 3 it can also be observed that IEEE 802.22 provides similar capabilities to those offered by IEEE 802.16/WiMAX. Therefore, it can be concluded that IEEE 802.22 presents a feasible solution for the implementation of rural telemedicine networks.

4.3 Telemedicine Services over BWA Technologies in Mexico

The deployment of WiMAX technology in Mexico has mainly focused on educational services. Since 2008, the Corporación Universitaria para el Desarrollo de Internet (CUDI) has developed several WiMAX projects. For instance, the proposal named "*Proyecto WiMAX Jalisco*" is related to the transmission of voice and data, interconnecting universities to extend the technological infrastructure in the Jalisco state.

As detailed in [59], the IEEE 802.22 standard is a suitable technology for the deployment of wireless rural networks in locations where population density is below 60 inhabitants/Km^2. However, to the best of our knowledge, no WRAN project has been proposed in Mexico yet. This, despite the fact that nearly 50% of Mexican cities fulfill this requirement [53]. In Mexico, regulation of the TVWS channels has been delayed. However, the objective of this proposal is to evaluate the suitability of using IEEE 802.22/WRAN technology to deliver telemedicine services in rural areas, despite the current status of the TVWS channels regulations.

5 Case Studies

Case 1. This chapter shows the throughput performance of the IEEE 802.22/WRAN standard for all available modulation and coding schemes based on 1/8 cyclic prefix mode. However, a total of four cyclic prefix modes are included in the IEEE 802.22/WRAN PHY layer definition. Hence, perform a calculation of the maximum data rate over one channel offered by the IEEE 802.22 for all available cyclic prefixes using the same system parameters considered in this chapter to observe the throughput capacity relation with the maximum system capacity.

Case 2. In section 4, a performance evaluation of telemedicine services bandwidth allocation capacity for IEEE 802.22/WRAN and IEEE 802.16/WiMAX standards based on the Chiapas example with a maximum of 16 municipalities included has been presented. However, there are two aspects of teleconsulting and telediagnosis (clinic/patient and clinic/ambulance traffic profiles, respectively) that have been studied independently. Hence, perform a calculation of the maximum number of teleconsulting and telediagnosis services overall traffic profiles (including the two sides on each case) that can be allocated in the IEEE 802.22/WRAN and IEEE 802.16/WiMAX for each UL/DL scheme based on QPSK-3/4 and 16-QAM-2/3 modulation-coding modes respectively.

6 IEEE 802.16/WiMAX and IEEE 802.22/WRAN Technologies Future Direction

Even though IEEE 802.16/WiMAX has been widely deployed to enable rural telemedicine applications, there are still important open research issues around this scenario that must be addressed, such as: end-to-end security and QoS for telemedicine traffic. Considering that medical information from patients is transmitted over wireless links, end-to-end security is an important issue, as data confidentiality must be observed overall the wireless telemedicine network. In this regard not only engineering issues must be addressed, but also legislative issues. Since telemedicine traffic profiles may define asymmetric traffic flows (i.e. telediagnosis), the study of QoS provisioning techniques is needed to properly implement wireless telemedicine services.

The IEEE 802.16 standard has been constantly updated to adjust its competencies considering the emerging needs. Hence, the following updates should consider using TVWS channels for its operation and include scheduling techniques to add QoS capabilities. Because spectrum license fees may represent an additional cost in the deployment of wireless rural telemedicine networks, IEEE 802.22 may present an alternative solution for the implementation of such networks. Nevertheless, the analog switch-off to enable TVWS channels worldwide is planned to finish at the end of 2020. Therefore, the deployment of IEEE 802.22/WRAN technology will depend on the portion of TVWS channels regulated as unlicensed according to each country's normativity. Since IEEE 802.22 includes cognitive radio capabilities, the study of primary user detection techniques should be developed. Similarly, coexistence engineering issues should be addressed to guarantee no-interference operation among IEEE 802.22 networks nearby located.

In the future, the deployment of IEEE 802.16 and IEEE 802.22 networks will enable the implementation of mHealth applications for underserved areas. Over this assumption, highly populated continents (i.e. Africa and Asia) with large rural regions can greatly benefit from the deployment of wireless rural telemedicine networks.

7 Conclusions

In this chapter, a wireless telemedicine solution for the delivery of telemedicine services over rural regions has been proposed. In order to assess the suitability of IEEE 802.16/WiMAX and IEEE 802.22/WRAN for the implementation of rural telemedicine networks, a rural telemedicine model that includes traffic profile descriptions and a deployment scenario was introduced. The traffic profile descriptions were telediagnosis, teleconsunting, telemonitoring, and teleeducation services. The system performance evaluation was carried out considering asymmetric data bandwidth allocations for the UL and DL of both technologies. By considering the overhead percentage factor, the maximum number of active telemedicine services was calculated for each standard. The presented results showed that the IEEE 802.22/WRAN standard is fully capable of supporting telediagnosis, teleconsulting, telemonitoring, and teleeducation services in rural deployments. Moreover, the IEEE 802.22/WRAN and IEEE 802.16/WiMAX standards showed similar performance regarding the maximum number of active telemedicine services. Consequently, it can be concluded that IEEE 802.22/WRAN presents a viable solution for the implementation of wireless rural telemedicine networks.

The rural telemedicine model used in this chapter included a deployment scenario where the case of the State of Chiapas, Mexico was used as an example. The aim was to assess the potential benefits that the deployment of wireless rural telemedicine networks can offer. Based on the results presented in this chapter, it can be said that it is feasible to deploy wireless rural telemedicine networks within Mexico by using IEEE 802.22/WRAN technology. Considering that IEEE 802.22/WRAN has the potential to operate in unlicensed TVWS channels, and its maximum transmission range doubles the one offered by IEEE 802.16e/WiMAX, then the deployment of IEEE 802.22/WRAN rural telemedicine networks has the potential to offer lower operating and implementation costs compared to WiMAX or another rural connectivity solutions (e.g. BSC). Since as many as 50% of Mexican rural regions may present favorable conditions for the deployment of IEEE 802.22/WRAN networks, we consider that its utilization for telemedicine applications can improve the provisioning of healthcare services in Mexico.

References

[1] Malindi, P.: QoS in Telemedicine. In: Graschew, G., Rakowsky (eds.) Telemedicine Techniques and Applications, pp. 119–120. InTech, Rijeka (2011)

[2] World Health Organization, Atlas eHealth country profiles based on the findings of the second Global survey on eHealth. World Health Organization, Geneva, Switzerland (2010)

[3] Vergados, D.J., Vergados, D.D., Maglogiannis, I.: NGL03-6: Applying Wireless DiffServ for QoS Provisioning in Mobile Emergency Telemedicine. presented at the Global Telecommunications Conference, GLOBECOM 2006, pp. 1–5. IEEE (2006)

[4] Grigsby, J., Barton, P.L., Kaehny, M.M.: Analysis of Expansion of Access to Care Through Telemedicine, Report 4, Study Summary and Recommendations for Further Research. Center for Health Policy Research, Denver, CO (1994)

[5] Dussault, G., Franceschini, M.C.: Not enough there, too many here: understanding geographical imbalances in the distribution of the health workforce. Hum. Resour. Health 4(1) (2006)

[6] Siddiqua, P., Awal, M.A.: A portable Telemedicine system in the context of rural Bangladesh. presented at the 2012 International Conference on Informatics, Electronics & Vision (ICIEV), pp. 608–611 (2012)

[7] Kuran, M.S., Tugcu, T.: A survey on emerging broadband wireless access technologies. Comput. Netw. 51(11), 3013–3046 (2007)

[8] 802.16-2009 (May 2009)

[9] So-In, C., Jain, R., Tamimi, A.-K.: Scheduling in IEEE 802.16e mobile WiMAX networks: key issues and a survey. IEEE J. Sel. Areas Commun. 27(2), 156–171 (2009)

[10] Li, B., Qin, Y., Low, C., Gwee, C.: A Survey on Mobile WiMAX Wireless Broadband Access. IEEE Commun. Mag. 45(12), 70–75 (2007)

[11] 802.22-2011 (July 2011)

[12] Niyato, D., Hossain, E., Diamond, J.: IEEE 802.16/WiMAX-based broadband wireless access and its application for telemedicine/e-health services. IEEE Wirel. Commun. 14(1), 72–83 (2007)

[13] Komnakos, D., Vouyioukas, D., Maglogiannis, I., Constantinou, P.: Performance Evaluation of an Enhanced Uplink 3.5G System for Mobile Healthcare Applications. Int. J. Telemed. Appl. 2008, 1–11 (2008)

[14] Chorbev, I., Mihajlov, M.: Building a wireless telemedicine network within a WiMax based networking infrastructure. presented at the IEEE International Workshop on Multimedia Signal Processing, MMSP 2009., Rio de Janeiro, pp. 1–6 (2009)

[15] Chorbev, I., Mihajlov, M., Jolevski, I.: WiMAX supported telemedicine as part of an integrated system for e-medicine. presented at the 30th International Conference on Information Technology Interfaces, ITI 2008, pp. 589–594 (2008)

[16] Zhang, Y., Ansari, N., Tsunoda, H.: Wireless telemedicine services over integrated IEEE 802.11/WLAN and IEEE 802.16/WiMAX networks. IEEE Wirel. Commun. 17(1), 30–36 (2010)

[17] Alinejad, A., Philip, N., Istepanian, R.S.H.: Mapping of multiple parameter m-health scenarios to mobile WiMAX QoS variables. presented at the 2011 Annual International Conference of the IEEE Engineering in Medicine and Biology Society, EMBC, pp. 1532–1535 (2011)

[18] Alinejad, A., Istepanian, R.S.H., Philip, N.: Dynamic subframe allocation for mobile broadband m-health using IEEE 802.16j mobile multihop relay networks. presented at the 2012 Annual International Conference of the IEEE Engineering in Medicine and Biology Society, EMBC, pp. 284–287 (2012)

[19] Neves, P., Simoes, P., Gomes, A., Mario, L., Sargento, S., Fontes, F., Monteiro, E., Bohnert, T.: WiMAX for Emergency Services: An Empirical Evaluation. presented at the 2007 International Conference on Next Generation Mobile Applications, Services and Technologies, NGMAST 2007, pp. 340–345 (2007)

[20] Mignanti, S., Castellano, M., Augusti, V., Martufi, G., Neves, P.M., Mambretti, C., Andreotti, F.: WEIRD - Real Use Cases and Applications for the WiMAX Technology. presented at the 5th IEEE Consumer Communications and Networking Conference, CCNC 2008, pp. 948–952 (2008)

[21] Lam, K.L., Tung, H.Y., Tsang, K.F., Ko, K.T.: WiMAX telemedicine system for emergence medical service. presented at the 7th International Conference on Information, Communications and Signal Processing, ICICS 2009, pp. 1–5 (2009)

[22] Ito, K., Tsuruta, K., Sugano, S., Iwata, H.: Evaluation of a wearable tele-echography robot system: FASTele in a vehicle using a mobile network. presented at the 2011 Annual International Conference of the IEEE Engineering in Medicine and Biology Society, EMBC, pp. 2093–2096 (2011)
[23] Debono, C.J., Micallef, B.W., Philip, N.Y., Alinejad, A., Istepanian, R.S.H., Amso, N.N.: Cross-Layer Design for Optimized Region of Interest of Ultrasound Video Data Over Mobile WiMAX. IEEE Trans. Inf. Technol. Biomed. 16(6), 1007–1014 (2012)
[24] Panayides, A., Antoniou, Z., Pattichis, M.S., Pattichis, C.S.: The use of H.264/AVC and the emerging high efficiency video coding (HEVC) standard for developing wireless ultrasound video telemedicine systems. presented at the 2012 Conference Record of the Forty Sixth Asilomar Conference on Signals, Systems and Computers (ASILOMAR), pp. 337–341 (2012)
[25] Panayides, A., Antoniou, Z.C., Mylonas, Y., Pattichis, M.S., Pitsillides, A., Pattichis, C.S.: High-Resolution, Low-Delay, and Error-Resilient Medical Ultrasound Video Communication Using H.264/AVC Over Mobile WiMAX Networks. IEEE J. Biomed. Health Inform. 17(3), 619–628 (2013)
[26] Pawar, P., Jones, V., van Beijnum, B.-J.F., Hermens, H.: A framework for the comparison of mobile patient monitoring systems. J. Biomed. Inform. 45(3), 544–556 (2012)
[27] Yen, Y.-S., Chiang, W.-C., Hsiao, S.-F., Shu, Y.-P.: Using WiMAX network in a telemonitoring system. presented at the 2011 3rd International Conference on Computer Research and Development (ICCRD), pp. 313–318 (2011)
[28] Ibikunle, F.A.: Comparative Analysis of Alternative Last Mile Broadband Access Technologies (Wi-Fi and WiMAX). Pac. J. Sci. Technol. 10(1), 280–285 (2009)
[29] Feng, S., Liang, Z., Zhao, D.: Providing telemedicine services in an infrastructure-based cognitive radio network. IEEE Wirel. Commun. 17(1), 96–103 (2010)
[30] Kumar, R., Adhikary, K., Vaid, R.: WiMAX Propagations. Int. J. Comput. Sci. Eng. Technol. 1(8), 480–483 (2011)
[31] Prasad, R.: OFDMA WiMAX Physical Layer. In: WiMAX Networks: Techno-Economic Vision and Challenges, pp. 63–135. Springer, Dordrecht (2010)
[32] WiMAX Network Capacity and Radio Resource Management. In: Radio Resource Management in WiMAX: From Theoretical Capacity to System Simulations, pp. 53–107. ISTE; Wiley, London, Hoboken (2009)
[33] Chakraborty, M., Bhattacharyya, D.: Overview of End-to-End WiMAX Network Architecture. In: Tang, S.-Y., Müller, P., Sharif, H.R. (eds.) WiMAX Security and Quality of Service, pp. 1–22. John Wiley & Sons, Ltd., Chichester (2010)
[34] Wickramasinghe, Y., Wijethilake, D., Vatsalan, D., Sudhahar, S., Runathilake, S., Seneviratne, G., Thilakarathna, K., Chapman, K., Mahanamahewa, P.: BudhuDas: An eHealth Business Model for Emerging Countries. presented at the IST-Africa, Africa (2010)
[35] Sudhahar, S., Vatsalan, D., Wijethilake, D., Wickramasinghe, Y., Arunathilake, S., Chapman, K., Seneviratna, G.: Enhancing Rural Healthcare in Emerging Countries through an eHealth Solution. presented at the Second International Conference on eHealth, Telemedicine, and Social Medicine, ETELEMED 2010, pp. 23–28 (2010)
[36] Whitacre, B.E., Hartman, P.S., Boggs, S.W.: The Economic Impact of Telemedicine Capability in a Rural Hospital. National Center for Rural Health Works, United States of America (December 2007)

[37] Lu, W.-P., Leung, H., Estrada, E.: Transforming telemedicine for rural and urban communities Telemedicine 2.0 - any doctor, any place, any time. presented at the 2010 12th IEEE International Conference on e-Health Networking Applications and Services (Healthcom), pp. 379–385 (2010)
[38] Polze, A., Tröger, P., Hentschel, U., Heinze, T.: A Scalable, Self-Adaptive Architecture for Remote Patient Monitoring. presented at the 2010 13th IEEE International Symposium on Object/Component/Service-Oriented Real-Time Distributed Computing Workshops (ISORCW), pp. 204–210 (2010)
[39] Ali, A., Faisal, A., Sorwar, G.: Telemedicine framework for Bangladesh. presented at the TENCON 2010 - 2010 IEEE Region 10 Conference, pp. 335–338 (2010)
[40] Bravo, S., Valero, M.A., Blandino, J.R., Ruiz, M.: A telematics solution for the remote follow-up of malnourished children in underserved areas. presented at the 2013 IEEE Point-of-Care Healthcare Technologies (PHT), pp. 65–68 (2013)
[41] Fong, B., Pecht, M.G.: Prognostics in wireless telecare networks: A perspective on serving the rural Chinese population. presented at the Prognostics and Health Management Conference, PHM 2010, pp. 1–6 (2010)
[42] Kun, L., Hao, W.: Remote medicine in Xinjiang province of China. presented at the 2010 International Conference on Computer Application and System Modeling (ICCASM), pp. V15-623–V15-626 (2010)
[43] Zhu, H., Dong, J.: Wearable ECG terminal and its telemedicine-health services. presented at the 2011 IEEE/SICE International Symposium on System Integration (SII), pp. 133–137 (2011)
[44] Mulvaney, D., Woodward, B., Datta, S., Harvey, P., Vyas, A., Thakker, B., Farooq, O.: Mobile communications for monitoring heart disease and diabetes. presented at the 2010 Annual International Conference of the IEEE Engineering in Medicine and Biology Society (EMBC), pp. 2208–2210 (2010)
[45] Meethal, S.P., Jyotish, J.: A Low Cost Connectivity Solution for Rural Mobile Telemedicine. presented at the 2011 IEEE Global Humanitarian Technology Conference (GHTC), pp. 506–511 (2011)
[46] Sharma, M.K., Vaisla, K.S.: E-health for rural areas of Uttarakhand under e-Governance service delivery model. presented at the 2012 1st International Conference on Recent Advances in Information Technology (RAIT), pp. 622–625 (2012)
[47] Kachieng'a, M.O.: Challenges in managing diffusion of telemedicine technology in South Africa. presented at the 2011 IEEE International Technology Management Conference (ITMC), pp. 121–125 (2011)
[48] Foster, K.: Telehealth in Sub-Saharan Africa: Lessons for Humanitarian Engineering. IEEE Technol. Soc. Mag. 29(1), 42–49 (2010)
[49] Husni, E., Waworundeng, J.: Telemedicine system based on Delay Tolerant Network. presented at the 2011 International Conference on Electrical Engineering and Informatics (ICEEI), pp. 1–5 (2011)
[50] Zambrano, A., Garcia Betances, R.I., Huerta, M.K., De Andrade, M.: Municipal Communications Infrastructure for Rural Telemedicine in a Latin-American Country. IEEE Lat. Am. Trans. 10(2), 1489–1495 (2012)
[51] Kapoor, L., Mishra, S.K., Singh, K.: Telemedicine: experience at SGPGIMS, Lucknow. J. Postgrad. Med. 51(4) (2005)
[52] Misic, J., Misic, V.: Bridging between ieee 802.15.4 and IEEE 802.11b networks for multiparameter healthcare sensing. IEEE J. Sel. Areas Commun. 27(4), 435–449 (2009)

[53] INEGI, Anuario estadistico de los Estados Unidos Mexicanos. Instituto Nacional de Estadistica y Geografia, Mexico (2011)
[54] Coneval, Anexo estadistico pobreza 2012. Consejo Nacional de Evaluacion de la Politica de Desarrollo, Mexico (2012)
[55] Kas, M., Yargicoglu, B., Korpeoglu, I., Karasan, E.: A Survey on Scheduling in IEEE 802.16 Mesh Mode. IEEE Commun. Surv. Tutor. 12(2), 205–221 (2010)
[56] Ergen, M.: Mobile broadband: including Wimax and LTE, 1st edn. Springer, New York (2009)
[57] Tarhini, C., Chahed, T.: On capacity of OFDMA-based IEEE802.16 WiMAX including Adaptive Modulation and Coding (AMC) and inter-cell interference. presented at the 15th IEEE Workshop on Local & Metropolitan Area Networks, LANMAN 2007, pp. 139–144 (2007)
[58] Nuaymi, L., Bouida, N., Lahbil, N., Godlewski, P.: Headers Overhead Estimation, Header Suppression and Header Compression in WiMAX. In: WiMob 2007 Third IEEE International Conference on Wireless and Mobile Computing, Networking and Communications, October 8-10. IEEE, New York (2007)
[59] Stevenson, C., Chouinard, G., Lei, Z., Hu, W., Shellhammer, S., Caldwell, W.: IEEE 802.22: The first cognitive radio wireless regional area network standard. IEEE Commun. Mag. 47(1), 130–138 (2009)

Questions

Q1. Because health specialists tend to concentrate in the main cities. What are some types of patients that may benefit with the deployment of rural telemedicine networks based on mobility issues?

Q2. The IEEE 802.16/WiMAX technology has been leading the deployment of rural telemedicine. What is the major benefit offered by the IEEE 802.22/WRAN technology based on its TVWS operating channels compared to the IEEE 802.16/WiMAX standard?

Q3. In Section 2, it has been established through a coverage radio overlapping that different modulation schemes can be delivered in the same location considering IEEE 802.16/WiMAX or IEEE 802.22/WRAN standards. What are the IEEE 802.22 WRAN modulation schemes offered overall IEEE 802.16/WiMAX standard considering QPSK, 16-QAM and 64-QAM schemes?

Q4. Based on the proposed wireless telemedicine model, what are the telemedicine services with asymmetric traffic profile consider in the bandwidth allocation capacity evaluation?

33. mHealth: WBANs' Issues and Challenges

Saadi Boudjit[1] and Hassine Moungla[2]

[1] 99, Avenue J-B Clément
93430 Villetaneuse, France
University of Paris 13
Sorbonne Paris Cité
boudjit@univ-paris13.fr
[2] 45, Rue des Saints Pères
75006 Paris, France
University of Paris Descartes
Sorbonne Paris Cité
hassine.moungla@parisdescartes.fr

Abstract. Recent advances in wireless networked systems, intelligent low-power sensors and medical sensors, have led to the development and emergence of new embedded networks in the last years known as Wireless Body Area Networks (WBANs). These WBANs carry the promise of expanding the quality of life and care across a large variety of healthcare applications. In this chapter, we will review two fundamental mechanisms of WBANs including data dissemination and sensor deployment. A bi-objective nonlinear non-convex model based on a Min-Max formulation is proposed for deployment issue. On the other hand, a trade-off between energy consumption and the number of hops in the network was proposed for the purpose of data dissemination. The common objective of these two main proposals is saving energy and hence increasing network lifetime.

1 Introduction

Telemedicine integrated wearable health monitoring system is a novel technology with aiming to support early detection of abnormal conditions and prevention of its serious consequences. These systems are based on information technologies and wireless communication for the benefit of those needing medical permanence. Such wireless health monitoring platforms aim to continuously monitor mobile patients needing permanent surveillance. Patients benefit from continuous ambulatory monitoring as a part of a diagnostic procedure, optimal maintenance of a chronic condition or during supervised recovery from an acute event or surgical procedure. However, to set up such platforms several issues along the communication chain should be resolved. The acquisition of medical information, the treatment and use of this information either by a local contractor equipment or offset after transfer in 3G and/or WiFi connection to a data server, confidentiality

S. Adibi (ed.), *Mobile Health*,
Springer Series in Bio-/Neuroinformatics 5, DOI: 10.1007/978-3-319-12817-7_33

and data security, QoS over heterogeneous wireless networks, ...etc. are some of the important challenges that should be considered.

The corner stone of such systems is the wireless body area network where patient' medical information are acquired via a set of wireless sensors embedded in the patient himself (Figure1.a). The devices used in WBANs have limited energy resources available and consequently the computational power and available memory of such devices are limited. This poses a number of challenges on the design of communication protocols and deployment patterns for WBANs. In this chapter, we discuss the problem of optimal power consumption in wireless body area networks taking into account two main constraints: optimal routing protocols and optimal sensors deployment.

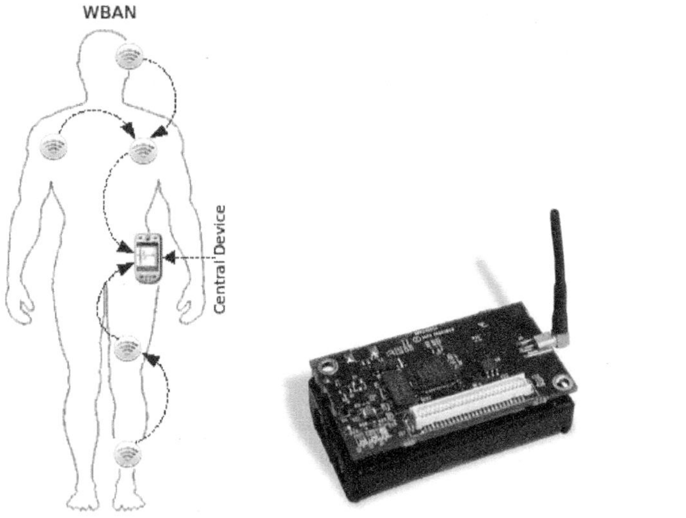

(a) Wireless Body Area Network (b) Mica2 sensor mote MPR2400CA

Fig. 1 A WBAN Platfom

Several studies have shown that multi-hop routing in WBANs result in non-negligible lifetime increase of sensors as compared to direct communication between sensors and the central device [3]. In this case, information dissemination between sensors and the central device requires an ad hoc routing protocol. Some other studies, however, argue contrarily. They show that direct communication between sensors and the central device considerably increase the lifetime of the wireless body area network [4]. They show that executing a multi-hop adhoc routing protocol on embedded sensors consumes more energy. In this chapter we are exploring both information dissemination techniques in a WBAN scenario. Energy consumption for various scenarios is evaluated through simulations. In addition, in order to have an insight view of the energy consumption in various working modes, the power consumption values of Mica2 sensor motes from

Crossbow (Figure 1.b) have been used. Our objective here is to find a trade-off between the number of hops in the network and the energy consumption.

Moreover, the effectiveness of Wireless Body Sensor Networks(WBSNs) is often determined to a large extent by the sensing coverage provided by the actual node deployment. The locations of the sensor nodes strongly affect energy consumption, operational lifetime and sensing coverage [11][12][13]. Maximizing lifetime, sensing coverage and other constraints are conflicting objectives and thus warrant a trade-off [14]. Thereby, the problem should be formulated as a Multi-objective Optimization Problem (MOP); since all objectives are considered and there is not a single solution to optimize them at the same time (the set of Pareto optimal or non-dominated solutions is usually called Pareto Front (PF) [15]). In this chapter, we focus on two important objectives that have to be taken into account in sensor deployment problem. The energy consumption is minimized using a MinMax function and the coverage area is maximized by minimizing the distance between sensors. This latter objective, deals with maintaining connectivity between sensor nodes, in the case of multi-hop communication. We propose a multi-objective optimization model and solve it using the open source software Couenne.

The remainder of this chapter is as follows: in section 2, we introduce data dissemination problem in WBANs and discuss two main directions suggested in literature. Section 3 presents our optimal deployment pattern exploration and a mathematical formulation using non-convex optimization. Examples and experimental comparative studies will be discussed in each section. Section 4 concludes the chapter by presenting the future trends of our approaches.

2 Data Dissemination in WBANs

2.1 Problem Analyses

Power consumption for transceivers is different in different communication states i.e. transmission/reception/idle/sleep. Transmission energy depends upon the power with which signal is propagated to attain longer ranges while consumption in other states is less variable. Unlike other technologies, reception/idle state consumption for low power, limited range WSN is not negligible as compared to transmission state power consumption. Thus global energy minimization requires optimum selection of transmission range. Latré et. al. in [3] showed that utilizing multi-hop communication by reducing transmission power in WBAN reduces overall energy consumption. This is true but authors in [3] ignored increased energy consumption due to multiple receptions. Wang et. al. in [4] proposed a realistic power consumption model for WSN. Authors in [4] showed that multi-hop communication is more energy efficient when destination is out of reach. That is when destination cannot be reached in single hop. They show that multi-hop communication by controlling transmission power does not necessarily result in energy gain. This is due to non-negligible energy consumptions in reception and idle modes.

In sensor motes energy consumption in reception and idle mode is relatively high. It can be equal to or greater than transmission energy consumption for low power transmissions. For low transmission ranges as in WBAN, utilizing single hop data delivery to gateway node while other sensors are put in idle/sleep state might be more energy efficient. Real experiments conducted in [5] showed that reception and idle listening consume a considerable amount of energy. Especially for sensor nodes, very low power transmissions for ranges as short as in WBAN consume lesser power than reception. For micaZ motes [6], reception energy is higher than transmission energy even with maximum power transmission. Above stated variations in literature, motivated us to better understand this energy consumption scenario and choose a better communication strategy for our future test bed implementation for WBANs.

2.2 Channel Model and Energy Consumption

Let P_t and P_r be the transmission and reception signal power respectively, where P_r is equal to receive sensitivity of mica2 node (-98 dBm). Let d be the communicating nodes' inter-node distance and, L the system loss, then our WSN model can be represented by the well known *TwoRayGround* radio model as depicted in NS2 simulator [7]:

$$P_r = ReceiveSensitivity = P_t\, G_t\, G_r h_t h_r / d^4 L \quad (1)$$

Where $G_t = G_r = 1.2$ are antenna transmission and reception gains respectively. $h_t = h_r = 16$cm are transmission and receptor's antenna heights. Transmission power P_t is varied according to desired range. Equation (1) can be rearranged to determine d for given values of P_t as follows:

$$d = [P_t\, G_t\, G_r h_t h_r / P_r L]^{1/4} \quad (2)$$

We need minimum P_t that ensures successful reception of packet at destination with $P_r >$ *ReceiveSensitivity*. For given values of P_t, approximate range values are obtained from radio model given in equation (2). The actual power consumed by mica2 motes while transmitting with permissible power P_t is obtained from CC1000 data sheet [8]. The range results for given P_t in *Table1* conform to the mica2 data sheet and are experimentally obtained range values.

Table 1 Mica2 Mote Power Consumption and Range Values

T_x power (dBm)	Power consumed (mW)	R_x power / Idle power (mW)	T_x range (meter)
05	76.2	30	60
-07	32.4	30	30
-14	27.9	30	20
-20	25.8	30	10

Energy consumption of sensor nodes in various states can be obtained by the following equations:

$$E_{tx} = P_{tr} \text{ x } T_{tx}$$
$$E_{rx} = P_{rec} \text{ x } T_{rx}$$
$$E_{idle} = P_{idle} \text{ x } T_{idle}$$

Where P_{tr}, P_{rec}, P_{idle} are the powers consumed by the Mica2 mote's CC1000 transceiver in transmission, reception and idle mode respectively, and T_{tx}, T_{rx}, T_{idle} are times spent in each mode. Time for transmitting a packet of size *b* bits is equal to *b*/*R* where *R* is the data rate. Total energy consumed by the network is given by:

$$E_{total} = E_{tr} + E_{rec} + E_{idle}$$

2.3 Simulations

Network Simulator 2 (NS2) is used for performing simulations. We performed simulations utilizing the actual power consumption values of sensor motes in various states. In order to have an insight view of the energy consumption in various working modes, the power consumption values of Mica2 [9] sensor motes have been used. Mica2 has an on-board CC1000 transceiver for communication. Thus power consumption values for communication have been taken from CC1000 data sheet [8]. Deployed topology consists of a set of seven equidistant nodes $\{n_s, n_2, n_3, n_4, n_5, n_6, n_d\}$ deployed linearly with adjacent inter-node distance of 10 meters. This accounts to maximal source-destination ($n_s \rightarrow n_d$) distance of 60 meters. Source n_s generates packets at regular intervals and transmits them towards sink n_d. Simulations are performed with four different relaying scenarios as detailed in table 2 and illustrated in figure 2.

Table 2 Simulated scenarios

Scenario	Hops	Distance	Nodes
1	1	60	$n_s \rightarrow n_d$
2	2	30	$n_s \rightarrow n_4 \rightarrow n_d$
3	3	20	$n_s \rightarrow n_3 \rightarrow n_5 \rightarrow n_d$
4	6	10	$n_s \rightarrow n_2 \rightarrow n_3 \rightarrow n_4 \rightarrow n_5 \rightarrow n_6 \rightarrow n_d$

$n_s \rightarrow n_d$ transmission is varied from single hop to a maximum of six hops. In all the scenarios source node n_s sends 50 bytes packets towards desination node n_d. MAC layer issues like collisions, retransmissions and scheduling are not taken into account. This is realistic assumption as the goal here is to analyze only the energy consumption effect with varying hop distances.

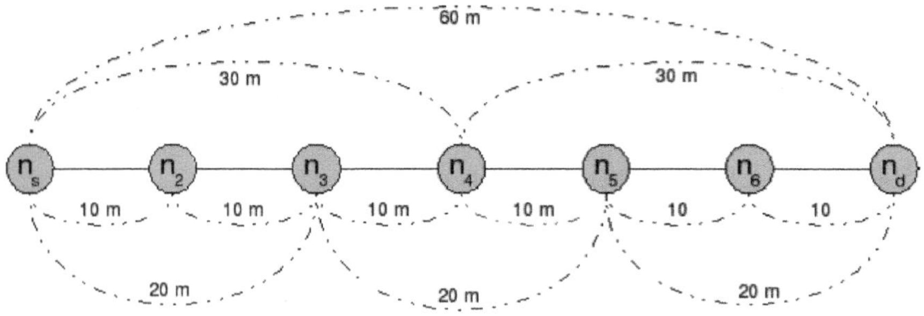

Fig. 2 Simulation scenarios

Figure 3 shows total network energy consumption when radio is always active i.e. either in Tx/Rx state or in idle state. In this case energy consumption reduces a bit with increasing number of hops.

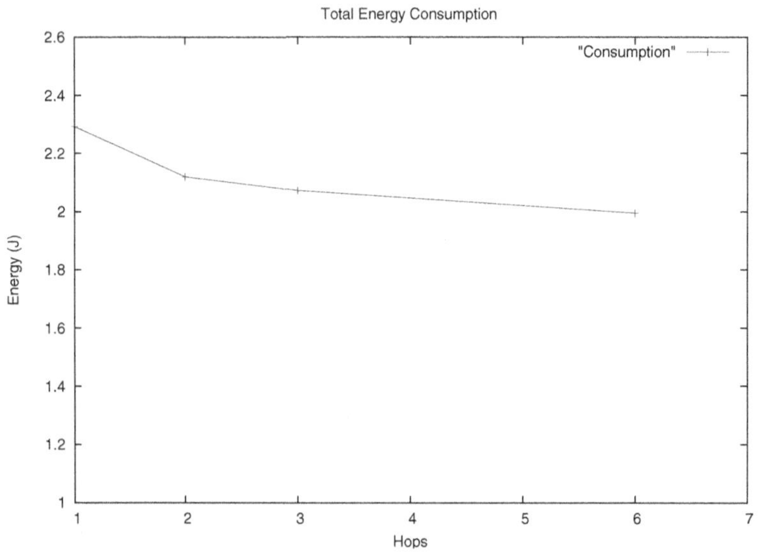

Fig. 3 Total energy conusmption when radio always ON/IDLE

Note that for this case $P_{rec} = P_{idle} = 30$mW. This is because current draw through mica2 is same when it is either receiving a packet or performing idle listening. Power transmission values P_{tr} for each case are obtained from table 1. Energy consumption for direct $n_s \rightarrow n_d$ communication is 2.29J. For two hops $n_s \rightarrow n_4 \rightarrow n_d$ and three hops $n_s \rightarrow n_3 \rightarrow n_5 \rightarrow n_d$ scenarios total energy consumption slightly reduces to 2.12J and 2.07J respectively. Results are not very appealing. With least transmission power of -20dBm, that draws 25.8mW from mica2, minimum total consumption of 1.99J is acheived.This accounts to maximum energy saving of 13% as compared to direct $n_s \rightarrow n_d$ communication. At -20dBm, power consumed $P_{tr} < P_{rec}$ (25.8 < 30mW). As compared to this, multiple WBAN sensors are very close to each other (tens of centimeters) and to the gateway node e.g. PDA. In such scenario further reduction in transmission power will increase multiple receptions/Listening dominancy over transmission energy. Thus multi-hop option is not practical in small WBAN. Heinzelman et. al. in [10] propose a WSN based energy model. They also conclude that multi-hopping is energy efficient when destination cannot be reached in a single hop. A contrary argument to this point could be that transmission range reduction is neccessary to avoid collisions and large number of overhearings. This is true, but in small scale WBANs, source to sink synchronized single hop communications while keeping other nodes' transceivers off would be more optimal.

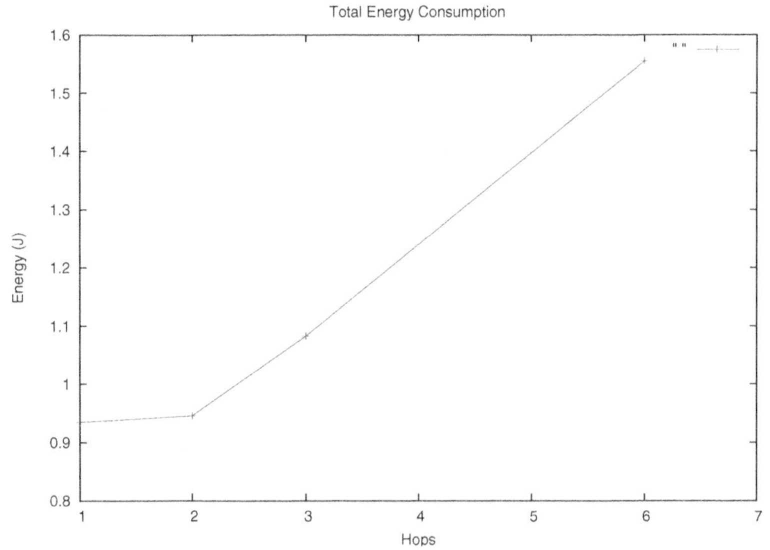

Fig. 4 Total energy conusmption when radio ON only for transmission

Figure 4 shows network energy consumption when nodes switch on their radio only when they need to transmit their information. Energy consumption for direct $n_s \rightarrow n_d$ communication is 0.93J. For two hops $n_s \rightarrow n_4 \rightarrow n_d$ and three hops $n_s \rightarrow n_3 \rightarrow n_5 \rightarrow n_d$ scenarios total energy consumption increases to 0.946J and 1.083J respectively. With least transmission power of -20dBm, that draws 25.8mW from mica2, minimum total consumption of 1.55J is acheived. This accounts to 40% increase in network's total energy consumption as compared to direct $n_s \rightarrow n_d$. Clearly, reducing transmission range to perform data relaying does not seem to be good option in WBAN. Energy efficiency gain in WSN is maximum when sensor nodes periodically go to deep sleep mode. We are not considering this case here as turning off health sensing equipment may prove fatal to patient's life. Patient's health needs to be regularly monitered and transmitted to the concerned data center. Although the periodicity of monitering depends upon the nature of observation and patients condition. It may not be necessary to transfer health update to data center. This could be utilized only under emergency condition to trigger the call for medical assistance. From this discussion, it is concluded that, unless really required, unneccessary relaying should be avoided. Some real experiments show high path atenuation values in WBAN with the atenuation factor α as high as 5.8. If such condition arrives the deployed routing mechanism should be able to adapt itself for multi-hop communication.

2.4 *Discussion*

From the aforementioned discussion, it is concluded that unless really required, unnecessary multi-hop relaying in case of WBANs should be avoided. However, some real experiments show high path attenuation values in WBAN with the attenuation factor α as high as 5.8. This may happen, for instance, if a patient is sitting on a chair, and has one sensor installed on his back and the central device is in his front pocket. In this case, the central device may not be able to receive a packet transmitted by the sensor as the signal gets absorbed by the human body and the chair on his back. If such a condition arrives, the data dissemination mechanism in use should be able to transmit data to the personal coordinator via another health sensor installed on the body; the sensor on the back could transmit data to the central device via another sensor installed on for example patient's shoulder as this would allow nodes to communicate with the central device over two hops, when required. The assumption of using maximum two hops seems logically correct as this option will be used in only special scenarios described above. In addition, our assumption of considering two hops has also been enforced by the IEEE 802.15.6 WPAN Task Group 6 [20] which recently recommended to use two hops in case of WBANs.

3 Sensors Deployment in WBANs

WBAN nodes are mainly battery operated and have limited energy and processing capabilities. Unlike the traditional networks, WSNs differs in several aspects. In order for large deployments to be cost-effective, nodes are resource-constrained in terms of energy capacity, radio transmissions and processing capabilities.

Depending on the application field, the reliable data transfer and energy efficiency play a very important role in WSN research. Many protocols have been proposed to provide energy efficiency and reliable data transfer of packets in WSNs. These protocols are broadly classified as (a) Protocols that provide Reliability and (b) Protocols that provide Energy Efficiency.

In these mission-critical applications, a set of parameters requirement on network performance and constraints should be satisfied. This poses a number of challenges on the design and analysis of WBANs. Wireless body sensors nodes have a severe limitation in their storage, radio communication, capabilities, bandwidth, and energy, the most important design consideration for WBAN is to extend their operational lifetime by minimizing power consumption. This is the reason why, protocols with low energy consumption has been an important research orientation in this field. Maximizing lifetime and other constraints are conflicting objectives and thus warrant a trade-off. Thereby, the problem should be formulated as a Multi-objective Optimization Problem (MOP).

The effectiveness of WBANs is often determined to a large extent by the sensing coverage provided by the actual node deployment. The locations of the sensor nodes strongly affect energy consumption, operational lifetime and sensing coverage [11][12][13]. Maximizing lifetime, sensing coverage and other constraints are conflicting objectives and thus warrant a trade-off [14]. Thereby, the problem should be formulated as a Multi-objective Optimization Problem (MOP); since all objectives are considered and there is not a single solution to optimize them at the same time (the set of Pareto optimal or non-dominated solutions is usually called Pareto Front (PF) [15]).

In this chapter, we focus on two important objectives that have to be taken into account in sensor deployment problem. The energy consumption is minimized using a MinMax function and the coverage area is maximized by minimizing the distance between sensors. This latter objective, deals with maintaining connectivity between sensor nodes, in the case of multi-hop communication. We propose a multi-objective optimization model and solve it using the open source software Couenne.

3.1 *Preliminary*

In this section, we present the preliminaries for our optimal deployment pattern exploration. When considering wireless sensor transmission around, and on the body for healthcare applications, one of important issues are radiation absorption and heating effects on the human body.

Like cells phone radio frequency (RF) heating of body tissues and possible others alteration (mutation) happens to 100% of the people exposed to RF [16]. The amount of heating is determined by a combination of the four factors below:

–Frequency - Certain frequencies are absorbed in the body more than others.

–Duration - How long you are exposed to the radiation, or how long the transmitter is "on."

–Distance- How close you are to the antenna. Energy levels decrease with the square of the distance.

–Power Level- What the strength of the signal is. This is measured in microwatts, milliwatts and watts.

Modern medical science knows there are limits to how much of an assault on the body. The immune system can deal with to reduce tissue heating. The radio's transmission power can be limited or traffic control algorithms can be used.

In [16] rate control is used to reduce the bioeffects. Adapt a transmission and respectively reception range of all the nodes is taking into account in order to reduce tissue heating the radio's transmission power, there are no clear studies on the amount of energy, frequency are absorbed in the body to reduce the bioeffects.

We mean, that electric and magnetic fields, often referred to as electromagnetic fields, occur naturally and as a result of the power generation, power Transmission, power distribution and use of electric power. This energy increases significantly and proportionally when distance between sensors increase. High voltages cause electrostatic effects, where as short circuit currents and line loading currents are responsible for electromagnetic effects. The effect of these electrostatic fields is seen prominent with living things like humans, plants, animals along with vehicles.

3.2 System Model

Sensor deployment aims to find the best location of a set of sensors in a given region, taking into account several constraints. We consider in this chapter n sensors to deploy on a square region network of length A (figure 5). Note that we can equivalently consider rectangular or circle regions. We consider a two dimensional continuous space, each sensor $i \in \{1,\ldots,n\}$ is located at $X_i = (x_i, y_i)$, where x_i refer to the x-coordinate of the sensor location and y_i refer to the y-coordinate, respectively. The function of each sensor i is to collect data from a region, defined as a disc of center X_i and radius R_{cov_i} (called radius coverage). These data are sent to the sink of coordinates $sink = (x_{sink}, y_{sink})$ which receives informations from all sensors and treat them. The sink can detect and communicate with sensor i if the sum of the communication radio range of the sink $R_{com_{sink}}$ and the communication radio range of node i R_{com_i} is lower than a given upper bound $R = R_{com_i} + R_{com_{sink}}$. We use the Euclidean distance $\|X_i - X_j\| = \sqrt{(x_i - x_j)^2 + (y_i - y_j)^2}$ to express the distance between i and j, $i, j \in \{1,...,n\} \cup \{sink\}$. Under these assumptions, we formulate the problem of finding an optimal sensors deployment by minimizing the energy consumption and by maximizing the region coverage. We consider a *many-to-many* configuration in which each sensor can communicate with either another sensor or the sink node.

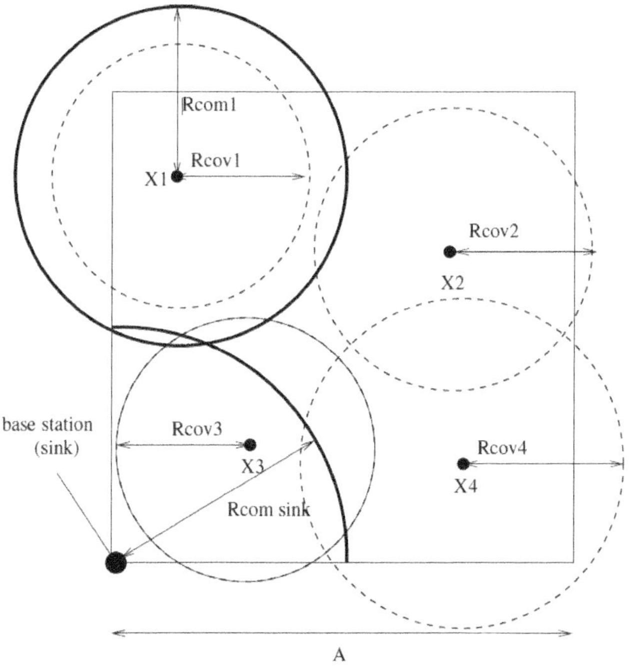

Fig. 5 An example of sensors deployment system with four sensors

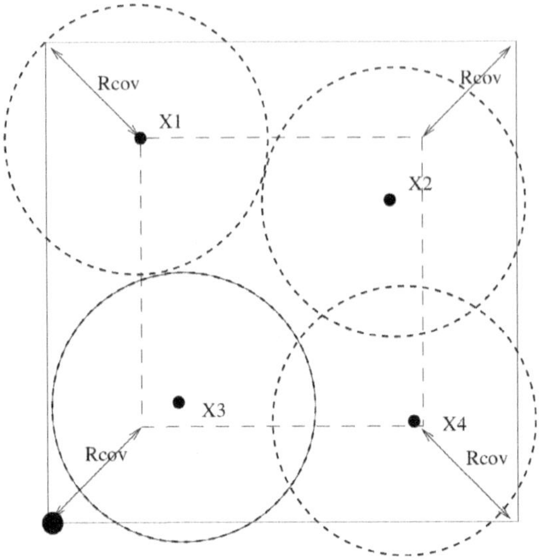

Fig. 6 Feasible region

Fig. 7 Degree of Irregularity : DOI=0, DOI=0.003 and DOI=0.01 respectively

3.3 WBAN Radio Model Parameters

The human body has a severe influence on radio wave propagation, and so it affects the connectivity and the topology of the WBAN. Thus it is very important to take the body effect on propagation loss and link quality in various postures into consideration. As the effect of the body strongly depends on the relative position of nodes and the body situation, it is very useful to include that in the WBAN deployement model. Based on the radio propagation model which is used for network simulation [17], Radio Irregularity Model (RIM) proposed here is an extension to isotropic radio models. It enhances isotropic radio models by approximating three main properties of radio signals: non-isotropic, continuous variation and heterogeneity. The RIM model is motivated by a simple DOI (Degree of Irregularity) model briefly mentioned in the localization work [17]. The DOI is used to denote the irregularity of the radio pattern. Provided that, a path-loss model is used, for instance, we specify for every pair of nodes (i, j) the link loss coefficient.

3.4 Objectives of Many-to-Many Model

We consider a many-to-many model, where each sensor can communicate with another one. Our objectives aim to minimize the the consumed ernegy, maximize the covered space and the number of connected sensors.

1. ***Minimize the Transmission Energy***
   Energy optimization is a significant component of WBANs. It is well known that energy increases linearly with transmission distance [18]. Minimizing the transmission distance also has to be performed to avoid a strong radiation caused by long emission. We consider here a MinMax formulation which consists of minimizing the maximum distance between the sinkand all the sensors. This objective is given as follows:

   $$\min \max_{i=1}^{n} \|X_i - X_{sink}\|$$

   This objective can be formulated as follows:

   $$\min \quad Z$$
   $$Z^2 \geq \|X_i - X_{sink}\|^2, \quad \forall i \in \{1,\ldots,n\}$$

2. ***Maximize the Covered Space***
   It is important in a deployment strategy to maximize the coverage area of WSN. For this purpose we choose to maximize the distance between each pair of sensors. This can be given by $n \times (n-1)$ objectives:

   $$\max \|X_i, X_j\|, \forall i, j \in \{1,\ldots,n\}, i \neq j.$$

   These objectives can be formulated as follows:

   $$\max D, \quad \|X_i - X_j\|^2 \geq D^2, \quad \forall i, j \in \{1,\ldots,n\}, i \neq j.$$

   Note that this objective function permits also the minimization of overlapping regions. In addition to the coverage objective, the number of connected sensors has to be maximized.

   Lets C_{ij} be a binary variable equal to 1 if sensor i can communicate with sensor j and 0 otherwise. The objective function is:

   $$\max \sum_{ij} C_{ij}$$

3. ***Feasible Deployment Region***
   A feasible region consists of all points in the space where a sensor can be deployed. In figure 5 for example, X_i is a feasible location of sensor i if and only if $x_{sink} \leq x_i \leq (x_{sink} + A)$ and $y_{sink} \leq y_i \leq (y_{sink} + A)$.

Where A is the width of the given square area. When a sensor is located on the border of the square region (for example $X=(x_{sink}+A,y_{sink}+A)$ or $x=(0,y_{sink}+A)$), the sensor can cover only a half of its capacity coverage the feasible region. To avoid these bad positions, we consider another feasible region called the restricted feasible region. For sake of simplicity, we consider that $R_{cov_i}=R_{cov},\forall i\in\{1,...,n\}$. The restricted feasible region satisfy the following constraints (see figure 6):

$$\forall i\in\{1,...,n\},\ \left\{\begin{array}{l} x_{sink}+R_{cov}\le x_i\le x_{sink}+A-R_{cov} \\ y_{sink}+R_{cov}\le y_i\le y_{sink}+A-R_{cov} \end{array}\right\}$$

4. ***Connectivity Constraints***

Taking into account the link loss coefficient between two sensors, i and j, are able to communicate with each other when the Euclidean distance between them is less than or equal to the sum of their communication radius, denoted by $R_{com_{i,j}}=R_{com_i}+R_{com_j}$. This can be formulated by the following constraint:

$$C_{ij}\left[\left\|X_i-X_j\right\|^2-R^2_{com_{i,j}}\right]\ge 0,\quad \forall i,j\in\{1,...,n\},i\ne j.$$

When $\left\|X_i-X_j\right\|^2-R_{com_{i,j}}\ge 0$, the variable C_{ij} takes the value 1 (see the objective function) and when $\left\|X_i-X_j\right\|^2-R_{com_{i,j}}<0$, the variable C_{ij} takes the value 0.

5. ***The Optimization Problem***

We obtain the following bi-objective nonlinear non-convex programming problem:

$$\left\{\begin{array}{l} \max \beta D+\gamma\sum_{ij} C_{ij} \\ \left\|X_i-X_j\right\|^2\ge D^2,\quad \forall i,j\in\{1,...,n\},i\ne j \\ C_{ij}\left[\left\|X_i-X_j\right\|^2-R^2_{com_{i,j}}\right]\ge 0,\quad \forall i,j\in\{1,...,n\},i\ne j \\ x_{sink}+R_{cov_i}\le x_i\le x_{sink}+A-R_{cov_i},\forall i=1,...,n \\ y_{sink}+R_{cov_i}\le y_i\le y_{sink}+A-R_{cov_i},\forall i=1,...,n \\ C_{ij}\in\{0,1\},\quad \forall i,j\in\{1,...,n\},i\ne j. \end{array}\right.$$

and we consider the single objective version by aggregating the two objectives functions as follows: $\max \beta D + \gamma \sum_{ij} C_{ij}$ with β and γ suitably choosed such as $\beta + \gamma = 1$.

We consider also the feasible deployment region constraints (section III-D-3).

3.5 Simulations

As our models are nonlinear non-convex programs, we solve them using Couenne software (Convex Over and Under ENvelopes for Nonlinear Estimation).

Couenne (http://www.coin-or.org/Couenne/) is an Open Source branch&bound algorithm for solving Mixed-Integer Nonlinear Programming (MINLP) problems. It implements linearization, bound reduction, and branching methods within a Branch&Bound framework.

This is particularly required for the the many-to-many model. We Consider in this section a test instance with $n = 13$, $A = 50\,cm$ $A = 50$, $R_{cov_i} = 7, \forall i \in \{1,...,13\}$, $R_{com} = 50\sqrt{2}\,cm$ and $R_{com_{i,j}} = 35cm, \forall i, j \in \{1,...,13\}$.

Our objective is the deployment of sensors using the considered models.

Authors in [17] explored the RIM model's applicability across platforms and radios, they repeat the experiments with MICAZ [CROSSBOW] motes, each of which has a whip antenna of 1 /4 radio wavelength. As Figure 7 illustrates, the RIM model applies in MICAZ devices dtermine that the DOI model assumes an upper and lower bound on signal propagation. Beyond the upper bound, all nodes are out of communication range; and within the lower bound, every node is guaranteed to be within the communication range. The Degree of Irregularity and consequently the variation of the link loss coefficient used in our experimentations was between 0.001 and 0.003.

Figures 11,and 12 present the sensors deployment on the considered area based on different weights β and γ.

Figures 8 and 9 present the sensors deployment on the considered area for different weights β and γ.

The distribution obtained with parameters $(\beta, \gamma) = (0.8, 0.2)$ ensure a good connectivity (the number of possible connexions between sensors is $\sum_{ij} C_{ij} = 78$). With parameters $(\beta, \gamma) = (0.9, 0.1)$ the coverage is better, but with less connectivity ($\sum_{ij} C_{ij} = 76$).

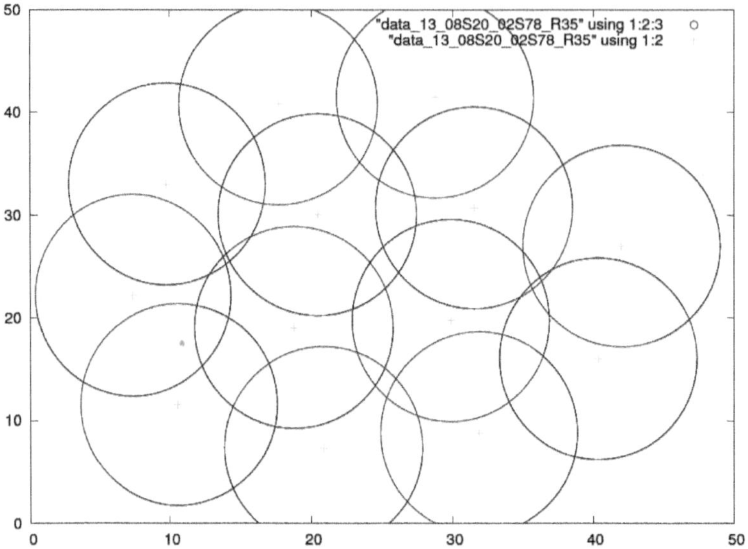

Fig. 8 $(\beta, \gamma) = (0.8, 0.2)$, $D = 11.10$, $\sum Cij = 78$

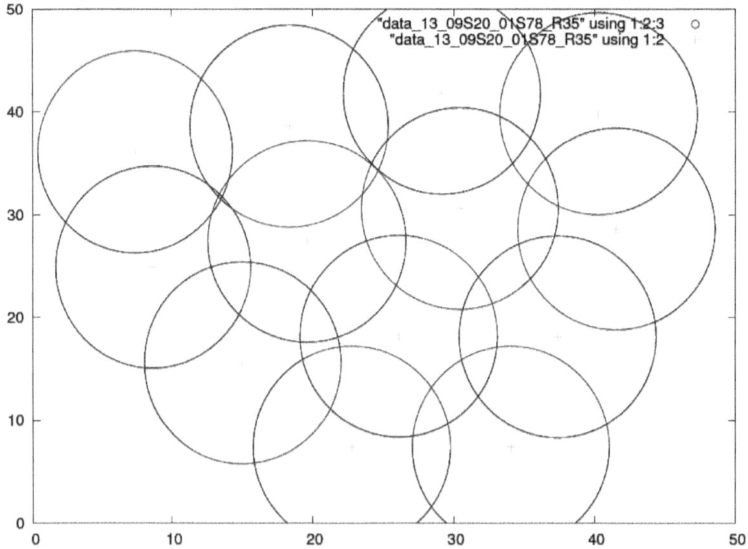

Fig. 9 $(\beta, \gamma) = (0.9, 0.1)$, $D = 11.28$, $\sum Cij = 76$

We observe that the distribution with the lower maximum distance between sensors (with lower energy consumption) is obtained by varying the two ponderation. Some distributions ensure worst coverage. The optimal coverage was obtained with a greater energy consumption.

The best compromise between energy consumption and coverage should be configured based applications types needs.

3.6 Evaluation

In order to evaluate our solution's performance, we implemented and compared maximum area coverage and network connectivity with the approximative strategy deployment (which mean that sensors will be deployed with approximative precision in the interested area).

We compared the proposed solutions to the related approximative deployment strategies. The maximum number of sensors that could be deployed was set to fourteen, as obtained with the proposed method. We can notice that our method proposal ensures both connectivity and maximum coverage.

We evaluate also the accuracy of our computations for the relay-sensor ratio for many-to-many configuration. Figure 10 compares the practical relay-sensor ratio with the theoretical results. The relay-sensor ratio should be not sensitive to the network density which is shown in figure 13. Considering the effect from remedy relays in practice, the practical relay-sensor ratio is larger than the theoretical one but when the density is increased to 0.9 they will quickly converge. Another important observation is even when the network density is as low as 0.3, the Wireless Body relay-sensor ratio is just around 0.04 and from more than 20 sensors just a few relays are needed to forward the packets.

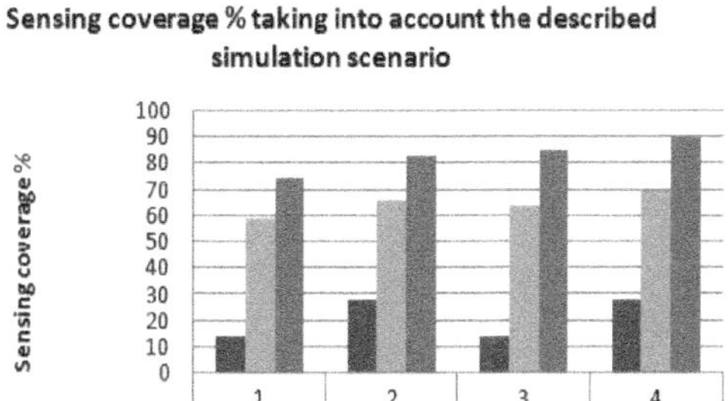

	1	2	3	4
Deployed Sensor node number	14	28	14	28
Approximative deployment %	59	66	64	70
Optimal deployment %	75	83	85	90

Fig. 10 WBAN relay numbers vs. sensor numbers

We compared the proposed solutions to the related approximative deployment strategies in many-to-many configuration using OPNET simultor. The maximum number of sensors that could be deployed was set to fourteen, as obtained with the proposed method. We can notice that our method proposal ensures both connectivity and maximum coverage. To compare the energy consumption (lifetime).

Note that in existing literature, we found many different definitions of WBAN lifetime. Most of the research considered the network to be dead when the first sensor fails. However, in our work, we considered the network dead when no sensors were able to communicate with the sink.

Our proposed deployments already leads to strong energy savings of 23% respectively for both configurations many-to-one and for many-to-many and to slightly better sensing coverage as we can see in the figure11.

We observe that our optimal deployment models clearly outperform the approximative model. Significant improvements of the sensing coverage can be achieved if the other objective will be relaxed. According to the objective that we would to perform we could meet to adjust the weights β and γ to improve the sensing coverage or reduce the energy consumption or/and tissue heating.

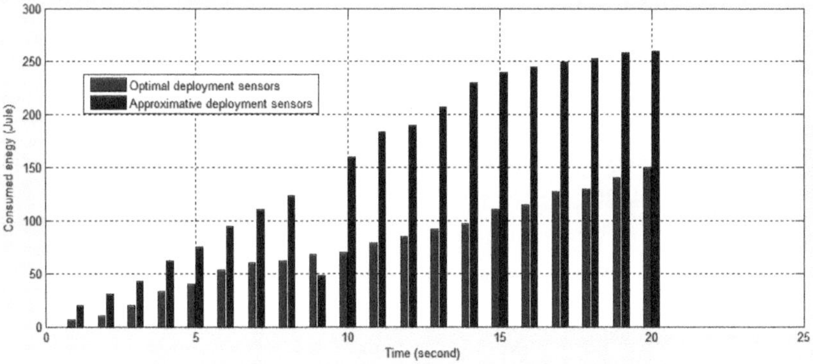

Fig. 11 Many-to-Many WBAN energy consumption comparison between an approximative and the optimal sensor deployment

4 Conclusion

We are addressing the issues regarding minimum energy consumption, reliable routing , maximum area coverage and optimal deployment in a WBAN network. A bi-objective nonlinear non-convex model based on a Min-Max formulation is proposed for deployment issue. We also integrate the effect of radiation in wireless medium, and try to devise many-to-many model that minimizes the bioeffects throughout the body and reduces tissue heating. Topologies generated from our proposed model exhibit minimum energy consumption at maximum sensing coverage. Our optimal deployment permit optimal data dissemination solution which maximizes the network lifetime, by minimizing the maximum energy consumption of sensor nodes in the network and maximizing the sensors connectivity. The obtained results show that the models propose good solutions, with suitable objective weights. Finally, we highlight the relationship between an optimal deployment and the connectivity quality. In fact, we show in our experimentations that an optimal deployment permits the network life time improvement.

On the other hand, an energy consumption comparison for various communication scenarios has been made through simulation. It has been shown that at very low ranges, transceivers consume almost equal or more power on receiving packets than on transmitting them. Thus, deliberate reduction of transmission range to induce multi-hop scenario is not energy efficient. Though this is device dependent but general characteristics of very low power transceivers seem to show the same results.

Our future work is to propose an energy efficient, reliable routing architecture keeping in view the results obtained through this study. The objective is to adopt a simple routing protocol that uses a smart neighbor discovery mechanism. This could be achieved by a simple modification of the 2-hop neighbor discovery algorithm of OLSR [19] and would allow nodes to communicate with sinks over two hops if required.

References

[1] Patel, M., Wang, J.: Applications, challenges and prospective in emerging body area networking technologies. IEEE Wireless Communications 17(1), 80–88 (2010)

[2] Ullah, S., et al.: A comprehensive survey of Wireless Body Area Networks. Journal Med. Syst. (published online, August 2010)

[3] Latre, B., Moerman, I., Dhoedt, B., Demeester, P.: Networking in Wireless Body Area Networks. In: 5th FTWPHD Symposium (2004)

[4] Wang, Q., Hempstead, M., Yang, W.: A Realistic Power Consumption Model for Wireless Sensor Network Devices. In: IEEE Comm. Society Conference on Sensor, Mesh and Ad hoc Communications and Networks (2006)

[5] Anastasi, G., et al.: Performance Measurements of Motes Sensor Networks. In: International Workshop on Modeling Analysis and Simulation of Wireless and Mobile Systems (2004)

[6] Chipcon AS. CC2420 2.4 GHz IEEE 802.15.4/ZigBee-ready RF Transceiver (2007), http://inst.eecs.berkeley.edu/cs150/ (retrieved October 2009)

[7] Network Simulator 2 (n.d.), http://www.isi.edu/nsnam/ns/

[8] Chipcon AS. CC1000 Single Chip Very Low Power RF Transceiver (2007), http://focus.ti.com/lit/ds/symlink/cc1000.pdf (retrieved October 2009)

[9] Crossbow Technology MICA2 Motes, http://www.xbow.com (retrieved June 2009)

[10] Heinzelman, W.R., Chandrakasan, A., Balakrishnan, H.: Energy-efficient communication protocol for wireless microsensor networks. In: Proceedings of the 33rd Hawaii International Conference on System Sciences (2000)

[11] Yang, S., Li, M., Wu, J.: Scan-Based Movement-Assisted Sensor Deployment Methods in Wireless Sensor Networks. IEEE Trans. on Parallel and Distributed Systems 18(8), 1108–1121 (2007)
[12] Senouci, M.R., Abdelhamid, M., Assnoune, K.: Localized Movement-Assisted Sensor Deployment Algorithm for Hole Detection and Healing. IEEE Transactions on Parallel and Distributed Systems (May 29, 2013)
[13] Oldewurtel, F., Riihijarvi, J., Mahonen, P.: Impact of Correlation in Node Locations on the Performance of Distributed Compression. In: Proceedings of the WONS, Snowbird, USA, pp. 135–142 (2009)
[14] Özdemir, S., Attea, B.A., Khalil, Ö.A.: Multi-Objective Evolutionary Algorithm Based on Decomposition for Energy Efficient Coverage in Wireless Sensor Networks. Wireless Personal Communications 71(1), 195–215 (2013)
[15] Deb, K.: Multi-Objective Optimization Using Evolutionary Algorithms. Wiley and Sons (2002) ISBN: 978-0-471-87339-6
[16] Ren, H., Meng, M.Q.H.: Rate control to reduce bioeffects in wireless biomedical sensor networks. In: 3rd Annual International Conference on Mobile and Ubiquitous Systems - Workshops, San Jose, CA, pp. 1–7 (July 2006)
[17] Zhou, G., He, T., Krishnamurthy, S., Stankovic, J.A.: Models and solutions for radio irregularity in wireless sensor networks. ACM Trans. Sen. Netw. 2(2), 221–262 (2006)
[18] Ramos, H.S., Zhang, T., Liu, J., Priyantha, B., Kansal, A.: LEAP: A Low Energy Assisted GPS for Trajectory-Based Services. In: 13th ACM International Conference on Ubiquitous Computing (UbiComp), September 17. ACM (2011)
[19] Jacquet, P., Muhlethaler, P., Minet, P., Qayyum, A., Laouiti, A., Clausen, T., Viennot, L., Adjih, C.: Optimized Link State Routing Protocol - Internet Engineering Task Force – Manet Group, RFC 3626 (2003)
[20] IEEE 802.15.6WPAN Task Group 6 (TG6) – Body Area Networks, http://www.ieee802.org/15/pub/TG6.html

34. Motion Capture: From Radio Signals to Inertial Signals

Matteo Giuberti[1,*] and Gianluigi Ferrari[2]

[1] Xsens Technologies B.V.
Enschede, The Netherlands
matteo.giuberti@xsens.com
[2] Department of Information Engineering
University of Parma
Parma, Italy
gianluigi.ferrari@unipr.it

Abstract. The study of the motion of individuals allows to gather relevant information on a person status, to be used in several fields (e.g., medical, sport, and entertainment). Over the past decade, the research activity in motion capture has benefited from the progress of portable and mobile sensors, paving the way toward the use of motion capture techniques in mHealth applications (e.g., remote monitoring of patients, and telerehabilitation). Indeed, even if the optical motion capture, which typically relies on a set of fixed cameras and body-worn reflecting markers, is generally perceived as the standard reference approach, other motion capture techniques, such as radio and inertial, are attracting an increasing attention because of their suitability in remote mHealth applications.

Moreover, several hybrid approaches have been studied and proposed in order to overcome the limitations of component technologies considered independently. In this chapter, we present an overview of possible integration strategies between radio and inertial motion capture techniques. We start by investigating a radio-based approach, based on the fingerprinting radio localization technique. Then, the previous approach is improved by integrating inertial measurements: namely, accelerometers are used to provide an estimate of the nodes' pitches. Finally, the radio signals are abandoned in favor of only inertial measurements (obtained through accelerometers, gyroscopes, and magnetometers). The advantages and limitations of all approaches are discussed in a comparative way, characterizing the similarities and differences between the various approaches.

Keywords: Motion capture, radio fingerprinting, inertial sensors, remote monitoring, body area network, body sensor network, sensor fusion.

[*] University of Parma while performing this work.

S. Adibi (ed.), *Mobile Health*,
Springer Series in Bio-/Neuroinformatics 5, DOI: 10.1007/978-3-319-12817-7_34

1 Introduction

The health of an individual is mostly assessed through the evaluation of specific biophysiological signals. In particular, the evolution over time of such signals may give the doctors (or caregivers) very relevant information about the evolution of the pathology itself. However, this typically requires a frequent, or sometimes constant, monitoring of the patient and, thus, a large use of human resources.

Parameters such as blood-pressure, heart-pulse-rate, and respiration are common examples of this class of biophysiological signals. Besides them, it has been shown that the motion of individuals allows to gather relevant information on a person status, especially in many healthcare scenarios characterized by patients affected by motion disorders (e.g., Parkinson's Disease or Stroke) or, more simply, by patients who need a constant control of their movements (e.g., post-rehabilitation after surgery or monitoring of activities of elderly people).

Furthermore, over the past decade, the progress of portable and mobile sensors has paved the way toward the use of these sensors in mHealth applications, allowing a remote evaluation of different biophysiological signals with economic savings (in terms of specialized human resources) and improved patient comfort.

Even if optical motion capture, which often relies on a set of fixed cameras and body-worn reflecting markers, is generally perceived as the standard reference approach for the evaluation of the motion of individuals, other motion capture techniques, such as radio and inertial, are attracting an increasing attention because of their suitability in remote mHealth applications. Moreover, several hybrid approaches have been studied and proposed in order to overcome the limitations of component technologies considered independently. More specifically, in the literature, two main approaches have been mainly considered for motion capture: optical and inertial.

Concerning *optical motion capture*, the widely used technology is optoelectronic (e.g., Vicon system [3]). Optoelectronic systems require the user to wear reflective markers and to move in a space completely visible by a set of cameras. Because of their accuracy, these systems are often used as ground truth references for other motion capture systems. On the other hand, their use is mostly limited to clinical environments or specialized laboratories, due to their high cost and complexity. Other optical motion capture systems comprise the class of markerless systems. Microsoft Kinect [2], which makes use of an RGB camera and a depth sensor (composed of an infrared camera and projector), can be considered one of the most significant examples of markerless systems. Its low cost (with respect to systems like Vicon), along with its still quite good performance, has made it a widely used solution. Typically, optical systems suffer from problems related to different lighting conditions and markers' occlusion. Moreover, the user movements must be limited to the area captured by the cameras. The reader is referred to [8, 22] for accurate surveys on optical motion capture.

Concerning *inertial motion capture*, inertial sensors are used to estimate the orientation of rigid body segments and, thus, to capture the motion of a user. One of the most successful and complete commercial products is the Xsens MVN [27], which comprises 17 inertial sensors (equipped with triaxial accelerometers, gyroscopes, and magnetometers) attached to the body of the user by a Lycra suit. The major advantage of this technology, with respect to optical systems, is that the user is completely free to move everywhere because no camera is needed. Moreover, the visibility of the nodes placed on the user body is not an issue. However, the accuracy of these systems is typically lower than that of optical systems and the cost, particularly for systems which rely on a large number of nodes and types of sensors (as Xsens MVN), is not significantly lower than that of optoelectronic systems. Finally, especially when used for a long time, a significant drift in the sensors' measurements can be observed, leading to a performance degradation.

When the motion of the whole body needs not to be tracked, besides complex systems like Xsens MVN, simpler configurations of nodes can be used. To this end, many other subsystems have been studied and developed in order to provide simpler solutions to simpler scenarios. These systems still rely on the use of accelerometers, gyroscopes, magnetometers, and/or combinations of them for different types of biomedical applications [7, 18, 29, 38, 42].

Finally, solutions based on the joint use of different technologies, designed in order to tackle and overcome the limitations which characterize component technologies considered independently, have also been investigated. For instance, inertial/radio, inertial/GPS, inertial/optical, and inertial/acoustic joint measurements are considered in [11, 12, 21], in [6], in [32], and in [36], respectively. In particular: in the inertial/radio approach, radio signal strengths are used to determine (after a preliminary training phase) a first rough estimate of the user's pose, eventually corrected through inertial measurements. In the inertial/GPS approach, nodes equipped with GPS are also used to track the user, providing better performance (especially) in outdoor scenarios (e.g., sport sessions, such as skiing). In the inertial/optical approach, inertial sensor nodes and cameras are jointly used, leading to a decreased user's freedom of movement but also to an improved robustness to occlusions. In the inertial/acoustic approach, nodes equipped with microphones and speakers are used in addition to inertial sensor nodes, providing better performance (thanks to the estimates of relative distances between nodes) but suffering from "acoustic occlusions."

Since motion capture applications typically rely on the use of a centralized base station, which collects all the data generated from the sensor networks, they are very suitable to be used in mHealth scenarios where, for instance, a mobile phone can directly communicate with the base station to get the processed data and send them to some specialized personnel. This is also reflected by various efforts already spent in developing mHealth applications based, for instance, on the use of inertial sensors [5, 23, 24, 25, 37, 41].

This chapter aims at presenting a comparative overview of the possible integration between radio and inertial approaches, with specific focus on the motion capture of an arm. In Sect. 2, the considered approaches are properly introduced and described; the mathematical background behind each approach is also briefly presented along with some performance results. In Sect. 3, a comparative overview between the selected approaches is given: to this end, the advantages and limitations of every approach are discussed, trying to motivate the transitions between different approaches and to highlight their relationships. In Sect. 4, two application case studies are presented. In Sect. 5, future directions of the presented techniques and technologies are discussed. Finally, conclusions are drawn in Sect. 6.

2 Motion Capture: From Radio Signals to Inertial Signals

In this section, different approaches to (arm) motion capture are described. The mathematical background behind each approach is briefly presented and some performance results are also discussed. In particular, we start by investigating a radio-based approach (Sect. 2.1), based on the fingerprinting radio localization technique. Then, the previous approach is improved by integrating inertial measurements (Sect. 2.2): namely, accelerometers are used to provide an estimate of the nodes' pitches. Finally, the radio signals are abandoned in favor of only inertial measurements obtained through accelerometers, gyroscopes, and magnetometers (Sect. 2.3).

2.1 The Radio Fingerprinting-Based Approach

Fingerprinting is a robust localization technique for indoor scenarios, which are typically characterized by reflections, multipath, and fading [34]. We now provide some intuition on the fingerprinting technique—the interested user can find more details in [11, 12, 21]. Fingerprinting requires three kinds of nodes: target nodes, anchor nodes, and a base station. Target nodes have to be localized, whereas anchors nodes have fixed known positions and are used to generate a reference system. Finally, the base station is the processing center. Two phases are involved: a training phase, during which a radio map of fingerprints is generated; and an online phase, during which localization is performed.

During the training phase, the target node continuously broadcasts packets to be received by the anchor nodes. The latter, upon reception of the packets sent by the target, measure the Received radio Signal Strengths (RSSs) and relay this information to the base station. The base station collects the RSS values and generates a "fingerprint," i.e., a vector containing the average RSSs measured by the anchors. Different positions of the target (i.e., fingerprint positions) in the monitored area (properly chosen depending on the considered application) correspond to different fingerprints. The set of fingerprints created in the training phase represents the "radio map" of the environment. The major strength of fingerprinting consists of the fact that the fingerprint vectors implicitly take into account the impact of reflections and multipath on the RSSs, i.e., it is "tailored" to the specific indoor environment. This makes fingerprinting virtually insensitive to indoor propagation limitations—provided that the propagation environment remains quasi-static. The radio map (obtained from the average RSSs) can be used to run either deterministic (based on simple comparisons between the RSSs and the fingerprints) or probabilistic (based on a more accurate statistical characterization of the RSS) localization algorithms. To this end, during the training phase the entire Probability Mass Functions (PMFs) of the RSS from each anchor need to be also computed and stored.

After the training phase is completed, the online phase starts (following the same operations of the training phase, but now building a so-called "online" vector) and the radio map (deterministic approach) or the PMFs (probabilistic approach) created in the training phase are used to localize the target. Note that, at this stage, the target node can move freely and should not necessarily be placed in the fingerprint positions. Given the measured "online" vectors, different algorithms can be used to estimate the target positions.

One of the simplest *deterministic* fingerprinting algorithm is the Nearest Neighbor (NN) algorithm, whose generalization is known as kNN [9]. It consists of the computation of a specific distance metric between the received online vector and every fingerprint. According to the kNN algorithm, by applying the Shepard method [30], the estimated target position $\hat{\mathbf{s}} = (x, y, z)$ is given by

$$\widehat{\mathbf{s}} = \sum_{i=1}^{k} \frac{w_i}{\sum_{j=1}^{k} w_j} \cdot \widehat{\mathbf{s}}_i \tag{1}$$

where $\{\widehat{\mathbf{s}}_i\}_{i=1}^{k}$ are the positions of the k best neighbors (i.e., those with shortest distances from the target) and

$$w_i \triangleq \frac{1}{d_i^{p_s} + 0.0001} \tag{2}$$

where d_i is the distance computed between the i-th neighbor and the target (defined in the space of RSS vectors) and p_s is an integer larger than 0. The term 0.0001 is used to prevent a division by zero if the online vector is equal to one of the fingerprints. Various definitions of distance can be considered, such as Euclidean and Manhattan [9].[3] Observe that, when $k = 1$ (i.e., with the NN algorithm), Eq. 1 reduces to the coordinates of the closest fingerprint position and, then, p_s has no influence on the system. Finally, note that, due to the interpolation between fingerprint positions in Eq. 1, the estimated position $\widehat{\mathbf{s}}$ may (likely) differ from any of the considered fingerprint position.

Unlike the deterministic approach, in the *probabilistic* approach (straightforwardly called p-kNN) the RSSs measured at the anchor nodes are characterized, using the samples received in the training phase, through their entire PMFs. More precisely, the estimated target position can be expressed as

[3] Other distance definitions can be applied, e.g., the Mahalanobis distance, which takes also into account the contribution of covariance matrix computed for every fingerprint [9]. Little performance differences are, however, observed.

$$\widehat{\boldsymbol{s}} = \sum_{i=1}^{k} \frac{P\left(\widehat{\boldsymbol{s}}_i|\boldsymbol{r}\right)}{\sum_{j=1}^{k} P\left(\widehat{\boldsymbol{s}}_j|\boldsymbol{r}\right)} \widehat{\boldsymbol{s}}_i \tag{3}$$

where: $\boldsymbol{r}$ is the "online" vector; $P\left(\widehat{\boldsymbol{s}}_i|\boldsymbol{r}\right)$ is the a-posteriori probability of the i-th (out of k) best neighbor; and the k best neighbors are chosen so that the corresponding a-posteriori probability is maximized. Note again that, when $k = 1$ (and thus reducing to a so-called p-NN), Eq. 3 returns exactly the coordinates of the closest fingerprint position.

Even if in common practice fingerprinting is often used to localize subjects (with accuracy in the order of a few meters) in large (indoor) areas [4, 13, 31], it can be also exploited to estimate the positions of (i.e., localize) multiple target nodes, properly placed on a user arm (e.g., one on the upper arm and one on the forearm), and straightforwardly derive the whole arm posture (and motion). To this end, a fixed origin must be properly chosen (e.g., the shoulder) and the user must try to keep this origin fixed during the evaluation. In such scenario, the fingerprint positions are predefined postures of the arm that must be held by the user during the *training* phase. During the following *online* phase, the user can instead move his/her arm freely (always recalling not to move the reference system origin, i.e., his/her shoulder).

In [11, 21], a detailed experimental analysis was carried out for the motion capture of the arm of an individual. Specifically, 3 SunSPOTs, acting as targets, were placed on the right arm of the user: the first node was on the hand, the second on the forearm, and the third on the upper arm. In the surroundings of the user, $N = 8$ anchor nodes were considered. An illustrative representation of the overall experimental set-up is shown in Fig. 1. Four postures of the arm (on the basis of a priori known movements of the arm) were trained leading to the computation of 12 fingerprint positions (4 positions per target node) and the origin of the reference system was fixed on the shoulder. In the online phase, the system performance was tested asking the user both to replicate trained postures and to perform different (intermediate) ones.

In Fig. 2, the precision is shown, as a function of the accuracy, for both the kNN and the p-kNN algorithms.[4] Either fingerprint or intermediate positions are considered during the online phase. The results show that the system performance, when estimating intermediate positions, is remarkably worse than that in correspondence to the estimation of the fingerprint positions. Note also that the kNN algorithm performs better than the p-kNN algorithm when the target is positioned in the intermediate positions. On the other hand, when the target is in one of the fingerprint positions, the p-kNN algorithm has a better performance. This is due to the fact that the probabilistic approach tends to give more weight to the fingerprint positions.

[4] Note that the performance graph can be read as follows: percentage of correct position estimates (i.e., precision) as a function of a given maximum tolerable error (i.e., accuracy).

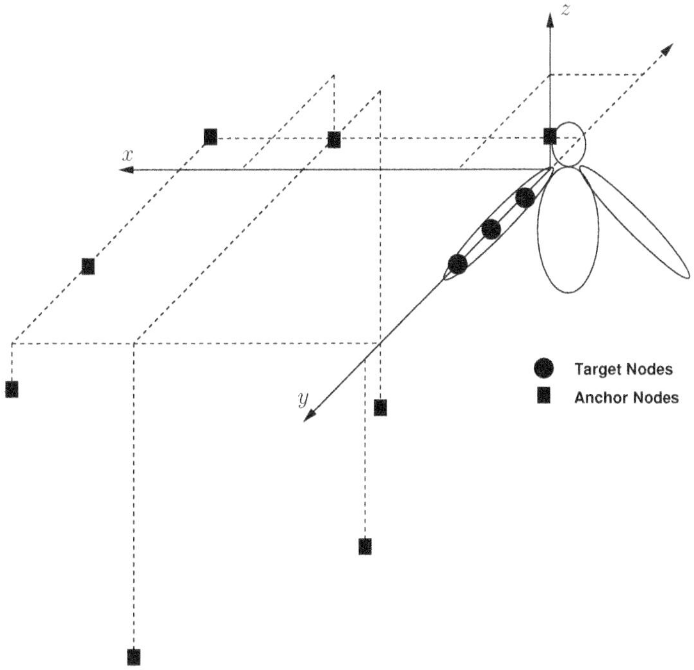

Fig. 1 Considered motion capture set-up: placement of 3 SunSPOTs (i.e., target nodes) on the user arm and overall scenario

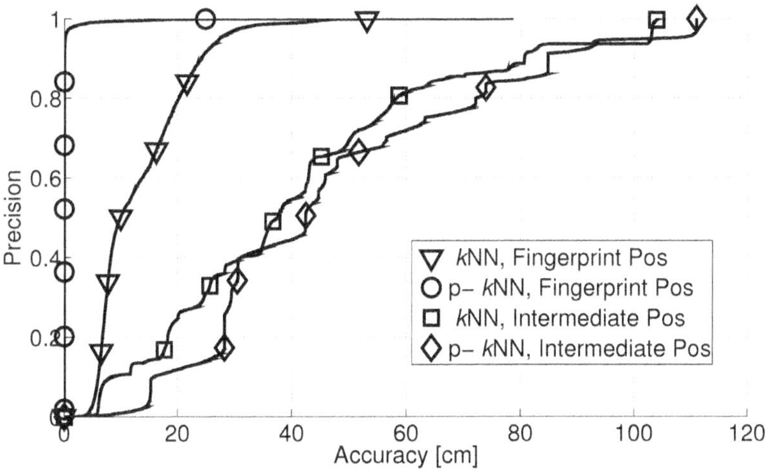

Fig. 2 Precision, as a function of the accuracy, for both the *k*NN and the p-*k*NN algorithms (using the optimized values $p_s = 1$ and $k = 2$). The system performance is shown for the estimation of both fingerprint and intermediate positions.

2.2 The Radio/Accelerometric Approach

As previously observed, the radio fingerprinting approach described in Sect. 2.1 may introduce errors, especially when a target node is in a position that differs from the trained fingerprint positions (i.e., when the arm is in an untrained posture). A possible way to improve the system performance is to estimate the arm orientation by making use of other inertial sensors (e.g., accelerometers, gyroscopes, and/or magnetometers), which the target nodes can be equipped with. In particular, considering proper combinations of these sensors (e.g., an accelerometer and a gyroscope), the orientation of a device (and, thus, of the arm) can be estimated [20].

Generally, the orientation of a device can be described by three parameters: yaw (or heading), pitch (or elevation), and roll (or bank) [16]. In order to define them, one must consider two reference systems: (i) the device coordinate system (that can be considered equal to the coordinate system of the inertial sensor mounted on it) and (ii) the Earth coordinate system. It is known that a rigid body can be arbitrarily rotated by first rotating it around its z axis by an angle ψ (the yaw), then around its y axis by an angle θ (the pitch), and finally around its x axis by an angle ϕ (the roll) [16]. It can be shown that, using just an accelerometer (in order to minimize the system cost), the pitch of a *still* device can be estimated as follows [40]:

$$\theta = \arcsin \overline{a}_{\mathrm{arm}} \tag{4}$$

where $\overline{a}_{\mathrm{arm}}$ is the normalized acceleration (in g units) measured along the arm axis (which is the device axis that points toward the direction of the user arm, as shown in Fig. 3). Note that, according to Eq. 4, the pitch belongs to $[-\pi/2, \pi/2]$.

As the above approach is valid only if the device is still, static (as the constant force of gravity) and dynamic (as the movements or vibrations of the accelerometer itself) accelerations need to be discriminated. This problem is generally solved by applying a low-pass filter to the acceleration outputs or by taking into account only acceleration measurements with amplitudes in $[g-\xi, g+\xi]$, where g is the gravity acceleration (i.e., 9.81 m/s^2) and ξ needs to be properly chosen with respect to the application context (i.e., considering only data segments in which the user is not moving and, thus, $\xi \cong 0$ m/s^2).

Furthermore, if a device is placed on a user body, its position can be estimated more accurately by taking into account the implicit constraints of the human body. The following assumptions are expedient to derive useful human body constraints:

- the pitch of each device is known (e.g., using Eq. 4);
- the lengths of the body segments on which the devices are mounted (e.g., the length of the arm between two joints) are known;
- preliminary estimates of the positions of the devices are available (e.g., from radio localization).

Focusing on the motion capture of a user arm, assume that two nodes are placed on the user arm: the first on the upper arm and the second on the forearm. For ease of clarity, let us denote their physical positions as $\{\mathbf{D}_i\}_{i=1}^{2}$, where $\mathbf{D}_i = (x_{\mathrm{D}_i}, y_{\mathrm{D}_i}, z_{\mathrm{D}_i})$.

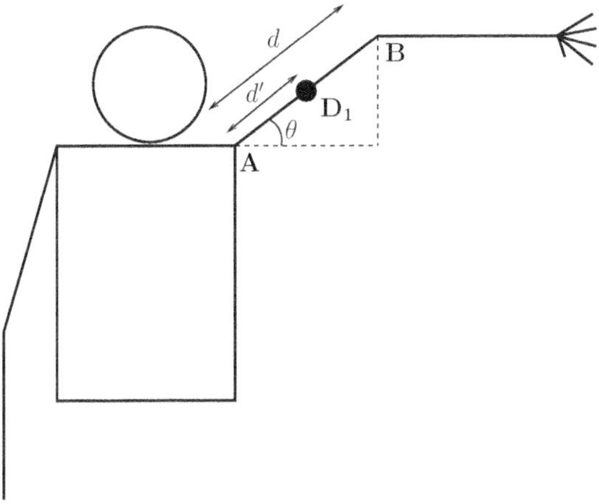

Fig. 3 Arm constraints: key points and distances used in the algorithm, focusing on the first step of the proposed approach. In particular: **A** and **B** are the three-dimensional coordinates of the joints of the arm segment considered in the first step of the proposed approach (i.e., shoulder and elbow); $\mathbf{D}_1$ indicates the three-dimensional coordinates of target 1; d is the actual length of the considered arm segment (in this case, the upper arm); and d' is the distance between a joint and the next (starting from the shoulder) sensor device (in this case, the distance between the shoulder and target 1).

The origin $(0,0,0)$ will be the shoulder. Another reference point will be the joint of the arm between the upper arm and the forearm, namely the elbow.

First, the "reference pitch" of each device has to be defined. In our case, we define it as the pitch that every device has when the user raises his/her arm in a direction perpendicular to the vertical axis of his/her body. As shown in Fig. 3, at the first step the arm reference point, corresponding to the shoulder, is denoted as point $\mathbf{A} = (x_\mathrm{A}, y_\mathrm{A}, z_\mathrm{A})$ and the nearest joint (i.e., the elbow) as point $\mathbf{B} = (x_\mathrm{B}, y_\mathrm{B}, z_\mathrm{B})$. We then consider the first device, denoted as point $\mathbf{D}_1$ and positioned in the middle of the upper arm, i.e., between **A** and **B**. Its initial position estimate (namely, from radio localization) is denoted as $\widehat{\mathbf{D}}_1$ and will be refined using the pitch information (recovered through Eq. 4). In order to do this, we define the two values d' and d as the actual distances between **A** and $\mathbf{D}_1$ and between **A** and **B**, respectively. Knowing the device pitch θ, it can be shown (through geometric considerations relative to the scenario in Fig. 3) that the adjusted position estimate of the first device, denoted as $\overline{\mathbf{D}}_1$, can be expressed as

$$\overline{\mathbf{D}}_1^\mathrm{T} = \begin{pmatrix} x_{\overline{\mathrm{D}}_1} \\ y_{\overline{\mathrm{D}}_1} \\ z_{\overline{\mathrm{D}}_1} \end{pmatrix} = \begin{pmatrix} x_\mathrm{A} + R\left(x_{\widehat{\mathrm{D}}_1} - x_\mathrm{A}\right) \\ y_\mathrm{A} + R\left(y_{\widehat{\mathrm{D}}_1} - y_\mathrm{A}\right) \\ z_\mathrm{A} + d' \sin\theta \end{pmatrix} \tag{5}$$

where $(\cdot)^\mathrm{T}$ denotes vector transposition and

$$R = \frac{d' \cos\theta}{\sqrt{\left(x_{\mathrm{A}} - x_{\widehat{\mathrm{D}}_1}\right)^2 + \left(y_{\mathrm{A}} - y_{\widehat{\mathrm{D}}_1}\right)^2}}. \quad (6)$$

On the basis of the estimate $\overline{\mathbf{D}}_1$ and taking into account the coordinates of $\mathbf{A}$ and the pitch θ, the coordinates of $\mathbf{B}$ (namely, the elbow) can be expressed as follows:

$$\mathbf{B}^{\mathrm{T}} = \begin{pmatrix} x_{\mathrm{A}} + K\left(x_{\overline{\mathrm{D}}_1} - x_{\mathrm{A}}\right) \\ y_{\mathrm{A}} + K\left(y_{\overline{\mathrm{D}}_1} - y_{\mathrm{A}}\right) \\ z_{\mathrm{A}} + d\sin\theta \end{pmatrix} \quad (7)$$

where $K \triangleq d/d'$.

The above estimation procedure can be repeated to estimate the position of the second device $\mathbf{D}_2$, positioned in the middle of the forearm, between point $\mathbf{B}$ (the elbow) and point $\mathbf{C}$ (the wrist), as shown in Fig. 3. To this end, Eqs. 5 - 6 are used again considering proper values of θ, d', and d (namely, the ones related to the arm segment between $\mathbf{B}$ and $\mathbf{C}$), and substituting $\mathbf{A}$ and $\mathbf{D}_1$ with $\mathbf{B}$ and $\mathbf{D}_2$, respectively.[5] Note that as the position of $\mathbf{B}$ is estimated (i.e., $\mathbf{B}$ is not a reference point as $\mathbf{A}$), it is likely that the estimate of the coordinates of $\mathbf{D}_2$ will be less accurate than that of the coordinates of $\mathbf{D}_1$.

According to the described hybrid approach, the z coordinate is estimated relying only on the pitch estimation (which depends on the accelerometric measurements), whereas x and y coordinates are first coarsely estimated through the radio localization and then their estimates are refined (more precisely, rescaled) according to the pitch estimation. Alternatively, we can interpret our motion capture system as a system able to estimate the user arm heading (i.e., the direction of the arm in a horizontal plane) and pitch (i.e., the inclination of the arm). According to this point of view, the roles of the two component techniques in our hybrid approach (i.e., radio localization and accelerometric measurements) are even more definite. Indeed, it can be observed that the arm pitch estimation is carried out exclusively by using the accelerometric measurements, whereas the arm heading is estimated only through radio localization.

Finally, we also remark that the integration of the accelerometers into the system does not have any impact on the training phase. Indeed, all the acceleration measurements are taken into account only during the online phase.

In [12], this hybrid motion capture system was then tested for a scenario similar to the one described in Sect. 2.1. In particular, a body area network with target nodes on the user arm was still considered, but now the anchor nodes were in part placed (and fixed) in the surroundings of the user and in part attached on his/her body. This was obtained through the use of (i) a home-made t-shirt with folders where anchors can be placed and (ii) a hat with an anchor attached to it. Though the presence of some anchor nodes at fixed (outside the body) positions still forces the user to

[5] Observe that this process can be iteratively repeated, should more consecutive body segments be considered (e.g., by adding a device for the hand in our testbed).

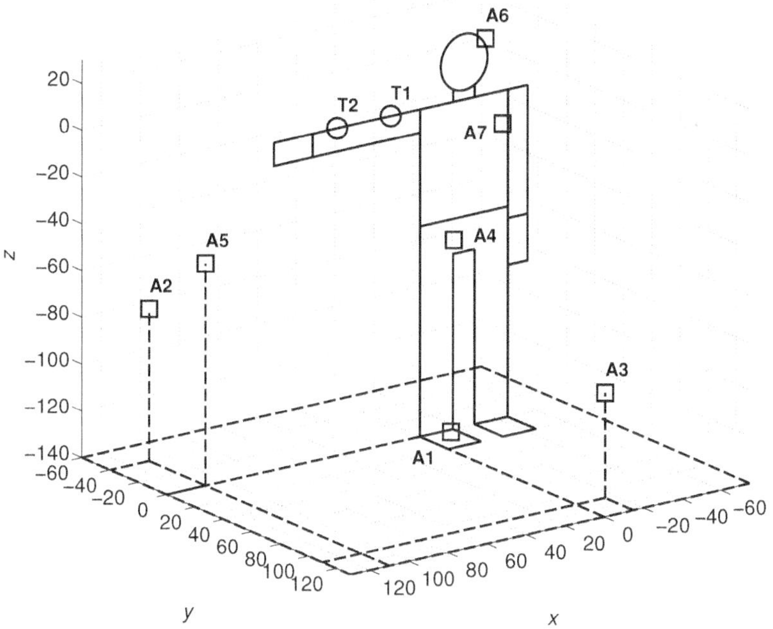

Fig. 4 Experimental motion capture set-up for the radio/accelerometric approach. Note that A2, A3, and A5 are fixed in the surroundings of the user, whereas A1, A4, A6, and A7 are attached on his/her body.

remain in its initial position in the room (in order to make the fingerprinting technique work consistently), the considered testbed was a first step toward a fully portable arm motion capture system. Furthermore, a noisier (and, thus, more realistic) environment was there taken into account (e.g., with more than one people moving in the room while testing the system), whereas in [11, 21] the system performance was analyzed in more controlled (interference-free) scenarios.

In the experimental testbed, only 2 SunSPOTs, acting as targets (i.e., target 1 and target 2), were placed on the right arm of the user: the first node (i.e., target 1) is on the upper arm and the second (i.e., target 2) on the forearm. $N = 7$ anchor nodes were considered: 4 of them were placed on the user body, whereas the remaining 3 were placed in its proximity. For ease of clarity, an illustrative representation of the overall experimental set-up is shown in Fig. 4, where a map of the positions of the nodes is shown. Still fixing the origin of reference system on the shoulder (and thus forcing the user to move arm and keep the shoulder still), 5 fingerprint positions per target node were considered, corresponding to 5 simple arm postures. The choice of the 5 fingerprint positions was made in order to cover as uniformly as possible the surroundings of the user arm.

During the online phase, the user was first asked to replicate the 5 training fingerprint positions (in order to test the system in *static* conditions). Furthermore, a

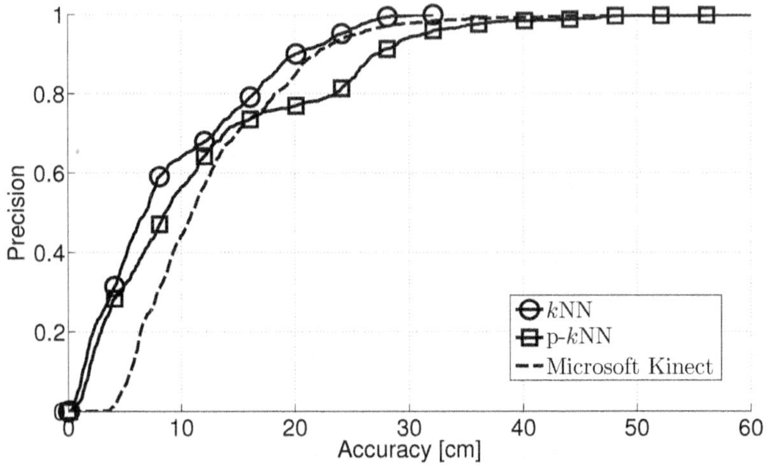

Fig. 5 Precision, as a function of the accuracy, for both deterministic and probabilistic approaches. A comparison with the Kinect performance is also offered.

few transitional movements, which start and end at two of the trained fingerprint positions, were also executed (in order to test the system in *dynamic* conditions). For the purpose of comparison, a Microsoft Kinect system [2] was also used to perform arm motion capture.

In Fig. 5, the performance of the hybrid radio/accelerometric approach is shown, considering both deterministic and probabilistic approaches, along with that of Microsoft Kinect. It can be seen that the deterministic approach slightly outperforms the probabilistic one and the Kinect system, whereas the Kinect system outperforms the probabilistic approach for accuracy values higher than 15 cm (i.e., if a position error larger than 15 cm is accepted the Kinect system is more suitable than the probabilistic approach of our system).

2.3 The Inertial Approach

If one wants to avoid the use of fingerprinting and, thus, of radio signals, the only use of accelerometers is not sufficient to provide unambiguous information about the motion capture of the arm. Instead, a complete Inertial Measurement Unit (IMU) is generally used, which typically comprises a triaxial accelerometer and a triaxial gyroscope. Furthermore, a triaxial magnetometer can be included to measure the direction of the magnetic field in the surroundings of the sensor device (as a digital compass).

The inertial approach mainly relies on the use of an orientation estimation filter which must be able to continuously estimate and track the orientation of the sensor device, starting from the raw measurements of its acceleration (from the accelerometer), angular velocity (from the gyroscope), and sensed magnetic field (from the magnetometer). The use of different sensors in the same device is mainly motivated by the fact that component sensors, if used independently, are prone to "intrinsic" (and unavoidable) problems. More precisely: accelerometers measure gravity acceleration, which is hardly discriminated from the linear acceleration of the device; gyroscopes can lead to a drift in the measurements due to implicit bias errors; magnetometers are very susceptible to magnetic disturbances of the local magnetic field (e.g., due to ferromagnetic materials in the proximity of the sensor).

The orientation of a device can be expressed, alternatively, in terms of: Euler angles (i.e., yaw, pitch, roll); rotation matrices; and quaternions [16]. Even though Euler angles give a more direct indication of how the device is rotated, quaternions are mostly used due to their robustness to unwanted phenomena (e.g., gimbal lock [16]) that can lead to singularities and ambiguities in the orientation estimation. In the last 20 years, many studies related to motion capture have led to the development of several types of orientation estimation filters. Furthermore, even if solutions based on Kalman filtering [14] have been mainly adopted in such studies [10, 19, 26, 28], in the last years many solutions have focused on complementary filtering techniques due to their simplicity and low computational requirements [20, 39, 40].

Provided that (i) each device is properly placed on each rigid body segment and (ii) its estimated orientation is correct, a kinematic chain thus appears. If a fixed origin is then chosen (e.g., the pelvis for full body motion capture, or a shoulder for arm motion capture), the positions of the extremes of all body segments can be easily derived from the estimated segments' orientations, leading to the reconstruction of the full posture of the user (or part of it in the case one wants to focus on a specific part). For ease of clarity, in Fig. 6 an illustrative representation of the reconstruction of the posture of the user, starting from the orientations of inertial devices attached to the user body segments, is shown only focusing on the upper body of the user.

The performance of the approach outlined above has been shown to be very accurate. Indeed, the estimated orientations of the body segments typically differ from the actual orientations of less than a degree in static conditions and of at most a few degrees in dynamic conditions, often leading to position errors in the order of a few centimeters.

3 A Comparative Overview

As previously discussed in Sect. 1, most of current state-of-the-art solutions for motion capture rely on optoelectronic or inertial approaches. However, the costs of an optoelectronic system and of a complete inertial suit are both very high. The only difference between these systems consists in the trade-off between accuracy and freedom of movement. Indeed, optoelectronic systems make errors in the order of a few millimeters, whereas errors with inertial systems are in the order of a few centimeters. Nevertheless, optoelectronic systems (and, more generally speaking, optical systems) are very sensitive to body occlusions and force the user to move in a specific area. On the contrary, inertial systems allow users to move freely and do not suffer from occlusion problems.

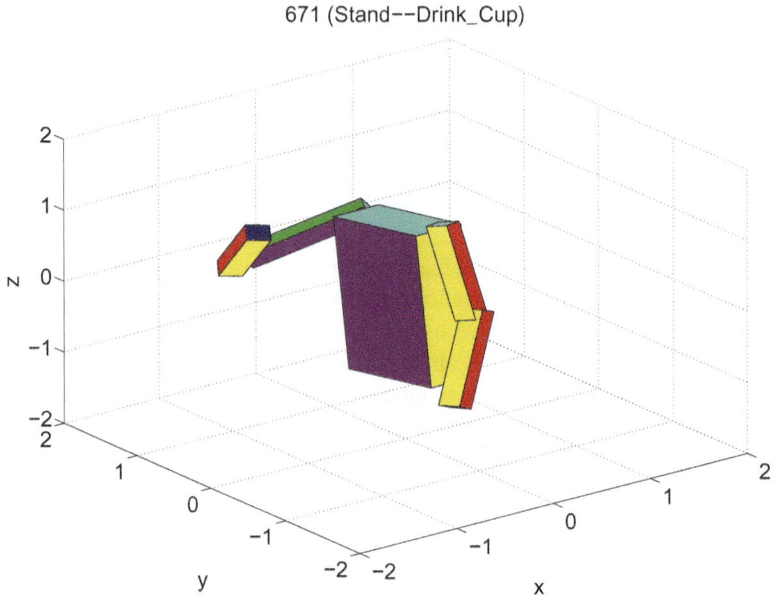

Fig. 6 Illustrative representation of the reconstructed posture of the user upper body. As shown in the label, the user was drinking something from a cup.

Another major advantage of inertial systems is that they scale better in cost when the whole body motion capture is not necessary. Indeed, the cost of optoelectronic systems is mainly associated to high quality cameras whose use cannot be avoided, whereas inertial systems cost depends on the number of inertial devices that the user wants to use (and, thus, the number of body segments that the user wants to monitor). Furthermore, the cost of an inertial device mostly resides in the sensors cost. Therefore, the less sensors are required, the lower the cost is.

From the previous considerations, it is easy to understand that the transition between the three approaches described in Sect. 2 mainly owes to cost motivations. A device with only radio communication hardware (Sect. 2.1) costs less than the same device comprising also an accelerometer (Sect. 2.2), which, in turns, costs less than a device supplied with an accelerometer, a gyroscope, and a magnetometer (Sect. 2.3). However, one can also observe that the same transition allows to move toward systems with increasingly higher accuracy. Indeed, the radio-based approach is the less accurate, whereas the inertial approach is able to achieve the best performance. Similar considerations can be carried out with reference to the computational efforts and compactness of the various approaches. In particular, the radio-based approach requires limited computational efforts, whereas the radio/accelerometric hybrid approach and the inertial approach require—in this order—increasingly higher computational efforts. However, the inertial approach is self-contained and does

Table 1 Comparative overview of the characteristics of the three presented approaches

	Component Technology		
	Radio	**Radio/Accelerometric**	**Inertial**
Accuracy	Medium	Medium/High	High
Cost	Low	Medium	Medium/High
Computational Efforts	Low	Low/Medium	Medium
Compactness (Self-containedness)	Low	Low	High

Table 2 Overview of the advantages/disadvantages of the three presented approaches

	Component Technology		
	Radio	**Radio/Accelerometric**	**Inertial**
Advantages	Very low cost	Low cost	Very accurate and independent of any external infrastructure
Disadvantages	External nodes are required (limited user motion freedom) and the estimate of the posture of the user is qualitative	External nodes are required (limited user motion freedom)	Higher cost and unable to provide the absolute position of the user

not require any external infrastructure. Comparative overviews of the characteristics and of the advantages/disadvantages of the three approaches are summarized in Table 1 and Table 2, respectively.

Going deeper into the working principles of the three approaches, it can be of interest to highlight differences and similarities of the considered approaches. Concerning the radio approach (Sect. 2.1), it can be observed that the focus is on giving a qualitative (rather than quantitative) evaluation of the arm motion. Indeed, even if the interpolation between known trained postures allows to capture the arm motion in a continuous way, this approach can be perhaps more suitable to posture *classification*. The hybrid approach (Sect. 2.2), placing itself in the middle between the other two approaches, relies less on the interpolation. Indeed, through the use of a single sensor (i.e., the accelerometer), the hybrid approach makes use of a more reliable estimation of the device inclination and uses the radio signals component just to estimate the heading of the device. Finally, the inertial approach (Sect. 2.3) completely abandons the contribution of the radio signals and estimate the position of the considered body segments starting only from the knowledge of the three-dimensional orientation of the device (estimated through a joint use of accelerometer, gyroscope, and magnetometer). This is also the reason why the first two approaches (Sect. 2.1 and Sect. 2.2) require the user not to move during the evaluation of the arm motion (or, at least, not to move the shoulder, being the origin of the reference system). Indeed, a relative difference between the position of the user and the positions of anchor nodes in his/her surroundings would invalidate the evaluation of the arm position. To this end, in [12] a hybrid placement of the anchor nodes also on the body user has been investigated, as an intermediate step toward the implementation of a fully portable system. Potentially, a future system design could exploit radio localization to directly estimate distances between (mobile) nodes placed only on the user body, possibly removing the distinction between anchor and target nodes (i.e., every node could be at the same time an anchor and a target node).

Concerning the use of the described approaches in specific applications, the performance results have proved that the inertial approach (Sect. 2.3) can be applied to a wide range of biomedical applications due to its high accuracy with respect to the others. The other approaches (Sect. 2.1 and Sect. 2.2) can instead be considered for health applications where it is relevant to discriminate discrete cases of arm motion (e.g., in rehabilitation after a surgery, if arm can be raised half-way or all the way).

Finally, a further development for the first two approaches (Sect. 2.1 and Sect. 2.2) can be that of exploiting the available environment infrastructure. For instance, if some fixed base stations (e.g., routers) are available in the room/building, they could be used as anchor nodes to capture the motion of the user in that room/building.

4 Case Studies

The previously presented approaches can be applied, for instance, in the following two case studies that will be straightforwardly referred to as *posture classification* and *motion capture*.

In *posture classification* applications, the aim is to determine the posture (i.e., the joints' positions) of a user body part (e.g., the arm), choosing between a predefined set of possible postures. In particular, the term *classification* specifically refers to a choice among a *discrete* set of postures. Therefore, it is not of interest to precisely estimate the positions of the joints, but a rough estimate is typically sufficient. This, however, typically implies that the system should be properly trained to recognize the user postures and discriminate between them. Finally, since the aim of posture classification is just to determine the posture of a user choosing among a discrete set of previously trained postures, posture classification could be considered in scenarios where a few cases need to be discriminated (e.g., while monitoring the recovery improvement after an arm surgery, it can be of interest to just evaluate whether the arm can be raised half-way or all the way).

On the other hand, in *motion capture* applications, the goal is to monitor and evaluate the precise movements of a user body part (e.g., the arm). This means recognizing and estimating the posture of the considered body part, without prior assumptions of possible postures that the body part can assume. In this case, the system does not need to be trained. However, since the problem is "unconstrained," it is not always easy to solve it and, in order to accurately estimate the body part posture, advanced signal processing techniques may be required.

At the end of this chapter, we will try to answer some common questions that can arise if one wants to apply the previously described techniques in the context of these two case studies.

5 Future Directions

Due to the continuous advancements in the production of small and portable sensors, able to measure a wide variety of biophysiological signals, motion capture techniques and applications are probably experiencing, in these years, one of their most successful periods in terms of development and interest from a user perspective. The technology is almost mature and the first commercial applications are becoming month by month more known and used by the end users. The commercial cost, which now may still not be affordable for everyone, is continuously decreasing [1].

Especially in the gaming and biomedical fields, many applications based on motion capture and inertial sensors are already attracting the attention of gamers and doctors, respectively. In particular, in 5-10 years, the gaming scenario will probably be dominated by games where the user will interactively control virtual characters by using some easily wearable devices (at most 4-5 on the whole body, e.g., possible locations could include feet, ankles, wrists, and head [15, 17, 33]). The medical scenario will instead begin to adopt wearable sensors able to assess health of the patients by remotely measuring vital signs/parameters and allowing patients to freely stay at their homes still keeping a direct connection with their doctors [5, 7, 18, 23, 24, 25, 29, 37, 38, 41, 42].

Concerning the technologies described in this chapter, the living environments in the next years will be covered even more by radio devices able to communicate with sensors and devices placed on the user body, facilitating the application of the techniques previously described [35]. Also inertial and magnetic sensors will be further developed in order to become even more small, light, and less bulky. These advancements will then increase the "usability" gap between optical and inertial systems, making the latter more appealing for their flexibility and ease of use.

Finally, the technological developments foreseen above are also paving the way to the use of body sensor networking for more fundamental research activities, concerning the design of advanced biophysiological signal processing techniques able to detect, from posture and movement signals, the status of a person's pathological condition. In other words, the technological development of advanced posture recognition techniques could allow to tackle relevant interdisciplinary (engineering/medical) research questions.

6 Conclusions

In this chapter, the importance and role of motion capture in mHealth applications has been investigated. To this end, three different motion capture approaches (i.e., radio-based, inertial, and radio/accelerometric) have been presented and described, in terms of performance, cost, and complexity. Furthermore, the advantages and limitations of all approaches have been discussed in a comparative way, characterizing their similarities and their differences and trying to highlight the most suitable application of each approach. Furthermore, some possible appealing extensions of the considered approaches have been also proposed.

References

[1] Invensense, http://www.invensense.com/
[2] Kinect: markerless optical motion capture system from Microsoft, http://www.microsoft.com/en-us/kinectforwindows/
[3] Optoelectronic motion capture system from Vicon, http://www.vicon.com/
[4] Bahl, P., Padmanabhan, V.N.: RADAR: an in-building RF-based user location and tracking system. In: Proc. IEEE Conf. on Computer Commun (INFOCOM), Tel Aviv, Israel, vol. 2, pp. 775–784 (2000)
[5] Bonato, P.: Wearable sensors and systems. IEEE Eng. Med. Biol. Mag. 29(3), 25–36 (2010)
[6] Brodie, M., Walmsley, A., Page, W.: Fusion motion capture: a prototype system using inertial measurement units and GPS for the biomechanical analysis of ski racing. Sports Technology 1(1), 17–28 (2008)
[7] Chen, B., Patel, S., Buckley, T., Rednic, R., McClure, D., Shih, L., Tarsy, D., Welsh, M., Bonato, P.: A web-based system for home monitoring of patients with Parkinson's Disease using wearable sensors. IEEE Transactions on Biomedical Engineering 58(3), 831–836 (2011)
[8] Chen, L., Wei, H., Ferryman, J.: A survey of human motion analysis using depth imagery. Pattern Recognition Letters 34(15), 1995–2006 (2013)
[9] Duda, R.O., Hart, P.E., Stork, D.G.: Pattern Classification and Scene Analysis, 2nd edn. Wiley-Interscience, New York (2000)
[10] Favre, J., Jolles, B.M., Siegrist, O., Aminian, K.: Quaternion-based fusion of gyroscopes and accelerometers to improve 3D angle measurement. Electronics Letters 42(11), 6125–6614 (2006)
[11] Giuberti, M., Martalò, M., Ferrari, G.: Fingerprinting-based wireless 3D localization for motion capture applications. In: Proc. 1st ACM MobiHoc Workshop on Pervasive Wireless Healthcare (MobileHealth), Paris, France, pp. 6.1–6.8 (2011)
[12] Giuberti, M., Martalò, M., Ferrari, G.: A hybrid radio/acclerometric approach to arm posture recognition. Journal of Ambient Intelligence and Smart Environments (2014); Under revision. Available upon request
[13] Kaemarungsi, K.: Design of indoor positioning systems based on location fingerprinting technique. PhD thesis, University of Pittsburgh, Pittsburgh (2005), http://etd.library.pitt.edu/ETD/available/etd-02232005-235903%/unrestricted/dissertation28Feb05.pdf

[14] Kalman, R.E.: A new approach to linear filtering and prediction problems. Trans. of the ASME—Journal of Basic Eng. 82(Series D), 35–45 (1960)
[15] Kim, J., Seol, Y., Lee, J.: Realtime performance animation using sparse 3D motion sensors. In: Kallmann, M., Bekris, K. (eds.) MIG 2012. LNCS, vol. 7660, pp. 31–42. Springer, Heidelberg (2012)
[16] Kuipers, J.B.: Quaternions and Rotation Sequences: A Primer with Applications to Orbits, Aerospace and Virtual Reality. Princeton University Press, Princeton (1999)
[17] Liu, H., Wei, X., Chai, J., Ha, I., Rhee, T.: Realtime human motion control with a small number of inertial sensors. In: Proc. of Symposium on Interactive 3D Graphics and games (I3D), San Francisco, CA, USA, pp. 133–140 (2011)
[18] Lo, G., Suresh, A.R., Stocco, L., Gonzalez-Valenzuela, S., Leung, V.C.M.: A wireless sensor system for motion analysis of Parkinson's disease patients. In: 2011 IEEE International Conference on Pervasive Computing and Communications Workshops (PERCOM Workshops), Seattle, WA, USA, pp. 372–375 (2011)
[19] Luinge, H.J., Veltink, P.H., Baten, C.T.M.: Ambulatory measurement of arm orientation. Journal of Biomechanics 40(1), 78–85 (2007)
[20] Madgwick, S.O.H.: An efficient orientation filter for inertial and inertial/magnetic sensor arrays. Tech. rep., Department of Mechanical Engineering, University of Bristol, Bristol, UK (2010), `http://www.x-io.co.uk/res/doc/madgwick_internal_report.pdf`
[21] Martalò, M., Giuberti, M., Ferrari, G.: Experimental investigation of wireless sensor networks for fingerprinting-based posture recognition. In: Riunione Annuale 2011 del Gruppo Nazionale Telecomunicazioni e Teoria dell'Informazione (GTTI), Messina and Taormina, Italy (2011)
[22] Moeslund, T.B., Hilton, A., Krüger, V.: A survey of advances in vision-based human motion capture and analysis. Computer Vision and Image Understanding 104(2), 90–126 (2006)
[23] Patel, S., Park, H., Bonato, P., Chan, L., Rodgers, M.: A review of wearable sensors and systems with application in rehabilitation. Journal of NeuroEngineering and Rehabilitation 9(21), 1–17 (2012)
[24] Rahimi, F., Duval, C., Jog, M., Bee, C., South, A., Jog, M., Edwards, R., Boissy, P.: Capturing whole-body mobility of patients with Parkinson disease using inertial motion sensors: expected challenges and rewards. In: Proc. of the 33rd Annual Int. Conf. of the IEEE Eng. in Medicine and Biology Society (EMBS), Boston, MA, USA, pp. 5833–5838 (2011)
[25] Rodriguez-Silva, D.A., Gil-Castineira, F., Gonzalez-Castano, F.J., Duro, R.J., Lopez-Pena, F., Vales-Alonso, J.: Human motion tracking and gait analysis: a brief review of current sensing systems and integration with intelligent environments. In: Proc. of IEEE Int. Conf. on Virtual Environments, Human-Computer Interfaces, and Measurement Systems (VECIMS), Instanbul, Turkey, pp. 166–171 (2008)
[26] Roetenberg, D., Slycke, P.J., Veltink, P.H.: Ambulatory position and orientation tracking fusing magnetic and inertial sensing. IEEE Trans. on Biomedical Eng. 54(5), 883–890 (2007)
[27] Roetenberg, D., Luinge, H.J., Slycke, P.J.: Xsens MVN: Full 6DOF human motion tracking using miniature inertial sensors. Xsens Technologies B V (2009)
[28] Sabatini, A.M.: Quaternion-based extended Kalman filter for determining orientation by inertial and magnetic sensing. IEEE Trans. on Biomedical Eng. 53(7), 1346–1356 (2006)

[29] Salarian, A.: Ambulatory monitoring of motor functions in patients with Parkinson's disease using kinematic sensors. PhD thesis, École polytechnique fédérale de Lausanne, Lausanne, Switzerland (2006), http://biblion.epfl.ch/EPFL/theses/2006/3475/EPFL_TH3475.pdf

[30] Shepard, D.: A two-dimensional interpolation function for irregularly spaced data. In: ACM National Conf. (1968)

[31] Swangmuang, N., Krishnamurthy, P.: Location fingerprint analyses toward efficient indoor positioning. In: IEEE Int. Conf. Pervasive Comp. and Commun. (PERCOM), Hong Kong, China, pp. 100–109 (2008)

[32] Tao, Y., Hu, H., Zhou, H.: Integration of vision and inertial sensors for 3D arm motion tracking in home-based rehabilitation. The International Journal of Robotics Research 26(6), 607–624 (2007)

[33] Tautges, J., Zinke, A., Krüger, B., Baumann, J., Weber, A., Helten, T., Müller, M., Seidel, H., Eberhardt, B.: Motion reconstruction using sparse accelerometer data. ACM Trans. Graph. 30(3), 18:1–18:12 (2011)

[34] US Wireless Corporation. RadioCamera Wireless Caller-Location System, http://www.directionsmag.com/companies/us-wireless-corp/93474

[35] Viani, F., Robol, F., Polo, A., Rocca, P., Oliveri, G., Massa, A.: Wireless architectures for heterogeneous sensing in smart home applications: Concepts and real implementation. Proceedings of the IEEE 101(11), 2381–2396 (2013)

[36] Vlasic, D., Adelsberger, R., Vannucci, G., Barnwell, J., Gross, M., Matusik, W., Popović, J.: Practical motion capture in everyday surroundings. ACM Trans on Graphics 26(3), 35.1–35.9 (2007)

[37] Wong, W.Y., Wong, M.S., Lo, K.H.: Clinical applications of sensors for human posture and movement analysis: a review. Prosthetics and Orthotics International 31(1), 62–75 (2007)

[38] Ying, H., Schlösser, M., Schnitzer, A., Schäfer, T., Schläfke, M.E., Leonhardt, S., Schiek, M.: Distributed intelligent sensor network for the rehabilitation of Parkinson's patients. IEEE Trans. on Information Technology in Biomedicine 15(2), 268–276 (2011)

[39] Young, A.D.: Wireless realtime motion tracking system using localised orientation estimation. PhD thesis, University of Edinburgh, Edinburgh, UK (2010), http://www.xsens.com/images/stories/PDF/young_phd.pdf

[40] Yun, X., Bachmann, E.R., McGhee, R.B.: A simplified quaternion-based algorithm for orientation estimation from earth gravity and magnetic field measurements. IEEE Trans. on Instrumentation and Measurement 57(3), 638–650 (2008)

[41] Zheng, H., Black, N.D., Harris, N.D.: Position-sensing technologies for movement analysis in stroke rehabilitation. Medical and Biological Engineering and Computing 43(4), 413–420 (2005)

[42] Zwartjes, D.G.M., Heida, T., van Vugt, J.P.P., Geelen, J.A.G., Veltink, P.H.: Ambulatory monitoring of activities and motor symptoms in Parkinson's Disease. IEEE Trans. on Biomedical Eng. 57(11), 2778–2786 (2010)

Questions

Case Study #1: "Posture Classification"

As described in Sect. 4, in this scenario it is of interest to determine which posture is assuming a user body part (e.g., the arm), choosing between a predefined discrete set of possible postures.

Q1. Which approach would be preferable to use (among those presented in the chapter), in order to minimize the cost of the system but allowing a minimum level of acceptable performance?

Q2. Which kind of arm postures could be considered for the classification? How many devices would be required to be placed on the user body?

Q3. In order to classify the posture of a leg, how should the system be changed?

Case Study #2: "Motion Capture"

As described in Sect. 4, in this scenario it is of interest to continuously monitor and evaluate the movements of a user body part (e.g., the arm).

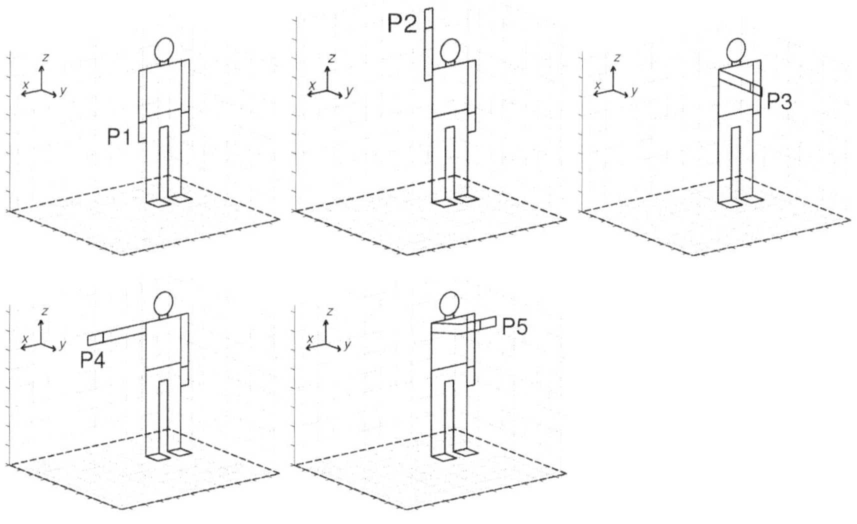

Fig. 7 Representation of the 5 fingerprint positions (i.e., P1, P2, P3, P4, and P5) considered in the training phase

Q1. Which approach would be preferable to use (among those presented in the chapter), in order to minimize the cost of the system but allowing a minimum level of acceptable performance?

Q2. Which kind of arm movements could be evaluated using the previous approach? How many devices would be required to place on the user body?

35. Design and Evaluation of Near Field Communication (NFC) Technology Based Solutions for mHealth Challenges

Jürgen Morak and Günter Schreier

AIT Austrian Institute of Technology GmbH, Safety & Security Department, Graz, Austria

Keywords: mHealth, Telehealth, Near Field Communication Technology, Usability, Security, Internet of Things.

1 Introduction

1.1 mHealth-Based Telehealth

Today's healthcare systems are facing major challenges generated by the increasing prevalence of chronic diseases. Telehealth services appear as promising approaches to attenuate those challenges by changing the way of delivering healthcare services. Telehealth is based on the inclusion of the last unutilized resource that has to be brought into the process of healthcare, the patient him/herself [1]. Information is an essential asset in the field of medicine, and telehealth utilizes information and communications technologies (ICT) to include patients into data and information flow, particularly in treating chronic diseases. An appropriate ICT infrastructure provides a virtual communication link between patients and their caregivers to allow for just-in-time intervention instead of routinely consulting a medical practitioner or the outpatient clinic in a just-in-case manner. mHealth-based telehealth services include mobile devices and services to this infrastructure in order to enable ubiquitous availability.

Mobile phones play an eminent role in today's information and communication technology-dominated world. Over the past decades, since the roll-out of mobile telephone services, the mobile communications industry has presented drastic technology advances [2]. Digital mobile networking technologies advanced from Global System for Mobile communications (GSM) over 2G and 3G networks to 4G networks with data rates of 50Mbit/s and beyond [3]. In parallel to that, the end user devices have evolved from simple voice telephony devices to feature-rich electronic devices. While in the first years these devices only allowed for a limited set of functions, today's smartphones are versatile mobile computing platforms providing personal information management functions, Web browsing, entertainment, and other forms of personal computing [4]. With respect to user

S. Adibi (ed.), *Mobile Health*,
Springer Series in Bio-/Neuroinformatics 5, DOI: 10.1007/978-3-319-12817-7_35

interfaces, sensors and wireless interfaces, mobile phone technology has demonstrated a high level of utility. Thus, devices already may feature high-resolution displays with multi-finger touch recognition and/or a full QWERTY layout keyboard, high-resolution cameras with video recording capabilities, sensors for measuring acceleration, rotation, ambient light, and even barometric pressure. In order to provide geolocation services and communication with devices in their surrounding environment, they provide a GPS receiver paired with a digital compass sensor and wireless networking transceivers for Wi-Fi, Bluetooth, and ANT+ connectivity [5, 6, 7, 8].

Using a mobile phone as the patient terminal device for telehealth scenarios is one of the key scenarios in mHealth. During the last years, the term “mHealth” has been most succinctly defined as services related to health and wellness delivered via mobile communications devices [9]. However, instead of using the mobile phone’s basic communication features such as voice communication, text messaging and using a Web browser, in typical mHealth-based telehealth applications the mobile phone acts as data hub to electronically collect and forward data from medical sensor devices in an automatic or at least semi-automatic way.

Nonetheless, handling a mobile phone on the one hand, and its interplay with medical sensor devices on the other, still remains a challenge. The majority of mobile phones on the market today are sophisticated devices with relatively complicated user interfaces. Mastery of their functions can sometimes take a long time [10]. Additionally, chronic diseases are predominantly affecting elderly people, thus, providing telehealth systems, and the patient terminal in particular, with a barrier-free user interface design is a must-have requirement. Usability and Plug-and-Play interoperability are essential and determine whether the patient is able and willing to use it in daily routine.

1.2 NFC as the Enabling Technology

A wireless technology that became available in mobile phones during the last decade is the Near Field Communication (NFC) technology. NFC evolved from Radio Frequency Identification (RFID) technology that additionally uses features of common wireless technology like Bluetooth. Definition of standards to ensure interoperability as well as certification of conformity is managed by the NFC Forum [11]. NFC operates in the industrial, scientific and medical (ISM) band of 13.56 MHz and supports communication within a range of a few centimeters (<10) at a data rate of up to 424 Kbit/s [12]. The short operating range is the major advantage of this technology since two NFC-enabled items start communicating with each other automatically once they are brought close to each other.

The intention behind developing NFC technology was its integration into mobile phones to enable this kind of ubiquitous accessory, acting as a virtual smart card for ticketing, loyalty and payment applications by waving it in front of a reader terminal [13, 14]. The major advantages of such a virtual card are remote deployment and maintenance by means of a mobile Internet connection. Having the mobile phone acting as a contactless smartcard is the so-called NFC card-emulation mode.

In addition, NFC supports the card-reader mode, which enables the mobile phone to act as reader device for reading passive RFID tags and cards being compliant with certain RFID standards. The third operation mode is called the peer-to-peer (p2p) mode and supports communication between two active devices (e.g. two mobile phones exchanging electronic business cards, pictures or music files) [15].

NFC is supposed to change the way we interact with the physical world in a subtle, but pervasive way and provides an intuitive method for linking any object from the physical world into the virtual world of the Internet, hence, it is conceived as "the" enabler for the "Internet of things" [16, 17].

In the field of mHealth, this technology might be a game-changer as well. As mentioned before, the short operation range leads to an automatic data exchange when objects are brought together. Using a gesture of pointing the NFC-enabled mobile phone towards a specific object, communication with this object starts immediately. An application running in the mobile phone or the firmware itself could behave distinctively, depending on the data received from this object and hence, perform a specific action. This feature enables the user to express the willingness to perform a specific action simply by touching the corresponding object with the mobile phone.

1.3 NFC vs. Bluetooth

Bluetooth is currently the most popular wireless technology for linking a mobile phone acting as an application hosting device (AHD) to one or more medical sensor devices. Bluetooth operates in the ISM frequency band of 2,4 GHz and allows for setting up a master-slave oriented wireless network with an operation range of up to 100 m. Different versions of the Bluetooth specifications have been published by the Bluetooth Special Interest Group (SIG) [18] over the last years and implemented by the industry. These specifications include the Health Device Profile (available in Bluetooth V2.1 and V3.0), which is based on the IEEE 11073 standards framework, as well as some Bluetooth Low Energy profiles, which focus on healthcare applications. These interoperability approaches by the SIG makes Bluetooth the most prominent wireless personal area network (PAN) interface promoted by the Continua Health Alliance (CHA) [19]. This industrial consortium tries to promote and establish industry standards and guidelines in order to create interoperable telehealth devices and services.

Although the feasibility of using Bluetooth technology in mHealth applications could be demonstrated in many research and telemonitoring pilot projects, this technology still shows some drawbacks when it comes to the need of balancing usability and security on an adequately high level.

One thing is the need of setting up a paired connection between the AHD and each sensor device in order to make them communicate. But, so far, performing this setup procedure below Bluetooth V4.0 is not very intuitive. On a mobile phone, this procedure includes several manual steps such as opening the Bluetooth menu, enabling Bluetooth and starting the device and service discovery. Once a device and a service have been selected, the user is asked to enter a PIN code to finalize the pairing process [20] that guarantees encrypted data transmission. This configuration procedure (device/service discovery and pairing) requires some basic technical understanding and can be performed by technically skilled persons, but may overstrain especially elderly patients [21]. Therefore, the most common way is to provide patients with a prepaired patient terminal. This approach, however, also means a more complex replacement procedure if one or more devices need to be exchanged (a technician who replaces them and does the pairing again).

Having a sensor device paired with the AHD means that only this mobile phone is able to receive the data from this sensor device. Additionally, in most cases these data are transmitted automatically leading to the situation that the application may forward the data to the backend system automatically. Without handling these data, which implies further action to be taken by the patient, wrong values could be automatically forwarded to the backend system (e.g. the guest problem: a guest uses the weight scale in the bathroom).

What if it is intended to share one or more sensor devices like a blood pressure meter or a weight scale among several persons living in the same household? This kind of multi-user scenario cannot be covered by using Bluetooth technology since a sensor device is paired and, therefore, linked to one specific AHD usually belonging to one person. This can only be handled by means of different patient accounts on a single AHD, which is atypical when using a mobile phone.

Using NFC, many process steps to be taken during a data acquisition workflow could be simplified and optimized at the same time. It may provide patients with full authority to acquire their data without being bypassed through automatic data transmission, even in a multi-user environment, and provides potential for usability as well as security, both on a high level. In order to improve the level of security while keeping the process simple, the user can be authenticated by means of a token-based identification approach as known from RFID instead of remembering and always typing a username/password combination.

1.4 Continua Health Alliance Goes NFC

Beginning in Q3 of the year 2011, the Continua Health Alliance also considered NFC for acting as a potential linking technology between AHD and agents. Therefore, the CHA started collaboration with the NFC Forum in order to facilitate the development of related Continua specifications for NFC-enabled personal healthcare devices [22]. Next to Bluetooth, the CHA's predominant wireless PAN interface, and ZigBee, the most popular wireless Local Area Network (LAN) technology, NFC was selected as the linking technology to create a Touch Area Network (TAN). It was added to the range of existing technologies in order to cover the "tap-and-go" use cases defined by a sub-team of CHA's technical work group.

In Q4 2012, a version of the Personal Health Device Communication (PHDC) specification has been published by the NFC Forum as a candidate technical specification [23]. After final approval of this specification, it can be expected that several products will appear on the market in a little while. In parallel to appearance of new devices, the release of this specification will simplify development of mobile applications. Based on this, mobile applications will communicate exclusively with these devices by means of standards. These standards will be implemented on the operating system level, thus, developers will not have to take care of them anymore. In Android OS, the Bluetooth Health Device Profile (HDP) is already implemented. Thus, it can be expected that the NFC PHDC specification will be considered in Android OS as well, as soon as it is ready for integration.

Although NFC was selected to become a Continua-compliant transport technology in 2011, it was already addressed as a linking technology for mHealth applications much earlier. Thus, many solutions may be compliant to some low-level specifications by the NFC Forum, but transmitted data are still in a proprietary format due to the lack of an existing standard.

One of these solutions is the "Keep In Touch" (KIT) telehealth system of the AIT Austrian Institute of Technology, which is presented in the following. It is based on mobile handsets acting as patient terminals and medical sensor devices with NFC technology. After a few years of development and evaluation the system was deployed to several telehealth projects focusing on clinical research and chronic disease management in routine-like operation.

2 Methods – System Components

2.1 *Mobile Client Application*

Mobile phones with NFC capabilities were used as AHDs. Either a Java 2 Micro Edition (J2ME, Oracle, Redwood City, CA) [24] software application or an Android® (Google Inc., Mountain View, CA) app was developed to run on different NFC-enabled handsets. The application called KIT - MobileMonitor was developed in a generic design to be deployed to dedicated telehealth scenarios after specific customization. Therefore, it was based on a configurable workflow engine to provide an individual data acquisition procedure depending on the scenario's use cases, the needs of the user group as well as the range and combination of data to be collected.

The application was launched after touching a patient-specific RFID token (see section patient ID card). The graphical user interface based on icons guided the patient through the data acquisition process. Following this, the patient had to enter a set of health parameters which was carried out by touch-based interaction with KIT-enabled sensor devices and smart icons (see below). Additional quantitative data could be entered manually (insulin dosage, bread units, comments, etc.).

Acquired data were aggregated and stored on the handset's memory. They were synchronized with a remote telehealth system via a mobile Internet connection. Depending on the configured workflow, the application synchronized the data automatically after the patient had performed the last step or waited until he/she had touched the RFID token again, which triggered this process at any time.

2.2 Patient ID Card

A Mifare Classic 1K (NXP Semiconductors, Gratkorn, Austria) contactless smart card was used as a security token to identify the patient and authenticate access to the remote Web-based telehealth platform. The card was programmed with a special record according to the NFC Data Exchange Format (NDEF) specified by the NFC Forum. Touching this card to the mobile phone resulted in an automatic launch of the installed J2ME or Android application. In case of using a smartphone running Android OS, the card also kept an application record which pointed the handset to the right application at Google Play for downloading if it was not yet installed.

The ISO7816 format compliant plastic card (Figure 1) with an inlaid RFID transponder was printed with a project-specific logo, user ID phrase, and two fields to which the patient's first and last name could be written manually.

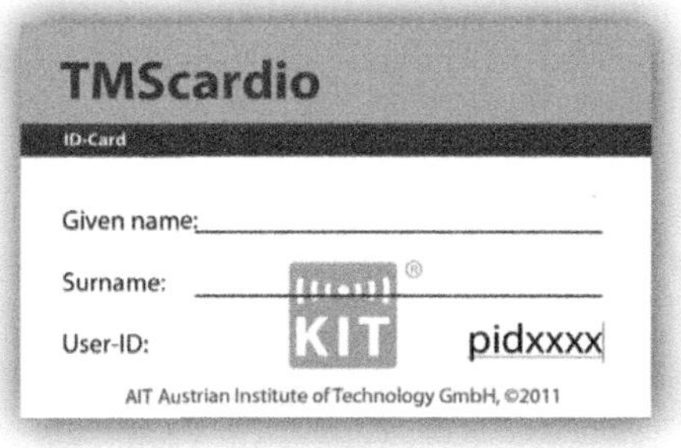

Fig. 1 A patient ID card

2.3 KIT Sensor Devices

Starting with a blood pressure meter, several medical sensor devices have been equipped with NFC technology in order to allow for simple data transmission to an NFC-enabled handset by simply bringing them close together. For this purpose, an NFC module was developed in a generic design to be integrated into a device or hooked up to a wired interface of a meter device. Based on this CE-marked NFC module, an industrial medical device manufacturer (A&D Medical, Tokyo, Japan) commercialized the world-wide first NFC-enabled medical device, by integrating it into a common blood pressure meter. After a successfully performed measurement, the data could be read out immediately by an application running on an NFC-enabled mobile phone (Figure 2).

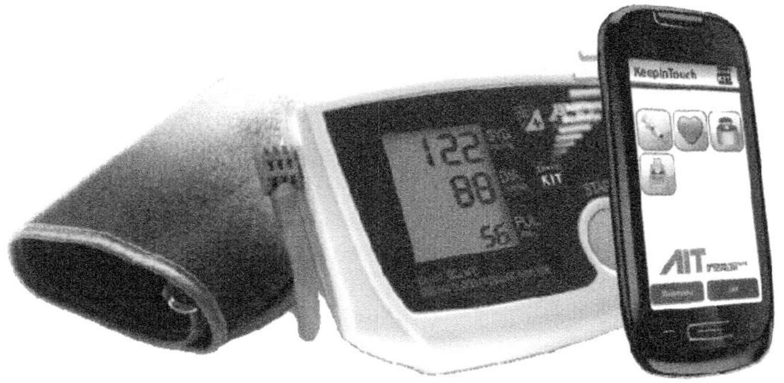

Fig. 2 A blood pressure meter with NFC-based data transfer to a mobile handset

2.4 KIT Box – The NFC Interface

Further devices like blood glucose meters and body weight scales were equipped with NFC by means of connecting an external interface device (KIT box) to the meter's serial data port. Therefore, the NFC module was put into a separate enclosure with an integrated battery compartment. In case of the body weight scales (UC-321PL, A&D Medical, Tokyo, Japan), only a single measurement value (the one from the last measurement) can be transmitted immediately after weighing. The KIT box for the blood glucose meter needed to be activated by means of pressing a button to get the values from the meter. It was able to store and transmit up to several measurement values collected over a period of at least one week. The KIT box for the body weight scale as well as for the blood glucose meter connected to the respective device is shown in Figure 3.

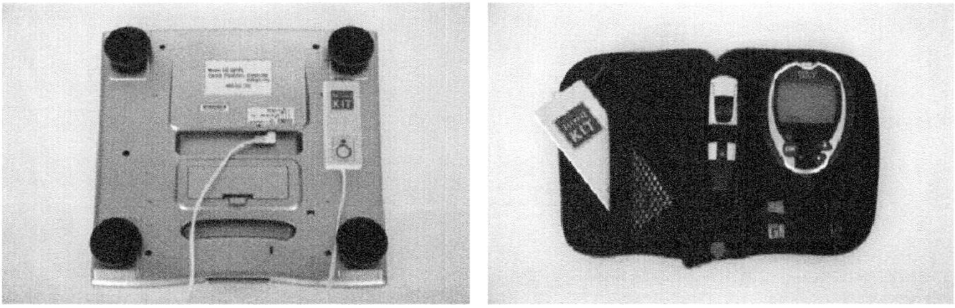

Fig. 3 Body weight scale (left) and blood glucose meter (right) with a connected KIT box

2.5 KIT 2.0 – A Combination of NFC and Bluetooth

In order to provide wireless connectivity beyond the range and data rate of NFC, a further communication concept called KIT 2.0 was designed, which enabled acquisition of huge amounts of data over a larger distance. The intention was to initiate the transmission process in the same simple way as known from pure NFC by bringing both devices close together. In this concept, the NFC and Bluetooth technologies were combined in a way that NFC was used to activate Bluetooth and set up the connection link by exchanging the required pairing information. After this process, both devices can be separated again and communicate with each other via the Bluetooth link that was established automatically.

This communication concept was used as an interfacing technology between a handset and a prototype of a blood pressure meter with an integrated electro cardiogram (ECG) recorder. This device allowed for intuitive acquisition of ECG data while performing a blood pressure measurement. By means of touching the device with the mobile phone at the position where the NFC antenna was located, a Bluetooth link between these two devices was set up automatically and the application running on the handset was able to control the device, i.e. starting and stopping the blood pressure measurement as well as recording a one-channel ECG signal. Therefore, the patient had to touch three metallic electrodes on the rear side of the enclosure (see Figure 4).

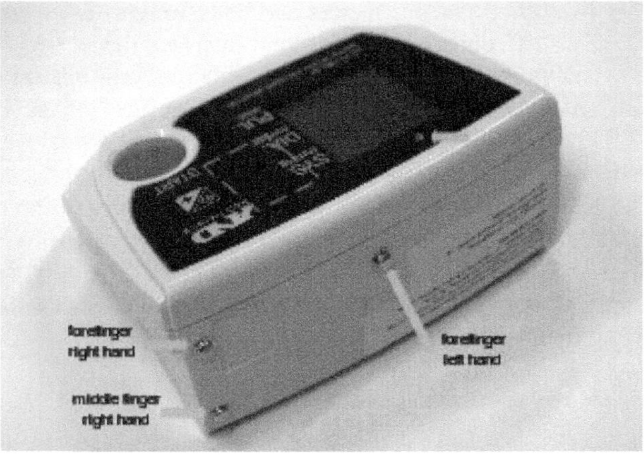

Fig. 4 A blood pressure meter with external electrodes and an integrated ECG recorder

2.6 *The NFC-Enabled Electronic Medication Blister*

Electronic medication blisters designed and developed by our partners (OtCM, DSM TCG B.V., Mauritslaan, The Netherlands) were added to the patient terminal platform in order to objectively track the dosage and timing of medication intake. These blisters were based on standard medication blisters in combination with some electronic parts and a coin-cell battery. A printed circuitry label was applied to the aluminium foil cover of the blister (bottom side) and linked to a small printed circuit board (keeping the electronic parts). By taking out a pill, the conductive track inside the label directly underneath the respective position was broken. This event was recognized at the input pin of a microcontroller which stored the corresponding data (position and timestamp) in its internal memory. In the following, these data could be interrogated wirelessly by an NFC-enabled mobile phone. This interface was also used to program and configure each individual blister. Figure 6 (left) shows an electronic medication blister read out by an application running on an NFC-enabled handset (Nokia 6131, Nokia, Espoo, Finland).

2.7 The RFID Dialogue Book

A foldable icon table (Figure 5) was designed to realize a physical symbol-based interaction method for answering a questionnaire displayed on the handset. Therefore, the user had to touch the appropriate symbol on the icon table which was underfed with an RFID tag that electronically stored the meaning of the respective icon. This table was printed with icons to answer various types of questions (yes/no/more/less, good/medium/bad, scale: very low/low/medium/high/very high) or, in specific cases, contained some activity icons (running, hiking, swimming, etc.) to document physical activities.

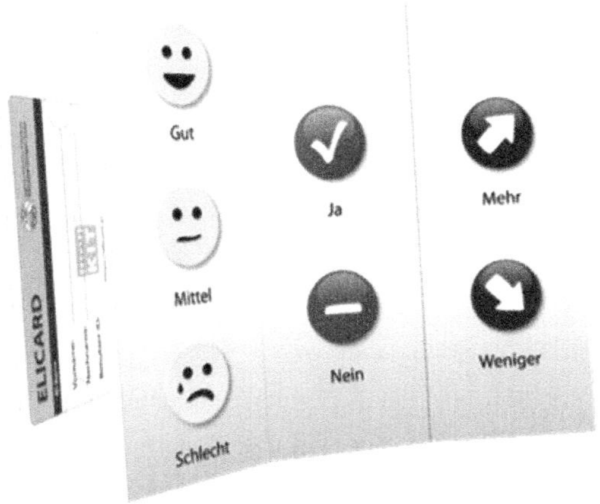

Fig. 5 A dialogue book with icons (good/medium/bad and yes/no/more/less)

All components of this patient terminal concept based on mobile phones and NFC technology were designed and developed in parallel in order to meet the requirements of high usability and security. This patient terminal allowed for intuitive interaction and acquisition of health parameters simply by touching objects and sensor devices. All acquired data were synchronized with a project-specific remote telehealth system which was customized separately to fit the requirements of each telehealth project (proof-of-concept or clinical trial). Several of them and the specific application of the patient terminal solution are presented in the following.

3 Selected Telehealth Projects

3.1 An NFC-Based mHealth Solution for Chronic Disease Management

3.1.1 ELICARD

One telehealth system was set up to run a proof-of-concept at the Elisabethinen general hospital in Linz, Upper Austria. This system, called ELICARD [25], was used to support professionals from the outpatient clinic for cardiovascular disease in treating their patients suffering from Chronic Heart Failure (CHF) and Pulmonary Arterial Hypertension (PAH). The application was registered at the Austrian data protection committee. AIT acted as a technical service provider and also provided the helpdesk.

Patients visiting the outpatient clinic as well as those with inpatient treatment who were eligible for participation were registered after signing informed consent. In the course of the registration, the attending physician entered anamnesis data, selected and configured monitoring parameters, and defined the medication to be monitored by the system. A doctor's letter was generated and printed to inform the general practitioner about the treatment supported by the telemonitoring system. Before discharge, the data acquisition procedure using the KIT-based telemonitoring equipment was presented to the patient by the physician. The equipment was comprised of the following components:

- a patient ID card
- an NFC-enabled mobile phone (Nokia 6131)
- a blood pressure monitor
- a body weight scale with a KIT box
- a pulse oximeter (only for PAH patients)
- a dialogue book to document well-being (good, medium, bad), medication intake and to answer further yes/no questions

All components were configured and prepared (including a graphical user manual) to be sent to the patient's home via mail service. After an additional explanation by the helpdesk via phone (if necessary) the patient acquired and transmitted his/her health data on a daily basis according to a strictly defined workflow consisting of two separate paths. One contained the data from the weight scale (placed in the bathroom) and the second was from the blood pressure meter together with the answered well-being and medication intake questions. Patients suffering from PAH had to answer additional questions regarding:

- difficulties with breathing
- whether oxygen saturation was below a certain threshold
- whether a laboratory test had already been done

- whether a new prescription was necessary.

Data were synchronized automatically after touching the KIT box of the weight scale (path 1) or after the last question was answered (path 2). A sound and a specific icon indicated that data were transmitted successfully or will be transmitted next time due to temporary network problems.

In case of malfunction of any system component, the helpdesk at AIT Austrian Institute of Technology was called either by the physicians or the patients themselves. If the problem could not be solved remotely, the faulty device was replaced via mail service.

ELICARD started operation at the Elisabethinen general hospital, Linz, on March, 1st 2009. At the end of July 2013, after a period of 51 months, 41 patients from Upper Austria were registered and equipped with the patient terminal equipment to be monitored by their attending physicians. 24 CHF patients (six females, median age 73.6 ± 11.3 years) were consecutively admitted to the service. All of them had been hospitalized because of cardiac compensation and were registered before discharge. After the system had been updated to meet the requirements for monitoring patients suffering from PAH, physicians started to also admit this patient group into the program. From August 2010, 17 PAH patients (11 females, median age 68.4 ± 8.3 years) were consecutively included to be monitored by their attending physicians. Thirteen patients were admitted after visiting the outpatient clinic, while four patients had in-hospital treatment prior to that.

During this period, a total of 134,282 data items were received in the course of 36,812 data transmission sessions in both patient groups. The cumulative monitoring period of the CHF patient group was 17,886 days (median 716, IQR [262-1,376] per patient). Data were transmitted in at least 14,670 days, which resulted in a patient adherence rate of 82.0%.

The PAH patient group was monitored over a cumulative period of 11,205 (median 763, IQR [250-1,039] per patient) days. A total of in 7,914 (median 519.0, IQR [210-681] per patient) data transmission sessions were observed. Considering six requested transmissions per week, this patient group showed an overall adherence rate of 83.7%.

During the 51-month operation period, four CHF patients (2 f) stopped using the system, because they didn't want to handle the devices every day. Three of them stopped within the first week and one person returned the devices after a period of one month. In the PAH patient group, no person dropped out from monitoring.

In 136 cases, the helpdesk was called due to problems with devices. In more than 65% of those cases, the problem could be resolved remotely or by giving a further explanation of how to handle the device in the current state. In the remaining cases one or more devices had to be replaced using mail service. In six home visits by the helpdesk, patients received an additional individual training session on how to handle the devices.

3.2 DiabMemory

The second proof-of-concept medium-scale telehealth project utilizing the KIT-based patient terminal was done in cooperation with the Austrian Health Insurance Company for Railway Workers and Miners (VAEB) that operated its own rehabilitation facility. During their stay at this healthcare institution, patients were asked to participate in a health program called "Gesundheitsdialog Diabetes". Patients who had a high risk for diabetes or already suffered from diabetes received the opportunity to get in close contact with their caregiver and general practitioner to permanently receive reliable and individual medical attention. The remote telehealth system called DiabMemory [26] was used as a therapy management component for this project. It was provided and operated by AIT and included helpdesk service.

Persons who were willing to participate signed informed consent and were able to choose between a manually-operated Web browser based user interface using a PC or the KIT-based patient terminal consisting of the following components:

- a patient ID card
- an NFC-enabled mobile phone (Nokia 6212, Samsung GT-S5230N, Nokia 701, or LG Optimus L5)
- a blood glucose meter (One Touch Ultra II) with a KIT box
- a blood pressure monitor (UA-767Plus NFC), (optional)
- a body weight scale (UC-321PL) with a KIT box, (optional)
- a dialogue book for documenting well-being (good, medium, bad), activity, and activity intensity

The components were preconfigured by a technically skilled person and handed out to the patients after the first week of their three weeks stay at the health resort. How to handle the devices was explained in the course of an initial training session immediately after handout. Subsequently, patients were asked to start acquiring their data and had to continue after discharge from the health resort.

Depending on the patient's underlying disease, an individual data acquisition and transmission scheme was defined. Blood glucose meters were given to all, while blood pressure meters and body weight scales were only given to selected patients. Patients suffering from Type I Diabetes were asked to perform three blood sugar measurements per day. Type II Diabetes patients had to perform at least three blood sugar measurements per week. In both groups, blood pressure and body weight had to be measured only once per week (if the respective devices were handed out).

The workflow of the MobileMonitor version for this project was designed in a more flexible way. The procedure of data acquisition was both started and finished by touching the patient ID card. When the application was running, it asked for a few data measurements (blood sugar, blood pressure, body weight, well-being, activity and its intensity). The user interface provided an input field to enter free-text comments. Patients were able to decide whether they wanted to acquire only a single parameter per session or more, by touching up to all items one after each other. The patient decided when to close the data acquisition session by touching the ID card again. This action triggered the MobileMonitor to aggregate, locally store, and upload the data. In case of a mobile network blackout, the data were transmitted within the next session. In case of any problem related to software or device malfunction, patients could call the helpdesk at the health resort or a technically skilled contact person at AIT.

The DiabMemory telehealth system started operation in May 2010. After a period of 39 months, a total of 542 persons (90 f) with a median age of 56.4 ± 10.4 years were registered. Most of these patients were suffering from type II diabetes (473 in total). The remaining persons were suffering from type I diabetes (30) or had been classified with an increased diabetes risk (9). More than four fifths of all patients, 437 in total (85 f, median age 57.2 ± 10.6 years) chose the KIT-based patient terminal to collect their health data and communicate with their caregivers. Seventeen of these 437 patients only received a mobile phone and an ID card, since they wanted to keep their existing blood pressure and blood glucose meters. Thus, they typed in their measurements manually. The remaining 105 patients (5 f, median age 53.5 ± 9.2 years) used their personal computers to access the Web system for entering all data manually.

Less than four percent of all mobile users, 17 in total (3 f), were classified as never-beginners, since they were registered, equipped and trained, but did not start to acquire their data autonomously. 32 (8 f) patients started with telemonitoring, but stopped participation within the first week after leaving the health resort.

During 39 months of operation, a total of 294,538 data transmission sessions were performed (median 271, IQR [131-503]) by the remaining 388 patients with mobile access. All data carried by these sessions consisted of a total of 454,160 (median 799, IQR [352-1553]) measurement values from NFC-enabled sensor devices, 133,053 (median 155, IQR [21-415]) data items indicating well-being conditions or physical activities, and 12,408 manually written comments. The major part of transmitted data items was represented by blood glucose values (57.3%). The remaining items comprised data related to blood pressure, body weight, activity, well-being, and comments with a proportion of 18.2%, 4.6%, 10.7%, 7%, and 2.2%, respectively. An analysis of time-related participation (duration between first and last transmission) of these 388 patients revealed a cumulative monitoring period of 229,012 days (median 566, IQR [296-890] days per patient).

During the 39 months of operation, a total of 52 mobile phones had to be replaced by mail service due to problems that couldn't be solved remotely. In 34% of those cases, the application running on the mobile phone could not be fixed by sending a reset SMS to the MobileMonitor. Only five blood pressure monitors had to be replaced due to problems with the internal NFC interface. In all cases, the problems could be traced to a loose plug connection inside the device's enclosure. After re-fixing the plug, the devices were sent back to the health resort to be handed out again. In case of the KIT box for the blood glucose meter, the broad application in this large population revealed a mechanical design problem of the first version. Due to manufacturing by manual work, the KIT box was fragile and broke apart in both enclosure components when the user tried to replace batteries. Thus, starting in January 2012, most of those devices were replaced by an entirely new version based on an industrially manufactured enclosure.

Recently it could be shown that the compliance of patients in a telehealth diabetes management program revealed a significantly higher compliance rate for patients using a mobile phone and NFC-based application user interface to acquire and transmit their health data than for patients submitting data predominantly via Web browser. Hence, different types of data acquisition technologies may have an important effect on patients' willingness to participate in telehealth programs in the long-term [27].

3.3 An NFC-Based Solution for Medication Management

The KIT-based telehealth solution was used for a medication compliance management study conducted at the diabetes outpatient clinic, Health Center South, Vienna, Austria. This randomized single-blinded (doctor-blinded), controlled, single-center study with cross-over design included participants with the presence of at least two of three health risks: Type II diabetes, hypertension, and hypercholesterolemia. Adherence to medication prescription was measured based on four defined standard drugs for treating diabetes (metformin 1,000 mg, Arcana), hypercholesterolemia (simvastatin 40 mg, Bayer, and rosuvastatin 10 mg, AstraZeneca), and hypertension (ramipril 5mg, Sandoz).

Study physicians accessed the Web-based study system by means of a Web browser for registering new patients. During registration, the physician had to prescribe one or more medications and set the timeframe and dosage of pills to be taken.

In the monitoring phase of the study, the following equipment was handed out to the participants:

- a patient ID card
- an NFC-enabled mobile phone (Nokia 6131)
- up to three different medications packaged in smart medication blisters

A special NFC-enabled terminal device located at the study site was used to connect to the database and to receive the prescription data. This device was used during the blister handout process to initialize each unit and program it with the patient and prescription data. After receiving the entire equipment, the patients had to take their pills according to the schedule communicated by the physician and transmitted each take-out event to the remote study system. Therefore, the patients had to launch the mobile application by touching the user ID card followed by an interrogation of the smart blister (Figure 6, left). This procedure had to be repeated for each blister after taking out one or more pills.

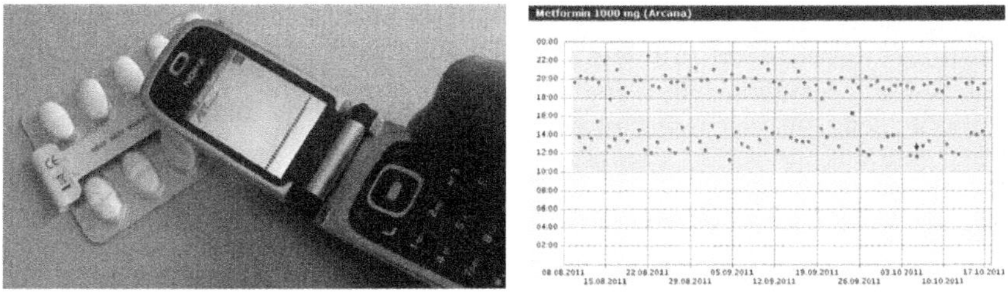

Fig. 6 Nokia 6131 interrogating a smart medication blister (left) and a representation of the adherence data (right)

In case of not transmitting all expected data, the patient received an SMS on the following day as a reminder to transmit the blister data. Received data were analyzed in real-time to check whether the patient was adherent. Adherence data, as represented in Figure 6, right, were regularly observed by a study coordinator who contacted all subjects with minor adherence (<70% of prescribed drugs taken) via phone call once per week, trying to provide adherence motivation.

Once the patient had consumed all pills, he/she was requested to visit the study site again. All blisters were put onto an NFC reader to finally read all takeout events stored in the blisters' memory. These data were sent to the remote study system to be merged with the self-transmitted data already available in the central database. New blisters were handed out as described above and the patient continued with medication adherence monitoring. Changes in dosage or schedule were communicated by phone or during consultations with the physician.

The study was started in September 2010 and ran for a period of 13 months until the study sponsor, i.e. the company that provided the electronic blisters, decided to terminate the study prematurely. During this period, 77 patients had been registered to the study system. 24 participants categorized as drop-outs have been counted at the time of termination. They stated different reasons. Only a single participant stopped using the system because of operating difficulties.

The remaining 53 participants had entered the monitoring phase before the study was terminated. Their mean age at registration was 69.4 ± 4.75 years. All of them were equipped with a mobile phone, a user ID card, and electronic blisters to acquire and transmit their medication adherence data. These 53 participants received a total of 1,654 electronic blisters, with an average of 30.6 ± 13.1 blisters per patient. With a proportion of 57.8% (957 blisters), metformin was prescribed the most. The other three drugs were handed out less frequently with a quantity of 332 (ramipril), 193 (simvastatin), and 172 (rosuvastatin) blisters. The cumulative monitoring period of these 53 participants was 6,788 days with a mean period of 125.7 ± 27.4 days per individual. 14,203 pill take-out events were recognized by the electronic blisters and have been transmitted by a total of 10,359 data transmission sessions. 627 of these 14,203 recorded events did not fall into the defined time range and have been labeled as "non-adherent". Thus, the time adherence of all taken pills was calculated to be 95.59%. Each participant in the monitoring phase recorded and transmitted a mean quantity of 263.0 ± 123.4 pill take-out events. Analysis of the take-out event and transmission timestamps revealed that only 56% of the detected events were transmitted immediately after the pill was taken out (< 5 minutes). 88% of the data were transmitted within 24 hours after the takeout event.

Further details have been published in [28].

3.4 An NFC-Based mHealth Solution for Biomedical Research

The first employment of a KIT-based telemonitoring system in the field of biomedical research was in the course of a clinical pilot trial. In this case, it was linked to the telehealth platform acting as an electronic data capture (EDC) system to evaluate the feasibility of a new haemodynamic sensor device for monitoring various patient groups with specific cardiac disorders. This non-invasive sensor device uses a conventional upper-arm blood pressure cuff to measure the patient's systolic and diastolic blood pressure and heart rate using the oscilometric method [29]. In the course of a second measurement cycle, the device records several peripheral pulse-waves and uses a neural network-based mathematical algorithm to calculate the aorta pulse wave, stroke volume, total peripheral resistance and augmentation index. CHF patients who were hospitalized were asked to participate in this trial. Their cardiovascular parameters were measured using this sensor device. A study physician performed the measurements twice a day during the in-hospital stay. Data were acquired and transmitted to the telehealth platform using a Nokia 6131 NFC with a preinstalled MobileMonitor. Additionally, data were also documented manually in a patient-specific case report form (CRF) for later comparison. To identify each person, the study physician as well as every patient were equipped with an ID card to be read by the mobile phone during the data acquisition process as summarized in Table .

Table 1 Data acquisition workflow

Step	Action by the Study Physician	Mobile Application Behavior
1	Touch physician's ID card	MobileMonitor launch, authentication of physician Next: ask for sensor data (symbol-based)
2	Put on cuff and start measurement	
3	Touch CardioMon to interrogate the data	Read data and check the CardioMon symbol Next: ask for patient ID
4	Touch patient's ID card	Read ID information and start sending data Next: ask for sensor data
5	Manually document the measured data in a patient-specific CRF	
6	Repeat the procedure (step 2-5) Quit by pressing the hang up button	Same as above Close MobileMonitor

New patients were registered automatically after touching a new patient ID card. Full registration had to be completed afterwards using the study physician interface (PC with a Web browser).

In case of failure during data transmission, the data were stored locally to be transmitted with the next acquired record. However, all errors related to the data acquisition and transmission process as well as those related to the measurement device were documented to determine the system's reliability.

This EDC system [30] has been evaluated at the Department of Internal Medicine - Division of Cardiology of the Medical University of Graz. During a total a period of 77 days, 25 patients (14 f, mean age 77.5 ± 10.0 years) suffering from CHF were observed with the sensor device during in-hospital clinical treatment. Observation was performed by a study physician and the mean period per patient was 6.0 ± 5.0 days. From each CHF patient participating in this clinical trial, a mean of 13.2 ± 10.6 data sets were acquired during the hospital stay after an acute cardiac decompensation. Each patient was measured twice a day.

The study physician manually documented 331 (164x2 + 3x1) successfully performed measurements. At the backend system, a total of 323 data transmission sessions were recognized. These transmission sessions carried a total of 331 acquired data sets which were stored in the database. A correlation of the manually documented and electronically acquired data sets revealed a 100 % concordance and proved the system's accuracy in terms of linking medical data to the corresponding patients.

Comparing the timetable with the timestamps of interrogating the device with the NFC handset and the timestamps of the records being stored in the central database revealed that in eight of the 331 instances the MobileMonitor had to store the data locally due to temporary GPRS/EDGE network blackouts. In all of those cases, the locally buffered record was eventually successfully transmitted, either with the subsequent record (seven cases) or on the next day (one case).

3.5 An mHealth Solution Combining NFC and Bluetooth

The KIT 2.0 communication platform and the developed blood pressure/ECG monitor were evaluated in the course of a clinical pilot trial. The study was conducted at the Medical University of Graz and was approved by the corresponding ethics committee (21-283 ex 09/10). Patients have been recruited by study physicians at the Division of Cardiology and signed informed consent, which was obtained from all participating patients. During the registration process, the study physician entered the past medical history data and performed an ECG recording. Afterwards, the patient received the telemonitoring equipment consisting of:

- a patient ID card
- an NFC-enabled mobile phone (Nokia 6212 classic)
- a KIT 2.0-enabled ECG/blood pressure sensor device
- a body weight scale (UC-321PL) with a KIT box

- an icon table for documenting well-being (good, medium, bad) and medication intake
- a step-by-step user manual

Patients were instructed in using the system by initially observing the trainer operating the system followed by one assisted data acquisition by the patients themselves. After handout of patient terminal equipment and successful training, patients were asked to use the system for a period of one week two times a day according to the workflow illustrated in Table with the following two slightly different usage scenarios:

- morning: sensor device had to be applied to the chest so that the three electrodes contacted the naked skin
- evening: electrodes had to be touched with the fingertips.

Table 2 Data acquisition workflow including the KIT 2.0-enabled ECG/blood pressure sensor device

Step	Interaction by the patient	MobileMonitor Behavior
1	Touch ID card	MobileMonitor launch, user authentication Next: ask for touching sensor device
2	Put on cuff Touch sensor device	Establishes a connection to sensor device Next: ask to contact the electrodes
3	Contact electrodes (fingertips or chest)	Triggers sensor device to perform measurement → displays ECG signal while measurement runs Next: ask for body weight
4	Step on body weight scales and touch NFC interface	Read body weight Next: ask for CHF medication
5	Touch one of four icons to confirm intake of CHF drugs (up to four times)	Read the tag data of the touched icon (up to four times) Next: ask for well-being
6	Touch one of three smiley icons to indicate current well-being state	Read the tag data of the touched icon, upload all data and close application

Medical intervention was not intended during the observation period. In case of questions or technical problems, patients could contact the study physician or a technical helpdesk. After one week the patients returned the system and attended a final examination where an ECG (12-lead) was acquired using a clinical ECG recorder. To assess individual patient satisfaction, the patients were asked to fill out a standardized usability questionnaire.

The results of this clinical pilot trial [31] focusing on technical feasibility, patient adherence, general user acceptance, and usability are listed in the following. Patient characteristics and monitoring results are summarized in Table . A total of 21 patients (54.1 ± 14.7 years, 4 f) were enrolled into the clinical trial and have been observed over a period of ten weeks. All patients suffered from CHF with a median CHF severity of NYHA II. In addition to CHF, three patients had already suffered from stroke and two patients suffered from diabetes mellitus or COPD, respectively.

Table 3 Patient and transmission characteristics

Parameter	Total	Mean (SD)
Patients (female) [n]	21 (4)	
Age [years]		54.1 (14.7)
Patients in NYHA class (I / II / III / IV) [n]	9 / 7 / 5 / 0	
Patients with stroke / diabetes mellitus / COPD	3 / 2 / 2	
Drop-outs (female) [n]	1 (1)	
Patients with at least one transmission per day [n]	18	
Monitoring period [days]	140	
Transmissions [n]	211	10.6 (3.8)
Adherence rate [%]	82.2	
Transmissions of subgroup with 18 patients [n]	209	11.6 (2.1)
• Adherence rate [%]	89.3	
Received data items [n]	1501	
• ECG signals [n]	211	10.6 (3.8)
• Blood pressure [n]	198	9.9 (3.8)
• Body weight [n]	211	10.6 (3.8)
• Well-being		
o Good [n]	161	8.1 (4.1)
o Medium [n]	46	2.3 (3.2)
o Bad [n]	4	0.2 (0.5)
• Medication [n]	673	33.7 (15.3)

A total of 262 data sets were finally stored in the system's database. 41 data sets were acquired during the training session at the hospital and 211 data sets were acquired in real-life settings at the patients' homes by themselves. One female participant didn't start to acquire and transmit her data and was thus rated as a drop-out. In case of two patients, only a single data set could be transmitted due to technical problems (weak battery condition and defective body weight scales). The cumulative monitoring period of all active patients (N=20) was 140 days. By expecting 13 autonomously acquired data sets per patient, the overall adherence of all 20 participants was calculated at 82.2%. In consideration of two participants with technical problems, the remaining 18 patients showed a patient adherence rate of 89.3%.

Individual patient satisfaction was analyzed based on 20 completed questionnaires. 19 patients reported that they already owned a mobile phone, while only two of them already used mobile Web services. Patients stated that they had no problems in reading the display or understanding the voice command given by the mobile application. The whole data acquisition procedure including the six steps specified in Table 2 took 3.9 minutes on average. The individual statements about expenditure of time ranged from 1 to 10 minutes. 17 of 20 patients rated the idea for this service as very good. All of them would recommend this service to other patients, while only ten patients would continue to use this service. Seven of them would be willing to pay for it.

4 Discussion

4.1 NFC Bridges the Gap between Usability and Security

The intuitive solution of touch-based data acquisition given by NFC provides a high level of usability, which, among others, is essential for gaining user acceptance. User acceptance is one of the most important prerequisites when it comes to elderly and technically unskilled persons. The process of data acquisition is easy to use, as it can be conducted without a single keypress either for digging into a menu to start the application or for configuring the communication link to sensor devices.

High usability is also given by the fact that the NFC-based KIT solution provides only a single and, hence, consistent solution for acquiring all kinds of data. Table 4 shows an overview how different data can be acquired using KIT technology.

Table 4 List of linking technologies and data types to be transmitted

Link	Technology	Data volume	Data type
KIT 1.0	RFID	few bits or bytes (static)	User and object identification, symptomatic data, compliance data, etc.
KIT 1.0	NFC	bytes to kilobytes (dynamic)	Vital signs: blood pressure, body weight, blood sugar, O2 saturation, step counts, etc.
KIT 2.0	NFC+Bluetooth	kilobytes to Megabytes (dynamic)	Streamed or stored (large volume) biosignal data: ECG records, 3D accelerometer data, plethysmography, etc.

The incidental coupling of devices allows for a Plug-and-Play-like expansion of existing patient terminals and usage in multi-user environments. These features also contribute to this high level of usability, which has an enormous influence on the patient training effort as well as on the patient adherence.

In terms of security, the KIT solution provides an easy-to-use mechanism to secure all data. The ID card is not only used to launch the application, but also to store user credentials. These data are used to login to the application and to authenticate the patient against the backend system. Although the MobileMonitor can be started manually in the application menu, the ID card is absolutely necessary to log in. In addition to that, all data stored locally are encrypted based on these credentials. Thus, an attacker needs to get both, the mobile phone as well as the ID card in order to hack the stored data. If the card gets lost, the patient needs to contact the helpdesk to receive a new one. But this ID card does not work until the helpdesk has sent a new configuration SMS to the phone number that is assigned to this patient.

Data reliability also belongs to the field of security. As demonstrated in the clinical pilot trial with the new haemodynamic sensor device, the KIT approach allows for an error-proof assignment of measurement values to patient records. Not only for clinical research, but also for routine telemonitoring applications, the automated data acquisition procedure prevents intended or unintended entry of wrong or faked data.

Due to the fact that devices do not have to be paired, they can be simply replaced without any configuration in case of damage or malfunction. Thus, neither the patient or a relative nor a technician is required to add a new device to an existing patient terminal setup. This is valid even for mobile phones and KIT 2.0 devices with Bluetooth technology. This circumstance allows adding or replacing devices via mail service, which is less time-consuming and more cost-effective.

4.2 Tailored Solutions for Different Indications and Age Groups

A comparison of CHF and PAH patients with diabetes patients - all of them equipped with the KIT patient terminal - shows a different distribution in acquired data. In CHF and PAH patients, the transmitted blood pressure, body weight and well-being data are uniformly distributed, while in diabetes patients, blood glucose data were acquired more often. Compared to blood glucose data, the remaining parameters were less often documented at different frequencies.

This is due to the fact that vital signs are of different importance depending on the disease. The patient terminal needs to be designed considering this situation in addition to other circumstances like the patient's age or life situation. With KIT technology, these conditions were considered by means of a process-oriented approach with workflow support.

Patients suffering from CHF or PAH mostly stay at home and need to collect the relevant data every day or every other day. Since these patients are generally of higher age [32], the workflow of data acquisition needs to be designed in an appropriate way in order to avoid intentionally or unintentionally leaving out crucial data. Additionally, there should be no chance of getting lost in a submenu. Thus, in this case, the MobileMonitor requests all necessary data in a kind of dialogue workflow. One device or item has to be touched after another in order to follow the displayed instruction or answer a question. In order to consider the situation where sensor devices may be distributed in the home (having the weight scale placed in the bathroom, while operating the blood pressure meter mainly in the kitchen or living room where the medications are taken), data acquisition is separated in two paths. One is the acquisition of the body weight, while the other path combines the remaining data (blood pressure and questions). Once the last data item of each path was acquired, the application transmitted the data immediately and visualized feedback indicating that the transmission was successful. This feedback was found to be essential for the patients, as they need to be assured that their data can now be seen by the doctor. Otherwise, as observed during the MOBITEL trial [33], they were unsure and sometimes contacted the doctor in order to check whether the data had been received.

In telemonitoring of diabetes patients, use cases are totally different to those as given for CHF and PAH patients. Diabetes patients are also of younger age, are in working life and are usually mobile. The most crucial parameter to be monitored is the blood glucose level. It has to be documented regularly, sometimes even several times a day, partially in combination with intake of bread units and insulin. Other vital signs, such as blood pressure or body weight, are of less importance as reflected by a sample rate of only once per week. Documentation of well-being and physical activities is done occasionally. Thus, it is not feasible to combine different data and create a predefined path that has to be run through. This resulted in a fully open data acquisition without a predefined workflow. The patient has full authority in terms of timing and data to be acquired. He/she is able to document only a single parameter at a time at exactly the time when it occurs or he/she gets the possibility to acquire more parameters at a time and it is up to the patient to leave out any parameter. Since this approach always allows for adding another parameter, the application doesn't recognize automatically whether the last parameter was entered. Thus, the patient has to provide this information in order to start the transmission process. In this case, he/she touches the ID card to indicate that the data acquisition process has finished.

These two different workflows were deployed by means of individual compilation of the MobileMonitor application. But in order to configure the scope of parameters to be acquired and the sequence of a predefined path, the MobileMonitor could be configured remotely. This allowed for an individual setup for each patient definable by the doctor.

4.3 Outlook: Future Potential of Smartphone Platforms with NFC

In 2013 the diversity and regional availability of NFC-enabled mobile phones has changed dramatically compared to the early years. The term mobile phone now has to be split into feature-phone with an ordinary firmware and smartphone with a sophisticated operating system. Starting in December 2010, NFC-enabled smartphones based on operating systems like Android OS (Google Inc, Mountain View, CA), Blackberry OS (RIM, Waterloo, ON) or Windows Phone (Microsoft, Redmond, WA) appeared on the marked. In March 2013, more than a hundred individual devices – manufactured by different companies like Samsung, Sony, Nokia, HTC, LG and so on – were available. When these operating systems appeared, development of mobile applications also became much simpler and, hence, more popular due to easy-to-handle development tools and powerful APIs. Additionally, development of so-called "apps" became popular due to a new sort of distribution by means of dedicated application stores like the PlayStore (Google Inc, Mountain View, CA) or the App Store (Apple Inc, Cupertino, CA). This concept allows for simple installation of any software application by searching a Web-based catalog with any keyword and selecting the desired app from the result list. By the end of 2012, tens of thousands of apps related to health, fitness and wellness were available in different application stores.

In order to take advantage of this broad variety and availability of NFC-enabled smartphones, the KIT-based patient terminal was also ported to the Android OS. Besides simplifying the deployment process (by using the PlayStore), this transaction also was beneficial in respect of NFC functionalities, push features, security and maintainability.

Next to mobile phones and tablet computers, Android is also a popular operating system of smart television (TV) devices. These are connected to the Internet as well and able to run the same apps as known from mobile devices. In January 2013, Sony demonstrated a remote control featuring an NFC interface [34] that could be used in combination with an Android-based smart TV in order to serve the same functionality as demonstrated by the KIT system.

In addition to mobile phones and other consumer electronic products enabled with NFC technology, a few medical sensor devices started to appear, which also featured touch-based interrogation by means of NFC. Most of these devices originated from Japanese companies, as they are based on the Japanese FeliCa technology, which was developed by Sony. FeliCa also belongs to the NFC standard and is compatible with almost any NFC-enabled smartphone. The line-up of these FeliCa-enabled sensor devices ranges from pedometer or activity monitors, over body weight scales and blood pressure monitors, up to blood glucose sensors and body compensation monitors. Since these devices are based on the "FeliCa Plug" interface that offers memory for storing measurement values to be interrogated by a reader device, each company using this interface may implement its own proprietary data format. Therefore, Sony has implemented a so-called NFC healthcare library [35] that takes care of these different proprietary implementations in order to ease application development.

This new and continuously growing ecosystem of smartphones as well as sensor devices featuring an NFC interface will contribute to the ramp-up of development and deployment of easy-to-use mHealth applications.

5 Conclusion

A patient terminal concept based on mobile phones and NFC technology was designed and implemented to establish a versatile and easy-to-use link between patients suffering from chronic diseases and their attending physicians. It was used in a number of proof-of-concept telehealth projects in cooperation with hospitals (treatment of heart failure patients) and health insurance companies (disease management for diabetes) as well as clinical trials. During these projects – most of which are still running in a routine-like operation – we gained comprehensive experience in terms of how to operate mHealth-based telehealth in a real world scenario in general, and with NFC technology in particular. The results obtained so far are promising. They indicate that NFC may act as the enabling technology with potential for reducing the complexity of telehealth applications to a minimum. Additionally, it keeps security on a high level, which is a basic prerequisite, since healthcare deals with sensitive personal data. However, there is still a necessity and a huge potential for future developments and further enhancements, e.g. towards including environmental information. Such next-generation mHealth systems are expected to utilize concepts like "The Internet of Things" for medical applications, where we see NFC as a key enabling technology again.

References

[1] Kinsella, K., He, W.: An Aging World, `http://www.census.gov/prod/2009pubs/p95-09-1.pdf` (cited October 29, 2013)

[2] Hsiao-Chen, M., Che, J.L., Shihmin, L., Sher, P.J.: Does technological standardization enhance or inhibit firm's capabilities? Research on dynamic capabilities in the mobile communication market. In: Proceedings of PICMET 2010: Technology Management for Global Economic Growth (PICMET), pp. 1–8 (2010)

[3] Katz, D.L., Nordwall, B.: Novel Interactive Cell-Phone Technology for Health Enhancement. J. Diabetes Sci. Technol. 2(1), 147–153 (2008)

[4] Hedtke, P.A.: Can wireless technology enable new diabetes management tools? J. Diabetes Sci. Technol. 2(1), 127–130 (2008)

[5] Agarwal, Y., Chandra, R., Wolman, A., Bahl, P., Chin, K., Gupta, R.: Wireless wakeups revisited: energy management for voip over wi-fi smartphones. In: Proceedings of the 5th International Conference on Mobile Systems, Applications and Services, San Juan, Puerto Rico, pp. 177–191 (2007)

[6] Johnson, D.A., Trivedi, M.M.: Driving style recognition using a smartphone as a sensor platform. In: Proceedings of 14th International IEEE Conference on Intelligent Transportation Systems (ITSC), October 5-7, pp. 1609–1615 (2011)

[7] Oshin, T.O., Poslad, S., Ma, A.: Improving the Energy-Efficiency of GPS Based Location Sensing Smartphone Applications. In: Proceedings of IEEE 11th International Conference on Trust, Security and Privacy in Computing and Communications (TrustCom), June 25-27, pp. 1698–1705 (2012)

[8] Palit, R., Singh, A., Naik, K.: Enhancing the capability and energy efficiency of smartphones using WPAN. In: Proceedings of IEEE 22nd International Symposium on Personal Indoor and Mobile Radio Communications (PIMRC), September 11-14, pp. 1020–1025 (2011)
[9] Vital Wave Consulting. mHealth for Development: The Opportunity of Mobile Technology for Healthcare in the Developing World, http://unpan1.un.org/intradoc/groups/public/documents/unpan/unpan037268.pdf (cited October 29, 2013)
[10] Holman, H., Lorig, K.: Patients as partners in managing chronic disease - Partnership is a prerequisite for effective and efficient health care. BMJ 320(7234), 526–527 (2000)
[11] NFC Forum http://www.nfc-forum.org/home/ (cited October 29, 2013)
[12] Rinner, D., Witschnig, H., Merlin, E., Broadband, N.F.C.: - A system analysis for the uplink. In: Proceedings of 6th International Symposium on Communication Systems, Networks and Digital Signal Processing, CNSDSP 2008, July 25-25, pp. 292–296 (2008)
[13] Yang, D., Wang, Q.: The study on the application of RFID- based mobile payment to the Internet of Things. In: Proceedings of 2011 International Conference on Multimedia Technology (ICMT), July 26-28, pp. 908–911 (2011)
[14] Ozdenizci, B., Coskun, V., Aydin, M.N., Kerem, O.: NFC loyal: A beneficial model to promote loyalty on smart cards of mobile devices. In: Proceedings of 2010 International Conference for Internet Technology and Secured Transactions (ICITST), November 8-11, pp. 1–6 (2010)
[15] Dobrigkeit, P., Flach, T., Faymonville, P., Mueller, J.: Exchange of contact data between mobile phones using NFCIP. In: Proceedings of 2008 4th European Workshop on RFID Systems and Technologies (RFID SysTech), June 10-11, pp. 1–9 (2008)
[16] Garrido, P.C., Miraz, G.M., Ruiz, I.L., Gómez-Nieto, M.A.: A Model for the Development of NFC Context-Awareness Applications on Internet of Things. In: Proceedings of 2010 Second International Workshop on Near Field Communication (NFC), April 20, pp. 9–14 (2010)
[17] Semiconductors, N.X.P.: NFC is the double-click in the internet of the things. In: Proceedings of 2007 3rd European Workshop on RFID Systems and Technologies (RFID SysTech), June 12-13, pp. 1–5 (2007)
[18] Bluetooth Special Interest Group, https://www.bluetooth.org/en-us (cited October 29, 2013)
[19] The Continua Health Alliance, http://www.continuaalliance.org/ (cited October 29, 2013)
[20] Sharmila, D., Neelaveni, R., Kiruba, K.: Bluetooth man-in-the-middle attack based on secure simple pairing using out of band association model. In: Proc. Int. Conf. Control, Automat., Commun. Energy Conservation, pp. 1–6 (June 2009)
[21] Istepanian, R.S., Zitouni, K., Harry, D., Moutosammy, N., Sungoor, A., Tang, B., Earle, K.: Evaluation of a mobile phone telemonitoring system for glycaemic control in patients with diabetes. J. Telemed. Telecare 15(3), 125–128 (2009)
[22] NFC Forum. Continua Health Alliance and the NFC Forum Collaborate to Expand Connectivity, Simplify Data Exchange in Healthcare IT, http://www.nfc-forum.org/news/pr/view?item_key=0986047c50bdb4848314f1427210ce91a8ff78aa (cited October 29, 2013)

[23] NFC Forum. NFC Forum Candidate Technical Specifications, `http://www.nfc-forum.org/specs/candidate_spec_list/` (cited October 29, 2013)

[24] Oracle. Java ME and Java Card Technology, `http://www.oracle.com/technetwork/java/javame/index.html` (cited October 29, 2013)

[25] Kastner, P., Morak, J., Modre, R., Kollmann, A., Ebner, C., Fruhwald, F.M., et al.: Innovative telemonitoring system for cardiology: from science to routine operation. Appl. Clin. Inf. 1(2), 165–176 (2010)

[26] Morak, J., Schreier, G.: mHealth based on NFC technology – Preliminary results from medium scale proof of concept projects. In: Proceedings of the eHealth 2012, Vienna, Austria, May 10-11, pp. 131–137. OCG (2012)

[27] Schreier, G., Eckmann, H., Hayn, D., Kreiner, K., Kastner, P., Lovell, N.: Web versus App - compliance of patients in a telehealth diabetes management programme using two different technologies. J. Telemed. Telecare 18(8), 476–480 (2012)

[28] Brath, H., Morak, J., Kästenbauer, T., Modre-Osprian, R., Strohner-Kästenbauer, H., Schwarz, M., Kort, W., Schreier, G.: Mobile health (mHealth) based medication adherence measurement - a pilot trial using electronic blisters in diabetes patients. Br. J. Clin. Pharmacol. 76(suppl. 1), 47–55 (2013)

[29] Pacheco, M., Jiménez, M.J., Ramírez, L., Nalda, M.: Clinical study of a new monitor for the determination and recording of arterial pressure by a non-invasive electronic-oscilometric method. Rev Esp Anestesiol Reanim 31(3), 99–104 (1984)

[30] Morak, J., Schwetz, V., Hayn, D., Fruhwald, F.M., Schreier, G.: Electronic data capture platform for clinical research based on mobile phones and near field communication technology. In: Conf. Proc. IEEE Eng. Med. Biol. Soc., pp. 5334–5337 (2008)

[31] Morak, J., Kumpusch, H., Hayn, D., Leitner, M., Scherr, D., Fruhwald, F.M., Schreier, G.: Near Field Communication-based telemonitoring with integrated ECG recordings. Appl. Clin. Inf. 2(4), 481–498 (2011)

[32] Boonman-de Winter, L.J., Rutten, F.H., Cramer, M.J., Landman, M.J., Liem, A.H., Rutten, G.E., Hoes, A.: High prevalence of previously unknown heart failure and left ventricular dysfunction in patients with type 2 diabetes. Diabetologia 55(8), 2154–2162 (2012)

[33] Scherr, D., Kastner, P., Kollmann, A., Hallas, A., Auer, J., Krappinger, H., Schuchlenz, H., Stark, G., Grander, W., Jakl, G., Schreier, G., Fruhwald, F.M.: Effect of home-based telemonitoring using mobile phone technology on the outcome of heart failure patients after an episode of acute decompensation: randomized controlled trial. J. Med. Internet Res. 11(3), e34 (2009)

[34] The Verge. Sony adding Miracast to all high-end TVs, introduces NFC remote control for Xperia screen sharing, `http://www.theverge.com/2013/1/7/3848362/sony-mirasol-nfc-smartphone-tv-screen-sharing` (cited October 29, 2013)

[35] Sony. NFC Healthcare Library, `http://www.sony.net/Products/felica/business/products/NFC_healthcare.html` (cited October 29, 2013)

36. RFID in Healthcare – Current Trends and the Future

Saravanan Sundaresan*, Robin Doss, and Wanlei Zhou

School of Information Technology, Deakin University, Burwood, Australia

Keywords: RFID Healthcare Systems, RFID Tags with Sensor Capabilities, EMI Caused by RFID, Security and Privacy Issues of RFID.

1 Introduction

Radio Frequency Identification (RFID) enables automatic identification of objects using radio waves. The identified objects can be in and out of the line of sight and there is no need for physical contact with them. RFID technology is deployed in a wide range of industries such as supply chain management, inventory control, farming (to track animals), e-Passports, the tracking of humans (in prisons and hospitals) and in healthcare [1]. The three key elements of an RFID system are the tags, readers and the backend server. Tags are devices physically attached to objects and readers (wired or mobile) recognize the presence of objects in its range. The server maintains all the information about IDs for the tags and readers, their respective secrets and any information about the object attached to the tag. A tag is typically made up of an antenna for receiving and transmitting a radio-frequency (RF) signal and an integrated circuit for modulating and demodulating the signal and storing and processing information. There are three types of RFID Tags: active tags, semi-active tags and passive tags [2]. Active tags have their own battery to power its internal circuitry and transmission components; semi-active tags also have their own power source which is used only for powering the internal circuitry but not for transmission; passive tags have no internal battery to power themselves and they use the electromagnetic signal from the reader as the power source. This makes passive tags highly cost-effective thereby enabling large-scale applications. However, they are highly resource constrained and have very limited capabilities in terms of functionalities. The total volume of tags used worldwide was estimated to be 10.6 billion in 2011 of which 80% were low-cost UHF passive tags such as the EPC Class-1 Generation-2 (C1G2) tags [3].

The ability for readers and tags to be mobile and wireless in an RFID system opens up a world of possible applications in healthcare. In this chapter, we provide extensive coverage of how RFID impacts the healthcare arena. In depth analysis is

* Corresponding author.

S. Adibi (ed.), *Mobile Health*,
Springer Series in Bio-/Neuroinformatics 5, DOI: 10.1007/978-3-319-12817-7_36

provided covering areas such as RFID based healthcare applications (including tags with sensor capabilities and implantable tags), a study on Electro Magnetic Interference (EMI) caused by RFID, security and privacy issues of RFID in healthcare, secure RFID protocols for healthcare, the ethical implications of deploying RFID in healthcare, barriers impacting their deployment, benefits reaped from successful implementation, and finally we provide future research directions. By taking a holistic approach, we aim to assist common healthcare users who wish to gain an understanding of the impact of RFIDs in healthcare and provide guidance to Information Technology professionals to design innovative applications and further research in this field. Table 1 gives a brief list of some commercially available RFID healthcare products and Table 2 shows some of the experimental studies conducted on RFID based healthcare deployments and the results obtained from these studies. The following statistics emphasize the need for deploying RFID in healthcare not only to improve services but to save precious human lives.

1.1 Healthcare Statistics

In 1999, the Institute of Medicine (IOM) estimated approximately 98,000 deaths would occur each year in the US due to preventable medical errors, costing $37.6 billion [4], and becoming the sixth biggest killer in the US [5]. Five years later, a study released by Health Grades [6] reported the same number of deaths due to a failure to improve patient safety. One in five Americans (nearly 22% of the population) say they or a family member have faced a medical error of some kind [7], while McGann [8] argues the percentage of Americans who believe they have personally experienced a medical error could be as high as 42%. Graban [9] states that about 7,000 people are estimated to die each year due to medication errors costing USD $4 billion, and that the US is not alone in experiencing these problems. The 2008 British Parliamentary report estimated over 11,000 cases of serious injury or death due to medical errors, while the Health Select committee estimates the number to be as high as 72,000. In Canada, the estimate is 9,000 to 24,000 deaths each year, while in Germany the estimate is 17,000. In Australia 11% of all deaths are attributed to medical error, while an in depth analysis may conclude it to be as high as 19%, which is 1 in every 5 deaths. A report on Sentinel and Serious Untoward Events from Hong Kong (cited in [10]) states that complaints about retained instruments after surgery constituted about 36.4% while the 3rd highest medical error happens to be wrong site/patient surgery which constitutes 15.2%. A 2009 report [11] says western medical practice in countries such as Europe, USA, Canada, Australia and New Zealand has killed more than 5 million in the past decade; the economic impact of which is estimated to be over USD $1 trillion. McGee (2005) (cited in [12]) states that in the US, 5 to 8 wrong-site surgeries occur every month. Pappu *et al.* (cited in [12]) state the cost to hospitals in the US is $3.9 billion every year due to theft of equipment and supplies. Wurster *et al.* (2009) (cited in [13]) state that healthcare expenses in the US are expected to go up by nearly 20% by 2017.

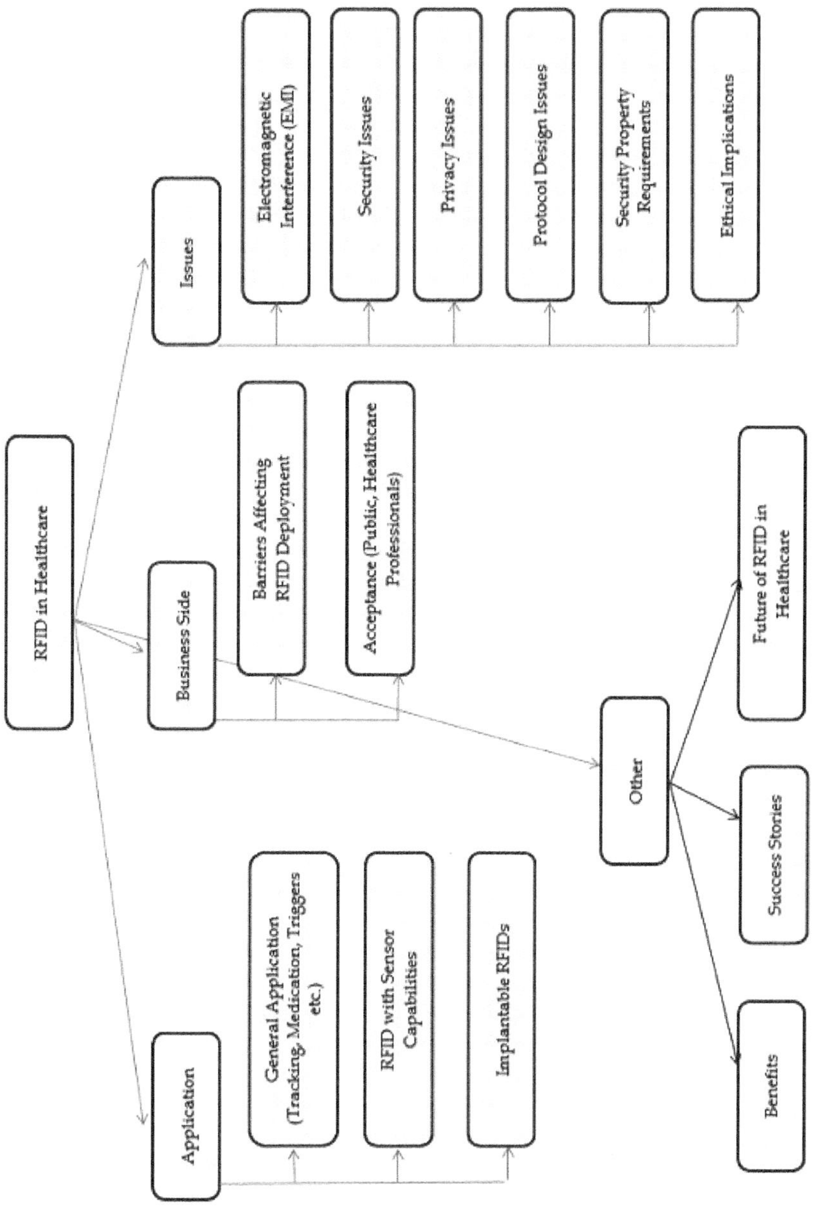

Fig. 1 Impact of RFID in Healthcare - An Overview

2 RFID'S Impact in Healthcare

From the above statistics, it is evident there are many convincing reasons to adopt Information and Communication Technology (ICT). The goal of this chapter is to explore how RFID contributes to improving healthcare services thereby helping to increase public confidence in the system. In healthcare, RFID is widely deployed in the context of patient identification [14] and realtime location tracking systems (RTLS) to track healthcare professionals and patients [14, 15, 16, 17, 18, 19] and assets [20]. It is also used for inventory control of equipment, medical supplies and medicine, and tracking the expiry dates of medicines [21]. RFID is expected to solve some of the major problems in healthcare such as reducing medical errors [22, 23, 24, 25] thereby improving patient-safety [13, 26, 27, 28, 29], improving work efficiency [10, 30, 31], producing cost savings [12], and reducing litigation from incorrect treatment, surgery or medication. Barua *et al.* (2006) (cited in [32]) state the financial benefits of RFID in retail and healthcare touch the $40 billion mark with an estimated Return on Investment (ROI) of 900%. We now explore in detail the impact of RFID in healthcare and Figure 1 reflects the same.

2.1 RFID Based Healthcare Applications

Mowry [12] highlights several RFID based healthcare applications. The deployment of active tags with thermometers to combine patient location tracking and real time patient symptom tracking was used by Taipei Medical University Hospital to contain a SARS outbreak. A RFID based Real Time Location System (RTLS), eShephard from Exavera Technologies, allows doctors to correctly identify patients thereby assuring correct treatment, reducing medical errors and increasing operational efficiency. RFID tagged implant objects introduced by implant manufacturer Biomet in 2006 have assisted in identifying items used from a kit so they could quickly and accurately invoice the hospital. RFID enabled smart cabinets track inventories of medical devices to ensure patient safety, improve billing efficiency and inventory management, and reduce the error-prone manual labor involved in the process. RFID based system SurgiChip assists doctors to prevent wrong-site and wrong-procedure surgery by correctly identifying the patient, the exact surgical site and other vital information. ClearCount that uses RFID tagged surgical sponges ensures sponges are not left inside patients after surgery. Macario *et al.* [33] proposed a similar system to detect commonly used surgical gauze sponges tagged with RFID. 8 patients undergoing abdominal and pelvic surgery participated in the evaluation. 8 untagged sponges and 28 tagged sponges were placed in patients. The result of the study showed the RFID reader detected all tagged sponges correctly (100% accuracy) in less than 3-seconds with no false-positives or false-negatives. A context-aware middleware for mobile healthcare services has been proposed by Song *et al.* [34] which focusses on preventive medicine. Information about a patient is gathered using middleware and the Bayesian probability formula is used to determine the high-frequency context to manage risk situations. The middleware also provides real-time analysis of information, calculating risk frequency via a real-time sensor and RFID.

Martinez *et al.* [15] proposed an application that uses RFID for tracking patients and Near Field Communication (NFC) for drug administration. It was argued RFID is not yet compatible with universal devices such as cell phones and this was the reason why NFC was used. This argument is not true, as RFID enabled mobile phones are already in use [35, 36]. A similar patient tracking system was proposed in [21, 22]. Lin *et al.* [37] proposed a remote healthcare system (RHS) using RFID with an emergency monitoring and reporting mechanism to protect the safety of senior patients who live on their own. Though the application has potential use, there is a major weakness in the design. The system requires the patient to be at home to scan his RFID tag to the reader connected to the home computer. The patient ID is sent to the server in the clinic which automatically completes the authentication and authorization. The system measures the physiological signals of the patient and sends it to the clinic. The downside is that, if a patient's tag were lost or stolen, an adversary can use the publicly available system and compromise the safety and privacy of the patient. This is because the system does not verify the person scanning the tag is actually the patient and it simply performs the authentication and authorization once it receives the request. Similar applications are discussed in [38, 39].

Chen and Collins [16] proposed a Real-Time Tracking System to optimize healthcare facility operations, throughput time, room utilization and patient flow. The system expects to increase visibility within the facility by allowing medical staff to find where their co-workers and patients are at any given time. The system also records the amount of time nurses and doctors spend performing their tasks. While these two features help the administrative side of healthcare, in our opinion, this system poses some privacy issues for both staff and patients. Kim *et al.* [40] proposed an indoor Ubiquitous Healthcare (U-Healthcare) system using RFID to locate and track the elderly. Sensors were used in U-Healthcare to record information about patients such as heart rate, blood pressure, pulse oximeter readings and electrocardiogram data. When an abnormal condition was detected, the system alerted healthcare providers/family members so immediate action could be taken. The system uses modules such as 1) a Tag Sensing Module to determine tag location, 2) a Location Sensing and Tracking Module that uses the tag's location to know the patient's location, 3) a Monitoring Module to monitor the patient's health status, 4) an Alarm Module to alert concerned people when certain events occur, and 5) a Health Status Information Analysis Module which analyzes the data gathered. The system uses encryption techniques such as a cryptographic hash function to secure the communication, and employs timestamps and a MAC algorithm to defend against spoofing attacks.

An evaluation was done at a medical center to see if RFID along with infrared signals would increase equipment tracking/utilization [20]. Results showed the system performed well in localizing equipment to a ward with over 80% accuracy compared to a specific room (between 60 - 80% accuracy). Accurate designation of beds to patients assisted in capturing charges accurately and improving equipment utilization/availability-status. Despite some concerns, the system was fairly well received from the focus group. Polycarpou *et al.* [14] presented design, implementation and testing of a RFID system in healthcare aimed at error-free patient identification, RTLS and drug-inventory management. The application was tested in a hospital environment. The results showed tag readability was in the order of 88% for vertically oriented tags, which is considered satisfactory. The down side of the system was that while it worked well in powder-based drugs, there were readability issues when the tag was in direct contact with liquid bottles as the reflection coefficient at the chip's terminal was severely affected thus reducing the input power to the chip. To improve readability, it was suggested near-field tags with near-field antennas could be used. The patient identification system worked well and proved quite useful to doctors.

Symnonds *et al.* [41] discuss the challenges and solutions involved in implementing RFID for assisted living especially in cases where people experience short term memory loss. Factors such as providing a means of locating objects, being unobtrusive, being usable by older people and its cost effectiveness were considered in designing the system. The system also accounted for broken tags (when an object is dropped for example), with the loss of a signal from the tag time stamped, the nearest landmark noted and an alert stating where the object was last located and the time it was located. Unluturk and Kurtel [27] proposed an RFID and Web Service based Assisted Living System for elderly people. The system aimed to reduce medical costs and the complexity of services, improve patient safety and quality of care, and provide timely decision making capabilities.

A High Efficiency Particulate Air (HEPA), an Air Ventilation System using RFID was developed by Lin *et al.* [42]. The system aims to control the quality of air in hospital operating rooms to prevent hospital-acquired infection. The system is composed of RFID tags for staff and patients, sensors and a reacting module that receives information about the number of staff, the status of the surgery and the volume of air in the room. Of the two control mechanisms that were deployed, one was time-based where the system will keep the volume of airflow to a minimum when the operating theater is not in use. The second control mechanism was controlled by RFID which monitors the number of people in the room and regulates the airflow accordingly. By installing two readers in the room, the system achieved a 99% read rate. While the initial setup cost may be high, results show that an energy savings of 50% and 40% were achieved thus providing more life for the air-filter. It is important to note the system not only aids in preventing hospital-acquired infection but other occupational diseases from spreading.

The European Agency (2000) (cited in [43]) states that over 5720 fatal work-related accidents occur in the European Union. This prompted the development of the Human Condition Monitoring System in Hazardous Locations using RFID [43]. The application uses a mobile system based on energy-efficient wireless sensor networks (WSNs) and active RFID to achieve ubiquitous positioning and monitoring of people working in hazardous locations. The system uses an algorithm called Time Division Double Beacon Scheduling (TDDBS) to increase operation time and data transmission rates of nodes in the system. The implementation results show the system was able to pinpoint the location and state of an individual, and also provide analysis of potential risks at any given time. The system uses radio communication between transmission points which form a network of routers sending and receiving information, and acting as signal repeaters. When a lack of signal is detected (in a mine for example) using an acoustic warning signal, the individual can take a few steps backwards to activate and deposit a new repeater. The ZigBee communications network is used among these repeater nodes and a RF signal is also used from the tagged bracelet to the nodes. The system can also detect emergency situations such as a fall or loss of consciousness to trigger an alarm.

McCall *et al.* [44] proposed a RFID based Medication Adherence Intelligence System (RMAIS) that enables patients to manage their medications in terms of taking the right dosage at the right time in an automated fashion. The system also provides medication reminders and sends alerts to care-givers about non-compliance. The National Council Report (2007) (cited in [44]) states there is compelling evidence to show patients are not taking their medicines as prescribed for one or more of the following reasons: a) patients are not convinced about the need for treatment, b) fear of adverse effects, and c) difficulty in managing the medication. The proposed system addresses the last concern - medication management. Medicine bottles fitted with RFID tags enables the system to read the medicine information, and the built-in scale tells if the patient has taken the right dosage, while the motorized rotation platform positions the correct medicine bottle in front of the patient and the built-in network interface card enables the system to alert medical professionals when necessary. A similar application was proposed by Sun *et al.* [45]. and a medicine manufacturing system using RFID and Service Oriented Architecture (SOA) was proposed by Deng *et al.*

Ting and Tsang [10] investigated the potential use of RFID in surgery management specifically targeting hospitals where the nurse to patient ratio is much lower. This Surgery Management System addresses issues such as: a) identifying/tracking patients, b) eliminating manual redundant checks to increase work-efficiency, c) automating materials and equipment management, d) retention of instruments inside a patient's body, and e) wrong site/person surgery. Critical elements to be considered when implementing RFID are discussed: i) technical dimensions covering encryption for security; EMI interference and reliability of communication in terms of reducing false reads and misreads by deploying anti-collision/error-correcting codes, ii) a management dimension that includes gaining top-level support and training, iii) capital costs, and iv) privacy issues. Yamashita *et al.* [46] studied data management of surgical instruments using RFID. A tag was designed and attached to surgical instruments used in the experimental study. The tag was cylindrical in shape with a diameter of 6mm, a thickness of 2mm, weighing 1 gram, a storage capacity of 128bytes and consisting of an original loop antenna. The internal chip was manufactured by Philips (ICODE SLI) and a frequency of 13.56MHz was used. The focus was on the effect of the tag being repeatedly sterilized, its durability to contamination tests during the washing process and finally its impact resistance against being dropped. The results of the study showed no tags were incapable of communicating or had any damage after high-pressure sterilization (50 times for 30 minutes on 30 tags), the high-heating challenge test (200 degree centigrade for 300 hours on 20 tags), the contamination test (77 surgical instruments for 31 cycles using a washer disinfector), high-pressure steam sterilization (68 surgical instruments for 15 cycles using an autoclave process) and were resistent to repeated impact. The failure rate was remarkably 0%.

Wang *et al.* [47] conducted a case study on RFID integration in a Taiwan hospital. This RFID infrastructure consisted of 163 field generators, 41 RFID readers, 27 Yagi antennas, 15 programming stations and one system API. The project was a huge success motivating the hospital to develop other medical applications based on this to track equipment, auditing of medicine, and drug control. Some of the key findings from the project were: a) RFID was considered a part of the entire IT infrastructure rather than an add-on, b) device management is highly environment and context dependent. In hospitals, an active RFID solution is more appropriate, and frequency spectrum management is vital along with a combination of tags with sensors that help to provide better care for patients.

Stahl *et al.* [48] conducted a study to determine if a RFID based indoor positioning system (IPS) could objectively and unobtrusively capture outpatient clinic behavior. The key performance parameters were flow-time, wait-time and patient to doctor face-time. The tags in the system generated events based on the current location and time. The data stream provided an indication of how long a tag was in a given location. Thus, the system allowed the tracking of not only the path the patients took but also how long they spent in each location. Face-time was calculated when the doctor's tag and patient's tag were together in the same place at the same time. The study was conducted in a primary clinic and an urgent care clinic. The results showed the flow time and wait time were similar in both places but face time was significantly longer in the primary clinic. It thus proved that a) the use of RFID based IPS had strong potential to provide insight into the behavior of tightly coupled clinical systems enabling them to identify bottlenecks; and b) it can help identify room utilization patterns to enhance scheduling processes. Similar work to track cardiac patient flow was proposed by Baarah *et al.* [49] and Unluturk [50] proposed a system called Supervised Event Executive (SEE) to improve nurse-patient communications. The system aims to provide a common, convenient and reliable way to transfer nurse call system events.

Shim *et al.* [30] proposed a specimen management system and architecture with RFID for a clinical laboratory to solve the known problems of specimen management using barcodes. In order to avoid interference with other electrical systems in the lab, the antenna's reading range was adjusted to 10cms. The following observations were made: a) the location of tags and the antenna should be parallel to decrease false recognition; b) the reading ability decreases if the tags overlap each other; and c) there has to be some distance between tags to increase reading ability. The advantages of the system include an expectation that overall clinical laboratory work processes will improve and work efficiency will increase. From sampling to analysis, the number of contacts between a person and the container reduced by at least 1 from 4 or more.

The numerous applications of RFID in healthcare make it evident that RFID has become an integral part of healthcare. We now explore RFID tags with sensor capabilities and their impact in healthcare.

2.2 *RFID Tags with Sensor Capabilities*

Tags can be used to collect sensor-derived data by extending the chip's interface capability to a sensor [51]. Combining the tag's sensor capabilities with sensor nodes has provided new dimensions for interaction and operation with each other [43].

Pletervsek *et al.* [52] proposed a hand disinfection control system using RFID named Automatic Disinfectant Tracker (ADT). Experimental results show 100% efficiency in detecting propanol and ethanol concentration using the static disinfectant control method. Ethanol sensors were used with the IDS-SL900A RFID data-logger which can be used to record both body temperature as well as readings from an external sensor. The 8Kb EEPROM memory was used to store the level of alcohol concentration with a timestamp of hand rubbing. The reader detects the presence of an individual then records the time period of hand scrubbing, the temperature and concentration of the applied ethanol. When the reader detects the ethanol from a tag-sensor that exceeds a certain level, it updates the nearby person's tag with the disinfection information. If the desired level of disinfection is not achieved, the system alerts the personnel. Thus, ADT allows the quality of hand rubbing to be monitored via a clean and contact-less method.

Hung *et al.* [53] proposed a Blood Pressure Measurement (BPM) with ease of use, low cost and wireless connectivity. The authors note that recording blood pressure changes over a period of time will help doctors control cardiovascular conditions. A condenser microphone is used to listen to a weak heartbeat and amplify it. The Micro Controller Unit (MCU) listens to the heartbeat until the sound disappears and calculates the time difference between start and end. This allows the system to calculate the Systolic Blood Pressure (SBP), Diastolic Blood Pressure (DBP) and heart rate. Functions such as measuring the BPM signal, detecting the position, intercepting the tag identifier information and analyzing/transferring the BPM signals form the core components of the system. The RFID module is added for convenience and security reasons in the sensor because it is used to identify the patient, thus eliminating the need for manual entry of the ID and other personal information.

Vecchia *et al.* [54] proposed intelligent infrastructure for smart hospitals to optimize medical staff to patient interaction and to provide ubiquitous and transparent access to medical data. The system uses RFID and photosensor technologies to accomplish these tasks. The purpose of the photosensor is to trigger RFID antennas to on/off to reduce RF exposure and power consumption. The architecture is three-layered: application layer, service layer and data layer. RFID and photosensor systems are part of the data layer. The system is event driven and service oriented, aimed to shorten the existing gap between healthcare professionals and patients. Opasjumruskit *et al.* [55] proposed an RFID based temperature sensor.

Though the design is aimed at implantable chips for application in animal healthcare, it does raise some interesting points. The RFID chip has five parts: a) analog front, b) digital controller, c) charge pump d) EEPROM, and e) temperature sensor. The metal-oxide semiconductor temperature sensor is suggested for passive RFID since it consumes less power allowing for ease of integration. The microchip in the tag can operate and communicate in ice or water-filled packages, or food containers without any problems. When a reader sends a command to the tag using a 100% Amplitude Shift Keying (ASK), the demodulator in the tag converts this RF signal to digital data. If the sensor unit received a "read temperature" command, it converts the temperature to 16bit output data. During this phase, care is taken not to modulate the RF field to avoid power supply fluctuation that would deteriorate the accuracy of the measurement. The authors note that embedding RFID with other sensory circuitries such as a humidity sensor, pressure gauge or a chemical detector, will dawn a new age in measurement techniques.

Other notable sensor-attached RFID based applications include Barman *et al.*'s [56] sensor enabled RFID for monitoring arm activity after a stroke. Catarinucci *et al.* [57] proposed a label-type sensor tag that uses as input a generic sensor and transmits the measured data. The sensor tag is based on four optimized antennas and it takes into account the size constraints and mutual coupling effects. Experimental results have shown the antenna design and their optimized mutual position reduced coupling/parasitic effects by 44%. Romain *et al.* [58] proposed an implantable RFID based pressure sensor that could be used in network configuration. The goal is to develop a communicating endo-prosthesis which includes integrated sensors made of a pressure transducer. RFID used in recording and transmitting recordings of pressure in the aneurysmal sac during post-surgical consultations provides a means to monitor the condition of Abdominal Aortic Aneurysm (AAA).

The above RFID sensor tag based applications support the statement that chemical sensing can aid in advanced medical monitoring [59]. We now explore the on-body/implantable RFID tags and their impact in healthcare.

2.3 On-Body/Implanting RFID in Humans

Aubert [60] conducted an indepth analysis of RFID implants on humans for locating/tracking and remotely controlling human biological functions. Miniature implantable RFID devices assist in biomedical/therapeutic applications such as blood-sugar sensors. Implantable chips also assist in monitoring brain functions via implanted probes that communicate via a transponder embedded within a skull. The author notes that RFID implants are not mature enough to communicate with GPS satellites to facilitate real-time tracking of a person however the remote control of human biological functions are more realistic. Some of the technical requirements for RFID human implants include: a) implants have to be as small as possible; b) they have to be passive because using a battery to power the device is impractical; c) selecting the human biological functions that will be remote controlled; d) using bio-compatible material for fabricating the device to avoid tissue reactions; e) secure and reliable wireless transmission of data; f) complying with human tissue exposure IEEE standards for safety reasons related to implanted antennas, and g) feasibility to remotely control at a bio-molecular scale level. Two wireless techniques for powering passive RFID-implant devices are: 1) inductive coupling is where the power required to operate the device is supplied by an interrogator via inductive (near field) coupling, and 2) electromagnetic coupling which is based on an electromagnetic field that uses electromagnetic waves propagating from the antenna in the far field region to power the passive tag.

Grosinger and Fischer [61] evaluated backscatter RFID systems on the human body at two different frequencies (900 MHz and 2.45 GHz). The necessity to investigate the on-body RFID propagation channel, including the effects of antennas to understand the robust power transfer and communication in such systems is emphasized. It is noted mutual influence between the antenna and the human body should be low. The system uses a monopole antenna/patch antenna and five on-body tags at various locations in a human-body to measure the physiological parameters of a person with the reader placed on the stomach. The study concluded that patch antennas were suitable for on-body applications and the outage probabilities at 900 MHz were lower than the probabilities at 2.45 GHz due to increased energy absorption in human tissues at higher frequencies.

Occhiuzzi *et al.* [62] proposed a design for implanting RFID tags for passive sensing of the human body. The purpose is to sense the evolution of human physiological and pathological processes involving a local change for effective permittivity inside the body. Medical devices such as prosthesis, sutures, and orthopedic fixings can be fitted with battery-less RFID radio-sensors to acquire data at varying times. This helps to record the geometrical/chemical features of the tissues to observe the healing process, and identify possible complications. It is noted that biological processes differ in terms of timeline and amplitude, and hence the measurable response of RFID radio-sensors needs to be properly correlated to the process under observation. This design particularly focuses on the monitoring of biological ducts that host a stent implant. To achieve the best result from the sensors, careful attention was given to the shape of the antenna in the RFID tag, the power threshold and input impedance of the Integrated Circuit (IC) and lastly, the state of the biological process where the antenna was forced to have the best impedance to match the IC. Analysis and experimentation suggest the possibilities of sensing some inner biological processes using implanted RFID tags. It is noted this design technique could be extended to other implanted devices mentioned earlier where long-term monitoring may provide added value.

Thevissen *et al.* [63] proposed the use of implanted RFID tags into a human tooth to handle mass deaths that arise from diaster situations such as tsunamis and hurricanes. Tags used by veterinarians were modified for this purpose. During the modification process, 10% of the tags failed due to the unavoidable manipulation of the fragile connections. To minimize the loss, naked tags were protected with polymerized composite bonding. It was shown that dental x-rays could be taken without disturbing the normal activity of the tags. A similar application to identify disaster victims was proposed in [64], the difference being the placement of the chip inside cranial structures. The maxillary sinus seemed most suited for the purpose and the majority of chips were placed in that area for identifying the 2004 tsunami victims. Madrid *et al.* [65] have also studied the reliability of RFID in the identification of dentures in long-term care facilities. The tag was programmed and inserted in the dentures, with only one tag unreadable and the rest still operational at the end of three months. At six months, 100% of 50 initially placed tags were readable proving the reliability of the technology for tracking and identifying dentures.

Other applications that use implantable RFID technology include the following. Burke *et al.* (cited in [60]) discuss the possibility of using RFID in the remote activation/deactivation of biochemical activity in a single living cell. The miniature nanotechnology FM-radio by Jensen *et al.* (cited in [60]) could be inserted into a living human cell to give a subcellular remote and real-time controlled interface. Berger (2008) (cited in [12]) proposes an RFID enabled orthopedic implant to assess the function of the implanted device and the surrounding tissue. VeriMed application from VeriChip (cited in [12]) uses a small glass capsule about the size of a grain of rice (FDA approved device) that is implantable in humans. This capsule assists in patient identification when they arrive at a hospital unconscious or unresponsive. Zhou *et al.* (cited in [12]) propose a human implantable RFID enabled sensor tag that could help diabetic patients monitor their blood sugar levels and save them from having to prick themselves several times a day to check their blood-sugar levels.

2.4 Insight into EMI Caused by RFID

Varying results can be seen from the extensive research carried out to measure the Electro Magnetic Interference (EMI) caused by RFID. Here, we analyze some of the important studies conducted in this area.

Houliston *et al.* [66] conducted a study to replicate EMI due to RFID with a drug infusion device. Infusion pumps known to have failed due to EMI were placed in RF fields with varying strengths created using the following: varying numbers of RFID readers; high-gain antennas; varying distance between readers to infusion pump; and the presence of an RFID tag on the infusion pump. The results concluded the infusion pump was not affected by low-power RFID readers even when they were in direct contact. However, the pump attached to a tag was disrupted when a high-power reader was used at a distance of 10cms. Similar interruptions were noticed when high-power and low-power readers were used at a distance of 10cms. It was noted that once the pump failed due to EMI, it subsequently failed at other times even when no RFID readers were active. This points out the unreliability of a device once it has failed due to EMI, thus emphasizing the importance of thoroughly resetting medical devices after failure. It was concluded that EMI incidents increase due to: higher RF field strengths determined by high output power from the RFID system; a shorter distance between the reader and the medical device, and the presence of a tag.

Fiocchi *et al.* [67] conducted an EMI study of a RFID-based mother-newborn identification system with some interesting results. The study included testing the compliance of the EMF exposure guidelines in terms of the magnetic field threshold and the maximum distance of the reader to wake the tag. The results indicate the need to identify the technical specifications of the optimal reader-tag for this type of application by performing an accurate exposure assessment investigation particularly for newborns. The study used two realistic Virtual Family Models based on high-resolution Magnetic Resonance (MR) images of healthy volunteers. The system used passive inductive coupling RFID at a frequency of 13.56MHz. Experiments showed higher levels of EMF for both the mother and newborn in the regions of the body that were closer to the reader, with a trend for higher values in the newborn compared to the mother for all reader positions. In addition, the highest exposure value for the newborn was not only restricted to the proximal limb but also extended to the trunk, genital organs and head/neck regions. Hence, it is recommended the tag be positioned on the ankle instead of the wrist so the reader is at a greater distance from the sensitive organs.

Kapa *et al.* [68] evaluated the likelihood of EMI when using RFID. 32 medical devices were tested with a maximum magnetic output field ranging from 8 to 800 mgauss. Results show light-to-hazardous incidents occurred in 8 devices, hazardous EMI in 2 devices, significant EMI in fluoroscopy and echo-cardiograms, and 1 out of 5 cardiac monitors plus light EMI in a defibrillator and cardiac monitor. It is important to note that no EMI was recorded in any device at distances greater than 4 ft. The authors recommend rigorous on-site testing before implementing such devices. Kuo *et al.* (cited in [69]) observed that a shift in resonance frequency causes a degraded reading range for RFID. A thin EMI sheet made of a magnetic composite absorber placed between the tag and the metal recovered nearly 60% of the reading range. An EMI study by Ting and Tsang [10] recommends that RFID systems be positioned at least 30cm away from a medical device to avoid problems of interference.

Togt *et al.* [70] conducted a study to assess the EMI interference from RFID. The study was conducted in a controlled environment without patients using active 125KHz and passive 868 MHz tags. 41 medical devices under 17 different categories from 22 different manufacturers were tested. The results showed that out of 123 EMI tests, 34 EMI incidents were induced by RFID, of which 22 were classified as hazardous, 2 as significant and 10 as light. The 868 MHz passive tags induced 26 incidents in 41 EMI tests and its counterpart induced only 8 incidents. The passive tags induced EMI in 26 medical devices, 8 of which were also affected by active tags. The authors concluded RFID potentially induced hazardous incidents in medical devices and the implementation of RFID in healthcare should require on-site EMI testing and that international standards be followed. As Payne [71] quotes *"..these findings sparked a flurry of media attention and panic regarding the safety of RFID.."* Bacheldor [72] quotes John Collins (Director of Engineering and Compliance with the American Society for Healthcare Engineering) who says *"There needs to be a growing awareness among hospitals that problems can occur and it is difficult to say if something such as an RFID system will interfere with critical care equipments unless a study is conducted"*. Collins did deem the study well conducted. Following this, several reports came out to address consumers' concerns. Payen's [71] report states the following: a) the research has not been in line with most current hospital RFID deployments where passive RFID is the mainstream technology; b) most UHF deployments call for a 1-watt or less reader but the study used 3-watt readers which are far more powerful; c) this, combined with the gain of antenna used, indicate that the study was highly unlikely to be compliant with FCC regulations; and d) the most commonly used frequency (13.56 MHz) in hospital settings was not mentioned in the study and hence does not truly reflect the reality of most current RFID deployments. The author does note the value of the study as it would serve as a warning for future RFID deployments based on UHF tags and emphasizes the need for standards in relation to RFID in hospitals. Roberti's [1] report states the following: a) no known cases of patients being harmed have been reported due to RFID, and b) every speaker who participated in the *RFID in Healthcare* event [1] stated they have not seen RFID equipment affect medical devices of any kind. Researchers from Indiana University and Purdue University, Indianapolis, USA found no incidents of EMI from passive UHF RFID systems [73]. The first test used 868 MHz tags and an interrogator to study the effect on 41 medical devices. The second test used 902-928 MHz tags and readers on 25 medical devices. Iadanza *et al.* [74] conducted a similar study by testing 16 medical devices from 5 different categories in a children's intensive care unit using active RFID systems with low power and a relatively high operating frequency. The results do not report any malfunction in any devices.

Cheng and Chai [69] proposed an experimental design using simulation to measure the efficiency and reliability of implementing RFID in a MRI department which is a major source for EMI. Cheng's experimental results show the MRI machine did not cause any interference on RFID readability. The tags were also functioning normally without any data loss after the MRI scanning, and the tags did not lose any identifying function in the MRI environment. In addition, patients did not have to remove their tags during the scan and doctors did not have blurred images during diagnosis. The study concluded there was no evidence to prove interference between the two technologies.

Other noteworthy studies/results in this area include the following. Macario *et al.* [33] note RFID did not exhibit any EMI while using their RFID wand device to detect surgical sponges. Rieche [75] proposed a novel modulation and transmission technique named the Gaussian Derivative Frequency Modulation (GDFM) to mitigate EMI interference. Censie *et al.* [76, 77] analyzed EMI of RFID regarding medical devices and recommended standards of electromagnetic immunity for devices, and standards for the electromagnetic compatibility (EMC) of RFID systems. Ogirala [78] mathematically analyzed the EMI of RFID specific to Cardiac Rhythmic Monitoring Devices (CRMD) and proposed mitigation strategies. The study concluded that the power level of RF alone does not contribute to the EMI but the rate of change of the power output does play a significant role in the interference. The use of Pitt interference as a mitigation technique has been experimentally shown to solve the interference issue.

As clearly seen above, results vary and it highlights the need for continuing research in this area to ensure RFID technology is safely deployed in healthcare.

Table 1 Commercial RFID Healthcare Products and their Applications

Product Name	**Product Category**	**Application**	**Pros & Cons**
eShepherd	Real Time Location Tracking (RTLS)	Patient, staff and equipment tracking	Initial investment of USD$400,000 could potentially save USD 4 million annually
Biomet	Supply Chain	Knee implant surgical kits	Enables effective tracking, invoicing and replenishing
Mobile Aspects, TAGSYS RFID and Terso Solutions	Inventory	Smart Cabinets to inventory medical devices	Costs about USD$25,000. Improves patient safety, billing efficiency and inventory management
SurgiChip	Patient/Site/Surgery verification	Helps to identify patient and surgical site	Prevents wrong site and wrong procedure surgery
ClearCount	Detection	Detects surgical sponges and other surgical devices placed in a patient	Prevents leaving surgical devices inside patients; cost-savings; prevents litigations
VeriChip	Implantable Device	Patient identification	Ethical and privacy issues; associated health hazards; helps identify patients and retrieve patient medical history in emergency situations
Medication Adherence Intelligence System (RMAIS)	Medication Management	Enables patients to take medications in an automated fashion	Provides medication reminders and alerts care-givers
Specimen Management System	Specimen management	To manage specimens in clinical laboratories	Improved work flow and specimen management
Blood Pressure Management	Monitoring	Records blood pressure changes over a period of time	Helps doctors to control cardiovascular conditions; low-cost

2.5 Secure RFID Based Protocol Designs for Healthcare

Several protocol designs have been proposed over the years, specific to RFID based healthcare applications. Here, we analyze some of them, highlight their strengths and weaknesses and identify areas of improvement.

Rrub *et al.* [79] proposed a secure RFID model based on smartcard RFID design. The model adopts the technology in a contactless payment module used in credit cards to design a secure, practical and cost-effective solution for the healthcare sector. The proposed solution is based on Mifare cards DesFire EV1 and is

Table 2 Deployment of RFID Healthcare Products and their Effects

Product Description	Where Deployed	Environment	Issues	Results
Patient Identification	Chang-Gung Memorial Hospital, Taiwan	13.56MHz tags embedded in wrist-bands; HP iPAQ PDS equipped with RFID interrogators	None Identified	Hospital saved 4.3 minutes per patient in identification and verification and achieved 100% identification accuracy.
Patient Tracking	Eastern Asia	-	None Identified	Helped contain SARS outbreak
Real Time Location Tracking (RTLS)	Virginia Hos-pitals, USA	303MHz tags, 100s of readers	None Identified	Net Savings of USD 3 million in the first three years
Implantable sensor tag to monitor blood sugar levels	Inside hu-man body	N/A	Body's reaction to foreign objects; pri-vacy issues;	Enables diabetics patients to accu-rately measure blood-sugar levels painlessly
High Efficiency Particulate Air (HEPA) Air Ventilation System	Taiwan Hospital	434MHz active readers (model: ARUnew01); Tags (model: ACUnew01-Passlt); 1500 bed hospital; 20 staffs wearing tags	Reader cost $600; tag cost $20USD, software $50K	Two readers en-sured 99% read rate; 50% energy savings; 40% more filter life; pre-vents occupational diseases
Indoor Positioning System (IPS) to evaluate patient flow-time, wait-time and face-time	Primary & Urgent Care Clinics	-	None Identified	Showed flow-time and wait-time to be the same in both clinics but face-time was significantly longer in primary clinic, thus en-abling bottle-neck identification.
Automatic Disinfectant Tracker (ADT)	Slovenia	Clinical Hospital; Ethanol sensors used with IDS-SL900A RFID data logger	None Identified	The system suc-cessfully alerted when disinfec-tion has not been achieved to the desired level
EMI Study 1 - Mother-Newborn Identity Reconfir-mation	Virtual Family Model / Hospital in Italy	13.56MHz reader/tag; ISO 15693 Tags	None Identified	Results showed for both mother and newborn the regions closer to the reader were exposed to higher levels of EMF. It highlights the need for optimal reader-tag technical specifications
EMI Study 2 - Passive UHF RFID Systems	Purdue Uni-versity, USA	868MHz and 902 – 928MHz tags	None Identified	No EMI incidents were reported

ISO/EPC-14443A compliant. The tags used in this model have more storage capacity and processing power than passive tags. The model leverages these abilities to perform complex security operations to protect the security and privacy of the tag holder. The advantages are strong security options and fast communication ability. However, the downside is the 10cm maximum reading distance that requires the tag to be close enough to the reader so it can be read. In terms of security, it uses Message Digest to generate a digital signature, along with random numbers and hashing algorithms as part of the communication protocol. Yu and Hou [80] introduced the common criteria security evaluation methodology known as ISO/IEC 15408 for security evaluation of RFID tags and proposed a framework as a minimal requirement for RFID tags to improve security assurance. The framework consists of a) identifying information value and threats; b) defining the evaluation assurance level; c) deriving the security objective; d) deriving security functional requirements and e) performing rational mapping.

Min and Yih [81] proposed a fuzzy-logic based RFID monitoring system for an outpatient clinic. The motivation was to collect raw data from a RFID system containing noise and missing reads that prevented the identification of the exact tag location. Fuzzy-logic algorithms were used to interpret the raw RFID data to extract accurate information. Tag location was determined by evaluating the possibility of presence/absence. Their experimental study in an eye exam room with two readers showed significant false-positives and false-negatives, thus making unreliable RFID reads as one of the main obstacles in its implementation. It was noted the average read rate in real-world RFID applications is only 60-70% which is consistent with this study. Smoothing methods (static and dynamic) have been used to overcome these issues but they don't entirely solve the problem. The static smoothing method uses a fixed window size that cannot be effective for all applications. The dynamic smoothing method is based on the assumption that the read rate of RFID decreases proportionally to the distance from the antenna, however in a small enclosed area, the effect of distance from an antenna on the read rate is not significant. Information on the amount of time a RFID tag is detected or not detected motivates the use of fuzzy-logic based possibility-theory in the protocol instead of probability-theory. Two algorithms were proposed in the protocol. The first algorithm determines the presence of a RFID tag by simply comparing the possibility of presence over the possibility of absence. If the possibility of presence is greater than the possibility of absence, then the algorithm considers the tag to be present. The second algorithm not only considers the current possibility when the tag is detected or not detected but it also considers the previous decision. If the current decision is consistent with the previous one then it definitely confirms the presence/absence of a tag. If the current decision is not consistent then the maximum possibility is taken for the final decision. Numerical experiments show both the proposed algorithms perform better than the static smoothing method by minimizing false positives/negatives. Tu *et al.* [82] also conducted a similar study in healthcare, along with the dynamics of locating and identifying the presence of a tag and have proposed algorithms to accomplish the same. It was noted that both false-negatives and false-positives can be a problem in RFID embedded systems particularly when tags are blocked by impenetrable objects. The algorithms aims to reduce this problem through a variation of triangulation where the presence/absence of a related tag is considered. The results of the study indicate the location accuracy of tags can be improved through the proper collection of data and the use of algorithms for inference of data.

Many yoking-proof/grouping-proof RFID protocols have been proposed over the years, however in this chapter we limit them to the healthcare context. The purpose of the yoking-proof protocol is to build a validation chain to sequentially verify each tag [83]. Juels (2004a) cited in [83] presented a classic example of the application of this protocol in a hospital setting. US FDA regulations require pharmacists to attach leaflets that describe the side-effects and dosage instructions of medications. When a tag is attached to the medicine container and to the leaflet, the yoking-proof can verify if the pharmacist has complied with the requirement by validating the presence of the tags together. Huang and Ku [83] proposed a light-weight grouping-proof protocol for EPC C1G2 tags using CRC and PRNG operations. The authors also presented a practical application of the protocol in a hospital environment to enhance patient medication safety. The nurse giving medication to a patient would scan all drugs from the medicine trolley and the pallet tag from the patient's bracelet to verify if the drugs were consistent with the prescription of the patient. The association rule would be derived from the medical record of the patient and hence the protocol would ensure that nothing more or nothing less was given to the patient. The authors state that more focus should be given to the integrity and accuracy of the data rather than security strength or issues regarding malicious attacks. In our opinion, this is incorrect as security is just as critical in healthcare as the application of RFID is in any other field. Otherwise, attackers could easily compromise the system causing severe risks to patient safety (for example, giving incorrect medication or performing the wrong surgery on the wrong patient). Chein *et al.* [84] proposed online and offline verifier yoking-proof protocols similar to [83], however the difference is that tags are authenticated in an independent way. Wickboldt and Piramuthu [85] on the other hand identified vulnerabilities in both protocols: [83] is vulnerable to DoS attack and patient privacy/security is compromised because the EPC of the tag is sent in plaintext; [84] is also prone to tracking/tracing attack due to the same issue. Yu *et al.* [86] proposed a binding-proof protocol using AND, XOR and ADD logic gates for low-cost passive tags to reduce medication errors. This protocol is based on the Lightweight Mutual Authentical Protocol (LMAP) from Peris *et al.* (2006). Wu *et al.* [87] highlight that LMAP has been broken by Barasz *et al.* (2007) and argue that [86] is vulnerable to the same type of attacks. An improved version of the protocol is presented in [87]. More recent developments in grouping proofs by Sundaresan *et al.* [88] and Sundaresan *et al.* [89] cover all the unique design requirements of grouping proofs and have been shown to be resistent to the known attacks on RFID systems. The protocols are light-weight in terms of applying only XOR, PRNG and MOD operations on the tag which make them compliant with EPC C1G2 standard and enable large-scale implementations.

Other significant contributions in this area include Chen *et al.*'s [90] proposal for an RFID based tamper resistent prescription access control protocol with an authentication and authorization mechanism. The protocol claims to protect against attacks such as MITM, and spoofed-reader/tag, and provides mutual authentication and location privacy. The protocol aims to reduce forged/altered prescriptions and deter drug use. The downside is its use of complex symmetric encryption scheme and a hash function. Their implementation in low-cost passive tags is not practical as these tags are highly resource constrained in terms of storage and computation capabilities. Zhou *et al.* [91] proposed a multi-level RFID tag ownership transfer protocol for healthcare environments based on symmetric-key cryptography. Three different scenarios were considered: a) where tagged-items in the healthcare supply chain had different levels of ownerships; b) seamless ownership incorporation by a third party logistics provider; and c) where tags simultaneously possess a temporary and a permanent key for communicating with a temporary and a permanent owner respectively. Yeh and Wang [92] proposed an e-Health system based on RFID that addresses patient privacy and increases the efficiency of the out-patient clinic procedure. The protocol uses lightweight functions such as XOR and PRNG, making it possible for large scale implementation using low cost passive tags. However, an in-depth security analysis is missing from this study to prove the robustness of the design which is an important factor for implementation.

2.6 Security and Privacy Issues of RFID in Healthcare

Gasson [93] conducted an interesting experiment with an implantable RFID tag infected with a virus. Through the experiment, successful infection of computer systems via the spread of the infection from the tag was demonstrated. It was claimed the author may have been the first human to be infected with a computer virus. A glass capsule sized tag HITAG S 2048 infected with a malicious code was implanted in the hand of Gasson, and his mobile phone was RFID reader enabled. The malicious code was written in such a way so instead of simply reading the data, it also executed some SQL injection code which successfully damaged the user's profile information with further payloads. Through this experiment, the author highlights that implantable devices have evolved to the point where they should be treated as small computers and close attention should be paid to security issues.

Thiesse *et al.* (cited in [32]) pointed out some security/privacy issues that may arise from the use of RFID. These issues are injection attack, eavesdropping and DoS attacks that could compromise the security of the system, and the ability to permanently save and link information about individuals through temporal and spatial extension of data collection activities representing the privacy aspect. Parks *et al.* [32] applied the Fair Information Practice (FIP) principles such as *Notice, Choice, Access, Security and Enforcement* to address the privacy issues in RFID. It was noted that the *Health Insurance Portability and Accountability Act* (HIPAA) privacy and security requirements, such as Security Breach Notification Requirements and Patient Rights, can be addressed by applying FIP principles in RFID development and deployment. The technical approach takes two dimensions: 1) the *physical solution* comprises options like faraday-cage and the active jamming method;

2) *logical solution* comprises a destruct-tag using a kill command, a control tag response using blocker tag and agents, use of encryption schemes such as hash-lock, re-encryption, silent tree walk and zero knowledge. Kim [94] addresses the security and privacy issues suggesting that confidentiality can be achieved by authentication and encryption, and data privacy and availability can be achieved through authorization, encryption and data backups. The author also adopted the four security layers proposed by Ko *et al.* (2010): mobile layer, network, application and database. The mobile layer is where components such as RFID and NFC are incorporated.

Hawrylak *et al.* [95] address several security risks associated with using RFID in healthcare, such as threats to patient privacy and safety based on interception of messages, interruption of communication, modification of data, and fabrication of messages and devices. Attacks are categorized as interception, interruption, modification and fabrication. Interception can be carried out by wireless sniffing (eavesdropping) and man-in-the-middle (MITM) attacks. The former is where an attacker simply listens to the communications between tags and the reader attempts to retrieve the information. MITM attacks are even more difficult to detect because they overcome a number of security issues. In a relay attack, the attacker uses a fake reader and tag to extend the range of a legitimate reader and tag. Several distance bounding protocols have been proposed to solve relay attacks but they do not offer complete protection as they can be defeated using a modified transmitter. Interruption attacks are accomplished by jamming the network and blocking reader-tag communication. Some methods include interference of RF noise with any RFID signals using rogue devices, killing tags and energy draining attacks. Currently, only physical security offers the best defense, however monitoring for abnormal conditions does add some value but does not entirely solve the problem. Modification attacks focus on maliciously modifying the information in the RFID system by performing injection attacks. Countermeasures such as mutual authentication, encryption, and challenge-response methods are suggested. Fabrication attacks use a separate device to inject false information into the system. Cloning is one type of attack where an attacker reads the data from the legitimate tag and writes the data to a counterfeit tag. As a countermeasure, strong authentication schemes, and two-factor authentication schemes are suggested. The authors note that while countermeasures do exist, some are too costly to be implemented in low-cost systems.

The required security and privacy properties common to most RFID systems as noted in [96] [97], can be applied in the context of healthcare and can be summarized as follows:

- *Tag Anonymity*: The protocol should protect against information leakage that can lead to the disclosure of a tag's real identifier. This is important as an attacker may be able to clone a valid tag.
- *Tag Location Privacy*: The protocol should ensure the message contents is sufficiently randomized so they cannot be used to track the location(s) of the tags and thereby glean social information about the wearer of the tag.
- *Forward Secrecy*: The protocol should ensure that on compromise of the internal secrets of a tag, its previous communications cannot be traced by an attacker.

This requires previous messages be independent of current resident data on the tag.
- *Forward Untraceability*: The protocol should ensure the old owner is unable to trace or communicate with the tag post-ownership transfer. This means the old owner is unable to learn or compromise any new secret(s) that have been established between the tag and the new owner.
- *Replay Attacks*: The protocol should be able to resist compromise by an attacker through the replay of messages that have been collected by an attacker during previous protocol sequences. This requires protocol messages in each round of the protocol are unique.
- *Denial of Service (DoS)*: The protocol should be able to recover from incomplete protocol sequences that can occur due to an attacker selectively blocking messages. Importantly, such blocking of messages by an attacker should not lead to desynchronisation between the tag and the servers.
- *Server Impersonation*: The protocol should ensure the new server cannot be impersonated by an attacker. This requires the tag to challenge a server requesting ownership transfer to prove its legitimacy.

2.7 Ethical Implications of RFID in Healthcare

Lewan (2007), cited in Katz and Rice [98], notes that two states in the US, Wisconsin and North Dakota, have passed legislation proscribing mandatory microchip implantation in humans, and other states like Oklahoma are considering implanting chips in certain violent criminals upon their release from prison. The latter is definitely worthy of consideration. Under normal circumstances, every human and even animals deserve complete freedom. But when freedom is abused to the extent of harming themselves and others, then it calls for restrictions. People who have a history of harming others, inevitably lose that freedom because of the choices they make. The Government and the Justice System have a duty to protect the common person and if technology can be exploited to do this, then why not use it?

Though the effects of RFID in healthcare have been explored from various perspectives, the ethical implications have not been significantly addressed. Monahan and Fisher [99] conducted a study in 23 US hospitals that use an RFID based patient-identification system. The study used semi-structured interviews to find the social and ethical implications of such systems and some important findings were discovered. These included a) unfair prioritization of patients; b) diminished trust of patients; and c) endangerment of patients. More detailed and commendable work in this area has been done by Foster and Jaeger [100]. Key issues such as ethics, concerns about privacy and potentially forced implantations of RFID are discussed emphasizing the need for a national discussion to identify the limits regarding implantable RFID tags in humans before it is too late to prevent the misuse of this useful but ethically problematic technology. The following implantable RFID systems are quoted in [100]: a) VeriMed from the VeriChip corporation for use by patients who present themselves to healthcare facilities unresponsive; b) implantable tags in dentures/human molars by Thevissen (2006); c) edible RFID tags from Eastman Kodak Company to monitor internal bodily events following the ingestion of medication (Spoonhower 2007); d) ClearCount Solution to detect implanted surgical sponges; e) VeriGuard from VeriChip to control individuals accessing secure areas; f) the proposal from VeriChip to implant chips into immigrants and guest workers. Colombian President Alfaro Uribe apparently welcomed the idea and was quoted as telling a US senator he would get Colombian citizens to agree to being implanted with RFID chips if they were granted permission to work in the US. VeriChip is also trying to convince the US military to replace metal "dog tags" worn by soldiers with implantable RFID tags, and g) Professor Kevin Warwick from the University of Reading, UK implanted an RFID chip as part of his work *Cyborg I project*. He was able to walk the halls of his university, operating doors and lights without lifting a finger. He had a microelectrode array implanted in the median nerve fibers of his left arm to allow him to send signals back and forth between his nervous system and a computer. Using these examples as a baseline, Foster and Jaeger [100] reported the following implications:

1. **Possible Medical Risks:** In its ruling, the US Food and Drug Administration (FDA) classified VeriChip as a class II device and listed the potential hazards as: *adverse tissue reaction, migration of implanted transponder, compromised information security, failure of implanted transponder, failure of inserter, failure of electronic scanner, electromagnetic interference, electrical hazards, magnetic resonance imaging incompatibility, and needle stick (US FDA 2004)*. Controversy also arose about possible carcinogenic effects of implanted chips. Blanchard *et al.* (1999), Elcock *et al.* (2001) and Tillmann *et al.* (1997) (cited in [100]), state that it has been known for implanted RFID tags, similar to the ones used in the VeriMed system, to be associated with tumors in rodents, mostly sarcomas. Vascellari *et al.* (2006) (cited in [100]), mentioned tumors in dogs implanted with microchips.
2. **Ethical Issues:** A lot of literature can be found about privacy and security issues of RFID and activists' websites also discuss possible harm from implanted RFIDs, which included the tracking of individuals to mind control. The latter being in the realm of science fiction at least for the foreseeable future.
3. **Disclosure of Risks:** It is only ethical to disclose adverse effects of a treatment and individuals have a right to know about the effects. It was noted that VeriChip failed to disclose the rodent studies until the issue was raised by activists. In addition, the carcinogenic effect of an implant in rodents suggests the possibility of a similar effect in humans. A cancer researcher states - "*I think the evidence from the animal studies is indeed alarming and one should refrain from chipping people unless the mechanisms and long-term effects are known*". Foster [100] suggests cancer be added to the list of adverse side effects provided by the FDA.
4. **Truth in Advertising:** The above-mentioned occurrences call for an independent assessment of risks and benefits using implanted RFID tags so they can be advertised truthfully.

5. **Coercion:** Cottle (2006) (cited in [100]) stated the *New Republic Online* defended a proposal to implant RFID tags in sex offenders because such people are already subject to extensive restrictions and tracking via chips might be preferable. However, fundamentalist Christian groups in the US vehemently objected to such ideas as "marks of the beast". Judaism and Islam also prohibit tattoos and their religious heads may forbid implanted tags for similar reasons. The proposal from VeriChip to implant immigrant/guest workers indirectly forces some people to accept the chip because of poverty at home and the greater likelihood of a job in the US. This would be deeply offensive on the grounds of human rights and would be in violation of Article 3 of the Universal Declaration of Human Rights (1948) which guarantees right to "life, liberty and security of person". It would also be a violation of the privacy provision of the International Covenant on Civil and Political Rights (1966) to which the US is a party.

Foster and Jaeger [100] conclude that technologies such as fingerprint and retinal scans allow reliable identification of individuals without the need to compromise bodily integrity. They also quote Anderson and Labay's remark - "*a decision about where to draw the line of acceptable use must be made soon, before the technology becomes rampant and it becomes too late to prevent misuse*".

2.8 Barriers Impacting Adoption of RFID in Healthcare

The key barriers to adopting RFID include high up front costs and the global healthcare industry's main objection to adopting ICT solutions in general because paper is more reliable even though it is harder to manage [101]. In addition, barriers such as technological limitations, interference concerns, lack of global standards, privacy concerns, and social and legal issues also slow down the adoption of RFID in healthcare [51].

Fisher and Monahan [102] evaluated several RTLS systems currently in use in US hospitals. The results of their study showed most RTLS systems have substandard functionality, including material and organizational constraints that pose major obstacles to effective deployment of systems, not to mention the best use of RTLS is asset-tracking. Some reasons for the failure of RTLS systems are identified as: short battery life of tags; failure to achieve room-level accuracy when it comes to tracking which reveals the technological incapabilities of RTLS; the difference in the physical structure of every hospital may pose major deployment challenges; and the organizational culture of hospitals - territoriality and poor division of labor may result in limiting the scope of deployment. It was also noted that using RTLS to track people yielded worse results due to resistance from staff.

Katz and Rice [98] note that many object to the technology on religious grounds and that many web sites refer to the biblical chapter of Mark as justification for their opposition. Verse 16-17 states the Antichrist will "*causeth all*" to "*receive a mark in their right hand or in their foreheads and that no man might buy or sell, save he that had the mark*". This is quoted as a prophetic warning to resist RFID as the mark of the devil. However, experts in the industry are confident that despite these barriers, RFID in healthcare will be where wireless technology ends up.

2.9 Benefits, Success Stories and Future of RFID in Healthcare

Here, we discuss some of the benefits and successful implementations of RFID in healthcare and also briefly address the industry forecast concerning future prospects.

Improved patient safety, reduced medical errors, time and cost savings, and improved medical processes are some of the major benefits identified by Yao *et al.* [51]. ClearCount, (cited in [12]), states that nurses spend 15 to 30 minutes counting surgical sponges and other instruments after every surgery, costing more than $1 billion every year. When the count does not match, X-rays are taken to look for items inside patients costing $375 million, and any resulting litigation costs an additional $125 million. The ClearCount application that helps identify RFID tagged surgical sponges, improves patient safety, work efficiency and saves the image of healthcare professionals and hospitals by reducing the number of litigations. Chang-Gung Memorial Hospital in Taipei, Taiwan, achieved time savings of 4.3 minutes per patient to perform a patient identification/verification procedure with an accuracy level of 100% for patient identification [12]. Virginia Hospital realized a net savings of USD$3 million over the first 3 years after implementing RTLS from Agility Healthcare Solutions [12]. The system was used to track over 10,000 pieces of equipment. The RFID Lab Specimen Tracking System claims to achieve an error-rate of less than 1%, which is far below average lab error rates [26].

Saarijarvi [101] predicts a promising future for RFID in Healthcare. It is noted that "*RFID tops the chart in medicine with vastly increased safety, efficiency and ROI*". For example, staff can quickly identify an optimal bag of blood from a blood bank freezer close to its expiry by using RFID that can read as many as 400 bags per second. Needless to say, with barcodes this is an impossible task to carry out in freezers that maintain subarctic temperatures. After implementing an RFID/infrared tracking system in 2010, Mission Hospital in California posted an ROI of 186% within a year. The system also boosted device utilization rates, eliminated equipment rentals/thefts, and equipment shrinkage dropped from $150,000/yr to zero. When trauma cases came to the hospital, equipment was ready and waiting because of the ability to locate it within seconds. After their huge success with RFID in other areas, Mission Hospital is planning on expanding the system to prevent cross-contamination (hand washing for example, the number one preventer of cross-infection in hospitals). Healthcare professionals will receive alerts on their mobile devices should they forget/neglect to wash their hands after moving in and out of patient rooms. The forecast is that in 15 years, all healthcare processes will be electronically managed because RFID enables hospitals to do things they have never done before.

3 Future Direction

Here we present some research directions for RFID in healthcare so it can be safely and securely deployed.

- Most of the literature addresses the application of RFID in healthcare in one form or another. But not many address the security and privacy issues RFID may cause in healthcare, even though there are few security protocols specifically designed for this sector. It is vital this gap be filled. Recent developments by Sundaresan *et al.* [88] and Sundaresan *et al.* [89] address this gap in the context of grouping proofs and this can be extended to other areas.
- As seen in Section 2.4, several contradicting reports about electromagnetic interference due to the use of RFID. There is a definite need for further research to qualitatively and quantitatively measure this factor. Ensuring RFID is a safe and reliable technology to be used in healthcare is vital to increase consumers' confidence in the technology as well as in the healthcare system.
- Implantable RFIDs and the impact it could have on humans is still among the major unknowns. Though there are some reports that make speculations, further in-depth research needs to be undertaken to clearly identify the pros and cons of such systems before large-scale implementations are encouraged.
- The medical industry does not have any universal standard RFID products or practices [12]. Hosaka [103] recommends following the use of anti-medical tags in healthcare, reducing Q (the quality factor) in the resonance circuit of tags to improve the communication distance, and adhereing to 125KHz and 2.4GHz bands as the communication frequency. The need for universal standards for RFID enabled healthcare products is apparent.
- While the benefits of RFID in terms of work efficiency, and cost savings are discussed, statistics on how RFID has helped to reduce medical errors (or even the number of deaths) were not found. It would be interesting to find out if the implementation of RFID has improved the situation. A detailed survey of the hospitals before and after RFID would provide a good answer. As seen, the statistics shown in Section 1.1 depict the dark side of the medical industry. A thorough research on the after-effects of deploying technologies such as RFID would not only help improve the consumers' confidence in the medical system but definitely help improve the image of RFID and promote the technology further.

4 Summary

In this chapter, we provided extensive coverage of RFID in the healthcare context, in order to gain a clear understanding of current trends and to explore future prospects. Analysis included areas such as RFID based healthcare applications (including sensor based and implantable tags); Electro Magnetic Interference (EMI) and RFID; healthcare specific security protocols; security and privacy issues posed by RFID in healthcare; ethical implications; factors acting as barriers in the adoption of RFID in healthcare; benefits reaped from the system; and finally, the future of RFID in healthcare. This holistic approach aims to help the common healthcare users who wish to gain an understanding of RFID's impact in healthcare, and serve as a guide for Information Technology professionals to design innovative applications and motivate further research in the area.

References

1. Roberti, M.: Does RFID Create Negative Effects in Hospitals? RFID Journal (2008), http://www.rfidjournal.com/article/view/5264
2. Moessner, M., Khan, G.N.: Secure authentication scheme for passive C1G2 RFID tags. Computer Networks 56(1), 273–286 (2012)
3. EPCGlobal, EPC Radio-Frequency Identity Protocols,Class-1 Generation-2 UHF RFID Protocol for Communications at 860MHz-960MHz Version 1.2.0 (2008)
4. Kohn, L.T., Corrigan, J.M., Donaldson, M.S.: To Err Is Human: Building a Safer Health System. The National Academies Press (2000), http://www.nap.edu/openbook.php?record_id=9728
5. Preventable Medical Errors – The Sixth Biggest Killer in America, American Association for Justice, http://www.justice.org/cps/rde/justice/hs.xsl/8677.htm
6. H. G. Inc., Patient Safety in American Hospitals (2004), http://www.healthgrades.com/media/english/pdf/HG_Patient_Safety_Study_Final.pdf
7. Patient Safety: Current Statistics, Patient Safety Focus, http://www.patientsafetyfocus.com/patient-safety-current-st.html
8. McGann, E.: Medical Error Index. Human Factors MD (2010), http://www.humanfactorsmd.com/our-medical-error-index/
9. Graban, M.: Statistics on Healthcare Quality and Patient Safety Problems – Errors & Harm (2012), http://www.leanblog.org/2009/08/statistics-on-healthcare-quality-and/
10. Ting, S.L., Tsang, A.H.C.: Development of an RFID-based Surgery Management System: Lesson Learnt from a Hong Kong Public Hospital. Journal of Healthcare Engineering 3(3), 347–372 (2012)
11. Drugs and Medical Errors Killing One of Every Five Australians. Best Health (2009), http://besthealth.com.au/drugs-and-medical-errors-killing-one-of-every-five-australians/
12. Mowry, M.: A Survey of RFID in the Medical Industry with emphasis on Applications to Surgey and Surgical Devices, pp. 1–19 (2008), http://www.winmec.ucla.edu/rfid/course/2008s/Final_project_Mike_Mowry.pdf
13. Wamba, S.F.: RFID-enabled Healthcare Applications, Issues and Benefits: An Archival Analysis (1997-2011). Journal of Medical Systems, 3393–3398 (2012)
14. Polycarpou, A.C., Dimitriou, A., Bletsas, A., Polycarpou, P.C., Papaloizou, L., Gregoriou, G., Sahalos, J.N.: On the Design, Installation and Evaluation of a Radio Frequency IDentification System for Healthcare Applications. IEEE Antennas and Propagation Magazine 54(4), 255–271 (2012)

15. Martínez, M., Fontecha, J., Vizoso, J.R., Bravo, J., Cabrero-Canosa, M.J., Martín, I.: RFID and NFC in hospital environments: Reaching a sustainable approach. In: Bravo, J., López-de-Ipiña, D., Moya, F. (eds.) UCAmI 2012. LNCS, vol. 7656, pp. 125–128. Springer, Heidelberg (2012), http://dx.doi.org/10.1007/978-3-642-35377-2_17
16. Chen, J.C., Collins, T.J.: Creation of a RFID Based Real Time Tracking (R-RTT) System for Small Healthcare Clinics. Journal of Medical Systems 36, 3851–3860 (2012)
17. Musa, A., Yusuf, Y., Meckel, M.: A Hospital Resource and Patient Management System Based on Real-Time Data Capture and Intelligent Decision Making. In: International Conference on Systems and Informatics, pp. 776–779 (2012)
18. Yeung, C.L., Kwok, S.K., Mui, H.C.: An Investigation of an RFID-based Patient-tracking and Mobile Alert System. International Journal of Engineering Business Management 3(1), 50–56 (2011)
19. Laskowski, M., Demianyk, B., Friesen, M.R., Mcleod, R.D.: Uncertainties Inherent in RFID Tracking Systems in an Emergency Department. In: 2010 IEEE Workshop on Health Care Management (WHCM), pp. 1–6 (2010)
20. Ostbye, T., Lobach, D.F., Cheesborough, D., Lee, A.M.M., Krause, K.M., Hasselblad, V., Bright, D.: Evaluation of an Infrared / Radiofrequency Equipment-Tracking System in a Tertiary Care Hospital. Journal of Medical Systems 27(4), 367–380 (2003)
21. Polycarpou, A.C., Gregoriou, G., Dimitriou, A., Bletsas, A., Sahalos, I.N., Papaloizou, L., Polycarpou, P.: Design and Implementation of a Radio Frequency IDentification (RFID) System for Healthcare Applications. In: Medicon 2010 IFMBE Proceedings, vol. 29, pp. 897–900 (2010)
22. Perez, M.M., Cabrero-Canosa, M., Hermida, J.V., Garcia, L.C., Gomez, D.L., Gonzalez, G.V., Herranz, I.M.: Application of RFID Technology in Patient Tracking and Medication Traceability in Emergency Care. Journal of Medical Systems 36, 3983–3993 (2012)
23. QIN, Z., Huang, G.Q., DAI, Q.: No Room for Error: RFID-enabled Smart Point-of-Care Medication Process in Hospital Wards. In: IEEE International Conference on RFID-Technologies and Applications, pp. 353–358 (2011)
24. Xu, S., Duan, H., Lu, X., Huang, Z.: A RFID-based Tracking System of Endoscopes. In: 4th International Conference on Biomedical Engineering and Informatics (BMEI), pp. 1909–1913 (2011)
25. Chen, P.-J., Chen, Y.-F., Chai, S.-K., Huang, Y.-F.: Implementation of an RFID-Based Management System for Operation Room. In: Proceedings of the 8th International Conference on Machine Learning and Cybernetics, pp. 2933–2938 (2009)
26. Wisdom, R.: Health Care Compliance - Automated through RFID (2012), http://blog.odintechnologies.com/bid/93002/Health-Care-Compliance-Automated-through-RFID
27. Unluturk, M.S., Kurtel, K.: Integration of RFID and Web Service for Assisted Living. Journal of Medical Systems 36, 2371–2377 (2012)
28. Akiyama, M., Koshio, A.: It can improve healthcare management for patient safety - minimizing risk of blood transfusion with point-of-act-system. In: 2011 IEEE International Conference on Industrial Engineering and Engineering Management (IEEM), pp. 979–984 (2011)
29. Jeong, B.H., Cheng, C.Y., Prabhu, V., Yu, B.J.: An rfid application model for surgery patient identification. In: IEEE Symposium on Advanced Management of Information for Globalized Enterprises, pp. 1–3 (2008)
30. Shim, H., Uh, Y., Lee, S.H., Ro, Y.Y.: A New Specimen Management System Using RFID Technology. Journal of Medical Systems 35, 1403–1412 (2011)

31. Liu, C.C.H., Chang, C.-H., Su, M.-C., Chu, H.-T., Hung, S.-H., Wong, J.-M., Wang, P.-C.: RFID-initiated workflow control to facilitate patient safety and utilization efficiency in operation theater. Computer Methods and Programs in Biomedicine 104(3), 435–442 (2011)
32. Parks, R., Chu, C.-H., Xu, H.: RFID Privacy Issues in Healthcare: Exploring the Roles of Technologies and Regulations. Journal of Information Processing Systems 61, 1–24 (2004), `http://faculty.ist.psu.edu/xu/papers/jips.pdf`
33. Macario, A., Morris, D., Morris, S.: Initial Clinical Evaluation of a Handheld Device for Detecting Retained Surgical Gauze Sponges Using Radiofrequency Identification Technology. American Medical Association 141, 659–662 (2006)
34. Song, J.-G., Jung, S., Kang, B.H., Hwang, D.-J., Kim, S.: Design of context-aware middleware for mobile healthcare services. Journal of the Chinese Institute of Engineers 35(5), 535–545 (2012)
35. Zetter, K.: RFID-Enabled Phones Could Let Credit Card Companies Track Users (2009), `http://www.wired.com/threatlevel/2009/06/rfid-enabled-phones-would-let-credit-card-companies-track-users/`
36. NFC: RFID enabled smartphones and mobile devices are coming," (2010), `http://intrepidusgroup.com/insight/2010/12/nfc-rfid-enabled-smartphones-and-mobile-devices-are-coming/`
37. Lin, S.-S., Hung, M.-H., Tsai, C.-L.: Development of an Ease-of-Use Remote Healthcare System Architecture Using RFID and Networking Technologies. Journal of Medical Systems 36, 3605–3619 (2012)
38. Shih, S.-T., Hsu, C.-M., Chao, C.-Y.: Self Healthcare Management System based on RFID Technology. In: 2011 7th International Conference on Networked Computing and Advanced Information Management (NCM), pp. 252–255 (2011)
39. Zhang, C., Gruver, W.A.: Distributed Agent System for Behavior Pattern Recognition. In: Proceedings of the 9th International Conference on Machine Learning and Cybernetics, pp. 204–209 (2010)
40. Kim, S.-C., Jeong, Y.-S., Park, S.-O.: RFID-based indoor location tracking to ensure the safety of the elderly in smart home environments (2012)
41. Symonds, J., Parry, D., Briggs, J.: An RFID-based system for assisted living: Challenges and Solutions. Studies in Health Technology and Informatics, 127–138 (2007)
42. Lin, J., Pai, J.-Y., Chen, C.-C.: Applied Patent RFID Systems for Building Reacting HEPA Air Ventilation System in Hospital Operation Rooms. Journal of Medical Systems 36, 3399–3405 (2011)
43. Escribano, J.G., García, A.: Human condition monitoring in hazardous locations using pervasive RFID sensor tags and energy-efficient wireless networks. Journal of Zhejiang University 13(9), 674–688 (2012)
44. Mccall, C., Maynes, B., Zou, C.C., Zhang, N.J.: An automatic medication self-management and monitoring system for independently living patients. Medical Engineering & Physics, 1–10 (2012)
45. Sun, P.R., Wang, B.H., Wu, F.: A New Method to Guard Inpatient Medication Safety by the Implementation of RFID. Journal of Medical Systems 32, 327–332 (2008)
46. Yamashita, K., Kusuda, K., Iwakami, Y., Tanaka, K., Tokuda, Y., Tanaka, S., Kuraoka, K., Honda, H., Komino, M., Komatsu, H., Shimada, S., Obayashi, T., Saito, Y., Ino, S., Ifukube, T., Okubo, T.: Development of System for Managing Individual Surgical Instruments by RFID. In: 5th European IFMBE Conference, IFMBE Proceedings, vol. 37, pp. 725–728 (2011)

47. Wang, S.-W., Chen, W.-H., Ong, C.-S., Liu, L., Chuang, Y.-W.: RFID applications in hospitals: a case study on a demonstration RFID project in a Taiwan hospital. In: Proceedings of the 39th Hawaii International Conference on System Sciences, pp. 1–10 (2006)
48. Stahl, J.E., Holt, J.K., Gagliano, N.J.: Understanding Performance and Behavior of Tightly Coupled Outpatient Systems Using RFID: Initial Experience. Journal of Medical Systems 35, 291–297 (2011)
49. Baarah, A., Mouttham, A., Peyton, L.: Improving Cardiac Patient Flow Based on Complex Event Processing. In: IEEE Jordan Conference on Applied Electrical Engineering and Computing Technologies, pp. 1–4 (2011)
50. Unluturk, M.S.: SEE: Improving Nurse-Patient Communications and Preventing Software Piracy in Nurse Call Applications. Journal of Medical Systems 36, 1909–1916 (2012)
51. Yao, W., Chu, C.-H., Li, Z.: The Adoption and Implementation of RFID Technologies in Healthcare: A Literature Review. Journal of Medical Systems 36, 3507–3525 (2011)
52. Pletervsek, A., Sok, M., Trontelj, J.: Monitoring, Control and Diagnostics using RFID Infrastructure. Journal of Medical Systems 36, 3733–3739 (2012)
53. Hung, C.-H., Bai, Y.-W., Tsai, R.-Y.: Design of Blood Pressure Measurement with a Health Management System for the Aged. In: 2012 IEEE International Conference on Consumer Electronics (ICCE), pp. 619–625 (2012)
54. Vecchia, G.D., Gallo, L., Esposito, M., Coronato, A.: An infrastructure for smart hospitals. Multimed Tools Application 59, 341–362 (2012)
55. Opasjumruskit, K., Thanthipwan, T., Sathusen, O., Sirinamarattana, P., Gadmanee, P., Pootarapan, E., Wongkomet, N., Thanachayanont, A., Manop, T.: Self-Powered Wireless Temperature Sensors Exploit RFID Technology. In: IEEE CS and IEEE ComSec, pp. 54–61 (2006)
56. Barman, J., Uswatte, G., Sarkar, N., Ghaffari, T., Sokal, B.: Sensor-Enabled RFID System for Monitoring Arm Activity in Daily Life. In: 33rd Annual International Conference of the IEEE EMBS, pp. 5219–5223 (2011)
57. Catarinucci, L., Colella, R., Tarricone, L.: Optimized Antennas for Enhanced RFID Sensor Tags. In: Proc. of IEEE Antennas and Propagation Society, AP-S International Symposium (Digest), pp. 1433–1436 (2011)
58. Romain, O., Mazeyrat, J., Garda, P., Talleb, H., Lautru, D., Wong, M.F., Wiart, J., Hanna, F.V., Lagree, P.Y., Bonneau, M., Kang, C., Fernandex, M., Gerbeau, J.-F., Deplano, V., Berthier, B., Legallais, C., Leprince, P.: Rfid implantable pressure sensor for the follow-up of abdominal aortic aneurysm stented. In: 2011 6th International Conference on Design Technology of integrated Systems in Nanoscale Era (DTIS), pp. 1–6 (2011)
59. Yao, W., Chu, C.-H., Li, Z.: The Use of RFID in Healthcare: Benefits and Barriers. In: 2010 IEEE International Conference on RFID-Technology and Applications (RFID-TA), pp. 128–134 (2010)
60. Aubert, H.: RFID technology for human implant devices. Comptes Rendus Physique 12(7), 675–683 (2011)
61. Grosinger, J., Fischer, M.: Evaluating On-Body RFID Systems at 900 MHz and 2. 45 GHz. In: Fourth International EURASIP Workshop on RFID Technology, pp. 52–58 (2012)
62. Occhiuzzi, C., Contri, G., Marrocco, G.: Design of implanted rfid tags for passive sensing of human body: The stentag. IEEE Transactions on Antennas and Propagation 60(7), 3146–3154 (2012)

63. Thevissen, P.W., Poelman, G., Cooman, M.D., Puers, R., Willems, G.: Implantation of an RFID-tag into human molars to reduce hard forensic identification labor. Part I: Working principle. Forensic Science International, 33–39 (2006)
64. Meyer, H.J., Chansue, N., Monticelli, F.: Implantation of radio frequency identification device (RFID) microchip in disaster victim identification (DVI). Forensic Science International 157, 168–171 (2006)
65. Madrid, C., Korsvold, T., Rochat, A., Abarca, M.: Radio frequency identification (RFID) of dentures in long-term care facilities. The Journal of Prosthetic Dentistry 107(3), 199–202 (2012)
66. Houliston, B., Parry, D., Webster, C.S., Merry, A.F.: Interference with the operation of medical devices resulting from the use of radio frequency identification technology. Journal of the New Zealand Medical Association 122 (2009)
67. Fiocchi, S., Parazzini, M., Paglialonga, A., Ravazzani, P.: Computational Exposure Assessment of Electromagnetic Fields Generated by an RFID System for Mother and Newborn Identity Reconfirmation. Journal of Bioelectromagnetics 32, 408–416 (2011)
68. Kapa, S., Pierce, T., Hayes, D.L., Holmes Jr., D.R., Asirvatham, S.J.: Electromagnetic interference of magnetic field based auto identification technologies in healthcare settings. International Journal of Medical Informatics 80(4), 239–250 (2011)
69. Cheng, C.-Y., Chai, J.-W.: Deployment of RFID in Healthcare Facilities - Experimental Design in MRI Department. Journal of Medical Systems 36, 3423–3433 (2012)
70. Van Der Togt, R., Van Lieshout, E.J., Hensbroek, R., Beinat, E., Binnekade, J., Bakker, P.: Electromagnetic Interference From Radio Frequency Identification Inducing Potentially Hazardous Incidents in Critical Care Medical Equipment. Journal of the Americal Medical Association 299(24), 2884–2890 (2008)
71. Payne, M.: Dutch RFID Interference Study Is a Worst-Case Test. RFID Journal (2008), http://www.rfidjournal.com/article/view/4192
72. Bacheldor, B.: Researchers Warn RFID May Disrupt Medical Equipment. RFID Journal (2008), http://www.rfidjournal.com/article/view/4166
73. Bacheldor, B.: New RFID Study Finds No Interference With Medical Devices. RFID Journal (2008), http://www.rfidjournal.com/article/view/4186
74. Iadanza, E., Dori, F., Miniati, R., Corrado, E.: Electromagnetic Interferences (EMI) from Active RFId on Critical Care Equipment. In: MEDICON 2010. IFMBE Proceedings, vol. 29, pp. 991–994 (2010)
75. Rieche, M., Komensky, T., Husar, P.: Radio Frequency Identification (RFID) in medical environment: Gaussian Derivative Frequency Modulation (GDFM) as a novel modulation technique with minimal interference properties. In: 33rd Annual International Conference of the IEEE EMBS, pp. 2180–2183 (2011)
76. Censi, F., Calcagnini, G., Mattei, E., Triventi, M., Bartolini, P.: RFID in Healthcare Environment: Electromagnetic Compatibility Regulatory Issues. In: 32nd Annual International Conference of the IEEE EMBS, pp. 352–355 (2010)
77. Censi, F., Calcagnini, G., Mattei, E., Triventi, M., Bartolini, P.: Electromagnetic Compatibility of Portable RF Emitters in Uniquos Health Environment: Regulatory Issues. PIERS Online 7(5), 456–460 (2011)
78. Ogirala, A., Stachel, J.R., Mickle, M.H., Fellow, L.: Electromagnetic Interference of Cardiac Rhythmic Monitoring Devices to Radio Frequency Identification: Analytical Analysis and Mitigation Methodology. IEEE Transactions on Information Technology in Biomedicine 15(6), 848–853 (2011)

79. Rrub, J.A., Al-Jabi, J., El-Khatib, K.: Security Model for Real Time Tracking System (RTLS) in the Healthcare Sector. In: 2nd International Conference on Communications and Information Technology (ICCIT): Wireless Communications and Signal Processing, pp. 369–373 (2012)
80. Yu, Y.-C., Hou, T.-W.: Utilize Common Criteria Methodology for Secure Ubiquitous Healthcare Environment. Journal of Medical Systems 36, 1689–1696 (2012)
81. Min, D., Yih, Y.: Fuzzy Logic-Based Approach to Detecting a Passive RFID Tag in an Outpatient Clinic. Journal of Medical Systems 35, 423–432 (2011)
82. Tu, Y.-J., Zhou, W., Piramuthu, S.: Identifying RFID-embedded objects in pervasive healthcare applications. Decision Support Systems 46(2), 586–593 (2009)
83. Huang, H.-H., Ku, C.-Y.: A RFID Grouping Proof Protocol for Medication Safety of Inpatient. Journal of Medical Systems 33, 467–474 (2009)
84. Chien, H.-Y., Yang, C.-C., Wu, T.-C., Lee, C.-F.: Two RFID-based Solutions to Enhance Inpatient Medication Safety. Journal of Medical Systems 35, 369–375 (2011)
85. Wickboldt, A.-K., Piramuthu, S.: Patient Safety Through RFID: Vulnerabilities in Recently Proposed Grouping Protocols. Journal of Medical Systems 36, 431–435 (2012)
86. Yu, Y.-C., Hou, T.-W., Chiang, T.-C.: Low Cost RFID Real Lightweight Binding Proof Protocol for Medication Errors and Patient Safety. Journal of Medical Systems 36, 823–828 (2012)
87. Wu, S., Chen, K., Zhu, Y.: A Secure Lightweight RFID Binding Proof Protocol for Medication Errors and Patient Safety. Journal of Medical Systems 36, 2743–2749 (2012)
88. Sundaresan, S., Doss, R., Zhou, W.: Grouping Proof with Forward Security for Gen 2 RFID Tags. In: 2013 IEEE RFID TA (2013)
89. Sundaresan, S., Doss, R., Zhou, W.: Offline Grouping Proof Protocol for RFID Systems. In: The 9th IEEE International Conference on Wireless and Mobile Computing, Networking and Communications, WiMob 2013 (2013)
90. Chen, Y.-Y., Huang, D.-C., Tsai, M.-L., Jan, J.-K.: A Design of Tamper Resistant Prescription RFID Access Control System. Journal of Medical Systems 36, 2795–2801 (2012)
91. Zhou, W., Yoon, E.J., Piramuthu, S.: Simultaneous multi-level RFID tag ownership & transfer in health care environments. Decision Support Systems (2012)
92. Yeh, K.-H.: A Patient Privacy-aware e-Health System based on Passive RFID. In: 9th International Conference on Ubiquitous Intelligence and Computing & Autonomic and Trusted Computing, pp. 967–972 (2012)
93. Gasson, M.N.: Human Enhancement: Could you become infected with a computer virus? In: IEEE International Symposium on Technology and Society, pp. 61–68 (2010)
94. Kim, J.T.: Privacy and Security Issues for RFID Healthcare System in Wireless Sensor Networks. In: Lee, G., Howard, D., Ślęzak, D., Hong, Y.S. (eds.) ICHIT 2012. CCIS, vol. 310, pp. 594–601. Springer, Heidelberg (2012)
95. Hawrylak, P.J., Schimke, N., Hale, J., Papa, M.: Security Risks Associated with Radio Frequency Identification in Medical Environments. Journal of Medical Systems 36, 3491–3505 (2012)
96. Duc, D., Kim, K.: Defending RFID authentication protocols against DoS attacks. Computer Communications (2010)
97. Lopez, P., Ofila, A., Castro, J., van der Lubbe, J.: Flaws on RFID grouping proofs. Guidelines for future sound protocols. Journal of Network and Computer Applications (2010)
98. Katz, J.E., Rice, R.E.: Public views of mobile medical devices and services: A US national survey of consumer sentiments towards RFID healthcare technology. International Journal of Medical Informatics 78, 104–114 (2009)

99. Monahan, T., Fisher, J.A.: Implanting inequality: Empirical evidence of social and ethical risks of implantable radio-frequency identification (RFID) devices. International Journal of Technology Assessment in Healthcare 26(4), 370–376 (2010)
100. Foster, K.R., Jaeger, J.: Ethical Implications of Implantable Radiofrequency Identification (RFID) Tags in Humans. American Journal of Bioethics 8(8), 44–48 (2008)
101. Saarijarvi, M.: RFID: The Medical Miracle. RFID Arena (2011), `http://www.rfidarena.com/2011/12/15/rfid-the-medical-miracle.aspx`
102. Fisher, J.A., Monahan, T.: Evaluation of real-time location systems in their hospital contexts. International Journal of Medical Informatics 81(10), 705–712 (2012)
103. Hosaka, R.: An Analysis for Specifications of Medical Use RFID System as a Wireless Communication. In: Proceedings of the 29th Annual International Conference of the IEEE EMBS, pp. 2795–2798 (2007)

Questions

Q1. Discuss the Role Played by RFID with Sensor Capabilities

Q2. Comment on EMI Effects due to the Use of RFID in Health Care

Q3. Discuss Any Five Health Care Applications That Uses RFID

Q4. Comment on On-Body /Implanting RFID in Humans

Q5. Comment on Security and Privacy Requirements of RFID in Healthcare

Q6. Discuss about Ethical Implications of RFID in Healthcare

Part VII

Communication Network and Systems

Chapter 37: Wireless Body Area Networks in mHealth

Chapter 38: An Integrated Wireless Communication Platform for End-to-End and Automatic Wireless Vital Sign Capture Using Personal Smart Mobile Devices

Chapter 39: Energy Analysis and QoE of Wireless Sensor Networks

Chapter 40: Quality of Service in Wireless Technologies for mHealth Service Providing

Chapter 41: mHealth over "Medical Grade" High Quality of Service Mobile Networks

37. Wireless Body Area Networks in mHealth

Garth V. Crosby[1], Craig A. Chin[2], Tirthankar Ghosh[3], and Renita Murimi[4]

[1] Department of Technology, Southern Illinois University Carbondale, Illinois, USA
garth.crosby@siu.edu
[2] Department of Electrical & Computer Engineering Technology,
Southern Polytechnic State University, Georgia, USA
[3] Department of Computer Science & Information Technology,
St. Cloud State University, Minnesota, USA
[4] College of Business, Oklahoma Baptist University, Oklahoma, USA

Abstract. Wireless body area networks (WBANs) are emerging as important networks, applicable in various fields. In this chapter a comprehensive survey of WBANs that are designed for applications in healthcare is presented. The survey consists of stand-alone sections focusing on important aspects of WBANs. Topics covered are: monitoring and sensing, power efficient protocols, physical layer, MAC protocols, routing, system architectures and security. The chapter concludes with a discussion of open research issues, their potential solutions and future trends.

Keywords: Wireless body area network, body sensor network, healthcare, WBAN survey.

1 Introduction

The increase in average lifespan and health cost in many developed nations are catalysts to innovation in health care. These factors along with the advances in the miniaturization of electronic devices, sensing, battery and wireless communication technologies have led to the development of Wireless Body Area Networks (WBANs). WBANs consist of smart miniaturized devices (motes) that are able to sense, process and communicate. They are designed such that they can be worn or implanted, and they can be used to monitor a variety of physiological signals as well as signals associated with the location and posture of an individual. The captured signals are transmitted to specialized medical servers without much interference to the daily routine of the patient.

The general architecture of a WBAN is shown in Figure 1, and the position of a WBAN within the more general architecture of a Wireless Health Monitoring System (WHMS)[1] is shown in Figure 2. These figures illustrate the necessity of WBANs for continuous/vital sign monitoring in mHealth applications.

S. Adibi (ed.), *Mobile Health*,
Springer Series in Bio-/Neuroinformatics 5, DOI: 10.1007/978-3-319-12817-7_37

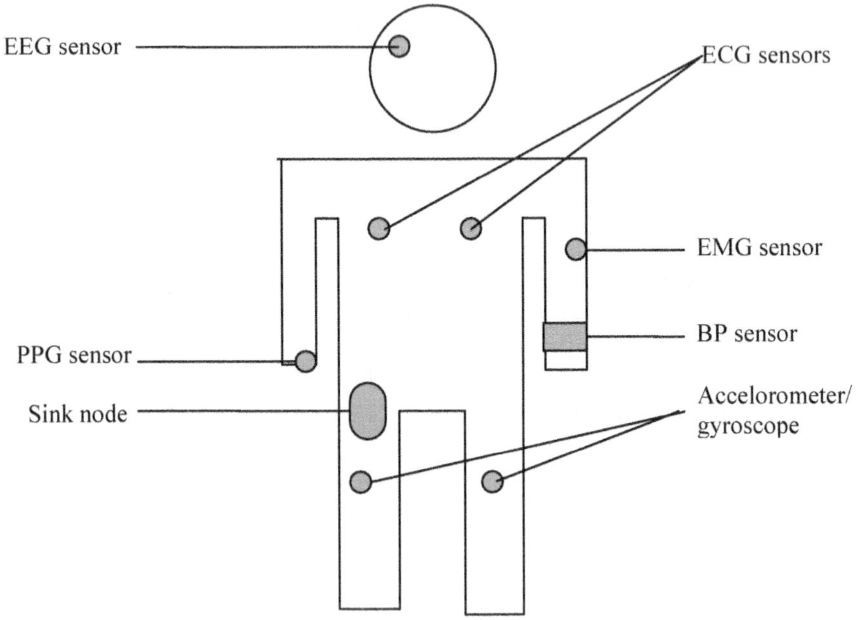

Fig. 1 – Architecture of a WBAN

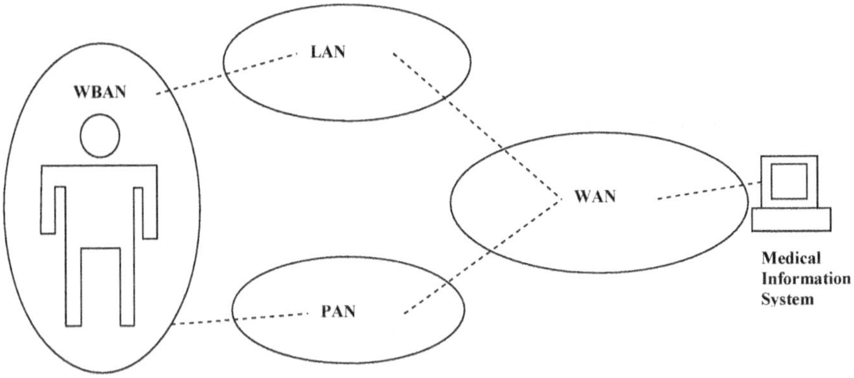

Fig. 2 – Communications Architecture for WHMS

2 Monitoring and Sensing

2.1 Sensor Taxonomy

Sensors can be classified in different ways [2]. If we consider the deployment position of the sensors, sensors may be classified as wearable (on the body) and implantable (within the body). Based on the transmission medium used for communication between sensors, there are wireless sensors, wired sensors and human body communication (HBC) sensors. Sensors may also be categorized according to the frequency of samples required for monitoring purposes. There are signals that require continuous monitoring (e.g. ECG, EEG, EMG, accelerometers, gyroscopes, etc.), and those that require occasional (a time-scale of minutes to hours) readings (e.g. blood pressure, glucose and temperature). We may also classify sensors based upon their adaptation ability. Self-adapting sensors can automatically adjust various processing parameters based on changes observed in measured data in order to optimize some performance measure. Non-self-adapting sensors do not possess this capability.

2.2 Sensor Examples

Some examples of sensors currently used in the construction of WBANs is reviewed in this section.

Pulse Oximetry

A pulse oximeter is a medical device that indirectly measures the oxygen saturation levels (SpO_2) in an individual's blood as well as the changes in blood volume in the skin that coincide with the cardiac cycle. Typically, a pulse oximeter is attached to a finger or an earlobe, and it consists of red and infra-red light-emitting diodes (LEDs) and a photodetector. The photodetector measures the amount of red and infra-red light that is transmitted through or reflected by the body part, which is partially dependent on the amount of light absorbed by the blood that perfuses the body part. The relative absorption of red and infra-red light by the blood is related to the ratio of oxygenated hemoglobin to deoxygenated hemoglobin, and this serves as the basis of the SpO_2 measurement. The overall amount of light absorption varies as the pulsatile volume of blood within the body part changes with time. This quasi-periodic signal is called a photoplethysmograph (PPG), and can be used to determine heart-rate.

A wearable PPG biosensor in the form of a ring has been developed by Yang and Rhee [3]. As an article of clothing, a ring is more likely to be worn continuously, making it suitable for continuous monitoring applications. Asada *et. al.* have further refined the design of the ring sensor to ensure that the PPG signal output is more resistant to noise components due to motion artifacts and changes in ambient light levels [4]. Also, they have sought to reduce power consumption by using a high frequency, low duty cycle modulation scheme. A picture of the ring sensor is illustrated in Figure 3.

Shnayder, *et al.* have designed a pulse oximeter that integrates a BCI micro power oximeter [5] with a Mica2 [6] or MicaZ [7] wireless sensor platform. This pulse oximeter serves as a node in a medical sensor network platform developed by a research group at Harvard University called *CodeBlue* [8]. Anliker *et. al.* have also designed a pulse oximeter sensor as a part of their wearable multiparameter medical monitoring and alert system called AMON (Advanced care and alert portable telemedical MONitor) [9]. The pulse oximeter sensor is integrated into a single wearable wrist device with the other sensor components of this system (Figure 4).

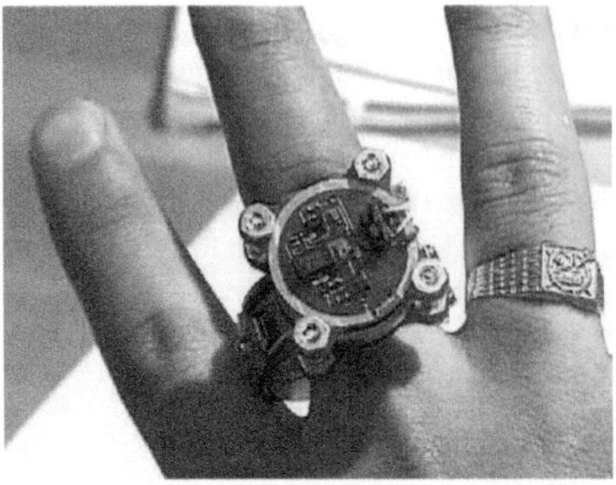

Fig. 3 Photograph of PPG Ring Biosensor Prototype (Courtesy of Asada *et. al* [4])

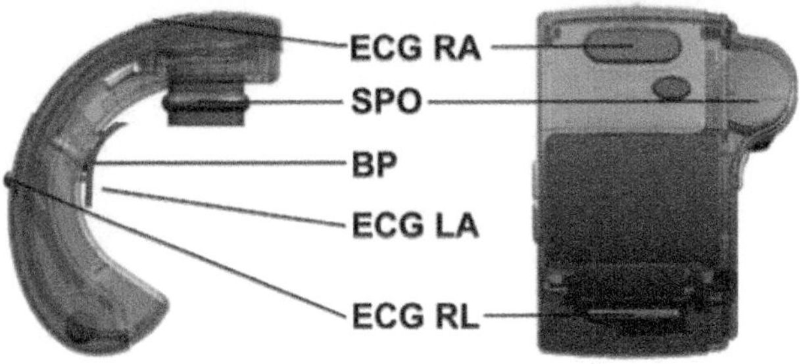

Fig. 4 Photograph of AMON Sensors: ECG, SPO_2, and Blood Pressure (Courtesy of Anliker *et. al* [9])

Wireless PPG devices may be used in applications that require data on the cardiovascular (CV) state of an individual [4].

Electrocardiography (ECG)

The electrocardiogram (ECG) is a waveform that represents the propagation of electric potentials through the heart muscle with respect to time. The propagation of these potentials results in the quasi-periodic contraction of the heart muscle. Therefore, the ECG waveform provides a non-invasive means for investigating heart function. Standard ECG measurements utilize twelve leads or 'views' of the electrical activity of the heart. However, ECG measurements using wireless sensors are generally for ambulatory applications and will typically utilize a subset of these leads. Figure 5 illustrates the basic features and intervals pertaining to ECG waveforms.

Fulford-Jones *et. al.* designed an ECG sensor that is supported by a Mica2 mote hardware platform [10]. The sensor utilized two electrodes to produce a single ECG signal. This sensor is another node in the *CodeBlue* medical sensor network platform.

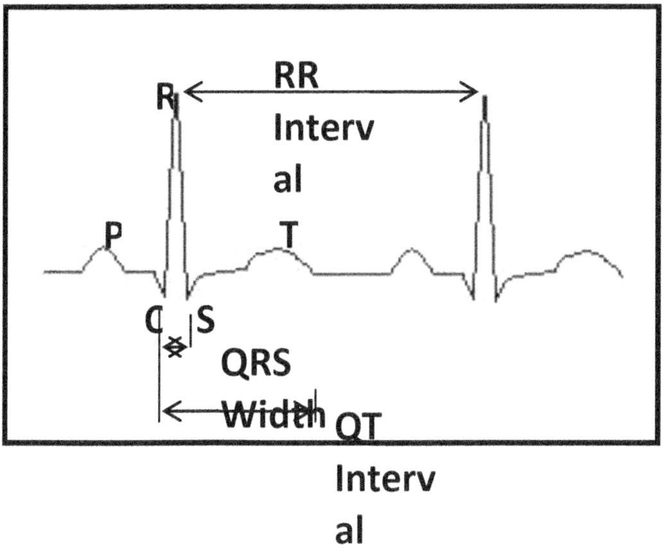

Fig. 5 Basic ECG Features and Intervals

The AMON system also includes an ECG sensor that produces a single-lead ECG signal. The sensor also utilizes an algorithm that calculates QRS pulse width, RR distance and QT interval from the ECG waveform. The Human++ research program has also developed a wireless single-lead ECG patch sensor, where the ECG signal is measured across two Ag/AgCl electrodes [11]. Cao et. al. have created a wireless three-pad ECG system (W3ECG)[12, 13], A pad is a self-contained board with two/three electrodes and front-end analog amplification circuitry. The advantage of this system over the others mentioned is that it enables a user to synthesize conventional 12-lead ECG signals from the three leads being measured.

In addition to the application areas mentioned for pulse-oximeters, the ECG signal can also be used to detect/classify cardiac arrhythmias [14].

Blood Pressure
A blood pressure (BP) reading is a measure of the force exerted by circulating blood on the walls of blood vessels. BP varies between a maximum (systolic) and a minimum (diastolic) pressure during a cardiac cycle. It has been observed that ambulatory BP is more closely related to target organ damage and cardiovascular events than BP readings taken in a clinical environment [15]. This fact provides the motivation for the creation of wireless BP sensors. The AMON system has a BP sensor that uses an inflatable cuff around the wrist and obtains systolic and diastolic readings via the oscillometric method [16]. Though this method can used to obtain ambulatory BP readings, it cannot monitor BP variations continuously and the cuff-based measurement may cause user discomfort. These issues are

remedied by Poon et. al. in the creation of a cuff-less BP watch sensor, based on the pulse transit time (PTT) method for measuring BP [17] (Figure 6).

Electromyography (EMG)

Electromyography is the study of muscle function through the monitoring of the electrical signals emitted by the muscle [18]. When a surface electrode is placed on the skin above a superficial muscle while it is contracting, it will receive electrical signals emanating from several muscle fibers associated with different motor units. The spatio-temporal summation of these electrical signals results in what is called an electromyogram (EMG) signal. Therefore, the EMG signal provides an effective means of monitoring muscle activity.

Bonato et. al. describe a data-mining technique (self-organizing maps) that is used to cluster EMG data recorded over a 24-hour period of daily activity [19]. These clustered data are used to determine neuromuscular mechanisms associated with specific biomechanical activities. One possible application of this work is to use EMG signals to investigate mechanisms associated with daily physical activities that induce pain, with the intention of devising treatment regimes.

Activity/Motion Detection

The level of activity or the nature of motion of an individual can be detected by a system that combines an accelerometer with a gyroscope. An accelerometer is a sensor that measures acceleration with respect to gravity, and can be used to determine the orientation of a body part in the absence of movement. A gyroscope is a sensor that measures angular velocity and can be used to determine the orientation of a moving body part as a function of time.

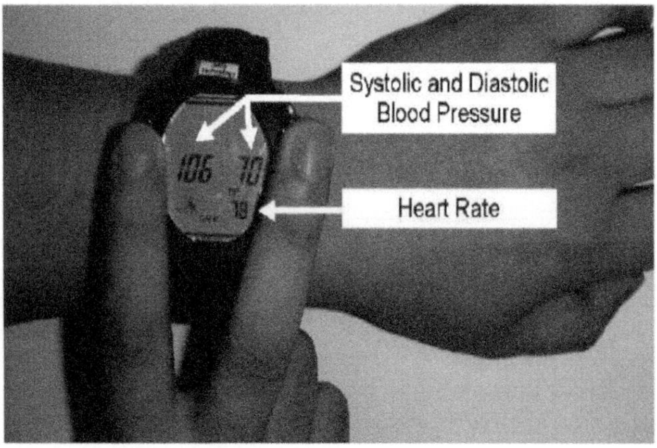

Fig. 6 Cuff-less Blood Pressure Watch Prototype (Courtesy of Poon *et. al.* [17])

Examples of 3-axis accelerometers can be found in [20, 21], and example of an integrated accelerometer/gyroscope in [8]. Kusserow et. al. presented a wireless sensor node design called BodyANT, which is equipped with an accelerometer[22].

Their investigations revealed the effectiveness of this device in long-term activity monitoring/recognition applications. One potential application of these motion detectors is to monitor the occurrence of uncontrolled body movements, called dyskinesias, experienced by sufferers of Parkinson's disease while undergoing a levodopa (the generic name for a drug used in the treatment of Parkinson's) treatment regime [23]. These investigations may lead to a more effective control of levodopa intake in order to reduce the occurrence of these dyskinesias.

Electroencephalography (EEG)
Electroencephalography (EEG) is a representation of the electrical activity of the brain. Currently, ambulatory EEG (AEEG) recordings have been shown to have great value in the diagnosis of epilepsy and in the monitoring of patient response to therapy [24]. Much of the information obtained from AEEG recordings may not be obtained during a routine, 20-min EEG test. This serves as motivation to develop wireless EEG sensors that will make the recording of AEEG signals during daily activities less obtrusive and less cumbersome.

Jovanov *et. al.* have proposed using their Wireless Intelligent SEnsor (WISE) for EEG signal acquisition applications [25]. This sensor is a microcontroller-based system capable of data acquisition, analog signal conditioning, low-level real-time signal processing, and wireless communication. Farshchi *et. al.* have also introduced a wireless neural interface, using Mica2 and Mica2dot [26] systems as the wireless sensor platforms, which is capable of acquiring two channels of EEG data [27].

Glucose Monitoring
It has been shown in [28] that real-time continuous blood glucose data will assist in reducing hyperglycemic excursions for individuals with type 1 diabetes, while lowering the risk of episodes of hypoglycemia caused by the administered levels of insulin being too high. Continuous monitoring is enabled by placing an implantable sensor covered with a multilayered membrane in the subcutaneous tissue of the abdomen. Glucose levels were determined every 30 s and radio transmission of the glucose data occurred every 5 minutes. If this sensor were to be combined with a implantable drug delivery system, such as the one in [29] a closed feedback loop would be formed for the control of blood glucose levels via the delivery of variable amounts of insulin.

Implantable Neural Stimulators
Implantable neural stimulators send electrical impulses into the brain or spinal cord for the treatment of Parkinson's disease, intractable epilepsy and chronic pain. An example of such a device is given in [30].

2.3 Sensor Node Architecture

The sensor itself is only one of number of components required to build a sensor node. A typical architecture for a sensor node is shown in Figure 7.

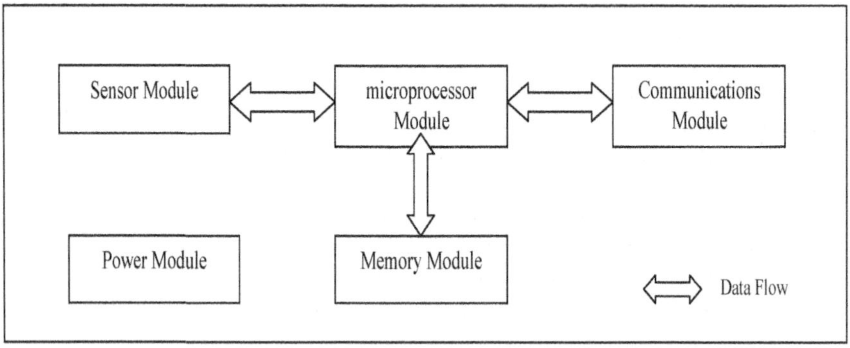

Fig. 7 Block Diagram of Sensor Node Architecture

The function of the sensor module is to acquire the relevant signal data and convert these into analog electrical signals. This module may also convert the analog data into digital format for interface with the microprocessor module. From a data flow perspective, the microprocessor module is located at the intersection of the sensor, communications and memory modules. It is responsible for executing data processing algorithms that are energy/data efficient. The communications module consists of the components required to transmit/receive processed data between nodes in the WBAN. The memory module stores data and program information for the node. The power module supplies power to all the modules in the node, and is typically a battery.

2.4 Sensor Node Design Considerations

Key considerations for sensor node design are wearability and low power consumption. Wearability refers to the ability of a user to wear the sensor with minimal impact on the user's ability to perform daily activities. Therefore, miniaturization and small form factor are common design goals for improving the wearabilty of the sensor node. Low power consumption refers to utilizing various sensor node design techniques that will reduce the amount of energy expended by the sensor node in the performance of its operations. Sensor node low power consumption design areas include: low-power architecture design, low-power processor design, low-power transceiver design and energy harvesting [2]. Low power consumption is essential because the desired battery lifetime of theses nodes is on the order of weeks to years, and battery replacement is not easily performed for implantable sensor nodes. The size of the battery represents an area where the design goals of wearability and low power consumption are in conflict. Wearbability design requirements seek to minimize the size of the battery, while low power consumption would seek to increase battery size so as to improve sensor node lifetime. The health impact of implantable sensors is another design consideration. Tissue damage due implantation and the dissipation of heat into surrounding tissue due to the transmission of RF electromagnetic waves are among the primary concerns. Tissue damage may involve tissue distortion and occlusion of blood vessels. To minimize tissue damage, it is recommended that the sensor should be blunt and rounded instead of sharp [31]. This reduces the probability that tissue will be sliced. Heat dissipation into tissue will impact the amount time allocated for transmission.

3 Physical Layer Advances and Research

Advances and research at the physical layer can be categorized in two areas: layer specification and channel characterization.

3.1 Layer Specification

The IEEE 802.15.6 standard [32] provides three physical layer specifications for narrowband (NB), ultra wideband (UWB) and human body communications (HBC) operation. Each specification covers PHY frame format, frequency bands of operation, modulation schemes, transmission/ reception specifications, timing/synchronization, etc. The frequency bands supported by each physical layer are shown in Figure 8.

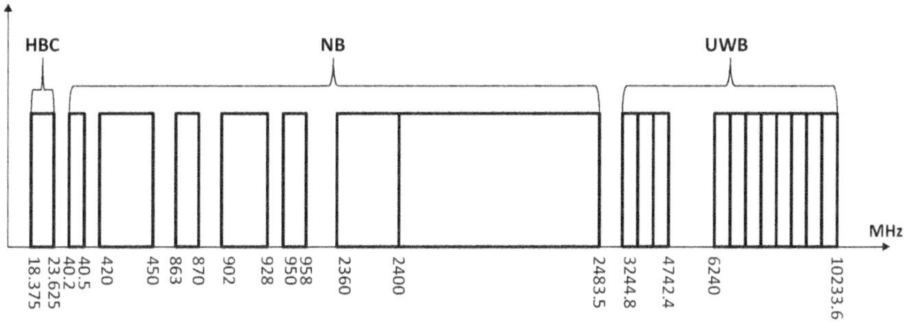

Fig. 8 IEEE 802.15.6 Frequency Bands

Human Body Communication (HBC) band uses the human body as the transmission medium and utilizes electric field coupling (galvanic and capacitive) as the mechanism for communication [33, 34]. Communication in this band can provide an enhanced degree of security compared to the other bands, because the human body transmission medium is generally not readily available to attackers.

The NB frequency bands specified can satisfy the data rate requirements of many WBAN applications. Interference with other systems is primary concern when using these bands, especially the 2400 – 2483.5 MHz band that corresponds to one of the Industrial, Scientific and Medical (ISM) bands. UWB has the advantages of high data rate and low interference over a short range and can be utilized WBAN applications that have a high data rate requirement.

3.2 Channel Characterization

Channel characterization is useful for estimating the delay and path loss associated with specific channels and can be used to evaluate and improve the performance of WBAN application. Three common channel characterization methods are: measurement-based methods, simulation-based methods, and combination of simulation-based and measurement-based methods [2].

3.2.1 RF Communication

RF communication can occur using NB or UWB signals and propagation can place inside the body or along the body. A review of RF communication findings applicable to WBANs follows.

3.2.1.1 Inside the Body

The path loss that occurs during the propagation of electromagnetic RF waves in human tissue is due mainly to absorption by the tissue, which is dissipated as heat. The Specific Absorption Rate (SAR) can be used to determine the amount of power lost due to heat dissipation, which can be in turn used to create a propagation model for the human body. An SAR-based propagation loss model (PMBA) for homogeneous tissue bodies is presented in [35]. This model indicates that path loss for RF propagation within the human body can be as much 30 – 35 dB more than free space propagation.

3.2.1.2 Along the Body

RF propagation along the body may be categorized as line-of-sight (LOS) and non-line-of-sight (NLOS) situations.

Channel models have been developed for LOS propagation in [36, 37] for UWB and [38, 39] for NB. The path loss exponents (η) obtained for these models are found to be in the range of 3 to 4, and antenna height was found to have a major influence on path loss.

NLOS occurs primarily as a creeping wave that diffracts around the surface of the body [40, 41]. Investigations into UWB, NLOS propagation around the body in [37, 42] resulted in a channel model with path loss exponents in the range of 5 to 7.

Taken together, these studies show that the path losses incurred due to LOS and NLOS propagation is significantly greater than that experienced in free space

($\eta = 2$), making single –hop communication unfeasible at times and necessitating the utilization of multi-hop communication [43].

3.2.1.3 Body Movement

Limb movements that cross the path between transmitter and receiver have a significant impact on received signal strength [37]. Attenuation losses of up to 20 dB [44] have been observed. In order to mitigate the path loss effects due to body movement. The nature of the wireless link between two sensor nodes is highly dynamic due in part to body movement. Therefore, a dynamic link power assignment mechanism is required to improve energy management. In [45], a dynamic power assignment scheme based on received signal strength indicator (RSSI) and packet drops is proposed. This scheme is found to provide better packet delivery and energy performance compared to a range of static and other dynamic assignment strategies.

3.2.2 HBC Communication

In [33], the human body is characterized as the transmission medium for galvanic coupling between on-body sensors by means of numerical simulations and measurements. Finite-element simulations revealed that an increase in distance of 5 cm between transmitter and receiver at the arm increases the attenuation by 6 to 9 dB. It was also found that joints can increase the attenuation by up to 8 dB. Measurements from clinical trials showed that the thorax exhibited excellent transmission characteristics with a typical SNR of 20 dB, while the extremities and joints lowered the quality of transmission over larger distance. Also, improvements in transmission characteristics were observed during body activity. These results Therefore, galvanic coupling is a potential communication technology for sensor application on the thorax and for short distances on the extremities.

In [34], a capacitive coupling technique was used to investigate the body channel characteristics based on measurements made both in the frequency domain and in the time domain. measurement of channel attenuation between two electrodes at various locations across the whole body revealed that the body maintained a stable attenuation vs. frequency characteristic across locations. the lowest attenuation observed is approximately -15 db at 28mhz. joints such as the elbow, knee and wrist affected the channel attenuation characteristic by leading to higher attenuation of about 2 db. also, experimental results illustrate that fat content in the body can cause increased attenuation in the channel characteristic by 4 db. the attenuation curves showed no significant change due to natural movements such as swinging hands, walking and breathing. in fact, running causes lower attenuation from 1 mhz to 30 mhz, since the value of capacitance and resistance between electrodes and moist skin was different from dry skin, and moist skin induces better coupling.

4 Power Efficient Protocols

Wireless body area sensor nodes, due to their size, use miniaturized batteries. The battery life of a sensor node is proportional to the battery size. Hence, WBANs must sense, process and communicate data in a power efficient manner. Power efficiency, therefore, is of utmost importance and is a key emphasis of design efforts for WBAN protocols.

4.1 Multi-layer Protocol Solutions

The majority of work in this area has been on developing energy-efficient medium access control (MAC) protocols. However, there has been some research in implementing energy-efficient methodologies in the higher layers of the TCP/IP model. One such work proposed a robust protocol stack for multi-hop WBANs [46]. A multi-hop architecture is utilized because of the possibility of low quality links that may arise due to the movement of on-body nodes. Link degradation may be large enough to prevent a sensor node from having a direct wireless link to the gateway of the network. The proposed protocol uses a gossiping strategy for data routing between the sensor and the gateway and a time division multiple access (TDMA)-based MAC protocol. Simulation results show that power consumption is reduced compared to a TDMA-based star network.

Another research effort proposed an energy efficient computing model, called wireless device driver for low-duty peripherals for WBANs [47]. The wireless device driver, which is part of the host's operating system, performs the control operations for a wireless peripheral device. Also, the wireless device driver relies on wireless communication protocols for a reliable connection with the peripheral. For energy-efficiency reasons, a wireless peripheral device is not always connectable, but switches its state between connectable and idle continuously. These techniques serve to achieve minimum connection latency without changing power consumption. The experimental results using BlueTooth and ZigBee show that the techniques are effective in reducing energy consumption while meeting connection latency requirements.

Shankar *et. al.* analyzed the energy efficiency of two protocols designed for wireless communication in biosensor networks [48]. TDMA is used for medium access in both protocols. The first protocol is applied to a cluster-based topology, where the network is divided into clusters consisting of sensors and a leader node. Energy saving is accomplished in this protocol by requiring only the leader nodes to make the long distance transmission to the base station. The other protocol is applied to a tree-based topology. In this protocol, a node transmits data to the base station via a tree hierarchy of parents and children with the base station as the root. Energy saving is obtained in this protocol, because no node is required to make a long distance transmission. Simulation results show that the cluster-based approach is more energy efficient that the tree-based approach, because the tree-based approach incurs transmit/receive costs at each level of the tree hierarchy.

4.2 MAC Protocol Solutions

The main sources of energy waste in the design of a MAC protocol for a WBAN have been identified as collision, overhearing, control packet overhead, and idle listening [49, 50]. Currently, there are two main schemes used for MAC protocols of sensor networks. Contention-based MAC protocols, such as carrier sense multiple access/collision avoidance (CSMA/CA), have their nodes contend for channel access prior to transmission. The advantages of these protocols are scalability, adaptability to network changes and no time synchronization constraint. In schedule-based MAC protocols, such as TDMA, access to the channel is divided into time slots that are of fixed or variable duration. Each node is assigned a time slot(s) by a controller, and it will only transmit within that time window. TDMA-based protocols eliminate collision, overhearing and idle listening, and are typically utilized in some form in energy-efficient MAC protocols.

IEEE Standard 802.15.4 defines the physical layer (PHY) and medium access control (MAC) sublayer specifications for low-data-rate WPAN (LR-WPAN), typically operating in the personal operating space (POS) of 10 m [51]. IEEE 802.15.4 has two operational modes: a beacon-enabled mode and a non-beacon enabled mode. In the beacon-enabled mode, the network is controlled by a coordinator, which is responsible for device synchronization. Optionally, the superframe can be subdivided into active and inactive periods. The coordinator may enter sleep mode during the inactive period. The active period contains three components: a beacon, a Contention Access Period (CAP), and a Contention Free Period (CFP). During the CAP period, devices contend for channel access using slotted CSMA-CA. It has been found that IEEE 802.15.4 does not meet all the energy efficiency requirements for WBAN applications [52-54]. It is because of these shortcomings that the IEEE 802.15 Task Group 6 (Wireless Body Area Networks) began developing a communication standard optimized for low power devices that operate on, in or around the human body (but not limited to humans) to serve a variety of applications including medical, consumer electronics / personal entertainment and others [32]. This standard was approved in February 2012.

While this standard was being developed, several energy efficient MAC protocols specifically designed for WBAN applications were proposed and/or implemented. An energy-efficient MAC protocol was presented in [55]. This protocol was designed specifically for a star topology that consists of clusters with a central master node plus slave sensor nodes. The network access is obtained using clear channel assessment and collision avoidance with time division multiplexing (CCA/TDMA). This network access scheme significantly reduces the likelihood of collision and idle listening, leading to power savings. The hardware implementation of the MAC incorporates cross-layer optimization by performing some functions from the session layer in the physical layer, reducing the power overhead of software implementations. A list of proposed power efficient MAC protocols is presented in Table I, while Table II presents their respective advantages and disadvantages.

Table 1 Power Efficiency of MAC Protocols for WBAN

Protocol Description	Power Consumption Measure	Measurement Conditions	Reference
CCA/TDMA with cross-layer optimization	$P_{AVG} = 3$ mW	Protocol implemented on Sensium SOC WBASN ASIC for temperature sensing and ECG applications.	Omeni *et al.*[55]
Low-duty cycle TDMA	$P_{AVG} = 2.04$ mW at 3 VDC	Protocol implemented using Analog Devices 70XMBZ2 platform with ADF7020 RF transceivers.	Marinkovic *et al.*[56]
HMAC: TDMA with heartbeat rhythm synchronization	Increased network lifetime 15% - 300% compared to 2 WSN MACs	OMNet++ simulation, cardiac rhythm obtained from MIMIC database, Tmote Sky specification	Li *et al.* [57]
BodyMAC: Energy efficient TDMA-based MAC	$P_{AVG} = 3$ mW (sleep mode) $P_{AVG} = 6$ mW (no sleep mode)	Simulation uses single star topology and MICAz mote specification.	Fang *et al.* [58]
MedMAC: Adaptive TDMA-based MAC protocol	$P_{AVG} \approx 1$ mW (EEG) $P_{AVG} = 1.81$ mW (Respiration) $P_{AVG} = 0.2$ mW (Temperature) $P_{AVG} = 0.2$ mW (Pulse monitor)	Opnet simulation model uses star topology for two applications: EEG (up to 24 nodes at 86.4 kbps/node) and health/fitness (respiration node at 640 bps + pulse at 8 bps + temperature at 16 bps). MICAz mote specification.	Timmons *et al.* [59]
DQBAN (Distributed Queuing Body Area Network): Cross-layer fuzzy logic scheduling mechanism	All sensors transmit at lowest possible level specified in 802.15.4. (i.e. -25 dBm)	Simulation uses a star-based WBAN. Matlab simulations using the CC2420 transmitter-receiver benchmark used two scenarios: • Homogeneous: 5-35 wireless ECG sensors • Heterogeneous: 10-35 wireless ECG sensors along with 4 other sensors for clinical doctor PDA, respiratory rate, blood pressure and endoscope imaging	Otal *et al.* [60]

Marinkovic et. al. [56] present an energy-efficient, low duty cycle MAC protocol based on TDMA. The MAC protocol enables access to the physical layer for a hierarchical topology consisting of nodes communicating with master nodes, which in turn communicate with the monitoring station. The hierarchy removes the need for sensors to expend power by transmitting to the monitoring station. Also, the use of TDMA ensures collision-free transfer and minimization of idle-listening. The protocol is implemented using the Analog Devices 70XMBZ2 platform with ADF7020 RF transceivers. Measurements reveal that the protocol is energy-efficient for streaming and short burst data communications.

A novel TDMA-based protocol for BSNs, called H-MAC, is proposed in [57]. This protocol improves energy-efficiency by using the heartbeat rhythm to perform TDMA synchronization, avoiding energy consumption associated with transmitting time synchronization beacons. Power efficiency is also accomplished in H-MAC as a TDMA-based protocol assigns time slots to each biosensor to guarantee collision-free transmission. Simulations show that H-MAC prolongs the network life of sensors dramatically.

In [58] a TDMA-based MAC protocol called BodyMAC is proposed. Three types of bandwidth allocation schemes are devised to cope with different types of data communications, such as periodic data sensing and important event allocation. In conjunction with bandwidth allocation, a sleep mode mechanism is introduced, which turns off a node's radio during beacon, uplink and downlink periods, as much as possible. Simulations results show superior performance of BodyMAC compared to that of IEEE 802.15.4 MAC.

Timmons et. al. [59] introduce an adaptive TDMA-based MAC protocol called MedMAC. MedMAC incorporates a novel adaptive TDMA synchronization mechanism in which only a multi-superframe beacon has to be listened to by the nodes. An optional contention period is also available for low-grade data, emergency operation and network initialization procedures. Simulations show that MedMAC consumes less power than IEEE 802.15.4 for two classes of medical applications.

Table 2 Pros and Cons of Power Efficient MAC Protocols for WBAN

Protocol Description	Advantages	Disadvantages
CCA/TDMA with cross-layer optimization	Reduced likelihood of collision and idle listening, leading to power savings.	Limited cluster size. Communication has to be initiated by the master node in the cluster.
Low-duty cycle TDMA	The MAC protocol allows collision-free transfer, robustness to communication errors, longer sleep times, and real-time patient monitoring.	Sensor resynchronization overhead is reduced using a rigid structure of TDMA slot assignment for predetermined sensor functions
HMAC: TDMA with heartbeat rhythm synchronization	Implements TDMA without distributing synchronization information.	Complex. Single point of failure problem with the biosensors.
BodyMAC: Energy efficient TDMA-based MAC	Accommodates on-demand traffic using downlink subframes.	Uplink frames are energy-intensive.
MedMAC: Adaptive TDMA-based MAC protocol	Developed for ultra-low power WSN applications for monitoring applications that cannot be operated under the standard 802.15.6 WBAN standard. Consumes less power than IEEE 802.15.4 for two classes of medical applications.	Not suitable for high data rate applications.
DQBAN (Distributed Queuing Body Area Network): Cross-layer fuzzy logic scheduling mechanism	Distributed queuing eliminates back-offs and collisions.	End-to-end packet delivery time increases with the number of sensors.

In [61] a power efficient MAC protocol is proposed for WBANs. This work presented a traffic-based wakeup mechanism that utilizes the three categories of traffic patterns of the body sensor nodes, namely: normal traffic, on-demand traffic and emergency traffic. The wakeup patterns of all body sensor nodes are organized into a table called traffic-based wakeup table. The table is maintained and modified by a network coordinator according to the application requirements. Based on the body sensor node's wakeup patterns, the network coordinator can also calculate its own wakeup pattern. During normal traffic, both the body sensor nodes and the network coordinator send data based on the traffic–based wakeup table. The network coordinator sends a wakeup radio signal to body sensor nodes, which wake up in response to these signals, during on-demand traffic period. During emergency traffic period, the body sensor nodes send a wakeup radio signal to the network coordinator, which responds to the wakeup radio signal.

Otal and Alonso proposed an energy-saving MAC protocol, DQBAN (Distributed Queuing Body Area Network) for WBAN in [60] as an alternative to the 802.15.4 MAC protocol which suffers from low scalability, low reliability and limited QoS in real-time environments. The proposed DQBAN is a combination of a cross-layer fuzzy-logic scheduler and energy-aware radio-activation policies. The queuing of access packets and data packets is determined by fuzzy-logic rules, which permit body sensors to find out '*how favorable*' or '*how critical*' their situation is in a given time-frame. The fuzzy-logic scheduling algorithm is shown to optimize QoS and energy-consumption by considering cross-layer parameters such as residual battery lifetime, physical layer quality and system wait time. The authors tested their proposed protocol on two scenarios: a homogenous scenario of a body sensor network with 5 – 35 homogenous ECG wireless sensors and a heterogeneous scenario of a body sensor network with wireless ECG sensors and four other sensors for clinical doctor PDA, respiratory rate, blood pressure and endoscope imaging. Further specifications are shown in Table 1. The authors showed that the DQBAN protocol has higher reliability, while fulfilling certain battery limits and latency demands and displays energy-saving behavior compared to most MAC implementations.

Another cross-layer protocol, CICADA (Cascading Information retrieval by Controlling Access with Distributed slot Assignment) for low-delay and energy efficiency has been proposed in [62]. The authors used a spanning tree approach to determine presence/absence of next-hop nodes, and hence the data transfer schedule. By leveraging the network topology, CICADA enforces a downward motion of control information and upward information of data along the spanning tree. This allows for simplified routing, lower interference and idle listening; all of which contribute to energy-efficiency for body area networks.

4.3 IEEE 802.15.6

The purpose of this standard is to enable short-range wireless communication in the vicinity of, or inside, a human body (but not limited to humans) [32]. It is designed to support quality of service (QoS), extremely low power, and data rates up to 10 Mbps, while simultaneously complying with strict noninterference guidelines where needed.

The standard supports a one-hop star BAN network topology consisting of a single hub communicating with nodes. It also supports a two-hop extended star BAN where the hub and a node may exchange frames optionally via a relay-capable node.

The hub is specified to operate in one of three access modes: beacon mode with beacon periods (superframes), non-beacon mode with superframes, and non-beacon mode without superframes. The arrangement of access phases in a beacon period for beacon mode is shown in Figure 9.

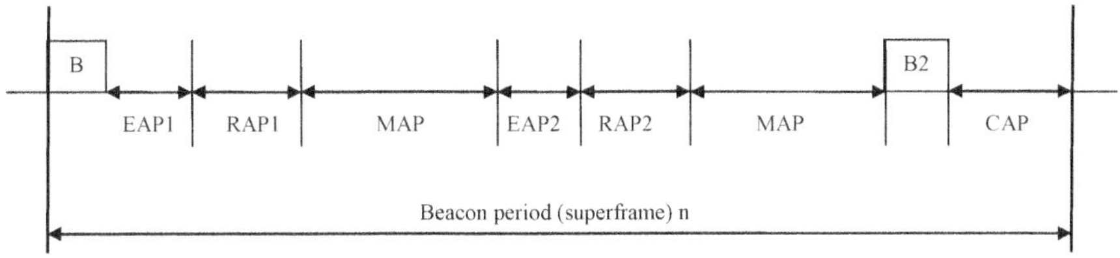

Fig. 9 Layout of Access Phases in a Beacon Period (superframe) for Beacon Mode

The hub arranges the access phases in the following order: exclusive access phase 1 (EAP1), random access phase 1 (RAP1), managed access phase (MAP), exclusive access phase 2 (EAP2), random access phase 2 (RAP2), another managed access phase (MAP), and contention access phase (CAP). Access phases EAP1, RAP1, EAP2, RAP2 and CAP are contention-based allocations that may use CSMA/CA or slotted Aloha as the access schemes. The MAP access phase is the only phase that permits scheduled, unscheduled and improvised access intervals. The exclusive access phases are designated to be used for high priority/emergency communications. The hub may set any of the access phases to zero length, under the constraint that RAP1 cannot end before the guaranteed earliest time as communicated in Connection Assignment frames sent to nodes that are still connected with it. The hub must transmit a B2 frame prior to implementing a non-zero length CAP.

When operating in non-beacon mode with superframes, a hub may have only a managed access phase (MAP) with the same access interval allocations specified for MAP previously.

In non-beacon mode without superframes, a hub may provide unscheduled bilink allocation intervals. Also, after determining that the hub for the next frame exchange is operating in non-beacon mode without superframe boundaries, a node may treat any time interval as a portion of EAP1 or RAP1 and employ CSMA/CA based random access to obtain a contended allocation.

The standard also allows a node to implement power efficiency mechanisms at the MAC layer by enabling node hibernation, that is, a node can assume an inactive state across its non-wakeup beacon periods. It also enables sleep, that is, a node can assume an inactive state over some time intervals in its wakeup beacon periods.

5 Routing

Routing and related issues have not been sufficiently addressed, while a lot of attention has been paid on designing MAC layer, energy and power constraints, and related Physical layer characteristics.

The principal characteristics of a Wireless Body Area Network (WBAN) that necessitate the design of a routing protocol are frequent network partitioning due to postural mobility of the on-body sensors, high propagation loss across the human body, low transmission power of the sensors, and low reliability of end-to-end path from source to sink. Link layer behavior of WBANs at 2.4 GHz have been studied in [63], where the authors observed the following:

(i) Average packet delivery ratio (PDR) increases with increase in transmission power.
(ii) Increasing transmission power at regions with low multipath increases PDR even more.
(iii) Environments do have an impact on PDR. In a lab setting more than 70% of links have PDR 90% or more; while in an open setting (on the roof) about 50% of links have 90% or more PDR.

The authors in [63] also found that channel symmetry is better in environments having more reflective surfaces (more multipath). Average channel symmetry in a lab environment was found to be 3.9% as compared to 7.3% on a roof. Sensor placements on the body also have an impact on the network performance. While average PDR was found to be 94.9% between nodes on the left hand side and 94.4% between nodes on the right hand side, significant drop in PDR was observed between the left-hand side and right-hand side nodes.

A fundamental question arises on whether there is a need to design a routing protocol on top of the link layer protocol already in place. Some preliminary studies [64] have shown that there are clear tangible benefits of multihop communication between on-body sensors. In an open environment with low transmission power, there has been significant improvement in PDR with multihop communications. Moreover, as the propagation loss around the human body is high with high signal attenuation, direct communication between the sensors is energy-expensive. Multihop communication is therefore a more viable alternative among on-body sensors.

Traditionally, there have been two approaches to routing in BANs. One approach is to integrate the routing functions with the MAC layer, with a fundamentally cross-layer approach. The other is to design a routing layer on top of the MAC layer, where link qualities are measured based on selected parameters, and taken into path computation.

The first approach has been studied and proposed in [65-67]. The authors in [66] have proposed a cross-layer CICADA protocol that sets up a spanning tree and uses time slots for controlling each node's transmission and reception cycles. Each node informs its children about their turns for sending data. Data transfer takes place as a sequence of cycles: a control cycle and a data cycle. In the control cycle all nodes are informed about the order in which they can transmit. When all nodes receive their control schemes, that data cycle starts. In the data cycle each data scheme has two parts: a data period (α), and a waiting period (β). The data period also provides a contention slot to allow nodes to join the tree. This also provides mobility support for the network where nodes may get disconnected due to postural mobility. The authors have also discussed the energy efficiency of the algorithm, which depends on the network topology. As the nodes have to spend time on idle listening and overhearing during the control cycle, depth of the tree plays a significant role in controlling the energy efficiency of the protocol. However, the efficiency of the protocol needs to be evaluated based on frequent node disconnection and network partitioning due to postural mobility. The protocol also lacks an approach to define link metrics and use those metrics for finding effective multihop routes. This can be a compelling argument for necessitating a separate routing layer on top of MAC that would provide the basis for computing efficient multihop paths based on link quality metrics.

The second approach has been investigated in [71], where the authors have proposed a probabilistic packet routing protocol, PRPLC, using a stochastic link cost. The experimental topology is being built in the laboratory with on-body sensor nodes using 900 MHz Mica2Dot Motes running TinyOS. The motes have Chipcon's SmartRF CC1000 radio chips and MTS 510 sensor cards from CrossBow Technologies. The radio chips' transmission powers are reduced to set the transmission range between 0.3 to 0.6 meters. The proposed protocol, based on postural link cost formulation, uses time-varying costs formulated for each link based on the locality in the connectivity patterns of the links. The protocol uses postural link costs to compute probabilistic forwarding of data packets. The authors have defined the Link Likelihood Factor (LLF), $P_{i,j}^{t}$, which represents the probability of the link $L_{i,j}$being connected at discrete time interval, t. The LLF is dynamically adjusted at each discrete time slot by a constant factor, ω, such that $P_{i,j}^{t}$ increases fast and decreases slowly for a historically good link, and decreases fast and increases slowly for a historically bad link. When a node *i* wants to forward a packet to node *d*, it will send the packet to an intermediate node *j* if $P_{i,d}^{t} < P_{j,d}^{t}$. The authors have reported a significantly low end-to-end packet delay using PRPLC with adaptive ω, as the protocol can successfully capture the locality in postural movements. The worst case scenario in the number of transmissions per packet for PRPLC is comparable to other store and forward protocols. While most of the packets are transmitted in two hops, some also take three and four hops to reach the destination. Packet delivery ratio for PRPLC is also comparable to other store and forward protocols.

However, the proposed protocol computes links cost based only on the connectivity pattern of the links, and lacks consideration of transmission power and residual energy of the nodes in the link cost computation. An alternative approach would be to consider transmission power of forwarding packets from node *i* to node *j* at time *t*, $E_{i,j}^{t}$. The point to consider is if $P_{i,d}^{t} >= P_{j,d}^{t}$, node *i* will send the packets directly to node *d*, but if $E_{i,d}^{t} > E_{i,j}^{t}$, it will require more transmission power to forward the packets. Another approach would be to consider the residual energy at each node, along with the link cost, in making the forwarding decision. All nodes would send their residual energy at that time instant to all their neighbors by piggybacking with the Hello messages. For example, if R_i^t denotes the residual energy at node *i* at time instant *t*, and *i* wants to forward packets to node *d*, it will forward the packets only to an intermediate node *j*, only if the following two conditions satisfy:

$$P_{i,d}^t < P_{j,d}^t$$

and, $R_j^t > R_i^t$

A challenge that will arise in this context is the channel asymmetry and instability when nodes transmit data. Suitable metrics need to be incorporated in the route and next hop selection for wireless link reliability using channel asymmetry and instability.

Reliability, as it is defined, is also addressed at the MAC layer in the IEEE Standards Association's IEEE802.15.6. The standard has provision of reducing interference by shifting each beacon transmitted by a known time period. In addition, dynamic channel hopping is also used to avoid narrowband interference. Four MAC techniques to increase reliability have been proposed in [68]. The first among them is dynamic slot allocation, where knowledge of future link state is used to dynamically allocate slots to sensor nodes. The authors based link allocation on a two-state Markov model – whether the link is good or bad. Each link is characterized by two parameters – stability, S and volatility, V. S is a measure of long term average probability of a link to be in a good state, while V is a measure of how frequently a link transitions between good and bad. A flipping algorithm is proposed by the authors where all bad links of the previous superframe are allocated last in reverse order, and all good links are allocated first in the reverse order. The second method that the authors proposed to increase reliability is an adaptive retransmission technique that adaptively allocated retransmission slots to anywhere in the superframe based on channel conditions. They showed that adaptive retransmission strategy has a better performance in recovering packets than fixed retransmission under an energy-constrained situation. The authors have also proposed use of relay nodes to achieve better reliability. This is a precursor to multi-hop communication in WBAN using appropriate routes.

Another approach is to design the routing protocol based on the mobility pattern of the nodes. This approach has been investigated in [69], although not in context of WBANs, where a generic algorithm has been presented based on the use of high-dimensional Euclidian space. This is somewhat difficult to accomplish in WBANs, where the only mobility of the nodes is postural mobility. A possible approach can be using temporal mobility patterns of nodes based on historical data of link disconnections. A temporal pattern of link disconnection data can be used to select links for computing a probabilistic end-to-end path. However, extensive evaluations of these protocols need to be conducted based on prototype implementation in laboratories.

Mobility of sensor nodes gives rise to another challenge of ensuring connectivity at all times. If sensor nodes are communicating with multiple base stations, the underlying network needs maintain an uninterrupted state, which in turn demands that the sensor nodes are always connected to at least one base station at any point in time. Authors in [70] have discussed several handover mechanisms where nodes are handed over to appropriate base stations during intra-mobility. Received Signal Strength Indicator (RSSI) is the dominant parameter used to make handover decisions. There are two approaches to make a decision: choosing the best RSSI, and having a threshold comparison. In the later, the sensor nodes monitor the received signal strength from the base station that they are attached to, and start searching for other base stations when the signal strength falls below a predefined threshold. In addition to RSSI, other parameters that are taken into consideration while making a handover decision are Link Quality Indicator (LQI), Speed and direction of node movement, and node position. The authors in [70] have surveyed different handover techniques, and have compared them in a table, details can be found in their paper.

In summary, following challenges need to be addressed while designing a multi-hop communication algorithm:

- Suitable next hop selection (and probable route selection) based on wireless channel asymmetry and instability
- Route selection based on residual power on each node
- Route selection based on required transmission power

6 System Architectures

In this section, we survey the architectures of WBANs with respect to network topologies, application-specific network design, and topologies with various physical locations.

6.1 Network Architectures

The network architecture of WBANs can be broadly classified into two major categories: flat architectures and multi-tier architectures. Flat architectures comprise of a single data-gathering unit that sends its data to a personal computer or a personal server application running on a PDA. An example of this is described in [25]. Here the authors describe the use of wireless intelligent breathing telemetry sensor (WIBRATE) to study breathing patterns. A thermistor-based breathing sensor is equipped with on-board data processing mechanisms. This device is connected to a PC using a wireless link. This is an example of a point-to-point network of a single client and server. Data upload to a personal server is facilitated whenever a wireless link is available. A flat architecture of WBAN is illustrated in Figure 10.

Multi-tier architectures, as shown in Figure 11, are widely used to achieve large data gathering of multiple physiological signals using multiple nodes in the base tier, a gateway at the second tier that acts an interface between first tier and a server at the third tier. For example, the architecture described in [71] is a prototypical WBAN. In [71], the authors have studied the performance of a

WBAN for remote patient monitoring through an OPNET simulation study. The multi-tier architecture used for the WBAN in [49] is the widely used three-tier architecture. A WBAN is connected through a voice and data gateway to the hospital core-switch, which in turn is connected to the IP phone. The voice and data gateway is also connected to the wireless access point which provides access to the patient's and doctor's PDAs. Finally the core switch is connected to the hospital's ZigBee network which records the data from multiple patient WBANs.

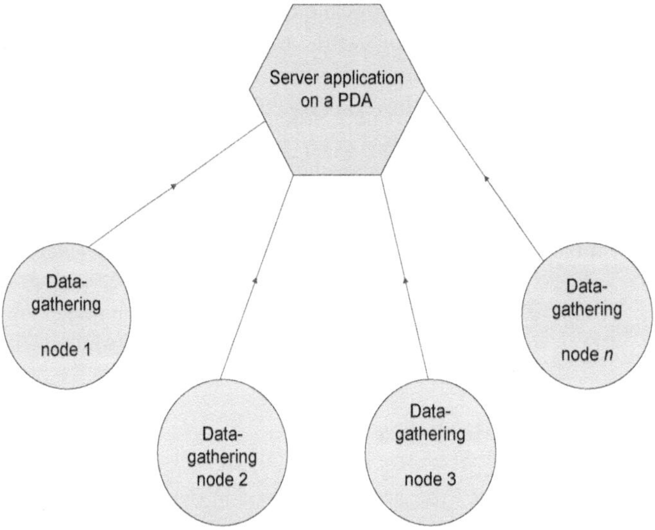

Fig. 10 Flat architectures in WBANs

The rest of this section focuses on various multi-level architectures with an emphasis on specific topologies for parameter optimization of security, energy efficiency, node locations and the emerging use of ultra-wideband transceivers for WBANs.

Security-Centric Architectures

In [72], the authors have studied the design of WBAN architecture with an emphasis on security and cross-layer operation. The paper studies CICADA-S, a secure multi-hop protocol that places stringent security requirements on the existing CICADA (Cascading Information retrieval by Controlling Access with Distributed slot Assignment) protocol. The main components of the WBAN are the network of nodes, external networks and back-end server. Each patient has their own unique gateway that securely processes patient data and sends it to the back-end server. The CICADA protocol addresses both MAC and routing layer by ensuring collision-free medium access, low interference and avoidance of idle listening. Routing is performed along a distributed spanning tree. Security is enforced during node initialization using a secure key which is generated by back-end servers uniquely for each node in each session. When a node leaves or rejoins the WBAN, it triggers a network topology update which generates a new key for each node. This ensures that nodes which leave the network cannot read/modify the patient data. In addition, node membership in WBAN networks is mutually unique and location privacy has been enforced through anonymity in messages shared between the back-end server and the WBAN.

Energy Efficient Architectures
In [73], the authors described an energy efficient TDMA-based MAC protocol for a multi-tier architecture of a WBAN. The multi-tier architecture is that of the widely used sensor nodes at the first tier, a set of master nodes that collect data from first tier nodes and a monitoring station in the highest (third) tier that encompasses the monitoring and data access functionalities at the user level.

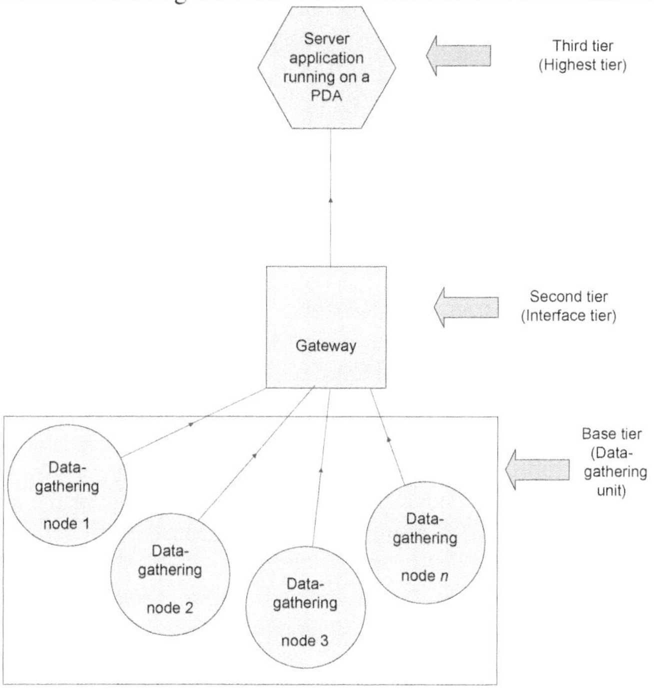

Fig. 11 Multi-tier architectures in WBANs

By exploiting the features of a fairly stationary network of first tier nodes with fixed data collecting functions (as opposed to dynamic ad-hoc networks or mesh networks with dynamic topologies of wireless sensor networks (WSNs)), and smaller distances between the first tier nodes and second tier master nodes, the authors develop a MAC protocol that is energy efficient and uses concise synchronization messages to co-ordinate data transfer on the network.

In [74], the authors presented a WBAN that measures stress through the heart-rate variability parameter. The BAN is organized as a network of (**W**ireless **I**ntelligent **Se**nsor) WISE sensors (clients) that connect to a personal server (PS).

In addition to data processing, the PS also facilitates synchronization of clients. In order to reduce the power consumption of the wireless transceiver on the PS the authors introduce the concept of a mobile gateway (MOGUL) that establishes wireless communication with the PS and downloads data. Energy efficiency is implemented using energy efficient communication protocols and low-energy radio layer 900 MHz RF modules. Increased on-board data processing at sensor nodes reduces the size of the transmitted packets to the PS. The system architecture is a master-slave architecture, where the MOGUL acts as the master and downloads data from WISE sensors when it is in within its communication range. The architecture allows for multiple sessions of communication with WISE sensors. All data obtained from these sensors is aggregated in a central device that is connected to the multiple mobile gateways generating these sessions.

In [25] , the authors present the prototype implementation of a network of WISE sensors. To minimize power consumption in communications, the communication range of a sensor is limited. The architecture features sensors that perform data collection and on-board signal processing. The multiple WISE clients communicate with a single server. A remote intelligent control system determines when a new measurement is needed and signals the sensor to do so or requests that the patient do it. The collected data is transmitted to the databases on the personal server.

Ultra-wideband (UWB) transceiver design

In [75], the authors proposed the use of a UWB transmitter for energy-efficient operation of WBANs. Due to the high interference generated by the human body and its environment, one solution to develop low power-output transceivers for radios in the sensor nodes is to optimize the air interface of the network. By creating architectures that exploit the features of robust nodes and energy-constrained nodes, low latency and simple network topologies, the authors show that an UWB based architecture is advantageous over narrowband radio communication.

The use of UWB has been further explored in [76]. Here the authors described the use of UWB for an in-body WBAN application of capsule endoscopy. Within the 402-405 MHz frequency band, allowed by the FCC for in-body communication systems, UWB communication has shown to be most effective for integrating in-body and on-body medical sensors into a single system. With the help of radio channel simulation results, the authors presented the link capacity, signal power spectral density and interference mitigation. Finally a WBAN coordinator acts as an interface between the network of sensors and a medical server. The current state of this work is documented under the MELODY Project [77, 78].

Application-Specific Network Design

WBANs can have unique architectures to tailor them for various applications. In [78], the authors described the implementation of a WBAN for bio-telemetry for elderly people. Although ZigBee and Bluetooth are widely used in WBANs due to their low-power simple topology operation, they can be inefficient due to the high data volume sensing and communication operations required in monitoring physiological parameters. The network architecture is a scalable architecture of bio-sensors and motion sensors at the first tier, a gateway at the second tier that interfaces between the first tier nodes and the monitoring station and medical station at the third tier.

In [79], the authors described the implementation of a WBAN for hip rehabilitation called HipGuard. This implementation uses the ANT WSN technology since it enables repetitive measurements with low latency requirements. The ANT architecture uses a master-slave framework, where a central unity is connected with sensor units. The focus of this paper is on network issues with the HipGuard implementation and the authors study various performance parameters such as human body interference, throughput and latency in this context.

In [80], the authors presented general system architecture for WBANs. They propose a system for activity sensing (Actis) that is a multi-tier system architecture of intelligent physiological sensors at the lowest level, a personal server running on a PDA in the second tier, which is connected to a medical server or the Internet in the third tier.

A re-configurable sensor architecture for enabling a range of detection of physiological signals is presented in [81]. The authors develop this architecture for the application of pulse oximetry and use a reconfigurable platform that can be self-initiated or request-triggered. The advantages of such architecture include reusability of hardware and code, precise time synchronization and cost efficiency. Run-time configuration adapts readily to change in patient environments and supports updation of software parameters for data processing, thus customizing the medical application.

In [82], the authors described the architecture and implementation for two types of tier -1 sensors- Actis and eActis, where eActis is the ECG and tilt sensor enabled framework for Actis. The architecture uses the ZigBee star network topology to minimize power consumption, a Flooding Time Synchronization Protocol (FTSP) for accurate synchronization of WBAN nodes, and power-saving sleep modes for the sensor nodes. The architecture allows for scenarios where raw data can be sent over a link to a personal server. This approach allows for precise localization of events leading to a specific occurrence in the patient's health, as opposed to transmission of processed data that loses some information in the data aggregation process.

Node Locations

WBAN architectures can also be classified based on their location in the human vicinity. In particular, these architectures can be wearable or implanted on/within the human body. Capsule endoscopy discussed in [76] is an example of implantable WBAN architecture. In [83], the author reports various implementations of body area networking. One such implementation is that of a pill camera which the patient ingests. The camera takes pictures and transmits them to a recorder. The camera is scheduled for an operational time of up to eight hours. Another application is that of using signals from implants to stimulate muscles to allow limb movements.

Most of the WBAN architectures and applications discussed in this section are wearable architectures. An additional example is [84]. Here, the authors described wearable WBAN (WWBAN) for health monitoring. The proposed WWBAN architecture is a multi-tier architecture, where first tier comprises of sensors, second tier is an application-specific layer and third tier includes data accessibility through servers connected to the Internet. Specifically, the sensors in first tier monitor physiological signals such as ECG and blood pressure. In second tier, a personal server (PS) running on a PDA acts as an interface between the user, medical server connected to the Internet (third tier) and the sensors in first tier. Hence the server application in second tier is responsible for network configuration of the sensors, secure communications and data processing for data obtained from first tier sensors. In the event that the second tier server is unable to connect a third tier server, the proposed architecture enables local data storage. Finally third tier enables data access by healthcare providers. The proposed WWBAN architecture can be configured for use at home, work or in a medical setting, where the architecture ties in seamlessly by virtue of the connectivity between second tier and third tier servers. The authors developed a WWBAN prototype for monitoring ECG, trunk and ankle motion. The network coordinator manages synchronization, ensures secure links for data transmission and data aggregation before forwarding it to the second tier PS application. Using a star topology of the ZigBee architecture, a master node (network coordinator) periodically transmits beacons to its slave nodes and allows for energy conservation through the use of power-saving sleep modes.

In the area of on-body communications, Sensium [85] is an application for monitoring physiological signals. It comprises of a '*digital plaster*' worn on the body to measure parameters such as heart rate. Multiple Sensium devices report to a base station Sensium plugged into a PDA or a smartphone. This is a multi-tier architecture with the base-tier of Sensium devices that collect data, a base station (second tier) that collects data from (first tier) nodes, and the highest tier is a server that acts as a local database and can also be connected to the Internet.

In [86], the authors described a specific routing architecture for the MAC protocol for body sensor networks. The authors propose the *data-pulling* paradigm for body sensor networks (BSNs) as opposed to the *data-pushing* model adopted by nodes in current WSN architectures. The advantages of using a *data-pulling* model in a BSN with resource constrained nodes and a robust base station (BS) are that it allows for sensors that support high throughput over the single hop to the base station while eliminating collisions and overhead due to synchronization and timing. The BS is a robust data processing unit that has a larger battery and larger processing power that enables it to carry on data processing and network configuration. The data collection is achieved through a *data-pulling* model, where the BS requests (pulls) data from a set of nodes with the help of simple commands, thus eliminating packet collision and channel contention issues. It uses the same model to manage the nodes in power-saving models and for network configuration. The sensor nodes are however configured as a thin server architecture where nodes collect data about the physiological signals and store them locally. Since the nodes are lighter, smaller and more resource-constrained than the BS, they have lower processing power which also enables wearability. These characteristics call for *thin servers* that are smaller data collection entities in the BSN network. The authors use this architecture and the resulting MAC protocol for a wearable ECG monitoring system and show that this architecture achieves high channel utilization by eliminating power consumed in the time synchronization and avoids the use of complex MAC protocols.

7 Security

7.1 *Security Requirements of WBAN*

The WBAN and supporting infrastructure must implement security operations that guarantee the security, data integrity, privacy and confidentiality of the patients' medical records. In addressing privacy issues it must be ensured that the Health Insurance Portability and Accountability Act of 1996 [87] is observed. The following security requirements must be attained:

Authentication: This is necessary to enable the WBAN to validate network nodes and thus prevent network compromise and/or node impersonation.

Data Integrity: this is needed to prevent the altering of data traversing the communication paths between nodes, and to prevent replay attacks.

Confidentiality: the network should be able to guarantee the secrecy of message exchange among nodes.

Availability: Since this network carries highly sensitive, important and potentially life-saving information, it is of utmost importance that the network resources are available at all times.

Privacy: The patients' data should not be disclosed to unauthorized entities (persons). Medical information is one of the most sensitive forms of personal data. The system must employ mechanisms that are able to adequately address current and potential laws in the future, governing the privacy of medical data.

7.2 *Proposed Security Solutions*

We now turn our attention to proposed security approaches in WBANs.

TinySec
TinySec is proposed in [88] as a security solution in biomedical sensor network to achieve link-layer encryption and data authentication. TinySec [89] is a software based security architecture that implements link-layer encryption. It is a

component of the official TinyOS release. TinySec is very popular in the wireless sensor community and has even been implemented on a variety of custom hardware.

TinySec encrypts the data packet with a group key common to the sensor nodes and computes a message authentication code (MAC) for the entire packet including the header. This group key is shared network-wide and manually programmed into the nodes prior to deployment. This network-wide key presents a single point of vulnerability. TinySec does not protect against node capture. If a node is compromised and keying material revealed the entire network can be compromised.

IEEE 802.15.4 Security

Several security suites can be implemented under the IEEE 802.15.4. The IEEE 802.15.4 security suite modes can be classified into two basic modes: unsecured mode and the secured mode. The unsecure mode simply means no security suite has been selected. The standard defines 8 distinct security suites (see Table III). The first of these is the Null suite that provides no security. The others can be further classified based on the security properties they provide. There is encryption only AES-CTR (counter mode of cryptographic operation with AES), authentication only (AES-CBC-MAC), and encryption & authentication (AES-CCM). Encryption is performed using AES encryption [5], which consumes comparatively less energy than other algorithms. Authentication is achieved using the cipher block chaining with message authentication code (CBC-MAC). A detailed description can be found in [90].

Table 3 III IEEE 802.15.4 Security Suites

Name	Description
Null	No security
AES-CTR	Encryption only. This provides access control, data encryption, and optional sequential freshness.
AES-CBC-MAC-128 AES-CBC-MAC-64 AES-CBC-MAC-32	Authentication only allowing flexibility by the selection of different MAC lengths: 32, 64, 128 bits.
AES-CCM-128 AES-CCM-64 AES-CCM-32	This provides authentication and encryption allowing flexibility by the selection of different MAC lengths: 32, 64, 128 bits

ZigBee Security Services

ZigBee is a consortium of industry players which came together to define a new standard for ultra-low power wireless communication [91]. The ZigBee network layer (NWK) is designed to operate on top of the IEEE 802.15.4 defined PHY and

MAC layers. The ZigBee standard defines extra security services including processes for key exchange and authentication, in addition to the security services of IEEE 802.15.4, upon which it is built.

The ZigBee standard specifies a "Trust Center". Usually the function of the Trust Center is performed by the ZigBee coordinator. The Zigbee coordinator is responsible for allowing nodes to join the network and for the distribution of keys. The roles specified for the Trust Center are: (1) trust manager- responsible for authenticating nodes requesting to join the network, (2) network manager-responsible for key maintenance and distribution, (3) configuration manager-responsible for ensuring end-to-end security[92]. More information can be found in [91].

Hardware Encryption
Instead of using software encryption as done in TinySec, hardware encryption can be implemented utilizing the ChipCon 2420 ZigBee compliant RF Transceiver. The CC2420 is able to execute IEEE 802.15.4 security operations with AES encryption using 128-bit keys. These operations include the counter (CTR) mode encryption and decryption, CBC-MAC authentication and CCM encryption plus authentication.

Hardware encryption has been implemented in a WBAN project with off-the-shelf ZigBee platform [93]. In this project it was determined that the hardware encryption does not significantly increase power consumption on the sensor platform. This was attributed to the efficient on-chip hardware support for encryption on the wireless controller and the dominant power consumption of the radio frequency (RF) unit when compared to the processing circuitry. However, the drawback of this method is that it is dependent on the specific sensor platform. Not all sensor node hardware offers hardware encryption support.

Elliptic Curve Cryptography
Elliptic curve cryptography (ECC) has emerged as a viable option for public key cryptography in wireless sensor networks. The main reason for this is its comparatively fast computation, small key size and compact signatures. There have been several noteworthy contributions in the past few years. One of the earliest works utilizing ECC in sensor networks was done by Malan *et al.* [94]. In this work, a public key infrastructure, using ECC, was implemented and evaluated on a Mica2 sensor mote platform supported by TinyOS. Uhsadel *et al.* [95] proposed an efficient implementation of ECC. Liu *et al.* [96] proposed TinyECC, which is another variation of ECC designed for TinyOS environment. As stated by the developers, "TinyECC is a configurable library for ECC operations in wireless sensor networks. The primary objective of TinyECC is to provide a ready-to-use, publicly available software package for ECC-based PKC operations that can be flexibly configured and integrated into sensor network applications"[96]. Recently, Szczechowiak *et al.* proposed NanoECC[97], which executes comparatively faster than existing ECC implementations but typically requires significant amount of ROM and RAM.

Although ECC has been successfully implemented in several variations it is still not a top choice for WBAN. This is because its energy requirements are still significantly higher than symmetric systems. This being the case, others have proposed that ECC be implemented only for infrequent and security-sensitive operations such as key establishment during the initial setup of the network or code updates. In line with this thinking, Malasri *et al.* [98] proposed a solution for medical sensor networks that uses: (i) an ECC-based secure key exchange protocol to set up shared keys between sensor nodes and base stations, (ii) symmetric encryption and decryption for protecting data confidentiality and integrity, and (iii) an authentication scheme for verifying data source.

Identity-Based Encryption

Oliveira *et al.*[99] proposed TinyTate, a lightweight Identity-Based Encryption (IBE) security solution for traditional wireless sensor networks. Tan *et al* [100] proposed an Identity-Based cryptographic security solution for WBAN. In their work, the sensor nodes compute public keys by applying a hash function on an arbitrary number of application dependent self-generated keys. These keys are stored on their flash memory and are used to execute elliptic curve encryption/decryption using Elliptic Curve Digital Signature Algorithm (ECDSA). This approach has several drawbacks: higher execution time, greater energy consumption due to increased computational overhead, and higher storage requirement for flash ROM as a result of the public key storage. Sankaran *et al.* [101] proposed IDKEYMAN, an identity-based key management scheme for wireless body area networks. IDKEYMAN is designed for a publisher-subscriber architecture like that of CodeBlue[102]. It uses IBE to set up pair-wise symmetric keys to preserve data confidentiality and integrity. IBE is only used to exchange pair-wise symmetric keys between publishers and subscribers. To reduce the computational overhead on the publisher, the symmetric keys are used in all communications subsequent to setup. IDKEYMAN takes advantage of the superior security strength of public key cryptography while minimizing energy consumption by only utilizing IBE in the bootstrapping phase.

Biometrics

Biometrics has emerged as a useful mechanism to use in the key establishment and authentication of body sensor nodes [103-106]. This method uses measurement of physiological characteristics of the body itself as an important parameter in a symmetric key management system. While the measurement of several physiological signals can be used for biometrics, the ECG (electrocardiogram)[106], and the timing information of heart beats, that is, interpulse interval (IPI)[103], are among the most appropriate since they exhibit proper time variance and randomness. The following are necessary characteristics for a useful biometric physiological value [103, 107]:

- Universal: possessed by most patients
- Distinctive: sufficiently different in any two patients
- Collectable: easily measured and collected

- Effective: able to implement a relatively secure biometric system within the constraints of processing, computing and power of the body sensor nodes
- Acceptable: adoption by the public
- Invulnerable: difficult to compromise
- Random: difficult to guess
- Time variance: changes over time

Currently there exist low cost sensor devices for medical applications that can record suitable biometric physiological signals [103, 104, 108, 109]. This could mean that in some current and future WBANs the additional system requirement for implementing a biometric based system would be almost negligible.

8 Discussion: Open Research Problems and Future Trends

In this section, we provide an overview of open research problems in WBANs and suggest some potential solutions. We also discuss future trends in the area of WBANs especially as it relates to healthcare applications.

Extended Power Supply Lifetimes
Sensor node lifetimes on the order of several days to years are required in order to ensure user acceptance of wearable sensors and because it is infeasible to replace batteries in the case of implantable sensors. Fuel cell technology provides a potential alternative to conventional batteries [110]. Micro-fuel cells provide high energy efficiency and density and refueling simply requires a cartridge replacement. These characteristics make fuel cells attractive for portable applications such as WBANs. Energy scavenging of solar, heat or vibration from the ambient environment also has the potential to extend the life of the power supply. Leonov et. al. have designed a self-powered wireless sensor node that is powered solely by human body heat [111]. The thermoelectric generator (TEG) converts temperature difference into electrical energy by the Seebeck effect [112], and the power generated is used to support the acquisition and transmission of battery voltage, temperature and light intensity data.

Low Power Consumption
The majority of the power consumption budget is dedicated to wireless communication. Possibilities for reducing communication-based power consumption are the use of ultra wideband (UWB) transceivers, because of the high data rates and low power consumption they provide [113]. Methods that seek to minimize the amount of redundant data being transmitted by each sensor will also reduce power consumption. Use of energy-efficient data compression algorithms to the data obtained from the A/D [114] would reduce the number of bits that would need to be transmitted by the transceiver. Power consumption may also be reduced by organizing the sensor data into packets and transmitting the data in short communication bursts. In between bursts, the transceiver is off. This

is called data-level duty cycling and can be combined with signal duty cycling to reduce power consumption [11]. Compressed sensing (CS) is a signal processing technique that takes advantage of the sparcity of an acquired signal and its incoherence with the sensing modality to enable signal recovery with much fewer samples compared to traditional Nyquist-based methods [115]. Mamaghanian et. al. [116] have shown that the lower complexity and CPU execution time exhibited by a CS-based ECG compression algorithm compared to DWT-based ECG compression, enables it to outperform DWT-based ECG compression in terms of overall energy efficiency. Methods that take advantage of the redundancy found in data from different sensors also have the potential for reducing power consumption. An example of this is the use of distributed source coding (DSC) to obtain compression of the signals acquired from a cluster of sensors that have correlated signals [117].

WBAN Usability

As WBAN systems transition from research laboratory prototypes to commercial products, greater emphasis must be placed on the ease of use of these systems considering the target population of users. One area that is must be considered is sensor wearability, that is, on-body sensors must be small enough to not impede a user's daily activities, and they must be resistant to common causes of wear in order to prevent frequent replacement. One way to make sensor nodes small enough for wearability specifications is to find an energy efficient way to reduce battery size (e.g. adopting fuel cell technology). Also, the utilization of ASIC technology will produce greater levels of integration than would be obtained by adapting commercial off-the-shelf (COTS) motes to a specific sensor application, and this would result in a reduction in sensor node size. Research into how sensor material is impacted by human activities, such running, walking and bathing needs to performed and sensor node construction must be modified accordingly. In addition to these issues, the set-up or replacement of a sensor node should be such that a non-technical user can accomplish these tasks.

Biocompatibility

Biocompatibility refers to "the ability of a biomaterial to perform its desired function with respect to a medical therapy, without eliciting any undesirable local or systemic effects in the recipient or beneficiary of that therapy, but generating the most appropriate beneficial cellular or tissue response in that specific situation, and optimizing the clinically relevant performance of that therapy [118]." In the context of implantable sensors, biocompatibility encompasses the reactions the sensor undergoes once being placed within the body (sensocompatibility) and the reaction the body experiences in response to the sensor. One aspect of a sensor's reaction to being placed within the body is called biofouling, which refers to the accumulation of proteins, cells and other unwanted biomaterials on a surface. If biofouling occurs on the active surface of an implantable sensor, it will result in a fall in the sensor current and may eventually result in sensor failure. Wisniewski et al. present nine sensor modifications to mitigate the effects of biofouling [119].

Physical Layer Modeling
A significant amount of research has been performed in the area of modeling the physical channel for RF propagation. More work has to be done in incorporating the impact of body movement on the channel model, and the modeling of the HBC channel needs to be investigated more extensively. In the future, model development for both the RF and HBC forms of propagation should focus on the bands specified in the 802.15.6 standard.

Optimization of network resources
Since WBANs are either wearable or implanted and are developed primarily for the study of physiological parameters, a key concern is the development of network protocols that use ultra-low radio power levels for transmission and reception that are safe for human use. With increasing number of WBANs, the radio layers used by the 802.15.4 networking standard need to be optimized to increase the throughput and minimize interference between multiple networks and users of the frequency bands.

Security
As computing and technology inches closer to the human body (WBANs being the latest forefront), it is important to protect the privacy of the data collected and disseminated by these networks. Complex security mechanisms require more computational and power resources, and optimizing the tradeoff between these is crucial for the widespread use of WBANs.

As mentioned earlier, there exists a number of emerging security approaches to WBAN. Novel methods of achieving security to satisfy low resource consumption should be explored. The IEEE 802.15 Task Group 6 proposal for a standard for WBAN was approved in February 2012. This new standard, IEEE 802.15.6, outlines a path for further research in the security operations of WBAN and has defined important security metrics for the development of secure WBAN.

Preventative Healthcare
Present WBAN technology is mostly developed on-demand; WBANs are developed in response to specific physiological requirements. Using dynamic programming environments and cognitive interfaces, we should be able to measure multiple parameters in the human body and use this data to aid in preventive medicine and diagnosis.

Computational and Economic Perspective
Since WBANs comprise of networks of sensors performing specialized detection of physiological data, the cost and size of these devices impose limitations on their use. Creating networks that seamlessly interface with the human environment will be an interesting area of research.

Routing
The main challenges in designing any efficient routing protocol would be to address network partitioning with postural mobility, design an energy-constrained

protocol, and ensure end-to-end path reliability. Some research questions that need to be addressed are:

- Can an effective multihop communication protocol be designed taking into consideration constraints of wireless channels and power constraints of on-body sensor nodes?
- Can the mobility pattern of on-body sensor nodes be effectively designed to assist in designing a mobility pattern-based communication protocol?

Emerging Markets and Future Applications
Although WBANs have been developed for medical applications, they can be easily tailored for smart environments that combine sectors such as business, entertainment and education for a heightened seamless experience. The recent explosion of personal computing devices into the consumer market that combine social networking applications can be boosted with the introduction of WBANs without active involvement of the individual at the center of the networks. This kind of passive involvement in data transfer can ease the cognitive burden on the individual and result in more unobtrusive computing applications. It also has the potential for breakthroughs in the study of medicine, ecology and other civilian and military applications. A major concern underlining these advances will be that of privacy and security of data and the applications running on them, since this kind of passive involvement by the individual might result in the use of default settings for data privacy and might be vulnerable to malice by unintended users of the data. WBANs will also need evolving standards for co-existence and data transfer with other ad hoc networks, mobile networks and the Internet. The success of WBANs and their eventual widespread acceptance will be accelerated by merging it with existing technologies and creating frameworks with ease of data transfer and data access.

References

[1] Custodio, V., Herrera, F.J., López, G., Moreno, J.I.: A Review on Architectures and Communications Technologies for Wearable Health-Monitoring Systems. Sensors 12 (2012)

[2] Lai, X., Liu, Q., Wei, X., Wang, W., Zhou, G., Han, G.: A Survey of Body Sensor Networks. Sensors 13, 5406–5447 (2013)

[3] Yang, B.H., Rhee, S.: Development of the ring sensor for healthcare automation. Robotics and Autonomus Systems 30, 273–281 (2000)

[4] Asada, H.H., Shaltis, P., Reisner, A., Rhee, S., Hutchinson, R.C.: Mobile Monitoring with Wearable Photoplethysmographic Biosensors. IEEE Engineering in Medicine and Biology Magazine, 28–40 (2003)

[5] Perrig, A., Szewczyk, R., Wen, V., Culler, D., Tygar, J.: SPINS:Security Protocols for Sensor Networks. In: Seventh Annual ACM International Conference on Mobile Networking and Computing (Mobicom 2001), Rome, Italy (2001)

[6] Callaway, E.: Secure Low-Power Operation of Wireless Sensor Networks. Sensors Online Magazine (January 2004), http://www.sensorsmag.com/articles/0104/22/main.shtml
[7] Giovanni, C.: Topology for Denial of Service. Endeavour Systems Inc., White Paper (July 12, 2000)
[8] Shnayder, V., Chen, B., Lorincz, K., Fulford-Jones, T.R.F., Welsh, M.: Sensor Networks for Medical Care. Technical Report TR-08-05, Division of Engineering and Applied Sciences, Harvard University (2005)
[9] Anliker, U., Ward, J.A., Lukowicz, P., Tröster, G., Dolveck, F., Baer, M., Keita, F., Schenker, E., Catarsi, F., Coluccini, L., Belardinelli, A., Shklarski, D., Alon, M., Hirt, E., Schmid, R., Vuskovic, M.: AMON: A Wearable Multiparameter Medical Monitoring and Alert System. IEEE Transactions on Information Technology in Biomedicine 8, 415–427 (2004)
[10] Fulford-Jones, T.R.F., Wei, G.-Y., Welsh, M.: A Portable, Low-Power, Wireless Two-Lead EKG System. In: 26th Annual International Conference of the IEEE EMBS, San Francisco, CA, USA, pp. 2141–2144 (2004)
[11] Penders, J., Gyselinckx, B., Vullers, R., Nil, M.D., Nimmala, V.S.R., van de Molengraft, J., Yazicioglu, F., Torfs, T., Leonov, V., Merken, P., Hoof, C.V.: Human++: from technology to emerging health monitoring concepts. In: 5th International Workshop on Wearable and Implantable Body Sensor Networks, The Chinese University of Hong Kong, HKSAR, China, pp. 94–98 (2008)
[12] Cao, H., Li, H., Stocco, L., Leung, V.C.M.: Demonstration of A Novel Wireless Three-pad ECG System for Generating Conventional 12-lead Signals. In: 5th Annual International ICST Conference on Body Area Networks, Corfu Island, Greece (2010)
[13] Cao, H., Li, H., Stocco, L., Leung, V.C.M.: Design and Evaluation of a Novel Wireless Three-pad ECG System for Generating Conventional 12-lead Signals. In: 5th Annual International ICST Conference on Body Area Networks, Corfu Island, Greece (2010)
[14] Liszka, K.J., Mackin, M.A., Lichter, M.J., York, D.W., Pillai, D., Rosenbaum, D.S.: Keeping a Beat on the Heart. IEEE Pervasive Computing 3, 42–49 (2004)
[15] Kario, K., Yasui, N., Yokoi, H.: Ambulatory Blood Pressure Monitoring for Cardiovascular Medicine. IEEE Engineering in Medicine and Biology Magazine, 81–88 (2003)
[16] Carr, J.J., Brown, J.M.: Introduction to Biomedical Equipment Technology, 4th edn. Prentice-Hall (2001)
[17] Poon, C.C.Y., Wong, Y.M., Zhang, Y.-T.: M-Health: The Development of Cuff-less and Wearable Blood Pressure Meters for Use in Body Sensor Networks. In: Proceedings of IEEE/NLM Life Science Systems and Applications Workshop (2006)
[18] Basmajian, J.V., Deluca, C.: Muscles Alive, 5th edn. Williams & Wilkins, Baltimore (1985)
[19] Bonato, P., Mork, P.J., Sherrill, D.M., Westgaard, R.H.: Data Mining of Motor Patterns Recorded with Wearable Technology. IEEE Engineering in Medicine and Biology Magazine, 110–119 (2003)
[20] Tapia, E.M., Marmasse, N., Intille, S.S., Larson, K.: MITes: Wireless Portable Sensors for Studying Behavior. In: Proceedings of Extended Abstracts Ubicomp 2004: Ubiquitous Computing (2004)

[21] Park, C., Chou, P.H.: Eco: Ultra-Wearable and Expandable Wireless Sensor Platform. presented at the Third International Workshop on Body Sensor Networks, BSN 2006 (2006)
[22] Kusserow, M., Amft, O., Troster, G.: BodyANT: Miniature wireless sensors for naturalistic monitoring of daily activity. In: 4th Annual International ICST Conference on Body Area Networks, pp. 1–8 (2009)
[23] Keijsers, N.L.W., Horstink, M.W.I.M., Gielen, S.C.A.M.: Online Monitoring of Dyskinesia in Patients with Parkinson's Disease. IEEE Engineering in Medicine and Biology Magazine, 96–103 (2003)
[24] Waterhouse, E.: New Horizons in Ambulatory Electroencephalography. IEEE Engineering in Medicine and Biology Magazine, 74–80 (2003)
[25] Jovanov, E., Raskovic, D., Price, J., Chapman, J., Moore, A., Krishnamurthy, A.: Patient Monitoring Using Personal Area Networks of Wireless Intelligent Sensors. Biomedical Sciences Instrumentation 37, 373–378 (2001)
[26] Intanagonwiwat, C., Govindan, R., Estrin, D.: Directed diffusion:a scalable and robust communication paradigm for sensor networks. In: ACM MobiCom 2000, Boston, MA, pp. 56–57 (2000)
[27] Farshchi, S., Mody, I., Judy, J.W.: A TinyOS-Based Wireless Neural Interface. In: Proceedings of the 26th Annual International Conference of the IEEE EMBS (2004)
[28] Garg, S.K., Schwartz, S., Edeleman, S.V.: Improved Glucose Excursions Using an Implantable Real-Time Continuous Glucose Sensor in Adults With Type 1 Diabetes. Diabetes Care 27, 734–738 (2004)
[29] Maloney, J.M., Santini, J.J.T.: Implantable Microchips for Controlled Drug Delivery. In: Proceedings of the 26th Annual International Conference of the IEEE EMBS, San Francisco, CA, USA, pp. 2668–2669 (2004)
[30] Ghovanloo, M., Beach, K., Wise, K.D., Najafi, K.: A BiCMOS Wireless Interface Chip for Micromachined Stimulating Microprobes. In: Proceedings of IEEE-EMBS Special Topic Conference on Microtechnologies in Medicine and Biology, pp. 277–282 (2002)
[31] Patel, B.A., Anastassiou, C.A., O'Hare, D. (eds.): Biosensor Design and Interfacing (Body Sensor Networks). Springer (2006)
[32] S. Association. 802.15.6-2012 IEEE Standards for Local and Metropolitan Area Networks–Part 15.6: Wireless Body Area Networks (2012), `http://standards.ieee.org/findstds/standard/802.15.6-2012.html`
[33] Wegmueller, M.S., Kuhn, A., Froehlich, J., Oberle, M., Felber, N., Kuster, N., Fichtner, W.: An Attempt to Model the Human Body as a Communication Channel. IEEE Transactions on Biomedical Engineering 54, 1851–1857 (2007)
[34] Wang, W., Nie, Z., Guan, F., Leng, T., Wang, L.: Experimental Studies on Human Body Communication Characteristics based upon Capacitive Coupling. In: International Conference on Body Sensor Networks, pp. 180–185 (2011)
[35] Gupta, S., Lalwani, S., Prakash, Y., Elsharawy, E., Schwiebert, L.: Towards a Propagation Model for Wireless Biomedical Applications. In: IEEE International Conference on Communication (ICC), pp. 1993–1997 (2003)
[36] Zasowski, T., Althaus, F., Stage, M., Wittneben, A., Troste, G.: UWB for noninvasive wireless body area networks: channel measurements and results. In: IEEE Conference on Ultra Wideband Systems and Technologies, pp. 285–289 (2003)

[37] Fort, A., Desset, C., Ryckaert, J., Doncker, P.D., Biesen, L.V., Wambacq, P.: Characterization of the Ultra Wideband Body Area Propagation Channel. In: IEEE international Conference on Ultra-Wideband, pp. 22–27 (2005)
[38] Roelens, L., Van den Bulcke, S., Joseph, W., Vermeeren, G., Martens, L.: Path Loss Model for Wireless Narrowband Communication Above Flat Phantom. Electronics Letters 42, 10–11 (2006)
[39] Reusens, E., Joseph, W., Vermeeren, G., Martens, L., Latré, B., Moerman, I., Braem, B., Blondia, C.: Path Loss Models for Wireless Communication Channel along Arm and Torso: Measurements and Simulations. In: IEEE Antennas and Propagation Society International Symposium, Honolulu, HI, pp. 336–339 (2007)
[40] Ryckaert, J., Doncker, P.D., Meys, R., Hoye, A.D.L., Donnay, S.: Channel Model for Wireless Communication Around Human Body. Electronics Letters 40, 2–3 (2004)
[41] Zasowski, T., Meyer, G., Althaus, F., Wittneben, A.: Propagation Effects in UWB Body Area Networks. In: IEEE International Conference on Ultra-Wideband, pp. 16–21 (2005)
[42] Fort, A., Ryckaert, J., Desset, C., Doncker, P.D., Wambacq, P., Biesen, L.V.: Ultra-Wideband Channel Model for Communication Around the Human Body. IEEE Journal on Selected Areas in Communications 24, 927–933 (2006)
[43] Braem, B., Latré, B., Moerman, I., Blondia, C., Reusens, E., Joseph, W., Martens, L., Demeester, P.: The Need for Cooperation and Relaying in Short-Range High Path Loss Sensor Networks. In: First International Conference on Sensor Technologies and Applications (SENSORCOMM 2007), Valencia, Spain, pp. 566–571 (2007)
[44] Renzo, M.D., Buehrer, R.M., Torres, J.: Pulse Shape Distortion and Ranging Accuracy in UWB–based Body Area Networks for Full–Body Motion Capture and Gait Analysis. In: IEEE Global Telecommunications Conference (Globecom 2007), pp. 3775–3780 (2007)
[45] Quwaider, M., Rao, J., Biswas, S.: Transmission Power Assignment with Postural Position Inference for On-Body Wireless Communication Links. Transactions on Embedded Computing Systems (TECS) 10, 1–27 (2010)
[46] Nabi, M., Basten, T., Geilen, M., Blagojevic, M., Hendriks, T.: A Robust Protocol Stack for Multi-hop Wireless Body Area Networks with Transmit Power Adaptation. In: 5th Annual International ICST Conference on Body Area Networks, Corfu Island, Greece (2010)
[47] Yan, L., Zhong, L., Jha, N.K.: Energy Comparison and Optimization of Wireless Body-Area Network Technologies. In: The Proceedings of the ICST 2nd International Conference on Body Area Networks(BodyNets 2007), Florence, Italy, pp. 1–8 (2007)
[48] Shankar, V., Natarajan, A., Gupta, S.K.S., Schwiebert, L.: Energy-Efficient Protocols for Wireless Communication in Biosensor Networks. In: IEEE Personal, Indoor and Mobile Radio Communications Conference, San Diego, CA, pp. D-114–D-118 (2001)
[49] Ullah, S., Bin, S., Islam, S.M.R., Khan, P., Saleem, S., Kwak, K.S.: A Study of MAC Protocols for WBANs. Sensors 10, 128–145 (2009)
[50] Ye, W., Heidemann, J., Estrin, D.: An Energy-Efficient MAC protocol for Wireless Sensor Networks. In: IEEE Infocom, New York City, NY, USA, pp. 1567–1576 (2002)

[51] Wireless Medium Access Control (MAC) and Physical Layer (PHY) Specification for Low-Rate Wireless Personal Area Networks (LR-WPANs), IEEE 802.15.4. IEEE, New York (2003)
[52] Cavalcanti, D., Schmitt, R., Soomro, A.: Performance Analysis of 802.15.4 and 802.11e for Body Sensor Network Applications. In: 4th International Workshop on Wearable and Implantable Body Sensor Networks Aachen, Germany (2007)
[53] Timmons, N.F., Scanlon, W.G.: Analysis of the Performance of IEEE 802.15.4 for Medical Sensor Body Area Networking. In: IEEE Annual Conference on Sensor, Mesh and Ad Hoc Communications and Networks (SECON 2004), Santa Clara, CA, USA, pp. 1266–1275 (2004)
[54] Li, C., Li, H.B., Kohno, R.: Performance Evaluation of IEEE 802.15.4 for Wireless Body Area Network (WBAN). In: IEEE International Conference on Communications Workshops, Dresden, Germany, pp. 1–5 (2009)
[55] Omeni, O., Chi, A., Wong, W., Burdett, A.J., Toumazou, C.: Energy Efficient Medium Access Protocol for Wireless Medical Body Area Sensor Networks. IEEE Transactions on Biomedical Circuits and Systems 2 (December 2008)
[56] Marinkovic, S.J., Popovici, E.M., Spagnol, C., Faul, S., Marnane, W.P.: Energy-Efficient Low Duty Cycle MAC Protocol for Wireless Body Area Networks. IEEE Transactions on Information Technology in Biomedicine 13 (November 2009)
[57] Li, H., Tan, J.: Heartbeat-Driven Medium-Access Control for Body Sensor Networks. IEEE Transactions on Information Technology in Biomedicine 14 (January 2010)
[58] Fang, G., Dutkiewicz, E.: BodyMAC: Energy efficient TDMA-based MAC protocol for Wireless Body Area Networks. In: 9th International Symposium on Communications and Information Technology (ISCIT 2009), pp. 1455–1459 (2009)
[59] Timmons, N.F., Scanlon, W.G.: An Adaptive Energy Efficient MAC Protocol for the Medical Body Area Network. In: 1st International Conference on Wireless Communication, Vehicular Technology, Information Theory and Aerospace (Wireless VITAE 2009), Aalborg, Denmark, pp. 587–593 (2009)
[60] Otal, B., Alonso, L.: Highly Reliable Energy-Saving MAC for Wireless Body Sensor Networks in Healthcare Systems. IEEE Journal on Selected Areas in Communications 24, 553–565 (2009)
[61] Ullah, S., An, X., Kwak, K.S.: Towards Power Efficient MAC Protocol for In-Body and On-Body Sensor Networks. In: Håkansson, A., Nguyen, N.T., Hartung, R.L., Howlett, R.J., Jain, L.C. (eds.) KES-AMSTA 2009. LNCS, vol. 5559, pp. 335–345. Springer, Heidelberg (2009)
[62] Latré, B., Braem, B., Moerman, I., Blondia, C., Reusens, E., Joseph, W., Demeester, P.: A Low Delay Protocol for Multihop Wireless Body Area Networks. In: The 2007 Fourth Annual International Conference on Mobile and Ubiquitous Systems: Networking & Services (MobiQuitous), Philadelphia, PA, USA, pp. 1–8 (2007)
[63] Natarajan, A., de Silva, B., Yap, K.-K., Motani, M.: Link Layer Behavior of Body Area Networks at 2.4 GHz. In: The 15th Annual International Conference on Mobile Computing and Networking (MobiCom 2009), Beijing, China, pp. 241–252 (2009)
[64] Natarajan, A., Motani, M., de Silva, B., Yap, K.-K., Chua, K.C.: Investigating Network Architecture for Body Sensor Networks. In: The 1st ACM SIGMOBILE International Workshop on Systems and Networking Support for Healthcare and Assisted Living Environments, San Juan, Puerto Rico, USA, pp. 19–24 (2007)

[65] Braem, B., Latré, B., Moerman, I., Blondia, C., Demeester, P.: The Wireless Autonomous Spanning Tree Protocol for Multihop Wireless Body Area Networks. In: 3rd Annual International Conference on Mobile and Ubiquitous Systems - Workshops, San Jose, CA, USA, pp. 1–8 (2006)

[66] Latré, B., Braem, B., Moerman, I., Blondia, C., Reusens, E., Joseph, W., Demeester, P.: A Low Delay Protocol for Multihop Wireless Body Area Networks. In: The 2007 Fourth Annual International Conference on Mobile and Ubiquitous Systems: Networking & Services (MobiQuitous), pp. 1–8 (2007)

[67] Singelée, D., et al.: A Secure Cross-Layer Protocol for Multi-hop Wireless Body Area Networks. In: Coudert, D., Simplot-Ryl, D., Stojmenovic, I. (eds.) ADHOC-NOW 2008. LNCS, vol. 5198, pp. 94–107. Springer, Heidelberg (2008)

[68] Boulis, A., Smith, D., Miniutti, D., Libman, L., Tselishchev, Y.: Challenges in Body Area Networks for Healthcare: the MAC. IEEE Communications Magazine 50, 100–106 (2012)

[69] Leguay, J., Friedman, T., Conan, V.: Evaluating Mobility Pattern Space Routing for DTNs. In: The 25th IEEE International Conference on Computer Communications (INFOCOM 2006), Barcelona, Spain (2006)

[70] Caldeira, J., Rodrigues, J., Lorenz, J.: Intra-Mobility Support Solutions for Healthcare Wireless Sensor Networks – Handover Issues. IEEE Sensors Journal PP, 1–12 (2013)

[71] Khan, J.Y., Yuce, M.R., Karami, F.: Performance evaluation of a wireless body area sensor network for remote patient monitoring. In: 30th Annual International Conference of the IEEE Engineering in Medicine and Biology Society (EMBS), pp. 1266–1269 (2008)

[72] Singelée, D., et al.: A Secure Cross-Layer Protocol for Multi-hop Wireless Body Area Networks. In: Coudert, D., Simplot-Ryl, D., Stojmenovic, I. (eds.) ADHOC-NOW 2008. LNCS, vol. 5198, pp. 94–107. Springer, Heidelberg (2008)

[73] Marinkovic, S., Spagnol, C., Popovici, E.: Energy-Efficient TDMA-Based MAC Protocol for Wireless Body Area Networks. In: Third International Conference on Sensor Technologies and Applications, pp. 604–610 (2009)

[74] Jovanov, E., Lords, A., Raskovic, D., Cox, P., Adhami, R., Andrasik, F.: Stress Monitoring Using a Distributed Wireless Intelligent Sensor System. IEEE Engineering in Medicine and Biology Magazine, 49–55 (May/June 2003)

[75] Ryckaert, J., Desset, C., Fort, A., Badaroglu, M., Heyn, V.D., Wambacq, P., Van der Plas, G., Donnay, S., Poucke, B.V., Gyselinckx, B.: Ultra-wideband Transmitter for Low-Power Wireless Body Area Networks: Design and Evaluation. IEEE Transactions on Circuit and Systems - I: Regular Papers 52, 2515–2525 (2005)

[76] Santiago, R.C., Khaleghi, A., Balasingham, I., Ramstad, T.A.: Architecture of an Ultra Wideband Wireless Body Area Network for Medical Applications. In: IEEE 2nd International Symposium on Applied Sciences in Biomedical and Communication Technologies (ISABEL), pp. 1–6 (2009)

[77] MELODY Project (2010), `http://www.melody-project.info/`

[78] Saeed, A., Faezipour, M., Nourani, M., Banerjee, S., Lee, G., Gupta, G., Tamil, L.: A Scalable Wireless Body Area Network for Bio-Telemetry. Journal of Information Processing Systems 5, 77–86 (2009)

[79] Soini, M., Nummela, J., Oksa, P., Ukkonen, L., Sydänheimo, L.: Wireless Body Area Network for Hip Rehabilitation System. Ubiquitous Computing and Communication Journal 3, 1–7 (2008)

[80] Jovanov, E., Milenkovic, A., Otto, C., de Groen, P.C.: A wireless body area network of intelligent motion sensors for computer assisted physical rehabilitation. Journal of NeuroEngineering and Rehabilitation 2 (March 1, 2005)
[81] Jovanov, E., Milenkovic, A., Basham, S., Clark, D., Kelley, D.: Reconfigurable Intelligent Sensors for Health Monitoring: A Case Study of Pulse Oximeter Sensor. In: The 26th Annual International Conference of the IEEE Engineering in Medicine and Biology Society, pp. 4759–4762 (2004)
[82] Otto, C., Milenkovic, A., Sanders, C., Jovanov, E.: System Architecture of a Wireless Body Area Sensor Network for Ubiquitous Health Monitoring. Journal of Mobile Multimedia 1, 307–326 (2006)
[83] Bindra, A.: Medical info-communications signals an era of body area networking (2008), http://www.rfdesign.com
[84] Milenkovi, A., Otto, C., Jovanov, E.: Wireless sensor networks for personal health monitoring: Issues and an implementation. Computer Communications 29, 2521–2533 (2006)
[85] Manjeshwar, A., Agrawal, D.P.: APTEEN: A Hybrid Protocol for Efficient Routing and Comprehensive Information Retrieval in Wireless Sensor Networks. In: The 2nd International Workshop on Parallel and Distributed Computing Issues in Wireless Networks and Mobile Computing, Fort Lauderdale, FL (April 2002)
[86] Yoo, S., Chen, C., Chou, P.H.: Low-Complexity, High-Throughput Multiple-Access Wireless Protocol for Body Sensor Networks. In: The Sixth International Workshop on Wearable and Implantable Body Sensor Networks (BSN), pp. 109–113 (2009)
[87] The Health Insurance Portability and Accountability Act of 1996 (HIPAA), Ed: Centers for Medicare and Medicaid Services (1996)
[88] Marci, S.S., Meingast, Roosta, T.: Security and Privacy Issues with Health Care Information Technology. In: 28th Annual International Conference of the IEEE Engineering in Medicine and Biology Society (EMBS 2006), New York City, NY, USA, pp. 5453–5458 (2006)
[89] Karlof, C., Sastry, N., Wagner, D.: TinySec:A Link Layer Security Architecture for Wireless Sensor Networks. In: Second ACM Conference on Embedded Networked Sensor Systems (SenSys 2004), Baltimore, Maryland, USA, pp. 162–175 (2004)
[90] Tanenbaum, A.S.: Computer Networks, 4th edn. Prentice Hall, New Jersey (2003)
[91] ZigBee Specification v1.0, ZigBee Alliance, San Ramon, CA, USA (2005)
[92] Boyle, D., Newe, T.: Securing Wireless Sensor Networks:Security Architectures. Journal of Networks 3 (January 2008)
[93] Warren, S., Lebak, J., Yao, J., Creekmore, J., Milenkovic, A., Jovanov, E.: Interoperability and Security in Wireless Body Area Network Infrastructures. In: IEEE Engineering in Medicine and Biology 27th Annual Conference, Shanghai, China (2005)
[94] Malan, D.J., Welsh, M., Smith, M.D.: A Public Key Infrastructure for Key Distribution in TinyOS Based on Elliptic Curve Cryptography. In: IEEE Sensor and Ad Hoc Communications and Network (SECON 2004), Santa Clara, California, USA (2004)
[95] Uhsadel, L., Poschmann, A., Paar, C.: Enabling Full-Size Public-Key Algorithms on 8-Bit Sensor Nodes. In: Stajano, F., Meadows, C., Capkun, S., Moore, T. (eds.) ESAS 2007. LNCS, vol. 4572, pp. 73–86. Springer, Heidelberg (2007)

[96] Liu, A., Ning, P.: TinyECC: A Configurable Library for Elliptic Curve Cryptography in Wireless Sensor Networks. In: International Conference on Information Processing in Sensor Networks (IPSN 2008), St. Louis, Missouri, USA (2008)
[97] Szczechowiak, P., Oliveira, L.B., Scott, M., Collier, M., Dahab, R.: NanoECC: Testing the Limits of Elliptic Curve Cryptography in Sensor Networks. In: Verdone, R. (ed.) EWSN 2008. LNCS, vol. 4913, pp. 305–320. Springer, Heidelberg (2008)
[98] Malasri, K., Wang, L.: Addressing Security in Medical Sensor Networks. In: ACM SIGMOBILE International Workshop on Systems and Networking Support for Healthcare and Assisted Living Environments (HealthNet 2007), San Juan, Puerto Rico, USA (2007)
[99] Oliveira, L.B., Aranha, D., Morais, E., Daguano, F., Lopez, J., Dahab, R.: TinyTate:Identity-Based Encryption for Sensor Networks. Cryptology ePrint Archive, vol. 2007/020 (2007)
[100] Tan, C.C., Wang, H., Zhong, S., Li, Q.: Body Sensor Network Security: An Identity-Based Cryptography Approach. In: Frist ACM Conference on Wireless Network Security (WiSec 2008), Alexandria, Virginia, USA (2008)
[101] Sankaran, S., Husain, M.I., Sridhar, R.: IDKEYMAN:An Identity-Based Key Management Scheme for Wireless Ad Hoc Body Area Networks. In: The 5th Annual Symposium on Information Assurance (ASIA 2009), Albany, New York, USA (2009)
[102] Malan, D., Fulford-Jones, T., Welsh, M., Moulton, S.: Codeblue:An Ad hoc Sensor Network Infrastructure for Emergency Medical Care. In: MobiSys 2004 Workshop on Applications of Mobile Embedded Systems (WAMES 2004), Boston, Massachusetts, USA (2004)
[103] Poon, C.C.Y., Zhang, Y.-T., Bao, S.-D.: A Novel Biometrics Method to Secure Wireless Body Area Sensor Networks for Telemedicine and M-Health. IEEE Communications Magazine (April 2006)
[104] Cherukuri, S., Venkatasubramanian, K.K., Gupta, S.K.S.: BioSec: A Biometric Based Approach for Securing Communication in Wireless Networks of Biosensors Implanted in the Human Body. In: Workshop on Wireless Security and Privacy (WiSPr), International Conference on Parallel Processing Workshops (2003)
[105] Bao, S.D., Zhang, Y.T., Shen, L.F.: A New Symmetric Cryptosystem of Body Area Sensor Networks for Telemedicine. In: The Japan Society of Medical Electronics & Biological Engineering, Japan (2005)
[106] Bui, F.M., Hatzinakos, D.: Biometric Methods for Secure Communications in Body Sensor Networks: Resource-Efficient Key Management and Signal-Level Data Scrambling. EURASIP Journal on Advances in Signal Processing 8, 1–16 (2008)
[107] Uludag, U., Pankanti, S., Jain, A.K., Prabhakar, S., Jain, K.: Biometric Cryptosystems: Issues and Challenges. In: The IEEE, pp. 948–960 (2004)
[108] Venkatasubramanian, K.K., Gupta, S.K.S.: Security for Pervasive Health Monitoring Sensor Applications. In: 4th International Conference on Intelligent Sensing and Information Processing (ICISIP 2006), Bangalore, India, pp. 197–202 (2006)
[109] Kannavara, R., Bourbakis, N.G.: Iris Biometric Authentication based on Local Global Graphs: An FPGA Implementation. In: IEEE Symposium on Computational Intelligence for Security and Defense Applications (CISDA), Ontario, CA (2009)
[110] Heinzel, A., Hebling, C., Muller, M., Zedda, M., Muller, C.: Fuel cells for low power applications. Journal of Power Sources 105, 250–255 (2002)

[111] Leonov, V., Torfs, T., Fiorini, P., Hoof, C.V.: Thermoelectric Converters of Human Warmth for Self-Powered Wireless Sensor Nodes. IEEE Sensors Journal 7, 650–657 (2007)
[112] Herwaarden, A.W.V., Sarro, P.M.: Thermal Sensors Based on the Seeback Effect. Sensors and Actuators 10, 321–346 (1986)
[113] Alomainy, A., Hao, Y., Hu, X., Parini, C.G., Hall, P.S.: UWB on-body radio propagation and system modelling for wireless body-centric networks. IEE Proceedings Communications 153, 107–114 (2006)
[114] Schwiebert, L., Gupta, S.K.S., Weinmann, J.: Research challenges in wireless networks of biomedical sensors. In: 7th Annual International Conference on Mobile Computing and Networking Rome, Italy, pp. 151–165 (2001)
[115] Candès, E.J., Wakin, M.B.: An Introduction to Compressive Sampling. IEEE Signal Processing Magazine 25, 21–30 (2008)
[116] Mamaghanian, H., Khaled, N., Atienza, D., Vandergheynst, P.: Compressed Sensing for Real-Time Energy-Efficient ECG Compression on Wireless Body Sensor Nodes. IEEE Transactions on Biomedical Engineering 58, 2456–2466 (2011)
[117] Xiong, Z., Liveris, A.D., Cheng, S.: Distributed Source Coding for Sensor Networks. IEEE Signal Processing Magazine 21, 80–94 (2004)
[118] Williams, D.F.: On the mechanisms of biocompatibility. Biomaterials 29, 2941–2953 (2008)
[119] Wisniewski, N., Reichert, M.: Methods for Reducing Biosensor Membrane Biofouling. Colloids and Surfaces B: Biointerfaces 18, 197–219 (2000)

Questions

Problems:

Q1. What are the characteristic of WBAN that contribute to the need of designing a routing protocol?

a. frequent network partitioning of the on-body sensors
b. high propagation loss across the human body
c. low transmission power of the sensors
d. all of the above

Q2. What are the benefits of having a multihop communication between the on-body sensors?
a. significant improvement in packet delivery ratio in an open environment with low transmission power
b. energy-efficient end-t-end communication
c. both a and b
d. None of the above

Q3. Ultra-wideband radio (UWB) is a good candidate for high data rate applications like wireless video. The Federal Communications Commission (FCC) defines UWB as any radio technology that has a bandwidth > 500

MHz or larger than 20% of the center frequency. How can the Shannon-Hartley theorem be used to explain two methods to support UWB? Comment on the feasibility of the two methods.

(Hint: Recall that the Shannon Hartley theorem is mathematically given by: $C = W \log_2\left(1+\frac{S}{N}\right)$, where C is the capacity, W is the bandwidth and the ratio $\frac{S}{N}$ is the signal-to-noise ratio, commonly referred to as the SNR.)

Q4. Compare the following:

a. Flat architectures versus multi-tier architecture

Flat architectures comprise of a multiple data-gathering units in the base tier that sends their data to a personal computer or a personal server application running on a PDA (upper tier).

Multi-tier architectures comprise of multiple nodes in the base tier, one or more gateway(s) at the next tier(s) that acts an interface between the base tier and a server at the highest tier.

b. Data-pulling versus data pushing models of information dissemination

Case Study One:

Q5. Consider the case of energy-generating floors. Energy-generating floors are made of piezoelectric tiles that use the kinetic energy produced by stepping on them to power other applications. These floors are being tested for their energy-production applications in the Tokyo subway, dance clubs and shopping centers. Construct a sample network architecture for a voice-activated energy-generator that can be used to supply power to a grid for other applications. Comment on the choice of architecture (flat or multi-tier), node locations, security and energy efficiency, information dissemination model (data pulling versus data pushing) as well as possible applications of this technology.

Case Study Two:

Q6. Consider a situation where the on-body sensors need to find energy-efficient and reliable end-to-end paths to other on-body sensors. Design the selection criteria based on the following parameters:

a. Link likelihood factor (LLF), i.e., the probability of link $L_{i,j}$ being connected between nodes i and j at discrete time interval, t

b. Residual energy of the sensors (nodes)

What is the role of choosing the weightage factor, w, with LLF?

38. An Integrated Wireless Communication Platform for End-to-End and Automatic Wireless Vital Sign Capture Using Personal Smart Mobile Devices

Quang-Dung Ho and Tho Le-Ngoc

Department of Electrical and Computer Engineering
McGill University, 3480 University Street, Montreal, Quebec, H3A 2A7, Canada
{quang.ho,tho.le-ngoc}@mcgill.ca

Keywords: Smartphone, vital signs, bluetooth low energy (BLE), machine-to-machine (M2M) communications, Wi-Fi networks, cellular networks, long term evolution (LTE), prototype, experiment.

1 Wireless Vital Sign Capture: A Review

In recent years, in addition to fast development and maturity of information and communication technology, personal smart mobile devices (e.g., smartphones, phablets, tablets) have been becoming more and more popular. According to statistics in 2013, more than 50% of mobile consumers in United States and Canada have smartphones [1]. This number keeps increasing due to the fact that smartphones have been equipped with more powerful microprocessors that are capable of supporting more advanced operating systems and applications, larger and higher-resolution touch displays, higher capacity batteries, etc., yet at more affordable prices. Therefore, these devices are expected to have a wider adoption in electronic healthcare (eHealth). They can be integrated into patient interactions, electronic health record (EHR), and many other aspects of medicine. In other words, mobile-health (mHealth) has emerged as an important segment of eHealth. By definition, it is the delivery of healthcare services via mobile communication devices.

Mobile systems have been used to monitor, record, and relay a variety of patient vital signs (VSs) such as body temperature (TEMP), non-invasive blood pressure (NIBP), heart rate (HR), blood oxygen saturation level (SpO2), respiration rate (RR), electrocardiographic (ECG), etc., and disease-related conditions (e.g., blood sugar level). Other health-related activities such as weight tracking, cardiovascular monitoring, and patient location tracking are also monitored using mHealth facilities. The work in [2] presents a comprehensive overview and classification of a broad

S. Adibi (ed.), *Mobile Health*,
Springer Series in Bio-/Neuroinformatics 5, DOI: 10.1007/978-3-319-12817-7_38

range of mHealth solutions that have been implemented for elderly patients with chronic conditions. In [3], a number of biomedical applications and various low-power wireless sensor and link technologies for mHealth are surveyed. Platforms developed for monitoring ECG in clinical and domestic environments have been demonstrated in [4, 5]. Personal mobile health systems aimed to support individuals suffering from chronic illness such as diabetes, obesity, cardiovascular disease, etc., for long-term management and treatment have been introduced in [6-10]. In [11-13], mobile applications for fitness, rehabilitation and individual activity monitoring are investigated. An important observation is that most existing systems focus on long-term patient monitoring for a number of specific chronic diseases in clinical and home scenarios and/or for fitness applications. mHealth systems designed for large-scale use in hospitals or for emergency medical situations (e.g., car accidents, natural disasters) have not been receiving sufficient attention. As a representative example, nowadays, in most hospitals, medical staff (e.g., nurses, physicians) capture patient VSs "manually": they attach probes to patients, perform measurements, read the results and note them down by hand writings. Then, the data is manually input into EHR using a workstation in the hospital. Inherently, this workflow cannot ensure the accuracy and reliability of collected data and thus the quality of patient care. It is also time-consuming and there is a significant delay to the availability of information in EHR. Usually, hand-written data recorded by a medical staff cannot be input into EHR until he/she finishes a whole round of visits/measurements for many patients and can find an available workstation in his/her proximity. The incurred delay can be several hours or even days.

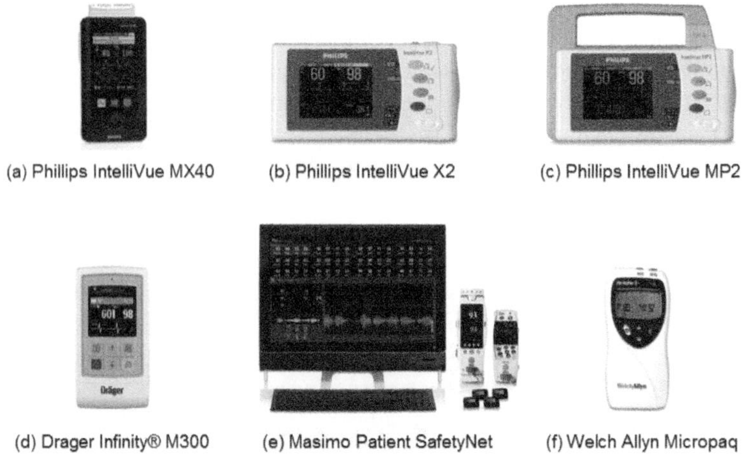

Fig. 1 Example commercial wireless patient monitors

In fact, there are a number of commercial products that support automatic data capture and connectivity to EHR. Most of them rely on wired Ethernet connection for transporting data between patient monitors and EHR servers. Due to the wiring requirement, they are typically used in emergency rooms and intensive care units for patients who are in critical conditions and generally do not move around. There are also products that can support wireless data transport to allow patient mobility. A number of examples of such kind of products are presented in Fig. 1. The IntelliVue MX40 wearable patient monitor (Fig. 1a) manufactured by Phillips Healthcare [14] can continuously traces ECG, SpO2, and other vital signs of ambulatory patients in real-time when they are on the move. IntelliVue smart-hopping wireless medical telemetry service network is used to transport data from this monitor to a gateway which connects to the medical data center. IntelliVue X2 and MP2 portable patient monitors (Figs. 1b and 1c, respectively) can measure ECG, SpO2 and NIBP and transmit data wirelessly to the information center over Intellivue clinical IEEE 802.11/WiFi-based network. The Infinity®M300 patient-worn monitor (Fig. 1d) manufactured by Draeger [15] utilizes the Infinity IEEE 802.11/Wi-Fi-based network for wireless telemetry. Patient SafetyNet (PSN) from Masimo [16] is a communication platform integrated with Masimo SET pulse oximetry to support continuous surveillance monitoring for patients and automatic alerts when a clinician is not at the bedside. It employs WiFi technology for data connectivity. Micropaq manufactured by Welch Allyn [17] is another wearable patient monitor. It reports ECG and SpO2 measurements to central monitoring stations using FlexNet Wi-Fi based wireless network.

2 Contribution and Outline of This Chapter

For commercial products surveyed in the previous section, a very important observation is that proprietary communication protocols, technologies and software have to be purchased and deployed. This makes these products quite expensive and it is almost impossible to customize and develop them for specific needs. Besides, proprietary concerns limit the inter-operability of devices and software products from different vendors. Maintenance, extension and upgrade of this kind of products have to be loyal to a given manufacturer/vendor. In order to address these limitations and issues, this chapter presents the design and implementation of an intelligent VS capture platform, namely iVS, where wireless communication technologies are used jointly with personal smart mobile devices that act as the primary point of interactions for medical staff. The platform provides an end-to-end solution for fully-automatic patient VS capture in order to mitigate mistakes, reduce latencies and simplify associated workflows. Information is transferred from patient monitors to medical staff's personal mobile devices, then to end-users and vice versa with minimal manual operations. Additionally, by using personal smart mobile devices, the designed platform offers medical staff full mobility and convenience. These devices are small and can fit into pockets or hand bags and thus allow medical staff access to EHR at anytime and from anywhere for fast and efficient responses to emergency situations. Another beneficial feature of this platform is that it attempts to provide connectivity to various EHR software platforms and to give a unified user interface to patient monitors from different manufacturers. This helps to save time and costs in system integrations and to boost the efficiency of clinical and administrative workflows.

The rest of this chapter is organized as follows. Section 3 presents the overall architecture of the communication network enabling mHealth applications. Technical details of various wireless technologies that can be employed to build up the mHealth communication network are also discussed. Sections 4-6 present technical details of our designed and implemented mHealth wireless communication platform, iVS, for capturing patient VSs and making them accessible to authorized medical staff everywhere. Specifically, Section 4 first presents the network architecture and system design of iVS. Next, Section 5 elaborates machine-to-machine (M2M) communication protocols used for reliable data exchanges between involved devices and sub-systems in iVS. Workflows associated with iVS for VS capture and patient file access are then presented in Section 6. Experiment results related to the performance of iVS are also summarized and discussed. Section 7 presents two case studies. Future directions on communication technologies and services for mHealth are then addressed in Section 8. Section 9 concludes this chapter.

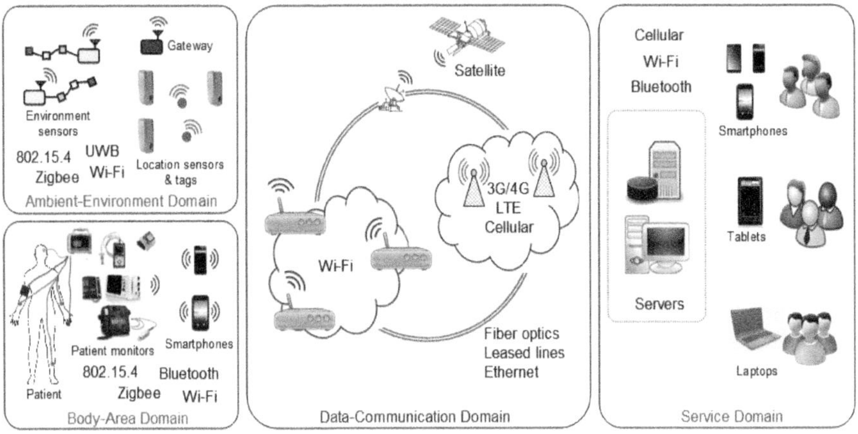

Fig. 2 Overall communication network architecture for mHealth

3 Communication Infrastructure for mHealth: An Overview

3.1 Network Architecture

A typical mHealth system is an integration of four domains: the body-area domain, ambient-environment domain, data-communication domain and service domain, as illustrated in Fig. 2. The following section explains in details the components and the functional roles of each domain as well as the interaction among them.

3.1.1 Body-Area Domain

The body-area domain is defined by a number of wireless body-area networks (WBANs), each corresponding to a patient. A WBAN is composed of a few sensors and a gateway. Body sensors are the essential parts of the WBAN. They are placed in, on, or around a patient's body for monitoring of physiological conditions such as heart rate, blood pressure, body temperature, respiration rate. Depending on patient medical condition or nature of the disease, the monitoring could be performed continuously or on a regular basis. A smartphone can be used as a gateway which relays sensor information between WBAN and other domains. The primary role of this domain is to supply the information about what is happening with the patient at specific time instances. From this information, the trend of VSs over time can be obtained to diagnose some specific disease (e.g., a curve of heart rates over a period of 24 hours can give a detailed understanding of the patient heart condition) or to measure the response or reaction of the patient to a treatment plan.

3.1.2 Ambient-Environment Domain

This domain consists of a number of sensors and a gateway that are deployed in patient's surroundings to monitor the environmental conditions such as air temperature, humidity, brightness, etc. Indoor real-time locating system (RTLS) used to locate users and assets can be also included in this domain. The data provided by the ambient sensors are important. Coupled with patient's physiological conditions measured by the body sensors, they provide valuable supplementary information for medical diagnosis and treatment as well as contexts for emergency responses. For example, RTLS can pinpoint the current location of the patient who is having a stroke so that emergency nurses can find him/her quickly for an efficient and timely life-saving intervention.

3.1.3 Data-Communication Domain

Data from body and ambient sensors needs to be relayed to servers for processing and then displaying to end-users (physicians, nurses, patients) or automating some programmed routines. This domain is mainly responsible for transporting such data from gateways in body-area and ambient-environment domains to servers in the service domain, usually over a large geographical area. The communications in the opposite direction also needs to be supported by this domain, e.g., a control message issued by an end-user can be transported over the communication network to a given sensor device to configure the data sampling rate or start/stop a measurement. Besides, this domain provides services to end-users by enabling two-way information exchange between servers to end-users' mobile devices in the service domain.

3.1.4 Service Domain

Numerous servers for managing medical/ambient sensor data and various types of information related to patients, prescriptions, medications, etc., processing data/information, providing user interfaces, etc., are major components of this domain. Those servers can be hosted in the same machine in a specific location or in a number of networked machines in a cloud-based data center. This domain is also responsible for providing services to end-users for real-time or near-real-time information access for diagnostic, treatment, emergency responses, etc, through various types of communication technologies and networks.

3.2 Communication Technologies

Various types of communication technologies can be employed to provide data connectivity enabling mHealth applications. Those technologies can be categorized depending on their supporting data rate, communication range, power consumption, etc. Following is a summary of a number of representative technologies that can be used to implement mHealth communication infrastructure.

3.2.1 Short-Range Radios

IEEE 802.15.1/Bluetooth [18] is a very well known technology for communications between closely located devices as a substitute for data transfer cables. This standard defines the physical (PHY) and medium access control (MAC) layers for wireless communications over a short range (typically, a few tens of meters) and with low power consumption (from less than 1 mW up to hundreds of milliwatts). It operates in the industrial, scientific and medical (ISM) frequency band from 2.4 to 2.48 GHz and can support high data rates of a few Mbps.

Bluetooth low energy (BLE) [19], marketed as Bluetooth Smart, is intended to provide considerably reduced power consumption and lower cost, compared to classic BT. It aims to facilitate a wide array of emerging applications with smaller form factor devices. By optimizing the protocol design for low-duty-cycle types of applications, BLE consumes a fraction of the power of classic BT. The connection setup time of BLE is only a few milliseconds, compared to several seconds of BT. BLE is designed for transporting small bursts of data and it has a data rate of 1 Mbps.

Near field communication (NFC) [20] is a set of very short-range radio transmission standards for smart phones and similar devices to establish communication with each other by touching them together or bringing them into close proximity. It operates at 13.56 MHz on ISO/IEC 18000-3 air interface and at rates ranging from 106 to 424 Kbps. Applications of NFC include contactless transactions, data exchange, and simplified setup of more complex communications such as Wi-Fi and BT/BLE.

3.2.2 Wireless Sensor Networks (WSNs)

IEEE 802.15.4 is a commonly-used standard for WSNs that enable monitoring and control applications [21, 22]. It operates on 860 MHz, 900 MHz and 2.4 GHz frequency bands that can support data rates of up to 250 kbps. The MAC layer uses carrier sense multiple access with collision avoidance (CSMA-CA) so that multiple nodes can share the same channel. Low cost, low power consumption and small form factor are main advantages of this technology. However, the typical radio communication range is only 10 to 75 m and nodes can only communicate in two simple models: point-to-point or multiple-point-to-point (star). Therefore, the coverage and flexibility of IEEE 802.15.4-based networks are quite limited. For more sophisticated communication models, a network layer which enables data routing between nodes is required. ZigBee [22], wireless highway addressable remote transducer protocol (WirelessHART) [23] and IPv6 over low power personal area network (6LoWPAN) [24] are technologies that supply network and higher layers running on top of IEEE 802.15.4 PHY and MAC layers.

ZigBee enhances the IEEE 802.15.4 standard by additionally including authentication with valid nodes, encryption for security, and mesh routing layer. Nodes can communicate with any other node even in cases they are not within the range of each other by relaying the transmission through multiple additional nodes. The network can thus spread out over a larger area and it is also resilient to link/node failures. Another compelling feature of ZigBee is the availability of profiles developed for various applications.

3.2.3 Wireless High-Speed Local Area Networks (WLANs)

IEEE 802.11/Wi-Fi [25] is the most popular and successful technology for wireless local access that allows portable and stationary consumer electronics devices to exchange data over a computer network. It is based on IEEE 802.11 standards that specify over-the-air interfaces between a wireless client and an access point (AP) or between two wireless clients. While IEEE 802.15.1 and 802.15.4 standards are oriented towards low-power and low-data rate applications, IEEE 802.11 standard is primarily designed for broadband connections, as an extension of or substitution for cabled LANs.

There are several specifications in the 802.11 family. The original 802.11 provides 1 or 2 Mbps transmission in the 2.4 GHz band. 802.11a/b/g/n have been introduced to boost the rates. For example, 802.11n can support a few hundreds of Mbps. 802.11e adds quality-of-service (QoS) features and multimedia support. 802.11ac is built upon the 802.11n standard to deliver data rates of 433 Mbps per spatial stream, or 1.3 Gbps in a three-antenna design. 802.11ad operating in the 60 GHz frequency band is under development and it has a theoretical maximum transfer rate of up to 7 Gbps. Since the communication range of Wi-Fi devices is short, typically from 30 m to 100 m, WiFi networks in some cases are configured in a mesh topology to extend their coverage as specified in IEEE 802.11s standard.

3.2.4 Wide Area Networks (WANs)

IEEE 802.11 Wi-Fi mesh can be deployed to cover large geographical areas for WANs. In fact, it has been a widely used large-scale back-haul mesh networking solution. A mesh is resilient to network element failures and requires low maintenance. In addition, this kind of network overcomes the line-of-sight issues that may occur when a space is crowded with buildings or industrial equipment.

Cellular communications [26] is the most successful and widely used wireless communication technologies. It has experienced exponential growth in the past two decades with global service coverage and billions of users. Although, cellular communications was originally developed for voice services, it has been evolving and therefore nowadays it has been extensively used for a wide range of data services. Most cellular systems operate on the licensed frequency spectrum of 824-894 MHz/1900 MHz. Data transmission rate of this technology varies from a few tens of kbps to hundreds of Mbps, depending on technologies, communication distances and deployment scenarios.

Long term evolution (LTE) [27], the fourth-generation (4G) cellular, has introduced a number of new technologies when compared to the previous cellular systems. These technologies include frequency-division multiplexing (OFDM), orthogonal frequency-division multiple access (OFDMA), single-carrier frequency-division multiple access (SC-FDMA), multiple-input and multiple-output (MIMO), system architecture evolution (SAE), etc. They significantly improve the system performance. Peak data rates for downlink and uplink that LTE can achieve are 100 Mbps and 50 Mbps, respectively. Latency of round trip packet transmission is around 10 ms. Towards the truly 4G cellular technology, LTE advanced (LTE-A) has been developed. MIMO evolves further and more advanced techniques with additional antennas to enable additional paths are used. Compared to LTE, LTE-A exhibits a number of advantages. Its peak data rates for downlink and uplink are 1 Gbps and 500 Mbps, respectively. Its spectrum efficiency is three times greater than that of LTE. LTE-A is capable of supporting scalable bandwidth use and spectrum aggregation where non-contiguous spectrum needs to be used. The latency is less than 5 ms one way for individual packet transmission.

It is noted that LTE is not the only candidate 4G technology. Mobile worldwide interoperability for microwave access (WiMAX) [28] is also promising since it can offer very high data rates and high levels of mobility. However it seems less likely that WiMAX will be adopted, as compared to LTE. One of the main reasons why LTE is appearing to be better positioned is that LTE has been designed to be well integrated with existing cellular technologies for smoother and more economical transitions.

Besides, satellite communications [29] provides a microwave radio relay technology complementary to that of communication cables. They are also used for mobile applications such as communications to ships, vehicles, planes and hand-held terminals. Dependent on the type of satellite communication system and the orbits used, it is possible to provide complete global coverage. As a result, satellite communication systems are used for providing communication capabilities in many remote areas where other technologies would not be viable. This technology can support data rates ranging from 9.6 Kbps to a few tens of Mbps. However, very high cost and significant communication latency are the two key disadvantages of this technology.

3.3 Candidate Technologies for Each mHealth Domain

Fig. 3 summarizes and compares the above-mentioned communication technologies from different perspectives including supporting data rate, coverage, mobility, indoor/outdoor use, power consumption and operating frequency band. For body-area domain, short-range communication technologies including NFC, BT/BLE, IEEE 802.15.4 and its derivatives are found to be the most suitable. These technologies support low data rates over short distances, however, they are optimized for ultra low or low power consumption which is the key requirement for this domain where devices are battery-powered and cannot be recharged often. Multiple technologies may need to be integrated together to enable mHealth applications that offer mobility, ubiquity and convenience to end-users. IEEE 802.15.4-based technologies and infrastructure Wi-Fi can be used for the ambient environment to give connectivity to the back-haul. When larger areas need to be covered for the data-communication domain, various technologies appear to be good candidates. Wi-Fi mesh can establish a network which can support relatively high speed at a reasonable cost. LTE/LTE-A can offer global broadband coverage, however, costs related to frequency spectrum and system operation and maintenance are much higher than those of Wi-Fi mesh. WiMAX is another solution for long-range and high-speed communications but it is less likely to be widely deployed and used due to its high costs and it cannot be smoothly integrated with previous cellular networks. In some special scenarios, e.g., in remote areas, airplanes, ships, satellite communications is the only feasible technology even though it is apparently quite expensive and communication latency could be an issue.

It is noted that, regardless of the fact that cable communication technologies (e.g., Ethernet, digital subscriber line, leased line, fiber optics, etc.) are not mentioned in this section, they should be employed whenever they are accessible (except the cases when portability or mobility is required) due to the fact that in general they can offer higher capacity and higher levels of stability, reliability and security as compared to their wireless counterparts. For example, new buildings/institutions often have gigabit fiber links that can be used for data back-haul.

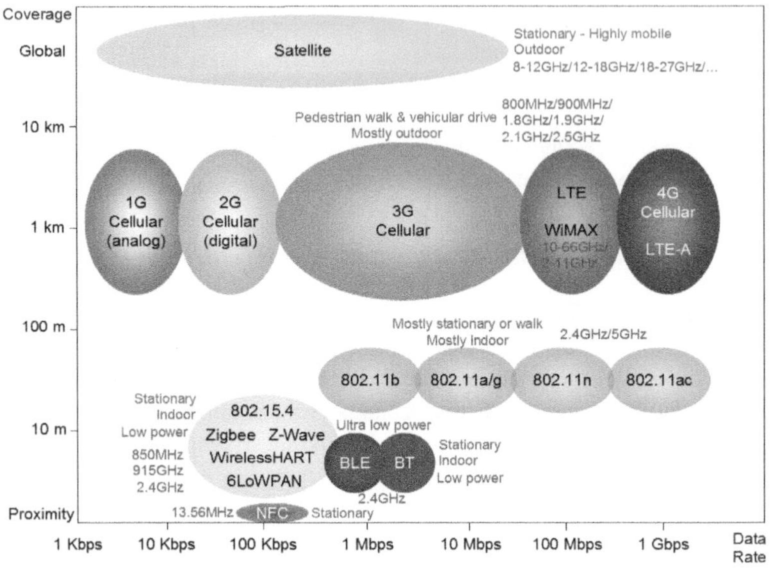

Fig. 3 An overview on wireless communication technologies

4 Network Architecture and System Design of iVS

4.1 System Requirements

As mentioned in Section I, the main objective of iVS is to transform the process of monitoring patient VSs from manual to automatic operations in order to avoid mistakes, enhance reliability, reduce delays and thus improve healthcare service quality in clinical settings. Therefore, iVS should be designed with following primary requirements. In the first place, iVS must reliably and quickly capture patient VSs with minimal manual operations. Besides, from an economical perspective, conventional patient monitors currently in use should be integrated with iVS and continue their services since they are quite expensive (several thousand dollars per unit) and their service lifetime is typically 5 to 10 years. In order to further reduce the capital expenditure, hardware design and fabrication required for iVS should be minimized. Also, in order to simplify the workflow associated with iVS (which in turn reduces training, maintenance and operation expenses), a universal user interface that allows iVS to communicate with different conventional patient monitors made by different manufactures should be developed. Furthermore, the time spent by a medical staff to monitor each patient by using iVS should be similar to or faster than that by using manual operations.

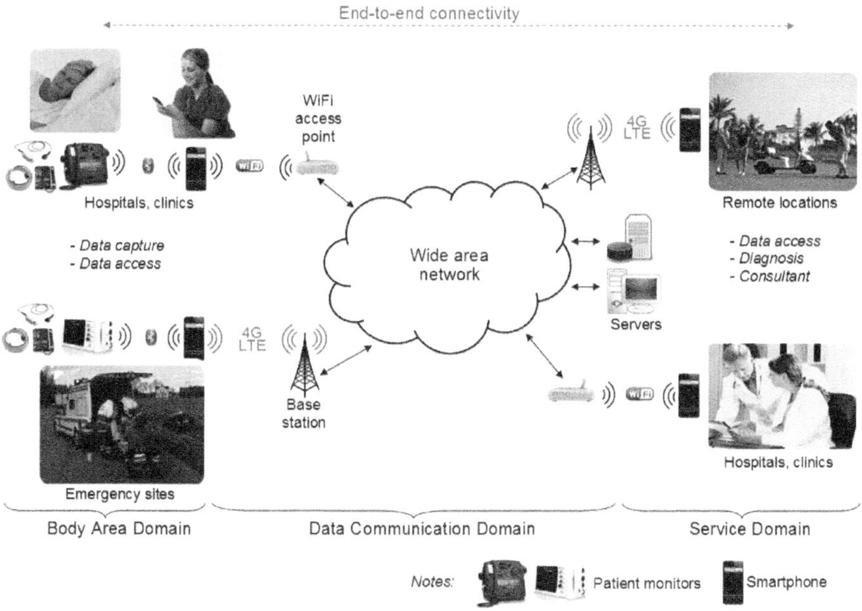

Fig. 4 iVS overall network architecture

4.2 *System Components and Network Architecture*

As sketched in Fig. 4, iVS encompasses the following components: patient monitors, personal smart mobile devices, EHR and web servers. Patient monitors are medical devices that can measure one or all of the following VSs: TEMP, NIBP, HR, SpO2, RR, and so on. These monitors are conventional devices currently used in hospitals, clinics and ambulances. General Electric (GE) DINAMAP PRO 400 [30] is one of the monitors used for development and preliminary experiments. However, iVS is designed to be able to work with a wide range of devices from different manufacturers. Personal smart mobile devices can be smartphones, phablets, tablets and so on that are generally built with powerful micro-processors, high-quality screens and various communication interfaces including BT, Wi-Fi and 3G/4G cellular radios. In order to simplify the presentation and explanation, the smartphone is mentioned throughout this chapter since it is the most representative and widely-used personal smart mobile device. EHR server is basically a database server for EHR where all information (demographic data, VS measurements, medication/treatment records, etc.) related to patients is managed. Web server is connected to EHR database and provides interactive web interfaces that allow users to query or update data for e-health applications. These two servers are hosted in a computing cloud. It is noted that smartphones provide primary communication and application interfaces that allow medical staff to interact with patient monitors and servers.

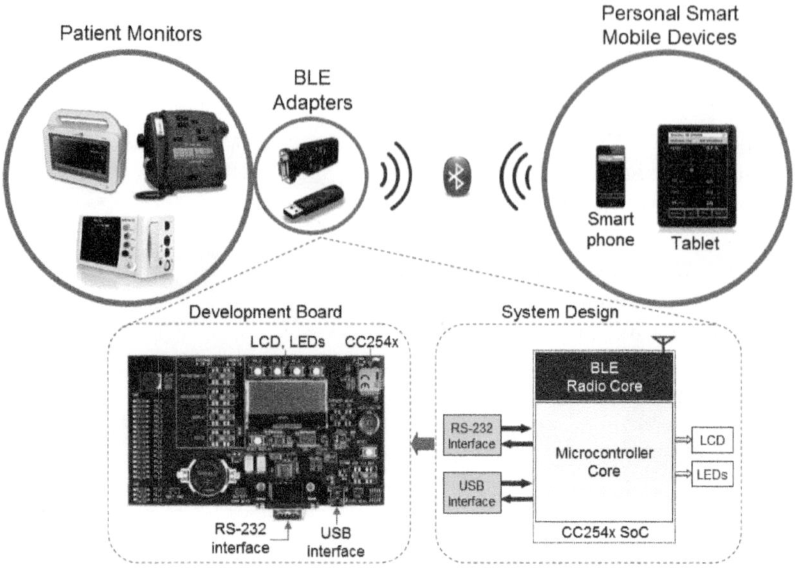

Fig. 5 Communications between a patient monitor and a smartphone and the implementation of BLE adapters

Basically, patient VSs measured by a patient monitor are wirelessly captured by a smartphone which in turn records them in EHR server. In order to ensure reliability of the system, medical staff at first need to explicitly pair the smartphone with the intended patient monitors (on the patient bedside) and then to validate and verify every measurement captured from the paired patient monitor. In the opposite direction, the smartphone can accept control information from medical staff and send respective queries and instructions to patient monitor. Once the information is validated by medical staff, the smartphone transmits this information to the patient file in EHR server and the recorded information must be verified again by the medical staff in order to avoid any mistake or error due to data transmission/recording. Information in patient files is available to any authorized medical staff by using a smartphone accessing EHR server via user web interfaces supplied by web server.

Fig. 4 illustrates that VS capture routine can be carried out in various settings including hospitals, clinics, or even in arbitrary emergency sites outside of hospitals. Data captured from the patients is updated to EHR immediately, and thus far-away physicians can access their patients' medical conditions in a near-real-time fashion. This results in life, time and cost savings. There are various local wireless technologies (as investigated in subsection 1.3.2) that could be used for establishing the short-range communications between a patient monitor and a smartphone. Classic BT and BLE are among the best candidate technologies for this link. As presented in Section II, BLE is superior to classic BT due to the fact it is specifically designed for extremely low-power consumption and low-data-rate applications. Especially, the pairing protocol for BLE is also simplified and optimized in such a way that the connection setup time is much faster than that of classic BT, i.e., only a few milliseconds compared to several seconds. These advantages of BLE match well with the various iVS requirements that are addressed in the previous subsection. Additionally, using BLE technology, there is no need to have any extra hardware on smartphones since BLE interface is built in a large number of smartphones. This is not possible with other short-range communication technologies including ZigBee, RF4CE, WirelessHAR and 6LoWPAN. As a result, BLE technology is selected to realize the communications between patient monitors and smartphones. For the communications between smartphones and EHR server, Wi-Fi and 3G/4G cellular technologies are employed. This selection is quite straightforward since hardware interfaces for these two technologies are always integrated in smartphones. Wi-Fi is a well-known and matured technology for high-speed wireless local-area connectivity while 3G/4G cellular offers global coverage, high data rate and reliable wide-area communications.

4.3 System Design of BLE Adapter

The communication link between a pair of patient monitor and smartphone is zoomed in Fig. 5. Since almost all patient monitors currently used in hospitals are not equipped with BLE radio communication interface, a BLE adapter is developed to bridge patient monitors and smartphones. There are two interfaces integrated in each BLE adapter: a serial interface (either RS-232 or USB) and a BLE interface to enable the communications between a patient monitor and the adapter and that between the adapter and a smartphone, respectively. System design and development board of BLE adapters are presented in the lower part of Fig. 5. As can be seen, the main component of each BLE adapter is a Texas Instruments CC2540/CC2541 system-on-chip (SoC) which consists of a 8051-compatible micro-controller and a BLE transceiver. To simplify the design and operation of the adapter, power supply for the adapter is obtained directly from 5-volt power pins driven by patient monitor serial interface. For development purpose, a liquid-crystal display (LCD) and other peripherals are also included in the development board. However, the final packaging of BLE adapters can be as small as a RS-232/USB dongle which is plugged into the serial host connector in the back of each patient monitor. There are two light-emitting diodes (LEDs) showing status of the adapter: the red one for power supply ("off" means no power supply, "solid red" means power supply is alright) and the green one for BLE connectivity with smartphone ("off" means no connection, "solid green" means the connection pair is established, "flashing green" indicates transmission/reception of information).

5 M2M Communication Protocols of iVS

The iVS involves different M2M communications between various devices to enable end-to-end connectivity between patient monitors and servers and vice versa. As presented in the previous section, the connectivity between patient monitor and smartphone is established with BLE adapter as a relay and therefore it can split into two segments: (i) patient monitor-BLE adapter over RS-232 or USB serial link using host communication protocol specified by patient monitor and (ii) BLE adapter-smartphone over BLE link using BLE Generic Attribute Profile (GATT), namely iVS GATT, designed in this work. The connectivity between smartphone and (EHR server, web server) traverses various physical networks including Wi-Fi, 3G/4G cellular, Ethernet, Internet, etc. However, only one application data exchange protocol is required. In iVS, health level-7 (HL7) protocol specified by the Health Level Seven International [31] is employed for this connectivity.

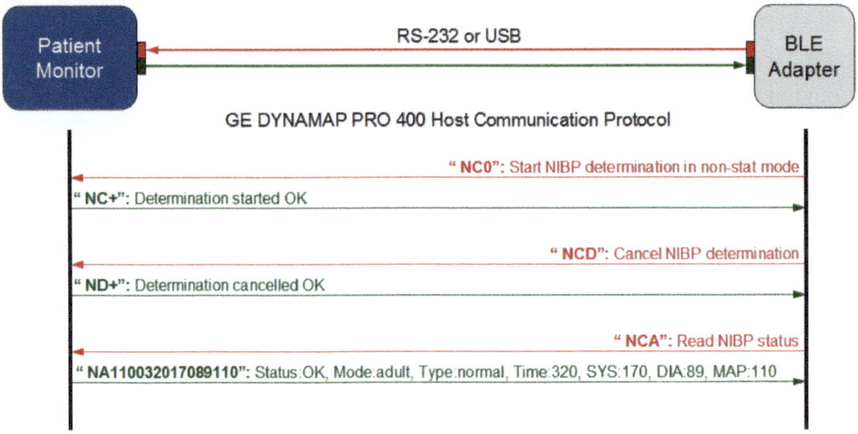

Fig. 6 Host communication protocol of patient monitors

5.1 *Patient Monitor Host Communication Protocol*

Each patient monitor is integrated with a serial port (could be either RS-232 or USB) which allows data communications with a host. A host communication protocol is specified by the manufacturer to ensure that the patient monitor and the host can understand each other. Each device manufacturer has their own proprietary host communication protocols; however, they usually share a few design principles. First, communications is based on a query-reply model and text/binary messaging. Second, there are three typical types of messages: sensor control (e.g., start/stop NIBP determination), data query/reply (e.g., read results of the latest NIBP measurement) and device information/configuration/status (e.g., read device manufacturer, read device model, setup serial data rate). As an illustrative example, Fig. 6 shows a number of sample messages related to NIBP sensor in the GE DYNAMAP PRO 400 patient monitor [32]. The host in the figure is actually BLE adapter in iVS. In the first example, BLE adapter may start an NIBP determination in non-state mode by sending "NC0" text message to the monitor which then replies with "NC+" text message indicating that NIBP pump is started and determination is in progress. In the second example, BLE adapter may cancel an in-progress NIBP determination by sending "NCD" text message to the monitor which then replies with "ND+" text message after the NIBP pump is stopped and determination is aborted. In the last example, after an NIBP determination is done properly, the measurement results can be queried by sending "NCA" text message to the monitor, then various parameters are returned from the monitor: status, type of measurement, time stamp, values of systolic, diastolic and mean arterial pressure (MAP). In the iVS prototype, a smartphone communicates with BLE adapter to allow medical staff to interact with patient monitors.

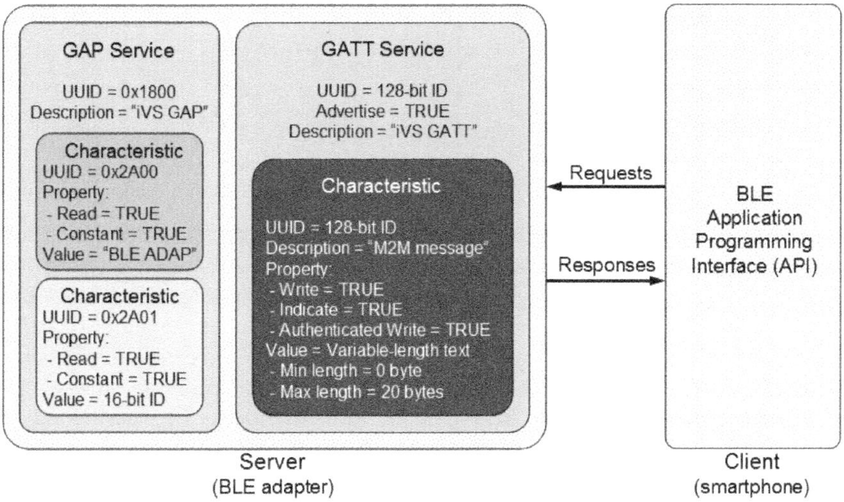

Fig. 7 GATT and GAP for the BLE link

5.2 *BLE iVS Generic Attribute Profile*

As addressed in subsection 1.4.1, one of the main requirements of BLE technology is device inter-operability. In order to assure this, a standardized wireless protocol to transfer bytes of information must be used, and data representation levels must be shared as well. In other words, the two devices that want to exchange information have to send or receive data in the same format using the same data interpretation based on intended device functionality. The communications between BLE devices is done via Generic Attribute Profile (GATT) which is a bridge between the wireless protocol stack and the application and functionality of the device. GATT defines basic elements including profiles, services and characteristics. The top level of the GATT hierarchy is a profile that is composed of one or more services. A service contains a number of characteristics or references to other services. Each characteristic is described by a value and optional descriptors about the value. The service, characteristic and the components of the characteristic (i.e., value and descriptors) are all stored in an attribute server.

BLE adapter is designed to work as a relay between patient monitors and smartphones. In order to enable smartphones to communicate with various types of

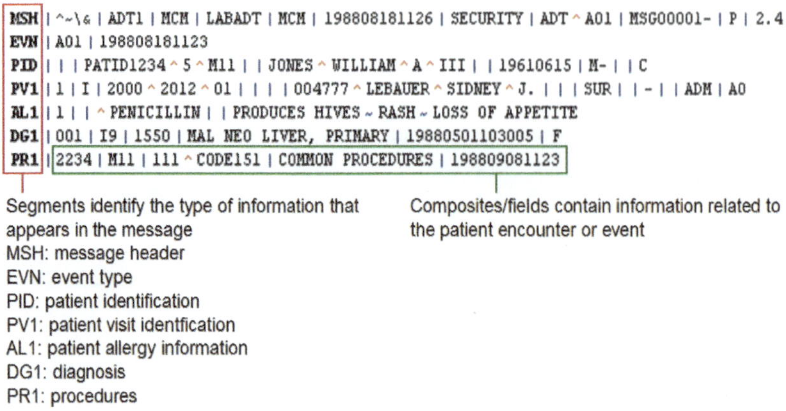

Fig. 8 An example of HL7 message

patient monitors that have different data communication protocols, BLE adapter is transparent to data that it relays. Semantics of exchanged information are only translated and understood by patient monitors and smartphones. As a result, GATT for BLE link is quite simple. First a 128-bit universally unique identifier (UUID) is defined to specify BLE Medical Device service. This UUID is randomly generated according to RFC 4122 [33] so that it would be hard for unauthorized persons to connect to the device. It could be interpreted as an extra layer of security: only the application which was made specifically with these smartphones will be able to connect to them. The service is also included in the advertisement packet when the device is advertising. This enables smart-phones to find a particular BLE adapter faster. There is only one characteristic included in this service which contains the actual data to be exchanged between two devices. "Write" property is enabled to allow the characteristic on one device to be written by the other device. "Indicate" property is enabled so that the value of this characteristic on one device is actively reported to the other device whenever it is changed. Therefore, there is no need for periodic polling and this gives significant power saving for smartphones. Indication also provides acknowledgements of data; hence the data is always transmitted reliably.

The GATT designed for BLE link in iVS is summarized in Fig. 7. The value of characteristic has a variable length of up to 20 bytes which is the maximum data length in BLE characteristics with indication. GATT employs client-server communications to allow an attribute server to expose a set of characteristics and their associated values to an attribute client. These characteristics can be discovered, read, and written by peer devices. The servers have data and clients request data to/from servers where data is encapsulated in "services" and exposed in "characteristics". In iVS, BLE adapter and smartphone work as GATT server and client, respectively. Fig. 7 also shows the generic access profile (GAP) designed for BLE communications. The most important field in the GAP is a 16-bit identification which allows iVS application on the smartphone to recognize each detected adapter in the pairing procedure.

5.3 HL7 Protocol

Hospitals and other healthcare provider organizations typically have many different computer systems used for everything from billing records to patient tracking. HL7 provides a number of flexible standards, guidelines, and methodologies for data exchange to allow inter-operability between those systems. HL7 works in the application layer protocols and it is independent of lower layers. Since HL7 has become the dominant interface protocol in healthcare, it is selected to implement the communications between a smartphone and an EHR server.

A HL7 message is composed of a series of segments, each of which identifies the type of information the message contains (e.g., diagnosis, insurance, observation result). In turn, each segment includes one or more composites (or fields) that carry the actual information being conveyed (e.g., name, test value, dosage). In short, an HL7 message is defined by its structure as well as by the information that it contains. A HL7 message is exemplified in Fig. 8. The acknowledgment (ACK) protocol is a key element of HL7's utility in a healthcare setting. Once an HL7 message is received, the recipient system returns an ACK with message status (received and processed, error detected, or rejected). If the application that initiated the communication does not receive an ACK, it will continue to send the message. Only when an ACK message is received will it stop transmitting and take corrective action when necessary.

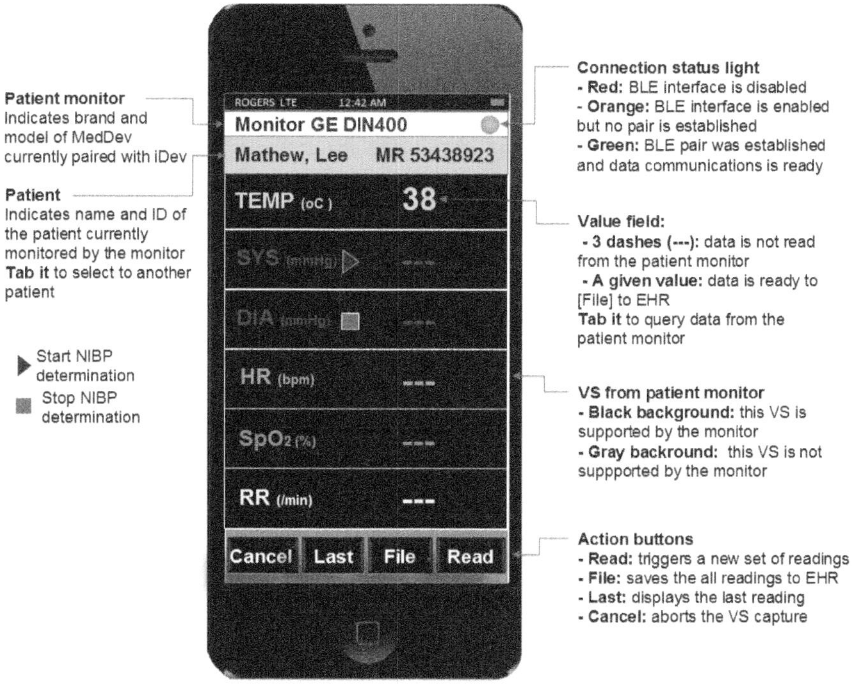

Fig. 9 The main GUI of iVS on the smartphone

6 iVS Workflows and Experiment Results

6.1 Capturing Patient VSs

iVS can be used in various scenarios in hospitals, clinics, emergency sites, etc. In this section, clinical workflow to capture VSs of a patient in the hospital is presented as an illustration. Similar workflows with minor modifications can be applied in other scenarios. Each iVS user initially needs to register and setup their own usernames and passwords that managed by the system administrator. They are granted a given type of system access privileges (e.g., read-only, read and write…, within given clusters of database and time slots) by the institution manager depending on the nature of their jobs and management policies. VSs of patients admitted to the hospital for observation or treatment need to be monitored periodically by nurses (e.g., every 2 or 8 hours depending on health conditions and diseases of the patients). The nurse is assumed to be at the bedside of the patient to be monitored. Upon launching the smartphone iVS application, the nurse logs into the system using his/her registered username and password. A data connection between iVS application and EHR server is then established. The iVS application downloads patient database from EHR server and allows the nurse to select the patient of interest from the GUI. The patients can be grouped by departments, first/last names, birth dates, and so on for quick searching and selection. Once the patient identification is confirmed, iVS application automatically attempts to pair the nurse's smartphone with the patient monitor which is currently assigned to the patient. The main GUI of iVS is then displayed to support VS capture for the selected patient as depicted in Fig. 9. After the nurse attaches all sensor probes properly to patients, measurements of a number of VSs (e.g., TEMP, SpO2, HR, etc.) are automatically initiated by patient monitor. NIBP determination needs to be started by the nurse by clicking NIBP [Start] button in the GUI. Once all VS measurements are performed properly, all corresponding readings can be captured by clicking [Read] button. VS measurements captured from patient monitor are instantly displayed in the GUI. After verification and validation, the nurse saves all readings to the file of patient in EHR server by clicking [File] button. The nurse needs to verify that data has been accurately recorded to the associated patient file. Each individual VS measurement can be measured and recorded separately when needed as explained in Fig. 9. Details of this workflow and related information exchanges are depicted in Fig. 10. It is noted that the same GUI and workflow can be used with a wide range of patient monitors from different manufacturers. This results in users' convenience, cost and time saving in system operation and maintenance.

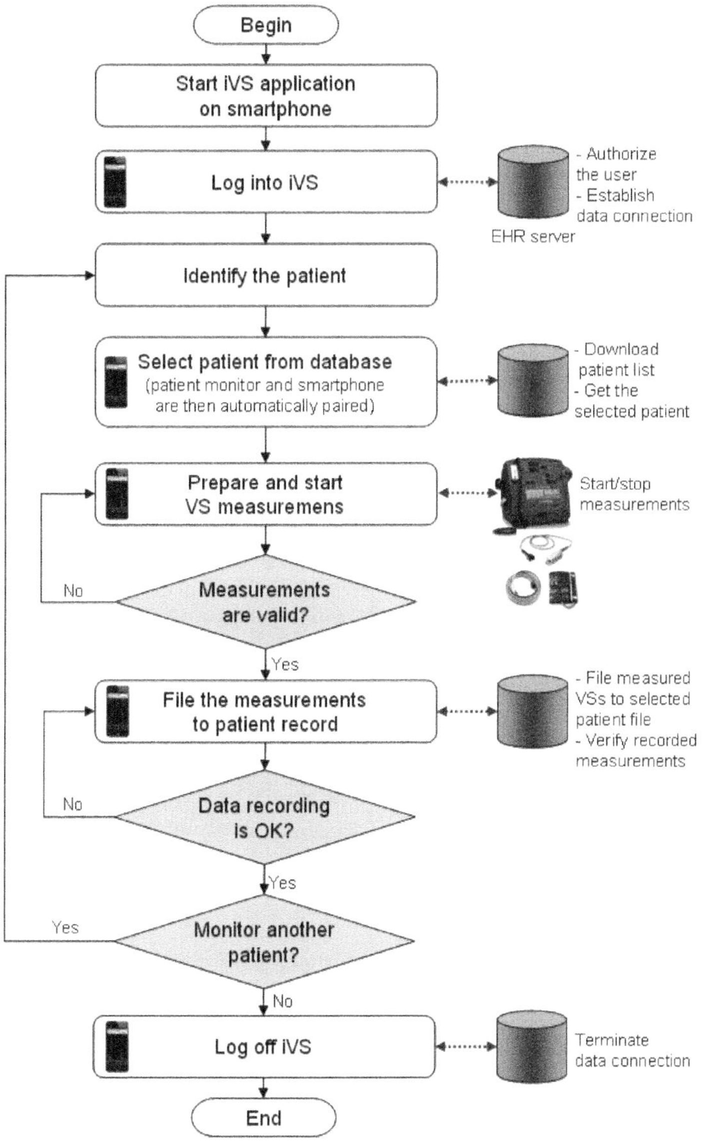

Fig. 10 Patient VS capture procedure and related information exchanges

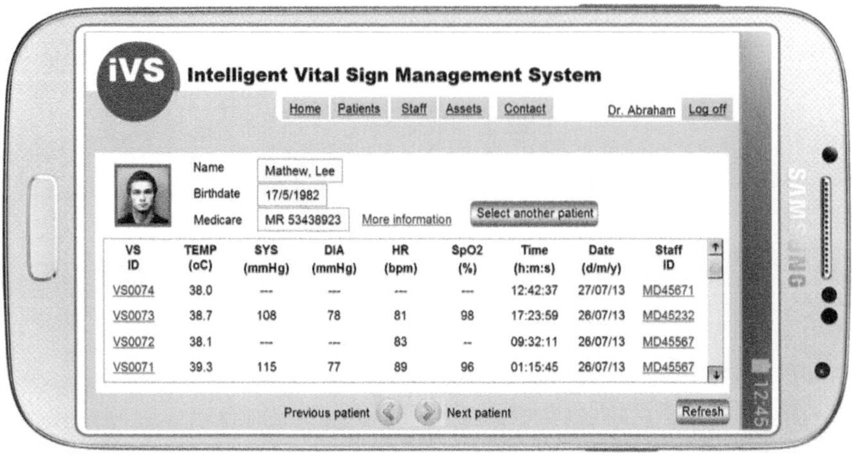

Fig. 11 The end-user GUI allows near-real-time patient monitoring

6.2 Workflow for Accessing Patient Files

Information captured by iVS is managed by EHR server and exposed by web server to physicians at any location with their smartphones. After opening the iVS web interface in smartphones, username and password are required to log into the system. The profile of each end-user determines corresponding privileges to access the EHR. The file of a patient can be pulled out quickly by browsing the list of accessible patients grouped by institutions, departments, wards, or rooms. A wealth of information related to the latest VS measurements, historical VS measurements with their trends, statistics, medical treatment histories, allergies, medications, etc., of the selected patient can be displayed in iVS web GUIs to help the end-users in making diagnosis, treatment plan, prescriptions, etc. A GUI displaying latest VS measurements of a given patient is demonstrated in Fig. 11.

6.3 Experiments

We have carried out experiments to investigate the user-friendliness of GUIs and performance of the communication link between a GE DINAMAP PRO 400 patient monitor [30] (supplemented with a BLE adapter) and an iPhone5 running iOS6. The iPhone application for iVS employs the Core Bluetooth framework [34] to access the BLE hardware built in the phone. The results show that by using intuitive GUI design and BLE technology, a patient of interest can be selected easily and then within a fraction of second the pair between iPhone and the corresponding patient monitor is automatically established. The reliability of this link is measured when the message rate is swept. It is observed that when there are no more than 10 message exchanges per second (the corresponding application data throughput is around 2.15 kbps), data transport over this link is very reliable (i.e., 100% of messages are successfully delivered). However, when higher message rates are attempted, BLE cannot ensure the delivery. This phenomenon can be explained by following two main reasons. First, iOS restricts the number of messages that can be sent in a single connection event (defined by I_{conn} in BLE protocol stack [35]) to 4 and I_{conn} is limited and cannot be controlled by user applications under the current iOS versions. These limitations are imposed since the BLE radio is shared with classic BT and Wi-Fi (for cost and space efficiency) and user applications are not allowed to consume all resources otherwise they disturb the behavior and hinder the performance of other applications. Second, the wireless channel between the two communicating devices is not always reliable considering the fact that frequency spectrum used for BLE radio (2.4 GHz) is unlicensed and quite crowded. This results in packet losses, packet re-transmissions or even pre-matured connection termination in case of the reception of two consecutive packets with bit errors or corruption of the access address field. Fortunately, for iVS applications, communication rates of less than 5 messages per second are always sufficient for all possible interactions between a patient monitor and a smartphone.

Once data is ready in the iPhone, it is recorded to EHR and available to end-users within a few seconds. Therefore, end-users can remotely monitor patients in a near-real-time fashion regardless their locations as long as they have Internet connection over Ethernet, WiFi or cellular networks with their computers, laptops, smartphones, or tablets. Server applications are also implemented with data analyzing modules that can intelligently detect abnormal VSs or trends with a patient and then automatically sends notifications (using mobile phone's short message service or email) to nurses and/or physicians responsible for that patient to trigger fast and efficient actionable responses.

7 Case Studies

Case Study 1: *Assuming that BLE radio is used to exchange application data messages between two devices (e.g., a patient monitor and a smartphone). Calculate the data rate at the BLE physical layer versus application data lengths and message rates.*

First, it is necessary to capture the BLE protocol stack and the structure of BLE messages used for application data transmission. As can be seen in Fig. 12, the BLE protocol stack is composed of two main parts: the controller and the host. The controller comprises physical and link layers, and is typically implemented as a system-on-chip with an integrated BLE transceiver. The host runs on an application processor and includes upper layer functionality, i.e., Logical Link Control and Adaptation Protocol (L2CAP), Attribute Protocol (ATT), Generic Attribute Profile (GATT), Security Manager Protocol (SMP) and Generic Access Profile (GAP). Communication between the host and the controller is standardized as the Host Controller Interface (HCI). Finally, Applications (Apps) is placed on top of the host.

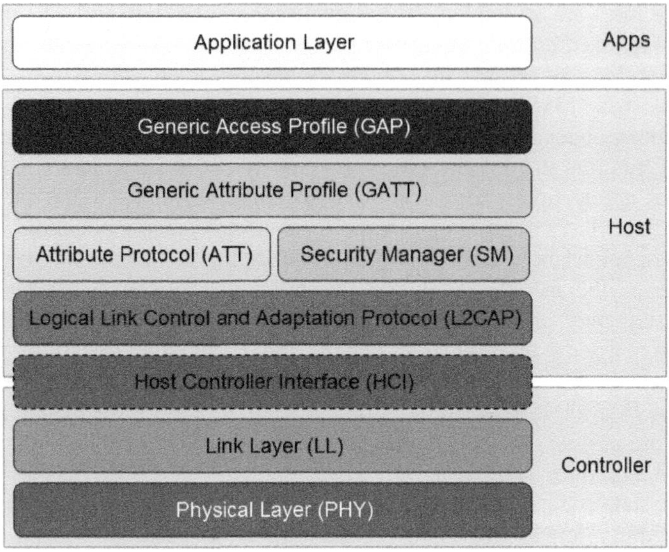

Fig. 12 BLE protocol stack

Next, Fig. 13 depicts the structure and size of each field contributed by each layer to a PHY data unit when application data is transmitted. The total overhead that each application data message carries is then (1 + 4 + 2 + 3 + 4 + 4 + 1) = 19 bytes.

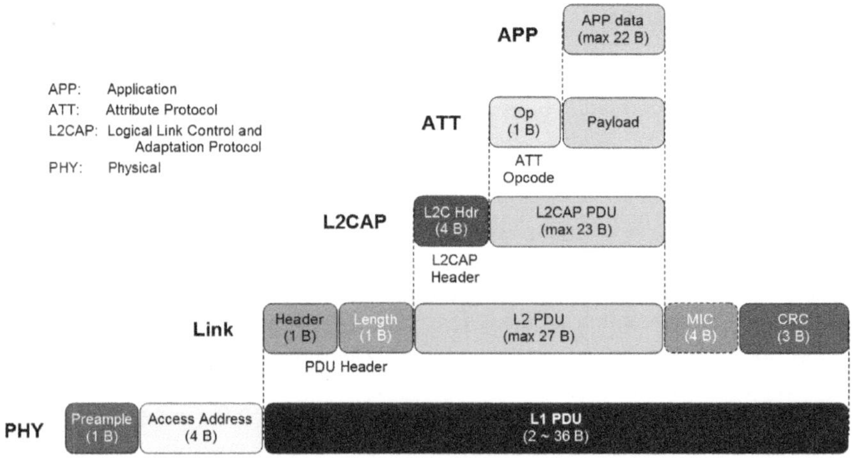

Fig. 13 The format of a BLE data message (note: the size of each field is expressed in bytes)

Finally, denote n (byte) and r (message/s) for the length of application data and the message rate at the APP layer, respectively. Data rate at the PHY layer (denoted by R) is calculated as follows:$R = (n + 19) \times 8 \times r$ (bps). Sampled calculations of R versus n and r are given in Table 1.

Table 1 PHY data rate R (bps) versus application data lengths and message rates

n \ r	0.1	0.5	1	2	3	4	5	6	7	8	9	10	20	30	40	50
1	16	80	160	320	480	640	800	960	1120	1280	1440	1600	3200	4800	6400	8000
2	16.8	84	168	336	504	672	840	1008	1176	1344	1512	1680	3360	5040	6720	8400
3	17.6	88	176	352	528	704	880	1056	1232	1408	1584	1760	3520	5280	7040	8800
4	18.4	92	184	368	552	736	920	1104	1288	1472	1656	1840	3680	5520	7360	9200
5	19.2	96	192	384	576	768	960	1152	1344	1536	1728	1920	3840	5760	7680	9600
6	20	100	200	400	600	800	1000	1200	1400	1600	1800	2000	4000	6000	8000	10000
7	20.8	104	208	416	624	832	1040	1248	1456	1664	1872	2080	4160	6240	8320	10400
8	21.6	108	216	432	648	864	1080	1296	1512	1728	1944	2160	4320	6480	8640	10800
9	22.4	112	224	448	672	896	1120	1344	1568	1792	2016	2240	4480	6720	8960	11200
9	22.4	112	224	448	672	896	1120	1344	1568	1792	2016	2240	4480	6720	8960	11200
10	23.2	116	232	464	696	928	1160	1392	1624	1856	2088	2320	4640	6960	9280	11600
11	24	120	240	480	720	960	1200	1440	1680	1920	2160	2400	4800	7200	9600	12000
12	24.8	124	248	496	744	992	1240	1488	1736	1984	2232	2480	4960	7440	9920	12400
13	25.6	128	256	512	768	1024	1280	1536	1792	2048	2304	2560	5120	7680	10240	12800
14	26.4	132	264	528	792	1056	1320	1584	1848	2112	2376	2640	5280	7920	10560	13200
15	27.2	136	272	544	816	1088	1360	1632	1904	2176	2448	2720	5440	8160	10880	13600
16	28	140	280	560	840	1120	1400	1680	1960	2240	2520	2800	5600	8400	11200	14000
17	28.8	144	288	576	864	1152	1440	1728	2016	2304	2592	2880	5760	8640	11520	14400
18	29.6	148	296	592	888	1184	1480	1776	2072	2368	2664	2960	5920	8880	11840	14800
19	30.4	152	304	608	912	1216	1520	1824	2128	2432	2736	3040	6080	9120	12160	15200
20	31.2	156	312	624	936	1248	1560	1872	2184	2496	2808	3120	6240	9360	12480	15600

Case Study 2: *Consider a scenario where a GE DINAMAP PRO 400 patient monitor (through a BLE adapter) periodically reports vital signs to a smartphone using BLE radio. Four vital signs are individually sent by the patient monitor in each report. Assume that BLE link is ideal (no error and no retransmission), calculate the data rate at the BLE physical layer versus report rates.*

Communication protocol of GE DINAMAP PRO 400 patient monitor is explained as follows. Each message is always started with one-byte start-of-text [STX], followed by the raw text, and then ended by 2-byte check sum [C1][C2] and one-byte end-of-text [ETX]. The format of raw texts is given in Table 2.

BLE "indication" messages with application data of up to 20 bytes are used to carry reports from the patient monitor. Therefore, in each report, the following BLE messages (19 bytes headers are included in each message as explained in Case Study 1) are sent to the phone (Table 3):

Table 2 The format of raw text representing vital signs reported by the patient monitor

Vital signs	Raw text	Notes
Non-invasive blood pressure (NIBP)	"uNAabcddddeeefffggg"	- "a": determination status - "b": adult/neonate status - "c": determination type - "dddd": time since last determination of systolic diastolic and mean arterial pressure - "eee": systolic pressure - "fff": diastolic pressure - "ggg": mean arterial pressure
Pulse oximeter	"uOAabbbcdd"	- "a": channel status - "bbb": oxygen saturation - "c" Signal strength - "dd": averaging interval
Heart/pulse rate	"uRAabbb"	- "a": heart/pulse rate source - "bbb": deart/pulse rate
Temperature	"uTAabbb"	- "a": display format - "bbbb": temperature value

Table 3 BLE messages sent in each report

Vital signs	Application Data	Data length (byte)	Total message length (byte)
NIBP	Message 1: [STX]"uNAabcdddd" Message 2: "eeefffggg"[C1][C2][ETX]	11 12	11+19 = 30 12+19 = 31
Pulse oximeter	[STX]"uOAabbbcdd"[C1][C2][ETX]	14	14+19 = 33
Heart/pulse rate	[STX]"uRAabbb"[C1][C2][ETX]	11	11+19 = 30
Temperature	[STX]"uTAabbb"[C1][C2][ETX]	11	11+19 = 30
	Total report length (byte)		154

As a result, the data rate at the PHY layer (denoted by R) is calculated as follows: $R = 154 \times 8 \times n$ (bps), where n is the report rate. Illustrated values of R versus report rates are given in Table 4.

Table 4 PHY data rate versus report rates

n (report/s)	1	2	3	4	5	6	7	8	9	10
R (bps)	1232	2464	3696	4928	6160	7392	8624	9856	11088	12320

8 Future Directions

With the emergence of various applications including traffic control and smart navigation, smart metering, electronic/mobile healthcare, etc., the number of devices/machines that need wireless connectivity is continuously increasing, changing the conventional perception of communications. Machine-to-machine (M2M) communications will be dominant in the future of Internet. The current cellular networks serving around 6.8 billions mobile-cellular subscriptions [36] might not be appropriate for future M2M applications due to many reasons. First, a vast number of devices, which is expected to be 50 billions [37], requires connectivity. Second, devices are heterogeneous in terms of computing capability, power type and capacity, communication technology and supportable data rate. Finally, M2M communications is characterized by short burst traffic and requires diverse quality of services in terms of reliability, latency and security. For example, in healthcare industry, one of the use cases is the combination of conventional medical telemetry and multi-point high-definition video conferencing. This allows doctors to view health-related data such as vital signs, weight, blood glucose level, etc., (requiring a low data rate but high reliability and security) while holding a video conference with the patient (requiring high bandwidth and low latency). To address challenges of M2M communications, new wireless communication/networking technologies and communication paradigms have been developed.

Long term evolution (LTE), one of the 4G mobile networking standards, is an advanced technology that has been introduced to accommodate future M2M communications. When compared to existing cellular systems, it has incorporated a number of new technologies such as OFDM, SC-FDMA, OFDMA, MIMO and so on. They enable LTE to operate more efficiently with respect to the use of spectrum, and also to provide much higher data rates. Besides, LTE system architecture evolution (SAE) transforms the system architecture to improve the performance. In this way data can be routed more directly to its destination and thus latency can be reduced [38].

While the 4G telecommunications systems is now starting to be deployed, research communities and industries are looking towards the development of the fifth-generation (5G) technology and services. 5G mobile systems are expected to provide better levels of connectivity and cover-age, as well as efficient support of machine-type devices. There are many new concepts that are being investigated and developed for this technology. They include pervasive networks, group cooperative relay, cognitive radio, wireless mesh networking and dynamic ad-hoc networking, smart antennas, wearable devices with artificial intelligence and so on. Currently, 5G cellular is very much in the early development stages and related standards have not been defined yet [39].

In addition to the evolution of cellular mobile networks, there is an ongoing effort to improve the performance of wireless local area network (WLAN) and wireless personal network (WPAN). One of the solutions is to migrate them toward less congested higher frequency unlicensed spectrum bands such as 60 GHz. Several standards have been drafted for the 60-GHz PHY and MAC protocols approaching Gbps range for service rate. These include the IEEE 802.11ad and wireless gigabit alliance (WiGig) for WLAN, IEEE 802.15.3c and wireless high-definition (WirelessHD) for short-range PANs. The IEEE 802.11ad draft standard introduces a new network architecture named personal basic service (PBSS). With PBSS, access points are no longer required and each station can serve the role of a central coordination point which supports a combination of random carrier sense multiple access (CSMA) and scheduled time division multiple access (TDMA) modes. The IEEE 802.15.3c standard defines a central controlled network topology and TDMA-based MAC protocol for 60-GHz wireless PANs [38]. Overall, 60-GHz technology is expected to mature during the next three to five years and will provide an important option for high-speed indoor connectivity associated with applications such as multi-point high-definition video conference for healthcare services.

In order to cope with network capacity scarcity, lack of coverage and dropped calls, heterogeneous network (HetNet) concept which is the use of multiple types of access nodes is also introduced. A HetNet interoperates macrocells and small cells (e.g., femtocells, picocells, etc.) to increase network capacity and/or to offer a wide variety of wireless coverage zones, ranging from an open outdoor environment to office buildings, homes, and underground areas [40]. Recently, Wi-Fi access is also integrated into cellular infrastructure with seamless handoff capability between network elements to enhance coverage and data offload. Small cells are an integral part of future LTE networks [41].

Finally, considering that the telecommunication business has become more competitive than ever, an increased synergy in terms sharing of resources and mainly considering the use of cloud system is required. Specifically, revenues from telecommunication service sector are not proportional to the sharp increase of traffic volume and more stringent quality of services demanded by emerging applications, therefore, network infrastructure needs to be more efficiently used. In order words, the cost per data bit needs to be lowered to maintain competitiveness. Cloud system embraces many concepts to achieve this. Cloud computing relies on sharing of resources (computing hosts or platforms, software, storage, etc.) to achieve coherence and economies of scale. It allows companies to avoid upfront infrastructure costs, and focus on projects that differentiate their businesses instead of infrastructure. Network virtualization focuses on the sharing of a network of hosts that allow multiple operators to provide services on top of a common physical network. Wireless access virtualization is subset of network virtualization. It is the abstraction and sharing of wireless communication resources to give multiple virtual owners the illusion of sole ownership. There is a wide range of wireless resource including wireless fabrics/devices (e.g., network interfaces, Wi-Fi access points, wireless routers, cellular base stations), physical wireless channels or spectrum, wireless links, address space, labeling tags and so on. Cloud-based services and infrastructures have appeared as one of the most promising solutions that can bring about numerous advantages for the telecom industry, such as reduced operational and capital expenditures, more dynamic and flexible service provisioning and lower energy consumption [42, 43]. This is believed to be the future of Internet that can efficiently and economically support emerging applications.

Tele-medicine, especially mHealth, is obviously playing a crucial role in human life and thus gaining more and more popularity in the years to come. It appears to be a very real world application that takes advantages of above-mentioned wireless communication technology and networking evolutions.

9 Conclusions

This chapter presents the overall architecture of an integrated communication network required to support mobile health (mHealth) applications. Various wireless communication technologies are comprehensively surveyed and candidate technologies for each domain of the mHealth communication network are then discussed. Next, this chapter addresses the need for an end-to-end and fully-automatic vital sign capture system in mHealth settings and presents the design and implementation of such a system, namely iVS, for real-life applications. Conventional patient monitors currently used in healthcare institutions are integrated into iVS by equipping each of them with a BLE adapter which wirelessly exchanges information with a smartphone in the vicinity. iVS employs smartphones as a gateway or an end-device for capturing patient's VSs or providing ubiquitous healthcare services to end-users. Machine-to-machine communication protocols for reliable and inter-operable data exchange among various components of iVS are also proposed and implemented. iVS can be used in many situations (e.g., hospitals, clinics, emergency sites and remote/isolated areas) since it ubiquitously provides reliable and fast data connectivity between patients and healthcare professionals. Especially, in a medical emergency, its capability to transmit information quickly and efficiently can make all the difference and results in time and life savings.

Experiments on the implemented iVS have been carried out to measure the time responsiveness of the system and the supportable throughput of the BLE link between patient monitors and smartphones. The results show that vital signs of the patient can be captured quickly and accurately thanks to extremely fast BLE connection setup time and end-to-end machine-to-machine communications between patient monitors and end-user devices. Measurements also reveal that the actual BLE throughput is much lower than the theoretical rate due to a variety of factors (e.g., signaling overheads, limitations on BLE chip implementation, firmware and operating system constraints, wireless signal propagation issues, etc.). However, the obtained throughput over the BLE link does not pose any challenge since, for iVS applications; this link is only required to deliver short text-based messages at a low rate.

Acknowledgment. This work was partially supported by the Natural Sciences and Engineering Research Council (NSERC) and industrial and government partners, through the Healthcare Support through Information Technology Enhancements (hSITE) Strategic Research Network.

References

1. Smartphone adoption tips past 50 percent in major markets worldwide. eMarketer (May 2013), http://www.emarketer.com
2. Chiarini, G., Ray, P., Akter, S., Masella, C., Ganz, A.: mHealth technologies for chronic diseases and elders: A systematic review. IEEE Journal on Selected Areas in Communications 31(9), 6–18 (2013)

3. Adibi, S.: Link technologies and BlackBerry mobile health (mHealth) solutions: A review. IEEE Transactions on Information Technology in Biomedicine 16(4), 586–597 (2012)
4. Huang, A., et al.: WE-CARE: An intelligent mobile telecardiology system to enable mHealth applications. IEEE Journal of Biomedical and Health Informatics (99), 1–10 (2013)
5. Saldarriaga, A., et al.: A mobile application for ambulatory electrocardiographic monitoring in clinical and domestic environments. In: 2013 Pan American Health Care Exchanges (PAHCE), pp. 1–4 (May 2013)
6. Zhao, Z.W., et al.: A machine-to-machine based framework for diabetes lifestyle management. In: The 10th IEEE International Conference on Networking, Sensing and Control (ICNSC), pp. 562–567 (April 2013)
7. Garibaldi-Beltran, J., Vazquez-Briseno, M.: Personal mobile health systems for supporting patients with chronic diseases. In: 2012 IEEE Ninth Electronics, Robotics and Automotive Mechanics Conference (CERMA), pp. 105–110 (November 2012)
8. Pan, Q., et al.: A mobile health system design for home and community use. In: 2012 IEEE International Conference on Biomedical and Health Informatics (BHI), pp. 116–119 (January 2012)
9. Lopes, I., et al.: A mobile health monitoring solution for weight control. In: 2011 International Conference on Wireless Communications and Signal Processing (WCSP), pp. 1–5 (November 2011)
10. Chomutare, T.F.: A context-sensitive framework for mobile terminals for assisting type 2 diabetes patients. Master thesis, Univ. Tromso, Tromso, Norway (June 2008)
11. Rozanowski, K., Piotrowski, Z., CiolekMilitary, M.: Mobile application for driver's health status remote monitoring. In: The 9th International on Wireless Communications and Mobile Computing Conference (IWCMC), pp. 1738–1743 (July 2013)
12. Milenkovic, A., Milosevic, M., Jovanov, E.: Smartphones for smart wheelchairs. In: 2013 IEEE International Conference on Body Sensor Networks (BSN), pp. 1–6 (May 2013)
13. Weghorn, H.: Applying mobile phone technology for making health and rehabilitation monitoring more affordable. In: 2013 Biosignals and Biorobotics Conference (BRC), pp. 1–5 (February 2013)
14. IntelliVue MX40 wearable patient monitor, Phillips Healthcare, `http://www.healthcare.philips.com`
15. Infinity M300 wearable patient monitor, Dragger, `http://www.draeger.ca`
16. Patient SafetyNet, Masimo, `http://www.masimo.com`
17. Micropaq wearable patient monitor, Welch Allyn, `http://www.welchallyn.com`
18. Miller, B.A., Bisdikian, C.: Bluetooth revealed. Prentice-Hall (September 2000)
19. Heydon, R.: Bluetooth low energy: The developer's handbook. Prentice Hall (October 2012)
20. Coskun, V., Ok, K., Ozdenizci, B.: Near field communication (NFC): From theory to practice. Wiley (February 2012)
21. Yick, J., Mukherjee, B., Ghosal, D.: Wireless sensor network survey. Computer Networks 52(12)
22. Tennina, S., et al.: IEEE 802.15.4 and ZigBee as enabling technologies for low-power wireless systems with quality-of-service constraints. Springer (May 2013)

23. Wireless HART, HART Communication Foundation, `http://www.hartcomm.org`
24. Shelby, Z., Bormann, C.: 6LoWPAN: the wireless embedded Internet. Wiley (December 2009)
25. Gast, M.: 802.11 wireless networks: The definitive guide. O'Reilly Media (April 2005)
26. Rao, G.S.: Mobile cellular communication. Pearson Education India (April 2012)
27. Dahlman, E., Parkvall, S., Skold, J.: 4G: LTE/LTE-Advanced for mobile broadband. Academic Press (December 2013)
28. Rao, G.S.: WiMAX: a wireless technology revolution. Auerbach Publications (September 2012)
29. Roddy, D.: Satellite communications. McGraw-Hill Professional (January 2006)
30. DINAMAP PRO series, GE Healthcare, `http://www.gehealthcare.com`
31. HL7 messaging, Health Level Seven International, `http://www.hl7.org`
32. DINAMAP PRO host communications manual, GE Medical Systems
33. A Universally Unique IDentifier (UUID) URN Namespace, Internet Engineering Task Force (IETF), RFC 4122 (July 2005)
34. Core Bluetooth framework reference, Apple Inc., `https://developer.apple.com`
35. Bluetooth specification version 4.0, Bluetooth SIG, `https://www.bluetooth.org`
36. The key 2006-2013 ICT data for the world, International Telecommunication Union (February 2013), `http://www.itu.int`
37. More than 50 billion connected devices, White Paper, Ericsson (February 2013)
38. Raychaudhuri, D., Mandayam, N.B.: Frontiers of wireless and mobile communications. Proceeding of IEEE 100(4)
39. The 5th generation wireless systems, Wikipedia, `http://www.wikipedia.org`
40. Damnjanovic, A., et al.: A survey on 3GPP heterogeneous networks. IEEE Wireless Communications (June 2011)
41. Cellular-WiFi integration, White Paper, InterDigital Inc. (June 2012)
42. Mosharaf, N.M., Chowdhury, K., Boutaba, R.: A survey on network virtualization. Journal of Computer and Telecommunications Networking (April 2010)
43. Wang, A., Iyer, M., Dutta, R., Rouskas, G., Baldine, I.: Network virtualization: Technologies, perspectives, and frontiers. Journal of Lightwave Technology (August 2012)

Questions

Q1. What Are the Key Attributes of Bluetooth Low Energy (BLE)?

Q2. Compare Bluetooth Classic (BT) and Bluetooth Low Energy (BLE)

Q3. List Some Use Cases Planned for Bluetooth Low Energy (BLE)?

Q4. Advantages of Long Term Evolution (LTE)?

Q5. Access the Relevance of LTE for M2M Communications?

39. Energy Analysis and QoE of Wireless Sensor Networks

Chenfu Yi, Da Xu, and Ye Li*

Research center for Biomedical and Information Technology,
Shenzhen Institutes of Advanced Technology, Chinese Academy of Sciences,
Shenzhen 518055, China
ye.li@siat.ac.cn

Abstract. With recent advances in wireless networking, easy-of-use sensors, low-power microelectronics, and embedded computing technologies, wireless sensor networks (WSNs) have many applications in mHealth, such as wireless body area networks (WBANs), and wireless body sensor networks (WBSNs), which are viewed as a vital sign perception and transmission unit for the health internet of things. However, as for the battery-powered wearable or implantable sensor nodes in the WBAN, energy consumption has been the biggest bottleneck in the long-term work. Because of the medical business diversity, service differentiation, and environmental dynamics, it would make the system energy consumption increase if the traditional fixed communication mode is still applied to WSNs. In this chapter, a system level energy consumption model is constructed and analyzed for WSNs.

On the other hand, Quality of Experience (QoE) is proposed to evaluate user's overall satisfaction with the network system and services. For mHealth systems, improving QoE means starting from user's view to make patients and doctors (the main users of mHealth system who are especially captious to services related to health) to achieve best satisfaction. In this chapter, we also proposed an improved model for QoE of mHealth systems based on energy consumption and information integrity received by the users, and then explore on how the users' satisfaction with mHealth systems' energy consumption and received information integrity by exponential formula, which is influenced by the quantity of the transmitted information. Finally, from the results of simulation, we concluded that the appropriate compression of information quantity is the best way to improve QoE performance of the mHealth system.

Keywords: wireless sensor networks, energy model, modulation, QoE, mHealth, information integrity.

1 Introduction

With the rapid development of wireless communication and low power embedded techniques, wireless sensor networks (WSNs) are widely used in many industrial

* Corresponding author.

S. Adibi (ed.), *Mobile Health*,
Springer Series in Bio-/Neuroinformatics 5, DOI: 10.1007/978-3-319-12817-7_39

and civilian application areas, including industrial process monitoring and control, machine status monitoring, environment and habitat monitoring, healthcare applications, home automation, and traffic control. For example, because of the strong requirements of medical healthcare, special populations care, and community healthcare service etc., WBSNs have become a focus of intensive research. As shown in Figure 1, by the worn and/or implanted sensors, the sensing physiological parameters, e.g., blood pressure, body temperature, body posture, cardiac rate, photoplethysmograph (PPG) and electrocardiogram (ECG) signals, are sent to the remote medical cloud server for the further real-time analysis and diagnosis. Then, the patient can experience a greater mobility and is no longer constraint to stay at hospital. This is called as mHealth, which is defined as a healthcare technique that uses mobile equipment (e.g., phone and/or pad). By using mHealth monitoring technology, people can observe their physiological signals anywhere at any time. This is very important for the health of people. In recent years, more and more mHealth applications appear around us.

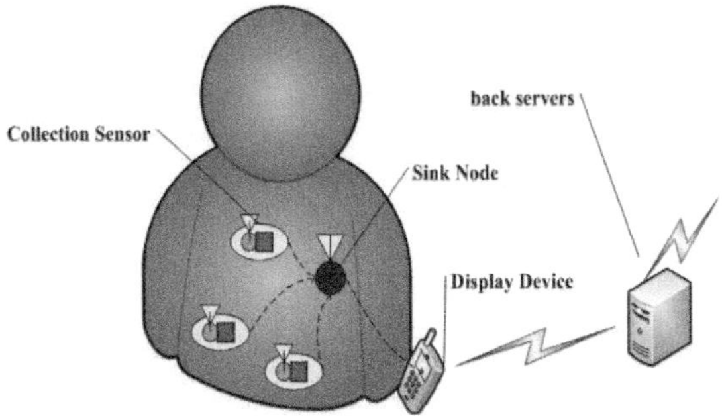

Fig. 1 Architecture for mHealth monitoring [1]

However, there are mainly two types of practical problems to be investigated in this chapter. On the one hand, a sensor node in WSNs is a tiny with power-operated battery and limited energy budget. Moreover, it is usually impracticable and inconvenient to recharge and/or replace a new battery for the drained sensor, since amount of sensor nodes are distributed in a hostile and/or unreachable circumstance. Therefore, we can draw a conclusion that, the greater the charge capacity of the battery, the longer the lifetime of network. However, because of the stringent constraints on the size and weight of sensor nodes, the battery with limited capacity must be relatively small and light. Moreover, the battery technology cannot keep pace with the development of the digital integration technology [2]. Consequently, the energy budget of WSNs is limited, and the energy efficiency in packet transmission is a crucial issue to be considered by the designer.

On the other hand, mHealth systems have not been prevalent as supposed. The main constraint is the experiences of existing mHealth monitoring services do not match users' expectation. For a long time, the main promoters of mHealth, telecommunication equipment manufactures and operators are using Quality of Service (QoS) to estimate the quality of mHealth monitoring system [3]. QoS indexes are delay, jitter and loss tolerance. QoS cannot evaluate users' feeling with services, so an mHealth monitoring system driven by QoS often cannot satisfy users' specific experience requirements. Quality of Experience (QoE) is proposed to measure user's synthetic feeling about the device, network, system and applications. The best way to popularize the mHealth applications is to raise users' satisfaction with them during operation. For mHealth system, QoE is a standard, reflecting the gap between existent service and users' expectation. People are always harsh and prudent for health-related services, so if they feel uncomfortable, they'll give up the idea of keeping on and distrust the data given by mHealth monitoring system. If we want to design mHealth monitoring system that users are willing to use, increasing QoE degree is crucial that we have to take QoE into account before we start designing mHealth monitoring systems.

2 Energy Analysis of WSNs

2.1 Background of Low Power Techniques for WSNs

In recent years, different aspects of low power techniques have been addressed for WSNs. These include delay controlled transmission scheme [4], energy efficient cooperative communication in WSNs [5], energy efficient modulation scaling [4][5], energy efficient routing schemes [6][7], power management-based task scheduling for digital communication processors [8], cross-layer energy optimization [10], etc. However, in most of the prior work, the power dissipated in the RF and analog circuit is often ignored or over simplified to constant values. This may lead to erroneous energy evaluations, since the RF and analog sections process analog signals with high frequency content and typically consume more energy compared to the digital part. From [10], about 75% of the total power is dissipated in the RF front-end. So, it is important to develop an accurate and comprehensive energy model for WSNs RF front-end. Such a model would definitely be helpful to develop a thorough tradeoff analysis between energy consumption and communication quality of WSNs.

Several system level energy models have also been proposed in recent years. For WSNs, the transceiver energy model considers the circuit start-up energy in addition to steady-state dissipated energy in [11]. Unfortunately, the energy consumption of every component is assumed to be constant, this is not suitable for real WSNs application. Another high level model was proposed in [12], which divides the circuitry power into two parts, one related to the instantaneous symbol rate and the other to the highest symbol rate. However, the dissipated circuit power is not only related to the symbol rate, but also to system parameters, such as signal Peak to Average Ratio (PAR), and signal bandwidth. In [9], a more comprehensive model is presented, however, most of the RF front-end components are still assumed to have constant power consumption, and the power model for power amplifier is gain dependent, and does not take into account signal characteristics for instance signal PAR in M-QAM modulation. In order to accurately evaluate the effects of different communication system parameters on RF front-end energy consumption, we build a system level RF front-end energy model with QAM modulation for Wi-Fi application in the former works [13], and tie the physical layer (PHY) parameters, such as bit error rate (BER), modulation level, bandwidth, and bit rate, to the RF circuit energy consumption. However, the prior work just considers one architecture or linear modulation scheme, and has not considered the nonlinear system. The results can just be applied to the Wi-Fi system instead of WSNs application. Because there are both linear and nonlinear RF front-ends architectures and modulation schemes widely used in WSNs. So, to design an energy efficient system, we should not consider just one architecture or only linear modulation. For various applications, the optimal architecture or modulation scheme may be different. For instance, it is shown in [14] that the energy efficiency for 4QAM is higher than 16QAM with respect to long distance transmission. But 4QAM still be superior compared with other modulation schemes such as OOK modulation, FSK modulation, and PSK modulation.

To design the energy efficient systems suitable for various WSNs applications, we must know the available RF front-end architectures used in WSNs nodes. In this chapter [15], we develop the energy models for four types of RF front-ends architectures with OOK, PSK, QAM and FSK modulation respectively, which are popular in WSNs nodes. We consider the effects of BER, PAR, modulation level, bandwidth, data rate and the transmission distance etc. in our energy models. And based on the energy models, we first theoretically compare the energy performance of different modulation schemes for different WSNs applications in system level. For instance, we compare the energy performance of OOK, BPSK and 2FSK modulation schemes for narrow bandwidth and low data rate application; QPSK, 4FSK and 8QAM modulation schemes for medium bandwidth and medium data rate application; and 16QAM, 32QAM, 64QAM modulation schemes are considered for wide bandwidth and high data rate application. At last, we give the most energy efficient modulation scheme for different scenarios and summarize how to design the energy efficient system.

2.2 Typical RF Architectures for WSNs Nodes

In order to design an energy efficient system, we must understand the RF architectures of the WSNs nodes. In this section, we choose four typical RF architectures with different classes of PA for WSNs nodes. There are two types of PA used in these architectures, which are the linear PA (Class A) and the nonlinear PA (Class E). The linear PA is adopted in linear modulation systems, such as OOK, PSK, and QAM, whereas the nonlinear PA would be employed in the nonlinear modulation systems, for example FSK modulation.

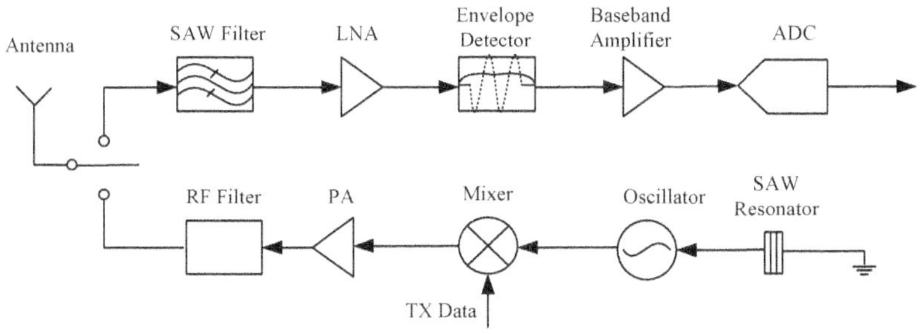

Fig. 2 OOK Transceiver [16][17]

For the linear modulation, the OOK transceiver has the simplest architecture, which is shown in Figure 2. Direct-conversion transmitter is used in the OOK transceiver architecture. Digital baseband signal is mixed with carrier generated by the crystal and up converted to RF band. Besides, in the receiver, the signal received from antenna is filtered by the SAW filter, amplified by low noise amplifier (LNA), demodulated by the envelope detector, and amplified by the baseband amplifier. Finally the original data are recovered by Analog-to-Digital Converter (ADC). Non-coherent demodulation is applied in the receiver, which eliminates the mixer and reduces the overall size and power consumption.

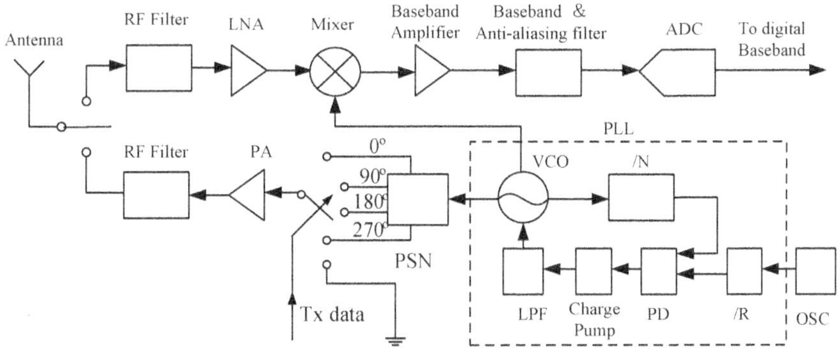

Fig. 3 PSK Transceiver [16][18]

The architecture of PSK transceiver is more complex than OOK transceiver, which is shown in Figure 3. The direct-conversion transmitter is also adopted in the PSK system, in which the phase modulation is accomplished by the Phase-shifting-network (PSN) controlled by digital baseband signal. The modulation signal is transmitted directly to the PA without crossing the mixer. The traditional heterodyne low-IF receiver is adopted in PSK systems. The main components in the receiver signal chain are the RF filter, LNA, down conversion mixer, baseband amplifier, baseband and anti-aliasing filter, and ADC.

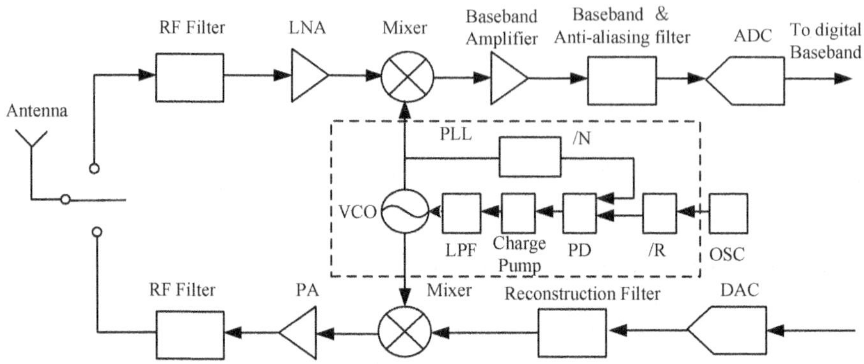

Fig. 4 QAM Transceiver [16][18]

Figure 4 describes the RF front-end of QAM system, which also employs the same receiver and direct-conversion transmitter. The main components of the analog signal chain of the transmitter are digital-to-analog converter (DAC), reconstruction filter, up conversion mixer, power amplifier (PA), and RF filter.

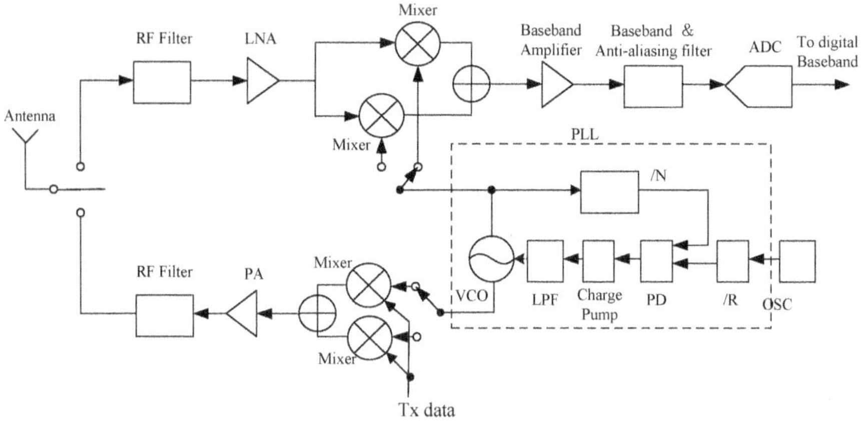

Fig. 5 FSK Transceiver [16][18][19]

Besides, for the nonlinear modulation, the architecture of FSK transceiver is shown in Figure 5. Direct-conversion transmitter is adopted in FSK system too. The carriers in different frequencies are also generated by Phase-Locked Loop (PLL) and controlled by the baseband signal. Then, the following processes are the same as the PSK system. The coherent demodulation is applied in the receiver. Note that there is just one pair of mixers in the transmitter and receiver active at the same time for the purpose of low power.

Generally speaking, the power efficiency of linear PA is lower than that of the nonlinear PA, while the bandwidth efficiency of linear modulation is higher than that of nonlinear modulation. For instance, according to [20], the bandwidth for PSK is

$$BW_{MPSK} = \frac{1+\alpha}{b} R_b \tag{1}$$

where α denotes the roll-off factor of the pulse-shaping filter [21], b is the modulation level, and R_b denotes data rate. The bandwidth of MQAM is equal to MPSK. However, the bandwidth of FSK modulation can be expressed as

$$BW_{\text{MFSK}} = \frac{2^b + 3}{2 \cdot b} \cdot R_{\text{b}} \tag{2}$$

Since $0 \le \alpha \le 1$ and $b \ge 1$, if the data rate is equal, it is obvious that $(1+\alpha)/b < (2^b + 3)/2b$. So we can conclude that for the same modulation level, PSK and QAM have higher bandwidth efficiency than FSK.

2.3 System Level Power Models

To minimize the total energy consumption of a transceiver, it is essential to develop accurate energy models for all the key signal processing blocks. In the former work [14], we develop the energy models of RF front-end for Wi-Fi application and give the power consumption of the main components. As an extension work, in this part, we establish energy models for WSNs nodes with both linear and nonlinear PAs. Since most power parameters of the analog blocks except PA are hard to be adjusted in WSNs for low power purpose, we assume that the power consumption of WSNs nodes except PA is constant. As the power consumption of PA is dominant [22], we focus on it first.

The power consumption of PA can be divided into two parts: signal transmission power which is delivered to the antenna and dissipated power which is consumed by the electronic circuits. The transmission power P_{out} is proportional to the detected signal power P_{detected} at the receiver, the antenna gain, and the propagation distance. Assuming free space propagation at distance d, the radiated power *P*out is given by

$$P_{\text{detected}} = \frac{P_{\text{out}} G_{\text{t}} G_{\text{r}} \lambda^2}{(4\pi)^2 d^2 L} \tag{3}$$

where G_{t}, G_{r}, λ, d are transmitter and receiver antenna gains, carrier wavelength and the transmission distance respectively. L is the system loss factor not related to propagation.

According to [23], the efficiency of Class A PA is defined as

$$\eta = \frac{P_{\text{out}}}{P_{\text{PA}}} = \frac{K}{PAR} \tag{4}$$

where P_{PA} is the total power consumed by PA, K is a constant, and we choose K=0.5. So, we can get

$$P_{\text{PA}} = \frac{P_{\text{out}}}{K} \cdot PAR \tag{5}$$

According to [20], the BER for OOK and QPSK modulation schemes are as follows

$$\text{OOK:} \quad BER_{\text{OOK}} = Q\left(\sqrt{\frac{P_{\text{detected}}}{2N}}\right) \tag{6}$$

$$\text{BPSK:} \quad BER_{\text{BPSK}} = Q\left(\sqrt{\frac{2P_{\text{detected}}}{N}}\right) \tag{7}$$

$$\text{QPSK:} \quad BER_{\text{QPSK}} = Q\left(\sqrt{\frac{2P_{\text{detected}}}{N}}\right) \tag{8}$$

$$\text{MPSK:} \quad BER_{\text{MPSK}} = \frac{2}{b}\text{Q}\left(\sqrt{\frac{4P_{\text{detected}}}{N}} \cdot \sin\left(\frac{\pi}{2^{b+1}}\right)\right) \tag{9}$$

where N is the noise power and b is the modulation level, $b = \log_2 M$, and $Q(\cdot)$ function is given by:

$$Q(\text{x}) = \frac{1}{\sqrt{2\pi}} \int_x^{\infty} \text{e}^{-y^2/2} dy \tag{10}$$

So, the power consumption for the class A PA for OOK, BPSK, QPSK and MPSK modulation can be written as

$$P_{\text{PA\_OOK}} = \frac{32\pi^2 d^2 L}{G_t G_r \lambda^2 K} \cdot N \cdot PAR \cdot \left[Q^{-1}(BER)\right]^2 \tag{11}$$

$$P_{\text{PA\_BPSK}} = \frac{8\pi^2 d^2 L}{G_t G_r \lambda^2 K} \cdot N \cdot PAR \cdot \left[Q^{-1}(BER)\right]^2 \tag{12}$$

$$P_{\text{PA\_QPSK}} = \frac{8\pi^2 d^2 L}{G_t G_r \lambda^2 K} \cdot N \cdot PAR \cdot \left[Q^{-1}(BER)\right]^2 \tag{13}$$

$$P_{\text{PA\_MPSK}} = \frac{4\pi^2 d^2 L}{G_t G_r \lambda^2 K} \cdot N \cdot PAR \cdot \left[Q^{-1}\left(\frac{b \cdot BER}{2}\right) \Big/ \sin\left(\frac{\pi}{2^{b+1}}\right)\right]^2 \tag{14}$$

where $Q^{-1}(\cdot)$ is the inverse function of $Q(\cdot)$, PAR=1 for PSK modulation.

In our former work [15], the power consumption of PA in MQAM systems has been proposed as

$$P_{\mathrm{PA\_MQAM}} = \frac{16\pi^2 d^2 L}{3G_t G_r \lambda^2 K}\left(2^b - 1\right)\cdot N \cdot \left[\mathrm{Q}^{-1}\left(\frac{1}{4}\left(1-\frac{1}{2^{b/2}}\right)^{-1}\cdot b\cdot BER\right)\right]^2 \cdot PAR_{\mathrm{MQAM}} \tag{15}$$

For the linear modulation, the BER expression for MFSK modulation is shown in [20],

$$BER_{\mathrm{MFSK}} = \frac{2^b - 1}{b} Q\left(\sqrt{\frac{b\cdot P_{\mathrm{detected}}}{N}}\right) \tag{16}$$

So, the power model for Class E PA [32] with FSK modulation is as follow

$$P_{\mathrm{out\_MFSK}} = \frac{16\pi^2 d^2 L}{G_t G_r \lambda^2}\cdot\frac{N}{b}\cdot\left[Q^{-1}\left(\frac{b\cdot BER}{2^b - 1}\right)\right]^2 \tag{17}$$

$$P_{\mathrm{PA\_MFSK}} = \left(1+\frac{R_{\mathrm{on}}}{V_{\mathrm{DD}}^2}\cdot\frac{Q_{\mathrm{Lf}}}{Q_{\mathrm{Lf}}-Q_{\mathrm{f}}}\cdot F\cdot P_{\mathrm{out\_MFSK}} + \frac{G}{Q_{\mathrm{Ls}}}\right)\cdot\left(1-\frac{Q_{\mathrm{f}}}{Q_{\mathrm{Lf}}}\right)^{-1}\cdot P_{\mathrm{out\_MFSK}} \tag{18}$$

where Q_{f} , Q_{Lf} are the quality factors of output LC filter with or without the load. Q_{L} is the quality factor of shunt inductor. R_{on} , V_{DD} are the switch resistance and supply voltage of the Class E PA. F , G are the power dissipation coefficients and $F = 2.3$, $G = 1.7$ are chosen in our simulation to make the PA operate at the highest efficiency [24]. $P_{\mathrm{out\_MFSK}}$ is the output power of Class E PA.

As the energy consumption for the RF front-end includes the energy consumption of PA and other blocks, the energy per bit can be expressed as:

$$E_{\mathrm{bit}} = P_{\mathrm{E}}/(b\cdot R_{\mathrm{s}}) + P_{\mathrm{PA}}/(b\cdot R_{\mathrm{s}}) \tag{19}$$

where E_{bit} denotes the energy per bit consumed by RF front-end, P_{E} is the power consumption of the RF circuits except PA, R_{s} denotes the symbol rate.

2.4 *Energy Evaluations for Different Applications*

In this part, we analyze and model the energy consumption of the four types of RF front-ends with different modulation schemes for WSNs applications, which is an extension of our former energy modeling work [14]. According to the data rate and bandwidth requirements, we divide the applications into six types, namely narrow bandwidth (less than 50KHz), low data rate (less than 50Kbps), medium bandwidth (from 50KHz to 500KHz), middle data rate (from 50Kbps to 500Kbps), wide bandwidth (above 500KHz) and high data rate (above 500Kbps). Assume the transmission distance is within 150 meters. So, the values of d^2BW and d^2R_s are less than 1.125×10^9 for narrow bandwidth and low data rate application, less than 1.125×10^{11} for medium bandwidth and middle data rate application, respectively, and we assume the values of d^2BW and d^2R_s are less than 1.125×10^{13} for wide bandwidth and high data rate application. In this chapter, we just investigate the OOK, BPSK and 2FSK modulation schemes for low bandwidth or low data rate communication system. Other modulation schemes can be adopted for middle bandwidth or middle data rate instance by the similar way. The related parameter values used in our simulation are listed in TABLE I [14].

Table 1 Simulation Parameters

f=2.4GHz	Q_f=10	$\alpha = 0.2$
BER=1e-3	Q_{Lf}=100	P_{E_OOK}=72.35e-3W
L=1.25	Q_{Ls}=10	P_{E_BPSK}=126.85e-3 W
G_r=G_t=1	R_{on}=0.05 Ω	P_{E_QPSK}=126.85e-3 W
K=0.5	V_{dd}=3.3V	P_{E_8PSK}=126.85e-3 W
N_0=2e-16W/Hz	F=2.3	P_{E_MFSK}=147.85e-3 W
PAR_{roll_off}=5dB	G=1.7	P_{E_MQAM}=165.75e-3 W

(Reproduced from Ye Li, Dengyu Qiao, et al., Energy-Model-Based Optimal Communication Systems Design for Wireless Sensor Networks, TABLE 1, International Journal of Distributed Sensor Networks Volume 2012, Article ID 861704, 11 pages doi:10.1155/2012/861704. © [2012] Hindawi Publishing Corporation. Reprinted, with permission.)

Since bandwidth (BW) is a precious resource in most communication system, in this chapter we assume it is constant and the data rate can be adjusted by changing the modulation level. Also we assume pulse shaping technique is used in the system to remove Inter-Symbol Interference (ISI) and pulse-shaping filter roll-off factor α equals 0.2. The power consumption values P_E for every modulation scheme are referred from the data sheets of the RF or analog components by TI or RFMD companies. Note that P_E can be set as different values but the trends of E_{bit} are similar.

2.4.1 Narrow Bandwidth System

We choose OOK, BPSK and 2FSK modulation for low bandwidth system and the relations between data rate and bandwidth are as follows:

$$R_{b\_OOK} = R_{b\_BPSK} = \frac{BW}{1+\alpha} \qquad \mathrm{R}_{b\_2FSK} = \frac{2}{5}BW \tag{20}$$

By (11-12), (18) and (19), energy per bit for OOK, BPSK, 2FSK can be expressed respectively as

$$E_{bit\_OOK} = \frac{(1+\alpha)\cdot P_{E\_OOK}}{BW} + \frac{32\pi^2(1+\alpha)\cdot L\cdot d^2\cdot BW\cdot N_0}{BW\cdot G_t\cdot G_r\cdot\lambda^2\cdot \mathrm{K}}\cdot PAR\cdot\left[\mathrm{Q}^{-1}(BER)\right]^2 \tag{21}$$

$$E_{bit\_BPSK} = \frac{(1+\alpha)P_{E\_BPSK}}{BW} + \frac{8\pi^2(1+\alpha)L\cdot d^2\cdot BW\cdot N_0}{BW\cdot G_t\cdot G_r\cdot\lambda^2\cdot \mathrm{K}}\cdot PAR\cdot\left[\mathrm{Q}^{-1}(BER)\right]^2 \tag{22}$$

$$E_{bit\_2FSK} = \frac{5P_{E\_2FSK}}{2BW} + \left(1-\frac{\mathrm{Q}_f}{\mathrm{Q}_{Lf}}\right)^{-1}\cdot\frac{40\pi^2 L\cdot d^2\cdot BW\cdot N_0}{BW\cdot G_tG_r\lambda^2}\cdot PAR\left[\mathrm{Q}^{-1}(BER)\right]^2\cdot\left(\begin{array}{c}1+\frac{R_{on}}{V_{DD}^2}\cdot\frac{Q_{Lf}}{Q_{Lf}-Q_f}\cdot F \\ \cdot\frac{16\pi^2 Ld^2N_0\cdot BW}{G_tG_r\lambda^2}\cdot PAR \\ \cdot\left[\mathrm{Q}^{-1}(BER)\right]^2+\frac{G}{Q_{Ls}}\end{array}\right) \tag{23}$$

1) BPSK vs OOK

Firstly, we compare the energy per bit of BPSK and OOK modulation scheme. Let

$$F_1(d^2,BW) = E_{bit\_BPSK} - E_{bit\_OOK}$$
$$= \frac{(1+\alpha)\cdot(P_{E\_BPSK}-P_{E\_OOK})}{BW} - \frac{24\pi^2(1+\alpha)\cdot L\cdot d^2\cdot BW\cdot N_0}{BW\cdot G_t\cdot G_r\cdot\lambda^2\cdot \mathrm{K}}\cdot PAR\cdot\left[\mathrm{Q}^{-1}(BER)\right]^2$$

If $F_1(d^2,BW)<0$, it means $E_{bit\_BPSK} < E_{bit\_OOK}$, by solving the inequation, we have

$$d^2\cdot BW > \frac{(P_{E\_BPSK}-P_{E\_OOK})\cdot G_t\cdot G_r\cdot\lambda^2\cdot K}{24\pi^2 L\cdot N_0\cdot PAR\cdot\left[\mathrm{Q}^{-1}(BER)\right]^2} \tag{24}$$

We set the values of the parameters as listed in TABLE I, then

$$d^2\cdot BW > 2.1844\mathrm{e}+8 \tag{25}$$

If $F_1(d^2,BW)=0$, it means OOK and BPSK modulation scheme have the same energy efficiency. We can also get the solution as $d^2\cdot BW = 2.1844\mathrm{e}+8$.

If $F_1(d^2,BW)>0$, it means $E_{bit\_BPSK} > E_{bit\_OOK}$. Similarly, we can get the solution as $d^2\cdot BW < 2.1844\mathrm{e}+8$.

From the analysis above and the restriction of $d^2 \cdot BW < 1.125e+9$, we can conclude that when $1.125e+9 > d^2 \cdot BW > 2.1844e+8$, BPSK modulation scheme is more energy efficient than OOK modulation scheme, and when $d^2 \cdot BW < 2.1844e+8$, OOK performs better.

2) 2FSK vs OOK

Secondly, let's compare the energy per bit of 2FSK and OOK modulation. Suppose

$$\xi = \left(1 - \frac{Q_f}{Q_{Lf}}\right)^{-1} \cdot \frac{16\pi^2 L \cdot N_0}{G_t G_r \lambda^2} \cdot PAR \cdot \left[Q^{-1}(BER)\right]^2 \tag{26}$$

$$F_2\left(d^2, BW\right) = E_{bit\_2FSK} - E_{bit\_OOK} = \frac{\frac{5}{2} P_{E\_2FSK} - (1+\alpha) \cdot P_{E\_OOK}}{BW} \tag{27}$$

$$-\frac{2(1+\alpha)\xi \cdot d^2 \cdot BW}{BW \cdot K} \cdot \left(1 - \frac{Q_f}{Q_{Lf}}\right) + \frac{5}{2}\frac{d^2 \cdot BW}{BW} \cdot \xi \cdot \left(1 + \frac{R_{on}}{V_{DD}^2} \cdot F \cdot \xi \cdot d^2 \cdot BW + \frac{G}{Q_{Ls}}\right)$$

Let $A = \frac{5}{2}\frac{R_{on}}{V_{DD}^2} \cdot F \cdot \xi^2$, $B = \frac{5}{2}\xi \cdot \left(1 + \frac{G}{Q_{Ls}}\right) - \frac{2(1+\alpha) \cdot \xi}{K} \cdot \left(1 - \frac{Q_f}{Q_{Lf}}\right)$, and $C = \frac{5}{2} P_{E\_2FSK} - (1+\alpha) \cdot P_{E\_OOK}$, then

$$F_2\left(d^2, BW\right) = \frac{1}{BW} \cdot \left(A \cdot \left(d^2 \cdot BW\right)^2 + B \cdot \left(d^2 \cdot BW\right) + C\right) \tag{28}$$

Assume $F_2\left(d^2, BW\right) = 0$, which means 2FSK and OOK modulation scheme have the same energy efficiency, we can solve the above function and get the roots $d^2 \cdot BW = \frac{-B \pm \sqrt{B^2 - 4A \cdot C}}{2A}$. We set the values of the parameters as listed in TABLE I and calculate the root values of $d^2 BW$ as:

$$X_1 = \frac{-B - \sqrt{B^2 - 4A \cdot C}}{2A} = 6.2616e+8 \tag{29}$$

$$X_2 = \frac{-B + \sqrt{B^2 - 4A \cdot C}}{2A} = 2.5369e+13 \tag{30}$$

Since A>0, if $6.2616e+8 < d^2 \cdot BW < 2.5369e+13$, then $E_{bit\_2FSK} - E_{bit\_OOK} < 0$, and consider the restriction of $d^2 \cdot BW$, we can get the result that when $6.2616e+8 < d^2 \cdot BW < 1.125e+9$ 2FSK modulation scheme is more energy efficient than OOK modulation scheme. If $d^2 \cdot BW < 6.2616e+8$, then $E_{bit\_2FSK} - E_{bit\_OOK} > 0$, it means OOK modulation scheme is more energy efficient than 2FSK modulation scheme.

3) 2FSK vs BPSK

Similarly, we can compare the energy performance of 2FSK and BPSK modulation. If $2.2465e+9 < d^2 \cdot BW < 5.1002e+12$, then $E_{bit\_2FSK} - E_{bit\_OOK} < 0$, but considering the restriction of $d^2 \cdot BW$, we find $2.2465e+9 < d^2 \cdot BW < 5.1002e+12$ is not suitable for narrow bandwidth application, and have to be abandoned. If $d^2 \cdot BW < 2.2465e+9$ or $d^2 \cdot BW > 5.1002e+12$, then $E_{bit\_2FSK} - E_{bit\_OOK} > 0$, and with the consideration of the value restriction of $d^2 \cdot BW$, we can conclude when $d^2 \cdot BW < 1.125e+9$, OOK is more energy efficient than 2FSK.

4) Example and Discussion

To verify the above theoretical analysis, we choose BW=25KHz as an example and draw Figure 6. From Figure 6, we can see that the OOK modulation scheme has the lowest energy consumption when $d^2 \cdot BW$ is small. This is because when $d^2 \cdot BW$ is small, the analog circuit energy consumption is dominant instead of PA. The OOK modulation system is much simpler than others and thus OOK has the best energy performance. But, with the increase of transmission distance, PA consumes much more energy compared with other RF blocks. When $d^2 \cdot BW$ increases to a certain value, BPSK modulation becomes the most energy efficient. 2FSK has higher energy consumption than that of BPSK, this is because when the bandwidth is fixed, according to (1) and (2), the data rate of 2FSK modulation is much slower than BPSK. From Figure 6, we can also see that the energy consumption of 2FSK modulation scheme changes much slower than that of OOK and BPSK, this owes to the high efficiency of Class E PA used in 2FSK system.

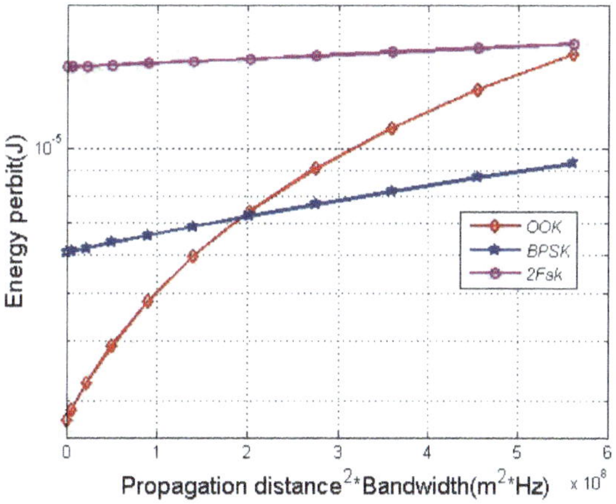

Fig. 6 Energy per bit for different RF front-end architecture in narrow bandwidth application, BW=25KHz, d=(0 150]m (*Reproduced from Ye Li, Dengyu Qiao, et al., Energy-Model-Based Optimal Communication Systems Design for Wireless Sensor Networks, Figure 7, International Journal of Distributed Sensor Networks Volume 2012, Article ID 861704, 11 pages doi:10.1155/2012/861704. © [2012] Hindawi Publishing Corporation. Reprinted, with permission.*)

From the above example and theoretical analysis, we can conclude that when $d^2 \cdot BW < 2.1844\mathrm{e}+8$, OOK modulation scheme is high energy efficient; when $2.1844\mathrm{e}+8 < d^2 \cdot BW < 1.125\mathrm{e}+9$, BPSK modulation scheme has the best energy performance.

2.4.2 Low Data Rate System

For low data rate application, the bandwidths of OOK, BPSK and 2FSK modulation can be expressed as follows:

$$BW_{\mathrm{OOK}} = BW_{\mathrm{BPSK}} = (1+\alpha)R_{\mathrm{b}} \qquad BW_{\mathrm{2FSK}} = \frac{5}{2}R_{\mathrm{b}} \tag{31}$$

So, the energy per bit for OOK, BPSK, and 2FSK can be expressed respectively as

$$E_{\mathrm{bit\_OOK}} = \frac{P_{\mathrm{E\_OOK}}}{R_{\mathrm{b}}} + \frac{32\pi^2 L \cdot d^2 \cdot N_0 \cdot (1+\alpha) R_{\mathrm{b}}}{R_{\mathrm{b}} \cdot G_{\mathrm{t}} \cdot G_{\mathrm{r}} \cdot \lambda^2 \cdot \mathrm{K}} \cdot PAR \cdot \left[\mathrm{Q}^{-1}(BER)\right]^2 \tag{32}$$

$$E_{\mathrm{bit\_BPSK}} = \frac{P_{\mathrm{E\_BPSK}}}{R_{\mathrm{b}}} + \frac{8\pi^2 L \cdot d^2 \cdot N_0 \cdot (1+\alpha) R_{\mathrm{b}}}{R_{\mathrm{b}} \cdot G_{\mathrm{t}} \cdot G_{\mathrm{r}} \cdot \lambda^2 \cdot \mathrm{K}} \cdot PAR \cdot \left[\mathrm{Q}^{-1}(BER)\right]^2 \tag{33}$$

$$E_{\mathrm{bit\_2FSK}} = \frac{P_{\mathrm{E\_2FSK}}}{R_{\mathrm{b}}} + \frac{Q_{\mathrm{Lf}}}{Q_{\mathrm{Lf}} - Q_{\mathrm{f}}} \cdot \frac{40\pi^2 L \cdot d^2 \cdot N_0 \cdot R_{\mathrm{b}}}{R_{\mathrm{b}} \cdot G_{\mathrm{t}} G_{\mathrm{r}} \lambda^2} \cdot \left[\mathrm{Q}^{-1}(BER)\right]^2 \cdot \left(1 + \frac{R_{\mathrm{on}}}{V_{\mathrm{DD}}^2} \cdot \frac{Q_{\mathrm{Lf}}}{Q_{\mathrm{Lf}} - Q_{\mathrm{f}}} \cdot F \cdot \frac{40\pi^2 L \cdot d^2 \cdot N_0 \cdot R_{\mathrm{b}}}{G_{\mathrm{t}} G_{\mathrm{r}} \lambda^2} \cdot \left[\mathrm{Q}^{-1}(BER)\right]^2 + \frac{G}{Q_{\mathrm{Ls}}} \right) \tag{34}$$

According to (32-34), Figure 6 shows the energy per bit with OOK, BPSK and 2FSK modulation when the data rate is 25Kbps. From Figure 7 we can see that for a given data rate, the OOK modulation scheme has the lowest energy consumption when the transmission distance is short. This is because when the transmission distance is short, which means $d^2 \cdot R_{\mathrm{b}}$ is small; the analog circuit energy consumption is dominant instead of PA. The OOK modulation system is much simpler than others and thus OOK has the best energy performance. But with the increase of $d^2 \cdot R_{\mathrm{b}}$, PA consumed much more energy compared with other RF blocks. When $d^2 \cdot R_{\mathrm{b}}$ increases to a certain value, BPSK modulation becomes the most energy efficient. However, when $d^2 \cdot R_{\mathrm{b}}$ is large enough, 2FSK modulation scheme becomes the best choice. This leads to the high efficiency of Class E PA used in 2FSK system.

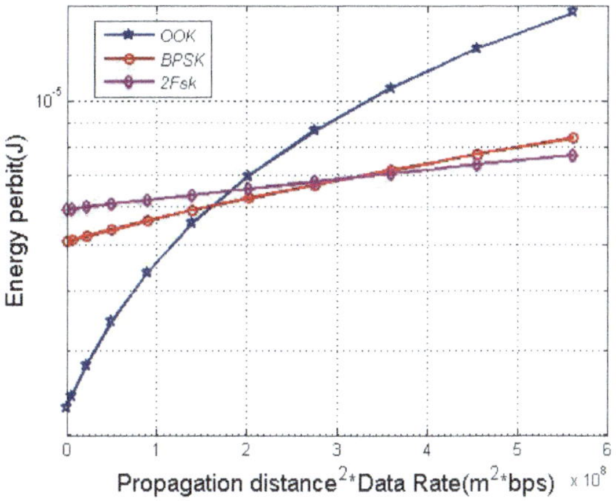

Fig. 7 Energy per bit for different RF front-end architecture in low data rate application, Rb=25Kbps, d=(0 150]m (*Reproduced from Ye Li, Dengyu Qiao, et al., Energy-Model-Based Optimal Communication Systems Design for Wireless Sensor Networks, Figure 8, International Journal of Distributed Sensor Networks Volume 2012, Article ID 861704, 11 pages doi:10.1155/2012/861704. © [2012] Hindawi Publishing Corporation. Reprinted, with permission.*)

With the similar method used to analyze the low bandwidth system, we can conclude that when $d^2 \cdot R_b < 1.8203e+8$, OOK modulation scheme is the most energy efficient; when $1.8203e+8 < d^2 \cdot R_b < 4.2214e+8$, BPSK modulation scheme has the best energy performance; if $4.2214e+8 < d^2 \cdot R_b < 1.125e+9$, 2FSK modulation scheme consumes the lest energy.

2.4.3 Summary and Design Steps for Energy Efficient System

In the former sections, we evaluate the energy performance of different modulation schemes under different WSNs applications. In reality, as to most of wireless communication systems, BW is fixed and cannot be adjusted. Then, in order to clearly illustrate the relationship among propagation distance, modulation scheme and energy consumption, BW is assumed to be constant during numerical analysis in the former subsections. According to the theoretical analysis results, if substituting the values of BW or R_b that we assumed, some recommended modulation schemes for designing an energy efficient system are given and summarized in TABLE 2 and TABLE 3. It is seen from TABLE 2 and TABLE 3 that under different application scenario, the optimal modulation scheme is different. So when designing an energy efficient system, we should firstly understand the application requirement, such as BW, data rate, and etc. Then, the theoretical analysis results in this section can be utilized to find out the optimal modulation scheme and choose the corresponding RF front-end architecture.

Table 2 Styles Energy Efficient System for Same Bandwidth

Modulation Scheme		d^2BW Large	Medium	Small
Bandwidth	***Wide***	16QAM	16QAM	64QAM, 32QAM
	Medium	4FSK, QPSK	QPSK	8QAM
	Narrow	BPSK	BPSK, OOK	OOK

Table 3 Styles Energy Efficient System for Same Data Rate

Modulation Scheme		d^2R_b Large	Medium	Small
Data Rate	***High***	16QAM	16QAM	16QAM
	Medium	4FSK	4FSK	QPSK
	Low	2FSK	BPSK	OOK

(Reproduced from Ye Li, Dengyu Qiao, et al., Energy-Model-Based Optimal Communication Systems Design forWireless Sensor Networks, TABLE 2-3, International Journal of Distributed Sensor Networks Volume 2012, Article ID 861704, 11 pages doi:10.1155/2012/861704. © [2012] Hindawi Publishing Corporation. Reprinted, with permission.)

3 Quality of Experience (QoE) in mHealth Monitoring System

3.1.1 Previous Work on QoE

Recently, some studies on QoE have been presented. Gong et al. proposed a pentagram model for measuring QoE degree [25]. They proposed 5 factors contributed to the QoE degree of a service. But they don't link these factors with objective measurable parameters. An adaptation scheme that is QoE-driven for optimizing content provisioning and network resource utilization for video applications over wireless networks has been proposed [26]. Unfortunately, this design is not for mHealth applications.

Among mHealth applications, energy consumption [27] affects QoE seriously. The battery capacity of mobile devices is usually limited, so if mHealth devices' energy consumption is too heavy, it will lead to interruptions of service and reduce QoE greatly as a result. Information integrity [28] received by users is also crucial because if users only receive part of the medical information, their experience for this monitoring service will be damaged.

In our research [29], the management of original medical information will influence energy consumption of mHealth systems as well as the information integrity. We proposed some models to describe the correlations among the sensor's management of original information, energy consumption, information integrity and QoE. Through theoretical derivations and simulations, we propose an optimal transmitting compress mechanism in wireless mHealth system considering health information integrality and energy consumption issues to achieve the optimum QoE performance.

3.2 System Model and Problem Formulation

3.2.1 QoE Model

In [25], Gong et al. proposed a pentagram model to measure QoE degree, which can be computed as:

$$QoE = \frac{1}{2}\sin(\lambda)(a \cdot b + b \cdot c + c \cdot d + d \cdot e + e \cdot a), \tag{35}$$

where λ represents the 72-degree angle between the two sides in the QoE pentagram model, and *a, b, c, d* and *e* denote service availability, service usability, service integrality, service retainability, and service instantaneousness, respectively.

The five parameters *a, b, c, d* and *e* reflect users' experience of different aspects of the service. For mHealth services, integrality (c) is used to measure how much information users get from mHealth monitoring service [30]. For example, a doctor hope to get patients' daylong body temperature data, but mHealth monitoring cluster node only transmit 18-hours data to him, which transmits 75% information of user's expectation, so user's experience for information integrality of this service would not be 100%. Here, we take user experience for mHealth monitoring service's information integrality as QoE for Information Integrity (QoE_Inf); retainability (d) describe user experience for how long the monitoring can last, so it can be regarded as QoE for Energy Consumption (QoE_Eng). So that

$$QoE = \frac{1}{2}\sin(\lambda)(a \cdot b + b \cdot QoE\_Inf + QoE\_Inf \cdot QoE\_Eng + QoE\_Eng \cdot e + e \cdot a), \tag{36}$$

3.2.2 System Model

Information Integrity Received By User: Users receive medical information from sink sensor node (cluster sensor) in monitoring system. The information integrity of one monitoring service is decided by the information quantity users receive from cluster sensor [31]. Cluster sensor manages original information collected by acquisition sensors and then transmits managed information to users.

In [32], the authors point out that user experience of a service follows the logarithmic laws, and QoE function can be modeled in the logarithmic form. However, the QoE in [31] only focuses on the aspect of information integrity, apart from the synthetic user experience for the whole system. So we adopt the QoE for Information Integrity as the logarithmic function between r_i and r in mHealth system, which is specified as

$$QoE\_Inf(r_i, r) = a_1 \ln \frac{a_2 r_i}{r}, \tag{37}$$

where the constant parameters a_1 and a_2 are both positive and vary according to the information type the system transmits. r_i is the actual transmit information quantity and r is the excepted information quantity.

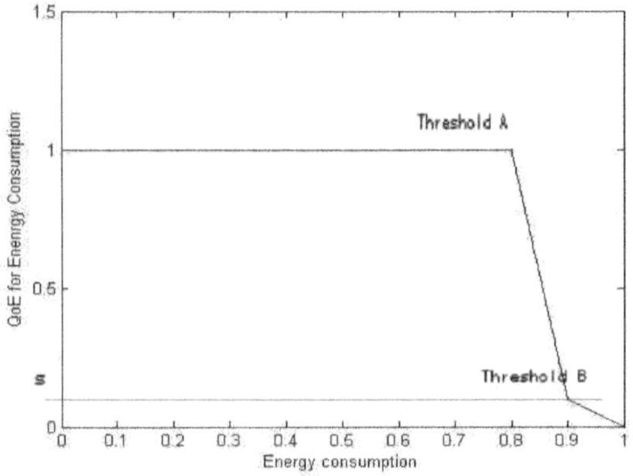

Fig. 8 Linear approximation for relationship between the QoE and remained energy (Reproduced from Da Xu, Chenguang He, and Ye Li, Improving Quality of Experience in M-health Monitoring System, Figure 2, proceedings of the 35th Annual International Engineering in Medicine and Biology Society (EMBC). © [2013] IEEE . Reprinted, with permission.)

Energy consumption influence on QoE Model: In mHealth monitoring system, the energy consumption is a key factor influences user experience that cannot be ignored [32]. QoE varies with energy consumption, and the trend map is given as Figure 8 [13].

We normalized QoE for Energy Consumption as 0~1, when $QoE\_Eng = 1$, it means users feel completely satisfied while $QoE\_Eng = 0$ means the user are completely unsatisfied. The bigger the value of QoE_Eng is, the more satisfied the user will be about the energy consumption performance of the system.

When the energy consumption is less than threshold *A* (means the whole monitoring time of the system is lasting long enough), *QoE* for Energy Consumption always stay at high level; on the contrary, if the energy consumption is more than threshold *A*, QoE_Eng will drop acutely. In addition, when the energy consumption is more than threshold *B* (means the monitoring service cannot last for a very short time), the user has accepted the fact that the system would power off soon. So, QoE for Energy Consumption will drop slowly but in a very low level. In summary, the relationship between the QoE for energy of the system and the remained energy is shown in Figure 8. Here, *s* is the value of QoE for Energy Consumption when the energy consumption of the system is *B*.

Mathematically, the relationship shown in Figure 8 is as follows:

$$QoE\_Eng = \begin{cases} 1 & (NEC < A) \\ \dfrac{1-s}{A-B}(NEC - A) + 1 & (A < NEC < B) \\ \dfrac{s}{B-1}(NEC - 1) & (B < NEC < 1), \end{cases} \tag{38}$$

where *NEC* means Normalized Energy Consumed in one monitoring service [14].

Problem Formulation: The QoE of one mHealth monitoring service are influenced by the energy consumption and information integrity. Information integrity and energy consumption are both determined by the information quantity the users receiving during the service. To achieve the highest *QoE*, the problem can be formulated mathematically as the following constrained optimization problem :

$$Max: QoE = \frac{1}{2}\sin(\lambda)(a \cdot b + b \cdot QoE\_Inf(r_i, r) + QoE\_Inf(r_i, r) \cdot QoE\_Eng + QoE\_Eng \cdot e + e \cdot a),$$

$$s.t.: \quad r_{min} < r_i < r_{max},$$

where r_{min} and r_{max} stand for the minimum and maximum of information quantity transmitted to users in a system, respectively. It is limited by the hardware configuration and software scheme of the mHealth monitoring system.

3.3 Optimal Transmission Driven By QoE

In this section, we will set some values to formulae to investigate the influence of transmitting medical information quantity on the QoE in an mHealth monitoring system.

The research problem is to calculate the appropriate information quantity transmitted to users in one monitoring service. First, we suppose r as 1, if the information quantity user received in line with his expectation, i.e. $r_i= 1$; if exceeds expectation, then $r_i> 1$; if it is below expectation, then $r_i< 1$. For ease of calculation, we take a_1 equals 1 and a_2 equals 10, so that when information quantity reaches user's expectation, the QoE for the information integrity equals 1 which represents users are completely satisfied with the information integrity of this service. So,

$$QoE\_Inf(r_i) = \ln 10 r_i. \tag{39}$$

The energy consumption for a completed service is positively related to the information quantity transmitted during the service. We propose

$$NEC = kr_i, \tag{40}$$

where k is the energy it consumes to transmit unit quantity of information. So QoE for Energy Consumption can be written as

$$QoE\_Eng = \begin{cases} 1 & (r_i < \frac{A}{k}) \\ \frac{1-s}{A-B}(kr_i - A) + 1 & (\frac{A}{k} < r_i < \frac{B}{k}) \\ \frac{s}{B-1}(kr_i - 1) & (\frac{B}{k} < r_i < \frac{1}{k}). \end{cases} \tag{41}$$

According to Figure 8, we take A=0.8, B=0.9, and s=0.1. Therefore, QoE for Energy Consumption can be written as

$$QoE\_Eng = \begin{cases} 1 & (r_i < 0.8) \\ -9r_i + 8.2 & (0.8 < r_i < 0.9) \\ -r_i + 1 & (0.9 < r_i < 1). \end{cases} \tag{42}$$

Substituting equation (5) and equation (8) into equation (2), the QoE of an mHealth system is given by

$$QoE = \begin{cases} \frac{1}{2}\sin(\lambda)(\ln 10r_i + a\ln 10r_i \\ +b + ad + bd) & (r_i < 0.8) \\ \frac{1}{2}\sin(\lambda)((8.2 - 9r_i)\ln 10r_i + a\ln 10r_i \\ +(8.2 - 9r_i)b + ad + bd) & (0.8 < r_i < 0.9) \\ \frac{1}{2}\sin(\lambda)((1 - r_i)\ln 10r_i + a\ln 10r_i \\ +(1 - r_i)b + ad + bd) & (0.9 < r_i < 1). \end{cases} \tag{43}$$

We suppose $a=b=d=1$, which means the availability, usability and instantaneousness of the system have reached users' expectation. So the overall *QoE* function is given by

$$QoE = \begin{cases} 0.4755(\ln r_i + \ln r_i + 3) & (r_i < 0.8) \\ 0.4755((8.2-9r_i)\ln r_i + \ln r_i \\ +(8.2-9r_i)+2) & (0.8 < r_i < 0.9) \\ 0.4755((1-r_i)\ln r_i + \ln r_i + (1-r_i) \\ +2) & (0.9 < r_i < 1). \end{cases} \tag{44}$$

Thorough calculation, we find that *QoE* reaches the peak when $r_i = 0.8$. It means that we only need to offer 80% information quantity of user's expectation to make the total QoE of mHealth system the maximum. When cluster node receives the information collected by collection sensors, it should do some data process. Through our model and derivations, the optimal size of the processed information has been given. It can be concluded that the QoE in mHealth monitoring services is promoted definitely by opting for the optimal information transmission mechanism of the system.

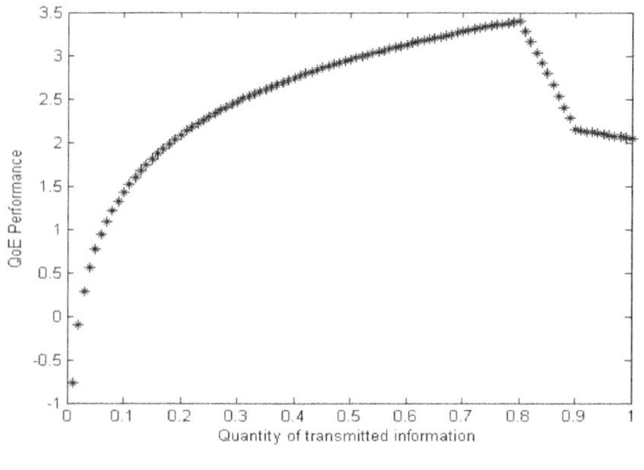

Fig. 9 Variation of QoE degree along with the information quantity (*Reproduced from Da Xu, Chenguang He, and Ye L, Improving Quality of Experience in M-health Monitoring System, Figure 2, proceedings of the 35th Annual International Engineering in Medicine and Biology Society (EMBC). ©[2013] IEEE . Reprinted, with permission.*)

From Figure 9, we conclude that for an mHealth monitoring service, when the information quality transmitted to users is low, energy consumed in transmission keeps low and the service can run for a long time with the information integrity increases quickly. So the whole QoE increase in the end. When the transmitted information quantity reaches a threshold *A*, energy consumption reaches the level that makes mHealth monitoring system to offer only limited service time, so user's experience for service retainability reduces more drastically than increase speed of QoE for information integrity, which leads to the reduction of synthetic QoE performance. The last curve after information quantity reaches threshold *B* indicates user's experience for service retainability is very low and even though the QoE for information integrity is very high, the synthetic QoE still keeps low. And this is coincident with our common sense.

4 Conclusions and Future Work

Wireless Sensor Networks (WSNs) have become an abstractive solution in many potential applications. However, energy consumption is a key issue since sensors nodes in WSNs is battery-operated with limited energy sources. Therefore, in this chapter, we firstly focus on how to minimize energy consumption to extend the life time for WSNs by designing the optimal energy-quality system in terms of various application scenarios. On the other hand, energy consumption is a QoS index, which cannot evaluate absolutely users' feeling with services. Therefore, we also investigated the problem of how to adapt the information collected by body sensors and transmit it to users for the highest QoE performance in mHealth system, and propose new model, and figured out the optimal information quantity transmitted to users by the cluster sensor. In the future work, we will investigate the energy efficiency and novelty applications for mHealth devices based on wireless sensor networks.

References

[1] He, C., Fan, X., Li, Y.: Toward ubiquitous healthcare services with a novel efficient cloud platform. IEEE Transactions on Biomedical Engineering 60(1), 230–234

[2] Tang, Q., Yang, L., Giannakis, G.B., Qin, T.: Battery power efficiency of PPM and FSK in wireless sensor networks. IEEE Transactions on Wireless Communications 6(4), 1308–1319 (2007)

[3] Chen, D., Varshney, P.K.: QoS support in wireless sensor networks: A survey. In: International Conference on Wireless Networks (ICWN 2004), Las Vegas, Nevada, USA (June 2004)

[4] Wang, X., Ren, Y., Zhao, J., et al.: Energy efficient transmission protocol for UWB WPAN. In: Proc. IEEE VTC 2004, USA, pp. 5292–5296 (2004)

[5] Zhou, Z., Zhou, S., Cui, S., Cui, J.: Energy-efficient cooperative communication in a clustered wireless sensor network. IEEE Transactions on Vehicular Technology 57(6), 3618–3628 (2008)

[6] Wang, T., Heinzelman, W., Seyedi, A.: Minimization of transceiver energy consumption in wireless sensor networks with AWGN channels. In: 46th Annual Allerton Conference on Communication, Control, and Computing, USA, pp. 62–66 (September 2008)

[7] Uppu, N., Subrahmanyam, B.V.S.S., Garimella, R.: Energy Efficient Routing Technique for Ad-Hoc Sensor Networks [EERT]. In: Proc. IEEE Sensors Applications Symposium, USA, pp. 228–232 (February 2008)
[8] Lahiri, K., Raghunathan, A., Dey, S.: Communication-based power management. In: Proc. IEEE Design Test Comput., pp. 118–130 (July 2002)
[9] Cui, S., Madan, R., Goldsmith, A.J., Lall, S.: Cross-layer energy and delay optimization in small-scale sensor networks. IEEE Transactions on Wireless Communications 6(10), 3688–3699 (2007)
[10] Srivatsava, M.: Power-aware communication systems. In: Rabaey, M. (ed.) Power-Aware Design Methodologies, ch. 11, Kluwer, Norwell (2002)
[11] Wang, A.Y., Chao, S., Sodini, C.G., Chandrakasan, A.P.: Energy efficient modulation and MAC for asymmetric RF microsensor system. In: International Symposium on Low Power Electronics and Design, USA, pp. 106–111 (2001)
[12] Schurgers, C., Srivastava, M.B., et al.: Modulation scaling for energy aware communication systems. In: Proc. ISLPED 2001, USA, pp. 96–99 (August 2001)
[13] Li, Y., Bakkalogu, B.: A system level energy model and energy-quality evaluation for integrated transceiver front-ends. IEEE Transactions on Very Large Scale Integr. (VLSI) Systems 15(1), 90–103 (2007)
[14] Li, Y., Reisslein, M., Chakrabarti, C.: Energy efficient video transmission over a wireless link. IEEE Transactions on Vehicular Technology 58(3), 1229–1244 (2009)
[15] Li, Y., Qiao, D., et al.: Energy-model-based optimal communication systems design for wireless sensor networks. International Journal of Distributed Sensor Networks 2012, Article ID 861704, 11 pages, doi:10.1155/2012/861704
[16] Razavi, B.: R. F. Microelectronics. Prentice-Hall, Englewood Cliffs (1998)
[17] Daly, D.C., Chandrakasan, A.P.: An energy-efficient OOK transceiver for wireless sensor networks. IEEE Journal of Solid-State Circuits 42(5), 1003–1009 (2007)
[18] Wang, A.Y., Sodini, C.G.: On the energy efficiency of wireless transceiver. In: IEEE International Conference on Communication, ICC 2006, pp. 3783–3787 (2006)
[19] Huang, K.H., Wang, C.K.: A cost effective binary FSK demodulator for low-IF radios. In: International Symposium on VLSI Technology, Systems, and Applications, pp. 133–134 (2001)
[20] Rappaport, T.S.: Wireless Communications: Principles and Practice, 2nd edn. Pearson Education Asia Limited and Publishing House of Electronics Industry (2004)
[21] Gustavsson, M., Wikner, J.J., Tan, N.N.: CMOS Data Converters for Communications. Kluwer Academic Publishers (2002)
[22] Cui, S., Goldsmith, A.J., Bahai, A., et al.: Energy constrained modulation optimization. IEEE Transactions on Wireless Communications 4(5), 2349–2360 (2005)
[23] Chen, P.F., Larson, L.E., et al.: High-efficiency power amplifier using dynamic power-supply voltage for CDMA applications. IEEE Trans. Microw. Theory Tech. 47(8), 1471–1476 (1999)
[24] Hasani, J.Y., Kamarei, M.: Analysis and optimum design of a class E RF power amplifier. IEEE Transactions on Circuits Systems I, Reg. Papers 55(6), 1759–1768 (2008)
[25] Gong, Y., Yang, F., Huang, L.S.S.: Model-based approach to measuring quality of experience. In: First International Conference on Emerging Network Intelligence, pp. 29–32 (2009)

[26] Khan, A., Sun, L., Jammeh, E., Ifeachor, E.: Quality of experience-driven adaptation scheme for video applications over wireless networks. Communications (IET) 4(11), 1337–1347 (2010)
[27] Ksentini, A., Hadjadj-Aoul, Y.: QoE-based energy conservation for VoIP over WLAN. In: IEEE Wireless Communications and Networking Conference, pp. 1692–1697 (2012)
[28] Liu, L., Zhou, W., Song, J.: Quantitative customer perception evaluation for telecommunication service. In: Third International Conference of Pervasive Computing and Applications (CPCA), pp. 912–915 (2008)
[29] Xu, D., He, C., Li, Y.: Improving Quality of Experience in m-health monitoring system. proceedings of the 35th Annual IEEE International Engineering in Medicine and Biology Society (EMBC)
[30] Lee, Y.W., Strong, D.M., Kahn, B.K., Wang, R.Y.: AIMQ: a methodology for information Quality assessment. Information & Management 40(2), 133–146 (2002)
[31] Reich, P., Tuffin, B., Schatz, R.: Logarithmic laws in service quality perception: where microeconomics meets psychophysics and quality of experience. Telecommunication Systems, 1–14 (2011)
[32] Eberle, W., Bougard, B., Pollin, S., Catthoor, F.: From myth to methodology: cross-layer design for energy-efficient wireless communication. In: 42nd Design Automation Conference, pp. 305–308 (2005)

Questions

Q1. If wireless sensor networks (WSNs) are used to healthcare and other engineering applications, what are the main problems to be concerned in the WSNs design? and why?

Q2. Please list the low power technologies from different aspects.

Q3. Consider medium bandwidth application. Assume that QPSK, 8QAM and 4FSK modulation schemes are used. Please give out the simulation results by MATLAB.

Q4. What is the difference between QoS and QoE?

40. Quality of Service in Wireless Technologies for mHealth Service Providing

M. Sanchez Meraz, A. Leyva Alvarado, and S. Gonzalez Ambriz

Telecommunications Department of the National Polytechnic Institute, Mexico
mmeraz@ipn.mx, {sgonzaleza0500,aleyvaa1100}@alumno.ipn.mx

Abstract. Wireless communication technologies are an essential part of the broadband infrastructure of countries. The advantages of the installation of such networks made them very attractive to deploy broadband services in regions where wired technologies do not provide coverage. These technologies have been very successful alternative mainly in developing countries and particularly to support remote health services. Additionally, the unceasing development of mobile applications, most of them using smartphones, is changing the way medical care services are provided. The use of wireless technologies is improving the operating way of health care services and the conditions for the users to receive them. Ability of wireless technologies to transmit differentiated services is essential for the deployment of networks for mHealth applications. The efficient operation of such networks requires identifying requirements of real-time applications such as biomedical signal transmission, and requirements of *store and forward* applications such as access to EMR (Electronic Medical Record). In the first case it is necessary to assign a high priority to the transmission, but also to meet strict parameters of Quality of Service (QoS) such as delay, delay variation and packet loss rate. In the second case the QoS requirements for transmission are less strict. This chapter presents an overview of the QoS mechanisms offered by current and future wireless mobile and wired IP technologies. The defined technical requirements of quality of service for the main mHealth applications are also presented. Finally a review of the capabilities of 4G technologies to support future mHealth applications is presented.

Keywords: mHealth, QoS, QoE, LTE Advanced, WiMAX 2.0, ITU QoS recommendations, IMT Advanced, mHealth QoS requirements.

1 Traffic Treatment in Mobile and Wired IP Networks

The capacity of every communication network is a limited resource so all their carried services have to compete for using this shared network resource. Because of the sustained increase in the traffic transported by the current communications networks, several mechanisms for traffic classification have been designed in order to assign different priority levels to the requests for network resources. In the case of mobile

S. Adibi (ed.), *Mobile Health*,
Springer Series in Bio-/Neuroinformatics 5, DOI: 10.1007/978-3-319-12817-7_40

communication networks and wired IP networks such traffic classification mechanisms have followed their own evolution. However the current technological convergence to *All-IP Networks* demands a transparent coexistence and integration of both mobile and wired networks. This interaction is necessary to ensure that traffic classification keeps on the same priority level through different networks.

Since its origins and during the first and second generation of mobile communication networks most traffic was voice traffic. At that stage of mobile networks standard mechanisms for traffic classification were not available because they transport one single traffic type. Under such scenario the congestion of a network was solved mainly by expanding the capacity of the network or through traffic rerouting policies.

Since the introduction of the third generation of mobile communication networks in the early 2000s, data traffic has had a fast growing. This behavior will be maintained in the future because of a technologic convergence to the so called *All-IP Networks*. Voice communication still is an important service but currently most traffic is generated by data and multimedia applications and this proportion will continue growing. Current 3G and 4G broadband wireless communication technologies like LTE and WiMAX own the capacity to provide differentiated services, i.e. assign different priorities to selected data flows or applications. Under this premise the management of shared network resources is done in order that services or applications operate with a specified quality level. With an increase in the network traffic load, prioritization determines which services have to maintain preferential transmission and performance characteristics. These traffic classification mechanisms are of capital importance in the case of network congestion events. By contrast when a network is operated with a traffic load far below the network´s capacity, the classification mechanisms are not useful because the network is over provisioned. In the case of wired IP networks most of the carried services are delivered in a quality level classification referred as *Best Effort*. Under this classification all carried services compete under equal conditions for the network resources. This is the main resource allocation mode defined for the Internet operation when this network was created. However due to the large increase in the traffic from multimedia applications and high data transfer services, like Peer to Peer (P2P) file sharing, it has become necessary to apply policies and control mechanisms for management of IP networks traffic. All these actions are intended to ensure defined quality levels for an adequate delivery of specific services.

Two of the most important mechanisms to support traffic classification and quality of service for IP networks are *DiffServ* [1] and *MPLS* [2]. *DiffServ* is a standard mechanism for traffic classification and management used to provide quality of service. *MPLS* on the other hand is a network protocol that specifies routing information to each IP packet and allows routers to assign explicit paths to several traffic classes. DiffServ is based on traffic conditioners located at the edge of the network to set requirements for each IP packet. In the case of *MPLS*, this protocol works over label-switching routers with the capability of reading header information and assigning packets to specific routes. Both mechanisms *DiffServ* and *MPLS* work together to provide quality of service over IP networks.

Since its creation the Internet has been transformed from being a communication network to support academic and research activities to be a global information network. This global network is based on the integration of infrastructure of network service providers (NSP) and services and applications from content service providers (CSP). In the case of the network infrastructure this is based on several technologies including optical like Optical Transport Network (OTN), wired like Ethernet, mobile like WiMAX and LTE and wireless like WiFi. Under this scenario with the coexistence of heterogeneous networks, the delivered services to an end user may require to be transported across different networks. In this case the service classification management becomes a major challenge to ensure that services are delivered with the required quality of service on an *end-to-end* basis.

Currently Internet has the capability to transport a wide variety of services and multimedia applications including *anywhere anytime* real time apps and it has become an integral part of the ordinary people´s daily life. In this sense for several decades health sector has maintained a constant research and development activity to solve a large number of needs in this area [3]. Recently these activities have been using the *Information and Communication Technologies* (ICT) as a principal support. New developed solutions include a variety of healthcare services based on the use of mobile networks and identified as mHealth services.

Most users of medical applications based on mobile technologies don´t have knowledge about technical requirements to receive their services with quality. But they want to take a quick and reliable access to their different multimedia applications on an *anywhere anytime* basis. QoS mechanisms provided by 3G and 4G technologies will be essential to support current and future applications of mHealth. It is important to notice that these services may require strict management of quality of service because life of people may be at risk.

This chapter presents a review of two of the main new generation mobile technologies, WiMAX and LTE, with a specific focus on their QoS capabilities. The purpose of this review is to evaluate the ability of such technologies to support mHealth services. With this target section 2 presents technical recommendations from the International Telecommunications Union (ITU) related to quality of service with application to mobile networks. Section 3 presents the technical requirements related to quality of service for providing the main mHealth applications. In section 4 it is presented a review of the LTE and WiMAX QoS capabilities to support the requirements of mHealth services. Finally a brief review of the pending challenges to provide quality of service on an *end to end* basis is presented.

2 ITU QoS Recommendations for Mobile Networks

Quality has been defined by several standardization organizations using their own contexts. Here is presented the definition of QoS from ITU, the United Nations agency responsible for the promotion of the international standardization of telecomm infrastructure. ITU defines QoS as the "Totality of characteristics of a telecommunications service that bear on its ability to satisfy stated and implied needs of the user of the service" [4].

Defining the quality of a service can be a complicated task because this definition includes the people perception and this is a subjective matter. However, there are several technical parameters that are used to qualify the performance of a communications network and they are included in the ITU recommendations as performance thresholds. The main technical parameters to assess the quality of service in a communication network are the following, including general definitions:

1. **Delay**. The elapsed time for a data packet generated by a source application, at one end of a link, to reach the destination application, at the other end of the link. This parameter is fundamental to define if a service can be provided on a real time basis.
2. **Delay variation**. The difference between one packet delay and other packet delay used as reference. This parameter determines the consistency in the delay value from the packets related to a specific application. Interactive services require a low level of delay variation.
3. **Packet loss rate**. Percentage of packet lost from the packets related to a specific application or service. This parameter is related with the quality of the service perceived by the final user, like sound or image quality.
4. **Bandwidth**. Throughput required in a link to transport a specific service with a defined quality. Although this is not a standard quality parameter it is usually defined to complete the specifications to provide a service with a defined quality level.

This is a set of standard technical parameters that allow objectively evaluate the performance of a network or link.

The following review includes a specific ITU recommendation for IP networks usually applied to backbone networks. A recommendation for wireless access networks is also presented. These two recommendations are considered because in a general scenario to deliver mHealth services it will be required to transport the end user traffic by these two types of networks: a wireless access network and an IP backbone network. Figure 1 shows the relationship between the reviewed ITU QoS recommendations that can be applied to the existing 3G and future 4G mobile networks. This figure shows the year when these recommendations were approved. It is important to notice that QoS technical requirements for several services have been well established for at least two decades. Under these known performance targets, new mobile network technologies fundamentally aim to expand their capacity, coverage and mobility characteristics. However these recommendations will remain as the mandatory references to assess their quality of service.

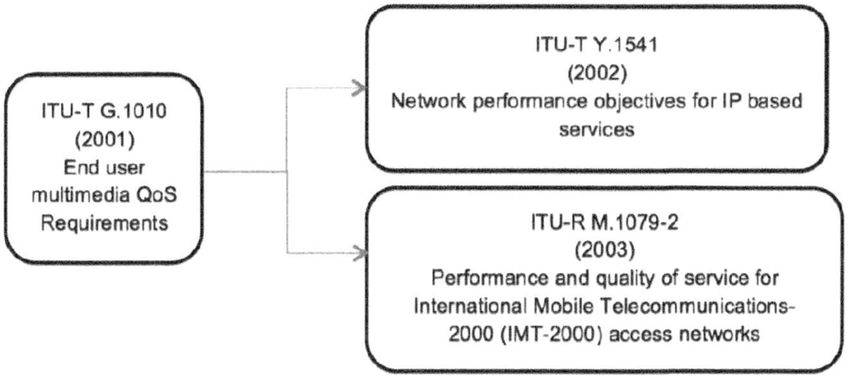

Fig. 1 Relationship between ITU QoS recommendations related to mobile networks

Relationship between ITU-T G.1010, ITU-T Y.1541 and ITU-M.1079-2 Recommendations. The main ITU recommendations related to the QoS issue on mobile networks are the ITU-T Rec. G.1010, ITU-T Rec. Y.1541 and ITU-R Rec. M.1079-2. Each of these recommendations have a different scope as illustrated in Figure. 2. *End-to-end* QoS implies ensuring quality requirements for an entire link, from the user terminal equipment, for example a smartphone, up to a second terminal equipment located for example at a medical centre. The scope of the ITU-T Rec. G.1010 is general and define *end-to-end* technical requirements to provide multimedia services. This recommendation is not associated with specific technologies or networks. The ITU-T Rec. Y.1541 has also an *end-to-end* scope and it is related to the characteristics of quality required by the services provided on IP networks. The ITU-R M.1079-2 defines QoS characteristics for the wireless radio access section of the network, which enables the link between the user terminal equipmet and the core network. The ITU-T Rec. Y.1541 is general covering the *end-to-end* link, while the ITU-R Rec. 1079-2 is more specific and it only covers the radio access network segment in an *end-to-end* path.

It is important to notice that in Figure 2 the recommendations do not cover the link between the sensor of the mHealth application and the terminal equipment of the user of this application. The most common technologies currently used for this link are WiFi and Bluetooth but these technologies are beyond the access network which is the border of the ITU QoS recommendations. So in the context of the ITU recommendations an *end-to-end* link refers to a communication link established between terminal equipments from different users, wich are served by any access network. Although WiFi is also considered an access technology, in this case the WiFi link is not used to connect the equipment to an access network but to create an ad hoc network to connect the smartphone and the mHealth sensors. So for this wireless link between the mHealth sensors and the user terminal equipment there are not specific ITU recommendations that address the issue of quality of service.

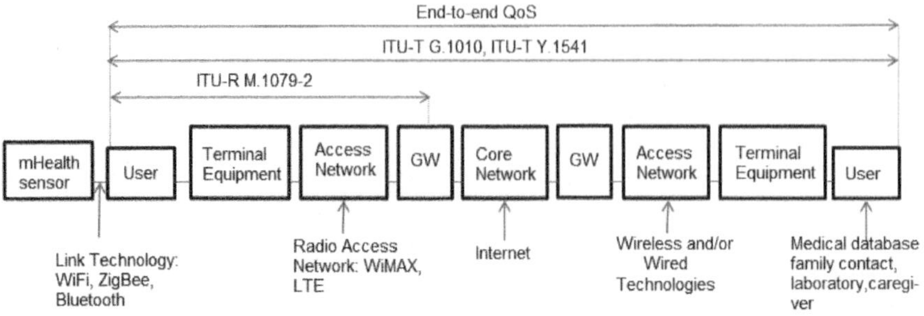

Fig. 2 *End-to-end* network architecture and ITU recommendations scope

ITU-T Recommendation G.1010. End-user Multimedia QoS Requirements. This recommendation defines the required performance of multimedia applications using several technical parameters, with a focus on the effects that the user can perceive from the services. This recommendation is independent of the network architecture on which the services are transmitted, so it can be used as a guide to set service requirements for different networks. The parameters considered are delay, delay variation (jitter) and information loss (expressed as Packet Loss Ratio). The quality characteristics presented in Table 1, allow a distinction between different application, e.g. by indicating the requirements for "urgent" applications like transactions, voice and video as well as "less urgent" applications like file download and messaging. The QoS values are guidelines for upper limits of technical characteristics for different applications in order to deliver a service perceived as acceptable by the user. When the upper limits are exceeded, the service is considered as not satisfactory. The application categories are intended to be used as a basis for delivering QoS classes and QoS mechanisms for the underlying transport networks [5].

ITU-T Recommendation Y.1541. Network Performance Objectives for IP-Based Services. This recommendation defines five different QoS classes for IP-based networks and the performance parameters to be used in the planning of international IP service offerings. The objective values are a result of the analysis of key applications such as conversational telephony, multimedia conferencing, reliable data exchange over TCP, etc. Each QoS class has a different specific combination of performance values as it is shown in Table 2. Any flow that meets all performance objectives is considered to meet with the class normative recommendation.

Also some applications are suggested to be used for each QoS class, for example: for QoS class 0 should be used real-time and jitter sensitive with high interaction applications, such as VoIP; for QoS class 1 should be used real-time, sensitive and interactive applications as video teleconference; for QoS class 2 should be used highly interactive applications as transaction data applications [6].

Table 1 Model for user-centric QoS categories (Source: ITU-T G.1010)

	Interactive (delay << 1s)	Responsive (delay ~ 2s)	Timely (delay ~ 10s)	Non-critical (delay >>10s)
Error tolerant	Conversational voice and video	Voice/video messaging	Streaming audio and video	Fax
Error intolerant	Command/Control (e.g. Telnet, interactive games)	Transactions (e.g. e-commerce, www browsing, email access)	messaging, downloads (e.g. FTP, still image)	Background

Table 2 IP network QoS class definitions and network performance objectives (Source: ITU-T Y.1541)

Network performance parameter	Nature of network performance objective	QoS Classes					
		Class 0	Class 1	Class 2	Class 3	Class 4	Class 5
IP Packet Transfer Delay	Upper bound on the mean IP transfer delay	100 ms	400 ms	100 ms	400 ms	1 s	Unbounded
IP Packet Delay Variation	Upper bound on the $1\text{-}10^{-3}$ quantile of IPTD minus the minimum IPTD	50 ms	50 ms	Unbounded	Unbounded	Unbounded	Unbounded
IP Packet Loss Rate	Upper bound on the packet loss probability	10^{-3}	10^{-3}	10^{-3}	10^{-3}	10^{-3}	Unbounded
IP Packet Error Rate	Upper bound	10^{-4}	10^{-4}	10^{-4}	10^{-4}	10^{-4}	Unbounded

ITU-R Recommendation M.1079-2. Performance and Quality of Service Requirements for International Mobile Telecommunications-2000 (IMT-2000) Access Networks. In this recommendation QoS is defined as "the collective effect of service performance which determines the degree of satisfaction of a user of a service". This takes into consideration elements as the following:

- **Speech quality:** Used to express the degree of the customer satisfaction with conversational speech transmission. This includes the whole speech path and is dependent on the handset and the transmission medium.
- **Connection performance:** Expressed in the ITU-T Recommendation E.770 as Grade of Service. Consists of the signaling delay for call set-up and call release, the probability of *end-to-end* blocking, the probability of unsuccessful handover, etc.

- **Service retainability performance:** Defined in ITU-T Recommendation E.800 as the probability that a service, once obtained, will continue to be provided for a communication under given conditions.

However, external bearers are out of the scope of the recommendation, thus only is indicated the IMT-2000 bearer to provide QoS within the IMT-2000 access network. The IMT-2000 bearer consists of two parts:

- **Radio access bearer service:** which provides confidential transport of signaling and user data between the mobile terminal and the core network radio access network (CN-RAN) interface with the appropriated IMT-2000 QoS negotiated or with a default QoS for signaling.
- **Core network bearer service:** which connects the IMT-2000 core network radio access network (CN-RAN) node with the core network (CN) gateway of the external network. This bearer controls and utilizes the backbone network to provide the contracted IMT-2000 bearer service.

The defined QoS classes consider the effects the user can perceive and they are independent of the network architecture or technology. Four QoS classes are defined for IMT-2000: Conversational class, interactive class, streaming class and background class. The conversational class includes very delay-sensitive applications, while background class includes delay-tolerant applications. These QoS classes are related with the services transported in the mobile access network and they do not include the performance of *end-to-end* bearers. However these requirements for the access networks are aligned with those in ITU-T G.1010 for *end-to-end* QoS evaluation. The M.1079-2 recommendation defines the QoS performance values the end user expects for the conversational, interactive and streaming services for IMT-2000. Table 3 shows the QoS requirements for the interactive class, that could be applied to mHealth services [7].

3 mHealth QoS Requirements

New immersive, pervasive and interactive mHealth applications have emerged in the recent years and it is expected to continue a great development activity of the same in the future. These applications can provide emergency medical services, self-diagnosis and preventive care, mobile patient monitoring, post-hospital care, collaborative teleconsult, medical information management, etc. All these services require specific data rates and QoS parameters to be properly delivered. However for the proper delivery of these applications with a suitable *end-to-end* QoS guaranty several factors are involved, such as the wireless mobile communication environments, available and/or assigned network resources, IP packet delay, delay variation, PLR, etc. This clearly implies the need of mechanisms and context-aware policies for the transmission of these applications through different mobile environments where their traffic may be transported [8][9].

Table 3 End-user performance expectations – Interactive services. (Source: ITU-R M.1079-2)

Medium	Application	Degree of symmetry	Data rate (kbps)	Key performance parameters and target values		
				One way delay	Delay variation (ms)	Information loss
Audio	Voice messaging	Primarily one way	4-32	<1 s for playback <2 s for record	< 1	< 3 % PLR
Data	Web brows-ing,-HTML	Primarily one way		Preferred: <2 s/page Acceptable: <4 s/page	Not appli-cable	Zero
Data	Transaction services, High priori-ty	Two way		Preferred: <2 s Acceptable: <4 s	Not appli-cable	Zero
Data	E-mail (server access)	Primarily one way		Preferred: <2 s Acceptable: <4 s	Not appli-cable	Zero

Table 4 presents a resume of the QoS requirements to the adequate delivery of mHealth services. For example, a three-lead ECG signal is sampled and quantized with 12 bits per sample and a strict delay requirement is set for this service in order to provide the capability of real-time detection of problems with the patient health status. Blood pressure presents a data rate less than 10 Kbps (low data rate) and requires similar delay values that those of ECG. Common image files (such as mammogram, MIR and ultrasound image) require the transfer of a large size files and also requires data integrity. While voice and video services are considered error tolerant, ECG and file transmission require data integrity [10]. Table 5 shows acceptable values for delay, PLR, and data rate for voice, video and ECG traffic, these values are presented according with the reviewed recommendations in Section 2.

The quality of service delivered by the mHealth service platform must match with the quality of service required by the medical professionals. The network must have the ability to provide an adequate QoS level defined, for example to serve an emergency application, which requires timely medical assistance to the patient. As can be seen, the QoS level for an emergency situation differs from that of a non-emergency situation. Therefore mHealth services platform must have knowledge of the current situation (i.e. health status) of the patient. This application-level knowledge is contextual information, and currently it is possible to integrate this information to the mHealth platform [8][11]. This contextual information may include the current situation of the patient and with this knowledge it is possible to assign specific QoS parameters to meet a service level in accordance with that patient situation. The complication arises when the data from a

Table 4 mHeatlh services requirements (Source: [9][10][11])

mHealth Service	Data Rate	QoS parameters	Temporal/spatial (number of samples/sec)	Contrast/resolution (bits per sample)
Electrocardiography (ECG) monitoring	12 Kbps/s 12 channels	Delay	1250	x12
Blood pressure monitoring	<10 Kbps	Delay	1	x16
Heart rate	~ 120 Kb/s	Packet loss, Delay PSNR, Frame size,	10000	×12
Region of Interest Image (e.g Mammogram)	15- 24 MB (Image size)	Packet loss, PSNR, Frame size,	4096 × 4096	×12
Radiology	~ 6 MB (Image size)	Packet loss PSNR, Frame size,	1024 × 1250	×12
Magnetic Resonance Imaging (MIR)	< 1MB (Image size)	Packet loss PSNR, Frame Rate,		
Ultrasound video streaming	250 Kb/s – 1.2 Mb/s (WMV2)	Frame size, Packet loss, Delay	512 × 512	×8

Table 5 Standard values for QoS parameters transmission from network access until the end point (Source: [8])

Services	Data rate	Maximum Delay	Packet Loss
Audio	4 – 25 kbps	150 – 400 ms	3%
Video	32 – 384 kbps	150 – 400 ms	1%
ECG	1 – 20 kbps	Approx. 1s	Zero
File Transfer	Not available	Not available	Zero

patient is transported on a mix of communication network infrastructure; i.e. there may be different QoS classifications in the different involved networks. Therefore to meet the required QoS, the mHealth platform must be able to acquire and use contextual information about the QoS offered through all the communication network infrastructures (WLAN/3G/4G/IP).

In an *end-to-end* sense, the biomedical data is generated by the sensors and transmitted to the processing unit (e.g. a smartphone) through a wireless link (e.g. using WiFi or Bluetooth). Then the biomedical information is stored, processed and/or transmitted to the core network. Therefore to maintain a global sense of QoS, a proper mapping has to take place between all network segments (sensor-to-smartphone, smartphone-to-access network, access network-to-core network and core network-to-terminal equipment) [12]. As can be seen it is fundamental to assure a specific performance for the *end-to-end* quality of service parameters in order to deliver adequately the mHealth applications. But the traffic transport over heterogeneous networks becomes this demand in a great challenge. In order to ensure this *end to end* quality of service there have been several proposals like that reported in [13]. This paper reports the development of a framework to evaluate the *end-to-end* quality of service regardless of the structure and network technologies that transport the mHealth services. The main idea of this framework is to perform a series of tests that allow to rate the quality level of the link. Based on the results of these tests it is possible to define what mHealth services can be provided appropriately in that specific network.

The current and future growth of data rate transmission of multimedia applications in the next years such as high-definition Video on Demand is expected to increase by 16-fold from 2012 to 2017, accounting for over 66 percent of total mobile data traffic by the end of the forecast period [14]. This trend also allows the growth of mHealth applications like managing patient care, the use of high definition 2D/3D imaging for diagnostic and multi-view, stereoscopic and holographic video communications for tele-diagnosis and remote operations [12]. This continuous traffic growing scenario predicts congestion problems, especially in the broadband wireless access segment where the 4G mobile networks will be, at that moment, in a wide deployment phase. In this sense the quality of service capabilities of these 4G technologies will be critical to support the future mHealth applications.

4 QoS in 3G and 4G Mobile Technologies as Support for mHealth Applications

The ITU International Mobile Telecommunications (IMT-2000) initiative arises in the mid-1980s, to be used as the third generation of mobile networks. Later in the early 2000s the ITU IMT Advanced initiative was proposed as the technical specification for the fourth generation of mobile networks. The main characteristic of that initiative was the definition of an *All-IP* network architecture. In the transition from the third to the fourth generation of mobile networks two main technologies have maintained their reconnaissance of approved technologies on both third and fourth generation: LTE and WiMAX. These are the two reviewed technologies in this section, with focus on their ability to manage quality of service.

In 2000 the technical specifications of the systems belonging to the IMT-2000 were approved. This standard consists of several recommendations that enable the convergence of fixed, mobile, voice, data, Internet and multimedia services, allowing full interoperability and interworking between mobile systems [15].

Initially there were approved five radio interfaces comprising the IMT-2000 technologies: Universal Mobile Telecommunication System (UMTS), EDGE, cdma2000, Time-Division CDMA, Time Division-Synchronous CDMA and Digital Enhanced Cordless Telecommunications [16]. In 2007 was added the OFDMA

TDD WMAN technology as part of the IMT-2000 radio interfaces. The minimum transmission rates were specified as 2 Mbps in fixed or building environments, 384 Kbps in pedestrian or urban environments and 144 Kbps in wide area vehicular environments. Besides high data rate, these systems also envisioned providing a better QoS control for a variety of applications [3]. These technologies were accompanied with different challenges, including indoor coverage and sustainable data rates for different mobility types [16]. Some technologies such as Mobile WiMAX and LTE overcome such technical challenges with their QoS management ability.

QoS in mobile WiMAX. The QoS in the IEEE 802.16 standards are based on service flows (SF). A SF is a logical unidirectional flow of packets between the access service network gateway (ASN-GW) and a mobile subscriber with a particular set of QoS attributes [17].

For mobile WiMAX there were defined five types of services to meet the characteristics of different applications according to their requirements of parameters such as delay, jitter and minimum and maximum transmission rates [18]. The different WiMAX service classes are shown in Table 6 and they are [19][20]:

- **UGS** (Unsolicited Grant Service): Is designed to support real-time service flows that generate fixed-size data packets on a periodic basis such as VoIP.
- **rtPS** (real time Polling Service): Is designed to support real-time services that generate variable-size data packets on a periodic basis, such as MPEG video.
- **ertPS** (extended real time Polling Service): Supports real-time traffic that generates variable-size data packets on a periodic basis with a sequence of active and silence intervals, such as VoIP with activity detection.
- **nrtPS** (non-real time Polling Service): Supports delay tolerant traffic that requires a minimum reserved rate, such as FTP.
- **Best Effort**: It is used for services that do not require guaranteed bandwidth and delay.

It should be noticed that WiMAX QoS classes indicate the recommended application to be used with, although typical performance values are not indicated.

QoS in LTE. In LTE networks QoS is established from the user terminal to the packet data network gateway (PDN-GW) in the core network using a bearer service. A bearer provides the QoS granularity in the LTE for different SF. Table 7 shows the QoS classes used for LTE. In order to support multiple QoS requirements, multiple bearers are set up within the Evolved Packet System. Bearers can be classified into two categories based on the nature of the QoS they provide [21]:

- **Minimum GBR** (Guaranteed Bit Rate): Bearers that can be used for applications such as VoIP. These bearers have an associated GBR value for which dedicated transmission resources are permanently allocated. Higher bit rates may be allowed for a GBR bearer if resources are available.

Table 6 Mobile WiMAX applications and QoS (Source: WIMAX FORUM)

	Maximum sustained traffic rate	Minimum reserved traffic rate	Low latency	Jitter tolerant	Example Services
UGS	X		X		VoIP
ertPS	X		X		VoIP with activity detection
rtPS	X	X	X	X	Audio/Video streaming
nrtPS	X	X		X	FTP
BE	X			X	Web navigation, e-mail

Table 7 Policy and Charging Control quality classes for LTE (Source: ETSI TR 103 122 v1.1.1 2012-11)

QoS Class Identifier	Resource type	Priority	Packet delay budget	Packet Error Loss Rate	Example Services
1	GBR	2	100ms	10^{-2}	Conversational voice
2	GBR	4	150ms	10^{-3}	Conversational voice (live streaming)
3	GBR	3	50ms	10^{-3}	Real-time gaming
4	GBR	5	300ms	10^{-6}	Non-conversational video (buffered streaming)
5	Non-GBR	1	100ms	10^{-6}	IMS signaling
6	Non-GBR	6	300ms	10^{-6}	Video (buffered streaming), TCP-based
7	Non-GBR	7	100ms	10^{-3}	Voice, video (live streaming), interactive gaming
8	Non-GBR	8	300ms	10^{-6}	Video (buffered streaming), TCP-based (e.g. www, e-mail, chat, ftp, p2p file sharing, progressive video, etc.)
9	Non-GBR	9	300ms	10^{-6}	

– **Non GBR**: These are bearers that do not guarantee any particular bit rate. These are used for applications such as web browsing or FTP transfer. For these bearers, no bandwidth resources are allocated permanently.

Although WiMAX and LTE provide QoS control mechanisms, it is important to notice that they are applicable only in the domain of their respective wireless access networks. Additional QoS mechanisms must be applied for each of the different network segments (for example the IP core network) included in a path to provide the full *end-to-end* QoS mHealth services require.

The fourth generation of mobile networks form part of the ITU IMT-Advanced initiative, this new generation represents an evolution of the IMT-2000 technologies to support new applications and services. Some of the new capabilities include the data rates support of up to approximately 100 Mbps for high mobility, such as mobile access, and up to 1 Gbps for low mobility such as nomadic or local wireless access. These systems will support a wide range of symmetrical (e.g. VoIP), asymmetrical (e.g. patients video monitoring) and unidirectional (e.g. SMS with medical indications to the patients) services and will provide management of different QoS levels based on different requirements of these packet-based services. Also they will be compatible within IMT and wired networks, and will interwork with different radio access systems [22].The general requirements for IMT-Advanced surpass the performance levels of the 3G networks and an enhanced support of basic services (conversational, interactive, streaming and background) is expected [3].

WiMAX release 2.0 supports the same classification of mobile WiMAX QoS categories and includes a new service: adaptive granting and polling (aGP) service. This service is used to face applications which require variable traffic patterns not only with variable packet sizes, but also with variable periodic intervals such as VoIP with activity detection. aGP supports the adaptation of the QoS parameters to serve the dynamic traffic characteristics of the applications. For LTE-Advanced most of the characteristics related to QoS are supported from LTE, assuring backwards compatibility. The QoS performance measures, classification, signaling bandwidth requests and grants are almost similar to those of LTE [16]. For example a new addition includes carrier aggregation, which includes two or more component carriers to provide larger bandwidths to meet the IMT-Advanced requirements [19].

4.1 4G Mobile Networks Support for mHealth Services

The 4G mobile networks that will be deployed in the future will allow handling different bandwidth links for the transmission of several mHealth applications. The tendency of these future mobile networks is to provide a higher transmission rates and in general deploy a network with a higher capacity. These characteristics will allow to support a greater number of mHealth applications and to reach to a greater number of users. Additionally the delivered services will operate with more powerful features, providing an acceptable QoS for the mHealth services users and for the remote health centers like hospitals and also the caregivers.

Table 8 presents a proposed mapping between the QoS categories of 4G mobile access technologies, the ITU recommendations revised in Section 2 as well as with the QoS requirements of mHealth services revised in Section 3. The 4G mobile access technologies allow adapting the type of service to be transmitted

Table 8 QoS mapping between ITU recommendations, wireless access technologies and suitable mHealth applications

ITU-T Rec Y.1541 (IP Network QoS classes)	ITU-T Rec. G.1010, ITU-R rec. M.1079-2	LTE (QCI)	WiMAX (QoS classes)	mHealth applications
0	Conversational – error intolerant, Conversational – error tolerant	1, 2	UGS, ertPS	ECG monitoring, heart rate (real time)
1		3, 4	rtPS	Blood pressure monitoring, ultra sound video streaming (real time)
2, 3	Interactive – error intolerant, Interactive – error tolerant	5, 6	nrtPS	Region of interest (e.g. mammogram), radiology, magnetic resonance imaging (non-real time)
4	Streaming – error intolerant, Streaming – error tolerant	7, 8		
5	Background – error intolerant, Background – error tolerant	9	BE	Medical reminder messages, access to administrative files and electronic health records (non-real time)

according to the importance of the service and to the requirements of delay, delay variation and packet loss required, also allowing the definition of the bandwidth required for each mHealth service. In this mapping it is considered that the listed services are using the recommended bandwidth for each specific mHealth service.

5 Pending Challenges on QoS and 4G Mobile Technologies for mHealth Support

As already indicated in the previous sections one of the most important challenges related to use of 4G mobile technologies for mHealth applications will be the definition of mechanisms to ensure the provision of quality of service on an *end-to-end* basis. This will be a task of great importance for the following years because the different technologies including wireless, mobile and IP networks manage their own classifications of services. It will be necessary to develop a framework for an efficient mapping of the QoS classifications between these different technologies. This mapping will ensure the QoS characteristics of the flows will be maintained as they pass through different and heterogeneous networks.

Another major challenge will be to define network architectures and mechanisms to operate networks with a high availability level and a short recovery time in the case of a disruption of services. For real-time interactive services such as the monitoring of patients in critical condition this characteristic of a network will be of capital importance. This capacity to recover the operating conditions of a network is named the network resilience. The integration of contextual information will be used to define the minimum conditions of the network resilience and the particular treatment of mHealth traffic flows as a function of the patient's health status. The operation of networks with these characteristics will ensure that data flows associated with critically ill patients will be granted with the greatest privileges of quality of service and the best possible network resilience.

Another pending task is to develop the definition of a framework for incorporating mHealth sensors within the chain for the provisioning of *end-to-end* quality of service. As mentioned in section 2 current ITU recommendations about quality of service assume that the end of the communication chain is at the terminal equipment connected to an access network, in this case the 3G and 4G mobile networks. But it will be necessary to consider processing delays in the mHealth sensors and the transmission characteristics of their data flows to the smartphone or communication device. This information should be included in the analysis of *end-to-end* QoS to provide a closer approximation about the quality of service measures in mHealth applications.

Another aspect to be developed with great intensity in the coming years is the concept of Quality of Experience (QoE). This concept represents an extension or evolution of the concept of quality of service. In this case the evaluation of the services includes subjective aspects to get a more precise assessment of the level of satisfaction the users experience about the services they are receiving. For example currently the ITU P.800 recommendation allows evaluating the voice quality perceived by the users of voice services [23]. In this way in the future there will be developed some other frameworks to extend the evaluation of the quality of service to the concept of QoE in other services and applications.

Finally it is very important to notice that one of the most significant technologies in the near future will be the so-called Internet of Things (IoT). Under this new paradigm of the Internet development a great number of applications are being proposed in a variety of areas. Such applications will operate automatically and the support of their access to the network resources will be of capital importance. The typical structure of these applications includes the monitoring of specific variables and the reporting of this data to intelligent management systems which could include some control mechanisms applied on the system of interest. This general structure of IoT applications fits perfectly to the development of mHealth applications that will allow automatic and real-time report of the condition of monitored patients. The data generated from these monitored patients will be processed by intelligent management systems and could perform automatic actions based on the result of their processing. For example it will be possible to automatically dispense the application of certain drugs to patients by using medical devices with access to the bloodstream, and to report permanently the patient health condition. In support to these applications the medical devices will require a communication link with granted QoS characteristics. All these operations could be performed without the direct intervention of patients. These applications would be very valuable in treating patients with certain physical disabilities or elderly patients. As can be seen, mHealth applications developed under the IoT paradigm will require intelligent management of their traffic and this can be achieved only through technologies with guaranteed QoS abilities as LTE Advanced and WiMAX 2.0.

6 Conclusions

Present and future mHealth applications will require the support of network infrastructure with large bandwidth and the ability for providing transparent traffic classification over heterogeneous networks. These capabilities will allow a differential treatment of the services based on their assigned priority. New wireless and wired technologies provide QoS management functionalities but it is necessary to develop policies and mechanisms to integrate the management of the QoS at the points of interconnection between different networks. This integrated management of QoS will be required in future applications to offer services with consistent QoS levels on an *end-to-end* basis. Under this scenario the *end-to-end* QoS will be ensured with independence of the networks where the traffic of mHealth applications will be transported. Technical specifications of QoS on IP networks and broadband wireless access networks have been established by the ITU recommendations for several years and remain as the mandatory reference for the design of different types of services.

LTE Advanced and WiMAX 2.0 are the two technologies recognized by ITU as the only 4G technologies and they will be used in future deployments of these type of networks. LTE advanced is a technology that currently enjoys a great popularity in the mobile operators market and the first commercial deployments of this technology are expected for this year. However the capabilities offered by WiMAX 2.0 are equivalent to those of LTE Advanced so it is possible that later in this decade some deployments of this technology will also appear. Capabilities offered by 4G mobile technologies for traffic classification and management of quality of service will be critical to support the deployment of future mHealth applications. Quality of service requirements defined for major mHealth applications can be supported by both the LTE Advanced and WiMAX 2.0 technologies.

References

1. Request for comments 2474 Definition of the Differentiated Services Field (DS Field) in the IPv4 and IPv6 Headers (1998), https://www.ietf.org/rfc/rfc2474.txt
2. Request for comments 3031 Multiprotocol Label Switching Architecture (2001), https://www.ietf.org/rfc/rfc3031.txt

3. Najah, A., Abd-Elhamid, M., Hossam, H.: Fundamentals of LTE. Pearson Education, United States of America (2011)
4. ITU-T Recommendation E.800 Definitions of terms related to QoS (2008), http://www.itu.int/rec/T-REC-E.800-200809-I (accessed September 1, 2013)
5. ITU-T Recommendation G.1010 End user multimedia QoS requirements (2001), http://www.itu.int/rec/T-REC-G.1010-200111-I (accessed September 1, 2013)
6. ITU-T Recommendation Y.1541 Network performance objectives for IP based services (2011), http://www.itu.int/rec/T-REC-Y.1541-201112-I (accessed September 1, 2013)
7. ITU-R Recommendation M.1079-2 Performance and quality of service requirements for international Mobile Telecommunications-2000 (IMT-2000) access networks (2003), http://www.itu.int/rec/R-REC-M.1079 (accessed September 1, 2013)
8. Takács, A., Bokor, L.: A Distributed Dynamic Mobility Architecture with Integral Cross-Layered and Context-Aware Interface for Reliable Provision of High Bitrate mHealth Services. In: Godara, B., Nikita, K.S. (eds.) MobiHealth. LNICST, vol. 61, pp. 369–379. Springer, Heidelberg (2013)
9. Alinejad, A., Philip, N., Istepanian, R.S.H.: Mapping of Multiple Parameter M-health Scenarios to Mobile WiMAX QoS Variables. Paper presented at the 33rd Annual International Conference of the IEEE EMBS, Boston, Massachusetts, USA, August 30-September 3 (2011)
10. Komnakos, D., Vouyioukas, D., Maglogiannis, I., Constantinou, P.: Performance Evaluation of an Enhanced Uplink 3.5G System for Mobile Healthcare Applications. International Journal of Telemedicine and Applications, 11 pages (2008), doi:10.1155/2008/417870
11. Skorin-Kapov, L., Matijasevic, M.: Analysis of QoS Requirements for e-Health Services and Mapping to Evolved Packet System QoS Classes. International Journal of Telemedicine and Applications, 18 pages (2010), doi:10.1155/2010/628086
12. Adibi, S., Mobasher, A., Tofigh, T.: LTE networking: extending the reach for sensors in mHealth applications. Transactions on Emerging Telecommunications Technologies (2013), doi:10.1002/ett.2598
13. Adibi, S.: Biomedical Sensing Anaizer (BSA) for Mobile-Health (mHealth)-LTE (2013), doi:10.1109/JBHI.2013.2262076
14. CISCO. Cisco Visual Networking Index: Global Mobile Data Traffic Forecast Update, 2012–2017, http://www.cisco.com/en/US/solutions/collateral/ns341/ns525/ns537/ns705/ns827/white_paper_c11-520862.html (accessed September 1, 2013)
15. International Telecommunication Union: About Mobile technology and IMT-2000 (2011), http://www.itu.int/osg/spu/imt2000/technology.html#Cellular%20Standards%20for%20the%20Third%20Generation (accessed September 4, 2013)
16. Hossam, H., et al.: LTE, LTE-Advanced and WiMAX: Towards IMT-Advanced Networks. John Wiley& Sons, United Kingdom (2012)
17. Rengaraju, P., Lung, C.-H., Yu, F., Srinivasan, A.: On QoE Monitoring and e2e service assurance in 4G wireless networks. IEEE Wireless Communications 19(4), 89–96 (2012), doi:10.1109/MWC.2012.6272428
18. WiMAX Forum (2001), http://www.wimaxforum.org (accessed September 2, 2013)

19. Yan, Z.: WiMAX Network Planning and optimization. CRC Press, United Sates of America (2009)
20. Alasti, M., Neekzad, B., Hul, J., Vannlthamby, R.: Quality of service in WiMAX and LTE networks. IEEE Communications Magazine, Topics in Wireless Communications, 108–111 (May 2010), doi: 10.1109/MCOM.2010.5458370
21. Sesia, S., Toufik, I., Baker, M. (eds.): Front Matter, in LTE - The UMTS Long Term Evolution: From Theory to Practice. John Wiley & Sons, Ltd., Chichester (2011)
22. ITU-R Recommendation M.1645 Framework and overall objectives of the future deployment of IMT-2000 and systems beyond IMT-2000 (2003), `http://www.itu.int/rec/R-REC-M.1645/en` (accessed September 1, 2013)
23. ITU-T Recommendation P.800Methods for subjective determination of transmission quality (1996), `http://www.itu.int/rec/T-REC-P.800-199608-I` (accessed September 1, 2013)

Questions

Q1. Why it is necessary to have QoS implemented for mHealth services delivery?

Q2. What are the QoS parameters considered to evaluate the performance of an *end-to-end* link?

Q3. What are some considerations to provide QoS on an *end-to-end* basis for mHealth applications?

Q4. What are the advantages offered by 3G / 4G mobile technologies for mHealth applications?

Q5. Compliance of the ITU recommendations like ITU-T G.1010, ITU-T Y.1541 and ITU-R M.1079-2 are enough to ensure the QoS required for the delivery of mHealth services?

41. mHealth over "Medical Grade" High Quality of Service Mobile Networks

Michael Kedar

Mobilexchange Ltd., 1131A Leslie Street, Suite 401, Toronto, Ontario (Canada) M3C 3L8
mike.kedar@mobilexchange.ca

Keywords: Mobile infrastructure, Broadband 700 MHz Mobile Networks, Ubiquitous Telemedicine, Public Real-Time Social Services, Interoperability, Industry Canada - IC, Federal Communications Commission - FCC, Quality of Service - QOS, Wireless Body Area Network-WBAN, Bring Your Own Device - BYOD.

1 Introduction

This chapter is focused on discussions of a number of issues and challenges associated with the underlying infrastructure of mHealth, including: regulation of medical devices/smartphones, network security and privacy, interoperability, scalability, ubiquity, data management and ownership, remote monitoring, and the importance of end-to-end Quality of Service (QoS) for full reliance on mHealth technology in medicine.

The rapidly accelerated adoption of broadband mobile technology in healthcare will have significant impact in transforming the industry; [4] in this chapter we will address the criticality to the medical profession of developing sufficient confidence in the accuracy and security of the underlying wireless infrastructures.

A pervasive expansion of mHealth is already occurring:

- Within hospitals and clinics, over unlicensed private Wi-Fi networks.
- Over ubiquitous wide area carriers' mobile networks owned by commercial public wireless networks.

We will review the evidence indicating that interfaces among these networks are commonly insecure. They are typically interconnected at the smartphone level, where medical patients' records and real-time, life-dependent monitoring data are interchanged from one network to another.

This chapter will show that the most important aspect of mHealth is the ability of the mobile/wireless infrastructure to reliably and accurately accommodate large amounts of end-to-end real-time data. Thus, dependability and end-to-end QoS of the underlying Radio Frequencies (RF) infrastructure are paramount.

> ***"As more and more wireless technologies and systems are introduced and rolled out in the hospital, proper management of this fundamental resource has never been more important."*** [3]

S. Adibi (ed.), *Mobile Health*,
Springer Series in Bio-/Neuroinformatics 5, DOI: 10.1007/978-3-319-12817-7_41

We will see how mHealth is situated at the intersection of Electronic Health (IT) and smartphone technology. The discussion will center on the infrastructure behind smartphone technology.

In order to evaluate the present and future QoS issues and challenges of mHealth, a review is included as to how mobile communications technologies and smartphone technology came to be used in healthcare.

Imagine the following scenario:

A major train derailment occurs, injuring hundreds of people spread over two city blocks. Within minutes first responders arrive and immediately use their two-way radios to call for all equipped ambulances within 30 kilometers to rush to the scene. Hundreds of local residents untouched by the disaster rush to the derailment site to help in the rescue. Suddenly, hundreds of cell phones are turned on, jamming the four-cell site serving the two affected city blocks. Police and fire crews, as well as all the ambulance paramedic teams, start attending to the injured while communicating over their own radios to their individual headquarter dispatchers.

No direct radio communication is possible between the police, fire and paramedic crews on the scene because each of their radio networks operate on a different frequency.

In the ensuing chaos, many of the injured require urgent medical attention. Paramedics do their best to attend to these cases. They use the EMS (Emergency Medical Service) protocol to deal with simple injuries, but to obtain instructions to deal with the most serious injuries; they need to connect to the local hospital emergency medical staff. As the cellular network is now jammed, they call their dispatchers, who in turn relay their requests to the hospital ER staff.

Now let's zoom in. A middle aged man is lying motionless without any signs of bleeding. He has a very weak pulse and shallow breathing. He appears to be having a cardiac arrest. The attending paramedic uses his portable radio and calls the dispatcher who is busy with many other calls. Eventually, he gets a response and reports, "We have a 50-something year old male, no signs of bleeding, weak pulse, shallow breathing and possible cardiac arrest, but we have no data on him. What is the safest approach to revive his heart in addition to providing CPR?" The dispatcher relays his interpretation of the hospital's recommendations, translated over the busy telephone. The message is, "Use Epinephrine 10 mg IV/IO every 3-5 minutes." "Are you sure?" asks the paramedic. "Yes" replies the dispatcher. The procedure is administered and the patient dies.

The amount of the Epinephrine should have been 1 mg and not 10 mg; however, due to the poor communication quality, the dosage was misunderstood and, ultimately, caused the patient to die.

This emergency event scenario shows how, in real-time situations, medical personnel are critically dependent upon dedicated, reliable, broadband, end-to-end communication infrastructure.

2 Brief History of Mobile Communications in Healthcare

Wireless communications came to the healthcare industry as an ad hoc phenomenon, introduced some twenty-five years ago along with the introduction of commercial cellular radios. Doctors and other healthcare providers, having taken personal subscriptions on public commercial cellular carrier networks, imported their devices into "work," thereby commencing mHealth through the back door.

The use of personalized wireless phones and other mobile devices at work is now called "BOYD" – Bring Your Own Device. This phenomenon presents serious headaches to hospital IT departments who have to accommodate devices operating on foreign networks totally out of their control. The initial BOYD deployment in healthcare was limited to voice communications and was managed through privacy hospital rules and procedures. That has changed, however, since the mid-1990 when data was added to these devices and hospitals and clinics started to deploy Wi-Fi networks: "wireless local area networks" (WLAN) (Reference #5). We will review the evolution of indoor and various wide area wireless networks and their "medical grade" suitability in this section.

Mobile communication over a wide area has been deployed by Public Real-Time Social Services (PRTSS) agencies such as police, fire, ambulance (EMS) and others since the invention of the two-way radio in the 1920s. Two-way communication devices, "radios", were installed in vehicles to allow instant communication from headquarters (dispatchers) to their operating fleets. Public safety and social service agencies have deployed private two-way radio systems for over eighty-five years. Mobile radios became a fundamental "tool of the trade" for the PRTSS industry.

Public safety, airlines, the military, and other real-time operations have customized and deployed mobile networks over dedicated Radio Spectrums licensed by governments, based on the unique requirements and specifications which met their voice communication needs.

These private mobile networks were independently deployed by each agency in municipalities across the country and all over the world. Deployments in big metropolitan cities and small rural areas were individually undertaken on a variety of narrow band mobile channels in the LB spectrum (L Band used by the military), VHF (Very High Frequency), and UHF (Ultra High Frequency). Diverse frequency bands and serendipitous governmental channel allocations created a series of mostly short-term patches rather than a consistent, integrated system. Visionary planning to coordinate statewide, countrywide and, ultimately, worldwide interoperability needs during major emergencies were completely overlooked.

3 The Mobile Communication Gap

Following the September 11, 2001 (9/11) catastrophe in the U.S.A., the FCC and Industry Canada set aside dedicated spectrum for first respondents that includes the fundamental broadband (voice data and video) design requirements to enable regional and countrywide co-ordination and interoperability.

Unfortunately, despite the availability today of advanced broadband mobile technologies, nothing has dramatically changed over the twelve years since 9/11. In many wide areas, natural or manmade disasters, of serious local, regional or multiple wide area situations, such as fire, flood, and terrorism attacks, the PRTSS agencies are still unable to perform "real-time end-to-end" broadband communications. Communication from victim field locations to hospital Emergency Rooms (ER) is still available only over commercial cellular networks.

Victims of major emergencies are attended to by paramedics who are unable to directly interface with hospital ER personnel. Typically, during emergencies, public cellular networks are totally jammed, preventing real-time access by rescue workers. Consequently, they must depend upon the only wireless communication available: The old narrow-band, two-way radios, capable of providing only voice and rudimentary data between them and their base dispatchers.

Hospitals are rarely equipped with PRTSS mobile base stations to accommodate such communications. So the link ends up flowing from the injured patient to the dispatcher to the dispatcher's landline (internet) to the hospital.

When local cellular networks are blocked or in failure mode, current radio systems cannot provide real-time, direct co-ordination between the field PRTSS personnel and the landline-connected institutions.

As private PRTSS radio systems evolved, their vehicular and portable radio devices, operating on different frequency bands, between different agencies, in different regions, created a chaotic mélange of products. In addition, as might be expected, unit cost structures soared. Volume production for these radios is in the thousands, unlike popular cellular phones, which are standardized all over the world and are produced in millions per annum, and have a significant lower unit cost. A police "walkie-talkie" could cost over $2,500.00 in comparison to a much more sophisticated "smart cellular phone" at less than $500.00.

As a result of budgetary and regulatory constraints that impede the allocation of dedicated public safety/healthcare spectrum and their subsequent deployment, many PRTSS agencies now rely upon commercial cellular mobile and local area Wi-Fi unlicensed networks for their operations. These include hospitals and telemedicine practitioners who employ commercial third-party networks not only for day-to-day correspondence but for emergency and life-dependent real-time operations as well.

4 Public Discussions

The growing reliance on commercial wireless networks is one of the most critical challenges of the burgeoning mHealth and its related "first responders" emergency operations.

To illustrate this critical point a CBSNEWS article from September 2011 and a Bloomberg BusinessWeek article from April 2013 have been reprinted below:

On September 11, 2011 CBSNEWS reported: [11]

WASHINGTON – Amid the chaos of the Sept. 11 attacks in 2001, emergency responders found they could not communicate with each other. That problem persists 10 years later, according to a review of the 9/11 Commission's recommendations.

A National Preparedness Group report released Wednesday concludes that the recommendation that a nationwide broadband network for emergency responders be created "continues to languish."

"Despite the lives at stake, the recommendation to improve radio interoperability for first responders has stalled because of a political fight over whether to allocate 10 MHz of radio spectrum … directly to public safety for a nationwide network, or auction it off to a commercial wireless bidder who would then be required to provide priority access on its network dedicated to public safety during emergencies," says the report, whose authors include 9/11 Commission chairmen Lee Hamilton and Thomas Kean.

Law enforcement and emergency responders around the country have long supported the creation of the communication network. In January, President Obama announced his support for allocating the radio space, known as the D-block spectrum, to police and other emergency workers.

Bills that would set aside the D-block and create a communication network have been introduced in both the House and the Senate this year, but so far have not been passed in either chamber.

Several big wireless carriers have supported auctioning off the airwaves to the wireless industry, a move that the government has estimated would raise about $3.1 billion. That could help pay to build a public safety communication network.

The Federal Communications Commission has tried to auction off the D-block before, but that 2008 auction required that the winning bidder help build a network to be shared by first responders and give them priority in an emergency. It did not attract any serious bidders.

The National Preparedness Group report stated, "…statewide communications, interoperability plans, and the creation of a national emergency communication plan have improved emergency co-ordination across different jurisdictions. More work needs to be done and the ultimate resolution is to follow the 9/11 Commission's recommendation and create a nationwide communication network."

A more recent article following the Boston Marathon bombing in April 2013; Bloomberg BusinessWeek (Technology by Brad Stone April 16, 2013) [8]

Toward the bottom of the list of disturbing aspects about Monday's bombing at the Boston Marathon was this: Cellular networks in the area almost immediately slowed down and, for a period of time, appeared to stop working altogether. Runners and their loved ones could not connect, and victims had trouble communicating with emergency responders.

That frustrating scene has become familiar, evoking disasters from the Sept 11th attacks in 2001 to Hurricane Sandy in 2012. We rely on cell phones to run our lives, but they tend to be useless-or at least far from useful-when we need them most.

Initial reports suggested that authorities deliberately shut down mobile networks in Boston to avoid the possibility that mobile devices could be used to trigger more bombs.

The Associated Press reported this and then quickly reversed itself. Spokespeople from the major wireless carriers denied it, saying their networks were simply suffering from congestion: Too many people were trying to use their phones at the same time in order to get help or tell family and friends they were O.K.

"Thinking of those at Boston Marathon," AT&T tweeted at around 5:30 p.m. East Coast time. "For those in the area, please use text & we ask that you keep non-emergency calls to a minimum."

A Verizon Wireless spokesperson also suggested that customers in the area use text messages or e-mail to free-up capacity for public safety officials.

The science behind these failures in wireless connectivity isn't complicated. In every city, each mobile carrier operates hundreds or thousands of cell towers, which route calls and data to the carrier's backbone network. Each tower is designed to accommodate a set number of calls per second, per a certain geographic area. In a crisis, when everyone naturally reaches for their phone, that limit is quickly surpassed and the radios on the tower get sluggish. Mobile analyst, Chetan Sharma, who estimates that a cell site can handle 150 to 200 calls per second per sector: "We've all had the experience of a fast-busy signal. That is the network telling you, "Sorry, but your cell is overloaded. There is no more space."

In these situations, sending texts or e-mail is a better option because messages can be queued up and delivered as capacity becomes available. Twitter is especially efficient because information can be spread quickly and widely. Until it was debunked, the rumor that police deliberately shut down cell networks was an intriguing one. This is almost unheard of. When it has occurred, the practice has been controversial. In April 2011, public transit authorities in the San Francisco Bay Area on the region's BART subway system used to prevent activists from coordinating their activities as they protested shooting by BART police. The shutdown led to criticism that police were hindering the activists' right to free speech, spurring a further round of protests. The police argued that they acted to protect the public safety.

Monday's situation was different, of course. ABC News reported in a tweet that police were asking people on the scene not to use cell phones because they might set off other explosive devices. Still, it does not appear there was any widespread directive from law enforcement to shut down cell networks, and authorities appear to believe that cell phones do more to help than hurt in an emergency. "The belief is that much more good comes from networks being available than the risk of them being utilized for bad intentions," says Charles Golvin, an analyst at Forrester Research.

Can anything be done to improve network performance during tragedies such as Monday's bombing? Probably not much. Carriers could engineer networks to handle these situations by adding radios and towers and amplifying connections from the towers to the network's backbone.

But a vast majority of that extra capacity would then be wasted. "Over the years, a lot of research has gone into analyzing the trade-off between cost and capacity, and today's networks are designed to provide enough capacity during the vast majority of 'normal' conditions," says Golvin.

Just as land line networks were never meant to allow every person with a phone to place a call at the same time, wireless networks are designed to align resources with normal peak call volumes in the most efficient manner possible. As a result, hearing a loved one's voice in the tense minutes after an emergency may always be difficult. Perhaps it's best to hope we have fewer and fewer of these kinds of tragedies in the first place.

5 The Urgent Need for Dedicated Broadband Mobile Networks

As seen from these articles, national dedicated networks have not been deployed for neither first responders in the "public safety" arena, nor for mHealth and Healthcare in general. All real-time broadband wireless transactions are still shared with non-urgent media and voice networks without due priorities.

Dedicated broadband wireless networks were designed exclusively and deployed for public safety/healthcare and other real-time services; PRTSS, will overcome the serious shortcomings of current commercial wireless services. These shortcomings include poor availability, inadequate coverage, and, above all, the lack of priority access during emergencies.

Network infrastructure requirements for PRTSS communications include:

- Broadband applications to facilitate real-time responses by providing simultaneous voice and data images, including video.

- Coverage of the entire country, ensuring service to meet PRTSS emergencies anywhere.
- Voice communication capabilities, to be used as backup to real-time critical voice channels on other frequencies, including one-to-one and one-to-many connectivity.
- Network software that provides traffic management services such as prioritizing. (If multiple networks were built separately and then linked together, interoperability and nationwide roaming would need to be ensured.)
- Mobile broadband applications designed for PRTSS. In particular, radio chipsets need to be developed for wireless devices that can connect to a Long Term Evolution (LTE) network.
- Secure authentication software to regulate access to the PRTSS networks; to achieve large volume, unit cost reductions. Such software must be incorporated into standard, commercially available "smart" phones.
- Network cell towers that are strengthened against natural hazards and furnished with back-up power supplies that can outlast extended power outages.
- Robust backhaul, high capacity, landline communication networks. Backhaul (which typically refers to the connections between the air interface access points between cell towers) is an essential component of wireless network infrastructure.

The increasing demand of data traffic on commercial wireless networks worldwide is fast exceeding governments' abilities to allocate more spectrums to public carriers as well as carriers' capabilities to build adequate capacity. [13, 15]

The result is already being noticed by consumers, as the frequency of dropped calls and delayed data transfer (emails) increases with ever more regularity. In other words, the carrier grade Quality of Service (QoS) of commercial cellular networks in large metropolitan areas is degrading. The existing commercial wireless infrastructures, which were originally designed for voice, are hard pressed to carry simultaneously vast amounts in real-time configurations of broadband mobile data and multimedia technology resulting from exponentially increasing usage.

A recent CTIA semi-annual data survey [14] shows more than 69% year-over-year growth in the amount of mobile data traffic (bytes) handled. The 2011 domestic traffic was more than 633 million Mbytes over a six month period ending in June 2012.

To put it into a slightly different perspective, in 2011, every person in the US was receiving more than 11 Mbytes/day. The well respected Cisco Global Mobile Traffic Data Forecast, 2011-2017, is predicting a 66% [15] compound annual growth rate for North America. This forecast would result in every American's receiving more than 200 Mbytes/day by 2017, nearly a twentyfold increase in just four years.

Weak RF signal levels coupled with this tremendous growth in demand will further diminish the reliability of these commercial wireless carriers. The variety and complexity of scenarios involving wireless devices will require thorough analysis. The variable signal strengths of disparate wireless sources, in conjunction with separate, co-existing wireless systems can clearly lead to potential interference and degradation of overall system performance.

The existing commercial wireless infrastructures, which were originally designed for voice alone, are hard pressed to carry increasingly vast amounts of broadband mobile and multimedia data, all simultaneously. Constant internet browsing, on a daily basis, increasing exponentially worldwide, is pushing the limits of current wireless infrastructures.

- **The Network requirements for mHealth**

The growth of mHealth diagnostic services radically increases the demand for wireless capacity. Sending real-time detailed images, such as MRIs, will require significant additional real-time digital bandwidth throughout, which the current carriers will not always be able to provide, particularly at peak hours.

Today's wireless network operators also do not provide priority assurances for real-time urgent communications. They provide only marketing Quality of Service (QoS) information about their networks. They refuse to give any detailed information to mobile users. Some mobile applications even use a network's signal strength to estimate QoS rather than capacity. Signal strength, of course, does not provide any information related to actual network congestion.

Taking into consideration that the mHealth users are moving in space and time, the task of managing the QoS by figuring out and trying to anticipate QoS levels of the underlying wireless networks on a continuous basis is impossible. Reliance on inappropriate metrics leads to worst case scenario assumptions and lowers the usability of real-time mHealth applications.

- **Real-Time, Real Life Example**

Let us take a real life situation, one of many thousands occurring every day, not in major emergencies involving large communities as we discussed earlier, but a typical day-to-day telemedicine/mHealth, doctor-patient remote interaction.

A 30-year old male is recovering from a surgical procedure and has developed an alarming infection. He is being treated by an antibiotic drip and is being monitored remotely by his Doctor who is presently working in another facility. The drip is being remotely operated; the patient's temperature is transmitted constantly and a camera is pointed to the infected area for monitoring on demand. The information is collected either continuously or on demand via a WBAN [16] (Wireless Body Area Network).

WBAN is a low-powered wireless communication and physiological sensor eco-system that has developed over FCC's specially assigned low power spectrum (40 MHz of spectrum at 2360-2400 MHz) using tiny, lightweight, ultra-low power monitoring devices. WBAN with sensors consuming extremely low power is used to monitor patients in critical conditions inside hospitals. Outside the hospital, the network can transmit patients' vital signs to their physicians over internet in real time.

In our example, the data is being collected wirelessly over the WBAN by a small computer connected to a cellular device: The "Collector/Operator". The Collector/Operator can be interrogated for information by the distant doctor to periodically check on the condition of his patient and remotely activate and change the drip rate of the antibiotic medicine. The system also allows for continuous monitoring on the distant doctor's smart phone screen.

Now, let's look at the wireless infrastructure and risk factors associated in this example of day-to-day telemedicine.

The information hub, the Collector/Operator is situated at the hospital within ten meters of the patient's bed, probably at the nursing station.

Cellular signal strength at that location is usually good; however, today, unannounced by the commercial cellular carrier to which the device is registered, a failure requiring maintenance has shut off the cell site covering the hospital.

Cellular commercial networks are consumer oriented operations with no direct interaction protocol between the users and the network operators, regardless of the situation.

So without notice, the distant doctor has lost his ongoing control and monitoring of his patient.

Very likely he will resort to a telephone to alert the hospital and to arrange alternative supervision, but the critical emphasis here is that there are no "medical grade" technology standards. Mobile networks are operated and managed by consumer oriented companies without any 'medical grade" QoS obligations.

- **The Network Regulatory Impairments**

The sad reality is that, in the twenty-first century, there are no end-to-end wireless standards for a "medical grade" mobile infrastructure. [10]

In "Medical Grade Wireless Utility Case Study for Hospitals – By Ed Cantwell- Oct 28, 2011" [12] various networks' are categorized by quality of service:

Network	Consumer Grade	Enterprise Grade	Medical Grade
Coverage The % of the "area" that has wireless signal	Best Effort 85%	Engineered 95%	Guaranteed 100% Clinical Space
Signal Strength Received at device in presence of RF interference	Nominal Will Vary	Engineered Prioritized	Engineered Guaranteed
Capacity Available to the "devices" within the area	Shared	Prioritized Not Dedicated	Engineered Guaranteed Per unit area
Security Physical, Over the Air, and Digital measures to protect	Best Effort	By Policy Engineered	Comprehensive
Certainty Ability of protocol to operate in the area's RF environment	Vary Widely	Engineered Prioritized	Very High Mission & Life Critical

Source: Gary & Mary West Wireless Health Institute

The FDA and its equivalents in many other countries have strict compliance rules regarding medical devices; similarly, the FCC and its global equivalents have strict regulations regarding spectrum utilization and management. **Yet there is no domestic or international organization that has the responsibility to establish end-to-end "medical grade" [5, 6, 9, 12] mobile network infrastructure standards.**

The FDA has begun to talk about the issue: "Radio Frequency Wireless Technology in Medical Devices - Guidance for Industry and Food and Drug Administration Staff" issued on August 13, 2013, includes a request for comments and recommendations on the QoS (section 3b.): [7]

3b. Wireless Quality of Service

Wireless Quality of Service (QoS) refers to the necessary level of service and performance needed for the wireless functions of the medical device. While the QoS of cellular telephone networks might be acceptable for voice communication, it might not be sufficient for certain medical functions. Connections lost without warning, failure to establish connections, or degradation of service can have serious consequences, especially when the medical device relies heavily on the wireless connection. Such situations can compromise the wireless transmission of high-priority medical device alarms, time-sensitive continuous physiological waveform data, and real-time control of therapeutic medical devices (such as wireless footswitches).

If a wireless medical device becomes part of a network, wireless QoS should be carefully considered in conjunction with the intended use of the wireless medical device. The following factors should be assessed when developing QoS standards: acceptable latency, acceptable level of probability for loss of information within the network, accessibility, and signal priorities of the network.

When the network is chosen or designated, FDA recommends use of a risk management approach to deployment, security, and maintenance of the network's QoS. Depending on the intended use of the device, additional failure modes may need to be considered. Once failure modes and associated risks are identified, we recommend a justification of acceptable risk, or testing or other measures to demonstrate appropriate risk mitigation.

6 Recommendations for Solutions – mHealth over Reliable Broadband Networks

The discussion so far has centered on the infrastructure that underlies smartphone technology, a key factor in establishing universal mHealth and telemedicine. Another essential issue that must be addressed is regulations, specifically regulation of medical devices/smartphone applications and technologies.

In Canada, a comprehensive approach to interoperability, ubiquitous scalability, data management and ownership, remote monitoring, dependability assessment and end-to-end QoS standardization, has been proposed to Industry Canada (equivalent to the FCC in the U.S.). **This wide-ranging solution can be achieved by including mHealth in the dedicated spectrum that has already been allocated to public safety.**

Leading the advocacy of this solution is the Canadian Advances Technology Association Campaign (CATAAlliance). The CATAAlliance's solution, "mHealth Real Time," forms an integral part of the industry's Innovation Nation program, a program, under the tutelage of Canada's leading entrepreneur, Sir Terence Matthews, which lays out what Canada must do as a nation to achieve digital leadership for the 21st Century.

According to CATAAlliance's CEO, John Reid, "Recent emergencies in public health and increased reliance on mHealth have demonstrated the urgency of getting the right information securely into the hands of those who need it the most, in real time, and across jurisdictional boundaries...Enhancing quality, improving convenience, extending reach and reducing costs of healthcare are the potential benefits of mHealth, should the underlying mobile infrastructure be designed to provide ubiquitous broadband 'medical grade' performance...Local governments, public safety and healthcare agencies across the country are evaluating the benefits of broadband wireless mobile networks for remote access to information, both mission critical and routine, to improve real-time awareness for better decision making and to enhance collaboration and information sharing."

The Canadian Federal Government has recognized the situation and has allocated mobile broadband (20 MHz under consideration) dedicated spectrum in the 700 MHz band to be shared by all public safety agencies operating in Canada under municipal, provincial and federal jurisdictions. This band is the same band that has been allocated by the FCC to its PRTSS communities. The spectrum must be cleared within 100 kilometers of the border, which includes the majority of the Canadian population.

EMS (Emergency Medical Services) is included in the definition of "Public Safety" and will have access to the dedicated spectrum. However, mHealth and Healthcare in general and Telemedicine in particular are not included, unless Industry Canada recognizes the needs of mHealth in its review.

Industry Canada is currently evaluating and considering, in a public consultation process (SMSE-007-12), [1, 2] this key issue which embodies unprecedented opportunities for Canada's Digital Leadership in the 21st Century.

The question is: "Who, besides Police, Fire, and EMS, should have access to the secure LTE dedicated broadband mobile network, once it is deployed across the country?"

The digital leadership answer is: m-Health/Healthcare and other Public Real Time Social Services.

The opportunities for Canada and the healthcare industry worldwide arise from the fact that the same 20 MHz spectrum that the FCC has allocated exclusively to Public Safety in the U.S. has also been recommended for allocation in Canada. But Canada, with 1/10 the population of the United States, has just 1/10 of the traffic demand. So, if it chooses to include healthcare in this dedicated spectrum, Canada can easily accommodate mHealth within its own borders. In contrast, Homeland Security and first respondents in the U.S. understandably argue they need the full 20 MHz just for emergency responses.

The efficiencies and cost reductions that will be generated by ubiquitous mHealth use would greatly help fund the building and the ongoing operation of the countrywide (province by province) deployment of public safety/medical grade broadband mobile networks. As we have discussed, these savings and the widespread adoption of mHealth by the medical profession will occur only if the underlying wireless/mobile infrastructure is fully trustworthy end-to-end, via "medical grade" dedicated mHealth networks.

In Canada such efficiency improvements may represent annual savings of 15 billion dollars.

The R.O.I (Return on Investments) associated with healthcare participation in the dedicated Public Safety spectrum would justify the building of ubiquitous networks for all mobile Public Real-Time Social Services' needs. Emergency and real-time access to services would be secure and reliable.

Without Provincial Healthcare participation, the municipal Police, Fire, and EMS authorities will not be able to fund and justify the operating cost from operational savings.

Spectrum availability by itself will not produce the full benefits to society; the Return on Investment will come from the construction of a dedicated, secure, broadband mobile infrastructure, available *all-the-time, and everywhere,* encompassing all Public Real-Time Social Services, including:

- Fire (Municipal and Provincial)
- Police (Municipal, Provincial, Territorial, and Federal)
- EMS (Municipal)
- Healthcare (Provincial and Federal)
- Utilities (Provincial and Federal)
- Other governmental Real-Time Social Services
- Jointly named: Government provided; Public Real-Time Social Services (PRTSS)

Governments need to develop strategic e-Health frameworks in order to provide access to dedicated mobile spectrum of which mHealth is an integral component.

Such frameworks will enable them to provide mHealth implementers with more guidance when deploying projects, to ensure alignment with their governments' e-Health information systems and other relevant policies to leverage technology for public good. Government involvement should include national health priorities, enterprise architecture for interoperability and data standards, and commonly used metrics for assessing the impact of m-Health.

CATAAlliance's CEO John Reid notes that, "If Canada is to become a competitive Innovation Nation, we've got to excel in key sectors such as healthcare where given the right conditions we can create new flagship enterprises and hundreds of SME's (Small Medium Enterprises) to supply domestic and global healthcare needs."

Including mHealth in public safety networks is once-in-a-lifetime opportunity for Canada to establish worldwide leadership in "medical grade" mobile networks. This inclusion would facilitate the deployment of such networks throughout the world for years to come.

Now, let's go back to the train disaster scenario. Imagine it all happening in the future with the benefit of mHealth; i.e., over a PRTSS dedicated broadband mobile LTE network.

The Command and Control Incident (CCI) field H.Q, is quickly established on the scene upon the arrival of the first responders - police, fire and ambulance (EMS). It establishes priorities by assigning responders to various rescue operations, based on the urgency of the situation in each case (all responders of all three agencies come under the control of the CCI). All communication is done over the data/video/voice network.

Full interoperability between the three first responder groups allows seamless coordination on the scene and with their respective home bases in a pre-rehearsed precise manner. Information from building layouts, such as water, gas and electricity distributions for the entire two affected city blocks, are downloaded instantly for viewing by all responders via their mobile devices. Simultaneous video, data and voice communications on split (real-time) screens make operations' support extremely efficient.

7 Achieving Quality of Service

Now let's zoom into the unfortunate, injured, middle aged man. Earlier, we left him lying motionless without any signs of bleeding, and with a very weak pulse and shallow breathing. His fate was sealed by poor communication quality among the dispatcher, the hospital, and the paramedic leading to a fatal overdose.

In a mHealth future, the paramedic in charge immediately presses the patient's thumb (to transmit his finger print) onto the glass front of his mobile device and instantaneously receives the man's picture, ID and a request for an access code to his medical EHR (Electronic Health Records).

The paramedic enters his authorization code and gets the patients' health records which indicate that a serious sensitivity to Epinephrine (the drug that killed him in our early scenario). The paramedic clicks on the "sensitivity' field on the screen and four alternative drugs are displayed, with proper application instructions. Needless to say, the patient would still be alive, under this mHealth operational scenario.

8 The Growing Demand for Reliable mHealth Transactions

In addition to ameliorating crisis management, as is widely discussed in the literature and in this book, reliable mHealth will have wide-reaching benefits in managing the health of aging populations worldwide. Accurate and reliable patient monitoring, facilitated by mHealth applications, will be key to the burgeoning "home care" transformation of the healthcare industry, as the percentage of the aging population continues to grow. Reliable, real-time monitoring over commercial, unsecure wireless networks, as is the case today, will simply not be possible because of the structural limitations of the commercial infrastructure we have discussed.

9 Conclusion

In this chapter we have discussed the critical need for the medical profession to operate with sufficient confidence in the reliability, accuracy and security of the underlying wireless infrastructures on which mHealth applications and transactions are being conducted.

We also discussed the fundamental point, that unless such confidence can be sustained the potential benefits of mHealth will remain unrealized.

> ***"There is a fear that oversight in the development of mobile medical apps will present risks to patients if the networks fail to work effectively."*** **[17]**

The recommended solution, currently being contemplated in Canada, is the integration of mHealth and healthcare in general, with EMS (Emergency medical Service) and other Public Real-Time Social Services (PRTSS), user group. The PRTSS user group will be licensed by the government as a dedicated RF (Radio Frequency) spectrum to facilitate the design, deployment and operation of an independent dedicated broadband wide area mobile infrastructure.

This solution will enable the medical profession to benefit from mHealth under similar reliability and security conditions, with due process priorities, as is being contemplated for "first responders" (police, fire and EMS) worldwide.

As this (Canadian) solution unfolds, further understanding and research of various alternatives should be undertaken, particularly if funding limitation were slow to the speed with which these networks are implemented and made available to the medical communities.

Acknowledgments. I am grateful to Dr. Eva Seidner and David Francis who have provided generous support in the preparation of this chapter.

References

[1] Gazette Notice SMSE-007-12 – Consultation on a Policy, Technical and Licensing Framework for Use of the Public Safety Broadband Spectrum in the Bands 758-763 MHz and 788-793 MHz (D Block) and 763-768 MHz and 793-798 MHz (PSBB Block) Comments by Mobilexchange Ltd., `http://www.ic.gc.ca/eic/site/smt-gst.nsf/vwapj/SMSE-007-12-Mobilexchange-Comments.pdf/FILE/SMSE-007-12-Mobilexchange-Comments.pdf`

[2] Gazette Notice SMSE-007-12 – Consultation on Policy, Technical and Licensing Framework for Use of the Public Safety Broadband Spectrum in the Bands 758-763 MHz and 788-793 MHz (D Block) and 763-768MHz and 793-798 MHz (PSBB Block) Reply comments by Mobilexchange Ltd., http://www.ic.gc.ca/eic/site/smt-gst.nsf/vwapj/SMSE-007-12-MobilExchange-ReplyComments.pdf/FILE/SMSE-007-12-MobilExchange-ReplyComments.pdf
[3] A special compilation of AAMI Wireless articles and resources Association for the Advancement of Medical Instrumentation, 4301 N. Fairfax Dr., Suite301, Arlington, VA 22203-1633 www.aami.org2013, http://www.aami.org/hottopics/wireless/AAMI/Going_Wireless_2013.pdf
[4] Mehregany, M., Saldivar, E.: Opportunities and Obstacles in the Adoption of mHealth, http://www.mhimss.org/sites/default/files/resourcemedia/pdf/CHAPTER%202_mHealth.pdf
[5] Baker, S.D., Hoglund, D.H.: Medical Grade, Mission-Critical Wireless Networks Designing an Enterprise Mobility Solution in the Healthcare Environment (March/April 2008), http://www.arubanetworks.com/pdf/technology/whitepapers/wp_Medical-grade_Life-critical_Wireless_Networks_SteveBaker.pdf
[6] FDA on Health IT Adverse Consequences: 44 Reported Injuries and 6 Deaths in Two Years, Probably Just Tip of Iceberg (February 2010), http://hcrenewal.blogspot.ca/2010/02/fda-on-health-it-adverse-consequences.html
[7] Radio Frequency Wireless Technology in Medical Devices - Guidance for Industry and Food and Drug Administration Staff issued on August 13, 2013, include a request for comments and recommendations on the QoS (section 3B), http://www.fda.gov/medicaldevices/deviceregulationandguidance/guidancedocuments/ucm077210.htm
[8] Boston Marathon bombing in April 2013, Bloomberg BusinessWeek (Technology by Brad Stone April 16, 2013)
[9] Cantwell, E.: Harnessing the Power of Mobility for Healthcare: Why We Need a "Medical Grade Wireless Utility". West Wireless Health Institute (May 2011), http://medsmagazine.com/2011/05/harnessing-the-power-of-mobility-for-healthcare-why-we-need-a-%E2%80%9Cmedical-grade-wireless-utility%E2%80%9D/
[10] Wireless Health – A Prescription for Regulatory Clarity (June 2011), http://s3.amazonaws.com/wwhi.org/documents/WWHI-RegulatoryIssuePaper-June2011.pdf
[11] CBSNEWS report- Sept. 11 attacks in 2001 (September 11, 2011)
[12] Cantwell, E.: Medical Grade Wireless Utility Case Study for Hospitals (October 28, 2011), http://wireless.fcc.gov/spectrum/Small_Cell/Lunch_speaker.pdf
[13] The Spectrum Crunch - Sorry, America: Your wireless airwaves are full – CNNMoney (February 21, 2012), http://money.cnn.com/2012/02/21/technology/spectrum_crunch/index.htm
[14] CTIA Year-End 2012 Semi-Annual Wireless Industry Survey, http://blog.ctia.org/2013/05/02/semi-annual-survey/
[15] Cisco Visual Networking Index: Global Mobile Data Traffic Forecast Update 2012–2017, http://www.cisco.com/en/US/solutions/collateral/ns341/ns525/ns537/ns705/ns827/white_paper_c11-520862.html

[16] Versel, N.: FCC finalizes MBAN rule, still must appoint coordinator - Mobihealthnews (September 13, 2012), `http://mobihealthnews.com/18431/fcc-finalizes-mban-rule-still-must-appoint-coordinator/`

[17] Barriers to Adopting – New intelligent data management systems combat mobile network congestion problems as demand for mHealth data services explodes (September-October 2012), `http://www.roke.co.uk/resources/articles/Barriers-to-Adopting-Wireless-Healthcare.pdf`

Additional Articles and Papers Recommended

1. Mar 1, 2012 - 9-1-1 Magazine: Pre-Hospital Care & 700 MHz Broadband for Mission Critical Public Safety Data – From Calling Rampart to Communicating with the Enterprise – Author: Pascal B. Rodier, Superintendent, British Columbia Ambulance Service *(http://www.9-1-1magazine.com/Rodier-700MHz-for-EMS)*
2. May 16, 2013 - Making Care Mobile: Transforming Health Care Delivery – PWC *(http://www.mowatcentre.ca/Fellows/Will-Falk/AppsforHealth%20May%2016th_Falk_vFinal.pdf)*
3. May 23, 2013- CIO Council – Government Mobile and Wireless Security Baseline *(https://cio.gov/wp-content/uploads/downloads/2013/05/Federal-Mobile-Security-Baseline.pdf)*
4. National Broadband Plan- Chapter 10: Healthcare *(http://www.broadband.gov/plan/10-healthcare/#s10-5)*
5. Thinking about EMS Telemedicine? By Curt Bashford (http://www.general-devices.com/files/learning_pdf/EMS_World_Article.pdf)
6. Communication Issues in Pervasive Healthcare Systems and Applications – Chapter 10 – By Demosthenes Vouyioukas and Ilias Maglogiannis *(http://www.icsd.aegean.gr/dvouyiou/dmdocuments/pubs/BOOK%20CHAPTERS/IGI/978-1-61520-765-7.ch010.pdf)*
7. Capturing Value in the mHealth Oasis – An Opportunity for Mobile Network Operators – By Arthur D. Little *(http://www.adlittle.com/downloads/tx_adlreports/ADL_mHealth.pdf)*
8. Wireless Sensor Networks for Personal Health Monitoring: Issues and an Implementation – By Aleksandar Milenković, Chris Otto, Emil Jovanov – Electrical and Computer Engineering Department – The University of Alabama in Huntsville*(http://citeseerx.ist.psu.edu/viewdoc/download?doi=10.1.1.92.1861&rep=rep1&type=pdf)*
9. FCC – The Federal Communications Commission – USA - FCC finalizes MBAN rule, still must appoint coordinator *(http://mobihealthnews.com/18431/fcc-finalizes-mban-rule-still-must-appoint-coordinator)*
10. Amcom Software – Ten Predictions for 2012 on Smartphones in Hospitals *(http://www.hahleq.com/stuff/3pT/WP-US-10-Predictions-2012-Smartphones-Hospitals.pdf)*
11. Medical Applications of Wireless Networks – By Tam Vu Ngoc (http://www.cse.wustl.edu/~jain/cse574-08/ftp/medical/)

12. Requirements and Design Spaces of Mobile Medical Care – Mobile Computing and Communications Review, Volume 11, Number 3 - By Prajakta Kulkarni and Yusuf Őztürk *(http://tomx.inf.elte.hu/twiki/pub/Team/Research/p12-kulkarni.pdf)*
13. Quality of Service, Adaptation, and Security Provisioning in Wireless Patient Monitoring Systems- InTechOpen *(http://www.intechopen.com/books/biomedical-engineering-trends-in-electronics-communications-and-software/quality-of-service-adaptation-and-security-provisioning-in-wireless-patient-monitoring-systems)*
14. mHealth in an mWorld – How Mobile technology is transforming health care – Deloitte Center for Health Solutions *(http://www.deloitte.com/assets/Dcom-UnitedStates/Local%20Assets/Documents/us_chs_2012_mHealth_HowMobileTechnologyIsTransformingHealthcare_032213.pdf)*
15. Why common carrier and network neutrality principles apply to the Nationwide Health Information Network (NWHIN) – Journal of the American Medical Informatics Association *(http://jamia.bmj.com/content/early/2013/07/08/amiajnl-2013-001719.full)*
16. FCC chief: We have a three-pronged push for mhealth *(http://www.mobilehealthlive.org/articles/fcc-chief-we-have-a-three-pronged-push-for-mhealth/20551/)*
17. NCBI Bookshelf - Improving Patient Safety with the Military Electronic Health Record - Advances in Patient Safety: From Research to Implementation *(http://www.ncbi.nlm.nih.gov/books/NBK20562/)*
18. Healthcare Wireless and Device Connectivity *(http://davidhoglund.typepad.com/integra_systems_inc_david/)*
19. BCIT - E-Health Accelerator links major references to regulation of "Medical Grade" communications *(http://www.bcit.ca/appliedresearch/ehealth/links.shtml)*
20. FDA and FCC Joint Hearing *(http://www.electricalpollution.com/FCCandFDAHearingJuly2010.html)*
21. Wireless Health Care - IEEE Spectrum *(http://spectrum.ieee.org/biomedical/devices/wireless-health-care/0)*
22. Ed Cantwell: Medical Grade Wireless Utility *(http://wireless.fcc.gov/spectrum/Small_Cell/Lunch_speaker.pdf)*
23. Regulations and Standards for Wireless Medical Applications – written by Salim A. Hanna, Industry Canada *(http://www.ismict2009.org/pdf/2D05.pdf)*
24. Philips suggests dedicated mHealth spectrum *(http://mobihealthnews.com/7178/philips-suggests-dedicated-mhealth-spectrum/)*
25. Fish & Richardson | News | President Signs into Law the 700 MHz Public Safety Bill and Other Spectrum Provisions *(http://www.fr.com/700mhz-public-safety-bill/)*
26. Secure Wireless Communications *(http://www.cewit.org/downfile/CEWITFaculty%20Research%20Long%20Version.pdf)*

27. FCC, FDA to work on effective mHealth regulation I mobihealthnews *(http://mobihealthnews.com/6945/fcc-fda-to-work-on-effective-mhealth-regulation/)*
28. Six hospitals deploy West's medical grade wireless I mobihealthnews - Ed Cantwell *(http://mobihealthnews.com/16369/six-hospitals-deploy-wests-medical-grade-wireless/)*
29. ITU – Standards and eHealth (ITU-T Technology Watch Report) *(http://www.scribd.com/doc/47248412/Standards-and-eHealth)*
30. National-Academics.org – To improve patient safety, health information technology needs better oversight, accountability *(http://www.gallitanooconnor.com/pdf/2011_11_08_Press_Release_IOM_Patient_Safety_National-Academies_org_Newsroom.pdf)*
31. OECD Health Policy Studies - Improving Health Sector Efficiency, The Role of Information and Communication Technologies *(http://ec.europa.eu/health/eu_world/docs/oecd_ict_en.pdf)*
32. Amednews.com – American Medical News, Smartphones blamed for increasing risk of health data breaches *(http://www.ama-assn.org/amednews/2011/12/19/bil21219.htm)*
33. FlexITy Healthcare, Medical Grade Network - Security, Reliability and Compliance *(http://healthcare.flexity.ca/medical-grade-network)*
34. Commentary – Why private HIEs are more sustainable than public exchanges *(http://www.govhealthit.com/news/commentary-why-private-hies-are-more-sustainable-public-ones)*
35. Federal Communications Commission (US) – FCC/FDA Joint meeting scheduled to streamline review process for life-saving wireless medical technology *(http://www.fcc.gov/events/fccfda-joint-meeting-life-saving-wireless-medical-technology-day-2)*
36. RP-25 - Policy Principles for Public Safety Radio Interoperability - Spectrum Management and Telecommunications *(http://www.ic.gc.ca/eic/site/smt-gst.nsf/eng/sf09554.html#a5.1)*
37. Convergence of IP and Wireless Networks for Public Safety, by Edward Kelly *(http://www.theiacp.org/LinkClick.aspx?fileticket=fwcmm9XkPog%3D&tabid=734)*
38. Healthcare Mobility Solutions, End-to-End Mobility for every healthcare workflow *(http://www.motorola.com/web/Business/Solutions/_ChannelDetails/_Documents/static%20files/HealthcareAAG_BRO_0211-8pg-layout-v2.pdf)*
39. Forbes – How Healthcare's Embrace of Mobility has turned dangerous *(http://www.forbes.com/sites/sap/2012/01/05/how-healthcares-embrace-of-mobility-has-turned-dangerous)*
40. mHealth: From Smartphones to Smart Systems *(http://www.mhimss.org/sites/default/files/resource-media/pdf/CHAPTER%202_mHealth.pdf)*

41. Vodafone Global Enterprise Evaluating mHealth Adoption Barriers: Privacy and Regulation *(http://mhealthregulatorycoalition.org/wp-content/uploads/2013/01/VodafoneGlobalEnterprise-mHealth-Insights-Guide-Evaluating-mHealth-Adoption-Privacy-and-Regulation.pdf)*
42. mHealth Task Force Findings and Recommendations – Improving care delivery through enhanced communications among providers, patients, and payers. *(http://www2.itif.org/2012-mhealth-taskforce-recommendations.pdf)*
43. IEEE Xplore - Worldwide Network-Centric Healthcare Operations and the Associated Telecommunications Infrastructure *(http://ieeexplore.ieee.org/xpl/login.jsp?tp=&arnumber=4265819&url=http%3A%2F%2Fieeexplore.ieee.org%2Fiel5%2F4265778%2F4265779%2F04265819.pdf%3Farnumber%3D4265819)*
44. Mendeley: From mobile phone to responsible devices (vol. 4, Issue 6, Pages 719-726) *(http://onlinelibrary.wiley.com/doi/10.1002/sec.218/abstract)*
45. Key Issues in Mobile Health, by Robyn Whittaker *(http://www.hinz.org.nz/uploads/file/2011conference/P13_Whittaker.pdf)*
46. Data Breach Press: Top 11 Trends for 2012 in Healthcare Data, According to Industry Experts (Portland, Oregon Jan 5, 2012) *(http://www2.idexpertscorp.com/press/top-11-trends-for-2012-in-healthcare-data-according-to-industry-experts/)*
47. The Institute for Policy Innovation Event Recap - Mobile Health: Innovations in Care and the Spectrum Challenge (PDF) *(http://www.ipi.org/docLib/20120313_MHealth_Event_Recap.pdf)*

Questions

Q1. How will medical staff access the PRTSS networks, and would they need a different smartphone or tablet?

Q2. In major emergencies, how will the PRTSS network be configured to allow various agencies (Police, Fire, EMS healthcare, etc…) to simultaneously access and convey large amount of information?

Q3. How will PRTSS networks be funded?

Q4. Will governments be willing to allocate cellular compatible “mobile spectrum” for mHealth rather than auction it to the higher bidder?

Q5. Why would public wireless mobile carriers not accommodate a priority secure and reliable service for PRTSS use?

Part VIII

Security and Privacy

Chapter 42: Mobile Data Collection: A Security Perspective

Chapter 43: Security, Reliability and Usability of mHealth Environments

Chapter 44: "Security and Privacy Issues for Mobile Health"

Chapter 45: Impact of Privacy Issues on User Behavioural Acceptance of Personalized mHealth Services

42. Mobile Data Collection: A Security Perspective

Samson Gejibo, Federico Mancini, and Khalid Azim Mughal

University of Bergen, Department of Informatics
{samson.gejibo,federico,khalid.mughal}@uib.no

Keywords: Data collection, secure mobile storage, cloud storage, information sharing, mobile security.

1 Introduction

Remote mobile data collection systems (MDCSs) are a combination of a client application running on mobile devices, wireless infrastructures and remotely accessible server databases. Most of the existing systems share common principles and guidelines to collect data remotely. As reported also in [3], MDCSs have been mostly aimed at projects with tight budgets, deployed in developing countries with sparsely populated areas, where low data rate and intermittent connectivity exist.

This resulted in the development of light mobile clients targeting the so called "feature phones", i.e., cheap phones with only basic functionality, but able to run simple Java applications. It is not surprising that most MDCSs until a couple of years ago would provide only Java ME (Java Mobile Edition)[1] based clients which could run on phones such as Nokia 2330c with very low specifications, i.e., a 4.7MHz processor, 4MB RAM, and 128KB persistent storage. At this point, security was not considered in any of MDCSs. In fact, the lack of support for cryptographic libraries in these feature phones, their limited computing capability and memory resources, and least privileges access for third-party application to the phone, made it extremely difficult to adopt the most basic and standard security solutions. Even using the HTTPS [12] protocol for secure communication could be a big problem because of the lack of a standard list of pre-installed Certificate Authorities (CAs) on the phones, the cost of these certificates and the poor implementation of the protocol itself on many devices [15].

However, in the past few years we observed a dramatic change in the mobile operating systems (Mobile OS) market share. Java ME, that dominated the scene for decades, has left the competitive era to the new born Android[2] and iOS[3]. We also

[1] http://www.oracle.com/technetwork/java/javame/index.html
[2] http://www.android.com
[3] http://www.apple.com/ios/

S. Adibi (ed.), *Mobile Health*,
Springer Series in Bio-/Neuroinformatics 5, DOI: 10.1007/978-3-319-12817-7_42

saw that novel mobile OSes such as Ubuntu Mobile[4] and Firefox OS[5] are trying to penetrate the market by targeting the emerging market. This OS explosion has brought new opportunities and challenges to the data collection systems community. The boost in device computation performance, longer battery life, larger screen, and natural way of communicating with the device are some of the benefits. On the other hand, cost of the device and maintaining multiple code base targeting different OS have been the major challenges. Besides, the skills needed to customise the code-base written in native languages require high level computing competence which is hard to find in low-and-middle income countries. A recently conducted survey by the OpenDataKit group [4] shows that most of their users are experiencing problems with customisation, forcing to spend extra resources on this issue. New security challenges are, of course, also another problem.

The server side of MDCSs has also undergone some evolution. Although the technologies used in the implementation might not have changed so drastically as the mobile counterpart, the way of deploying and offering services on the web has changed. It should suffice to mention cloud-based services.

As this book is about a technology roadmap for mHealth, in this chapter we will mainly focus on the latest technologies and the security challenges faced by MDCSs specifically when Android-based smart phones and cloud based servers are employed. However, some basic security issues are conceptually identical no matter the underlying system, and most of the proposed solutions can be applied in several contexts and Mobile platforms. We will also discuss the solutions we designed for older Java ME based devices, and how they are still relevant.

1.1 Typical Use Cases for MDCS in mHealth

MDCS have been used to collect different types of data in multiple sectors including agriculture, education, and Health. The Health sector in particular leverages MDCS to facilitate data gathering and information distribution from/to remote locations.Some of the use cases are: mobile monitoring and reporting of HIV related data collection programs in South Africa[6]; Home-based Counseling and Rapid HIV Testing program by USAID-AMPATH in Kenya [21]; Childhood Vaccinations program in Uganda [1]; and meeting community health worker needs for maternal health care service delivery using mobile technologies in Ethiopia [8].

The United Nations Foundation and Vodafone Foundation report [18] covers a comprehensive list of MDCS use cases in different parts of the world. Moreover, readers can get a list of MDCS use cases in[7] and[8].

---

[4] `http://www.ubuntu.com/phone`

[5] `http://www.mozilla.org/en-US/firefox/os/`

[6] `http://www.cell-life.org/projects/data-collection-for-community-media-trust/`

[7] `http://opendatakit.org/about/deployments/`

[8] `https://datadyn.zendesk.com/entries/21282536-Case-Studies-Who-Uses-Magpi-formerly-EpiSurveyor`

2 Typical MDCSs Architecture and Organisation

We start by describing more formally and in detail how MDCSs are organised, both from a software and a deployment point of view.

A typical mobile data collection project consists usually of the following entities:

1. *Project owner* or a *manager* which initiates and manages the research and has the highest privileges on the central server,
2. *Data collectors* who have responsibilities for downloading form definitions, collecting data and uploading collected data, and possibly the ability to download submitted data for further data manipulation.
3. *Data viewers* that are able to log onto the server and to export data onto their devices.
4. *Site administrator* – all the abilities of a project manager plus responsibilities for system setup, provides technical support including the ability to add users, set passwords, and grant these capabilities to other users.

This role overlapping creates problems when trying to define a security architecture where it must be clear who has the right to do what. Thus we define a semi-hierarchical structure that can model most systems, but where we clearly decouple data access from system administration.

1. *The site administrator*: is in charge of server maintenance, user management and data recovery. This administrator does not need to be a physical person, but just a role in the system that does not need to access form definitions and collected data.
2. *The form managers*: are in charge of creating form definitions and have access to the data collected using such forms. In practice, form managers are the data owners.
3. *The data collectors*: are trained specifically for a specific project and sent into the field to actually collect the data at different locations. They usually do not have access to the collected data after upload to the server.
4. *The local administrators*: provide technical support on site. Data collectors might forget their password or lose their phone. Since we are talking about remote data collection, in places where network connectivity is scarce and unreliable, each location might then have a local supervisor with some administrator rights on the system that can provide this type of support while coordinating the data collection process locally. They do not have direct access to any form definition or collected data.
5. *The data viewers*: are usually analysts that are only interested in reading the collected data for analysis purposes.

The collection process itself, schematically illustrated in Figure 1, starts by setting up the server with the list of predefined collectors, form managers, local administrators and data viewers, which all receive a user-name and password to get access to the server. The project organisation remains pretty much unchanged over time. The form managers design the form definitions to be used with each project, and the collectors can download these definitions on their mobile client to start collecting data. Data collectors must first of all install the application on their mobile devices, and then register to the server for the first time, where they are presented with the list of possible form definitions they can download. A predetermined work-flow might decide in which order the forms should be downloaded. The form definitions are then downloaded and the data collection starts. Once the forms are filled, they are regularly uploaded to the server and deleted from the phone. At this point the form managers and data viewers can synchronise with this central repository to download the most recent data to their personal machines for further analysis.

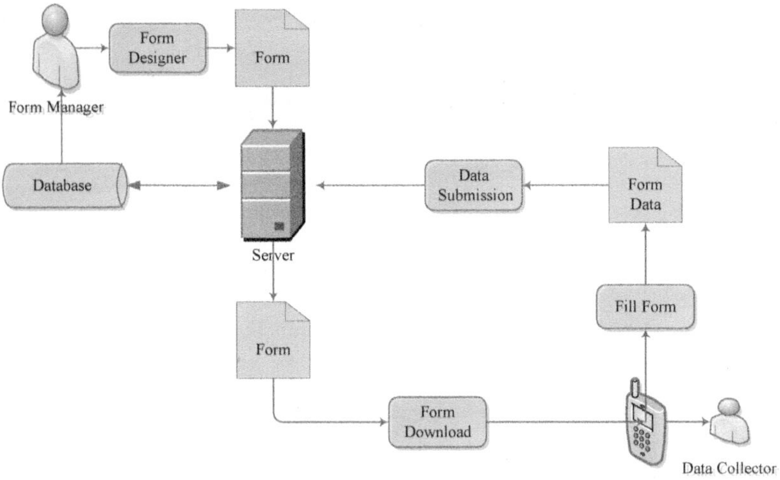

Fig. 1 Simple MDCS Data Flow

The electronic form definition, which is a replacement for the traditional paper form, is created through a form designer engine. There are several form designer tools including openXdata form designer[9], WEPI and Voozanoo 3 from Epiconcept[10], ODK Build and XLSform from OpenDataKit (ODK)[11], Vellum from Dimagi[12], and several others. The form designer assists to formulate relevant questions used to conduct particular research. Typical question types are text, number, date, time, single/multi-select fields, multimedia, and geo-tagging. After creating a form definition, an organisation or a research institute interested in conducting the research has to decide where to host these forms and the collected data. Generally, there are four alternatives:

---

[9] `http://www.openxdata.org/`
[10] `http://www.epiconcept.fr/en`
[11] `http://opendatakit.org/`
[12] `http://www.dimagi.com/`

1. **Private server:** demands substantial resources such as dedicated IP, hosting location, network connectivity, hardware, power, cooling system, and obviously backup and possible fail-over solution. However, it certainly provides a complete control over the collected data and form definitions.
2. **Third party server:** this solution requires low resources compared to the previous one and it is the easiest and cost-effective way to setup the hosting server within a short time. The only thing required by the project is to set up an account on the server, download the client on the mobile devices, and pay per number of users and data traffic, as in dedicated services like formHub[13], Commcarehq[14], Magpi formerly Episurveyor[15]. All the data stored on these type of servers will be however potentially exposed to third party entities who might not be authorised to access it.
3. **Private Server on Third Party Infrastructure:** As in the first case, a virtual server is offered by third-party as Infrastructure as service (IaaS) and the project administrator setups the server and load MDCSs server application. Amazon EC2 is a typical example of IaaS. Theoretically, this would allow the project to have full control over the server applications and collected data, but IaaS providers do not take any responsibilities related to security breaches and data loss on the server which potentially would still expose data stored on it to a third party.
4. **Server application on Platform as a Service:** In contrast to third party server, Platform as a Service (PaaS) provides a range of platforms and the project owner needs to deploy the server application onto the platform. Google App Engine is a typical PaaS and Aggregate server from ODK is developed to run on App Engine. It is believed that IaaS provides better data control over PaaS.

Other differences among different MDCS regard the mode of use of the client and the technology used. For example whether a mobile device is shared by multiple collectors or if a collector has the ability to modify form submissions, and which platform the client application has been developed, i.e., Java ME, Android, iOS or HTML5 browsers.

2.1 Functional Requirements

Another important aspect of concerns are the functional requirements we can expect from these kind of applications on the client side, as they also put some constraints on the security solutions that can be designed. After thorough discussion with ODK and openXdata community, we identified the following functional requirements:

- *Sharing:* The same mobile phone might be shared among multiple collectors, while the same collector might use more than one phone. The client must therefore support multiple users.

[13] `http://www.formhub.org/`
[14] `http://www.commcarehq.org`
[15] `https://www.magpi.com/`

- *Offline capabilities:* As often collectors might not have sufficient connectivity to connect to the server, the client application must also work off-line and for example store collected data on the mobile device until connectivity is available or data can be backup through a different connectivity such as USB and Bluetooth.
- *Rendering and storing media files:* As a part of a data collection process, different data types might have to be collected. The client must then be able to capture, store and display video, audio, text, geo-coordinates and so on.
- *Tight budgets:* Any solution that such system might have to adopt should not exceed the cost of what a typical project can afford.

2.2 MDCSs Overview

To conclude this section, we present a non-exhaustive list of current MDCSs on the market, classified by platform and business model, in Table 1. Many other systems exist, but they are mostly derivations of the ones listed here[16].

Table 1 List of MDCSs, with the platform they support, the type of server deployment and whether they are open source

MDCS	Client	Server	Open
Nokia Data Gathering[17]	Java ME/Windows	Private/Third Party	YES
openXdata	Java ME	Private	YES
EpiCollect[18]	Android/iOS	Third Party	YES
OpenDataKit	Android	Private/PaaS	YES
MagPi	Java ME/Android/iOS	Third Party	NO
Mobenzi Researcher[19]	Java ME	Third Party	NO
CommCareHQ[20]	Java ME/Android	Third Party	YES

3 Security Requirements and Threat Model

Among all MDCSs available today, there is none with a complete secure solution for the data handled either by the mobile client or sent and stored on the server, wherever it might be processed. We discuss what solutions are in place and whether they are adequate or not in Section 4, but first we identify the critical security aspects in MDCSs. Most of them are standard security requirements, but nevertheless, are often neglected. A good place to start is the OWASP Top Ten Mobile Risks[21], where

[16] `http://opendatakit.org/help/help-for-hire/`
[17] `https://projects.developer.nokia.com/ndg/wiki`
[18] `http://www.epicollect.net`
[19] `http://www.mobenzi.com/`
[20] `http://www.commcarehq.org/`
[21] `https://www.owasp.org/index.php/Projects/OWASP_Mobile_Security_Project_Top_Ten_Mobile_Risks`

the ten most common security risks for mobile devices and corresponding server-side infrastructure are classified. In particular we focus on the three of them we have worked on, which are in the top 5: Insecure Data Storage, Insufficient Transport Layer Protection and Poor Authorisation and Authentication. Besides these general security concerns, there are others that are more specific to MDCSs: for instance *Data recovery* and *Inter-process communication*.

Here is a brief description of the different security aspects we will consider:

- *Insecure Data Storage:* Insecure data storage is probably the most concerning aspect in mobile devices, especially when large amount of health related data are collected. The reason is that, unlike desktop computers, mobile devices are much more likely to be lost or stolen, or to be easily accessible in a short period of time if left unattended. It is enough that most application data is stored unprotected on the memory card, which can easily be taken out of the phone and read without problems on another device. This is why it is necessity to always encrypt sensitive data before storing them locally on a mobile device. With encryption, however, come also a lot of other unavoidable problems. How to generate strong cryptographic keys? How to store them in a secure way? Which algorithm to use? Does the device offer adequate support? Similar problems arise also on the server side, at least when the server is not under direct control, like for example in the cloud case.
- *Insufficient Transport Layer Protection:* Protecting the data in transit from the mobile devices to the server, and vice-versa, is another natural security requirement. When using wireless communication, eavesdropping the data traffic is not a difficult task, and if no encryption is in place, sensitive information can be easily stolen. However, only because the communication to the first wireless hotspot or router is encrypted, it does not mean that the entire communication up to the server is, and end-to-end encryption should be adopted. The most widespread solution to this problem is usually HTTPS, but we will argue later that better solutions might exist for MDCSs.
- *Poor Authorisation and Authentication:* An authentication is a major problem both on the client and on the server side. Especially considering that the same phone might be shared among collectors, there should be mechanisms that guarantee that collectors only have access to their own data, and this can be done only if some form of authentication and access control is in place on the client application. A similar problem is present when a user wants to transfer data from or to the server. Then it is essential to have mutual authentication, where the user knows that the server is the real one, and the server knows the identity of the user in order to enforce the right permissions. There exists various authentication protocols that allow for mutual authentication, but often the main problem is the generated traffic and the definition of a root of trust. For example, if we use a certificate to authenticate the server, we have to trust the certificate. This can be done by either buying one from a well known Certificate Authority or by provisioning it to the user through a secure side-channel. This secure side-channel becomes then the new problem: how do we know we can trust this channel? We

will discuss this issue in the Provisioning section 5.1. Similar problems might exist when generating and distributing user accounts.

- *Data recovery:* When data is going to be encrypted, it is important to have some recovery solution in case the collectors or even the project administrators lose their encryption keys or passwords. At the same time we do not want to have a recovery mechanism that can be easily exploited by an attacker or unauthorised entities.
- *Process to Process Communication and Separation:* Typically, when a collector needs to fill a form, it is not only text that is entered, but also GPS-coordinates, short movie clips, pictures and so on. On modern smart phones this can be done by any application by simply accessing the existing hardware on the phone (like the camera) through public APIs offered by the OS. The problem is that often the data captured in this way is not under the direct control of the application, which only gets a copy or a link to it, and it is therefore very difficult to secure it or make sure it is deleted from the phone memory.

3.1 Threat Model

In order to devise adequate security mechanisms, it is also important to define the possible threats to the system and what its critical parts are. We divide the sensitive parts into client, server, and communication channel.

3.1.1 Mobile Device

On the mobile device, the most likely threats in a Remote Mobile Data Collection scenario are the following:

- *The data is lost or stolen*: This can be due to losing or damaging the phone, in which case the data is most likely lost, or due to someone willingly stealing the phone or the memory card, so that data becomes actually available for an unauthorised party.
- *Data pollution*: a collector might voluntarily upload fake data and affect the results of a project.
- *Malicious applications*: a collector might download malicious apps on the mobile device that can interfere with the MDCS client and steal, read or hijack the collected data. The MDCSs application might be replaced with a malicious version at download time.

3.1.2 Server

On the server side we should distinguish between private and third party servers.

- *Private server:* In case the server is running on private machines and is operated by the project itself, then the most likely threats are standard security issues like wrong configuration settings and old applications that are not being patched. This issues are largely covered elsewhere. The other typical threat are insiders, like a malicious administrator. In this case we can compare the private server case to the third party server where we do not trust those who operate it.

- *Third party:* We focus on the cloud version of a third party server, and in this case we assume the adversary model of the "honest but curious server". That is, the cloud platform where our server is running might monitor our input and outputs, both from network and the local memory, but it will not compromise our application code or change the data stored in our database. In other words, assuming that what is stored on the cloud might be exposed to an attacker, we want to minimise the potential damage that such a leakage could cause. In particular, we want to protect the data in case the database is stolen, or accounts that have access to it are compromised.
- *Communication:* Here we can assume that eavesdropping the traffic from a mobile device might be within the reach even of an opportunistic attacker with the right equipment, especially if WiFi is used rather than data traffic based on the cellular network infrastructure. However, these days even the eavesdropping on cellular network is becoming a reality with individual affordable cost. The General Packet Radio Service (GPRS)[22] is one of the widely deployed cellular communication networks in low-income countries. GPRS provides only data confidentiality based on stream cipher between a mobile device to Serving GPRS Support Node (SGSN). These algorithms were kept secret for long time, but recently, researchers have reverse engineered most of these algorithms and demonstrate successful exploitations of GPRS systems across multiple networks in Germany and Italy. The researchers used a free and open source software, OsmocomBB[23] to intercept and decrypt GPRS network traffic within 5km radius.[24,25] . This implies the need of end-to-end security.

4 Current State of Security

Given the security concerns and the threat model, we can quickly review the existing MDCSs mentioned in Table 1 and evaluate them from a security point of view. Notice that the list given here is by no means exhaustive, since there exists many variations from the same original systems that might differ in many ways from the original. However in our experience, the majority of them focus on improving flexibility and usability rather than security and application logic. A more detailed analysis can be found in [9].

- *Secure Data Storage*: None of the MDCSs we considered had any kind of local encryption to secure stored data at rest on the mobile device. The only attempt was made by Nokia Data Gathering, where the user can choose a password that is used as encryption key. Unfortunately, also this solution had some weakness as the encryption key was also stored in clear on the client. The only other exception

[22] `http://www.etsi.org`
[23] `http://bb.osmocom.org/`
[24] `http://www.techadays.com/2011/08/gprs-is-too-easy-to-hack/`
[25] `http://events.ccc.de/camp/2011/Fahrplan/events/4504.en.html`

is ODK for Android, but, as we will explain later on, the encryption, although very strong, is very little flexible and quite heavy.

- *Authentication and access control*: Again, only very few of the systems we reviewed had any local authentication or multi-user support. Only Episurveyor, CommCareHQ and openXdata had a user authentication procedure. However, Episurveyor supports only remote authentication which was a problem for projects deployed in remote region without connectivity, and also had the option of saving the user credentials in clear on the phone in order to auto-complete the log-in form during form submission. On the other hand, CommCareHQ and openXdata provides local authentication by comparing entered password against the locally stored hash of the salted passwords. For remote authentication they both also support the standard digest based server authentication[26], which is not recommended without a secure underlying communication channel. OpenXdata also used to download the entire users table during server authentication and store it on the device storage for local authentication. This was later fixed.
- *Secure Communication*: Most MDCSs implemented HTTPS link for secure data transmission, but due to a range of issues such as Java ME weak random key generation, and multiple certificates needed to achieve HTTPS link on different device brands, and more, made establishing a secure transmission channel far from practical. The situation is better on Android based systems, but it is still up to the single projects to buy and setup the required certificates if a private server is used.
- *Data Recovery*: No notable mechanisms were in place anywhere to allow for password recovery or consistent password changes. No encryption was used on the client, so no mechanism for recovering encrypted data were devised either.
- *Provisioning*: Java ME applications need to be signed with a special code signing certificate in order to access some critical APIs on the phone. Without this signature the application can however still be modified into a malicious one to steal user data or upload data to the wrong server. None of the reviewed client apps were signed, and no alternative solution to signing was proposed either. Android based applications are usually self-signed as this is enough to publish them on the Google PlayStore, but this does not really give any security as we explain later on, unlike the iTune store for iOS apps.
- *Cloud*: For a few years now, cloud computing has been a hot research area in both industry and academia. As thoroughly described in Hassan et al. [7], security and privacy are still the major challenges for cloud computing. The only MDCS currently offering a server component thought to be deployed to the cloud is ODK. What we can do is to look at what security they offer and alternatively look at independent existing secure cloud storage solutions and discuss whether they would work well in a Mobile Data Collection context. What ODK offers, is to download a public key attached to the form definition, and then upload each single form to the cloud encrypted with a random symmetric key, which is in turn encrypted with the aforementioned public key. This assure very strong

[26] `http://www.ietf.org/rfc/rfc2617.txt`

data security both at rest on the client and on the server. Secure communication might even be superfluous since data is already encrypted. The downside is the overhead generated by all the encryption required for each form and the little flexibility the collector has, since forms cannot be edited after being encrypted. Besides only the owner of the public key can decrypt and possibly redistribute the collected data. When it comes to secure cloud storage providers, there is a comprehensive report from the Fraunhofer Institute for Secure Technology [3]. Of all the security features they look at, we are most interested in confidentiality and recovery i.e., data protection on the cloud and data recovery process when a user loses credentials. It is clear that the safest solution when it comes to confidentiality is to encrypt the data locally before upload with a key which is not shared with the cloud storage provider, as offered in TeamDrive, Wuala, Mozy and CrashPlan (to which we can add Mega[27] and nCrypted[28]). The main drawback of this approach is that if the user loses the encryption keys stored locally, also the data is lost.

A compromise is to derive the encryption keys from the user password, so that, even if the password is never sent to the provider, users can re-generate the keys as long as they remember the password. This however can compromise confidentiality itself if the password is not strong enough.

Hence the main problem to solve for secure cloud storage in MDCSs is the recovery of the encrypted data if password or keys are lost, without creating a security hole that an attacker can easily exploit. Sharing encrypted data is also a very difficult problem, but we will try to give a possible solution to this issue in the next section.

5 Proposed Solution

Although it is clear that the current security is inadequate, we do not think there is a simple and straightforward fix either. The reason is that standard security solutions are not always suitable for these systems. From a discussion based on the field experience of openXdata and ODK, it turned out that even the widely deployed HTTPS protocol used to protect data transmission in most systems, might not always be a viable solution. The working environment in which these MDCSs are deployed, the low budget settings of the projects using them, and other factors, impose heavy restrictions on their capabilities and reliability. Therefore, we want to discuss different approaches that can provide the needed security to mobile data collection.

At the time this writing, Android takes 70% of the global smartphone market share and it has become the platform of choice for most MDCSs. Thus, our main focus will be on the Android platform and its security model as far as the client is concerned. However, we provide an overview of the security features in iOS based devices, and recall same issues we faced in Java ME enabled phones.

---

[27] `https://mega.co.nz`

[28] `https://www.ncryptedcloud.com/`

We start with the problem of protecting the integrity of an application during distribution as the initial step for a secure MDCSs.

5.1 Secure Application Distribution

Keeping the integrity of an application during installation with its pre-set contents in a distributed environment is imperative to the rest of proposed secure solution. The major mobile OS providers enforce integrity checking through signature based system. However, application provisioning is one of the areas that the two rival platforms Andorid and iOS differ. The iOS built-in security model enabled Apple to distribute applications through a single channel.[29] The integrated security architecture of hardware and software within iOS devices relies on the public key infrastructure (PKI) with Apple as a top Certificate Authority (CA). Figure 2 shows the bottom-up signature based chain validation technique from initial boot-up to iOS Kernel loading which takes care of verifying third-party app signature before installing. This makes application distribution safer than most other solutions.

The iOS kernel make sure that apps signed with Apple issued certificate can run on iOS device and prevents unsigned apps or apps signed with self-signed certificate from installing. Apple issues a code signing certificate to an individual or organisation after verifying their identity through different ways. Then, application can be signed with issued certificate and distributed through the iTunes app store. There is an additional app checking process that the Apple store enforces before the app is shown to the public. The app passes through an automatic rigorous test following some check list. As of this writing, the check list and how it is enforced is not public, but through cumulative experiences from several developers, there are app performance, memory leak, malware, and signature checking. The overall built-in security and permission enforcement allows to build a trusted relationship between developers, distribution channel, and consumers.

The provisioning of Java ME applications reminds in part of iOS, where code must be signed by a valid CA in order to be installed on the phone. Actually, apps can still be unsigned, but then they would not get access to critical system APIs. In fact, different levels of code signing certificates are needed to allow an application to use different APIs offered by the mobile platform. However, no single distribution channel or appstore exists and applications can be provisioned freely Over The Air (OTA).

Android completely depart from the idea of bottom-up built-in security model, controlling, and validating app distribution channel. The main principle for departing from this idea is that with less restriction developers can build apps faster and put them into market in a shorter period of time. Practically, this principle does not seem to be working as expected. If we look at some statistics, by March 2009, Android Market started with hosting 2300 Android apps and on May 2013, the number of apps on Google Play was estimated to 1,000,000 with 48 billion downloaded

[29] `http://images.apple.com/iphone/business/docs/iOS_Security_Oct12.pdf`

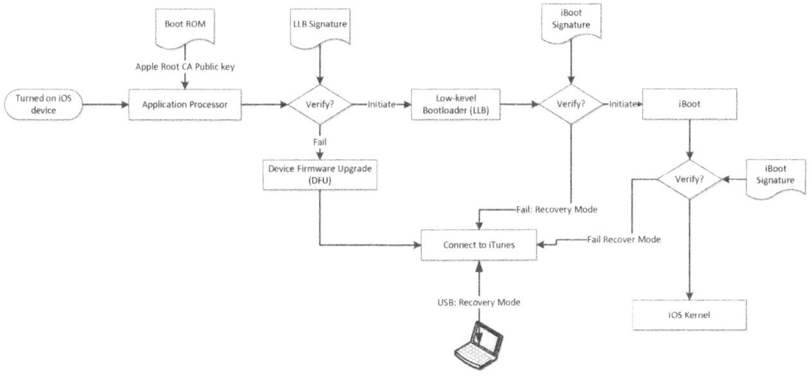

Fig. 2 iOS Security Model

to Android devices[30]. On the other hand, Apple started with hosting 500 apps in July 2008. As of July 2013, it was hosting 900,000+ apps with 50 billion download which is close to the number in Android[31]. In conclusion, Apple submission criteria and checking procedures did not affect the number of apps. Therefore, the premise from Android team's security trade off for faster app development principle, does not seem to be justified.

The Android security model spurred other app stores to host Android apps. At the beginning, Android departed from PKI based code signing architecture with a top CA and introduces a self-signed code signing mechanisms for application testing and release. The departure from PKI based architecture helped others to host Android apps in their store. As of this writing, there are more than thirty Android app stores. Some of them are Google play, Amazon App Store[32], Mikandi Adult App store[33], Samsung App Store[34], Android App Online[35], Aptoide[36], and Opera App Store[37].

Android app distribution starts with by digitally signing the application with a self-signed certificate (no CA required). The only requirement is that the code signing certificate should be valid for more than 25 years. The Android security model uses the application signature to ensure the authenticity of the app author at installation time and during app update, and to establish a trust between different apps that are signed with the same key. Applications signed with the same key share the same Linux kernel User Id, which helps to share resource, permissions, and

[30] http://www.youtube.com/watch?v=9pmPa_KxsAM
[31] http://www.apple.com/itunes/50-billion-app-countdown/
[32] https://developer.amazon.com/welcome.html
[33] http://mikandi.com/
[34] http://apps.samsung.com/
[35] http://www.appsgeyser.com/
[36] http://www.aptoide.com/
[37] http://apps.opera.com/

processes. David et al. [2] perform thorough analysis on collective Android apps from different app stores and have shown Android's current signing architecture does not encourage best security practices and propose some platform level mitigation methods. In order to upload the app to Google Play, Google requires an initial registration through a Google account and a payment of 25 USD paid through an individual credit card. Some of the third-party app stores are free or require an annual developer membership fee for uploading apps.

Reverse engineering has never been easier when we come to Android apps. There are tools available that make it very easy for an attacker to get into the source code and resource files. Once the installed app is extracted from the mobile device we can unzip the application file and all class files. Generally, it is recommended to obfuscate apps, but even then it is not so hard to get the source code. Traditionally, an attacker can manipulate the source files, add malware into it, re-package it, and sign it with his/her own key and finally publish to the app store. Rahul et al. [11] demonstrated this type of attack and proposed a three detection system level scheme that relies on syntactic fingerprinting to detect plagiarised applications under different levels of obfuscation used by the attacker. This kind of practice has been overwhelmingly high in Google Play before Google introduced the Google Bouncer[38] in 2011. The purpose of the bouncer is to scan apps for known malware, spyware, Trojans, and hidden and malicious behaviour (comparing it with previously analysed apps, and checking new developer accounts against previously known offenders) by using a simulated area inside Google's cloud. Google also introduced a remote kill-switch mechanisms to remove already installed apps remotely. As a result of the bouncer, the number of potential malicious apps has been reduced by 40%[39]. To further strengthen the application validation, Google introduced a client verification mechanism in Android 4.2 (Jelly bean), but latter incorporated it into Google play service app (which already exists from Android 2.3, a.k.a Gingerbread). In the end, there is no guarantee that the application being downloaded is genuine. Someone might have tempered with it, turned off its security features, or added a malware in it, and we do not have any way of detecting that for sure.

We propose three possible ways to alleviate this problem without having to change the whole system:

1. Manual installation is the ideal choice of secure app distribution. Apps delta update (aka a patch) can be pushed to client by the project server in coordination with Google Cloud Messaging system (GCM)[40]. The Android platform enforces the proper signature checking procedure during update. The main drawback of this approach is that it will be cumbersome installing the app at the initial stage of a large scale project. An alternative solution might be necessary for this type of scenario.

---

[38] http://jon.oberheide.org/files/summercon12-bouncer.pdf

[39] http://googlemobile.blogspot.no/2012/02/android-and-security.html

[40] http://developer.android.com/google/gcm/index.html

2. Bypassing the app stores altogether and distributing apps through the vendor websites. Application vendors such as ODK or openXdata act like a trusted central CA. The application is going to be signed, even self-signed, for integrity checking, but we are not relying on this signature to trust its authenticity. We are relying on the fact that we download the app from the developer site in a secure manner, and that we are actually connecting to the correct download site. The download link can be sent through a secure side channel as SMS or mail, and now the root of trust becomes the combination of this secure link and the SSL certificate of the application vendor signed by a trusted CA pre-installed on the device. The downside is that the application cannot be customised and no project specific configuration added to it, because then it would have to be signed again and downloaded possibly from another site we might not trust. The same problem as the appstore would arise again. However, the advantage is that the application is not compromised and it will behave as expected, which is the fundamental requirement to be able to enforce any other security mechanism later on.
3. A third option might be to have some MDCSs vendors become specialised appstores, where they guarantee for all customised versions of their systems. This requires to set up a model similar to that of the iTunes, but the chances of achieving this within the foreseeable future are slim.

5.2 Secure Communication

In some previous work [9, 5] we argued why HTTPS cannot always be considered as the best solution for secure communication in low-budget data collection projects. The reasons were mainly three: bad support/implementation in older phones, criticisms towards the Certificate Authority Trust Model (see for instance [13]) and the cost of the certificates. A typical feature phone would, in fact, have a limited list of root certificates which could not be modified by the user, and that was not standardised across models and manufacturer. This would make it difficult for a project to choose a CA from which to buy an SSL certificate that could be supported by all the handsets deployed in the project. On the other hand, not being able to modify the list of root certificate would give a guarantee that self-signed certificates could not be accepted, providing higher security. The result however was that most MDCSs would simply not use HTTPS at all. With smart phones, the support for HTTPS improved dramatically, but, especially in Android, self-signed certificates and CA signed certificates are treated in the same way, with the consequences discussed in the previous section on app distribution.

We looked then for other protocols that are not dependent on certificates, and realised that a password-based approach would be best. One of the most reliable such algorithms is the Secure Remote Password Protocol (SRP), a Password Authenticated Key Exchange (PAKE) protocol based on a pre-shared password for mutual authentication and key exchange, standardised in RFC 2945[41]. It allows mutual authentication and secure key exchange, while being resistant to on-line brute-force

[41] https://www.ietf.org/rfc/rfc2945.txt

and Man-in-the-Middle attacks (MITM). In addition, it can be used in conjunction with TLS[42] to create a secure transport layer without the need of certificates or new protocols. In previous MDCSs this approach would have been ideal since every collector would get a password based account to access the server anyway, but they would lacked the necessary APIs to support it. Besides, with new authentication technologies like OAuth, that can be used to authorise access to some server resources without sharing the user's credentials, a password based approach might not be necessary. ODK for example is using Gmail accounts to grant access to their servers, without having to issue a new user-name and password for each collector, as long as they have a mail account with Google.

Thus, we started looking at another strategy called certificate pinning. With the help of strict-host-key-checking feature, an OpenSSH client can be configured to grant or reject a connection request by verifying incoming request key against known keys list stored in the key store. This idea has been adopted in certificate or public key pinning. Pinning is an emerging concept of associating a given client with a list of known public keys or certificates. The client then verifies an incoming connection request against the list and grant or reject accordingly. After Man-in-the-Middle attack on Gmail's SSL through a compromised DigiNotar certificate (DigiNotar was one of well know CAs before bankrupting)[43], Google and other organisations have been actively engaged on making pinning as a part of their products. As a result, there are public APIs to implement pinning solution on mobile apps and web browsers. There is one fundamental problem with pinning though. There is no secure way of distributing the certificates or public keys to a client at the beginning of a project in a distributed environment. Basically, there are two ways to distribute the certificate or public key to the client. It can be incorporated with the application before distribution install or it can be fetched after the application is installed. OWASP recommends the former - before app is installed. However, if we choose the second distribution scheme proposed in the previous section, where no project specific configuration can be added to the app, we have to go for the later approach. That is, to fetch the keys and certificates after the application is installed. We found SRP protocol to be a potential candidate to accomplish this task.

Assuming we downloaded a genuine app, we propose to use SRP based on a one-time password delivered to the user by SMS or email, and use this protocol to authenticate the server for the first time and download its certificate. Successively, we use the certificate pinning strategy offered for example by openSSH and use HTTPS for future communications and perform the chosen user authentication on this secure channel.

In general, given that the application has been correctly and securely installed and configured, and the user has a unique identifier and a pre-shared secret, e.g. username and password or a one time password, we want to achieve the following requirements: Confidentiality; Mutual Authentication; Protection against reply and

---

[42] `http://tools.ietf.org/html/rfc5054`

[43] `https://www.owasp.org/index.php/Certificate_and_Public_Key_Pinning`

MITM attacks; Data Integrity; Key generation and management; and Compatibility with the existing applications.

SRP is currently resilient to many attacks and no vulnerable proof exist. SRP was proposed by Thomas Wu and its latest version is SRP6a. Due to the strength and the simplicity of the protocol, we can use SRP to fulfill all security requirements except data integrity, since it is used for authentication and not secure data transport. Compatibility is an issue as there are surprisingly few APIs offering the complete SRP protocol on both client and server side. We are working on this issue. The two SRP roundtrips needed to perform mutual authentication are shown in Figure 3. The final result is a shared symmetric key K, which can be later used in a TLS based session for example, or to encrypt the server certificate. According to the project setup, SRP might be used only once to download the server certificate, or every time in conjunction with TLS and the user password. Whether user authentication is based on[44] user certificates, passwords or one time codes is also open. In any case, when using TLS based communication, all our requirements would be met.

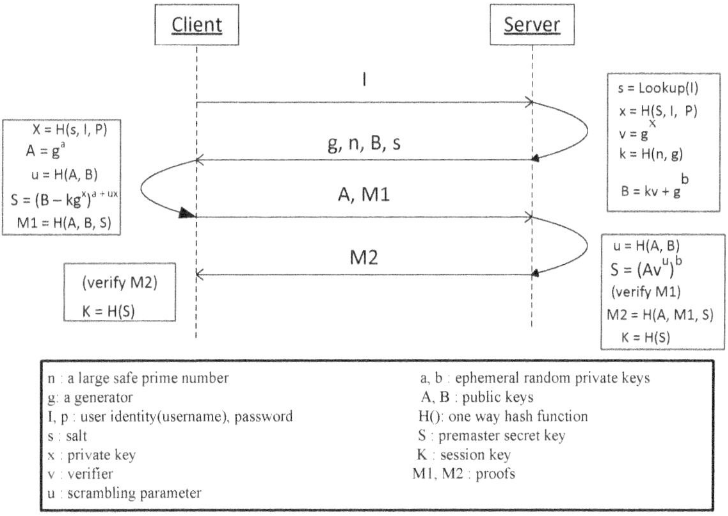

Fig. 3 SRP-6 Protocol

5.3 Secure Mobile Storage

When designing a solution for secure storage, we had to account for some typical scenarios in mobile data collection as described in Section 2.1. In particular that multiple users should allowed to use the same phone and that Internet access might not be always be available. This means that mobile devices can no longer be considered private or personal to a user and that most of the data collection might have

[44] `http://tools.ietf.org/html/rfc6749`

to be done offline. From a security perspective this translates into the following concerns:

1. Confidentiality (encryption)
2. Authorization (users can access only their own data)
3. (Off-line) authentication
4. Password and data recovery
5. Password changes should account for the possibility of using the same credentials to authenticate on different phones
6. Data should be reasonably protected also if all information stored on the phone is available to an attacker
7. Breaking the storage encryption should not compromise the rest of the system

In this section, we go through different possible solutions, discuss why they do not meet all the requirements, and gradually improve them until we get to a satisfactory one. Notice that, although at a high level we are guaranteed some security properties because we use standard algorithms which are proven to be secure, the security of actual implementations depends on the correct implementation of the crypto libraries used, the correct usage of the algorithms and the security model of the platform. Such libraries might be Bouncy Castle[45] for Java ME, Spongy Castle[46] for Android or OpenSSL[47], just to mention some freely available APIs. We will also report on our experiences with the actual implementation of this solution on different platforms.

Here we assume that user credentials consist of a unique user name and a password, and that to each user (on a given device) is assigned a different encryption key, so that a successful authentication grants direct access to this key and the data encrypted with it, but nothing else.

- *Straightforward solution*: The user is authenticated by computing a hash of the password and comparing it to the hash stored in clear on the phone. The same hash is also used as the user's encryption key (or to derive it). This allows separate encryption for each user (requirements 1 and 2), off-line authentication (requirement 3), but the authorisation mechanism prevents unauthorised access only as long as data are accessed through the application. If an attacker can copy the phone memory, then the stored hash can be used to decrypt the data directly, just as it can be used to recover the data if the password is lost (requirement 4).
- *Slightly better solution*: An easy improvement to the above solution is not to store the hash of the password. Authentication is then performed by verifying that the key derived from the password can correctly decrypt the data. This can be done by employing some kind of integrity check on the data, like appending a digest to the data before encryption, or adopting algorithm that can reliably detect errors caused by the use of a wrong key. But the safest and the cleanest way is to compute a keyed-hash message authentication code (HMAC). This approach

[45] `http://www.bouncycastle.org/`

[46] `http://rtyley.github.io/spongycastle/`

[47] `http://www.openssl.org/`

would satisfy also requirement 6 if the key is generated with a reasonable number of iterations and therefore more time-consuming to break, but makes the recovery of the password and the data impossible unless a copy of the derived key or the password itself is stored somewhere and it is accessible through some other form of authentication.

- *Complete solution with a private server*: Recovering the password or the derived key is not a trivial problem. This information would have to be stored somewhere and the user should have an alternative authentication mechanism to access it, creating a circular problem. In any case, when the encryption key is derived from the password, even though we had a way to recover the data, a password change would require re-encrypting the whole storage since also the encryption key must be changed. This problem can be solved by creating an encryption key that is not directly derived from the password, but is instead encrypted separately with a password based derived key. Combining the previous solutions we would satisfy almost all our requirements, except for 4, 5 and 7, as long as the same password is used both for accessing the phone and the server. In fact changing a password on the server from one phone, would mean that the old password must still be used to login locally on other phones, creating potential synchronisation problems. Also, if one phone encryption is broken and the password recovered, the attacker could impersonate the user to the server and gain access to the rest of the system. One approach that could give a satisfactory solution from a security perspective, is to separate completely the local authentication from the server authentication. In this way, requirement 4 would be satisfied by storing a copy of the encryption key on the server, and if the mobile password is forgotten, the user could login on the server, retrieve the key and reset the mobile password. Requirement 7 would be satisfied since no trace of the server password could be found on the mobile phone, neither explicitly (the hash of the password), nor implicitly (some keys are still derived from the password, so it is still possible to guess the password, generate a key and verify whether it is correct just as the authentication mechanism does).

Figure 4 shows a solution that satisfies all given requirements based on the previous discussion. The user will have to register on the phone the first time, and this requires remote authentication on the server with the server password. If the authentication is successful, then key to encrypt stored data ($STKey$ in the figure) can be created for the specific user, either by the server or the mobile phone itself, and the user may be asked to choose a new mobile password to access the encrypted storage. A copy of user master key is kept on the server in clear, since this is a private server we trust, and allows for recovery. The master key MK is created from the mobile password and used to encrypt the MK. A Hash-based Message Authentication Code (HMAC) is generated by combining MK and the encrypted $STKey$. Local authentication can now be performed using this HMAC.

How difficult it is to break the data encryption depends on the strength of the user password and the key derivation algorithms used. There is not much that can be done to enforce a very strong password while keeping it secure. It is common knowledge that users will just write it down somewhere or forget it. A good compromise could

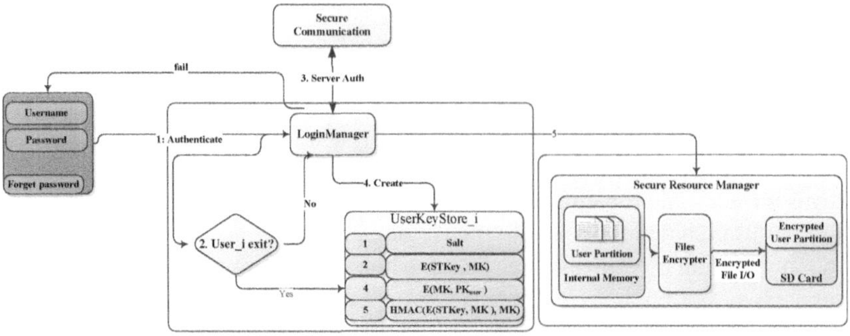

Fig. 4 Local Authentication. The notation $E(Data, Key)$ means that the *Data* is encrypted with *Key*.

be to require an alphanumerical password of a given minimum length, or a pass phrase, while using algorithms that could slow down as much as possible a brute force attack, but without compromising usability. For this purpose, simple MD5 hashes or even a salted SHA-1 hash might not be enough, so we recommend using a password based key derivation function such as PBKDF2 (Password Based Key Derivation Function)[48] and scrypt[49], with as many iterations as possible, based on the computation power of the phone.

5.3.1 Secure Storage Implementation in openXdata (Java ME)

Although the previous solution satisfies all the requirements we defined, it does not mean that it is the best in practice. From a usability point of view it is very heavy for a user to remember two different passwords, and from an implementation point of view, it might mean to re-design the whole application. Hence, in collaboration with openXdata, we designed a simplified solution that uses only one password and still satisfies most requirements except for the complete decoupling between client and server authentication. However, this was a risk that they were willing to take in order not to compromise usability. The details can be found in [6] and we will not report them here as they deal with many technicalities specific to the Java ME platform, which is not the focus of this chapter.

5.3.2 Secure Storage Implementation Experience on Android Platform

The solution sketched out for openXdata is general enough to be easily ported also to the Android platform, but with different implementation challenges that we will discuss here. The first and foremost task during secure storage implementation for Android platform is to figure out what Android platform security model provides us

[48] `http://www.ietf.org/rfc/rfc2898.txt`

[49] `http://www.tarsnap.com/scrypt.html`

to facilitate smooth implementation. In this section, we cover the following issues and recommend possible remedies.

1. Does Android platform package provides built-in Crypto APIs?
2. Can Android provide truly random key generation tool on the device?
3. Is there a secure place to store keys and user credentials?
4. Are there any Android features to handle multi-user support per phone?

Android platform includes two major open source crypto library providers, OpenSSL (C and Assembly implementation) and Bouncycastle (Java implementation). Bouncy castle has been shipped with Android platform in a cut-down and crippled state due to device memory constraints. As a result of this, the BouncyCastle API has been re-packaged and re-named as Spongycastle to avoid class loader conflicts with the old built-in Bouncycastle API. The Spongy Castle must be included with an Android project as external library, while OpenSSL is supported and maintained as a native library. OpenSSL is also available in iOS which make solution porting easier. Developers must however know how to use these libraries properly, or security might be compromised[50].

The other issue with Android is where to store keys and credentials in a secure manner in the device. The SIM card (Subscriber Identity Module) is a type of smart card and can be considered as a secure key store. Android provides APIs to access to the SIM element but, as of this writing, there is no Android crypro API exposed to perform cryptographic operations in the SIM. Basically, the SIM element is owned by the network operator and they have all privilege and control over the SIM. An application signed with operator signing key may have access to the SIM card and use it as secure element. Another way to store keys and credentials in Android is through its built-in keystore daemon. Since Android 1.6 (Donut), the keystore daemon was introduced for maintaining cryptographic keys for system-level apps such as VPN and Wifi connection. Despite the fact that Keystore deamon provides a secure store, it doesn't maintain multi-user credentials. As a result, we don't recommend use of keystore if a project is looking for multi-user support per device. But, it is worth mentioning that third-party application can leverage system level protection for its credentials through keystore and enhance application usability by introducing a single login interface for accessing the device. A limited built-in multi-user support per device was introduced in Android 4.2 (a.k.a Jelly bean). Therefore, as long as a multi-user environment is not fully supported, our proposed solution is still a valid alternative to handle the phone sharing problem. A secure ODK client based on this scheme is being implemented and the technical details will on be published.

5.3.3 Secure Cloud Storage

This section involves both the secure storage on the mobile device and the server counter-part. As we mentioned above, the recovery procedure we proposed involves storing a copy of data encryption key on the server, but this scenario works only if

[50] `http://android-developers.blogspot.com.au/2013/08/some-securerandom-thoughts.html`

the data is hosted at a privately owned server with exclusive control. If the data is hosted in a cloud or third party server, which are becoming increasingly popular, the proposed solution does not work because it completely exposes the collected data to the hosting organisation. If we look at today's trend, most mobile data collection systems are hosted in the third party server including Formhub[51] , DiMagi, Google App Engine, and Amazon EC2 and many more. Here we change slightly the way encryption keys are stored both on the client and on the server to guarantee an acceptable data security. We build on the current ODK solution mentioned in Section 4, and a more detailed explanation can be found in [5].

The main idea for sharing is the same as for many cloud storage providers and ODK: encrypt the master key used to encrypt the submitted data with the public keys of those who can read the data. However, unlike user-centric models, in a data collection project we know in advance who are the data consumers are. This allows the public keys to be distributed before data collection even starts, and therefore to enable sharing for multiple data readers in an asynchronous way. In addition we will use the same approach to publish securely other information that will allow for data recovery, and discuss some variations on this model to increase efficiency. A summary of the overall solution is showed in Figure 5.

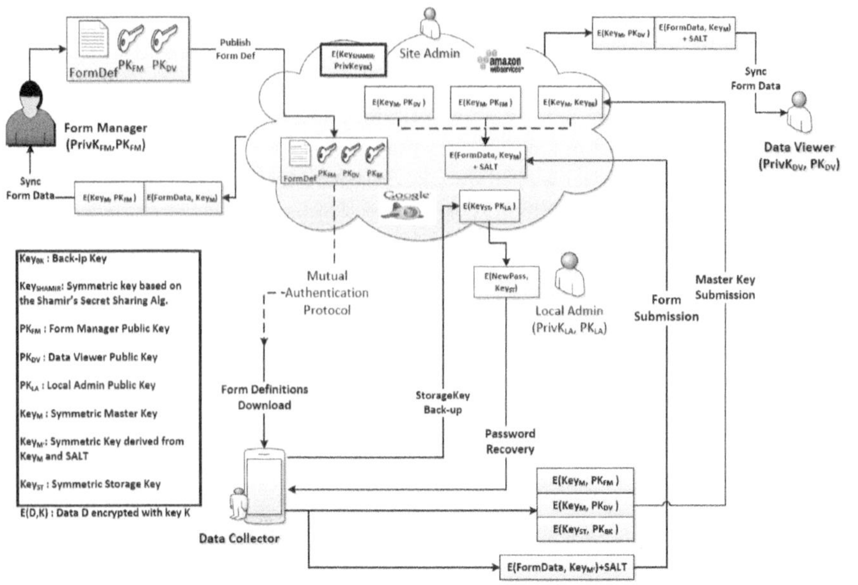

Fig. 5 Proposed Solution for Secure Cloud Storage for RMDC

Once the local authentication in subsection 5.3 and mutual authentication in subsection 5.2 are in place, the shared symmetric key created during SRP handhake can be used to encrypt or sign the public keys and form definitions downloaded from

[51] `http://www.formhub.org/`

the server, so that their authenticity is verified also without certificates. Such keys can then be securely stored on the phone encrypted with the user storage key. The public keys can subsequently be used also to verify the integrity of form definitions or other data loaded on the phone through other channels, but signed by their source.

At set-up time, a special public/private key pair is created and published on the server: the *back-up key*. This key pair is used solely for back-up purposes. All the master encryption keys used in the collection process will always be encrypted with the back-up public key, while the private back-up key will be encrypted with a strong symmetric key and stored on the cloud. This symmetric key will then be split up and distributed to a number of participants in the project (most likely the project organisers), according to Shamir's Secret Sharing algorithm [14].

The main differences with the current ODK approach lie also in the way forms are encrypted. Rather than using heavy public key encryption for each form, one symmetric master key is created when the form definition is downloaded. This key is subsequently uploaded to the cloud encrypted with all the public keys that where attached to the form definition, and stored locally encrypted with another key derived from the collector's local password. After this is done, every new form can be encrypted simply with a symmetric key derived from the master key and directly uploaded together with the value used to derive the new key. In this way anyone who can access the master key, can also derive the keys for each form. Besides, the collectors can locally edit forms also after having encrypted them, and possibly even download them again without having to have a public/private key pair. Multiple public keys enable also more flexible sharing, although limited to selected number of data viewers.

The other big difference is the possibility of recovery in most situations, while not compromising data security. Site and local administrators cannot see the data unless their public keys are attached to the form definitions, but they still have control over the server so that they can create accounts and enable and disable various features. This includes the recovery of user passwords. These passwords are needed to access the system and upload and download data, but are in no way connected to the encryption. In the catastrophic event that the owner of the public key which is used to encrypt all master password loses the corresponding private key, we still have the back-up key on the server. However, a collaborative effort is needed to decrypt and use this key, so that a single attacker or malicious admin needs a lot of resources in order to obtain it, thanks to Shamir's Secret Sharing scheme.

A downside of this approach is that the heavy cryptography limits the possibilities to collaborate on-line and the number of possible data viewrs, but it might be an appropriate solution for small research projects that place a premium on privacy.

6 Discussion and Related Work

In this chapter we have tried to summarise the experiences we had while working with MDCSs, and discussed the most critical security concerns around these systems. We have followed their development from the first Java ME based systems until the latest cloud-based solutions. Unfortunately, security was and still is a major issue in many of these systems. We have designed some solutions in close collaboration with the main players in this field, like openXdata and ODK, and have tried to find solutions that was both secure and user-friendly, and that could easily be integrated in a transparent way in their systems. The secure client storage has been the area where we were able to make most progress, and now both openXdata and ODK have a secure client version in production. We are actively working on the communication part, where a SRP based solution is being implemented and tested. This would not have been possible on low-end phone running Java ME applications, with limited or non-existing crypto APIs and low CPU power. Secure provisioning and cloud storage remain the most challenging parts, and we still do not have a working solution in these areas. The idea we proposed here for the secure cloud storage is probably good enough for small projects, but would probably prevent good usability in large scale projects. Notice also that most of what we discussed here can easily be applied to most mobile client-server applications, although some specific requirements are typical of MDCSs.

When it comes to related works, we are not aware of any security initiative specifically aimed at a comprehensive solution for remote mobile data collection, but there is a vast literature about the many individual problems we have discussed. The most relevant ones, however, have already been discussed throughout the chapter and we will therefore not repeat them here. Let us also point out that we considered mHealth systems where data is entered by human collectors, and we are not considering the situation where a sensor is feeding data to a remote server or device, or a monitoring program is installed on the mobile device and their correct operation must be guaranteed at all times as in [17]. The collected data is also not thought to be readily accessible to health-care providers, or in general to be offered as a service. These are different area of mHealth where much research is ongoing, but not directly applicable to our case. For instance, in [20] a solution for secure data storage is suggested, and they also propose to encrypt the data when it is not in use. However, this solution is designed to protect patient records that are retrieved to be read and should not be edited or distributed. Clearly this cannot be adopted in a data collection scenario. The same applies to solutions like those proposed in [16, 10] or [19] for secure patient's record retrieval.

7 Case Studies

We have had a close collaboration with these two organizations:

- openXdata is a free and open-source mobile data collection system used to run health related projects in low-income countries and has been deployed in different countries such as South Africa, Pakistan, and Uganda[52]
- OpenDataKit (ODK) is a free and open-source set of tools which help organizations author, field, and manage mobile data collection solutions. ODK has been deployed in several countries around the world[53]

[52] `http://www.openxdata.org/`

[53] `http://opendatakit.org/`

We developed secure storage and secure communication solutions based on their field experience and functional requirements. Our solution is currently being integrated in their clients.

8 Conclusion and Future Directions

Our final goal is to develop generic APIs that can easily be integrated in existing mobile applications and provide adequate security at all levels of the project deployment. As technology advances it becomes easier to adopt more sophisticated solutions, but at the same time new security challenges arise due to the complexity of the new systems.

In the future we see more and more on-line collaboration between collectors and data viewers, and scenarios where health personnel will need to access collected data on a real-time basis. Cloud resources will be used to analyse the huge amounts of data on a central server, so that anyone who needs it will be able to simply query the server without having to download all the data on some local machine and perform the analysis locally. This implies that our current solution for encryption of the data is not portable for big collaborative projects and new solutions must be devised. A possible compromise we are thinking about, is to leverage the form definition structure to define which information is sensitive and which can be processed without privacy concerns, so that a selective encryption can be performed and only privacy-critical fields will be protected with strong cryptography. Authentication mechanisms to access the on-line data without having to maintain huge user databases with passwords and access control lists are starting to get mature and widely used. OAuth authorization framework[54] is an example which is already being adopted by ODK. User-names and passwords might soon become obsolete, and one single account might be enough for a variety of services. This entails some challenges when mobile devices are involved, and even shared as in Mobile Data Collection. Having the device, might be enough to access a service without further authentication. However, mobile devices are evolving continuously and we have already mentioned that Android has recently introduced a multi-user environment. So, many challenges we have today might be solved by the natural evolution of mobile devices, but not all. The compromise between security and flexibility usually leads to problems that are very difficult to solve. An example is the open distribution model Android adopted, that make it almost impossible to know whether we are installing a genuine application or not. The same goes for applications that wants to use resources present on the phone, but do not have control over the way the data is transferred or stored during the inter-process communication.

On the client side we are observing a specific trend, where many MDCSs are being implemented as a hybrid application which is built from HTML5/CSS/JavaScript and native apps. Native apps are still proven to be the best in terms of performance and security but they are specific to a given platform and requires multiple code bases to run on multiple platforms. HTML5/CSS/JavaScript is promising to bridge

[54] `http://oauth.net/`

the gap and run on cross-platform with better user interface but still has limitation on accessing device hardware such as camera and geolocation. Therefore, the current solution leverages native functionality by embedding HTML5/JavaScript within the web-browser container such as Webkit. This architecture is known to be vulnerable to a range attacks such as Cross-Site-Scripting and needs in-depth security analysis before adopting it as a solution.

In general, mobile data collection might evolve in order to make the collected data available automatically available to other services in real time, so that decision might be taken based on the latest information collected also in remote locations and be as precise and updated as possible. Privacy issues might then become even more pressing, and privacy preserving and secure data retrieval techniques developed for other areas of mHealth might become relevant for data collection as well.

References

[1] Babirye, J.N., Engebretsen, I.M.S., Makumbi, F., Fadnes, L.T., Wamani, H., Tylleskar, T., Nuwaha, F.: Timeliness of childhood vaccinations in kampala uganda: A community-based cross-sectional study. Plos One: The Open Access Journal 7(4) (2012)

[2] Barrera, D., Clark, J., McCarney, D., van Oorschot, P.: Understanding and improving app installation security mechanisms through empirical analysis of android. In: 2nd Annual ACM CCS Workshop on Security and Privacy in Smartphones and Mobile Devices (SPSM), pp. 81–92 (2012)

[3] Borgmann, M., Hahn, T., Herfert, M., Kunz, T., Richter, M., Viebeg, U., Vowe, S.: On the security of cloud storage services. Technical report, Fraunhofer Institute for Secure Information Technology SIT, Darmstadt, Germany (2012)

[4] Brunette, W., Sundt, M., Dell, N., Chaudhri, R., Breit, N., Borriello, G.: Open data kit 2.0: expanding and refining information services for developing regions. In: Agarwal, S., Varshavsky, A. (eds.) HotMobile - 14th Workshop on Mobile Computing Systems and Applications, p. 10. ACM (2013)

[5] Gejibo, S., Grasso, D., Mancini, F., Mughal, K.A.: Secure cloud storage for remote mobile data collection. In: Proceedings of the Second Nordic Symposium on Cloud Computing and Internet Technologies, NordiCloud 2013, pp. 8–14. ACM (2013)

[6] Gejibo, S., Mancini, F., Mughal, K.A., Valvik, R., Klungsøyr, J.: Secure data storage for mobile data collection systems. In: Kacprzyk, J., Laurent, D., Chbeir, R. (eds.) MEDES 2012 - International Conference on Management of Emergent Digital EcoSystems, pp. 131–144. ACM (2012)

[7] Joshi, J.B.D., Takabi, H., Ahn, G.: Security and privacy challenges in cloud computing environments. IEEE Security & Privacy 14(3), 24–31 (2010)

[8] Little, A., Medhanyie, A., Yebyo, H., Spigt, M., Dinant, G.-J., Blanco, R.: Meeting community health worker needs for maternal health care service delivery using appropriate mobile technologies in ethiopia. Plos One: The Open Access Journal 8(10) (2013)

[9] Mancini, F., Gejibo, S., Mughal, K.A., Valvik, R., Klungsøyr, J.: Secure mobile data collection systems for low-budget settings. In: ARES 2011 - The Seventh International Conference on Availability, Reliability, pp. 196–205. IEEE Computer Society (2012)

[10] Mirkovic, J., Bryhni, H., Ruland, C.M.: Secure solution for mobile access to patient's health care record. In: 2011 13th IEEE International Conference on e-Health Networking Applications and Services (Healthcom), pp. 296–303 (2011)
[11] Potharaju, R., Newell, A., Nita-Rotaru, C., Zhang, X.: Plagiarizing smartphone applications: Attack strategies and defense techniques. In: Barthe, G., Livshits, B., Scandariato, R. (eds.) ESSoS 2012. LNCS, vol. 7159, pp. 106–120. Springer, Heidelberg (2012)
[12] Rescorla, E.: SSL and TLS: Designing and Building Secure Systems. Addison-Wesley Professional (2000)
[13] Roosa, S.B., Schultze, S.: The "Certificate Authority" Trust Model for SSL: A Defective Foundation for Encrypted Web Traffic and a Legal Quagmire. Intellectual Property & Technology Law Journal 22(11), 3–7 (2010)
[14] Shamir, A.: How to share a secret. Commun. ACM 22(11), 612–613 (1979)
[15] Simonsen, K.I.F., Moen, V., Hole, K.J.: Attack on sun's MIDP reference implementation of SSL. In: Proceedings of NORDSEC 2005 - 10th Nordic Workshop on Secure IT Systems, Tartu, Estonia (2005)
[16] Singh, K., Zhong, J., Mirchandani, V., Batten, L., Bertok, P.: Securing data privacy on mobile devices in emergency health situations. In: Schmidt, A.U., Russello, G., Krontiris, I., Lian, S. (eds.) MobiSec 2012. LNICST, vol. 107, pp. 119–130. Springer, Heidelberg (2012)
[17] Sorber, J., Shin, M., Peterson, R.A., Kotz, D.: Plug-n-trust: practical trusted sensing for mhealth. In: Davies, N., Seshan, S., Zhong, L. (eds.) MobiSys, pp. 309–322. ACM (2012)
[18] Vital Wave Consulting. mHealth for Development: The Opportunity of Mobile Technology for Healthcare in the Developing World. Washington, D.C. and Berkshire, UK: UN Foundation-Vodafone Foundation Partnership (February 2009)
[19] Weerasinghe, D., Muttukrishnan, R.: Secure trust delegation for sharing patient medical records in a mobile environment. In: 2011 7th International Conference on Wireless Communications, Networking and Mobile Computing (WiCOM), pp. 1–4 (2011)
[20] Weerasinghe, D., Rajarajan, M., Rakocevic, V.: Device data protection in mobile healthcare applications. In: eHealth, pp. 82–89 (2008)
[21] Rajput, Z., Mbugua, S., Viola, C., Saleem, J., Anokwa, Y., Hartung, C., Borriello, G., Mamlin, B., Ndege, S., Were, M.: Evaluation of an android-based mhealth system for population surveillance in developing countries. In: AMIA 2011 Annual Symposium (2011)

Questions

Q1. Why is security important in Mobile Data Collection?

Q2. Are current data collection systems secure?

Q3. Is mobile data collection an important application area in mHealth?

Q4. What is the future of mobile data collection?

Q5. How far are we from a secure solution for these systems?

Q6. What kind of technology will influence data collection in the future?

Q7. Why not to use standard security solutions also in mobile data collection?

Q8. Does any Mobile Data Collection System interested in complying with some regulatory security and privacy requirements such as HIPAA (The Health Insurance Portability and Accountability Act)?

43. Security, Reliability and Usability of mHealth Environments

Martin Gerdes, Yohanes Baptista Dafferianto Trinugroho,
Mari Næss, and Rune Fensli

University of Agder, Grimstad/Norway
{Martin.Gerdes,Dafferianto.Trinugroho,Rune.Fensli}@uia.no

Keywords: privacy, identification, access control, relationship management, reliability, redundancy, self-monitoring, user interface, portable communication gateway, integration platform.

1 Introduction

Mobile technologies confer mobility and autonomy on patients with the advantage of access to home care and health care services on demand. However, these benefits impose challenges to the future health care services. For instance, computation capacity of a conventional smartphone provides applications and services with sufficient power of calculation and automation to assist in daily life activities and medical purposes. Combined with a user-friendly interface, mobile technologies can be an easy and efficient manner to help people who are in a condition of cognitive deterioration or have a chronic disease which demands a close connection to near family members and/or to health care services. These scenarios require a technical solution guaranteeing that the user will always be "on-line" and connected to the mobile telecommunication network. In addition, to be able to share personal information and specifically for sharing medical information with health care services, the current solutions have to take into consideration both security and reliability requirements. The security issues will be those of standard computer network security measures, because there will be a need for secured user authentication and data integrity. This means that the data must not be corrupted or modified during the transference from the user's mobile to the health care service. Finally, to comply with data protection regulations data will only be available to people who have been granted authorisation to access the relevant information in specific cases to provide necessary follow-up and support.

The need for system's reliability implies the use of technology proven against mobile network or computer failures. In addition, emergency alarms and their data must be ranked with the highest priority to ensure immediate transmission to rescuing personnel.

S. Adibi (ed.), *Mobile Health*,
Springer Series in Bio-/Neuroinformatics 5, DOI: 10.1007/978-3-319-12817-7_43

In this chapter, we will focus both on the security issues and reliability precautions, including the safety aspects and user-friendliness for the end-user. In addition, we highlight actual use-case scenarios where future mobile technology solutions can provide relevant health and care services to elderly people and patients. Based on these scenarios, we will discuss some important user requirements to be considered by the system designers and technical developers, and also by the support chain of services that has to be established by private health care and security companies, close relatives, and public health care services such as telecare and telehealth functions.

1.1 Examples of Actual User Scenarios as Study Cases

A) Elderly Person at an Early Stage of Dementia

A typical situation for an elderly person at an early stage of dementia can be that she/he is living alone at home. For the upcoming generations, most elderly people are born after World War II and an increasing number of them are used to modern ICT solutions at work. Thus, they are likely to be familiar to use smartphones and tablet PCs on a daily basis, in contrast with today's elderly generation. Even with gradual decline of cognitive abilities by aging, some smartphone functions can usefully be used for assisting elderly people in daily life activities, such as in situations where they experience difficulties and need remote assistance by their close relatives. This could lead to improvements in situations as:

- Elderly person does not remember appointments and intake of medication
  - ➢ Support from a memory planner and electronic pill dispenser
- Elderly person does not find her/his way back home
  - ➢ Support from a traveller guide with GPS positioning alarm
- Loneliness and lack of social contacts
  - ➢ Use of social media functions such as Skype and Facebook
- Need for permanent availability of electronic assistance
  - ➢ Hand-watch display for user-friendly solution
- Easy connection to family members
  - ➢ Smartphone with video and voice over IP system (e.g., "Face time") call-up functions

B) Mobile Patient with Chronic Disease

In several European projects, there has been a focus on medical support for patients at home. When living with a chronic disease, it can be important to be able to make daily measurements of vital signs recordings. This could in particular be the case for patients with chronic diseases, such as diabetes, chronic heart failure (CHF) and chronic obstructive pulmonary disease (COPD). To date, experiences from the Renewing Health project[1] reveal that chronic disease patients need to go to hospital for treatment less frequently and they feel more secure being at home with remote home monitoring possibilities.

In general, current technology should allow recordings of the medical data at patient's home, and when traveling on vacation, etc., and to upload the data to a secure server storage. This storage should preferably be available within the computer network of the health care services, from where the authorized medical personnel taking remotely care of the patient can access the recordings of specific patients. In principle, such a database for the patient's medical information will be a Personal Electronic Health Record (PEHR). The Markle Foundation (2003) has defined a PEHR as:

> *"An electronic application through which individuals can access, manage and share their health information, and that of others whom they are authorized, in a private, secure, and confidential environment."*

This leads to the requirements of authentication and authorization, i.e., based on the authenticated identity of the patient, they should be able to administer access authorization to this database and give privileges to the health care personnel for access to the actual information. This relates to the requirements of privacy protection, which can be defined as:

> *"The right of individuals to determine for themselves when, how and to what extent information about them is communicated to others"* (Agrawal et al., 2002).

Enabling patients to manage their own medical recordings taken from devices in their home can be achieved by a Patient administered Personal electronic Health Record (PaPeHR) as described by (Fensli et al., 2011).

There might also be situations where close relatives need access to the vital signs recorded. As an example, this can be the case when children carry out their regular measurements of blood glucose, and want to forward this information to their parents to get advice about the correct insulin dosage to be given (Gammon et al., 2005).

For patients at home a diary of their medical symptoms and feelings can be relevant for the correct treatment and corresponding follow-up. A Spanish multicenter study with COPD patients at home concluded that a report on symptoms in a diary carried out by the patients seems to be a valuable tool in primary care for at-home-treatment and rehabilitation (Llor et al., 2012).

In this scenario, several aspects need to be considered:

- Remote monitoring
    - ➢ Secure storage in PaPeHR solution

---

[1] `http://www.renewinghealth.eu/en/`

- Children with close supervision by parents
  - Secure transmission of medical data to close relatives
- Trend curves for medical measurements
  - Enable visual view of changes in medical conditions
- Daily writing on a medical diary
  - Electronic storage of the patient's day-by-day experienced medical situation
- Personal Health Record securely stored
  - Administered by the patient, with access for authorized personnel

C) Rehabilitation and Physical Training

Patient treatments at hospitals are highly time efficient. The average length of stay (ALoS) has, according to OECD indicators, fallen over the past decade in nearly all OECD countries (OECD, 2011). At the same time, the number of people receiving long-term-care is rising. OECD reasons for this trend the population's aging and the implementation of home-based-care.

In a review of wearable sensors and systems with application in rehabilitation, Patel et al. concluded that there was a focus towards health and wellness in the use of wearable sensors, and towards the area of remote rehabilitation (Patel et al., 2012). In a study of home-based versus in-hospital based cardiac rehabilitation after cardiac surgery, Scalvini et al. found that the tele-rehabilitation program was effective and comparable to in-hospital rehabilitation (Scalvini et al., 2013). However, it will be necessary to implement functional telemedicine services with rehabilitation equipment and training records deployed as a normal service from the physiotherapists. An interesting on-going project in North-Norway is using video support for at-home exercise where the patients have installed a treadmill and have to fill in a regular training report to the physiotherapist[2]. On the market for fitness products, numerous *apps* are available for recording of physical exercises together with pulse activity, step counters and GPS data in order to track the training route. In addition, game controllers, such as Wii and Xbox ones, create opportunities for home-based training guided by an avatar on the screen[3]. The Kinect 3D camera will track body movements and limb positions; those functions can be useful in systematic training of motoric functions, as can be the case for rehabilitation after cerebral stroke.

For the home-rehabilitation use-case, important aspects will be:

- Training activities
  - Step-counter, GPS tracking, etc.
- Personal training guide with reminders and stimuli
  - Patient identification
  - Training log and remote video follow-up by physiotherapist

---

[2] `http://telemed.no/chromed.5125023-247951.html`
[3] `http://www.youtube.com/watch?v=2v9ZtD6oVJY`

- Avatar guidance
  - ➢ Training programs in game-controllers (e.g., Wii, Xbox)
- Integration of medical sensors
  - ➢ Pulse watch, respiration etc.

2 Reference Architecture

mHealth environments are characterized by the mobility of patients, their health and care service providers and other supporters, such as relatives. On the one hand information about the medical status of patients and their context (e.g., the physical location) shall be gathered and made available as soon as possible when needed for health and care service providers. On the other hand, patients shall receive information and support as fast as possible from the remote service providers and relatives when needed, in particular in certain emergency situations.

The fundamental requirement of gathering medical and environmental information from mobile patients on the one side and making it available to doctors, nurses and other people involved in the health and care on the other shall be made independently from their physical location. This requirement determines the specification of devices, services, and the underlying information and communication technology (ICT) infrastructures. Fig. 1 illustrates the high-level reference architecture, comprising the main domains of a mHealth system, namely the *Point-of-Care environment*, the *Health & Care Services environment*, and the *Health Information Services (HIS) environment.* This reference architecture covers all user scenarios described above, and will support the studies and discussion of technical requirements and solutions needed to realize the use cases, with particular focus on security, reliability and usability aspects.

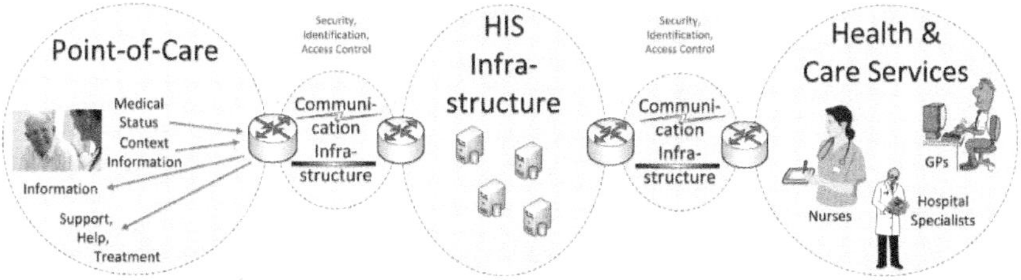

Fig. 1 High-level Reference Architecture

The main architecture domains comprising components and solutions from a variety of stakeholders are shortly described next.

Point-of-Care

The Point-of-Care (PoC) environment describes the immediate environment of the patient, see Fig. 2 for more details.

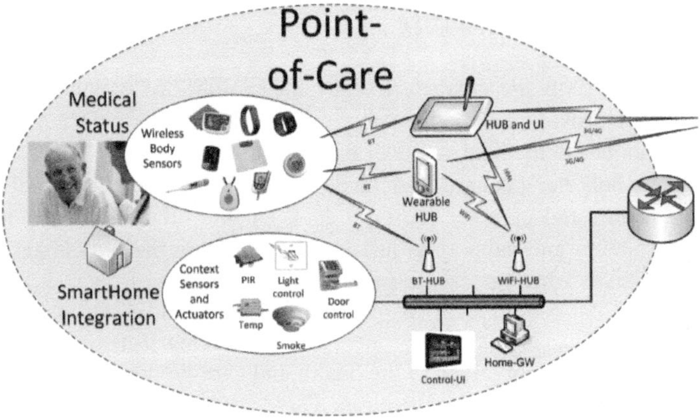

Fig. 2 Point-of-Care Environment

It consists of a manifold mixture of wireless body sensors capturing medical, health and (potentially) other kinds of information from the patient's environment. Furthermore, multiple kinds of personal user devices (for example, smartphones and tablet PCs) carry out the support applications and services for the patient, while they may, at the same time, act as communication gateways between the Wireless Body Sensors and the HIS infrastructure. Besides that, the point-of-care environment also includes content sensors and actuators, communication as well as control application devices, all belonging to the facility infrastructure the patient gets connected to temporarily, e.g., his own SmartHome network, or a care home network.

Health and Care Services

The Health & Care Services environment (see Fig. 3) is an inhomogeneous world of various kinds of medical and care service providers and their systems, which are partly based on standards and partly are proprietary, including a significant base of legacy systems. A common demand of health care experts is to have access to latest patient data, either from the patient's devices in the point-of-care environment or from other health care expert, in order to carry out remote diagnoses, and to provide information and remote medical and care support to the patient. From a mHealth perspective, this demand applies in particular to nurses and doctors visiting the patient at home, or outside in case of emergency.

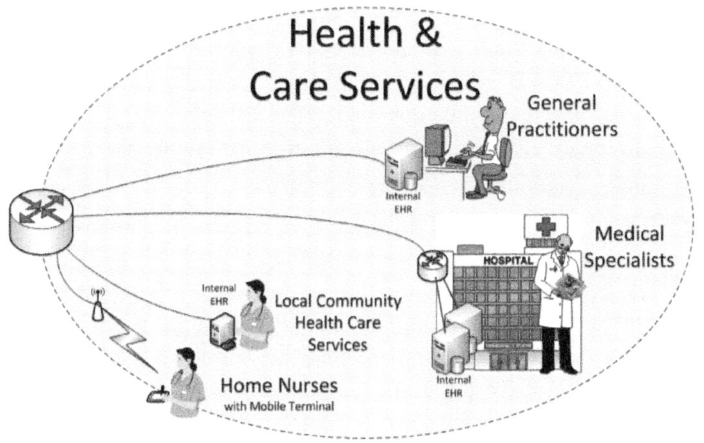

Fig. 3 Health & Care Services Environment

<u>Health Information Services (HIS) Infrastructure</u>

Information from patients shall be available for medical and care specialists and vice versa. To achieve that, a heterogeneous Health Information Services (HIS) infrastructure (Fig. 4) is continuously being developed, which contains a multitude of Electronic and Personal Health Record (EHR, PHR, xHR) systems from commercial and public stakeholders, based on both closed, proprietary and on open standardized technologies. Through means of authentication, authorization and accounting (AAA), it interconnects private, commercial and open systems, including point-of-care infrastructures, with closed Health Information Infrastructures of national or international health systems.

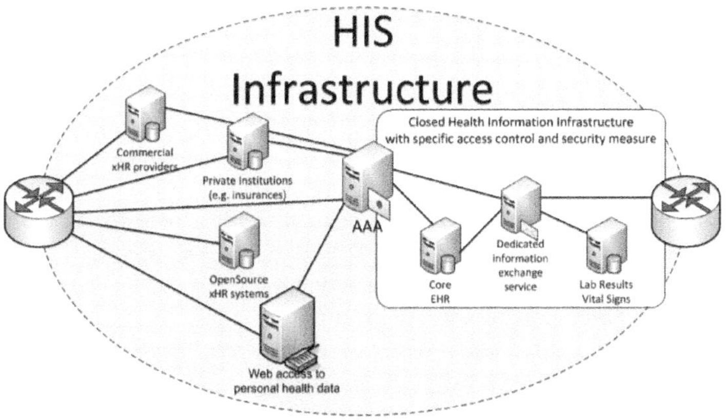

Fig. 4 Health Information Services (HIS) Infrastructure

From a mHealth perspective, the main role of the HIS infrastructure is to enable information exchange between (potentially) mobile patients (=information sources) and (optionally) mobile health service specialists (=information receivers) in a secure, privacy protecting, and efficient manner, while allowing high flexibility with regards to the different types of information, devices and services.

3 Requirements

In order to realize the anticipated mHealth use cases, various ICT components within the different domains of the reference architecture have to fulfil certain requirements. Within this section the technical requirements are explained, which are based on an analysis of the user scenarios described in the introduction (section 1) above. Particular focus is put on those requirements that relate to the different security, reliability and usability aspects of devices, applications and services that have to be considered for the development of corresponding solutions.

3.1 Security

One of the major challenges of ICT systems are scenarios where different individuals use different connected devices to access different resources (= data and services) in a shared service and information infrastructure. This applies in particular to the user scenarios B) and C). Different people, identifiable by a certain individual identifier (e.g. a generic personID, a patientID, or a healthExpertID) can use certain devices (as personal smartphones, tablet PCs, etc.), which come with device-specific identifiers (deviceID), to get connectivity through certain communication gateway devices (as "Wearable HUB" or "HUB and UI" in Fig. 2), having certain gatewayIDs and network-specific addresses, and to access certain data and services. On top of that an infrastructure with multiple communication gateways has to be considered, together with multiple users (i.e., patients) per communication gateway, multiple personal medical devices (PMDs) per person, per gateway, and per application/service, multiple patient applications (in the point-of-care environment), and multiple health and care related services (in the health & care services environment).

The mapping of the different identifiers to each other requires a flexible and at the same time safe, reliable and trusted *relationship management*. Furthermore, smart and user friendly applications and services are needed for the configuration, provisioning and enforcement of such identity mappings.

Hence, *Security* in this context refers to the need for *access control* and *privacy protection*, based on unique *identification*, *authentication* and *authorization*.

These logical needs lead to the following functional requirements:

Table 1 Security Requirements

Identification	Technical IDs for PMDs (sensorID, ...), patient devices (deviceID, ...), point-of-care applications (appID, ..), communication gateways (commGwID, ...), patients, health & care experts (nurseID, doctorID, ...), health & care experts services (serviceID, ...), health & care experts devices (deviceID, ...), HIS infrastructure services (xhrID)
Relationship Management	Mapping of PMDs, patient devices, point-of-care applications and communication gateways to a specific patient or point-of-care user (temporary, at the moment of use)
	Mapping of health & care experts devices and health & care experts services to a specific health & care expert (nurse, doctor, ...)
Access Control	Authorize or deny access attempts from identified and authenticated users (point-of-care users, health and care experts) to any specific resource (data, service) in the mHealth infrastructures
Privacy Protection	Protection of access to transmitted and stored data and information about a specific patient, based on individual access grants specified by the patient, and corresponding access control

3.2 Reliability

The availability of mHealth services and components can be of vital importance, in particular when it comes to the fast detection of and appropriate reaction on potentially life threatening situations or medical conditions of the patient. This applies in particular to the user scenario A), and also to scenario B). The transmission of vital signs from the patient together with additional context information, such as the location while being outdoors, to the HIS infrastructure and the availability of that information to health and care experts in critical situations (e.g., critical vital signs recordings, or loss of orientation of a dementia patient) has to be ensured.

The design guidelines for typical consumer ICT equipment do not include high guaranteed availability requirements. Hence, the reliability of mHealth solutions corresponds on the one side to specific availability requirements for the involved equipment, but also to solutions for a centralized or de-centralized self-monitoring of all involved HW and SW components, and measures to react on detected malfunctions in order to restore the functionality.

The following technical requirements can be identified:

Table 2 Reliability Requirements

Self-Monitoring	Monitoring of the operation status of HW and SW components (i.e., applications and services, in particular communication services)
HW redundancy	Provision of redundant HW (e.g., second communication gateway), that gets in operation triggered by Self-Monitoring component
Power fall-back	Switch to fall-back power (e.g., battery) in case of power breakdown, automatically or triggered by monitoring component
Connectivity fall-back	Switch to fall-back connectivity (e.g., fixed to cellular, cellular operator-1 to operator-2, cellular to WiFi, …) in case of connectivity loss, automatically or triggered by monitoring component
Capacity Adaptation	Automatic adaptation of processing and storage capacities of HIS Infrastructure services according to the load caused by growing number of active mHealth users (patients as well as health and care experts)

3.3 Usability

The capabilities and functionalities of a mHealth system as a whole can only be fully utilized, if its applications and services can be easily and efficiently accessed and used by the different users. The means to utilize certain functionalities of a system differ depending on the target users, the purpose of the applications and services, and the specific usability features of the user devices that are involved. A patient for example shall be able to use a tablet PC type of device, to check certain personal medical information at their current location, and to get in contact with a nurse or a doctor. Relatives or a nurse of the patient might utilize Web-technology based services on their smartphone or table PC device to access certain care and personal data via a dedicated portal service, independently from their own physical location. A doctor, visiting patients at their point-of-care, will use a laptop or tablet device with specific applications and services to access patients' medical and other data from the health information services infrastructure. Corresponding requirements apply to all three study cases from the Introduction section.

One requirement area concerning the usability of the applications and services comprises the various aspects of the user interface (UI, also called Human-Machine-Interface HMI), i.e., how the different user groups can be supported to optimally interact with the system. The UI design has to consider multiple factors, such as the needs, capabilities and limitations of different user groups and individual users: a doctor has UI requirements that may differ from a nurse's, a blind patient has obvious sensory limitations and a dementia patient cognitive impairments.

Beside those requirements on the visual and interaction design, there are usability-related requirements on the underlying mHealth system infrastructure. The same medical, health and care and context data from a patient shall be, under protection of defined privacy rules, available for different user groups of the mHealth system and their specific applications and services. The vital signs recordings of the mobile COPD patient (user scenario B) shall be available for the patient for information, for the nurse for regular control, and for a doctor to support the diagnosis. Though using the same data, the user interfaces realized in the different applications and services have to be adapted according the specific needs of the users with regards to content, visual presentation, and used HW devices. The underlying requirement is a decoupling of the applications and services that gather patients' data and information, and transmit them into the HIS infrastructure, from the applications and services that present selected patients' data through dedicated UIs for doctors, nurses, and the patients.

Furthermore, the portable personal user device of the patient has to be able to carry out certain local logic based on data gathered from connected sensors and PMDs, in order to be usable for certain emergency detection and care support features of all described user scenarios.

This leads to the following overview of functional requirements related to usability:

Table 3 Usability Requirements

Adaptive UI design	User interfaces (UIs) shall allow easy and user-friendly interaction with services and applications, while supporting personalization according to the patient's needs and preferences
Adaptation to PoC environment	User interfaces of point-of-care applications and services shall allow easy or automatic adaptation to the specific involved PMD(s) the user has to involve in an user scenario
Patient-friendly utilization of PoC data to support multiple support scenarios	Utilization of data gathered one time from PMDs in the point-of-care environment for different applications and services, locally in the PoC environment, and for health and care experts
Easy identification and relationship management	Identification of information (patientID, deviceIDs, ...) shall be easily be configurable or seamlessly provisioned and propagated to support the relationship management (see Security requirements above)

4 Solutions

The solutions described in this section are those that address the three requirement areas: security, reliability and usability. Furthermore, it is explained how the solutions can be applied to realize the user scenarios from the Introduction section.

4.1 Security

Some of the main challenge areas of multi-sensor, multi-device, multi-patient, multi-gateway environments as described in the reference architecture are security management, access control and privacy protection.

User authentication is very important in healthcare information systems as it is the first step in the entire health information access process, and mHealth systems are not an exception. It is primarily used to assure that the user is indeed who they claim to be and, therefore, some mechanisms are needed to verify the user's identity. These mechanisms are commonly known as authentication factors. There are mainly three categories of authentication factors:

1. Something the user knows (e.g., password, pass phrase, PIN).
2. Something the user possesses (e.g., smart card, security token, cell phone).
3. Something the user is or does (e.g., fingerprint, retina pattern, signature, voice pattern).

There are previous works that focus on authentication of users for accessing healthcare information systems via mobile devices. Poon and Zhang (Poon et al., 2006) explored security mechanism of body area sensor network (BASN) by means of biometrics that uses intrinsic characteristics of the human body as the authentication identity. The system requires biometric sensors around the user's body for gathering the necessary human body characteristics, and thus, it is rather complex to be used in a multi-user environment. A wireless application protocol (WAP)-based telemedicine system was proposed and developed which utilizes WAP devices as mobile access terminals for general inquiry and patient monitoring services (Hung and Zhang, 2003). The users authenticate themselves by introducing username and password through a wireless markup language (WML) page loaded by the mobile devices. Marti et al. (Martí et al., 2004) presented security services required mobile eHealth applications with the MobiHealth project described as an example. The presented MobiHealth system makes use of mobile devices (e.g., smartphone) as gateways for body are network (BAN) sensors to communicate with back-end servers. However, multi-user environment where the sensors and mobile devices are used by more than one user was not described. A wearable mHealth system using a wearable wrist-worn platform called *Amulet* was proposed in (Sorber et al., 2012). The *Amulet* acts as a gateway between various sensors and the user's mobile device, which is connected to back-end servers in the cloud. The *Amulet* also acts as the user's identity proxy, so that the sensors can be used by other users, and different mobile devices can also be used as its gateway. The main issue of this approach is that every user should have his/her own *Amulet* which cannot be exchanged with other users.

To build up the authentication level, multi-factor authentication scheme is normally used by combining different authentication mechanisms. In Norway, for example, four security levels for authentication and non-repudiation are officially defined, and any use of electronic identification to authorize access to health information is subject to strong security with strong authentication. At level 4, two-factor authentication is required where one factor should be based on qualified certificates.

Authentication can be conducted by using digital signature as the electronic ID, which can be used to log into different systems. For the initial certification and secure distribution of required public-private key pairs, existing and widely accepted solutions for Public Key Infrastructure (PKI) shall be deployed, which is outside the scope of this chapter. On reception of a log in request, the receiving system responses with a random number (nonce) or a string text. Encryption is then carried out on the device of the user who wants to log in using their private key, resulting in a "digital signature". The receiving system then checks this "digital signature" with the user's public key. The private key can be stored in smart cards, key fobs, universal subscriber's identity module (USIM) or any device that has a computer chip. The common technical solution to access the private key is by inserting a card to a card reader and entering a PIN. With key fobs, the user normally has to enter a PIN, and then the device will display a number, also known as Temporary Authorization Number (TAN), which can be used to log in. In order to enable multiple users to use one mobile gateway device (communications device) for sending measurements to back-end servers (hosting various services such as the PHR) as well as for accessing/reading measurement values from those back-end servers, a solution based on digital certificate is proposed to be used as shown in Fig. 5.

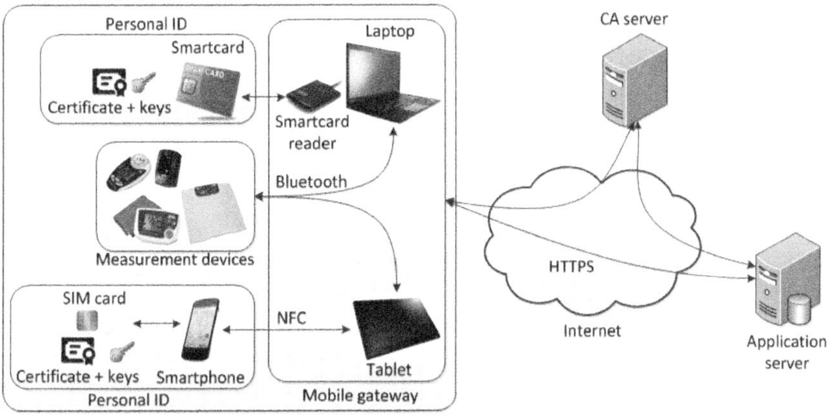

Fig. 5 Proposed architecture for multi-user mHealth environment

Each user of the system needs to identify themselves to the back-end servers hosting various services via the mobile gateway (e.g., tablet, laptop). To achieve this, each user should have their own personal digital certificate. Such a certificate can be issued, e.g., by a health service provider, by a health insurance or by the national health system administration/authorities. It must be certified by a Certification Authority (CA), and it is stored in their personal identification device, such as smartcard or SIM card connected to a smartphone. The application in the gateway device is then responsible to send the user's credentials to the back-end server for authentication. To ensure that both gateway device and server trust each other, 2-way mutual authentication is proposed to be utilized (i.e., user's digital certificate to authenticate the user and server's digital certificate to authenticate the server). A secure connection between the gateway device and the server is then established for transmitting measurements data based on SSL technology.

4.2 *Reliability*

In this section, solution approaches will be described on conceptual level to increase the reliability of the mHealth infrastructure components, in order to safeguard the operation of (potentially) vital monitoring devices and alarm services even in the situation of outages of electrical power, loss of connectivity to communication networks or their complete failure, or malfunction of certain mHealth devices. The solutions address the identified requirements, and focus on the mobility of the patient and the health and care experts, and on the corresponding portability of devices, in particular within the Point-of-Care environment.

Self-monitoring

Self-monitoring refers to the identification of failures, in order to trigger applicable recovery or fallback actions, or to initiate alarms to local users or remote support staff. A self-monitoring function can either be deployed on the to-be-monitored device itself, in order to monitor the status of interfaces and SW components on that device, and to trigger a re-initialization or re-start if required. Alternatively, the monitoring function can be deployed on a separate, high-availability device, to avoid the impact of malfunctions on the monitoring functionality itself. However, that requires dedicated interfaces between the monitoring device and the to-be monitored device.

The main risks for the patients in all three user scenarios when being outdoor on the move is the loss of connectivity to the HIS infrastructure, or with one or multiple wireless body sensors. A monitoring function on the "Wearable HUB" (e.g., a smartphone) or on the "HUB and UI" (e.g., a tablet PC) continuously controls the connectivity with the wearable body sensors and with the HIS infrastructure via mobile communication, such as via a cellular communication network or via a Wireless LAN (WLAN) infrastructure. On detection of a connectivity loss, the monitoring function will initiate a connectivity recovery function that tries to re-establish the connectivity. If connectivity recovery to either the wearable sensor/s or the HIS infrastructure is not possible, the patient will be provided with corresponding information and recommendations, under consideration of their usability and interaction needs.

HW Redundancy

In order to take care of the availability of essential mHealth solutions also in the case of HW failures, critical components can be duplicated. In case of outage of a component, the stand-by device (so far redundant) with replicated application and service SW takes over the functionalities.

Two modes of operation are proposed: (1) a "master-slave operation", and (2) a "parallel operation" mode. In mode (1), a self-monitoring component (see above) activates a "slave" device on detection of a malfunction of the "master" device. The slave device carries out the same applications and logical functionalities as the master device, and the monitoring function also takes care for the capturing and provisioning of operational status information from the master device, in order to continue the operation on the slave device as seamless as required and possible. In mode (2) the critical SW applications and functionalities are deployed and executed on two identical, separate HW devices in parallel. If one HW device stops working, the other one continues with seamless operation. A monitoring function detects the malfunction of one device, and initiates a notification in order to request a HW replacement.

The utilization of HW redundancy is applicable to all three user scenarios. The communication functionality of a tablet PC can be taken over by a mobile phone or portable embedded device with mobile communication support, in case of failure of the tablet PC. Though the support applications are potentially reduced in functionality and usability on a smaller or even headless (i.e., without display) device, the transmission of potentially vital life sign recordings from the patient to the health & care experts via the HIS infrastructure continues. Wearable sensor devices for the measurement of important health data can be duplicated, in order for the data to be available also in the case one sensor stops working.

Power Fallback

One aspect of the "power-fallback" is the continuation of operation of devices in the patient's point-of-care environment at home in the case of a failure of the public power supply. In this case the role and functionalities of a device with power supply from the grid, e.g., a stationary "Home-Gateway", is taken over by a battery-powered device as the "Wearable HUB" (e.g., smartphone) or a "HUB and UI" (e.g., tablet PC). This is of particular relevance for the communication functionality. The detection of the power-failure and the handover to the alternative hardware is supported by the (battery-backed) monitoring function.

Another aspect is the reliability and availability of functionalities of the portable devices the patient has with them when being outdoors on the move. A power-monitoring functionality monitors the power status of the devices (in particular of the communication and application device (e.g., tablet PC), reduces the power consumption at certain charging levels (e.g., by reducing communication frequency or screen brightness), generates power alarms in form of notifications to the patient and to remote personnel in order to request a re-charge, and initiates a handover to a second redundant device that still has power.

Connectivity Fallback

The availability and reliability of the communication functionalities is of essential importance for all user scenarios that include the measuring of vital signs from patients on the move, and the transmission of those data to HIS infrastructure services, from where they are utilized by remote health and care experts or for the information of relatives. Hence, in case of breakdown of one underlying communication service or certain involved connectivity components, the communication continues through another available communication link or technology.

In case the fixed broadband connectivity of a Home-gateway in the patient's PoC environment stops working, the monitoring function (see above) initiates the handover of the communication to any available portable device with cellular connectivity, such as a smartphone or cellular-enabled tablet PC.

In case the cellular connectivity of the "wearable HUB" (e.g., the smartphone) of the patient stops working due to, e.g., loss of coverage or network failures while the patient is on the move, the monitoring functionality initiates the re-connection to any alternative cellular or WLAN network, or informs the patient about recommended actions.

Capacity Adaptation

Within the context of reliability of mHealth services there is also the availability of underlying infrastructure components and services. In order to adapt to increasing processing and storage requirements caused by an increasing number of service and infrastructure users, also the corresponding capacities of the involved components have to be adapted. The basic concept is to apply cloud technologies, such as load balancing on virtual servers to infrastructure components as the proposed Information Integration Platform (IIP). However, the discussion of those technologies is out of the scope of this chapter.

The prototypical verification of the proposed solution concepts is part of ongoing work (see also the outlook section).

4.3 Usability

The importance of a user-friendly design needs to be taken into account when developing new mHealth services. The combination of user-specific requirements and limited relative small-screen interfaces as on smartphones creates design challenges in usability, which includes learnability, efficiency and satisfaction as the general requirements described by (Nielsen, 1993). In a systematic approach to design principles, Su et al. showed how a mobile nursing information system can be developed when taking care of the usability aspects and the user's cognitive model, and not only the technology aspects (Su and Liu, 2012).

It will also be important to perform adequately usability testing in a realistic scenario and environment with a high level of realism, as this will influence the actual use, as highlighted by (Svanæs et al., 2010). They have proposed a model for usability tests for mobile health care systems where the physical environment will play an important role, looking at ergonomic aspects such as the ability to have both hands free. Those principles can also be applied to the health and care support for a mHealth patient. How the actual devices and sensors can be used in a daily situation will be of outmost importance for how those systems will be used. This is also according to the Technology Acceptance Model as described by Davis (1989), where the combination of easiness in use and usefulness of the solution can give an indication of future use of a system.

In order for elderly people to become familiar with the usage of new mobile technology solutions, it will be relevant for developers to gain interactive experience of how elderly feel when using the mobile technology. The functions they are provided with should not be limited to assist the user but also to let the user interact with the system, as described by Isaacs et al. when addressing issues of surrounding accessibility to mobile devices and services (Isaacs et al., 2013). Design requirements for mHealth solutions will need to take care of the different users and their needs, with a special focus on the elderly person who is supposed to give inputs from vital signs measurements, understand information and warnings given from sensors and reminder functions, as well as to interact with health care workers and close relatives at distance by video conference and social media solutions.

Traditionally, the patient care and daily support will be the responsibility of the actual hospital, the local doctor (GP) or the municipality home health care services. As the access to a patient's health record normally will be restricted to health care personnel employed at the actual organization, the case for our mHealth patient will be a seamless health care service across organizational borders. However, several obstacles both or regulatory issues and of technical interoperable issues will be a challenge in defining a reference architecture. The focus on shared access to medical information will be important for inter-organizational collaboration when introducing new care pathways based on the mHealth technology options. McGinn et al. stressed the importance of understanding the user needs and perspectives when implementing multi-organizational care services because of the complexity and focus by different health care professionals (McGinn et al., 2012). Especially, the team communication will require efficient information flow (Kuziemsky et al., 2009, Borycki et al., 2010). However, limited interoperability and inflexible workflow processes were discovered to be major obstacles when introducing new health care services for the community care in Australia (Clark and Lewis, 2012), and this is probably a general problem in most countries. Seamless healthcare will require access to medical information across organizational boundaries when the patient's travel between different health care services will have a need of co-ordinated follow-up, as described in a report from NHS in England (Eason et al., 2012).

Beside the User Interface (UI) and interaction aspects of applications and services the patient (or user in more general terms) is confronted with when using any mHealth device or application. There are aspects in the underlying communication infrastructure related to the distribution of measurements and other patient data to multiple services, which also have a significant impact on the overall usability of mHealth environments.

As stated in the usability requirements above, the patient data measured in the PoC may be needed for more than just one service (e.g., remote health monitoring service, PHR service, hospital EHR service, alarming service, and social media service). In the traditional siloed approach, each measurement device sends its measurements to a specific application running on the mobile gateway device, which then forwards the data to the corresponding back-end server hosting a specific service. The usability of the system is sacrificed since the user has to interact with different applications running on the mobile gateway device, or even with different dedicated devices for each specific application. Furthermore, running many applications on a mobile device will also consume more battery power, which will reduce the lifetime of the device. One solution is to deploy one application on the mobile gateway device which is responsible for handling measurement data from all measurement devices, and sends the data to different services. The drawback of this approach is that the mobile device still needs to send measurement data to multiple destinations (i.e., multiple services), which will also affect the power consumption negatively. To tackle this issue, an Information Integration Platform (IIP) is proposed to act as aggregation and distribution point for measurement data, so that the mobile device only needs to send the measurements once. The IIP is responsible to deliver the measurement data to services that use it. Publish/subscribe messaging pattern is utilized by the IIP so that different services can subscribe to different measurement data channel and get notified whenever new data is available. Fig. 6 shows the deployment of an IIP within the proposed reference architecture. A more detailed description of the IIP's functionalities is available in (Trinugroho et al., 2013).

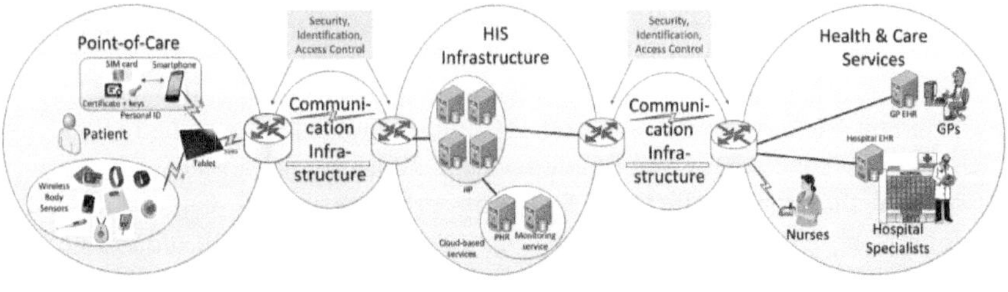

Fig. 6 End-to-End Architecture with Information Integration Platform (IIP)

5 Experiences and Results from Prototypes

Several prototypes have been designed and developed, covering the essentials of the proposed solutions for the mHealth ecosystem. In order to enable measurement data to be reused by different services, the IIP has been developed and deployed in a lab environment. As mobile devices play an important role within the mHealth ecosystem in supporting mobile patient, their lifetime operation can be very crucial for the safety and health of the patient. By utilizing the IIP, which is deployed in the cloud, mobile devices used by the patient only need to send one message for each measurement to the IIP. The IIP is then responsible to distribute this message containing the measured data to different services (e.g., alarming service, remote health monitoring dashboard service, PHR service, and hospital EHR service). This will reduce the mobile devices' power consumption while also eliminating the requirement of the patient to do multiple measurements with different redundant devices for different services, which can be seen as a serious usability drawback. The IIP itself makes use of publish/subscribe messaging pattern which enables various services to subscribe to different data and get notifications whenever new data is available. More detailed functionalities of the IIP are described in a separate article written by the authors in (Trinugroho et al., 2013). Fig. 7 depicts two deployed prototype services, namely remote health monitoring service and SOS service by utilizing the IIP as service broker.

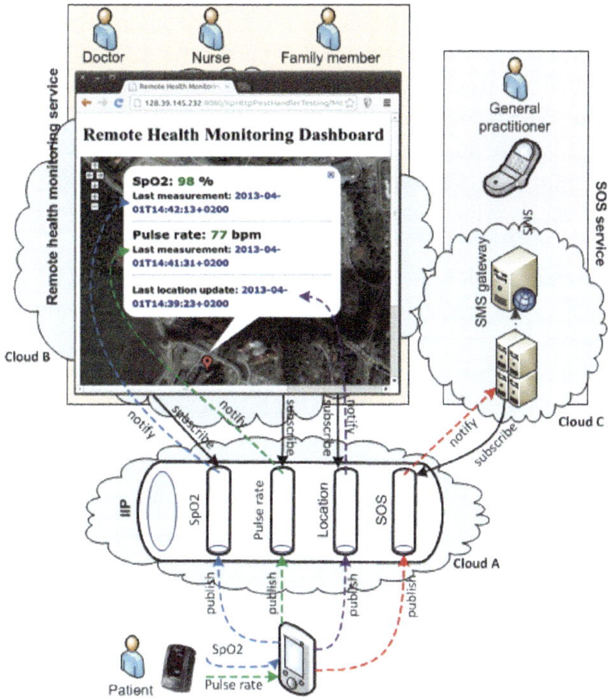

Fig. 7 Remote health monitoring and SOS services through IIP (Trinugroho et al., 2013)

Digital certificates play a central role in Public Key Infrastructure (PKI), deployed for example on electronic ID cards which provide means of proving an identity in electronic transactions (Adams and Lloyd, 1999, Housley and Polk, 2001). It is also used to verify that a public key belongs to a specified identity. A digital certificate is issued by a certification authority (CA), which is a third party entity whom both the user of the certificate and the serving party, such as a Web site, trust. The signature on a certificate is testament by the certificate signer that the identity information and the public key belong together. PKI forms a hierarchy where root CA may issue certificates directly to an entity or delegate this process to subordinate CAs, which results in a chain of certificates. The root CA and the subordinate CAs along the chain must have the trust of the verifiers to verify an end-host certificate. In some cases, the CA incorporates the role of a registration authority (RA). Where a separate RA is used, the RA is a trusted End-Entity certified by the CA, acting as a subordinate server of the CA. The CA can delegate some of its management functions to the RA. For example, the RA may perform personal authentication tasks, report revoked certificates, generate keys, or archive key pairs.

The CA is responsible for issuing and verifying certificates. The CA can be a unit within an organization or a company, but also an independent entity. The CA signs the public key certificate to prevent modification or falsification. The Root

CA's certificate forms the root of all certificate validation. The validation of certificates occurs using the appropriate public certificate residing within the root CA list. When a client has established a trust to a CA, they can trust certificates issued by that CA. The clients are ensured that the certificate is genuine by verifying the signature using the CA's public key. The infrastructure consists of a CA with Certificate Revocation List (CRL) publication and Online Certificate Status Protocol (OCSP). The CRL is a list of revoked certificates, while OCSP is a standard protocol to verify certificates online.

A prototype that enables the mobile patient to do measurements and to send the data to a remote PHR server has also been implemented, incorporating mutual authentication between the mobile device and remote server based on digital certificates to comply with the Norwegian level 4 security requirements. The prototype supports multiple users to use one mobile device, with a simplification that all client certificates are stored in the mobile device itself, and different users (e.g., patients, nurses) have to authenticate themselves to the mobile application by using passwords. The password used for logging in to the mobile application must match the password for the client certificate in order to establish a secure connection. If the password does not match any stored client certificate, the interaction is cancelled. The prototype consists of a server-side and a client-side application. A Web server and the CA are hosted in the server, and the Web server mainly hosts a PHR application with a database. The client-side is an Android application that is responsible for sending measurement data to the server and for displaying a Web page containing information from the server-side PHR application. Both the client and the server have their own certificates issued from the CA. Before any data gets exchanged, a secure connection has to be established.

Fig. 8 shows a sequence diagram of a patient sending a measurement data to the PHR application through a mobile device, and Fig. 9 depicts a sequence diagram of a patient accessing their health record via a mobile device.

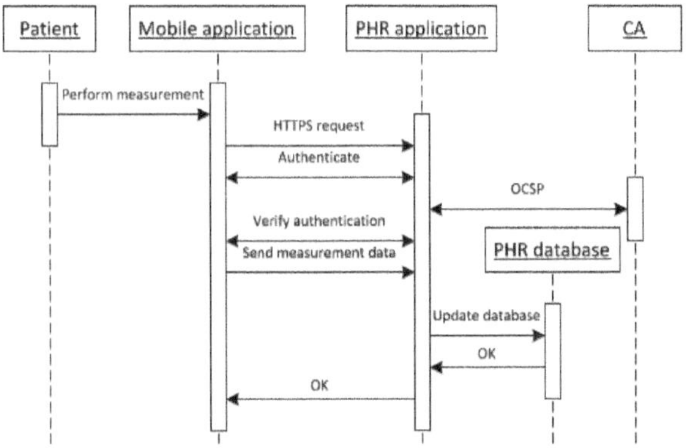

Fig. 8 Sequence diagram of sending measurement data

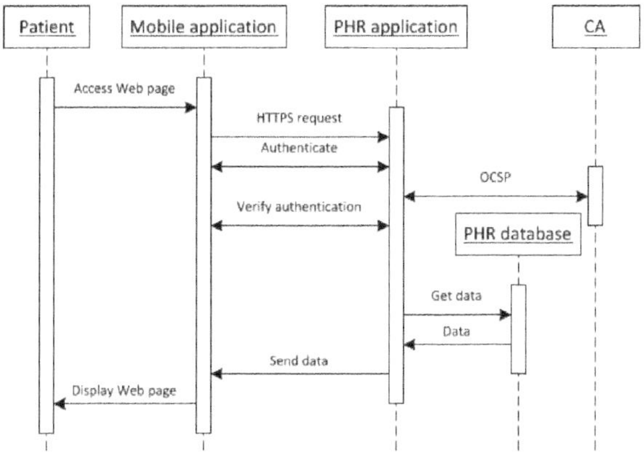

Fig. 9 Sequence diagram of accessing measured data

6 Conclusions

Within this chapter, security, reliability, and usability aspects within mHealth environments have been addressed. Three user scenarios have been described to act as guiding case studies for the analysis of requirements and solutions, and a generic reference architecture for the mHealth ecosystem has been described, encompassing three main entities, namely the point-of-care environment, the HIS infrastructure, and the health and care services. Based on the concrete user scenarios, a requirements analysis has been carried out, determining requirements that general mHealth solutions put on the various components of the mHealth reference architecture, with focus on the main study aspects.

Addressing the requirements, a number of solutions have been proposed, including state-of-the-art solutions, prior-art technologies, and concepts for further study and verification. The proposed solutions have been partly realized and verified as proof-of-concept implementations.

One learning from the realization of the proposed security solutions is the principle trade-off between security and usability. A secure communication channel and a reliable access control and privacy protection are very important for exchanging and storing health-related data, and user authentication is the first and utmost important step. The safer the user authentication process should be, the more authentication factors are required to be used. However, security levels are usually increased in proportion to the increased difficulty level of efforts that the user has to perform, which leads to significant usability drawbacks. Considering disabilities and limitations of typical mHealth user, the level of security has to be compromised with the required ease of use.

Another major result is the developed general reference design, which follows the reference architecture and contains solution components as a portable communication gateway in the PoC environment and the Information Integration Platform (IIP) in the HIS infrastructure, which have been developed using open standard technologies. As a conclusion from the prototyping and verification, we propose to use a general open infrastructure for the development of new mHealth solutions and services, and recommend following the approach selected for the described reference design.

7 Outlook

The currently implemented prototype for the verification of the authentication solution for multiple users using one mobile device to send measurement data to back-end servers is simplified to utilize username and password as credentials, which are then mapped to corresponding digital certificates stored in the mobile device. A dedicated personal identification device is under development, including smartcards and SIM cards. The digital certificates which are issued by the CA are stored in these personal identifiers. The utilization of USIM cards, which carry personal certificates for identification and authentication purposes, is also being studied.

The proposed solution concept for reliability and availability of connectivity including monitoring components is under preparation.

Furthermore, there is an ongoing work on a UI adaptation component to adapt UI and interaction characteristics from the perspective of usability and user needs.

References

Adams, C., Lloyd, S.: Understanding the Public-Key Infrastructure: Concepts, Standards and Deployment Considerations. Sams Publishing (1999)

Agrawal, R., Kiernan, J., Srikant, R., Xu, Y.: Hippocratic databases. VLDB Endowment 154 (2002)

Borycki, E., Kushniruk, A., Kuo, M.-H., Kuwata, S.: Requirements for Communication and Exchange of Health Information: A Socio-Technical Approach to Understanding Healthcare Information Exchange. In: 2010 International Conference on Advanced Information Technologies (AIT), Taichung County, Taiwan, April 23-24 (2010)

Clark, A., Lewis, J.: The challenges in developing a seamless model of aged and disability community care in Australia within a mixed economy of service providers. In: 2012 IEEE Conference on Technology and Society in Asia (T&SA), October 27-29, pp. 1–5 (2012)

Davis, F.D.: Perceived Usefulness, Perceived Ease of Use, and User Acceptance of Information Technology. MIS Quarterly 13, 319–340 (1989)

Eason, K., Dent, M., Waterson, P., Tutt, D., Hurd, P., Thornett, A.: Getting the benefit from electronic patient information that crosses organisational boundaries. Final report. NIHR Service Delivery and Organisation Programme, Department of Health, London (2012)

Fensli, R., Oleshchuk, V., O'Donoghue, J., O'Reilly, P.: Design requirements for a Patient Administered personal Electronic Health Record. In: Laskovski, A.N. (ed.) Biomedical Engineering, Trends in Electronics, Communications and Software, InTech, Austria (2011)

Gammon, D., Årsand, E., Walseth, O.A., Andersson, N., Jenssen, M., Taylor, T.: Parent-child interaction using a mobile and wireless system for blood glucose monitoring. Journal of Medical Internet Research 7 (2005)

Housley, R., Polk, T.: Planning for PKI: best practices guide for deploying public key infrastructure. John Wiley & Sons, Inc. (2001)

Hung, K., Zhang, Y.-T.: Implementation of a WAP-based telemedicine system for patient monitoring. IEEE Transactions on Information Technology in Biomedicine 7, 101–107 (2003)

Isaacs, J., Martinez, S., Scott-Brown, K., Milne, A., Evans, A., Gilmour, D.: Mobile Technology and E-Inclusion. In: Stephanidis, C., Antona, M. (eds.) UAHCI 2013, Part III. LNCS, vol. 8011, pp. 626–635. Springer, Heidelberg (2013)

Kuziemsky, C., Borycki, E., Purkis, M., Black, F., Boyle, M., Cloutier-Fisher, D., Fox, L., Mackenzie, P., Syme, A., Tschanz, C., Wainwright, W., Wong, H., Team, I.P.: An interdisciplinary team communication framework and its application to healthcare 'e-teams' systems design. BMC Medical Informatics and Decision Making 9, 43 (2009)

Llor, C., Moragas, A., Miravitlles, M.: Usefulness of a patient symptom diary card in the monitoring of exacerbations of chronic bronchitis and chronic obstructive pulmonary disease. International Journal of Clinical Practice 66, 711–717 (2012)

Markle Foundation 2003. Connecting for Health. The personal health working group final report (2003)

Martí, R., Delgado, J., Perramon, X.: Network and application security in mobile e-health applications. In: Kahng, H.-K., Goto, S. (eds.) ICOIN 2004. LNCS, vol. 3090, pp. 995–1004. Springer, Heidelberg (2004)

Mcginn, C., Gagnon, M.-P., Shaw, N., Sicotte, C., Mathieu, L., Leduc, Y., Grenier, S., Duplantie, J., Abdeljelil, A., Legare, F.: Users' perspectives of key factors to implementing electronic health records in Canada: a Delphi study. BMC Medical Informatics and Decision Making 12, 105 (2012)

Nielsen, J.: Usability engineering. Academic Press, Boston (1993)

OECD 2011: Health at a Glance 2011. OECD Publishing (2011)

Patel, S., Park, H., Bonato, P., Chan, L., Rodgers, M.: A review of wearable sensors and systems with application in rehabilitation. Journal of Neuroengineering and Rehabilitation 9, 1–17 (2012)

Poon, C.C., Zhang, Y.-T., Bao, S.-D.: A novel biometrics method to secure wireless body area sensor networks for telemedicine and m-health. IEEE Communications Magazine 44, 73–81 (2006)

Scalvini, S., Zanelli, E., Comini, L., Dalla Tomba, M., Troise, G., Febo, O., Giordano, A.: Home-Based Versus In-Hospital Cardiac Rehabilitation After Cardiac Surgery: A Nonrandomized Controlled Study. Physical therapy (2013)

Sorber, J., Shin, M., Peterson, R., Cornelius, C., Mare, S., Prasad, A., Marois, Z., Smithayer, E., Kotz, D.: An amulet for trustworthy wearable mHealth. In: Proceedings of the Twelfth Workshop on Mobile Computing Systems & Applications, vol. 7. ACM (2012)

Su, K.-W., Liu, C.-L.: A Mobile Nursing Information System Based on Human-Computer Interaction Design for Improving Quality of Nursing. Journal of Medical Systems 36, 1139–1153 (2012)

Svanæs, D., Alsos, O.A., Dahl, Y.: Usability testing of mobile ICT for clinical settings: Methodological and practical challenges. International Journal of Medical Informatics 79, e24-e34 (2010)

Trinugroho, Y.B.D., Gerdes, M., Amjad, M.M.M., Reichert, F., Fensli, R.: A REST-Based Publish/Subscribe Platform to Support Things-to-Services Communications. In: 19th Asia-Pacific Conference on Communications, Bali, Indonesia, August 29-31, pp. 296–301. IEEE (2013)

Questions

Q1. What are the main scenarios / case studies for mHealth solutions?

Q2. How can mHealth solutions support the life of patients in an early dementia stadium?

Q3. What are the key requirements for mHealth solutions as discussed in this chapter?

Q4. What are the main domains of an end-2-end mHealth reference architecture?

Q5. How can a patient be reliably identified and authenticated in a mHealth environment in a user-friendly way, i.e., without much user involvement?

Q6. Why can self-monitoring, application recovery and fallback actions be very important for mHealth services?

Q7. What are crucial usability aspects for mHealth solutions?

Q8. What are the common authentication factors and their examples? What are the main advantage and drawback of having multiple authentication factors?

44. "Security and Privacy Issues for Mobile Health"

Patricia A.H. Williams[1] and Anthony J. Maeder[2]

[1] eHealth Research Group, School of Computer and Security Science, Edith Cowan University, Joondalup, Western Australia, Australia
trish.williams@ecu.edu.au
[2] Telehealth Research and Innovation Laboratory, School of Computing, Engineering and Mathematics, University of Western Sydney, Campbelltown, New South Wales, Australia
A.Maeder@uws.edu.au

Keywords: security, privacy, standard, protocol, user.

1 Introduction

Security and privacy have been well established as major considerations in health informatics generally (Rindfleisch 1997). A challenge for healthcare innovation is to embrace the potential of mobile health creatively within the healthcare system and not to merely replicate current technologies into a parallel wireless environment. In addressing this challenge, the complexities of securing health information along a composite clinical information pathway and in each situation of use must be defined.

This chapter, as the title reflects, is targeted at identifying the issues in mobile health security and privacy rather than solving them. The reason for this is that these are human and organisational concerns, and cannot be addressed simply by adopting a purely technical perspective. This chapter expands on the issues associated with security and privacy rather than the solutions because, whilst the solutions are included and described in necessarily broad terms, each mhealth situation is unique. Indeed, to date one of the particular problems facing mhealth, in its relative infancy, is the ability to apply consistency in standardisation and interoperability of mhealth solutions (Williams & McCauley, 2013). Currently the development of each mhealth solution is distinctive, and the specific nature of these designed systems limits their wider applicability. This is not unusual in the early stages of development of new technology. Whilst it is possible to provide a list of generic security and privacy solutions, such as the parameters of access control, encryption, and public key infrastructure (PKI), each generic solution is a major topic in its own right and is well described in literature elsewhere. The purpose of this chapter is to provide guidance on what to address, given the breadth of the topic of mhealth in an environment characterised by rapid evolution.

S. Adibi (ed.), *Mobile Health*,
Springer Series in Bio-/Neuroinformatics 5, DOI: 10.1007/978-3-319-12817-7_44

Mobile health has been recognised as one of the fastest growing and most rapidly diversifying technological areas in healthcare, with growth rates of 10% (compounded) per annum reported (Dunbrack 2011) and consequential diversification of users and settings for which it is deployed. The same report identifies security as the dominant issue for concern in the market, rated top by over 50% of stakeholders surveyed. This is reinforced by current health services management thinking, where reluctance to adopt mobile technologies has been associated with inadequacy of addressing security and privacy concerns (Morgan & Agee 2012). Confounders in the security arena are numerous and include prolific private ownership of mobile devices with consequential uncontrolled access to public telecommunications infrastructure, widespread use of unencrypted data transmission protocols, variations in jurisdictionally localised regulatory compliance regimes, ease of covert and unconsented use of mobile devices within clinical workflow fuelled by increasing societal acceptance of sharing of information captured on mobile devices (e.g. in social networking).

For the purposes of this chapter, it is helpful to categorise mobile health solutions into several groups, characterised by different kinds of healthcare function. These groups embrace substantially different issues in security and privacy, according to their properties in terms of clinical, technical and user aspects. Four such distinctive groups can be identified:

- delivery of routine clinical care in controlled clinical settings (e.g. teleconsultation using video; store-and-forward reporting of dermatology);
- remote access to health data from patient records or clinical databases by health care professionals (e.g. electronic medical records for hospital admissions; individual care plan monitoring and adjustment; messaging between health services);
- collection of patient-specific data through wearable or ambient wireless sensors (e.g. vital signs monitoring; smart home health surveillance);
- patient-initiated consumer-services oriented transfer and sharing of health information, outside a formal care setting and potentially connecting to the Internet (e.g. health diaries and promotional aids, online health coaching or opinion seeking).

In each case, the main healthcare "function of interest" within a group would be undertaken using some form of wireless transmission medium, ranging from wide range connection of laptop and tablet computers and smart phones using mobile telephony or IEEE 802.11x Wi-Fi, to mid range connection of peripheral devices such as wearable vital signs and movement detectors using IEEE 802.15 with Zigbee or Bluetooth, to close range or surface connection devices such as Smart Cards or RFID. The wide variety of these ICT infrastructure choices introduces a confounding situation, because no one solution for privacy or security can span such a wide field, nor could a single solution possibly be implemented in harmony with the all the different technical standards that determine how these ICT components are used.

Furthermore, any mobile health activity undertaken by one or more of a variety of human actors (e.g. doctors, nurses, carers, patient) aiming at a specific type of clinical process (e.g. diagnosis, treatment, procedure, followup) and relying on the support of external agencies (e.g. call centres, patient monitoring companies, telecommunications service providers). The activity may be compounded by involvement of further human participation or assistance, as aspects not normally regarded as part of the main activity (e.g. technicians, family, consultant clinicians), who may in turn introduce further complexity in the telecommunications aspects. This diversity creates a further confounding effect, as different risks and solutions exist in each case based on the selection of these aspects, and are consequently difficult to map or generalise to other situations.

This chapter initially reviews examples of mobile health security and privacy from current common clinical practice. Subsequently, to provide a picture of the constraints on the operational mhealth environment, the legislative requirements and standards pertaining to security and privacy in e-health are discussed, explaining where and how these are also pertinent and applicable to mobile health. To support the underlying security and privacy principles, basic security and risk measures specifically applicable to mobile health are outlined. This is logically followed by a discourse on how software standards play a key role in mobile application development and contribute to addressing security and privacy. To consolidate these vitally important aspects, which shape the development and use of mobile health, there follows a generic description of the mobile health ICT ecosystem which informs a high level view of the security and privacy aspects that need consideration. These concepts are explained in more detail using two case studies of mobile health security and privacy violations, together with possible solutions.

2 Review of Existing Common Practice

The effect of confounders identified above means that it is difficult to provide a comprehensive template or "recipe" approach to identify and overcoming security and privacy issues in mobile health (FDA 2011, Brusco 2012). Past attempts have typically concentrated either on the telecommunications aspects (e.g. Ryoo et al 2009; Osunmuyiwa & Ulusoy 2012) or device aspects (e.g. Markovic et al 2006; Takahashi et al 2008). Kotz (2011) has proposed a "threat" based taxonomy for mobile health with three dimensions: identity, access and disclosure. Other studies treat mobile health as a subset of e-Health and recommend similar approaches be adopted (e.g. Appari & Johnson 2010), or emphasise the personal care aspects as paramount in determining appropriate practice (e.g. Avancha et al 2012). It is not feasible to synthesize all these different perceptions into one composite methodology, so here the approach taken instead is to review exist practices in the four healthcare function groups identified above, and to provide some benchmark examples which can be used to inform and scaffold further solutions.

In teleconsultation situations, there are two schools of thought as to how much control is needed over the telecommunications mechanism. Hospital and health system providers often maintain strict in-house procedures such as use of proprietary encrypted video systems and restriction of endpoint equipment to their own devices. These approaches can in principle be expanded to include approved mobile devices, but often the proprietary systems were not designed for lower compute power or display resolution associated with mobile devices, and useability is poor. However, the inbuilt security features and privacy protocols which are part of these procedures, extend naturally to the mobile situation if it is part of the controlled environment. The second school of thought is typically promoted by general practitioners and other primary care clinicians: it espouses use of commercial, openly available solutions in both video conferencing and image transfer, on the grounds that the risk of violation is no greater than if a wired telephone connection was being used, which is already well established as an accepted mechanism for remote connection between clinicians and their patients, or other clinicians. This same dichotomy is also apparent in store-and-forward situations, however the opportunities for violation of security and privacy requirements are higher due to the non-immediacy of the traffic. The landscape in this group has been surveyed (Garg & Brewer 2011) and the results serve to underline the diversity of thinking and practice that exists.

For remote data access, it is often not feasible to run a reduced scope version of a major vendor EHR management product on a mobile device. Such systems are frequently built assuming substantial enterprise architectures and supporting complex query and update topologies. Mobile health applications must then be satisfied with a summary or subset "snapshot" type of record access, with salient field information preselected and automated resolution of queries to feed a "back door" patched mobile solution. Nevertheless, the information is equally affected by security and privacy issues, and so additional protection is needed if it is being accessed in this way outside of the commercial enterprise system. In situations where multiple parties must access shared information on a patient, such as care management records for a chronic disease patient, the data may be protected with different levels of severity according to roles and types of carers, e.g. doctors may have less stringent constraints on what is accessible compared with community support carers. The development of integrated services (e.g. Marti et al 2004) has been advocated as a promising approach in this area.

Home monitoring systems are typically the least carefully considered mobile health environments from security and privacy perspectives. This may be a reflection of lower levels associated with other in-home systems such as cordless phones, home wi-fi, smart home environmental and security system controls. Basic access security measures such as password entry available for computer connection wireless networks are often deemed sufficient. While the incentives for "snooping" in a home setting may be less compelling than in a hospital, the damage that could be done by compromised data accuracy and privacy violations are no less severe. Some vendors offer systems which have inbuilt proprietary encryption and audit capabilities, but these are usually not testable against open standards. Implementation of conventional data protection protocols may be computationally expensive and detract unfavourably from time and power budgets for small devices. A multi-layered approach (e.g. Sahoo 2012) may offer the most expedient solution to this situation.

The overall space of Personal Health Record and related publicly accessed online health information presents particular security and privacy issues (Tang et al 2006, Win et al 2006). Consumer data provision and management in third party data storage environments, such as cloud storage (e.g. Microsoft Healthvault), or social networking health promotion sites (e.g. physical activity logging sites), can be compromised by any break-in to those environments. In mobile settings, opportunities for break-in are expanded, e.g. through the subversion of hand-over or channel allocation strategies. Also, the wider set of consumer and public participants who use the systems and their data have differing attitudes to security and privacy behaviour and indeed may seek to operate outside of normal clinical sector guidelines and requirements. This trend may well have the result of relaxing norms and community expectations on health data security and privacy in the future, but so far have not prompted change.

Across the four groups identified in our introduction, clearly a diversity of security and privacy practices exists. Some overarching unity is imposed on these by international best practice pressures, as well as local jurisdictional and operational requirements. From a data protection perspective, boundaries that influence common practice are driven by regulation, by risk management, and by standardization. These drivers are discussed in the next sections.

3 Legislative Requirements

The mobile health space has been approached primarily from a technology perspective understandably because it is incorporates medical devices which are strictly regulated and controlled by international standards (Silberman & Clark, 2012). Thus, there needs to be symbiotic relationship between standards, legislation and best practice. National and jurisdictional privacy laws have driven development of frameworks to assist in the interpretation and practical application of privacy measures (Avancha et al 2012), however these fall short of providing a "cross spectrum flow of mobile healthcare" (Williams & Maeder, 2013). Whilst a full review of international legislation in not included here, the following overview provides a perspective on what should be considered relevant in legislation, with selected examples of regulations.

Whilst the use of mobile devices in healthcare has seen some legal and regulatory requirements put in place, these requirements are not specifically related to the mobile healthcare environment. For instance the US Health Insurance Portability and Accountability Act (HIPAA) legislation, focusing on the protection of health information, specifies that policies and procedures should be in place to ensure that a record of mobile equipment, including backup media, is maintained, and to ensure that secure access to health information, regardless of its movement, is maintained. Indeed, HIPAA has stringent requirements for the security and privacy protections associated with sharing of patient information. It provides a much needed standardization for the healthcare industry and includes consideration of new technologies and new uses of the technologies. However, due to its stringent nature, healthcare organisations have found it challenging to meet its requirements and the privacy aspects are now being challenged as they limit healthcare research (Steinberg & Rubin, 2009).

In addition, HIPAA has a requirement for protection using encryption, which is further supported by the Health Information Technology for Economic and Clinical Health (HITECH) legislation in its requirements for data breach. The HITECH Act, part of the American Recovery and Reinvestment Act of 2009, has impacted sections of HIPAA, particularly in regard to the encryption of patient information and the impact on mandatory data breach notification to patients where encryption of data is not used. Such specification highlights the importance that awareness of national legislation plays in the collection and storage of patient information.

In Australia, there is no equivalent legislation to the US HIPAA, however the driving force has been through the use of the Privacy Act and privacy principles. The Australian Privacy Principles (OAIC, 2013), which are an amalgamation of the previous Information Privacy Principles and National Privacy Principles, are fundamental to the protections required by Australian legislation. Significantly, Principle 1 impacts mobile health as it requires that clearly expressed policy on the management of health information. Further, 'Principle 11 – Security of personal information' states that reasonable steps must be taken to protect information and to destroy it securely, and 'Principle 12 – Access to personal information'- presents significant challenges in the distributed mobile environment.

Similarly, in the European Union, there are broad data-protection and privacy laws, but no law or EU directives specifically for the privacy of health information (Thomson Reuters Foundation, 2013). It is important to note that in addition to national regulations many countries such as the US and Australia, have additional jurisdictional legislation which supersede or add additional requirements to national legislation. In addition, codes of conduct, professional standards and accreditation standards should also be considered.

To put effective security in place to support and comply with legislation and regulations, and to embrace fundamental underlying security and privacy principles, basic security and risk measures specifically applicable to mobile health need understanding and elucidation, as in the next section.

4 Basic Security Measures and Risk Considerations

The management and implementation of security in the mobile device environment presents numerous challenges. The increased security and privacy risks to the data, together with the increased opportunity for theft or loss of devices, create the potential for data breaches impacting patient privacy. To compensate for this and to provide a minimum of basic level of security, best practice measures should be followed. This includes:

- monitored, enforceable organisational policy in relation to the parameters for using mobile devices in the healthcare setting and in particular the use of personal mobile devices. This would include clear guidance for the appropriate use of mobile devices, together with reporting mechanisms for incidents that may result in data loss or breach;
- avoid storing patient healthcare information on mobile devices;
- enforce encryption on all devices;
- implement mobile device management processes which may include features such as an ability to remotely wipe the mobile device in the event that it is lost or stolen;
- regular security maintenance such as scanning to ensure anti-malware controls are functioning effectively; and
- appropriate physical security measures such as secure storage of mobile devices when not in use and staff training on best practice security measures, including appropriate use of passwords.

Further consideration should be given to the potential security issues related to hidden data stores that mobile devices can facilitate. This sort of potential vulnerability should be identified through effective risk assessment processes. The discussion on basic security practice would not be complete without some reference to standards in risk assessment and governance compliance. The accepted control standards include:

- ISO/IEC 27000:2012 Information technology -- Security techniques -- Information security management systems -- Overview and vocabulary
- ISO 27799:2008 Health informatics -- Information security management in health using ISO/IEC 27002. This standard specifies a set of detailed controls for managing health information security and provides health information security best practice guidelines
- ISO/IEC 27002:2005 Information technology -- Security techniques -- Code of practice for information security management
- ISO/IEC 27005:2011 Information technology -- Security techniques -- Information security risk management
- ISO/TS 25238:2007 Health informatics -- Classification of safety risks from health software
- ISO 31000:2009 Risk management -- Principles and guidelines
- COBIT. This is an IT governance framework. It is designed to assist in development and implementation of measures for governance in IT particularly in relation to compliance. (COBIT, 2013).

Whilst examining mobile health from each of its constituent parts affords a certain levels of protection, i.e. at each device and each data store, what also needs to be considered is that mobile health is generally part of a bigger process of capturing and using data to provide healthcare services. Therefore, mobile health security must also be considered and analysed from the whole mobile health workflow process perspective. To date, this has not been considered in any depth as demonstrated by the lack of standards to address this.

Further, a key aspect of providing mobile solutions is the development of software to international standards to ensure both interoperability and reliability. A detailed review of the relevant mobile standards is provided in the following section.

5 Software Standards

In a review of the standards it is evident that there is significant contention between the definition of health software and the delineation between software that is embedded in medical devices. This section provides an overview of the relevant standards in this complex area in relation to software and to mobile device workflow. It does not cover standards and regulations in relation to medical devices themselves. However, ISO/TC 215 Health Informatics Technical Committee has been providing standards for mobile medical devices for a number of years as part of the *ISO/IEEE11073* family of standards. These include a specialised Bluetooth for health standard, which incorporates enhanced security, as well as sensor data acquisition standards. These standards have subsequently been leveraged to develop a range of mobile devices and ehealth platforms that communicate securely, for instance using the Integrating the Health Enterprise (IHE) XDR profile. The following standards are also utilised for this purpose:

- ISO 13485:2003 Medical devices – Quality management systems – Requirements for regulatory purposes
- ISO 14969:2004 Medical devices – Quality management systems – Guidance on the application of ISO 13485: 2003
- IEC 62304:2006 Medical device software – Software lifecycle processes
- ISO 14971:2007 Medical devices – Application of risk management to medical devices

These and other similar standards demonstrate the focus on the development aspects of software in medical (mobile) devices, and highlight the lack of attention to the use of such devices in a broader networked environment. Subsequently, collaboration between IEC 62A and ISO TC215 has published the initial standard that sets basic requirements for setup and management of safe medical networks incorporating medical devices including mobile devices *IEC 80001-1:2010 Application of risk management for IT networks incorporating medical devices, Part 1 – Roles, responsibilities and activities.*

The key security related standards applying to mobile health include the following major items.

ISO/DTR 17791 Health Informatics – Guidance on Standards for Enabling Safety in Health Software (In Development)

The intent of this standard is to provide a guide to a coherent set of international standards needed for the patient-safe development, implementation and use of health software. Its scope is to provide guidance on which standards are applicable to enabling safety in health software. It takes a risk management approach to this task and identifies where gaps exist in relation to health software safety issues. Of significance to the mobile setting, is the omission of any standards that relate to the clinical process and human factors.

ISO/TR 27809:2007 Health Informatics – Measures for Insuring Patient Safety of Health Software

This standard was developed to ensure patient safety in regards to health software products that are not covered by other medical device regulations. The standard includes consideration of the associated standards and controls that are applied to medical devices and suggest how these might be adapted.

ISO/IEC 82304-1 Healthcare Software Systems – Part 1: General Requirements (In Development)

This standard, (under development at time of printing), defines the requirements to enable safety in ehealth software which will be applicable to mobile devices. It applies to the safety of health software that is designed to operate on general purpose IT platforms, which includes mobile devices. The standard covers the entire lifecycle including design, manufacture, validation, installation, maintenance, and disposal of health software and applies to Health Software (e.g. Medical Apps, Health Apps) intended to be used in combination with mobile computing platforms.

IEC/TR 80002-1:2009 Medical Device Software – Part 1: Guidance on the Application of ISO 14971 to Medical Device Software (ISO 14971:2007 Medical Devices – Application of Risk Management To Medical Devices)

This standard specifies a process for the application of the ISO14971 standard, to assist manufacturers identify hazards associated with medical devices. As with other standards in space it is concerned practical assistance to risk management practitioners who need to perform risk management when software is included in the medical device/system, and at software engineers who need to understand how to fulfil the requirements for risk management addressed in ISO 14971. ISO 14971, recognized worldwide by regulators, is widely acknowledged as the principal standard to use when performing medical device risk management. However, this standard also has applicability to any safety risk management process for software in the healthcare environment independent of whether it is classified as a medical device, which includes mobile health devices.

ISO 27799:2008 Health Informatics – Information Security Management in Health Using ISO/IEC 27002

This standard defines guidelines to support the interpretation and implementation in health informatics of ISO/IEC 27002 and is a companion to that standard. ISO 27799 specifies a set of detailed controls for managing health information security and provides best practice guidelines and mandatory requirements to protect the confidentiality, integrity and availability of personal health information. These include requirements for design, development, implementation and maintenance of information systems. Whilst not mobile health specific, this standard's general principles apply equally to the mobile situation as the more controlled network environment.

ISO 27789 Health Informatics – Audit trails for Electronic Health Records

The standard is not directly able to be applied to the mobile environment, however it provides a common framework for audit trails for electronic health records (EHR), in terms of audit trigger events and audit data. This is important because in the mobile environment it is necessary to keep the complete set of personal health information auditable across information systems and domains. It is applicable to systems processing personal health information which, complying with ISO 27799, create a secure audit record each time a user accesses, creates, updates or archives personal health information via the system.

The tension created by the integration of mobile technology with medical device functionality has seen the landscape for standards forced to embrace this, to address these tensions. For instance, where an application effectively transforms the mobile platform such as a smart phone or tablet computer, into a role with clinical applicability such as an electronic stethoscope, the device together with the application becomes a regulated medical device (Ardito, 2011). Consequently, new international standards such as ISO/IEC 82304-1, are emerging to address the safety concerns associated with healthcare software systems and their crossover with medical devices. Whilst the domains of medical devices and of healthcare software have to date been considered separate, as the medical device typically contains proprietary embedded software, when you transfer the functionality of medical device to a more accessible and open platform such as a smartphone, the delineation between specific device functionality and the device itself becomes blurred. Clearly a medical device and a smartphone are different types of architecture designed for different purposes: hence the tension is evident in trying to use a telephony tool for something other than for what it was originally designed using software applications to facilitate this. Standards such as ISO/IEC 82304 are addressing such issues.

To draw the multiple aspects that need consideration in mobile health solutions (as described in the previous sections) together, the following section defines the design factors that influence security and privacy, from a systems perspective. This is vitally important to understand when designing solutions, as the three factors that impact any 'clinical information system', namely technology, clinical and human, must achieve synergy to ensure effective and safe outcomes.

6 ICT Ecosystem Design Factors

There are various common design considerations impacting upon security and privacy that need to be considered when adopting or incorporating use of mobile health. These include the understanding and addressing of the following factors:

- capability and potential of mobile devices for the intended purpose, both by the clinical community and by consumers of healthcare;
- ease and accuracy of data capture / storage / transfer, and integration of data into existing databases and health record systems;
- review and management requirements for such data including integration into existing workflow and decision-making processes;
- equitable and ubiquitous access to the technology across a broad range of connectivity and user settings; and
- realistic usability of the devices and applications in situ.

Ultimately the clinical continuity of care and the end-to-end use of information, from the information gained by remote monitoring or consultation through to the use of this information for clinical decision support, demands a comparable level of integrity in mobile health solutions as for any e-Health system.

A frequent criticism of the implementation of health information systems is the lack of conceptualisation of the practical relevance and lack of a foundational conceptual design framework outside the adoption of the technology itself. The authors have proposed such a framework (Williams & Maeder 2013), summarising the practical design considerations to address those aspects particularly relevant to mobile health situations. This framework characterizes the ICT ecosystem according to 3 dimensions of "design tensions": human factors (e.g. ease of use), clinical factors (e.g. integration into workflow) and technology factors (e.g. functional capacity of the platform).

A comprehensive identification of the relevant factors for each of the framework categories is provided below in this section. Details of the interplays which exist between factors from each category are also explained, based on the healthcare function groups identified in the Introduction section. In the following section several case studies are analysed, illustrating some specific mobile health security and privacy design considerations. These examples show how analysis and adaption of security risk mitigations can be developed for practical application, using the framework.

Technology Factors: These exist across the spectrum of *hardware, software, communications and data.* Each of these aspects has its own security characteristics and implications for protection of healthcare information. In essence, this covers the factors of software incorporation into personal device function, application development, data capture, acquisition and transmission, and physical device security. Common technology issues include choice of methodologies for securing information, software-as-medical-device, bring-your-own-device (BYOD), seamless migration between personal and corporate use, management and transfer of personal data, wireless attacks and device durability. Some frequent major options to be considered in mobile health situations are:

- mobile platform choice, including bandwidth, display capabilities, software environment
- connectivity media and channels, data/telecommunication transfer mechanisms
- technology persistence and battery life
- integrity of the ICT environment, detectability of violations through audit or scanning.

Clinical Factors: These cover clinical processes such as *diagnosis and treatment,* as well as related systems elements such as *EHR, workflow, clinical need and benefits.* Whilst there is a distinct and rapidly growing presence of mobile applications for preventative health for personal use, there are relatively few which have been developed for direct clinical applicability. However, the demand on mobile health technologies by healthcare providers is reportedly set to increase rapidly over the next few years. Such development aspirations need however to be tempered with concerns regarding data security. Typical clinical issues here include healthcare process workflow, clinical interactions between parties, new models of care and applications (e.g. home patient monitoring; preventative health), stakeholder impacts (e.g. data safety monitoring; health software certification). In mobile health settings, some major recurring options are:

- availability of clinical support tools such as EHR access, CDS systems, guidelines
- integration of clinical activities into longitudinally managed episodes of care
- assuring and combining of clinical information on the patient from multiple sources.

Human Factors: These address utility aspects such as *human computer interaction and usability,* as well as organisational aspects such as *business case and user adoption.* One of the major benefits of mobile health is the accessibility to accurate personal healthcare information in emergency situations where critical decisions need to be made rapidly. This is one area in which an understanding of the capabilities and potential by consumers is necessary, which requires education and engagement. However, as in any healthcare innovation, consideration of the equitable access to the level of care provided must be included. This may incorporate the cost of devices and communications coverage, a particular issue in rural and remote communities as well as developing world settings. Similarly, the usability of devices and applications is an area of consideration studied as human-computer interaction, and technology acceptance or adoption modelling. The major options for this category are:

- useability and intelligability of the user interface, which may differ from web or desktop
- replicability and scaleability of the mobile solution, depending on platform and software
- technology adoption and change management processes associated with the solution.

Mobile health is a compound synergy of components embracing many of the above factors, which can present disruption to conventional healthcare processes in part due to the shift in responsibility and placement of diagnosis in the healthcare process workflow, and an increased focus on patient self-management. This disruption can also be attributed to the vast amount of data that can be recorded in a variety of media using single convergent technology devices. From a security perspective these two disruptive aspects create multiple complexities and require an understanding of the context of use, as well as the technological and workflow elements, in order to provide effective security of data and assurances to patients and healthcare providers participating in the healthcare delivery.

Another major consideration is the proof of clinical usefulness and efficacy in the use of the mobile healthcare delivery continuum, applying appropriate evaluation methods to support an acceptable level of clinical evidence. This issue is not restricted to the domain of mobile health, and has had much discussion in the associated area of telehealth. Indeed, the need for business case sustainability as well as assessment of clinical and economic benefit is a major consideration in the clinical adoption and use of health technologies generally. The impact on, and by, security or privacy requirements and mechanisms must be considered as part of this.

From a risk perspective, each element in the end-to-end delivery of healthcare using mobile devices can be independently assessed and mitigations devised. However, this is not necessarily the best or most effective perspective with which to look at the security solutions. The frequency and severity of threats and vulnerabilities, and the ability and feasibility to mitigate against these threats, also need to be considered. Therefore, applying the same approach to each element independently, for instance where human factors are high risk and technology factors are low risk, may result in ineffective solutions and unnecessary and costly mitigation strategies which could compromise the effectiveness of the clinical process. An understanding of the co-dependency of the risk factors needs to be present and the risk mitigations tailored accordingly.

In order to see how the ICT ecosystem is applied in practice, the following case studies highlight the security and privacy issues that exist in real world mobile health environments, illustrating numerous concepts discussed above.

7 Case Studies

In this section, two case studies are presented, for which we identify various security and privacy violations that may occur in the chain of events and suggest possible approaches to avoid these.

Scenario A – Assessing Burns Patients Remotely

A patient presents at small Regional Hospital Emergency Department with severe burns on both forearms and lower torso, resulting from ignition of a flammable liquid in a workplace materials handling accident.

1. The attending clinician seeks advice from an associated Tertiary Hospital using the dedicated State Health Service telehealth system, which provides proprietary videoconferencing. Views of the patient burns are obtained by zooming and panning of the dedicated video camera from the videoconferencing system.

2. The associate clinician at the Tertiary Hospital decides he needs additional advice, so he contacts a plastic surgeon who is on call for specialist consultation, and who can receive video streamed to his private tablet computer. The associate clinician initially attempts to establish the video stream using the camera and video transmission capability of his mobile phone, by simply holding it up to the screen of the videoconferencing unit (as seen by him at the Tertiary Hospital).

3. As the visual quality is not sufficient for the specialist consultant, a nurse suggests they try and use output from the open source screen-slaving software which the Tertiary Hospital has installed on all its screens to take advantage of internal hospital-wide wireless connectivity. One of the hospital tablet computers is used to receive the image stream from the videoconferencing unit screen, and then stream it to the consultant's tablet computer by Skype screen sharing.

4. The consultant provides advice but requests that higher quality images of the burns in their current initial state should be acquired by the Regional Hospital clinician prior to treatment and covering, to allow for reference during recovery. The consultant had also saved several screen shots on his tablet while the video streaming was in progress, on his tablet computer. The Regional Hospital clinician captures a number of images of the burns with his mobile phone, and emails these to both the associate clinician at the Tertiary Hospital, and the specialist consultant, using a secure e-mail environment provided for use by all associated doctors within the State Health Service.

5. Both recipients store the images in their personal filing systems on their State Health Service owned laptops. The consultant also copies these to his private tablet computer, as part of his strategy of holding copies of current cases to allow maximum accessibility. Unknown to the consultant, the file store of the tablet computer has been configured to mirror files into a public cloud site for automated backup purposes.

6. The patient is admitted for a one week period in the Regional Hospital. At the end of the week, dressings are changed and further images are

acquired. In response to the consultant's request, the images are emailed to him using a conventional public email service, but invoking an open source wrapper service that encrypts the email contents.

Notes for Scenario A

1. A Dedicated State Health Service telehealth system with proprietary videoconferencing would normally be setup as using encrypted transmission for video and voice. It is reasonable to assume that the proprietary system would have these facilities. From a security perspective this would only be potentially insecure if the proprietary protocol has been compromised (by hacking) and that the transmissions were being monitored between the two end-points. Since this communication would be streamed and not sent as a discrete file, the potential to exploit such a vulnerability would require ongoing monitoring and capture of the packets of data during transmission for reassembly. Hence the risk would be assessed as low. The impact of this would be a potential breach of privacy for the patient. Video in telehealth consultation is not generally recorded - however if it is it must become part of the patient record and protected as for any other data file.

2. Using the mobile phone camera and video transmission (as a file not streaming) is inherently insecure as its transmission would not have been encrypted. This also raises the question of the video being retained on the mobile phone, in an unencrypted manner. This has several impacts not only on the privacy of the patient in accordance with clinical information retention as well as potential sharing and loss of loss of this video, as it is reliant on the clinician who owns the phone to protect the video, not share it, and delete it securely from the device at some point in time but also the compliance with hospital policy and records retention policy. There is also the consideration that the video once taken should become part of the patient record. There is a responsibility that once recorded any record of the consultation, in whatever form, should be recorded for future reference and form part of the electronic health record for the patient. This is also important that there needs to be a record of any information that a diagnosis and clinical decisions were based upon.

3. The screen slaving solution tried was an innovative idea however there are several aspects to consider here. Firstly, similar to point 2 above, the deployment of the open source screen slaving software assumes that the software has been deemed to be secure for use in this environment. Despite being open source it means that the onus is on the user of the software to ensure that it functions in a secure and protected manner. This is clearly outside the capability and scope of what would be expected of clinicians to assess. Trusted open source software is a technical matter for the IT and security expertise within an organisation, however unless this team has such expertise the inadvertent vulnerability of backdoors and

access portals is a possibility. The use of Skype is a protective step as Skype uses a proprietary encryption methodology which protects video and voice conversations. The exception to this is where voice calls are made to landlines or mobile phones. However, as for any proprietary solution there is a great deal of trust placed in the third party provider of the service in relation to the encryption protocol, and there is no control over or knowledge of the offshore transfer and archiving.

4. The recording of video is not by default set using Skype. However the screen shots recorded by the consultant on his tablet are a vulnerability if these images are not protected by encryption on the tablet computer. In addition, it should be noted that at this point there are now several copies of the video in different locations, all of which need to be protected. The additional photos captured by the regional clinician are left on the mobile device unencrypted and without any knowledge of the protection installed or used on the mobile phone. Retention as well as the storage of these images is of concern, although depending on the image not identifiable for the patient. There are also issues with meeting or breaking hospital policy in regards to the images on the phone being taken outside the hospital environment.

5. The transfer of the images taken on the mobile phone via a secure email application is a positive protection method. Unfortunately this chain of protection is broken once the consultant also records these images on his private 'uncontrolled' tablet computer for convenience of access the images at a later date in the patient's treatment. This highlights the balance required between accessibility and security. This is further compounded by the automatic mirroring of the tablet computer for backup purposes – which of itself is a good security measure, but there is a need to ensure the protection of this information in the cloud is commensurate with privacy law.

6. The transfer of the images, one week post treatment, via a public email system, without secure transfer encryption is of concern particularly of the email may be encryption but the attachments are not.

Scenario B – Home Monitoring of Rehabilitation Patient

A patient has been discharged after heart surgery with a requirement to log daily activity and ECG/vital signs data using two small wearable devices, one on the wrist and one on a waistband.

1. The devices transmit data in burst mode via Bluetooth connection to the patients' mobile phone, which has had an app installed to encode and relay the data to a third party commercial monitoring service provider using a conventional mobile phone connection, established automatically by the app.

2. The monitoring service provider is responsible for aggregating and analysing the raw data, providing a daily report to the hospital by secure email, and generating real-time alerts during the course of the day if any exceptional conditions occur. Depending on the severity of the alerts, the monitoring service provider may contact the patient directly to check on them verbally (on the same mobile phone used to transfer the data) or may contact the hospital monitoring call-centre to escalate the level of intervention. The patient may also call the monitoring service provider or the hospital monitoring call-centre directly to discuss any concerns or queries.

3. Unfamiliar with the nature of the app, the patient accidentally reconfigures it when switching on the phone in the morning, changing the transmission characteristics to send the data messages out locally on Bluetooth and Wi-Fi as well as making the automated phone calls. A neighbouring resident detects the routine Bluetooth to mobile phone data transmissions and being uncertain of their provenance, attempts to analyse their content. Not realising they are patient data, the neighbour attempts to block the signals in case they are hostile to the local in-house Bluetooth environment. Later realising they are personal data, the neighbour decides to save them in case there is some follow-up incident where he might be able to assist.

4. A passing market survey scanning vehicle detects the phone connection and transmission messages between the patient's mobile phone and the monitoring service provider. In an attempt to push marketing information to the mobile phone, it interrupts the data sending protocol from the mobile phone.

5. The nurse at the hospital call-centre needs a quick look at the patient's data from the past 2 hours, and to expedite that he/she requests the monitoring service provider to send a screen shot of the visualization of the data directly to his/her personal tablet computer, which is connected to the mobile phone network.

6. The monitoring service provider is showing its operations centre to a party of potential investors and clients, including viewing of the incoming "live data streams" from multiple patients on their big screen. The screen is an intelligent client able to be linked to any of the operator screens using Wi-Fi. One of the party has a tablet computer which captures Wi-Fi transmissions and so is able to save copies of the screen transmissions for further inspection later, with the intention to contact some of the patients to ask about their experience with the service.

Notes for Scenario B

1. The initial problem is the lack of security controls on the app in being able to accidentally reconfigure the transmission characteristics. Good application security would protect against this occurring.

2. The security and privacy regime adopted by the monitoring service provider should have been subject to validation and audit processes if they are working in the health sector, as part of their data management compliance policy.

3. The subsequent capture of the data by the neighbour is a privacy concern as whilst the data may be encoded it is not encrypted in transfer between devices nor in transmission via the phone connection. The retention of the information by the neighbour is another vulnerability, not only in regard to privacy of the patient but also because it too is likely to be unprotected from further third party access.

4. The passing market survey scanning vehicle presents a security issue in that it interrupts the data transmission and information may then not be sent. This has the potential to be highly dangerous if the data transmissions that are interrupted are abnormal or life threatening and do not reach the monitoring service or are delayed.

5. The screen shot transmission is sent unencrypted from the monitoring service to the personal computer and the screen shot recorded by the nurse on their personal tablet are a vulnerability if these images are not protected by encryption on the tablet computer. Deletion of the screen shots is an issue as is the uncontrolled sharing of this information. Further, if the nurse is using this information for the patient's care it should also become part of the patient record. There is a responsibility that once recorded any record of the consultation, in whatever form, should be recorded for future reference and form part of the electronic health record for the patient. This is also important that there needs to be a record of any information that a diagnosis and clinical decisions were based upon.

6. The operations centre role in the scenario has several issues. Firstly, the information captured from the live streams is not unsecured (as they are a screen capture) and most likely without patient consent, which is required as the use of this information is not for the provision of healthcare directly to the patient and under the privacy principles contravenes use of that information for viewing by the visiting party and for the later surveying of patients about the monitoring service.

Learnings from the Case Studies

- Any situation using mobile technology and services is complex and made up of many, mostly discrete parts. Each discrete entity must be assessed and the risks defined in order to understand the overall risk involved.
- Security in mobile health requires a multi-faceted approach as it includes (as most security does) technology and people. It is often the socio-technical aspects that create the vulnerabilities in security and therefore the technological protection controls should be as strong as possible to minimise these vulnerabilities.
- There is a significant amount of assumed trust in the solutions provided to support health, and rightly it is not up to the clinician to also be a security expert and be mindful of all these aspects when trying to undertake their primary role of treating patients.
- The support and configuration of mobile solutions must consider the environment and situations under which they are operated and required, as this case study clearly demonstrates.

8 Conclusion and Future Directions

Security presents a gamut of issues that need addressing in the use of mobile devices in the healthcare setting. Common mobile devices have "hidden" data stores and logs which can be a source of compromise of patient/provider confidentiality. Support for standard PKI certificates routinely used for ehealth encryption and authentication, and their associated software libraries are not generally available. It is generally considered that 256 bit encryption should be a requirement in the mobile environment, but there are restrictions on export of such applications in some jurisdictions. Hence, local data storage, even if encrypted, may be at risk. However, implementation of sophisticated encryption algorithms and secure network protocols, such as VPN, lead to complexity, slow performance and reduced battery life that with current devices may discourage use.

Ultimately the continuum of data collection, information transfer and storage in the mobile environment has to be a balance between access and appropriate security to protect patient information and patient and healthcare provider privacy. Data collected via mobile applications or devices becomes part of the patient record if that information is used to make decisions regarding provision of care of a patient, and all the subsequent legal ramifications this places on the information. The challenge in this environment is the tracking and management of the information that is stored in a diversity of places and devices. At present the solution should, as with all security solutions, be layered and consisting of multiple protections. Firstly, policies should be established and monitored to guide users in how mobile devices should be used in the specific healthcare environment. Secondly, where technical solutions, such as mobile device management (MDM), are available they should be employed.

As with other types of security, it is vital that mobile security is also protected using a multi-faceted and layered approach. It is not solely about having multiple methods for specific protection objectives, but of realising the mobile security environment encompasses the application level security, the mobile operating system software, the device hardware, the wireless network and service. This is in addition to the equipment and software used to transmit, process and store the data. It is clear that managing the security and thus privacy of data in this environment is highly complex.

Future trends affecting the mobile environment offer a new set of challenges in security and privacy. The virtualisation and distribution of data and services enabled by Cloud Computing and Software as a Service approaches provide more opportunities for violations and lack of control across multiple points in the ICT ecosystem. This may prompt development of specific solutions for such environments (e.g. Lin et al 2013). At the same time, the number of participants in data collection and exchange within the Health ecosystem is increasing, leading to the notion that data control may in the future need to be undertaken in a dynamic adaptive approach (e.g. Evesti et al 2013).

References

Appari, A., Johnston, M.: Information security and privacy in healthcare: surrent state of research. Int. J. Internet and Enterprise Management 6(4), 279–314 (2010)

Ardito, S.: Mobile apps for the health professional. Searcher 19(6), 46–50 (2011)

Avancha, S., Baxi, A., Kotz, D.: Privacy in mobile technology for personal healthcare. ACM Comput. Surv. 45(1)3, 1–54 (2012)

Brusco, J.: Mobile health application regulations and compliance review. AORN Journal 95(3), 391–394 (2012)

COBIT, COBIT 4.1: Framework for IT Governance and Control (2013), http://www.isaca.org/knowledge-center/cobit/Pages/Overview.aspx

Dunbrack, L.A.: The Second Wave of Clinical Mobility: Strategic Solution Investments for Mobile Point of Care. IDC Health Insights, 29 (December 2011)

Evesti, A., Suomalainen, J., Ovaska, E.: Architecture and knowledge-driven self-adaptive security in smart space. Computers 2, 34–66 (2013)

FDA. Draft Guidance for Industry and Food and Drug Administration Staff - Mobile Medical Applications (2011), http://www.fda.gov/medicaldevices/deviceregulationandguidance/guidancedocuments/ucm263280.htm

Garg, V., Brewer, J.: Telemedicine security: a systematic review. Journal of Diabetes Science and Technology 5(3), 768–777 (2011)

Kotz, D.: A threat taxonomy for mHealth privacy. In: 2011 Third International Conference on Communication Systems and Networks (COMSNETS), pp. 1–6. IEEE (January 2011)

Lin, H., Shao, J., Zhang, C., Fang, Y.C.: Cloud-assisted privacy preserving mobile health monitoring. IEEE Trans Information Forensics and Security 8(6), 985–997 (2013)

Markovic, M., Savic, Z., Kovacevic, B.: Secure mobile health systems: principles and solutions. In: Istepanian, R., Laxminarayan, S., Pattichis, C.S. (eds.) M Health: emerging mobile health systems, pp. 81–106. Springer, Berlin (2006)

Martí, R., Delgado, J., Perramon, X.: Security specification and implementation for mobile e-health services. In: 2004 IEEE International Conference on e-Technology, e-Commerce and e-Service, pp. 241–248 (March 2004)

Morgan, S.A., Agee, N.: Mobile healthcare. Frontiers of Health Services Management 29(2), 3–10 (2012)

NIST (n.d.) Guide to Storage Encryption Technologies for End-User Devices. National Institutes of Science and Technology Special Publication 800-111, http://www.csrc.nist.gov/publications/nistpubs/800-111/SP800-111.pdf

OAIC 2013, Australian Privacy Principles Guidelines (draft), Office of the Australian Information Commissioner 2013 (2013), http://www.oaic.gov.au/privacy/privacy-engaging-with-you/current-privacy-consultations/draft-australian-privacy-principles-guidelines/draft-app-guidelines

Osunmuyiwa, O., Ulusoy, A.: Wireless security in mobile health. Telemed J. EHealth 18(10), 810–814 (2012)

Rindfleisch, T.: Privacy, information technology, and health care. Comm. ACM 40(8), 92–100 (1997)

Ryoo, J., Choi, Y.B., Oh, T.H.: Security and Privacy in Mobile Telemedicine. Mobile Telemedicine: A Computing and Networking Perspective, 175–193 (2008)

Sahoo, P.K.: Efficient security mechanisms for mHealth applications using wireless body sensor networks. Sensors 12, 12606–12633 (2012)

Silberman, M.J., Clark, L.: M-Health: The union of technology and healthcare regulations. J. Med. Pract. Manage. 28(2), 118–120 (2012)

Steinberg, M.J., Rubin, E.R.: The HIPAA Privacy Rule: Lacks Patient Benefit, Impedes Research Growth. Association of Academic Health Centers (2009)

Takahashi, D., Xiao, Y., Hu, F.: A survey of security in telemedicine with wireless sensor networks. Mobile Telemedicine: A Computing and Networking Perspective, 209–235 (2008)

Tang, P.C., Ash, J.S., Bates, D.W., Overhage, J.M., Sands, D.: Personal health records: definitions, benefits, and strategies for overcoming barriers to adoption. JAMIA 13(2), 121–126 (2006)

Thomson Reuters Foundation. Patient privacy in a mobile world: A framework to address privacy law issues in mobile health (2013), http://www.mhealthalliance.org/images/content/trustlaw_connect_report.pdf#sthash.vpNtzBwS.dpuf

Williams, P.A.H., Maeder, A.: A conceptual framework for secure mobile health. J. Int. Soc. Telemed. eHealth 1(1), 44–51 (2013)

Williams, P.A.H., McCauley, V.B.: A rapidly moving target: Conformance with e-health standards for mobile computing. In: 2nd Australian eHealth Informatics and Security Conference, pp. 40–49 (2013), http://ro.ecu.edu.au/cgi/viewcontent.cgi?article=1013&context=aeis (retrieved)

Win, K.T., Susilo, W., Mu, Y.: Personal health record systems and their security protection. Journal of Medical Systems 30(4), 309–315 (2006)

Questions

Q1. Why are security and privacy issues important in mobile health, and why are they difficult to achieve?

Q2. Describe four distinctive groups of service types in mobile health and the settings in which they occur, which require different approaches to security and privacy solutions.

Q3. Discuss the security and privacy issues of private vs public ICT environments for teleconsultation using mobile health solutions.

Q4. Describe the framework of three different fundamental "design tensions" used for analysing security and privacy issues in mobile health.

Q5. Identify the different settings for security or privacy violation threats encountered in connection with in-home collection of patient data for analysis by a monitoring service, retained by and reporting to a health service.

45. Impact of Privacy Issues on User Behavioural Acceptance of Personalized mHealth Services

U.S. Premarathne, Fengling Han, Haibin Liu, and Ibrahim Khalil

RMIT University, VIC 3001, Australia
{uthpalasubodhani.premarathne,fengling.han,
haibin.liu,ibrahim.khalil}@rmit.edu.au

Abstract. Health can provide efficient and convenient personalized healthcare services to a variety of patients with diverse medical needs. Existing security vulnerabilities, such as identity theft, loss or theft of mHealth devices and health information raise grave concerns in preserving privacy as well as in promoting user acceptance. With the advent of location dependent personalized mHealth services, access control coupled with cryptographic techniques are seen as pragmatic solutions in preserving privacy of mHealth data in addition to the formal regulatory requirements. In this view, we discuss the impact of privacy threats on user acceptance of personalized mHealth services. As well as the implications of enforcing privacy preserving enforcements at mHealth device level, network level and regulatory measures for sustainable and wide-spread use of personalized mHealth services in future.

Keywords: mHealth, security, mobile devices, access control, user adoption, security threats.

1 Introduction

mHealth service systems are potential strongholds for providing more social inclusion and efficient healthcare service facilitation to a wide variety of communities. Recently, mHealth Alliance [3] was formed by the United Nations Foundation to promote mHealth services to the developing countries. mHealth systems have also been used for supporting and diagnosis and treatment in hospitals. For example, the diagnosis and treatment support system, in Alfred hospital in Melbourne Australia [40], uses mHealth based devices (e.g. RFID and sensors) to collect data, medications and to monitor the health status. Personalized mHealth services cater to the specific healthare needs of a patient and delivers health services more focused on these specific needs. Examples of such personalized mHealth services include healthcare information services to specific patient categories (e.g. diabetes [2]) or personalized physical activities and dietary information services (e.g. GPSexergame [17]).

S. Adibi (ed.), *Mobile Health*,
Springer Series in Bio-/Neuroinformatics 5, DOI:10.1007/978-3-319-12817-7_45

As evidenced in a recently published report [7], there are over 70% patients and physicians that want to get personalized mHealth services via mobile devices. However, the potential privacy related security issues, such as identity theft, theft of medical records and unlawful publications of personal health information are heavy deterrents for proliferation of personalized mHealth services. As revealed by a recent survey [6], over 78% of the participants worry about the information leakage and misuse of their personal health information. Based on user behavioral intentions, such perceived risks associated with the privacy issues thus have become significant deterrents to wide acceptance of personalized mHealth services [22][61][38]. Privacy preserving techniques such as mHealth device level security features (e.g. device lock mechanisms) cryptographic techniques are robust security enforcements in facilitating secure mHealth services. In addition, there are regulatory enforcements that foster the privacy preservation requirement in mHealth systems. Therefore, it is interesting to understand the impact of such privacy preserving security enforcements on improving the user acceptance as well as the user-centric support to mitigate potential privacy violations in the context of personalized mHealth services.

In this chapter we cover the i) privacy issues of personalized mHealth services, ii) user-centric best practices to reduce the impact of privacy-centric attacks, iii) existing privacy preserving techniques, strategies and law enforcements and to iv) review the behavioral intentions for adopting security measures by users and the usability of these features in view of mHealth services.

2 mHealth Services and Privacy Issues

2.1 *Personalized mHealth Services*

In this section we discuss two main categories of personalized mHealth services: i) remote monitoring services and ii) location-dependent mHealth information services) and the related privacy issues. Next, in view of privacy centric threats, we cover the mHealth device level security features.

2.1.1 Real-Time Monitoring mHealth Services

Ubiquitous health monitoring and personal health management are promising mHealth applications due to recent advancements in miniature biosensing devices, smart textiles and wireless communications [49][14]. An example of a remote monitoring mHealth service is shown in Figure 1. Such mHealth remote monitoring solutions are low-cost unobtrusive alternatives beneficial to a wide range of patients with diverse medical conditions (e.g. elderly, patients with cardiac diseases). In Australia, cardiovascular diseases are the leading causes of deaths in 2011 [1], out of which most deaths are preventable with timely medical services. Wearable or mHealth monitoring systems can facilitate monitoring of various bio-signals (e.g. ECG, blood glucose level, body temperature). Depending on the health condition of the patient, the severity of losing or receiving false

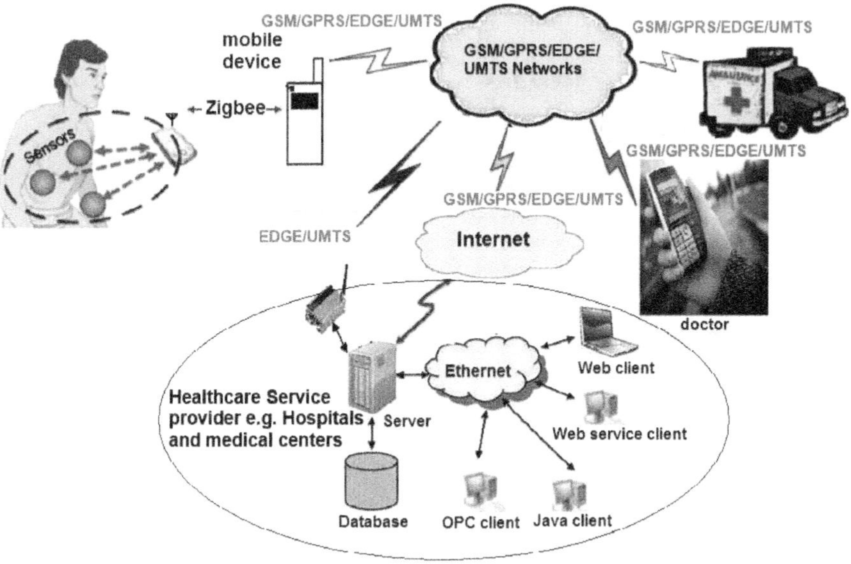

Fig. 1 mHealth Service System Overview

information by the mHealth monitoring systems can lead to disastrous situations, causing mis-diagnosis, prescribing erroneous medication dosages, or even deaths.

mHealth devices, such as sensors, implantable sensor based devices and RFID tags, are vital components in real-time monitoring mHealth services. These mHealth devices sense, collect, log and transmit vital bio-signals (e.g. ECG) and health status indicator (e.g. blood glucose level). Privacy issues (e.g. data leakages, malicious signal interceptions and interruptions) related to using these devices are found to be of significantly strong deterrents in real-time monitoring mHealth services.

2.1.2 Location Dependent Personalized mHealth Services

Location based health information services have emerged as a new paradigm for delivering personalized health information via GPS enabled mobile devices (e.g. smartphones). These mHealth information services, focus on health behaviors by providing information and services via personalized messages on to the mobile devices. The information conveyed include medical appointment and medication reminders, self-tracking tools, educational resources, lab and clinical result delivery.

Personalized mHealth information services are of great advantage over traditional informational health campaigns as it provides a more effective media to reach a broader audience engagement and direct interaction [8]. Examples of location-dependent health information services (HIS) include multiaccess services for Telematic management of Diabetes Mellitus [2] and MobiHealth [4]. Both these systems can send information to mobile devices of the users irrespective of any knowledge of its current location. A more intelligent location based health information service was tested using wearable sensors and GPS equipment, to detect and identify the current location of the users' mobile device [54].

Moreover, there is a growing body of evidence that supports the value of mHealth applications for health promotion, behaviour change (e.g., smoking cessation), chronic disease management, medication adherence and management, prenatal care, maintaining a healthy lifestyle, persistently encouraging weight loss and physical activity [8]. For example, GPSexergames are attractive new location dependent HIS applications [17] that aims to promote physical activities for children and adolescents. The applications are video games that can adaptively change their game based on the GPS location information (e.g. actual street names) based on the current location of the user. In addition, these mobile applications can adaptively change according to the health status of the user to give appropriate messages and warnings (e.g. warns when the user play for prolonged time without taking an adequate rest).

Also, preference-based personalized mHealth consultation services have also proposed as emerging personlized mHealth servies [52][51]. As evidenced in revcent reviews [25], in order to have a wider social inclusion, it is anticipated to use soial networks (e.g. Facebook, Twitter) to provide the location based information on the available mHealth services. For example for mHealth consultation services if a user is in location L1, then the social network site can provide available mHealth consultation service centers in close proximity to L1.

2.2 Privacy Issues and User-Centric Best Practices

In this section we discuss the privacy issues related to the mHealth services mentioned above and discuss the user-centric best practices to reduce the impact of potential malicious privacy-centric attacks.

2.2.1 Location Privacy

Location privacy and health information privacy are grave concerns in delivering location dependent personalized health services. In order to preserve information privacy, proper disclosure of sensitive information to authorized entities is vital. Since mobile devices are equipped with location tracking capabilities (e.g. GPS), location dependent mHealth applications are good venues to gather location information of participating users. This is a sever privacy breach. For example in GPSexergame applications, other players can identify the fellow players and their locations [17].

Privacy preserving features at application level are seen as a strong security feature in such personalized mHealth services. In addition, user enabled privacy settings are also significant determinants in preserving privacy of location privacy.

2.2.2 Lost or Theft of Health Information

When a mobile device is lost more than the cost of the device and the cost of losing valuable data is high. Such sensitive data can breach the confidentiality and the privacy of the user. It is also important to note that securing a mobile device is the sole responsibility of the user. The manufacturer or the mobile service provider has minimum share in securing such devices.

Best practices in minimizing the impact of losing sensitive information due to a stolen or lost mobile device include physical protection measures (e.g. locks, alarms), avoid carrying the device unless absolutely necessary, to avoid sharing the device with others or leaving the device unattended [59][62]. For wearable mHealth devices these practices provide a limited practical use. However all these avoidance practices imply that the users need to be vigilant about their personal mHealth devices. Remedial counter-measures refer to the actions that can help to cope with the consequences of losing a mobile device. One of the fore-most remedial measure is to report to an authorized body the incident of losing the mobile device [59][62]. This may help to trace the device or increase the chances to find it. Another potentially safety measure is to get an appropriate comprehensive insurance coverage. This depends on the type of data and applications being used in the device. Also, to maintain an up-to-date backup of the data in a secure storage can reduce the impact of losing sensitive information [59][62]. For example most service providers have secure Cloud storages where a user can store data (e.g. Google).

2.2.3 Identity Theft

Identity theft is a serious privacy issue as a malicious attacker can impersonate legitimate user by exploiting the true identification credentials. Identity theft can be performed by eavesdropping by intercepting the signal, snooping, phishing attacks via SMS, stealing the mobile device containing confidential information. An example of a severe privacy breach was reported in Norway. A very intensive privacy violation occurred when hackers stole confidential data on 60,000 Norwegians (1.3 % of the country's population) [24]. The vulnerability was on the web site of the mobile network operator Tele2 and stole the national personal identity numbers and addresses of these users. The information was thought to be exploited to maliciously intercept mail and other information from the users' accounts by changing certain credentials (e.g. address).

Best practices against identity theft are avoiding storing sensitive data without any security support (e.g. encryption, password or biometrics based protection), limit the remote access based mobile services and applications and to use the service provider services for anti-theft mechanisms (e.g. monitoring, memory wipeout, password resets, device disable or blocking remotely) [24]. These safe practices implies that a user should be assertive and mindful about the device capabilities, security services offered by the service providers and in using applications and the storing data.

2.2.4 Denial of Service (DoS) Attacks

DoS can have severe impact on the real-time bio-signal monitoring mHealth services. This is due to the possibility of battery drain attacks, near field communications based attacks (especially on Bluetooth) and frequency jamming attacks [18]. When the signal is jammed the mHealth devices are completely cut off from sending the vital health information. Implications of these attacks are disastrous consequences. For example, a real time health monitoring or data collecting application for remote diagnosis would be completely disturbed leaving the patients' life in danger.

Best practices include use of secure managed wireless networks and to disable the near field communications unless absolutely necessary [59][62].

2.2.5 Eavesdropping Attacks

These attacks can be launched by extracting information (e.g. username and password) by intercepting transmitted signals and by analyzing the incoming and outgoing network traffic from mHealth devices [42]. Personal area networks (e.g. bluetooth and Wi-Fi) as well as near field network connections can be used to launch these attacks [42]. Particularly in mobile wireless platforms this is a critical aspect that directly impacts the privacy and confidentiality of health information.

Robust countermeasures that can significantly reduce the possibility of such attacks are packet level encryption, physical and link layer fingerprinting [13]. The fingerprinting techniques can also avoid potential man-in-the-middle attacks, where an adversary can inject frames or packets maliciously to disturb accurate communication. Identity based encryption techniques have also been proposed to accurately and safely identify devices [60]. In addition the best practices recommended for users would be to minimize the usage of unmanaged and untrusted networks for private and confidential data and communication needs. Security features are aimed at providing certain prevention and detection techniques to safeguard against known attacks. However, the flexibility offered to the mobile device users to enable and use these security features depends on the users' ability and familiarization of technology. Therefore, it is important to consider the usability aspects of security features in mobile devices that can support mHealth services.

3 Privacy Preserving Techniques

In delivering personalized mHealth services, there are four main types of information: i) continuously monitored raw data (e.g. electrocardiogram (ECG), blood glucose level), ii) electronic medical records (EMR), iii) mobile personal health records (PHR) and iv) other related information sources (eg. exercise routines, diets). EMRs are generally controlled by the healthcare provider while the mobile PHRs are more user centric. There are number of PHR service providers that help the user to store and manage their PHRs (e.g. Google Health, Microsoft Health Vault). Mobile PHRs have evolved as an option with high availability where the users can store their health information in their own mobile devices. These mHealth data are accessed by patients and healthcare professionals (e.g. doctors, specialists, care givers, physiotherapists, dietitians) for various purposes (e.g. lab reports for updating health status, review, medication updates, follow-up reporting)[8].

Example: A remote ECG monitoring service, alerts a doctor of an anomaly in ECG data. The doctor may need to see the entire medical history and the associated test results and health records for treatment and diagnosis in order to ascertain the observed anomaly not as a false alarm and to provide necessary emergency care advice. A patient will need to access their personal health records to view, review, update their health status, medication management, education and self-management of personal health. An insurer would also want to know the up-to-date health profile of an insured as well as the history of medical insurance claims.

Based on the above discussion we discuss the following in this section.

- Mobile device level and network level privacy preserving security enforcements.
- Law enforcements, standards and regulations for imposing control on information access and management in mHealth service systems.

3.1 Security Features of Mobile Devices

In this section we discuss the security enforcements that can prevent and detect privacy breaches at mobile device level. Most commonly the mobile phones are used for collecting, transmitting and relaying confidential health information.

3.1.1 Intrusion Detection

The well-researched technique for detecting security threats in mobile devices is host-based intrusion detection [58]. The intrusion detection can be done at three levels, i) location-based detection, ii) traffic anomaly detection and iii) anomalous behaviors of mobile phone users. These detection methods can be implemented as rule based, feature based or both.

The authors in [65] have proposed a rule-based approach with an implementation on a Symbian operating system based Nokia mobile phone to detect unauthorized attempts to create Short Message Service (SMS) by using a simulated Trojan horse. This is applicable to a single solution. In [21], agent based detection architecture is proposed for multiple smartphones. The agents interact with a backend proxy to deliver the collected information about the devices. The information is processed and analyzed by the server. The solution was proved to be a viable detection mechanism by analysing the experimental data of SMS traces collected over three weeks by simulating a virus outbreak scenario. In the solution proposed in [56], the features describing the state of the mobile device are forwarded to a remote service. This solution analyses the information and detects the anomalies remotely. Energy-depletion related anomaly detection based on the generated power consumption signatures by observing the power consumption patterns was proposed by several researchers [58].

Temporal feature based intrusion detection framework was proposed in [16] using temporal logic approach for detecting Trojan horse and worm activities. The first stage is signature based mapping of malicious behavioural signatures. Then a supervised learning based classification was performed to identify the monitored events. The accuracy of classifying new events were over 96%. Many other researchers have proposed intrusion detection systems based on temporal features represented as temporal probabilistic tree [57], frequency table storing sequence [29], Allens' temporal relations [12] and to use chronical formalism [47]. All these methods consider the time dependent behaviour of features that represents the security threats in mobile devices and to consider time as a dominant determinant of the behavioural inferences in these systems.

3.1.2 Encryption Based Security Features

Secure data transfer is essentially has three phases. Encryption is required for preserving confidentiality of data. Digital signatures are required for preserving integrity and authenticity. Authentication is required to assure legitimate access to data. Though encryption is a viable solution, it does not guarantee the privacy preservation of signals. As evidenced in [33], such encrypted signals (e.g. GSM encryption algorithm) are susceptible to strategic attacks. Encryption based solutions for securing data in smartphones have been proposed in [27]. The authors propose a cryptographic protocol that enables personal health records to be viewed under specific scenarios in a controlled manner using multi-biometric feature based authentication. The secret key is managed according to a hierarchical tree. However, the way in which the scenario and validation of its credibility is not demonstrated adequately.

EMR stored in mobile devices ensures more availability in events of online only access control policies, high risk EMR transportation and complex access control requirements. In this view, health information need to be securely stored in users' mobile devices. Attribute based encryption (ABE) have been proposed as a viable protection mechanism [10]. In ABE, each user is given a key which can decrypt ciphertext if and only if certain attributes match. These attributes are specified at the time when the ciphertext is created. ABE is public key encryption scheme. The private keys are generated by a trusted third party. The attributes are embedded onto the users' private key.So there are two policies. First policy binds the entitled parties for decryption using boolean expressions. The second policy inverts the relationship between the ciphertext and the key. This scheme ensures only the entitled users can get access to the data.

In conclusion, above mentioned security features can greatly enhance the secure usability of personalized mHealth services on mobile devices if the users are knowledgeable of such security features. Therefore, mHealth service providers, mobile devices and application developers need to educate and inform users about the privacy-preserving security features of mobile devices.

3.2 Location Privacy Preservation Techniques

Location privacy is an important aspect in delivering personalized mHealth services. Recent publication [28] provides an excellent review of the state-of-the-art location privacy protection mechanisms. As identified in the early works [30], location privacy of location-dependent services include location tracking, position awareness, location-based queries as well as the privacy level of disclosure of location information. In addition to the above mentioned aspects, personalized mHealth services need to consider the situation dependency coupled with the privacy of disclosure of location information. For example, in remote monitoring mHealth services during an emergency situation, even if the disclosure risk is high, location information will have to be revealed to get timely care.

Most widely adopted and largely researched location privacy techniques include anonymization techniques, cryptographic techniques, geometric transformation techniques and differential privacy [28].

Anonymization techniques are broadly categorized as two-tier and three-tier transformations. Two-tier transformation techniques (e.g. Hilbert curve mapping, dummy locations based mapping, feature based obfuscation maps) do not require any trusted third party, and the query anonymization is performed by the mobile user itself. Three-tier transformation techniques (e.g. spatial k-anonymity, cloaking techniques) require a trusted third-party anonymizer server as a more robust strategy to mitigate background knowledge attacks. However, three-tier transformations are more computationally expensive and incur more runtime overheads. Cryptography based protocols (e.g. private information retrieval (PIR), computation PIR) have been proposed to allow a client to privately retrieve information from a database, without the database server learning what particular information the client has requested. These protocols are of significant use to the personalized location-dependent mHealth service providers. Privacy-preserving trajectory publication techniques are important for emerging personalized mHealth services (e.g. physical activity monitoring and encouraging applications such as GPSexergame applications). Privacy of trajectories can be preserved by using existing publishing techniques: reveal independent location samples or individual trajectories with distorted location samples at each time instance. Differential privacy allows users to interact with data through statistical queries. The amount of perturbation is added by means of random noise addition. Differential privacy can guarantee much higher level of privacy, since an attacker cannot be certain whether the target data is present by using the statistical queries. Essentially mHealth related information are often accessed for a variety of purposes (e.g. clinical research, diagnosis and treatment plans, long-term care plans) demand more robust measures to preserve privacy. In this view, differential privacy seems to be more applicable in mHealth services.

Implication of the above mentioned location privacy preserving mechanisms are largely applicable to personalized mHealth services. However, it is also necessary to have such location privacy preserving enforcements at the application level in mobile devices. This means that more computationally efficient techniques are vital as the mobile devices are resource constrained low computational devices. In addition, certain controls on setting the privacy levels should also be available to the users. Largely, it is the mHealth service providers' responsibility to preserve the location-sensitive information of their clients. However, in addition to the responsibility of mHealth service providers, regulatory enforcements should also be in place to preserve the location privacy of personalized mHealth services.

3.3 Cryptographic Techniques

Cryptographic techniques have been increasingly seen as promising solutions for securing health information.

In [66], high availability and interoperability of the electronic health record content protection system uses a personalized media recorder with symmetric Advanced Encryption Standard (AES) encryption. In [43], a more comprehensive smartcards based cryptographic key management scheme is proposed for securing the health information. Both these systems are patient centric content protection systems. In [43], user consents are exchanged using session based keys. At the healthcare authority, each user is issued a master key which is stored in a smartcard. At each access, this master key is used to generate the session key and encrypts the information. In order to decrypt, the information, the patient again has to reproduce the session key generated based on the particular master key. This method is attractive as it ensure non-repudiation and privacy. In [35], instead of session based keys, contract oriented scheme is proposed based on hybrid public key infrastructure (PKI). In this scheme, the security management is off-loaded to a trusted security service provider as a more scalable solution. Recently, a PKI based security system is proposed for large-scale distributed healthcare information networks [46]. Also, [63] analyses the privacy issues in exchanging health information in peer-to-peer systems. The authors propose a consent management scheme for a centralized health information distribution architecture using access control and PKI encryption based message exchange.

Recently, steganographic techniques have been proposed to secure EMRs. Authors in [36] propose a wavelet based steganography technique to protect patient confidential data in remote cardiac monitoring systems. The proposed method combines encryption and scrambling technique and uses ECG signal to hide its corresponding patient confidential data and other physiological information. Steganographic techniques are being proposed as it is more efficient in computation and sorage compared to standard cryptographic techniques which are more computationally expensive withy more overheads [36].

Cloud based systems are sought after alternatives for economical and easily manageable health information storages. mHealth services offered based on cloud based health information systems are seen increasingly viable economical solution. Therefore, it is important to understand the security and privacy issues related to cloud that can affect the privacy of mHealth information services. Standard encryption techniques (e.g. symmetric key , asymmetric key encryption, ciphertext-policy attribute based encryption) have been proposed for cloud based EMR systems. In [44], a patient centric cloud based PHR framework with semitrusted servers is proposed. This system use ABE technique to encrypt each patients' PHR file which is stored in a cloud storage. Security issues in hybrid cloud EMR systems when the information is shared by various entities have been studied in [20]. Cloud systems can be private, public or hybrid (i.e. a combination of private and public clouds). The access control policies for EMR information in such systems are complex due to the existence of various policies. These complex policies can significantly affect the utility and availability of EMR information. Therefore, mHealth services that make use of cloud based services for health information storage and distribution need to be aware of these complexities in designing appropriate security policies for access control in service facilitation.

3.4 Regulations and Policies

Many countries (e.g. USA, Australia) have specific regulations in handling health information [5][45]. These law enforcements are necessary to conform different health service providers to engage in common security assured practices. For example the Australian privacy acts are declared for each state in handling health related information [5]. In addition to the rule and regulations there are guidelines and recommendations for security risk analyses for continuous secure operations of health information. Such privacy preserving practices are vital for mobile healthcare service platforms as the inherent vulnerabilities of wireless communications and mHealth devices.

Privacy regulations are formal specifications of patients' rights for controlled use of Protected Health Information (PHI) such as patients' name, information on medical records. Health Insurance Portability and Accountability Act (HIPAA) is the United States law enforcement for securing the privacy and confidentiality of electronic health data. HIPAA regulates the use, transmission and disclosure of PHI [45]. It also covers audits and validation systems to be implemented in managing [45]. In addition, there are other alliances formed by commercial organizations for managing security enforcements in handling health related information [45].

In order to maintain the security of health data it is vital to perform security risk analysis of the information systems holding the health data. Recently, the Office of the National Coordinator (ONC) for Health Information Technology (HIT) have suggested five necessary steps to perform a security risk analysis [15], i) review current security on health information, ii) identify potential threats and the vulnerabilities, iii) assess the likelihood of the threats, iv) prevent and mitigate the security risks and v) monitor and review the results.

In addition to the above mentioned regulations, XML (eXtensible Markup Language - XML) standards have also been put in place to preserve the privacy of EMR. Continuity of Care Record (CCR) and Continuity of Care Document (CCD) are two such standards for protecting EMRs [10]. The main drawback of these standardizations is that there is not enough information on the extent of the privacy preserved by using these guidelines. Therefore, additional network level and device level protective measures are vital in mHealth services.

In view of emerging new mHealth information services, existing health information privacy laws and regulations need to adaptively change in order to accommodate the growing privacy and security concerns.

3.5 mHealth Provenance

Provenance is defined as "lineage, pedigree and filiation of a resource (or data item)" [19]. Provenance data are represented as directed acyclic graphs (DAG). An example of provenance data DAG representation is shown in Figure2. This example shows open provenance model (OPM). OPM use three types of entities: artefacts, processes and agents as well as annotations to describe the causal relationships between the entities. Artefacts represent the information resources. Processes represent the actions performed on the artefacts. The users that have performed the respective actions are denoted by entities.

Recently, provenance of mHealth data is being viewed as an increasingly important aspect due to the use of third party services such as cloud based services for mHealth data storage and utility. In [50] a provenance framework for mHealth devices is presented. This framework aims to help the data consumer to confirm the validity of mHealth data by verifying certain provenance properties. These provenance properties vary based on the mHealth service. Thus, domain expert knowledge is mandatory in configuring these verifiable provenance properties.

Access control is a strong privacy preserving mechanism that can limit the access to confidential EMR information. Since provenance data of EMR can be used for i) user authentication, ii) action validation and iii) policy retrieval, provenance based access control is seen apt for more comprehensive authorization. Provenance based access control models [48] have been proposed in order to facilitate more reliable fail-proof authorization in accessing confidential health information.

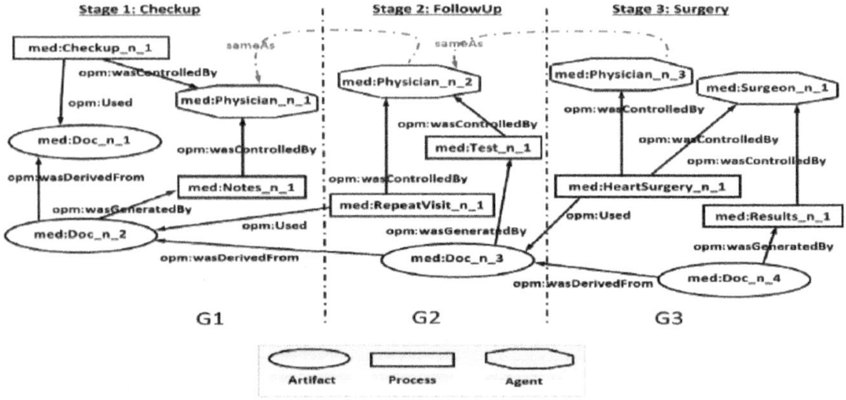

Fig. 2 Provenance Data Representation as a Directed Acyclic Graph [19]

4 Privacy Issues and User Behavioral Acceptance

In personalized mHealth service systems, users can access, exchange and distribute these information via secured Internet networks, unmanaged wireless networks, mobile cellular networks or through personal or near field communication networks.

Proliferation and acceptance of personalized mHealth services depend on various factors [53]. As shown in Figure 3, these multiple predictors are shown to have significant associations with mHealth usage intentions. The authors further postulate that in order to gain wide acceptance of personalized mHealth services, it is vital to have a strategic plan in promoting these value-oriented services in a more personalized manner.

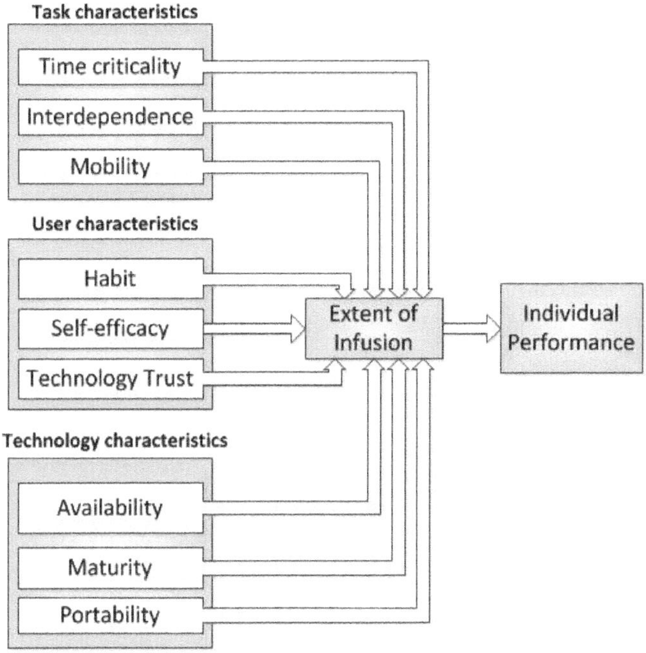

Fig. 3 mHealth Success Determinants [53]

4.1 Determinants of User Acceptance of mHealth Services

Acceptance by users (e.g. patients, healthcare professionals) is a vital determinant of the proliferation of mHealth devices and health information systems usage. A recent empirical study [26] further emphasizes that the social context is not fully explored but it is still an open research problem that can provide valuable inputs to successful technology integrations into healthcare domain. Social influence, perceived security, self-efficacy, persuasive communications (e.g. alerts, fear appeals, warnings) and the information security literacy are among the main determinants of adopting and continuous usage assurance for healthcare information management systems and applications [37] [34]. The success of achievement of security objectives of a distributed healthcare information management system or an application is dependent on the computer security related actions and user behavioral intention for adoption [37]. Theoretical basis for such understanding is the protection motivation theory [55].

According to [41][38], among the many reasons of wide spread usage and user acceptance of RFID devices is the perceived risk due to security threats. As revealed in [41] the privacy concern with the existing security and privacy threats on RFID is a growing concern for large scale deployments of healthcare applications. In addition the trust in using these devices is also a major concern [41] in data accessing, sharing and storing. The survey conducted on the public perceptions on mHealth applications and devices, the results reveal that the users have a positive attitude towards mHealth applications while the security issues do pose a negative perception due to the associated privacy and security risks associated [38]. The possible solutions that can improve the user adoption and wide-spread deployment are, to impose policies and regulations on RFID based mHealth device usage and management, up-to-date user education and awareness, appropriate training and education on the functionalities and technical information in using and managing these devices.

An important recommendation [31] related to enforcement of certain security measures on implantable devices is that these should not overly complicate the device usage to the users (i.e. both the patient and the healthcare providers such as doctors) due to: a) the necessity of quick decision making during emergency situations and b) long term sustainability of the device usage.

In [34], privacy issues that determine the user acceptance of healthcare information systems has been studied using the theory of acceptance and use of technology (UTAUT). There are two very important results that have direct impact on the acceptance and proliferation of a healthcare information system. The first is that the people with higher information security literacy would accept a healthcare management service only if they are made well aware of the security measures sufficient protection of privacy of information. The second is that word-of-mouth type peer-influences through social media can have significant impact on the service proliferation. This is a significant observation as the positive comments can enhance the proliferation while negative comments can retard the adoption proliferation.

Potential user-behaviour oriented techniques to motivate the users to use security features and also to conduct safe practices include fear appeals, warnings, disclaimers, formal policies and guidelines. The effect of fear appeals on behavioural intentions have been studied in [37][64]. Witte et al [64] states that effective fear appeals implementation should start with increasing the perceived threat and then highlighting the value of the recommended response. Johnston improves this approach by introducing persuasive communications to make the above two factors more effective. In smartphones certain applications provide warnings prior to any interaction. However, recent studies reveal that often these warnings do not convey the intended message fully to the users [24]. According to their survey based empirical study, users mostly prefer warnings on application actions for these risks, i) premium rate SMS messaging, ii) deleting all the contacts, applications, settings or logs and when iii) using peripherals to get data (e.g. using radio of the mobile phone to collect credit card number from the wallet). Such studies provide more information as to how to best construct applications with effective security enforcements that will help the users to use the mobile applications securely. This is particularly important for the acceptance and foster proliferation of mobile personal health record applications.

Any security threat coping feature will only be successful if the user willingly adopts it. The behavioural intention to take actions can be self-motivated or explicitly gained externally. For example warnings can compel a user to close a session of EMR web page upon logout. Various empirical studies [32] reveal the positive and negative determinants of coping behavioural intention to adopt a security feature. The strong positive determinants are, i) perceived risk, ii) self-efficacy, iii) perceived usefulness and iv) perceived ease of use.. Perceived risk can be significantly increased by informing and motivating the users to be educated about the current security threats and the potential harmful consequences. So a person with a higher perceived risk is more likely to adopt. And also, higher the self confidence in potential actions provide less likelihood for adopting certain security features whereas for other features the probability of adoption is high. Similarly, higher perceived usefulness is a strongly determinant of behavioural intention to motivate users to make use of an existing security feature. Also, majority of users are willing to adopt a security feature if the effort is less and the responses for actions are promptly revealed. Thus it is important to design security features that are easy to use and understand. Privacy related concerns are the main negative impact on the behavioural intention of adopting a security feature. Persuasive communications via social media through trusted sources is a viable approach that can potentially lower this negative effect.

Thus, in order to improve the user acceptance and to assure long-term sustainability, it is important to reduce the risk of privacy centric information threats as well as improve the understanding of the potential benefits of mHealth services.

4.2 Case Studies

Following case studies further confirms the above inferences.

4.2.1 Security Features of Mobile PHR Applications

The empirical study conducted by Kharrazt et al [39] uses 19 stand-alone mobile PHR applications in three different platforms: iOS, Blackberry and Andriod. The analysis was conducted to reveal the variety of features present in these based on the i) product characteristics (e.g. price, number of records), ii) data (e.g. types of data) and iii) application based characteristics - data security (e.g. application level password).

The application level security features that were recommended were mobile phone (or device) level password, application level password, ability to disclose limited data depending on the receiving entity and the support for emergency access to health data.

Based on the results, the important inferences in relation to the information security are, mobilty pose greater vulnerabilities in preserving privacy of mHealth services and associated health information, skill and knowledge of the users are main determinants of successfully operating the available security enforcements in mobile devices, regulations and laws are vital in standardizing and enforcing privacy preservation methods in mHealth services.

4.2.2 Promoting User Acceptance of Personalized mHealth Services

Recent empirical study [11] on user perceived quality of mHealth was aimed to find the determinants of success of mHealth services. They postulate the assessment of service quality of a mHealth system can in turn help in promoting mHealth services.

The results reveal that reliability, privacy and efficiency to be of significance impact on the quality evaluation. These factors have a significant impact on the initial adoption as well as the progressive sustainability of user acceptance. Also, the results imply that in the information security perspective, a user would want to adopt a mHealth service as a quality service is significantly determined by the sufficient security and privacy preserving features are available in the mHealth service system.

Based on the above discussion, it is evident that identification of privacy issues and mitigation of possible privacy centric threats are vital for ensuring long term sustainability of mHealth services. Access control enforcements are widely used privacy preserving security enforcements. The access control policies ensures that only authorized entities have access to confidential information resources while unauthorized entities are not allowed to access these resources. In addition, with the advent of mobile personal health record services, it is vital to ensure device level security enforcements for mitigating privacy centric attacks on health information stored in mobile devices.

5 Future Directions

According to [23], the future directions for behavioural information security research need to deal with several aspects. Firstly, investigations on the insider attack using real-world data sets. This research direction focuses to identify the misbehaviour patterns. Secondly, behaviour based studies of hackers of both types: ethical and unethical. Research on this aspect require to focus on the behavioural patterns and the personality traits of these two groups. Thirdly, fear based persuasive communications to increase security compliances and the temporal features of fear-based behavioral patterns. The implications of these in relation to mHealth are evident. The first, second and third research focuses can provide valuable information for developing more secure mHealth services with stronger security features. The pattern based studies in relation to the personality traits can provide valuable insight in understanding the areas that needs more focus to improve the user compliance and acceptance. mHealth is useful for all communities and mostly in the developing nations [18]. Most of these countries literacy and awareness of information technology and the security issues needs more improvement to reap the benefits of mHealth services. Thus the relevance of these research themes are timely and deserves significantly more attention.

In addition, more robust and computationally efficient location preserving mechanisms at application level is seen as a future research direction. Since proliferation of mobile devices (e.g. smartphones) are evident, people are willing to get access to mHealth services via their mobile devices. So it is vital to enhance location preserving mechanisms in the context of mHealth personalized services.

6 Conclusions

In conclusion, personalized mHealth services have great potential in delivering healthcare services more effectively and conveniently to patients with diverse healthcare needs. However, the privacy of personal health information is a strong determinant in proliferating personalized mHealth services. In order to improve the user acceptance robust privacy preserving security enforcements at i) mHealth device and application level, ii) service provider end and iii) sufficient regulatory and law enforcements to guide the service facilitators and users are vital. Existing privacy enforcement mechanisms are seen to be limited in terms of applicability to personalized mHealth services. In order to overcome the limitations future research focus should be on developing more user friendly security features at mHealth device and application level, more robust and fail-proof personal health information and location privacy preserving mechanisms at service-provider end and also guidelines and regulatory enforcements to cover the privacy issues of emerging personalized mHealth services.

References

1. http://www.abs.gov.au/ausstats/abs@.nsf/Lookup/3303.0Chapter42011
2. M2dm, http://www.labmedinfo.org/research/m2dm
3. mhealth alliance, http://www.mhealthalliance.org/about/founding-partners
4. Mobihealth, http://www.mobihealth.org/
5. The regulation of health information privacy in australia (2004), http://www.nhmrc.gov.au/filesnhmrc/publications/attachments/nh53.pdf
6. http://healthcaremgt.net/blog/2011/08/are-you-educating-patients-on-ehr/ (2011)

7. http://www.research2guidance.com/us-1.3-billion-the-market-for-mhealth-applications-in-2012 (2012)
8. Abebe, N.A., Capozza, K.L., Des Jardins, T.R., Kulick, D.A., Rein, A.L., Schachter, A.A., Turske, S.A.: Considerations for community-based mhealth initiatives: Insights from three beacon communities. Journal of Medical Internet Research 15(10) (2013)
9. Adesina, A.O., Agbele, K.K., Februarie, R., Abidoye, A.P., Nyongesa, H.O., Cape, W., Adesina, A.: Ensuring the security and privacy of information in mobile health-care communication systems. S. Afr. J. Sci. 107(9/10), 26–32 (2011)
10. Akinyele, J.A., Pagano, M.W., Green, M.D., Lehmann, C.U., Peterson, Z.N., Rubin, A.D.: Securing electronic medical records using attribute-based encryption on mobile devices. In: Proceedings of the 1st ACM Workshop on Security and Privacy in Smartphones and Mobile Devices, pp. 75–86. ACM (2011)
11. Akter, S., DAmbra, J., Ray, P.: Development and validation of an instrument to measure user perceived service quality of mhealth. Information & Management (2013)
12. Naldurg, P., Sen, K., Thati, P.: A temporal logic based framework for intrusion detection. In: de Frutos-Escrig, D., Núñez, M. (eds.) FORTE 2004. LNCS, vol. 3235, pp. 359–376. Springer, Heidelberg (2004)
13. Avancha, S., Baxi, A., Kotz, D.: Privacy in mobile technology for personal healthcare. ACM Computing Surveys (CSUR) 45(1), 3 (2012)
14. Baig, M.M., Gholamhosseini, H., Connolly, M.J.: A comprehensive survey of wearable and wireless ecg monitoring systems for older adults. Medical & Biological Engineering & Computing, 1–11 (2013)
15. Blumenthal, D., Tavenner, M.: The meaningful use regulation for electronic health records. New England Journal of Medicine 363(6), 501–504 (2010)
16. Bose, A., Hu, X., Shin, K., Park, T.: Behavioural detection of malware on mobile handsets. In: Proceedings of the Sixth International Conference on Mobile Systems, Applications and Services (2008)
17. Boulos, M.N.K., Yang, S.P.: Exergames for health and fitness: the roles of gps and geosocial apps. International Journal of Health Geographics 12(1), 18 (2013)
18. Briggs, J., Adams, C., Fallahkhair, S., Iluyemi, A., Prytherch, D.: M-health review: joining up healthcare in a wireless world (2012)
19. Cadenhead, T., Khadilkar, V., Kantarcioglu, M., Thraisingham, B.: Trasforming provenance using redaction. In: SACMAT, pp. 93–102. ACM (2011)
20. Chen, Y.Y., Lu, J.C., Jan, J.K.: A secure ehr system based on hybrid clouds. Journal of Medical Systems 36(5), 3375–3384 (2012)
21. Cheng, J., Wong, S., Yang, H., Lu, S.: Smartsiren: virus alert and detection for smartphones. In: Proceedings of the Fifth International Conference on Mobile Systems, Applications and Services (2007)
22. Chin, E., Felt, A.P., Sekar, V., Wagner, D.: Measuring user confidence in smartphone security and privacy. In: Proceedings of the Eighth Symposium on Usable Privacy and Security, p. 1. ACM (2012)
23. Crossler, R.E., Johnston, A.C., Lowry, P.B., Hu, Q., Warkentin, M., Baskerville, R.: Future directions for behavioral information security research. Computers & Security 32, 90–101 (2013)
24. Felt, A., Finifter, M., Chin, E., Hanna, S., Wagner, D.: A survey of mobile malware in the world. In: Proceesings of the First ACM Workshop on Security and Privacy in Smartphones and Mobile Devices, pp. 3–14 (2011)

25. Fernandez-Luque, L., Karlsen, R., Bonander, J.: Review of extracting information from the social web for health personalization. Journal of Medical Internet Research 13(1) (2011)
26. Fisher, J.A., Monahan, T.: Tracking the social dimensions of rfid systems in hospitals. International Journal of Medical Informatics 77(3), 176–183 (2008)
27. Gardner, R., Garera, S., Pagano, M., Gren, M., Rubin, A.: Securing medical records on smartphones. In: Proceedings on SPIMAC 2009 (2009)
28. Ghinita, G.: Privacy for location-based services. Synthesis Lectures on Information Security, Privacy, & Trust 4(1), 1–85 (2013)
29. Ghosh, A., Swartzbard, A., Schatz, M.: Using problem behaviour profiles for intrusion detection. In: Proceedings of the First USENIX Workshop on Intrusion Detection and Network Monitoring (1999)
30. Gruteser, M., Liu, X.: Protecting privacy, in continuous location-tracking applications. IEEE Security & Privacy 2(2), 28–34 (2004)
31. Halperin, D., Kohno, T., Heydt-Benjamin, T.S., Fu, K., Maisel, W.H.: Security and privacy for implantable medical devices. IEEE Pervasive Computing 7(1), 30–39 (2008)
32. Herath, T., Chen, R., Wang, J., Banjara, K., Wilbur, J., Rao, H.R.: Security services as coping mechanisms: an investigation into user intention to adopt an email authentication service. Information Systems Journal (2012)
33. Hoover, J.: Black hat conference: Security researchers claim to hack gsm calls (2008), `http://www.informationweek.com/news/mobility/security/showArticle.jhtml?articleID=206800800&cid=RSSfeed-IWK-All`
34. Hsu, C.-L., Lee, M.-R.: User acceptance of a community-based healthcare information system preserving user privacy. In: Stephanidis, C., Antona, M. (eds.) UAHCI 2013, Part III. LNCS, vol. 8011, pp. 453–462. Springer, Heidelberg (2013)
35. Hu, J., Chen, H.H., Hou, T.W.: A hybrid public key infrastructure solution (hpki) for HIPAA privacy/security regulations. Computer Standards & Interfaces 32(5), 274–280 (2010)
36. Ibaida, A., Khalil, I.: Wavelet based ecg steganography for protecting patient confidential information in point-of-care systems. IEEE Transactions on Bio-medical Engineering 60(12), 3322–3330 (2013)
37. Johnston, A.C., Warkentin, M.: Fear appeals and information security behaviors: an empirical study. MIS Quarterly 34(3), 549–566 (2010)
38. Katz, J.E., Rice, R.E.: Public views of mobile medical devices and services: A us national survey of consumer sentiments towards rfid healthcare technology. International Journal of Medical Informatics 78(2), 104–114 (2009)
39. Kharrazi, H., Chisholm, R., VanNasdale, D., Thompson, B.: Mobile personal health records: an evaluation of features and functionality. International Journal of Medical Informatics (2012)
40. Khosla, R., Chowdhury, B.: Real-time RFID-based intelligent healthcare diagnosis system. In: Zhang, D. (ed.) ICMB 2008. LNCS, vol. 4901, pp. 184–191. Springer, Heidelberg (2007)
41. Kuo, C.H., Chen, H.G.: The critical issues about deploying rfid in healthcare industry by service perspective. In: Proceedings of the 41st Annual on Hawaii International Conference on System Sciences, pp. 111–111. IEEE (2008)
42. La Polla, M., Martinelli, F., Sgandurra, D.: A survey on security for mobile devices (2012)

43. Lee, W.B., Lee, C.D.: A cryptographic key management solution for hipaa privacy/security regulations. IEEE Transactions on Information Technology in Biomedicine 12(1), 34–41 (2008)
44. Li, M., Yu, S., Zheng, Y., Ren, K., Lou, W.: Scalable and secure sharing of personal health records in cloud computing using attribute-based encryption (2013)
45. Manion, F.J., Robbins, R.J., Weems, W.A., Crowley, R.S.: Security and privacy requirements for a multi-institutional cancer research data grid: an interview-based study. BMC Medical Informatics and Decision Making 9(1), 31 (2009)
46. Mantas, G., Lymberopoulos, D., Komninos, N.: Pki security in large-scale healthcare networks. Journal of Medical Systems 36(3), 1107–1116 (2012)
47. Morin, B., Debar, H.: Correlation of intrusion symptoms: an application of chronicles. In: Proceedings of Recent Advances in Intrusion Detection Symposium (2003)
48. Nguyen, D., Park, J., Sandhu, R.: Dependency path patterns as the foundation of access control in provenance-aware systems, pp. 1–4 (2013)
49. Pantelopoulos, A., Bourbakis, N.G.: A survey on wearable sensor-based systems for health monitoring and prognosis. IEEE Transactions on Systems, Man, and Cybernetics, Part C: Applications and Reviews 40(1), 1–12 (2010)
50. Prasad, A., Peterson, R., Mare, S., Sorber, J., Paul, K., Kotz, D.: Provenance framework for mhealth (2013)
51. Premarathne, U.S., Han, F., Khalil, I., Tari, Z.: Fuzzy logic based load balancing for an online medical consultation system. In: 2013 8th IEEE Conference on Industrial Electronics and Applications (ICIEA), pp. 284–289. IEEE (2013)
52. Premarathne, U.S., Han, F., Khalil, I., Tari, Z.: Preference based load balancing as an outpatient appointment scheduling aid. In: 2013 35th Annual International Conference of the IEEE Engineering in Medicine and Biology Society (EMBC), pp. 1298–1301. IEEE (2013)
53. Rai, A., Chen, L., Pye, J., Baird, A.: Understanding determinants of consumer mobile health usage intentions, assimilation, and channel preferences. Journal of Medical Internet Research 15(8) (2013)
54. Rocha, A., Martins, A., Freire Junior, J.C., Kamel Boulos, M.N., Vicente, M.E., Feld, R., van de Ven, P., Nelson, J., Bourke, A., 'OLaighin, G., et al.: Innovations in health care services: The caalyx system. International Journal of Medical Informatics (2011)
55. Rogers, R.W.: A protection motivation theory of fear appeals and attitude change1. The Journal of Psychology 91(1), 93–114 (1975)
56. Schimdt, A., Peters, F., Lamour, F., Scheel, C., Camtepe, S., Albyrak, S.: Monitoring smartphones for anomaly detection. Mobile Networks and Applications 14(1), 92–106 (2009)
57. Seleznyov, A., Maazhelis, O.: Learning temporal patterns for intrusion detection. In: Proceedings of the 17th ACM Symopsium on Applied Computing (2002)
58. Shabati, A., Kanonov, U., Elovici, Y., Glezer, C., Weiss, Y.: Andromaly: A behavioural malware detection framework for android devices. Journal of Intelligent Information Systems 38(1), 161–190 (2012)
59. Stanford University: Guidelines for securing mobile computing devices, `http://www.stanford.edu/group/security/securecomputing/mobiledevices.html`
60. Tan, C.C., Wang, H., Zhong, S., Li, Q.: Ibe-lite: a lightweight identity-based cryptography for body sensor networks. IEEE Transactions on Information Technology in Biomedicine 13(6), 926–932 (2009)

61. Ting, S., Kwok, S.K., Tsang, A.H., Lee, W.: Critical elements and lessons learnt from the implementation of an rfid-enabled healthcare management system in a medical organization. Journal of Medical Systems 35(4), 657–669 (2011)
62. University of Michigan: Mobile computing guidelines for traveling abroad, `http://www.safecomputing.umich.edu/MDS/overseastravel.html`
63. Weber-Jahnke, J.H., Obry, C.: Protecting privacy during peer-to-peer exchange of medical documents. Information Systems Frontiers 14(1), 87–104 (2012)
64. Witte, K.: Putting the fear back into fear appeals: The extended parallel process model. Communications Monographs 59(4), 329–349 (1992)
65. Yap, T.S., Ewe, H.T.: A mobile phone malicious software detection model with behavior checker. In: Shimojo, S., Ichii, S., Ling, T.-W., Song, K.-H. (eds.) HSI 2005. LNCS, vol. 3597, pp. 57–65. Springer, Heidelberg (2005)
66. Yu, W., Chekhanovskiy, M.A.: An electronic health record content protection system using smartcard and pmr. In: 2007 9th International Conference on e-Health Networking, Application and Services, pp. 11–18. IEEE (2007)

Questions

Q1. What are the main types of security threats affecting the use of mHealth devices for personalized mHealth applications?

Q2. What are the user adoption related implications for developing more robust security features in mHealth devices?

Q3. What are the challenges seen for progressing research on the impact of security issues on user acceptance of personalized mHealth applications?

Q4. What are the implications of provider-end security enforcements in fostering personalized location-based mHealth services?

Q5. How important it is to have formal legal regulations for preserving the privacy features of health information involved in mHealth applications?

Answers

2. mHealth Sensors, Techniques, and Applications for Managing Wandering Behavior of People with Dementia: A Review

A1. Clinically, wandering is a form of meandering, aimless or repetitive locomotion that exposes a person to harm and is incongruent with boundaries, limits or obstacles. According to the clinical definition, wandering is characterized by (1) frequent or continuous movement from place to place, often revisiting the same destinations; (2) persistent locomotion in search of "missing" or unattainable people or places; (3) haphazard locomotion; (4) locomotion in unauthorized or private spaces; locomotion resulting in unintended leaving of a premise; (5) long periods of locomotion without an apparent destination; (6) fretful locomotion or pacing; (7) inability to locate significant landmarks in a familiar setting; (8) locomotion that cannot be easily dissuaded or redirected; (9) following behind or shadowing a caregiver's locomotion; (9) trespassing; hyperactivity; scanning, seeking, or searching behaviors; (9) periods of locomotion interspersed with periods of nonlocomotion (e.g., sitting, standing, sleeping); (10) getting lost.

Scientifically, wandering is a syndrome of dementia-related locomotion behavior having a frequent, repetitive, temporally-disordered, and/or spatially disoriented nature that is manifested in lapping, random, and/or pacing patterns, some of which are associated with eloping, eloping attempts, or getting lost unless accompanied. According to the scientific definition, wandering is measured in terms of *frequency* (a sum, mean, or rate of wandering episodes per unit time); *repetitiveness*(characterized by 2 items: wanderer goes repeatedly to the same location; wanderer travels repeatedly the same route while walking); *temporal distribution* (wandering duration between lunch and dinner; between breakfast and lunch; between dinner and bedtime; between waking and breakfast); *spatial disorientation* (characterized by 5 items: wanderer cannot locate dining room/own room/bathroom without help; wanderer bumps into obstacles or other people while walking alone; wanderer walks about aimlessly; wanderer gets lost); *geographical patterns* (wandering in pacing, lapping, random patterns); *eloping behavior* (characterized by 4 items: wanderer runs off; wanderer enters unauthorized area; wanderer attempts to leave authorized area; wanderer is returned to authorized area after leaving unnoticed); *negative outcomes* (characterized by: fatigue; inadequate food intake; falls and injuries; exit attempts despite caregivers' efforts to keep wanderer in a location or home).

A2. The proposed system architecture is shown in Fig. 1. It consists of four main components: the client devices, the wireless access network, the Internet and its hosts, and the server side that accommodates the database and application-specific servers that are required in our system.

There is a location module at the client devices. The module handles the end-system device hardware and returns the requested user location data to the application, regardless of the type of technology or method of implementation that actually calculates the user's location. In this application, GPS location technology is selected for the requirements of services that take place outdoors. By using Web service technology, the location module can be queried by a Web application to obtain the location of a handheld device without a requirement to install third-party software on the client devices. When queried by a Web application, the location module returns the location of the handheld device in an extensible markup language (XML) format to the Web server that initiated the request.

The wireless access network serves as Transmission Control Protocol/Internet Protocol (TCP/IP) pipes over the underlying infrastructure. For example, a carrier network can use high-speed, downlink-packet access (HSPDA) or general-packet radio service (GPRS) to facilitate data traffic.

The Internet component consists of third party application servers such as a Simple Mail Transfer Protocol (SMTP) mail server, voice over IP (VoIP) server, and Map server. Map server provides map Web service that processes the user's location and displays it on a geographical map viewed on a Web page. There are publicly available map servers that can be utilized and that provide well-defined, albeit proprietary, application programming interfaces (APIs). The application servers in the Internet component also can be deployed on the server side.

There are five modules on the server side: the Web server, location database, profile repository, matching logic, and privacy control. The Web server that runs the HTTP protocol is the data bearer that handles requests from the user and provides responses to it. Web services on the Internet, such as Google Maps, can be mashed up with the application specific services on the server side. The location database is responsible for storing access point (AP) and base station (BS) location coordinates so that the client devices can use them to estimate their current locations based on the beacons from nearby APs/BSs. Profile repository stores all the profiles of the users and caregivers. Matching logic enables the system to decide who to contact when a user asks for help. Factors like the availability of the caregivers and their distance to the user are required to design the matching logic. Privacy control is an essential function that is required to ensure that the process of interacting with others is secure. This is to protect both caregivers and people with MCI from malicious users.

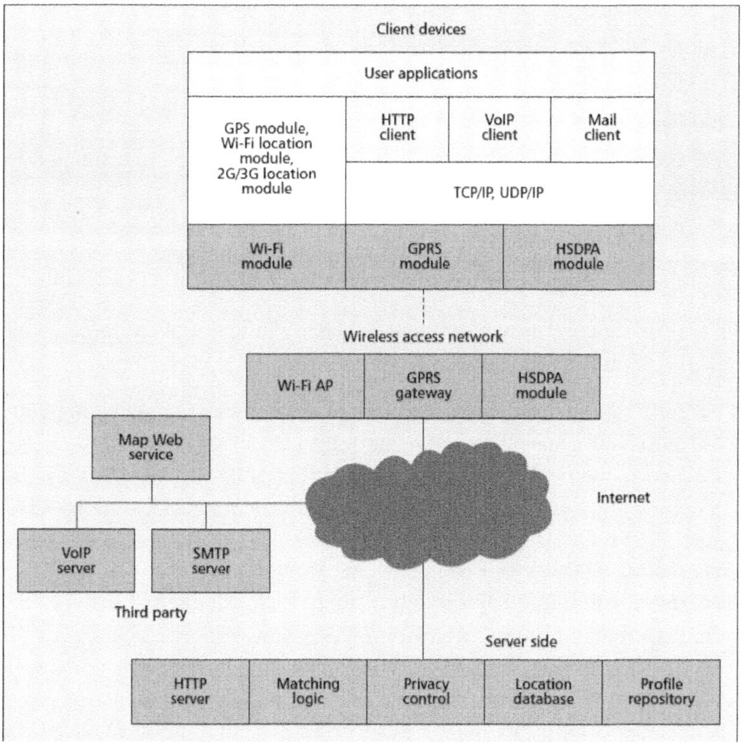

Fig. 1 System architecture

In the figure on the right, A represents the estimated location, B is the true location. AB represents the location error estimated. $C_1, C_2, \ldots, C_k$ are the nearest k caregivers that the application produces using the estimated location. We need to prove that if AB satisfies the constraint, then $C_1, C_2, \ldots, C_k$ are indeed the nearest k caregivers to B.

We use contradiction approach. Suppose there exists a caregiver C, who is closer to the person with MCI (i.e. B) than one of the caregivers $C_1, C_2, \ldots, C_k$, then we need to prove that $AB > (r'_{k+1} - r'_k)/2$.

Without loss of generality, assume that $BC < BC_i$ (C_i is the caregiver that satisfies the condition C is closer to C_i than B, $1 \leq i \leq k$).

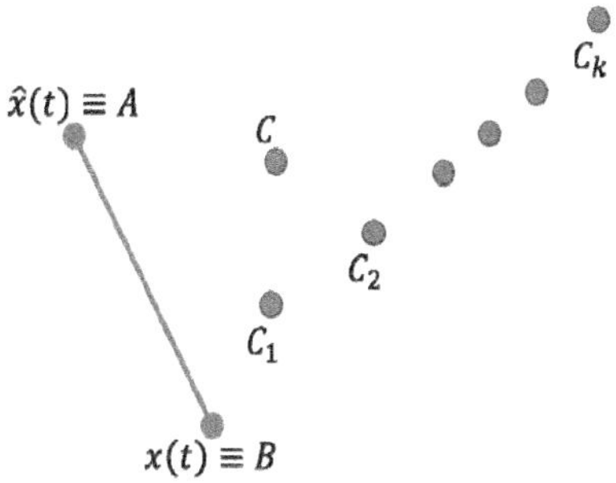

Then,
In triangle ABC: $AC - AB < BC$
In triangle ABC_i: $BC_i < AB + AC_i$
Hence, $AC - AB < AB + AC_i$. Or, $> \frac{AC-AC_i}{2}$.
Since C is not the one produced by the application using the estimated location A, $AC \geq AC_{k+1}$.
And because $1 \leq i \leq k$, $AC_k \geq AC_i$ (by definition).
Thus, $AB > \frac{AC-AC_i}{2} > \frac{AC_{k+1}-AC_k}{2}$. Or, $AB > (r'_{k+1} - r'_k)/2$.

A4. The factors that are most important or desirable to be taken into consideration when selecting sensors are:

Cost. Currently, the cost of location systems and devices for managing wandering is high for personal users. Typically, the total cost of an intelligent safety phone with integrated elopement prevention is 950USD (including installation). A UWB-RFID system with 4 sensors can cost 7000USD in the USA and 16000USD in Asia markets [98]. Devices to monitor activities such as wearable inertial sensors range from 1300USD to 3500USD per item. Though such cost is high for personal users, it is more affordable if group users share the systems and services together. For example, a dementia care centre purchases a system and charges its clients a fee for using the devices with the service. The cost per person will be reduced significantly. A previous study reported that the typical rental cost of the devices without other services is about 13USD to 65USD per month. Devices are most effective when they are used in large numbers, since they provide a comprehensive map or a rich and dense set of information about an environment or health condition that had been unobservable before. However, these applications will be widely deployed only if sensors are affordable.

Size. The size of a sensor is important in wandering management applications. In general, the smaller the size, the more acceptable they appear to be to the user. A result from a study on attitudes and expectations of technologies to manage wandering behavior in PWD revealed that nursing home residents are more concerned with the weight of tracking device placed on them than privacy issue. The study found that "a device should weigh no more than 113g if attached to an arm or leg, but it could weigh 227g if the device was worn on a belt" (Kearns, Rosenberg, West, Applegarth, 2007). Some advantages of small and lightweight sensors include ease of attachability to the body, non-intrusiveness, and convenience to users.

Power Consumption. Most sensors have to use a battery as their only source of power. It is essential for sensors to consume as little energy as possible so that the battery lasts as long as possible – ideally, for the life of the sensors. Some sensors may be able to harvest power themselves; however, power-harvesting devices carry with them other unwanted maintenance problems.

Mobility. For wandering management applications, it is highly desirable that sensors be able to communicate with other sensors or control devices while in motion, as they are often attached to a mobile user.

Processing Capability. It is often desirable for a sensor to perform some simple data processing such as filtering, removing additive noise before sending useful data to a data collector, etc. This is done to reduce data redundancy and minimize the cost (power) of data transmission.

Storage Capability. It is often desirable for a sensor to possess a small amount of storage for critical (or preprocessed) data prior to being passed on to a data collector. For example, the communication channel may be temporarily unavailable or interrupted.

Other factors may include ease of use, clinical approval, durability, and accuracy of measurement, social and fashion concerns.

A5. A tri-axis accelerometer streams the movement data using 3 channels. Therefore, the bandwidth required is:

Bandwidth = [Sampling rate] x [Number of bytes per sample] x [Number of channels] = [25, 100] x 2 x 3 = [150 B/sec, 600 B/sec].

3. Ubiquitous mHealth Approach for Biofeedback Monitoring with Falls Detection Techniques and Falls Prevention Methodologies

A1. As mobile health applications evolve, more sophisticated and advanced health applications also emerge, raising concerns about mobility, facility to access anytime and anywhere, and considering extreme conditions for their usage. Mobile technology has been registered a progressive evolution due to the growing ubiquity of those devices in contrast to desktop computing. These devices can be used in any location with many different formats. Mobile devices aims to provide users many services available anytime and anywhere in a transparent manner, focusing on the information perceived by the user instead of the technology behind responsible for its operation.

A2. Biofeedback is the process of gaining greater awareness of many physiological functions of human being using instruments that provide information about electrical activity. Many biological processes in a patient can be measured. The measurement of this process produces a physiological signal that gives an idea of what is happening with a person in a precise moment of time.

A3.For biofeedback monitoring in fall prevention and detection system five physiological sensors and one inertial sensor to extract precious information from patient's body are used. The sensors considered are the Electrocardiography (ECG), the Electromyography (EMG), the Blood Volume Pulse (BVP), the Electro Dermal Activity (EDA) or Galvanic Skin Response (GSR), the Respiration Sensor, and the Accelerometer.

A4. The most important signal is Electrocardiography (ECG). ECG monitors the heart with three sensors placed on body. Each sensor consists of an electrode which records the electrical activity in a different section of the heart. Several detection algorithms allow health professionals to find heart diseases and other interrelated heart problems.

A5. The system is responsible for receiving data from sensors in real time and sends them to a smartphone or tablet. It should serve as a support platform for medical health monitoring or to identify several occurrences, such as a stroke, a fall, and other incidents that endanger patient lives (users). The application also offers visual tools through the device screen for monitoring the related bio-signals using the data received from sensors that can be processed and observed in real time. Biofeedback monitoring with fall detection is a solution that enrich the knowledge of bio-signals before, after, and during a fall. Another objective of this system is the identification of fall characteristics collection for fall prevention strategies through biofeedback. Unfortunately, the available technology cannot detect all the falls but it was shown that not all the falls are preventable but many of them can be avoided with this system.

4. A Review of Methods to Characterize and Classify Sleep, Depression and Schizophrenia Disorders

A1. Comorbidity is termed as the occurrence of at least two different disorders in the same patient. Although, there is no consensus about the cause-effect relationship of one condition triggering another. Several studies have emphasised the comorbid nature of insomnia and psychiatric diseases. Moreover, there is rich evidence that links the chronic sleep disruption experienced by insomniac patients, could play a causal role in the emergence of depressive symptoms.

A2. Several studies have hypothesised the appearance of abnormalities between autonomous nervous system and autonomous cardiac activity during clinical charts of primary and chronic insomnia. The synchronised/desynchronised interaction of thalamic, thalamocortical and cortical neurons are associated to the generation of Slow Wave Sleep (SWS) during deep sleep and arousals episodes. Likewise, the interdependence of cardiac cycle is observed within the cells of thalamus. In insomniacs, it has been observed diminished correlation between cardiac vagal and neuronal low frequency bands (e.g. delta and theta), in comparison to healthy individuals.

A3. A significant amount of investigations have posited beta and gamma power bands as suitable mechanisms to describe abnormal integration of spatially distributed brain activities. More precisely, deficits in autonomous and voluntary actions could be produced by a communication malfunction between frontal and temporal lobes; such disturbances are observed in schizophrenic patients as an expression of central nervous system inability to inadequately interconnect neural processes.

A4. Inability the central nervous system to adequately bind neural processes segregated across distributed areas of the brain.

A5. Sleep disturbance (characterised by the loss of stage 4 sleep and difficulty staying asleep), increased resting heart rate, hyperactivity of the sympathetic nervous system, hypothalamic pituitary overactive state.

A6. Primary insomnia is used to describe a subtype of insomnia that constitutes the disorder itself and is not a consequent to any other disorder. Secondary insomnia is known as comorbid insomnia, which is usually associated with new episodes of depression.

5. A New Era in Sleep Monitoring: The Application of Mobile Technologies in Insomnia Diagnosis

A1. Commercially available sleep monitoring systems monitor sleep using a range of different modalities including the following:

i) Complete polysomnography
ii) EEG brain signals
iii) Sleep activity using wearable accelerometers
iv) Sleep activity using accelerometers or pressure sensors on the bed
v) Heart rate using a chest band or watch
vi) Respiration using a chest band
vii) Blood oxygen levels using a finger cuff
viii) Skin conductance and temperature using a watch or wearable band
ix) Sleep diary

A2

i) Non-invasive recording of physiological signals using minimal sensors that maximize patient comfort and do not disrupt sleep,
ii) Electronic diary to collect subjective measures from patients about the sleep quality,
iii) Transportation of collected data to the clinical back-end system using secure and efficient protocols,
iv) Accurate sleep analysis to classify each epoch into one of the sleep stages,
v) Detailed reporting to the clinician to assist in accurate diagnosis.

A3

i) The real-time availability of accurate sleep data
ii) Scalability to effectively enable the remote monitoring of a large number of in-home patients and to accommodate the collection of a large amount of real-time sleep recordings.
iii) Keeping the collected patient's health information security and private
iv) Ensuring the system is reliable, resilient and has good quality of service to provide an end-to-end connectivity between in-home patients and their medical centers.
v) Maintaining a high level of usability to allow communications errors to be recovered without disturbing patients or insomnia monitoring operations.

A4. The automatic sleep staging procedure consists mainly of three steps: preprocessing, features extraction and classification.

A5. A sleep onset index (SOI) which is measured from objective data (EEG, EOG, ECG) can assist in accurate insomnia detection. The index would need to include various feature extractions, such as measures to characterize the degree of EEG/EOG signal order/disorder, coherence and power of EEG frequency bands, various spectral edge frequencies and measures to detect micro and macro-structure of sleep EEG using a finer time scale.

A6. An arousal index can be used to quantify sleep fragmentation, derived as arousal number per hour of sleep. There is a need to develop an arousal index (AI) which will include the features, such as: the dynamic measure of multiple phases during the sleep onset; % of NREM N1 transitional sleep; number of transient arousals; exclusion of sleep spindles; and an intrusion of alpha wave activity into SWS (deeper) sleep.

A7

i) spontaneous arousal – an EEG arousal not associated with respiratory or limb movement event;
ii) limb movement – an EEG arousal that results from limb movement (e.g. periodic leg movements);
iii) respiratory arousal – an EEG arousal resulting from the occurrence of an apnea/hypopnea.

A8. The time in bed, sleep onset latency, total sleep time, mean motor activity (number of movement in one minute), number of awakenings longer than 5 minutes (NA>5), wake after sleep onset, and sleep efficiency

A9. Arousals increase the sympathetic nervous system and therefore increase blood pressure, decrease PTT, and change position of the dicrotic notch in PPG. It is suggested that recurrent arousals or awakenings from sleep (which reduce sleep efficiency), increase BP, and lead to sleep fragmentation associated with low apnea-hypopnea index (AHI).

6. Current Status and Future Trends of Wireless and Mobile Health Technologies in Sleep Medicine: Insomnia Case Study

A1. Based on International Classification of Sleep Disorder, Second Edition(ICSD-2) criteria, insomnia is defined as complaint of difficulty either initiating or maintaining sleep despite adequate opportunity, increased awakening, non-refreshing sleep and Low sleep efficiency for at least one month

A2. There are three different factors which have role in causing insomnia. Predisposing factors are group of factors which present before starting insomnia. Examples include personality trait, genetic predisposition, physiological hyperarousal and tendency to worry or ruminate. Precipitating factors are acute stressors which trigger starting the insomnia. The most common precipitating factors are drug/medication use, medical condition with nocturnal symptom, environmental factors, psychiatric disorders, primary sleep disorder, stressful events and changes in useful habits. Perpetuating factors contribute to the persistence of insomnia and include excessive time in bed, irregular sleep/wake schedule, daytime napping, caffeine, nicotine, alcohol, drugs, and anxiety about daytime consequences.

A3. Cognitive Behavioral therapy (CBT) is considered as the first line of treatment for chronic insomnia. CBT, with the focus on recognizing and changing pattern of thoughts and behaviors that worsen sleep helps to overcome the underlying causes of sleep problems. It also unlike medication does not cause side effects and dependency. CBT is as effective as pharmacologic therapy in the short term duration and has better long-term outcomes. CBT for insomnia contains sleep education, cognitive control and psychotherapy, sleep restriction, stimulus control therapy, sleep hygiene, relaxation training, paradoxical intention, sleep diary and phototherapy. The most effective treatment approach may combine several of these methods.

A4. Describe what information can be extracted from each graph.

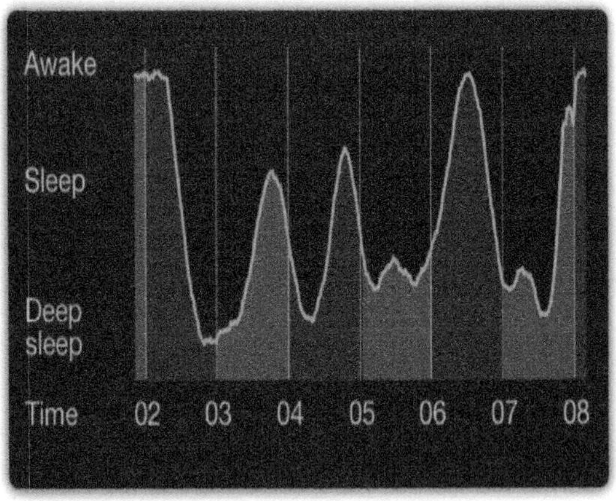

Fig. 1 Sleep stages during sleep [1]

The graph in Figure 1 is a normal sleep graph. Each cycle of sleep lasts for about 90 minutes, these cycles go from deep sleep to shallow sleep and dreaming. The peaks on the graph are about 90 minutes apart – these are the sleep cycles of this person. For every cycle the person goes into less deep sleep, this is typical for a normal sleep graph.

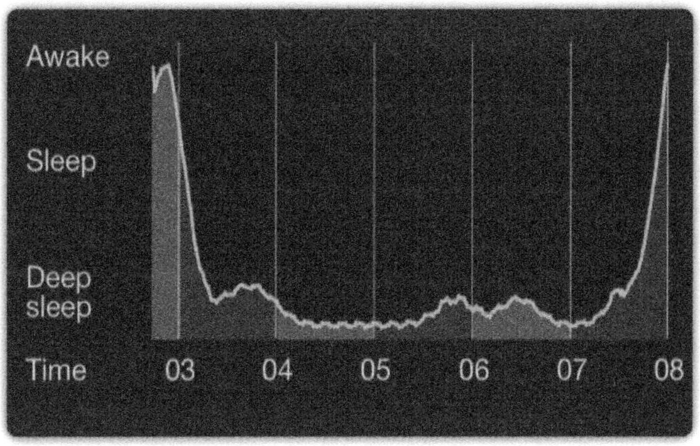

Fig. 2 Sleep stages during sleep [1]

Figure 2 shows a scenario when the phone has trouble sensing movements. For this graph, you should make sure the placement test in sleep cycle works well.

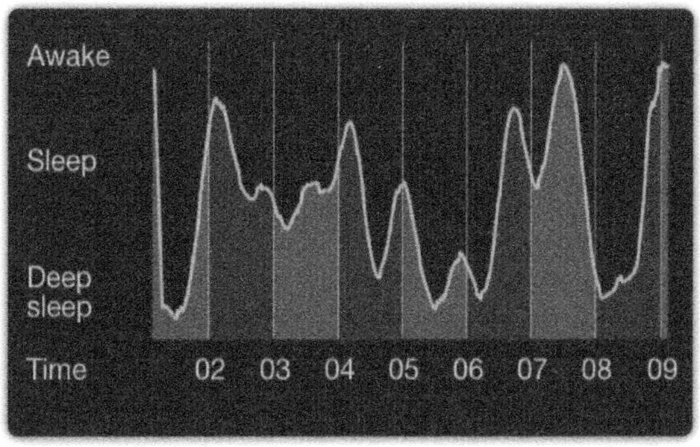

Fig. 3 Sleep stages during sleep [1]

The example shown in Figure 3 shows an irregular sleep. The user did probably not sleep as well as in the first example.

[1] http://www.sleepcycle.com/examples.html Accessed on June 21, 2014.

7. Accelerometer-Based Human Activity Recognition in Smartphones for Healthcare Services

A1. In order to reduce the cost and guarantee a good performance of a given solution, factors such as sensor sampling rate, processing time, memory size and power consumption, are particularly important.

A2. It is an orientation-independent measurement that combines the measures of three independent axes and easy activity recognition.

A3. Filtering for gravity and noise removal, as well as interpolation to reduce the effect of the non-regular sampling rate usually found in resource-constrained mobile platforms.

A4. Time-domain features are less computational expensive than frequency-based features. Furthermore, they are more appropriate to analyze short-time events.

A5. Reduce the number of features of interest into a smaller set of components with less redundancy. As a consequence the complexity of classifiers might be reduced.

A6. Mobile platforms capabilities are enhanced by computation and data offloading in Cloud computing since it has better processing and storage capabilities. On the other hand, the use of push notifications as a mechanism to receive information or feedback without explicitly pooling for updates through active transactions initiated by the remote system.

8. A Formal Approach for a Dependability Assessment of a mHealth Monitoring System

A1. A formal approach is particularly useful to assess the dependability of a mHealth monitoring system since formal reasoners generate event flows; they are useful for analysis in that the behavior of this system, and in particular of the BAN, is characterized in terms of events (e.g. a node turns on, a packet is sent, a node stops due to failure, etc.). In fact, the narrative, generated by a formal reasoner, helps to calculate the dependability metrics over time.

However, this kind of approach is not yet widely considered due to the inadequate communication between system engineers and formal methods experts and the need to re-adapt a formal specification to different design choices. Also, the definition of the general correctness specifications is not a trivial matter and therefore their conception and testing can take considerable time.

A2. The narrative will be composed of the sequence of events and fluents reported in Listing 6.

Listing 6. Listing for the narrative generated by the tool (question n.2)

```
1
2
Happens(Stop(2), 2).
3
-IsAlive(2).
Happens(Isolate(5), 3).
4
-IsReachable(5).
Happens(Stop(3),4).
5
-IsAlive(3).
Happens(Isolate(6),5).
6
-IsReachable(6).
Happens(Isolate(8), 6).
Happens(Isolate(9), 6).
7
-IsReachable(8).
-IsReachable(9).
8
9
10
P
```

A3. The approach concentrates on the following three main critical events: an isolation event, a packet loss event and battery exhaustion.

A4. In the chapter the following dependability metrics are defined : *Coverage*, being the time interval in which the BAN can operate, preserving a given number of nodes connected to the sink, *Connection Resiliency*, being the number of node failures and disconnection events that can be sustained preserving a given number of nodes connected to the sink, *Data Delivery Resiliency*, being the number of node failures and disconnection events that can be sustained preserving a given number of correct packets delivered to the sink and *Power Consumption*, being the battery power consumed by each sensor, useful to estimate the expected lifetime of the BAN.

A5. A typical mHealth monitoring system is usually composed of a number of sensors (the BAN), a gateway device and a medical station; typical communication means are bluetooth/zigbee (for Intra-BAN communication), and WiFi and cellular (for Extra-BAN communication).

A6. By means of an FMEA study.

A7. In the general correctness specification, all the events and fluents are identified and defined using event calculus formalism; they are also valid for every BAN. The general correctness specifications are complemented by structural correctness specifications. In particular, the topology of the BAN and the other parameters regarding the target BAN are specified. Indeed, the structural correctness specifications vary depending on the basis of the characteristics of the target BAN.

9. Wireless Monitoring System for Wheelchair Users with Severe Mobility Impairment

A1. No, this system does not connect any sensor or cable to the patient. EMFi sensors for ballistocardiogram measurements can produce Heart rate and Respiration rate data after processing. Other signals are obtained from sensors connected directly to the wheelchair.

A2. The system uses 802.11b WiFi to communicate with the server in the nursing home. The system can be upgraded with an EDGE or GPRS mobile system to communicate with central server from anywhere with mobile data coverage. This way, people who live independently could also benefit from enhanced physiological monitoring.

A3. EMFi sensors have a bigger area than FSR sensors. This characteristic of EMFI sensors allows the measurement of small changes on pressure generated by the natural movement of the body. In fact, EMFi sensors are more sensitive than FSR to detect Heart Rate and Respiratory Rate. FSR are used for detecting changes in position of the patient and in estimating patient activity.

A4. Using a Raspberry PI central unit to calculate and communicate with the central server, the power consumption of the system is very low and does not affect the autonomy of the wheelchair.

A5. Some authors use accelerometers to measure the natural movement of the body. To this end, they put the accelerometers directly on the body. In this project, no devices can be in contact with the patients. The accelerometer is attached to the wheelchair structure to measure the tilt angles and whether the wheelchair is moving.

A6. The monitoring system generates alarms to help patients locally. At the same time, the system sends information to a central server to store information and to alert the caregiver that something is wrong with the patient.

A7. The actual version of the system does not have information about the patient position. However, a GPS or an indoor positioning system can be included to the system for outdoor or indoor position reporting respectively.

10. Remote Health/Vital Sign Monitoring

A1. Due to Nyquist, the channel capacity C depends on the channel bandwidth B and the number of signal levels N that is represented by word with log_2N bits. Therefore, with multilevel signaling the Nyquist formulation becomes:

$$C = 2B * log_2 N$$

If each signal element (level) is encoded by one $n=4$-bit word, the $N=2^n=2^4=16$ levels of signal are used. Therefore,

for BTv1.2: $B = \dfrac{C}{2log_2 N} = \dfrac{1.2*10^6}{2*log_2 16} = 0.15(\,MHz\,)$ and

for BTv3.0+HS: $B = \dfrac{C}{2log_2 N} = \dfrac{24*10^6}{2*log_2 16} = 3.0(\,MHz\,)$

A2. The Shannon formula defines the maximum channel capacity in bits per second according to the following expression:

$$C = 2Blog_2(1+SNR)$$

In this particular case of ECG acquisition system:
$SNR=3dB$ and $B=500Hz$, therefore:

$$C = 2*500*log_2(1+3) = 1000*2 = 2Kbps$$

A3. The time between two peaks (period) of ECG wave necessary for heart rate computing in the worst case of tachycardia with 150 heartbeats per minute is:

$$t_{period} = \frac{1\ min}{150\ peaks} = 0.4s$$

For each period 360 samples per second are used. Thus, the required time for processing each sample is

$$t_{sample} = \frac{0.4s}{360\ samples} = 0.0011s = 1.1ms$$

Before processing ECG sample, it must be received by cell phone (the transmission time is $t_{transm}=0.3ms$) and storage to DB requires $t_{store}=0.3ms$. Therefore, the required sample processing time t_{proces} may be computed as:

$$t_{proces} = t_{sample} - (\,t_{transm} + t_{store}\,) = 1.1ms - 0.3ms - 0.3ms = 0.5ms$$

A4. A 4 On the horizontal axis: 5 peaks of bradycardia with 50 heartbeats per minute require

$$t_{of\,5\,peaks} = 5*\frac{60\ s}{50\ peaks} = 6s$$

$t_{of\,5\,peaks} = 6s$. This value corresponds to 360 pixels. It means that $1s$ of ECG wave is displayed on

$$Number_{of\ \ pixels\ \ per1s} = \frac{1\,s*360pixels}{6s} = 60\,pixels$$

If *1mm* of ECG grid corresponds to *0.04* seconds therefore, in pixels *1mm* is equal to

$$Number_{of\ \ pixels\ \ for1mm} = \frac{360pixels}{1s} * 0.04s = 12\,pixels$$

On the vertical axis: *1mm* corresponds to *0.1mV*. Thus, for presentation of ECG signal with upper limit of amplitude of *3mV* the

$$Number_{of\ \ mm\ \ per\ \ 3mV} = \frac{1mm * 3mV}{0.1mV} = 30mm\,.$$

3mV is the maximum amplitude of signal and it corresponds to *30mm* of ECG grid. On display we have available *200* pixels on vertical axis. Therefore, *1mm* of grid may be displayed using

$$Number_{of\ \ pixels\ \ for1mm} = \frac{1mm * 210pixels}{30mm} = 7\,pixels$$

A5. According to Eq. 2 presented in this chapter, the relative error $\delta(\%)$ of computing heart rate is equal to

$$\delta(\%) = \frac{\left|N_{max/\,min} - N_{real}\right|}{N_{real}} * 100(\%)\,,$$

where $\left|N_{max/\,min} - N_{real}\right|$ is the absolute error of measuring real value N_{real} of heart rate with respect to the maximum or minimum values $N_{max/\,min}$ computed and displayed on cell phone.

In Fig 13b the heart rate is equal *83* bpm on *200* sps, and it is equal to *71* bps on *360* sps. Therefore, in this particular case the increment of precision is

$$\delta(\%) = \frac{|83-71|}{71} * 100(\%) = 16.9\%$$

11. Operation, Analysis and Optimization of Wireless Sensor Devices in Health Oriented Monitoring Systems

A1. Deployment can be done by using Public Land Mobile Networks (2G GSM/IS-95, 3G UMTS and 4G-LTE) Wireless Local Area Networks or Wireless Personal Area Networks (mainly Bluetooth and ZigBee). A possible combination can be Bluetooth to connect acquisition sensors and a PLMN mobile terminal as a gateway.

A2. We can employ Personal Health Sensors and Context Sensors. In the case of Personal Sensors, we have ECG monitor, ECG recorder, Impedance cardiograph, Weigh scale, Blood pressure monitor, Pulse oximeter, Blood glucose meter and Peak flow meter. In the case of Context Sensors, we have Temperature monitor, Humidity monitor, Motion detector, Water consumption meter, Open/close door detector, Gas detector, Flood detector and On/off electrical appliance.

A3. The surrounding environment determines wireless channel losses, due to absorption, diffraction and scattering. Therefore, the topology and morphology play a key role in overall wireless system performance.

A4. Due to the limitations in size, power consumption and frequency allocation, adequate analysis of the impact of the environment on radio channel performance must be considered. This will lead to the adequate choice of transmit power, frequency band and antenna selection. Also, the required channel bandwidth must be determined, as it impacts on receiver sensitivity.

A5. The use of wireless systems allows mobility, real time monitoring and user interaction. The use of widespread elements such as Bluetooth devices and mobile terminals adds an additional degree of flexibility, due to its popularity and potential ease in adoption.

12. mHealth Sensors and Applications for Personal Aid

A1. Analysts forecast over 7 billion mobile subscribers and more than 1 billion smartphones in 2014. Smartphones are now part of peoples' lives and so are smartphone applications. Currently, there are more than 500 mHealth projects worldwide and approximately 40,000 medical apps available for smartphones and tablets.

Most health applications focus on prevention, and they are targeted for individuals interested in solution to manage their health, and for health professionals, to train, monitor remote patients, and manage health records.

By 2015, nearly 500 million users around the world will been using health applications enabling people to reduce the number of doctor visits, reduce hospital stays length and improve patient prescription tracking.

A2. The mobile application using data from a GPS system and a physiological data give a geolocalization of physiological data. This kind of solution allows us to obtain a variety of data about the individual person in a particular site in real time. Some changes such as the cardiac frequency and electrodermal activity can give important information about the emotional state (particularly stress) of the person and using GPS to locate where these changes occurred. With the accelerometer is possible to monitor what kind of activity the person held at the time of these changes, for example, if stopped, dropped or has accelerated the pace.

A3. For instance some equipment can be controlled remotely using other devices, saving time for its users; it can help keep track of patients, medical staff and equipment within a controlled environment; patients can monitor several physiological signals to help prevent further clinical complications; it can help medical professionals keep up with hygiene standards; and sensors can be used to monitor if a patient leaves the position he should keep.

A4. Using a simple smartphone, the jump is perceived by the user as a recreational activity, and as the same time the user is enjoying himself, is also practicing some physical activity, and is more motivated to see that he is also burning calories.

A5. As statistics refer (Index Mundi, 2013) nowadays almost every person in the world owns a cell phone and Portugal is not an exception (Index Mundi, 2013) with 12.33 million registered cell phone subscriptions in 2010 against 10.28 million persons living in Portugal (INE, 2013), which represents a rate of 120% of population that owns a cell phone, therefore representing more than 1 cell phone per person.

Following this statistic it can be concluded that the majority of the Portuguese and World population, carry all day long at least one cell phone with them, and for sure that people can easily handle a cell phone and send SMS messages, and that is the main reason why the SMS technology may be a strong ally in this type of treatments. Another reason for choosing SMS technology lies in the fact that not every person that carries a cell phone carries in fact a smartphone, and so there are no exclusions from this type of programs.

Index Mundi: Mobile cellular subscriptions, Index Mundi, http://www.indexmundi.com/facts/indicators/IT.CEL.SETS (last access 13 Sep 2013)
Instituto Nacional de Estatística: População residente em Portugal, Instituto Nacional de Estatística. http://www.ine.pt/xportal/xmain?xpid=INE&xpgid=ine_indicadores&indOcorrCod=0005888&contexto=pi&selTab=tab0 (last access 13 Sep 2013)

14. mHealth Monitoring System for Hospitalised Older Adults – Current Issues and Challenges

A1. Developments in wireless communication technologies and trend towards using hand-held mobile devices are forcing a re-evaluation of existing technology infrastructures within healthcare.

A2. Clinician's incorporation, patient's (user) centric design and lack of real-time clinical testing are the three main barriers identified that prevent worldwide adoption of mHealth applications.

A3. Major challenges identified are reliability, efficiency, mobile phone platform variability, cost effectiveness, energy usage, user interface, quality of medical data and security and privacy.

A4. Vital sign(s) monitoring applications are one of the fastest emerging areas of mHealth application in the clinical environment.

A5. Incorporation of clinician's feedback and user's involvement from the early stages of mHealth application development is considered as a key success for such applications.

15. Ubiquitous Health Monitoring: Integration of Wearable Sensors, Novel Sensing Techniques, and Body Sensor Networks

A1. Any three of the following: ECG, PPG, PCG, images of the face

A2. Body temperature, respiration activity, body movement

A3. First, research work on sensor design focuses on the wearability of sensors, energy consumption and sensor deployment. Second, research work on data fusion focuses on the techniques of denoising, feature extraction, data classification, data compression, and other technologies relating to processing of sensor signals. Third, research work on network communications focuses on the problems of channel characteristics, network topology, media access protocols, and routing algorithms.

A4. The first one is about monitoring and eavesdropping on patient vital signs. Second, there exist some threats to data integrity when WBSN nodes are sending information. Third, there are various threats relating to unguarded rely nodes in wireless networks.

16. Interoperability and mHealth – Precondition for Successful eCare

A1. A transformation where the main focus is shifting from acute care to managing chronic diseases.

A2. Regulatory reform driven by demographics changes, industrialization of the healthcare sector, and personalized medicine.

A3. Interoperability.

A4. Frameworks, theories and models.

A5. Enterprise, Organizational, Semantic, and Technical Interoperability.

A6. Data and representation schemas, XML standards, common ontologies, metadata sets, unified code lists, core components, data elements, and glossaries. Examples: SNOMED, HL7 CDA R2, ICD-10, LOINC …

A7. They come above the vocabularies and code sets and below the content structures. We express clinical content which we map to vocabularies and code sets, and express in different forms – e.g. HL7 CDA XML form.

A8. It is care process model that is common to all interventions that are supported on the eCare platform. It represents a reusable, extendable, and scalable component.

A9. Architecture homogeneity and shared ontology systems.

A10. It is the only ontology based approach of knowledge representation in the field of healthcare.

17. Recent Advances in mHealth: An Update to Personal Health Device Interoperability Based on ISO/IEEE11073

A1. The X73PHD consists of an Optimized Exchange Protocol (11073-20601) and a set of device specializations (11073-104xx). Currently the X73PHD counts with more than 11 specializations that have been successfully published by the Personal Health Device Working Group (PHD WG).

A2. The use cases represent the evolution of existing specializations. They are proposals that include new features that increase the functional characteristics and give an evolutive characteristic to the existing specializations.

A3. X73PHD identifies the information through a device, and models the information as a set of objects with certain attributes. Changes in the values of those attributes are transmitted to other peer using announcement services. In addition, the other peer can execute a remote procedure call to initiate a predefined operation.

A4. The standard has been developed to work in personal health environments, such as personal area networks, and currently is introducing its application to bodily health environments (Body Area Networks) for mobile Health applications.

A5. Yes, the PHDs can also being used for wellness, sports and independent living. We can find other applications in different research areas as data collection devices for further analysis.

A6. The updated Standard X73-PHD remains constant from PH-DWG work, which is responsible for keeping the standard and its specializations up-to-date. Their developer members will also propose new updates and standardizations to the current version.

A7. Command and control are specific mechanisms that allow us to change and modify the operational parameters of an agent from the manager. At this moment, the author is working to include few more characteristics from previous proposals to create a better model that enables the service of remote control in the Optimized Exchange Protocol part 20601.

A8. The main objective of this work is firstly to show the high capability of a set of standards called X73PHD which allow the interoperability of PHDs in multiple scenarios regardless of the transport technology used, and, secondly, if is implemented within mHealth scenarios, to increase the quality of the service providing significant benefits to their users and manufacturers.

18. Design Guidelines for Wireless Sensor Network Architectures in mHealth Mobile Patient Monitoring Scenarios

A1. The main challenges are: lower power consumption (for autonomous systems), miniaturization of the sensors, information security and lower cost. Moreover, there are some non-technical factors for the success of WBANs such as affordability, legality, regulatory and ethical issues, comfort, and user friendliness. Also, in order to become a long-term technology WBANs need the widespread acceptance of key collaborators in the medical community.

A2. Homogeneous WSN architectures are best suited for low patient mobility and low data rate applications or when each patient carries his/her own information sink. In this sense, by adequately implementing the homogeneous architecture, energy consumptions savings can be achieved since there is no need for a bridge device to retransmit the information through another wireless technology. On the other hand, homogeneous WSN architectures can be seriously limited when there is a high density of users connected to the patient information sink.

Heterogeneous WSN architectures can be used when the density of patients to be served within the same coverage area is a critical design parameter. Interference between patients' sensors can be mitigated by decreasing the transmission power of the sensors to WBAN limits. Compared to homogeneous architectures, by selecting a bridge-to-sink communication technology that provides a higher transmission data rate than the one present in the WBAN technology, a significant increment on the number of supported users can be achieved. Nevertheless, the energy consumption in this architecture may be higher compared to a homogeneous architecture since there is a bridge device retransmitting the information from one network to another.

A3. For fixed sensor data rates, when using technologies based on the CSMA-CA MAC mechanism, the number of samples transmitted per packet can define boundaries for network performance in terms of packet loss rates and number of supported users.

A4. The most significant application requirements are: patient mobility range, transmission requirements of bio-physiological variables to monitor, reliability of information, network scalability (i.e., number of concurrent users being served), required security, and the distance between the patient(s) and the information sink.

A5. It is believed that the future of wireless sensor communication for patient monitoring will be dominated by open-standard technologies, since these technologies provide a better platform for interoperability between different implementations and manufacturers.

19. Evaluation of Health Care System Model Based on Collaborative Algorithms

A1. COHESY is deployed over three basic usage layers. The first layer consists of the bionetwork (implemented from various body sensors) and a mobile application that reads sensor data and allows the users to specify the activities they perform. The second layer is presented by the social network implemented as a web portal which enables different collaboration within the end user community. The third layer enables interoperability with the primary/secondary health care information systems which can be implemented in the clinical centers and different policy maker institutions.

A2. We separate the health parameters into two groups: descriptive and control parameters. Descriptive parameters express the factors that are important for the people's health (for example age or place of living) and that cannot be affected. Control parameters express the characteristics of the people's health (for example blood pressure or blood sugar level) that can be affected. The algorithm tries to change the only the control parameters towards desirable state, but both types of parameters are used in the phase of the algorithm where we select the users most similar to the active user.

A3. The recommendation algorithm in COHESY gives recommendations about physical activities that should be performed by the patients in order to improve their health condition.

A4. The recommendation algorithm uses the history of performed activities and the history of measurements of all users in order to generate recommendations.

A5. The combination of the user's values of her/his health parameters represents his health profile.

A6. The algorithm consists of four phases: categorization of the users according to their diagnosis, selection of the users most similar to the active user, calculating the usefulness of the activities and generation of recommendations.

A7. First two phases are not necessary, but they contribute to improving the quality of the recommendations by selecting the most relevant data.

A8. We chose the most recent measurement made before the activity and we choose the measurement made after the activity with has the biggest validity (relevance) according to a validity function. We cannot assume that immediately after the completion of some activity there will be a change on the parameter value because some time should pass until the effect of the activity could be visible. On the other side, we cannot assume that the results of a single activity will be visible after a long time period (few months). That is why this validity function gives smaller relevance to measurements made immediately after the activity and to measurements made very long time after the activity.

20. Software Architecture for Emergency Remote Pre-hospital Assistance Systems

We provide answer guidelines for some of the questions stated.

A1. Golden hour refers to the qualified medical attention that must be giving with in the first hour of an accident event. The metric is an estimated value an actually refers more to a concept than to a scientific metric.

A2. Restricted bandwidth affects the communication links and lost of data becomes a problem. One alternative is to limit the number of active sensors, restrict the use of video (considering that a parmedic is present) and use alternative (traditional) communication via such as radio civil band.

A3. Open question, one strategy could be use just one type of smart phone-operating system and develop hardware specific solution, then interoperability is scarified.

A4. Interoperability, modifiability, usability, performance, and availability.

A6. Consider measures for interoperability, modifiability, usability, performance, and availability. As an example: number of platforms supported for interoperability of number of modules or software layers for modifiability.

A8. Body sensor network and communication network paramedic-medical center.

A9. Timely and quality information.

22. A Product Line Architecture for Mobile Patient Monitoring System

A1. **Body Area Network** - The BAN is defined as a network of communicating devices worn on or around the body which is used to acquire health related data to provide mobile health services to the patient. The BAN consists of a Mobile base Unit (MBU) and a set of BAN devices such as sensors, actuators or other wearable devices used for medical purpose. The sensors may directly transmit the Biosignals data to the MBU or do it via the Sensor Front-End (SFE).

Mobile Base Unit – Mobile base unit (MBU) acts as an intermediately unit between BAN and BESys. The MBU mainly gathers the Biosignals acquired by the BAN and transmits it to the BESys via a wireless network. MBU also provides initial Biosignal interpretation and data processing facilities.

Back end Systems - The Back end Systems (BESys) is the back bone of the mobile patient monitoring systems. All key functions like data warehousing, data mining, decision support, threshold generation, Biosignal interpretation are performed by BESys. It mainly comprises of two components: Back-end server to which the MBU transmits Biosignal data and Clinical back-end server which may host custom healthcare applications.

A2. The generic architecture of MPMs is broadly classified into two components named: Body Area Network (BAN) and a Back-End System (BESys) as shown in Figure 1. The BAN shown in the figure is mainly consist of communicating devices used to acquire health related data such as blood pressure, oxygen saturation level, blood glucose, respiration, Electrocardiogram etc. to provide mobile health services to the patient. The BAN consists of a Mobile base Unit (MBU) and a set of BAN devices such as sensors, actuators or other wearable devices used to acquire these biosignals and deliver them to the BESys. The sensors may directly transmit the biosignals data to MBU or do it via the SFE. The BESys further processes the Biosignals acquired by the BAN for all type of abnormalities and informs the caregiver or doctor if any irregularity is discovered in the Biosignals. The BESys typically comprises of a back-end server to which the MBU transmits biosignals data and clinical back-end server which hosts the custom health-care applications for Biosignal processing.

From process perspective, we can represent the mHealth architecture into five steps as: (i) Data acquisition (ii) Data Transmission (iii) Bio-signal Interpretation and Processing (iv) Notification (v) Intervention. BAN acquires the physiological data actively or passively by interacting with the patient. Data can comprise of vital signs, blood glucose levels, responses to specific health conditions and general health questions as well as patient location. Data is then packaged, transferred and delivered to the caregivers and third parties via different modes of communication. Algorithms and/or healthcare workers interpret and review the data and determine if there are any areas for concern. If needed clinicians, family, caregivers, third parties or the patients themselves are notified of a potential problem through either the device itself or an intermediary health-care worker. They can then take an action when intervention is needed.

In software engineering Product is something we get after going through a process and that product is used as part for carrying out the Process. In the above perspective we have defined process as a series of steps and products is the mean to accomplish these steps. This separation between product and process perspective will help the system modelers to individually focus on structural and functional elements of MPMS facilitating systematic development of MPMS.

A3. The FODA is a method of requirements analysis which defines both the products and the processes of domain analysis. In FODA, a feature is a prominent or distinctive user-visible aspect, a quality, or a characteristic which can be identified as mandatory, optional, or alternative property for a system. In FODA, domain analysis is defined as a process of collecting, identifying, representing and organizing the relevant information in a domain based on the study of existing systems and their expansion histories, knowledge taken from domain experts, emerging technology, and underlying theory within an application domain. FODA mainly involves 2 steps know as Context Analysis and Domain Modeling.

Context Analysis: The purpose of context analysis is to define the boundaries of a domain by using the domain data and constraints. This phase establishes a proper scope inside which the domain analysis is performed. In context analysis, relationships of the candidate domain with peripheral elements are analyzed, and variabilities are evaluated. The output of context analysis is described through structure and context diagrams.

Structure Diagram: The diagram covers complete application domain and it consists of subsystems necessary for realizing a full functional mobile patient monitoring system.

Context Diagram: The Context Diagram in FODA is a data-flow diagram showing data flows between a generalized application within the domain and the other entities and abstractions from system environment with which it communicates.

Domain Modeling: It provides a better understanding of the systems by systematically describing functions, objects, data, and relationships. The domain modeling phase mainly consists of three phases, namely, feature modeling, entity relationship modeling and operational modeling. The output produced by each phase describes what a system is, what the functions to be performed by an application are and how they perform it.

Feature analysis: The feature model captures the capabilities of a system from end users point of view. The analysis is mainly focused on the functionalities of the system, performance parameters, and operating environment in which the system application run.

Entity Relationship Model: The ER model captures and defines the domain knowledge essential for implementing applications in the system domain.

Operational model: The operational model provides an immediate understanding of the system to the user by describing that how the system works. The model captures functional and behavioral relationships from feature model and entity relationship model. The operational model mainly provides the data and control flow of the system.

A4. Various pros and cons of FODA are given below:

- To reduce the development cost and improves the quality due to reuse of well tested components in various systems
- To lower hardware costs.
- To achieve faster integration and development due to focused application specific development.
- FODA does not provide any systematic method to use the assets Developed during analysis.

A5. fodaA is a formal language that provides translation of any FODA Diagram into a quite formal natural language using the classical Process Algebras notations. The following Representations have been used to convert a FODA diagram into a fodaA grammar:

Table 1 FODA Diagram to fodaA grammar

	FODA Diagram	fodaA term
Feature		*F*; ✓
Mandatory	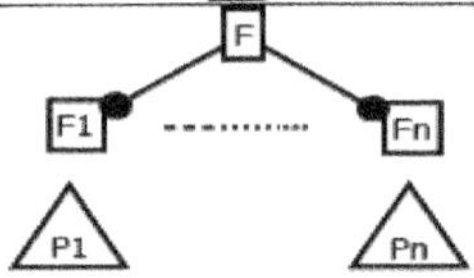	F; (F1; ✓ ∧ $\triangle P$ .... ∧ Fn; ✓ $\triangle Pn$)
Optional	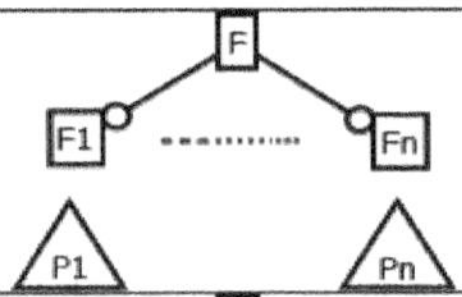	F; ($\overline{F1}$; ✓ $\triangle P$ ∧ ...... ∧ $\overline{Fn}$; ✓ $\triangle Pn$)
Choose 1	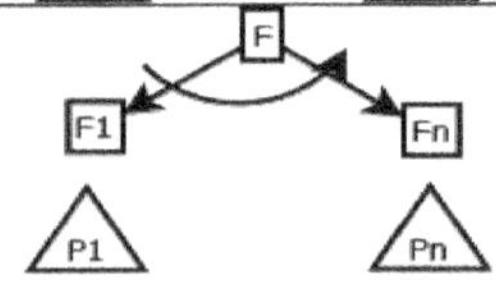	F; (F1; ✓ □ $\triangle P$ □ Fn; ✓ $\triangle Pn$)
Requires	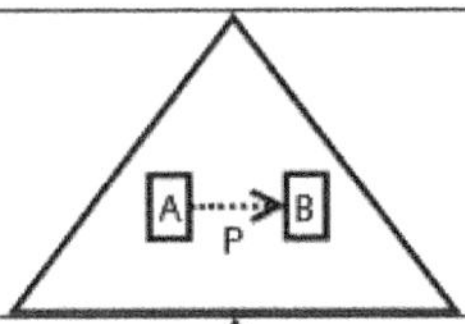	A ⇒ B in $\triangle P$
Excludes	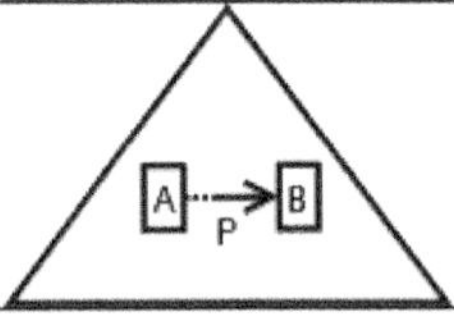	A ⇏ B in $\triangle P$

A6. A Software Product Line (SPL) is a set of software-intensive systems that shares a common, managed set of features satisfying the specific needs of a particular market segment or mission and that are developed from a common set of core assets in a prescribed way [58]. These SPL systems are developed in such a prescribed way to adopt the variabilities according to the customer needs.

Core assets are those reusable artifacts and resources that form the basis for the software product line. Core assets often include, but are not limited to, the architecture, reusable software components, domain models, requirements statements, specifications and process descriptions [58]. The architecture is a key among the collection of core assets.

A software product line offers the following benefits:

- Increased quality
- Decreased cost
- Improved productivity
- Decreased labor
- Decreased time to market

23. The Design of Integrated Circuit for Biomedical and mHealth Application

A1. This chapter focuses on introducing some recent IC design technologies which are especially suitable for mHealth applications.

A2. Portable, wearable and networked medical devices for the diagnosis, monitoring, and treatment of diseases are vital and desired for future mHealth applications. For the pursuit of convenient, low power and low cost, the system of mHealth instrumentations should be developed as small as possible and the power consumption should be the lower the better, so integrated circuits (ICs) designed for biomedical applications are necessary.

A3. These technologies are suitable for biomedical and mHealth applications include the design of analog circuits such as amplifier and filter, digital circuits , and radio-frequency (RF) circuits which are widely used in physiological signal acquisition, signal processing in signal transmission respectively.

A4. The key challenge in designing biomedical ICs is the pursuit of low frequency, low noise, and low power design.

A5. Because: 1) biomedical signals are usually low frequency, week and noisy, and 2) the wearable and portable application oriented.

24. A Secured Hybrid Cloud Architecture for mHealth Care

A1. Privacy & Confidentiality: means that cloud providers do not learn about client's data and integrity means that users can detect any modification happened to their data.

Integrity: means that data and processes must be modified only in authorized ways and by authorized users.

Authentication & Authorization: enables a party to ensure that he/she is communicating with the correct party on the other side. Without authentication, unauthorized access to resources and confidential information could be gained by an unauthorized party.

A2. The feel of safety and comfort as the patient knows that his medical records are kept confidential, his privacy is maintained, it is not possible for his medical records to be viewed or modified except by authorized people and his records are reviewed and audited only by authorized bodies. Doctors feel confident that they are dealing with the right patient with the correct medical file. They feel relieved when they know that their patients are monitored 24 hours per day and their patients can reach them at any time especially in case of emergency.

A3

1. Mobile devices can easily be misplaced, damaged, or stolen.
2. Authenticating and encrypting new mobile devices that are used to transmit confidential information could be a long process and time consuming. Help can be provided to the user as required.
3. Data can be hacked, manipulated, or destroyed by internal or external users, so security measures and ongoing educational programs must include all users.
4. Some security measures that protect data integrity include firewalls, antivirus software, and intrusion detection software must be continuously upgraded.
5. A full security program must be in place to maintain the integrity of the data, and a system of audit trails must be operational.
6. A security officer must be designated to regularly evaluate security measures and identify the security weaknesses and threats; assign a risk or likelihood of security concerns in the organization and address them.
7. Physicians mistakenly enter data on the wrong patient when working with an electronic list of patients and physicians have to choose a patient from an extensive listing.

A4. mHealth speed up and improve medical services. With mHealth patients and their families can get health care service anywhere, anytime or anyhow. Doctors can keep monitoring their patients who in critical situations and give them medical advices. mHealth devices can help elderly people by tracking their whereabouts and activities and allow them to call for help in case of emergency situations. mHealth devices are smart enough to alert the patients against critical situations such as environments that may trigger allergy attack and can give real-time advice to them about treatments via mobile health applications.

A5

a) Advantages:

1. Scalability: A scalable system is one that can serve a large number of users and scale up operations while maintaining an acceptable response time, with constant or slowly degrading performance with little overheads. mHealth systems based on hybrid cloud platforms achieve scalability through using the abundant resources of the public cloud platform. This means that hybrid cloud mHealth system can handle many patients simultaneously without any failure noticed and without any crashing due to resources exhausting. Given the pay-as-you-go situation the computational cost will grow slowly as the number of patients increase.
2. Flexibility: Doctors using mHealth systems based on cloud platform will be flexible to access their patients' medical records and providing them with medical advices and recommendations from any place they are and at any time. Hospitals and healthcare givers will be more flexible using IT resources available in the cloud, they can double their patients in half an hour. They can scale up and down depending on the number of patients using their mHealth system.
3. Monitoring: One of the hot issues in cloud computing is monitoring the use of the data by the cloud providers. In our proposed hybrid cloud mHealth architecture a third party will act as an auditor. Audit means recording all activities on a patient's medical record in chronological order.
4. Robustness and availability: By using the technology of virtualization, resiliency, redundancy, data restoration and disaster recovery the cloud environment are highly tolerant against many failure scenarios which make the proposed cloud mHealth system architecture robust against service disruptions due to power outages, denial of service attack, hardware failures, and system upgrades. This guarantees the availability and utility of mHealth services at all times.
5. Security through cryptography: Our proposed design uses both symmetric and asymmetric cryptography to secure, protect patients' data and to control access to vital information. Each doctor must sign the patient's ID in order to verify that the retrieved medical file belongs to the required patient. Each patient has a symmetric key which is also called a secret key for the encryption and decryption of his medical record. The hospital or emergency center itself has a public key to double encrypt patients' medical records.
   Authentication and authorization: Mobile devices authenticate themselves to the crypto-coprocessor before accessing the cloud storage. Also, patients and doctors who want to access the cloud storage should authenticate themselves to the cloud provider using the role based access control ($RBAC$). The access of medical information is limited by the access right $uRight$ recorded in the

license ACL. These rights are enforced using the$RBAC$. Also, other authentication functions can be integrated in our design to solve the private and public cloud authentication.

6. Ownership of information: In the mHealth system, medical information is created by the hospital and managed by the patients. Medical records are encrypted and decrypted using patients' secret keys. The corresponding access license is encrypted with the patient's public key. The encrypted medical file can be only decrypted by the secret key pSk embedded in the license $eACLci$. To access the patient's medical record the patient should provide the clinician or physician with the secret key needed to decrypt the medical file and the private key to decrypt the license. Notify the data owner, the patient and the hospital, before accessing the patient's data. Based on such concept, the patient and the hospital private cloud are both involved in each electronic medical record access in our design. Even in the cross hospital situation, the similar access process is implemented to avoid any unauthorized access or misuse of patient's medical information.
7. Integrity and confidentiality of data: Confidentiality can be achieved by access control and encryption techniques in mHealth systems. Integrity is achieved by encrypting all the medical file using the patient's secret key sPK which is recorded in the license ACL which is double encrypted by the patient's public key and then by the hospital's public key. It can only be decrypted by the hospital private key and then by the patient private key. Hospital digital signature is used to verify that medical information was not changed by an attacker while it was being stored in the cloud storage.
8. Non-repudiation: Non-repudiation is achieved by creating a signature $ScID = E_{h_{public}}(ID, time)$ with each query of the patient's medical record. The signature contains the patient's ID and the time of request which is encrypted by the hospital's public key. This signature can be used in the auditing process.
9. Auditing: is particularly useful to identify suspicious accesses and common access.
10. Availability: The proposed mHealth system is highly available all year 24 x 7 because it uses the cloud technology of virtualization, resiliency, redundancy, data restoration and disaster recovery. This means that no service disruptions due to hardware failure, power outage or system upgrades or denial of service attacks can happen.

b) Privacy and Confidentiality

Privacy and confidentiality are improved by:

1. Encrypting mobile devices that are used to transmit confidential information.

2. Using crypto-coprocessors to authenticate mobile devices that want to communicate with cloud providers.
3. Using the RBAC to control users' access which limits who can see what. The practice administrator identifies the users, determines what level of information is needed. For example, the physician, nurse and receptionist in a hospital, do not access the same information because they have different responsibilities and tasks.
4. Using IBE to generate patient's secret key, public and private keys which helps avoid keys collision.
5. Using symmetric and asymmetric cryptography improves data privacy and confidentiality.
6. Using the secured cloud storage to securely access the medical information in the cloud storage.

27. Mobile Health Improves Healthcare Delivery

A1. Electronic health refers the healthcare practice supported by electronic processes and communication.

A2. Mobile health refers to electronic health supported by mobile technologies such as smart phones and tablets.

A3. To improve healthcare delivery in terms of availability of health information, containment of health costs, and reaching populations in underserved regions of the world.

A4. An electronic health record is health information relating to a patient which is generated, managed, and used by authorized healthcare providers across several medical facilities. It is a patient-centric longitudinal electronic record containing health information from multiple electronic medical record systems.

A5
1. Healthcare professionals' resistance to change
2. Perceived security and privacy risks

A6
1. Perceived security and privacy risks
2. Bluetooth attacks
3. Internet bandwidth disparities

29. Developing Multi-agent Systems for mHealth Drug Delivery

A1. The home healthcare setting is a dynamic environment characterized with many uncertainties. For instance, the preferences or expectations of care workers, patients, and the management are ever-changing. These dynamic changes have to be taken into account when making task allocation decisions.

A2. mHealth technology facilitates the efficient real-time information update on such issues as nursing staff workloads, resource availabilities, upcoming patient requirements, re-scheduling, and management choices. In other words, data collection is enhanced.

A3. Multi-agent systems provide effective reasoning and planning mechanisms that can make decisions in the absence of the human decision maker. In addition, the response time is significantly reduced, as compared to the human decision maker. Moreover, multi-agent systems tend to be more consistent and accurate in the operation than human decision makers.

A4. Some of the disadvantages are: (i) the use of multi-agent systems increased the complexity of the healthcare system, demanding more expertise and training, (ii) oftentimes, patients and care givers are concerned about the weakening of patient-physician relationship, and (iii) there are possibilities of technology perversion, which may raise ethical and security concerns.

A5. The satisficing approach is a tradeoff between conflicting objectives; therefore, it seeks to satisfy all the concerned players, i.e., the patients, the workers, and the management. Moreover, the approach is a heuristic, which is less complex and more flexible and user-friendly than other methods such as mathematical programming.

30. Wireless Networks in Mobile Healthcare

A1. Answer: LTE (Long Term Evolution) is a standard for wireless data communications technology and an evolution of the GSM/UMTS standards. The LTE was achievement of 4G that is the advanced personal mobile communication technology. Currently, by the time of writing this chapter, LTE-A, LTE-B and LTE-C respectively.

A2. Some communication technologies are suitable for healthcare systems while others are not well suitable for such systems. Healthcare systems are delivering biomedical signals and other healthcare information over a long distance using communication technologies. Same services can be delivered anytime from anywhere using mobile communication systems with high-speed and high-reliability services. Such mobile communication services can be applied to a mobile hospital (M- hospital), emergency ambulance service, and early warning systems for diseases, illness rehabilitation, and general healthcare and so on. As a pre-hospital medical care, videos, images, electrocardiogram (ECG), pulse biomedical signals of the patient can be transmitted via Global System for Mobile communication (GSM), General Packet Radio Service (GPRS), and third generation (3G) mobile cellular systems, fourth generation (4G), WiMAX, and mobile satellite communication (MSC) systems (Pattichis et al., 2002, Rashvand et al., 2008). In this section we summarized main wireless technologies that are used in healthcare systems such as GSM, GPRS, 3G, 4G, LTE, Wi-Fi, WiMAX, satellite, WAP, wireless IP, wireless LAN, Wireless Personal Area Networks (WPAN), Wireless Sensor Networks, Wireless Body Area Network (WBAN).

A3. The author have classified four main groups consist of (1) diagnostic healthcare systems (Tele-cardiology and Tele-radiology), (2) emergency healthcare systems (Consultation in Healthcare Systems), (3) remote monitoring systems (Maritime Healthcare System and Healthcare Systems for Diabetes), and (4) location-based healthcare systems that was summarized in Figure 3 of this chapter.

A4. While monitoring, consultation and emergency system may requires pictures or multimedia video in order to evaluate condition of patients. Multimedia audio or video requires strict QoS requirement such as packet loss, jitter, out-of-order-delivery, and importantly bandwidth. Weak wireless signals will directly impacts on QoS parameters resulting disconnections, video missing or incomplete video transmission.

The QoS parameters are as follows:

Low Throughput: The threshold of the resource utilization in the wireless network gets maximize, the due to limited bandwidth users will compete to get fulfills their application requirements. In case if patient resides in a wireless network where the wireless throughput is low or limited then the data transmission will be effects especially for video and audio-based contents.

Packet Loss: Packet loss can happen due to limited resource in the wireless network or through congestion where data is corrupted or packet drop on the router due buffer overflow. Patient important information might be lost in effect to packet loss in remote monitoring and tele-consultation service.

Error: The information can be transmitted in case of loss of packet, but sometimes the error arises with this phenomena might not be detected. Imagine, patient's data is corrupted and transmission sent to the healthcare unit. With wrong data, transmitted, healthcare unit will unable to know about patient's original situation that is potentially dangerous.

Latency: Sometimes due to congestion in network, packet might reach to its destination late. Latency is much important when an audio or video transmission takes place. That might be cautious in the emergency case of health management system.

Jitter: The variation of delay might seriously affects the quality of video and audio as much of the information gets lost. A potentially dangerous for healthcare services that require audio and video transmission.

A5. Healthcare solutions have quite larger classification ranged from patient diagnostic methods, emergency system, and consultancy with physician or specialist doctor and remote monitoring of the patients. Quite recently, the development in location-based services in healthcare not only help in remote monitoring of the patient but also in the emergency management system. Mobile device should always be near to the patient as Bluetooth have specific data transmission coverage. GPS will be used in case if patient is residing outside, while indoor location coordinates will be calculated using in a different way. The information data about patient health in the form of text information does not have hard quality of service requirement and can be transmitted with even low signal coverage. While monitoring, consultation and emergency system may requires pictures or multimedia video in order to evaluate condition of patients. Multimedia audio or video requires strict QoS requirement that comprises of throughput, packet loss, error, latency, and jitter respectively.

A6. Evolution of wireless technological advancement and rapid improvement of application solutions for mobile healthcare facilitates the remote patient. The coverage of wireless and cellular communication becomes vastly available even in the remote location that compliments the healthcare solutions for the patients anywhere anytime. Researchers and Industry experts are working on the improvement of wireless infrastructure, algorithm, and architecture suitable for healthcare solutions in even unstable wireless networks. Evolution of these technologies will evolve further with advancement of application solutions that will require more requirements to facilitate the end user. With advancement of technologies, some technologies such as GSM, GPRS WiMAX, and 3G are already obsolete in healthcare. Recently progressed communication standard of LTE-A, B, C have more bandwidth and data coverage that will complements the healthcare sector. These enabling technologies will be highly efficient and cost effective with increase of flexibility and integration that promises to enhance the level, features and experience of future mobile healthcare systems.

32. Telemedicine Services over Rural Broadband Wireless Access Technologies: IEEE 802.22/WRAN and IEEE 802.16 WiMAX

A1. Post-operative, re-convalescing, and elderly patients

A2. The IEEE 802.22/WRAN offers a maximum transmission range up to 100 kms compared to the IEEE 802.16/WiMAX maximum transmission range of up to 50 kms.

A3. Within the 64-QAM modulation coverage radius based on IEEE 802.16/WiMAX there is no different modulation scheme from the IEEE802.22/WRAN standard. Within the 16-QAM modulation coverage radius based on IEEE 802.16/WiMAX a 16-QAM and 64-QAM modulation schemes are offered by the IEEE802.22/WRAN standard. Within the QPSK modulation coverage radius based on IEEE 802.16/WiMAX a QPSK and 16-QAM modulation schemes are offered by the IEEE802.22/WRAN standard.

A4. Telemonitoring and Telediagnosis.

34. Motion Capture: From Radio Signals to Inertial Signals

Case Study #1: "Posture Classification"

A1. If the cost of the system should be minimized, the approach based on radio fingerprinting could be sufficient for the arm posture classification. Furthermore, the use of *k*NN and p-*k*NN with $k = 1$ should also be sufficient for such scenario. Indeed, it has been shown in the chapter that the estimation of the only fingerprint positions is very accurate also using the simplest version of radio fingerprinting. Inertial sensors are not required here.

A2. The choice of the arm postures can be chosen by the doctor, recalling to properly train the system on the fingerprint positions associated to the chosen postures. Adjacent arm postures, however, should be chosen at a sufficiently large (spatial) distance in order to minimize the estimation error when classifying two different but close postures. For instance, in [11, 12, 21] the arm postures shown in Fig. 7 have been considered, leading to good classification performance. Finally, for these postures, even if the choice of two devices (i.e., on the upper arm and on the forearm) could be a more robust choice, one can consider to use only the node on the forearm, which is already associated with very distinct fingerprint positions (and, thus, arm postures). Observe that, if one is only interested in classifying different inclinations of the user arm (and its heading is not of interest), the best choice would be to use a device supplied with only an accelerometer.

A3. To this end, the devices should be placed on the user leg (e.g., one on the thigh and one on the shank) and during the session the user should pay attention not to move his/her hip (which will be now the origin of the reference system).

Case Study #2: "Motion Capture"

A1. Even if almost every approach should be theoretically able to reconstruct the continuous movements of the arm (just discarding versions of *k*NN and p-*k*NN with $k = 1$), if one wants to obtain a minimum performance level, the approach based on the use of inertial and magnetic sensors would be the preferable. Indeed, in the chapter it has been mentioned that such approach could lead to errors in the order of only a few centimeters.

A2. Every possible movement of the user arm could be evaluated. However, it must be clear that if one wants to estimate the motion of a specific arm segment a sensor device must be placed on that segment *and* on all the arm segments between it and the origin of the system (i.e., the shoulder). For instance, if one wants to estimate the motion of the hand, three sensor devices are necessary: one on the hand, one on the forearm, and one on the upper arm. Nevertheless, here the user is free to move in the room and no preliminary system training is needed.

36. RFID in Healthcare -- Current Trends and The Future

A1. RFID tags can be used to collect sensor-derived data by extending the chip's interface capability to a sensor [51]. Combining the tags' sensor capabilities with sensor nodes has provided new dimensions for interaction and operation with each other [43]. In addition, chemical sensing can support advanced medical monitoring [59]. A hand disinfection control system using RFID named Automatic Disinfectant Tracker (ADT) has been proposed in [52]. Experimental results show 100% efficiency in detecting proponal and ethanol concentration using the static disinfectant control method. Hung *et al.* [53] proposed a Blood Pressure Measurement (BPM) with ease of use, low cost and wireless connectivity. The authors note that recording blood pressure changes over a period of time will help doctors control cardiovascular conditions. Vecchia *et al.* [54] proposed an intelligent infrastructure for smart hospitals to optimize medical staff to patient interaction and to provide ubiquitous and transparent access to medical data. The system uses RFID and photosensor technologies to accomplish these tasks. The purpose of the photosensor is to trigger RFID antennas to on/off to reduce RF exposure and power consumption. Other notable sensor-attached RFID based applications include the label-type sensor tag proposed by Catarinucci *et al.* [57] which takes as input a generic sensor and transmits the measured data. Romain *et al.* [58] proposed an implantable RFID based pressure sensor that could be used in network configuration. The goal is to develop a communicating endo-prosthesis which includes integrated sensors made of a pressure transducer. Thus, it is evident that RFID with sensor capabilities plays a vital role in healthcare.

A2. Kuo *et al.* (cited in [69]) observed that a shift in resonance frequency causes a degraded reading range for RFID, and as a result they suggest placing a thin EMI sheet made of a magnetic composite absorber between the tag and the metal which recovers nearly 60% of the reading range. Ting and Tsang [10] recommended RFID systems be positioned at least 30cm away from a medical device to avoid problems of interference, based on their findings from an EMI study. Houliston *et al.* [66] conducted a study to replicate EMI due to RFID with a drug infusion device. Infusion pumps known to have failed due to RFID EMI were placed in RF fields with varying strengths created by varying numbers of RFID readers, high-gain antennas and varying distance between readers to infusion pump. The results concluded the infusion pump was not affected by low-power RFID readers even when they were in direct contact. However, the pump attached to a tag was disrupted when a highpower reader was used at a distance of 10cms. It was noted that once the pump failed due to EMI, it subsequently failed at particular moments even when no RFID readers were active. This points out the unreliability of a device once it has failed due to EMI, thus emphasizing the importance of thoroughly resetting medical devices after failure.

Togt *et al.* [70] conducted a study to assess the EMI interference from RFID. The study was conducted in a controlled environment without patients using active 125KHz and passive 868 MHz tags, and testing 41 medical devices under 17 different categories from 22 different manufacturers. The results of the study show that out of 123 EMI tests, 34 EMI incidents were induced by RFID, of which 22 were classified as hazardous, 2 as significant and 10 as light. The 868 MHz passive tags induced 26 incidents in 41 EMI tests and its counterpart induced only 8 incidents. The passive tags induced EMI in 26 medical devices, 8 of which were also affected by active tags. The authors concluded that RFID potentially induced hazardous incidents in medical devices and the implementation of RFID in healthcare should require on-site EMI testing and follow international standards. However Payen's [71] report argues the above conclusion is incorrect due to the following: a) the research has not been in line with most current hospital RFID deployments where passive RFID is the mainstream technology; b) most UHF deployments call for a 1-watt or less reader but the study used 3-watt readers which are far more powerful; c) this, combined with the gain of antenna used, means the study was highly unlikely to be compliant with FCC regulations; and d) the most commonly used frequency (13.56 MHz) in hospital settings was not mentioned in the study and hence does not truly reflect the reality of most current RFID deployments.

The varying results indicate that EMI effects of RFID in healthcare requires further indepth research in order for informed decisions to be made.

A3. A High Efficiency Particulate Air (HEPA) Air Ventilation System using RFID has been developed by Lin *et al.* [42]. The aim of this system is to control the quality of air in hospital operating rooms to prevent hospital-acquired infection. The system is composed of RFID tags for staff and patients, sensors and a reacting module that receives information about the number of staff, the status of the surgery and the volume of air in the HEPA Air Ventilation System. Two control mechanisms are deployed. One is time-based where the system will keep the volume of airflow to a minimum when the operating theater is not in use. The second controlmechanism is controlled by RFID which monitors the number of people in the room and regulates the airflow accordingly. By installing two readers in the room, the system achieves a 99% read rate. While the initial setup cost may be high, estimated results show an energy saving of 50% and 40% for the air-filter was achieved. It is important to note the system not only aids in preventing hospital-acquired infection but other occupational diseases from spreading.

Yamashita *et al.* [46] studied data management of surgical instruments using RFID. A tag was designed and attached to surgical instruments used in the experimental study. The tag was cylindrical in shape with a diameter of 6mm, a thickness of 2mm, weighing 1 gram, a storage capacity of 128bytes and consisting of an original loop antenna. The internal chip was manufactured by Philips (ICODE SLI) and the frequency used was 13.56MHz. The focus was on the effect of the tag being repeatedly sterilized, its durability to contamination tests during the washing process and its impact resistance against being dropped. The results of the study showed no tags were incapable of communicating or had any damage after high-pressure sterilization (50 times for 30 minutes on 30 tags), the high-heating challenge test (200 degree centigrade for 300 hours on 20 tags), the contamination test (77 surgical instruments for 31 cycles using a washer disinfector, high-pressure steam sterilization with 68 surgical instruments for 15 cycles using an autoclave process) and were resistent to repeated impact. The failure rate was remarkably 0%.

The deployment of active tags with thermometers to combine patient location tracking and real time patient symptom tracking was used by Taipei Medical University Hospital to contain a SARS outbreak. eShephard, a RFID based Real Time Location System from Exavera Technologies, allows doctors to correctly identify patients thereby assuring correct treatment, reducing medical errors and increasing operational efficiency. RFID tagged implant objects introduced by implant manufacturer Biomet in 2006 have assisted in identifying items used from a kit so they could quickly and accurately invoice the hospital. RFID enabled smart cabinets track inventories of medical devices to ensure patient safety, improve billing efficiency and inventory management, and reduce error-prone manual labor involved in the process. RFID based system SurgiChip assists doctors to prevent wrong-site and wrong-procedure surgery by correctly identifying the patient, the exact surgical site and other vital information. ClearCount that uses RFID tagged surgical sponges ensures sponges are not left inside patients after surgery.

A4. Aubert [60] has conducted an indepth analysis of RFID implants on humans for locating/tracking and remotely controlling human biological functions. The author notes that RFID implants are not mature enough to communicate with GPS satellites to facilitate real-time tracking of a person however the remote control of human biological functions are more realistic. Burke *et al.* (cited in [60]) discuss the possibility of using RFID in the remote activation/deactivation of biochemical activity in a single living cell. The miniature nanotechnology FM-radio by Jensen *et al.* (cited in [60]) could be inserted into a living human cell to give a subcellular remote and real-time controlled interface. The application VeriMed from VeriChip uses a small glass capsule about the size of a grain of rice (FDA approved device) that is implantable in humans. This capsule assists in patient identification when they arrive at a hospital unconscious or unresponsive [12]. Zhou *et al.* (cited in [12]) propose a human implantable RFID enabled sensor tag that could help diabetic patients monitor their blood sugar levels and save them from having to prick themselves several times a day to check their blood-sugar levels. Berger (2008) (cited in [12]), proposes an RFID enabled orthopedic implant to assess the function of the implanted device and the surrounding tissue. Occhiuzzi *et al.* [62] proposed a design of implanting RFID tags for the passive sensing of the human body with the purpose of sensing the evolution of human physiological and pathological processes involving a local change for effective permittivity inside the body. Thevissen *et al.* [63] proposed the use of implanted RFID tags into a human tooth to handlemass deaths that arise from diaster situations such as tsunamis and hurricanes.

A5. The required security and privacy properties for RFID protocols in healthcare can be summarized as follows [96] [97]:

- *Tag Anonymity*: The protocol should protect against information leakage that can lead to the disclosure of a tag's real identifier. This is important as an attacker may be able to clone a valid tag.
- *Tag Location Privacy*: The protocol should ensure the message contents is sufficiently randomized so they cannot be used to track the location(s) of the tags and thereby glean social information about the wearer of the tag.
- *Forward Secrecy*: The protocol should ensure that on compromise of the internal secrets of a tag, its previous communications cannot be traced by an attacker. This requires previous messages be independent of current resident data on the tag.
- *Forward Untraceability*: The protocol should ensure the old owner is unable to trace or communicate with the tag post-ownership transfer. This means the old owner is unable to learn or compromise any new secret(s) that have been established between the tag and the new owner.
- *Replay Attacks*: The protocol should be able to resist compromise by an attacker through the replay of messages that have been collected by an attacker during previous protocol sequences. This requires protocol messages in each round of the protocol are unique.
- *Denial of Service (DoS)*: The protocol should be able to recover from incomplete protocol sequences that can occur due to an attacker selectively blocking messages. Importantly, such blocking of messages by an attacker should not lead to desynchronisation between the tag and the servers.
- *Server Impersonation*: The protocol should ensure the new server cannot be impersonated by an attacker. This requires the tag to challenge a server requesting ownership transfer to prove its legitimacy.

A6. Foster and Jaeger [100] report the following ethical implications of RFID in Healthcare:

1. **Possible Medical Risks:** The US Food and Drug Administration (FDA) classified VeriChip as a class II device and listed some potential hazards as: adverse tissue reaction, electromagnetic interference, electrical hazards, magnetic resonance imaging incompatibility, etc. Several researchers argue that implanted RFID tags, similar to the ones used in the VeriMed system, have been associated with tumors in rodents, mostly sarcomas. Vascellari *et al.* (2006) (cited in [100]), mentioned tumors in dogs implanted with microchips.
2. **Ethical Issues:** A lot of literature can be found about privacy and security issues of RFID and activists' websites also discuss possible harm from implanted RFIDs, which include the tracking of individuals to mind control. The latter being in the realm of science fiction at least for the foreseeable future.
3. **Disclosure of Risks:** It is only ethical to disclose adverse effects of a treatment and individuals have a right to know about the effects. It was noted that VeriChip failed to disclose the rodent studies until the issue was raised by activists. In addition, the carcinogenic effect of an implant in rodents suggests the possibility of a similar effect in humans. A cancer researcher states - "*I*

think the evidence from the animal studies is indeed alarming and one should refrain from chipping people unless the mechanisms and long-term effects are known". Foster [100] suggests cancer be added to the list of adverse side effects provided by the FDA.

4. **Truth in Advertising:** The above-mentioned occurrences call for an independent assessment of risks and benefits using implanted RFID tags so they can be advertised truthfully.
5. **Coercion:** Cottle (2006) (cited in [100]) stated the *New Republic Online* defended a proposal to implant RFID tags in sex offenders because such people are already subject to extensive restrictions and tracking via chips might be preferable. However, fundamentalist Christian groups in the US vehemently objected to such ideas as "marks of the beast". Judaism and Islam also prohibit tattoos and their religious heads may forbid implanted tags for similar reasons. The proposal from VeriChip to implant immigrant/guestworkers indirectly forces some people to accept the chip because of poverty at home and the greater likelihood of a job in the US. This would be deeply offensive on the grounds of human rights and would be in violation of Article 3 of the Universal Declaration of Human Rights (1948) which guarantees right to "life, liberty and security of person". It would also be a violation of the privacy provision of the International Covenant on Civil and Political Rights (1966) to which the US is a party.

Foster and Jaeger [100] conclude that technologies such as fingerprint and retinal scans allow reliable identification of individuals without the need to compromise bodily integrity. They also quote Anderson and Labay's remark - "*a decision about where to draw the line of acceptable use must be made soon, before the technology becomes rampant and it becomes too late to prevent misuse*".

37. Wireless Body Area Networks in mHealth

A1. d

A2. c

A3. From the equation of the Shannon Hartley theorem, $C = W \log_2\left(1 + \frac{S}{N}\right)$, there are two ways to support high capacity C:

a. Expand the bandwidth W and/or
b. Increase the SNR (by using a better constellation in the modulation process)

Of these two methods, it is easier to expand the bandwidth. Most indoor environments support high SNR, and in these environments, UWB can be implemented as the radio technology for the network architecture.

A4. In the *data-pulling* model for sensor networks, the server or base station (BS) requests data from a single node or groups of nodes. In the data-pushing model, the nodes transmit their data to the server. The advantages of using a *data-pulling* model in a network with resource constrained nodes and a robust base station are that it allows for nodes that support high throughput over the single hop to the base station while eliminating collisions and overhead due to synchronization and timing. The BS is a robust data processing unit that has a larger battery and larger processing power that enables it to carry on data processing and network configuration. In the data pulling model, the BS requests (pulls) data from a set of nodes with the help of simple commands, thus eliminating packet collision and channel contention issues. The data pushing model can help in real-time data processing, since it enables a node to transmit its data immediately after collecting and processing rather than having to wait for the BS to request it.

Case Study One:

A5. The reader is encouraged to be creative in this solution, while keeping the constraints of WBANs in mind. One possible solution to this problem is detailed below:

Vibration energy from the vocal cords of a human can be used to provide energy for applications. It should be noted however, that this energy is extremely small. A recent article by M. Wischke, et al. "Vibration harvesting in traffic tunnels to power wireless sensor nodes." Smart Mater. Struct. 20 (2011) 085014 (8pp).explores this idea of harvesting energy from train vibrations for applications in tunnels. However, there are scenarios where voice-activated energy could be used:

i. In classrooms where instructors are delivering information by talking (as opposed to project-based or lab-based classes), voice activated energy could be used to supply power into the grid for other applications.
ii. Music instruction classes that focus on voice lessons.
iii. Concerts and games, where the audiences generally indulge in voice-based communication at a higher intensity than usual communication.

The architecture can be a node that comprises of a sensor that collects vibrational energy and a capacitor that collects the energy. The node is worn on the user's body, either on fabric or on the skin, as close to the vocal cords as possible. The vibrational energy is harvested in a capacitor/stored in a battery that can be disconnected from the node. A flat architecture is adequate for this scenario. Based on the utility of power generation, additional wireless capabilities can be built into the node. For example, user A's node might have a lot of harvested energy, which can be lent or sold to another user. To facilitate this, a node can have a transceiver that broadcasts the state of its battery (e.g., full or halfway), and other nodes can respond. This scenario can facilitate "power-sharing" for mobile applications.

Security aspects will focus on the tracking technology that can be facilitated from the wireless transmissions. A user's node might transmit that its battery level is halfway, and such transmissions can be used for localization. This raises security and privacy concerns. (Other security issues can be discussed in class.)

As previously mentioned, the vibrational energy harvested from vocal cord vibrations is generally small, so the node design has to be energy efficient to justify its operations. For the "power-sharing" concept that can be enabled by this technology, both data-pulling and data-pushing are applicable modes of data transmission. Further research and a test bed could validate the choice of one over the other.

Case Study Two:

A6. When a node i wants to forward a packet to node d, it will send the packet to an intermediate node j if $P_{i,d}^{t} < P_{j,d}^{t}$.

If R_i^t denotes the residual energy at node i at time instant t, and i wants to forward packets to node d, it will forward the packets only to an intermediate node j, only if the following two conditions satisfy:

$$P_{i,d}^{t} < P_{j,d}^{t}$$

and, $$R_j^t > R_i^t$$

The LLF is dynamically adjusted at each discrete time slot by a weightage factor, ω, such that $P_{i,j}^{t}$ increases fast and decreases slowly for a historically good link, and decreases fast and increases slowly for a historically bad link

38. An Integrated Wireless Communication Platform for End-to-End and Automatic Wireless Vital Sign Capture Using Personal Smart Mobile Devices

A1. One of the key attributes of BLE is the extremely fast response time, a crucial requirement in real-time human interface applications. Additionally, it demands very low power consumption (e.g., its maximum and sleep currents are 15 mA and a few μA, respectively). In many cases, BLE products will be able to operate more than a year on a button cell battery without recharging. BLE is the first low-power standard to be adopted by the mobile phone community which opens up many scenarios that are previously impossible.

A2. Table 1 shows the comparison between BT and BLE

Table 1 Technical comparison between BT and BLE

Technical Specification	Bluetooth Classic	Bluetooth Low Energy
Distance/Range	~ 100 m	~ 50 m
Over the air data rate	1 ~ 3 Mbps	1 Mbps
Application throughput	~ 0.5 ~ 2.0 Mbps	~ 0.2 Mbps
Active slaves	7	Not defined; implementation dependent
Security	56/128-bit and application layer user defined	128-bit AES with Counter Mode CBC-MAC and application layer user defined
Robustness	Adaptive fast frequency hopping, FEC, fast ACK	Adaptive frequency hopping, Lazy Acknowledgement, 24-bit CRC, 32-bit Message Integrity Check
Set-up time	A few hundreds of ms	A few ms
8-octet packet transmission time	214 μs	144 μs
ACK packet transmission time	126 μs	80 μs
Network topology	Star and piconet	Pure star
Power consumption	1 as the reference	0.01 to 0.5 (depending on use case)
Peak current consumption	< 30 mA	< 15 mA
Max transmission power	+20 dBm (class 1) +4 dBm (class 2)	+10 dBm

A3. BLE is expected to enable various applications in control, fitness, healthcare and so on. Here are some typical use cases: proximity, time, emergency, network availability, personal user interface, simple remote control, browse over Bluetooth, temperature sensor, humidity sensor, HVAC control, generic I/O (automation), battery status, physical activity monitor, cycling sensors, blood glucose monitor, vital sign (pulse oximeter, body temperature,) monitor.

A4. LTE is a fourth-generation (4G) mobile networking standard. It can support much faster data transmission, i.e., 50 times faster than 3G networks. Peak upload and download rates could be up to 500 Mbps and 1 Gbps, respectively. LTE is aimed at simplifying the architecture of the system as it transits from the existing circuit and packet switching combined networks to an all-IP flat architecture system. It also supports more data capacity because it focuses on voice over Internet protocol (VoIP). Low data transfer latencies (sub-5 ms latency for small IP packets in optimal conditions), lower latencies for handover and connection setup are other advantages of LTE, compared to previous radio access technologies. Moreover, LTE also support voice and SMS text messaging using existing networks via generic access, and also allows wireless broadband provides to transition to this new technology without rebuilding entire networks from the ground up. This makes upgrading more affordable for service providers.

A5. There are several reasons to promote the use of LTE for M2M communications. First, for high-performance M2M applications (e.g., real-time monitoring and control, video streaming, etc.), very high-speed and low-latency data transmission of LTE will provide immediate benefits. LTE also introduces more sophisticated resource scheduling and quality of service provisioning mechanisms. This makes it suitable for critical M2M applications in healthcare, safety and security sectors. Unfortunately, low-speed M2M applications may find it difficult to justify investing in LTE. However, in the long term, LTE exhibits many advantages. The life cycle of existing cellular networks is unpredictable, and the cost of upgrading deployed technologies can be high. LTE, on the other hand, is expected to support a large number of connected devices in the future at potentially lower service costs. LTE is very much spectrum-efficient and less expensive to operate and maintain. Besides, LTE is designed to support IPv6, making it scalable when a huge number of devices require network connectivity.

39. Energy Analysis and QoE of Wireless Sensor Networks

A1. Two important problems should be concerned in the WSNs design, i.e., energy consumption and QoE. As we known, sensors in WSNs are operated with energy-limited battery. Moreover, it is usually impracticable and inconvenient to recharge and/or replace a new battery for the drained sensor, since mount of sensor nodes are distributed in a hostile and/or unreachable circumstance. But, the battery technology cannot keep pace with the development of the digital integration technology. Therefore, energy consumption is one of the key issues. On the other hand, along with high requirements, the conventional QoS (Quality of Service) indexes cannot reflect the real experience of users. Thus, QoE (Quality of Experience) is another important concern when we design a WSN application.

A2. delay controlled transmission scheme, energy efficient cooperative communication in WSNs, energy efficient modulation scaling, energy efficient routing schemes, power management-based task scheduling for digital communication processors, cross-layer energy optimization, etc.

A3

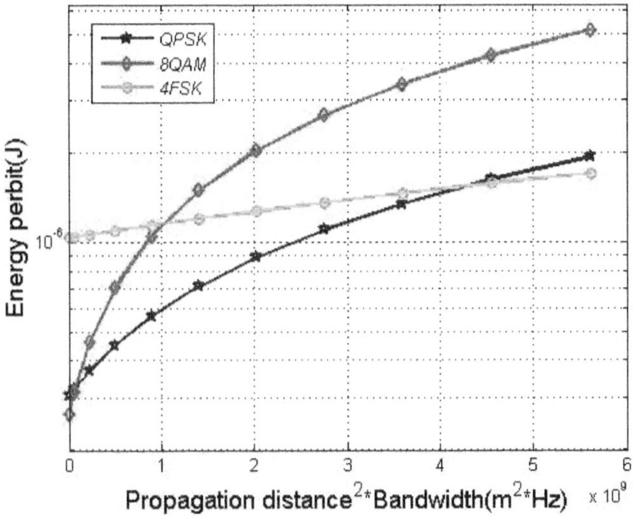

Fig. 1 Energy per bit for different RF front-end architecture in medium bandwidth application, BW=250KHz, d=(0 150]m

According to (1) (13) (15) (18) and (19), the above Figure shows the energy per bit with QPSK, 8QAM and 4FSK modulation when the bandwidth BW is 250KHz. From this Figure we can see when the transmission distance is short, which means $d^2 \cdot BW$ is small. 8QAM modulation scheme has the lowest energy consumption. This is because 8QAM has the highest data rate and saves the active time. With the increase of the transmission distance, which means the value of $d^2 \cdot BW$ will become more and more large, QPSK modulation scheme performs best. On the other hand, 4FSK modulation system has the highest energy consumption for small $d^2 \cdot BW$, which comes from the fact that for a given transmission bandwidth, the data rate of 4FSK modulation is lower than QPSK and 8QAM. But with the increase of $d^2 \cdot BW$, the energy consumption of 4FSK increase much slower than the other two modulation schemes, this is because the nonlinear PA used in 4FSK modulation system has higher efficiency. For large $d^2 \cdot BW$, the advantage of the high efficiency of nonlinear PA is obvious.

With the similar method to analyze the low bandwidth system, we can compare the energy performance of 4FSK, QPSK and 8QAM modulation. When $d^2 \cdot BW < 5.8472e+7$, 8QAM modulation scheme is the most energy efficient. When $5.8472e+7 < d^2 \cdot BW < 5.2965e+9$, QPSK modulation scheme is the best choice. If $5.2965e+9 < d^2 \cdot BW < 1.125e+11$, then 4FSK is superior.

A4. As we known, QoS is a traditional index. Its main concern focuses on the quality of the network itself. For example, if a network has a relative short delay-time, or low energy, so we can say that it is a good network, because it has high QoS indexes. While QoE is concerned about the experience of users, like service availability, service usability, service integrality, service retainability, and service instantaneousness. Therefore, with the development of technology and the requirements of people, we not only care about QoS indexes, but also QoE.

40. Quality of Service in Wireless Technologies for mHealth Service Providing

A1. With the extensive use of several services in the network, some of them require more bandwidth than others, and also requires satisfying some technical parameters such as packet delay or packet loss rate for the user to be satisfied with the service. When the network traffic load increases it is possible that the network not fully comply with these requirements. When using differentiated services with a defined QoS, services classified as real-time or urgent will be delivered before services classified as non-real time or less urgent. This allows that mHealth services such as emergency calls or blood pressure monitoring to be transmitted with priority, as they are emergency services and requires a real-time treatment.

A2. The common features of QoS are based on technical parameters such as delay, delay variation, packet loss rate and the ability of the system to provide the requested bandwidth for the service. However, it could also be included some additional characteristics such as the speech quality and connection performance to obtain a broader evaluation about the quality of the service.

A3. The provision of *end-to-end* QoS for mHealth applications involves several aspects such as the mobile communication environment, the availability and allocation of network resources, the applied policies and context- aware transmission. It is also important to ensure that the QoS requirements for a service can be supported by several different technologies, so the service will not be degraded during its transmission over heterogeneous networks.

A4. 3G and 4G mobile technologies have many advantages, including a high transmission rate required for some mHealth applications, such as image transmission and video streaming. They also allow the assignment of bandwidth according to the data rate required for the service such that no resources are wasted. These systems also enables service differentiation and QoS assignment for delay, delay variation and packet loss rate according to whether the service is delivered in real time or non-real time, ensuring the transmission will comply with the requirements of the mHealth applications.

A5. Not completely. These ITU recommendations cover the IP core network and the wireless access networks. However they do not specify QoS characteristics for the Personal and Body Area Networks the mHealth applications use to collect data from the sensors the patient use or wear.

41. mHealth over "Medical Grade" High Quality of Service Mobile Networks

A1. Standard commercial devices will be downloaded with the PRTSS software, including the authorization access routine. The BOYD practice will be enhanced to allow both routine transactions over commercial networks and priority real-time broadband communications over the secure PRTSS networks.

A2. The PRTSS network architecture, will be based on real-time capacity allocation based on operational needs. During the networks deployment and on an ongoing basis the operational needs are being reviewed by the participant PRTSS agencies. For example, patient monitoring during a catastrophic emergency would be discontinued until capacity demands are back to day-to-day normal levels.

A3. It will be funded by including mHealth and healthcare in general, into the PRTSS group of public safety agencies and by engaging the provincial health authorities (in the Canadian situation). An approach known as PPP (Public- Private- Partnerships) will allow the private sector, in conjunction with the Provincial and Municipal governments, to jointly fund and benefit from the projected R.O.I (Return on Investments) of mHealth over secure "medical grade" mobile networks.

A4. Spectrum assignments exclusively for mHealth or healthcare are rare. But as proposed in the Canadian context, spectrum is being assigned, where healthcare is proposed to be combined with other PRTSS, such as public safety and security. In the U.S and Canada up to 20 MHz in the 700 MHz band have been allocated for public safety and by extension to PRTSS agencies.

A5. Governments rarely dictate networks' specifications in licensing terms and conditions. Without such regulatory requirements there are no economic incentives for the carriers to meet public safety or healthcare "Security/Medical Grade" QoS facilities and services.

42. Mobile Data Collection: A Security Perspective

A1. The data collected is often very sensitive, like for instance medical records, and unlike personal computers, it is very easy to access data stored on a mobile phone, which is also more likely to be lost, stolen or damaged.

A2. Not at all. There is absolutely no security in the majority of these systems, especially on the mobile side. Data are stored in clear on the phone memory card and are accessible by anyone with physical access to the device.

A3. Yes, it is completely replacing paper based data collection in developing countries where the mobile phone penetration is much higher than personal computers, and it is used for anything from vaccination trial, to deforestation monitoring and for conducting social-demographic surveys.

A4. Right now data collection is a static activity, where information flows from collectors to data viewers and data is used mostly for statistical purposes and monitoring the progress of different activities. In the future the collected data might be used in real time and the type and modality of the collection might change dynamically over time based on previously accumulated data.

A5. We are not too far from a good solution to protect the data collected and stored on the mobile devices and to protect the communication with the server. However protecting the data on a server which is not under our direct control is a big open problem. Sharing the data to allow on-line collaboration while protecting sensitive information is also a big challenge.

A6. Sensors in particular Bio-sensors will advance today's data collection system to another level. Bio-sensor based mHealth can improve remote diagnostics and health-care information. It will improve the quality of remote diagnostic both for ambulatory and continuous monitoring of chronic diseases.

A7. Because of the mode of deployment and organization of the projects involved with data collection. Standard solutions can be adopted, but need to be tailored for this specific area of application. What is good enough for a normal mobile app with a single user might not fit in a collaborative system.

A8. Yes, in the past two years the data collection systems communities have shown enormous interest in being compliant with regulatory requirements. However, the regulatory bodies are not ready to deliver a comprehensive list of requirements related to mHealth privacy and security.

43. Security, Reliability and Usability of mHealth Environments

A1. Support for patients in early dementia stage living in their own home; allow patients with chronic diseases to live at their own home and keep mobile as much as they can; "home-based-care" = support the remote treatment of patients at home or on the move to achieve rehabilitation and physical training, to avoid long-term and expensive hospitalization.

A2. remind of regular tasks, appointments, medication, etc; support for orientation and positioning, including tracking in case of emergency; keep social contacts and initiate audio-visual-communication.

A3. Security (involving access control and privacy protection, based on identification, authentication, authorization); Reliability (including self-monitoring, redundancy/fall-back solutions); Usability (including accessibility).

A4. Point-of-Care environment; Health & Care Services environment; Health Information Services (HIS) environment.

A5. With an electronic, personal user ID, in form of a private, certified key stored on a personal device as a smartcard, key fob, or SIM/USIM card attached to a personal device, such as a smartphone or tablet PC.

A6. In order to safe-guard the operation of (potentially) vital monitoring devices and alarm services even in the situation of outages of electrical power, loss of connectivity to communication networks or their complete failure, or malfunction of certain mHealth devices.

A7. Adaptation of user interfaces to the needs, limitations, and preferences of patients; adaptation to the specific point-of-care environment and its characteristics with regards to specific medical devices, treatments; compromise between security and usability.

A8. Something the user knows (e.g., password), something the user possess (e.g., smart card), something the user is or does (e.g., fingerprint). The more authentication factors being used, the more secure a system is. However, multiple authentication factors can lead to usability issues as it becomes more complicated to use the system.

44. "Security and Privacy Issues for Mobile Health"

A1. There is a high level of public expectation that individual privacy and security will be protected in health, widely underpinned by legislation and standards internationally. The ubiquity and consumer oriented nature of mobile devices and services leaves them exposed to a higher degree of vulnerability than typical "closed" ICT systems in health settings.

A2

Delivery of Routine Clinical Care: occurs in controlled clinical service provision circumstances such as teleconsultation.

Remote Access to Health Data Records: occurs in accessing information systems used by health care professionals, and messaging between systems.

Collection of Patient-Specific Data: occurs in external or home settings using proprietary devices and consumer telecommunications technologies.

Patient-Initiated Health Information Transfer: occurs outside a formal care setting and on Internet with uncontrolled involvement of third parties.

A3. Strict in-house procedures such as use of proprietary encrypted video systems and restriction of endpoint equipment may be applied to privately controlled computing devices and connectivity infrastructure, and can be expanded to include approved mobile devices albeit with lower compatibility and useability, as the inbuilt security features and privacy protocols extend naturally to the mobile device situation in the controlled environment. Alternatively, the use of commercial, publicly available solutions in video/audio conferencing can be justified on the basis that the risk of violation is no greater than if a wired telephone connection was being used, and that such systems usually provide internal encryption and access or connectivity controls.

A4

Technology Factors: e.g. software incorporation into personal device function, application development, data capture, acquisition and transmission, and physical device security.

Clinical Factors: e.g. healthcare process workflow, clinical interactions between parties, new models of care and applications, stakeholder impacts.

Human Factors: e.g. cost of devices, communications coverage, usability of devices and applications, human-computer interaction, technology acceptance or adoption modelling.

A5

Within-Home: accidental disclosure by local close-range logging systems or covert surveillance of data streams from collection devices.

Outside-Home: interception of or interference with wi-fi or mobile phone service transmissions.

Mobile-Carrier: attack on carrier protocol, downloading of messages from protected carrier storage, denial of service or physical equipment attacks.

Monitoring-Service: accidental or intentional disclosure by employees or visitors, snooping and covert surveillance of client data, unauthorised copying or backup.

Health-Service: accidental or intentional disclosure by employees or visitors, snooping and covert surveillance of client data, unauthorised copying or backup.

45. Impact of Privacy Issues on User Behavioural Acceptance of Personalized mHealth Services

A1. Various personalized mHealth applications use different devices such as sensors, RFID, implantable devices and special devices for specific purposes. These mHealth devices are vulnerable to loss or theft of devices, eavesdropping and denial of service attacks.

A2. The new privacy preserving security features should be,

- easily understood by users with different cognitive abilities.
- easy to use (e.g. minimum number of page/window navigations).
- Effective user awareness and necessary education on how to use new features when these are made publicly available.

A3

- Lack of real-world applications based data sets.
- Lack of empirical studies on different types of mHealth applications.
- Lack of large scale user groups based empirical studies.

A4. Service-provider end security enforcements for secure information storage, access and location-privacy preserving mechanisms are vital in personalized location-based mHealth services. Regulatory guidelines and law enforcements on the security measures that are mandatory to be implemented at the service-provider end are also seen as crucial elements in preserving privacy of personal health information.

A5. Regulations can provide formal control on health information of mHealth services. This is seen as important because various mHealth service providers can be guided to follow the same mHealth information management practices.

Author Index

GPSR Compliance

The European Union's (EU) General Product Safety Regulation (GPSR) is a set of rules that requires consumer products to be safe and our obligations to ensure this.

If you have any concerns about our products, you can contact us on ProductSafety@springernature.com

In case Publisher is established outside the EU, the EU authorized representative is:

Springer Nature Customer Service Center GmbH
Europaplatz 3
69115 Heidelberg, Germany

Batch number: 10761746

Printed by Printforce, the Netherlands